ANNALS OF THE NEW YORK ACADEMY OF SCIENCES

Volume 600

EDITORIAL STAFF
Executive Editor
BILL M. BOLAND
Managing Editor
JUSTINE CULLINAN
Associate Editor
JOYCE HITCHCOCK

The New York Academy of Sciences
2 East 63rd Street
New York, New York 10021

THE NEUROPHARMACOLOGY OF SEROTONIN

ANNALS OF THE NEW YORK ACADEMY OF SCIENCES
Volume 600

THE NEUROPHARMACOLOGY OF SEROTONIN

Edited by Patricia M. Whitaker-Azmitia and Stephen J. Peroutka

The New York Academy of Sciences
New York, New York
1990

The art on the softcover edition of this volume has been modified from the papers by E. C. Azmitia *et al.* (brain) and C. H. Vanderwolf *et al.* (tracings) and the originals appear on pages 358 and 370, respectively.

Library of Congress Cataloging-in Publication Data

The Neuropharmacology of serotonin / edited by Patricia M. Whitaker-Azmitia and Stephen J. Peroutka.
718 p. cm.—(Annals of the New York Academy of Sciences, ISSN 0077-8923; v. 600)
Result of a conference held in New York City, July 10–13, 1989.
Includes bibliographical references.
Includes index.
ISBN 0-89766-609-7 (cloth: alk. paper).—ISBN 0-89766-610-0 (pbk.: alk.)
1. Serotonin—Congresses. 2. Serotonin—Receptors—Congresses. 3. Serotoninergic mechanisms—Congresses. I. Whitaker-Azmitia, Patricia M. II. Peroutka, Stephen J. III. New York Academy of Sciences. IV. Series.
[DNLM: 1. Nervous System—drug effects—congresses. 2. Serotonin-pharmacology—congresses. W1 an626YL v. 600 / QV 126 N4931 1989]
Q11.N5 vol. 600
[QP364.7]
500 s—dc20
[615'.78]
DNLM/DLC
for Library of Congress

SP
Printed in the United States of America
ISBN 0-89766-609-7 (cloth)
ISBN 0-89766-610-0 (paper)
ISSN 0077-8923

ANNALS OF THE NEW YORK ACADEMY OF SCIENCES

Volume 600
October 15, 1990

THE NEUROPHARMACOLOGY OF SEROTONIN[a]

Editors and Conference Organizers
PATRICIA M. WHITAKER-AZMITIA AND STEPHEN J. PEROUTKA

CONTENTS

[a]This volume is the result of a conference sponsored by the New York Academy of Sciences entitled **The Neuropharmacology of Serotonin**, which was held in New York City on July 10–13, 1989.

Part III. Expression of Serotonergic Function

Session Chair: J. P. Green

Part IV. Serotonin in Development and Aging

Session Chair: Robert Y. Moore

Part V. Role of Serotonin in Clinical Issues
Session 1 Chair: Hermann Van Praag

Session 2 Chair: Max Fink

Part VI. MDMA and Related Compounds
Session Chair: Lewis S. Seiden

Funding was provided by:

- AMERICAN CYANAMID COMPANY
- BRISTOL-MYERS COMPANY
- CIBA-GEIGY PHARMACEUTICALS
- GLAXO GROUP RESEARCH LIMITED
- ICI PHARMACEUTICALS GROUP
- INSTITUT DE RECHERCHES INTERNATIONALES SERVIER
- JANSSEN RESEARCH FOUNDATION
- LILLY RESEARCH LABORATORIES (A DIVISION OF ELI LILLY AND COMPANY)
- MEAD JOHNSON PHARMACEUTICALS
- MERCK SHARP & DOHME RESEARCH LABORATORIES
- MERRELL DOW RESEARCH INSTITUTE
- MILES INC.
- NATIONAL INSTITUTE OF MENTAL HEALTH/NIH
- NATIONAL SCIENCE FOUNDATION
- PARKE-DAVIS PHARMACEUTICAL RESEARCH DIVISION, WARNER LAMBERT COMPANY
- PFIZER CENTRAL RESEARCH
- SANDOZ PHARMACEUTICALS
- WYETH-AYERST RESEARCH
- SYNTEX RESEARCH

Preface

This volume of the *Annals of the New York Academy of Sciences* is the result of a meeting held in New York City in July of 1989. To organize that meeting, we selected several themes which we considered to be the most important in understanding the current status of the neuropharmacology of serotonin. Thus, we chose receptor pharmacology, expression of function such as second messengers and behavior, serotonin in development and aging, clinical issues, and finally, studies on MDMA. For each theme, we invited speakers and session chairs whose expertise in their particular fields was well known and whose contributions could be both an update for other experts and an introduction for the novice serotonin researcher. We are deeply indebted to these scientists. Their efforts and contributions to this meeting and to the field are evident in this volume, which we believe will become an essential part of many researchers' libraries.

In addition to the written chapters from each participant, we have included transcripts of the question period after each presentation. We hope that many of the memorable moments of the meeting are evident, so that those who could not attend will still appreciate the convivial spirit and sense of uniqueness which all of those present felt.

We wish to express our gratitude to the companies and organizations who contributed to the financial support of this meeting. In addition to those, we wish also to acknowledge the staff at the New York Academy of Sciences, whose experience in organizing meetings of this calibre made our task possible.

In conclusion, we are proud to be the editors of this collection of highly important scientific articles. We hope the reader will find them stimulating and useful.

PATRICIA M. WHITAKER-AZMITIA　　　　STEPHEN J. PEROUTKA

Photo by Joseph Queli.

THE NEUROPHARMACOLOGY OF SEROTONIN
New York, New York
July 10–13, 1989

1. G. K. Aghajanian
2. B. L. Jacobs
3. H. Tamir
4. M. Teitler
5. G. A. Ricaurte
6. R. W. Fuller
7. L. S. Seiden
8. E. C. Azmitia
9. P. M. Whitaker-Azmitia
10. S. J. Peroutka
11. D. P. Taylor
12. A. Frazer
13. S. Maayani
14. D. L. Murphy
15. T. R. Insel
16. M. B. Tyers
17. G. R. Heninger
18. H. Y. Meltzer
19. D. S. Charney
20. G. Curzon
21. C. J. Schmidt
22. B. P. Richardson
23. A. Sjoersdma
24. A. J. Cross
25. M. Fink
26. D. L. Nelson
27. D. C. Jimerson
28. E. B. De Souza
29. D. N. Middlemiss
30. J. M. Palacios
31. A. Wauquier
32. J. J. Frost
33. D. Hoyer
34. P. P. A. Humphrey
35. P. J. Cowen

History of Serotonin and Serotonin Disorders

ALBERT SJOERDSMA[a] AND MICHAEL G. PALFREYMAN

Merrell Dow Research Institute
Cincinnati, Ohio 45215-6300

Recurring cycles of scientific interest in serotonin over past decades, with concomitant waves of new findings, reports, symposia and reviews, literally defy the imagination and challenge description. We have elected to divide this saga of serotonin into four fairly distinct periods (Eras I–IV) and by necessity have been selective in our choices of subjects and when convenient have risked being hoist with our own petard.

ERA I: THE PRIMORDIAL PERIOD (— TO 1953).

This period is obscure in its beginnings but probably lasted a century or so. It ended abruptly in the years 1948–1953 with the isolation, identification and chemical synthesis of 5-hydroxytryptamine (5-HT). Primacy goes to Page's group (Rapport, Green and Page) which isolated and correctly identified the serum vasoconstrictor factor, released from platelets during clotting of blood, as 5-hydroxytryptamine. Quite appropriately, they (with Corcoran) named this widely encountered serum tonic factor "serotonin."[1,2] Independently, Erspamer[3] in Italy had been demonstrating a widespread distribution in nature of a smooth muscle contracting substance which he called "enteramine," because large amounts of it were stored in the enterochromaffin cell system of the gastrointestinal tract. In 1952, he and Asero reported that serotonin and enteramine were identical. In 1953, Twarog and Page,[4] using a sensitive bioassay (mollusk heart), detected serotonin in extracts of brain. Thus, by 1953, the three key locations (i.e. platelets, gut, and brain) of serotonin in animals were known.

Even before its detection in brain, it seemed that serotonin must have had something to do with brain function, as evidenced by the obsession with which these early pioneers extracted the substance from the serum of almost two tons of beef blood, 30,000 (30 Kg) octopuses' salivary glands (*Octopus vulgaris*) and 1,020 amphibian skins (*Discoglossus pictus*) in order to obtain a few crystals of serotonin (as creatinine sulfate or picrate). A similar obsession appears to have gripped our "modern" molecular biologists as they seek to identify various serotonin receptors by screening millions of clones and injecting thousands of oocytes.

ERA II: THE RENAISSANCE (1953 TO 1970)

This era was heralded by the availability of serotonin creatinine sulfate to a small group of interested scientists at the National Institute of Health (NIH) and elsewhere, courtesy of the Upjohn and Abbott companies. A research renaissance ensued, made

[a]Address correspondence to: Albert Sjoerdsma, M.D., Ph.D., Merrell Dow Research Institute, 2110 E. Galbraith Road, Cincinnati, Ohio 45215-6300.

possible by development of new analytical methods, delineation of pathways for synthesis and degradation and elaboration of drugs interacting with serotonin-related processes, especially enzymes and receptors.[1,5–9] One of us (AS) was among the first few scientists at the NIH to work on serotonin, initially because its known pressor action in man suggested possible involvement in hypertension and new approaches to development of antihypertensive agents. This interest soon evolved into an extensive and lengthy collaboration with Udenfriend who was then responsible for much of our knowledge of the biochemistry of serotonin.[5] By 1954 he had developed sensitive methods of chemical assay for various 5-hydroxyindoles, which were readily applied to numerous studies of serotonin metabolism *in vitro* and *in vivo*. One of the first clinical applications and one which exemplified the excitement which was beginning to surround serotonin, was discovery of the carcinoid syndrome and the possible role of serotonin in its pathophysiology. We had been tipped off by Erspamer that carcinoid tumors might produce excess enteramine (serotonin) because they were derived from the enterochromaffin cells of the gut mucosa. We immediately began to search for patients with such tumors in order to check their urine for elevated excretion of the acid metabolite of serotonin, 5-hydroxyindoleacetic acid (5-HIAA). Shortly (summer, 1954), we were referred the first U.S. case (38 y/o woman) of carcinoid syndrome by Mattingly and established the diagnosis by a simple color test for excess 5-hydroxyindoles in urine, the still preferred method of diagnosis. Wooley and Shaw and others had begun to talk about a central role for serotonin in mental illness at this time, chiefly because of psychotomimetic activity of serotonin analogues and of lysergic acid diethylamide (LSD), a serotonin antagonist on smooth muscle. Serotonin was not thought to readily enter brain, but the very aberrant mental state and behavior of our first carcinoid patient, along with production by her tumor of several hundred mgs of serotonin per 24 hours, was suggestive of a serotonin-mediated mental disturbance. In many subsequent patients, no such association appeared. Furthermore, LSD was shown to produce its full psychologic response in two carcinoid patients at a normal dose.

This early period was also marked by the finding by NIH colleagues working with Brodie that reserpine depleted serotonin in brain whereas monoamine oxidase inhibitors (MAOI) increased serotonin content. Concomitant effects of these and other drugs on the catecholaminergic and serotonergic systems in the brain sparked enormous activity around the world as well as excitement and endless debate as to the roles of the two systems in mental function and in mechanisms of drug action. Toward the end of the era, *p*-chlorophenylalanine (PCPA, fenclonine) was elaborated by Pfizer scientists as a means of inhibiting serotonin biosynthesis at the tryptophan hydroxylase step.[8]

Throughout the late 1950s and 1960s, patients with carcinoid syndrome served as valuable models for study of the biochemistry and pharmacology of excess circulating serotonin in man.[7,8] Neither hypertension nor migraine (nor mental illness, as noted) was associated with the condition. Drugs studied in these patients included new ergot alkaloids, serotonin receptor blockers, inhibitors of aromatic L-amino acid decarboxylase (AADC) and tryptophan hydroxylase. In an unusual series of studies with PCPA,[8] it was shown that PCPA treatment of 16 carcinoid patients generally reduced 5-HIAA excretion by 70–80 per cent, improved diarrhea but not flushing, caused a reduction in REM sleep which could be reversed by DL-5-hydroxytryptophan (5-HTP), and produced symptoms in a subset of patients consistent with a role for serotonin in mental depression. At the same time, PCPA was shown by Gessa to increase mounting behavior in male rats and rabbits, considered by some to be an aphrodisiac effect of sorts.

The richness of this era is amply illustrated by the number of new drug entities appearing, related to serotonin (TABLE 1). A more detailed description of many of these drugs, as well as those interacting with other monoaminergic transmitter systems, was presented by Pletscher in his remarks at the dedication of our Strasbourg Research Center in 1973. A written version[9] appeared as the "Keynote Address" in the Proceedings of The Third International Catecholamine Symposium held in Strasbourg, France, May 20–25, 1973.

TABLE 1. Serotonin Drugs of the Renaissance Era (1953–1970)

Drugs	Actions	Uses
1. Reserpine (Serpasil)	Release/depletion	Hypertension, psychosis
2. MAO Inhibitors	Increase amine content	
a. Pargyline[a] (Eutonyl)		a. Hypertension
b. Phenelzine (Nardil)		b. Depression
b. Tranylcypromine (Parnate)		
b. Isocarboxazid (Marplan)		
3. Ergots	Receptor antagonism	
a. LSD		Investigational, recreational
b. Methysergide (Sansert)		Migraine, carcinoid
4. Cyproheptadine (Periactin)	Receptor antagonism	Eating disorders, allergy, carcinoid
5. Methyldopa[a] (Aldomet)	Decarboxylase inhibition	Hypertension, carcinoid
6. Carbidopa[a] (+ L-dopa = Sinemet)	Decarboxylase inhibition	Parkinson's disease
Benserazide (+ L-dopa = Madopar)	Decarboxylase inhibition	Parkinson's disease
7. Parachlorophenylalanine[a] (fenclonine)	Tryptophan hydroxylase inhibition	Carcinoid, investigational
8. Tryptophan[a] (Trofan)	Precursor, mechanism unclear	Sleep
9. Tricyclic Antidepressants Imipramine (Tofranil) Desipramine (Norpramin) Amitriptyline (Elavil) Nortriptyline (Aventyl) Protriptyline (Vivactil)	5-HT and/or norepinephrine uptake inhibition	Depression
10. Metyrosine[a] (Demser)	Tyrosine hydroxylase inhibition	Pheochromocytoma, investigational

[a]First or early clinical studies performed in Experimental Therapeutics Branch of the National Heart Institute.

ERA III: THE DROUGHT (1970 TO 1979)

In the early 1970s, one of us (AS) became disenchanted with the importance of serotonin in the cardiovascular system and moved on to develop a major interest in polyamines. Meanwhile, the other author (MGP), having recently graduated in the U.K., became interested in the evolving European movement aimed at establishing a role for serotonin in depression.[10] Although this period was relatively barren in serotonin-related drug discoveries, it was shaped by three important observations: 1) several authors suggested that the serotonin precursors L-tryptophan and 5-hydroxytryptophan had antidepressant effects, whereas, the norepinephrine precursor L-3,4-

dihydroxyphenylalanine (L-dopa) did not;[11] 2) Shopsin and his colleagues [12,13] showed that PCPA but not α-methyl-p-tyrosine reversed the antidepressant effects of imipramine and tranylcypromine and perceptively commented, "Implications are that serotonergic mechanisms are likely involved in the antidepressant effects of imipramine in man," and finally; 3) a group of Scandinavian scientists inspired by Carlsson and Corrodi[14] developed selective inhibitors of serotonin uptake and showed them to be effective antidepressants. It is noteworthy that this era was marked by an apparent dichotomy between Europe and the United States. European scientists were strong advocates of a pivotal role for serotonin in depression, whereas, most in the U.S. supported a catecholamine hypothesis. As an aside, zemelidine[15] and fluoxetine[16] were discovered at almost the same time in Europe and the U.S., respectively, and yet the former was recognized to be therapeutically effective several years before the latter. The lack of effect of 5-HT uptake inhibitors on the widely used reserpine-induced ptosis and hypothermia models for antidepressants no doubt contributed to scientific and managerial skepticism. This third era was finally closed out by the application of receptor-binding techniques to the question of 5-HT receptor diversity.

ERA IV: THE NEW WAVE (1979 TO 1989)

A classic paper by Peroutka and Snyder[17] defining "multiple" 5-HT receptors ushered in the modern era. Their classification of 5-HT_1 and 5-HT_2 receptors, as defined by radioligand binding in brain membranes, did not correspond to the "renaissance" classification of "M" and "D" tryptamine receptors by Gaddum and Picarelli,[18] defined by the actions of 5-HT antagonists on guinea pig ileum. Subsequently, the brain 5-HT_3 receptor(s) was defined following upon the development of potent radiolabeled 5-HT_3 antagonists. Anecdotally, for those who find it useful to do so, "M" may be thought of as 5-HT_3 and "D" as 5-HT_2, but perhaps not, and maybe not for long.

Much of the work to discover potent 5-HT_3 antagonists was initiated by Fozard, first at the University of Manchester and then at the Merrell Dow Research Institute in Strasbourg. By marrying the 5-HT_3 antagonist properties of cocaine and metoclopramide into one molecule, the medicinal chemist, Gittos, and Fozard, produced the first potent and selective 5-HT_3 antagonist, MDL 72,222.[19] The early 1980s were also characterized by further subdivision of the 5-HT_1 receptor of Peroutka and Snyder into subtypes A through D, to which Fozard, Middlemiss, Tricklebank and Hibert at our Strasbourg center contributed much.[20] The fruits of their labor, and that of Carr, Dudley, Freedman, Ursillo and Palfreyman of our Cincinnati research center, are exhibited in FIGURE 1. The early '80s also saw the development of the "selective" 5-HT_2 antagonist, ketanserin, and renewal of the theses that 5-HT was involved in blood pressure control and migraine.[21,22]

In parallel with the receptor work, a long standing goal of AS to develop a brain selective MAOI was achieved. This compound, MDL 72,394 (FIG. 1), is a tyrosine analogue which makes use of the amino acid transporter that carries tyrosine and tryptophan into the brain. Therein, the amino acid analogue is decarboxylated by AADC to yield the potent irreversible inhibitor of MAO-A, (E)-β-fluoromethylene-m-tyramine.[23] The 5-OH-indole analogue, (E)-β-fluoromethylene-5-hydroxytryptophan, MDL 73,095, has recently been synthesized and also promises to have interesting central properties.[24] (FIG. 1).

Consequent upon greater understanding of 5-HT receptor subtypes, a plethora of new and unique drugs have been designed or identified to have 5-HT receptor selectivity. These new compounds, several of which are now in clinical trials or already marketed, underscore the importance of serotonin and of 5-HT receptor subtypes in a

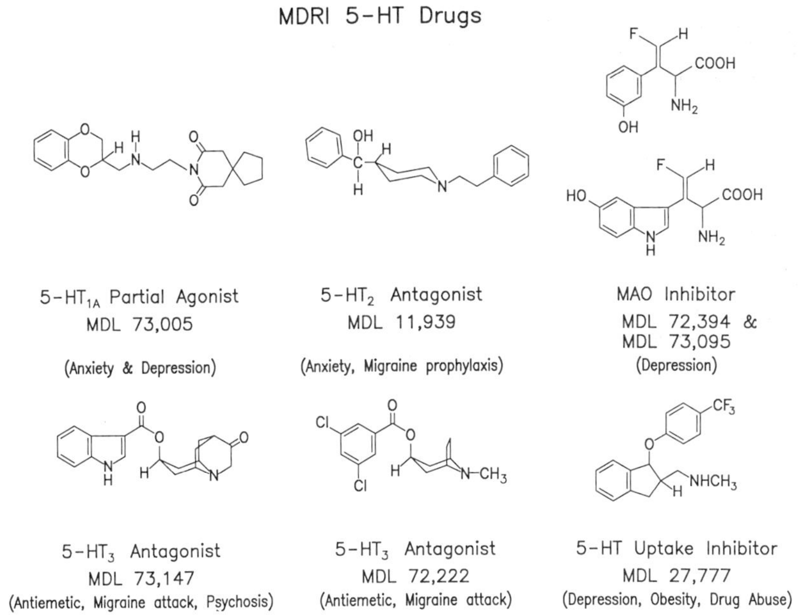

FIGURE 1. Structures, types of action and proven/probable clinical indications for Merrell Dow serotonin drugs.

variety of central nervous system disorders (TABLE 2). As the serotonin drugs have evolved, a number of simple tests have been established which permit an early assessment of drug effect in man. Several of these tests and drug classes to which they apply are shown in TABLE 3. Some of those listed were elaborated during the early renaissance period and still stand in good stead (e.g. platelet 5-HT and MAO).

TABLE 2. Serotonin Targets, Disorders, and Drugs

Targets	Disorders	Drugs
5-HT uptake site	Depression, obesity, substance abuse	Fluoxetine, fluvoxamine sertraline
5-HT_{1A}	Anxiety, depression	Buspirone, ipsapirone, gepirone
5-HT_{1B}	Only found in rodents	
5-HT_{1C}	Migraine prophylaxis ?	
5-HT_{1D}	Migraine attack	Sumatriptan
5-HT_2	Anxiety, sleep, schizophrenia, migraine prophylaxis, hypertension ?	Ketanserin, ritanserin
5-HT_3	Chemotherapy-induced vomiting, migraine attack, anxiety, schizophrenia, carcinoid syndrome[a]	MDL 72222, granisetron, odansetron ICS 205–930, zacopride
MAO-A	Depression	MDL 72394

[a]Somatostatin analogue, SMS 201-995 (octreotide), very effective against flushing and diarrhea.

If we have anything to learn from the history of serotonin, it is surely that this substance is of fundamental biological significance. It is a compound which has refused to die and nowhere is this more apparent than in its roles in "deep-seated" functions (e.g. affect, appetite, sex, sleep) of the nervous system, alterations thereof, and drugs which impinge thereon. Now that the molecular biologist has also entered the 5-HT receptor arena with his/her bag of tricks, and indeed already identified and sequenced three of the 5-HT receptors,[25] even more surprises in the area of neuropharmacology can be anticipated.

A final clue may be suggested from the striking efficacy of the stable somatostatin analogue, octreotide (Sandostatin), in treatment (reversal and control) of the humoral manifestation of the carcinoid syndrome, especially cutaneous flushing (hypotension), bronchoconstriction, and diarrhea.[26,27] This compound has supplanted serotonin antagonists and apparently acts by preventing release from the tumor of vasodilator peptides of the tachykinin family, plus other mediators including serotonin. Because of the often coincident distribution of serotonin and tachykinin peptides (e.g. substance P) in brain and spinal cord, studies of serotonergic-peptidergic interactions centrally and peripherally should provide new additions to the serotonin saga.

TABLE 3. Simple and Useful Serotonin Tests

Test	Indication
↑ 5-H1AA in urine	Carcinoid, eating bananas
↓ 5-H1AA in urine	MAOI, PCPA
↓ MAO in platelet	MAO-B inhibitors
↓ 5-HT content of platelet	5-HT uptake inhibitors
↓ [^{3}H]5-HT uptake into platelet	5-HT uptake inhibitors
↑ in serum prolactin	5-HT$_{1A}$ partial agonist ?
↓ 5-HT enhanced platelet aggregation	5-HT$_2$ antagonists
↓ 5-HT induced skin flare	5-HT$_3$ antagonists

REFERENCES

1. Page, I. H. 1968. Serotonin. Chicago: Year Book Medical Publishers, Inc.
2. Page, I. H. 1976. The discovery of serotonin. Perspect. Biol. Med. **20:** 1–8.
3. Erspamer, V. 1963. 5-Hydroxytryptamine. *In* Comparative Endocrinology. U. S. von Euler and H. Heller, Eds. Vol. **2:** 159–181.
4. Twarog, B. M. 1988. Serotonin: History of a discovery. Comp. Biochem. Physiol. **91:** 21–24.
5. Udenfriend, S. 1959. Biochemistry of serotonin and other indoleamines. Vitam. Horm. **17:** 133–151.
6. Garratini, S. & L. Valzelli. 1965. Serotonin. New York: Elsevier Publishing Co.
7. Sjoerdsma, A. 1959. Medical Progress—Serotonin. New Engl. J. Med. **261:** 181–188, 231–237.
8. Sjoerdsma, A., W. Lovenberg, K. Engelman, W. T. Carpenter, R. J. Wyatt & G. L. Gessa. 1970. Serotonin now: Clinical implications of inhibiting its synthesis with para-chlorophenylalanine. Ann. Int. Med. **73:** 607–629.
9. Pletscher, A. 1973. The impact of monoamine research on drug development. *In* Frontiers in Catecholamine Research. E. Usdin and S. H. Snyder, Eds. Oxford: Pergamon Press, pp. 27–37.
10. Clark, J. A., M. S. G. Clark, E. S. Palfreyman & M. G. Palfreyman. 1975. The effect of tryptophan and a tryptophan/5-hydroxytryptophan combination on indoles in the brain of rats fed a tryptophan deficient diet. Psychopharmacology **45:** 183–188.
11. Quadbeck, M., E. Lehmann & J. Tegeler. 1984. Comparison of the antidepressant action

of tryptophan, tryptophan/5-hydroxytryptophan combination and nomifensine. Neuropsychobiology **11:** 111–115.
12. SHOPSIN, B., S. GERSHON, M. GOLDSTEIN, E. FRIEDMAN & S. WILK. 1975. Use of synthesis inhibitors in defining a role for biogenic amines during imipramine treatment in depressed patients. Psychopharmacol. Commun. **1:** 239–249.
13. SHOPSIN, B., E. FRIEDMAN & S. GERSHON. 1976. Parachlorophenylalanine reversal of tranylcypromine effects in depressed patients. Arch. Gen. Psychiat. **33:** 811–819.
14. CARLSSON, A., H. CORRODI, K. FUXE & T. HOKFELT. 1969. Effect of antidepressant drugs on the depletion of intraneuronal brain 5-hydroxytryptamine stores caused by 4-methyl-α-ethyl-meta-tyramine. Eur. J. Pharmacol. **5:** 357–366.
15. ROSS, S. B., S. JANSA, L. WETTERBERG, B. FYRO & B. HELLNER. 1976. Decreased blood level of 5-hydroxytryptamine by inhibitors of the membranal 5-hydroxytryptamine uptake. Life Sci. **19:** 205–210.
16. WONG, D. T., J. S. HORNG, F. P. BYMASTER, K. L. HAUSER & B. B. MOLLOY. 1974. A selective inhibitor of serotonin uptake: Lilly 110140, 3-(*p*-trifluoromethylphenoxy)-*N*-methyl-3-phenylpropylamine. Life Sci. **15:** 471–479.
17. PEROUTKA, S. J. & S. H. SNYDER. 1979. Multiple serotonin receptors: Differential binding of [^{3}H]5-hydroxytryptamine, [^{3}H]lysergic acid diethylamide and [^{3}H]spiroperidol. Mol. Pharmacol. **16:** 687–699.
18. GADDUM, J. H. & Z. P. PICARELLI. 1957. Two kinds of tryptamine receptors. Br. J. Pharmacol. **12:** 323–328.
19. FOZARD, J. R. 1984. MDL 72,222: A potent and highly selective antagonist at neuronal 5-hydroxytryptamine receptors. Naunyn-Schmiedeberg's Arch. Pharmacol. **326:** 36–44.
20. FOZARD, J. P. 1987. 5-HT: The enigma variations. Trends Pharmacol. Sci. **8:** 501–506.
21. JANSSEN, P. A. J. 1983. 5-HT_2 receptor blockade to study serotonin-induced pathology. Trends Pharmacol. Sci. **4:** 198–206.
22. HOUSTON, D. S. & P. M. VANHOUTTE. 1986. Serotonin and the vascular system. Role in health and diseases, and implications for therapy. Drugs. **31:** 149–163.
23. PALFREYMAN, M. G., I. A. MCDONALD, P. BEY, P. J. SCHECHTER & A. SJOERDSMA. 1988. Design and early clinical evaluation of selective inhibitors of monoamine oxidase. Prog. Neuropsychopharm. & Biol. Psychiatry **12:** 967–987.
24. PALFREYMAN, M. G., M. ZREIKA & I. A. MCDONALD. 1989. Tyrosine and tryptophan analogues as dual enzyme-activated inhibitors of monoamine oxidase. *In* Enzymes as Targets for Drug Design. M. G. Palfreyman, P. P. McCann, W. M. Lovenberg, J. G. Temple, Jr. and A. Sjoerdsma, Eds. New York: Academic Press, pp. 139–156.
25. HARTIG, P. R. 1989. Molecular biology of 5-HT receptors. Trends Pharmacol. Sci. **10:** 64–69.
26. KVOLS, L. K., C. G. MOERTEL, M. J. O'CONNELL, A. J. SCHUTT, J. RUBIN & R. G. HAHN. 1986. Treatment of the malignant carcinoid syndrome. Evaluation of a long-acting somatostatin analogue. New Engl. J. Med. **315:** 663–666.
27. OATES, J. A. 1986. The carcinoid syndrome. New Engl. J. Med. **315:** 702–704.

DISCUSSION OF THE PAPER

LASZLO GYERMEK *(UCLA)*. I would like to congratulate Dr. Sjoerdsma on a very lucid and comprehensive summary of the history of serotonin, but I would like to also point to two, I think, significant cornerstones of serotonin research which developed in the so called renaissance era and of course many of us know about this. First there was the discovery of Gaddum and Picarelli in 1957 which was the first physiological/pharmacological characterization of different possible serotonin receptor sites by using the unspecific agents, morphine and dibenzyline, to differentiate between so called M and D receptor sites.

The other discovery came from our laboratory in the early 1960s which described for the first time antagonists to serotonin—some indol compounds, indol acetamidines and alkylated compounds of serotonin. So there was evidence in the middle 1950s and early 1960s of different receptor sites.

SJOERDSMA: We have included much of what you said in our manuscript, and cite specifically the classic work of Gaddum and Picarelli.

MANFRED GÖTHERT *(University of Bonn):* You didn't mention the serotonin neurotoxins 5,7 DHT and 5,6 DHT, and I think the names of Baumgarten and Björklund should be mentioned in this context. I think these compounds actually played a major role in research into the anatomy of the 5-HT system and also for physiology and behavioral studies. I think the New York Academy of Sciences has devoted a whole conference to this topic [Ann. N.Y. Acad. Sci. **305**].

SJOERDSMA: Well, I have to admit that the last time I was here was in 1963, so perhaps you ought to give the history of serotonin next year.

Anatomy of the Serotonergic System[a]

ISTVAN TÖRK

School of Anatomy
University of New South Wales
Kensington, NSW 2033, Sydney, Australia

INTRODUCTION

The serotonergic (5-hydroxytryptaminergic, 5-HT) system is one of the diffusively organized projection systems of the brain. Although the presence of serotonin itself was demonstrated in the brain more than thirty years ago[1,2] the full description of its neuronal localization and projections has only become possible with the availability of the histofluorescence[3,4] and immunohistochemical[5] techniques. Indeed it was only by the use of antibodies against serotonin that the detailed anatomy and topography of the serotonergic cells and their projections throughout the brain could be described. Therefore, the description of the 5-HT system below is based entirely on immunohistochemical data although some reference is made to other anatomical observations based on traditional neurohistological and histochemical techniques.

The 5-HT system consists of a morphologically diverse group of neurons the cell bodies of which are located in the brainstem raphe nuclei and some regions of the reticular formation, and complex axonal systems which innervate virtually all regions of the central nervous system, but with particular density the cerebral cortex, limbic structures, basal ganglia, many regions of the brainstem and the gray matter of the spinal cord.[5–10] This review discusses, from a structural as well as hodological viewpoint, the organizational principles of the 5-HT system. Since several detailed reviews are available on the serotonergic system, the repetition of all data is largely avoided. Rather, a more salient description of the general features of the brainstem 5-HT neurons and their terminal fields is attempted.

All investigators of the mammalian brain have agreed on the finding that the 5-HT system consists of two distinct subdivisions: a rostral division, with cell bodies localized in the midbrain and rostral pons and providing projections to the forebrain, and a caudal division, located primarily in the medulla oblongata, with major descending projections directed principally to the spinal cord.[11,12] Target areas within the brainstem itself and in the cerebellum are shared by these two divisions. The description given below is in part based on original experimental material (with particular reference to the human brain) prepared to illustrate the general morphology of the serotonergic neurons. For this reason, a brief description of the methodology used in our laboratory for the demonstration of the 5-HT neurons is given below.

EXPERIMENTAL PROCEDURES

For the study of the 5-HT system we have used immunohistochemical methods, in conjunction with two antibodies: (i) a commercial monoclonal antibody against

[a]This research was supported by grants from the National Health and Medical Research Council, the National Heart Foundation and the Asthma Foundation of New South Wales.

5-HT/bovine serum albumin conjugate (by Sera-Lab, UK)[13] and (ii) an antibody raised against monkey phenylalanine hydroxylase which in the brain demonstrates tryptophan hydroxylase (PH8[14,15]). The latter antibody is particularly useful in the study of the serotonergic system in post-mortem specimens of the human brain, since serotonin itself is eliminated from the cells and terminals within 30 minutes after death[16] while its rate-limiting biosynthetic enzyme, tryptophan hydroxylase, being a protein, remains within the cells for longer periods of time. In our experiments the serotonergic structures were demonstrated with the application of the biotin–avidin (ABC) method and the detailed technology is described elsewhere.[17–19] Briefly, in animal experiments, deep pentobarbital anesthesia was applied and the brain was fixed by perfusion through the heart with phosphate buffered saline (0.05 M phosphate buffer, pH7.4 in 0.9% NaCl, PBS) followed by 4% paraformaldehyde in 0.1M phosphate buffer, and the tissues were postfixed by immersion overnight. Human post-mortem tissue was fixed in buffered formaldehyde (4%) by immersion at the time of autopsy, and the blocks remained in the fixative for several months. Before sectioning the tissue blocks were washed in PBS for 1–2 days. Sections were obtained either with a Vibratome or by using a freezing microtome. Tissues which were cut on the freezing microtome (most human tissue) were presoaked in 30% sucrose in phosphate buffer for up to 7 days for cryoprotection.

Free-floating tissue sections, 50 μm thick, were preincubated with 1–5% normal serum of the host animal of the secondary antibody, followed by incubation in the primary antiserum for 2–7 days. If long incubation periods were used, 0.1% sodium azide was added to the primary antiserum. The monoclonal anti-serotonin antibody was diluted 1:250, while the PH8 antibody was diluted 1:1500. The secondary antibody was applied at a dilution of 1:100, for 1 h, followed by the ABC reagent, at a dilution of 1:80. All antibody solutions were made with PBS containing 0.1% Triton X-100, and the same PBS–Triton solution was used for washing sections between each reagent for 3 × 10 minutes. For electron microscopic examinations 0.2% glutaraldehyde was added to the perfusion fixative but not to the postfixing solution and the Triton X-100 was omitted from the incubation solutions. The peroxidase label in the ABC reagent was demonstrated by incubation in 0.05% diaminobenzidine (DAB) solution in 0.1M Tris–HCl buffer containing 0.01% hydrogen peroxide. Preceding the immunohistochemical reaction, human tissue sections were pretreated with 3% hydrogen peroxide in 50% methanol, to eliminate endogeneous peroxidase activity, primarily present in red blood cells.

GENERAL TOPOGRAPHY AND MORPHOLOGY OF SEROTONERGIC CELL BODIES IN THE BRAIN

The serotonin-containing neurons form a morphologically heterogeneous population of cells located in the brainstem.[7–9,19–24] The serotonergic cells throughout the brainstem are multipolar cells, however their size and orientation is remarkably different in different locations. While the cells located in the midline regions of the raphe nuclei are small to medium-sized, and are oriented with most of their dendrites along the mid-sagittal plane, cells that are found more distant from the midline send dendrites in all directions. Also, the midline cells usually exhibit only 2–3 dendrites, and in fact many cells appear bipolar. The strong alignment of the serotonergic cells in the midline is a characteristic feature of at least some of the cells in all raphe nuclei, but it is particularly striking in the case of the caudal linear nucleus and in the rostral part of the median raphe nucleus. The largest serotonergic cells are found in the lateral parts of the periaqueductal grey matter, in the reticular formation: in general in the peripheral zones of the serotonergic nuclei.

The cell bodies of most serotonergic neurons are found largely within the boundaries of the raphe nuclei of the brainstem. Because of their prominence, high serotonin content, and extensive projections these nuclei attracted some interest in the past and their topography has been the subject of several histological and histochemical studies.[7–10,19,20] The terminology to be used here follows that of recent reports of the raphe system in the rat,[18] primates[9,10] and in the human[19,21–23,25] and recent works on the organization of the brainstem.[26–28] Apart from the raphe nuclei serotonergic neurons are contained within regions of the reticular formation. Two areas are particularly prominent. In the rostral serotonergic system there are numerous cells located in the nucleus pontis oralis and in a region adjacent to the medial lemniscus (supralemniscal serotonergic cell group, or B9 cell cluster). In the medulla oblongata serotonergic cells are found near the ventrolateral surface lateral to the inferior olive. These cells are located within regions usually referred to as the rostral and caudal ventrolateral medulla (RVL and CVL, respectively).

The topography of the raphe nuclei and other serotonin containing cell groups located in the reticular formation are schematically depicted in FIGURE 1.

At the time of the discovery of the serotonergic neuronal cell groups in the brainstem, Dahlström and Fuxe[3] introduced a code system for the serotonergic cell groups found in the rat brainstem. According to this system the serotonergic cell clusters were coded B1–B9, B1 being the most caudal cell cluster. The B serotonergic cell groups mostly overlap with the cell masses of the raphe nuclei but there are some exceptions, i.e., some cytoarchitectonic entities belong to more than one B cluster or conversely one B cluster may cover more than one nucleus or only vaguely identifiable cytoarchitectonic entities. TABLE 1 presents the correlation between the B groups and the nuclear cell groups of the brainstem, representing the most frequently used terminology at the present time.

The 5-HT cells of the raphe express their neurotransmitter phenotype early in development[11,12,29]: in the rat the first 5-HT containing cells appear at E(embryonic day)12, and additional serotonergic neurons appear for at least 3 more days. Already at the time of early development the 5-HT system in the brain demonstrates its remarkable polarity. The 5-HT cells appear in two distinct groups, a caudal one in the medulla oblongata and in the caudal half of the pons and a rostral one in the midbrain and rostral pons. Immediately after their first appearance, the 5-HT cells develop axons which from the rostral group ascend towards the forebrain and from the caudal group descend to structures of the brainstem and to the spinal cord (FIG. 2). This high degree of polarity of the serotonergic projections is unique to the serotonergic system and is retained throughout life.

THE ROSTRAL SEROTONERGIC SYSTEM

The rostral part of the serotonergic raphe system consists of the caudal linear nucleus located almost entirely in the midbrain, the dorsal raphe nucleus (which in the human has also been termed the nucleus supratrochlearis[30]), extending from the caudal midbrain to rostral pontine levels, and the median raphe nucleus (also called the superior central raphe nucleus) which is found largely in the rostral pons. In addition, the rostral system contains the extensive B9 group and in some species, particularly in primates, a large dispersed group of cells located in the nucleus pontis oralis.[9,10,19–21]

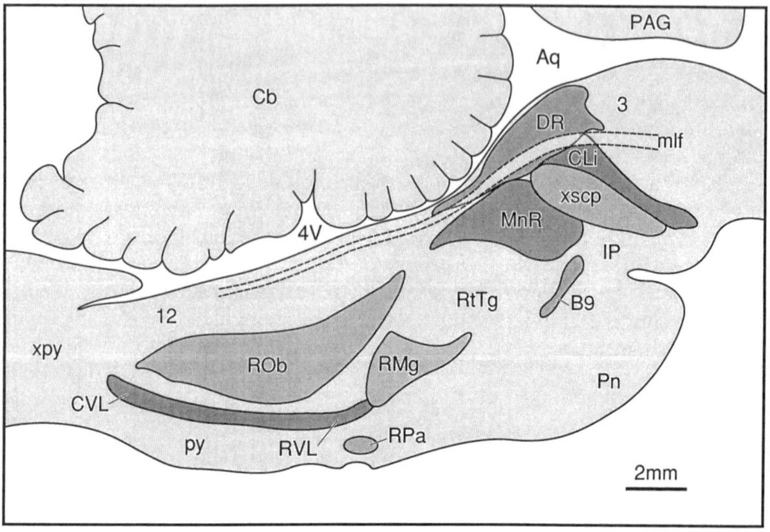

FIGURE 1. Schematic representation of the serotonergic cell groups in the raphe nuclei and in the reticular formation, as they would appear on a sagittal projection through the cat brainstem. The rostral (on the right) and caudal divisions of the serotonergic system appear almost contiguous, but really clearly separate. Note the close relationship of the caudal linear nucleus (CLi), dorsal raphe nucleus (DR) and the median raphe nucleus (MnR). The supralemniscal serotonergic cell group (B9) group, located in the reticular formation, appears at the same level as the MnR. Some scattered serotonergic cells present in the rostral pontine reticular formation are not illustrated. In the medulla, the most extensive nucleus is the raphe obscurus (ROb) preceded rostrally by the raphe magnus (RMg) nucleus, at the junction to the pons. The small raphe pallidus nucleus (RPa) is located near the ventral surface of the medulla between the two pyramidal tracts. Near the lateral surface of the medulla the reticular formation contains many serotonergic cells in the regions described as the rostral and caudal ventrolateral medulla (RVL and CVL). 3: oculomotor nucleus, 12: hypoglossal nucleus, 4V: 4th ventricle, Aq: cerebral aqueduct, Cb: cerebellum, IP: interpeduncular nucleus, mlf: medial longitudinal fasciculus, PAG: periaqueductal grey matter, Pn: pontine nuclei, py: pyramid, RtTg: reticular tegmental nucleus of pons, xpy: pyramidal decussation, xscp: decussation of superior cerebellar peduncles.

The Caudal Linear Nucleus (CLi)

This nucleus is relatively small and is located on the midline in the mesencephalon forming the rostral boundary of the decussation of the superior cerebellar peduncles. It is part of the ventromedial mesencephalic tegmentum, and contains a mixture of serotonergic and dopaminergic cells.[31] However, the serotonergic cells predominate in it and demonstrate a characteristic alignment with the midline. The serotonergic cells here are small, with round bodies and possess only two or three dendrites that extend in rostrocaudal directions. The caudal end of the nucleus is a contentious issue since dorsally the nucleus is closely related to the dorsal raphe nucleus the ventral part of which comes in close contact with the caudal linear through the narrow gap between the two medial longitudinal fasciculi. The ventrocaudal boundary of the nucleus is relatively sharply defined by the crossing fibers of the superior cerebellar peduncles, although a thin stream of serotonergic cells does appear in the midline within the decussation. In the human, for example, this thin streak of cells can be followed throughout the decussation to the rostral end of the median raphe nucleus.[19]

The Dorsal Raphe Nucleus (DR)

By contrast to the caudal linear nucleus, the dorsal raphe nucleus is the most prominent member of the brainstem serotonergic nuclei. It is located in the ventral part of the periaqueductal gray matter of the midbrain, but its caudal section extends well into the periventricular gray matter of the rostral pons. It demonstrates a highly characteristic bilateral symmetry, and in cross sections of the midbrain the appearance of the nucleus creates the outline of a fountain.[7–10,19,20,24,32] The neuronal population in the nucleus is quite diverse, ranging from small through medium sized to large cells. The smallest cells are found on or near the midline, while the largest ones are found primarily in the most lateral and dorsal regions of the nucleus.

The dorsal raphe nucleus is composed of several subregions[8,32,33] which can be distinguished not only because of their different cell densities and differential cell morphology but also because of their differential projections.[34–36] The following subnuclei can be distinguished in most species: interfascicular, ventral (ventromedial), ventrolateral (lateral), dorsal (dorsolateral and dorsomedial), and caudal.[8,9,21,33]

The interfascicular subnucleus is located between the two medial longitudinal fasciculi, and is made of small spindle-shaped cells. The density of neurons in this region of the dorsal raphe nucleus is high. Dorsally, but still close to the midline this cell group is replaced by the ventral group which is a somewhat more loosely organized cell cluster, made of bipolar and multipolar cells. The ventrolateral group, although present in the rat,[7,8] is most prominent in primates[9,10] and in the human.[19,20,25,32,33] The dorsal subgroup extends well into the lateral regions of the periaqueductal gray matter and contains the largest serotonergic neurons of the dorsal raphe nucleus (FIG. 3a). The caudal subnucleus forms the tail of the dorsal raphe nucleus (FIG. 4) extending as far caudally as the level of the motor nucleus of the trigeminal nerve. The serotonergic

TABLE 1. Anatomical Designation of the Serotonergic Cell Groups in the Raphe Nuclei and in the Brainstem Reticular Formation and the Corresponding Classification into the B Groups, According to Dahlström and Fuxe[3]

5-HT–Containing Cells in Structure	Designated as (or Anatomical Part of) 5-HT Cell Cluster
Raphe pallidus nucleus	B1
Raphe obscurus nucleus	B2
Raphe obscurus, dorsolateral part	B4
Raphe magnus nucleus	B3
Rostral ventrolateral medulla	B3
Lateral paragigantocellular reticular nucleus	B3
Caudal ventrolateral medulla	B1
Median raphe nucleus, caudal part	B5
Median raphe nucleus, rotral main part	B8
Dorsal raphe nucleus, caudal part	B6
Dorsal raphe nucleus principal, rostral part	B7
Caudal linear nucleus	B8
Nucleus pontis oralis	B8/B9
Supralemniscal region	B9

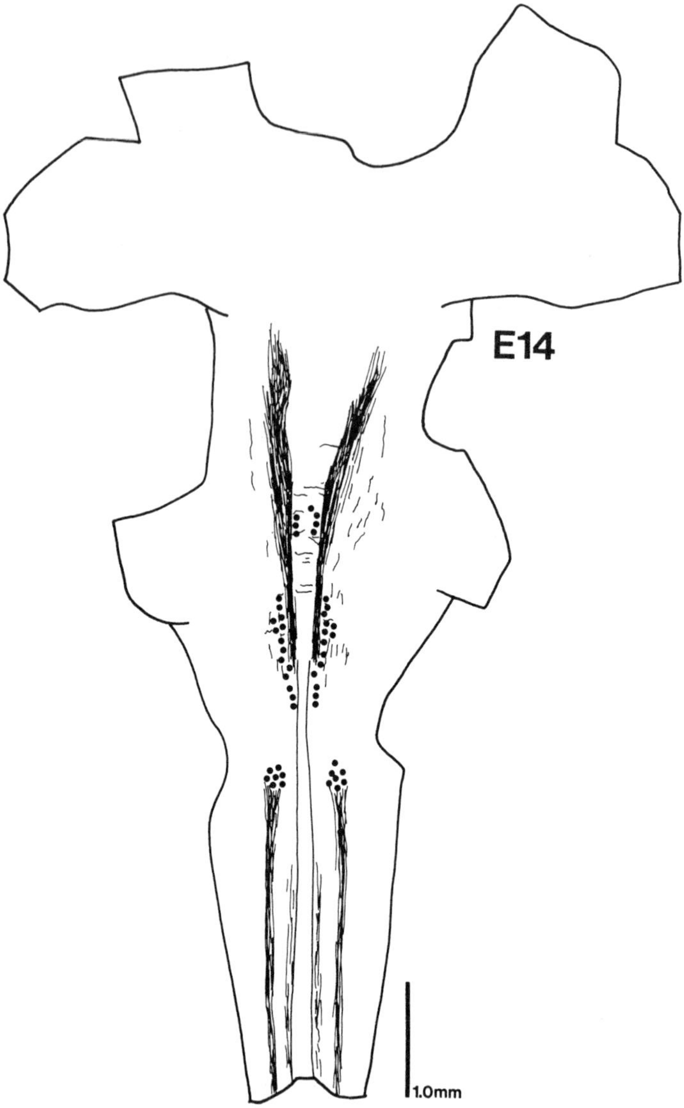

FIGURE 2. Camera lucida drawing of the rostral and caudal serotonergic system in the embryonic rat brainstem. The brain of the rat at embryonic day 14 (E14) was dissected and fixed as a wholemount preparation before immunohistochemistry for serotonin. The position of the two principal serotonergic cell groups is indicated by solid black circles, while the ascending and descending axons arising from them can be followed to the upper and lower limits of the brainstem. (From reference 12.)

cells in this subnucleus are arranged in two parallel streams of cells located just lateral to the midline in the floor of the fourth ventricle.

The number of serotonergic neurons is the largest in DR of all the raphe nuclei, 24,250 ± 1,600 in the cat,[6] and 165,000 ± 34,000 in the human.[25]

The Median Raphe Nucleus (MnR)

The serotonergic cells in the median raphe nucleus are arranged in two adjacent regions. On the midline there are numerous small cells, with relatively short dendrites, many of which are aligned parallel to the mid-sagittal plane. Outside the midline region there are many additional cells loosely arranged and showing no particular orientation. The lateral boundary of the nucleus is not sharply defined and many cells, particularly in primates extend into the adjacent pontine reticular formation. In primates and particularly in the human, a large number of serotonergic cells are found in the nucleus pontis oralis, in which they form a loose stream of cells extending dorsolaterally, just ventral to the central grey matter of the pons. The serotonergic cell cluster in the MnR forms the second largest group of serotonergic neurons in the brainstem and consists of 7,400 ± 780 in the cat,[6] and 64,300 PH8-immunoreactive neurons in the human.[21]

Serotonergic Neurons in the Pontomesencephalic Reticular Formation

In most species the reticular formation of the pons and midbrain contains a limited number of serotonergic neurons. Most of these cells are located adjacent to the medial lemniscus, forming a distinct layer of strongly immunoreactive cells which extend their dendrites in between the fiber bundles of the medial lemniscus. This supralemniscal serotonergic cell group has been originally termed group B9 and although in general these B codes are not used frequently today, the label B9 is commonly used. The position of the B9 cluster is slightly different in various species. In the rat and cat it is located largely in the midbrain while in primates it extends caudally into the pontine tegmentum. A recent description of the B9 cluster in the human indicated that it is located entirely in the pons and contained 28,980 serotonergic neurons.[21]

In addition to the supralemniscal cell cluster there are additional serotonergic cells in the pontine reticular formation. These cells, as mentioned above are in the dorsal parts of the nucleus pontis oralis. This group consists of large multipolar neurons, many of them bearing the characteristic features of neurons of the reticular formation, having an isodendritic dendritic pattern. Few data are available on this cell group separately, but in the human the group is very extensive and contains about 30,000 serotonergic neurons.[21]

THE CAUDAL SEROTONERGIC SYSTEM

The rostral limit of the caudal serotonergic cell clusters is not clearly defined. In most species the serotonergic cells extend into the pontine tegmentum but at midpontine levels there is always a region which is fully devoid of serotonergic neurons. During the early stages of development the two cell groups are distinctly separate from each other but due to their growth they approximate each other very closely. Nevertheless

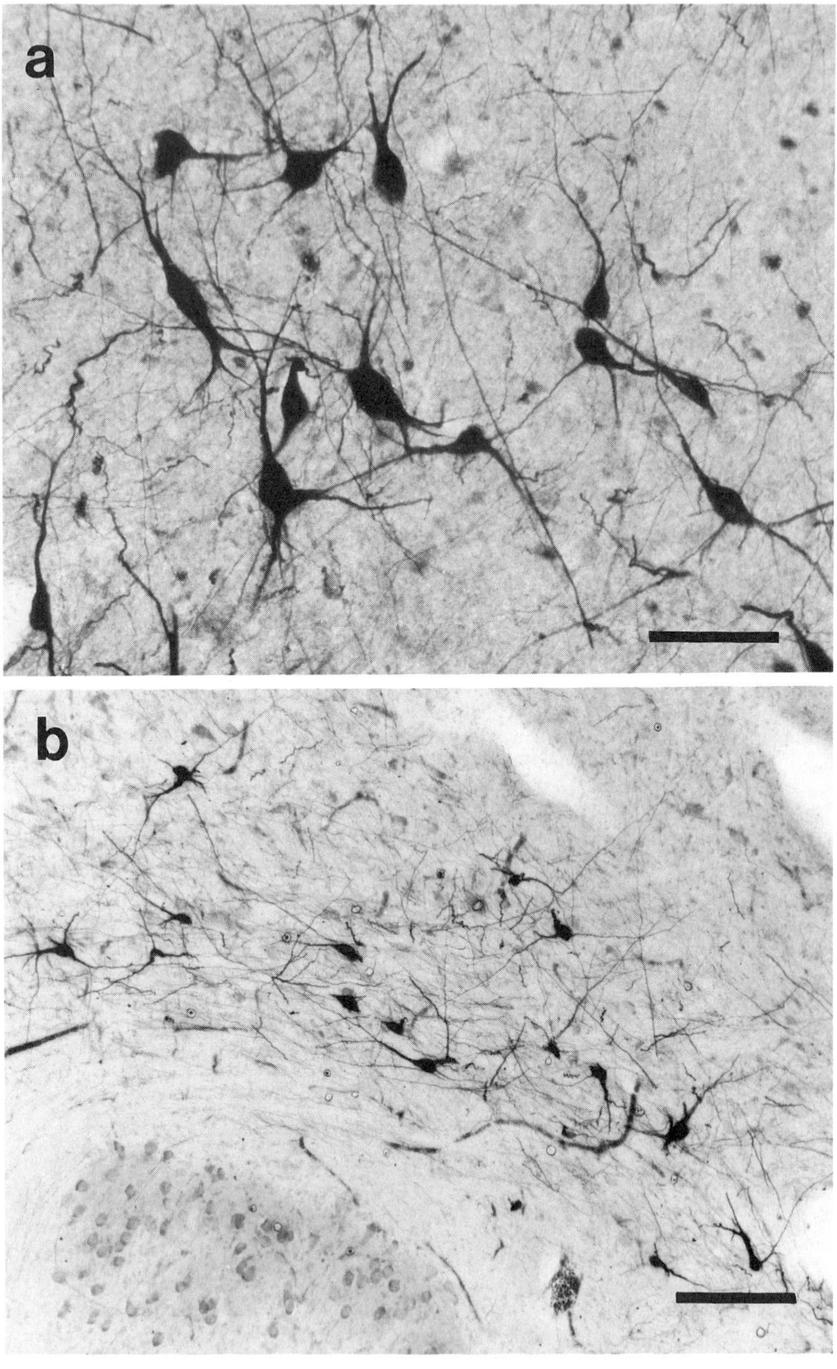
a
b

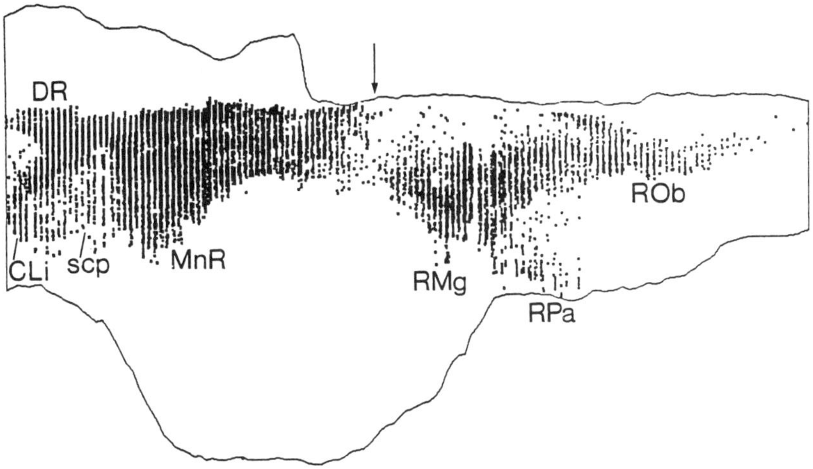

FIGURE 4. The raphe nuclei of the human brainstem on a computer-generated sagittal section. Rostral is to the left. The arrow on top illustrates the boundary between the rostral and caudal serotonergic cell groups. All abbreviations used for the different cell groups are the same as in FIGURE 1., scp: superior cerebellar peduncle. (Reproduced from reference 19.)

the differentiation of the two divisions of the serotonergic system is not difficult since the morphology of the cells found in them is remarkably different.

Raphe Magnus Nucleus (RMg)

This nucleus consists of large multipolar cells arranged on or near the midline at the junction of the pons and medulla. In transverse sections of the brainstem, the nucleus has a typical shape of an equilateral triangle, the apex of which points dorsally.[7,8,19] Most of the cells in this nucleus are dispersed among the fibers of the neighboring reticular formation and the most dorsal bundles of the medial lemniscus. Because of this prominent transverse orientation and the lateral extension of the dendrites the raphe magnus nucleus seems to extend laterally almost to the ventral surface of the medulla.[7–9] Nevertheless, the cells located laterally are not considered part of the raphe magnus nucleus. Rather they are counted to be part of the lateral paragigantocellular reticular nucleus (FIG. 3b), or even more laterally, the rostral ventrolateral medulla. This latter region is of considerable physiological interest because of its role in cardiovascular control functions.

In their cytoarchitectonic atlas of the human brainstem, Olszewski and Baxter[30] identified the region of the raphe magnus nucleus as the raphe pallidus nucleus.

FIGURE 3. Examples of serotonergic neurons, demonstrated in the human brain by immunocytochemistry of tryptophan hydroxylase. **a:** large multipolar serotonergic cells in the dorsolateral region of the dorsal raphe nucleus. These cells are among the largest serotonin-containing neurons in the brain; they have extensive dendrites. **b:** Serotonergic cells in the rostral lateral reticular formation of the medulla of a human infant located just dorsal to the inferior olive: the dendrites of these cells have a specific medio-lateral orientation. Scale bars: **a:** 100 μm, **b:** 200 μm.

However, our recent immunohistochemical investigations[19] and detailed studies on the human homologs of nuclei identified in animals[27] clearly identified this region of the brainstem as the human homolog of the raphe magnus nucleus identified in animals (FIG. 4).

Raphe Pallidus Nucleus (RPa)

A relatively small component of the medullary raphe system is the raphe pallidus nucleus. Typically the cells of this nucleus are closest to the ventral surface of the medulla and are located in between the pyramids found there. In many species the 5-HT cell of the pallidus continue uninterrupted dorsally and become contiguous with the cell cluster of the raphe obscurus nucleus. However, there are remarkable morphological differences: the cells in the raphe pallidus nucleus are slender and dorsoventrally oriented and are located very close or right on the midline; the nucleus itself rides on the midline without any significant interruption formed by the midline 'seam' of the brainstem. In the human, the raphe pallidus is only present in the rostral medulla,[19] but it is quite distinct from the raphe magnus and raphe obscurus nuclei.

Raphe Obscurus Nucleus (ROb)

The medullary 5-HT cell clusters are dominated by the cells located on or near the midline. Partially intermingled with the fibers of the medial lemniscus that in the medulla skirts the midline, the cells of the raphe obscurus form a continuous stream of cells extending from the caudal pons to the level of the pyramidal decussation (FIG. 4). The cells are closely aligned with the midline although not on it, and as a result they form two distinct sheets parallel to each other and to the midline. Most of the dendrites of the 5-HT cells are also aligned parallel with the midline, but very distinctly many of them send out processes at right angle to the mid-sagittal plane: these dendrites extend far laterally into the neighboring reticular formation.[6–10] A small number of serotonergic cells are also located dorsal and lateral to the medial longitudinal fasciculi and these cell clusters are considered to be part of the raphe obscurus nucleus.[29]

5-HT Neurons in the Medullary Reticular Formation and Other Non-raphe Regions

Beyond the lateral extensions of the raphe magnus nucleus there is a cluster of neurons extending over the lateral aspect of the inferior olivary nucleus, and many cells reaching the ventrolateral medullary surface.[7–10,19,23,24] These cells are dominant in the rostral parts of the medulla although numerous cells extend caudally and are present in this region along the whole length of the medulla oblongata. The lateral medullary serotonergic cell group is prominently present in the lateral paragigantocellular reticular nucleus,[19,22] a nucleus in the rostral half of the medulla. This serotonergic cell cluster is very well developed in primates including humans,[9,10,19,22] but is also well developed in sheep.[24]

Beyond the raphe nuclei and the reticular formation there is a limited number of serotonergic neurons within the solitary nuclear complex.[22,24,37] These neurons are mostly small, bipolar or spindle-shaped and seem to extend beyond the limits of the solitary complex into the area postrema and neighboring reticular formation. Their exact topography in relation to the functional subdivisions of the solitary nuclear complex is not yet known, although they seem to be concentrated in the medial part of the solitary nucleus.[22] In addition to the solitary complex there is a small number of 5-HT positive cells in the area postrema.[7,8,22]

OVERVIEW OF THE RAPHE NUCLEI

The serotonergic cells of the raphe nuclei are best seen in sagittal and parasagittal sections. However the majority of the cells in several raphe nuclei are not exactly on the midline, but rather slightly lateral to it (e.g., DR, ROb). Thus a single 50 μm section would not clearly illustrate the positions of the nuclei. Recently we have prepared a computer reconstruction of all the raphe nuclei[19] found in the human brainstem by plotting the serotonergic neurons from serial sections and entering them into a graphic computer. A midline "slab" of tissue has been created illustrating only the cells on or near the midline while more lateral cells have been ignored. The reconstruction (FIG. 4) clearly indicates the topography and relationships of the raphe nuclei and shows the higher density and larger extent of the serotonergic cell groups in the rostral raphe system in comparison to the caudal one. The computer reconstruction also shows the well defined boundary between the rostral and caudal parts of the serotonergic system.

PROJECTIONS OF BRAINSTEM SEROTONERGIC CELL GROUPS

Ascending Projections

The serotonergic projections arising from the nuclei described above are schematically illustrated in FIGURE 5. The principal target regions of the serotonergic projections are in the forebrain and in the spinal cord. Additional target regions located in the brainstem are not illustrated.

The ascending projections are very extensive and contain many collaterals innervating diverse regions of the cerebral cortex, basal ganglia, limbic system and diencephalon.[5,9,11,18,24,38–40] Initially the ascending projections travel in the medial forebrain bundle and many axons extend from this to different other fiber pathways which they follow to their target areas. All areas of the forebrain demonstrate a specific innervation pattern. In the cortex, the fiber density is different in the different layers and it is most prominent in layer I.[18,39,41–45] In primates the visual cortex receives a particularly dense innervation in layer IV.[42–44] In the striatum a very dense innervation can be observed.[5,46] The hypothalamus and septal area receive extremely strong serotonergic input and also some regions of the thalamus.[47]

Most authors studying the serotonergic innervation of the forebrain have observed that there were at least three types of serotonergic axons present in the forebrain.[18,48,49] The most common type is thin, with small (less than 1 μm in diameter) varicosities that are fusiform in shape (FIG. 6a). These axons are by far the most numerous and extend profusely over large areas of the forebrain. They also tend to branch frequently and the emerging branches have the same structure as the parent fibers. A second type of fiber is characterized by large round, or oval varicosities which stain very strongly with the immunocytochemical methods, and which are on very thin axons that themselves react rather weakly for serotonin (FIG. 6a, 8b). A third type of fiber, which is relatively thick (about 1 μm) and non-varicose, is also found in smaller numbers and usually in or close to the major tract systems from which serotonergic fibers emerge. Anterograde tracing studies using Phaseolus vulgaris lectin[49] have shown that the two principal types of axons found in the target regions of the brain have differential origins: the fine varicose

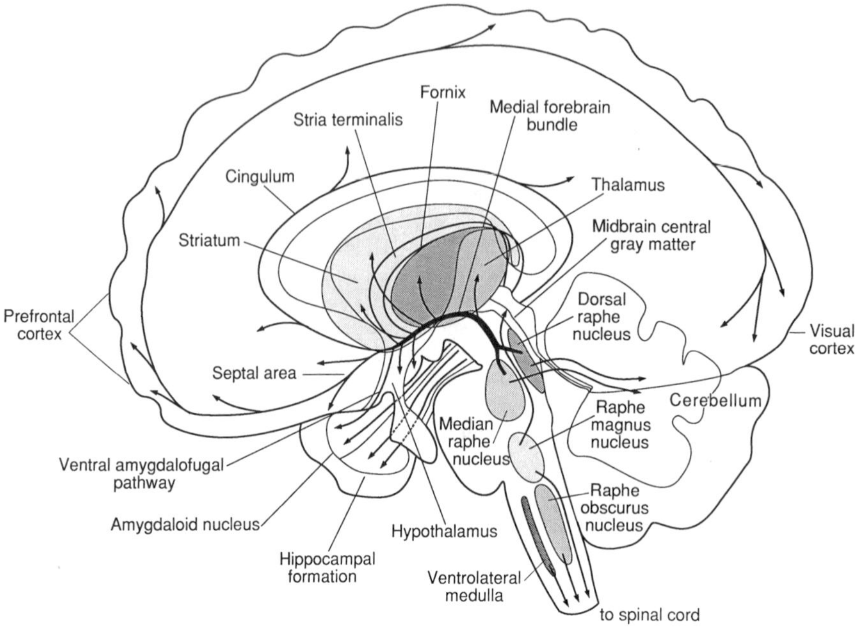

FIGURE 5. Schematic representation of the serotonergic projections of the brainstem serotonergic cell group to the forebrain, brainstem and spinal cord, projected onto the outline of the human brain. The ascending projections arise from the dorsal and median raphe nuclei and travel through the diencephalon in the median forebrain bundle. From there they enter several other fiber systems to reach their target regions, such as the limbic system, hypothalamus, striatum and cerebral cortex. The descending projections arise mainly from the raphe magnus and obscurus nuclei and also from the serotonergic cells located in the ventrolateral medulla. The projections of these nuclei are principally to the spinal cord. (Reproduced from reference 75.)

axons arise from the DR while the beaded axons arise from the MnR. In another study, in the cat, serial section analysis of the two fiber types has also indicated that the fine varicose axons and the beaded varicose axons form two independent fiber systems, although it has also been shown that the beaded axons arise from the thick non-varicose fibers.[18]

FIGURE 6. Light and electron microscopic appearance of serotonergic axons. In **a,** the two principal types of serotonergic axons have been demonstrated with serotonin immunocytochemistry. The small arrows indicate the fine varicose axons: these have small fusiform varicosities and are found in large numbers in all areas of the brain. In the center, (at hollow arrows) a beaded varicose axon is seen. The varicosities of this axon type are round or oval, and with a usual diameter of 3–4 μm. **b:** Electron micrograph of a varicosity of a beaded axon forming a pericellular array. Part of the cell body of the target neuron occupies the left lower half of the picture (S) and the beaded axon forms an axo-somatic synaptic contact with this cell (between arrows). The varicosity is densely packed with synaptic vesicles the interior of which is not immunoreactive, however the cytoplasm of the varicosity is heavily reactive for serotonin except for areas of mitochondria or larger vesicles. Scale bars: **a:** 10 μm, **b:** 1 μm.

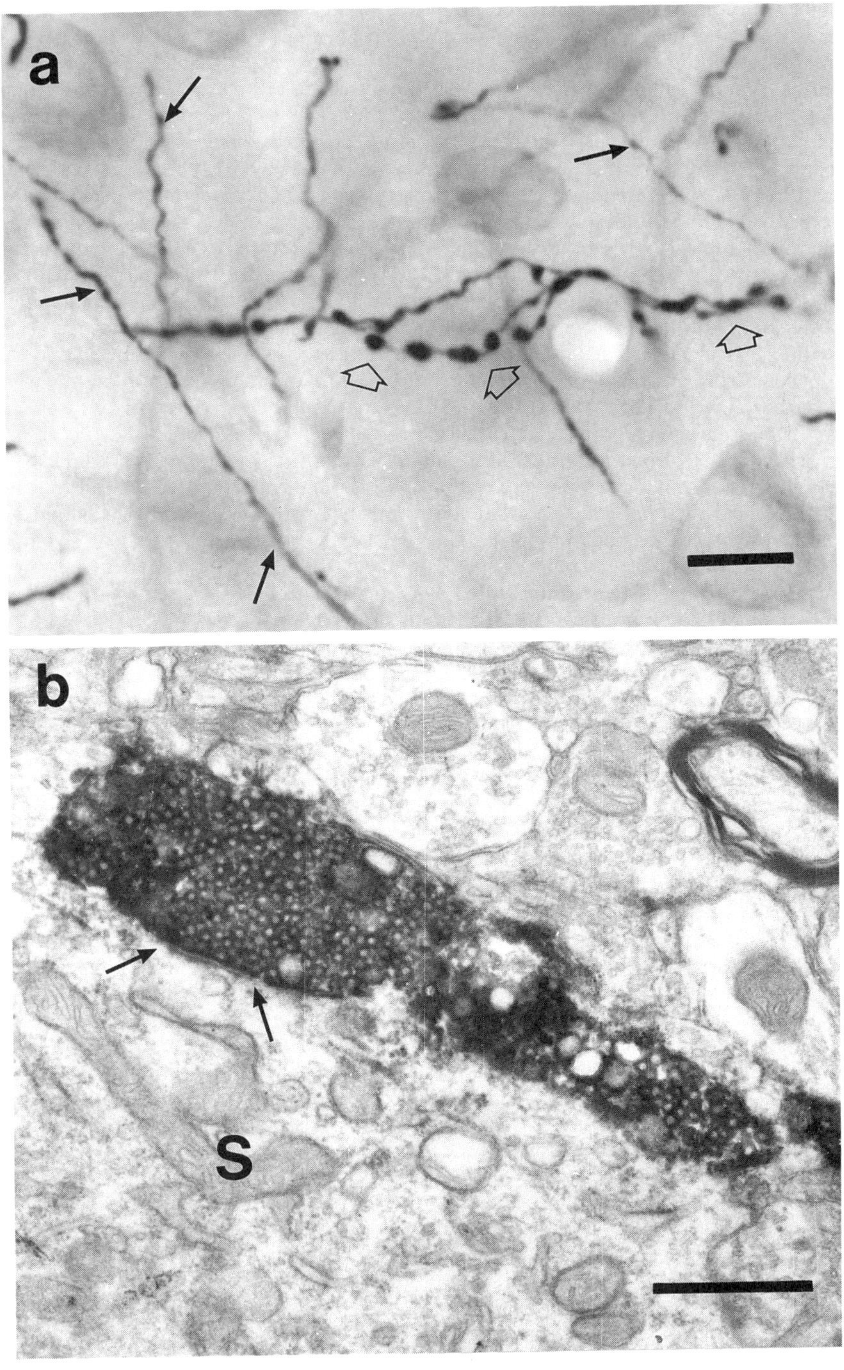
a
b
S

The aforesaid indicate that the ascending serotonergic systems form at least two parallel projection pathways which may also have distinct functions. There are several interesting observations in this regard. Firstly, many beaded axons in the cerebral cortex, septum and hippocampus form distinct pericellular arrays.[18,50,51] In the septum these have been demonstrated in the rat, cat, monkey, but in the cortex only in the cat and monkey. Septal pericellular arrays of the cat are shown in FIGURE 7 and examples of the cortical arrays ('5-HT baskets') are shown in FIGURE 8. The pericellular arrays in the cortex are often so extensive as to describe the morphology of the target cell.[18] In FIGURE 8a the outline of the target cell is that of a non-spiny stellate neuron of the cortex. In other cases horizontal and short axon cells of layer I of the cortex,[18] or bipolar and bitufted cells of the cortex could be seen innervated.[51] The pericellular arrays also occur in the hippocampus,[51,52] and the evidence is strong that it is the non-pyramidal neurons which are the targets of dense serotonergic innervation by the beaded axons. The target cells of the serotonergic axons in the cortex and in the hippocampus are consistently immunoreactive with antibodies against glutamic acid decarboxylase indicating that they are GABA-ergic (inhibitory) neurons.[53]

The forebrain and particularly the cortical serotonergic projections have been the target of many ultrastructural studies, initially using the re-uptake of ^{3}H-5-HT[54] for labeling the terminals and more recently using immunocytochemistry.[55–60] The results of these investigations are somewhat controversial: early studies have concluded that most varicosities or boutons of serotonergic axons do not form classical synaptic contacts.[54] More recent studies using immunocytochemistry appeared to be more successful, although the demonstrated synaptic densities in many cases seem to be rather small and often equivocal. The identification of small synaptic contacts formed by serotonergic axons which have been labeled with immunoperoxidase techniques is difficult because the usually strong immunoperoxidase reaction tends to penetrate into neighboring immuno-negative profiles, mimicking the appearance of postsynaptic densities. The presence of such densities is the only clearly visible sign of the synaptic contact since any specialization of the presynaptic membrane is usually hidden by the strong immunoreaction present in the presynaptic, serotonergic terminal. For this reason the older autoradiographic technique (where the label does not mask the ultrastructural features of a terminal) could be the method of choice except that it is more difficult to obtain and prepare larger samples of serial sections for autoradiography than it is using immunocytochemical methods. A certain advantage of the immunocytochemical approach is that it enables one to study the ultrastructure of light-microscopically identified serotonergic axons, providing the possibility for precise correlative light and electron microscopic analysis.

Recent electron microscopic studies of the synaptic connections of serotonergic axons in the cortex and in forebrain indicate that there is a remarkable variety of synaptic incidence that could be demonstrated. In the neocortex the percentage of serotonergic varicosities making classical synaptic connections was found to be as much as 80% in the rat[60] or as little as 3% in the monkey.[57] In the cortex and subcortical structures it appears that the varicosities of the fine fibers rarely make synaptic contacts while larger varicosities found on the beaded fibers tend to show distinct synaptic specializations, not only with the dendrites but also the somata of postsynaptic

FIGURE 7. Pericellular arrays formed by serotonergic axons in the septal region of the cat (at arrows). In all species so far studied the septal area is densely innervated by serotonergic axons which form distinct pericellular arrays around a small subpopulation of septal neurons. Scale bar: 50 μm.

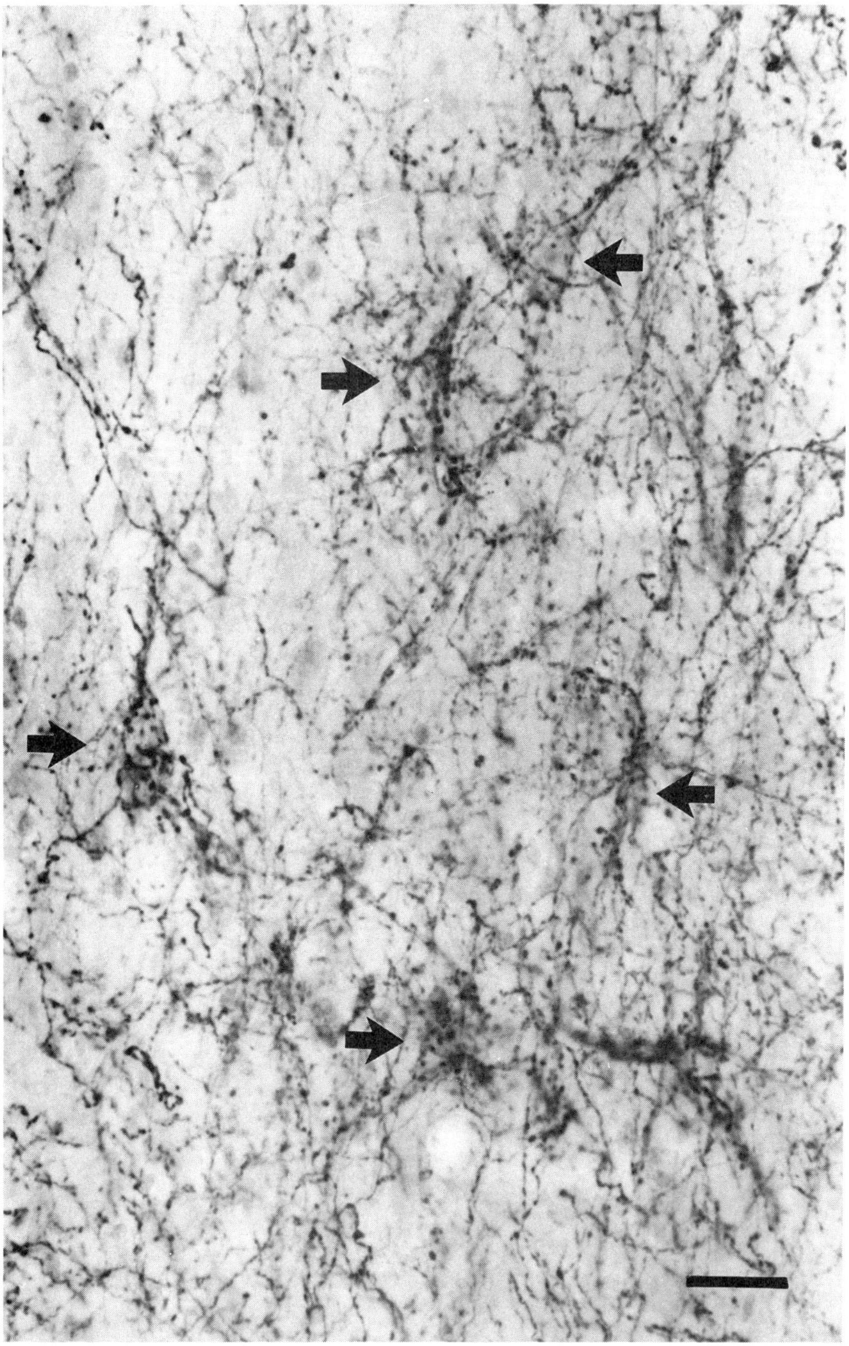

neurons.[53] An example of an axosomatic synaptic contact made by a beaded serotonergic axon on the soma of a neuron in the cortex is shown in FIGURE 6b.

Of the subcortical regions of the forebrain, the thalamus, hypothalamus and the basal ganglia have been studied. In the hypothalamus, just like in the septal region many cells are directly innervated by serotonergic axons and the serotonergic axons form direct synaptic contact with neuroendocrine cells.[61,62] On the other hand, synaptic connections in the thalamus[56] and even more so in the striatum appear to be rather infrequent.[63]

The differential distribution of axons with and without synaptic contacts seems to support the concept that the ascending serotonergic projections form a dual system[18] which would be different not only in their light microscopic but also in their electron microscopic features. Since most regions of the cerebral cortex contain both fiber systems, quantitative studies of the synaptic connections of varicosities may yield differential results depending on the relative density of the two fiber systems. On the other hand, we have found that in the cat the neostriatum, although densely innervated by serotonergic axons, contains relatively few beaded varicose axons (Török, unpublished observations). Thus in the striatum a low incidence of serotonergic synapses could be expected. Indeed, recently it was reported that only about 10–13% of striatal serotonergic varicosities from conventional synaptic contacts.[63]

A high density of beaded varicose axons can be observed in the hippocampal formation, particularly in the entorhinal cortex, CA3 area and in the polymorph layer of the dentate gyrus.[51,52] In the latter the density of pericellular arrays is also remarkably high.

The differential distribution of beaded and fine varicose axons can be dramatically demonstrated by the administration of certain amphetamine derivatives to experimental animals before fixation of brain tissue (for details see the chapter by Molliver in this volume). While the fine varicose axons degenerate as a result of treatment with PCA or MDMA the beaded varicose fibers survive and their most prominent positions can be easily seen in such preparations.[64,65]

The concept of the dual ascending serotonergic projection system is summarized in FIGURE 9. The two ascending systems have a differential origin, morphology and partially differential distribution. It is likely that most of the unambiguous synaptic contacts described in the literature are those formed by the beaded axons, which mostly form asymmetrical contacts. The origin of the beaded fibers is the median raphe nucleus, while the diffuse, fine varicose axons system arises from the dorsal raphe nucleus. Many of the findings described above support the concept illustrated in this figure, but clearly more exact data are required on the distribution and development of the two fiber systems; only then would it be possible to draw definite functional conclusions on the significance of the existence of the two parallel serotonergic systems in the forebrain.

FIGURE 8. Pericellular arrays of serotonergic axons in the cerebral cortex of the cat. **a:** Densely innervated cortical non-pyramidal neuron from the lateral suprasylvian cortex. Notice that several beaded axons run along the length of the principal dendrites and the soma of the cell which, although itself not stained is well described by the fibers. **b:** small number of beaded serotonergic axons making contact with the surface of the soma of a small neuron in layer I of the suprasylvian cortex. The soma is demonstrated with a counterstain. The electron micrograph of a varicosity shown in FIGURE 6b, is from a similar pericellular array. Scale bars: **a:** 50 μm, **b:** 10 μm.

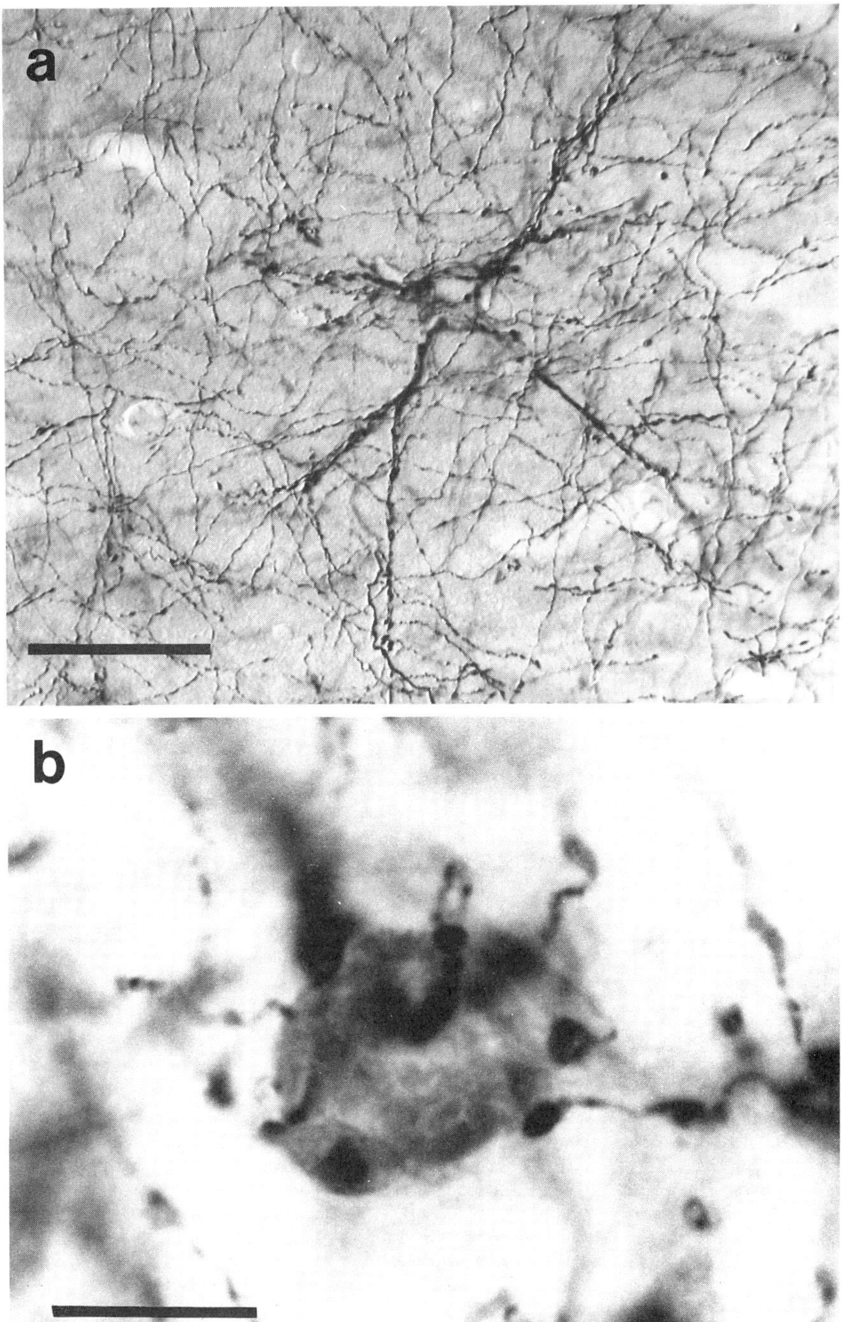
a
b

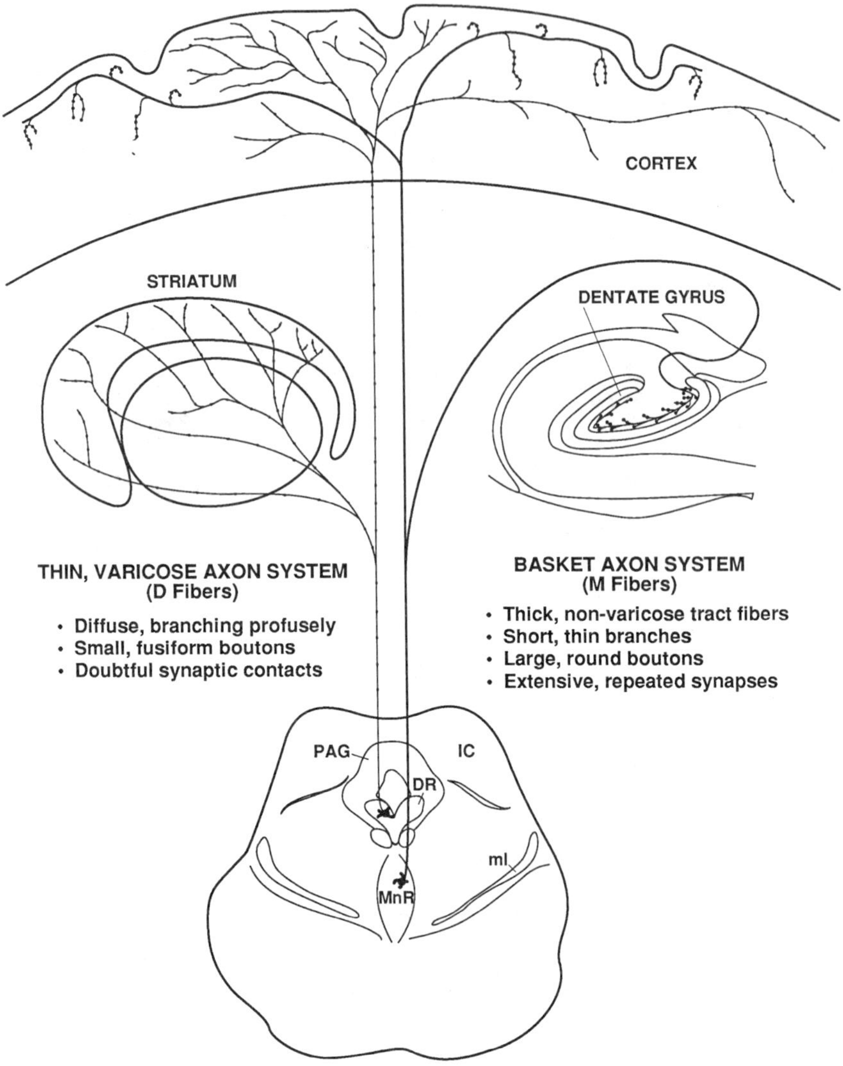

FIGURE 9. Simplified summary diagram of the main features of the dual serotonergic system innervating the forebrain. The fine varicose axons system (D-fibers) arises from the dorsal raphe nucleus with fibers that branch profusely in their target areas. It is difficult to demonstrate the synaptic connections of these fibers and therefore the incidence of synapses on these fibers is controversial. The second system—also called basket axons—with large, round or oval varicosities (M-fibers) arises from the median raphe nucleus with thick non-varicose axons, giving rise to branches with the characteristic beaded varicose axons. We have demonstrated that these large terminals make well-defined, asymmetrical synapses with the somata and dendrites of target cells. The two systems probably coexist in all areas of the brain, but their ratio can vary considerably. The most common region of widespread coexistence is the cerebral cortex. The striatum, on the other hand, seems to receive an almost exclusive innervation by the fine varicose fiber system, while the granule and polymorph layers of the dentate gyrus receive a high concentration of basket fibers. Other well innervated regions of the forebrain have not yet been surveyed from this point of view, thus they are not illustrated here.

Descending Serotonergic Projections

The spinal cord receives a strong serotonergic innervation. Throughout the grey matter there is a remarkable density of serotonergic axons which form a feltwork of fibers, describing the boundaries of the grey matter rather well. Particular regions of the spinal grey matter stand out because of their strong serotonergic innervation[66]: these are the dorsal horn, particularly lamina I, and to a lesser extent lamina II, the ventral horn motor nuclei (laminae VIII and IX) and in the thoracic cord, the intermediolateral column. FIGURES 10 and 11 show serotonin immunoperoxidase reaction in the dorsal and ventral horns of the spinal grey matter, respectively. The

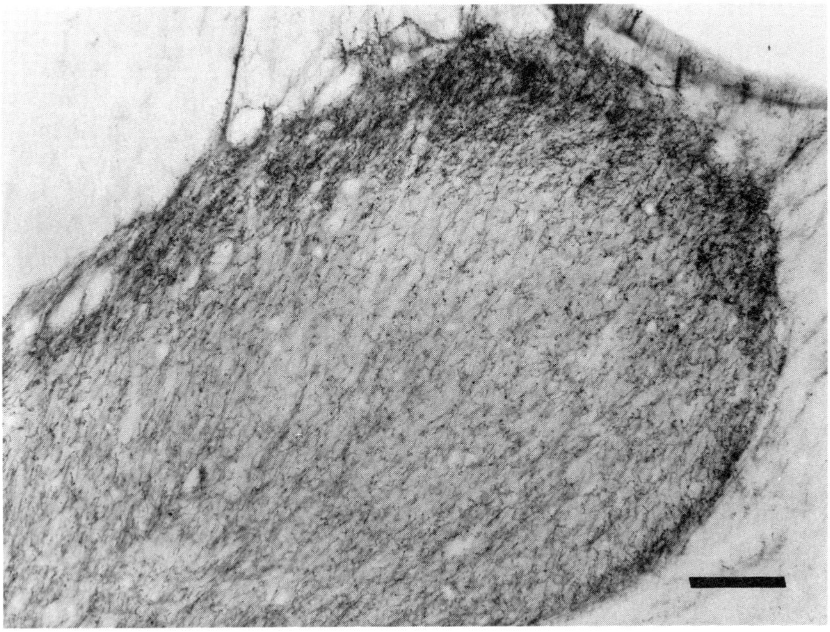

FIGURE 10. Distribution of serotonergic axons in the dorsal horn of the spinal cord. Although immunoreactive fibers are found throughout the dorsal horn the highest density is found closest to the dorsal surface. Scale bar: 100 μm.

fibers terminating in the dorsal horn are particularly dense in layer I (also called the marginal zone) and may terminate on the cell body and dendrites of spinothalamic tract neurons located there.[67] The origins of the serotonergic projections to the dorsal horn are principally the neurons of the raphe magnus nucleus and adjacent reticular formation,[66] and travel to the spinal cord in the dorsolateral fasciculus between the tip of the dorsal horn and the surface of the spinal cord.[68]

A second major target of the serotonergic projections in the spinal cord are the motoneurons, and a high density of serotonergic innervation can be observed in regions of lamina IX[69] (FIG. 11b). There is a particularly strong innervation to the phrenic motoneurons[70] in the cervical cord and the serotonergic axons innervating these neurons resemble the beaded varicose axons of the forebrain.[71] The serotonergic axons innervating the ventral horn are the termination of a second major descending system

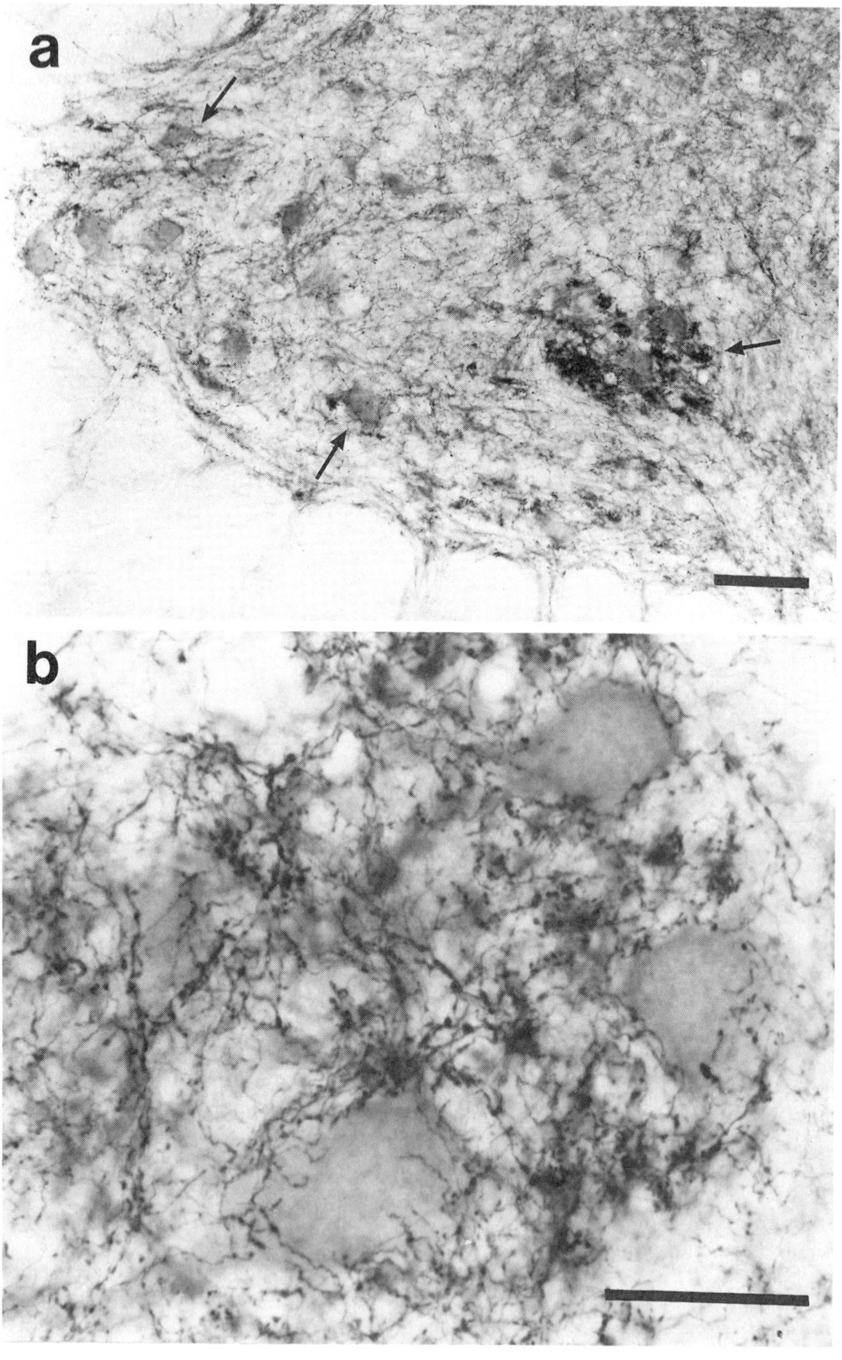
a
b

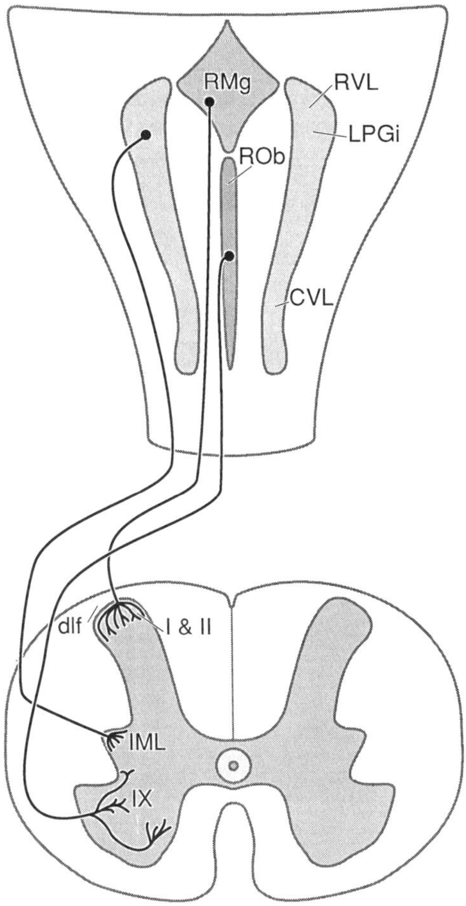

FIGURE 12. Schematic representation of the serotonergic projections from the brainstem to spinal cord. The figure emphasizes the three principal targets of these projections: the dorsal horn (Laminae I and II), the intermediolateral cell column (IML) and the motor columns (Lamina IX). The origin of the projections to the dorsal horn is the raphe magnus nucleus (RMg) and to the ventral horn the raphe obscurus (ROb) nucleus. The IML receives projections principally from the rostral ventrolateral medulla (RVL and LPGi). The projections of the serotonergic cells in the caudal ventrolateral medulla (CVL) are not known. dlf: dorsolateral fasciculus.

which arises primarily from the raphe obscurus and raphe pallidus nuclei.[66] Many neurons of these raphe nuclei have been shown to co-localize serotonin with substance P or thyrotropin-releasing hormone,[72] and as a result co-existence of serotonin with peptides in the ventral horn can also be observed.[69]

The preganglionic sympathetic neurons of the intermediolateral column in the thoracic cord receives a strong serotonergic input[66] and the cells are densely surrounded by serotonergic axons. Although there is evidence that some raphe magnus cells project to the intermediolateral cell column, the most significant source of these serotonergic projections is the ventrolateral medulla.[73] Indeed stimulation of the lateral

---

FIGURE 11. Serotonergic innervation of the ventral horn of the upper cervical spinal cord of the cat. **a:** low power view showing that the entire ventral horn contains serotonergic axons although there is higher density around the large motoneurons (at arrows). Ventral is to the left. **b:** higher power view of a few motoneurons showing the high density of serotonin-immunoreactive axons in their vicinity, many of them bearing large varicosities. Scale bars: **a:** 100 μm, **b:** 50 μm.

serotonergic regions of the medulla causes an increase of blood pressure and this effect can be abolished by the previous lesioning of the 5-HT cells with the use of the neurotoxin, 5,7-dihydroxytryptamine.[74]

The principal descending serotonergic pathways and their targets in the spinal grey matter are schematically illustrated in FIGURE 12.

There are some additional regions of the spinal cord which receive dense serotonergic input: thus the central gray, adjacent to the central canal is prominent, and also the ependyma lining the central canal, which is covered by a serotonergic nerve plexus on its surface. However little or no information is available on the origins of these serotonergic projections.

Notwithstanding that the most significant cells of origin of serotonergic afferents to the spinal cord are in the medullary raphe nuclei, additional minor projections have also been shown to exist. The additional descending projections arise from scattered cells of the B9 group in the pontomesencephalic reticular formation and from cells that are located within the medial longitudinal fasciculus.[66]

ACKNOWLEDGMENTS

I wish to thank the many collaborators and associates who have contributed to our ongoing studies of the serotonergic system: Adrian Aitken, Kathleen Mulligan, Glenda Halliday, Jean-Pierre Hornung, Kerry Baker, Deborah McRitchie, Glenys Rikard-Bell, Dzung Vu, Paul Halasz and Richard Cotton. The excellent technical assistance of Sharlane Hamblet, Mignon Wegner and Mirella Fabbri, and the graphic artistic assistance of Alicia Fritchle is gratefully acknowledged.

REFERENCES

1. BOGDANSKI, D. F., A. PLETSCHER, B. N. BRODIE & S. UDENFRIEND. 1956. Identification and assay of serotonin in brain. J. Pharmacol. Exp. Ther. **117:** 82–88.
2. COSTA, E. & M. H. APRISON. 1958. Studies on the 5-hydroxytryptamine (serotonin) content in human brain. J. Nerv. Ment. Dis. **126:** 289–293.
3. DAHLSTRÖM, A. & K. FUXE. 1964. Evidence for the existence of monoamine-containing neurons in the central nervous system. I. Demonstration of monoamines in cell bodies of brain stem neurons. Acta Physiol. Scand. Suppl. **232:** 1–55.
4. UNGERSTEDT, U. 1971. Stereotaxic mapping of the monoamine pathways in the rat brain. Acta Physiol. Scand. Suppl. **367:** 1–48.
5. STEINBUSCH, H. W. M. 1981. Distribution of serotonin-immunoreactivity in the central nervous system of the rat. Neuroscience **4:** 557–618.
6. WIKLUND, L., L. LEGER & M. PERSSON. 1981. Monoamine cell distribution in the cat brainstem. A fluorescence histochemical study with quantification of indoleaminergic and locus coeruleus cell groups. J. Comp. Neurol. **203:** 613–647.
7. STEINBUSCH, H. W. M. & R. NIEUWENHUYS. 1983. The raphe nuclei of the rat brainstem: A cytoarchitectonic and immunohistochemical study. *In* Chemical Neuroanatomy. P. C. Emson, Ed. New York: Raven Press, pp. 131–207.
8. TÖRK, I. 1985. Raphe nuclei and serotonin containing systems. *In* The Rat Nervous System. G. Paxinos, Ed.: Vol. 2: 43–78. Sydney: Academic Press.
9. AZMITIA, E. C. & P. J. GANNON. 1986. The primate serotonergic system: A review of human and animal studies and a report on Macaca fascicularis. Adv. Neurol. **43:** 407–468.
10. HORNUNG, J.-P. & J.-M. FRITSCHY. 1988. Serotonergic system in the brainstem of the marmoset: A combined immunocytochemical and three-dimensional reconstruction study. J. Comp. Neurol. **270:** 471–487.

11. LIDOV, H. G. W. & M. E. MOLLIVER. 1982. Immunohistochemical study of the development of serotonergic neurons in the rat CNS. Brain Res. **78:** 45–56.
12. AITKEN, A. R. & I. TÖRK. 1988. Early development of serotonin-containing neurons and pathways as seen in wholemount preparations of the fetal rat brain. J. Comp. Neurol. **274:** 32–47.
13. CONSOLAZIONE, A., C. MILSTEIN, B. WRIGHT & A. C. CUELLO. 1981. Immunocytochemical detection of serotonin with monoclonal antibodies. J. Histochem. Cytochem. **29:** 1425–1430.
14. HAAN, E. A., I. G. JENNINGS, A. C. CUELLO, H. NAKATA, H. FUJISAWA, C. W. CHOW, R. KUSHINSKY, J. BRITTINGHAM & R. G. H. COTTON. 1987. Identification of serotonergic neurons in human brain by a monoclonal antibody binding to all three aromatic amino acid hydroxylases. Brain Res. **426:** 19–27.
15. COTTON, R. G. H., W. MCADAM, I. JENNINGS & F. J. MORGAN. 1988. A monoclonal antibody to aromatic amino acid hydroxylases. Biochem. J. **255:** 193–196.
16. JOYCE, D. 1962. Changes in the 5-hydroxytryptamine content of rat, rabbit and human brain after death. Brit. J. Pharmacol. **18:** 370–380.
17. HSU, S.-M., L. RAINE & H. FANGER. 1981. Use of avidin–biotin peroxidase complex (ABC) in immunoperoxidase techniques: A comparison between ABC and unlabeled antibody (PAP) procedures. J. Histochem. Cytochem. **29:** 577–580.
18. MULLIGAN, K. A. & I. TÖRK. 1988. Serotoninergic innervation of the cat cerebral cortex. J. Comp. Neurol. **270:** 86–110.
19. TÖRK, I. & J.-P. HORNUNG. 1990. Raphe nuclei and serotonin containing systems. *In* The Human Nervous System. G. Paxinos, Ed. San Diego, CA: Academic Press, pp. 1001–1022.
20. TAKAHASHI, H., S. NAKASHIMA, E. OHAMA, S. TAKEDA & F. IKUTA. 1986. Distribution of serotonin-containing cell bodies in the brainstem of the human fetus determined with immunohistochemistry using antiserotonin serum. Brain Dev. **8:** 355–365.
21. BAKER, K. B., G. M. HALLIDAY, P. HALASZ, J.-P. HORNUNG, L. B. GEFFEN, R. G. H. COTTON & I. TÖRK. 1990. Cytoarchitecture of serotonin-synthesizing neurons in the pontine tegmentum of the human brain. Synapse. In press.
22. HALLIDAY, G. M., Y. W. LI, T. H. JOH, R. G. H. COTTON, P. R. C. HOWE, L. B. GEFFEN & W. W. BLESSING. 1988. Distribution of monoamine-synthesizing neurons in the human medulla oblongata. J. Comp. Neurol. **273:** 301–317.
23. JACOBS, B. L., P. J. GANNON & E. C. AZMITIA. 1984. Atlas of serotonergic cell bodies in the cat brainstem: An immunohistochemical analysis. Brain Res. Bull. **13:** 1–31.
24. TILLET, Y. 1987. Immunocytochemical localization of serotonin-containing neurons in the myelencephalon, brainstem and diencephalon of the sheep. Neuroscience **23:** 501–527.
25. BAKER, K. G., G. M. HALLIDAY, J.-P. HORNUNG, L. B. GEFFEN, R. G. H. COTTON & I. TÖRK. 1990. Distribution, morphology and number of monoamine-synthesizing and substance P-containing neurons in the human dorsal raphe nucleus. Submitted for publication.
26. PAXINOS, G. & C. WATSON. 1986. The Rat Brain in Stereotaxic Coordinates, 2nd edit. Sydney: Academic Press.
27. PAXINOS, G., I. TÖRK, G. M. HALLIDAY & W. R. MEHLER. 1990. Human homologs to brainstem nuclei identified in other animals as revealed by acetylcholinesterase activity. *In* The Human Nervous System. G. Paxinos, Ed. San Diego, CA: Academic Press, pp. 149–202.
28. PAXINOS, G., I. TÖRK, L. TECOTT & K. VALENTINO. 1990. Atlas of the Developing Rat Brain. San Diego, CA: Academic Press. In press.
29. OLSON, L. & Å. SEIGER. 1972. Early prenatal ontogeny of central monoamine neurons in the rat: Fluorescence histochemical observations. Z. Anat. Entwickl.-Gesch. **137:** 301–316.
30. OLSZEWSKI, J. & D. BAXTER. 1982. Cytoarchitecture of the Human Brain Stem, 2nd edit. Basel: Karger.
31. TÖRK, I. G. M. HALLIDAY, T. SCHEIBNER & S. TURNER. 1984. The organization and connections of the mesencephalic ventromedial tegmentum (VMT) in the cat. *In* Modula-

tion of Sensorimotor Activity during Alterations in Behavioral States. R. Bandler, Ed. New York: Alan R. Liss, pp. 39–73.
32. BRAAK, H. 1970. Über die Kerngebiete des menschlichen Hirnstammes II. Die Raphekerne. Z. Zellforsch. **107:** 123–141.
33. BAKER, K. G., G. M. HALLIDAY & I. TÖRK. 1990. Cytoarchitecture of the human dorsal raphe nucleus. J. Comp. Neurol. In press.
34. WATERHOUSE, B. D., G. A. MIHAILOFF, J. C. BAACK & D. J. WOODWARD. 1986. Topographical distribution of dorsal and median raphe neurons projecting to motor, sensorimotor, and visual cortical areas in the rat. J. Comp. Neurol. **249:** 460–476.
35. O'HEARN, E. & M. E. MOLLIVER. 1984. Organization of raphe-cortical projections in rat: A qualitative retrograde study. Brain Res. Bull. **13:** 709–726.
36. WILSON, M. A. & M. E. MOLLIVER. 1988. The anatomic organization of serotonergic projections to neocortex in the primate. Soc. Neurosci. Abstr. **14:** 210.
37. CALZA, L., L. GIARDINO, R. GRIMALDI, M. RIGOLF, H. M. W. STEINBUSCH & M. TEINGO. 1985. Presence of 5-HT-positive neurons in the medial nuclei of the solitary tract. Brain Res. **347:** 135–139.
38. FALLON, J. H. & S. E. LOUGHLIN. 1987. Monoamine innervation of cerebral cortex and a theory of the role of monoamines in cerebral cortex and basal ganglia. *In* Cerebral Cortex. A. Peters and E. G. Jones, Eds. New York: Plenum Press, pp. 41–127.
39. MOLLIVER, M. E., R. GRZANNA, H. G. W. LIDOV, J. H. MORRISON & J. A. OLSCHOWSKA. 1982. Monoamine systems in the cerebral cortex. *In* Cytochemical Methods in Neuroanatomy. V. Chan-Palay & S. L. Palay, Eds. New York: Alan R. Liss, pp. 255–257.
40. CROPPER, E. C., J. S. EISENMAN & E. C. AZMITIA. 1984. An immunocytochemical study of the serotonergic innervation of the thalamus of the rat. J. Comp. Neurol. **224:** 38–50.
41. LIDOV, H. G. W., R. GRZANNA & M. E. MOLLIVER. 1980. The serotonin innervation of the cerebral cortex in the rat—an immunohistochemical analysis. Neuroscience **5:** 207–227.
42. MORRISON, J. H., S. L. FOOTE, M. E. MOLLIVER, F. E. BLOOM & H. G. W. LIDOV. 1982. Noradrenergic and serotonergic fibers innervate complementary layers in monkey primary visual cortex: An immunohistochemical study. Proc. Natl. Acad. Sci. USA **79:** 2401–2405.
43. TAKEUCHI, Y. & Y. SANO. 1984. Serotonin nerve fibers in the primary visual cortex of the monkey. Quantitative and immunoelectronmicroscopical analysis. Anat. Embryol. **169:** 1–8.
44. BERGER, B., S. TROTTER, C. VERNEY, P. GASPAR & C. ALVAREZ. 1988. Regional and laminar distribution of the dopamine and serotonin innervation in the macaque cerebral cortex: A radioautographic study. J. Comp. Neurol. **273:** 99–119.
45. CAMPBELL, M. J., D. A. LEWIS, S. L. FOOTE & J. H. MORRISON. 1987. Distribution of choline acetyltransferase-, serotonin-, dopamine-b-hydroxylase-immunoreactive fibers in monkey primary auditory cortex. J. Comp. Neurol. **261:** 209–220.
46. SOGHOMONIAN, J.-J., G. DOUCET & L. DESCARRIES. 1987. Serotonin innervation in adult rat neostriatum. I. Quantified regional distribution. Brain Res. **425:** 85–100.
47. MORRISON, J. H. & S. L. FOOTE. 1986. Noradrenergic and serotonergic innervation of cortical, thalamic, and tectal visual structures in old and new world monkeys. J. Comp. Neurol. **243:** 117–138.
48. TAKEUCHI, Y. & Y. SANO. 1983. Immunohistochemical demonstration of serotonin nerve fibers in the neocortex of the monkey (Macaca fuscata). Anat. Embryol. **166:** 155–168.
49. KOSOFSKY, B. E. & M. E. MOLLIVER. 1987. The serotonergic innervation of cerebral cortex: different classes of axon terminals arise from dorsal and median raphe nuclei. Synapse **1:** 153–168.
50. GALL, C. & R. Y. MOORE. 1984. Distribution of enkephalin, substance P, tyrosine hydroxylase, and 5-hydroxytryptamine immunoreactivity in the septal region of the rat. J. Comp. Neurol. **225:** 212–227.
51. HORNUNG, J.-P., J.-M. FRITSCHY & I. TÖRK. 1990. Distribution of two morphological types of serotonergic axons in the cerebral cortex of the marmoset monkey. J. Comp. Neurol. **297:** 165–181.
52. TÖRK, I. & J.-P. HORNUNG. 1989. Serotonergic innervation of the cat hippocampus. Soc. Neurosci. Abstr. **15:** 1247.

53. TÖRK, I., J.-P. HORNUNG & K. A. MULLIGAN. 1990. Serotonergic innervation of cortical inhibitory interneurons. Submitted for publication.
54. DESCARRIES, L., A BEAUDET & K. C. WATKINS. 1975. Serotonin nerve terminals in adult rat neocortex. Brain Res. **100:** 563–588.
55. AZMITIA, E. C. & P. J. GANNON. 1983. The ultrastructural localization of serotonin immunoreactivity in myelinated and unmyelinated axons within the medial forebrain bundle of rat and monkey. J. Neurosci. **3:** 2083–2090.
56. DE LIMA, A. D. & W. SINGER. 1987. The serotonergic fibers in the dorsal lateral geniculate nucleus of the cat: Distribution and synaptic connections demonstrated with immunocytochemistry. J. Comp. Neurol. **258:** 339–351.
57. DEFELIPE, J. & E. G. JONES. 1988. A light and electron microscopic study of serotonin-immunoreactive fibers and terminals in the monkey sensory-motor cortex. Exp. Brain Res. **71:** 171–182.
58. DE LIMA, A. D., F. E. BLOOM & J. H. MORRISON. 1988. Synaptic organization of serotonin-immunoreactive fibers in primary visual cortex of the macaque monkey. J. Comp. Neurol. **274:** 280–294.
59. PARNAVELAS, J. G., H. C. MOISES & S. G. SPECIALE. 1985. The monoaminergic innervation of the rat visual cortex. Proc. R. Soc. Lond. B. **223:** 319–329.
60. PAPADOPOULOS, G. C., J. G. PARNAVELAS & R. BUIJS. 1987. Light and electron microscopic immunocytochemical analysis of the serotonin innervation of the rat visual cortex. J. Neurocytol. **16:** 883–892.
61. KISS, J. & B. HALÁSZ. 1986. Synaptic connections between serotoninergic axon terminals and tyrosine hydroxylase-immunoreactive neurons in the arcuate nucleus of the rat hypothalamus. A combination of electron microscopic autoradiography and immunocytochemistry. Brain Res. **364:** 284–294.
62. KISS, J., A. CSÁKY & B. HALÁSZ. 1988. Demonstration of serotoninergic axon terminals on somatostatin-immunoreactive neurons of the anterior periventricular nucleus of the rat hypothalamus. Brain Res. **442:** 23–32.
63. SOGHOMONIAN, J.-J., L. DESCARRIES & K. C. WATKINS. 1989. Serotonin innervation in adult rat neostriatum. II. Ultrastructural features: a radioautographic and immunocytochemical study. Brain Res. **481:** 67–86.
64. MAMOUNAS, L. A. & M. E. MOLLIVER. 1988. Evidence for dual serotonergic projections to neocortex: axons from the dorsal and median raphe nuclei are differentially vulnerable to the neurotoxin p-choloamphetamine (PCA). Exp. Neurol. **102:** 23–36.
65. WILSON, M. R., G. A. RICAURTE & M. E. MOLLIVER. 1989. Distinct morphological classes of serotonergic axons in primates exhibit differential vulnerability to the psychotropic drug 3,4-methylendioximethamphetamine. Neuroscience **28:** 121–137.
66. BOWKER, R. M., K. N. WESTLUND, M. C. SULLIVAN & J. D. COULTER. 1982. Organization of descending serotonergic projections to the spinal cord. Prog. Brain Res. **57:** 239–265.
67. HYLDEN, J. L. K., H. HAYASHI, M. A. RUDA & R. DUBNER. 1986. Serotonin innervation of physiologically identified lamina I projection neurons. Brain Res. **370:** 401–404.
68. BASBAUM, A. I., C. H. CLANTON & H. L. FIELDS. 1978. Three bulbospinal pathways from the rostral medulla of the cat: An autoradiographic study of pain modulating systems. J. Comp. Neurol. **178:** 209–224.
69. BOWKER, R. M. 1986. Serotonergic and peptidergic imputs to the primate ventral spinal cord as visualized with multiple chromagens on the same tissue section. Brain Res. **375:** 345–350.
70. HOLTMAN, J. R., W. P. NORMAN, L. SKIRBOLL, K. L. DRETCHEN, C. CUELLO, T. J. VISSER, T. HÖKFELT & R. A. GILLIS. 1984. Evidence for 5-hydroxytryptamine, substance P and thyrotropin-releasing hormone in neurons innervating the phrenic motor nucleus. J. Neurosci. **4:** 1064–1071.
71. LIPSKI, J., P. M. PILOWSKY, M. D. VOSS & D. DE CASTRO. 1988. Serotonin immunoreactive boutons make close contact with feline phrenic motoneurons. Soc. Neurosci. Abstr. **14:** 936.
72. JOHANSSON, O., T. HÖKFELT, B. PERNOW, S. L. JEFFCOATE, N. WHITE, H. W. M. STEINBUSCH, A. A. J. VERHOSTAD P. C. EMSON & E. SPEIDEL. 1981. Immunohistochemical support for three putative neurotransmitters in one neuron. Coexistence of 5-hydrox-

ytryptamine, substance P and thyrotropin releasing hormone-like immunoreactivity in medullary neurons projecting to the spinal cord. Neuroscience **6:** 1857–1881.
73. LOEWY, A. D. & S. MCKELLAR. 1981. Serotonergic projections from the ventral medulla to the intermediolateral cell column in the rat. Brain Res. **211:** 146–152.
74. HOWE, P. R. C., D. M. KUHN & J. B. MINSON, B. H. STEAD & J. P. CHALMERS. 1983. Evidence for a bulbospinal serotonergic pressor pathway in the rat brain. Brain Res. **270:** 29–36.
75. STONE, J., B. DREHER & I. TÖRK. 1990. The Neuroanatomists's Colouring Book. 3rd edit. Sydney: Maitland Publications.

DISCUSSION OF THE PAPER

LAURENT DESCARRIES (*Université de Montreal, Quebec, Canada*): I very much enjoyed your pictures of the cell groups. In the ultrastructural data you show this entirely from the cat? Because in the cortex of the rat I don't believe we have seen these pericellular arrangements. There were quite obvious in the septal area but not in the cortex and I was wondering if they are described, let's say, in monkey or in human, or in the monkey since there's not much data, ultrastructurally speaking, in the human. And the perisomatic arrangements, would they be peculiar to the cat?

TÖRK: We have conducted experiments in collaboration with Jean-Pierre Hornung and Jean-Marc Fritschy in the marmoset monkey and sectioned the cortex from the rostral end to the visual cortex and found that, even in the monkey, in certain areas of the cortex one can demonstrate these baskets, particularly in the prefrontal cortex and temporal cortex. But none, not a single one, was seen in the visual cortex. I think this is why in the monkey cortex these peculiar arrangements have not been found for long: 80% of the serotonin work in the monkey was done on the visual cortex. And it is really an extremely unique type of innervation the significance of which I can't explain, unfortunately.

We have also looked at the human cortex, in biopsy material. In such material of course you have difficulties in trying to catch serotonin in time, but one can also use a tryptophan hydroxylase antibody. And the finding is that the pericellular arrays are not so beautiful as they are in the cat or in the monkey, but one finds so-called hyperinnervated fields. That is made of a cluster of cells, say 5, 6, 7 or even more cells which are surrounded by a long stretch of fiber with large boutons. We haven't gone as far yet as to look at the synapses of these but extrapolating from the previous studies we are certain that they are also present. Wherever we find these fibers with large boutons there are synapses.

Now I can't explain why in the rat cortex there are less of these arrays but there are synapses there too, although in small numbers. And I believe that all that happens there that the basket fiber system is rather restricted. In fact this is borne out by the studies of Molliver and his group who used different drugs to deplete the fine fiber system, and then you can see, as you will hear it later on during his talk, that in fact the fine fiber system degenerates following the application of MDMA while the basket fiber system seems to be more resilient and resists this drug treatment: following which you can see where the basket fibers are.

If you were using this approach you might be able to actually find the part of the cortex that really contains higher concentrations of the boutons that do have synapses.

ALLEN MILLER (*Rockefeller University, New York City*): I may have missed this, but you said the, I believe, D fibers do not appear to make synaptic contacts. If not then what do you think they do?

TÖRK: Well there are lots of more qualified people here to talk about that. One of the things that they can do is to "drip-irrigate" the brain with serotonin, and of course this is, I do not like to use the word, what I believe. I'm just being convinced about the opposite of what I thought at the beginning when I believed that in fact all boutons make synapses, but we just find it difficult to actually identify them. In the electron microscope, the fine fibers are difficult to distinguish from the fibers with large boutons if you are just looking for them randomly in the cortex in thin sections, because their internal structure is not very much different and their size is not necessarily a good guide. What you need to do is to recover the fine fibers after you have embedded them flat in plastic and then cut thin sections through them. And when you've got 200 ultrathin sections to a fiber with 2 or 3 boutons perhaps there you still find no synapses.

Of course there are many more boutons and I've looked at maybe 10–20 only so that it may not be a representative sample. On the other hand people have been studying, for example, the striatum, like Laurent Descarries, who published not long ago a paper on the rat striatum and did not find all that many synapses either. This may be because they only find synapses on the M-fibers which are not very dense in the striatum.

Distribution of Serotonin Receptors

J. M. PALACIOS, C. WAEBER, D. HOYER, AND G. MENGOD

Preclinical Research
SANDOZ Ltd
CH-4002 Basel, Switzerland

During the last decade our knowledge of serotonin (5-hydroxytryptamine, 5-HT) receptors has undergone a dramatic change. The main reasons for this can be found in the introduction of radioligand binding techniques, the development of compounds acting selectively on receptor sub-populations and the utilization of anatomical techniques for identifying regions enriched in receptors subtypes. More recently, the application of molecular biology techniques has provided detailed insights into the protein structure of some 5-HT receptors.[1] While at the presynaptic level histofluorescence, autoradiography and immunohistochemistry have been the tools used to establish serotoninergic pathways,[2] autoradiography has been the exclusive way to visualize pre- and postsynaptic receptor sites for 5-HT until now. [^{3}H]LSD and [^{3}H]5-HT were the first serotoninergic ligands introduced. After the discovery of the serotoninergic component of several neuroleptics, [^{3}H]spiperone was also found to be a ligand for 5-HT receptors. The seminal work of Peroutka and Snyder let to the definition of 5-HT_1 and $5HT_2$ receptors and the use of [^{3}H]5-HT, [^{3}H]LSD and [^{3}H]spiperone for the selective labeling of these receptors.[3] Since then, many other ligands have been introduced for labeling more or less selectively 5-HT receptors subtypes. In the present paper we would like to review the results obtained with autoradiographic techniques in the analysis of 5-HT receptors in the brain and their contribution to our understanding of the multiplicity of receptors where 5-HT acts to produce its diverse effects.

TOOLS AND PROCEDURES

Receptor autoradiography is a technique based on the combination of receptor binding assays with autoradiography. Cryostat tissue sections mounted onto microscope slides are incubated with different radiolabeled ligands. Autoradiograms are generated by putting these labeled tissues in close contact with dry emulsions. These techniques are now extensively used in pharmacology and have provided a wealth of information (see reference 4 for a review).

TABLE 1 summarizes the radioligands used for the labeling of 5-HT receptors. The conditions for the selective labeling of 5-HT receptors with these ligands have been established after detailed analysis of the binding of their radioligands to mounted tissue sections (see refs. 5–7 for details). Because autoradiography can be performed quantitatively, the technique has not only been used to visualize the distribution 5-HT receptors sites, but also to quantify the densities of these receptors in different experimental situations and to carry out detailed pharmacological analysis of drug binding profiles in brain regions. This type of experiments has allowed the establishment of the experimental conditions in which the different subtypes can be labeled independently. As many of the ligands used in the visualization of 5-HT receptors are

TABLE 1. Radioligands used in the Autoradiographical Visualization of Serotonin Receptors

5-HT_{1A}	5-HT_{1B}	5-HT_{1C}	5-HT_{1D}	5-HT_2	5-HT_3
[^{3}H]5-HT	[^{3}H]5-HT	[^{3}H]5-HT	[^{3}H]5-HT	[^{3}H]Ketanserine	[^{3}H]ICS 205930
[^{3}H]8-OH-DPAT	[^{125}I]-CYP	[^{3}H]LSD		[^{3}H]Mesulergine	[^{3}H]Zacopride
[^{3}H]Ipsapirone		[^{125}I]LSD		[^{3}H]Spiperone	[^{3}H]GR65360
[^{3}H]LSD		[^{3}H]Mesulergine		[^{3}H]LSD	[^{3}H]BRL 43694
[^{125}I]-BH-8NeO-N-PAT		[^{125}I]MIL		[^{125}I]LSD	
		[^{3}H]DOB			
		[^{125}I]DOI			

not selective, unlabeled ligands are used together with the non-specific radiolabeled ligands to visualize separately the desired receptor populations.[5–7]

THE DISTRIBUTION OF 5-HT RECEPTORS IN THE BRAIN

5-HT is a sophisticated transmitter. It uses many different receptor classes to convey its messages. These classes are not only differentiated by their pharmacology, but also by their differential location in various brain regions and systems. Thus, 5-HT can act upon the functions of different brain systems via receptors subtypes which diverge in their drug binding profile and signal transduction mechanisms. In the following review we describe the distribution of the different receptor subtypes in the brain of several species as well as the attempts to visualize at a cellular level the location of these receptors. TABLE 2 summarizes the distribution of 5-HT receptors in the mammalian brain.

5-HT_{1A} Receptors—the Limbic Connection

When [^{3}H]5-HT is used to label all the populations of 5-HT_1 receptors in the rat brain, a number of regions stand out by their high densities of [^{3}H]5-HT binding sites[8] (see TABLE 2). The hippocampus is one of these regions (FIG. 1). Most of the 5-HT_1 sites present in this region belong to the 5-HT_{1A} class, as can be assessed by investigating the displacement of [^{3}H]5-HT labeling with 8-OH-DPAT (8-hydroxy-dipropylaminotetralin) or by direct labeling using [^{3}H]- or [^{125}I]labeled BH-8-NeO-N-

TABLE 2. Summary of the Distribution of the Different Subtypes of 5-HT Receptors in the Mammalian Brain

	5-HT_{1A}	5-HT_{1B}	5-H_{1C}	5-H_{1D}	5-HT_2	5-HT_3
Hippocampus	+++	+	+	+	+	+
Septum	+++	+	+	+	+	+
Amygdala	++	++	+	++	+	+
Hypothalamus	++	++	++	++	+	
Striatum		++	+	++	++	
Pallidum		+++	++	+++	+	
S. Nigra		+++	++	+++		
Neocortex	+++	++	+	++	+++	
Raphé	+++	+		+		
Choroid plexus			+++			
S. Gelatinosa	++					+++

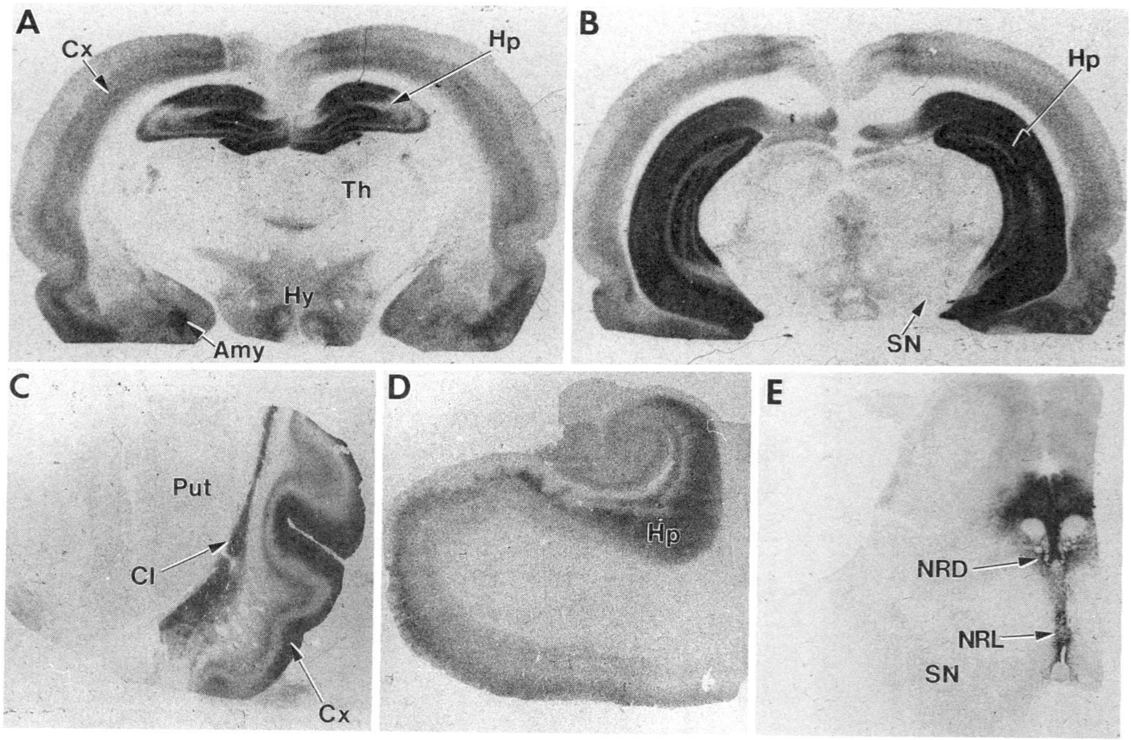

FIGURE 1. Autoradiographic visualization of 5-HT_{1A} receptors in the rat and human brain. The pictures are photographs from autoradiograms generated by putting a ^3H-sensitive photographic film in contact with microtome tissue sections where 5-HT_{1A} receptor binding sites were labeled by incubations with [^3H]8-OH-DPAT. Dark areas are rich in receptor binding. **A** and **B** are from two coronal sections of the rat brain illustrating the presence of high densities of 5-HT_{1A} receptors in the hippocampus (Hp) amygdala (Amy) hypothalamus (Hy) and cortex (Cx) particularly in the deep layers. Low or no binding is observed in the thalamus (Th) or substantia nigra (SN). **C, D,** and **E** show the binding of [^3H]8-OH-DPAT in the human brain. **C** is from a section at the level of the putamen (Put) which did not exhibit significant level of binding, while the claustrum (Cl) and cortex (Cx) were rich in 5-HT_{1A} receptors. **D** shows the presence of high levels of [^3H]8-OH-DPAT binding in the human hippocampus. **E** illustrates high densities of 5-HT_{1A} receptors in the human nucleus raphé dorsalis (NRD) and raphé linearis (NRL), while as in the rat brain the substantia nigra was characterized by the lack of 5-HT_{1A} sites.

PAT.[9,10] Several other brain areas are also enriched in 5-HT_{1A} binding. These are, for example, the septum, some of the amygdaloid nuclei and also the raphé nuclei, particularly the dorsal raphé nucleus. Many of these brain regions are components of the pathway involved in the modulation of emotion, the limbic system. High densities of 5-HT_{1A} binding have been found not only in the limbic system of the rat, but also of other mammalian species investigated, including humans[11,12] (FIG. 1). The predominance of 5-HT_{1A} receptors in this system suggests that the reported effects of 5-HT and serotoninergic drugs in emotional mechanisms could be mediated by 5-HT_{1A} receptors.[13] Furthermore, the presence of high densities of 5-HT_{1A} receptors in the raphé nuclei indicates that via these receptors 5-HT can modulate the activity of serotoninergic neurons. Other brain areas enriched in 5-HT_{1A} receptors are for example the neocortex, the hypothalamus and some parts of the spinal cord, particularly the substantia gelatinosa. The localization of 5-HT_{1A} receptors in these areas suggests that 5-HT_{1A} mechanisms could also be involved in the functions of the hypothalamus, in the regulation of propioception and in integrative functions of the neocortex.

In order to localize more precisely the neurons expressing 5-HT_{1A} receptors and because of the lack of cellular resolution of light microscopic autoradiography in the brain, lesion experiments were carried out. The result of the destruction of serotoninergic neuronal cell bodies by the selective neurotoxin 5,7-dihydroxy-tryptamine or using electrolytic lesions showed that indeed these cell bodies were the cellular elements carrying the receptor sites.[14] Interestingly no alterations of receptor binding sites were seen after these lesions in forebrain areas, including the neocortex and hippocampus. Although it is difficult to conclude definitively on the presence or absence of 5-HT_{1A} in terminals, the most conservative interpretation of these results is that the density of presynaptic 5-HT_{1A} receptors in forebrain regions is probably low. This is further supported by the effects of lesions of the target cells in areas such as the hippocampus. Injections of kainic acid into the hippocampus produces a selective degeneration of pyramidal cells and interneurons in this brain area (FIG. 2). This degeneration is accompanied by the loss of 5-HT_{1A} binding in the hippocampus. This is an indication of the predominant postsynaptic localization of 5-HT_{1A} receptors in this area. However, other experiments, for example the lesion of the pathway connecting the septum with the hippocampus, the fimbria, or the electrolytic destruction of the septum resulted in some changes in 5-HT binding in the hippocampus.[15]

In the human brain the localization of receptors can be investigated by examining the binding of radioligands in tissues obtained at autopsy from humans dying with neurological diseases where specific neuronal populations degenerate. An investigation of the presence of 5-HT_{1A} receptors in the hippocampus of Alzheimer's disease patients showed a loss of receptor binding sites which was correlated with the loss of pyramidal cells in the hippocampus (FIG. 2). When taken together with the experimental lesions in the rat brain, these results tend to suggest that 5-HT_{1A} in the hippocampus are most probably postsynpatic to serotoninergic afferents.[16]

5-HT_{1B} and 5-HT_{1D} Receptors—the Basal Ganglia Connection

Outside the limbic system, other brain areas highly labeled by [^{3}H]5-HT are the basal ganglia.[8] (TABLE 2). Particularly dense [^{3}H]5-HT binding can be seen in the globus pallidus and in the substantia nigra pars reticulata (FIG. 3). In the rat brain these receptors represent the characteristics of 5-HT_{1B} sites. The binding of [^{3}H]5-HT is selectively blocked by compounds such as RU 24969 or some β-adrenergic blockers.[8] In fact, in the rat brain [^{125}I]iodo-cyanopindolol labels the globus pallidus and the substantia nigra and this binding is preferentially blocked by 5-HT[17]. In species other than rat or mouse, the basal ganglia are also characterized by high densities of [^{3}H]5-HT binding sites (FIG. 3).

However, in contrast to the pharmacological characteristics of these sites in rats and mice, the [^{3}H]5-HT binding in the basal ganglia of the other mammalian species examined until now is not sensitive to β-blockers and presents a pharmacological profile different from the one characteristic of the 5-HT_{1B} sites. We observed first the lack of 5-HT_{1B} binding in the human brain.[11] We found that this binding displayed characteristics different from those expected for 5-HT_{1B}, particularly in some basal ganglia areas. Because of the unusual pharmacology of [^{3}H]mesulergine binding sites in these areas we defined these sites as atypical 5-HT_{1C}.[12] Later, it was realized by us

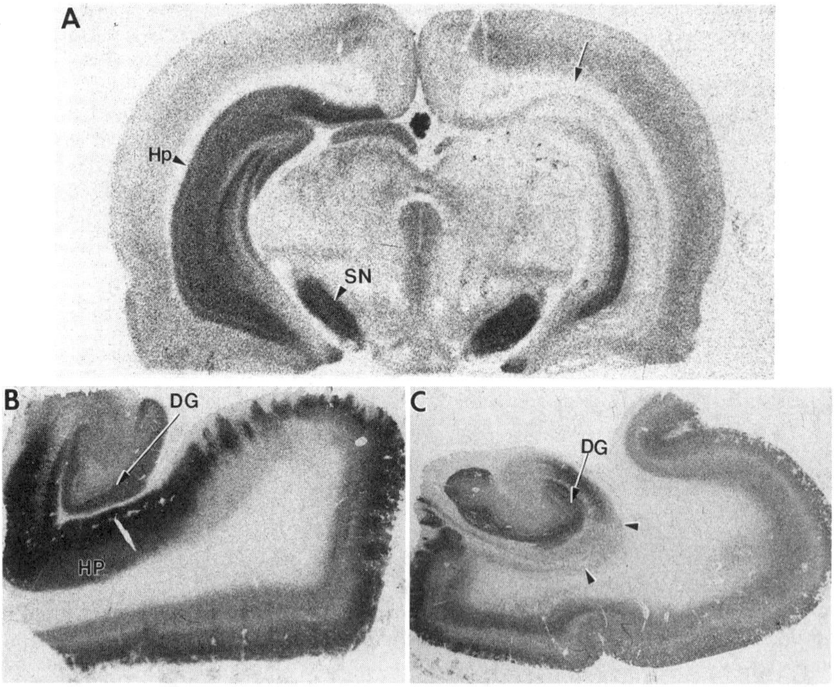

FIGURE 2. Effects of neuronal lesions on 5-HT receptors in the rat and human hippocampus. **B** is an autoradiogram obtained from a section of a rat injected with kainic acid in the right hippocampus and incubated with [^{3}H]5-HT to visualize 5-HT_1 receptor sites. Note the marked loss of binding in the lesioned hippocampus. **B** and **C** illustrate the distribution of 5-HT receptors in the hippocampus from a control human brain (**B**) and from an Alzheimer's disease (**C**) case. Note the loss of 5-HT receptor binding in the some hippocampal areas (*arrow heads*) of the Alzheimer's disease case associated with the loss of neurons in this area.

and others,[18,19] that in fact these sites belonged to a completely different class of 5-HT_1 receptors which has been called 5-HT_{1D}. Both in the rodents such as the rat or the mouse and in other mammalian species, including the guinea-pig, these 5-HT_{1B}/5-HT_{1D} receptors are also characteristic of presynaptic terminals of serotoninergic neurons.[20] The examination of the pharmacological profile of the receptor modulating the 5-HT release in the neo-cortex showed that this receptor belongs to the 5-HT_{1D} or 5-HT_{1B} class, depending on the species examined. Furthermore, it has been seen that the second messenger system used by these receptors is also similar, namely the inhibition of an adenylate cyclase (see Hoyer *et al.* in this volume). Thus, 5-HT_{1D} and 5-HT_{1B} receptors, although different pharmacologically, appear to serve the same type of functions and are localized in the same areas in the mammalian brain.

In the rat, lesions experiments have shown that these receptors could be presynaptically localized on the terminals of the striatal intrinsic neurons which innervate the substantia nigra pars reticulata, as destruction of caudate neurons (FIG. 4) or transection of the medial forebrain bundle results in a dramatic decrease of binding in the substantia nigra pars reticulata. On the other hand the lesion of dopaminergic neurons in the substantia nigra pars compacta does not induce a decrease of 5-HT_{1D}/5-HT_{1B} binding sites. It was interesting to examine pathologies of the human brain where the striatonigral pathway is known to be affected. For example in Parkinsons disease, the dopaminergic cells of the substantia nigra pars compacta die and this results in a denervation of the striatum. On the other hand, in Huntington's chorea, a large population of striatal neurons, particularly those projecting to the globus pallidus and substantia nigra, degenerate as a consequence of a genetic mutation. Examination of 5-HT binding in these brains showed a marked decrease in the caudate, the globus pallidus and the substantia nigra in the cases of Huntington's chorea, but no significant changes in Parkinson's disease cases. All these results taken together clearly indicate that in the human brain, as well as in the brains of the other mammalian species examined, 5-HT_{1B} and 5-HT_{1D} receptors are localized on cells which play an important role in the regulation of the activity of the basal ganglia. Furthermore these receptors do not appear to be under the influence of dopaminergic innervation.

The results obtained with 5-HT_{1B}/5-HT_{1D} receptors illustrate a general characteristic of 5-HT receptors, namely the possible presynaptic control of the release not only of 5-HT itself, but also of other transmitters. In other words, the 5-HT receptors can be presynaptic on 5-HT terminals, but also on non serotoninergic terminals: they can be autoreceptors or presynaptic heteroreceptors. It is interesting however, that the number of presynaptic heteroreceptors appears to be much larger than that of autoreceptors as lesion experiments do not seem to produce noticeable losses of 5-HT binding in most areas examined.

5-HT_{1C} Receptors—the Choroid Plexus and Much More

The presence of high densities of binding in the choroid plexus of the brain ventricles (TABLE 2) was already observed in the initial autoradiographic studies with serotoninergic ligands such as [^{3}H]LSD[22] or [^{3}H]5-HT.[23] As 5-HT_2 ligands did not label these tissues, it was assumed that these receptors in the choroid plexus belonged to the 5-HT_1 class. When [^{3}H]mesulergine became available it was initially described as a 5-HT_2 ligand.[24] It was then suprising to see on the first autoradiograms generated with this ligand that the choroid plexus was very intensively labeled by this ligand (FIG. 5). In agreement with the initial results these binding sites were more sensitive to 5-HT_1 than to 5-HT_2 ligands and our studies, both by autoradiography[8] and using membrane binding assays[25] showed that in fact these sites belonged to a new class of 5-HT receptors, which was named 5-HT_{1C} receptors.[25] 5-HT_{1C} sites have been visualized in the choroid plexus of all the mammalian species investigated until now. The properties of these receptors appeared to be homogenous throughout the species studies. High resolution autoradiography has shown that they are enriched on the epithelial cells of the choroid plexus.[26] As it is known that serotoninergic nerve terminals are present on the walls of the ventricles and that they could release 5-HT into the cerebro-spinal fluid (CSF) it was proposed that 5-HT could regulate the composition and volume of the CSF, acting via 5-HT_{1C} receptors.[25]

With the development of new 5-HT agents and the determination of the affinities of

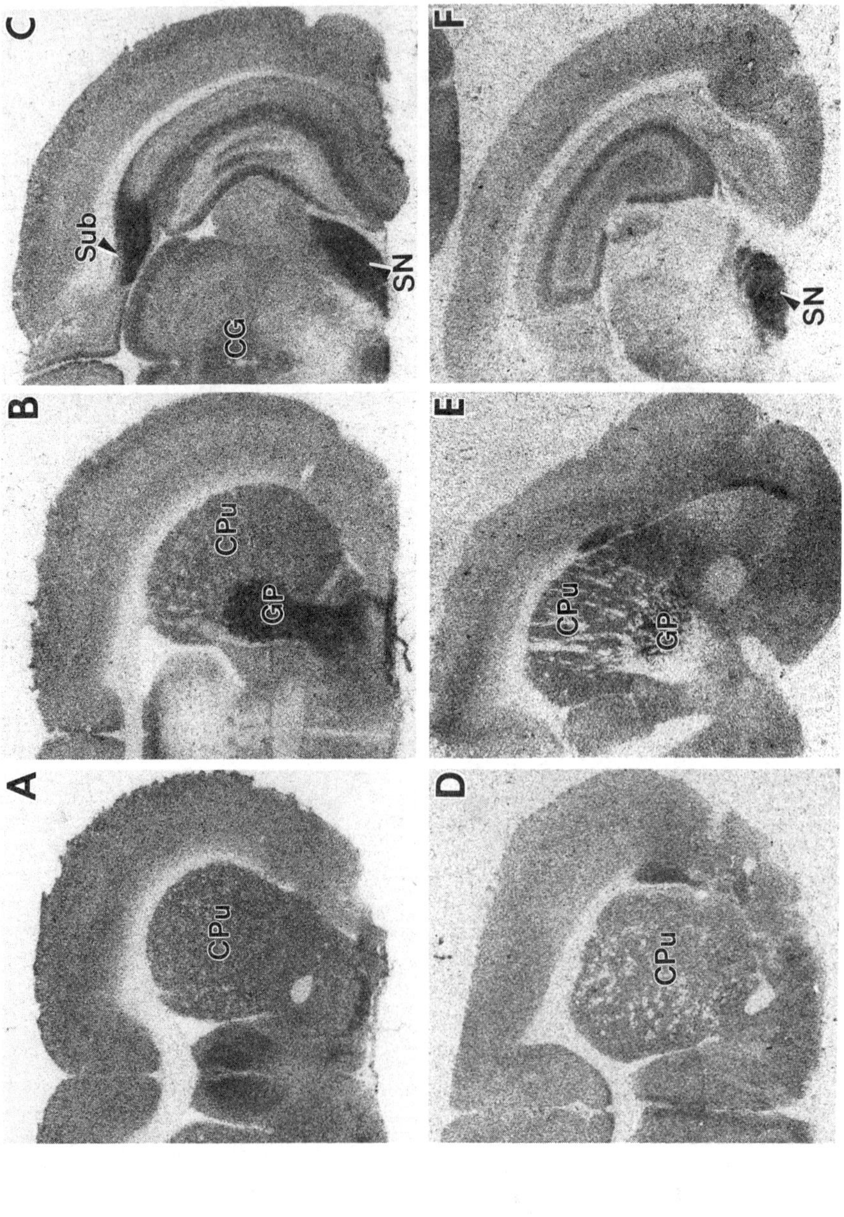
A
CPu
B
CPu
GP
C
Sub
SN
CG
D
CPu
E
CPu
GP
F
SN

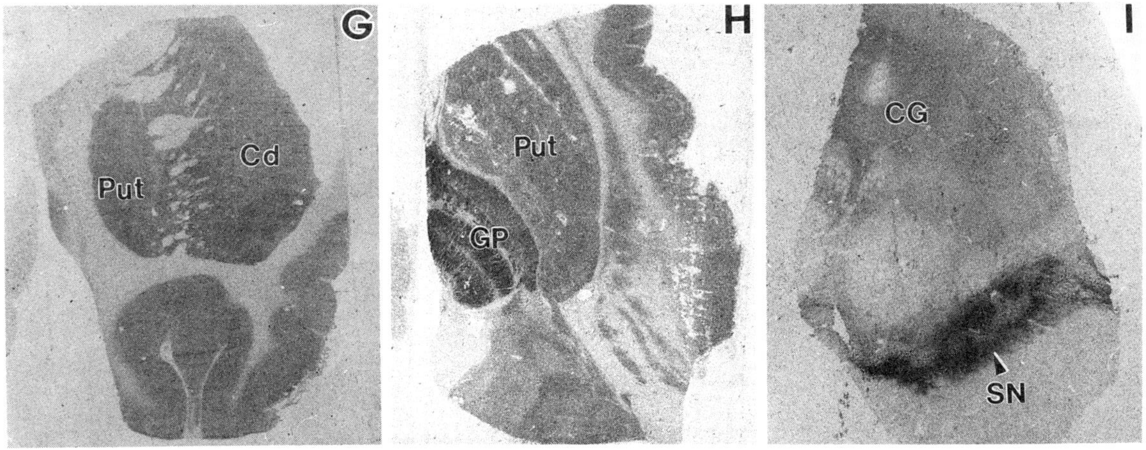

FIGURE 3. Visualization of $5\text{-}HT_{1B}/5\text{-}HT_{1D}$ receptor sites in the mammalian brain. The pictures show autoradiograms obtained using $[^3H]5\text{-}HT$ to label $5\text{-}HT_{1B}$ sites in the rat brain (**A,B,C**) and $5\text{-}HT_{1D}$ sites in the guinea-pig, (**D,E,F**) and human (**G,H,I**) brains. See text for details. **A,D** and **G** are sections at the level of the caudate putamen (CPu in rat and guinea pig, C and Put in human brain). **B,E** and **H** illustrate the high densities of $5\text{-}HT_{1B}/5\text{-}HT_{1D}$ binding in the globus pallidus (GP). **C,F** and **I** show these receptors enriched in the substantia nigra (SN) pars reticulata of the three species.

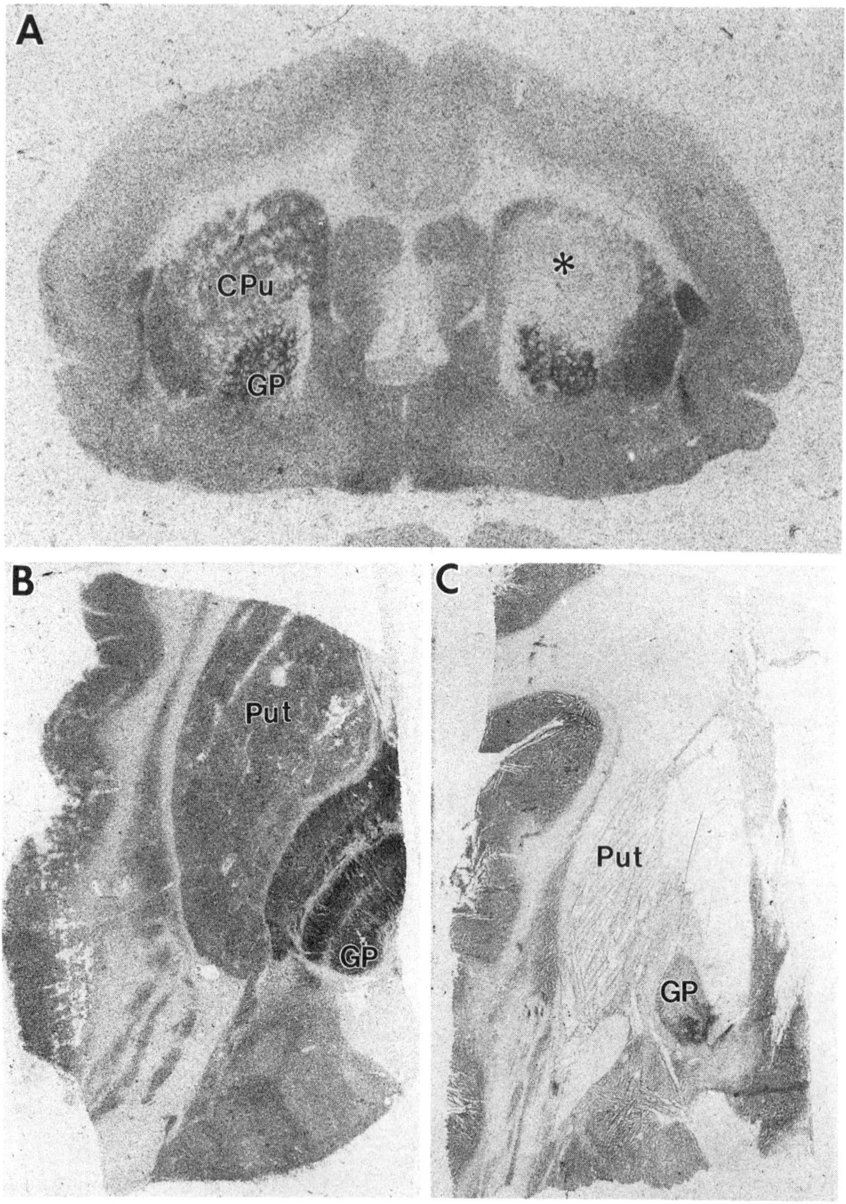

FIGURE 4. Effects of neuronal lesions in 5-HT_{1D} receptor binding. **A** is an autoradiogram showing the effects of a neurotoxin (quinolinic acid) unilateral injections (*) in the caudate putamen of a guinea-pig leading to a marked loss of 5-HT_{1D} binding. The decrease of 5-HT_{1D} binding sites in the basal ganglia of a Huntington's chorea is shown in **C** and compared to a control case (**B**).

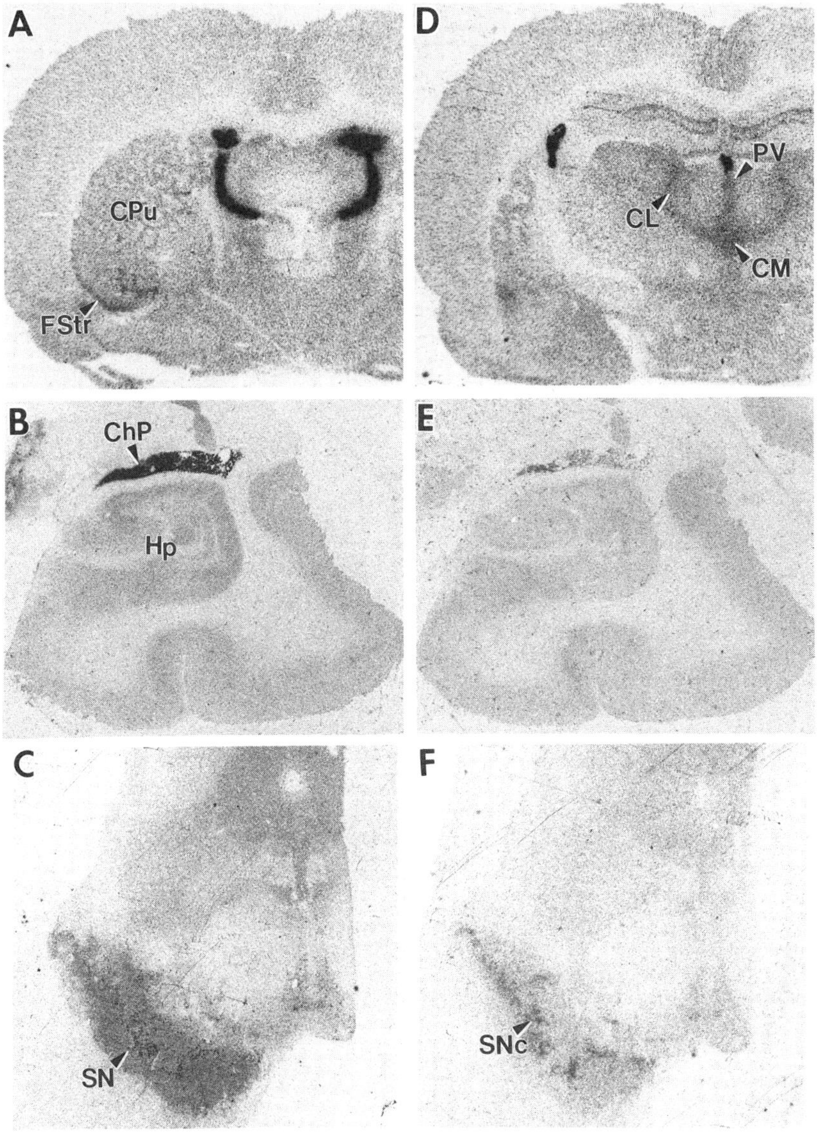

FIGURE 5. 5-HT_{1C} receptors in the rat and human brain. Autoradiograms from coronal rat brain sections incubated with [^{3}H]mesulergine in conditions in which 5-HT_{1C} binding is visualized (**A** and **D**). High densities of sites are seen in the choroid plexus (ChP). Other brain areas, such as the ventral striatum (FStr), hippocampus (Hp) and central medial (CM), centrolateral (CL) and paraventricular (PV) nuclei of the thalamus also present significant levels of binding. In the human brain high densities of 5-HT_{1C} binding sites were visualized in the choroid plexus (**B**) and substantia nigra (SN in **C**). Note that the levels of non-displaceable binding are important in some regions of the human brain (**E** and **F**), particularly in the substantia nigra pars compacta (SNc).

these compounds for different receptor subtypes, it was found that compounds such as *m*CPP (*m*-chloro-phenyl-piperazine) or TFMPP (trifluoro-methyl-phenyl-piperazine) presented some selectivity for 5-HT_{1C} sites. The administration of these compounds to laboratory animals and also to humans has shown that they produce effects that are probably mediated by brain areas different from the choroid plexus.[27] For instance, they can modify locomotor activity or feeding behavior. This has led us to re-evaluate the distribution of 5-HT_{1C} in the brain, and we have found that 5-HT_{1C} receptors are in fact very widespread throughout the mammalian brain. Particularly enriched, although in concentrations much lower than those seen in the choroid plexus, are the regions of the limbic system and those associated with motor behavior.[8] As we shall see later, the recent cloning of the gene for 5-HT_{1C} receptor has confirmed and extended these results.

Interestingly 5-HT_{1C} sites appear to be more abundant in the basal ganglia of humans, particularly the globus pallidus substantia nigra (FIG. 5).[12] All these data taken together indicate that 5-HT_{1C} receptors could be very important not only in regulating the volume and composition of the CSF, but also in particular functions of regions such as the human basal ganglia.

5-HT_2 Receptor—the Cortical Connection

When [^{3}H]spiperone was used to visualize receptors in the rat brain it was found that in addition to D_2 receptors this neuroleptic labeled high densities of sites in the neocortex, which were shown to present the characteristics of 5-HT_2 receptors.[28] This discovery led to the development of the first selective 5-HT_2 antagonist ketanserine.[29] Studies with [^{3}H]ketanserine confirmed these preliminary results with [^{3}H]spiperone and clearly showed that 5-HT_2 receptors are particularly enriched in many areas of the cortex in all the mammalian species examined until now.[30–32] (TABLE 2). In the neocortex, these sites are mainly concentrated in laminae I and V, but 5-HT_2 sites are also found in other brain areas (FIG. 6). They are particularly concentrated in the claustrum, a region which is connected to the visual cortex, some components of the limbic system, particularly the olfactory nuclei and parts of the basal ganglia. Attempts to characterize the cells expressing 5-HT_2 receptors in the neocortex using lesion experiments have suggested that these receptors are localized on the processes of intrinsic cells, for deafferentation and many other types of lesion do not result in changes in 5-HT_2 receptor densities. On the other hand, cortical [^{3}H]ketanserin binding sites have been reported to be decreased in senile dementia of the Alzheimer type, paralleling the loss of somatostatin immunoreactivity and the decrease of GABA concentration in this region.[33] This suggests that 5-HT_2 receptors could be located on intrinsic GABA- or somatostatin-containing neurons, which are known to be present in the cortex. As we shall show later, in situ hybridization studies have confirmed this assumption and demonstrated that cells intrinsic to the neocortex express mRNA coding for 5-HT_2 receptors.

5-HT_3 Receptors—the Link between the Periphery and the Central Nervous System. A New Limbic Connection?

The presence of excitatory 5-HT receptors in peripheral neurons has been established since the 1950s (for review see ref 34). The development of compounds such as ICS 205-930, MDL 72222 and GR 65630 in the last five years has allowed thorough characterization of these receptors, which were named 5-HT_3. These sites have been directly labeled in peripheral and central tissues using several tritiated ligands (see

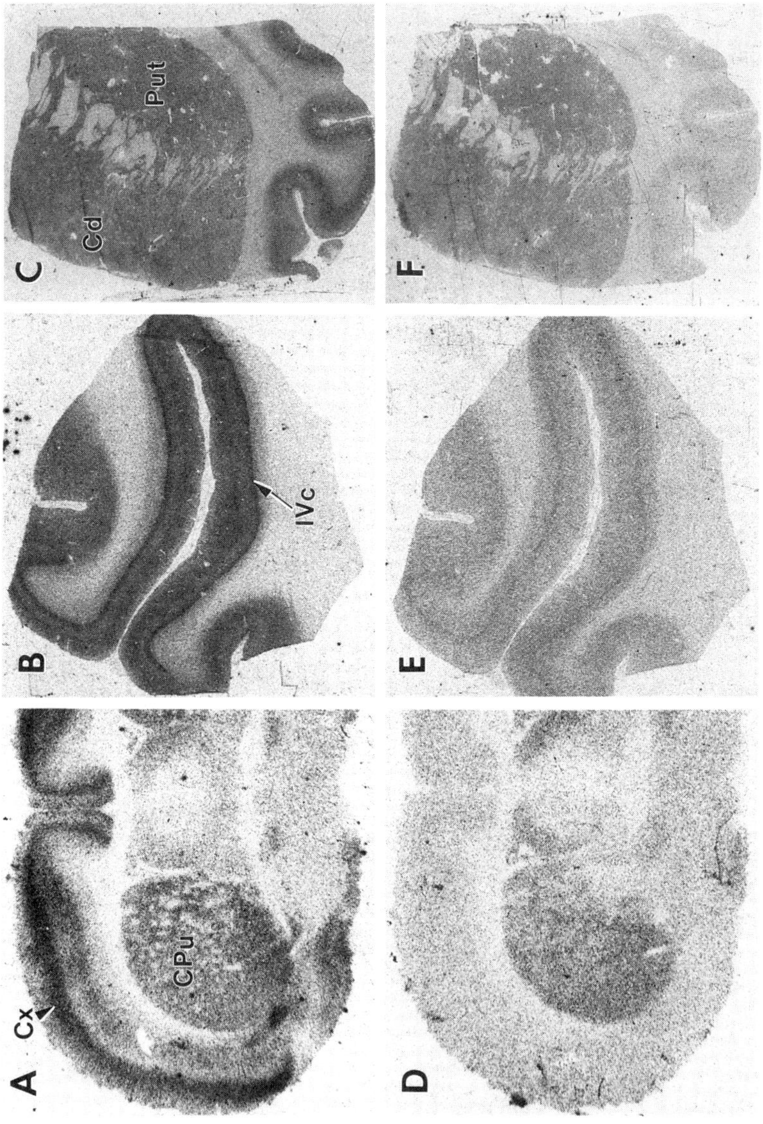

FIGURE 6. Visualization of 5-HT_2 sites in rat and human brain. Sections were labeled with [^{3}H]ketanserine. Total binding is illustrated in **A, B, C**. Non-displaceable binding is shown in **D, E, F**. Note high densities of specific binding in several layers of the cortex in rat (**A**) and man (**B** and **C**). Part of the binding to the caudate and putamen (CPu, **C**; Put in **A** and **C**) was, however, non-specific (**D** and **F**). High levels of non-specific binding were also observed in some layers (IVc) in the human visual cortex (**E**).

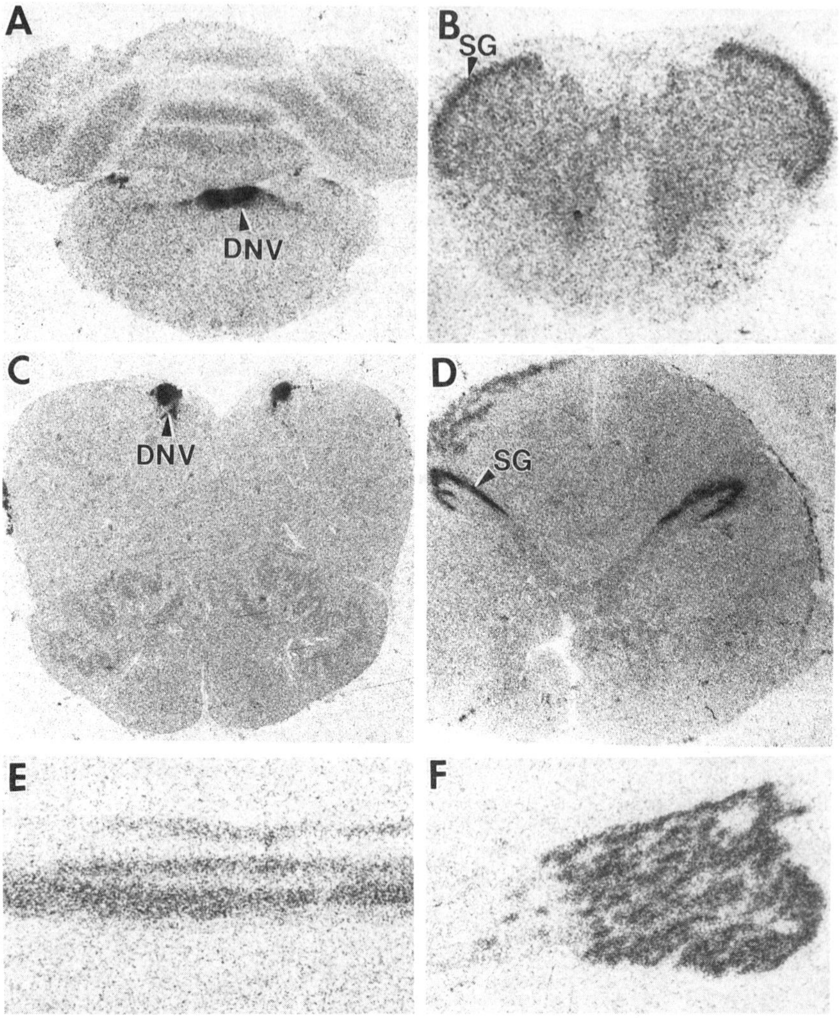

FIGURE 7. 5-HT_3 receptors visualized in rat, cat and human tissues. Autoradiograms were obtained using [^{3}H]ICS 205-930 as a ligand. In the rat high densities of binding were seen in the dorsal nucleus of the vagus (DNV in **A**) and in the substantia gelatinosa of the spinal cord (SG in **B**). In the human medulla (**C**) and spinal cord (**D**) the same regions were labeled. The cat nodose ganglion (**F**) and vagus nerve (**E**) were also rich in 5-HT_3 sites.

TABLE 1). We have used [^{3}H]ICS 205-930 to visualize 5-HT_3 sites in the brain and peripheral ganglia of several mammals, including man.[7,35] Other autoradiographical studies have been performed using [^{3}H]GR65630[36] and [^{3}H]zacopride.[37] FIGURE 7 illustrates the localization of 5-HT_3 receptors in the rat and human CNS as well as the cat nodose ggnglion and vagus nerve. The highest densities of 5-HT_3 receptors were found in discrete brainstem areas, such as the nucleus tractus solitarius, the dorsal nucleus of the vagus nerve, the substantia gelatinosa of the spinal trigeminal nucleus and the area postrema, as well as in the spinal cord, particularly in the substantia gelatinosa (TABLE 2). Examination of peripheral ganglia such as the superior cervical ganglion showed a very high density of these binding sites in the cell bodies of the ganglionic neurons, but also an intense labeling in the vagus nerve.[38] This suggests that these receptors are synthesized in the ganglionic cell bodies and are axonally transported to the terminals, both in the CNS and in the periphery. In forebrain areas, low densities of [^{3}H]ICS 205-930 binding sites were found in the hippocampal formation of mouse, rat and human brain.

The presence of 5-HT_3 binding sites in the peripheral ganglia and nerves as well as the substantia gelatinosa of the trigeminal nerve and spinal cord is compatible with a presynaptic localization of these sites on the terminals of the ganglionic afferents. In agreement with this hypothesis, selective destruction of these afferents with capsaicin results in a partial loss of [^{3}H]zacopride binding in the dorsal horn of the spinal cord.[37] The cellular localization of the remaining binding sites still has to be established. Independently of the pre- or postsynaptic localization of these sites, their enrichment in the mentioned spinal cord and medullary areas suggests that 5-HT could modulate nociceptive mechanisms via 5-HT_3 receptors. The particularly high densities of [^{3}H]ICS 205-930 binding sites in the area postrema provides an anatomical substrate for the well-known anti-emetic effects of 5-HT_3 antagonists.[39]

Other central effects of 5-HT_3 receptor antagonists have been described. For instance it has been proposed that these compounds could have anxiolytic effects.[40] However, with the ligands available until now it has been difficult to demonstrate a selective localization of 5-HT_3 sites to the limbic system. The development of more sensitive ligands with higher specific activity is necessary to further investigate the distribution of 5-HT_3 receptors throughout the forebrain.

Visualization of Cells Expressing mRNAs Coding for 5-HT Receptors Using in Situ Hybridization Technique

As already mentioned, light microscopic autoradiography does not allow the unequivocal establishment of the cellular localization of receptors. The recent molecular cloning of cDNAs or genes encoding three 5-HT receptors[1] has enabled us to develop synthetic oligonucleotides complementary to selected regions of the sequences of mRNAs coding for these proteins. By means of in situ hybridization histochemistry, we have visualized the cells expressing 5-HT receptors. 5-HT_2 and 5-HT_{1C} mRNAs were found localized to different cellular populations in the rat brain. High densities of 5-HT_{1C} receptor were seen not only in the choroid plexus but also in several brain areas and nuclei (FIG. 8). Particularly rich in 5-HT_{1C} transcripts were the olfactory nucleus, cingulate cortex, lateral habenula and subthalamic nucleus. 5-HT_2 receptor mRNA was mainly expressed in cells localized in the neocortex, the olfactory bulb and cortex and the claustrum. In general the localizations of mRNAs was compatible with the distribution of binding sites, although discrepancies were found in the relative densities of mRNA and binding sites. This could suggest, for example, that receptors, or their mRNAs, have different turnover rates in different cell populations.

CONCLUSION

The elucidation of the anatomical, and in some cases cellular, localization of the different subtypes of 5-HT receptors has significantly contributed to their pharmacological characterization and to the understanding of the role of 5-HT mechanisms both in the central and peripheral nervous system. The predictible development of new tools

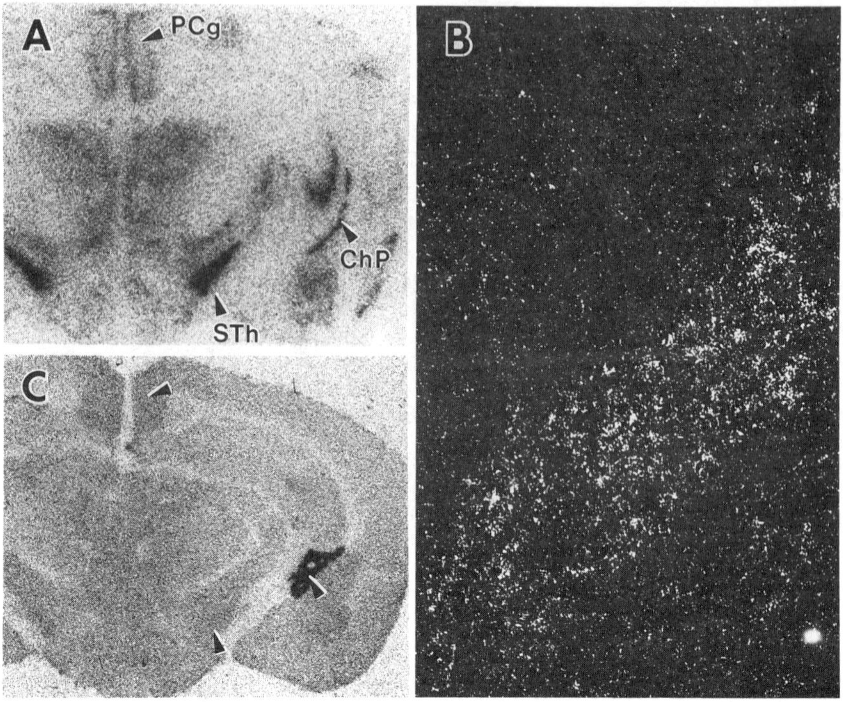

FIGURE 8. 5-HT_{1C} receptors mRNA containing cells visualized by in situ hybridization. **A** is a film autoradiogram showing the distribution of hybridization signal in a section of the mouse brain at the level of the subthalamic nucleus (STh), which is intensely labeled. Very high levels of hybridization signal are also seen in the choroid plexus (ChP). Also labelled are the CA3 area of the hippocampus (CA3), some amygdaloid nuclei and the posterior cingulate cortex (PCg). **B** is a higher magnification picture of a liquid emulsion autoradiogram illustrating intense hybridization signal over cell bodies of the subthalamic nucleus which appear as white dots under dark-field illumination. **C** shows the distribution of 5-HT_{1C} receptor binding sites in a section close to the one shown in **A**. Note the presence of high levels of binding in the choroid plexus, but not in the subthalamic nucleus.

such as more selective ligands, the cloning of other 5-HT receptors and the production of antibodies against receptor proteins will contribute to answering some of the remaining questions on their cellular and subcellular localization and on their regulation.

ACKNOWLEDGMENT

We wish to thank Dr. N. Venable for critically reading the manuscript.

REFERENCES

1. Hartig, P. R. 1989. Molecular biology of 5-HT receptors. TIPS **10:** 64–69.
2. Takeuchi, Y. 1988. Distribution of serotonin neurons in the mammalian brain. *In* Neuronal Serotonin. N. N. Osborne and M. Hamon, Chichester: John Wiley & Sons, pp. 25–56.

3. PEROUTKA, S. J. 1988. 5-Hydroxytryptamine receptor subtypes: Molecular biochemical and physiological characterization TINS **11:** 496–500.
4. PALACIOS, J. M. 1988. Mapping brain receptors by autoradiography. ISI Atlas of Science: Pharmacology **2:** 71–77.
5. PAZOS, A., D. HOYER, M. M. DIETL & J. M. PALACIOS. 1988. Autoradiography of serotonin receptors. *In* Neuronal Serotonin. N. N. Osborne and M. Hamon, Eds. Chichester: John Wiley & Sons, pp. 507–543.
6. WAEBER, C., M. M. DIETL, D. HOYER, A. PROBST & J. M. PALACIOS. 1988. Visualization of a novel serotonin recognition site ($5\text{-}HT_{1D}$) in the human brain by autoradiography. Neurosci. Lett. **88:** 11–16.
7. WAEBER, C., K. DIXON, D. HOYER & J. M. PALACIOS. 1988. Localisation by autoradiography of neuronal $5\text{-}HT_3$ receptors in the mouse CNS. Eur. J. Pharmacol. **151:** 351–352.
8. PAZOS, A. & J. M. PALACIOS. 1985. Quantitative autoradiographic mapping of serotonin receptors in the rat brain. I. Serotonin-1 receptors. Brain Res. **346:** 205–230.
9. GOZLAN, H., S. EL MESTIKAWY, L. PICHAT, J. GLOWINSKI, & M. HAMON. 1983. Identification of presynaptic serotonin autoreceptors using a new ligand: [^{2}H]PAT. Nature **305:** 140–142.
10. PONCHANT, M., J.-P. BEAUCOURT, A. VANHOVE, G. DAVAL, D. VERGE, M. HAMON & H. GOZLAN. 1988. ^{125}I-BH-8-NeO-N-PAT, un nouveau ligand pour l'étude des récepteurs $5\text{-}HT_{1A}$ dans le système nerveux central. C. R. Acad. Sci. Paris, **306(III):** 147–152.
11. HOYER, D., A. PAZOS, A. PROBST & J. M. PALACIOS. 1986. Serotonin receptors in the human brain. I. Characterization and autoradiographic localization of $5\text{-}HT_{1A}$ recognition sites. Apparent absence of $5\text{-}HT_{1B}$ recognition sites. Brain Res. **376:** 85–96.
12. PAZOS, A., A. PROBST & J. M. PALACIOS. 1987. Serotonin receptors in the human brain. III. Autoradiographic mapping of serotonin-1 receptors. Neuroscience **21:** 97–122.
13. IVERSEN, S. D. 1984. 5-HT and anxiety. Neuropharmacology **23:** 1553–1560.
14. VERGǴE, D., G. DAVAL, M. MARCINKIEWICZ, A. PATEY, S. EL MESTIKAWY, H. GOZLAN & M. HAMON. 1986. Quantitative autoradiography of multiple $5\text{-}HT_1$ receptor subtypes in the brain of control or 5,7-dihydroxy-tryptamine-treated rats. J. Neurosci. **6:** 3474–3482.
15. PALACIOS, J. M. & M. M. DIETL. 1989. Autoradiographic studies of serotonin receptors. *In* The Serotonin Receptors. E. Sanders-Bush, Ed. Clifton, N.J.: The Humana Press, pp. 89–140.
16. CROSS, A. J., T. J. CROW, I. N. FERRIER, J. A. JOHNSON, S. R. BLOOM & J. A. N. CORSELLIS. 1984. Serotonin receptor changes in dementia of the Alzheimer type. J. Neurochem. **43:** 1574–1581.
17. PAZOS, A., G. ENGEL & J. M. PALACIOS. 1985. β-Adrenoreceptors blocking agents recognize a subpopulation of serotonin receptors in brain. Brain Res. **343:** 403–408.
18. HEURING, R. E. & S. J. PEROUTKA. 1987. Characterization of a novel ^{3}H-5-hydroxytryptamine binding site subtype in bovine brain membranes. J. Neurosci. **7:** 894–903.
19. WAEBER, C., P. SCHOEFFTER, J. M. PALACIOS & D. HOYER. 1988. Molecular pharmacology of $5\text{-}HT_{1D}$ recognition sites: Radioligand binding studies in human, pig and calf brain membranes. Naunyn-Schmiedebergs Arch. Pharmacol. **337:** 595–601.
20. HOYER, D. & D. MIDDLEMISS. 1989. Species differences in the pharmacology of terminal 5-HT autoreceptors in mammalian brain. TIPS **10:** 130–132.
21. WAEBER, C. & J. M. PALACIOS. 1989. Serotonin-1 receptor binding sites in the human basal ganglia are decreased in Huntington's chorea but not in Parkinson's disease: A quantitative in vitro autoradiography study. Neuroscience **32:** 337–347.
22. MEIBACH, R. C., S. MAAYANI & J. P. GREEN. 1980. Characterization and radioautography of [^{3}H]LSD binding by rat brain slices in vitro: The effect of 5-hydroxytryptamine. Eur. J. Pharmacol. **67:** 371–382.
23. YOUNG, W. S. & M. J. KUHAR. 1980. Serotonin receptor localization in rat brain by light microscopic autoradiography. Eur. J. Pharmacol. **62:** 237–239.
24. CLOSSE, A. 1983. [^{3}H]Mesulergine, a selective ligand for serotonin-2 receptors. Life Sci. **32:** 2485–2495.
25. PAZOS, A., D. HOYER & J. M. PALACIOS. 1985b. The binding of serotoninergic ligands to porcine choroid plexus: Characterization of a new type of serotonin recognition site. Eur. J. Pharmacol. **106:** 539–546.

26. YAGALOFF, K. A. & P. R. HARTIG. 1985. ^{125}I-LSD binds to a novel serotonergic site on rat choroid plexus epithelial cells. J. Neurosci. **5:** 3178–3183.
27. KENNETT, G. A. & G. CURZON. 1988. Evidence that hypophagia induced by mCPP and TFMPP requires 5-HT_{1C} and 5-HT_{1B} receptors; hypophagia induced by RU24969 only requires 5-HT_{1B} receptors. Psychopharmacology **96:** 93–100.
28. PALACIOS, J. M., D. L. NIEHOFF & M. J. KUHAR. 1981. [^{3}H]Spiperone binding sites in brain: Autoradiographic localization of multiple receptors. Brain Res. **213:** 277–289.
29. LEYSEN, J. E., C. J. E. NIEMEGEERS, J. M. VAN NUETEN & P. M. LADURON. 1982. [^{3}H]Ketanserin (R 41 468), a selective [^{3}H]-ligand for $serotonin_2$ receptor binding sites, Mol. Pharmacol. **21:** 301–314.
30. PAZOS, A., A. PROBST, J. M. PALACIOS. 1987. Serotonin receptors in the human brain. IV. Autoradiographic mapping of serotonin-2 receptors. Neuroscience **21:** 123–139.
31. PAZOS, A., R. CORTES & J. M. PALACIOS. 1985. Quantitative autoradiographic mapping of serotonin receptors in the rat brain. II. Serotonin-2 receptors. Brain Res. **346:** 231–249.
32. HOYER, D., A. PAZOS, A. PROBST & J. M. PALACIOS. 1986. Serotonin receptors in the human brain. II. Characterization and autoradiographic localization of 5-HT_{1C} and 5-HT_2 recognition sites. Brain Res. **376:** 97–107.
33. CROSS, A. J. 1988. Serotonin in neurodegenerative disorders. *In* Neuronal Serotonin. N. N. Osborne and M. Hamon, Eds. Chichester: John Wiley & Sons, pp. 231–253.
34. RICHARDSON, B. P. & G. ENGEL. 1986. The pharmacology and function of 5-HT_3 receptors. TINS **9:** 424–428.
35. WAEBER, C., D. HOYER & J. M. PALACIOS. 1989. 5-Hydroxytryptamine-3 receptors in the human brain: Autoradiographic visualization using [^{3}H]ICS 205-930. Neuroscience **31:** 393–400.
36. KILPATRICK, G. J., B. J. JONES & M. B. TYERS. 1988. The distribution of specific binding of the 5-HT_3 receptors ligand [^{3}H]GR65630 in rat brain using quantitative autoradiography. Neurosci. Lett. **94:** 156–160.
37. HAMON, M., M. C. GALLISSOT, F. MENARD, H. GOZLAN, S. BOURGOIN, & D. VERGE. 1989. Biochemical and autoradiographic evidence of 5-HT_3 receptor binding sites on capsaicin-sensitive fibres in the rat spinal cord. Eur. J. Pharmacol. In press.
38. HOYER, D., C. WAEBER, A. KARPF, H. NEIJT & J. M. PALACIOS. 1989. [^{3}H]ICS 205-930 labels 5-HT_3 recognition sites in membranes of cat and rabbit vagus nerve and superior cervical ganglion. Naunyn-Schmiedeberg's Arch. Pharamcol. In press.
39. BORISON, H. L. 1984. History and status of the area postrema. Fed. Proc. **43:** 2937–2940.
40. JONES, B. J., B. COSTALL, A. M. DOMENEY, M. E. KELLY, R. J. NAYLOR, N. R. OAKLEY & M. B. TYERS. 1988. The potential anxiolytic activity of GR38032F, a 5-HT_3-receptor antagonist. Br. J. Pharmacol. **93:** 985–993.

DISCUSSION OF THE PAPER

JOSEE LEYSEN (*Janssen Research Foundation, Beerse Belgium*): You did not show the localization in the mesolimbic region particularly the olfactory tubercle, where there is a very high density—nearly as high as in the frontal cortex. You can see this with homogenate binding and with autoradiography. When you did the messenger RNA study did you see the expression of the receptors also there?

PALACIOS: A little bit, not as remarkable as it is in the cortex and the olfactory bulb.

LEYSEN: So there is a discrepancy between the messenger RNA and the number of sites which you can label?

PALACIOS: Yes, probably.

Serotonin-Storing Secretory Vesicles[a]

HADASSAH TAMIR[b,c] AND MICHAEL D. GERSHON[b]

[b]*Department of Anatomy and Cell Biology*
Columbia University College of Physicians & Surgeons
New York, New York 10032

[c]*Division of Neuroscience*
New York State Psychiatric Institute
New York, New York 10032

INTRODUCTION

Serotonergic neurons appear to play a critical role in the maintenance of mental health. For example, drugs that potentiate the action of serotonin (5-HT) by blocking its re-uptake alleviate depression[1] and agonists at 5-HT_{1A} receptors are potent anxiolytics.[2] Other drugs that act on 5-HT_2 receptors are hallucinogens.[3,4] Additional evidence has implicated serotonergic neurons in the control of mood, temperature regulation, pain perception, feeding, and sleeping behavior.[5] Not only are serotonergic neurons important in brain function, but peripheral serotonergic neurons in the gastrointestinal tract also appear to be vital to the normal function of the enteric nervous system (ENS).[6] It follows, therefore, that an understanding of the cellular biology of serotonergic neurons, including knowledge of how the transmitter is stored and re-cycled will be of great importance to understanding the role of this subset of neurons in health and disease. In addition this knowledge can be expected to foster the design of new psychotherapeutic and gastrointestinal drugs. A great deal of useful knowledge has already been learned about serotonergic mechanisms, including the location in the brain and PNS of 5-HT-containing neurons and their projections, the biosynthesis and metabolism of 5-HT, the mechanism of uptake of the amine from an ambient medium, and the identification of many subtypes of pre- and post-junctional 5-HT receptor (for a review see ref. 7).

Much less is known about the subcellular storage vesicles in which 5-HT is contained than about other components of serotonergic neurons. It is clear that these vesicles resemble in some respects vesicles that store catecholamines (CA). For example, reserpine and tetrabenazine, which inhibit the action of the transporter molecule in the vesicular membrane, deplete 5-HT as well as CA from the brain.[8] This indicates that the same or a similar transporter is present in both types of synaptic vesicle and that 5-HT, like CA, is dependent upon vesicular storage to protect intracellular amine from catabolism by monoamine oxidase (MAO).[9] Moreover, isolated amine-containing vesicles from several sources take up 5-HT as well as CA.[10] On the other hand, considerable evidence indicates that the synaptic vesicles of serotonergic neurons are also quite different from those of catecholaminergic neurons in a number of important respects. Some of this evidence has been derived from studies of the CNS; however, because neither serotonergic neurons, nor 5-HT- containing synaptic vesicles can be obtained from the brain uncontaminated by other types of neuron or other types of vesicle, respectively, studies of serotonergic vesicular properties have been supplemented by investigations of peripheral serotonergic neurons and paraneurons (thyroid parafollicular cells.[11]) These peripheral sources of 5-HT storage

[a]This work was supported in part by grants NIMH 37575, NS 12969, and NS15547.

organelles have been valuable models for the study of serotonergic mechanisms in much the same way that research done on chromaffin granules of the adrenal medulla has been helpful in revealing properties of CA storage organelles.[12] Peripheral serotonergic neurons are found in the gut in all vertebrates.[6] and serotonergic paraneurons are found in the thyroid gland of a number of mammalian species.[13,14] Platelets, which like serotonergic neurons, take up and store 5-HT have also been used as surrogates for neurons in the study of serotonergic mechanisms[15,16]; however, in contrast to central and peripheral serotonergic neurons and parafollicular cells, which are embryologically derived from neurectoderm, platelets are mesodermal derivatives. Moreover, platelets do not synthesize 5-HT as do the 5-HT-storing cells of neurectodermal origin, and platelets do not express the specific high affinity 5-HT binding protein (SBP) that characterizes all of the neurectodermal 5-HT-storing cells.[17,18] Furthermore, platelets are not whole cells, but cell fragments; therefore, the amount of cell biology that can be learned from the study of platelets is necessarily limited. In looking for a model relevant to central serotonergic neurons, therefore, it would appear that a neurectodermal serotonergic cell would be preferable to platelets.

PARAFOLLICULAR CELLS OF THE THYROID

During mammalian and avian development cells migrate from the region of the rhombencephalic (vagal) neural crest to the neck where they contribute to the formation of the ultimobranchial bodies.[19] In avians the ultimobranchial bodies persist, but in mammals they become incorporated into the developing thyroid gland. In both the avian ultimobranchial body and the mammalian thyroid, the crest-derived cells give rise to parafollicular or "C" cells (FIG. 1), which store 5-HT in concentrations that vary between species.[14] The serotonergic neurons of the gut are derived from the same region of the neural crest as are the parafollicular cells.[19] Parafollicular cells store calcitonin as well as 5-HT.[20] Recent EM immunocytochemical studies have revealed that calcitonin (FIG. 2) and 5-HT are co-stored in parafollicular cells (FIG. 3) in the same subcellular vesicles.[21] The relationship of parafollicular cells to neurons remains evident, not only in the structure and properties of the cells, but in their response to nerve growth factor (NGF) and microenvironmental conditions.[22]

Sheep parafollicular cells, which are particularly rich in 5-HT[23], have been isolated.[24] When cells from dissociated thyroid glands are treated with thyrotropin (TSH) follicular cells become phagocytic. TSH-activated cell suspensions can then be passed through columns to which thyroglobulin has been coupled. The follicular cells "attempt" to phagocytize the beads and are retained on the columns, while the parafollicular cells pass through. This procedure of "phagocytic chromatography," coupled with gradient centrifugation, results in a final preparation 97% of which is composed of parafollicular cells. When purified parafollicular cells are cultured in the presence of NGF they extend neurites.[22] More significantly, these neurites express neurofilament proteins and synthesize calcitonin gene–related peptide (CGRP), instead of calcitonin. CGRP is produced by alternative splicing of mRNA coded by the calcitonin gene in neurons, while calcitonin is the product of the same gene expressed in endocrine cells. These observations indicate that parafollicular cells can be induced to change their phenotype from that of an endocrine cell to that of a neuron. Within 24–48 h following exposure of parafollicular cells to NGF, the cells that have extended neurites die. In contrast, if isolated parafollicular cells are co-cultured with segments of aneuronal chick hindgut, those parafollicular cells that reach the region where the myenteric plexus of the bowel would normally form (in the external muscular layer) survive indefinitely and even give rise to a neural-appearing plexus.[22] These observa-

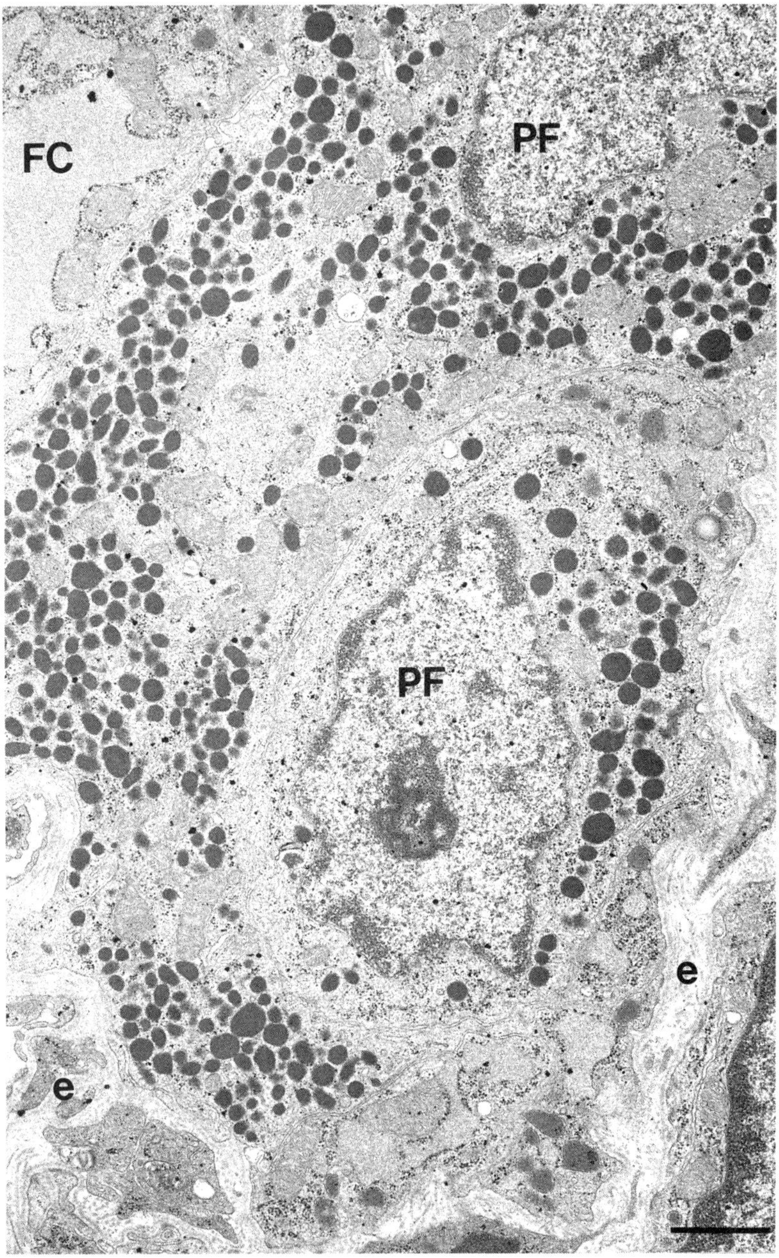

FIGURE 1. An electron micrograph showing the morphology of typical mammalian parafollicular cells. The illustration was obtained from a bat. Note the large number of dense cores secretory vesicles. PF = parafollicular cells; FC = follicular cells; e = extracellular space; The marker = 1.0 μm.

tions have led to the hypothesis that there may be a primitive serotonergic precursor cell that gives rise to serotonergic neurons in the gut and to parafollicular cells in the thyroid or ultimobranchial bodies. In any case, because of its strong resemblance to a neuron, the parafollicular cell has been called a paraneuron.[11] It follows that the properties of the 5-HT storage organelles of parafollicular cells are likely to have a great deal in common with the properties of 5-HT-storing synaptic vesicles.

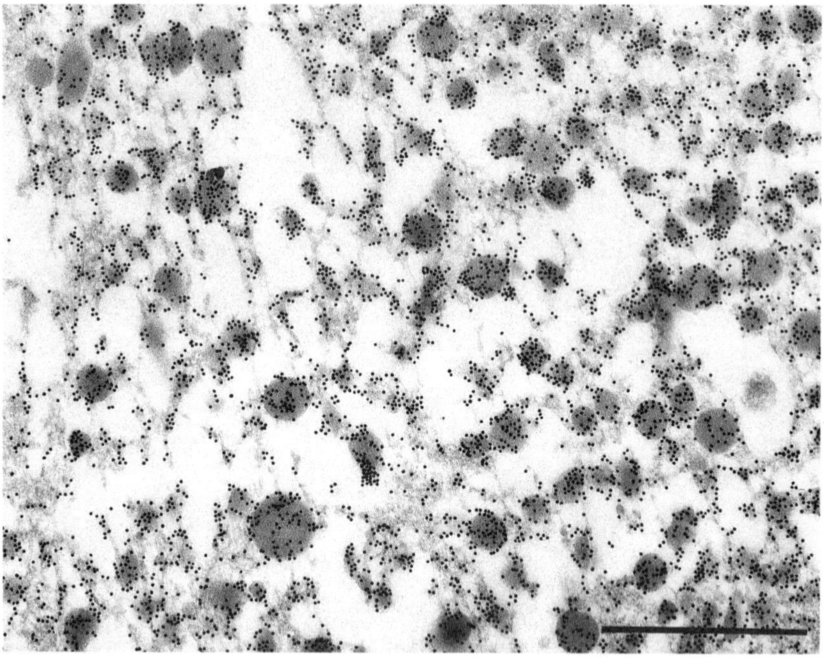

FIGURE 2. A portion of a resting parafollicular cell. The material has been fixed with aldehydes and embedded without osmication in LR White. The sections were incubated with anti-calcitonin sera and sites of immunoreactivity were visualized with a secondary antibody labeled with 5 nm particles of colloidal gold. Virtually every granule shows calcitonin immunoreactivity. The marker = 1.0 μm.

PROPERTIES OF 5-HT- AND CATECHOLAMINE (CA)-STORING VESICLES

5-HT-containing secretory vesicles have been isolated from sheep thyroid glands by a two-step process involving separation of subcellular particles, first by size, and then by density.[21] Storage organelles for 5-HT appear to differ from those which contain CA in ATP content, regulation of intravesicular pH, and in their complement of specific intravesicular proteins.[21,25] For example, ATP cannot be demonstrated histochemically in vesicles in known serotonergic axons, such as supraependymal fibers,[26] or in the 5-HT-storing parafollicular cells of sheep.[21] On the other hand, ATP is readily demonstrated in adrenal chromaffin granules.[27–29] Moreover, synaptic vesicles isolated from 5-HT-containing synaptosomes are poor in ATP[30] and parafollicular cell granules

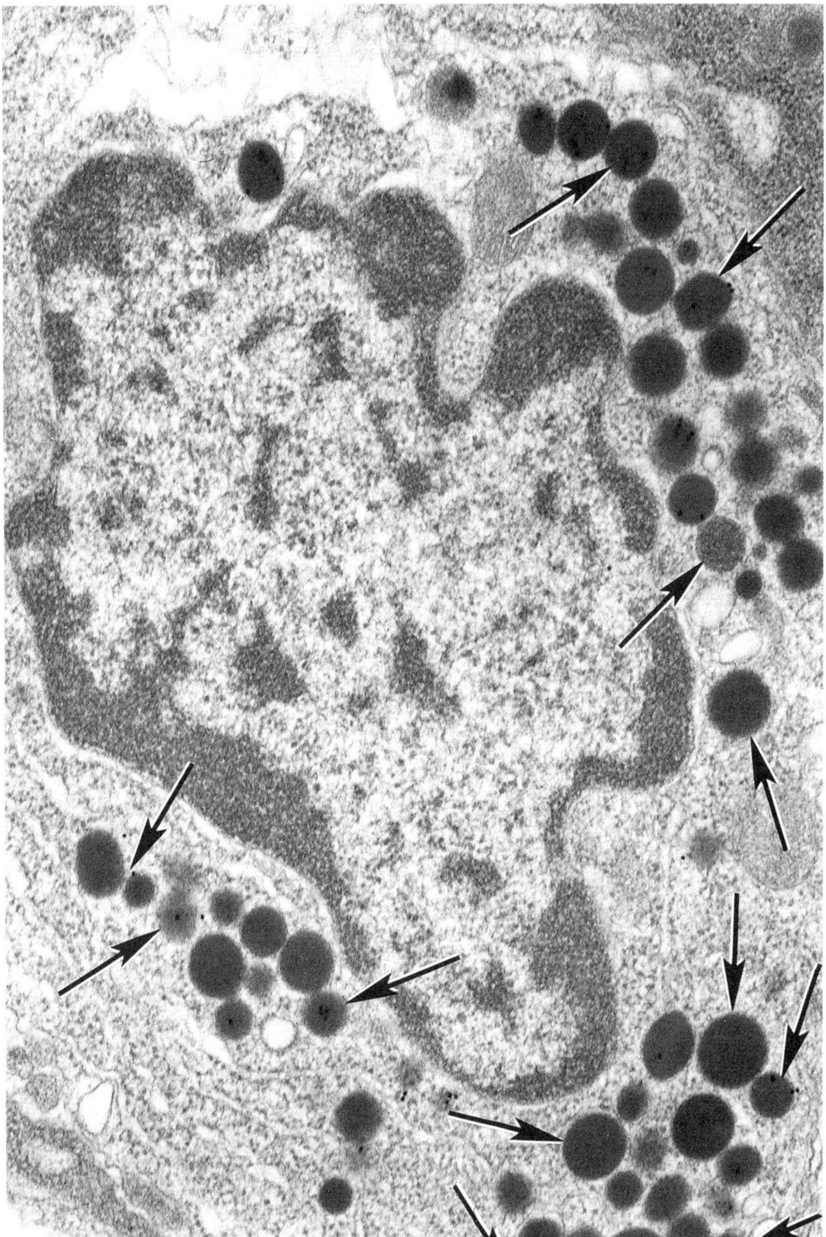

FIGURE 3. An electron micrograph of a parafollicular cell. The tissue was fixed with paraformaldehyde (4%) and glutaraldehyde (0.1%) and was lightly osmicated prior to embedding in Epon 812. The sections were picked up on nickel grids and exposed to anti-5-HT sera. Sites of immunoreactivity were visualized with a secondary antibody labeled with 5 nm particles of colloidal gold. About 50% of granules are labeled (→). The marker = 1.0 μm.

contain virtually none.[21] Additional properties of 5-HT-storing vesicles (isolated from sheep thyroids) which are shared with chromaffin granules include the presence in the vesicles of the same or a similar vacuolar H^+ATPase. Antibodies to each of the 4 subunits of the ATPase[31] cross react with proteins in parafollicular vesicles and the 72 kDa subunit can be demonstrated in the parafollicular granules of intact cells by EM immunocytochemistry (FIG. 4). The vesicles also contain synaptophysin and chromogranin-B[32].

The uptake of CA by adrenal chromaffin granules appears to depend upon the chemiosmotic mechanism.[28,29] The granular membrane contains a H^+-translocating ATPase, which pumps protons into the granules. The activity of this enzyme causes the formation both of a ΔpH and a potential difference ($\Delta\Psi$; inside positive) across the granular membrane. The proton gradient drives the transport of CA through an exchange of H^+ for CA. The membranes of isolated chromaffin granules are sufficiently permeable to Cl^- that ATP-dependent acidification of the interior of granules can occur and not be prevented by the development of a limiting $\Delta\Psi$. In contrast, ATP-dependent acidification of the interior of isolated serotonergic vesicles from parafollicular cells does not occur unless the vesicles are exposed to an ionophore, such as valinomycin (FIG. 5), which dissipates a $\Delta\Psi$[25]. The Cl^- conductance of these vesicles is low. These observations suggest that H^+ transport by isolated serotonergic vesicles is

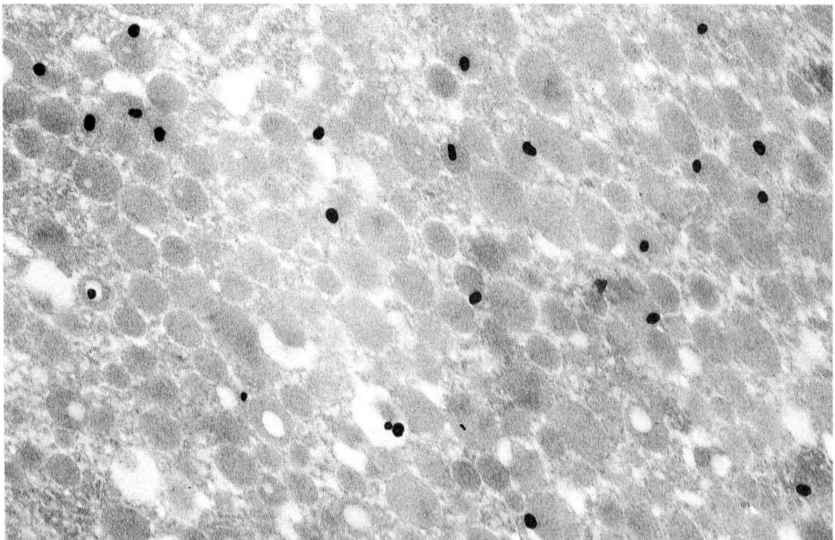

FIGURE 4. A portion of a resting parafollicular cell. The material has been fixed with aldehydes and embedded without osmication in LR White. The sections were incubated with antibodies to the 72 kDa subunit of the vacuolar ATPase and sites of immunoreactivity were visualized with a secondary antibody labeled with 40 nm particles of colloidal gold. Many granules are immunolabeled. The marker = 1.0 μm.

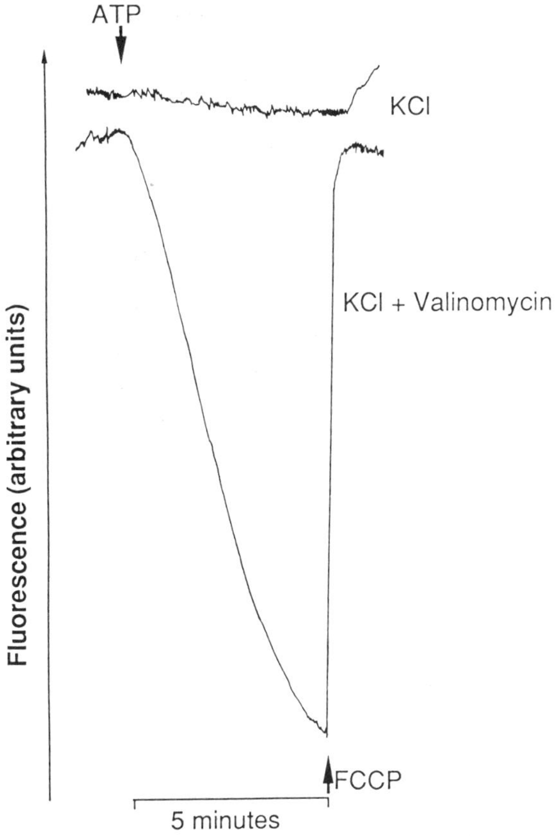

FIGURE 5. Uptake of acridine orange into purified parafollicular cell granules measured by dual wavelength spectroscopy. A downward deflection reflects trapping of acridine orange inside granules and thus H^+ transport. An upward deflection indicates movement of the dye out of granules. In the control trace little inward movement of acridine orange is initiated by the addition of ATP (↓); however, subsequent addition of the H^+ ionophore, FCCP (↑), drives acridine orange out of the granules. In the superimposed tracing (KCl + valinomycin), the addition of ATP, induces inward movement of acridine orange. The addition of FCCP causes rapid efflux of acridine orange from the granules.

electrogenic. With intact resting parafollicular cells the ΔpH across the membranes of individual granules is highly variable; however, most of the granules acidify greatly when the cells are stimulated to secrete by the addition of secretogogues, such as TSH or increased $(Ca^{2+})_e$. Acidification of vesicles is associated with a rise in a $(Ca^{2+})_i$ and does not occur in Cl^--free media.[25] These observations have led to the hypothesis that the internal pH of serotonergic vesicles may be regulated by a Ca^{2+}-dependent Cl^- channel in the vesicular membrane, the opening of which dissipates the $\Delta\Psi$ and permits further inward transport of H^+.

We have recently observed that isolated parafollicular vesicles, and ghosts derived from them, take up ^{3}H-5-HT. This uptake is linear for about 4 min and saturates within 8 min. The uptake of ^{3}H-5-HT by parafollicular vesicles or their ghosts resembles the uptake of ^{3}H-CA by chromaffin granules in that the uptake is inhibited by agents that dissipate the ΔpH across the membranes of the granules. These observations suggest that the transport of 5-HT into vesicles may be regulated by parafollicular cells as they modify the internal pH of their 5-HT storage organelles. Conceivably 5-HT may be loaded into vesicles in preparation for exocytosis when the cells are stimulated by a natural secretogogue.

SEROTONIN-BINDING PROTEIN

Serotonergic synaptic vesicles, but not chromaffin granules, appear to contain a specific type of protein that binds 5-HT with high affinity under conditions that approximate the intracellular milieu.[33] This protein, called SBP, disappears from the forebrain when lesions are made in the nuclei of the median raphe, where the cell bodies of serotonergic neurons are concentrated, and from the distal spinal cord when the cord is transected.[34,35] These observations indicate that SBP is contained within serotonergic neurons. Confirmation of this localization has recently been obtained. SBP has been purified and two forms of the protein have been found.[36] These differ from M_r; one form of SBP is 45 kDa and the other is 56 kDa. Antibodies have been raised against each. Many of these antibodies do not distinguish one form of the protein from the other; however, some antibodies specifically recognize only the 45 kDa or the 56 kDa protein.[36,37] Immunostaining of the rat brain and spinal cord with 45 kDa or 56 kDa SBP-specific antisera has revealed that the immunoreactivities of each is co-localized in cell bodies and processes with the immunoreactivity of 5-HT.[37] SBP has also been detected in peripheral cells that store 5-HT; however, the protein is found only in those which are derived from neurectoderm, such as the serotonergic neurons of the gut and parafollicular cells,[24,38,39] but not in 5-HT-containing cells derived from other germ layers, such as platelets, mast cells, or enterochromaffin cells.[18,33] Moreover, SBP accumulates above a spinal cord transection at a rate consistent with the proximo-distal movement of the protein by fast axonal transport.[35] Fast transport, in turn, is suggestive of an intracellular location in vesicles destined for the synaptic apparatus.[40] This suggestion has been supported by the observations that SBP is concentrated in synaptosomes isolated from the brain and especially in vesicles liberated from these synaptosomes in K^+-containing media.[30] SBP has also been shown to be released along with 5-HT from stimulated enteric neurons by a Ca^{2+}-dependent mechanism[41] and 45 kDa SBP has been localized chemically and immunocytochemically in the 5-HT storage vesicles of parafollicular cells.[21] There is evidence that the 45 kDa form of SBP may predominate in synaptic vesicles, while the 56 kDa SBP is more concentrated in cell bodies and preterminal axons.[35,42] Newly taken up ^{3}H-5-HT forms a complex mainly with 45 kDa SBP and during ontogeny the 45 kDa material appears later than 56 kDa SBP, but coincidentally with the period of synaptogenesis.[43] It has been proposed that the 56 kDa form of SBP is a precursor of the 45 kDa material.[36] Each of the 12 monoclonal antibodies that have been raised against SBP recognize both 45 and 56 kDa SBP, as does an anti-idiotypic antibody produced from affinity purified antibodies to 5-HT. Moreover, mild trypsinization of the 56 kDa SBP/5-HT complex yields a complex of 45 kDa SBP/5-HT. Finally, peptide maps derived from 45 kDa or 56 kDa SBP labeled with a photoaffinity probe for the 5-HT binding site are similar.[44] Fifty-six kDa SBP can be phosphorylated, which reduces its affinity for 5-HT, while 45 kDa SBP is not subject to phosphorylation.[45] Conceivably, phosphorylation prevents premature binding of 5-HT during the transport of the protein to terminals. This idea remains to be tested.

MEDULLARY THYROID CARCINOMA (MTC) CELLS

Difficulties arise in characterizing SBP when only the brain is investigated. The complexity of the system and its limited ability to survive *in vitro* complicate studies of the biosynthesis and intracellular transport and processing of the molecule. On the other hand, the purification of SBP and the development of monospecific anti-SBP sera make it possible to investigate these issues. Recently we have characterized a serotonergic cell line derived from parafollicular cells that can be exploited to gain insight into the questions of where and in what form SBP is synthesized, whether 56 kDa SBP is converted to 45 kDa SBP, and if so, where in cells this conversion occurs. This cell line is a human medullary carcinoma of the thyroid (MTC cells).[46] MTC cells produce 5-HT, both 56-kDa and 45 kDa, and peptides, including calcitonin, calcitonin gene related peptide (CGRP), somatostatin.[46–49] The cells actively take up ^{3}H-5-HT by a process that can be inhibited by fluoxetine and zimelidine, and their ability to store 5-HT is blocked by reserpine.[46] Like parafollicular cells, MTC cells secrete 5-HT in response to the secretogogues, TSH (50 mU) or elevated $(Ca^{2+})_e$ (7.0 mM). The effect of TSH is Ca^{2+}-dependent and is associated with the presence on these cells of TSH receptors, which can be demonstrated immunocytochemically by using a monospecific anti-TSH receptor serum (FIG. 6) The response of MTC cells to secretogogues may be coupled to cAMP as a second messenger. MTC cells can be induced to secrete 5-HT by addition of dibutyryl-cAMP (1.0 mM) or forskolin (1μM). It is also known that dibutyryl-cAMP and forskolin induce the secretion of calcitonin from MTC cells.[50] The coincident secretion of 5-HT and calcitonin is to be expected, since EM immunocytochemistry reveals that in MTC cells as well as in parafollicular cells 5-HT and calcitonin and co-stored in the same secretory vesicles (FIG. 7). In many respects, therefore, the MTC cells resemble neurons and in fact they can be induced to extend neuritic processes and express neurofilament proteins by manipulating the growth media.[46] Unlike parafollicular cells, however, MTC cells do not express receptors for NGF and do not respond to this growth factor. The existence of these cells provides a model that can be used to characterize the synthesis, processing, and putative release and re-cycling of SBP. What is learned from MTC cells can be verified in later studies of serotonergic neurons of the gut and brain.

SUMMARY

Advances have been made in the characterization of 5-HT-storing organelles of neurectodermal cells. The parafollicular cell of the thyroid has been used as a model. This cell stores 5-HT, shares many properties with neurons, and can be induced to change its phenotype from endocrine to neuronal by exposure *in vitro* to NGF. The membranes of isolated parafollicular 5-HT storage vesicles appear to contain a chloride channel that is gated in response to stimulation of the cells by secretogogues. Opening of this channel permits the interior of the vesicle to acidify in response to the action of a H^+ATPase in the vesicular membrane. Development of a $\Delta\Psi$ appears to limit acidification of the vesicular interior when the chloride conductance is low. Transmembrane transport of ^{3}H-5-HT into parafollicular vesicle is inhibited by dissipating the ΔpH across the granular membranes. The physiological significance of the ability of parafollicular vesicles to modify the internal pH of their 5-HT-storing organelles remains to be determined. Like the synaptic vesicles of central and peripheral serotonergic neurons parafollicular vesicles contain a specific 5-HT binding protein, SBP. 5-HT storage organelles and SBP have been found in medullary thyroid

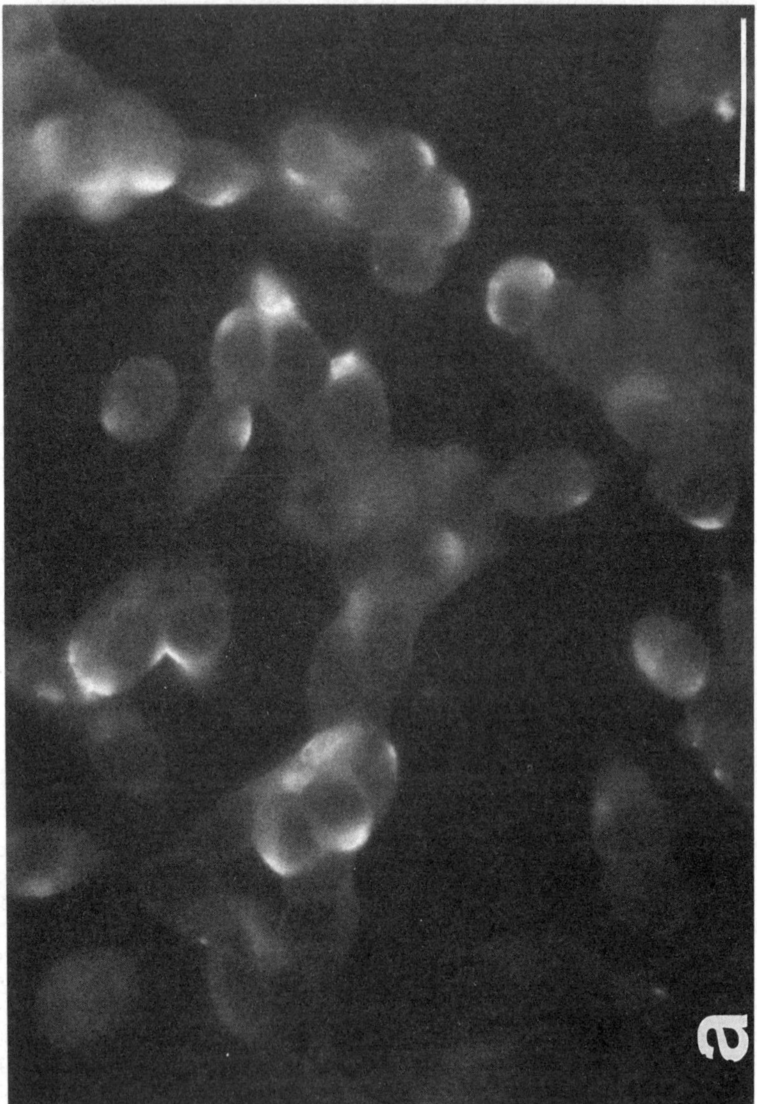
a

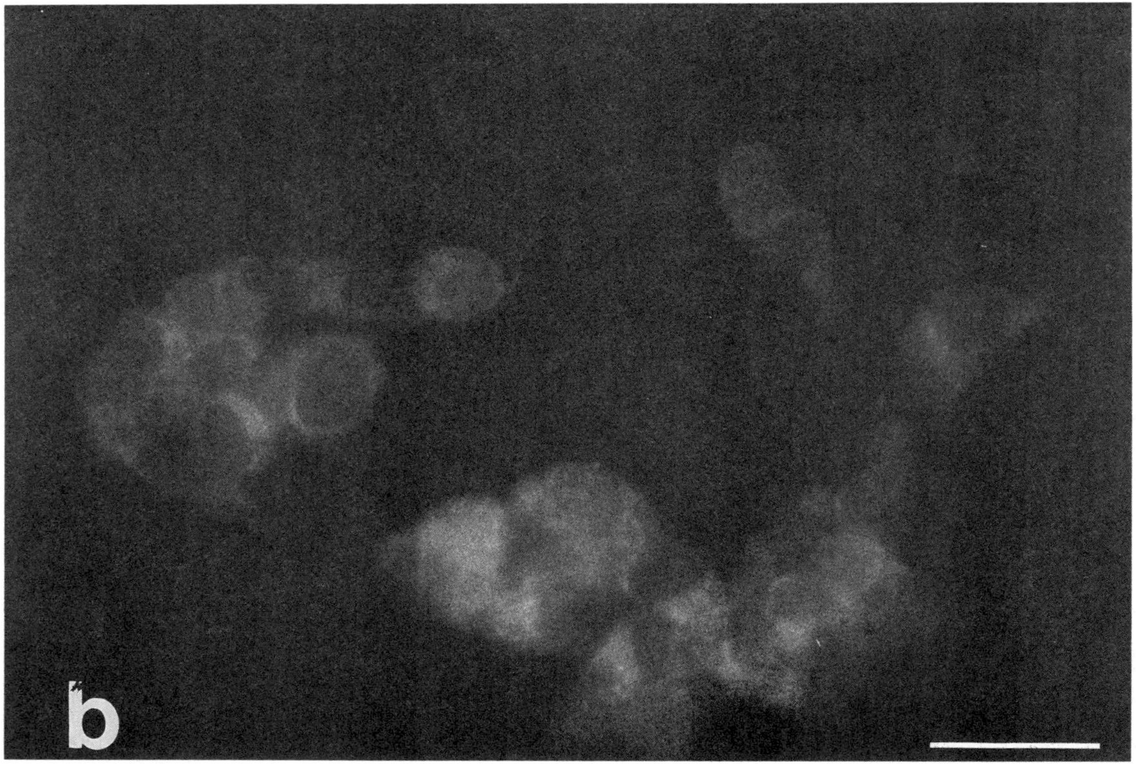

FIGURE 6. Immunofluorescence of MTC cells following exposure to monoclonal antibody that reacts with TSH receptors. **a.** Specific immunofluorescence. Note the bright patches of immunoreactivity at cell surfaces. **b.** Control cells exposed to preimmune sera. Only background fluorescence can be discerned. The marker = 30 μm.

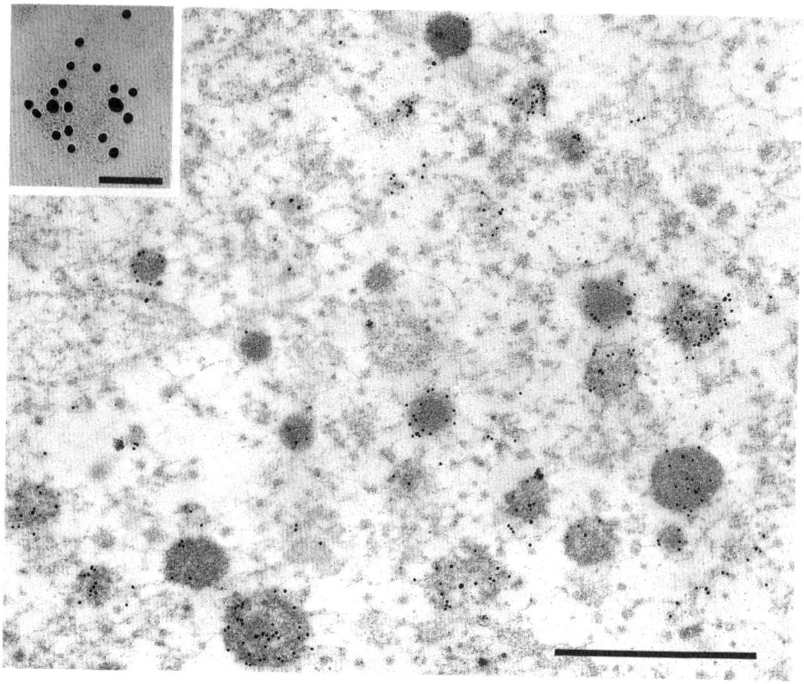

FIGURE 7. An EM showing a portion of an MTC cell. The preparation has been exposed to anti-calcitonin sera and processed as in FIG. 2. Virtually every granule displays calcitonin immunoreactivity. The inset shows the colocalization of 5-HT (5 nm gold) and calcitonin (15 nM gold) immunoreactivites in the same granule of an MTC cell. The marker = 1.0 μm.

carcinoma (MTC) cells, a tumor line derived from parafollicular cells. The cell biology of SBP is now under study utilizing the MTC cells.

REFERENCES

1. CURZON, G. 1988. Serotonergic mechanisms of depression. Clin. Neuropharmacol. **11:** S11–S20.
2. PEDIGO, N. J. W., H. I. YAMAMURA & D. L. NELSON. 1981. Discrimination of multiple (^{3}H) 5-hydroxytrptamine binding sites by the neuroleptic spiperone in rat brain. J. Neurochem. **36:** 220–226.
3. PEROUTKA, S. J. & S. H. SNYDER. 1979. Multiple serotonin receptors: differential binding of (^{3}H-)5-hydroxytryptamine, (^{3}H)lysergic acid diethylamide and (^{3}H)spiroperidol, Mol. Pharmacol. **16:** 687–699.
4. JACOBS, B. L. 1983. Mechanism of action of hallucinogenic drugs: focus upon postsynaptic serotonergic receptors. In Psychopharmacology, 1, D. G. Grahame-Smith, Ed. Excerpta Medica, pp. 344–76.
5. VOGT, M. 1982. Some functional aspects of central serotonergic neurons. In Biology of Serotonergic Transmission N. N. Osborne, Ed. New York: John Wiley & Sons, pp. 299–315.
6. GERSHON, M. D., G. M. MAWE & T. M. BRANCHECK. 1989. 5-hydroxytryptamine and enteric neurons. *In* Peripheral Action of 5-Hydroxytryptamine. J. R. Fozard, Ed. Oxford: Oxford University Press. pp. 247–265.

7. Sanders-Bush, E., Ed. 1988. The Serotonin Receptors. Clifton, NJ: The Humana Press.
8. Brodie, B. B., D. F. Bogdanski & L. Bonomi. 1964. Formation, storage and metabolism of serotonin(5-hydroxytryptamine) and catecholamine in lower vertebrates. *In* Comparative Neurochemistry. D. Richter, Ed. Oxford: Pergamon Press. pp. 367–378.
9. Ross, S. B. 1982. The characteristics of serotonin uptake system. *In* Biology of Serotonergic Transmission. N. N. Osborne, Ed. New York: John Wiley and Sons, pp. 160–195.
10. Slotkin, T. A., F. J. Seidler, W. L. Withmore, C. Lau, M. Salvaggio & D. F. Kirksey. 1978. Rat brain synaptic vesicles. Uptake and specificities of (^{3}H)norepinephrine and (^{3}H)serotonin in preparation of whole brain and brain regions. J. Neurochem. **31:** 961–968.
11. Fujita, T. 1977. Concept of paraneurons. Arch. Histol. Japan. **40** Suppl. 1–12.
12. Douglas, W. W. 1977. Secretory control of adrenal medullary secretion: Membrane and ionic events in stimulus-secretion coupling. *In* Handbook of Physiology and Endocrinology 6. H. Blaschko, G. Sayers, and S. D. Smith, Eds. Washington, DC: Am. Physiol. Soc. pp. 367–388.
13. Falck, B. & C. Owman. 1968. 5-hydroxytryptamine and related amines in endocrine system. Adv. Pharmacol. **6A:** 211–231.
14. Nunez E. A. & M. D. Gershon. 1978 Cytophysiology of thyroid parafollicular cells. Int. Rev. Cytol. **52:** 1–.
15. DaPrada, M. & A. Pletscher. 1968. Isolation of 5-hydroxytryptamine organelles of rabbit blood platelets: Physiological properties and drug induced changes. Br. J. Pharmacol. **34:** 591–597.
16. Fishkes, H. & G. Rudnick. 1982. Bioenergetics of serotonin transport by membrane vesicles derived from platelet dense core granules. J. Biol. Chem. **257:** 5671–5677.
17. Tamir, H., R. F. Payette, Y. L. Huang, K. P. Liu & M. D. Gershon. 1985. Human serotonectin: A blood glycoprotein that binds serotonin and is associated with platelets and white blood cells. J. Cell Sci **73:** 187–206.
18. Gershon, M. D. & H. Tamir. 1984. Serotonectin and the family of proteins that bind serotonin. Biochem. Pharmacol. **33:** 3115–3118.
19. Le Douarin, N. M. 1982. The neural crest. London: Cambridge University Press.
20. Bussolati, G., & A. G. E. Pearse. 1967. Immunofluorescent localization of calcitonin in the C-cells of pig and dog thyroid. J. Endocrinol. **37:** 205–209.
21. Barasch, J. M., H. Tamir, E. A. Nunez & M. D. Gershon. 1987. Serotonin-storing secretory granules from thyroid parafollicular cells. J. Neurosci. **7:** 4017–4033.
22. Barasch, J. M., H. Mackey, H. Tamir, E. A. Nunez & M. D. Gershon. 1987. Induction of neuronal phenotype in a serotonergic endocrine cell derived from the neural crest. **7:**(9): 2874–2883.
23. Atack, V. C., L. E. Ericson & A. Melander. 1972. Intracellular distribution of amines and calcitonin in the sheep thyroid gland. J. Ultrastruct. Res. **41:** 484–498.
24. Bernd, P., M. D. Gershon, E. A. Nunez & H. Tamir. 1981. Separation of dissociated thyroid follicular and parafollicular cells: association of serotonin binding protein. J. Cell Biol. **88:** 499–508.
25. Barasch, J. M., M. D. Gershon, E. A. Nunez, H. Tamir & Q. Al-Awqati. 1988. Thyrotropin induces the acidification of the secretory granules of parafollicular cells by increasing the chloride conductance of the granular membrane. J. Cell Biol. **107:** 2137–2147.
26. Richard, G. Ultrastructural visualisation of biogenic monoamines. *In* Handbook of Chemical Neuroanatomy. A. Bjorklund and T. Hokfelt, Eds. New York: Elsevier Science Publishers, pp. 122–146.
27. Carty, S., R. G. Johnson & S. Scarpa. 1981. The isolation of intact chromaffin granules using isotonic Percoll density gradient. Anal Biochem. **106:** 438–445.
28. Johnson, R. G. 1988. Accumulation of biological amines into chromaffin granules: A model for hormone and neurotransmitter transport. Physiol. Rev. **68:** 232–307.
29. Njus, D., P. M. Kelley & G. J. Harnadek. 1985. The chromaffin vesicles: A model secretory organelle. Physiologist. **28:** 235–241.
30. Tamir, H. & M. D. Gershon. 1979. Storage of serotonin and serotonin binding protein in synaptic vesicles. J. Neurochem. **33:** 35–44.

31. Cidon, S. & N. Nelson. 1986. Purification of N-ethylmaleimide-sensitive ATPase from chromaffin granule membrane. J. Biol. Chem. **261:** 9222–9227.
32. Winkler, unpublished observation.
33. Gershon, M. D. & H. Tamir. 1985. Peripheral sources of serotonin and serotonin binding proteins. *In* Serotonin and Cardiovascular System. P. M. Vanhoutte, Ed. New York: Raven Press, pp. 15–26.
34. Tamir, H. & M. J. Kuhar. 1975. Association of serotonin binding protein with projections of the midbrain raphe nuclei. Brain Res. **83:** 169–172.
35. Gershon, M. D., K. P. Liu., S. E. Karpiak & H. Tamir. 1983. Storage of serotonin in vivo as a complex with serotonin binding protein in central and peripheral serotonergic neurons. J. Neurosci. **3:** 1901–1911.
36. Liu, K. P., M. D. Gershon & H. Tamir. 1985. Identification, purification, and characterization of two forms of serotonin binding protein from rat brain. J. Neurochem. **44:** 1289–1301.
37. Kirchgessner, A. L., M. D. Gershon, K. P. Liu & H. Tamir. 1988. Co-storage of serotonin binding protein with serotonin in the rat CNS. J. Neurosci. **8:** 3879–3890.
38. Jonakait, G. M., H. Tamir., M. M. Rapport & M. D. Gershon. 1977. Detection of a soluble serotonin binding protein in the mammalian myenteric plexus and other peripheral sites of storage. J. Neurochem. **28:** 277–284.
39. Bernd, P., E. A. Nunez, M. D. Gershon & H. Tamir. 1979. Serotonin binding protein: Characterization and localization in the parafollicular cells of the sheep thyroid. Anat. Rec. **193:** 257–268.
40. Grafstein, B. 1977. Axonal transport: The intracellular traffic of the neuron. *In* Handbook of Physiology: The Nervous System. J. M. Brookhard, V. B. Mountcastle, E. R. Kandel, and S. R. Geiger, Eds. Bethesda, MD: Am. Physiol. Soc., pp. 691–717.
41. Jonakait, J. M., H. Tamir, A. R. Gintzler & M. D. Gershon. 1979. Release of serotonin and its binding protein from enteric neurons. Brain Res **174:** 55–69.
42. Adlersberg, M., H. Tamir & K. P. Liu. 1984. Photoaffinity labeling of 45 and 56Kd forms of serotonin binding protein(SBP); Evidence that 45 Kd SBP is stored in synaptic vesicles. Abs. **10:** 364.
43. Liu, K. P., H. Tamir, S. H. Hsiung, M. Adlersberg & M. D. Gershon. 1987. Prenatal development of serotonin binding protein in relation to other transmitter-related characteristics of central serotonergic neurons. Dev. Brain Res. **32:** 31–41.
44. Liu, K. P., H. Tamir & M. Adlersberg. 1990. Photoaffinity labeling of two forms of serotonin binding protein: Peptide mapping of the binding sites. J. Neurochem. **54:** 963–970.
45. Adlersberg, M., K. P. Liu, Y. Ehrlich & H. Tamir. 1987. A Ca^{2+} dependent protein kinase activity associated with serotonin binding protein. J Neurochem. **49:** 1105–1115.
46. Tamir, H., K. P. Liu, R. F. Payette, S. H. Hsiung, M. Adlersberg, E. A. Nunez & M. D. Gershon. 1989. Human medullary thyroid carcinoma: Characterization of the serotonergic and neuronal properties of a neuroectodermally derived cell line. J. Neuroscience. **9:** 1199–1212.
47. Leong, S. S., J. S. Horoszewicz, K. Shimaoka, M. Friedman, E. Kawinski, M. J. Song, R. Ziegel, T. M. Chu, S. Baylin, & E. A. Mirand. A new cell line for study of human medullary thyroid carcinoma. *In* Advances in Thyroid Neoplasia. M. Andreoli, F. Monaco, and J. Robbins, Eds., Rome: Field Education, Italia, pp. 95–108.
48. Gkonos, P. J., W. Born, B. N. Jones, J. B. Petermann, H. T. Keutmann, R. S. Birnbaum, J. A. Fischer & B. A. Roos. 1986. Biosynthesis of calcitonin gene-related peptide and cacitonin by a human medullary thyroid carcinoma cell line. J. Biol. Chem. **261:** 14386–14391.
49. Kameda, Y., M. Ito, & K. Ogawa. 1985. Immunoreactive somatostatin in thyroid C cells. An immunohistochemical study. Virchow's Arch. (A) **410:** 97–101.
50. deBustros, A., S. B. Baylin, M. A. Levine & B. D. Nelkin. 1986. Cyclic AMP and Phorbol esters separately induced growth inhibition, calcitonin secretion, and the calcitonin gene transcription in cultured human medullary thyroid carcinoma. J. Biol. Chem. **261:** 8036–8041.

DISCUSSION OF THE PAPER

QUESTION: Does the serotonin vesicle differ in different locales or species?

TAMIR: It varies with the species. Some species contain serotonin, but if they do have serotonin they will have it all along. In humans the embryo has serotonin and then later on in life the amount diminishes. But we haven't seen changes from serotonin to yet another phenotypic expression. In terms of locale, for example from gut to brain, we are just beginning to look at that.

QUESTION: What is the size of the serotonin vesicles? Is there only one kind or are there two kinds?

TAMIR: You see only one kind. They look like the dense core vesicles, about a 0.1 μm, that is quite large, and they all contain calcitonin. Almost all of them contain serotonin, whereas serotonin binding protein is seen in about half of them.

Serotonin Uptake and Serotonin Uptake Inhibition

RAY W. FULLER AND DAVID T. WONG

Lilly Research Laboratories
Eli Lilly and Company
Lilly Corporate Center
Indianapolis, Indiana 46285

INTRODUCTION

After serotonin is released from nerve terminals and has acted on post-synaptic receptors, it is transported out of the synaptic cleft back into the nerve terminal via the action of a specific membrane uptake carrier. During the past fifteen years, several compounds have been described that inhibit this serotonin uptake carrier with high selectivity, and their availability has helped in understanding the physiologic role of the uptake carrier. Administration of these uptake inhibitors to animals does not result in depletion of brain serotonin content, and in fact leads to a *reduction* in brain serotonin formation and release, suggesting that the physiological role of the neuronal uptake carrier is not to conserve serotonin but instead to inactivate serotonin by removing it from the synaptic cleft.

In the brain, serotonin uptake carriers also occur on glial cells, but the physiologic roles of these uptake carriers and their possible influence on synaptic function are still not well understood.[1] Blood platelets also have a serotonin uptake carrier, which is the means by which platelets acquire serotonin, since they do not synthesize it.[2] Inhibition of the serotonin uptake carrier on blood platelets over a prolonged time results in depletion of platelet (and whole blood) concentrations of serotonin.[3]

This chapter will focus on brain serotonin neurons and the ways in which serotonin uptake inhibition and the functional results of that inhibition have been studied.

CHEMICAL CLASSES OF SEROTONIN UPTAKE INHIBITORS

The chemical structures of several selective inhibitors of serotonin uptake are shown in FIGURE 1. The compounds cover a wide range of chemical types, with little similarity among them overall. All have primary or substituted amino groups separated from an aromatic nucleus by cyclic, branched or straight-chain connecting groups. Several, but not all, have halogen substituents on the aromatic nucleus.

SELECTIVITY OF SEROTONIN UPTAKE INHIBITORS *IN VITRO*

Many of the earlier uptake inhibitors, especially the tricyclic antidepressant drugs, were relatively nonselective in blocking the uptake of neurotransmitter monoamines. They inhibited norepinephrine and serotonin uptake, and some antidepressant drugs such as nomifensin also inhibited dopamine uptake.[4] The compounds in FIGURE 1 are

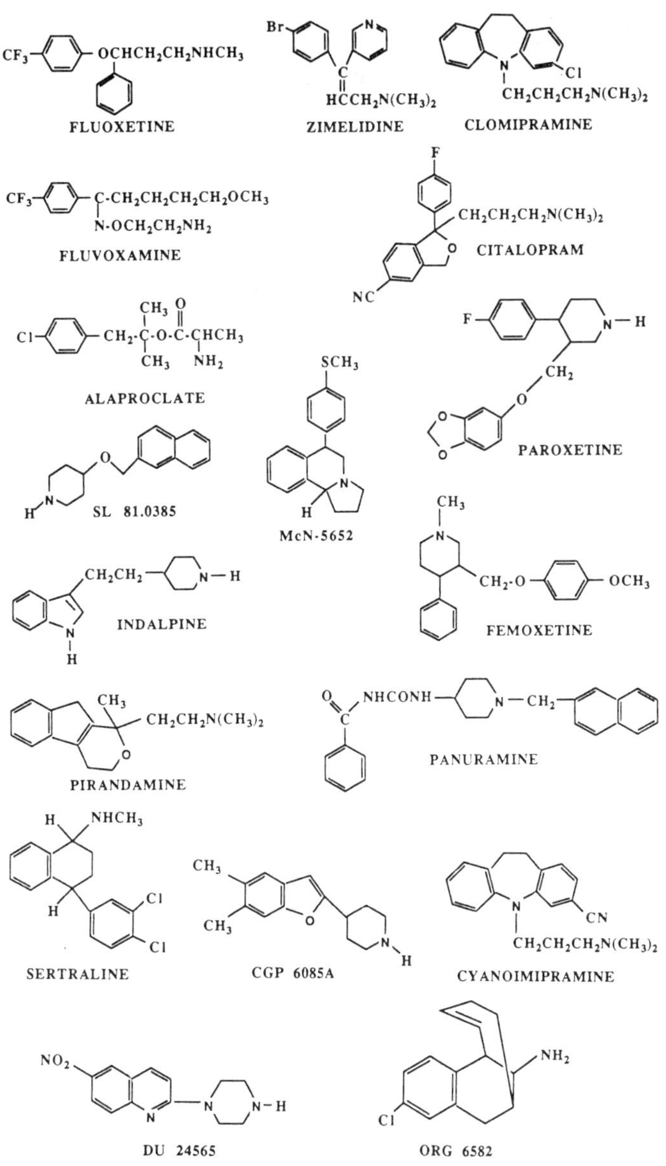

FIGURE 1. Chemical structures of some selective inhibitors of serotonin uptake.

all highly selective in blocking the uptake of serotonin, having effects on norepinephrine and dopamine uptake only at much higher concentrations *in vitro*. In several cases, the substituent on the nitrogen is an important determinant of selectivity. The removal of a methyl group from clomipramine results in a selective inhibitor of norepinephrine uptake, not serotonin uptake.[4] On the other hand, the removal of a methyl group from zimelidine increases potency of serotonin uptake inhibition.[5] The substituent on the aromatic nucleus also is an important determinant of selectivity. The absence of the chlorine atom in clomipramine reduces the potency and selectivity of serotonin uptake inhibition.[4] Replacement of the *p*-trifluoromethyl substituent on the phenoxy nucleus of fluoxetine, a selective inhibitor of serotonin uptake,[6] with an *o*-methyl or *o*-methoxy group results in selective inhibitors of norepinephrine uptake.[7,8]

DEMONSTRATION OF SEROTONIN UPTAKE INHIBITION *IN VIVO*

Serotonin uptake inhibition can be demonstrated *ex vivo* by measuring serotonin uptake *in vitro* in brain slices or synaptosomes prepared from animals treated with the uptake inhibitor *in vivo.* The percentage inhibition measured is probably not a true reflection of the percentage inhibition that occurred in the intact brain, since this type of experiment is to some extent a bioassay of the amount of drug present in the brain preparation used to measure uptake. Such experiments are especially useful to demonstrate the ability of serotonin uptake inhibitors to reach the brain, their potency and selectivity relative to one another, and their duration of action.[9]

A less direct method for evaluating uptake inhibition in brain, but one that measures an effect in the intact brain as opposed to *in vitro,* is the antagonism of brain serotonin depletion by a drug whose depleting action requires the membrane uptake carrier. Four such serotonin-depleting drugs whose antagonism by serotonin uptake inhibitors has been used as a measure of uptake inhibition *in vivo* are *p*-chloromethamphetamine,[10] *p*-chloroamphetamine,[11] fenfluramine,[12] and H75/12.[13] The accumulation of these drugs into serotonin neurons via the membrane uptake carrier apparently is required for their depletion of brain serotonin, since that depletion is blocked in a dose-dependent manner by serotonin uptake inhibitors.

Inhibition of the serotonin uptake carrier on blood platelets can also be demonstrated by measuring uptake *in vitro* after administration of the uptake inhibitor *in vivo.* This technique can be applied in human studies, since blood platelets are readily accessible, as well as in animal studies. Serotonin becomes depleted from blood platelets after continued administration of a serotonin uptake inhibitor, because uptake is the means by which platelets acquire serotonin. Inhibition of platelet serotonin uptake *ex vivo* and reduction of platelet serotonin content have been demonstrated in humans given selective inhibitors of serotonin uptake.[14]

UPTAKE INHIBITORS AS RADIOLABELS FOR SEROTONIN UPTAKE CARRIERS

Radioactive forms, especially tritiated forms, of several selective inhibitors of serotonin uptake have been used as radioligands to label the serotonin uptake carrier. Radiolabels used for *in vitro* binding studies have included imipramine,[15] nitroimipramine,[16] fluoxetine,[17] citalopram,[18] norzimelidine,[19] and paroxetine.[20] Radiolabels used for quantitative autoradiography *in vivo* or in brain slices include imipramine,[21] nitroimipramine,[22] paroxetine,[23] citalopram,[18] indalpine,[24] and cyanoimipramine.[25]

EVIDENCE FOR INCREASED SEROTONIN CONCENTRATIONS IN THE SYNAPTIC CLEFT AFTER SEROTONIN UPTAKE INHIBITION

As a consequence of administration of a serotonin uptake inhibitor, serotonin concentration in the synaptic cleft is expected to increase since it cannot be removed from the synaptic cleft by the membrane uptake carrier. Although the technology for measuring serotonin concentration directly in the synaptic cleft does not exist currently, several experimental approaches have documented increased extracellular concentrations of serotonin after uptake inhibition. Geyer *et al.*[26] presented histofluorescence evidence for increased extraneuronal concentrations of serotonin in raphe regions of rat brain after fluoxetine administration. Marsden *et al.*[27] showed that *in vivo* voltammetric signals believed to be serotonin increased in rat brain after fluoxetine administration. Guan and McBride[28] reported several-fold increases in serotonin concentration in extracellular fluid obtained from push-pull cannulae in the nucleus accumbens of rats after administration of fluoxetine. Perhaps the most extensive evidence for increased serotonin concentrations in the synaptic cleft and increased activation of synaptic receptors for serotonin comes from the numerous and varied functional effects that have been produced by serotonin uptake inhibitors.

PHARMACOLOGIC EFFECTS OF SEROTONIN UPTAKE INHIBITORS IN ANIMALS

Serotonin uptake inhibitors produce various functional effects on laboratory animals alone or in combination with 5-hydroxytryptophan (5HTP). These are thought to be due to the enhanced activation of synaptic receptors for serotonin, either presynaptic autoreceptors or postsynaptic receptors.

Neurochemical Effects

Serotonin uptake inhibitors cause a decrease in serotonin turnover, which has been measured in various ways, including a decrease in incorporation of radioactive tryptophan into 5-hydroxyindoles, a decrease in the accumulation of 5HTP after decarboxylase inhibition, a decrease in the steady state concentration of 5-hydroxyindoleacetic acid (5HIAA) and in the accumulation of 5HIAA after administration of probenecid to block its efflux from brain.[29] The decrease in serotonin turnover occurs in parallel with a decrease in firing of serotonin neurons measured electrophysiologically.[30]

The decrease in serotonin turnover may occur partly or entirely via activation of autoreceptors on serotonin neurons that regulate serotonin synthesis and release. Other neurochemical effects of serotonin uptake inhibitors probably result from increased activation of postsynaptic receptors by serotonin that has accumulated in the synaptic cleft. For example, serotonin uptake inhibitors potentiate the increase in striatal dopamine metabolites caused by neuroleptic drugs.[31] They also increase hypothalamic norepinephrine metabolites,[32] likely due to facilitatory input to noradrenergic neurons,[33,34] especially when combined with 5HTP.[35]

Neuroendocrine Effects

Serotonin uptake inhibitors increase serum corticosterone concentration in rats[36] by increasing corticotropin-releasing hormone release and increasing plasma adrenocorticotrophin (ACTH).[37] Pretreatment of rats with serotonin uptake inhibitors antagonizes the elevation of serum corticosterone concentration by *p*-chloroamphetamine[38] or fenfluramine[39] by blocking the carrier-mediated release of serotonin by these agents, but potentiates the elevation of serum corticosterone concentration by L-5-hydroxytryptophan (L-5HTP)[36] by inhibiting the inactivation of serotonin by the uptake carrier. Serotonin uptake inhibitors alone do not increase serum prolactin concentration in rats,[40] but they antagonize the elevation of serum prolactin by fenfluramine[41] and potentiate the elevation of serum prolactin by L-5HTP.[30]

Behavioral Effects

Animals treated with serotonin uptake inhibitors look normal in gross appearance, but subtle behavioral effects can be detected in suitable experimental paradigms. For example, serotonin uptake inhibitors inhibit muricidal behavior in rats[42] and improve passive avoidance behavior in some but not all behavioral tests in bulbectomized rats.[43] Serotonin uptake inhibitors potentiate behavioral effects of 5HTP, such as head twitch in mice[44] and the discriminative cue stimulus property of 5HTP in rats.[45]

Effect on Food Intake

Serotonin uptake inhibitors decrease food intake after single doses in animals.[46–48] The decrease in food intake persists after daily treatment, and body weight or weight gain is decreased after repeated administration to normal or obese rats and mice.[49,50]

Other Effects

Selective inhibitors of serotonin uptake have been shown to reduce rapid eye movement sleep in rats and cats,[51,52] to enhance memory processing in mice,[53] and to produce analgesic effects or potentiate opioid-induced analgesia.[54,55] The multiplicity of effects that can be elicited by enhancing serotonin function through uptake inhibition results from the widespread distribution of serotonergic projections to many different brain regions involved in various functions controlled by the brain.[56]

THERAPEUTIC EFFECTS OF SEROTONIN UPTAKE INHIBITORS IN HUMANS

Serotonin uptake inhibitors are used or have shown promise for use in treating numerous psychiatric illnesses and other conditions. Some of these therapeutic uses are outlined below.

Depression

There is now extensive literature on the efficacy of selective inhibitors of serotonin uptake in treating mental depression.[57,58] A major clinical advantage over most tricyclic antidepressant drugs is the lack of side effects related to anticholinergic, antihistaminergic, or anti-adrenergic actions, since most serotonin uptake inhibitors have little affinity for muscarinic, histamine H1, alpha-1 adrenergic or other neurotransmitter receptors.[59] Selective inhibitors of serotonin uptake generally lack direct effects on the heart, in contrast to many tricyclic antidepressant drugs.[60] Weight gain, a relatively common side effect of earlier antidepressant drugs, is also not a problem with selective inhibitors of serotonin uptake used to treat depression.[61] Zimelidine was the first selective inhibitor of serotonin uptake to be marketed for the treatment of depression worldwide, but it had to be removed from clinical use due to some infrequent but severe side effects including the Guillain-Barre syndrome. The adverse effects were not thought to be related to serotonin uptake inhibition but were somehow due to the particular chemical structure of zimelidine. Fluoxetine was the first selective inhibitor of serotonin uptake to be introduced as an antidepressant drug in the U.S. market. Other compounds in this group are in clinical use outside the U.S. and their development is continuing.

Postanoxic Intention Myoclonus

This frequently severe movement disorder that can follow brain injury by anoxia is thought to be related to impaired serotonergic neurotransmission and has been treated with L-5HTP.[62] Fluoxetine alone had minimal benefit in these patients, but the combination of fluoxetine and a low dose of L-5HTP (given with carbidopa to block its decarboxylation in peripheral tissues) resulted in enhanced antimyoclonic activity with fewer side effects.[63]

Obsessive-Compulsive Disorder

White *et al.*[64] have described a patient with severe obsessive-compulsive disorder refractory to treatment with a tricyclic antidepressant or a monoamine oxidase inhibitor who responded dramatically and repeatedly to treatment with two serotonin uptake inhibitors, clomipramine and citalopram. Turner *et al.*[65] had reported that fluoxetine was effective in patients with obsessive-compulsive disorder.

Panic Disorder

Zimelidine, a selective inhibitor of serotonin uptake, was superior to placebo in most rating scales in the treatment of patients with agoraphobia with panic attacks.[66] In this study, imipramine was not shown to be superior to placebo. Fluvoxamine has also decreased panic attacks in a study in which a norepinephrine uptake inhibitor did not.[67] Fluoxetine has caused complete cessation of panic attacks in some patients.[68]

Obesity

Serotonin uptake inhibitors decrease food intake in rats, acting synergistically with L-5HTP.[46] Unlike anorectic drugs acting via non-serotonergic mechanisms, serotonin uptake inhibitors have been shown in animal studies to suppress stress-induced eating,[69] to selectively suppress carbohydrate intake,[70] and to suppress insulin-induced hyperphagia as well as 2-deoxyglucose-induced hyperphagia.[71] Therapeutic utility of serotonin uptake inhibitors in the treatment of obesity in humans has been reported for zimelidine,[72] femoxetine,[73] fluvoxamine[74] and fluoxetine.[75]

Bulimia

Bulimia is an eating disorder with episodes of binge eating followed by self-induced vomiting. Recent evidence indicates that serotonin uptake inhibitors are effective in treating bulimia.[76]

Alcoholism

In rats, selective inhibitors of serotonin uptake including fluoxetine, zimelidine, fluvoxamine and sertraline, have reduced ethanol drinking in rats selected or bred to prefer alcohol when given a choice of drinking fluids.[77–79] In humans, zimelidine decreased the number of drinks consumed and increased the number of abstinent days in nondepressed heavy drinkers.[80,81] Similar effects have been reported with another serotonin uptake inhibitor, citalopram.[82]

Other Disorders

Serotonin uptake inhibitors including fluoxetine and clomipramine have been reported to be effective in treating cataplexy, sudden brief paralysis of voluntary movement and loss of muscle tone following a momentary decrease in alertness;[83] fluoxetine had earlier been effective in treating cataplexy in dogs.[84] There is extensive evidence that central serotonergic pathways are involved in the reception and processing of pain messages,[85] and serotonin uptake unhibitors may be useful in treatment of some types of pain, such as chronic pain.[86] It is possible that depletion of platelet serotonin by serotonin uptake inhibitors will have therapeutic benefit in reducing migraine attacks.[87]

SUMMARY

Serotonin uptake carriers occur on serotonin neurons, on glial cells and on blood platelets. The uptake carrier on serotonin neurons inactivates serotonin that has been released into the synaptic cleft by transporting it back into the nerve terminal. The serotonin uptake carrier is the means by which blood platelets acquire serotonin, since they do not synthesize it. The function of the serotonin uptake carrier on glial cells is poorly understood. Selective inhibitors of serotonin uptake enhance neurotransmission via serotonergic neurons and have been useful pharmacologic tools for studying physiologic roles of serotonin neurons. Some serotonin uptake inhibitors are finding therapeutic uses in mental depression and other psychiatric disorders and in treating obesity and bulimia; other therapeutic applications continue to be evaluated.

REFERENCES

1. Kimelberg, H. K. 1986. Occurrence and functional significance of serotonin and catecholamine uptake by astrocytes. Biochem. Pharmacol. **35:** 2273–2281.
2. Pletscher, A. 1987. The 5-hydroxytryptamine system of blood platelets: physiology and pathophysiology. Int. J. Cardiol. **14:** 177–178.
3. Petersen, E. N., E. Bechgaard, R. J. Sortwell & L. Wetterberg. 1978. Potent

depletion of 5HT from monkey whole blood by a new 5HT uptake inhibitor, paroxetine (FG 7051). Eur. J. Pharmacol. **52:** 115–119.
4. HYTTEL, J. 1982. Citalopram—pharmacological profile of a specific serotonin uptake inhibitor with antidepressant activity. Prog. Neuro-Psychopharmacol. Biol. Psychiat. **6:** 277–295.
5. ROSS, S. B. & A. L. RENYI. 1977. Inhibition of the neuronal uptake of 5-hydroxytryptamine and noradrenaline in rat brain by (Z)- and (E)- 3-(4-bromophenyl)-N,N-dimethyl-3-(3-pyridyl-allylamines and their secondary analogues. Neuropharmacology **16:** 57–63.
6. WONG, D. T., J. S. HORNG, F. P. BYMASTER, K. L. HAUSER & B. B. MOLLOY. 1974. A selective inhibitor of serotonin uptake: Lilly 110140, 3-(*p*-tri-fluoromethylphenoxy)-*N*-methyl-3-phenylpropylamine. Life Sci. **15:** 471–479.
7. WONG, D. T., J. S. HORNG & F. P. BYMASTER. 1975. dl-N-Methyl-3-(o-methoxy-phenoxy)-3-phenylpropylamine hydrochloride, Lilly 94939, a potent inhibitor for uptake of norepinephrine into rat brain synaptosomes and heart. Life Sci. **17:** 755–760.
8. WONG, D. T., P. G. THRELKELD, K. L. BEST & F. P. BYMASTER. 1982. A new inhibitor of norepinephrine uptake devoid of affinity for receptors in rat brain. J. Pharmacol. Exp. Ther. **222:** 61–65.
9. WONG, D. T., F. P. BYMASTER, J. S. HORNG & B. B. MOLLOY. 1975. A new inhibitor for uptake of serotonin into synaptosomes of rat brain: 3-(trifluoromethylphenoxy)-N-3-phenylpropylamine. J. Pharmacol. Exp. Ther. **193:** 804–811.
10. MEEK, J. L., K. FUXE & A. CARLSSON. 1971. Blockade of *p*-chloromethamphetamine induced 5-hydroxytryptamine depletion by chlorimipramine, chlorpheniramine and meperidine. Biochem. Pharmacol. **20:** 707–709, 1971.
11. FULLER, R. W., K. W. PERRY, H. D. SNODDY & B. B. MOLLOY. 1974. Comparison of the specificity of 3-(*p*-trifluoromethylphenoxy)-*N*-methyl-3-phenylpropylamine and chlorimipramine as amine uptake inhibitors in mice. Eur. J. Pharmacol. **28:** 233–236.
12. GHEZZI, D., R. SAMANIN, S. BERNASCONI, G. TOGNONI, M. GERNA & S. GARATTINI. 1973. Effect of thymoleptics on fenfluramine-induced depletion of brain serotonin in rats. Eur. J. Pharmacol. **24:** 205–210.
13. CARLSSON, A., H. CORRODI, K. FUXE & T. HOKFELT. 1969. Effect of antidepressant drugs on the depeletion of intraneuronal brain 5-hydroxytryptamine stores caused by 4-methyl-α-ethyl-meta-tyramine. Eur. J. Pharmacol. **5:** 357–366.
14. LEMBERGER, L., H. ROWE, R. CARMICHAEL, R. CRABTREE, J. S. HORNG, F. BYMASTER & D. WONG. 1978. Fluoxetine, a selective serotonin uptake inhibitor. Clin. Pharmacol. Ther. **23:** 421–429.
15. BRILEY, M. 1985. Imipramine binding: its relationship with serotonin uptake and depression. *In* Neuropharmacology of Serotonin. A. R. Green, Ed. Oxford: Oxford University Press, pp. 50–78.
16. REHAVI, M., H. TRAZER, K. RICE, P. SKOLNICK & S. M. PAUL. 1983. [3H]2-Nitroimipramine: a selective "slowly-dissociating" probe of the imipramine binding site ("serotonin transporter") in platelets and brain. Life Sci. **32:** 645–653.
17. WONG, D. T. & F. P. BYMASTER. 1983. Serotonin (5HT) neuronal uptake binding sites in rat brain membranes labeled with [3H]fluoxetine. Fed. Proc. **42:** 1164.
18. D'AMATO, R. J., B. L. LARGENT, A. M. SNOWMAN & S. H. SNYDER. 1987. Selective labeling of serotonin uptake sites in rat brain by [3H]citalopram contrasted to labeling of multiple sites by [3H]imipramine. J. Pharmacol. Exp. Ther. **242:** 364–371.
19. HALL, H., S. ROSS, S. O. OGREN & L. GAWELL. 1982. Binding of a specific 5-HT uptake inhibitor, [3H]norzimelidine, to rat brain homogenates. Eur. J. Pharmacol. **80:** 281–282.
20. MELLERUP, E. T. & P. PLENGE. 1986. High affinity binding of 3H-paroxetine and 3H-imipramine to rat neuronal membranes. Psychopharmacology **89:** 436–439.
21. DAWSON, T. M. & J. K. WAMSLEY. 1983. Autoradiographic localization of [3H]imipramine binding sites: Association with serotonergic neurons. Brain Res. Bull. **11:** 325–334.
22. RAINBOW, T. C. & A. BIEGON. 1983. Quantitative autoradiography of [3H]nitroimipramine binding sites in rat brain. Brain Res. **262:** 319–322.
23. DE SOUZA, E. B. & B. L. KUYATT. 1987. Autoradiographic localization of 3H-paroxetine-labeled serotonin uptake sites in rat brain. Synapse **1:** 488–496.

24. Benavides, J., H. E. Savaki, C. Malgouris, C. Laplace, C. Margelidon, M. Daniel, J. Courteix, A. Uzan, C. Gueremy & G. Le Fur. 1985. Quantitative autoradiography of [3H]indalpine binding sites in the rat: I. Pharmacological characterization. J. Neurochem. **45:** 514–520.
25. Kovachich, G. B., C. E. Aronson, D. J. Brunswick & A. Frazer. 1988. Quantitative autoradiography of serotonin uptake sites in rat brain using [3H]cyanomipramine. Brain Res. **454:** 78–88.
26. Geyer, M. A., W. J. Dawsey & A. J. Mandell. 1978. Fading: A new cytofluorimetric measure quantifying serotonin in the presence of catecholamines at the cellular level in brain. J. Pharmacol. Exp. Ther. **207:** 650–667.
27. Marsden, C. A., J. Conti, E. Strope, G. Curzon & R. N. Adams. 1979. Monitoring 5-hydroxytryptamine release in the brain of the freely moving unanaesthetized rat using in vivo voltammetry. Brain Res. **171:** 85–99.
28. Guan, X.-M. & W. J. McBride. 1988. Fluoxetine increases the extracellular levels of serotonin in the nucleus accumbens. Brain Res. Bull. **21:** 43–46.
29. Fuller, R. W. & D. T. Wong. 1977. Inhibition of serotonin reuptake. Fed. Proc. **36:** 2154–2158.
30. Clemens, J. A., B. D. Sawyer & B. Cerimele. 1977. Further evidence that serotonin is a neurotransmitter involved in the control of prolactin secretion. Endocrinology **100:** 692–698.
31. Waldmeier, P. A. & A. A. Delini-Stula. 1979. Serotonin-dopamine interactions in the nigrostriatal system. Eur. J. Pharmacol. **55:** 363–373.
32. Scatton, B., Y. Claustre, D. Graham, T. Dennis, A. Serrano, S. Arbilla, C. Pimoule, H. Schoemaker, D. Bigg & S. Z. Langer. 1988. SL 81.0385: A novel selective and potent serotonin uptake inhibitor. Drug Develop. Res. **12:** 29–40.
33. McRae-Degueurce, A., T. Dennis, L. Leger & B. Scatton. 1985. Regulation of noradrenergic neuronal activity in the rat locus coeruleus by serotoninergic afferents. Physiol. Psychology **13:** 188–196.
34. Feuerstein, T. J. & G. Hertting. 1986. Serotonin (5-HT) enhances hippocampal noradrenaline (NA) release: Evidence for facilitatory 5-HT receptors within the CNS. Naunyn-Schmiedeberg's Arch. Pharmacol. **333:** 191–197.
35. Smythe, G. A., R. M. Gleeson & G. M. Stead. 1988. Mechanisms of 5-hydroxy-L-tryptophan-induced adrenocorticotropin release: A major role for central noradrenergic drive. Neuroendocrinology **47:** 389–397.
36. Fuller, R. W. 1981. Serotonergic stimulation of pituitary-adrenocortical function in rats. Neuroendocrinology **32:** 118–127.
37. Gibbs, D. M. & W. Vale. 1983. Effect of the serotonin reuptake inhibitor fluoxetine on corticotropin-releasing factor and vasopressin secretion into hypophysial portal blood. Brain Res. **280:** 176–179.
38. Fuller, R. W. & H. D. Snoddy. 1980. Effect of serotonin-releasing drugs on serum corticosterone concentration in rats. Neuroendocrinology **31:** 96–100.
39. McElroy, J. F., J. M. Miller & J. S. Meyer. 1984. Fenfluramine, *p*-chloroamphetamine and *p*-fluoroamphetamine stimulation of pituitary-adrenocortical activity in rats: Evidence for differences in site and mechanism of action. J. Pharmacol. Exp. Ther. **228:** 593–599.
40. Krulich, L. 1975. The effect of a serotonin uptake inhibitor (Lilly 110140) on the secretion of prolactin in the rat. Life Sci. **17:** 1141–1144.
41. Van de Kar, L. D., J. H. Urban, K. D. Richardson & C. L. Bethea. 1985. Pharmacological studies on the serotoninergic and nonserotonin-mediated stimulation of prolactin and corticosterone secretion by fenfluramine. Neuroendocrinology **41:** 283–288.
42. Molina, V. A., S. Gobaille & P. Mandel. 1986. Effects of serotonin-mimetic drugs on mouse-killing behavior. Aggress. Behav. **12:** 201–211.
43. Joly, D. & D. J. Sanger. 1986. The effects of fluoxetine and zimeldine on the behavior of olfactory bulbectomized rats. Pharmacol. Biochem. Behav. **24:** 199–204.
44. Ortmann, R., P. C. Waldmeier, E. Radeke, A. Felner & A. Delini-Stula. 1980. The

effects of 5-HT uptake- and MAO-inhibitors on L-5-HTP- induced excitation in rats. Naunyn-Schmiedeberg's Arch. Pharmacol. **311:** 185–192.
45. Barrett, R. J., M. A. Blackshear & E. Sanders-Bush. 1982. Discriminative stimulus properties of L-5-hydroxytryptophan: Behavioral evidence for multiple serotonin receptors. Psychopharmacology **76:** 29–35.
46. Goudie, A. J., E. W. Thornton & T. J. Wheeler. 1976. Effects of Lilly 110140, a specific inhibitor of 5-hydroxytryptamine uptake, on food intake and on 5-hydroxytryptophan-induced anorexia. Evidence for serotonergic inhibition of feeding. J. Pharm. Pharmacol. **28:** 318–320.
47. Dumont, C., J. Laurent, A. Grandadam & J. R. Boissier. 1981. Anorectic properties of a new long acting serotonin uptake inhibitor. Life Sci. **28:** 1939–1945.
48. Lucki, I., M. S. Kreider & K. J. Simansky. 1988. Reduction of feeding behavior by the serotonin uptake inhibitor sertraline. Psychopharmacology **96:** 289–295.
49. Yen, T. T., D. T. Wong & K. G. Bemis. 1987. Reduction of food consumption and body weight of normal and obese mice by chronic treatment with fluoxetine: A serotonin reuptake inhibitor. Drug Develop. Res. **10:** 37–45.
50. Wong, D. T. & R. W. Fuller. 1987. Serotonergic mechanisms in feeding. Int. J. Obesity **11**(Suppl. 3): 125–133.
51. Slater, I. H., G. T. Jones & R. A. Moore. 1978. Inhibition of REM sleep by fluoxetine, a specific inhibitor of serotonin uptake. Neuropharmacology **17:** 383–389.
52. Pastel, R. H. & J. D. Fernstrom. 1987. Short-term effects of fluoxetine and trifluoromethylphenylpiperazine on electroencephalographic sleep in the rat. Brain Res. **436:** 92–102.
53. Flood, J. F. & A. Cherkin. 1987. Fluoxetine enhances memory processing in mice. Psychopharmacology **93:** 36–43.
54. Messing, R. B., L. Phebus, L. A. Fisher & L. D. Lytle. 1975. Analgesic effect of fluoxetine hydrochloride (Lilly 110140), a specific inhibitor of serotonin uptake. Psychopharmacol. Commun. **1:** 511–521.
55. Sugrue, M. F. 1979. On the role of 5-hydroxytryptamine in drug-induced antinociception. Br. J. Pharmacol. **65:** 677–681.
56. Azmitia, E. C. & P. J. Gannon. 1986. The primate serotonergic system: A review of human and animal studies and a report on Macaca fascicularis. Adv. Neurol. **43:** 407–468.
57. Lemberger, L., R. W. Fuller & R. L. Zerbe. 1985. Use of specific serotonin uptake inhibitors as antidepressants. Clin. Neuropharmacol. **8:** 299–317.
58. Asberg, M., B. Eriksson, B. Martensson, L. Traskman-Bendz & A. Wagner. 1986. Therapeutic effects of serotonin uptake inhibitors in depression. J. Clin. Psychiatr. **47**(suppl. 4): 23–35.
59. Richelson, E. & A. Nelson. 1984. Antagonism by antidepressants of neurotransmitter receptors of normal human brain *in vitro*. J. Pharmacol. Exp. Ther. **230:** 94–102.
60. Pary, R., C. R. Tobias & S. Lippmann. 1989. Antidepressants and the cardiac patient. Postgrad. Med. **85:** 267–276.
61. Russ, M. J. & S. H. Ackerman. 1988. Antidepressants and weight gain. Appetite **10:** 103–117.
62. Van Woert, M. H. & D. Rosenbaum. 1979. L-5-Hydroxytryptophan therapy in myoclonus. *In* Cerebral Hypoxia and Its Consequences. S. Fahn, J. N. Davis and L. P. Rowland, Eds. New York: Raven Press, pp. 107–122.
63. Van Woert, M. H., I. Magnussen, D. Rosenbaum & E. Chung. 1983. Fluoxetine in the treatment of intention myoclonus. Clin. Neuropharmacol. **6:** 49–54.
64. White, K., P. E. Keck, Jr. & J. Lipinski. 1986. Serotonin-uptake inhibitors in obsessive-compulsive disorder: A case report. Comprehensive Psychiatry **27:** 211–214.
65. Turner, S. M., R. G. Jacob, D. C. Beidel & J. Himmelhoch. 1985. Fluoxetine treatment of obsessive-compulsive disorder. J. Clin. Psychopharmacol. **5:** 207–212.
66. Evans, L., J. Kenardy, P. Schneider & H. Hoey. 1986. Effect of a selective serotonin uptake inhibitor in agoraphobia with panic attacks. A double-blind comparison of zimeldine, imipramine and placebo. Acta Psychiatrica Scand. **73:** 49–53.

67. DEN BOER, J. A. & H. G. M. WESTENBERG. 1988. Effect of a serotonin and noradrenaline uptake inhibitor in panic disorder; A double-blind study with fluvoxamine and maprotiline. Int. Clin. Psychopharmacol. **3:** 59–74.
68. GORMAN, J. M., M. R. LIEBOWITZ, A. J. FYER, D. GOETZ, R. B. CAMPEAS, M. R. FYER, S. O. DAVIES & D. F. KLEIN. 1987. An open trial of fluoxetine in the treatment of panic attacks. J. Clin. Psychopharmacol. **7:** 329–332.
69. ANTELMAN, S. M., N. ROWLAND & C. KOCAN. 1981. Anorectics: Lack of cross tolerance among serotonergic drugs and sensitization of amphetamine's effect. *In* Anorectic Agents: Mechanisms of Action and Tolerance. S. Garattini & R. Samanin, Eds. New York: Raven Press, pp. 45–62.
70. WURTMAN, J. J. & R. J. WURTMAN. 1977. Fenfluramine and fluoxetine spare protein consumption while suppressing caloric intake by rats. Science **198:** 1178–1180.
71. CARRUBA, M. O., S. RICCIARDI, P. SPANO & P. MANTEGAZZA. 1985. Dopaminergic and serotoninergic anorectics differentially antagonize insulin- and 2DG-induced hyperphagia. Life Sci. **36:** 1739–1749.
72. SIMPSON, R.J., D. J. LAWTON, M. H. WATT & B. TIPLADY. 1981. Effect of zimelidine, a new antidepressant, on appetite and body weight. Br. J. Clin. Pharmacol. **11:** 96–98.
73. SMEDEGAARD, J., P. CHRISTIANSEN & B. SKRUMSAGER. 1981. Treatment of obesity by femoxetine, a selective 5HT reuptake inhibitor. Int. J. Obesity **5:** 377–378.
74. ABELL, C.A., D.L. FARQUHAR, S. M. GALLOWAY, F. STEVEN, A. E. PHILIP & J. F. MUNRO. 1986. Placebo controlled double-blind trial of fluvoxamine maleate in the obese. J. Psychosom. Res. **30:** 143–146.
75. LEVINE, L. R., S. ROSENBLATT & J. BOSOMWORTH. 1987. Use of a serotonin re-uptake inhibitor, fluoxetine, in the treatment of obesity. Int. J. Obesity **11**(Suppl. 3): 185–190.
76. FREEMAN, C. P. L. & M. HAMPSON. 1987. Fluoxetine as a treatment for bulimia nervosa. Int. J. Obesity **11**(Suppl. 3): 171–177.
77. ROCKMAN, G. E., Z. AMIT, Z. W. BROWN, C. BOURQUE & S.-O. OGREN. 1982. An investigation of the mechanisms of action of 5-hydroxytryptamine in the suppression of ethanol intake. Neuropharmacology **21:** 341–347.
78. GILL, K., Z. AMIT & B. K. KOE. 1988. Treatment with sertraline, a new serotonin uptake inhibitor, reduces voluntary ethanol consumption in rats. Alcohol **5:** 349–354.
79. MURPHY, J. M., M. B. WALLER, G. J. GATTO, W. J. MCBRIDE, L. LUMENG & T.-K. LI. 1985. Monoamine uptake inhibitors attenuate ethanol intake in alcohol-preferring (P) rats. Alcohol **2:** 349–352.
80. NARANJO, C. A., E. M. SELLERS, C. A. ROACH, D. V. WOODLEY, M. SANCHEZ-CRAIG & K. SYKORA. 1984. Zimelidine-induced variations in alcohol intake by nondepressed heavy drinkers. Clin. Pharmacol. Ther. **35:** 374–381.
81. AMIT, Z., Z. BROWN, A. SUTHERLAND, G. ROCKMAN, K. GILL & N. SELVAGGI. 1985. Reduction in alcohol intake in humans as a function of treatment with zimelidine: Implications for treatment. *In* Research Advances in New Psychopharmacological Treatments for Alcoholism. C. A. Naranjo and E. M. Sellers, Eds. Amsterdam: Elsevier Science Publishers BV, pp. 189–198.
82. NARANJO, C. A., E. M. SELLERS, J. T. SULLIVAN, D. V. WOODLEY, K. KADLEC & K. SYKORA. 1987. The serotonin uptake inhibitor citalopram attenuates ethanol intake. Clin. Pharmacol. Ther. **41:** 266–274.
83. LANGDON, N., J. SHINDLER, J. D. PARKES & S. BANDAK. 1986. Fluoxetine in the treatment of cataplexy. Sleep **9:** 370–371.
84. BABCOCK, D. A., E. L. NARVER, W. C. DEMENT & M. M. MITLER. 1976. Effects of imipramine, chlorimipramine, and fluoxetine on cataplexy in dogs. Pharmacol. Biochem. Behav. **5:** 599–602.
85. ROBERTS, M. H. T. 1984. 5-Hydroxytryptamine and antinociception. Neuropharmacology **23:** 1529–1536.
86. GOURLAY, G. K., D. A. CHERRY, M. J. COUSINS, B. L. LOVE, J. R. GRAHAM & M. O. MCLACHLAN. 1986. A controlled study of a serotonin reuptake blocker, zimelidine, in the treatment of chronic pain. Pain **25:** 35–52.
87. SYVALAHTI, E., P. KANGASNIEMI & S. B. ROSS. 1979. Migraine headache and blood serotonin levels after administration of zimelidine, a selective inhibitor of serotonin uptake. Curr. Ther. Res. **25:** 299–310.

DISCUSSION OF THE PAPER

T. MENNINI (*Milan*): I would like just to add a brief comment because in my opinion the pharmacology of serotonin uptake inhibitors is not always so simple and it probably cannot be generalized. It is important to characterize the pharmacological and biochemical effects of serotonin uptake inhibitors by seeing if they are antagonized by serotonin antagonists in order to establish a mechanism of action. For example, indalpine is a good inhibitor of serotonin uptake but does not decrease food intake in rats. So this is just a comment in order to remind us that the pharmacology of serotonin uptake can not be oversimplified.

FULLER: Thank you.

QUESTION: After treatment with serotonin uptake inhibitors to reduce body weight or to decrease alcohol consumption, do the effects persist or does the weight go back to normal and does the alcohol intake go back to normal?

FULLER: I don't think there is much data available for alcoholism because of limited clinical studies. Naranjo and Sellers have shown that serotonin uptake inhibitors decrease alcohol intake significantly, but I don't think they have studied whether the effect persists.

There's also limited clinical data available on weight reduction at this point. Longer term studies are being done with fluoxetine and other selective inhibitors of serotonin uptake. Even if weight loss is maintained for as long as the drug is administered, I doubt that weight would stay down after the drug is discontinued unless other factors such as dietary intake and exercise patterns are changed.

EFRAIN AZMITIA (*New York University, New York City*): You suggested that uptake inhibitors such as fluoxetine block the pharmacological effects of a serotonin releaser, *p*-chloroamphetamine, by preventing the *p*-chloroamphetamine from getting into the terminal. Is there strong evidence supporting that, or is that just an hypothesis?

FULLER: It is a hypothesis that, in my judgment is the best one to fit all of the available data. There are no direct data to show that compounds which release serotonin are selectively accumulated in serotonin nerve terminals. But many effects which they produce are blocked by uptake inhibitors, including not only short-term effects whether the blocks might be explained in some other way but also long-term effects which I find difficult to explain any other way. It is a hypothesis that I favor but which is not yet proven.

GEORGE ANDERSON (*Yale Medical School, New Haven, CT*): Could you comment on whether the uptake inhibitors are acting directly at the 5-HT uptake site or at a second regulatory site?

FULLER: Dr. David Wong wants to respond to that question.

WONG: Most of the selective inhibitors of serotonin uptake show complete competitive inhibitors of serotonin uptake and they competitively inhibit the binding of radioligands to the uptake carrier, suggesting they act directly on the serotonin uptake site. There have been suggestions of a regulatory site, but I think the existence of such a site remains to be established.

ANDERSON: There may be differences among uptake inhibitors.

WONG: Some of the evidence for regulatory sites has been based on biphasic inhibition of tritiated imipramine binding. The low affinity binding site for tritiated imipramine may not be associated with the serotonin uptake carrier.

R. MURPHY (*New York University, New York City*): I'm just curious about the anorectic activity of fluoxetine in the rat. How robust is the effect in deprived vs. non-deprived rats or in rats fed different diets?

FULLER: Anorexic effects with serotonin uptake inhibitors have been shown by many investigators in various types of experiments. Goudie *et al.* [J. Pharm. Pharmacol. **28:** 318, 1976] were the first to report anorectic effects of fluoxetine in rats fed a single daily meal. Anorectic effects have also been demonstrated in rats fed different diets or induced to eat in some way. For example, Wurtman and Wurtman [Curr. Med. Res. Opin. **6**(Suppl. 1): 28, 1979] have shown that serotonin uptake inhibitors and other serotonergic drugs selectively suppressed carbohydrate consumption in rats given a choice between high protein and low protein diets. Rowland and others [Life Sci **36:** 2295, 1985] have shown anorectic effects of serotonin uptake inhibitors in various models and after chronic administration. But Dr. Mennini pointed out that not all serotonin uptake inhibitors have shown anorectic effects in all types of experiments, and the exceptions are not well understood.

J. CARUSO (*SUNY, Stony Brook, NY*): Have the glial uptake sites been characterized and are they the same as the 5-HT sites that are located on the 5-HT neuron?

FULLER: I haven't been involved directly with this study of glial uptake of serotonin, as some other people in the room, such as Dr. Whitaker-Azmitia, have been, but my reading of the literature is that to the extent uptake sites on glia have been characterized they appear to be similar in their specificity to the uptake sites on neurons.

Morphology of Central Serotonin Neurons

Brief Review of Quantified Aspects of Their Distribution and Ultrastructural Relationships[a]

LAURENT DESCARRIES, MARIE ANDRÉE AUDET, GUY DOUCET, SYLVIA GARCIA, SHARON OLESKEVICH, PHILIPPE SÉGUÉLA, JEAN-JACQUES SOGHOMONIAN, AND KENNETH C. WATKINS

Département de physiologie
Centre de recherche en sciences neurologiques
Faculté de médecine, Université de Montréal
Montréal, Québec, Canada H3C 3J7

INTRODUCTION

Having recently been shown to be present in neurons of the cnidarian Renilla koellikeri,[1] serotonin (5-hydroxytryptamine, 5-HT) could be the most ancient, phylogenetically, of currently known transmitter molecules. Interestingly, 5-HT neurons probably constitute also the most widely and ubiquitously distributed transmitter-defined system in vertebrate brain, since no region of the central nervous system appears to lack a 5-HT innervation, including each of the circumventricular organs (brain-blood interface) and the cerebroventricular cavity.[2,3] Owing to the various methodological approaches applicable to the identification and examination of central 5-HT neurons at the light and electron microscopic level (for recent reviews, see refs. 2 and 4), much has been learned about various aspects of their distribution and cytological features, particularly in the adult rat brain (e.g. refs. 4–6). In fact, of all transmitter-defined systems in brain, 5-HT must be the one about which the most is known in terms of ultrastructural relationships.

Yet, irrespective of the descriptive level envisaged, anatomical, histological or ultrastructural, most of this morphological information has remained essentially qualitative, or semi-quantitative at best, precluding close correlation with other measurable parameters of 5-HT function. In particular, even though the number of 5-HT nerve cell bodies in the brain of several mammalian species has now been accurately determined, few anatomical regions have actually been investigated for the number as well as distribution of their 5-HT axonal varicosities (density of innervation), and for the frequency with which these terminals establish structurally differentiated synaptic contacts (synaptic incidence). The present report focuses precisely on these two topics, thanks to recent methodological advances made in our laboratory, and reviews such data for three regions of the adult rat brain: neostriatum, cerebral cortex and hippocampus. It is thus hoped to exemplify the usefulness of such quantitative information and draw meaningful conclusions from the comparison between regions.

[a]This work was supported mainly by grant MT-3544 (to L. Descarries) from the Medical Research Council of Canada, and also, in parts, by the Fonds de la recherche en santé du Québec (Equipe de recherche sur les neurotransmetteurs du cerveau).

QUANTIFICATION OF THE 5-HT INNERVATION IN NEOSTRIATUM, CEREBRAL CORTEX, AND HIPPOCAMPUS

The method developed for estimating the actual number of monoamine axon varicosities per mm^3 of rat brain tissue has already been described and discussed in detail, using the cerebral cortex as an example.[7] It is based on the uptake labeling of monoamine terminals of a given type in whole hemisphere slices (200-μm-thick) of the rat brain incubated with the respective tritiated amine. The incubations are carried out in the presence of appropriate blockers of the eventual cross specific uptake of the tracer into other types of monoamine neurons, and of an inhibitor of monoamine oxidase to prevent its metabolic degradation after intraneuronal accumulation. In the case of serotonin, the incubations are made in 1 μM [^{3}H]5-HT, in the presence of a tenfold higher concentration of non-radioactive noradrenaline, to avoid uptake into catecholamine neurons. After fixation with glutaraldehyde, postfixation with osmium tetroxide and flat-embedding in Epon, 4-μm-thick sections of entire slices are cut with a microtome for large, hard objects, and radioautographed by dipping in nuclear emulsion.[8] Controls for labeling specificity include incubations carried out in presence of a highly selective inhibitor of 5-HT uptake, such as fluoxetine or citalopram, and experiments following selective lesioning of 5-HT neurons with cytotoxic dihydroxytryptamine.

The labeled varicosities, visualized as clusters of silver grains, are counted directly from the light microscope radioautographs using a microcomputer-based image analysis system linked to the microscope via a video-camera. In each counting window, the optical image transferred to the video-screen is digitized into a binary image and adjusted by grey-scale discrimination to match the distribution pattern observed in the light microscope. The number of separate features in the binary image is then counted and, whenever necessary, corrected for feature fusion on the basis of histograms of their size distribution. These initial results are expressed in number of labeled sites per surface unit (mm^2) of radioautograph.

In every region examined, three sets of complementary data are gathered for purposes of expressing the data in number of varicosities per mm^3 of tissue. To estimate the proportion of labeled varicosities detected at the time of the counts, small areas are radioautographed in serial sections exposed for increasing periods of time, and the number of labeled sites is plotted as a function of duration of exposure. To determine the proportion of labeled varicosities visualized in sections 2 or more μm in thickness, and therefore partially opaque to tritium, counts are obtained from semi-thin sections alternately cut at thicknesses of 0.5 μm and 2 μm and exposed for a fixed period of time. Lastly, the average diameter of the labeled varicosities is measured in thin sections from the same material and processed for electron microscope radioautography.[8] The initial counts may then be converted to number of varicosities per mm^3 of tissue using the stereological formula $N = n'' \times 1000/(t + d - 2h)$, where n'' is the doubly transformed value per mm^2 of radioautograph, t the 0.5 μm tissue thickness allowing full detection of labeled varicosities, d their mean diameter as measured in electron microscope radioautographs, and h the height of their lost caps, i.e. top portions insufficiently large to induce a detectable signal in light microscope radioautographs (0.1 μm according to Lapierre *et al.*[9]).

Neostriatum

The distribution of neostriatal 5-HT varicosities was quantified at three transverse levels equally spaced across the rostral two thirds of this anatomical region in five rats.[10] At each level, counts were obtained from 12 pairs of counting windows disposed in a checkered pattern to cover the whole medial to lateral and ventral to dorsal extent of neostriatum. As evaluated in number of axon terminals per mm^3 of tissue, the regional density of neostriatal 5-HT innervation ranged from 1.5 to 4.8, averaging 2.6×10^6 varicosities per mm^3 of tissue. It was generally higher ventrally than dorsally, medially than laterally, and caudally than rostrally. There was a close correlation ($r = 0.76$) between the number of axonal varicosities in each neostriatal sector and corresponding measurements of the length of the 5-HT axon network visualized after peroxidase-anti-peroxidase immunostaining with a 5-HT antiserum.

As indicated in TABLE 1 and explained in Soghomonian *et al.*,[10] correlation with biochemical measurements of neostriatal 5-HT yielded a mean endogenous content of 0.2 fg per varicosity. Based on an estimate of the total number of neurons in adult rat neostriatum, it could be calculated that the overall density of 5-HT innervation represents an average of 100–200 5-HT varicosities per neostriatal neuron. Assuming that one third of the 5-HT nerve cell bodies in the nucleus raphe dorsalis of adult rat project to neostriatum, it could also be inferred that each of these neurons issue an average of at least 60,000 axonal varicosities destined to this part of brain.

TABLE 1. Quantitative Data on 5-HT Neurons Innervating Neostriatum, Cerebral Cortex and Hippocampus[a]

Parameter	Neostriatum	Cerebral Cortex	Hippocampus
N varicosities per mm^3 of tissue	2.6×10^6	5.8×10^6	2.7×10^6
5-HT content per varicosity (concentration)	0.2 fg (1800 μg/g)	0.05 fg (535 μg/g)	0.06 fg (450 μg/g)
N varicosities per target neuron	100–200	145–230	20–130
N varicosities per cell body of origin	60,000	500,000	150,000
Frequency of junctional varicosities	~20%	30–40%	very low

[a]Values reported in Soghomonian *et al.*,[10,14] Audet *et al.*,[11] Séguéla *et al.*, [15] and Oleskevich and Descarries[12] and in preparation.

Cerebral Cortex

In this study,[11] counts were obtained at two transverse levels across the anterior half of the hemisphere, from seven distinct cytoarchitectonic areas in which every cortical layer could be sampled by aligning rows of counting windows perpendicular to the pial surface. As estimated in six rats, the mean regional density of 5-HT innervation amounted to 5.8×10^6 varicosities per mm^3, ranging from 4.4 (Cg2) to 7.1×10^6 (AIDr). Some cortical regions (e.g. Cg2, Fr1, Par1) showed significantly lower average densities than others (Cg3, AIDr and AIDc, Pir).[b] The highest laminar density was that of layer I in all regions except Pir, but several regions showed a distinct pattern of 5-HT innervation. In Par1, for example, there was a band of predilection at the height

[b]Nomenclature of cortical areas according to Zilles.[42] Cg2, cingulate cortex, area 2 (medial prefrontal, supragenual); AIDr, rostral agranular insular cortex, dorsal part; Fr1, frontal cortex, area 1 (primary motor); Par 1, parietal cortex, area 1 (primary somatosensory); Cg3, cingulate cortex, area 3 (medial prefrontal, pregenual); AIDc, caudal agranular insular cortex, dorsal part; Pir, prepiriform cortex (primary olfactory).

of layer Va, and in Pir, a strong innervation of the deep polymorph as well as the superficial plexiform layer contrasted with a very low number of 5-HT varicosities in the pyramidal cell body layer.

On the basis of these figures, it was possible to extrapolate a mean endogenous 5-HT content of 0.05 fg per varicosity, and a mean number of 5000–6000 recognition sites per varicosity for reuptake blockers such as paroxetine, citalopram or cyano-imipramine. Using current figures on the total number of cells and axon terminals in the rat cerebral cortex, the average number of 5-HT varicosities per cortical neuron was evaluated at 145–230, and their overall incidence among all cortical terminals at 1 in 200. Assuming that one third of the 15,000 5-HT cell bodies in the dorsal and median raphe nuclei project to the cortex, each of these neurons might thus be sending well over 500,000 axon terminals to this brain region (TABLE 1).

Hippocampus

The regional and laminar density of hippocampal 5-HT innervation was similarly quantified with rows of counting windows set in representative sectors of the subiculum (SUB), Ammon's horn (CA1, CA3a, CA3b) and dentate gyrus (DG-medial blade, crest and lateral blade).[12] Counts were obtained from six rats, at three horizontal levels across the ventral two thirds of the hippocampus. The mean regional density of 5-HT innervation was 2.6×10^6 varicosities per mm^3, and it appeared significantly higher in SUB (3.6×10^6) and Ammon's horn (3.1×10^6) than DG (2.2×10^6). SUB and DG-crest had the highest and lowest regional densities, respectively. There was a marked heterogeneity also in terms of laminar distribution. For example, the stratum moleculare of SUB and CA1, and the stratum oriens of CA3 (5.2×10^6 varicosities in CA3-a) showed much higher values than in pyramidal cell layer (1.3×10^6 on the average). Similarly, in DG, the compact granular layer (0.9×10^6) had a much lower density than the molecular (2.8×10^6) and the polymorph (2.4×10^6) layer.

In this region, extrapolation of the mean 5-HT content per varicosity yielded a value similar to that in the cerebral cortex, 0.06 fg, in spite of the lower average density of 5-HT innervation (TABLE 1). The number of 5-HT varicosities per target neuron ranged from 20–130, in keeping with the greater regional and laminar heterogeneity of distribution. Assuming once again that one third of the 5-HT neurons in the nucleus of origin (mostly raphe medianus) project to hippocampus, the number of hippocampal 5-HT varicosities per cell body of origin was 150,000. No data are currently available to indicate whether some of these cell bodies also innervate the cerebral cortex.

ULTRASTRUCTURAL RELATIONSHIPS OF SEROTONIN AXON TERMINALS IN NEOSTRIATUM, CEREBRAL CORTEX AND HIPPOCAMPUS

Both radioautography after [^{3}H]5-HT uptake labeling *in vivo* and 5-HT-immunocytochemistry have been used to characterize the ultrastructural features and intercellular relationships of central 5-HT neurons (for reviews, see refs. 4 and 6). The immunocytochemical approach is technically simpler, compatible with broader anatomical sampling, and more convenient for the electron microscopic examination of serial thin sections. Owing to the availability of antisera directed against 5-HT-glutaraldehyde-protein conjugate,[13] it can be applied to tissue primarily fixed with relatively high concentrations of glutaraldehyde and hence optimally preserved at the ultrastructural level, particularly in the absence of Triton X-100. The integrity of the tissue then allows for a reliable assessment of the presence or absence of junctional membrane specializations on immunostained varicosities and facilitates the identification of elements in their immediate vicinity (microenvironment).

Neostriatum

In this brain region, we have applied the two methods in parallel.[14] The varicosities radioautographed with [^{3}H]5-HT were examined in single thin sections from a paraventricular sector, whereas their 5-HT-immunostained counterparts were observed in serial thin sections from the same paraventricular plus a dorsal neostriatal sector. The two approaches yielded complementary results. The neostriatal 5-HT terminals from both sectors were relatively small (0.5–0.6 μm in average diameter). Their internal content, best viewed in radioautographs, consisted of small pleomorphic clear vesicles associated with occasional larger granular vesicles and mitochondria. Junctional 5-HT varicosities synapsed exclusively, and with a comparable frequency, on dendritic spines or shafts, almost always with asymmetrical membrane differentiations. The proportion of junctional varicosities was very low, however, in serial (immunocytochemistry) as well as single thin sections (radioautography). The population of serially sectioned immunostained varicosities (n = 109) showed a synaptic incidence of 10% in an average of 6.7 thin sections per varicosity. By way of comparison, 70% of 200 non-immunostained varicosities randomly selected from the same material and followed in an average of 5.8 sections exhibited a junctional complex. Of 484 single profiles of labeled varicosities, only 6% were seen in synaptic junction, yielding a stereologically extrapolated synaptic incidence of 12.8% for whole varicosities. Thus, it could be safely concluded that less than 20% of these 5-HT terminals are endowed with a synaptic specialization. A variety of structures comprising mostly other axon terminals, dendritic spines, and dendritic shafts, and rarely neuronal somata, were found to be closely apposed to neostriatal 5-HT varicosities.

Cerebral Cortex

The ultrastructural features of cortical 5-HT varicosities were examined immunocytochemically in a superficial (layers I-II) and a deep sector (layer VI) of primary motor cortex (Fr1-Fr2), and in the upper layers (I-II) of primary somatosensory (Par1) and of primary visual (Oc1M-Oc2) cortex. One hundred and forty-five immunostained varicosities from Fr1-Fr2 (54 from layers I-II; 91 from layer VI) and 97 each from Par1 and OcM1-Oc2 were scrutinized in serial thin sections. These terminals were of comparable shape and size in the four cortical sectors examined, and averaged 0.6–0.7 μm in mean diameter. The proportion of varicosities making a synaptic junction was evaluated by linear transformation of the relationship between the frequency of observed junctions and the number of thin sections available for examination. Reliability of this sampling was evidenced by a high coefficient of correlation ($r > 0.95$) in each cortical sector. The synaptic incidence thus extrapolated for whole varicosities ranged from 28% in layer VI of Fr1-Fr2 to 46% in Par1, without statistically significant differences between the 4 sectors examined. The mean synaptic incidence was therefore in the order of 30 to 40%. A comparable analysis of non-immunostained varicosities in the same material showed a junctional frequency of 98%. All synaptic 5-HT varicosities formed asymmetrical junctions, which were exclusively found on dendritic spines and shafts, and appeared more frequent on spines than shafts in the deep frontal and the upper occipital cortex. In every sector, dendritic shafts, dendritic spines, and axonal varicosities were the most frequently encountered elements in the immediate microenvironment of immunostained varicosities. It was concluded that the cortical 5-HT innervation is predominantly non-junctional throughout the neocortex of adult rat, reinforcing earlier views of a highly divergent afferent system probably capable of widespread, global and/or sustained influences in this part of the brain.

Hippocampus

To date, there has been only one published report on the ultrastructural features of 5-HT axon terminals in the hippocampus of adult rat.[16] In this immunocytochemical study carried out in single thin sections, 5-HT-immunostained varicosities from normal dentate gyrus ($n = 74$) were classified as either large (1–2 μm) or small (0.5–1 μm). The large varicosities ($n = 33$) were never observed in synaptic contact, but a number of the small ones (6 out of 41) were reported to show an asymmetrical pre- and post-synaptic membrane specialization "on what appeared to be dendritic spines or small dendritic branches."

We have also begun to examine the ultrastructure of 5-HT-immunostained varicosities from both a dentate gyrus (DG-crest) and an Ammon's horn (CA3) sector of adult rat hippocampus. Until now, among a total of 230 varicosities visualized in single sections (130 in DG and 100 in CA3-a), only one, from DG, has been found with synaptoid features (FIG. 1). All others failed to show any indication of a synaptic membrane differentiation, including streaming of the synaptic vesicles against a limited portion of their membrane (FIG. 2). It may therefore be concluded that the synaptic incidence of hippocampal 5-HT varicosities is exceedingly low. Interestingly, Anderson *et al.*[16] had also found a low proportion of junctional 5-HT varicosities after innervation of the dentate gyrus by an embryonic raphe transplant placed a few days after a neonatal lesion of the fimbria fornix and entorhinal cortex.

CONCLUDING REMARKS

These results indicate that the 5-HT innervation in three anatomical regions of adult rat telencephalon (neostriatum, cerebral cortex and hippocampus) is predominantly non-junctional. Perhaps this characteristic is related to the phylogenetic anciency of the 5-HT system. A similar lack of synaptic differentiation has recently been demonstrated for the 5-HT innervation in the spinal cord of the lamprey.[17]

Some of the biological implications of this finding have been discussed at length elsewhere.[18–20] It was then suggested that monoamine varicosities deprived of attach-

---

FIGURES 1 and 2. 5-HT-immunostained axonal varicosities from adult rat hippocampus. The synaptoid profile in FIGURE 1 is from the crest of the dentate gyrus (DG-c). In accordance with earlier observations by Anderson *et al.*,[16] most 5-HT varicosities visualized in this hippocampal region failed to show any sign of a synaptic membrane differentiation. In this regard, the above example was a rare exception. Note the unusual appearance of its junctional complex (between arrows), with a heavier deposit of electron dense material on the pre-synaptic (immunostained) side. The non-junctional 5-HT varicosity in FIGURE 2 is from Ammon's horn (CA3-a). As in DG, most if not all 5-HT varicosities from this region lacked a junctional complex. There were no indications of vesicular clustering or streaming against any portion of their membrane. $\times$36,000. Calibration bar: 0.5 μm.

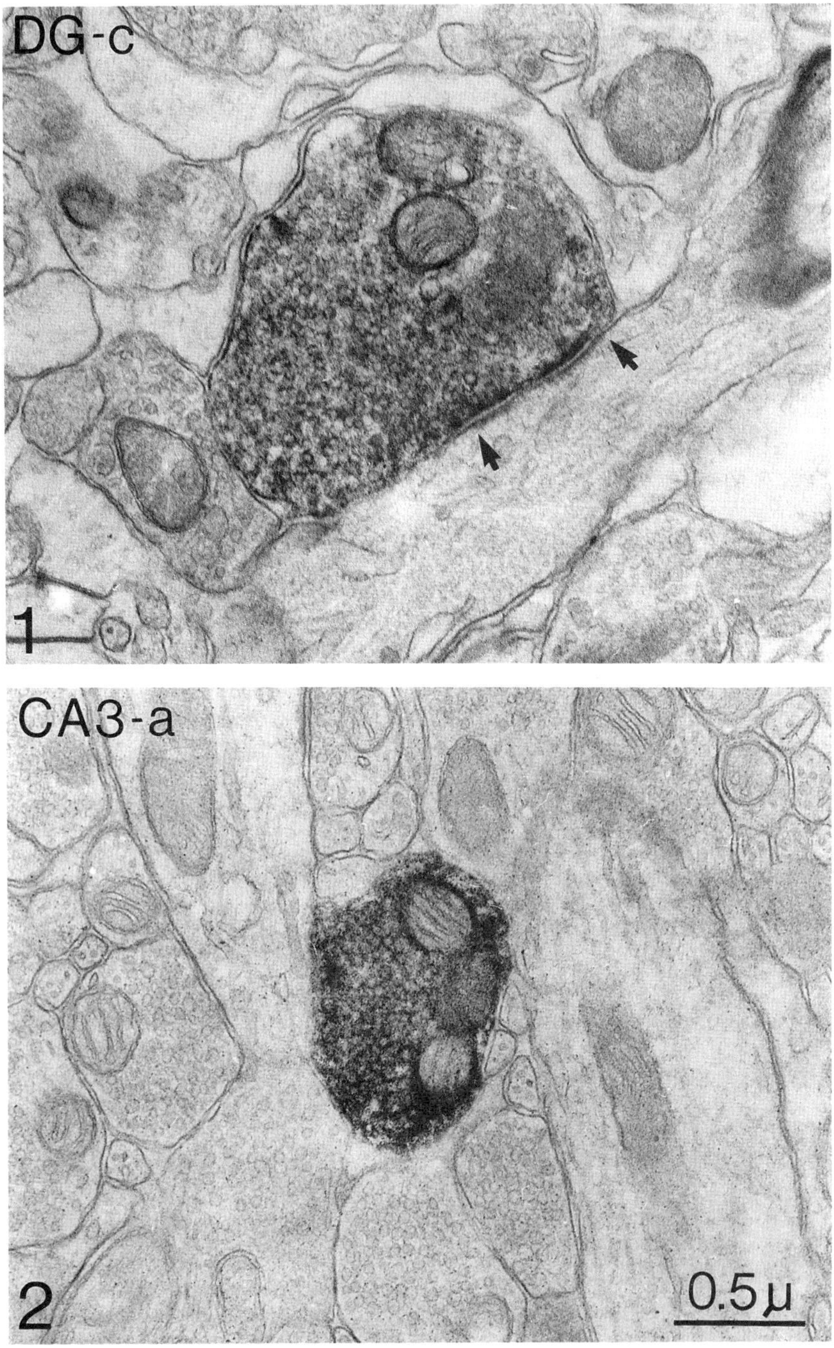

FIGURES 1 and 2

ment sites might undergo incessant movements of translocation and reshaping along their parent fiber and that, in the absence of synaptic junctions, their released amine might diffuse in the intercellular space to act on relatively distant cellular targets. Considering the average density and ubiquitous distribution of the 5-HT innervation in the three regions here examined, it is conceivable that most of their cellular constituents, neuronal, glial or endothelial, are within reach of the transmitter. Since various receptor subtypes have now been characterized for 5-HT and only the 5-HT2 binding site has a distribution resembling the regional and/or laminar density of 5-HT innervation in some parts of rat brain,[21] it seems likely that many of the receptors involved in the physiological effects of 5-HT are neither located on the 5-HT terminals themselves, nor on their occasional synaptic interfaces, nor even restricted to the immediate vicinity of these varicosities. Presently available data do not allow to relate specific mechanisms of action of 5-HT to the activation of receptor sites preferentially situated in extrajunctional as opposed to synaptic locations. Nevertheless, in view of an increasing amount of biochemical and electrophysiological evidence on other monoamines (e.g. ref. 22), it may be postulated that 5-HT will produce different effects by acting in different concentrations in various cellular locations, and notably at extrajunctional versus synaptic sites.

It is also clear that the incidence of junctional 5-HT varicosities varies between regions, as here reported for hippocampus, neostriatum, and cerebral cortex. While the lower frequency in hippocampus and neostriatum might again be indicative of an evolutionary trend, it suggests, in addition, local, epigenetic influences of the innervated region on the synaptic differentiation of its 5-HT input. Electron microscope radioautographic studies in rat agranular cerebellum have already indicated that manipulations of the cellular milieu can induce dramatic modifications in the incidence of junctional 5-HT varicosities.[23] In contrast, as already mentioned, 5-HT varicosities issued from a neonatal graft into the deafferented hippocampus do not appear more frequently synaptic than in normal tissue. In this respect, it will be of particular significance to analyze the 5-HT hyperinnervation which is observed in adult rat neostriatum following a neonatal 6-hydroxydopamine destruction of the nigrostriatal dopamine system,[24] and to relate its synaptic features to those of the missing innervation. Another intriguing possibility is that, in normal tissue, the junctional incidence of 5-HT varicosities might be correlated with the overall density of 5-HT innervation. While this appears consistent with the relatively higher synaptic incidence in the neocortex than hippocampus or neostriatum, it will be of even greater interest to compare sectors with higher and lower innervation densities within a given anatomical region, as in different layers of cerebral cortex or hippocampus.

Considering the ultrastructural and distributional data together, the 5-HT innervation of neostriatum appears ideally built to exert its effect on a multiplicity of cellular target sites in addition to the restricted number of dendritic spines and shafts synaptically contacted by this type of monoamine terminal. The relatively high endogenous concentration of 5-HT inside the terminals of this region might be consistent with this general hypothesis. We have also seen that the neuronal elements most frequently found in direct apposition with neostriatal 5-HT varicosities were other axon terminals. Through such axo-axonic appositions, 5-HT might exert presynaptic influences on other afferent systems, such as inhibition of dopamine[25] or acetylcholine release.[26]

The 5-HT innervation of the cerebral cortex should also be capable of diverse and widely distributed actions, even though these may remain difficult to characterize in such a vast and highly differentiated region of the brain (but see ref. 27). The cellular organization of this 5-HT input is beginning to be unraveled in higher species. Greater differences in the distribution of the 5-HT innervation have been observed in primates than in the rat, not only between different cortical regions but also within given cytoarchitectonic fields such as the motor and the striate cortex (e.g. refs. 28–30). These data have been taken as an indication that, coincident with the extensive phylogenetic development of the neocortex, there was a parallel elaboration and differentiation of its 5-HT input. Further examination of the topographic and ultrastructural features[31,32] of this innervation, along with the localization of its various receptors,[33] should provide important clues as to its role in cortical function.

Owing to the hippocampal slice preparation, much more is being known about 5-HT effects within this brain region. In Ammon's horn particularly, a considerable amount of electrophysiological and pharmacological data may be correlated with the heterogeneous regional and laminar distribution of the 5-HT innervation. For example, the relatively high laminar density in the stratum radiatum and lacunosum-moleculare of CA1 suggest a predilection of 5-HT for the distal apical dendrites of these pyramidal cells.[34] This predilection could in turn account for the high concentration of post-synaptic 5-HT1A receptors[35] mediating the strong inhibitory effect of 5-HT on CA1 pyramidal cells.[36,37] Two other effects of 5-HT on CA1 pyramidal cells have been reported to follow their initial hyperpolarization by 5-HT: a longer lasting depolarization and a decrease in the calcium-activated afterhyperpolarization.[36] The receptor(s) mediating these late effects have not yet been identified. Yet, in view of our data showing a distinct laminar distribution in CA1, these various effects might correspond to 5-HT actions on different parts of CA1 pyramidal neurons. It remains to be determined whether such actions are primarily exerted on the somata or apical dendrites of these neurons, but, judging from the small number of 5-HT terminals in the pyramidal cell body layer of CA1, the dendrites are likely to be favored.

In CA3-a and CA3-b, an even greater difference in 5-HT innervation density is found between the pyramidal and adjacent layers. As previously noted by others,[34,38] the lack of 5-HT varicosities at the level of these pyramidal cell bodies extends along the proximal portion of their apical dendrites in the narrow band corresponding to the stratum lucidum. Moreover, at variance with CA1, the greater number of 5-HT varicosities in CA3 is found in the region of the basal rather than apical dendrites of these neurons. The respective distributions of 5-HT1A and 5-HT2 receptors appear to match this particular pattern of 5-HT innervation, both types of receptors being found in greater concentration in the oriens layer of CA3 than in the subjacent pyramidal layer.[39,40] Disinhibition of CA3 neurons causes epileptiform activity in CA1, which is prevented by destruction of the CA3-CA1 connection.[41] Since disinhibitory effects of 5-HT have been demonstrated in hippocampus,[37] the dense 5-HT innervation of CA3 suggests that inhibitory actions of 5-HT in the dendritic domain of these pyramidal cells could play an important role in influencing neuronal activity throughout Ammon's horn. In the absence of synaptic specializations formed by a great majority of the 5-HT terminals in CA3 (as in DG), eventual knowledge of the ultrastructural localization of the various types of 5-HT receptors in this region should again be of crucial importance for understanding the role of 5-HT in brain function.

ACKNOWLEDGMENTS

The authors are indebted to Dr. Michel Geffard (Bordeaux, France) for his generous gift of 5-HT antiserum. They also thank Daniel Cyr for photographic work, and Hélène Collerette for secretarial assistance.

REFERENCES

1. UMBRIACO, D., M. ANCTIL & L. DESCARRIES. 1990. Serotonin-immunoreactive neurons in the cnidarian *Renilla koellikeri*. J. Comp. Neurol. **291:** 167–178.
2. STEINBUSCH, H. V. M. 1984. Serotonin-immunoreactive neurons and their projections in the CNS. *In* Handbook of Chemical Neuroanatomy, Vol. 3. Classical Transmitters and Transmitter Receptors in the CNS, Part II. A. Björklund, T. Hökfelt and M. J. Kuhar, Eds. Amsterdam: Elsevier, pp. 68–125.
3. TAKEUCHI, Y. 1988. Distribution of serotonin neurons in the mammalian brain. *In* Neuronal Serotonin. N. N. Osborne and M. Hamon, Eds. Chichester: John Wiley & Sons, pp. 25–56.
4. BEAUDET, A. & L. DESCARRIES. 1987. Ultrastructural identification of serotonin neurons. *In* Monoaminergic Neurons: Light Microscopy and Ultrastructure. H. M. W. Steinbusch, Ed. Chichester: John Wiley & Sons, pp. 265–313.
5. BEAUDET, A. & L. DESCARRIES. 1981. The fine structure of central serotonin neurons. J. Physiol. (Paris) **77:** 193–203.
6. SOGHOMONIAN, J.-J., A. BEAUDET & L. DESCARRIES. 1988. Ultrastructural relationships of central serotonin neurons. *In* Neuronal Serotonin. N. N. Osborne and M. Hamon, Eds. Chichester: John Wiley & Sons, pp. 57–92.
7. DOUCET, G., L. DESCARRIES, M. A. AUDET, S. GARCIA & B. BERGER. 1988. Radioautographic method for quantifying regional monoamine innervations in the rat brain. Application to the cerebral cortex. Brain Res. **441:** 223–259.
8. DESCARRIES, L. & A. BEAUDET. 1983. The use of radioautography for investigating transmitter-specific neurons. *In* Handbook of Chemical Neuroanatomy, Vol. I, Methods in Chemical Neuroanatomy. A. Björklund and T. Hökfelt, Eds. Amsterdam: Elsevier, pp. 286–364.
9. LAPIERRE, Y., A. BEAUDET, N. DEMIANCZUK & L. DESCARRIES. 1973. Noradrenergic axon terminals in the cerebral cortex of rat. II. Quantitative data revealed by light and electron microscope radioautography of the frontal cortex. Brain Res. **63:** 175–182.
10. SOGHOMONIAN, J.-J., G. DOUCET & L. DESCARRIES. 1987. Serotonin innervation in adult rat neostriatum. I. Quantified regional distribution. Brain Res. **425:** 85–100.
11. AUDET, M. A., L. DESCARRIES & G. DOUCET. 1989. Quantified regional and laminar distribution of the serotonin innervation in the anterior half of adult rat cerebral cortex. J. Chem. Neuroanat. **2:** 29–44.
12. OLESKEVICH, S. & L. DESCARRIES. 1990. Quantified distribution of the serotonin innervation in adult rat hippocampus. Neuroscience **34:** 19–33.
13. GEFFARD, M., J. DULLUC & A. M. ROCK. 1985. Antisera against the indole-alkylamines: Tryptophan, 5-hydroxytryptophan, 5-hydroxytryptamine, 5-methoxytryptophan, and 5-methoxytryptamine tested by an enzyme-linked immunosorbent assay method. J. Neurochem. **44:** 1221–1228.
14. SOGHOMONIAN, J.-J., L. DESCARRIES & K. C. WATKINS. 1989. Serotonin innervation in adult rat neostriatum. II. Ultrastructural features: A radioautographic and immunocytochemical study. Brain Res. **481:** 67–86.
15. SÉGUÉLA, P., K. C. WATKINS & L. DESCARRIES. 1989. Ultrastructural relationships of serotonin axon terminals in the cerebral cortex of adult rat. J. Comp. Neurol. **289:** 129–142.
16. ANDERSON, K. J., V. R. HOLETS, P. C. MAZUR, R. LASHER & C. W. COTMAN. 1986. Immunocytochemical localization of serotonin in the rat dentate gyrus following raphe transplants. Brain Res. **269:** 21–28.
17. CHRISTENSON, J., S. CULLHEIM, S. GRILLNER & T. HÖKFELT. 1989. 5-hydroxytryptamine immunoreactive varicosities in the lamprey spinal cord have no synaptic specializations—An ultrastructural study. Brain. Res. In press.
18. DESCARRIES, L., A. BEAUDET, A. & K. C. WATKINS. 1975. Serotonin nerve terminals in adult rat neocortex. Brain. Res. **100:** 563–588.
19. BEAUDET, A. & L. DESCARRIES. 1978. The monoamine innervation of rat cerebral cortex: Synaptic and non-synaptic axon terminals. Neuroscience **3:** 851–860.

20. DESCARRIES, L., K. C. WATKINS & Y. LAPIERRE. 1977. Noradrenergic axon terminals in the cerebral cortex of rat. III. Topometric ultrastructural analysis. Brain Res. **133:** 197–222.
21. PEROUTKA, S. J. 1988. 5-hydroxytryptamine receptor subtypes. Ann. Rev. Neurosci. **11:** 45–60.
22. MUELLER, A. L., B. J. HOFFER & T. V. DUNWIDDIE. 1981. Noradrenergic responses in rat hippocampus: Evidence for mediation by a and b receptors in the in vitro slice. Brain Res. **214:** 113–126.
23. BEAUDET, A. & C. SOTELO. 1981. Synaptic remodeling of serotonin axon terminals in rat agranular cerebellum. Brain Res. **206:** 305–329.
24. STACHOWIAK, M. K., J. P. BRUNO, A. M. SNYDER, E. M. STRICKER & M. J. ZIGMOND. 1984. Apparent sprouting of striatal serotonergic terminals after dopamine-depleting brain lesions in neonatal rats. Brain Res. **291:** 164–167.
25. ENNIS, C., J. D. KEMP & B. COX. 1981. Characterization of inhibitory 5-hydroxytryptamine receptors that modulate dopamine release in the striatum. J. Neurochem. **36:** 1515–1520.
26. GILLET, G., S. AMMOR & G. FILLION. 1985. Serotonin inhibits acetylcholine release from rat striatum slices. Evidence for a presynaptic receptor-mediated effect. J. Neurochem. **45:** 1687–1691.
27. DAVIES, M. F., R. A. DEISZ, D. A. PRINCE & S. J. PEROUTKA. 1987. Two distinct effects of 5-hydroxytryptamine on single cortical neurons. Brain Res. **423:** 347–352.
28. MORRISON, J. H., S. L. FOOTE, M. E. MOLLIVER, F. E. BLOOM & H. G. W. LIDOV. 1982. Noradrenergic and serotoninergic fibers innervate complementary layers in monkey primary visual cortex: An immunohistochemical study. Proc. Natl. Acad. Sci. USA **79:** 2401–2405.
29. TAKEUCHI, Y. & Y. SANO. 1983. Immunohistochemical demonstration of serotonin nerve fibers in the neocortex of the monkey (Macaca fuscata). Anat. Embryol. **166:** 155–168.
30. BERGER, B., S. TROTTIER, C. VERNEY, P. GASPAR & C. ALVAREZ. 1988. Regional and laminar distribution of the dopamine and serotonin innervation in macaque cerebral cortex. A radioautographic study. J. Comp. Neurol. **273:** 99–119.
31. DEFELIPE, J. & E. G. JONES. 1988. A light and electron microscopic study of serotonin-immunoreactive fibers and terminals in the monkey sensory-motor cortex. Exp. Brain Res. **71:** 171–182.
32. DE LIMA, A. D., F. E. BLOOM & J. H. MORRISON. 1988. Synaptic organization of serotonin-immunoreactive fibers in primary visual cortex of the macaque monkey. J. Comp. Neurol. **274:** 280–294.
33. LIDOW, M. S., P. S. GOLDMAN-RAKIC, D. W. GALLAGHER & P. RAKIC. 1989. Quantitative autoradiographic mapping of serotonin 5-HT_1 and 5-HT_2 receptors and uptake sites in the neocortex of the rhesus monkey. J. Comp. Neurol. **280:** 27–42.
34. MOORE, R. Y. & A. E. HALARIS. 1975. Hippocampal innervation by serotonin neurons of the midbrain raphe in the rat. J. Comp. Neurol. **154:** 171–184.
35. KÖHLER, C. 1984. The distribution of serotonin binding sites in the hippocampal region of the rat brain. An autoradiographic study. Neuroscience **13:** 667–680.
36. ANDRADE, R. & R. A. NICOLL. 1987. Pharmacologically distinct actions of serotonin on single pyramidal neurones of the rat hippocampus recorded in vitro. J. Physiol. **394:** 99–124.
37. ROPERT, N. 1988. Inhibitory action of serotonin in CA1 hippocampal neurons in vitro. Neuroscience **26:** 69–81.
38. KÖHLER, C. 1982. On the serotonergic innervation of the hippocampal region: An analysis employing immunohistochemistry and retrograde fluorescent tracing in the rat brain. *In* Cytochemical Methods in Neuroanatomy. V. Chan-Palay and S. L. Palay, Eds. New York: Alan R. Liss, pp. 387–405.
39. PAZOS, A. & J. M. PALACIOS. 1985. Quantitative radioautographic mapping of serotonin receptors in the rat brain. I. Serotonin-1 receptors. Brain Res. **346:** 205–230.
40. PAZOS, A., R. CORTÉS & J. M. PALACIOS. 1985. Quantitative autoradiographic mapping of serotonin receptors in the rat brain. II. Serotonin-2 receptors. Brain Res. **346:** 231–249.

41. Wong, R. K. S., R. D. Traub & R. Miles. 1986. Cellular basis of neuronal synchrony in epilepsy. *In* Advances in Neurology, Vol. 44. A. V. Delgado-Escueta, A. A. Ward Jr., D. M. Woodbury and R. J. Porte, Eds. New York: Raven Press, pp. 583–592.
42. Zilles, K. 1985. The Cortex of the Rat. A Stereotaxic Atlas. Berlin: Springer-Verlag.

DISCUSSION OF THE PAPER

Manfred Göthert (*University of Bonn*): As far as I understood, you have now revised the original figures indicating that 5-HT varicosities rarely formed true synapses in the cerebral cortex. You have also shown that these 5-HT varicosities were in frequent appositional contact with other cortical terminals. My question is whether you have any idea about the nature of these apposed terminals. I would like to point out that, in my laboratory, we have already described three types of presynaptic heteroreceptors on 5-HT terminals of the cerebral cortex: alpha-2 adrenergic, GABA-b and histamine H-3 receptors.

Descarries: Our initial radioautographic study (Descarries *et al.,* 1975), mainly carried out in single thin sections, revealed a 5% proportion of synaptic 5-HT profiles. As extrapolated stereologically to whole varicosities, this figure indicated a synaptic incidence of 15–20%. From our recent extensive and thorough immunocytochemical analysis in serial thin sections (Séguéla *et al.,* 1989), we estimate the synaptic incidence at a higher value, between 30 and 40%. The main point remains that this innervation is predominantly non junctional. We have not yet made any attempt to determine the ultrastructural localization of 5-HT receptors, nor the transmitter identity of the elements in the immediate microenvironment of junctional or non junctional 5-HT varicosities. This should certainly be of considerable interest but will require difficult combinations of cytochemical techniques at the electron microscopic level.

M. Kalia (*Philadelphia, PA*): I have two comments. First of all, in immunocytochemically stained material, the reaction product is often so dense that it becomes very difficult to distinguish pre- and post-synaptic thickenings. Secondly, in electron micrographs of the hippocampus, it is quite common to find presynaptic thickenings as described by Peters, Palay, and Webster. These should not come as a surprise. In the picture you showed us, there was also a widening of the synaptic cleft.

Descarries: I must agree that some of the terminals in my presentation looked rather dark. This was when the slides were illustrating several sections from a varicosity and had to be made from photographs rather than directly from the negatives. However, most of the pictures we had in hand for the analysis were from material that had indeed been lightly immunostained on purpose. By the way, this also seemed to improve the ultrastructural preservation. As for the presumed junctional varicosity from the hippocampus, I must emphasize that it was the only one in more than 230 single profiles, which led me to think that it was somewhat unusual.

Electrophysiology of the Central Serotonin System: Receptor Subtypes and Transducer Mechanisms[a]

G. K. AGHAJANIAN,[b,c,d] J. S. SPROUSE,[b] P. SHELDON,[c]
AND K. RASMUSSEN[b]

Departments of Psychiatry[b] and Pharmacology[c]
Yale University School of Medicine
Connecticut Mental Health Center and
the Abraham Ribicoff Research Facilities
New Haven, Connecticut 06508

INTRODUCTION

More than two decades ago, single-cell recordings from neurons in the cerebral cortex,[1] and later in the brain stem,[2] gave the first indication that multiple serotonin (5-HT) receptors existed in the central nervous system. These early studies showed that peripheral 5-HT antagonists such as methysergide and cinanserin are able to block excitatory but not inhibitory effects of 5-HT applied by microiontophoresis, suggesting that at least two types of 5-HT receptors may be present in brain, one for excitation and one for inhibition. Subsequently, multiple 5-HT binding sites were detected in brain homogenates by radioligand binding techniques. The neuroleptic agent [^{3}H]spiperone was found to label 5-HT-displaceable binding sites in frontal cortex[3–5] that exhibited features distinct from sites labeled by [^{3}H]5-HT itself.[6] These results suggested the existence of two distinct populations of 5-HT receptor sites, termed 5-HT$_1$ for the [^{3}H]5-HT-labeled site and 5-HT$_2$ for the [^{3}H]spiperone-labeled site.[6] Recently, binding techniques have disclosed a multiplicity of 5-HT$_1$ sites (e.g., 5-HT$_{1A}$. 5-HT$_{1B}$, 5-HT$_{1C}$, 5-HT$_{1D}$, etc.) as well as the presence of a 5-HT$_3$ site in the brain (for reviews, see Bradley *et al.*[7] and Peroutka[8]).

Within the past year, the existence of distinct 5-HT receptor subtypes has been confirmed by cloning techniques, unequivocally demonstrating that 5-HT$_1$ and 5-HT$_2$ receptors are different gene products (see Hartig[9]). Nevertheless, 5-HT$_1$ and 5-HT$_2$ receptors show a significant degree of sequence homology, both belonging to the G protein-coupled receptor superfamily whose members share the common structural feature of seven membrane spanning segments. The combined evidence from radioligand binding and molecular studies constitutes a powerful argument for the existence of distinct 5-HT$_1$, 5-HT$_2$, and 5-HT$_3$ molecular entities. This new knowledge about 5-HT receptor subtypes has had a profound impact upon research on the electrophysiology of 5-HT. The perplexingly diverse electrophysiological actions displayed by 5-HT in the central nervous system can now be organized within the framework of 5-HT receptor subtypes and their associated transducer mechanisms (i.e., G proteins, second messengers, and ion channels).

[a]Supported by USPHS Grants MH 17871, MH 15642, and the State of Connecticut.

[d]Address for correspondence: Dr. George Aghajanian, Yale University School of Medicine, Connecticut Mental Health Center, 34 Park St., New Haven, CT 06508.

PHYSIOLOGICAL ACTIONS MEDIATED BY 5-HT$_1$ RECEPTORS

Many of the electrophysiological studies on the 5-HT$_1$ receptor subtype have been conducted in the dorsal raphe nucleus and the hippocampal pyramidal cell layer, two areas where there are dense concentrations of 5-HT$_{1A}$ binding sites.[10] Some of these studies will be reviewed in the following sections.

Dorsal Raphe Nucleus: 5-HT$_{1A}$ Identity of Somatodendritic Autoreceptors

It has been shown that serotonergic dorsal raphe neurons can be inhibited by their own transmitter, 5-HT; the receptor mediating this effect has been termed an "autoreceptor." The somatodendritic 5-HT autoreceptor appears to mediate collateral inhibition with the raphe nuclei.[11] Lysergic acid diethylamide (LSD) and other indoleamine hallucinogens are powerful agonists at the somatodendritic 5-HT autoreceptor.[12] Intracellular recordings *in vivo* have shown that the inhibition of dorsal raphe neurons by LSD is caused by a hyperpolarization accompanied by an increase in ionic conductance.[13] Studies in the brain slice preparation have revealed that the ionic basis for the autoreceptor-mediated hyperpolarization, either by 5-HT or LSD, is an opening of K channels.[14,15] It has been proposed that the somatodendritic autoreceptor of dorsal raphe neurons is of the 5-HT$_{1A}$ subtype. This conclusion is based in part on studies with 8-hydroxy-2-(di-n-propylamino)tetralin (8-OH-DPAT), a novel agonist with 5-HT-like actions (but devoid of an indole nucleus) which possesses an almost 1000-fold selectivity for the 5-HT$_{1A}$ binding site.[16] In microiontophoretic, studies 8-OH-DPAT has been shown to exert a depressant effect on serotonergic neurons comparable to that of LSD.[17] Furthermore, buspirone, a non-benzodiazepine anxiolytic which displays 5-HT$_{1A}$ binding properties, has also been shown in extracellular recordings to potently inhibit the firing of 5-HT dorsal raphe neurons when administered systemically, microiontophoretically, or added to media bathing brain slices.[18]

A direct comparison of the effects of 5-HT$_{1A}$ and 5-HT$_{1B}$ selective compounds on the spontaneous firing rate of dorsal raphe neurons has disclosed striking differences between the two ligand subtypes. The *intravenous* administration of ipsapirone or PAPP, two long-chain substituted piperazines with high affinity for 5-HT$_{1A}$ sites,[19–22] potently inhibit raphe cell firing in a dose-dependent manner.[23] In contrast, TFMPP and mCPP, two short-chain substituted piperazines with relatively low 5-HT$_{1A}$ binding affinities and some degree of selectivity for the 5-HT$_{1B}$ site,[19,21] display much weaker or irregular actions. Raphe cell firing is also suppressed by the *microiontophoretic* application of ipsapirone and PAPP, indicating a direct action on 5-HT neurons. Microiontophoretic dose-response curves for these two 5-HT$_{1A}$-selective compounds are virtually indistinguishable from that of 5-HT itself. However, the two 5-HT$_{1B}$-selective compounds, TFMPP and mCPP again appear to be only weak autoreceptor agonists. These electrophysiological findings are consistent with results from binding studies obtained in conjunction with the use of selective neurotoxins which indicate that the 5-HT receptor binding on serotonergic cell bodies and/or dendrites (but not terminals) is of the 5-HT$_{1A}$ subtype.[24]

The efficacy of 5-HT$_{1A}$ ligands in depressing 5-HT cell firing has been further demonstrated by intracellular recordings of dorsal raphe neurons in brain slices. Ipsapirone and PAPP fully mimic the effects of 5-HT in hyperpolarizing the cell membrane and decreasing input resistance.[23] While the classical 5-HT antagonists have proven ineffective in blocking the electrophysiological effects of 5-HT at the autoreceptor, binding and behavioral studies have suggested that (−)-propranolol may possess 5-HT$_{1A}$ antagonistic properties.[25–27] Accordingly, electrophysiological studies have shown that low microiontophoretic currents of (−)- but not (+)-propranolol effectively block the suppressant effects of ipsapirone and 8-OH-DPAT on raphe cell firing.[28] These results fit with recent cloning data[29] which show a remarkable degree of sequence homology between the β_2-adrenoceptor and the 5-HT_{1A} receptor (especially in the membrane spanning segments), providing a basis for the interaction between β-adrenoceptor antagonists and 5-HT_{1A} ligands. Interestingly, (-)-propranolol is a relatively weak antagonist of 5-HT itself. It has been proposed that the seven transmembrane segments of G protein-coupled receptors (including 5-HT_1 receptors) form a relatively hydrophilic central cavity with ligand binding sites deep within the receptor's transmembrane domain (see Hartig[9]). Such a configuration may underlie the differential interactions of the β-antagonists with the 5-HT_1 ligands as compared to 5-HT itself.

5-HT_{1A}-Mediated Responses in Hippocampal Pyramidal Cells

In contrast to their full agonist activity in the dorsal raphe, 5-HT_{1A} agonists (e.g., 8-OH-DPAT, ipsapirone and PAPP) are only partially effective in suppressing the spontaneous activity of hippocampal pyramidal cells (CA1). The 5-HT_{1B} selective compounds, TFMPP and mCPP, also exhibit only weak effects.[30] Intracellular recordings from hippocampal pyramidal cells (CA1) in rat brain slices have shown that 5-HT produces a membrane hyperpolarization and reduction in input resistance due to an opening of potassium channels.[31] The receptors mediating these events appear to be of the 5-HT_{1A} subtype as 8-OH-DPAT and buspirone elicit similar albeit weaker changes in membrane potential and input resistance.[32,33] In addition, 8-OH-DPAT has been shown to occlude the hyperpolarization produced by 5-HT and unmask an increase in input resistance. Taken together, these studies support the notion of a partial agonist action for selective 5-HT_{1A} agonists in the hippocampus.

5-HT_1 Receptors: Electrophysiological Evidence for G-protein Coupling

GTP and GDP are active at inhibiting binding to the 5-HT_{1A} receptor.[34] Theoretically, the effects of guanine nucleotides involve binding to a G protein, resulting in a decrease in agonist affinity for the receptor. There is evidence that the opening of K channels via 5-HT_{1A} receptors in dorsal raphe neurons is mediated by a pertussis toxin-sensitive G protein. Pertussis toxin is known to catalyze the ADP ribosylation of the alpha subunit of certain G proteins (e.g., G_i and G_o), causing an uncoupling of the G protein from its receptor. Intracellular experiments in the dorsal raphe nucleus have shown that a 48 hour preinjection with pertussis toxin (local or intracerebroventricular) causes an almost total blockade of the hyperpolarizing effect of 5-HT.[15,35] Furthermore, the intracellular injection of GTPγS, a non-hydrolyzable analogue of GTP which induces an irreversible activation of G proteins, mimics and is non-additive with the hyperpolarizing action of 5-HT.[35] Consistent with their $5HT_{1A}$ binding properties, the inhibitory effects of ipsapirone and LSD in the dorsal raphe are also blocked by pertussis toxin.[36]

As in dorsal raphe nucleus, the possibility that the electrophysiological effects of 5-HT in the hippocampus may be mediated through a G protein has been evaluated by pretreating animals with pertussis toxin. In rat hippocampal slices, applications of 5-HT cause large hyperpolarizations which can be antagonized by spiperone.[37] Intraventricular injections of pertussis toxin attenuate or eliminate the 5-HT response. Although 5-HT is negatively coupled to adenylate cyclase via a pertussis toxin-sensitive G protein, the hyperpolarizing effect of 5-HT on membrane potential does not appear to involve cAMP as neither bath application of the membrane-soluble analogue of cAMP, 8-Br-cAMP, nor intracellular injection of cAMP reduces the response. A similar lack of effect of cAMP analogues has been found in the dorsal raphe nucleus.[35] On this basis, it has been suggested that hyperpolarizing response mediated by 5-HT_{1A} receptors may not involve second messenger systems but that there may be direct coupling by G-proteins to a K channel.

FIGURE 1 presents an hypothetical scheme for G protein/ion channel coupling mechanisms for the 5-HT_1 receptor. Note that there are two parallel G protein-coupled pathways depicted, one in which there is direct coupling to a K channel via a G protein (G″) and another where there is coupling to adenylate cyclase (AC) via another G protein (G′). It remains to be determined whether there are any modulatory interactions between the two pathways via a cyclic AMP-dependent protein kinase (PKA).

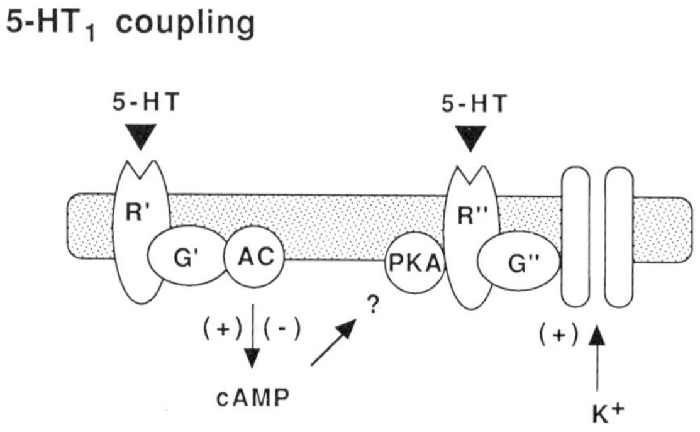

FIGURE 1. Hypothetical scheme for G protein/ion channel/second messenger coupling of 5-HT_1 receptors via two parallel pathways: 1) direct coupling of 5-HT_1 receptors (R″) to the opening of K channels via a G protein (G″); 2) coupling of 5-HT_1 receptors to adenylate cyclase (AC) via another G protein (G′); note that the model allows for the possiblity of a modulatory interaction between the two pathways via cAMP-dependent protein kinase (PKA). As yet it is unclear whether the two postulated pathways are subserved by the same or different types of 5-HT_1 receptors.

PHYSIOLOGICAL ACTIONS MEDIATED BY 5-HT_2 RECEPTORS

Quantitative autoradiographic studies show high concentrations of 5-HT_2 sites in the forebrain including the neocortex (layers IV/V), piriform cortex, claustrum, and olfactory tubercle.[38] With few exceptions (e.g., facial nucleus, n. tractus solitarius), relatively low concentrations of 5-HT_2 receptors are found in the brain stem. Single-cell studies aimed at examining the physiological role of 5-HT_2 receptors are discussed in the following sections.

Facilitatory Actions Mediated by 5-HT_2 (or 5-HT_{1C}) Receptors in the Facial Nucleus

Generally, while there are very few 5-HT_2 binding sites in the brainstem, one exception being the facial motor nucleus.[38] In this nucleus the microiontophoretic application of 5-HT (or norepinephrine) does not by itself induce firing in quiescent facial motoneurons but does facilitate the subthreshold and threshold excitatory effects of iontophoretically applied glutamate.[39] Activation of 5-HT receptors on facial motoneurons has since been shown to cause a slow depolarization, increased input resistance, and increased excitability, probably through a decrease in resting membrane conductance to potassium.[40,41]

The action of iontophoretically applied 5-HT in the facial nucleus can be blocked by the classical 5-HT antagonists metergoline, methysergide, cyproheptadine and cinanserin.[42] However, as all of these antagonists have been shown to interact with both 5-HT_1 and 5-HT_2 receptors,[43] it is conceivable that the action of 5-HT in the facial motor nucleus involves both 5-HT_1 and 5-HT_2 receptors. In an effort to resolve this question, we have recently reexamined the pharmacological characteristics of 5-HT responses in facial motoneurons both *in vivo* and in brain slices.[44] We have found that the 5-HT_2 antagonist ritanserin[45] is able to selectively block (vis-à-vis norepinephrine) the excitatory effects of 5-HT. On the other hand, the selective 5-HT_{1A} agonist 8-OH-DPAT, although it increases facial motoneuron excitability when given systemically, fails to produce excitation when applied locally either by microiontophoresis or bath application in brain slices. Thus, this selective 5-HT_{1A} ligand does not appear to have any direct excitatory) effect on motoneurons either in the brain stem or spinal cord. However, 5-carboxamidotryptamine (5-CT), a broad spectrum 5-HT_1 agonist,[46] is quite effective in enhancing facial motoneuron excitability.[44] Surprisingly, ritanserin is able to block the effects of 5-CT as well as those of 5-HT. Since ritanserin has extremely low affinity for all 5-HT_1 receptors except 5-HT_{1C}, it is possible that it is the 5-HT_{1C} component of 5-CT's receptor profile that is responsible for its effect of facial motoneurons. In this regard, it is interesting that the 5-HT_{1C} receptor has a high degree of homology with the 5-HT_2 receptor (see Hartig[9]), which may account for why it is so difficult to distinguish the two sites pharmacologically.

Recently, a large number of studies, using behavioral as well as binding techniques, have shown that indoleamine and phenethylamine hallucinogens share the property of interacting with 5-HT_2 receptors.[47–51] Therefore, one would expect members of both classes of hallucinogens to have direct effects in the facial motor nucleus. Thus, it is of interest that the iontophoretic administration of LSD, mescaline, or psilocin, although having relatively little effect by themselves at low doses, enhance the facilitation of facial motoneuron excitation produced by iontophoretically applied 5-HT and norepinephrine[52]; curiously, the enhancement can persist for several hours after only a single application of a hallucinogen. Intracellular studies in brain slices show that the enhancement is due to a small but persistent depolarizing effect of the hallucinogens.[44] The action of 5-HT and the hallucinogens in the facial motor nucleus appears to be a complex one, possibly involving both 5-HT_{1C} and 5-HT_2 receptors. A determination of the exact nature of this interaction awaits experiments employing more selective 5-HT_2 and 5-HT_{1C} agonists and antagonists.

Electrophysiology of 5-HT_2 Receptors in the Cerebral Cortex

The electrophysiological effects of serotonin have been studied in several cortical regions. Pyramidal cells in cortical brain slices have been found to respond to serotonin by either a small hyperpolarization, depolarization, or no change in potential.[53,54] The depolarizations appear to be mediated by 5-HT_2 receptors since they can be blocked by 5-HT_2 antagonists. Recently, we have observed a novel effect of serotonin on pyramidal cells of piriform cortex: an elicitation of IPSPs.[54] The IPSPs are blocked by the GABA antagonist bicuculline, suggesting that a GABAergic interneuron is excited by serotonin. Accordingly, a subpopulation of nearby interneurons (at the border of layers II and III) has been found that is excited by serotonin. Excitation of these interneurons (as well as associated IPSPs in pyramidal cells) is blocked by 5-HT_2 antagonists. Interestingly, the hallucinogens LSD and DOM behave as a partial agonists in this system, producing a modest activation by themselves but occluding the full effect of serotonin. Thus, the effects of serotonin and hallucinogens in the cerebral cortex are complex, involving actions upon interneurons as well as pyramidal cells.

Effects of 5-HT_2 Agonists and Antagonists in the Locus Coeruleus

Another example of a facilitatory effect mediated by 5-HT_2 receptors can be detected in the locus coeruleus (LC). In anesthetized rats, systemically administered mescaline or LSD induces a decrease in the spontaneous activity of noradrenergic cells in the LC but, paradoxically, facilitates the activation of these cells by tactile stimuli.[55,56] That the effects of LSD and mescaline (and other phenethylamine hallucinogens) on LC neurons are mediated by 5-HT_2 receptors is suggested by the fact that they can be reversed by selective 5-HT_2 antagonists such as ritanserin and LY-53857.[56] Antipsychotics with affinity for 5-HT_2 binding sites[3,57] are also able to reverse the actions of hallucinogens in the locus coeruleus independently of their actions at dopamine and adrenergic receptors.[58] The relative potencies of hallucinogens in their action on LC neurons[59] correlates with their affinity for 5-HT_2 receptors.[49] However, the effects of hallucinogens in the LC are not direct as they are not mimicked by local, iontophoretic application onto LC cell bodies (Aghajanian, unpublished observations). Moreover, the systemic administration of mescaline or LSD does not enhance the excitation of LC neurons evoked by microiontophoretically applied acetylcholine, glutamate or substance P.[55] These results imply that the hallucinogens are acting indirectly, perhaps via afferents to the LC.

5-HT_2 Receptors, G Proteins and PI Turnover: Physiological Correlates

Serotonin stimulates phosphatidylinositol (PI) hydrolysis in brain through 5-HT_2 and 5-HT_{1C} receptors.[60] LSD and DOM act as partial agonists of this effect.[61] PI hydrolysis yields at least two major second messengers: inositol trisphosphate and diacylglycerol (DAG). The latter activates protein kinase C (PKC) thereby potentially affecting many long term cellular responses through protein phosphorylation. In preliminary studies, we have tested the effect of protein kinase inhibitors on the response of facial motoneurons to serotonin (Aghajanian, in preparation). Two protein kinase inhibitors with different mechanisms of action, 1-(5-isoquinolylsulfonyl)-2-methylpiperazine (H7), a non-selective protein kinase inhibitor,[62] and sphingosine, a selective PKC inhibitor,[63] in concentrations that have no effect of their own (100 and 10 mM, respectively), both markedly enhance and prolong the excitation of facial motoneurons induced by 5-HT. Conversely, phorbol esters that are known to activate PKC reduce the excitatory effect of serotonin. These results suggest that activation of PI turnover, perhaps through receptor phosphorylation, has a negative feedback effect on 5-HT-induced excitations in the facial nucleus. Similar findings have also been obtained in preliminary studies in the cerebral cortex.[64] These observations are consistent with results from studies on 5-HT_2 receptor-mediated responses in rat aorta which indicate that the activation of PKC desensitizes 5-HT_2 receptors.[65]

A schematic representation of an hypothesized negative feedback regulation of 5-HT_2 responses via PI and PKC activation is presented in FIGURE 2. The essence of this model is that there are two subtypes of 5-HT_2 receptors: a subtype coupled to PI turnover (R′) via one type of G protein (G′) and another subtype coupled to the closing of K channels (R″) via another G protein (G″). A negative modulation of R″ via the PI pathway is depicted (−). Mediation by a G protein is suggested by the fact that the depolarizing response to 5-HT in the facial nucleus becomes irreversible after intracellular injection of GTPγS (Aghajanian, in preparation). An interesting implication of the negative feedback model is the possibility that the partial agonist[61] or even antagonist[66] properties of hallucinogens with respect to 5-HT-stimulated PI hydrolysis, may contribute to rather than interfere with their electrophysiological actions.

ELECTROPHYSIOLOGY OF 5-HT_3 RECEPTORS

Rapid neuronal responses to 5-HT have been described in the periphery which are mediated by 5-HT_3 receptors.[7] Recently, 5-HT_3 receptors have been demonstrated in

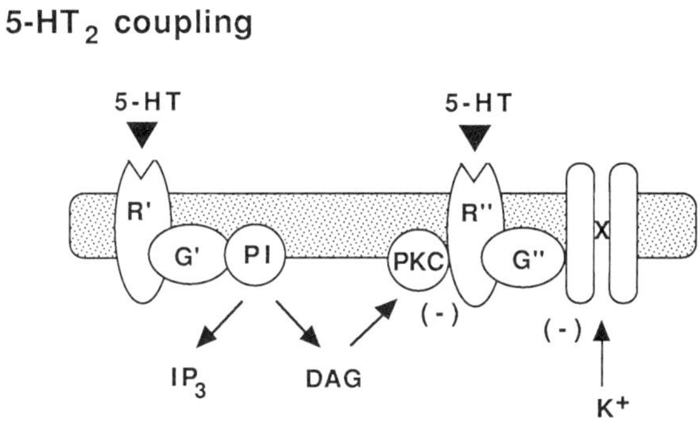

FIGURE 2. Hypothetical scheme for G protein/ion channel/second messenger coupling of 5-HT_2 receptors via two parallel pathways: 1) direct coupling of 5-HT_2 receptors (R″) to the closing of K channels via a G protein (G″); 2) coupling of 5-HT_2 receptors to the activation of phosphoinositide (PI) turnover; note that the PI pathway is depicted as negatively modulating the closing of K channels via diacylglycerol (DAG), a product of PI turnover which activates protein kinase C (PKC).

the brain[67] and rapid excitatory responses to 5-HT have been found in cultured mouse hippocampal and striatal neurons which have many of the characteristics of 5-HT_3 responses in the periphery: rapid onset, rapid desensitization, and blockade by 5-HT_3 antagonists.[68] It has been suggested that 5-HT_3 receptors, unlike other 5-HT receptors, are directly coupled to an ion channel and that the response is not mediated by a second messenger.[68] Nevertheless, cAMP may exert a modulatory influence on 5-HT_3 responses since rates of desensitization are accelerated by the adenylate cyclase activator forskolin.

SUMMARY AND CONCLUSIONS

It is becoming evident that the diverse electrophysiological actions of 5-HT in the central nervous system can be best formulated in terms of receptor subtypes and their respective effector mechanisms. Based on the findings described in this review, the following pattern of central 5-HT electrophysiology is emerging: 1) inhibitory effects are mediated by 5-HT_1 receptors linked to the opening of K channels via pertussis-toxin sensitive G proteins: 2) facilitatory effects are mediated by 5-HT_2 receptors and involve the closing of K channels, an effect which appears to be negatively modulated by activation of the PI second messenger system: 3) fast excitations are mediated by 5-HT_3 receptors, most likely involving a direct interaction with an ion channel rather than through coupling with a G protein or a second messenger. Further studies will be required in a wider range of brain areas to establish the generality of these conclusions.

REFERENCES

1. ROBERTS, M. H. T. & D. W. STRAUGHAN. 1967. Excitation and depression of cortical neurones by 5-hydroxytryptamine. J. Physiol. **193:** 269–294.
2. HAIGLER, H. J. & G. K. AGHAJANIAN. 1974. Peripheral serotonin antagonists: Failure to antagonize serotonin in brain areas receiving a prominent serotonergic input. J. Neural Trans. **35:** 257–273.
3. LEYSEN, J. E. & P. M. LADURON. 1977. A serotonergic component of neuroleptic receptors. Arch. Int. Pharmacodyn. Ther. **230:** 337–339.
4. CREESE, I. & S. H. SNYDER. 1978. ^{3}H-spiroperidol labels serotonin receptors in rat cerebral cortex and hippocampus. Eur. J. Pharmacol. **49:** 201–202.
5. LEYSEN, J. E., C. J. E. NIEMEGEERS, J. P. TOLLENAERE & P. M. LADURON. 1978. Serotonergic component of neuroleptic receptors. Nature **272:** 68–171.
6. PEROUTKA, S. J. & S. H. SNYDER. 1979. Multiple serotonin receptors: Differential binding of ^{3}H-serontonin, ^{3}H-lysergic acid diethylamide and ^{3}H-spiroperidol. Mol. Pharmacol. **16:** 687–699.
7. BRADLEY, P. B., G. ENGLE, W. FENIUK, J. R. FOZARD, P. P. A. HUMPHREY, D. N. MIDDLEMISS, E. J. MYLECHARANE, B. P. RICHARDSON & P. R. SAXENA. 1986. Proposals for the classification and nomenclature of functional receptors for 5-hydroxytryptamine. Neuropharmacology **25:** 563–576.
8. PEROUTKA, S. J. 1988. 5-hydroxytryptamine receptor subtypes: Molecular, biochemical and physiological characterization. TINS **11:** 496–500.
9. HARTIG, P. R. 1989. Molecular biology of 5-HT receptors. TIPS **101:** 64.
10. PAZOS, A. & J. M. PALACIOS. 1985. Quantitative autoradiographic mapping of serotonin receptors in the rat brain. I. Serontonin-1-receptors. Brain Res. **346:** 205–230.
11. WANG, R. Y. & G. K. AGHAJANIAN. 1977. Antidromically identified serotonergic neurons in the rat midbrain raphe: Evidence for collateral inhibition. Brain Res. **132:** 186–193.
12. AGHAJANIAN, G. K. & H. J. HAIGLER. 1975. Hallucinogenic indoleamines: Preferential action upon presynaptic serotonin receptors. Psychopharmacol. Commun. **1:** 619–629.
13. AGHAJANIAN, G. K. & C. P. VANDERMAELEN. 1982. Intracellular recordings from serotonergic dorsal raphe neurons: Pacemaker potentials and the effect of LSD. Brain Res. **238:** 463–469.
14. AGHAJANIAN, G. K. & J. M. LAKOSKI. 1984. Hyperpolarization of serotonergic neurons by serotonin and LSD: Studies in brain slices showing increased K^+ conductance. Brain Res. **305:** 181–185.
15. WILLIAMS, J. T., W. F. COLMERS & Z. Z. PAN. 1988. Voltage- and ligand-activated inwardly rectifying currents in dorsal raphe neurons *in vitro*. J. Neurosci. **8:** 3499–3506.
16. MIDDLEMISS, D. N. & J. R. FOZARD.1983. 8-Hydroxy-2-(di-n-propylamino)tetralin discriminates between subtypes of the 5-HT_1 recognition site. Eur. J. Pharmacol. **90:** 151–153.
17. DEMONTIGNY, C., P. BLIER & Y. CHAPUT. 1984. Electrophysiologically-identified serotonin receptors in the rat CNS. Neuropharmacology **23:** 1511–1520.
18. VANDERMAELEN, C. P., G. K. MATHESON, R. C. WILDERMAN & L. A. PATTERSON. 1986. Inhibition of serotonergic dorsal raphe neurons by systemic and iontophoretic administration of buspirone, a non-benzodiazepine anxiolytic drug. Eur. J. Pharmacol. **129:** 123–130.

19. ASARCH, K. B., R. W. RANSOM & J. C. SHIH. 1985. 5-HT_{1a} and 5-HT_{1b} selectivity of two phenylpiperazine derivatives: evidence for 5-HT_{1b} heterogeneity. Life Sci. **36:** 1265–1273.
20. DOMPERT, W. U., T. GLASER & J. TRABER. 1985. ^{3}H-TVX Q 7821: Identification of 5-HT_1 binding sites as target for a novel putative anxiolytic. Naunyn-Schmiedebergs Arch. Pharmacol. **328:** 462–465.
21. HAMON, M., J. M. COSSERY, U. SPAMPINATO & H. GOZLAN. 1986. Are there selective ligands for 5-HT_{1A} and 5-H_{1B} receptor binding sites in brain? TIPS **7:** 336–338.
22. RANSOM, R. W., K. B. ASARCH & J. C. SHIH. 1986. [^{3}H]1-[2-(4-Aminophenyl)ethyl]-4-(3-trifluoromethylphenyl)piperazine: A selective radioligand for 5-HT_{1A} receptors in rat brain. J. Neurochem. **46:** 68–75.
23. SPROUSE, J. S. & G. K. AGHAJANIAN. 1987. Electrophysiological responses of serotonergic dorsal raphe neurons to 5-HT_{1A} and 5-HT_{1B} agonists. Synapse **1:** 3–9.
24. VERGE, D., G. DAVAL, A. PATEY, H. GOZLAN, S. EL MESTIKAWAY & M. HAMON. 1985. Presynaptic 5-HT autoreceptors on serotonergic cell bodies and/or dendrites but not terminals are of the 5-HT_{1A} subtype. Eur. J. Pharmacol. **113:** 463–464.
25. GREEN, A. R. & D. G. GRAHAME-SMITH. 1976. (−)-Propranolol inhibits the behavioural responses of rats to increased 5-hydroxytryptamine in the central nervous system. Nature **262:** 594–596.
26. MIDDLEMISS, D. N. 1984. Stereoselective blockade of ^{3}H5-HT binding sites at the 5-HT autoreceptor by propranolol. Euro. J. Pharmacol. **101:** 289–293.
27. TRICKLEBANK, M. D., C. FORLER & J. R. FOZARD. 1985. The involvement of subtypes of the 5-HT_1 receptor and of catecholaminergic systems in the behavioural response to 8-hydroxy-2-(di-n-propylamino)tetralin in the rat. Eur. J. Pharmacol. **106:** 271–282.
28. SPROUSE, J. S. & G. K. AGHAJANIAN. 1986. (-)-Propranolol blocks the inhibition of serotonergic dorsal raphe cell firing by 5-HT_{1A} selective agonists. Eur. J. Pharmacol. **128:** 295–298.
29. FARGIN, A., J. R. RAYMOND, M. J. LOHSE, B. K. KOBILKA, M. G. CARON & R. J. LEFKOWITZ. 1988. The genomic clone G-21 which resembles a β-adrenergic receptor sequence encodes the 5-HT_{1A} receptor. Nature. **335:** 358–360.
30. SPROUSE, J. S. & G. K AGHAJANIAN. 1988. Responses of hippocampal pyramidal cells to putative serotonin 5-HT_{1A} and 5-HT_{1B} agonists: A comparative study with dorsal raphe neurons. Neuropharmacology **27:** 707–715.
31. SEGAL, M. 1980. the action of serotonin in the rat hippocampal slice preparation. J. Physiol. (Lond.) **303:** 423–439.
32. ANDRADE, R. & R. A. NICOLL. 1987. Pharmacologically distinct actions of serotonin on single pyramidal neurones of the rat hippocampus recorded *in vitro*. J. Physiol. (Lond) **394:** 99–124.
33. COLINO, A. & J. V. HALLIWELL. 1986. 8-OH-DPAT is a strong antagonist of 5-HT action in rat hippocampus. Eur. J. Pharmacol. **130:** 151–152.
34. SCHLEGEL, J. R. & S. J. PEROUTKA. 1986. Nucleotide interactions with 5-HT binding sites directly labeled by ^{3}H-8-hydroxy-2-(di-n-propylamino)tetralin (^{3}H-8OH-DPAT). Biochem. Pharmacol. **35:** 1943–1949.
35. INNIS, R. B., E. J. NESTLER & G. K. AGHAJANIAN. 1988. Evidence for a G protein mediation of serotonin- and $GABA_B$-induced hyperpolarization of rat dorsal raphe neurons. Brain Res. **459:** 27–36.
36. INNIS, R. B. & G. K. AGHAJANIAN. 1987. Pertussis toxin blocks 5-HT_{1A} and $GABA_B$ receptor mediated inhibition of serotonergic neurons. Eur. J. Pharmacol. **143:** 195–204.
37. ANDRADE, R., R. C. MALENKA & R. A. NICOLL. 1986. A G protein couples serotonin and $GABA_B$ receptors to the same channels in hippocampus. Science **234:** 1261–1265
38. PAZOS, A., R. CORTES & J. M. PALACIOS. 1985. Quantitative autoradiographic mapping of serotonin receptors in the rat brain. II. Serotonin-2 receptors. Brain Res. **346:** 231–249.
39. MCCALL, R. B. & G. K. AGHAJANIAN. 1979. Serotonergic facilitation of facial motoneuron excitation. Brain Res. **169:** 11–27.
40. VANDERMAELEN, C. P. & G. K. AGHAJANIAN. 1980. Intracellular studies showing modulation of facial motoneurone excitability by serotonin. Nature **287:** 346–347.
41. VANDERMAELEN, C. P. & G. K. AGHAJANIAN. 1982. Serotonin-induced depolarization of

rat facial motoneurons *in vivo:* Comparison with amino acid transmitters. Brain Res. **239:** 139–152.
42. McCall, R. B. & G. K. Aghajanian. 1980. Pharmacological characterization of serotonin receptors in the facial motor nucleus: a microiontophoretic study. Eur. J. Pharmacol. **65:** 175–183.
43. Leysen, J. E., F. Awouters, L. Kennis, P. M. Laduron, J. Vandenberk & P. A. J. Janssen. 1981. Receptor binding profile of R 41 468, a novel antagonist at 5-HT_2 receptors. Life Sci. **28:** 1015–1022.
44. Rasmussen, K. & G. K. Aghajanian. 1990. Serotonin excitation of facial motoneurons: Receptor subtype characterization. Synapse. In press.
45. Leysen, J. E., W. Gommereu, P. Van Gompel, J. Wynants, P. F. M. Janssen & P. M. Laduron. 1985. Receptor-binding properties in vitro and in vivo of ritanserin a very potent and long acting serotonin-S_2 antagonist. Mol. Pharmacol. **27:** 600–611.
46. Hoyer, D. 1988. Functional correlates of serotonin 5-HT1 recognition sites. J. Receptor Res. **8:** 59–81.
47. Glennon, R. A., R. Young & J. A. Rosecrans. 1983. Antagonism of the effects of the hallucinogen DOM and the purported 5-HT agonist quipazine by 5-HT_2 antagonists. Eur. J. Pharmacol. **91:** 189–196.
48. Glennon, R. A., M. Titeler & J. D. McKenney. 1984. Evidence for 5-HT_2 involvement in the mechanism of action of hallucinogenic agents. Life Sci. **35:** 2505–2511.
49. Heym, J., K. Rasmussen & B. L. Jacobs. 1984. Some behavioral effects of hallucinogens are mediated by a postsynaptic serotonergic action: Evidence from single unit studies in freely moving cats. Eur. J. Pharmacol. **101:** 57–68.
50. Colpaert, F. C., T. F. Meert, C. J. E. Niemegeers & P. A. J. Janssen. 1985. Behavioral and 5-HT antagonist effect of ritanserin: A pure and selective antagonist of LSD discrimination in rat. Psychopharmacology. **86:** 45–54.
51. Mokler, D. J., K. W. Stoudt & R. H. Rech. 1985. The 5-HT_2 antagonist pirenperone reverses disruption of FR-40 by hallucinogenic drugs. Pharmacol. Biochem. Behav. **22:** 677–682.
52. McCall, R. B. & G. K. Aghajanian. 1980. Hallucinogens potentiate responses to serotonin and norepinephrine in the facial motor nucleus. Life Sci. **26:** 1149–1156.
53. Davies, M. F., R. A. Deisz, D. A. Prince & S. J. Peroutka. 1987. Two distinct effects of 5-hydroxytryptamine on single cortical neurons. Brain Res. **423:** 347–352.
54. Sheldon, P. W. & G. K. Aghajanian. 1989. Serotonin (5-HT) induces IPSPs in pyramidal layer cells of rat piriform cortex: Evidence for the involvement of a 5-HT_2 activated interneuron. Brain Res. In press.
55. Aghajanian, G.K. 1980. Mescaline and LSD facilitate the activation of locus coeruleus neurons by peripheral stimuli. Brain Res. **186:** 492–498.
56. Rasmussen, K. & G. K. Aghajanian. 1986. Effects of hallucinogens on spontaneous and sensory-evoked locus coeruleus unit activity in the rat: Reversal by selective 5-HT_2 antagonists. Brain Res. **385:** 395–400.
57. Peroutka, S. J. & S. H. Snyder. 1980. Relationship of neuroleptic drug effects at brain dompaine, serotonin, α-adrenergic, and histamine receptors to clinical potency. Am. J. Psychiat. **137:** 1518–1522.
58. Rasmussen, K. & G. K. Aghajanian. 1988. Potency of antipsychotics in reversing the effects of a hallucinogenic drug on locus coeruleus neurons correlates with 5-HT_2 binding affinity. Neuropsychopharmacology **1:** 101–107.
59. Rasmussen, K., R. A. Glennon & G. K. Aghajanian. 1986. Phenethylamine hallucinogens in the locus coeruleus: Potency of action correlates with rank order of 5-HT_2 binding affinity. Eur. J. Pharmacol. **132:** 79–82.
60. Conn, P. J. & E. Sanders-Bush. 1986. Regulation of serotonin-stimulated phosphoinositide hydrolysis: relation to the serotonin 5-HT_2 binding site. J. Neurosci. **6:** 3669–3675.
61. Sanders-Bush, E., K. D. Burris & K. Knoth. 1988. Lysergic acid diethylamide and 2,5-dimethoxy-4-methylamphetamine are partial agonists at serotonin receptors linked to phosphoinositide hydrolysis. J. Pharmacol. Exp. Ther. **246:** 924–928.
62. Hidaka, H., M. Inagaki, S. Kawamoto & Y. Sasaki. 1984. Isoquinolinesulfonamides,

novel and potent inhibitors of cyclic nucleotide-dependent protein kinase and protein kinase C. Biochemistry **23:** 5036–5041.
63. HANNUN, Y. A., C. R. LOOMIS, A. H. MERRIL, JR & R. M. BELL. 1986. Sphingosine inhibition of protein kinase C activity and of phorbol dibutyrate binding *in vitro* and in human platelets. J. Biol. Chem. **261:** 12604–12609.
64. SHELDON, P. W. & G. K. AGHAJANIAN. 1989. Protein kinase C inhibitors potentiate serotonin-induced depolarizations in pyramidal layer cells of piriform cortex. Soc. Neurosci. Abs. In press.
65. ROTH, B. L., T. NAKAKI, D. M. CHUANG & E. COSTA. 1986. 5-Hydroxytryptamine$_2$ receptors coupled to phospholipase C in rat aorta: Modulation of phosphoinositide turnover by phorbol ester. J. Pharmacol. Exp. Ther. **238:** 480–485.
66. PIERCE, P. A. & S. J. PEROUTKA. 1988. Antagonism of 5-hydroxytryptamine$_2$ receptor-mediated phosphatidylinositol turnover by d-lysergic acid diethylamide. J. Pharmacol. Exp. Ther. **247:** 918–925.
67. KILPATRICK, G. J., B. J. JONES & M. B. TYERS. 1987. Identification and distribution of 5-HT$_3$ receptors in rat brain using radioligand binding, Nature **330:** 746–748.
68. YAKEL, J. L. & M. B. JACKSON. 1988. 5-HT$_3$ receptors mediate rapid responses in cultured hippocampus and a clonal cell line. Neuron **1:** 615–621.

DISCUSSION OF THE PAPER

QUESTION: Dr. Aghajanian, I would like to come back to your electrophysiological recordings where you showed in two of your slides that GABA is mimicking 5-HT in inhibiting cell firing. Could you inhibit that with the GABA-B receptor antagonist phaclofen? And my second question would be did you ever find the 5-HT$_{1a}$ somatodendritic autoreceptor in species other than the rat.

AGHAJANIAN: I'll get to your last question first. We haven't looked at any other species but the rat. It is necessary for this question to rely on other types of data, including data in humans, for the existence of the 5-HT$_{1A}$ receptor in the dorsal raphe of other species. As far as the GABA-B question is concerned, in the raphe and the hippocampus the GABA-B receptor seems to couple through the same type of G protein and induces the same kind of opening of potassium channels as does 5-HT. In addition, these GABA-B effects are also blocked by pertussis toxin.

E. BEKKER (*Bristol-Myers Holland, Weesp, the Netherlands*): What is the implication of this very detailed, very selective animal research for more general electrophysiological measures in human EEG. I studied the effects of ritanserin on the 24 hours EEG in normal volunteers, and it's difficult to find such systematic effects as you presented on electrophysiology.

AGHAJANIAN: It's a long way from single cells to EEG. I'll just mention one type of study with 5-HT$_2$ antagonists such as ritanserin. In a published study, it has been shown that these drugs prolong slow wave sleep. Thus, there are EEG studies with the more selective of the drugs that act at these various 5-HT receptors which show interesting changes.

Serotonin Receptor "Families" in the Central Nervous System: An Overview[a]

STEPHEN J. PEROUTKA,[b] ANNE W. SCHMIDT,
ANDREW J. SLEIGHT, AND MELISSA A. HARRINGTON

Department of Neurology
Stanford University
Stanford, California 94305

INTRODUCTION

The identification of multiple receptor subtypes for 5-hydroxytryptamine (5-HT) using radioligand binding techniques proliferated at a brisk rate in the 1980s. In the future, with the application of molecular biological techniques, the number of distinct 5-HT subtypes is likely to expand rather than contract. The evolving data suggest that 5-HT receptor subtypes can be categorized into 3 major "families." Each "family" consists of multiple receptor subtypes with similar molecular biological, pharmacological and biochemical properties. This brief review provides a summary of recent data as well as a framework for the classification of 5-HT receptor subtypes (TABLE 1).

THE 5-HT_1 RECEPTOR "FAMILY"

In 1979, radioligand binding techniques were used to differentiate two distinct subtypes of 5-HT recognition sites in the central nervous system.[1] One class, designated 5-HT_1 binding sites, was defined by nanomolar affinity for 5-HT and could be labeled with ^{3}H-5-HT. These same simple criteria might still be used as part of the definition of the 5-HT_1 "family" of 5-HT receptors (TABLE 1). The 5-HT_1 "family" includes at least three distinct subtypes: 5-HT_{1A}, 5-HT_{1B} and 5-HT_{1D} receptors.[2-4] The pharmacology and biochemistry of the 5-HT_{1C} receptor,[5,6] initially classified as a 5-HT_1 subtype, suggest that it should be more appropriately placed in the 5-HT_2 "family" of receptors.

Molecular Biology

The 5-HT_{1A} receptor was the first 5-HT receptor to be successfully cloned, although its identity remained unknown until mid-1988. The genomic clone G-21 was identified because of its cross-reactivity with a beta-adrenergic receptor sequence.[7] Radioligand binding studies using membranes from monkey kidney cells transfected with G-21 DNA revealed specific binding of ^{125}I-cyanopindolol, a radioligand which labels beta-adrenergic receptors with high affinity and 5-HT_{1A} receptors with slightly lower affinity. However, 5-HT was more potent than adrenergic agents in inhibiting ^{125}I-cyanopindolol binding to G-21 transfected cell membranes, and high affinity binding with the 5-HT_{1A} agonist ^{3}H-8-hydroxy-(2-di-*N*-propylamino)-tetralin (8-OH-

[a]This work was supported in part by the McKnight Foundation and NIH Grants NS 12151-14 and NS 23560-03.

[b]*Present address:* Genetech, Inc., 460 Point San Bruno Blvd., South San Francisco, CA 94080.

TABLE 1. Characteristics of 5-HT Receptor Subtypes

	5-HT$_1$ Receptors		
	5-HT$_{1A}$	5-HT$_{1B}$	5-HT$_{1D}$
Molecular Biology	Cloned	?	?
Potent Pharmacologic Agents (i.e. <10 nM)	Selective 8-OH-DPAT ipsapirone buspirone	Selective none available	Selective none available
	Non-selective 5-HT 5-CT RU 24969 d-LSD (−)pindolol	Non-selective 5-HT 5-CT RU 24969 (−)pindolol	Non-selective 5-CT Metergoline
Second Messenger	Inhibition of adenylate cyclase	Inhibition of adenylate cyclase	Inhibition of adenylate cyclase

5-HT$_2$ Receptors			5-HT$_3$ Receptors
5-HT$_{1C}$	5-HT$_{2A}$	5-HT$_{2B}$	5-HT$_3$
Cloned	?	Cloned	?
Selective none available	Selective DOB DOI	Selective none available	Selective GR 65630 Zacopride ICS 205-930 Granisetron Odansetron
Non-selective Mesulergine Metergoline Methysergide Mianserin	Non-selective Mesulergine Metergoline Ketanserin Spiperone	Non-selective Mesulergine Metergoline Methysergide Mianserin Ketanserin Spiperone	
PI Turnover	?	PI Turnover	?

DPAT) could also be detected in this transfected cell line. Competition experiments demonstrate that 5-HT$_{1A}$ selective agents such as buspirone and 8-OH-DPAT potently inhibit ^{3}H-8-OH-DPAT binding whereas dopamine, histamine and adrenergic agents weakly inhibit binding. Like other G-protein linked receptors, 5-HT$_{1A}$ receptors have seven membrane spanning units. The amino acid sequence of these membrane spanning domains shows the least variability in comparison to other cloned receptors of biogenic amines. The third cytoplasmic loop shows the greatest variability in its amino acid sequence and is believed to interact with the G-protein. More recently, additional receptor clones which display significant homology with the 5-HT$_{1A}$ receptor have been isolated, but not yet identified.[8]

Pharmacology

The 5-HT_1 "family" of receptors displays nanomolar affinity for 5-HT and 5-carboxyamidotryptamine (5-CT) and micromolar affinity for ketanserin, mesulergine, quipazine and ICS 205-930. The 5-HT_{1A} binding site was defined initially as the specific ^{3}H-5-HT binding in brain membranes which was sensitive to nanomolar concentrations of spiperone.[9] Subsequently, several radioligands have been developed which labeled 5-HT_{1A} binding sites more selectively such as ^{3}H-8-OH-DPAT[2,3] (TABLE 1). In general, the site displays high and selective affinity for certain aminotetralins (e.g., 8-OH-DPAT), pyrimidinylpiperazines (e.g., ipsapirone, buspirone) and benzodioxanes (e.g., WB 4101, spiroxatrine, MDL 72832, flesinoxan). Autoradiographic studies reveal a heterogeneous distribution of binding in the brain with highest densities of the 5-HT_{1A} site found in the CA1 region and dentate gyrus of the hippocampus and in the raphe nuclei.[10] Because of the large number of agents which display potent and selective affinity for the 5-HT_{1A} site, multiple correlations have been established between the pharmacological characteristics of this binding site and specific effects of 5-HT.[11]

Binding sites for 5-HT_{1B} have been directly labeled in rat brain with ^{125}I-cyanopindolol[3] and other radioligands. 5-HT_{1B} receptors display high affinity for 5-HT and RU 24969, and relatively low affinity for d-LSD and 8-OH-DPAT (TABLE 1). In the rat, autoradiographic studies with ^{3}H-5-HT reveal the highest density of binding sites with high affinity for 5-HT and RU 24969 in globus pallidus, dorsal subiculum, and substantia nigra.[10] Interestingly, the 5-HT_{1B} receptor binding site, as defined pharmacologically, appears to be species specific. The 5-HT_{1B} binding site is present in rat and mouse brain but not in guinea pig, cow, chicken, turtle, frog, or human brain membranes.[10,12]

Recently, the 5-HT_{1D} site, a third subtype of 5-HT_1 sites labeled by ^{3}H-5-HT, was identified in bovine brain membranes.[4] 5-HT_{1A} selective agents such as 8-OH-DPAT, ipsapirone and buspirone display micromolar affinities for these sites. RU 24969 and (−)pindolol are approximately two orders of magnitude less potent at these sites than at 5-HT_{1B} binding sites identified in rat brain. Agents which display nanomolar potencies for 5-HT_{1C} sites such as mianserin and mesulergine are two to three orders of magnitude less potent at ^{3}H-5-HT labeled 5-HT_{1D} binding sites in bovine caudate. In addition, both 5-HT_2 and 5-HT_3 selective agents are essentially inactive at these binding sites. These ^{3}H-5-HT sites display nanomolar affinity for 5-HT, 5-methoxytryptamine, metergoline and 5-HT (TABLE 1). Regional studies demonstrate that this class of sites is most dense in the basal ganglia but exists in all regions of bovine brain.[4] The 5-HT_{1D} binding site has been identified and characterized in multiple other species including bovine, pig, calf and human brain membranes.

Biochemistry

The 5-HT_1 receptor "family" has been linked to the modulation of adenylate cyclase via the pertussis-sensitive G_i protein. This G protein linkage is reflected in the regulation of agonist binding by guanine nucleotides. For example, GTP and GDP, but not GMP, inhibit the binding of ^{3}H-8-OH-DPAT to 5-HT_{1A} receptors in brain membranes.[2] Guanine nucleotides significantly reduce agonist potencies for ^{3}H-8-OH-DPAT binding sites whereas antagonist potencies are not affected by the addition of nucleotides. 5-HT, 8-OH-DPAT, and *d*-LSD all inhibit forskolin-stimulated adenylate cyclase activity in rat and guinea pig hippocampal membranes.[13,14] By contrast, spiperone is a competitive antagonist at this receptor whereas ketanserin, a 5-HT_2

antagonist, has no effect on 5-HT-induced inhibition of cyclase activity. Both (−)pindolol and (−)propranolol (10^5 M) reverse 8-OH-DPAT inhibition of forskolin-stimulated adenylate cyclase, suggesting that these drugs also act as antagonists at 5-HT$_{1A}$ receptors.[15]

Like 5-HT$_{1A}$ receptors, the 5-HT$_{1B}$ receptor mediates the inhibition of adenylate cyclase in rat substantia nigra. 5-HT, 5-CT, RU 24969 and CGS 12066B, reportedly a selective 5-HT$_{1B}$ agonist,[16] inhibit forskolin-stimulated adenylate cyclase, whereas 8-OH-DPAT and ipsapirone are ineffective.[17] Cyanopindolol, propranolol and metergoline reverse cyclase inhibition by 5-HT, whereas spiperone, mesulergine and ketanserin are ineffective. This pharmacological profile is consistent with the 5-HT$_{1B}$ receptor negatively coupled to adenylate cyclase.

The 5-HT$_{1D}$ receptor also appears to mediate the inhibition of forskolin-stimulated adenylate cyclase activity. In the presence of GTP, the 5-HT$_{1D}$ agonists 5-HT, 5-CT and 5-methoxy-tryptamine inhibit adenylate cyclase in the calf substantia nigra, whereas 8-OH-DPAT has no effect.[18] Cyanopindolol, spiperone and mianserin do not antagonize the inhibitory effects of 5-HT, while antagonism is observed with the non-selective 5-HT antagonist methiothepin. This pharmacological profile is consistent with a 5-HT$_{1D}$ receptor linked to inhibition of adenylate cyclase.

Interestingly, the 5-HT$_{1B}$ and the 5-HT$_{1D}$ sites may be analogous receptors found in different species. In addition to their pharmacological, anatomical, and biochemical similarities, both receptor subtypes appear to mediate the 5-HT "autoreceptor" in different species.[19] These observations suggest that the 5-HT$_{1B}$ receptor may be a relatively recent evolutionary descendant of the 5-HT$_{1D}$ receptor.

THE 5-HT$_2$ RECEPTOR "FAMILY"

The 5-HT$_2$ "family" of receptors displays relatively low affinity for 5-HT and nanomolar affinity for 5-HT antagonists such as ketanserin, mesulergine, metergoline and d-LSD. Recent pharmacological and molecular biological data suggest that subtypes exist within the 5-HT$_2$ receptor "family." For example, 5-HT$_{1C}$ receptors are presented here as part of the 5-HT$_2$ "family" based on their molecular biological, pharmacological and biochemical similarities to other 5-HT$_2$ receptor subtypes.

Molecular Biology

A functional 5-HT$_{1C}$ receptor cDNA encoding a protein of 460 amino acids was isolated by methods involving cloning in RNA expression vectors combined with a sensitive electrophysiological assay in *Xenopus* oocytes.[20] Application of 5-HT to *Xenopus* oocytes previously injected with cDNA generated from rat choroid plexus mRNA caused a detectable inward current. This inward current could be blocked by 1 μM mianserin, but not 1 μM spiperone, a pharmacological property that is typical of a 5-HT$_{1C}$ receptor. Radioligand binding studies on membranes from transfected mouse fibroblasts revealed high affinity binding of ^{125}I-LSD. Mianserin and 5-HT potently inhibited specific ^{125}I-LSD binding, whereas spiperone was a relatively weak competitor, as would be expected of a 5-HT$_{1C}$ receptor. *In situ* hybridization studies show that, in addition to choroid plexus, 5-HT$_{1C}$ receptor mRNAs are found in basal ganglia, hypothalamus, hippocampus and spinal cord.[21]

A functional 5-HT$_2$ receptor cDNA was isolated from a cloned cDNA library using probes generated from amino acid sequences 88–104 and 134–149 of the 5-HT$_{1C}$

receptor.[22] The 5-HT_2 receptor is encoded by 449 amino acids and shares a 51% homology with the 5-HT_{1C} sequence including identical fifth transmembrane domains. Radioligand binding studies on membranes from a transfected mammalian cell line reveal high affinity binding of ^{3}H-spiperone which is potently inhibited by ketanserin and mianserin. The 5-HT_{1A} agonist, 8-OH-DPAT, and the 5-HT_3 antagonist, MDL 72222, show low affinity against ^{3}H-spiperone binding to transfected cell membranes. The sequence of the human 5-HT_2 receptor appears to differ 10% from the sequence of the rat 5-HT_2 receptor,[23] although, the possible pharmacological and functional significance of this difference between species remains unknown.

Pharmacology

The 5-HT_{1C} binding site was first discovered as a result of autoradiographic analysis of 5-HT binding site subtypes in pig cortex and choroid plexus.[5,6] The site can be labeled by ^{3}H-5-HT, ^{3}H-mesulergine, ^{125}I-LSD or ^{3}H-SCH 23390. The 5-HT_{1C} binding site displays high affinity for 5-HT, methysergide and mianserin and relatively low affinity for 8-OH-DPAT, RU 24969 and spiperone (TABLE 1). The 5-HT_{1C} receptor has been identified and characterized in the rat, pig, and human brain membranes,[3,5,6] and the pharmacological profile and distribution of 5-HT_{1C} receptors appears similar for each of these species.

The 5-HT_2 class of binding sites has been extensively analyzed due to the availability of a large number of potent and selective antagonists. ^{3}H-spiperone, ^{3}H-LSD, ^{3}H-mianserin, ^{3}H-ketanserin, ^{3}H-mesulergine, ^{125}I-LSD and N_1-methyl-2-^{125}I-LSD can be used to label the 5-HT_2 binding site.[22] The 5-HT_2 binding site displays high affinity for a number of serotonergic antagonists while 5-HT and related tryptamines display markedly lower affinity (TABLE 1). 8-OH-DPAT, RU 24969, and 5-HT_3 selective agents display low affinity for this site. The highest density of 5-HT_2 sites is in the cerebral cortex and caudate with all other brain regions having substantially fewer sites.[24]

Recent studies have provided data which suggest that the putative hallucinogen ^{3}H-DOB (4-bromo-2,5-dimethoxyamphetamine) does not label an "agonist state" of the 5-HT_2 site,[25] but rather a distinct subtype of 5-HT_2 receptor. First, ^{77}Br-R(−)DOB–labeled sites comprise only 19% of the density of 5-HT_2 sites in rat cortex and are not eliminated by the addition of 10^{-4} M GTP.[26] In addition, 5-HT competition curves with ^{3}H-ketanserin in rat and human cortical membranes are shallow (Hill slopes = 0.67, 0.69 respectively), suggesting that ^{3}H-ketanserin labels more than one population of 5-HT_2 receptors, while 5-HT competition curves in bovine cortical membranes are steep (Hill slope = 0.97), indicating that a single site is labeled by ^{3}H-ketanserin. Furthermore, ^{3}H-DOB specific binding cannot be detected in bovine membranes. The high affinity component of ^{3}H-ketanserin binding (which can also be labeled with ^{3}H-DOB or ^{77}Br-(−)DOB in rat membranes) has been designated the 5-HT_{2A} binding site. ^{3}H-Ketanserin also labels a subpopulation of 5-HT_2 sites, (designated 5-HT_{2B} sites) in rat, human and bovine membranes.[27]

Biochemistry

Unlike other 5-HT_1 receptors, the 5-HT_{1C} receptor is linked to modulation of phosphoinositide (PI) turnover rather than adenylate cyclase. 5-HT stimulates PI turnover in choroid plexus, a brain area rich in 5-HT_{1C} receptors.[29] Mianserin, ketanserin and spiperone all inhibit the response, and their potencies correlate with their binding affinities at the 5-HT_{1C} receptor.

Phosphoinositide hydrolysis in rat cerebral cortex induced by 5-HT appears to occur as a result of 5-HT_2 receptor activation.[28] 5-HT stimulates PI turnover with an EC_{50} of 1 μM, and this 5-HT effect is potently blocked by ketanserin and other 5-HT_2 antagonists. The effectiveness of 5-HT_2 antagonists at blocking 5-HT stimulation of PI turnover correlates well with their binding affinities at ^{3}H-ketanserin labeled 5-HT_2 sites. Furthermore, tricyclic antidepressants decrease both 5-HT_2 binding sites and 5-HT-induced PI turnover.[28] Multiple other tissues such as the rat thoracic aorta, cultured bovine aortic smooth muscle cells and platelets contain a 5-HT receptor which appears to modulate phosphoinositide turnover via the 5-HT_2 binding site.

THE 5-HT_3 RECEPTOR "FAMILY"

Receptors of the 5-HT_3 class have been well characterized in the periphery where they mediate the excitatory effects of 5-HT. Selective 5-HT_3 antagonists such as odansetron (formerly called GR 38032F, zacopride, ICS 205-930, and MDL 72222 block these excitatory peripheral effects of 5-HT. It appears that heterogeneity may exist within the 5-HT_3 class of receptors. At least three distinct subtypes of 5-HT_3 receptors have been hypothesized, due to significant differences among the potencies of selective 5-HT_3 antagonists in various physiological systems.[30] Further definition of 5-HT_3 receptors and their subtypes awaits the development of more selective pharmacological agents.

Molecular Biology

At present, no information is available on the molecular biology of 5-HT_3 receptors.

Pharmacology

Recently, 5-HT_3 binding sites were identified in rat brain using ^{3}H-GR 65630[31] and a variety of other radioligands (TABLE 1). 5-HT_3 agonists 5-HT, 2-methyl-5-HT and phenylbiguanide have moderate affinity (K_i approximately 150 nM) for this site while antagonists such as granisetron (formerly called BRL 43694), zacopride and ICS 205-930 display subnanomolar affinity (K_i = 0.1–0.8 nM) for this site. 8-OH-DPAT, mesulergine and ketanserin display low affinity for 5-HT_3 binding sites in rat cortical membranes. The highest density of 5-HT_3 binding sites in rat brain is found in cortical areas particularly, entorhinal cortex and in the area postrema.

Biochemistry

5-HT_3 receptor activation appears to cause a release of dopamine from rat striatal slices.[32] 5-HT or 2-methyl-5-HT superfusion in rat striatal slices causes dopamine release which is potentially inhibited by ICS 205-930. By contrast, 5-HT_3 activation inhibits the release of acetyl choline in rat cortical slices, while treatment with 5-HT_3 antagonists facilitates acetyl choline release.[33] These biochemical results are consistent with the behavioral data that odansetron inhibits dopamine-mediated hyperactivity in rats.[34]

OTHER PUTATIVE 5-HT RECEPTORS

In addition to the 5-HT_1, 5-HT_2 and 5-HT_3 subtypes, a variety of other 5-HT receptor subtypes have been hypothesized. In general, the feature which distinguishes the following sites from the receptors listed above is that they have only been identified in a single laboratory. If their presence can be confirmed by independent investigators, then it is likely that each of the following sites will be added to the growing list of 5-HT receptor subtypes.

Gershon and colleagues have reported that a receptor designated 5-HT_{1P} exists in gut membranes.[35] This site can be labeled by ^{3}H-5-HT and displays a pharmacological profile which is distinct from all known 5-HT binding sites or receptors. Moreover, the pharmacological characteristics of this binding site correlate with the physiological effects of 5-HT in the gut.[35]

Recently, Bockaert and colleagues have described a "non-classical 5-HT_4" receptor which is positively coupled to adenylate cyclase.[36] In cell cultures generated from mouse embryo colliculi, 5-HT stimulates cAMP production in a dose-dependent manner. 5-Methoxytryptamine and 5-CT are full agonists, whereas tryptamine, bufotenine, and 2-methyl-5-HT are partial agonists. Spiperone, methiothepin, and ketanserin are ineffective in antagonizing 5-HT-induced increases in cAMP levels, whereas ICS 205-930, but not MDL 72222, weakly but competitively inhibited the effects of 5-HT ($K_i = 997$ nM). ICS 205-930 also competitively antagonizes 5-HT-induced increase in cAMP levels in adult guinea pig hippocampal membranes ($K_i = 454$ nM). This pharmacological profile is not consistent with any previously described 5-HT receptor "family." Further characterization of this site has revealed that 5-HT_3 antagonists of the benzamide chemical class (i.e. zacopride, cisapride, metoclopramide) act as agonists to stimulate cAMP levels in mouse embryo colliculi cell cultures, while ICS 205-930 competitively reversed the stimulation.[37]

FUTURE TRENDS

In this report, evidence has been provided from a variety of molecular, pharmacological and biochemical studies to suggest that multiple 5-HT receptors exist. At the present time, the differentiation of 5-HT receptors into at least 3 major "families" appears to be the most relevant classification system. Continued pharmacological progress will depend on the development of potent and selective agonists and antagonists for each 5-HT receptor subtype. However, a more important trend in the definitive classification of receptor subtypes is the advances that are currently being made in molecular biology. Progress in this area is likely to continue rapidly and yield detailed structural data on all 5-HT receptor subtypes in the very near future. The application of these techniques to the analysis of 5-HT receptors can be expected to lead to an expansion, rather than a contraction, in a number of 5-HT receptor subtypes. Once many of the sequences are determined, the current classification system based on radioligand binding and functional pharmacological studies may need to be revised. More likely, the current classification system of 5-HT receptors will be validated by the results of forthcoming molecular biological studies.

ACKNOWLEDGMENTS

We thank Debbie A. Bell for assistance in the preparation of the manuscript.

REFERENCES

1. PEROUTKA S. J. & S. H. SNYDER. 1979. Multiple serotonin receptors: Differential binding of ^{3}H-serotonin, ^{3}H-lysergic acid diethylamide and ^{3}H-spiroperidol. Mol. Pharmacol. **16:** 687–699.
2. GOZLAN H., S. EL MESTIKAWY, L. PICHAT, J. GLOWINSKY & M. HAMON. 1983. Identification of presynaptic serotonin auto-receptors using a new ligand: ^{3}H-PAT. Nature (London) **305:** 140–142.
3. HOYER D., G. ENGEL & H. O. KALKMAN. 1985. Molecular pharmacology of 5-HT$_1$ aad 5-HT$_2$ recognition sites in rat and pig brain membranes: Radioligand binding studies with ^{3}H-5-HT, ^{3}H-8-OH-DPAT, $(-)^{125}$I-Iodocyanopinolol, ^{3}H-mesulergine and ^{3}H-ketanserin. Eur. J. Pharmacol. **118:** 13–23.
4. HEURING R. E. & S. J. PEROUTKA. 1987. Characterization of a novel ^{3}H-5-hydroxytryptamine binding site subtype in bovine brain membranes. J. Neurosci. **7:** 894–903.
5. PAZOS, A., D. HOYER & J. M. PALACIOS. 1985. The binding of serotonergic ligands to the porcine choroid plexus: Characterization of a new type of serotonin recognition site. Eur. J. Pharmacol. **106:** 539–546.
6. YAGALOFF K. A. & P. R. HARTIG. 1985. ^{125}I-Lysergic acid diethylamide binds to a novel serotonergic site on rat choroid plexus epithelial cells. J. Neurosci. **5:** 3178–3183.
7. FARGIN A., J. R. RAYMOND, M. J. LOHSE, B. K. KOBILKA, M. G. CARON & R. J. LEFKOWITZ. 1988. The genomic clone G-21 which resembles a beta-adrenergic receptor sequence encodes the 5-HT$_{1A}$ receptor. Nature (London) **335:** 358–360.
8. LIBERT F., M. PARMENTIER, A. LEFORT, C. DINSART, J. VAN SANDE, C. MAENHAUT, M.-J. SIMONS, J. DUMONT & G. VASSART. 1989. Selective amplification and cloning of four new members of the G protein-coupled receptor family. Science **244:** 569–572.
9. PEDIGO, N. W., H. I. YAMAMURA & D. L. NELSON. 1981. Discrimination of multiple ^{3}H-5-hydroxytryptamine binding sites by the neuroleptic spiperone in rat brain. J. Neurochem. **36:** 220–226.
10. HOYER D., A. PAZOS, A. PROBST & J. M. PALACIOS. 1986. Serotonin receptors in the human brain. I. Characterization and localization of 5-HT$_{1A}$ recognition sites. Apparent absence of 5-HT$_{1B}$ recognition sites. Brain Res. **376:** 85–96.
11. PEROUTKA, S. J. 1988. 5-Hydroxytryptamine receptor subtypes. Annu. Rev. Neurosci. **11:** 45–60.
12. HEURING R. E., J. R. SCHLEGEL & S. J. PEROUTKA. 1986. Species variations in RU 24969 interactions with non-5-HT$_{1A}$ binding sites. Eur. J. Pharmacol. **122:** 279–282.
13. DE VIVO, M. & S. MAAYANI. 1986. Characterization of the 5-hydroxytryptamine$_{1A}$ receptor-mediated inhibition of forskolin-stimulated adenylate cyclase activity in guinea pig and rat hippocampal membranes. J. Pharmacol. Exp. Ther. **238:** 248–253.
14. BOCKAERT J., A. DUMUIS, R. BOUHELAL, M. SEBBERN & R. N. CORY. 1987. Piperazine derivatives including the putative anxiolytic drugs, buspirone and ipsapirone, are agonists at 5-HT$_{1A}$ receptors negatively coupled with adenylate cyclase in hippocampal neurons. Naunyn-Schmiedeberg's Arch. Pharmacol. **335:** 588–592.
15. OKSENBERG D. & S. J. PEROUTKA. 1988. Antagonism of 5-hydroxytryptamine$_{1A}$ (5-HT$_{1A}$) receptor-mediated modulation of adenylate cyclase activity by pindolol and propranolol isomers. Biochem. Pharmacol. **37:** 3429–3433.
16. NEALE R. F., S. L. FALLON, W. C. BOYAR, J. W. F. WASLEY, L. L. MARTIN, G. A. STONE, B. S. GLAESER, C. M. SINTON & M. WILLIAMS. 1987. Biochemical and pharmacological characterization of CGS 12066B, a selective serotonin-1B agonist. Eur. J. Pharmacol. **136:** 1–9.
17. BOUHELAL R., L. SMOUNYA & J. BOCKAERT. 1988. 5-HT$_{1B}$ receptors are negatively coupled with adenylate cyclase in rat substantia nigra. Eur. J. Pharmacol. **151:** 189–196.
18. SCHOEFFTER, P., C. WAEBER, J. M. PALACIOS & D. HOYER, 1988. The 5-hydroxy-

tryptamine 5-HT_{1D} receptor subtype is negatively coupled to adenylate cyclase in calf substantia nigra. Naunyn-Schmiedeberg's Arch. Pharmacol. **337:** 602–608.

19. HOYER D. & D. N. MIDDLEMISS. 1989. The pharmacology of the terminal 5-HT autoreceptors in mammalian brain: Evidence for species differences. Trends Pharmacol. Sci., **10:** 130–132.
20. JULIUS D., A. B. MACDERMOTT, R. AXEL & T. JESSEL. 1988. Molecular characterization of a functional cDNA encoding the serotonin lc receptor. Science **241:** 558–564.
21. PRITCHETT D. B., A. W. J. BACH, M. WOZNY, O. TALEB, R. DAL TOSO, J. C. SHIH & P. H. SEEBURG. 1988. Structure and functional expression of cloned rat serotonin 5-HT-2 receptor. EMBO J **7:** 4135–4140.
22. HARTIG P. R. 1989. Molecular biology of the serotonin receptor family. Abstracts of the International Symposium on Serotonin: From Cell Biology to Pharmacology and Therapeutics, **1.**
23. LEYSEN J. E., C. J. E. NIEMEGEERS, J. P. TOLLENAERE & P. M. LADURON. 1978. Serotonergic component of neuroleptic receptors. Nature (London) **272:** 168–171.
24. PAZOS A., R. CORTES & J. M. PALACIOS. 1985. Quantitative autoradiographic mapping of serotonin receptors in the rat brain. II. Serotonin-2 receptors. Brain Res. **346:** 231–249.
25. LYON R. A., K. H. DAVIS & M. TITELER. 1987. ^{3}H-DOB (4-bromo-2,5-dimethoxyphenylisopropylamine) labels a guanyl nucleotide-sensitive state of cortical 5-HT_2 receptors. Mol. Pharmacol. **31:** 194–199.
26. PEROUTKA S. J., A. HAMIK, M. A. HARRINGON, A. J. HOFFMAN, C. A. MATHIS, P. A. PIERCE & S.S-.H. WANG. 1988. (R)-(−)-[^{77}Br]4-Bromo-2,5-dimethoxyamphetamine labels a novel 5-hydroxytryptamine binding site in brain membranes. Mol. Pharmacol **34:** 537–542.
27. PIERCE P. A. & S. J. PEROUTKA. 1989. Evidence for distinct 5-$hydroxytryptamine_2$ binding site subtypes in cortical membrane preparations. J. Neurochem. **52:** 656–658.
28. SANDERS-BUSH E. & P. J. CONN. 1987. Neurochemistry of serotonin neuronal systems: Consequences of serotonin receptor activation. *In* Psychopharmacology: The third Generation of Progress, H. Y. Meltzer, Ed. New York: NY. Raven Press, 95–103.
29. LUBBERT H., T. P. SNUTCH, N. DASCAL, H. A. LESTER & N. DAVIDSON. 1987. Rat brain 5-HT_{1C} receptors are encoded by a 5–6 kbase mRNA size class and are functionally expressed in injected *Xenopus* oocytes. J. Neurosci. **7:** 1159–1165.
30. RICHARDSON, B. P., G. ENGEL, P. DONATSCH & P. A. STADLER. 1985. Identification of serotonin M-receptor subtypes and their specific blockade by a new class of drugs. Nature (London) **316:** 126–131.
31. KILPATRICK G. J., B. J. JONES & M. B. TYERS. 1987. Identification and distribution of 5-HT_3 receptors in rat brain using radioligand binding. Nature (London) **330:** 746–748.
32. BLANDINA P., J. GOLDFARB, & J. P. GREEN. 1988. Activation of a 5-HT_3 receptor releases dopamine from rat striatal slice. Eur. J. Pharmacol. **155:** 349–350.
33. BARNES J. M., N. M. BARNES, B. COSTALL, R. J. NAYLOR & M. B. TYERS. 1989. 5-HT_3 receptors mediate inhibition of acetyl choline release in cortical tissues. Nature (London) **338:** 762–763.
34. COSTALL B., A. M. DOMENEY, R. J. NAYLOR & M. B. TYERS. 1987. Effects of the 5-HT_3 receptor antagonist, GR38032F, on raised dopaminergic activity in the mesolimbic system of the rat and marmoset brain. Br. J. Pharmacol. **92:** 881–894.
35. MAWE G. M., T. A. BRANCHEK & M. D. GERSHON. 1986. Peripheral neural serotonin receptors: Identification and characterization with specific antagonists and agonists. Proc. Nat. Acad. Sci. USA **83:** 9799–9803.
36. DUMUIS A., R. BOUHELAL, M. SEBBEN, R. CORY & J. BOCKAERT. 1988. A nonclassical 5-hydroxytryptamine receptor positively coupled with adenylate cyclase in the central nervous system. Mol. Pharmacol. **34:** 880–887.
37. DUMUIS A., M. SEBBEN & J. BOCKAERT. 1989. Benzamide derivatives are agonists of nonclassical 5-HT receptors (5-HT_4) positive coupled with adenylate cyclase in the brain. *In* Abstracts of the International Symposium on Serotonin: From Cell Biology to Pharmacology and Therapeutics, 1989.

DISCUSSION OF THE PAPER

DUNCAN TAYLOR (*Bristol Myers, Wallingford, CT*): Just one comment Steve. The guinea pig is not a rodent, it's a lagomorph.

PEROUTKA: It's a what?

TAYLOR: It's a lagomorph.

PEROUTKA: But it says that it is a rodent in my biology book from college.

TAYLOR: The guinea pig is the same order as rabbits. It has four incisors back to back. Little piece of biological trivia for you.

PEROUTKA: All right, well I've got to go back, get my biology book out and write the editor (see editor's note below).

S.J. BOYSON (*University of Colorado HSC, Denver*): Are there drugs from other classes which commonly interact with the 5-HT_1 receptor?

PEROUTKA: Commonly is a good question. It seems as if the 5-HT_1 family is closely related to the adrenergic family. So there is considerable overlap. WB4101 and pindolol, for example, are adrenergic drugs which interact with the 5-HT_1 family. Also, I think you'll find the cholinergic family is going to look a little bit like the 5-HT_3 family. Atropine doesn't look really different from ICS 205-930. In fact, atropine and ICS 205-930 are very similar structures, but yet differ in their receptor profiles. So yes, there is obviously overlap between all receptor systems.

[**Editors' note:** Indeed, after further investigation, Dr. Peroutka's college biology textbook was correct. Dr. Taylor also investigated the matter and now agrees that the guinea pig is a rodent. Dr. Taylor apologized profusely to Dr. Peroutka, who graciously accepted the apology.]

The Central 5-HT$_{1A}$ Receptors: Pharmacological, Biochemical, Functional, and Regulatory Properties[a]

M. HAMON[b], H. GOZLAN, S. EL MESTIKAWY,
M. B. EMERIT, F. BOLAÑOS, AND L. SCHECHTER

INSERM U288
Neurobiologie Cellulaire et Fonctionnelle
Faculté de Médecine Pitié-Salpêtrière
91, boulevard de l'Hôpital
75634 Paris Cedex 13, France

INTRODUCTION

The first evidence for the existence of several classes of 5-HT$_1$ binding sites with high affinity (Kd ~ 1–5 nM) for serotonin (5-hydroxytryptamine, 5-HT) in brain membranes appeared in 1981 when Pedigo *et al.*[1] described the complex concentration-dependent inhibition of [^{3}H]5-HT specific binding to rat striatal and cortical membranes by spiperone. Two classes of binding sites were then postulated for explaining these data: one, called 5-HT$_{1A}$, with high affinity for both [^{3}H]5-HT and spiperone, and the other, designated 5-HT$_{1B}$, with high affinity for the indoleamine but low affinity for the neuroleptic.

Today, the situation is far more complex, since at least four distinct classes of 5-HT$_1$ receptors have been identified: 5-HT$_{1A}$, 5-HT$_{1B}$, 5-HT$_{1C}$ and 5-HT$_{1D}$, which all exhibit a nanomolar affinity for [^{3}H]5-HT[2]. Furthermore, the 5-HT$_1$ classes are not the only receptors for the indoleamine in the central nervous system (CNS) since other receptor types with lower affinity for 5-HT (Kd ~ 1–5 μM) have also been characterized in brain membranes: 5-HT$_2$,[2,3] 5-HT$_3$[2,3] and more recently 5-HT$_4$.[4]

Among the various classes of 5-HT$_1$ receptors, that designated "5-HT$_{1A}$" has been the most studied during recent years, for at least two reasons: 1) the development, in 1982, of a highly selective radioligand: [^{3}H]8-hydroxy-2-(di-*n*-propyl-amino)tetralin or [^{3}H]8-OH-DPAT, for the specific labeling of 5-HT$_{1A}$ sites in brain membranes,[5] and 2) the discovery that several 5-HT$_{1A}$ ligands (notably ipsapirone, buspirone, and gepirone) exert anxiolytic and antidepressant effects in relevant animal tests, and also in humans.[6,7]

The present review summarizes the main data on the pharmacological, biochemical, functional and regulatory properties of central 5-HT$_{1A}$ receptors, which have been obtained recently in this and other laboratories.

[a]The original data reported in the present review were obtained thanks to generous support from INSERM and BAYER-PHARMA.

[b]Address all correspondence to M. Hamon at the above address.

PROPERTIES OF CENTRAL 5-HT_{1A} BINDING SITES LABELED BY [^{3}H]8-OH-DPAT AND OTHER SELECTIVE RADIOLIGANDS

Soon after the discovery that 8-OH-DPAT is a central 5-HT agonist,[8] binding studies demonstrated that this drug differently interacts with [^{3}H]5-HT specific binding to membranes from various brain regions: whereas the concentration-dependent inhibition of [^{3}H]5-HT specific binding by 8-OH-DPAT was monophasic in the rat striatum,[9] clear biphasic curves were obtained in the hippocampus and cerebral cortex.[9,10] Furthermore in the two latter regions, inhibition curves became monophasic (as in the striatum) when 5-HT_{1A} sites have been occupied by spiperone, suggesting that 8-OH-DPAT normally interacts with spiperone-sensitive 5-HT_1 sites in the hippocampus and cerebral cortex. Such data led to conclude that: 1) 8-OH-DPAT is a high affinity ligand of 5-HT_{1A} sites in the rat brain,[9,10] and 2) central 5-HT_{1A} sites are heterogeneously distributed as they exist in the hippocampus and cerebral cortex but not in the striatum.[9]

The synthesis of [^{3}H]8-OH-DPAT as a possible radioligand for the selective labeling of central 5-HT_{1A} sites was then achieved in collaboration with the "Service des Molécules Marquées" of CEA (91191 Gif-sur-Yvette, France). Binding studies with this molecule fully confirmed the inferences from data with "cold" 8-OH-DPAT and allowed extensive characterization of the pharmacological properties and regional distribution of central 5-HT_{1A} sites.[5,11–13] More recently, other radioligands derived from the chemical structure of 8-OH-DPAT have been developed, which can be also used for the selective labeling of central 5-HT_{1A} receptors: [^{3}H]5-MeO-3-(di-n-propylamino) chroman,[14] and [^{125}I]BH-8-MeO-N-PAT[15] (TABLE 1). The latter molecule is particularly useful for autoradiographic studies since only two days (instead of 2–3 months with [^{3}H]8-OH-DPAT) are necessary for making autoradiograms from brain sections labeled with [^{125}I]BH-8-MeO-N-PAT.[15]

TABLE 1. Selective Radioligands, Preferential Agonists and Antagonists, and Main Locations of Central 5-HT_{1A} Binding Sites

Radioligands	Agonists	Antagonists	Main Locations
[^{3}H]8-OH-DPAT	DP-5CT	spiperone	–hippocampus (dentate gyrus, CA_1)
[^{3}H]5-MeO-3-DPAC	8-OH-DPAT	(—)propranolol	
[^{3}H]ipsapirone	ipsapirone	(—)pindolol	–lateral septum
[^{3}H]WB 4101	gepirone	penbutolol	–dorsal and median raphe nuclei
[^{3}H]spiroxatrine	buspirone	(8-MeO-2′-chloro-PAT)	–frontal and entorhinal cortex
[^{3}H]PAPP	PAPP		–central amygdala
[^{125}I]BH-8-MeO-N-PAT	CM 57493		

Particularly in case of antagonists, none of available drugs is really selective of central 5-HT_{1A} receptors, but exhibits high affinity also for dopamine receptors (spiperone) or β-adrenergic receptors (propranolol, pindolol, penbutolol). 8-MeO-2′-chloro-PAT is an alkylating agent producing irreversible blockade of 5-HT_{1A} and 5-HT_{1C} receptors.[23,24]

Abbreviations: 5-MeO-3-DPAC: 5-methoxy-3-(di-n-propylamino)chroman; WB 4101: 2-[(2′,6′-dimethoxy-phenoxyethyl)aminomethyl]-1,4-benzodioxan; PAPP: 1-[2-(4-aminophenyl)-ethyl]-4-(3-trifluoromethylphenyl)piperazine; BH-8-MeO-N-PAT: Bolton-Hunter-8-methoxy-2-(*N*-propyl-*N*-propylamino)tetralin; DP-5CT: *N*,*N*-dipropyl-5-carboxamidotryptamine; CM 57493: [4-(3-trifluoromethylphenyl)-1-(2-cyanoethyl)-1,2,3,6-tetrahydropyridine; 8-MeO-2′-chloro-PAT: 8-methoxy-2-(*N*-2′-chloropropyl,*N*-propyl)aminotetralin.

In addition to 8-OH-DPAT, several other molecules—often with a piperazine nucleus—have been shown to exhibit both a high affinity (Ki = 2–80 nM) and good selectivity for 5-HT_{1A} sites. Among them, ipsapirone, buspirone, and gepirone are of special interest since they are currently developed as anxiolytic and/or antidepressant drugs.[6,7] Furthermore, the radioactive derivatives of some of these piperazine agonists have also been successfully used for the selective labeling of 5-HT_{1A} binding sites in brain membranes and sections[16,17] (TABLE 1).

As expected for the interaction of an agonist with a receptor coupled to a G protein, the specific binding of [^{3}H]8-OH-DPAT is reduced in the presence of GTP, and can be enhanced by divalent cations, notably Mn^{2+}.[5,11] In addition, the curve of [^{3}H]8-OH-DPAT binding inhibition by increasing concentrations of agonists such as ipsapirone is shifted to the right by GTP, and to the left by Mn^{2+}.[18] In contrast, inhibition by antagonists (spiperone, propranolol, etc.) is unaffected by guanine nucleotides and divalent cations.[19] From these observations, a simple *in vitro* test has been proposed for a preliminary assessment of the agonist or antagonist properties of a given drug acting on 5-HT_{1A} receptors.[18,19] However, in all cases, conclusions from this test have to be confirmed by complementary experiments based on cellular and/or behavioral responses controlled by 5-HT_{1A} receptors (for instance the measurement of 5-HT_{1A}-dependent adenylate cyclase activity, the recording of 5-HT_{1A}-dependent inhibition of nerve impulse flow within serotoninergic neurons, etc; see ref. 19).

As for other receptors, the use of brain sections, instead of membranes, for the labeling of 5-HT_{1A} sites by selective radioligands allowed the description of their detailed regional distribution on the corresponding autoradiograms.[12,13,20] The highest density of 5-HT_{1A} sites is found in limbic areas, notably the hippocampus (dentate gyrus and CA1), the lateral septum, the frontal and entorhinal cortex and the central amygdala (TABLE 1). In addition, a high density of 5-HT_{1A} sites also exists in the anterior raphe nuclei (dorsal and median raphe, TABLE 1), where they correspond to the presynaptic autoreceptors located on the somas and/or dendrites of serotoninergic neurons.[20–22] Indeed the selective degeneration of these neurons by the intracerebral microinjection of 5,7-dihydroxytryptamine (5,7-DHT) resulted in a parallel decrease in the density of 5-HT_{1A} sites and of the activity of tryptophan hydroxylase (the enzyme responsible for the limiting step in 5-HT synthesis) within the nucleus raphe dorsalis in adult rats.[20–22] In all other regions, including those with relatively low densities of 5-HT_{1A} binding sites (hypothalamus, thalamus, etc.), the selective degeneration of serotoninergic fibers induced by this treatment did not affect the specific binding of [^{3}H]8-OH-DPAT, demonstrating that the latter concerns exclusively postsynaptic receptors.[20,21] Interestingly, 5-HT_{1A} sites are almost completely absent in extrapyramidal areas (caudate nucleus, putamen, globus pallidus, substantia nigra), where other 5-HT_1 receptor subtypes are particularly concentrated (notably 5-HT_{1B} and 5-HT_{1D} sites[2,13,20]).

Further progress in the knowledge on 5-HT_{1A} sites has been recently achieved by using new derivatives of 5-HT_{1A} ligands, notably an alkylating agent[23,24] and photosensitive probes.[25–27] In particular, the chloramine derivative of 8-OH-DPAT, 8-MeO-2′-chloro-PAT, has been successfully used for the *in vivo* irreversible blockade of 5-HT_{1A} receptors in discrete regions of the rat brain.[28] Thus, analysis of the time course of local recovery of 5-HT_{1A} binding sites after the microinjection of 8-MeO-2′-chloro-PAT into the ventral hippocampus led to an estimate of the receptor half-life in this structure: 2.3 days.[28] Similar studies at the level of the dorsal raphe nucleus indicated a half life value in the same range for the 5-HT_{1A} somato-dendritic autoreceptors (unpublished observations).

BIOCHEMICAL PROPERTIES OF THE 5-HT_{1A} RECEPTOR BINDING SUBUNIT

Radioactive photoaffinity derivatives of selective 5-HT_{1A} ligands such as [^{3}H]8-MeO-3′-nitro-azido-phenyl-PAT,[25] [^{3}H]p-azido-PAPP[26] and [^{125}I]azido-NAPS[27] have been used for the irreversible labeling of 5-HT_{1A} sites in brain membranes, and subsequent identification of the binding protein by sodium dodecylsulfate-polyacrylamide gel electrophoresis (SDS-PAGE) of solubilized membrane extracts. A single protein band of 63 kDa whose radioactive labeling could be prevented by "cold" 5-HT_{1A} ligands was thus extracted from membranes of brain regions enriched in 5-HT_{1A} receptors (notably the hippocampus).[25] In contrast, this 63 kDa protein was not found in the cerebellum where 5-HT_{1A} sites are hardly detected.[25]

Interestingly, previous determination of the molecular weights (MW) of various 5-HT binding sites using a radiation inactivation technique also gave a value close to 60 kDa for the 5-HT_{1A} subtype.[29] The convergence of MW values provided indirectly by the latter technique and by photoaffinity labeling led to the hypothesis of the monomeric nature of the 5-HT_{1A} receptor,[25] which could then be definitively demonstrated by the cloning and sequencing of the 5-HT_{1A} receptor binding subunit.[30,31] Indeed, the application of molecular biology techniques to the field of membrane-bound receptors allowed the identification of the 5-HT_{1A} receptor binding subunit as a 421 aminoacid chain with seven hydrophobic transmembrane domains,[30] like that observed for receptors functionally coupled to G proteins.[32] Glycosylated aminoacid residues appear to be located within the extracellular N terminal region of the receptor protein, whereas potential sites for phosphorylation by various kinases exist within the third intracellular loop,[30] the one probably involved in the interaction of the receptor binding subunit with its associated G protein.

In addition to the complete knowledge of the primary structure of the 5-HT_{1A} receptor binding subunit, information regarding its secondary and tertiary structures should be rapidly obtained as a two-step chromatographic procedure has been recently set up for the purification of the native receptor solubilized from rat hippocampal membranes.[33,34] Following membrane treatment with the detergent CHAPS, the soluble extract was first incubated with an affinity gel made of a 8-OH-DPAT derivative, 8-MeO-N-PBAT, coupled to Affigel 202. The adsorbed 5-HT_{1A} binding sites could then be eluted by 5-HT, and subsequently chromatographed onto a column of wheat-germ agglutinin (WGA) coupled to agarose.[34] Final elution with *N*-acetylglucosamine gave a fraction still able to bind [^{3}H]8-OH-DPAT and devoid of any trace of "cold" 5-HT used for the elution from the affinity column. SDS-PAGE of the eluted fraction with the highest binding activity revealed the presence of a single band with a MW of 63 kDa (FIG. 1), as expected for the purified 5-HT_{1A} receptor (see above).

As assessed by comparing [^{3}H]8-OH-DPAT high affinity binding activity in the WGA-agarose eluate with that in the starting CHAPS extract, the yield of this purification protocol was very low: 2–3%.[34] A possible proteolysis or denaturation of the 5-HT_{1A} receptor binding subunit during the purification procedure might account for this low yield. However, the addition of the wide spectrum protease inhibitor phenylmethyl-sulfonyl fluoride to the starting membrane material, and the use (throughout all the purification steps) of a buffer supplemented with several stabilizing agents (glycerol, dithiothreitol, CHAPS, Mn^{2+}, cholesteryl hemisuccinate, see ref. 34 and FIG. 1) made this explanation unlikely. Instead, it has to be emphasized that in addition to the 63 kDa-5-HT_{1A} receptor binding subunit, no other protein was detected by silver staining of the SDS polyacrylamide gel, which indicates that the purified receptor was dissociated from G proteins in the final WGA-agarose eluate. As

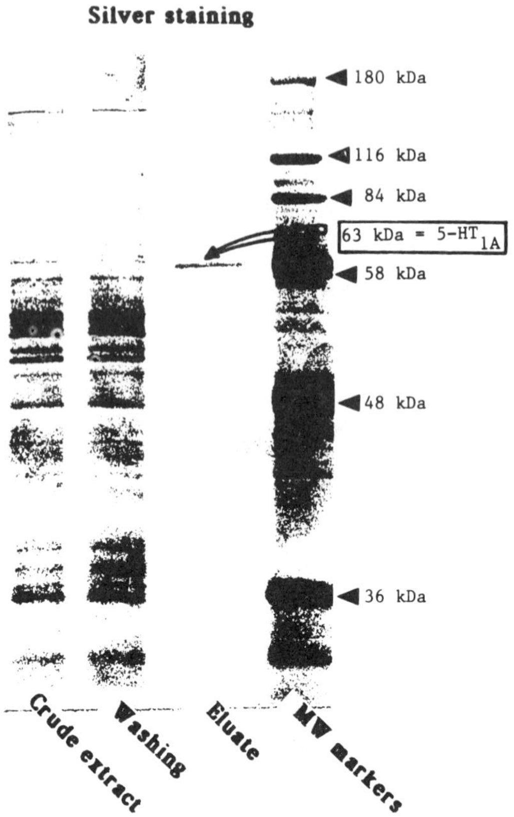

FIGURE 1. Sodium dodecylsulfate-polyacrylamide gel electrophoresis (SDS-PAGE) of the 5-HT$_{1A}$ receptor binding subunit purified from the rat hippocampus. Rat hippocampal 5-HT$_{1A}$ sites were solubilized by 10 mM CHAPS, and first adsorbed onto an affinity column made of 8-methoxy-2-[(*N*-propyl,*N*-butylamino)amino]tetralin (8-MeO-N-PBAT) coupled to Affigel 202.[34] After extensive washing with 50 mM Tris-HCl supplemented with 1 mM CHAPS, 50 μM dithiothreitol, 0.1 mM $MnCl_2$, 10% glycerol and 50 μg/ml of cholesteryl-hemisuccinate, 5-HT$_{1A}$ sites were eluted with 1 mM 5-HT and subsequently adsorbed onto a wheat germ-agglutinin (WGA)-agarose column. Final elution was achieved with 0.25 M *N*-acetylglucosamine, and aliquots of the starting crude extract, washing of the 8-MeO-N-PBAT-Affigel 202 affinity column, and WGA-agarose eluate were analysed by SDS-PAGE (4% stacking gel, 10.5% running gel). Silver staining revealed only one band in the WGA-agarose eluate, whose MW (63 kDa, by comparison with marker proteins) corresponds to that of the 5-HT$_{1A}$ receptor binding subunit previously determined by photoaffinity labeling[25] and a radiation-inactivation procedure.[29] Furthermore [^{3}H]8-OH-DPAT high affinity binding activity was measurable in the WGA-agarose eluate (see ref. 34).

dissociated 5-HT$_{1A}$ receptor binding sites lose their high affinity for agonists,[35] it is therefore not unexpected that only very low levels of [^{3}H]8-OH-DPAT high affinity binding could be recovered in the final eluate. A better estimate of the purification yield would be achieved using a radioactive antagonist for the selective labeling of 5-HT$_{1A}$ receptors, since their affinity for such a radioligand would be independent of the receptor coupling with G proteins.[19] Unfortunately, no radioactive 5-HT_{1A} antagonist has been developed so far.

Alternatively one could attempt the reconstitution of active Receptor-G protein (R-G) complexes for converting all uncoupled 5-HT_{1A} binding subunits into [^{3}H]8-OH-DPAT high affinity sites in the WGA-agarose eluate. Recently Emerit *et al.*[35] achieved the successful reconstitution of [^{3}H]8-OH-DPAT high affinity binding sites by incorporating soluble 5-HT_{1A} binding subunits (dissociated by GTP from G proteins) and a mixture of pure Gi and Go proteins into phospholipid vesicles made of azolectin plus octyl-β-D-glucoside. [^{3}H]8-OH-DPAT high affinity binding recovered under such conditions could be totally suppressed by the GTP analogue GppNHp, indicating it was produced by the functional coupling of the 5-HT_{1A} binding subunits with G proteins.[35]

At least two observations can be made from these data: 1) by converting all purified 5-HT_{1A} binding subunits into reconstituted high affinity sites for [^{3}H]8-OH-DPAT, the actual recovery of the purification procedure described above should be accurately estimated in future investigations; 2) the successful reconstitution of high affinity 5-HT_{1A} binding sites with a mixture of Gi and Go suggests that the endogenous G protein associated with 5-HT_{1A} receptors in brain membranes probably belongs to one of these G protein families. In line with the latter inference, we recently observed that *in vivo* inactivation of Gi and Go by an intracerebro-ventricular injection of Pertussis toxin produced a significant reduction of [^{3}H]8-OH-DPAT high affinity binding activity in the rat hippocampus (FIG. 2). Furthermore, this effect selectively concerned the fraction of [^{3}H]8-OH-DPAT binding which could be eliminated by guanine nucleotides (GppNHp, see FIG. 2), i.e. that depending on the functional coupling of 5-HT_{1A} sites with Gi/Go proteins.

One of our initial purposes when setting up the purification protocol described above was to use pure native 5-HT_{1A} receptors as antigens for the production of specific antibodies. Indeed, such antibodies would be particularly useful for the optical and ultrastructural visualization of the receptors, for their purification in large quantities (by immunoaffinity chromatography), etc., and should therefore be of considerable interest for thorough investigations on the 5-HT_{1A} subtype. However, the knowledge of the complete aminoacid sequence of the 5-HT_{1A} receptor from the rat brain (Civelli, personal communication) led us to choose another strategy for the production of specific antibodies. Indeed, the comparison of this sequence with those of other receptors coupled to G proteins, including the 5-HT_{1A} receptor from human tissues,[30,31] revealed that two domains located in the N terminal portion of the protein molecule and in the third intracellular loop are highly selective of the rat 5-HT_{1A} receptor. We thus synthesized a peptide corresponding to a 26 amino-acid sequence (from gly in position 243 to glu in position 268; see TABLE 2) within the third intracellular loop, for which homology with the corresponding region of the human 5-HT_{1A} receptor reaches only 58% (instead of 89% for the whole sequence). This peptide was covalently bound to a protein carrier (serumalbumin or keyhole limpet haemocyanin), and injected to rabbits for the production of polyclonal antibodies. This strategy revealed to be particularly efficient since antibodies immunoprecipitating 5-HT_{1A} receptors solubilized from rat hippocampal membranes could be detected in a rabbit serum as soon as after the second booster injection of the immunogen (TABLE 2). Furthermore, at a dilution of 1:1000, one antiserum could even be used for immunohistochemistry. As expected for the regional distribution of 5-HT_{1A} binding sites previously established by autoradiographic investigations,[12,13,20] positive immunostaining with the anti-peptide anti-bodies was notably found in the dentate gyrus and CA1 area of the hippocampus (onto pyramidal cells), in the lateral septum, and in the superficial layers of the dorsal horn within the spinal cord (unpublished observations).

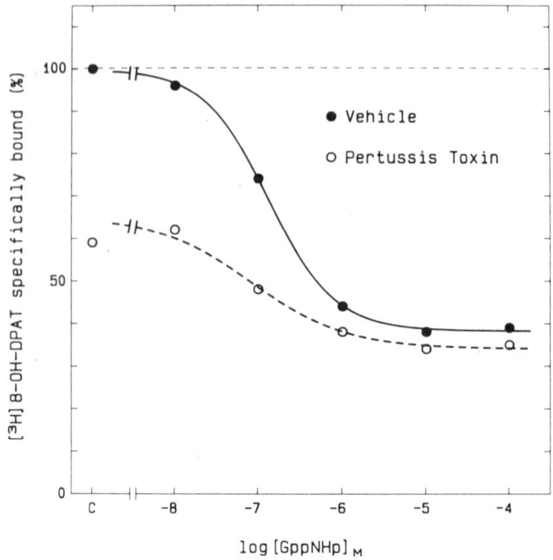

FIGURE 2. Effects of GppNHp on [^{3}H]8-OH-DPAT specific binding to 5-HT_{1A} sites solubilized from hippocampal membranes of control or Pertussis toxin-treated rats. Pertussis toxin (5 μg/rat) or saline (10 μl) was injected i.c.v. 48 h before sacrifice. Hippocampal 5-HT_{1A} sites were solubilized by 10 mM CHAPS,[33] and binding assays were carried out on 50 μl aliquots of soluble extracts, with 1.0 nM of [^{3}H]8-OH-DPAT and various concentrations of GppNHp (abscissa). The specific binding of [^{3}H]8-OH-DPAT is expressed as percentage of that found with soluble extracts from control rats, in the absence of GppNHp ("C" on abscissa, (●) 100% = 109 ± 5 fmol/mg protein). Each point is the mean of three independent determinations.

TABLE 2. Immunoprecipitation of [^{3}H]8-OH-DPAT High Affinity Binding Sites Solubilized from Rat Hippocampal Membranes by a Rabbit Antiserum Raised Against the 26 Amino-Acid Peptide Corresponding to aa 243–268 Sequence of the Rat 5-HT_{1A} Receptor Binding Subunit

Addition	[^{3}H]8-OH-DPAT Specifically Bound (fmol/ml)	(%)
None	335 ± 10	(100)
Control serum (1/200)	354 ± 8	(106)
Antiserum (1/200)	174 ± 4[a]	(52)[a]

The 26 amino-acid peptide (GTSSAPPPKKSLNGQPGSGDWRRCAE) corresponding to aa 243–268 of the 3rd intracellular loop of the rat 5-HT_{1A} receptor (Civelli, personal communication) was coupled to keyhole limpet haemocyanin by 1-ethyl-3-(3-dimethyl-aminopropyl)carbodiimide, and injected to a white male rabbit for raising antibodies. Serum was collected after the second booster injection, dialyzed and incubated for 12 h at 4°C for immunoprecipitating 5-HT_{1A} sites solubilized from CHAPS-treated hippocampal membranes.[33] Antigen-antibody complexes were adsorbed onto protein A-Sepharose 6MB, and binding assays were carried out using 1.5 nM [^{3}H]8-OH-DPAT on aliquots of clear supernatants from CHAPS soluble extracts incubated with control serum (1/200 final dilution), antiserum (1/200) or none. [^{3}H]8-OH-DPAT high affinity binding is expressed in fmol. recovered per ml of starting CHAPS soluble extract. Each value is the mean ± SEM of triplicate determinations in two separate experiments.

[a]Under such conditions, approximately half of soluble 5-HT_{1A} sites were immunoprecipitated by the antiserum at 1/200 final dilution.

TRANSDUCTION MECHANISMS ASSOCIATED WITH 5-HT$_{1A}$ RECEPTORS

Studies of [^{3}H]8-OH-DPAT high affinity binding already showed that 5-HT$_{1A}$ receptors are associated with G proteins belonging to Gi and/or Go families, suggesting a negative coupling with adenylate cyclase, and some coupling with ion channels in membranes. Direct demonstration of these inferences was subsequently made by several groups. In 1986, De Vivo and Maayani[36] and Weiss *et al.*[37] reported a negative influence of 5-HT$_{1A}$ agonists on adenylate cyclase activity in rat and guinea pig brain membranes treated with forskolin or VIP (vasoactive intestinal peptide), a finding subsequently confirmed by several groups.[18,19,38–40] Detailed pharmacological analyses then revealed that, in addition to 5-HT and 8-OH-DPAT, ipsapirone, gepirone and buspirone also inhibit the enzyme activity, whereas known (non-selective) 5-HT$_{1A}$ antagonists such as spiperone, penbutolol (FIG. 3), propranolol and pindolol[39] prevent this effect.

Surprisingly, Markstein *et al.*[41] described in contrast a stimulatory effect of 5-HT$_{1A}$ agonists on basal adenylate cyclase activity in rat hippocampal membranes, suggesting that 5-HT$_{1A}$ receptors might also be functionally associated with Gs in the rat brain. Recently, this problem has been extensively reinvestigated using mammalian cell lines (COS-7 and HeLa) transfected with the 5-HT$_{1A}$ receptor gene.[42] Only the negative coupling of 5-HT$_{1A}$ receptors with adenylate cyclase was evidenced in those cells.[42] However some stimulation by 5-HT of phospholipase C activity was also observed in HeLa cells expressing only the 5-HT$_{1A}$ type of 5-HT receptors, which may suggest that the 5-HT$_{1A}$ receptor can couple with multiple transduction pathways (i.e. adenylate cyclase activity and phosphoinositide turnover) via different G proteins (see ref. 43).

In addition to the coupling of 5-HT$_{1A}$ receptors with membrane-bound enzymes (adenylate cyclase and phospholipase C), evidence has been reported for a control of K^+ channels by these receptors in neurones, notably in the hippocampus[44] and the dorsal raphe nucleus.[45] In both regions, 5-HT$_{1A}$ agonists increase K^+ conductance via a receptor interaction with a G protein sensitive to Pertussis toxin.[44,45] Clearly this effect does not require second messengers such as cyclic AMP, diacylglycerol or inositol

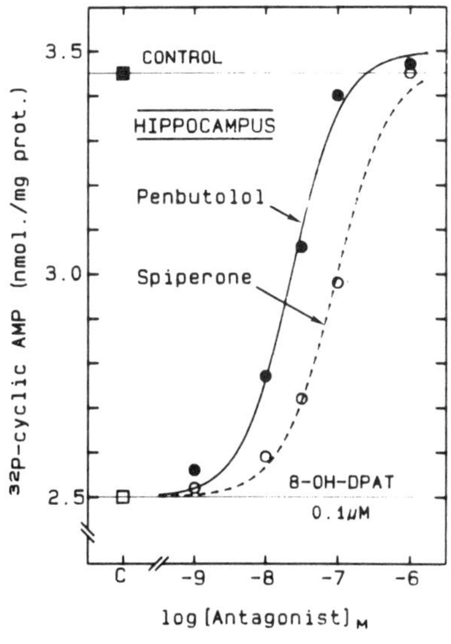

FIGURE 3. Antagonistic effects of spiperone and penbutolol on 5-HT$_{1A}$ receptors negatively coupled to adenylate cyclase in rat hippocampal membranes. Adenylate cyclase assays consisted of measuring the production of [^{32}P]cyclic AMP (in nmol./mg protein) from [^{32}P]α-ATP (0.5 mM) by hippocampal homogenates for a 20 min incubation at 30°C in the presence of 10 μM forskolin.[18] The inhibitory effect of 0.1 μM 8-OH-DPAT (−28%, □) was reversed in a concentration-dependent manner by penbutolol (IC_{50} = 15 nM) and spiperone (IC_{50} = 105 nM). Each point is the mean of triplicate determinations in three separate experiments.

trisphosphate,[44,45] leading to the concept of a direct control of K^+ channels by the Gi/Go protein associated with 5-HT_{1A} receptors.[46] Recently Lanfumey *et al.*[47] noted that 4-aminopyridine, a blocker of K^+ channels controlled by G protein(s), markedly reduced the negative influence of ipsapirone upon the nerve impulse flow within serotoninergic neurons of the dorsal raphe nucleus, as expected from the coupling of somato-dendritic 5-HT_{1A} autoreceptors with such K^+ channels.

As intracellular recordings were made for studies on hippocampal 5-HT_{1A} receptors coupled with K^+ channels,[44] it can be inferred that these receptors are located postsynaptically, on cell targets of serotoninergic projections, in agreement with data obtained from binding studies.[11,20,48] Recently, the pre- or post-synaptic location of 5-HT_{1A} receptors negatively coupled with adenylate cyclase was assessed by measuring this enzyme activity following the selective degeneration of serotoninergic fibers and terminals induced by an intracerebral microinjection of 5,7-DHT in rats. As shown in FIGURE 4, no change in the inhibitory effect of 8-OH-DPAT on forskolin-stimulated

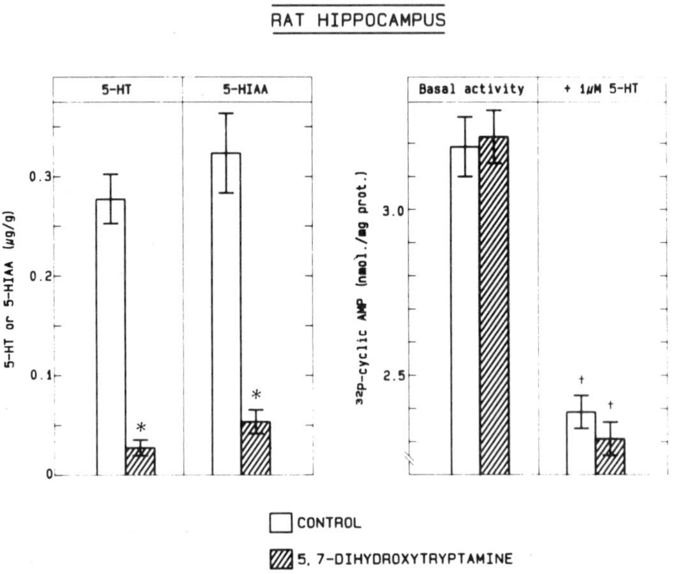

FIGURE 4. Effects of 5,7-DHT-induced degeneration of serotoninergic projections on the levels of 5-HT and 5-HIAA, and of forskolin-stimulated adenylate cyclase activity in the absence ("basal") or the presence of 1 μM 5-HT, in the rat hippocampus. The intracerebral injection of 5,7-dihydroxytryptamine (8 μg in 4 μl of isotonic NaCl containing 0.01% ascorbic acid[48]) or vehicle was made in desmethylimipramine (25 mg/kg i.p.)—pretreated rats, 18 days before sacrifice. 5-HT and 5-HIAA levels (in μg/g) were measured by HPLC coupled to electrochemical detection,[18] and forskolin (10 μM)—stimulated adenylate cyclase activity (with or without 1 μM 5-HT) was estimated by measuring the conversion of [^{32}P]α-ATP into [^{32}P]cyclic AMP (in nmol./mg protein) by hippocampal homogenates after a 20 min incubation at 30°C.[18] Each bar is the mean ± SEM of 6 independent determinations.

*****$P < 0.001$ when compared to respective control values.

† $P < 0.001$ when compared to [^{32}P]cyclic AMP production in the absence of 5-HT ("basal activity"). No significant difference between control and 5,7-DHT-treated rats was noted on forskolin-stimulated [^{32}P]cyclic AMP production in the absence as well as in the presence of 1 μM 5-HT.

adenylate cyclase activity could be observed in hippocampal membranes, in spite of a dramatic depletion of hippocampal 5-HT and 5-hydroxy-indole acetic acid (5-HIAA) levels in 5,7-DHT-treated rats. In contrast, the local microinjection of kainic acid to destroy intrinsic neurons completely suppressed the negative modulation by 5-HT_{1A} agonists of forskolin-stimulated adenylate cyclase activity in hippocampal membranes (unpublished observations). Accordingly, like the 5-HT_{1A} receptors coupled with K^+ channels, those coupled with adenylate cyclase are located on postsynaptic target cells of serotoninergic projections within the rat hippocampus. Whether both modes of coupling exist in the same cells or in different neuronal populations is yet an unsolved problem. In addition, further investigations will have to be done for assessing the relative functional importance for the target cells of 5-HT_{1A} receptors coupled with adenylate cyclase versus those controlling K^+ channels.

PHYSIOLOGICAL AND PHARMACOLOGICALLY INDUCED MODIFICATIONS IN CENTRAL 5-HT_{1A} RECEPTORS

Measurements of [^{3}H]8-OH-DPAT specific binding in various brain areas during the whole life have revealed significant changes for the early postnatal period,[49] and to a lesser extent for the aging process[50] compared to adult life in rats. In particular, 5-HT_{1A} binding sites are relatively abundant in the cerebellum of newborn rats,[49] whereas they are hardly detected in this structure in adult and senescent animals.[49,50] Careful examination of the time-course evolution of cerebellar 5-HT_{1A} sites during development indicates a progressive decrease in their density down to almost zero at the end of the second postnatal week.[49] Conversely, a progressive increase in 5-HT_{1A} receptor density was found for the same period in forebrain areas, notably the dentate gyrus of the hippocampus and the cerebral cortex.[49]

Already at birth, these binding sites correspond to functional receptors since 5-HT_{1A} agonists significantly decrease forskolin-stimulated adenylate cyclase activity in hippocampal, septal and cerebellar homogenates from newborn rats.[51] These observations raise the question of the possible role of 5-HT_{1A} receptors during CNS maturation, particularly in the cerebellum which develops mainly for the early postnatal period. It has been proposed that 5-HT may act as a "trophic" factor during brain maturation (see ref. 51), but further investigations are required before postulating the involvement of 5-HT_{1A} receptors in this action.

Recent studies in 22–24 month-old rats have shown a slight decrease in 5-HT_{1A} receptor density in various forebrain regions, possibly due to age-dependent degeneration of some 5-HT-target neurons, notably in the hippocampus.[50] However, the pharmacological properties of central 5-HT_{1A} sites in aged animals (as those of 5-HT_{1A} receptors during the early life period) are rigorously identical with those in adult rats.[50] For instance, the negative modulation of [^{3}H]8-OH-DPAT specific binding by GTP appeared to be as pronounced in senescent animals as in 3 month-old rats, suggesting that the coupling of 5-HT_{1A} receptors with G proteins is unaltered by aging.[50]

Changes in the responses to 5-HT_{1A} agonists have been reported after acute[52] or chronic[53–55] treatments with various drugs suggesting the existence of adaptive regulatory mechanisms of central 5-HT_{1A} receptors in adult rats. In particular electrophysiological recording of serotoninergic neurons within the nucleus raphe dorsalis and of postsynaptic 5-HT target neurons in the hippocampus have shown differential modifications of pre- versus post-synaptic 5-HT_{1A} receptors after the chronic administration of antidepressants.[56–58] For instance, desensitization of the somato-dendritic 5-HT_{1A} autoreceptors without any change of hippocampal postsynaptic 5-HT_{1A} receptors has been reported after a three week-treatment with the monoamine oxidase inhibitors phenelzine and clorgyline.[56] Similar observations have been made in rats treated chronically with selective 5-HT uptake inhibitors such as zimelidine[57] or citalopram,[58] or with the direct 5-HT_{1A} agonist gepirone.[59] In contrast the chronic administration of tricyclic antidepressants such as imipramine and desmethylimipramine did not alter the somato-dendritic 5-HT_{1A} autoreceptors but *increased* the sensitivity of hippocampal postsynaptic target cells to 5-HT.[60] However, the latter observation is in contradiction with biochemical data since a significant *decrease* in the inhibitory effect of 5-HT on 5-HT_{1A}-dependent adenylate cyclase activity was noted in hippocampal membranes from rats treated chronically with desmethylimipramine.[61] By recording electrical activity in hippocampal slices, Rowan and Anwyl[62] also concluded that the sensitivity of postsynaptic target cells to 5-HT was not enhanced, but was in fact significantly *reduced* after the chronic administration of tricyclic antidepressants to rats.

Because of these discrepancies regarding the adaptive regulations of postsynaptic 5-HT_{1A} receptors, we decided to re-examine also the possible changes in presynaptic 5-HT_{1A} autoreceptors under pharmacological conditions closely related to those previously used by Blier and De Montigny.[59] For this purpose, rats were treated for two weeks with the direct 5-HT_{1A} agonist ipsapirone (10 mg/kg i.p./day) (instead of gepirone[59]), and the sensitivity of somato-dendritic 5-HT_{1A} receptors was assessed by considering the negative effect of a further acute administration of this drug (5 mg/kg i.p. 24 h after completion of the chronic treatment) on 5-HT turnover in various brain regions. Indeed, previous studies[18,63,64] have clearly demonstrated that the inhibitory effect of 5-HT_{1A} agonists on central 5-HT turnover is a consequence of the stimulation of somato-dendritic 5-HT_{1A} autoreceptors by these drugs.

As shown in FIGURE 5, acute treatment with ipsapirone was equally effective to reduce 5-HT turnover (in the hippocampus as well as in other forebrain areas—not shown) in rats receiving ipsapirone or saline for the preceding two weeks. These data do not support that somato-dendritic 5-HT_{1A} autoreceptors become hyposensitive after long term exposure to a direct agonist (gepirone[59] or ipsapirone[18]), since, according to this hypothesis, a significant reduction in ipsapirone-induced inhibition of 5-HT turnover should have been observed in rats chronically treated with this drug. Data in FIGURE 5 also indicate that the stimulatory effect of acute ipsapirone treatment on dopamine turnover was unchanged by prior repeated administration of this drug. Accordingly, the 5-HT_{1A} receptors, probably located on *postsynaptic* target neurons,[18] which trigger this effect did not undergo adaptive modifications after a long term stimulation by an agonist. Similarly, measurements of [^{3}H]8-OH-DPAT binding and of 5-HT_{1A}-dependent adenylate cyclase activity confirmed that hippocampal postsynaptic 5-HT receptors were not altered after a two week treatment with ipsapirone (unpublished observations).

The striking discrepancies between electrophysiological and biochemical data on the adaptive regulatory properties of pre- and postsynaptic 5-HT_{1A} receptors may suggest that in fact, different 5-HT_{1A} receptor subtypes have been studied by these two approaches. Clearly, electrophysiological investigations concern 5-HT_{1A} receptors coupled with K^+ channels, whereas biochemical investigations may deal essentially with 5-HT_{1A} receptors coupled with adenylate cyclase, and even other —yet unknown— transduction mechanisms. Whether the nature of the transduction mechanism associated with 5-HT_{1A} receptors may influence their adaptive regulatory properties is therefore an interesting possibility.

CONCLUSIONS AND FUTURE TRENDS

Although considerable progress has been made recently in the knowledge of central 5-HT_{1A} receptors (since in particular their aminoacid sequence is presently known[30])

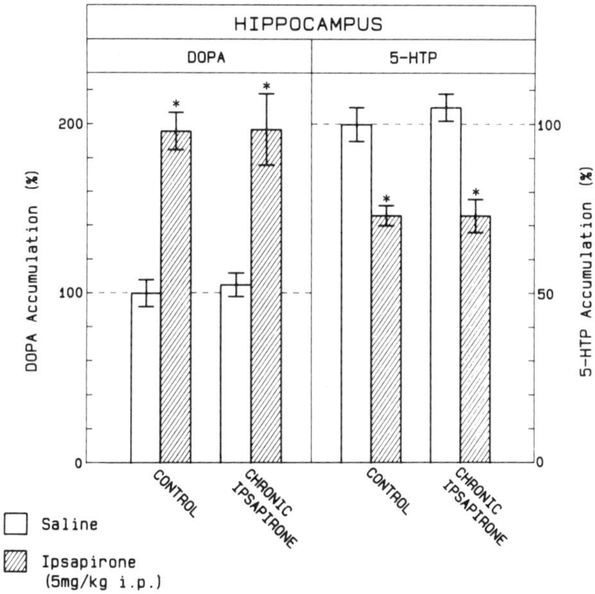

FIGURE 5. Effects of acute treatment with ipsapirone (5 mg/kg i.p.) on the synthesis of 5-HTP and DOPA in the hippocampus of rats pretreated for two weeks with saline ("control") or ipsapirone ("chronic ipsapirone"). "Chronic ipsapirone" treatment consisted of twice daily ip injections of ipsapirone (5 mg/kg at 9:00 A.M. and 6:00 P.M.) for 14 days. "Control" rats received saline instead of the drug. Twenty-four hours after the last injection, animals were treated with ipsapirone (5 mg/kg ip) or saline, followed by benserazide (800 mg/kg ip, an inhibitor of aromatic amino-acid decarboxylase) 30 min later. All rats were killed 30 min after the latter injection, and the levels of DOPA and 5-HTP accumulated in the hippocampus were measured by HPLC coupled to electrochemical detection.[18] Each bar is the mean ± SEM of 6 independent determinations (100% DOPA accumulation in control rats treated acutely with saline: 19.6 ng/g; 100% 5-HTP accumulation: 130.2 ng/g).

*$P < 0.05$ when compared to respective values in rats treated acutely with saline.

several important problems are yet unsolved. Among others, the following ones will require further extensive investigations:

1) Numerous 5-HT$_{1A}$ agonists are presently available, which exhibit potential therapeutic value for the treatment of anxiety and depression. In contrast, no selective 5-HT$_{1A}$ antagonist has been synthesized so far, although there is an urgent need of such a compound to really understand the physiological functions of central 5-HT$_{1A}$ receptors.

2) Although numerous data have demonstrated the functional coupling of central 5-HT$_{1A}$ receptors with G proteins, the exact nature (Gi, Go or even Gs?) of these proteins is yet unknown.

3) The regulatory properties of central 5-HT$_{1A}$ receptors appear to be extremely complex, since marked differences exist between pre- and postsynaptic 5-HT$_{1A}$ receptors, and possibly between those coupled with K^+ channels versus those coupled with adenylate cyclase or other transducing mechanisms. Maybe the nature of the G protein(s) associated with 5-HT$_{1A}$ receptors is of critical importance for such differential regulatory properties?

4) Since 5-HT$_{1A}$ receptors may function via a coupling with K^+ channels, ade-

nylate cyclase, *etc.*, do these various transducing mechanisms coexist in a single neuron? And what are the respective intracellular events finally triggered by the stimulation of 5-HT_{1A} receptors coupled with K^+ channels or adenylate cyclase in this neuron ?

5) Although the 5-HT_{1A} receptor protein is probably a substrate of protein kinase(s), the possible resulting alteration in 5-HT_{1A} receptor function by phosphorylation has not yet been explored. In particular, the possible involvement of 5-HT_{1A} receptor phosphorylation in the adaptive regulatory mechanisms described in this review should deserve further investigations.

6) The cloning and sequencing of the gene encoding the 5-HT_{1A} receptor[30] have made oligonucleotides and cDNA probes currently available for the visualization (*in situ* hybridization) and quantification of specific mRNAs. However, these approaches have not yet been introduced for a) the formal identification of the neurons which synthesize 5-HT_{1A} receptors in the CNS, and b) the demonstration of possible alterations in the synthesis of 5-HT_{1A} receptors associated with the adaptive regulatory mechanisms described above.

REFERENCES

1. PEDIGO, N. W., H. I. YAMAMURA & D. L. NELSON. 1981. Discrimination of multiple [^{3}H]5-hydroxytryptamine binding sites by the neuroleptic spiperone in rat brain. J. Neurochem. **36:** 220–226.
2. PEROUTKA, S. J. 1988. 5-hydroxytryptamine receptor subtypes: Molecular, biochemical and physiological characterization. Trends Neurosci. **11:** 496–500.
3. BRADLEY, P. B., G. ENGEL, W. FENIUK, J. R. FOZARD, P. P. A. HUMPHREY, D. N. MIDDLEMISS, E. J. MYLECHARANE, B. P. RICHARDSON & P. R. SAXENA. 1986. Proposals for the classification and nomenclature of functional receptors for 5-hydroxytryptamine. Neuropharmacology **25:** 563–576.
4. DUMUIS, A., R. BOUHELAL, M. SEBBEN, R. CORY & J. BOCKAERT. 1988. A nonclassical 5-hydroxytryptamine receptor positively coupled with adenylate cyclase in the central nervous system. Mol. Pharmacol. **34:** 880–887.
5. GOZLAN, H., S. EL MESTIKAWY, L. PICHAT, J. GLOWINSKI & M. HAMON. 1983. Identification of presynaptic serotonin autoreceptors using a new ligand: ^{3}H-PAT. Nature (Lond.) **305:** 140–142.
6. TRABER, J. & T. GLASER. 1987. The 5-HT_{1A} receptor-related anxiolytics. Trends Pharmacol. Sci. **8:** 432–437.
7. CERVO, L. & R. SAMANIN. 1987. Potential antidepressant properties of 8-hydroxy-2-(di-n-propylamino)tetralin, a selective $serotonin_{1A}$ receptor agonist. Eur. J. Pharmacol. **144:** 223–229.
8. HJORTH, S., A. CARLSSON, P. LINDBERG, D. SANCHEZ, H. WIKSTROM, L. E. ARVIDSSON, U. HACKSELL & J. L. G. NILSSON. 1982. 8-hydroxy-2-(di-n-propylamino)tetralin, 8-OH-DPAT, a potent and selective simplified ergot congener with central 5-HT receptor stimulating activity. J. Neural Transm. **55:** 169–188.
9. HAMON, M., S. BOURGOIN, H. GOZLAN, M. D. HALL, C GOETZ, F. ARTAUD & A. S. HORN. 1984. Biochemical evidence for the 5-HT agonist properties of PAT (8-hydroxy-2-(di-n-propyl-amino)tetralin) in the rat brain. Eur. J. Pharmacol. **100:** 263–276.
10. MIDDLEMISS, D. N. & J. R. FOZARD. 1983. 8-hydroxy-2-(di-n-proprylamino)tetralin discriminates between subtypes of the 5-HT_1 recognition site. Eur. J. Pharmacol. **90:** 151–153.
11. HALL, M. D., S. EL MESTIKAWY, M. B. EMERIT, L. PICHAT, M. HAMON & H. GOZLAN. 1985. [^{3}H]8-hydroxy-2-(di-n-propylamino)tetralin binding to pre- and postsynaptic 5-hydroxytryptamine sites in various regions of the rat brain. J. Neurochem. **44:** 1685–1696.
12. MARCINKIEWICZ, M., D. VERGE, H. GOZLAN, L. PICHAT & M. HAMON. 1984. Autoradiographic evidence for the heterogeneity of 5-HT_1 sites in the rat brain. Brain Research **291:** 159–163.

13. PAZOS, A. & J. M. PALACIOS. 1985. Quantitative autoradiographic mapping of serotonin receptors in the rat brain. I. Serotonin-1 receptors. Brain Research **346:** 205–230.
14. COSSERY, J. M., H. GOZLAN, U. SPAMPINATO, C. PERDICAKIS, G. GUILLAUMET, L. PICHAT & M. HAMON. 1987. The selective labelling of central 5-HT_{1A} receptor binding sites by [^{3}H]5-methoxy-3-(di-n-propylamino)chroman. Eur. J. Pharmacol. **140:** 143–155.
15. GOZLAN, H., M. PONCHANT, G. DAVAL, D. VERGE, F. MENARD, A. VANHOVE, J. P. BEAUCOURT & M. HAMON. 1988. ^{125}I-Bolton-Hunter-8-methoxy-2(N-propyl-N-propylamino)tetralin as a new selective radioligand of 5-HT_{1A} sites in the rat brain. In vitro binding and autoradiographic studies. J. Pharmacol. Exp. Ther. **244:** 751–759.
16. DOMPERT, W. U., T. GLASER & J. TRABER. 1985. ^{3}H-TVX Q 7821: Identification of 5-HT_1 binding sites as target for a novel putative anxiolytic. Naunyn-Schmiedeberg's Arch. Pharmacol. **328:** 467–470.
17. RANSOM, R. W., K. B. ASARCH & J. C. SHIH. 1986. [^{3}H]1-[2-(4-aminophenyl)ethyl]-4-(3-trifluoromethylphenyl)piperazine: a selective radioligand for 5-HT_{1A} receptors in rat brain. J. Neurochem. **46:** 68–75.
18. HAMON, M., C. M. FATTACCINI, J. ADRIEN, M. C. GALLISSOT, P. MARTIN & H. GOZLAN. 1988. Alterations of central serotonin and dopamine turnover in rats treated with ipsapirone and other 5-$hydroxytryptamine_{1A}$ agonists with potential anxiolytic properties. J. Pharmacol. Exp. Ther. **246:** 745–752.
19. ADRIEN, J., L. LANFUMEY, H. GOZLAN, C. M. FATTACCINI & M. HAMON. 1989. Biochemical and electrophysiological evidence for an agonist action of CM 57493 at pre- and postsynaptic 5-$hydroxytryptamine_{1A}$ receptors in brain. J. Pharmacol. Exp. Ther. **248:** 1222–1230.
20. VERGE, D., G. DAVAL, M. MARCINKIEWICZ, A. PATEY, S. EL MESTIKAWY, H. GOZLAN & M. HAMON. 1986. Quantitative autoradiography of multiple 5-HT_1 receptor subtypes in the brain of control or 5,7-dihydroxytryptamine-treated rats. J. Neurosci. **6:** 3474–3482.
21. VERGE, D., G. DAVAL, A. PATEY, H. GOZLAN, S. EL MESTIKAWY & M. HAMON. 1985. Presynaptic 5-HT autoreceptors on serotonergic cell bodies and/or dendrites but not terminals are of the 5-HT_{1A} subtype. Eur. J. Pharmacol. **113:** 463–464.
22. WEISSMAN-NANOPOULOS, D., E. MACH, J. MAGRE, Y. DEMASSEY & J. F. PUJOL. 1985. Evidence for the localization of 5-HT_{1A} binding sites on serotonin-containing neurons in the raphe dorsalis and raphe centralis nuclei of the rat brain. Neurochem. Int. **7:** 1061–1072.
23. EMERIT, M. B., H. GOZLAN, M. D. HALL, M. HAMON & A. MARQUET. 1985. Irreversible blockade of central 5-HT binding sites by 8-methoxy-2′-chloro-PAT. Biochem. Pharmacol. **34:** 883–892.
24. RADJA, F., G. DAVAL, M. B. EMERIT, M. C. GALLISSOT, M. HAMON & D. VERGE. 1989. Selective irreversible blockade of 5-HT_{1A} and 5-HT_{1C} receptor binding sites in the rat brain by 8-MeO-2′-chloro-PAT: a quantitative autoradiographic study. Neuroscience **31:** 723–733.
25. EMERIT, M. B., S. EL MESTIKAWY, H. GOZLAN, J. M. COSSERY, R. BESSELIEVRE, A. MARQUET & M. HAMON. 1987. Identification of the 5-HT_{1A} receptor binding subunit in rat brain membranes using the photoaffinity probe [^{3}H]8-methoxy-2-[N-n-propyl,N,3-(2-nitro-4-azidophenyl)aminopropyl] aminotetralin. J. Neurochem. **49:** 373–380.
26. RANSOM, R. W., K. B. ASARCH & J. C. SHIH. 1986. Photoaffinity labeling of the 5-$hydroxytryptamine_{1A}$ receptor in rat hippocampus. J. Neurochem. **47:** 1066–1072.
27. RAYMOND, J. R., A. FARGIN, M. J. LOHSE, J. W. REGAN, S. E. SENOGLES, R. J. LEFKOWITZ & M. G. CARON. 1989. Identification of the ligand binding subunit of the human 5-HT_{1A} receptor with [^{125}I]N_3-NAPS, a high affinity radioiodinated photoaffinity probe. Mol. Pharmacol. **36:** 15–21.
28. HAMON, M., H. GOZLAN, S. EL MESTIKAWY, M. B. EMERIT, J. M. COSSERY & O. LUTZ. 1988. Biochemical properties of central serotonin receptors. *In* Neuronal Serotonin. N. N. Osborne & M. Hamon, Eds. Chichester: John Wiley & Sons Ltd., pp. 393–422.
29. GOZLAN, H., M. B. EMERIT, M. D. HALL, M. NIELSEN & M. HAMON. 1986. In situ molecular sizes of the various types of 5-HT binding sites in the rat brain. Biochem. Pharmacol. **35:** 1891–1897.
30. KOBILKA, B. K., T. FRIELLE, S. COLLINS, T. YANG-FENG, T. S. KOBILKA, U. FRANCKE, R. J.

LEFOWITZ & M. G. CARON. 1987. An intronless gene encoding a potential member of the family of receptors coupled to guanine nucleotide regulatory proteins. Nature **329:** 75–79.

31. FARGIN, A., J. R. RAYMOND, M J. LOHSE, B. K. KOBILKA, M. G. CARON & R. J. LEFKOWITZ. 1988. The genomic clone G-21 which resembles a β-adrenergic receptor sequence encodes the 5-HT_{1A} receptor. Nature **335:** 358–360.
32. GOZLAN, H., S. EL MESTIKAWY & M. HAMON. 1989. Purification and cloning of central serotonin receptors. *In* Receptor Purification. G. Litwack, Ed. Clifton, NJ: Humana Press. In press.
33. EL MESTIKAWY, S., C. COGNARD, H. GOZLAN & M. HAMON. 1988. Pharmacological and biochemical characterization of rat hippocampal 5-HT_{1A} receptors solubilized by 3[3-(cholamidopropyl)dimethylammonio]-1-propane sulfonate (CHAPS). J. Neurochem. **51:** 1031–1040.
34. EL MESTIKAWY, S., D. TAUSSIG, H. GOZLAN, M. B. EMERIT, M. PONCHANT & M. HAMON. 1989. Chromatographic analyses of the serotonin 5-HT_{1A} receptor solubilized from the rat hippo-campus. J. Neurochem. **53:** 1555–1566.
35. EMERIT, M. B., S. EL MESTIKAWY, H. GOZLAN, B. ROUOT & M. HAMON. 1990. Physical evidence of the coupling of solubilized 5-HT_{1A} binding sites with G regulatory proteins. Biochem. Pharmacol. **39:** 7–18.
36. DE VIVO, M. & S. MAAYANI. 1986. Characterization of the 5-hydroxytryptamine$_{1A}$ receptor-mediated inhibition of forskolin-stimulated adenylate cyclase activity in guinea pig and rat hippocampal membranes. J. Pharmacol. Exp. Ther. **238:** 248–253.
37. WEISS, S., M. SEBBEN, D. KEMP & J. BOCKAERT. 1986. Serotonin 5-HT_1 receptors mediate inhibition of cyclic AMP production in neurons. Eur. J. Pharmacol. **120:** 227–230.
38. HAMON, M. 1987. Second messenger systems linked to different serotonin (5-HT) receptors. *In* Pharmacology. M. J. Rand & C. Raper, Eds. Amsterdam: Excerpta Medica , pp. 281–284.
39. OKSENBERG, D. & S. J. PEROUTKA. 1988. Antagonism of 5-hydroxytryptamine$_{1A}$ (5-HT_{1A}) receptor-mediated modulation of adenylate cyclase activity by pindolol and propranolol isomers. Biochem. Pharmacol. **37:** 3429–3433.
40. SCHOEFFTER, P. & D. HOYER. 1988. Centrally acting hypotensive agents with affinity for 5-HT_{1A} binding sites inhibit forskolin-stimulated adenylate cyclase activity in calf hippocampus. Br. J. Pharmacol. **95:** 975–985.
41. MARKSTEIN, R., D. HOYER & G. ENGEL. 1986. 5-HT_{1A}-receptors mediate stimulation of adenylate cyclase in rat hippocampus. Naunyn-Schmiedeberg's Arch. Pharmacol. **333:** 335–341.
42. FARGIN, A., J. R. RAYMOND, J. W. REGAN, S. COTECCHIA, R. J. LEFKOWITZ & M. G. CARON. 1989. Effector coupling mechanisms of the cloned 5-HT_{1A} receptor. J. Biol. Chem. **264:** 14848–14852.
43. HAMON, M., M. B. EMERIT, S. EL MESTIKAWY, M. C. GALLISSOT & H. GOZLAN. 1990. Regional differences in the transduction mechanisms of 5-hydroxytryptamine receptors in the mammalian brain. *In* Cardiovascular Pharmacology of 5-Hydroxytryptamine: Prospective Therapeutic Applications. P. R. Saxena, D. I. Wallis, W. Wouters & P. Bevan, Eds. Dordrecht: Kluwer Academic Publishers BV, pp. 41–59.
44. ANDRADE, R., R. C. MALENKA & R. A. NICOLL. 1986. A G protein couples serotonin and $GABA_B$ receptors to the same channels in hippocampus. Science **234:** 1261–1265.
45. INNIS, R. B., E. J. NESTLER & G. K. AGHAJANIAN. 1988. Evidence for G protein mediation of serotonin- and $GABA_B$-induced hyperpolarization of rat dorsal raphe neurons. Brain Res. **459:** 27–36.
46. NICOLL, R. A. 1988. The coupling of neurotransmitter receptors to ion channels in the brain. Science **241:** 545–551.
47. LANFUMEY, L., S. HAJ-DAHMANE, B. FRANC & M. HAMON. 1989. Potassium channels blockers and 5-HT_{1A} control of serotonergic neuronal activity in the dorsal raphe nucleus. Soc. Neurosci. **15:** 6.6.
48. NELSON, D. L., A. HERBET, S. BOURGOIN, J. GLOWINSKI & M. HAMON. 1978. Characteristics of central 5-HT receptors and their adaptive changes following intracerebral 5,7-dihydroxytryptamine administration in the rat. Mol. Pharmacol. **14:** 983–995.

49. DAVAL, G., D. VERGE, A. BECERRIL, H. GOZLAN, U. SPAMPINATO, & M. HAMON. 1987. Transient expression of 5-HT_{1A} receptor binding sites in some areas of the rat CNS during postnatal development. Int. J. Devl. Neurosci. **5:** 171–180.
50. GOZLAN, H., G. DAVAL, U. SPAMPINATO, D. VERGE, C. M. FATTACCINI, M. C. GALLISSOT, S. EL MESTIKAWY & M. HAMON. 1989. Differential alterations in central serotoninergic and dopaminergic neurotransmission in the aged rat. Neurobiol. Aging. In press.
51. HAMON, M. & M. B. EMERIT. 1989. Les facteurs chimiques impliqués dans la différenciation neuronale. *In* Médecine Périnatale (foetus et nouveau-né). J. P. Relier, J. Laugier & B. L. Salle, Eds. Paris: Médicine-Sciences. Flammarion, pp. 194–200.
52. KENNETT, G. A., M. MARCOU, C. T. DOURISH & G. CURZON. 1987. Single administration of 5-HT_{1A} agonists decreases 5-HT_{1A} presynaptic, but not postsynaptic receptor-mediated responses: relationship to antidepressant-like action. Eur. J. Pharmacol. **138:** 53–60.
53. GOODWIN, G. M., R. J. DE SOUZA & A. R. GREEN. 1985. Presynaptic serotonin receptor-mediated response in mice attenuated by antidepressant drugs and electroconvulsive shock. Nature **317:** 531–533.
54. WOZNIAK, K. M., C. S. AULAKH, J. L. HILL & D. L. MURPHY. 1988. The effect of 8-OH-DPAT on temperature in the rat and its modification by chronic antidepressant treatments. Pharmacol. Biochem. Behav. **30:** 451–456.
55. AULAKH, C. S., K. M. WOZNIAK, J. L. HILL & D. L. MURPHY. 1988. Differential effects of long-term antidepressant treatments on 8-OH-DPAT-induced increases in plasma prolactin and corticosterone in rats. Eur. J. Pharmacol. **156:** 395–400.
56. BLIER, P. & C. DE MONTIGNY. 1985. Serotoninergic but not noradrenergic neurons in rat central nervous system adapt to long-term treatment with monoamine oxidase inhibitors. Neuroscience **16:** 949–955.
57. BLIER, P. & C. DE MONTIGNY. 1983. Electrophysiological investigations on the effect of repeated zimelidine administration on serotonergic neurotransmission in the rat. J. Neurosci. **3:** 1270–1278.
58. CHAPUT, Y., C. DE MONTIGNY & P. BLIER. 1986. Effects of a selective 5-HT reuptake blocker, citalopram, on the sensitivity of 5-HT-autoreceptors: electrophysiological studies in the rat brain. Naunyn-Schmiedeberg's Arch. Pharmacol. **333:** 342–348.
59. BLIER, P. & C. DE MONTIGNY. 1987. Modification of 5-HT neuron properties by sustained administration of the 5-HT_{1A} agonist gepirone: Electrophysiological studies in the rat brain. Synapse **1:** 470–480.
60. BLIER, P., C. DE MONTIGNY & Y. CHAPUT. 1988. Electrophysiological assessment of the effects of antidepressant treatments on the efficacy of 5-HT neurotransmission. Clin. Neuropharmacol. **11** (suppl. 2): S1–S10.
61. NEWMAN, M. E. & B. LERER. 1988. Chronic electroconvulsive shock and desimipramine reduce the degree of inhibition by 5-HT and carbachol of forskolin-stimulated adenylate cyclase in rat hippocampal membranes. Eur. J. Pharmacol. **148:** 257–260.
62. ROWAN, M. J. & R. ANWYL. 1985. The effect of prolonged treatment with tricyclic antidepressants on the actions of 5-hydroxytryptamine in the hippocampal slice of the rat. Neuropharmacology **24:** 131–137.
63. HJORTH, S. & T. MAGNUSSON. 1988. The 5-HT_{1A} receptor agonist 8-OH-DPAT, preferentially activates cell body 5-HT autoreceptors in rat brain in vivo. Naunyn-Schmiedebergs Arch. Pharmacol. **338:** 463–471.
64. HIGGINS, G. A., A. J. BRADBURY, B. J. JONES & N. R. OAKLEY. 1988. Behavioural and biochemical consequences following activation of 5-HT_1-like and GABA receptors in the dorsal raphe nucleus of the rat. Neuropharmacology **27:** 993–1001.

DISCUSSION OF THE PAPER

FRANK YOCCA (*Bristol-Myers, Wallingford, CT*): The study that you did with the pertussis toxin where you looked at the GPPNHp inhibition of binding. How did you perform that experiment, was that infused into hippocampus or was it *in vitro* exposure?

HAMON: It was an intraventricular injection of pertussis toxin. 2.5 micrograms in 5 μl of saline were injected bilaterally, into each lateral ventricle, and the rats were killed two days after the treatment.

JOSEÉ LEYSEN (*Janssen Research Foundation, Beerse, Belgium*): Michel, when you add the antibody against the third intracellular loop of your receptor, does it affect in cyclase study the activity of the receptor? Do you have any effect on the coupling with the G protein?

HAMON: I don't know yet.

LEYSEN: You intend to do it though?

HAMON: Of course.

MENAHEM SEGAL (*The Weizmann Institute, Rehovot, Israel*): Those of us who have used 8-hydroxy-DPAT for physiology find quite consistently that in hippocampus it is more of an antagonist than a good agonist—unlike the raphe where 8-hydroxy-DPAT is clearly an agonist of 5-HT autoreceptors. Could you comment on this discrepancy between the biochemistry and some of the physiology?

HAMON: You raise a difficult point. However, I am not sure that everybody agrees about the antagonistic effect of 8-hydroxy-DPAT in the hippocampus.

SEGAL: But there are at least three or four reports showing that it is an antagonist. If anything it does not act like serotonin.

HAMON: Yes, I know. However evidence has been provided, notably by Andrade and Nicoll, that the 5-HT_{1A} receptor you are studying with electrophysiological recording techniques is not the one coupled to adenylate cyclase, but that which is directly coupled to a K+ channel via a G protein. This suggests that heterogeneity within the 5-HT_{1A} subclass probably exists, with 8-hydroxy-DPAT acting as an agonist on the 5-HT_{1A} receptor coupled to adenylate cyclase, and as an antagonist on that coupled to the K+ channel. However, as I mentioned before, there are conflicting reports on the latter point. Looking further to this possible heterogeneity of 5-HT_{1A} receptors, we recently compared the respective pharmacological profiles of 5-HT_{1A} binding sites located presynaptically on the somas and dendrites within the dorsal raphe nucleus, with those of the postsynaptic sites within the hippocampus. We thus incubated brain sections with 3H-8-hydroxy-DPAT plus increasing concentrations of various 5-HT_{1A} ligands, and made autoradiograms for quantitative analyses. A progressive decrease in optical density was found in sections exposed to increasing concentrations of 5-HT_{1A} agonists and antagonists allowing calculations of their IC50s. However, using this approach, we failed to see any significant differences between the pharmacological properties of 5-HT_{1A} binding sites in the dorsal raphe nucleus and those of 5-HT_{1A} sites in the hippocampus. In particular, the same IC50 values were found in the two areas for 8-hydroxy-DPAT itself, serotonin, buspirone, gepirone, ipsapirone, spiperone, methiothepin, and so on. Maybe the explanation of the possible opposite effects of 8-hydroxy-DPAT at the raphe level and at the hippocampal level stands in the nature of the G protein which is coupled to the 5-HT_{1A} receptor? Whether different G proteins are coupled to pre- and post-synaptic 5-HT_{1A} receptors, and whether they control the agonist or antagonist conformation of the receptor occupied by 8-hydroxy-DPAT are two important questions for which, I think, nobody has the answers yet.

QUESTION: 8-Hydroxy-DPAT has been used to label the serotonin transporter in brain and platelets. Can you comment on this and should it be a concern and is 8-hydroxy-DPAT a substrate for the transporter?

HAMON: 8-Hydroxy-DPAT is a competitive inhibitor of 5-HT uptake in synaptosomes with a Ki of 1. 4 μM under the conditions required for the measurement of 5-HT uptake, i.e. in Krebs medium–containing salts. However, under the conditions classically used for measuring the specific binding of 3H-8-hydroxy-DPAT, i.e. in Tris-HCl buffer in the absence of salts, the affinity of 8-hydroxy-DPAT for the 5-HT transporter is apparently much higher, with a Ki of approximately 20–30 nM. This observation has also been reported by Schoemaker and Langer almost three years ago. However, under the conditions we are using for the labeling of 5-HT_{1A} binding sites either in membranes or on brain sections, there is no significant binding of 3H-8-hydroxy-DPAT to the 5-HT transporter simply because the experiments are performed with 1–2 nM of this radioligand, i.e. a concentration in the range of the Kd of 5-HT_{1A} sites but far below the Ki of the transporter. And indeed, even in case of some binding of 3H-8-hydroxy-DPAT to the transporter, one can eliminate it quite easily by adding a 5-HT uptake inhibitor, for instance fluoxetine at 1 μM, into the assay mixture. Although I think it is not really needed, in fact we add systematically this drug into the incubating medium for the labeling of 5-HT_{1A} sites by 3H-8-hydroxy-DPAT on brain sections.

The 5-HT_{1B} Receptors

DEREK N. MIDDLEMISS AND PETER H. HUTSON

Merck Sharp and Dohme Research Laboratories
Neuroscience Research Centre
Terlings Park, Eastwick Road,
Harlow, Essex, CM20 2QR, England

INTRODUCTION

Central 5-hydroxytryptamine (5-HT) receptors were first described using radioligand binding techniques by Peroutka and Snyder in 1979 as consisting of two major subtypes, designated 5-HT_1 and 5-HT_2.[1] Subsequent studies have also identified a third major class of 5-HT receptor, the 5-HT_3 class,[2] and it is now accepted that both the 5-HT_2 and 5-HT_3 sites have many functional correlates in both the brain and the periphery.[3] Correlates of the 5-HT_1 site, however, have proved more difficult to delineate and indeed whether these sites actually represent functional receptors has been questioned. It is now evident that the description of the 5-HT_1 site as a single homogenous entity is an oversimplification and that at least four distinct subtypes of the 5-HT_1 recognition site may exist. These have been designated 5-HT_{1A}, 5-HT_{1B}, 5-HT_{1C} and 5-HT_{1D}.[4-6] The present paper will review the identification, location and functional correlates of one of these sites, the 5-HT_{1B} receptor.

RADIOLIGAND BINDING STUDIES

The key finding in the differentiation of the 5-HT_1 recognition site into its constituent components was the observation that the neuroleptics spiperone, and to a lesser extent (+)-butaclamol, can discriminate between two populations of [^{3}H]5-HT binding sites in rat brain.[4] Thus, spiperone shows some 3000-fold selectivity between its dissociation constants for the high and low-affinity sites which were designated 5-HT_{1A} and 5-HT_{1B}, respectively.

The significance of this observation was clarified by the discovery that the centrally active 5-HT receptor agonist, 8-hydroxy-2-(di-*n*-propylamino)-tetralin (8-OH-DPAT), also showed selectivity for the high affinity (5-HT_{1A}) component of the [^{3}H]5-HT recognition site with correspondingly low affinity for the 5-HT_{1B} site.[7] Radiolabeled 8-OH-DPAT[8] was subsequently shown to bind with high affinity and specificity to the 5-HT_{1A} receptor in the CNS[5,9], and many functional correlates of activation of this site have now been defined.[3,10] These studies with 8-OH-DPAT have led to a re-evaluation of the binding of [^{3}H]5-HT to rat brain membranes and further components of the binding, the so-called 5-HT_{1C} and 5-HT_{1D} sites, described.[5,6]

Each of these sites can be selectively labeled with radioligands. Specifically in the case of the 5-HT_{1B} sites, these can best be labeled with [^{3}H]5-HT in the presence of 8-OH-DPAT and mianserin to avoid binding to the 5-HT_{1A} and 5-HT_{1C} sites, respectively.[11-13] The potent β-adrenoceptor antagonist, [^{125}I] cyanopindolol, also binds to the 5-HT_{1B} recognition site when care is taken to avoid binding to β-adrenoceptors by the presence of isoprenaline in the incubation medium.[14-16] [^{125}I] Cyanopindolol used under these conditions is a particularly good ligand for autoradiographical studies in the rat, which have shown a particularly high concentration of sites in the striatum, substantia nigra, and dorsal subiculum.[15] A similar distribution of 5-HT_{1B} sites is seen with the ligand [^{3}H]5-HT in the presence of 0.1μM 8-OH-DPAT.[9] This distribution of sites measured autoradiographically broadly parallels receptor binding studies using homogenates of these rat brain areas. [^{3}H]Dihydroergotamine can also be used to label the 5-HT_{1B} receptor when binding to alpha and 5-HT_{1A} receptors is minimized by the use of phentolamine and 8-OH-DPAT respectively.[17]

It is not at present clear as to the pre- or postsynaptic localization of the 5-HT_{1B} sites since 5,7-dihydroxytryptamine lesions of 5-HT pathways in the rat have been reported to up regulate (substantia nigra,[18] caudate-putamen[16]), down regulate (substantia nigra[9]) or cause no change (substantia nigra,[16] cortex[16]) to 5-HT_{1B} recognition sites. The affinity of a range of drugs for the 5-HT_{1B} recognition site as measured using either [^{3}H]5-HT or [^{125}I]cyanopindolol are given in FIGURE 1. Also shown in FIGURE 1 are the activities of these drugs at other receptors which occur within a factor of 1.5 log units of the 5-HT_{1B} site affinity. It can be clearly seen that no drug shows any specificity or selectivity for the 5-HT_{1B} site.

Despite these problems, many authors have nevertheless argued that it is possible to establish the criteria necessary for definition of 5-HT_{1B}-mediated event and a consensus is now emerging which is summarized in TABLE 1. Thus the functional response should be stimulated by the 5-HT_1-like receptor agonists 5-carboxamidotryptamine (5-CT)[19] or 5-methoxy-3(1,2,3,6-tetrahydropyridin-4-yl-1H-indole (RU 24969)[20] and perhaps by 7-trifluoromethyl-4(4-methyl-1-piperazinyl-pyrolo[1,2-a]quinoxaline (CGS 12066B)[21] but not by sensible doses of the 5-HT_{1A} receptor agonist, 8-OH-DPAT[22] (for structures see FIG. 2). Conversely, the response should be antagonized by the 5-HT_1-like receptor antagonist metitepin (sometimes known as methiothepin)[2] and by the β-adrenoceptor antagonists, propranolol[23] or cyanopindolol[24,25] (FIG. 3). The response should not, however, be blocked by 5-HT_2 (e.g. ketanserin), 5-HT_{1C} (e.g. mesulergine) or 5-HT_3 (e.g. (3α-tropanyl)-1H-indole-3-carboxylic acid ester, ICS 205-930) antagonists.[2] These criteria will be used in the following sections to judge the validity of claims for 5-HT_{1B} receptor-mediated responses in a variety of *in vitro* and *in vivo* models.

IN VITRO FUNCTIONAL CORRELATES

The Terminal 5-HT Autoreceptor

Depolarization-evoked release of [^{3}H]5-HT from slices or synaptosomes prepared from the rat or mouse brain has been the subject of several studies in an attempt to define the receptor type which mediates the inhibitory actions of 5-HT receptor agonists. The first of these studies utilizing a wide range of agonists concluded that this receptor was of the 5-HT_1 subclass with a relatively low affinity for the 5-HT_1-like receptor antagonist, metitepin.[26] The low potency of metitepin was subsequently shown to be matched by the virtual inactivity of the 5-HT_{1A}-receptor agonist, 8-OH-DPAT.[24] Subsequent studies have indicated that 5-CT[28] and RU 24969[20,29–31] are potent agonists at this receptor mediating 5-HT release as is the putative 5-HT_{1B} receptor-selective agonist, CGS 12066B.[21] This receptor is blocked stereoselectively by metitepin[32] and by the β-adrenoceptor antagonists, propranolol,[23] isamoltane,[33] and cyanopindolol[24,25] which act as partial agonists at this site.[27] Cyanopindolol is a very powerful 5-HT autoreceptor antagonist with a pA_2 value of 8.3, a value identical to its affinity for the 5-HT_{1B} or the 5-HT_{1A} radioligand recognition sites. The latter site cannot however mediate the inhibition of release at the terminal 5-HT autoreceptor since 5-HT_{1A} receptor agonists such as 8-OH-DPAT, buspirone or ipsapirone are inactive in

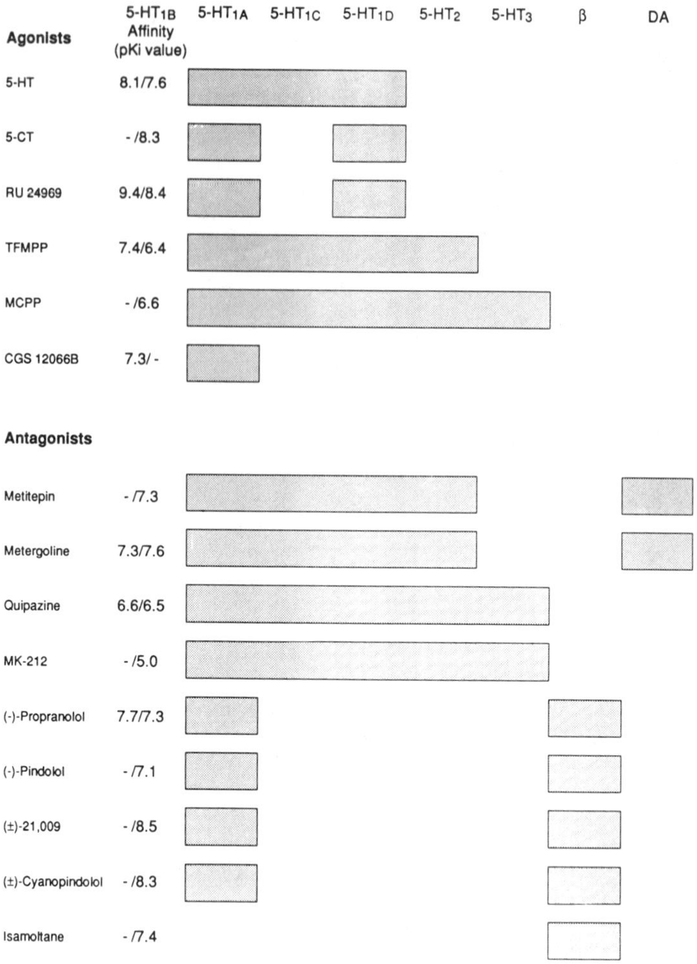

FIGURE 1. The affinities of compounds at the 5-HT_{1B} recognition site and their activities at other receptor sites. The values represent the pKi value against the 5-HT_{1B} recognition site measured using either [^{3}H]5-HT (1st number) or [^{125}I]cyanopindolol (2nd number). A box below any other receptor indicates that the compound possesses activity at that site within 1.5 log units of the 5-HT_{1B} activity and is taken to mean that the compound is non-selective. Data are taken from references 5, 6 10–14, 16, 21, 28, 33 & 85. **Note:** Sills *et al.*[84] have claimed that RU 24969 and TFMPP show a degree of selectivity for the 5-HT_{1B} site, but this is not a general finding.

inhibiting depolarization-evoked release of 5-HT.[28] That this receptor is of the 5-HT_{1B} subclass is further strengthened by studies carried out in the rat frontal cortex which indicate a strong correlation between the ability of a range of drugs to either stimulate or inhibit the terminal 5-HT autoreceptor and their affinity for the 5-HT_{1B} recognition site measured using [^{125}I]-cyanopindolol (FIG. 4). This study in the rat frontal cortex confirms and extends a seminal study of Engel *et al.* in the rat occipitoparietal cortex which reached a similar conclusion, that is that the terminal 5-HT autoreceptor is of the 5-HT$_{1B}$ subclass.[28]

The 5-HT Heteroreceptor on Cholinergic Terminals

The effects of 5-HT receptor agonists and antagonists on the release of [^{3}H] acetylcholine from rat hippocampal synaptosomes has been the subject of a careful pharmacological study which identifies this inhibitory receptor as being of the 5-HT$_{1B}$ subclass.[34] Thus K^+-evoked release of acetylcholine is inhibited by 5-HT and RU 24969 but not by 8-OH-DPAT. The inhibitory effects of 5-HT are blocked by metitepin or propranolol but not by the 5-HT$_{1A}$/5-HT$_2$ antagonist, spiperone or by the 5-HT$_{1C}$ antagonist, mesulergine. Interestingly, the same group have characterized a terminal 5-HT autoreceptor in the rat hippocampus as being of the 5-HT$_{1B}$ subclass,[35] indicating that both 5-HT autoreceptors located on 5-HT axon terminals and the 5-HT heteroreceptors located on cholinergic terminals in this brain area belong to the same (5-HT$_{1B}$) receptor subtype.

5-HT Mediated Inhibition of Adenylate Cyclase

The high concentration of 5-HT$_{1B}$ receptors in the rat substantia nigra prompted Bouhelal *et al.* to examine the effects of 5-HT receptor agonists for their ability to inhibit forskolin-stimulated adenylate cyclase activity in this brain region.[36] Adenylate cyclase activity was inhibited by 5-HT, RU 24969, 5-CT and CGS 12066B but not by 8-OH-DPAT or ipsapirone, an agonist pharmacological profile consistent with a 5-HT$_{1B}$-mediated effect. The inhibitory effects of 5-HT were not blocked by 5-HT$_{1A}$, 5-HT$_{1C}$ or 5-HT$_2$ receptor antagonists, but were antagonized by the 5-HT$_{1B}$ receptor

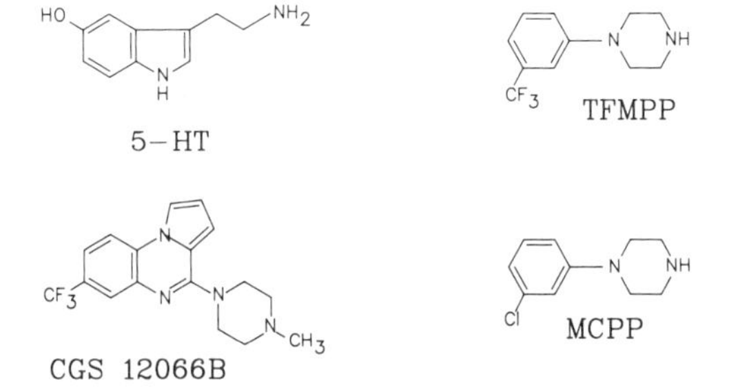

FIGURE 2. The structures of a range of 5-HT$_{1B}$ receptor agonists.

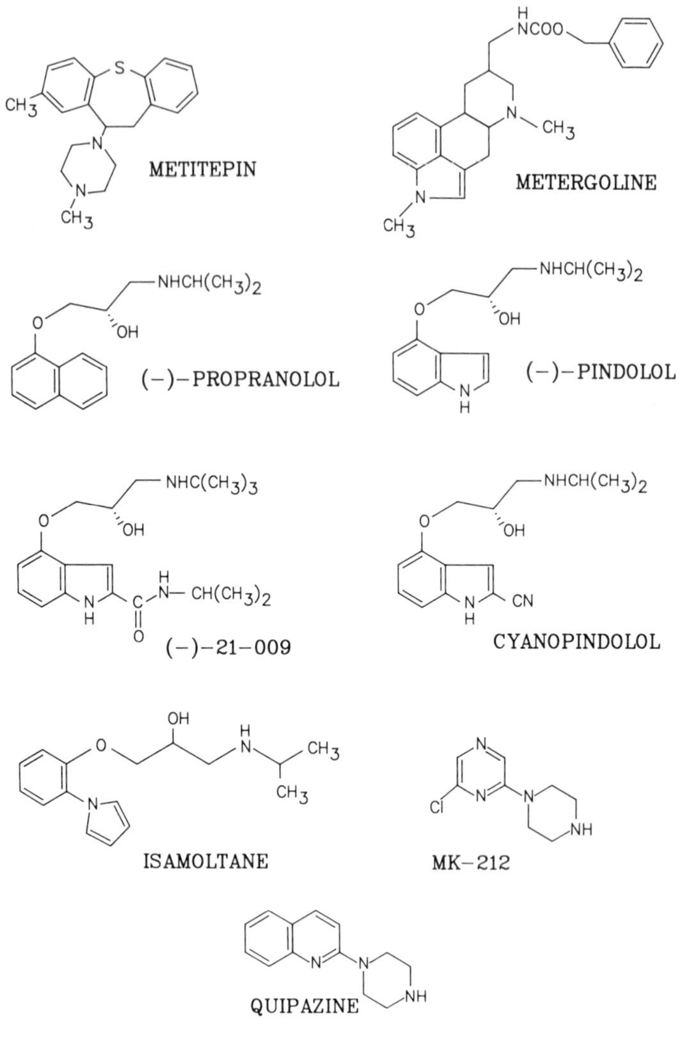

FIGURE 3. The structures of a range of 5-HT_{1B} receptor antagonists.

antagonists cyanopindolol, propranolol and to a lessor extent, metergoline. Thus it would seem that this receptor identified in the rat substantia nigra by radioligand binding techniques is functionally negatively linked to the adenylate cyclase second messenger system in this brain area. This inhibitory coupling of the 5-HT_{1B} receptor to adenylate cyclase has also been shown in a chinese hamster lung fibroblast cell line (CCL39 cells).[37] In this model, the stimulation of DNA synthesis by 5-HT (mitogenic effect) was shown to be mediated through 5-HT_{1B} receptors linked to inhibitory G proteins, indicating a role for this receptor in the control of cell proliferation. Further evidence for a G protein–linked 5-HT_{1B} receptor in the rat has been provided by experiments with *N*-ethylmaleimide which inactivates 5-HT_{1B} binding by an association with G proteins.[38]

The 5-HT Potentiation of Electrically Stimulated Contractions of the Mouse Urinary Bladder Strip.

The isolated mouse bladder strip, when contracted by a train of electrical pulses at 10 Hz, is sensitive to the potentiating effects of 5-HT at concentrations as low as 10nM, implicating an effect mediated by a 5-HT_1-like receptor.[39] The effect of 5-HT is mimicked by 5-CT but not by 8-OH-DPAT and is blocked by the 5-HT_{1B} receptor antagonists, propranolol[39] and cyanopindolol.[40] The latter compound displays a biphasic effect on the electrically stimulated mouse urinary bladder strip. Low concentrations, 10 or 100nM, antagonize the effects of 5-HT with a pA_2 of 8.3; higher concentrations, 1 or 10μM, actually stimulate the bladder. This latter effect may be due to actions on the tissue other than 5-HT blockade or may be related to the known partial agonist actions of cyanopindolol at 5-HT_{1B} sites.[26] Nevertheless the pharmacological characteristics of this mouse preparation are consistent with the potentiation of the contraction being mediated by a 5-HT_{1B} receptor.

Presynaptic 5-HT Receptors in Rat Vena Cava

Sympathetic nerve terminals of blood vessels are endowed with 5-HT receptors which serve to inhibit noradrenaline release when activated. An example of such a 5-HT receptor is that located on the cranial segment of the inferior vena cava of the rat and inhibition of the release of noradrenaline in this preparation can be followed using veins which have been preloaded with [^{3}H]-noradrenaline.[41] Electrically stimulated release of [^{3}H]noradrenaline from the rat vena cava was inhibited by a range of 5-HT receptor agonists and their potency was significantly correlated with their affinity for the 5-HT_{1B} recognition site. The effect of 5-HT was antagonized by metitepin and propranolol but not by ketanserin, a 5-HT_2 receptor antagonist or spiperone, a mixed

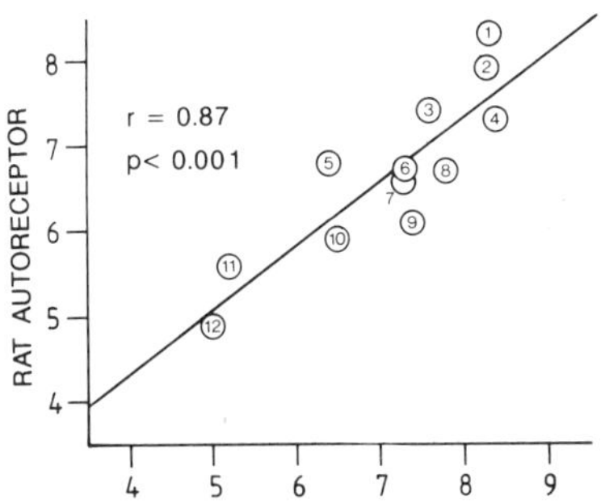

FIGURE 4. Correlation between the potency for agonists (pEC_{30}) or antagonists (pA_2) at the rat terminal 5-HT autoreceptor and their affinity against the 5-HT_{1B} receptor labeled with [^{125}I]cyanopindolol. Drugs used are as follows: 1. Cyanopindolol; 2. 5-CT; 3. RU 24969; 4. 5-HT; 5. 5-methoxytryptamine; 6. (-)-Propranolol; 7. Metitepin; 8. (-)-Pindolol; 9. Metergoline; 10. Quipazine; 11. Phenylpiperizine; 12. MK-212.

$5\text{-}HT_2/5HT_{1A}$ receptor antagonist; cyanopindolol acted as a partial agonist at this site. Thus this study indicates that the inhibitory presynaptic 5-HT receptors on the sympathetic nerve terminal of the rat vena cava belong to the $5\text{-}HT_{1B}$ subtype.

IN VIVO FUNCTIONAL CORRELATES

Effect of 5-HT Receptor Agonists on Brain 5-HT Metabolism

A decrease of brain 5-HT metabolism, synthesis or release is observed following administration of both direct and indirectly acting 5-HT receptor agonists. Consistent with 5-HT receptor agonist activity, RU 24969 (1.2–10mg/kg) caused a decrease of brain 5-HT synthesis (as determined by the rate of 5-HTP accumulation following decarboxylase inhibition[42]) and 5-HT turnover (as determined by a decrease of basal 5-HIAA concentration).[43] More recent studies using *in vivo* dialysis to monitor extraneuronal 5-HT concentrations in the rat cortex and hippocampus showed that RU 24969 (0.25–20mg/kg) also, in accordance with a 5-HT receptor agonist action, reduced 5-HT release *in vivo*.[44,45]

MCPP (10mg/kg) has also been shown to decrease brain 5-HT synthesis (5-HTP accumulation) although an action (direct or indirect) on catecholamine receptors was also indicated since brain-DOPA accumulation was decreased.[46] Agonist like effects were also described for TFMPP which dose-dependently (1.2–20mg/kg) decreased brain 5-HT turnover as determined by a reduction of basal 5-HIAA concentration and a reduced rate of 5-HIAA accumulation after probenecid.[47] More recently, studies with CGS 12066B, a novel $5HT_{1B}$ ligand, have indicated that this compound can decrease brain 5-HT synthesis (5-HTP accumulation) with an ED_{25} of 4.5 mg/kg. This is consistent with 5-HT receptor agonist activity but was found to be approximately seven times less potent than TFMPP (ED_{25} 2.9mg/kg).[21]

Neurochemical studies may provide evidence that drugs act as agonists *in vivo*. They do not however indicate whether this is an effect mediated directly via receptors or indirectly by decreasing 5-HT release. They also do not indicate whether the drug is acting pre- or post-synaptically since agonists may decrease 5-HT metabolism via presynaptic autoreceptors (terminal and somatodendritic) or via neuronal feedback loops following activation of post-synaptic receptors. Furthermore, as agonists of other 5-HT receptor subtypes e.g. 8-OH-DPAT also cause a reduction of 5-HT synthesis, such studies, performed in isolation, cannot be used to define which type of receptors are involved.[48] It is evident that careful antagonist studies along the lines outlined in TABLE 1 need to be carried out with each of the putative $5\text{-}HT_{1B}$ receptor agonists discussed above.

Neuroendocrine Studies with 5-HT Receptor Agonists

Elevation of plasma corticosterone and prolactin in response to 5-HT receptor agonists is thought to reflect activation of central 5-HT receptors. Thus, RU 24969 (1.2–10mg/kg), TFMPP (2.5–20mg/kg) and MCPP (2.5–20mg/kg) cause a marked increase of both hormones in rat plasma.[43,49–51] However, as 5-HT reuptake inhibiters and 5-HT agonists for other receptor subtypes also increase these hormones in plasma it cannot be considered an effect exclusive to the $5\text{-}HT_{1B}$ subtype.[52,53] This is apparent from studies in humans where MCPP (0.75mg/kg) was shown to increase plasma prolactin and cortisol which, although the effect was blocked by metergoline (4mg p.o),

TABLE 1. Drugs Useful in Defining 5-HT$_{1B}$ Receptor-Mediated Effects

	Concentration Range *In Vitro* (nM)	Dose Range *In Vivo* (mg/kg)
Agonists		
5-CT	1–100	0.0001–0.001[a]
RU 24969	1–100	2.5–10.0
CGS 12066B	300–3000	4.5
TFMPP	300–1000	1–20
MCPP	100–3000	1–20
8-OH-DPAT (5-HT$_{1A}$)	>1000	>1
Antagonists		
Metitepin	10–1000	0.2
(−)-Propranolol	100–1000	1–5
(±)-Cyanopindolol	1–100	4–8
Ketanserin (5-HT$_2$)	>1000	>2.5
Mesulergine (5-HT$_{1C}$)	>1000	0.2
ICS 205-930 (5-HT$_3$)	>1000	>1

[a]Profound cardiovascular effects which may not be 5-HT$_{1B}$ receptor mediated.[19]

cannot be interpreted as being mediated by a 5-HT$_{1B}$ receptor since 5-HT$_{1B}$ receptor binding sites have not been found in human post mortem brain tissue.[54]

Electrophysiological Studies with 5-HT Receptor Agonists

Studies with 5-HT receptor agonists have largely been confined to the dorsal and medial raphe nuclei which contain a high proportion of 5-HT cell bodies and a high density of 5-HT$_{1A}$ and 5-HT$_{1B}$ binding sites.

Whilst 5-HT$_{1A}$ receptor agonists potently inhibit dorsal raphe firing, the 5-HT$_{1B}$ receptor agonists MCPP and TFMPP do not reliably affect spontaneous firing of 5-HT neurones in the rat dorsal raphe nucleus when given either systemically or micro-iontophoretically.[21,55,56] However CGS 12066B, a 5-HT receptor agonist with a 17-fold selectivity for 5-HT$_{1B}$ versus 5-HT$_{1A}$ sites has been shown to inhibit dorsal raphe firing although this has not been characterized pharmacologically.[21,55] Interestingly, low doses of TFMPP and CGS 12066B were reported to increase the firing rate of 5-HT neurones in the medial raphe, although again, this effect has not been pharmacologically characterized.[55]

In contrast to the diverse effects of 5-HT$_{1A}$ and 5-HT$_{1B}$ receptor agonists on 5-HT cell body firing, hippocampal pyramidal cells were only weakly inhibited by the micro-iontophoretic application of either 5-HT$_{1A}$ or 5-HT$_{1B}$ agonists.[57] The response of post-synaptic hippocampal CA3 cells to electrical stimulation of ascending 5-HT pathways is an increase in post-synaptic inhibition and this effect has been used to assess the functional role of the terminal 5-HT autoreceptor *in vivo*. The autoreceptor antagonist metitepin was found to enhance the post-synaptic inhibition seen after stimulation of ascending 5-HT pathways and this has been interpreted as being mediated by the terminal 5-HT autoreceptor antagonist properties of this drug.[58] Further characterization of this effect using other 5-HT receptor antagonists is still lacking.

The Locomotor Response to 5-HT Receptor Agonists

Administration of the putative $5\text{-}HT_{1B}$ receptor agonists TFMPP (2–25mg/kg) or MCPP (2–20mg/kg) results in a decrease in locomotor activity in non-habituated rats which have been exposed to a novel environment.[59,60] This effect was blocked by the non-selective $5\text{-}HT_1$-like receptor antagonists metergoline (2.0mg/kg) and methysergide (10.0 mg/kg) but not by ketanserin (10.0mg/kg) thereby excluding a role for $5\text{-}HT_2$ receptors. MCPP-induced hypoactivity was found to be blocked by antagonists with high affinity for $5\text{-}HT_{1C}$ sites while antagonists at $5\text{-}HT_{1A}$, $5\text{-}HT_{1B}$, $5\text{-}HT_2$ and $5\text{-}HT_3$ sites were without effect.[61] These studies suggest, therefore, that MCPP and possibly TFMPP induced hypolocomotion is mediated by $5\text{-}HT_{1C}$ receptors and provide no evidence for an involvement of $5\text{-}HT_{1B}$ receptors in this particular behavioral response to these agonists.

The behavioral response seen with the $5\text{-}HT_1$ receptor agonist RU 24969 (0.5–36mg/kg), which was shown to cause an increase of locomotor activity, is similarly not mediated by $5\text{-}HT_{1B}$ receptors.[43,62,64] Thus, while the hyperactivity was enhanced in animals depleted of 5-HT with 5, 7 dihydroxytryptamine suggesting an involvement of post-synaptic 5-HT receptors, it would seem unlikely that this is due to activation of $5\text{-}HT_1$-like or $5\text{-}HT_2$ receptors as methysergide (5 and 10mg/kg) or metergoline (2.5 and 5mg/kg) did not block the hyperactivity.[63,64]

Regulation of Food Intake by 5-HT Receptor Agonists

Numerous studies with either non-selective 5-HT receptor agonists or drugs such as fenfluramine and fluoxetine, which indirectly enhance brain 5-HT availability, have demonstrated an inhibitory role for 5-HT in the regulation of food intake.[65]

Initial studies with TFMPP and MCPP supported this hypothesis and showed a marked hypophagia in both food-deprived and non-deprived rats, an effect blocked by metergoline.[50,51] More recent studies showed that both MCPP and TFMPP-induced hypophagia was blocked by 5-HT receptor antagonists with high affinity for $5\text{-}HT_{1C}$ sites i.e. metergoline (5mg/kg), mianserin (2 or 5mg/kg), mesulergine (0.2mg/kg) and also for $5\text{-}HT_{1B}$ sites i.e. cyanopindolol (8mg/kg) or (−) propranolol (16mg/kg). Antagonists with affinity for other 5-HT receptor sites i.e. ketanserin (0.2mg/kg), ritanserin (0.6mg/kg), ICS 205930 (1mg/kg) and idazoxan (1mg/kg) did not affect MCPP-induced hypophagia.[66]

Studies with RU 24969 (0.05–10mg/kg) also showed a dose-dependent hypophagia in non-deprived rats.[67] However, binding studies revealed that RU 24969 has high affinity for $5\text{-}HT_{1A}$ and $5\text{-}HT_{1B}$ sites but not $5\text{-}HT_{1C}$ sites. Agonists of $5\text{-}HT_{1A}$ receptors cause hyperphagia by activating somatodendritic autoreceptors.[68–70] It would seem likely therefore that RU 24969–induced hypophagia is mediated by an action at $5\text{-}HT_{1B}$ receptors, which are probably located postsynaptically as depletion of brain 5-HT by pCPA did not prevent the hypophagia.[67]

Antagonist studies revealed that RU 24969–induced hypophagia was blocked only by drugs with high affinity for $5\text{-}HT_{1B}$ sites i.e. cyanopindolol (8mg/kg), (−) pindolol (3mg/kg), and metergoline (5mg/kg). Antagonists at other 5-HT receptors i.e. mesulergine ($5\text{-}HT_{1C}$, 0.2mg/kg), mianserin ($5\text{-}HT_{1C}/5\text{-}HT_2$, 2 or 5mg/kg), ketanserin ($5\text{-}HT_2$, 2.5mg/kg), spiperone ($5\text{-}HT_2/5\text{-}HT_{1A}$, 0.05mg/kg), and haloperidol (DA, 0.1mg/kg) were without effect.[61,67]

As mentioned previously, TFMPP and MCPP cause pronounced hypolocomotion while RU 24969 induces hyperactivity. It is possible that the observed decrease of food intake is produced as a consequence of the change in locomotor activity, although this would seem unlikely. Thus, only the hypophagic response of MCPP was blocked by cyanopindolol and propranolol, whilst haloperidol blocked the hyperactivity but not the hypophagia produced by RU 24969.[67] Furthermore, local infusion of RU 24969 and TFMPP onto the paraventricular nucleus of the hypothalamus caused hypophagia without affecting locomotor activity.[71]

MISCELLANEOUS *IN VIVO* STUDIES

Penile Erection

In a recent study by Berendsen and Broekkamp it was suggested that penile erection in rats induced by serotonergic drugs is mediated by 5-HT_{1B} receptors.[72] They found that a number of serotonergic drugs including 5-HT reuptake inhibitors, fenfluramine, and the putative 5-HT_{1B} receptor agonist MCPP caused penile erections whereas 8-OH-DPAT and other 5-HT receptor agonists, including 5-MeODMT (5-methoxy-*N*,*N*-dimethyltryptamine) and LSD, did not cause this effect. RU 24969 also did not cause penile erection which is perhaps surprising given its activity at 5-HT_{1B} receptors. However, their conclusions were supported by antagonist studies which showed that penile erections induced by ORG 6997 (5α,8α,9α-2-chloro-5,8,9,10-tetrahydro-5,9-methanobenzocycloocten-8-amine HCl) were blocked by propranolol and (−)pindolol but not by the 5-HT_2 receptor antagonist, pirenperone or the 5-HT_{1A} receptor antagonist, spiperone.

Thermoregulation

Although 5-HT has been implicated in the regulation of body temperature, its effects are somewhat variable depending on species, dose, and ambient temperature. There has been much recent work on the hypothermic effects of 5-HT_{1A} receptor agonists[73] and there are now several reports that, in addition, 5-HT_{1B} receptors may be involved in thermoregulation. Thus, RU 24969 caused a small decrease of temperature in freely moving rats at doses of 0.3–3.0 mg/kg but an increase of temperature at 10 mg/kg. The hypothermic effects of RU 24969 were more pronounced in restrained rats although the hyperthermic effect was no longer apparent. Pre-treatment with (−)-pindolol blocked this hypothermia produced by RU 24969 suggesting an involvement of 5-HT_1 receptors.[64] Hyperthermia has also been observed in human subjects following MCPP, an effect which was blocked by the non-selective 5-HT receptor antagonist metergoline.[54] MCPP and TFMPP have also been shown to cause hypothermia in mice although significant effects were obtained only at the highest dose studied (20 mg/kg). Nevertheless antagonist studies indicated an involvement of 5-HT_{1B} as opposed to 5-HT_2 or 5-HT_3 receptors. Thus the hypothermia was blocked by pindolol, cyanopindolol and 21-009 whereas spiperone, mesulergine, cyproheptadine, mianserin, methysergide, ICS 205-930 and metoclopramide were without effect.[74]

Drug Discrimination

The use of drug discrimination and stimulus generalization studies may be of some value in the identification of receptor selective drugs. Animals are trained, using a two lever operant procedure, to discriminate the training drug from saline. In order to test for stimulus generalization, the drug under examination is then administered and if it is perceived by the animal as producing the same stimulus effects as the training drug the animal will respond on the lever associated with the training drug.

Putative 5-HT_{1B} receptor agonists have been examined in drug discrimination studies. Thus, rats trained to discriminate TFMPP (0.8–1.0 mg/kg) from saline showed stimulus generalization to fenfluramine (0.8–1.6 mg/kg), MCPP (0.1–1.6 mg/kg), and RU 24969 (0.1–1.6 mg/kg) and tetrahydro-β-carboline (20 mg/kg) but not to quipazine (0.2–3.2 mg/kg), the 5-HT_{1A} receptor agonist, 8-OH-DPAT (0.02–0.32 mg/kg) or the 5-HT_2 receptor agonist 1-(2,5 dimethoxy-4-methylphenyl)-2-aminopropane (DOM) (0.25–1.4 mg/kg).[75–77] Clearly, caution is required when interpreting such data as generalization does appear to occur between non-selective and selective drugs. An involvement of 5-HT_1 receptors in the TFMPP cue was suggested from rather limited antagonist studies which showed that the cue was attenuated by metergoline (0.4–6.4 mg/kg) and spiperone (0.08–1.28 mg/kg) but not by the 5-HT_2 receptor antagonists ketanserin (0.8–6.4 mg/kg) and pirenperone (0.08–0.64 mg/kg).[78] However, TFMPP stimulus generalization occurred with propranolol (1–14 mg/kg), mesulergine (0.6–4.0 mg/kg) and pindolol (5–22 mg/kg) indicating that some receptor antagonists may be capable of acting as agonists in this paradigm.[79] Further studies are needed in this area before a definitive classification as to the 5-HT receptor subtype involved in this behavior can be made.

Social Behavioral Deficits

Mice isolated for 7 or more days have been shown to display a social behavioral deficit in relation to their motivation to escape from an inverted beaker.[80] The putative 5-HT_{1B} receptor agonists MCPP, TFMPP and RU 24969 increase escape behavior and these effects are antagonized by the β-adrenoceptor antagonist penbutolol but not by 5-HT_2 or 5-HT_3 antagonists.[81] Taken together these studies may indicate a role for 5-HT_{1B} receptors in escape behavior in socially isolated mice.

SUMMARY

The 5-HT_{1B} receptors have been identified by radioligand binding techniques predominantly in the basal ganglia of the rat and mouse brain. A number of 5-HT receptor agonists have been shown to display high affinity but limited selectivity for the 5-HT_{1B} recognition site. These include 5-CT, 5-HT, RU 24969, TFMPP, MCPP, and CGS 12066B. Antagonists at the 5-HT_{1B} site include the drugs metitepin, metergoline, cyanopindolol, isamoltane, and propranolol but none of these drugs are selective for this receptor. Functional correlates of 5-HT_{1B} receptor activation have been most closely defined *in vitro*. These include inhibition of transmitter release, inhibition of forskolin-stimulated adenylate cyclase and actions on the mouse urinary bladder strip and the rat vena cava. Many functional correlates of 5-HT_{1B} receptor activation *in vivo* have been proposed, but convincing evidence from antagonist studies is generally lacking. The development of selective 5-HT_{1B} receptor agonists and antagonists will be a key step in defining the physiological role of this receptor site in the brain and periphery of the mouse and rat although it must be realized that these compounds, if they are developed, are unlikely to have functional effects in man since the 5-HT_{1B} recognition site is absent in the human CNS.[82] Nevertheless many of these studies on the 5-HT_{1B} receptor may aid the development of drugs acting at the 5-HT_{1D} site since this receptor has been identified as being the equivalent of the 5-HT_{1B} site in species other than the rat and mouse.[83]

ACKNOWLEDGMENTS

We are grateful to Mrs Eleonor Brawn for typing the manuscript and Roy Hammond for the artwork.

REFERENCES

1. PEROUTKA, S. J. & S. H. SNYDER. 1979. Multiple serotonin receptors: Differential binding of [^{3}H]5-hydroxytryptamine, [^{3}H]lysergic acid diethylamide and [^{3}]spiroperidol. Mol. Pharmacol. **16:** 687–699.
2. BRADLEY, P. B., G. ENGEL, W. FENIUK, J. R. FOZARD, P. P. A. HUMPHREY, D. N. MIDDLEMISS, E. J. MYLECHARANE, B. P. RICHARDSON & P. R. SAXENA. 1986. Proposals for the classification and nomenclature of functional receptors for 5-hydroxytryptamine. Neuropharmacology **25:** 563–576.
3. MIDDLEMISS, D. N., M. HIBERT & J. R. FOZARD. 1986. Drugs acting at central 5-hydroxytryptamine receptors. Ann. Rep. Med. Chem. **21:** 41–50.
4. PEDIGO, N. W., H. I. YAMAMURA & D. L. NELSON. 1981. Discrimination of multiple [^{3}H] 5-hydroxytryptamine binding sites by the neuroleptic spiperone in rat brain. J. Neurochem. **36:** 220–226.
5. HOYER, D., G. ENGEL & H. O. KALKMAN. 1985. Molecular pharmacology of 5-HT_1 and 5-HT_2 recognition sites in rat and pig brain membranes: Radioligand binding studies with [^{3}H]5-HT, [^{3}H]8-OH-DPAT, (−) [^{125}I] iodocyanopindolol, [^{3}H] mesulergine and [^{3}H]ketanserin. Eur. J. Pharmacol. **118:** 13–23.
6. HEURING, R. E. & S. J. PEROUTKA. 1987. Characterization of a novel ^{3}H-5-hydroxytryptamine binding site subtype in bovine brain membranes. J. Neurosci. **7:** 894–903.
7. MIDDLEMISS, D. N. & J. R. FOZARD. 1983. 8-Hydroxy-2-(di-n-propylamino)-tetralin discriminates between subtypes of the 5-HT_1 recognition site. Eur. J. Pharmacol. **90:** 151–153.
8. GOZLAN, H., S. EL MESTIKAWY, L. PICHAT, J. GLOWINSKI & M. HAMON. 1983. Identification of presynaptic serotonin autoreceptors using a new ligand: ^{3}H-PAT. Nature **305:** 140–142.
9. VERGE, D., G. DAVAL, M. MARCINKIEWICZ, A. PATEY, S. EL MESTIKAWY, H. GOZLAN & M. HAMON. 1986. Quantitative autoradiography of multiple 5-HT_1 receptor subtypes in the brain of control or 5, 7-dihydroxytryptamine-treated rats. J. Neurosci. **6:** 3474–3482.
10. HOYER, D. 1988. Functional correlates of serotonin 5-HT_1 recognition sites. J. Receptor Res. **8:** 59–81.
11. PEROUTKA, S. J. 1986. Pharmacological differentiation and characterization of 5-HT_{1A}, 5-HT_{1B}, and 5-HT_{1C} binding sites in rat frontal cortex. J. Neurochem. **47:** 529–540.
12. BLURTON, P. A. & M. D. WOOD. 1986. Identification of multiple binding sites for [^{3}H] 5-hydroxytryptamine in the rat CNS. J. Neurochem. **46:** 1392–1398.
13. ALEXANDER, B. S. & M. D. WOOD. 1987. Stereoselective blockade of central [^{3}H] 5-hydroxytryptamine binding to multiple sites (5-HT_{1A}, 5-HT_{1B} and 5-HT_{1C}) by mianserin and propranolol. J. Pharm. Pharmacol. **39:** 664–666.
14. HOYER, D., G. ENGEL & H. O. KALKMAN. 1985. Characterization of the 5-HT_{1B} recognition site in rat brain: Binding studies with (−) [^{125}I] Iodocyanopindolol. Eur. J. Pharmacol. **118:** 1–12.
15. PAZOS, A., G. ENGEL & J. M. PALACIOS. 1985. B-Adrenoceptor blocking agents recognise a subpopulation of serotonin receptors in brain. Brain Res. **343:** 403–408.
16. OFFORD, S. J., G. A. ORDUROY & A. FRAZER. 1988. Application of [^{125}I]Iodocyanopindolol

to measure 5-hydroxytryptamine$_{1B}$ receptors in the brain of the rat. J. Pharmacol. Exp. Ther. **244:** 144–153.

17. HAMBLIN, M. W., K. ARIANI, P. I. ADRIAENSSENS & R. D. CIARARELLO. 1987. [^{3}H] Dihydroergotamine as a high-affinity, slowly dissociating radioligand for 5-HT$_{1B}$ binding sites in rat brain membranes. Evidence for guanine nucleotide regulation of agonist affinity states. J. Pharmacol. Exp. Ther. **243:** 989–1001.
18. WEISSMANN, D., E. MACH, C. OBERLANDER, Y. DEMASSEY & J-F. PUJOL. 1986. Evidence for a hyperdensity of 5HT1B binding sites in the substantia nigra of the rat after 5, 7-dihydroxytryptamine intraventricular injection. Neurochem. Int. **9:** 191–200.
19. CONNOR, H. E., W. FENIUK, P. P. A. HUMPHREY & M. J. PERREN. 1986. 5-Carboxamidotryptamine is a selective agonist at 5-hydroxytryptamine receptors mediating vasodilatation and tachycardia in anaesthetised cats. Br. J. Pharmacol. **87:** 417–426.
20. MIDDLEMISS, D. N. 1985. The putative 5-HT$_1$ receptor agonist, RU 24969, inhibits the efflux of 5-hydroxytryptamine from rat frontal cortex slices by stimulation of the 5-HT autoreceptor. J. Pharm. Pharmacol. **37:** 434–437.
21. NEALE, R. F., S. L. FALLON, W. C. BOYAR, J. W. F. WASLEY, L. L. MARTIN, G. A. STONE, B. S. GLAESER, C. M. SINTON & M. WILLIAMS. 1987. Biochemical and pharmacological characterization of CGS 12066B, a selective serotonin-1B agonist. Eur. J. Pharmacol. **136:** 1–9.
22. MIDDLEMISS, D. N. 1984. 8-Hydroxy-2-(di-n-propylamino) tetralin is devoid of activity at the 5-hydroxytryptamine autoreceptor in rat brain. Implications for the proposed link between the autoreceptor and the [^{3}H]5-HT recognition site. Naunyn–Schmiedebergs Arch. Pharmacol. **327:** 18–22.
23. MIDDLEMISS, D. N. 1984. D. N. 1984. Stereoselective blockade at [^{3}H]5-HT binding sites and at the 5-HT autoreceptor by propranolol. Eur. J. Pharmacol. **101:** 289–293.
24. SCHLICKER, E., M. GÖTHERT & K. HILLENBRAND. 1985. Cyanopindolol is a highly potent and selective antagonist at the presynaptic autoreceptor in the rat brain cortex. Naunyn–Schmiedebergs Arch. Pharmacol. **331:** 398–401.
25. MIDDLEMISS, D. N. 1986. Blockade of the central 5-HT autoreceptor by B-adrenoceptor antagonists. Eur J. Pharmacol. **120:** 51–56.
26. ENGEL, G., M. GÖTHERT, E. MULLER-SCHWEINITZER, E. SCHLICKER, L. SISTONEN & P. A. STADLER. 1983. Evidence for common pharmacological properties of [^{3}H]5-hydroxytryptamine binding sites, presynaptic 5-hydroxytryptamine autoreceptors in CNS and inhibitory presynaptic 5-hydroxytryptamine receptors on sympathetic nerves. Naunyn-Schmiedebergs Arch. Pharmacol. **324:** 116–124.
27. MAURA, G., M. ULIVI & M. RAITERI. 1987. (−)-Propranolol and (±)-cyanopindolol are mixed agonists-antagonists at serotonin autoreceptors in the hippocampus of the rat brain. Neuropharmacol. **26:** 713–717.
28. ENGEL, G., M. GÖTHERT, D. HOYER, E. SCHLICKER & K. HILLENBRAND. 1986. Identity of inhibitory presynaptic 5-hydroxytryptamine (5-HT) autoreceptors in the rat brain cortex with 5-HT$_{1B}$ binding sites. Naunyn–Schmiedebergs Arch. Pharmacol. **332:** 1–7.
29. GÖTHERT, M., E. SCHLICKER, K. FINK & K. CLASSEN. 1987. Effects of RU 24969 on serotonin release in rat brain cortex: Further support for the identity of serotonin autoreceptors with 5-HT$_{1B}$ sites. Arch. Int. Pharmacodyn. **288:** 31–42.
30. BONANNO, G., G. MAURA & M. RAITERI. 1986. Pharmacological characterization of release-regulating serotonin autoreceptors in rat cerebellum. Eur. J. Pharmacol. **126:** 317–321.
31. RAITERI, M., G. MAURA, G. BONANNO & A. PITTALUGA. 1986. Differential pharmacology and function of two 5-HT$_1$ receptors modulating transmitter release in rat cerebellum. J. Pharm. Exp. Ther. **237:** 644–648.
32. HIBERT, M. & D. N. MIDDLEMISS. 1986. Stereoselective blockade at the 5-HT autoreceptor and inhibition of radioligand binding to central 5-HT recognition sites by the optical isomers of methiothepin. Neuropharmacology. **25:** 1–4.
33. WALDMEIER, P. C., M. WILLIAMS, P. A. BAUMANN, S. BISCHOFF, M. A. SILLS & R. F. NEALE. 1988. Interactions of isamoltane (CGP 361 A), an anxiolytic phenoxy propranolamine derivative, with 5-HT$_1$ receptor subtypes in the rat brain. Naunyn–Schmiedebergs Arch. Pharmacol. **337:** 609–620.

34. MAURA, G. & M. RAITERI. 1986. Cholinergic terminals in rat hippocampus possess 5-HT_{1B} receptors mediating inhibition of acetylcholine release. Eur. J. Pharmacol. **129:** 333–337.
35. MAURA, G., E. ROCCATAGLIATA & M. RAITERI. 1986. Serotonin autoreceptor in rat hippocampus: Pharmacological characterization as a subtype of the 5-HT_1 receptor. Naunyn–Schmiedebergs Arch. Pharmacol. **334:** 323–326.
36. BOUHELAL, R., L. SMOUNYA & J. BOCKAERT. 1988. 5-HT_{1B} receptors are negatively coupled with adenylate cyclase in rat substantia nigra. Eur J. Pharmacol. **151:** 189–196.
37. SENWEN, K., I. MAGNALDO & J. PONYSSEGUR. 1988. Serotonin stimulates DNA synthesis in fibroblasts acting through 5-HT_{1B} receptors coupled to a G_i-protein. Nature **335:** 254–256.
38. STRATFORD, C. A., G. L. TAN, M. W. HAMBLIN & R. D. CIARANELLO. 1988. Differential inactivation and G protein reconsititution of subtypes of [^{3}H]-hydroxytryptamine binding sites in brain. Mol. Pharmacol. **34:** 527–536.
39. HOLT, S. E., M. COOPER & J. H. WYLLIE. 1986. On the nature of the receptor mediating the action of 5-hydroxytryptamine in potentiating responses of the mouse urinary bladder strip to electrical stimulation. Naunyn–Schmiedebergs Arch. Pharmacol. **334:** 333–340.
40. WYLLIE, J. H. 1989. Blockade by cyanopinodolol of 5-HT receptors in mouse urinary bladder. *In* Serotonin from Cell Biology to Pharmacology and Therapeutics. Serotonin Club, Florence, Italy 56.
41. MOLDERINGS, G. J., K. FINK, E. SCHLICKER & M. GÖTHERT. 1987. Inhibition of noradrenaline release via presynaptic 5-HT_{1B} receptors of the rat vena cava, Naunyn–Schmiedebergs Arch. Pharmacol. **336:** 245–250.
42. CARLSSON, A., W. KEHR, M. LINDQUIST, T. MAGNUSSON & C. ATACK. 1972. Regulation of monoamine metabolism in the central nervous system. Pharmacol. Rev. **24:** 371–384.
43. EUVRARD, C. & J. R. BOISSIER. 1980. Biochemical assessment of the central 5-HT agonist activity of RU24969 (A piperidinyl indole). Eur. J. Pharmacol. **63:** 65–72.
44. BRAZELL, M. P., C. A. MARSDEN, A. P. NISBET & C. ROUTLEDGE. 1985. The 5-HT_1 receptor agonist RU 24969 decreases 5-hydroxytryptamine (5-HT) release and metabolism in the rat frontal cortex in vitro and in vivo. Br. J. Pharmacol. **86:** 209–218.
45. SHARP, T., S. R. BRAMWELL & D. G. GRAHAME-SMITH. 1989. 5-HT_1 agonists reduce 5-hydroxytryptamine release in rat hippocampus in-vivo as determined by brain microdialysis. Br. J. Pharmacol. **96:** 283–290.
46. ROKOSZ-PELE, A., L. ANTKIEWICZ-MICHALUK & J. VETULANI. 1980. 5-Hydroxytryptamine–like properties of M-chlorophenyl piperazine: Comparison with quipazine. J. Pharm Pharmac. **32:** 220–222.
47. FULLER, R. W., H. D. SNODDY, N. R. MASON & B. B. MOLLOY. 1978. Effect of 1-(m-trifluoro methyl phenyl)-piperazine on ^{3}H-serotonin binding to membranes from rat brain in-vitro and on serotonin turnover in rat brain in vivo. Eur. J. Pharmacol. **52:** 11–16.
48. HJORTH, S., A. CARLSSON, P. LINDBERG, D. SANCHEZ, H. WIKSTROM, L. E. ARVIDSSON, U. HACKSELL & J. L. G. NILSSON. 1982. 8-hydroxy-2-(di-n-propylamino) tetralin, 8-OH-DPAT, a potent and selective simplified ergot cogener with central 5-HT receptor stimulating activity. J. Neural Trans. **55:** 169–188.
49. VAN DE KAR, L. D., S. A. LORENS, J. H. URBAN & C. L. BETHEA. 1989. Effect of selective serotonin (5-HT) agonists and 5-HT_2 antagonists on prolactin secretion. Neuropharmacology **28:** 299–305.
50. FULLER, R. W., H. D. SNOODY, N. R. MASON, S. K. HEMRICK-LUECKE & J. A. CLEMENS. 1981. Substituted piperazines as central serotonin agonists: Comparative specificity of the postsynaptic actions of quipazine and m-trifluoromethyl phenylpiperazine. J. Pharmacol. Exp. Ther. **218:** 636–641.
51. FULLER, R. W., H. D. SNODDY, N. R. MASON & J. E. OWEN. 1981. Disposition and pharmacological effects of m-chlorophenylpiperazine in rats. Neuropharmacology **20:** 155–162.
52. KOENIG, J. I., G. A. GUDELSKY & H. Y. MELTZER. 1987. Stimulation of corticosterone and B-endorphin secretion in the rat by selective 5-HT receptor subtype activation. Eur. J. Pharmacol. **137:** 1–8.
53. CARLSSON, M. & E. ERIKSSON. 1986. A central serotonin receptor agonist, 8-hydroxy-2-(di-n-propylamino) tetralin, has different effects on prolactin secretion in male and female rats. Acta Pharmacol. Toxicol. **58:** 297–302.

54. MUELLER, E. A., D. L. MURPHY & T. SUNDERLAND. 1986. Further studies of the putative serotonin agonist, m-chlorophenylpiperazine: Evidence for a serotonin receptor mediated mechanism of action in humans. Psychopharmacology. **89:** 388–391.
55. SINTON, C. M. & S. L. FALLON. 1988. Electrophysiological evidence for a functional differentiation between subtypes of the 5-HT_1 receptor. Eur. J. Pharmacol. **157:** 173–181.
56. SPROUSE, J. S. & G. K. AGHAJANIAN. 1987. Electrophysiological responses of serotoninergic dorsal raphe neurons to 5-HT_{1A} and 5-HT_{1B} agonists. Synapse **1:** 3–9.
57. SPROUSE, J. S. & G. K. AGHAJANIAN. 1988. Responses of hippocampal pyramidal cells to putative serotonin 5-HT_{1A} and 5-HT_{1B} agonists: A comparative study with dorsal raphe neurones. Neuropharmacology **27:** 707–715.
58. CHAPUT, Y., P. BLIER & C. DE MONTIGNY. 1986. In vivo electrophysiological evidence for the regulatory role of autoreceptors on serotonergic terminals. J. Neurosci. **6:** 2796–2801.
59. LUCKI, I. & A. FRAZER. 1985. Changes in behaviour associated with serotonin receptors following repeated treatment of rats with antidepressant drugs. *In* Behavioural Pharmacology: Current Status. L.S. Seiden & R.L. Balster, Eds. New York: Alan Liss, 339–357.
60. LUCKI, I. & A. FRAZER. 1982. Behavioural effects of indole and piperazine-type serotonin receptor agonists. Soc. Neurosci Abst. **8:** 101.
61. KENNETT, G. A. & G. CURZON. 1988. Evidence that MCPP may have behavioural effects mediated by central 5-HT_{1C} receptors. Br. J. Pharmacol. **94:** 137–147.
62. GREEN, A. R., A. P. GUY & C. R. GARDINER. 1984. The behavioural effects of RU 24969, a suggested 5-HT_1 receptor agonist in rodents and the effect on the behaviour of treatments with antidepressants. Neuropharmacology **23:** 655–661.
63. OBERLANDER, C., B. BLAQUIERE & J. F. PUJOL. 1986. Distinct functions for dopamine and serotonin in locomotor behaviour: Evidence using the 5-HT_1 agonist RU 24969 in globus pallidus lesioned rats. Neurosci. Lett. **67:** 113–118.
64. TRICKLEBANK, M. D., D. N. MIDDLEMISS & J. NEILL. 1986. Pharmacological analysis of the behavioural and thermoregulatory effects of the putative 5-HT_1 receptor agonist, RU 24969, in the rat. Neuropharmacology **25:** 877–886.
65. SUGRUE, M. F. 1987. Neuropharmacology of drugs affecting food intake. Pharmacol Ther. **32:** 145–182.
66. KENNETT, G. A. & G. CURZON. 1988. Evidence that hypophagia induced by MCPP and TFMPP requires 5-HT_{1C} and 5-HT_{1B} receptors; hypophagia induced by RU 24969 only requires 5-HT_{1B} receptors. Psychopharmacology. **96:** 93–100.
67. KENNETT, G. A., C. T. DOURISH & G. CURZON. 1987. 5-HT_{1B} agonists induce anorexia at a post-synaptic site. Eur. J. Pharmacol. **141:** 429–435.
68. DOURISH, C. T., P. H. HUTSON & G. CURZON. 1985. Low doses of the putative serotonin agonist 8-hydroxy-2-(di-n-propylamino) tetralin (8-OH-DPAT) elicit feeding in the rat. Psychopharmacology **86:** 197–204.
69. HUTSON, P. H., T. P. DONOHOE & G. CURZON. 1986. Neurochemical and behavioural evidence for an agonist action of 1-[2-(4-aminophenyl)ethyl]-4-(3-trifluoro methyl phenyl)piperazine (LY 165163) at central 5-HT receptors. Eur. J. Pharmacol. **138:** 215–223.
70. HUTSON, P. H., C. T. DOURISH & G. CURZON. 1988. Neurochemical and behavioural evidence for mediation of the hyperphagic action of 8-OH-DPAT by 5-HT by 5-HT cell body autoreceptors. Eur. J. Pharmacol. **129:** 347–352.
71. HUTSON, P. H., T. P. DONOHOE & G. CURZON. 1988. Infusion of the 5-hydroxytryptamine agonists RU 24969 and TFMPP into the paraventricular nucleus of the hypothalamus causes hypophagia. Psychopharmacology **95:** 550–552.
72. BERENDSEN, H. H. G. & C. L. E. BROEKKAMP. 1987. Drug induced penile erections in rats: Indications of serotonin 1B receptor mediation. Eur. J. Pharmacol. **135:** 279–287.
73. GREEN, A. R. & G. M. GOODWIN. 1987. The pharmacology of the hypothermic response of rodents to 8-OH-DPAT administration and the effects of psychotropic drug administration on this response. *In* Brain 5-HT_{1A} Receptors Behavioural and Neurochemical Pharmacology. C. T. Dourish, S. Ahlenius & P. H. Hutson, Eds. Chichester: Ellis Horwood, pp. 161–198.
74. MAJ, J., E. CHOJNACKA-WÔJCIK, A. KLODZINISKA, A. DERÊN & E. MORYL. 1988. Hypothermia induced by m-trifluoro methylphenyl piperazine or m-chlorophenyl piperazine: an effect mediated by 5-HT_{1B} receptors? J. Neural Trans. **73:** 43–55.

75. McKenney, J. D. & R. A. Glennon. 1986. TFMPP may produce its stimulus effects via a 5-HT_{1B} mechanism. Pharmacol. Biochem. Behav. **24:** 43–47.
76. Glennon, R. A., J. D. McKenney & R. Young. 1984. Discriminative stimulus properties of the serotonin agonist 1-(3-trifluoromethylphenyl) piperazine (TFMPP). Life Sci. **35:** 1475–1480.
77. Schechter, M. D. 1986. Tetrahydro-β-carboline may produce its stimulus effects via 5-HT_{1B} receptors. Pharmacol. Biochem. Behav. **28:** 1–6.
78. Cunningham, K. A. & J. B. Appel. 1986. Possible 5-hydroxytryptamine, (5-HT_1) receptor involvement in the stimulus properties of 1-(m-trifluoromethylphenyl) piperazine (TFMPP). J. Pharmacol. Exp. Ther. **237:** 369–377.
79. Glennon, R. A., M. E. Pierson & J. D. McKenney. 1987. Stimulus generalisation of 1-(3-trifluoromethylphenyl)piperazine (TFMPP) to propranolol, pindolol and mesulergine. Pharmacol. Biochem. Behav. **29:** 197–199.
80. Frances, H., C. Lienard & J. Fermanian. 1990. Improvement of the isolation-induced social behavioural deficit involves activation of the 5-HT_{1B} receptors. Prog. Neuro-Psychopharmacol. & Biol. Psychiatry **14:** 91–102..
81. Frances, H. 1988. New animal model of social behavioural deficit: Reversal by drugs. Pharmacol. Biochem. Behav. **29:** 467–470.
82. Hoyer, D., A. Pazos, A. Probst & J. M. Palacios. 1986. Serotonin receptors in the human brain. I. Characterisation and autoradiographic localisation of 5-HT_{1A} recognition sites. Apparent absence of 5-HT_{1B} recognition sites. Brain Res. **376:** 85–96.
83. Hoyer, D. & D. N. Middlemiss. 1989. The pharmacology of the terminal 5-HT autoreceptors in mammalian brain: Evidence for species differences. TIPS **10:** 130–132.
84. Sills, M. A., B. B. Wolfe & A. Frazer. 1984. Determination of selective and nonselective compounds for the 5-HT_{1A} and 5-HT_{1B} receptor subtypes in rat frontal cortex. J. Pharmacol. Exp. Ther. **231:** 480–487.
85. Peroutka, S. J. & A. Hamik. 1988. [^{3}H]quipazine labels 5-HT_3 recognition sites in rat cortical membranes. Eur. J. Pharmacol. **148:** 297–299.

DISCUSSION OF THE PAPER

Manfred Göthert (*University of Bonn*): Derek I've just one comment and one question. The comment concerns the Ciba-Geigy compound CGS 12066B which you said is not very selective with respect to 5-HT-$_{1B}$ receptors. It's also not very selective with respect to 5-HT_{1D} receptors because in our pig autoreceptor model and in Daniel Hoyer's binding data it was active.

My question concerns your correlation. You know that we also try to correlate our rat data in cooperation with Dr. Hoyer with binding data. We found that there was not only a good correlation of the functional data at the rat 5-HT autoreceptor with the 5-HT_{1B} binding sites but also with 5-HT_{1A}. Did you also find such a good correlation with 5-HT_{1A}? We could only discriminate between both subtypes by utilizing 8-OH-DPAT and ipsaperone.

Middlemiss: In my own work in the rat frontal cortex there's no correlation with the 5-HT_{1A} receptor binding.

Göthert: So you had a very good correlation only with 5-HT_{1B}.

Middlemiss: Yes.

Question: Derek, one of the questions I had related to the use of the hypothermia model as a model for the 1B receptor, unless I'm mistaken I think also serotonin 1A agonists produce hypothermia as well. Is that not true?

Middlemiss: Yes, that's correct.

Comment: With regard to the species specificity of the 5-HT_{1B} receptor, I think I've heard also recently that it's present in the opossum as well as the rat and the mouse.

MIDDLEMISS: That is correct. There is a report by Murphy and Byland (JPET **249:** 535, 1989). It's supposed to be in a lung fibroblast cell line derived from the Chinese hamster as well. There was a recent paper which also indicated that there was a 5-HT_{1B} site there (Sewen *et al.,* Nature **335:** 254, 1988).

J. PALACIOS (*Sandoz, Basel*): About 5-HT_{1B} and 5-HT_{1D} receptors—we've been looking at a variety of species and could find 5-HT_{1B} receptors only in rat and mouse. Now we know that cell line from the opossum kidney is linked mainly to cyclase and I think the data are pretty convincing that this is a 5-HT_{1B} site. Also, in the hamster there is a hamster lung fibroblast, CCL39, in which 5-HT is showing mitogenic effects. This is mediated by inhibition of cyclase, and this is a 5-HT_{1B} receptor. Now we've looked at pigeons, and they are showing 5-HT_{1D}, no 5-HT_{1B} receptors; guinea pig has 5-HT_{1D} receptors, as do pig, cat, calf, human, monkey. Everything else we've looked at is showing 5-HT_{1D}. I don't mean to say that there are no 5-HT_{1B} receptors where we are finding 5-HT_{1D} receptors and vice versa, but most of the effects are mediated by one class of receptors in each species. Mind you this is in the brain, maybe in the periphery you still have some 5-HT_{1D} and 5-HT_{1B} receptors going in parallel.

MIDDLEMISS: What can I say?

The 5-HT_{1C} Receptor

PAUL R. HARTIG

Neurogenetic Corporation
215 College Road
Paramus, New Jersey 07652

BETH J. HOFFMAN

Laboratory of Cell Biology
National Institutes of Mental Health
Bethesda, Maryland 20892

MARC J. KAUFMAN AND FUSAO HIRATA

Department of Environmental Health Sciences
Johns Hopkins School of Hygiene and Public Health
Baltimore, Maryland 21205

INTRODUCTION

Sometimes the best model tissue for study of a neurotransmitter receptor is not found in the nervous system. Convenience, availability and receptor density may dictate the choice of a model tissue expressing the same protein in a more easily studied environment. Two of the most intensively studied neurotransmitter receptors, the beta adrenergic receptor and the nicotinic cholinergic receptor, have risen to prominence primarily as a result of two convenient non-neural model tissues, the turkey erythrocyte and the Torpedo electric organ. In both cases, the availability of the receptor in relatively high densities and pure form in a useful model tissue greatly facilitated the molecular characterization of the receptor protein.

A similar story has developed in the serotonin receptor field for the 5-HT_{1C} receptor. This receptor was discovered in the choroid plexus, a highly active secretory and transport tissue in the ventricles of the brain, as the result of two independent investigations. Pazos *et al.*[1,2] discovered this receptor in porcine choroid plexus by use of [^{3}H]mesulergine while Yagaloff and Hartig[3] were pursuing similar binding studies with [^{125}I]LSD in rat choroid plexus. Both rat and pig choroid plexus have provided excellent model systems for study of the 5-HT_{1C} receptor, owing to their lack of significant levels of other serotonin receptors and a receptor density which reaches the extraordinary level of 3100 fmol/mg protein in the rat choroid plexus.[3] Even higher 5-HT_{1C} receptor densities are found in certain transgenic mice which develop choroid plexus tumors of epithelial cell origin.[4] These tumors typically range from 20–200 mg in size and exhibit 5-HT_{1C} receptor densities of 6600 fmol/mg protein. In contrast, other brain tissues display multiple subtypes of serotonin receptors, and the highest serotonin receptor density found elsewhere in the brain is only 560 fmol/mg protein.[5,6] Attempts to use cortical or other tissues as 5-HT_{1C} models have led to quite different results in different laboratories. In the choroid plexus, by contrast, excellent signal to background ratios can be obtained in both binding and second messenger assays (see below). Therefore the choroid plexus is the tissue of choice for defining the properties of the 5-HT_{1C} site.

RADIOLIGAND BINDING STUDIES

At this time, three iodinated radioligands and five tritiated radioligands have been used to label the 5-HT_{1C} receptor (TABLE 1). The iodinated ligands are especially useful in characterizing the 5-HT_{1C} receptor owing to their high specific activities and thus high sensitivities in detecting the low levels of 5-HT_{1C} receptors found in most brain tissues. One of these iodinated ligands, [^{125}I]MIL, has been used to label the 5-HT_{1C} receptor in both membrane-bound and solubilized forms.[7]

The highest affinity radioligand for the 5-HT_{1C} receptor was recently reported[8] to be [^{125}I]SCH 23982, which was formerly thought to be a selective D1 dopamine receptor ligand.[9] This ability of a D1 receptor ligand to strongly cross-react with the 5-HT_{1C} receptor is a new, unexpected observation which raises intriguing questions about possible structural similarities between these two receptors. Unfortunately, little is known about the D1 dopamine receptor affinities of most other 5-HT_{1C} receptor ligands, so it is difficult to know if this cross reactivity will prove to be a general feature of 5-HT_{1C}-selective compounds.

TABLE 1. Affinities and Selectivities of Radioligands Commonly Used to Label the 5-HT_{1C} Receptor[a]

	Affinity (in nanomolar)			
Ligand	5-HT_{1C}	5-HT_2	Other Important Sites	References
[^{125}I]SCH23982	0.19		D1 dopamine: 0.7	8, 9
[^{125}I]MIL	0.41	0.14	D2 dopamine: 8.2	7, 63
[^{125}I]LSD	0.55–3.5	0.9–1.5	D2 dopamine: 9.1-40	3, 4, 11, 64–66
[^{3}H]mesulergine	1.1	1.9		1, 2
[^{3}H]serotonin	6.5	296	$5\text{-HT}_{1A,B,D}$: 2.2-11	1, 2, 67, 68
[^{3}H]LSD	9.7	2.5	5-HT_{1A}: 3.9	1, 2, 67, 68
[^{3}H]mianserin	1.0	2.4	H1 histamine: 3.3	69, 70
[^{3}H]SCH 23390	24.0	7.0	D1 dopamine: 0.4	12, 71

[a]All affinities are dissociation constants (K_d values) or inhibition constants (K_i values for the non-radioactive compound) at the indicated receptor subtype.

All radioligands which have been used to study the 5-HT_{1C} receptor, with the exception of the agonist [^{3}H]5-HT, exhibit strong cross-reactivity with the 5-HT_2 receptor (TABLE 1). Interestingly, this cross-reactivity is species dependent, since [^{3}H]mesulergine does not bind to porcine or human 5-HT_2 receptors but does strongly cross-react with 5-HT_2 receptors in rat brain tissue.[2] This cross-reactivity is a reflection of the many structural and functional similarities (see below) between the 5-HT_{1C} and 5-HT_2 receptor proteins. Discovery of selective, high affinity compounds which discriminate between the 5-HT_{1C} and 5-HT_2 receptors remains an elusive but important target for future research.

Since highly selective radioligands for the 5-HT_{1C} receptor are not available, investigators must rely on pharmacological profiles obtained with a variety of ligands in order to characterize this site. Among the most useful compounds for identifying "5-HT_{1C}-like" binding sites are the high affinity ligands: SCH 23982 ($pK_D = 9.73$), metergoline ($pK_D = 9.29$), mesulergine ($pK_D = 8.79$), ritanserin ($pK_D = 8.64$), and mianserin ($pK_D = 8.00$).[8,10,58] The observation that these high affinity ligands show potent effects in behavioral, physiological or binding assays while ligands such as 5-HT, spiperone and ketanserin show low to moderate affinities, provides at the present time the best evidence for the involvement of a "5-HT_{1C}-like" receptor in these actions.

Also, mCPP can serve as a useful agonist for the 5-HT$_{1C}$ receptor, when used with care to discriminate against its antagonist action at the 5-HT$_2$ receptor.[61] The difficult problem of separating 5-HT$_2$ from 5-HT$_{1C}$ receptor actions is best addressed by use of the most selective ligands currently available, 5-HT and spiperone, along with a full pharmacological profile using less selective, but high affinity antagonists. Special care must be exercised if mesulergine, methysergide or cyproheptadine are used in these experiments since these particular compounds show strong species differences in their affinities for 5-HT$_2$ and 5-HT$_{1C}$ receptors.[2,11] In addition, there is a large discrepancy between the pK$_D$ of SCH 23390 (the chlorinated form of SCH 23982) in 5-HT$_{1C}$ binding assays (pK$_D$ = 7.6)[12] verses response assays (pEC$_{50}$ = 5.7)[13] where it shows a much lower apparent affinity. The source of this affinity difference is unclear and deserving of further study.

SECOND MESSENGER COUPLING

Although a clear pharmacology for the 5-HT$_{1C}$ binding site soon emerged from early investigations, this site could not be considered biologically relevant until a biochemical or physiological response was found which coupled specifically to receptor activation. Early work by two groups suggested that serotonin might have a role in stimulating the formation of cAMP in choroid plexus. High concentrations of 5-HT evoked production of this second messenger in rabbit and bovine choroid plexus, although in bovine tissue cAMP production was blocked by histamine receptor antagonists.[14,15] However, involvement of the 5-HT$_{1C}$ site in this choroid plexus cAMP response was ruled out by the work of Palacios *et al.*[16] who demonstrated that no 5-HT$_{1C}$-dependent response could be detected.

Conn, *et al.*[17] chose to investigate a different potential second messenger coupling of the 5-HT$_{1C}$ receptor, namely stimulation of phosphoinositide hydrolysis. This group demonstrated that serotonin can stimulate a large increase in phosphoinositide (PI) turnover in rat choroid plexus, with pharmacological properties in agreement with the binding properties of the 5-HT$_{1C}$ receptor. The 5-HT$_{1C}$ receptor in porcine choroid plexus is also coupled to PI turnover, and the magnitude of the PI response is much greater than that observed in rat tissue.[13,18] PI turnover is not dependent on the activity of lipoxygenase or cyclooxygenase pathways, nor is it dependent on neurotransmitter release, suggesting that a direct coupling of the receptor to the enzyme phospholipase C (PLC) mediates the PI response.[19] In general, the relative efficacies of agonists at this site are indolalkylamines > phenylpiperazines > ergolines.[19,20] The ergot LSD is a partial agonist whose agonist activity is completely eliminated by halogenation at the two position,[18,20] while the hallucinogen DOM is a weaker, but more efficacious partial agonist.[21] The physiological function of PI turnover in this tissue is not clear, although it appears that the receptor and PI pathways are tonically activated by serotonin since lesioning of serotonergic fibers with the neurotoxin 5,7-dihdroxytryptamine results in a supersensitivity of the PI response.[22] The 5-HT$_{1C}$ receptor has a high affinity for the putative dopamine D1 ligand SCH 23390, and the hallucinogen DOM, both of which act as partial agonists for PI turnover.[21,13]

Kaufman *et al.*[22a] have recently demonstrated that the 5-HT$_{1C}$ receptor of porcine choroid plexus is also linked to another second messenger response, elevation of cGMP levels. As shown in FIGURE 1, guanylate cyclase activity reaches a peak of approximately 5-fold basal levels at 5 minutes following addition of 5-HT to porcine choroid plexus. The time course of IP$_3$ accumulation in response to serotonin (co-plotted in FIG. 1) is similar to the time course of cGMP production, suggesting a possible link between the two pathways. Cyclic GMP elevations in response to 1μM 5-HT are not mediated by muscarinic, adrenergic or dopaminergic receptors, as shown by the lack of effect of

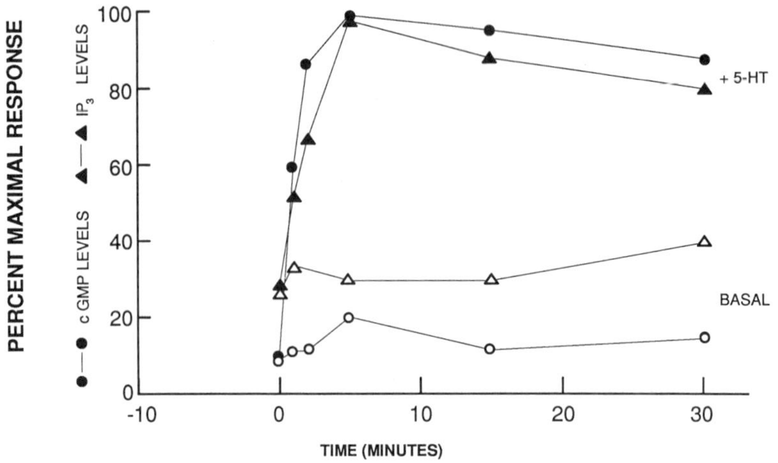

FIGURE 1. Time course of serotonin-stimulated cGMP and IP_3 formation in pig choroid plexus. Tissue was incubated in the presence (filled symbols) or absence (open symbols) of 1 μM serotonin for the indicated times and cGMP or IP_3 was assayed as previously described.[18,22a] Data are presented as mean values from 2–5 separate determinations, each in triplicate.

atropine, propranolol or low concentrations of spiperone in FIGURE 2; and the pharmacological properties of this response indicate that cGMP formation is mediated by the 5-HT_{1C} receptor[22a] (FIG. 2).

The phosphoinositide hydrolysis and guanylate cyclase activities of the 5-HT_{1C} receptor appear to be linked as sequential processes, with the linkage mediated by several pathways (FIG. 3). Cyclic GMP formation requires the presence of extracellular calcium[22a]; however, voltage-dependent calcium channels are not involved since the antagonists D-600 and nifedipine do not block cGMP formation.[22a] Extracellular calcium influx may be regulated by intracellular calcium (released by PI_3) as has been demonstrated in JURKAT cells.[23] The resulting intracellular calcium elevation may stimulate the enzyme phospholipase A2 (PLA2), which cleaves arachidonic acid from membrane lipids. Inhibition of PLA2 with *p*-bromophenacyl bromide substantially inhibits cGMP formation, suggesting that this phospholipase contributes to the formation of arachidonic acid, and to stimulation of the soluble form of guanylate cyclase (FIG. 3). It appears that arachidonic acid liberation is required for cGMP increases because blockade of the arachidonate metabolizing lipoxygenase pathway with BW 755C abolishes cGMP formation.[22a] Hydroxylated arachidonated (HETEs) and/or leukotrienes may be formed by this pathway and can stimulate the soluble form of guanylate cyclase to enhance cyclic GMP levels.[24]

The 5-HT_{1C} receptor now joins a growing family of receptors, including adrenergic and muscarinic receptors[25,26] which have been shown to couple to both PI turnover and guanylate cyclase activation. Kaufman *et al.*[22a] have provided several lines of evidence that PLA2-mediated arachidonic acid release contributes to cGMP production by the 5-HT_{1C} receptor. Other pathways might also contribute to arachidonate release, such as sequential activation of PLC and di- and mono-acylglycerol lipases. Preliminary analysis of release of arachidonate from lipids after serotonin stimulation suggests that both pathways contribute to arachidonate formation in the choroid plexus (unpublished data). This supports the hypothesis that multiple routes of liberation of arachidonate are involved in the 5-HT_{1C} receptor-mediated cyclic GMP responses.

PHYSIOLOGICAL FUNCTIONS IN THE CHOROID PLEXUS

The choroid plexus produces and secretes the majority of the cerebrospinal fluid (CSF) of the brain, and also acts as a bidirectional transport system between blood and brain compartments.[27] As shown in FIGURE 4, the secretory epithelial cells which are responsible for these activities form a monolayer at the surface of this tissue, with their apical surfaces in direct contact with the cerebrospinal fluid (CSF) of the venticles. The blood brain barrier is unusual in this tissue since the capillaries of the choroid plexus contain leaky junctions between individual capillary endothelial cells, which allow components of blood plasma to freely leak out and bathe the basal and lateral surfaces of epithelial cells. The epithelial cells, however, are joined by tight junctions which prevent transfer of plasma components directly to the CSF.[27] Thus, these tight junctions and the epithelial cells of the choroid plexus serve as a blood-CSF barrier in this brain region. Since this unique blood-CSF barrier is controlled primarily by the transport properties of these choroid plexus epithelial cells, determination of the cellular localization of the choroid plexus 5-HT$_{1C}$ receptor has been important for determining possible physiological roles of this receptor system.

High resolution autoradiographic labeling studies have shown that the 5-HT$_{1C}$ receptor is localized to the epithelial cell layer of this tissue.[3] This assignment has been confirmed by the localization of 5-HT$_{1C}$ receptors in choroid plexus tumors of epithelial cell origin,[4] by studies on isolated epithelial cells in culture,[28] and by '*in situ*' hybridization studies with 5-HT$_{1C}$ receptor mRNA.[29] Preliminary studies in which the choroid plexus was labeled by [^{3}H]5-HT from the blood side or from the CSF side,[30]

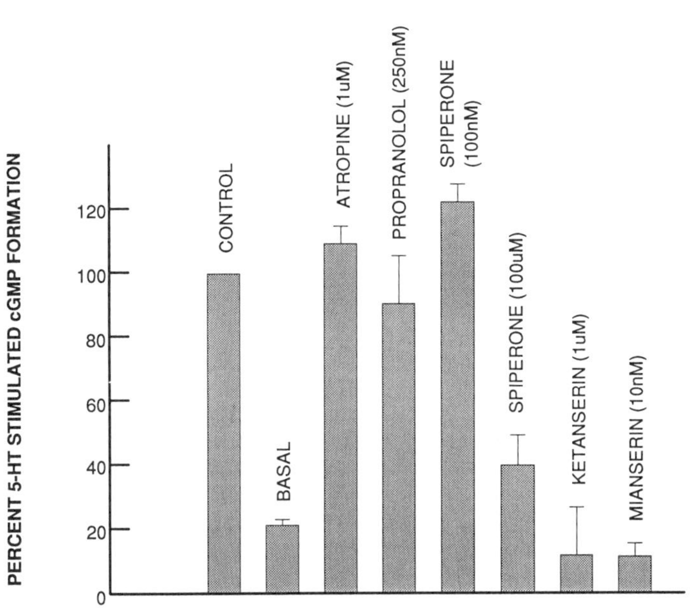

FIGURE 2. Effects of antagonists on cGMP formation by the pig choroid plexus. Samples were preincubated in the indicated concentration of each inhibitor for 20 minutes followed by addition of 1μM serotonin for 5 minutes. Data are presented as means and standard errors from 3–6 separate determinations.

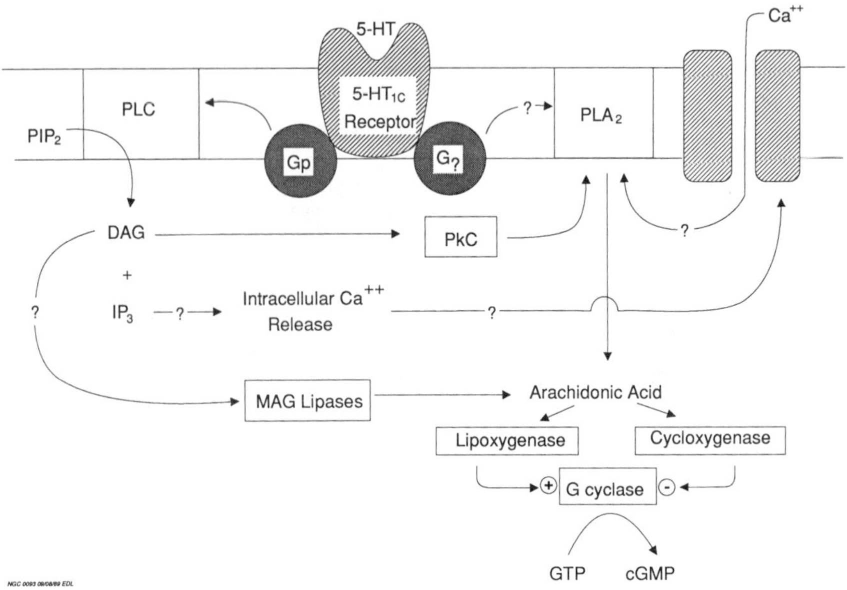

FIGURE 3. Proposed second messenger pathways of the 5-HT_{1C} receptor. Binding of 5-HT to the 5-HT_{1C} receptor activates G proteins (G_p and possibly others) which stimulates hydrolysis of phosphatidylinositol bisphosphate (PIP_2) by phospholipase C (PLC) and may directly activate phospholipase A_2 (PLA2). The phosphoinositide hydrolysis products diacylglycerol (DAG) and inositol triphosphate (IP_3) in turn activate protein kinase C (PkC) and stimulate intracellular calcium release. Intracellular calcium may trigger extracellular calcium influx, which along with PkC stimulates arachidonic acid release via PLA_2. DAG may also lead directly to arachidonic acid release via the action of DAG lipase. Arachidonic acid regulates the soluble form of guanylate cyclase (G cyclase) via both lipoxygenase and cyclooxygenase pathways.

suggest that the 5-HT_{1C} receptor is located on the apical face of this cell layer, in direct contact with CSF. Thus, this receptor appears to respond to CSF-borne rather than blood-borne serotonin. Since the serotonergic innervation of the choroid plexus is very sparse, with most fibers terminating in the vasculature (see ref. 3 for discussion), it would appear that serotonin is released into the CSF at a remote site, then carried to the choroid plexus. The dense supra-ependymal serotonergic fiber network, which does not form synaptic contacts with any known serotonin receptor sites, is a likely candidate for the origin of this CSF-borne serotonin.[3] In support of this view Conn *et al.*[22] have shown that intracerebroventricular injection of the serotonin neurotoxin 5,7 DHT (which would effectively lesion supra-ependymal fibers) causes supersensitivity of 5-HT_{1C} receptor-mediated second messenger responses in the choroid plexus.

The role of the 5-HT_{1C} receptor in modulating CSF production remains unclear and is complicated by the fact that 5-HT can exert strong effects on blood flow to the choroid plexus.[31] In choroid plexus tissue, agents which elevate cAMP levels tend to increase CSF formation[32] and the observation that 5-HT_{1C} receptor activation does not enhance cAMP levels[16] suggests that this receptor is not likely to play a role in enhancing CSF formation. Some evidence suggests that the 5-HT_{1C} receptor may, however, inhibit this pathway. Recent work in rabbits has shown that CSF-borne serotonin (which could directly activate the 5-HT_{1C} receptor) inhibits CSF formation.[33] Little information is available on the pharmacological profile of this response, and the optimal dose of serotonin is much higher than that of other 5-HT$_{1C}$ receptor-mediated responses. Another possible link between the 5-HT$_{1C}$ receptor and inhibition of CSF production comes from the guanylate cyclase activity of this receptor. Two other receptors which increase cGMP levels in choroid plexus (muscarinic and atrial natriuretic peptide receptors) also inhibit CSF formation.[34–37] Cyclic GMP inhibits $Na^+/K^+/2Cl^-$ cotransport in teleost intestine[38] and inhibits amiloride sensitive sodium flux in LLC-FK1 cells.[39] Protein kinase C (activated by the phosphoinositide hydrolysis pathway) has also been shown to modulate these two transport enzymes in other tissues.[40–42] Since sodium flux from blood to CSF is required for CSF formation and secretion, both cGMP and phosphoinositide second messengers might be able to inhibit CSF formation via modulation of these sodium conductive pathways.

The 5-HT$_{1C}$ receptor has recently been shown to regulate the synthesis and secretion of transferrin,[28] an iron carrier protein of the choroid plexus which transports this metal ion from blood into the CNS. Since iron is required for the growth and maintenance of neuronal and other cell types, and since the 5-HT$_{1C}$ receptor appears to be tonically activated by CSF-borne serotonin (above), this receptor may play an important role in the maintenance of brain cell function and growth. The fact that the 5-HT$_{1C}$ receptor shows a rapid decrease in density over the 2 days following birth of a rat pup, which is followed by a rapid rise to peak adult levels at 2 weeks[43] may provide a mechanism for regulating iron or other transport activities during the switch from fetal to adult hemoglobin. This developmental pattern differs substantially from that of other serotonin receptor subtypes[43] and may indicate a unique role of this receptor in the control of nervous system development.

MOLECULAR CLONING AND STRUCTURE

Numerous groups have used *Xenopus laevis* oocytes to study neurotransmitter receptors and ion channels. Following injection of mRNA from electrically excitable

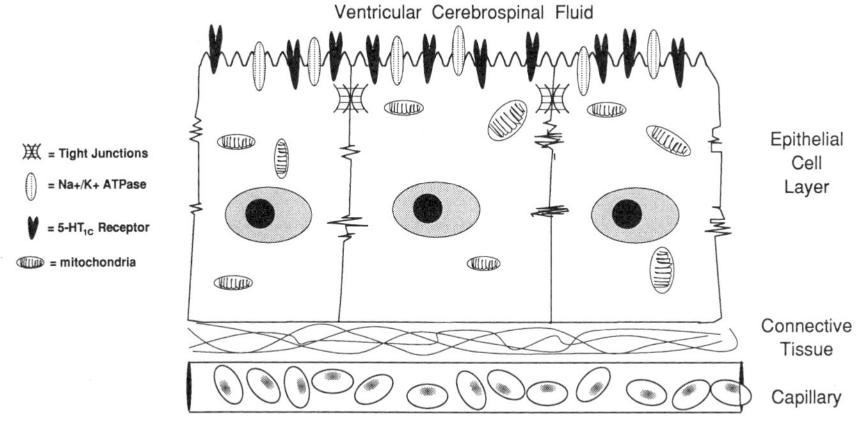

FIGURE 4. Structure of the choroid plexus. Epithelial cells of the choroid plexus form a monolayer joined by tight junctions. The apical face of these cells, which appears to contain both the Na^+/K^+ ATPase and 5-HT$_{1C}$ receptor, is in direct contact with cerebrospinal fluid, while the basal and lateral faces are bathed by plasma from blood capillaries. Thus, transport processes across these epithelial cells form critical components of the blood-CSF barrier.

tissue, membrane proteins are expressed on the oocyte surface and measured either electrophysiologically or biochemically. Gundersen, Parker and Miledi[44] have shown that rodent brain mRNA induces a 5-HT-mediated response in *Xenopus* oocytes. Lübbert and co-workers[45] characterized this 5-HT binding site as the 5-HT_{1C} receptor. This receptor couples to an oocyte guanine nucleotide binding protein (G-protein) and stimulates a Ca^{++}-activated Cl^- conductance increase via PI hydrolysis.[46] Interestingly, the oocyte serotonin response is sensitive to pertussis toxin treatment[46] but the PI response in rat choroid plexus is not.[20] A single 5 to 6 kilobase size class of rat brain mRNA is sufficient to generate the 5-HT response in oocytes and is particularly enriched in the choroid plexus, followed by hippocampus and striatum.[46–48] This oocyte expression system facilitated the cloning of the 5-HT_{1C} receptor by providing a sensitive bioassay.

Using this mRNA bioassay, methods were developed for the enrichment of mRNAs for low abundance receptors and ion channels, for the construction of libraries with cDNA directionally oriented in a single-stranded vector, and for the screening of libraries by hybrid depletion.[48] Using this methodology, partial cDNA clones for the 5-HT_{1C} receptor were isolated from a mouse choroid plexus tumor[4] cDNA library.[48] Using the same bioassay, Julius and colleagues[49] constructed a cDNA library such that a copy of the mRNA could be made *in vitro* and injected into oocytes. This approach requires a functional receptor, which facilitated the isolation of a rat choroid plexus cDNA with a complete protein coding region.[49] The amino acid sequence of this rat 5-HT_{1C} receptor is shown in FIGURE 5. This receptor is a member of the G protein-coupled receptor superfamily, with seven putative transmembrane domains. Other members of this receptor superfamily include β- and α- adrenergic, muscarinic, dopamine, substance K, and tachykinin receptors.[50] The two 5-HT_{1C} receptor clones reported so far, from rat[49] and mouse,[51] differ in length by one amino acid (460 and 459 amino acids, respectively) and share 97% homology at the amino acid level.

In general, the highest regions of homology between members of the G protein-coupled receptor superfamily are the transmembrane domains, which contain the ligand binding site of these receptors.[50] The amino- and carboxy- termini, and the large third cytoplasmic loop tend to be highly divergent. As shown in FIGURE 5, clusters of acidic and basic residues tend to mark the boundaries between transmembrane regions and cytoplasmic or extracellular loops of the 5-HT_{1C} receptor. These boundary charges may well interact with charged phospholipid head groups.[50] The third cytoplasmic loop and the carboxy-terminus, which are thought to be important for G-protein interactions,[50] contain several clusters of acidic and basic residues. These clusters may interact with phospholipid head groups or may form docking domains for the G-proteins.[52] The wasp venom mastoparan has similar clusters of positively charged residues, which can activate G proteins.[53]

The seven transmembrane domains of the 5-HT_{1C} receptor contain only two acidic and two basic residues (FIG. 5). The aspartic acid residue (Asp-118) of transmembrane region II (TM2) is in a position where it might form an ion-pair bond with either of the two basic residues, His-145 of TM3 or Lys-175 of TM4. The other acid residue, Asp-100 of TM3, does not appear to be in a position to neutralize its charge by bonding to a nearby basic residue. This residue (Asp-100) may well bind the amino group of serotonin and form one component of the ligand binding site, since this residue is conserved in all G protein-coupled receptors that bind monoamine ligands. Strader *et al.*[52] have proposed that this aspartic acid residue of TM3 serves just such a role in the adrenergic receptors.

Following the first structural description of the 5-HT_{1C} receptor, the amino acid sequences of two other 5-HT receptor subtypes, 5-HT_{1A} and 5-HT_2, have been described. Both of these subtypes were cloned by homology to other G-protein coupled receptors, the 5-HT_{1A} receptor using a probe from the β-adrenergic receptor,[54] and the

5-HT_2 receptor using oligonucleotides based on the 5-HT_{1C} receptor.[55] Comparison of the amino acid sequences of these three 5-HT receptors indicates that the 5-HT_{1C} receptor is more closely related to the 5-HT_2 than to the 5-HT_{1A} receptor (FIG. 6). Based on mutagenesis studies of the adrenergic receptors, the transmembrane domains are important in agonist and antagonist binding.[50,52] Since these hydrophobic regions are relatively conserved among members of this superfamily, homology in these domains may reflect similar function, i.e., drug binding.[56,50] The 5-HT_{1C} and 5-HT_{1A} receptors exhibit only 39% homology in the transmembrane domains. The highest degree of homology between the 5-HT_{1C} and 5-HT_{1A} receptors is in TM3 (16 of 25 residues) while the least conserved region, TM1, only shows 6 out of 25 amino acids conserved. The 5-HT_{1A} receptor is similar to the adrenergic receptor family in amino

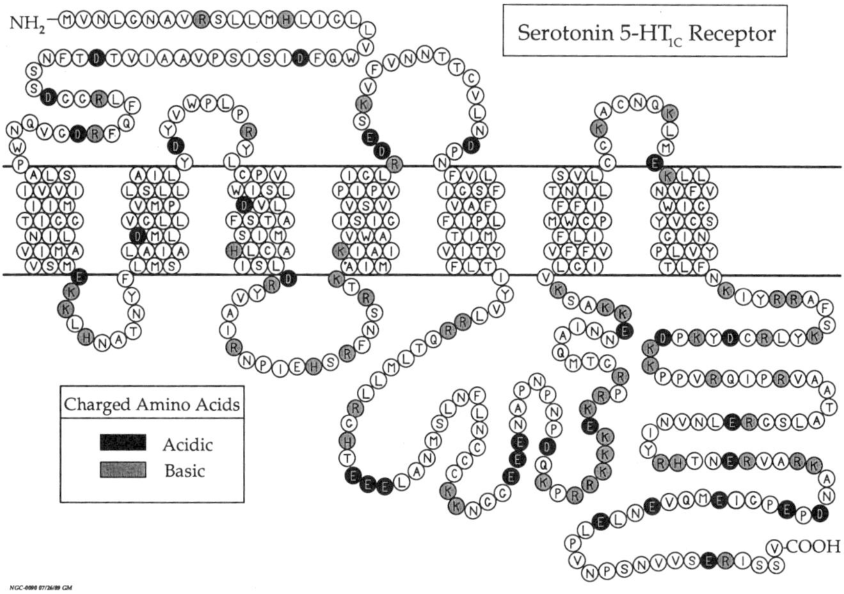

FIGURE 5. Amino acid sequence of the 5-HT_{1C} receptor. The sequence of the rat 5-HT_{1C} receptor[49] is shown with acidic and basic amino acid residues highlighted and proposed transmembrane segments indicated. Note the presence of only two acidic and two basic residues within the membrane bilayer, and the frequent presence of clusters of charged residues at the termination of transmembrane segments, within the third intracellular loop and within the COOH terminal tail.

acid sequence and in genomic structure, and in several respects resembles these receptors more than the serotonin 5-HT_2 or 5-HT_{1C} receptor.[54,57] The β-adrenergic, muscarinic and 5-HT_{1A} receptors are all characterized by their lack of introns in the coding region of the gene.

The 5-HT_{1C} and 5-HT_2 receptors are 51% homologous overall and show 80% identity in the transmembrane regions (FIG. 6). Within specific domains, TM5 is identical between 5-HT_{1C} and 5-HT_2 while TM3 shows the next highest degree of homology (22 of 25 amino acids) with the lowest homology in TM4 (15 of 24 residues). In addition to this amino acid homology, the 5-HT_{1C} and 5-HT_2 receptors share similar pharmacology (high affinity for many classical 5-HT antagonists and hallucinogens) and coupling to the same second messenger system (phosphoinositide hydrolysis).[58]

Also, both the 5-HT$_{1C}$ and the 5-HT$_2$ receptors have introns in the coding regions of their respective genes.[51] On the basis of these facts, we and others,[55,57,58] have suggested that the 5-HT $_{1C}$ receptor should be considered to be a member of the 5-HT$_2$ receptor family.

The rat 5-HT$_{1C}$ receptor cDNA, when expressed in a mammalian fibroblast cell line (NIH 3T3), exhibits the expected pharmacology and mediates a marked increase in intracellular free Ca^{++} in response to 5-HT.[49] Expression of functional 5-HT$_{1C}$

Transmembrane Segment Number	Receptor	Sequence
	5-HT$_2$:	S A L L T T V V I I L - T I A G N I L V I M A V S S
I	5-HT$_{1C}$:	P A L S - I V V I I I M T I G G N I L V I M A V S M
	5-HT$_{1A}$:	T S L L - L G T L I F C A V L G N A C V V A A I A L
	5-HT$_2$:	F L M S L A I A D M L L G F L V M P V S M L T I L
II	5-HT$_{1C}$:	F L M S L A I A D M L V G L L V M P L S L L A I L
	5-HT$_{1A}$:	L I G S L A V T D L M V S V L V L P M A A L Y Q V
	5-HT$_2$:	L C A I W I Y L D V L F S T A S I M H L C A I S L
III	5-HT$_{1C}$:	L C P V W I S L D V L F S T A S I M H L C A I S L
	5-HT$_{1A}$:	L C D L F I A L D V L C C T S S I L H L C A I A L
	5-HT$_2$:	A F L K I I A V W T I S V G I S M P I P V F G L
IV	5-HT$_{1C}$:	A I M K I A I V W A I S I G V S V P I P V I G L
	5-HT$_{1A}$:	A - L - I S L T W L I G P L I S I P - P M L G W
	5-HT$_2$:	F V L I G S - F V A F F I P L T I M V I T Y F L T I
V	5-HT$_{1C}$:	F V L I G S - F V A F F I P L T I M V I T Y F L T I
	5-HT$_{1A}$:	G Y T I Y S T F G A F Y I P L L L M L V L Y G R I F
	5-HT$_2$:	V L G I V F F L F V V M W C P F F I T N I M A V I
VI	5-HT$_{1C}$:	V L G I V F F V F L I M W C P F F I T N I L S V L
	5-HT$_{1A}$:	T L G I I M G T F I L C W L P F F I V A L V L P F
	5-HT$_2$:	L L N V F V - W I G Y L S S A V N P L V Y T L F
VII	5-HT$_{1C}$:	L L N V F V - W I G Y V C S G I N P L V Y T L F
	5-HT$_{1A}$:	L L G A I I N W L G Y S N S L L N P V I Y A Y F

FIGURE 6. Comparison of the transmembrane regions of the 5-HT$_{1A}$, 5-HT$_{1C}$, and 5-HT$_2$ receptors. Although the exact length of each transmembrane region is uncertain at this time, the length of each segment was chosed to be 24–26 amino acids based on hydrophobicity analysis, and gaps were inserted to maximize homologies.

receptors in NIH 3T3 cells resulted in transformed foci dependent on the continual presence of 5-HT.[59] Foci formation is blocked by the antagonist mesulergine. Introduction of cells from transformed foci into nude mice resulted in the generation of tumors.[59] Julius and co-workers concluded that the 5-HT$_{1C}$ receptor serves as a protooncogene in these fibroblasts. Future experiments will show whether other receptors which increase intracellular Ca^{++} and/or activate protein kinase C may also regulate the growth of NIH 3T3 cells. In CHO fibroblasts, increased PI turnover mediated by muscarinic receptors inhibits growth.[60] Thus, the same receptor may have different physiological effects in different environments.

DISTRIBUTION AND ROLES IN THE CNS

The distribution of 5-HT_{1C} receptors in rat and human brain has been determined by membrane binding assays and by receptor autoradiography. The 5-HT_{1C} receptors from both species exhibit similar pharmacology and anatomical distribution.[5,11] Choroid plexus has tenfold higher site density than other brain regions. Rat brain areas reported to have 5-HT_{1C} receptors are: claustrum, laminae III-IV of frontal cortex, lamina IV of the frontoparietal somatosensory cortex, lamina IV of the striate cortex, CA1 field of hippocampus, mammillary nucleus, central medial thalamic nucleus, lateral amydaliod nucleus, nucleus of the solitary tract, anterior olfactory nucleus and olfactory tubercle.

In addition to quantitating the location of 5-HT_{1C} receptor protein by ligand binding assays, the location of 5-HT_{1C} receptor mRNA (which should be found primarily in cell bodies) can be identified by RNA blot analysis following size fractionation of RNA extracted from brain regions (Northern blots). The choroid plexus has the highest levels of mRNA with lower levels in basal ganglia, pons-medulla, hippocampus and hypothalamus. Very little or no 5-HT_{1C} receptor mRNA was detected in cortex, olfactory bulb or cerebellum.[49]

In situ hybridization histochemistry (ISHH) is a technique which preserves the morphology of the brain and consequently provides cellular resolution and increased sensitivity not possible by RNA blot analysis. Using oligonucleotides specific for the 5-HT_{1C} receptor, Hoffman and Mezey[29] have described the distribution of this receptor mRNA in rat brain. Receptor mRNA is co-localized in most of the regions which also display specific binding, yet many more areas have 5-HT_{1C} transcripts than display detectible levels of receptor binding (TABLE 2). 5-HT_{1C} receptor mRNA is present in epithelial cells of the choroid plexus and in numerous neurons throughout the CNS. With the exception of the choroid plexus, the highest mRNA levels are in the limbic system: hippocampus, septum, amygdata, olfactory nuclei, endopyriform nuclei, cingulate cortex, and pyriform cortex. Receptor transcripts are also present in the hypothalamus, numerous motor nuclei and the subthalamus. In addition, 5-HT_{1C} receptor mRNA is found in catecholaminergic and serotonergic cell areas. Interestingly, the level of mRNA in many cells is comparable to levels found in choroid plexus epithelial cells. A comparison of 5-HT_{1C} receptor protein distribution (from ligand binding assays) and receptor mRNA distribution (from ISHH) is shown in TABLE 2. An exact match in these regional distributions would not be expected since cell bodies containing the receptor mRNA may be spatially separated from synaptic regions containing the receptor protein.

These observations on the regional distribution of the 5-HT_{1C} receptor suggest that it may mediate a number of the central effects of 5-HT. For instance, 5-HT_{1C} receptors in the limbic system may affect mood, behavior and hallucinogenesis while hypothalamic 5-HT_{1C} receptors may influence sleep, appetite, thermoregulation, sexual behavior and neuroendocrine function. In addition, 5-HT_{1C} receptors in neurons of motor nuclei may modulate motor activity. In fact, recent studies suggest that 5-HT_{1C} receptors mediate hypoactivity and decreased feeding in rats.[61] Also, a strong anxiogenic effect of 5-HT_{1C} receptor agonists was noted in recent rat social interaction tests.[62] Development of more specific 5-HT_{1C} receptor drugs as well as physiological manipulations of 5-HT_{1C} receptor mRNA should contribute greatly to our growing understanding of possible physiological roles of this receptor subtype.

TABLE 2. Regional Distribution of 5-HT_{1C} Receptor mRNA and Binding Sites[a]

	mRNA	Binding Site (fmol/mg)
Telencephalon		
Rhinencephalon		
Olfactory bulb (olfactory nuclei)	++	—
anterior olfactory nucleus	++	204
Olfactory tubercle	++	180
Nucleus of the diagonal tract	++	—
Hippocampus		
CA1	+	167
CA2	+++	—
CA3	+++	—
subiculum	+++	—
dentate gyrus	+++	—
Cerebral cortex		
Frontopolar cortex	+	—
frontal laminae III-IV	—	102
frontal parietal somatosensory lamina IV	—	84
Cingulate cortex	++	—
Pyriform cortex	++	—
Striate cortex lamina IV	—	51
Basal ganglia		
Caudate-putamen	single cells	—
Globus pallidus	single cells	—
Nucleus accumbens	++	—
Claustrum	+ to ++	185
Endopyriform nucleus	+++	—
Septum		
Dorsal nucleus	++	—
Lateral nucleus	++	—
Medial nucleus	+	—
Triangular nucleus	+	—
Subfornical organ	+++	—
Amydala		
Central amygdaloid nucleus	++ to +++	—
Basal amygdaloid nucleus	++ to +++	—
Lateral amygdaloid nucleus	+	179
All other amygdaloid nuclei	+	—
Bed nucleus of the stria terminalis	++	—
Diencephalon		
Thalamus	++	—
Periventricular nucleus	++	—

TABLE 2 Continued

	mRNA	Binding Sites (fmol/mg)
Reuniens nucleus	++	
Centromedial nucleus	++	134
Reticular formation	++	—
Epithalamus		
Habenula		
medial	+	—
lateral	+++	—
Metathalamus		
Lateral geniculate body	++	—
Medial geniculate body	++	—
Subthalamus		
Zona incerta	++	—
Entopeduncular nucleus	+++	—
Preoptic region		
Medial preoptic nucleus	+	—
Lateral preoptic nucleus	+	—
Hypothalamus		
Anterior hypothalamic nucleus	++	—
Suprachiasmatic nucleus	++	—
Paraventricular nucleus	++	—
Dorsomedial nucleus	+++	—
Median forebrain bundle	++	—
Medial mamillary nucleus	—	69
Mesencephalon		
Ventral tegmental area	+	—
Periaqueductal grey	+	—
Edinger-Westphal nucleus (oculomotor nucleus)	++	—
Dorsal raphe nucleus	++	—
Linear raphe nucleus	+++	—
Interpeduncular nucleus	+	—
Red nucleus	+	—
Cuneiform nucleus	+	—
Colliculus superior	++	—
Colliculus inferior	++	—
Substantia nigra	++	—
Catecholaminergic cell groups		
A7	++	—
A8	++	—

TABLE 2 Continued

	mRNA	Binding Sites (fmol/mg)
Pons		
Pontine nuclei	+	—
Pontine raphe	++	—
Locus coeruleus	++	—
Parabrachial nuclei	++	—
ventral	++	—
dorsal	++	—
Superior olive	+	—
Trigeminal motor nucleus	++	—
Cerebellum		
Cortex	+	—
Nuclei	++	—
Medulla oblongata		
Cochlear nuclei	+	—
Vestibular nuclei	+	—
Raphe obscurus	++	—
Raphe pallidus	++	—
Raphe magnus	++	—
Lateral reticular formation	++	—
Nucleus of the solitary tract		
commisuralis	++	—
commisuralis	+++	—
Cuneate nucleus	++	—
Choroid plexus		
Lateral ventricle	+++	1564
III ventricle	+++	—
IV ventricle	+++	1445

[a]Relative amounts of 5-HT_{1C} receptor mRNA have been estimated as low (+), intermediate (++) or high (+++) levels.[29] Binding data[5] is included for comparison. 5-HT_{1C} binding sites were defined, in reference 5, as ^{3}H-mesulergine binding displaceable by 100nM 5-HT.[5]

REFERENCES

1. PAZOS, A., D. HOYER & J. M. PALACIOS. 1984. Binding of serotonergic ligands to the porcine choroid plexus: Characterization of a new type of 5-HT recognition site. Eur. J. Pharmacol. **106:** 539–546.
2. PAZOS, A., D. HOYER & J. M. PALACIOS. 1984. Mesulergine, a selective serotonin-2 ligand in the rat cortex, does not label these receptors in porcine and human cortex: Evidence for species differences in brain serotonin-2 receptors, Eur. J. Pharmacol. **106:** 531–538.
3. YAGALOFF, N. A. & P. R. HARTIG. 1985. 125I-Lysergic acid diethylamide binds to a novel serotonergic site on rat choroid plexus epithelial cells, J. Neurosci. **5:** 3178–3183.
4. YAGALOFF, K. A., G. LOZANO, T. VAN DYKE, A. J. LEVINE & P. R. HARTIG. 1986.

Serotonin 5-HT_{1C} receptors are expressed at high density on choroid plexus tumors from transgenic mice. Brain Res. **385:** 389–394.

5. PAZOS, A. & J. M. PALACIOS. 1985. Quantitative autoradiographic mapping of serotonin receptors in the rat brain. I. Serotonin-1 receptors. Brain Res. **346:** 205–230.
6. PAZOS, A., A. PROBST & J. M. PALACIOS. 1987. Serotonin receptors in the human brain-III. Autoradiographic mapping of serotonin-1 receptors. Neurosci. **21**(1): 97–122.
7. YAGALOFF, K. A. & P. R. HARTIG. 1986. Solubilization and characterization of the serotonin 5-HT_{1C} site from pig plexus. Mol. Pharmacol. **29:** 120–125.
8. HOYER, D. & A. KARPF. 1988. [^{125}I]SCH23982, A 'selective' D-1 receptor antagonist, labels with high affinity 5-HT_{1C} sites in pig choroid plexus. Eur. J. Pharmacol. **150:** 181–184.
9. SIDHU, A., J. C. VAN OENE, P. DANDRIDGE, C. KAISER & J. W. KEBABIAN. 1986. [^{125}I]SCH 23982: The ligand of choice for identifying the D-1 dopamine receptor. Eur. J. Pharmacol. **128:** 213–220.
10. HOYER, D., G. ENGEL & H. O. KALKMAN. 1985. Molecular pharmacology of 5-HT_1 and 5-HT_2 recognition sites in rat and pig brain membranes: Radioligand binding studies with [^{3}H]5-HT, [^{3}H]8-OH-DPAT, (−)[^{125}I]iodocyanopindolol, [^{3}H]mesulergine and [^{3}H]ketanserin. Eur. J. Pharmacol. **118:** 13–24.
11. HOYER, D., S. SRIVATSA, A. PAZOS, G. ENGEL & J. M. PALACIOS. 1986. [^{125}I]LSD labels 5-HT_{1C} recognition sites in pig choroid plexus membranes. Comparison with [^{3}H]Mesulergine and [^{3}H]5-HT binding. Neuroscience Lett. **69:** 269–274.
12. NICKLAUS, K. J., P. MCGONIGLE & P. B. MOLINOFF. 1988. [^{3}H]SCH 23390 labels both dopamine-1 and 5-hydroxytryptamine$_{1C}$ receptors in the choroid plexus. J. Pharmacol. Exp. Ther. **247:** 343–348.
13. HOYER, D., C. WAEBER, P. SCHOEFFTER, J. M. PALACIOS & A. DRAVID. 1989. 5-HT_{1C} Receptor-mediated stimulation of inositol phosphate production in pig choroid plexus. Naunyn-Schmiedebergs Arch. Pharmacol. **339:** 252–258.
14. RUDMAN, D., B. M. HOLLINS, N. C. LEWIS & J. W. SCOTT. 1977. Effects of hormones on 3′,5′-cyclic adenosine monophosphate in choroid plexus. Am. J. Physiol. **232:** E353–357.
15. CROOK, R. B., M. B. FARBER & S. B. PRUSINER. 1984. Hormones and neurotransmitters control cyclic AMP metabolism in choroid plexus epithelial cells. J. Neurochem. **42:** 340–50.
16. PALACIOS, J. M., R. MARKSTEIN & A. PAZOS. 1986. Serotonin-$_{1C}$ sites in choroid plexus are not linked in a stimulatory or inhibitory way to adenylate cyclase. Brain Res. **380:** 151–154.
17. CONN, P. J., E. SANDERS-BUSH, B. J. HOFFMAN & P. R. HARTIG. 1986. A unique serotonin receptor in choroid plexus is linked to phosphatidylinositol turnover. Proc. Natl. Acad. Sci. USA **83:** 4086–4088.
18. HOFFMAN, B. J., P. R. HARTIG, P. J. CONN & E. SANDERS-BUSH. 1986. Thirty-fold stimulation of phosphatidylinositol hydrolysis by serotonin 5-HT_{1C} receptors in pig choroid plexus. Soc. Neurosci. Abst. **12:** 576.
19. CONN, P. J. & E. SANDERS-BUSH. 1986. Agonist-induced phosphoinositide hydrolysis in choroid plexus, J. Neurochem. **47:** 1754–1760.
20. KAUFMAN, M. J. 1989. Ph.D. Dissertation. Johns Hopkins University.
21. SANDERS-BUSH, E., K. D. BURRIS & K. KNOTH. 1988. Lysergic acid diethylamide and 2,5-dimethoxy-4-methylamphetamine are partial agonists at serotonin receptors linked to phosphoinositide hydrolysis. J. Pharmacol. Exp. Ther. **246:** 924–928.
22. CONN, P. J., A. JANOWSKY & E. SANDERS-BUSH. 1987. Denervation supersensitivity of 5-HT_{1C} receptors in rat choroid plexus. Brain Res. **400:** 396–398.

22a. KAUFMAN, M. J., P. R. HARTIG & B. J. HOFFMAN. 1990. The serotonin 5-HT_{1C} receptor of choroid plexus stimulates cyclic GMP formation. In preparation.

23. NG, J., B. B. FREDHOLM, M. JONDAL & T. ANDERSSON. 1988. Regulation of receptor-mediated calcium influx across the plasma membrane in a human leukemic T-cell line: Evidence of its dependence on an initial calcium mobilization from intracellular stores. Biochim. Biophys. Acta **971:** 207–214.

24. WALDMAN, S. A. & F. MURAD. 1987. Cyclic GMP synthesis and function. Pharmacological Rev. **39**(3): 163–196.
25. BURCH, R. M., A. LUINI & J. AXELROD. 1986. Phospholipase A_2 and phospholipase C are activated by distinct GTP binding proteins in response to alpha$_1$-adrenergic stimulation in FTRL5 thyroid cells. Proc. Natl. Acad. Sci. USA **83:** 7201–7205.
26. SEKAR, M. C. & L. E. HOKIN. 1987. Inhibitors of diacylglycerol lipase and diacylglycerol kinase inhibit carbamylcholine-stimulated responses in guinea pig pancreatic minilobules. Arch. Biochem. Biophys. **256:** 509–514.
27. JOHANSON, C. 1987. Choroid plexus. *In* Encyclopedia of Neuroscience, Vol. 1. G. Adelman, Ed. Boston: Birkhouser Publishers, pp. 236–239.
28. TSUTSUMI, M., M. K. SKINNER & E. SANDERS-BUSH. 1989. Transferrin gene expression and synthesis by cultured choroid plexus epithelial cells. J. Biol. Chem. **264:** 9626–9631.
29. HOFFMAN, B. J. & E. MEZEY. 1989. Distribution of serotonin 5-HT_{1C} receptor mRNA in adult rat brain. FEBS Lett. **247:** 453–462.
30. HARTIG, P. R. 1989. Serotonin 5-HT_{1C} receptors: What do they do? *In* Serotonin: Actions, Receptors, Pathophysiology. Proceedings of the 1987 IUPHAR Congress Satellite Meeting, Heron Island, Australia. E. J. Mylecharane, J. A. Angus, I. S. de la Lande, and P. P. A. Humphrey, Eds. England: Macmillan Press, pp. 180–187.
31. FARACI, F. M., W. G. MAYHAN & D. D. HEISTAD. 1989. Effect of serotonin on blood flow to the choroid plexus, Brain Res. **478:** 121–126.
32. FELDMAN, A. M., M. H. EPSTEIN & S. W. BRUSILOW. 1980. Role of cyclic AMP in cerebrolpinal fluid production. *In* Neurobiology of Cerebrospinal Fluid. J. H. Wood, Ed. New York: Plenum Publishing.
33. LINDVALL-AXELSSON, M., C. MATHEW, C. NILSSON & CH. OWMAN. 1988. Effect of 5-Hydroxytryptamine on the rate of cerebrospinal fluid production in rabbit. Neurology **99:** 362–368.
34. SCHINDLER, H., P. MAIER, E. SCHROTER & H. CRAMER. 1979. Stimulation of Cyclic GMP synthesis in the isolated choroid plexus of the rabbit by a muscarinic cholinergic mechanism and by sodium azide. Naunyn-Schmiedebergs Arch. Pharmacol. **3078:** R30.
35. LINDVALL, M., L. EDVINSSON & CH. OWMAN. 1978. Reduced cerebrospinal fluid formation through cholinergic mechanisms. Neurosci. Lett. **10:** 311–316.
36. TSUTSUMI, J., M. NIWA, T. KAWANT, M-A. IBARAGI, M. OZAKI & K. MORI. 1987. Atrial natriuretic polypeptides elevate the level of cyclic GMP in the rat choroid plexus. Neurosci. Lett. **79:** 174–178.
37. STEARDO, L. & J. A. NATHANSON. 1987. Brain barrier tissues: End organs for atriopeptins. Science **235:** 470–473.
38. O'GRADY, S. M., M. FIELD, N. T. NASH & M. C. RAO. 1985. Atrial natriuretic factor inhibits Na-K-Cl cotransport in teleost intestine. Am. J. Physiol. **249:** C531–534.
39. MOHRMANN, M., H. F. CANTIELLO & D. A. AUSIELLO. 1987. Inhibition of epithelial Na+ transport by atriopeptin, protein kinase C, and pertussis toxin. Am. J. Physiol. **253:** F372–376.
40. BERTORELL, A. & A. APERIA. 1988. Na+-K+-ATPase is an effector protein for protein kinase C in renal proximal tubule cells. Am. J. Physiol. **256:** F270–273.
41. LATTIMER, S. A., A. A. F. SIMA & D. A. GREEN. 1989. In vitro correction of impaired Na+-K+-ATPase in diabetic nerve by protein kinase C agonists. Am. J. Physiol. **256:** E264–269.
42. O'BRIEN, T. G., K. GEORGE & R. PRETTYMAN. 1988. Protein kinase C and membrane transport to exogenous diacylglycerol. Biochim. Biophys. Acta **945:** 41–50.
43. ZILLES, K., M. RATH, A. SCHLEICHER, T. GLASER & J. TRABER. 1986. Ontogenesis of serotonin (5-HT) binding sites in the choroid plexus of the rat brain. Brain Res. **380:** 201–203.
44. GUNDERSEN, C. B., R. MILEDI & F. R. S. PARKER. 1983. Serotonin receptors induced by exogenous messenger RNA in Xenopus oocytes. Proc. Royal Soc. London (Ser. B) **219:** 103–109.
45. LÜBBERT, H., T. P. SNUTCH, N. DASCAL, H. A. LESTER & N. DAVIDSON. 1987. Rat brain 5-HT_{1C} receptors are encoded by a 5–6 kilobase nRNA size class and are functionally expressed in injected Xenopus oocytes. J. Neurosci. **7:** 1159–1165.

46. DASCAL, N., C. IFUNE, R. HOPKINS, T. P. SNUTCH, H. LÜBBERT, N. DAVIDSON, M. I. SIMON & H. A. LESTER. 1986. Involvement of a GTP-binding protein in mediation of serotonin and acetylcholine responses in Xenopus oocytes injected with rat brain messenger RNA. Mol. Brain Res. **1:** 201–209.
47. LÜBBERT, H., T. P. SNUTCH, N. DASCAL, H. A. LESTER & N. DAVIDSON. 1986. The rat brain serotonin receptor functionally expressed in mRNA injected Xenopus oocytes is of the 5-HT$_{1C}$ type. Soc. Neurosci. Abst. **12:** 363.
48. LÜBBERT, H., B. J. HOFFMAN, T. P. SNUTCH, T. VAN DYKE, A. J. LEVINE, P. R. HARTIG, H. A. LESTER & N. DAVIDSON. 1987. cDNA cloning of a serotonin 5-HT$_{1C}$ receptor by electrophysiologial assays, of mRNA-injected Xenopus oocytes. Proc. Natl. Acad. Sci. USA **84:** 4332–4336.
49. JULIUS, D., A. B. MACDERMOTT, R. AXEL & T. M JESSELL. 1988. Molecular characterization of a functional cDNA encoding the serotonin 1c receptor. Science **241:** 558–564.
50. O'DOWD, B. F., R. J. LEFKOWITZ & M. G. CARON. 1989. Structure of the adrenergic and related receptors. Ann. Rev. Neurosci. **12:** 67–83.
51. HOFFMAN, B. J. 1988. Ph.D. Dissertation. Johns Hopkins University.
52. STRADER, C. D., I. S. SEGAL & R. A. F. DIXON. 1989. Structural basis of β-adrenergic receptor function. FASEB J. **3:** 1825–1832.
53. HIGASHIJIMA, T., S. UZU, T. NAKAJIMA & E. M. ROSS. 1988. Mastoparan, a peptide toxin from wasp venom, mimics receptors by activating GTP-binding regulatory proteins (G proteins). J. Biol. Chem. **263:** 6491–6494.
54. FARGIN, A., J. R. RAYMOND, M. J. LOHSE, B. K. KOBILKA, M. G. CARON & R. J. LEFKOWITZ. 1988. The genomic clone G-21 which resembles a beta-adrenergic receptor sequence encodes the 5-HT$_{1A}$ receptor. Nature **335:** 358–360.
55. PRITCHETT, D. B., A. W. J. BACH, M. WOZNY, O. TALEB, R. DAL TOSO, J. SHIH & P. H. SEEBURG. 1988. Structure and functional expression of cloned rat serotonin 5-HT$_2$ receptor. EMBO J. **13:** 4135–4140.
56. KOBILKA, B. K., T. S. KOBILKA, K. DANIEL, J. W. REGAN, M. G. CARON & R. J. LEFKOWITZ. 1988. Science **240:** 1310–1316.
57. HARTIG, P. R. 1989. Molecular biology of 5-HT receptors. Trends Pharmacol. Sci. **10:** 64–69.
58. HOYER, D. 1988. Molecular pharmacology and biology of 5-HT^{1C} receptors. Trends Pharmacol. Sci. **9:** 89–94.
59. JULIUS, D., T. J. LIVELLI, T. M. JESSELL & R. AXEL. 1989. Ectopic expression of the serotonin 1c receptor and the triggering of malignant transformation. Science **244:** 1057–1062.
60. CONKLIN, B. R., M. R. BRANN, N. J. BUCKLEY, A. L. MA, T. I. BONNER & J. AXELROD. 1988. Stimulation of arachidonic acid release and inhibition of mitogenesis by cloned genes for muscarinic receptor subtypes stably expressed in A9 L cells. Proc. Natl. Acad. Sci. USA **85:** 8698–8702.
61. KENNETT, G. A. & G. CURZON. 1988. Evidence that mCPP may have behavioral effects mediated by central 5-HT$_{1C}$ receptors. Br. J. Pharmacol. **94:** 137–147.
62. KENNETT, G. A., P. WHITTON, K. SHAH & G. CURZON. 1989. Anxiogenic-like effects of mCPP and TFMPP in animal models are opposed by 5-HT$_{1C}$ receptor antagonists. Eur. J. Pharmacol. **164:** 445–454.
63. HOFFMAN, B. J., U. SCHEFFEL, J. R. LEVER, M. D. KARPA & P. R. HARTIG. 1987. N1-Methyl-2-^{125}I-Lysergic acid diethylamide, a preferred ligand for in vitro and in vivo characterization of serotonin receptors. J. Neurochem. **48:** 115–124.
64. KADAN, M. J., A. M. KROHN, M. J. EVANS, R. L. WALTZ & P. R. HARTIG. 1984. Characterization of ^{125}I-LSD binding to serotonin receptors in rat frontal cortex. J. Neurochem. **43:** 601–606.
65. HARTIG, P. R., M. J. EVANS, A. M. KROHN, S. A. LEDER, P. C. SZE & D. A. STOFFERS. 1985. [^{125}I]LSD binding to serotonin and dopamine receptors in bovine caudate membranes. Neurochem. Int. **7:** 699–707.
66. ENGEL, G., E. MULLER-SCHWEINITZER & J. M. PALACIOS. 1984. 2-[125Iodo]LSD, a new ligand for the characterization and localization of 5-HT$_2$ receptors. Naunyn-Schmiedebergs Arch. Pharmacol. **325:** 328–336.

67. Peroutka, S. J. 1986. Pharmacologial differentiation and characterization of 5-HT_{1A}, 5-HT_{1B}, and 5-HT_{1C} binding sites in rat frontal cortex. J. Neurochem. **47:** 529–540.
68. Leysen, J. E., C. J. E. Neimegeers, J. M. Van Nueten & P. M. Laduron. 1982. [^{3}H]Ketanserin (R41 468), a selective ^{3}H-ligand for serotonin$_2$ receptor binding sites. Mol. Pharmacol. **21:** 301–314.
69. Sanders-Bush, E. & M. Breeding. 1988. Putative selective 5-HT_2 antagonists block serotonin 5-HT_{1C} receptors in the choroid plexus. J. Pharmacol. Exp. Ther. **247:** 169–173.
70. Peroutka, S. J. & S. H. Snyder. 1981. [^{3}H]Mianserin: Differential labeling of serotonin$_2$ and histamine 1 receptors in rat brain. J. Pharmacol. Exp. Ther. **216:** 142–148.
71. Hess, E. J., G. Battaglia, A. B. Norman, L. C. Iorio & I. Creese. 1986. Guanine nucleotide regulation of agonist interactions at [3H]SCH23390-labeled D1 dopamine receptors in rat striatum. Eur. J. Pharmacol. **121:** 31–38.

DISCUSSION OF THE PAPER

Question: Dr. Hartig, the work you're doing on the pig choroid plexus was presumably on whole choroid plexus, and if so are you sure that the cyclic GMP response occurs in the epithelium as opposed to endothelium or the vascular smooth muscle?

Hartig: I don't believe that we did any separations of the different tissue types because of the previous work indicating that the 1C receptor is not located on any tissue but the epithelial cells. So we don't have that specific answer, but I'd be extremely surprised if that turned out to be a problem because we just don't see the 1C pharmacology anywhere but the epithelial cells.

Sally J. Boyson (*University of Colorado Health Science Center, Denver*): I have evidence that the production of CSF can be inhibited in a whole animal by drugs which act at the 5-HT_{1c} receptor, I studied this in a rat model, I presented this at the neurology meetings a year ago. Unfortunately since it's given in a whole animal I can't guarantee that the only site it's acting at is at the 5-HT_{1c} receptor, but we do know that it doesn't decrease blood pressure and I'm currently looking into whether it decreases blood flow.

Hartig: I think there was a very nice study by Frank Faraci *et al.* (1989), showing that there were interesting effects of serotonin on blood flow in the choroid plexus that were quite different than on blood flow of other vascular beds. So I think that those of us who are working on that problem have to be very careful about its effects on blood flow because the choroid plexus shows unusual effects in response to serotonin.

J. Palacios (*Sandoz, Basel*): We did some studies on the *in vivo* binding of mesulergine many years ago that we never published. We saw that when you injected drug intravenously, you can label perfectly 1C receptors in the choroid plexus. So what I'm trying to say is that these receptors are accessible to drugs arriving via the intravenous supply.

Hartig: Does that surprise you? Isn't this a fairly lipophilic molecule compared to serotonin?

Palacios: What I'm trying to say is that probably these receptors don't only see serotonin coming from the brain.

Hartig: I have no debate with your results. But with mesulergine being a large organic molecule vs. serotonin, a highly water soluble, small molecule—that's why we didn't see serotonin labeling choroid plexus in our intravenous studies.

Palacios: My point is that receptors are accessible to drugs arriving via the blood supply.

HARTIG: To drugs, yes, that's a good point.

PALACIOS: The second point is about the presence of message for the 1C in the catecholamine cells in the brain. I think this is a very generalized statement that can be dangerous. In fact in the substantia nigra the receptors appear to be located between interneurons of the compacta which are probably GABA-ergic. And also in other brain areas, particularly in the subthalamic nucleus, these cells appear to be probably GABA-ergic rather than catecholaminergic.

HARTIG: First of all if I could stop you on the generalization—it was the correlation between location of the 1c mRNA and the general area containing catecholamine cell bodies that I wanted to point out. Secondly, the point about mismatch between what we see in "*in situ*" hybridization vs. receptor binding is going to be something that it will be important to work out. We of course don't know that there's direct linear correlation between the two.

Serotonin 5-HT_{1D} Receptors

DANIEL HOYER, PHILIPPE SCHOEFFTER, CHRISTIAN WAEBER, AND JOSE M. PALACIOS

Preclinical Research
386/525, Sandoz Ltd.
CH 4002 Basel, Switzerland

INTRODUCTION

The multiplicity of receptor subtypes through which the neurotransmitter serotonin (5-HT, 5-hydroxytryptamine)[a] exerts its various actions has been documented by abundant physiological, biochemical and pharmacological studies (for reviews, see refs. 1–3). True heterogeneity is supported by the recent determination of the different sequences of the genes or cDNA coding for at least three different subtypes of 5-HT receptors, 5-HT_{1A}, 5-HT_{1C} and 5-HT_2.[4–7] In order to reconcile the different existing nomenclatures, Bradley *et al.*[8] proposed the classification of 5-HT receptors into the "5-HT_1-like," 5-HT_2 and 5-HT_3 main types. Binding sites corresponding to each of these receptors have been labeled in the central nervous system. [^{3}H]Ketanserin is the most selective radioligand for 5-HT_2 recognition sites[9], while 5-HT_3 sites have been specifically labeled by compounds such as [^{3}H]GR 65630,[10] [^{3}H]zacopride,[11] [^{3}H]quipazine[12] or [^{3}H]ICS 205-930.[13,14]

[^{3}H]5-HT labels sites corresponding to "5-HT_1-like" receptors with nanomolar affinity.[15] However, these sites do not constitute a homogeneous population. To date, four subtypes of 5-HT_1 binding sites with distinct pharmacological profiles have been identified and successfully labeled in the central nervous system: 5-HT_{1A} labeled by [^{3}H]8-OH-DPAT,[16] 5-HT_{1B} labeled by [^{125}I]ICYP,[17] 5-HT_{1C} sites labeled by [^{3}H]mesulergine[18] and more recently 5-HT_{1D} sites labeled by [^{3}H]5-HT itself in the presence of selective blockers of the other subtypes.[19] It is established that 5-HT_{1A}, 5-HT_{1B}, 5-HT_{1C}, 5-HT_{1D} and 5-HT_2 receptors are pharmacologically different.[19–21]

The 5-HT_{1D} subtype has been characterized in pigeon, guinea-pig, bovine, porcine, and human brain membranes,[21–24] while binding sites with the pharmacological profile of 5-HT_{1B} sites have only been demonstrated in certain rodents (rat, mouse, hamster), but not in higher mammals, suggesting species differences in 5-HT recognition site subtypes.[3,20,25,26]

The distribution of the different classes of 5-HT binding sites has been examined by autoradiography (for review see ref. 27). Interestingly, 5-HT_{1B} sites in the rat brain[28] and 5-HT_{1D} sites in human brain showed similar distributions,[23] both subtypes being

[a] **ABBREVIATIONS:** 5-HT, 5-hydroxytryptamine (serotonin); 5-CT, 5-carboxamidotryptamine; 5-MeOT, 5-methoxytryptamine; DP-5-CT, *N,N*-dipropyl-5-carboxamidotryptamine; (−) α-Me-5-HT, (−) α-methyl-5-hydroxytryptamine; GR 43175, 3-(dimethylamino)ethyl-*N*-methyl-1H-indole-5 methane sulfonamide; RU 24969, 5-methoxy-3-(1,2,3,6-tetrahydro-4-pyridinyl)-1H indole; 8-OH-DPAT, 8-hydroxy-2-(di-*n*-propylamino)tetralin hydrobromide; PAPP, 1-[2-(4-aminophenyl)ethyl]-4-(3-trifluoromethylphenyl)-piperazine (LY 165163); CGS 12066, 7-trifluoromethyl-4-(4-methyl-1-piperazinyl)-pyrrolo[1,2-a] quinoxaline 1; ICS 205-930 ((1αH, 3α, 5αH-tropan-3-yl)1-H-indole-3-carboxylic acid ester); SDZ 21-009, 4[3-tert-butyl-amino-2-hydroxypropoxy]-indol-2 carbonic acid isopropylester; *m*CPP, *m*-chloro-phenyl-piperazine; TFMPP, trifluoromethyl-phenyl-piperazine.

enriched in the substantia nigra, the dorsal subiculum, and the basal ganglia. Moreover, both 5-HT_{1B} sites in the rat substantia nigra and 5-HT_{1D} sites in the calf substantia nigra are negatively coupled to adenylate cyclase.[29,30] These data will be briefly reviewed, and further potential functional correlates to 5-HT_{1D} binding will be presented.

PHARMACOLOGICAL PROFILE OF 5-HT_{1D} AND 5-HT_{1B} BINDING SITES

The affinity constants for various compounds for 5-HT_{1D} sites subtypes were determined in competition experiments using[^{3}H]5-HT in the presence of 100 nmol/1 8-OH-DPAT and 100 nmol/l mesulergine. TABLE 1 lists affinity values obtained in membranes from guinea-pig cortex, pigeon telecephalon and pig caudate. Under the conditions used, [^{3}H]5-HT labeled a site with high affinity for 5-CT and 5-HT. It has low affinity for drugs with high affinity and/or selectivity to 5-HT_{1A} (8-OH-DPAT, DP-5-CT), 5-HT_{1B} (SDZ 21-009) and 5-HT_{1C} or 5-HT_2 sites (mesulergine, mianserin). These sites have significant lower affinity for (−)propranolol, SDZ 21-009 and RU 24969 in comparison to 5-HT_{1B} sites. On the other hand, DP-5-CT, yohimbine and rauwolscine have a 100-fold higher affinity in pigeon, pig, calf, guinea pig and human brain membranes than at 5-HT_{1B} sites. The affinity values obtained in guinea-pig cortex, pigeon telencephalon, human and pig caudate membranes correlated highly significantly. Data from pig cortex (5-HT_{1A}), rat cortex (5-HT_{1B}) and pig choroid plexus (5-HT_{1C} sites) are listed for comparison. In contrast, no significant correlation was obtained between pK_D values at 5-HT_{1D} sites and the corresponding K_D values at 5-HT_{1B} sites in the rat brain (FIG. 1).

TABLE 1. Affinity Values of Various Drugs for Calf Caudate, Guinea-pig Cortex and Pigeon Telencephalon 5-HT_{1D} Sites Compared to 5-HT_{1A} (Pig Cortex), 5-HT_{1B} (Rat Cortex), and 5-HT_{1C} Sites (Pig Choroid Plexus)[a]

Drug	Calf 5-HT_{1D}	G-Pig 5-HT_{1D}	Pigeon 5-HT_{1D}	Rat 5-HT_{1B}	Pig 5-HT_{1A}	Pig 5-HT_{1C}
5-HT	8.44	8.51	8.46	7.61	8.51	7.69
5-CT	8.62	9.03	9.04	8.29	9.53	6.23
8-OH-DPAT	5.94	6.32	6.38	5.09	8.74	5.23
Mesulergine	5.20	5.93	6.19	4.88	6.23	8.79
SDZ 21-009	6.37	5.42	6.14	9.38	7.76	5.05
RU 24969	7.32	7.09	7.33	8.42	8.29	6.40
Yohimbine	7.12	6.64	6.62	5.46	6.86	4.36
Rauwolscine	7.65	6.10	6.85	5.25	6.86	5.83
Mianserin	6.37	6.43	6.27	5.33	5.94	8.00
DP-5-CT	6.92	6.53	6.68	4.88	9.54	4.51
(−)Propranolol	5.49	5.23	5.25	7.33	6.67	6.75
GR 43175	7.54	—	—	6.35	6.13	4.13
Ergotamine	7.82	—	—	8.69	8.31	7.32
Dihydroergotamine	7.86	—	—	8.36	8.56	7.26

[a]Affinity values are expressed as pK_D values (-log mol/l). Binding to 5-HT_{1D} sites was determined with [^{3}H]5-HT in the presence of 100 nmol/l 8-OH-DPAT and 100 nmol/l mesulergine. Binding to the other sites was determined as reported earlier.[3,18,20]

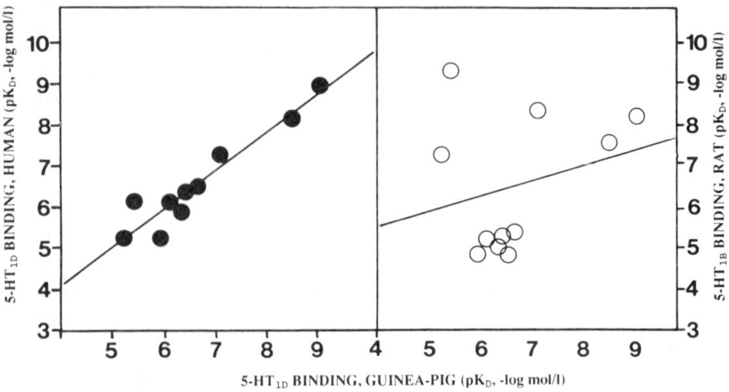

FIGURE 1. Correlation between 5-HT_{1D} binding in guinea-pig cortex versus human 5-HT_{1D} and rat 5-HT_{1B} binding. Comparison of the drug binding profiles of 5-HT_{1D} sites in guinea-pig cortex with human caudate 5-HT_{1D} (*left*) and rat cortex 5-HT_{1B} sites (*right*).

AUTORADIOGRAPHIC LOCALIZATION OF 5-HT_{1D} SITES IN THE VERTEBRATE BRAIN: COMPARISON WITH 5-HT_{1B} SITES

[^{3}H]5-HT binding was investigated by means of quantitative autoradiography in brain slices of a variety of species, including pigeon, rat, mouse, guinea-pig, cat, dog, monkey and humans.[23,24,27,28] High densities of non-5-HT_{1A} non-5-HT_{1C} [^{3}H]5-HT binding sites are present in the brain of all mammals investigated. These sites exhibit a similar regional distribution, particularly in the basal ganglia and associated structures, in spite of pharmacological differences. They correspond to the 5-HT_{1D} subtype in human, monkey, dog, cat and guinea-pig brain, whereas they are accounted for by 5-HT_{1B} sites in mouse and rat brain. Sites with a similar pharmacological profile are also present in the rest of the brain, although their anatomical distribution exhibits more pronounced species differences. Sites corresponding to the 5-HT_{1A} subtype have always been found in areas such as the hippocampus, the raphe nuclei and the cortex, while mesulergine-displaceable [^{3}H]5-HT binding sites are especially concentrated in the choroid plexus. Interestingly, the pigeon brain contains 5-HT_1 recognition sites corresponding to 3 subtypes found in higher mammals, the majority of them belonging to the 5-HT_{1D} subtype.

The autoradiographic data confirm results from radioligand binding and second messenger studies (see below). Using [^{125}I]cyanopindolol we demonstrated the presence of 5-HT_{1B} sites in rat brain,[17] but not in pig or human brain.[20,25] Similarly, species variations have been reported by Heuring *et al.*[26] Thus, although the pharmacological profile of non-5-HT_{1A} non-5-HT_{1C} [^{3}H]5-HT binding sites exhibits species differences, both 5-HT_{1B} receptors in the rat substantia nigra and 5-HT_{1D} receptors in calf and guinea-pig substantia nigra have been shown to be coupled to the same second messenger system, namely the inhibition of forskolin-stimulated adenylate cyclase (see below). The anatomical distribution of 5-HT_{1B} and/or 5-HT_{1D} sites has been conserved throughout the evolution of mammals, suggesting that 5-HT–mediated mechanisms in basal ganglia circuits are important enough to be preserved at different phylogenetic levels.

Data from lesion experiments and functional tests have revealed several possible synaptic localizations of 5-HT_{1B}/5-HT_{1D} receptors. A minor population appears to be localized on 5-HT terminals: the pharmacological profile of the modulation of 5-HT release corresponds to that of 5-HT_{1B} receptors in the rat cortex and that of 5-HT_{1D} receptors in the cortex of guinea-pigs, pigs and humans (see below). This population represents autoreceptors and probably does not account for all presynaptic receptors displaying the pharmacology of 5-HT_{1B} or 5-HT_{1D} recognition sites, for heteroreceptors seem to be present on the terminals of striatal neurons projecting to the substantia nigra.[31,32] In addition, the 5-HT_{1B}/5-HT_{1D} binding sites found in the striatum could be accounted for by receptors located postsynaptically to serotoninergic afferents.[31,32]

SECOND MESSENGER SYSTEMS LINKED TO 5-HT_{1D} RECEPTORS

By analogy to 5-HT_{1A} receptors, which are negatively coupled to adenylate cyclase activity in hippocampus,[33–35] the possibility of a coupling of 5-HT_{1B} and 5-HT_{1D} sites with adenylate cyclase was investigated in rat, guinea-pig, and calf substantia nigra.[29,30,36–39] Substantia nigra was chosen as the tissue source since this brain region contains a high density of 5-HT_{1B} sites in rat brain, and since the same was true for 5-HT_{1D} binding in guinea pig and a variety of other species (see above). Forskolin (10 μM) produced an increase in adenylate cyclase activity in calf and guinea-pig substantia nigra (see TABLE 2 and FIG. 2). The rank order of potency of agonists was comparable in membranes from both species: 5-CT > 5-HT > metergoline > 5-MeOT > methysergide > cyanopindolol > RU 24969 > yohimbine > DP-5-CT = SDZ 21-009 > buspirone. The rank order of potency for antagonists was: methiothepin > mianserin > corynanthine > spiperone > isamoltane. TABLE 2 compares the pEC_{50} or pK_B values obtained for inhibition of adenylate cyclase in calf hippocampus (5-HT_{1A}), rat (5-HT_{1B}), guinea-pig and calf substantia nigra (5-HT_{1D}).

COMPARISON OF THE 5-HT RECEPTORS MEDIATING INHIBITION OF ADENYLATE CYCLASE ACTIVITY IN GUINEA-PIG, CALF AND RAT SUBSTANTIA NIGRA

It is clear that the activities mediated by the three receptor subtypes are pharmacologically different. The profiles of the 5-HT receptors mediating inhibition of adenylate

TABLE 2. Comparison of Agonist pEC_{50} or Antagonist pK_B Values Obtained for Inhibition of Adenylate Cyclase Activity in Calf Hippocampus (5-HT_{1A}), Rat (5-HT_{1B}), Calf and Guinea-pig Substantia nigra (5-HT_{1D})[a]

Drugs	Guinea-pig 5-HT_{1D}	Calf 5-HT_{1D}	Rat 5-HT_{1B}	Calf 5-HT_{1A}
5-HT	7.62	7.62	7.79	7.83
5-CT	7.96	8.05	7.92	8.59
RU 24969	6.69	6.82	8.36	7.81
8-OH-DPAT	6.30	5.84	4.85*	8.22
SDZ 21-009	5.27	5.91	8.00*	8.30*
(−)Pindolol	4.90	4.90	6.78*	7.87*
Yohimbine	7.26*	6.93	6.14*	—
Isamoltane	5.22*	4.44*	7.28*	6.76
(−)Propranolol	4.34*	5.10	6.94*	6.60*

[a]Values for guinea-pig are taken from ref. 36, those for calf and rat from refs. 30,35,37–39. The asterisk (*) denotes an antagonist pK_B value; otherwise agonist pEC_{50} values are indicated.

cyclase activity in guinea-pig and calf substantia nigra are closely related (FIG. 3). They differ from the corresponding receptor of the rat, in that RU 24969, SDZ 21-009, (−)pindolol, isamoltane and (−)propranolol are more potent at the rat receptor (by at least 1.5 log units) whilst 8-OH-DPAT and yohimbine show a relatively lower potency at the rat than at the guinea-pig or calf receptor. A highly significant correlation was observed between guinea-pig and calf adenylate cyclase data, in contrast to the lack of

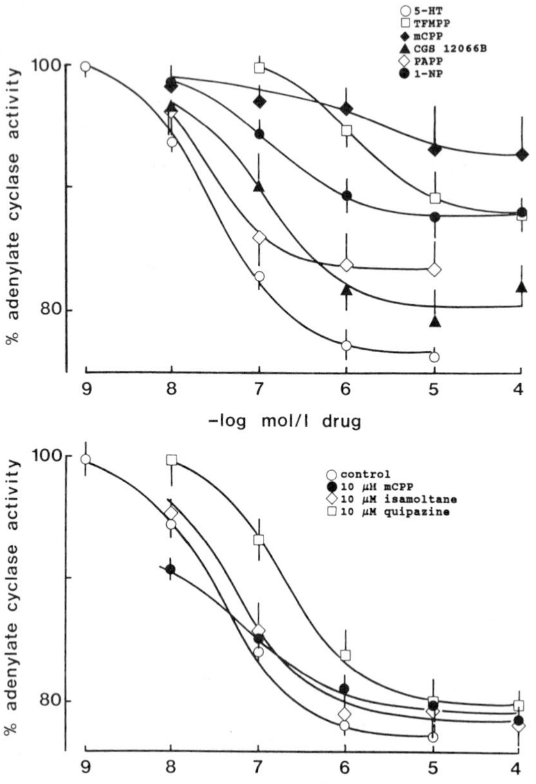

FIGURE 2. Concentration-effect curves of various agonists and antagonists for inhibition of forskolin-stimulated adenylate cyclase: *Top:* concentration-effect curves of 5-HT, TFMPP, mCPP, CGS 12066, PAPP and 1-NP for inhibition of forskolin-stimulated adenylate cyclase activity in calf substantia nigra. *Bottom:* antagonism of 5-HT-induced inhibiton of adenylate cyclase activity in calf substantia nigra. 5-HT concentration-effect curves are represented in the absence or in the presence of *m*CPP, isamoltane and quipazine.

correlation observed when comparing calf and rat cyclase data (as illustrated in TABLE 3 and FIG. 3).

Interestingly, 5-HT–mediated inhibition of adenylate cyclase with a 5-HT_{1B} pharmacological profile has also been described in cultures from hamster fibroblasts[40] and opossum kidney cells.[41]

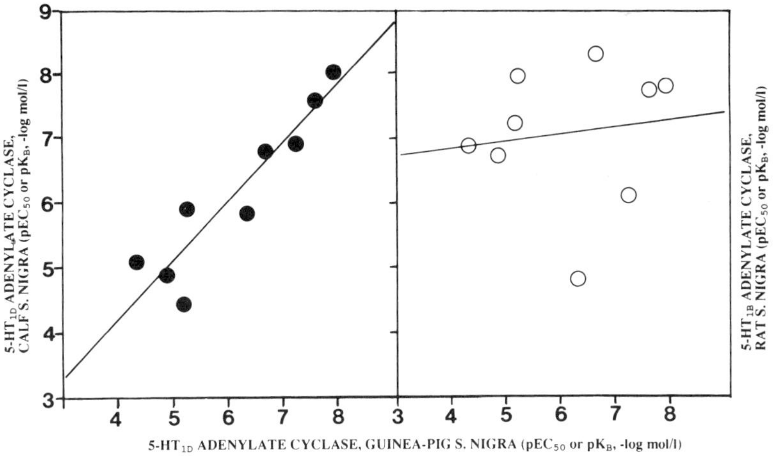

FIGURE 3. Correlation of adenylate cyclase data obtained in substantia nigra from guinea-pig with those from calf (*left*) or rat (*right*).

SELECTIVE LIGANDS FOR 5-HT_{1D} RECEPTORS

The complexity of the 5-HT_1 receptor family is certainly larger than originally expected. Pharmacologically, 5-HT_{1C} receptors resemble 5-HT_2 receptors,[2,3,42] and both differ markedly from the 5-HT_{1A}, 5-HT_{1B}, 5-HT_{1D} receptor group. Whereas a variety of rather selective 5-HT_{1A} ligands are available, this is not yet the case for 5-HT_{1B} and especially for 5-HT_{1D} receptors. A range of compounds has been described with "selectivity" for 5-HT_{1B} receptors, i.e. CGS 12066, isamoltane, TFMPP, *m*CPP

TABLE 3. Comparison of pEC_{50} or pK_B Values of Various Drugs at 5-HT_{1B} and 5-HT_{1D} Receptors Linked to Adenylate Cyclase[a]

Drug	5-HT_{1B}	5-HT_{1D}	Difference
Isamoltane	7.28	4.40	2.88
SDZ 21-009	8.00	5.91*	2.09
(−) Pindolol	6.78	4.90*	1.88
(−) Propranolol	6.94	5.10*	1.84
RU 24969	8.36*	6.82*	1.54
Cyanopindolol	8.24	6.82*	1.42
mCPP	6.46	5.08*	1.38
TFMPP	6.90	5.79	1.11
Quipazine	6.20*	5.70*	0.50
CGS 12066	7.56	7.11	0.45
Mianserin	5.95	6.49	−0.54
Yohimbine	6.14	6.91*	−0.77
DP-5-CT	5.76*	6.58*	−0.82
8-OH-DPAT	4.85	5.84*	−0.99
Rauwolscine	5.96	6.96*	−1.00
PAPP (LY 165163)	5.80*	7.59*	−1.79

[a]Compared are pEC_{50} (asterisks) or pK_B values determined in adenylate cyclase experiments performed in rate (5-HT_{1B}) and calf substantia nigra (5-HT_{1D}).

and others. However, a detailed analysis of the effects of these compounds on the various 5-HT_1 receptor subtypes has revealed that none of the compounds has actual selectivity for a single receptor.[37]

Recently, GR 43175 (3-[2-dimethylamino]ethyl-*N*-methyl-1H-indole-5 methane sulfonamide) has been described as a selective and potent agonist at 5-HT_1-like receptors present in the dog saphenous vein and carotid artery, cat arteriovenous anastomoses, and primate and dog basilar artery.[43–45] The effects of GR 43175 were mimicked by 5-HT and 5-CT (5-carboxamidotryptamine, the prototype 5-HT_1-like receptor agonist) and effectively antagonized by methiothepin, thus fulfilling the criteria for a 5-HT_1-like receptor agonist as defined by Bradley *et al.*[8] GR 43175 has aroused considerable interest, since it appears to be effective in the treatment of migraine.[46] However, up to now it has been unclear at which subtype of the 5-HT_1 receptor family GR 43175 is acting. The effects of the compound were investigated in radioligand binding studies and in functional models for 5-HT_1 receptor subtypes.[38] GR 43175 displayed the following order of affinity for 5-HT recognition sites (pK_D values, $-\log$ mol/l): 5-HT_{1D} (7.54) > 5-HT_{1B} (6.36) > 5-HT_{1A} (6.13) ≫ 5-HT_{1C} (4.13) > 5-HT_2 (3.72). The same order of potency was observed at functional 5-HT_1 receptors, at which GR 43175 acted as a full agonist, with the exception of the 5-HT_{1C} receptor, where the compound was a weak antagonist (pEC_{50} or pK_B values): 5-HT_{1D} (6.28) > 5-HT_{1B} (6.03) > 5-HT_{1A} (5.57) ≫ 5-HT_{1C} (4.25). Thus, GR 43175 appears to interact preferentially as an agonist with 5-HT_{1D} and 5-HT_{1B} receptors.

The effects of GR 43175 were also studied by autoradiography using [^{3}H]5-HT in rat, monkey and human brain slices.[47] There was no noticeable difference between the properties of GR 43175 in monkey and human brain. GR 43175 was effective at concentrations as low as 10 nM in regions rich in 5-HT_{1D} binding sites: the substantia nigra,[23] both parts of the globus pallidus, the caudate, and the putamen. In these areas, a micromolar concentration was able to displace [^{3}H]5-HT from more than 90% of its binding sites. In the raphe nuclei, the hippocampal formation, the claustrum, and layers I and II of the cortex, highly enriched in 5-HT_{1A} recognition sites, the lowest significantly effective GR 43175 concentration was 100 nM, the displacement extending to 70% at a micromolar concentration. In contrast, GR 43175 was unable to displace [^{3}H]5-HT binding even at the highest concentration (1 μmol/l) in the choroid plexus, where 5-HT_1 sites exclusively belong to the 5-HT_{1C} subtype.[27] In the rat brain, the displacing properties of GR 43175 were similar to those in human and monkey brain, in regions rich in 5-HT_{1A} and 5-HT_{1C} binding sites. However, at 5-HT_{1B} sites, enriched in the basal ganglia and the substantia nigra, GR 43175 was slightly less potent than in the corresponding 5-HT_{1D} site rich areas in higher mammals. The preferential affinity for the 5-HT_{1B} (in rat) and 5-HT_{1D} site subtypes (in monkey and man), is in agreement with the findings from radioligand binding and second messenger studies reported above. Thus, GR 43175 seems to be the first ligand with (limited) selectivity for 5-HT_{1D} recognition sites in the central nervous system of higher mammals.

THE TERMINAL 5-HT AUTORECEPTOR: A FUNCTIONAL CORRELATE TO 5-HT_{1D} BINDING IN SEVERAL SPECIES, COMPARISON TO 5-HT_{1B} AUTORECEPTORS

Terminal 5-HT autoreceptors mediating inhibition of 5-HT release have been identified in a variety of species and it is generally accepted that they belong to the 5-HT_1 receptor class.[48,49] Thus, these receptors are very sensitive to 5-CT, 5-HT and RU 24969, but much less to α-methyl 5-HT, (an agonist acting with some selectivity at

5-HT_{1C} and 5-HT_2 receptors), and virtually insensitive to the 5-HT_{1A} receptor agonist, 8-OH-DPAT. The non-selective 5-HT_1 receptor antagonist, methiothepin, and several β-adrenoceptor antagonists (e.g. SDZ 21009, cyanopindolol, pindolol and propranolol), show very potent antagonist activity at the rat terminal 5-HT autoreceptor, while spiperone, mesulergine and mianserin (antagonists at 5-HT_{1A}, 5-HT_{1C} and 5-HT_2 receptors) are ineffective. Several studies have established that the terminal 5-HT autoreceptors of the rat cortex are of the 5-HT_{1B} type.[50–53] There was a highly significant correlation between the potencies of drugs for the rat autoreceptors and their affinities at 5-HT_{1B} binding sites.[53]

It has become clear that 5-HT autoreceptors from other species (rabbit, guinea-pig, pig, humans) are not of the 5-HT_{1B} subtype.[54–56] These autoreceptors may instead be of the 5-HT_{1D} type[57–59] (see ref. 60 for a review). Indeed, 5-HT autoreceptors in guinea-pig and pig are minimally affected by β-adrenoceptor antagonists. On the other hand, 5-CT is a potent agonist and methiothepin a potent antagonist, indicating that 5-HT autoreceptors belong to the "5-HT_1-like" receptor family as originally defined by Bradley *et al.*[8] RU 24969, a potent agonist at the rat autoreceptor, is significantly less effective in other species. By contrast, 8-OH-DPAT, which is devoid of effect at the rat autoreceptor, is a weak agonist in the pig and perhaps the guinea-pig. Furthermore, yohimbine, a ligand with affinity and partial agonist activity at 5-HT_{1D} receptors, behaves either as an agonist (pig) or an antagonist (guinea pig) in the submicromolar concentration range at the terminal 5-HT autoreceptor.

The potencies of a variety of agonists and antagonists at the 5-HT_{1D} receptor mediating inhibition of adenylate cyclase correlated very significantly with their effects on [^{3}H]5-HT release in pig cortex slices[59] (see FIG. 4). The data showed clearly that the effects on 5-HT autoreceptors are not mediated by 5-HT_{1A}, 5-HT_{1B}, 5-HT_{1C}, 5-HT_2, or 5-HT_3 receptors. Preliminary results from experiments carried out with human brain cortex slices also suggest that the 5-HT autoreceptor in this model is of the 5-HT_{1D} type,[57] since the rank order of potency was: 5-CT > RU 24969 > 8-OH-DPAT for agonists and methiothepin > metergoline = methysergide > propranolol for antagonists. In the rabbit, the rank order of potency of agonists at the terminal 5-HT autoreceptor was similar: 5-CT > 5-HT > 8-OH-DPAT = methysergide > cyanopindolol > yohimbine = rauwolscine > corynanthine. Methiothepin was a very potent antagonist in this model, whereas propranolol was devoid of effects (Feuerstein, personal communication). It is tempting to speculate that rabbit 5-HT autoreceptors belong to the 5-HT_{1D} receptor category.

These studies strongly suggest that the terminal 5-HT autoreceptor is of the 5-HT_{1D} type in pig, guinea-pig, human and perhaps rabbit brain.

OTHER MODELS FOR 5-HT_{1D} RECEPTORS

As described above, GR 43175 acts as a full agonist at 5-HT_{1A}, 5-HT_{1B} and 5-HT_{1D} receptors, whereas its effects at 5-HT_{1C}, 5-HT_2 and 5-HT_3 receptors were negligible. GR 43175 showed selectivity for 5-HT_{1D} receptors. This compound has been described as a selective 5-HT_1-like receptor agonist in various peripheral systems.[44] It is not yet established whether the models used previously to characterize GR 43175, i.e., dog saphenous vein and basilar artery, and arteriovenous anastomoses in the cat (43–45), contain 5-HT_{1D} receptors: nevertheless, it can be ruled out that the dog saphenous vein contains 5-HT_{1A}, 5-HT_{1B} or 5-HT_{1C} receptors, since cyanopindolol (a mixed 5-HT_{1A} and 5-HT_{1B} receptor antagonist) and mesulergine (a mixed 5-HT_{1C} and 5-HT_2 receptor antagonist) failed to antagonize significantly the effects of GR 43175 and 5-HT or 5-CT. On the other hand, the rank order of potency of agonists (5-CT >

5-HT > GR 43175 > α-methyl-5-HT > 8-OH-DPAT; see ref. 45), the potent antagonism shown by methiothepin against GR 43175, and the partial agonist activity shown by cyanopindolol in the dog basilar artery and saphenous vein, are compatible with the presence of 5-HT_{1D} receptors in these tissues.

Finally, preliminary data suggest that the 5-HT receptor mediating endothelium-dependent relaxation in the pig coronary artery belongs to the 5-HT_{1D} type.[61,63] The potency of a variety of agonists and antagonists in this model correlated significantly with their affinity for 5-HT_{1D} sites.

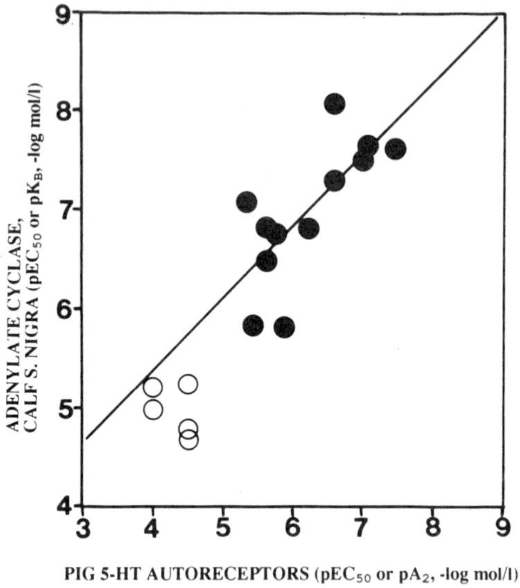

FIGURE 4. Correlation between 5-HT_{1D} adenylate cyclase values and effects of various compounds on terminal 5-HT autoreceptors in pig cortex slices: The figure compares the effects of agonists and antagonists (*closed circles*) on 5-HT_{1D} adenylate cyclase and terminal 5-HT autoreceptors in pig cortex (from ref. 59). *Open circles* represent compounds which were not active at the indicated concentrations. The correlation line was calculated without the inactive compounds.

CONCLUSION AND OUTLOOK

5-HT_{1D} recognition sites are present in the brain of several mammalian species where they display a regional distribution similar to that observed in the human brain. Interestingly, this is also the case in the guinea-pig and in the pigeon. It is established now that 5-HT_{1D} receptors are negatively coupled to adenylate cyclase activity in brain regions where high densities of 5-HT_{1D} sites have been documented. Further, it appears that presynaptic 5-HT autoreceptors which inhibit the release of 5-HT in the brain of a variety of species including man, belong to the 5-HT_{1D} receptor subtype. 5-HT_{1D} receptors could also be present in several vascular beds where they mediate either contraction or endothelium-dependent relaxation. There is an obvious parallel between 5-HT_{1B} receptors of the rat and 5-HT_{1D} receptors of other species in terms of distribution, second messenger coupling and function, both in the brain and the periphery.

With the application of molecular biological techniques to receptor studies, it should not be too long before the structure of 5-HT_{1B} and 5-HT_{1D} receptors is known. The cloning of 5-HT_{1A}, 5-HT_{1C} and 5-HT_2 receptors has already been reported.[4–7] All three receptors belong to the G-protein-coupled receptor family. There is little sequence homology between the 5-HT_{1A} receptor on the one hand, and 5-HT_{1C} and 5-HT_2 receptors on the other.[4,7] In contrast, the similarity between 5-HT_{1C} and 5-HT_2 receptors is striking, especially in their putative transmembrane spanning regions,[6,7] as already suggested by their similar pharmacology and their ability to stimulate inositol phosphate production and Ca^{++} mobilization.[42] Furthermore, 5-HT_{1C} and 5-HT_2 receptors expressed in Xenopus oocytes are both able to stimulate inositol phosphate production, Ca^{++} mobilization and Cl^- channels. Sequence homologies could be expected between 5-HT_{1B} and 5-HT_{1D} receptors, for their second messenger coupling (inhibition of adenylate cyclase), function (inhibition of 5-HT release) and distribution in the brain of various species (e.g. basal ganglia and substantia nigra) are very similar (see above). Finally, considering that 5-HT_{1B} and 5-HT_{1A} receptors share several features, with respect to second messenger coupling (inhibition of adenylate cyclase) and pharmacology (i.e. the high affinity of both receptors for a variety of β-adrenoceptor antagonists), it is tempting to predict significant sequence homologies between 5-HT_{1B} and 5-HT_{1A} receptors, and also β-adrenoceptors.[4,62] It will be interesting to see whether 5-HT_{1B} and 5-HT_{1D} receptors are products of separate genes (possibly by gene duplication). Also, it will soon be possible to assess whether both receptors are expressed in a single species, but with different levels of expression, or whether their presence is mutually exclusive, as suggested from biochemical studies.

SUMMARY

5-HT receptors are subdivided into 3 families, 5-HT_1, 5-HT_2 and 5-HT_3, of which subtypes have been described. The 5-HT receptor field has experienced over the last 10 years a revival due to the availability of new and more selective drugs and new techniques. This communication deals essentially with the biochemical approaches to characterize 5-HT_{1D} receptors, and their comparison with 5-HT_{1B} receptors. The methods used include radioligand binding, *in vitro* autoradiography, and second messenger studies.

5-HT_1 receptor subtypes are labeled with [^{3}H]5-HT and present a large heterogeneity: no less than 4 subtypes have been characterized: 5-HT_{1B} and 5-HT_{1D} are labeled respectively with [^{125}I]cyanopindolol, and [^{3}H]5-HT under appropriate conditions. Although some similarities are evident, the pharmacology of the two receptors is clearly different. Rat 5-HT_{1B} receptors recognize with high affinity a number of β-adrenoceptor antagonists, such as SDZ 21-009, cyanopindolol, pindolol, propranolol and isamoltane. In contrast, calf, pig or human 5-HT_{1D} receptors show significantly lower affinities for these drugs. 5-HT_{1D} receptors show high to intermediate affinities to compounds such as PAPP, DP-5-CT, 8-OH-DPAT, yohimbine and rauwolscine, whereas 5-HT_{1B} receptors have very low affinities for these compounds.

The presence of 5-HT_{1B} receptors has been documented convincingly only in rat, mouse and hamster. 5-HT_{1D} receptors have been demonstrated in pigeon, guinea-pig, cat, dog, pig, calf, monkey, and man. The distribution of 5-HT_{1B} and 5-HT_{1D} receptors in all species examined so far, is very similar: high concentrations of sites are found in the nigro-striatal pathway, caudate-putamen, globus pallidus and especially substantia nigra. The subicullum shows also high densities of sites.

Similar functional correlates have been proposed to 5-HT_{1B} and 5-HT_{1D} sites. Thus, 5-HT_{1D} receptors are negatively coupled to adenylate cyclase in guinea-pig and calf substantia nigra, and 5-HT_{1B} receptors are negatively coupled to adenylate cyclase in rat substantia nigra. Further, it is established that terminal 5-HT autoreceptors are of the 5-HT_{1B} type in rat cortex, and of the 5-HT_{1D} type in guinea-pig, pig, human and possibly rabbit cortex. In the rat saphenous vein, 5-HT_{1B} receptors mediate inhibition of noradrenaline release. Preliminary evidence suggests that the canine basilar artery and saphenous vein, described as models for "5-HT_1-like" receptors, could contain 5-HT_{1D} receptors. Indeed, the 5-HT_1 receptor agonist, GR 43175, is an agonist in these models and shows a certain degree of selectivity for 5-HT_{1D} receptors (determined in radioligand binding, autoradiographic and second messenger studies), and the pharmacological profile of the contractions induced by 5-HT receptor agonists in these models is compatible with that of a 5-HT_{1D} receptor.

5-HT_1 and 5-HT_2 receptors belong to the G-protein coupled receptor superfamily. 5-HT_{1A}, 5-HT_{1C} and 5-HT_2 have been cloned, sequenced and expressed, and present structural similarities with other receptors of this family. 5-HT_{1B} and 5-HT_{1D} receptors have not yet been cloned, but it is expected that both receptors will belong to the same family and present significant structural similarities to that of the 5-HT_{1A} receptor.

ACKNOWLEDGMENT

We thank John R. Fozard for carefully reading the manuscript.

REFERENCES

1. Peroutka, S. J. 1988. 5-Hydroxytryptamine receptor subtypes: Molecular, biochemical and physiological characterization. TINS **11:** 496–500.
2. Hoyer, D. 1988. Functional correlates to serotonin 5-HT_1 recognition sites. J. Rec. Res. **8:** 59–81.
3. Hoyer, D. 1989. Biochemical mechanisms of 5-HT receptor-effector coupling in peripheral tissues. *In* Peripheral actions of 5-HT. J. R. Fozard, Ed. Oxford: Oxford University Press, pp. 72–99.
4. Fargin, A., J. R. Raymond, M. J. Lohse, B. K. Kobilka, M. C. Caron & R. J. Lefkowitz. 1988. The genomic clone G-21 which resembles a β-adrenergic receptor sequence encodes the 5-HT_{1A} receptor. Nature **335:** 358–360.
5. Lübbert, H., T. Snutch, N. Dascal, H. A. Lester & N. Davidson. 1987. Rat brain 5-HT_{1C} receptors are encoded by a 5–6 kb mRNA size class and functionally expressed in injected Xenopus oocytes. J. Neurosci. **4:** 1159–1165.
6. Julius, D., A. B. MacDermott, R. Axel & T. M. Jessell. 1988. Molecular characterization of a functional cDNA encoding the serotonin 1c receptor. Science **241:** 558–564.
7. Pritchett, D. B., A. W. J. Bach, M. Wozny, O. Taleb, R. Dal Toso, J. C. Shih & P. H. Seeburg. 1988. Structure and functional expression of cloned rat serotonin 5-HT_2 receptor. EMBO J. **7:** 4135–4140.
8. Bradley, P. B., G. Engel, W. Feniuk, J. R. Fozard, P. P. A. Humphrey, D. N. Middlemiss, E. J. Mylecharane, B. P. Richardson & P. R. Saxena. 1986. Proposals for the classification and nomenclature of functional receptors for 5-hydroxytryptamine. Neuropharmacology **25:** 563–576.
9. Leysen, J. E., C. J. E. Niemegeers, J. M. Van Nueten & P. M. Laduron. 1982. [^{3}H]Ketanserin (R 41 468), a selective [^{3}H]-ligand for serotonin$_2$ receptor binding sites. Mol. Pharmacol. **21:** 301–314.
10. Kilpatrick, G. J., B. J. Jones & M. B. Tyers. 1987. Identification and distribution of 5-HT_3 receptors in rat brain using radioligand binding. Nature **330:** 746–748.

11. BARNES, N. M., B. COSTALL & R. J. NAYLOR. 1988. [^{3}H]Zacopride: Ligand for the identification of 5-HT_3 recognition sites. J. Pharm. Pharmacol. **40:** 548–551.
12. PEROUTKA, S. J. & A. HAMIK. 1988. [^{3}H]Quipazine labels 5-HT_3 recognition sites in rat cortical membranes. Eur. J. Pharmacol. **148:** 297–299.
13. HOYER, D. & H. C. NEIJT. 1988. Identification of serotonin 5-HT_3 recognition sites in membranes of N1E-115 neuroblastoma cells by radioligand binding. Mol. Pharmacol. **33:** 303–309.
14. WAEBER, C., K. DIXON, D. HOYER & J. M. PALACIOS. 1988. Localisation by autoradiography of neuronal 5-HT_3 receptors in the mouse CNS. Eur. J. Pharmacol. **151:** 351–352.
15. PEROUTKA, S. J. & S. H. SNYDER. 1979. Multiple serotonin receptors: Differential binding of [^{3}H]-5-hydroxytryptamine, [^{3}H]lysergic acid diethylamide and [^{3}H]-spiroperidol. Mol. Pharmacol. **16:** 687–699.
16. GOZLAN, H., S. EL MESTIKAWY, L. PICHAT, J. GLOWINSKI & M. HAMON. 1983. Identification of presynaptic serotonin autoreceptors using a new ligand: [^{3}H]PAT. Nature **305:** 140–142.
17. HOYER, D., G. ENGEL & H. O. KALKMAN. 1985. Characterization of the 5-HT_{1B} recognition site in rat brain: binding studies with (−)[^{125}I]iodocyanopindolol. Eur. J. Pharmacol. **118:** 1–12.
18. PAZOS, A., D. HOYER & J. M. PALACIOS. 1984. The binding of serotoninergic ligands to porcine choroid plexus: Characterization of a new type of serotonin recognition site. Eur. J. Pharmacol. **106:** 539–546.
19. HEURING, R. E. & S. J. PEROUTKA. 1987. Characterization of a novel ^{3}H-5-hydroxytryptamine binding site subtype in bovine brain membranes. J. Neurosci. **7:** 894–903.
20. HOYER, D., G. ENGEL & H. O. KALKMAN. 1985. Molecular pharmacology of 5-HT_1 and 5-HT_2 recognition sites in rat and pig brain membranes: Radioligand binding studies with [^{3}H]5-HT, [^{3}H]8-OH-DPAT, (−)[^{125}I]iodocyanopindolol, [^{3}H]mesulergine and [^{3}H]ketanserin. Eur. J. Pharmacol. **118:** 13–23.
21. HOYER, D., C. WAEBER, A. PAZOS, A. PROBST & J. M. PALACIOS. 1988. Identification of a 5-HT_1 recognition site in human brain membranes different from 5-HT_{1A}, 5-HT_{1B} and 5-HT_{1C} sites. Neurosci. Lett. **85:** 357–362.
22. WAEBER, C., P. SCHOEFFTER, J. M. PALACIOS & D. HOYER. 1988. Molecular pharmacology of 5-HT_{1D} recognition sites: Radioligand binding studies in human, pig and calf brain membranes. Naunyn-Schmiedeberg's Arch. Pharmacol. **337:** 595–601.
23. WAEBER, C., M. M. DIETL, D. HOYER, A. PROBST & J. M. PALACIOS. 1988. Visualization of a novel serotonin recognition site (5-HT_{1D}) in the human brain by autoradiography. Neurosci. Lett. **88:** 11–16.
24. WAEBER, C., M. M. DIETL, D. HOYER & J. M. PALACIOS. 1989. 5-HT_1 receptors in the vertebrate brain: regional distribution examined by autoradiography. Naunyn-Schmiedeberg's Arch. Pharmacol. **340:** 486–494.
25. HOYER, D., A. PAZOS, A. PROBST & J. M. PALACIOS. 1986. Serotonin receptors in the human brain: I Characterization and autoradiographic localization of 5-HT_{1A} recognition sites. Apparent absence of 5-HT_{1B} recognition sites. Brain Res. **376:** 85–96.
26. HEURING, R. E., J. R. SCHLEGEL & S. J. PEROUTKA. 1986. Species variations in RU 24969 interactions with non 5-HT_{1A} binding sites. Eur. J. Pharmacol. **122:** 279–282.
27. PAZOS, A., D. HOYER, M. M. DIETL & J. M. PALACIOS. 1988. Autoradiography of serotonin receptors. *In* Neuronal Serotonin. N. N. Osborne and M. Hamon, Eds. Winchester: John Wiley & Sons, pp. 507–543.
28. PAZOS, A. & J. M. PALACIOS. 1985. Quantitative autoradiographic mapping of serotonin receptors in the rat brain: I Serotonin-1 receptors. Brain Res. **346:** 205–230.
29. BOUHELAL R., L. SMOUNYA & J. BOCKAERT. 1988. 5-HT_{1B} receptors are negatively coupled with adenylate cyclase in rat substantia nigra. Eur. J. Pharmacol. **151:** 189–196.
30. SCHOEFFTER, P., C. WAEBER, J. M. PALACIOS & D. HOYER. 1988. The 5-hydroxytryptamine 5-HT_{1D} receptor subtype is negatively coupled to adenylate cyclase in rat substantia nigra. Naunyn-Schmiedeberg's Arch. Pharmacol. **337:** 602–608.
31. PALACIOS, J. M. & M. M. DIETL. 1988. Autoradiographic studies on 5-HT receptors. *In* The serotonin receptors. E. Sanders-Bush Ed. Clifton, NJ: Humana Press, pp. 89–138.

32. WAEBER, C. & J. M. PALACIOS. 1989. Serotonin-1 receptors in the human basal ganglia are decreased in Huntington's chorea but not in Parkinson's disease. Neuroscience. In press.
33. DE VIVO, M. & MAAYANI, S. 1986. Characterization of 5-hydroxytryptamine$_{1A}$-receptor-mediated inhibition of forskolin-stimulated adenylate cyclase activity in guinea-pig and rat hippocampal membranes. J. Pharmacol. Exp. Ther. **238:** 248–253.
34. BOCKAERT, J., A. DUMUIS, R. BOUHELAL, M. SEBBEN & R. N. CORY. 1987. Piperazine derivatives including the putative anxiolytic drugs, buspirone and ipsapirone, are agonists at 5-HT$_{1A}$ receptors negatively coupled with adenylate cyclase in hippocampal neurons. Naunyn-Schmiedeberg's Arch. Pharmacol. **335:** 588–592.
35. SCHOEFFTER, P. & D. HOYER. 1988. Centrally acting hypotensive agents with affinity to 5-HT$_{1A}$ binding sites inhibit forskolin-stimulated adenylate cyclase activity in calf hippocampus. Br. J. Pharmacol. **95:** 975–985
36. WAEBER, C., P. SCHOEFFTER, J. M. PALACIOS & D. HOYER. 1989. 5-HT$_1$ receptors in the vertebrate brain: Radioligand binding and biochemical studies. Naunyn-Schmiedeberg's Arch. Pharmacol. **340:** 479–485.
37. SCHOEFFTER, P. & D. HOYER. 1989. Interaction of arylpiperazines with 5-HT$_{1A}$, 5-HT$_{1B}$, 5-HT$_{1C}$ and 5-HT$_{1D}$ receptors: Do discriminatory 5-HT$_{1B}$ receptor ligands exist? Naunyn-Schmiedeberg's Arch. Pharmacol. **339:** 675–683.
38. SCHOEFFTER, P. & D. HOYER. 1989. How selective is GR 43175? Interactions with functional 5-HT$_{1A}$, 5-HT$_{1B}$, 5-HT$_{1C}$ and 5-HT$_{1D}$ receptors. Naunyn-Schmiedeberg's Arch. Pharmacol. **340:** 135–138.
39. SCHOEFFTER, P. & D. HOYER. 1989. 5-Hydroxytryptamine 5-HT$_{1B}$ and 5-HT$_{1D}$ receptors mediating inhibition of adenylate cyclase activity: Pharmacological comparison with special reference to the effects of yohimbine, rauwolscine and some β-adrenoceptor antagonists. Naunyn-Schmiedeberg's Arch. Pharmacol. **340:** 285–292.
40. SEUWEN, K., I. MAGNALDO & J. POUYSSÉGUR. 1988. Serotonin stimulates DNA synthesis in fibroblasts acting trough 5-HT$_{1B}$ receptors coupled to a G_i-protein. Nature **335:** 254–256.
41. MURPHY, T. J. & D. B. BYLUND. 1988. Oxymetazoline inhibits adenylate cyclase by activation of serotonin-1 receptors in the OK cell, an established renal epithelial cell line. Mol. Pharmacol. **34:** 1–7.
42. HOYER, D. 1988. Molecular pharmacology and biology of 5-HT$_{1C}$ receptors. TIPS **9:** 89–94.
43. FENIUK, W., P. P. A. HUMPHREY & M. J. PERREN. 1987. Selective vasoconstrictor action of GR 43175 on arteriovenous anastomoses in the anaesthetised cat. Br. J. Pharmacol. **92:** 756P.
44. HUMPHREY, P. P. A., W. FENIUK, M. J. PERREN, H. E. CONNOR, A. W. OXFORD, I. H. COATES & D. BUTINA. 1988. GR 43175—A selective agonist for the 5-HT$_1$-like receptors in dog isolated saphenous vein. Br. J. Pharmacol. **94:** 1123–1132.
45. CONNOR, H. E., W. FENIUK & P. P. A. HUMPHREY. 1989. Characterisation of 5-HT receptors mediating contraction of canine and primate basilar artery by use of GR 43175, a selective 5-HT$_1$-like receptor agonist. Br. J. Pharmacol. **96:** 379–387.
46. DOENICKE, A., J. BRAND & V. L. PERRIN. 1988. Possible benefit of GR43175, a novel 5-HT$_1$-like receptor agonist, for the acute treatment of severe migraine. The Lancet:1309.
47. WAEBER, C., D. HOYER & J. M. PALACIOS. 1989. GR 43175 displaces [^{3}H]5-HT from 5-HT$_{1D}$ sites in monkey and human brain: An autoradiographic study. Synapse. **4:** 168–170.
48. ENGEL, G., M. GÖTHERT, E. MULLER-SCHWEINITZER, E. SCHLICKER, L. SISTONEN, P. A. STADLER. 1983. Evidence for common pharmacological properties of [^{3}H]5-hydroxytryptamine binding sites, presynaptic 5-hydroxytryptamine autoreceptors in CNS and inhibitory presynaptic 5-hydroxytryptamine receptors on sympathetic nerves. Naunyn-Schmiedeberg's Arch. Pharmacol. **324:** 116–124.
49. MIDDLEMISS, D. N. 1988. Autoreceptors regulating serotonin release. *In* The Serotonin Receptors. E. Sanders-Bush, Ed. Clifton, NJ: Humana Press, pp. 201–224.
50. MIDDLEMISS, D. N. 1984. 8-hydroxy-2-(di-n-propylamino)-tetralin is devoid of activity at the 5-hydroxytryptamine autoreceptor and the [^{3}H]5-HT recognition site. Naunyn-Schmiedeberg's Arch. Pharmacol. **327:** 18–22.
51. MIDDLEMISS, D. N. 1985. The putative 5-HT$_1$ receptor agonist, RU 24969, inhibits the

efflux of 5-hydroxytryptamine from rat frontal cortex slices by stimulation of the 5-HT autoreceptor. J. Pharm. Pharmacol. **37:** 434–438.
52. MIDDLEMISS, D. N. 1986. Blockade of the central 5-HT autoreceptor by β-adrenoceptor antagonists. Eur. J. Pharmacol. **120:** 51–54.
53. ENGEL, G., M. GÖTHERT, D. HOYER, E. SCHLICKER & K. HILLENBRAND. 1986. Identity of inhibitory presynaptic 5-hydroxytryptamine (5-HT) autoreceptors in the rat brain cortex with 5-HT_{1B} binding sites. Naunyn-Schmiedeberg's Arch. Pharmacol. **332:** 1–7.
54. GALZIN, A. M., J. P. CHODKIEWICZ, M. F. POIRIER, H. LOO, F. X. ROUX, A. REDONDO, A. LISTA, R. RAMDINE, P. BLIER & S. Z. LANGER. 1988. Modulation of the electrically-evoked release of ^{3}H 5-HT from slices of human frontal cortex membranes. Br. J. Pharmacol. **93:** 14P.
55. SCHIPPER, J., J. A. M. VAN DER HEYDEN & B. OLIVIER. 1987. Species differences in serotonin autoreceptors. Soc. Neurosci. Abstr. **13(1):** 345.
56. MIDDLEMISS, D. N., M. E. BREMER & S. M. SMITH. 1988. A pharmacological analysis of the 5-HT receptors mediating inhibition of 5-HT release in the guinea-pig frontal cortex. Eur. J. Pharmacol. **157:** 101–107.
57. GALZIN, A. M., P. BLIER, J. P. CHODKIEWICZ, M. F. POIRIER, H. LOO, F. X. ROUX, A. REDONDO, A. LISTA, R. RAMDINE & S. Z. LANGER. 1988. Pharmacological characterization of the serotonin 5-HT autoreceptor modulating the electrically-evoked release of ^{3}H-5-HT from slices of human frontal cortex. Soc. Neurosci. Abstracts. **14:** 129.7
58. SCHIPPER, J. & M. T. M. TULP. 1988. Serotonin autoreceptors in guinea pig cortex slices resemble the 5-HT_{1D} binding site. Soc. Neurosci. Abstr. **14:** 221.8.
59. SCHLICKER, E., K. FINK, M. GÖTHERT, D. HOYER, G. MOLDERINGS, I. ROSCHKE & P. SCHOEFFTER. 1989. The pharmacological properties of the presynaptic 5-HT autoreceptor in the pig brain cortex conform to the 5-HT_{1D} receptor subtype. Naunyn-Schmiedeberg's Arch. Pharmacol.
60. HOYER, D. & D. N. MIDDLEMISS. 1989. The pharmacology of the terminal 5-HT autoreceptors in mammalian brain: Evidence for species differences. TIPS **10:** 130–132.
61. SCHOEFFTER, P. & D. HOYER. 1989. Is the sumatriptan (GR 43175)-induced endothelium-dependent relaxation of pig coronary arteries mediated by 5-HT_{1D} receptors? Eur. J. Pharmacol. **166:** 117–119.
62. KOBILKA, B. K., T. FRIELLE, S. COLLINS, T. YANG-FENG, T. S. KOBILKA, U. FRANCKE, R. J. LEFKOWITZ & M. CARON. 1987. An intronless gene encoding a potential member of the family of receptors coupled to guanine nucleotide regulatory proteins. Nature **329:** 75–79.
63. SCHOEFFTER, P. & D. HOYER. 1990. 5-Hydroxytryptamine (5-HT)-induced endothelium-dependent relaxation of pig coronary arteries is mediated by 5-HT receptors similar to the 5-HT_{1D} receptor subtype. J. Pharmacol. Exp. Ther. **252:** 387–395.

DISCUSSION OF THE PAPER

MANFRED GÖTHERT (*University of Bonn*): I noticed that several agonists were not able to fully relax the pig coronary arteries. Nevertheless, your functional data correlated with the results from radioligand binding experiments. Did you assume that these compounds were full agonists or did you calculate the EC_{50} on the basis of reduced intrinsic activity?

HOYER: The functional data were fitted using computer-assisted non-linear regression analysis (the same way binding data are being analyzed). The results are expressed as pEC_{50} and Emax values, which may be lower than the maximal effects obtained with 5-HT or 5-CT. No assumption is being made on whether the compounds act as partial or full agonists. Thus, we report true pEC_{50} values. Adenylate cyclase data are analyzed the same way and pEC_{50} values from both tests compared.

GÖTHERT: Did you find monophasic concentration response curves with agonists in the pig coronary arteries, or did you find, as we found in the pig coronary, bell shaped concentration response curves? We got a nice relaxation in the lower concentration range, but some agonists tend to reverse to contraction at higher concentrations.

HOYER: A few agonists will indeed produce bell shaped relaxations. The points used for the fit are those of the concentrations producing relaxation only, that is before reversal of the relaxing effect.

GÖTHERT: I noticed that most of the competition curves were monophasic, with the exception of 5-carboxamidotryptamine (5-CT). Would you speculate on this?

HOYER: I do not like to speculate on this matter. With the exception of 5-CT, most ligands produced steep (monophasic) competition curves in binding experiments. Heterogeneity has been suggested, and it seems that GR 43175 discriminates between two sites or affinity states (Br. J. Pharmacol, 1989, **98:** 2931); we could not reproduce such results in our test system, maybe because we use other membranes (calf compared to neonatal pig). Whatever the reasons (different affinity states or heterogeneity of binding sites), we obtained good correlations between data from radioligand binding inhibition of adenylate cyclase and relaxation of pig coronary arteries. This would speak against receptor heterogeneity, but it cannot be excluded.

GÖTHERT: Could you indicate the percentage of 5-HT_{1B} and 5-HT_{1D} binding in the calf caudate nucleus.

HOYER: We have not been able to identify 5-HT_{1B} sites in calf brain (or brain of any other species, except rat and mouse) in radioligand binding studies, autoradiography or inhibition of forskolin-stimulated adenylate cyclase. The same holds true for 5-HT_{1D} sites which could not be identified in rat brain: the brain of rat (or mouse) has high levels of 5-HT_{1B} binding which seems to be homogeneous. Further, adenylate cyclase in rat substantia nigra is characteristic of a 5-HT_{1B} receptor mediated effect, in contrast to substantia nigra from calf or guinea-pig which are of the 5-HT_{1D} type. These findings again suggest that the presence of these receptors is mutually exclusive. Nevertheless, the tools to demonstrate this convincingly are currently missing. Probably the final answer will come once both receptors will be cloned and their messenger RNAs detected.

GÖTHERT: Due to the good correlation between adenylate cyclase data and relaxation of pig coronary artery, do you believe that the cAMP dependent protein kinase is involved in the contraction?

HOYER: We have presently no evidence that these receptors are coupled to adenylate cyclase in the coronary, but we cannot exclude it either. I do not think that one can classify receptors on the basis of receptor effector coupling, although it may help. A given receptor can couple to different second messenger or effector systems.

GÖTHERT: May I add to this that there is a paper by Flavahan, Shimokawa and Vanhoutte (J. Physiol. 1989, **408:** 549–560) showing that pertussis toxin does abolish the relaxing effect produced by 5-HT in the pig coronary. These results would suggest that adenylate cyclase activity is involved in the relaxation.

HOYER: That pertussis toxin blocks the relaxant effects in this model does not help much. It is indeed possible that adenylate cyclase inhibition is involved, even more so since this is presently the only second messenger system described for 5-HT_{1D} receptors. But one should keep in mind that pertussis toxin interacts with a variety of G proteins, not only Ni.

5-HT_2 Receptors, Roles and Regulation

J. E. LEYSEN[a] AND P. J. PAUWELS

Department of Biochemical Pharmacology
Janssen Research Foundation
B-2340 Beerse, Belgium

INTRODUCTION

5-Hydroxytryptamine$_2$ (5-HT_2) receptors reveal a specific distribution in mesolimbic brain areas (frontal cortex, nucleus accumbens and tuberculum olfactorium) and occur on smooth muscle cells and platelets. They have been detected in radioligand binding studies using labeled 5-HT_2 antagonists and agonists[1]; the antagonist [^{3}H]ketanserin is the most widely used,[2] and recently [^{3}H]DOB was described as a useful agonistic ligand.[3] 5-HT_2 receptors operate through stimulation of the turnover of phosphatidyl inositol, in which diacylglycerol and inositol trisphosphate have second messenger functions.[4,5] In mammals, administration of a 5-HT_2 agonist causes detrimental effects such as head twitch response and clonic seizures in rodents, hallucinations in humans and vasoconstriction and platelet aggregation. The various effects are blocked by 5-HT_2 antagonists with potencies corresponding to their *in vitro* measured 5-HT_2 receptor binding affinities.[1-7] Administration of a 5-HT_2 antagonist by itself does not elicit directly observable neurological, behavioral or subjective effects in animals or humans.[6] Yet, increase in slow wave sleep[8,9] and improvement of blood circulation[10] were reported. In patients, the 5-HT_2 antagonist, ritanserin, revealed therapeutic action in dysthymia and generalized anxiety, and improved negative symptoms in schizophrenia.[11,12]

Studies on 5-HT_2 receptor regulation by long-term drug treatment in rodents, have shown that they are down regulated by treatment with various antidepressants, 5-HT antagonists and non-specific neuroleptics.[5,13,14] Several observations pointed to an anomalous regulation of these receptors: the down regulation by chronic treatment with antidepressants, thought to be a consequence of enhanced serotonergic transmission, still occurred in the absence of endogenous serotonin; chronic treatment with certain specific 5-HT-reuptake blockers did not produce receptor desensitization or down regulation; depletion or destruction of serotonergic neurons did not affect receptor numbers; treatment with various serotonin antagonists produced desensitization and down regulation instead of an expected supersensitivity and up regulation (for a review see ref. 5). Recently, we observed a rapid desensitization and down regulation of central 5-HT_2 receptors following repeated treatment of rats with DOM.[15] In cultured calf aorta smooth muscle cells, the 5-HT_2 receptor mediated phosphatidyl inositol hydrolysis became rapidly desensitized by treatment with 5-HT and the 5-HT_2 agonist, 1-(2,5-dimethoxy-4-methylphenyl)-2- aminopropane (DOM).[16] In this study we further elaborated the 5-HT_2 receptor down regulation *in vivo* and *in vitro* following agonist and antagonist treatment. A new hypothesis on the physiological versus pathological role of 5-HT_2 receptors is presented.

[a]Author to whom correspondence should be addressed.

MATERIALS AND METHODS

In Vivo *Experiments*

Male Wistar rats (200 g) were treated with DOM, ketanserin or ketanserin plus DOM with a 15 min interval between drugs. Drugs were given at a dose of 2.5 mg/kg s.c. (injection volume 1 ml/100 g body weight), repeated treatments were given every 8 h. Immediately after the injection, the head twitch response was scored for 1 h to 3 h. The animals were killed 24 h after the last injection and brain areas were immediately dissected. Tissue pools of 3 animals were made and either used freshly ([^{3}H]DOB binding) or frozen in liquid nitrogen and stored at −80°C. Tissues were homogenized in 50 volumes of the incubation buffer. The homogenate was centrifuged for 10 min at 30590 × *g* at 4°C in a refrigerated Sorvall RC-5B centrifuge. The pellet was homogenized in 10 ml buffer and incubated for 10 min at 37°C and centrifuged again. The homogenization, incubation, centrifugation procedure was repeated. The final pellet was suspended in the incubation buffer in a dilution of 100 or 200 volumes per g wet weight of tissue for [^{3}H]DOB or [^{3}H]ketanserin binding, respectively. Radioligand receptor binding assays were performed and data analyzed as previously described.[3,13,15]

In Vitro *Experiments on Cells*

Calf aorta smooth muscle cells were cultured in serum-supplemented DMEM (Dulbecco's modified Eagle medium) using 24-well tissue culture plates.[16]

Confluent cultures were kept for at least 24 h in serum-free medium. Cells were treated as described under results. The cells were loaded with [^{3}H]inositol (2.5 μCi/well) for 24 h. The cells were washed 5-times with DMEM at 37°C and then stimulated for 1 h with 10^{-5} M 5-HT in the presence of 10^{-3} M LiCl. [^{3}H]Inositol phosphates were extracted and separated from [^{3}H]inositol on Dowex AG1-X8 columns as described.[17]

RESULTS

In Vivo *Experiments*

Rats were treated 2, 4 and 10 times with DOM (2.5 mg/kg, s.c. every 8 h), the head twitch response was scored for 1 h immediately after the injections; 24 h after the last injection brain tissues were taken for radioligand receptor binding assays. When given acutely, 2.5 mg/kg DOM produced an intensive head twitch response which lasted for more than 3 h (see control in FIG. 2). Data, presented in FIGURE 1a, show that after 2 injections there was already a significant reduction in the head twitch response and in the total number of 5-HT$_2$ receptor binding sites labeled with [^{3}H]ketanserin. Near maximal reductions of both parameters were obtained after 4 injections. The K_D-values of [^{3}H]ketanserin binding were slightly increased in the rats treated 4 and 10 times. After 4 DOM injections the number of 5-HT$_2$ receptor sites in the frontal cortex, increased again slowly with prolonged drug-free periods; after 144 h they had returned to 84% of the control levels (see data in FIG. 1b). The receptor half-life was calculated to be 5 days in the 1 to 3 day drug-free period and 3 days in the 3 to 6 day drug-free period.

In the animals treated 4 or 10-times with DOM and killed 24 h later, radioligand

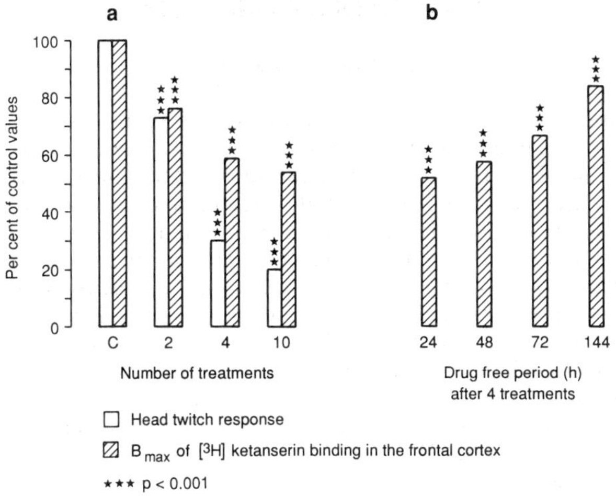

FIGURE 1. Head twitch response □ and B_{max}-value of [^{3}H]ketanserin binding to 5-HT$_2$ receptors in the frontal cortex ▨ in rats repeatedly treated with DOM (2.5 mg/kg, s.c. every 8 h). **a.** Rats were treated various times with DOM. The head twitch response was scored for 1 h immediately after each injection (control: 75 ± 33 twitches in 1 h, n = 96, mean value ± SD). [^{3}H]ketanserin in the frontal cortex was measured as previously described,[13,16] rats were killed 24 h after the last treatment (control B_{max} = 25.7 ± 3.2 fmoles/mg tissue, K_D = 0.47 ± 0.13 nM, n = 10, mean values ± SD). Presented values are means obtained in 10 independently assayed groups of rats. **b.** Rats were treated 4-times with DOM and killed at various times after the last injection. Values are means obtained in 5 independently assayed groups of rats.

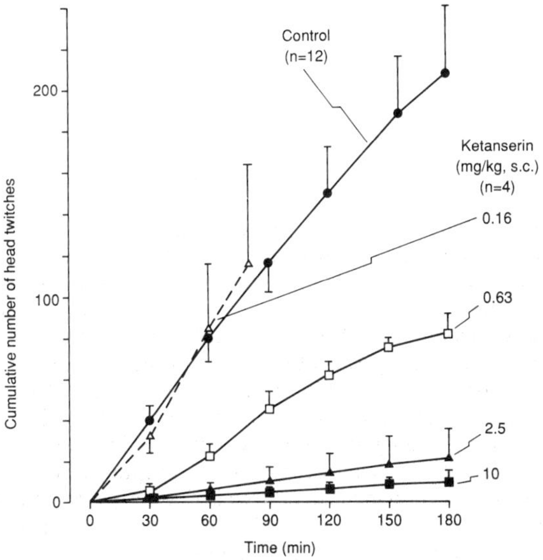

FIGURE 2. Head twitch response in rats treated with DOM (2.5 mg/kg, s.c.) (control) and with the indicated doses of ketanserin 15 min before DOM. Mean values ± SEM are presented.

binding to $5\text{-}HT_{1A}$ and $5\text{-}HT_{1B}$ sites in the hippocampus, to α_2- and β-adrenergic receptors in the cortex and to dopamine-D_2 receptors in the striatum was unaltered. The above study of agonist treatment was reported in detail previously.[15]

The antagonism between the $5\text{-}HT_2$ agonist, DOM and the $5\text{-}HT_2$ antagonist, ketanserin, was investigated. In an acute experiment various dosages of ketanserin were administered 15 min before DOM (2.5 mg/kg, s.c.) and the head twitch response was scored during 3 h. Data are presented in FIGURE 2. At a dose of 2.5 mg/kg, s.c. ketanserin was found to block completely the effect of DOM. Rats were then treated 4 times with either DOM, ketanserin plus DOM, or ketanserin; both drugs were given at a dose of 2.5 mg/kg, s.c. every 8 h; animals were killed 24 h after the last injection. Data on B_{max}- and K_D-values of [^{3}H]ketanserin binding in the frontal cortex are presented in TABLE 1. With all drug treatments the B_{max}-values were significantly reduced and K_D-values were unchanged.

In a last series of experiments, ketanserin (2.5 mg/kg, s.c. every 8 h) was given 10 times to rats. Twenty-four hours after the last injection both [^{3}H]ketanserin and [^{3}H]DOB binding were measured in the frontal cortex. Data are presented in TABLE 2. In the controls, B_{max}-values of [^{3}H]DOB binding were 3.55-times lower than those of [^{3}H]ketanserin. However, with both ligands a similar reduction by about 30% in B_{max}-values was seen in the ketanserin-treated animals.

In Vitro *Experiments on Vascular Smooth Muscle Cells in Culture*

In confluent cultures kept for at least 24 h in a serum-free medium, 1 h stimulation with 5-HT produced a concentration-dependent increase in inositol phosphate formation. Maximal stimulation was attained at 10^{-4} M 5-HT and amounted to 1200% of the basal level. The $5\text{-}HT_1$ agonists, *N,N*-dipropyl-8-hydroxy-2-aminotetraline (8-OHDPAT) and RU 24969, did not stimulate the [^{3}H]inositol phosphates formation up to 10^{-4} M. The 5-HT-stimulated [^{3}H]inositol phosphates formation gradually increased with time and was maximal after 1 h. The 5-HT-induced [^{3}H]inositol phosphates formation was inhibited at nM concentrations by various $5\text{-}HT_2$ antagonists (Pauwels *et al.*, in preparation).

Cells were pretreated with an agonist, 10^{-5} M 5-HT or DOM, for various time periods between 15 min and 24 h or with an antagonist, 10^{-7} M setoperone or ketanserin, for 15 min or 24 h. Drugs were given during the [^{3}H]inositol-loading period. Cells were washed 5 times and stimulated with 5-HT (10^{-5} M) for 1 h in the presence of 10^{-3} M LiCl. [^{3}H]Inositol phosphates were separated and quantified.

Pretreatment with 5-HT and DOM reduced the 5-HT–induced [^{3}H]inositol phosphates formation by 20% after 15 min and by 80% after 1 h pretreatment; the response

TABLE 1. [^{3}H]Ketanserin Binding to $5\text{-}HT_2$ Receptors in the Frontal Cortex of Rats 24 h after Repeated Drug Treatment (4 times 2.5 mg/kg s.c., every 8 h)

		Mean values ± SEM		
Treatment		B_{max} (fmoles/mg tissue)	Δ%	K_D (nM)
Control	n = 7	22.5 ± 0.7		0.43 ± 0.02
DOM	n = 5	13.6 ± 0.7***	−40	0.46 ± 0.05
Ketanserin + DOM	n = 5	15.4 ± 0.7***	−31	0.47 ± 0.05
Ketanserin	n = 5	18.2 ± 0.7***	−19	0.47 ± 0.04

***$P < 0.001$ significance of difference from controls according to Student's t-test.

TABLE 2. 5-HT_2 Receptor Binding in the Frontal Cortex of Rats 24 h after 10 Injections of Ketanserin (2.5 mg/kg s.c., every 8 h)

	Mean values ± SEM (n)					
	[^{3}H]Ketanserin Binding[a]			[^{3}H]DOB Binding[b]		
	K_D	B_{max} (fmoles/mg tissue)	Δ%	K_D	B_{max} (fmoles/mg tissue)	Δ%
Control	0.82 ± 0.09	21.7 ± 2.6 (11)		2.7 ± 0.3	6.1 ± 0.5 (6)	
Treated	0.96 ± 0.10	15.6 ± 1.6 (12)	−28	3.3 ± 0.4	4.2 ± 0.5 (6)	−31

[a] In this experiment the tissue was prepared in buffer containing 0.1% BSA in order to get a better desorption of residual drug from the tissue. 0.1% BSA was also present during the incubation with [^{3}H]ketanserin. This resulted in a 30% reduction of the non-specific binding, but also in a twofold increase in the K_D-value as compared to assays without BSA (see TABLE 2), probably due to a reduction in the free [^{3}H]ketanserin concentration because of binding of the ligand to the BSA.

[b] [^{3}H]DOB binding was measured in freshly prepared tissue in conditions as described by Titeler *et al.*[3] [^{3}H]DOB was used at 8 concentrations between 0.2 and 8 nM. Non-specific binding was measured in the presence of 1000-fold excess ketanserin. Data were analyzed in Scatchard plots as described.[13] The plots were linear over the entire concentration range, correlation coefficients of the lines calculated by the method of least squares were >0.9.

was completely abolished after 6 h pretreatment. In cells pretreated for 2 h and 24 h, respectively, with 10^{-5} M 5-HT, and then kept for various time periods in a 5-HT-free medium, the 5-HT–induced [^{3}H]inositol phosphates formation gradually returned to normal levels with half-times of 5 h and 12 h, respectively.

Following antagonist pretreatment, the 5-HT–induced inositol phosphate formation was reduced by about 50% with setoperone and by about 30% with ketanserin. Effects were similar with 15 min and 24 h pretreatment (see data in TABLE 3) (Pauwels *et al.*, in preparation).

DISCUSSION

As previously described,[15] the *in vivo* experiments with DOM treatment of rats, revealed a rapid desensitization and down regulation of the central 5-HT_2 receptors. This appeared from the reduction in the DOM-induced behavioral response, which was already significant after the second injection and maximal after 4 injections given in 24 h. Concomitantly, a substantial decrease (−24 and −41% after 2 and 4 injections) in the total number of 5-HT_2 receptor binding sites in the frontal cortex was found. *In vivo*, the 5-HT_2 receptor numbers recovered only slowly (receptor half-lives of 5 days initially to 3 days in the later drug-free period). Hence central 5-HT_2 receptors, seem to be rapidly degraded and a slow receptor resynthesis takes place. In accordance with our findings, a decrease in 5-HT_2 receptor binding in rat brain following 7 and 8 days treatment with two other 1-(2,5-dimethoxyphenyl)-2-amino-propane derivatives, DOB and DOI, was recently reported.[18,19] McKenna *et al.*[19] found a similar decrease of the receptor binding sites using the antagonist ligand [^{3}H]ketanserin and using the agonist ligand [^{125}I]R-DOI. Also the early noticed, rapid and long lasting tolerance to LSD-induced psychotic and behavioral effects in human[20] and laboratory animals[21-22] could probably be due to 5-HT_2 receptor down regulation.

The present investigations of antagonism of DOM induced effects by ketanserin, showed that the acute behavioral effect was dose-dependently antagonized. However, the DOM-induced down regulation of [^{3}H]ketanserin labeled 5-HT_2 receptor sites was apparently not prevented by co-treatment with ketanserin and DOM. Co-administration of both drugs produced a down regulation by 31%, which was 10% less than the down regulation by DOM alone. Ketanserin by itself, in a similar treatment schedule, produced a down regulation which was half the down regulation produced by DOM alone. The down regulation by the agonist and the antagonists were apparently not additive. However, it can be supposed that the agonist- and antagonist-induced down regulation occur via different mechanisms. It could be that the agonist-induced down regulation was (partially) prevented by ketanserin, but this effect was masked, by the down regulation via another mechanism, which was produced by ketanserin by itself. The presently observed 5-HT_2 receptor down regulation by ketanserin is in accordance with previous studies where a decrease in 5-HT_2 receptor numbers following treatment with various antagonists, such as ketanserin, setoperone, ritanserin, cyproheptadine, pizotifen, and mianserin was reported.[5,13] Finally we showed that repeated treatment with ketanserin alone produced the same degree of down regulation of 5-HT_2 receptor binding sites labeled with the antagonist [^{3}H]ketanserin and with the agonist [^{3}H]DOB. However, [^{3}H]DOB labeled a 3.55-times lower number of sites than [^{3}H]ketanserin. This was also apparent in the original study on [^{3}H]DOB binding[3] and was ascribed to the possibility that only part of the receptor binding sites exist in a high affinity agonist binding state. In spite of the differences in B_{max}-values, agonist and antagonist labeled 5-HT_2 receptor sites appear to undergo the same degree of down regulation *in vivo* following antagonist treatment, as seen in this study, and following agonist treatment, as reported by McKenna *et al.*[19]

In order to study 5-HT_2 receptor regulation on a more pure cellular system, we used cultured calf aorta smooth muscle cells. These cells show a 5-HT-induced phosphatidyl inositol hydrolysis, mediated by 5-HT_2 receptors.[16] The 5-HT–induced inositol phosphate formation was rapidly desensitized by agonist stimulation, nearly full desensitization was obtained after 2 h, but the response recovered readily with half-times of 5 h to 12 h. The rapid desensitization followed by a relatively rapid resensitization points to de-coupling/re-coupling or internalization/recycling phenomena. Probably little receptor degradation had occurred, which would require longer resynthesis times. In this regard the regulation of 5-HT_2 receptors on vascular smooth muscle cells differed from the regulation of central 5-HT_2 receptors *in vivo* where a rapid receptor degradation and slow resynthesis was observed.

After pretreatment of cells *in vitro* with the 5-HT_2 antagonists, setoperone and ketanserin, a decrease in the 5-HT-induced inositol phosphate formation was seen, but the effects were similar after short (15 min) and prolonged (24 h) pretreatment (TABLE 3). The most obvious interpretation would be that the drugs had been insufficiently removed from the cells. However, this is not expected according to receptor binding and physicochemical properties of the drugs. In *in vitro* radioligand binding experiments on rat frontal cortex, setoperone, and ketanserin displayed the same binding affinities (K_i: 0.40 nM), rapid dissociation from the receptor ($t_{1/2}$: 6.8 and 5.8 min), and fully competitive inhibition. Setoperone is less lipophilic and more basic than ketanserin (log P_{app} at pH 7.6: 1.60 for setoperone, 3.05 for ketanserin, pKa: 8.36 for setoperone, 7.5 for ketanserin).[23] Theoretically, the more lipophilic compound would be more difficult to wash out, yet the reduction in response was less with ketanserin than with setoperone. An alternative interpretation would be that the drugs have an immediate effect on the receptor-transducer coupling or receptor processing mechanism, leading to a reduction in the response. This latter explanation would be in accordance with the *in vivo* observed 5-HT_2 receptor down regulation by antagonists.[5,13]

The action of 5-HT_2 agonists and antagonists and the role and regulation of 5-HT_2 receptors reveals a number of remarkable facts. 5-HT_2 agonist elicit detrimental central and peripheral effects. In contrast, 5-HT_2 antagonist do not cause apparent neurological, behavioral or subjective effects, and they do not disrupt autonomic functions in laboratory animals or humans. The increase in slow wave sleep is considered to be beneficial, and thus far no unwanted side effects have been observed in the clinic. This is clearly distinct from antagonists at various other neurotransmitter receptors which cause disruption of autonomic functions (e.g. adrenergic and cholinergic antagonists), neurological side effects (dopamine antagonists), psychotic side effects (muscarinic antagonist), or sedation and depression (adrenergic antagonists). Moreover, for the known subtypes of adrenergic, dopaminergic and cholinergic receptors, chronic receptor blockade leads to receptor supersensitivity. This is also not the case with 5-HT_2 receptor blockade as described above. The 5-HT_2 receptor system is perhaps most comparable to the histamine-H_1 receptor system, where agonist stimulation is also most harmful, and where antagonists by themselves, besides occasional induction of somnolence, have little effect on normal functions. Recently we observed that persistent histamine-H_1 receptor blockade does not lead to receptor up regulation.

TABLE 3. 5-HT (10^{-5} M) Stimulation of Inositol Phosphate Formation in Calf Aorta Smooth Muscle Cells Pretreated with 5-HT_2 Antagonists (10^{-7} M) and Washed 5 Times

		Percent of Control Stimulus (mean ± SD, (n))[a]	
Drug\Pretreatment	Time	15 minutes	24 hours
Setoperone		46 ± 13 (6)	42 ± 7 (6)
Ketanserin		64 ± 18 (2)	74 ± 1 (2)

[a]Control values:
basal [^{3}H]inositol phosphates per well: (272 ± 90) DPM;
5-HT (10^{-5} M)-stimulated inositol phosphate per well: (2314 ± 1348) DPM;
ratio stimulated/basal: 8.71 ± 3.47; (mean values ± SD, n = 8).

The whole of observations on the 5-HT_2 receptor system, prompted a new hypothesis. 5-HT_2 receptors probably receive little impetus under normal physiological conditions. Therefore, acute blockade of the receptor by an antagonist, does not disrupt the normal situation, which explains the lack of observable effects by the 5-HT_2 antagonists by themselves. Moreover, with lack of sufficient stimulus, the 5-HT_2 receptors could normally exist in a supersensitive state, through which denervation or blockade cannot produce further supersensitivity. On the other hand, agonist treatment, which readily causes detrimental effects, induces a very rapid and, in the central nervous system, long lasting receptor desensitization. This looks as an effective protective mechanism against harmful stimulation.

The serotonergic system acting through central 5-HT_2 receptors, probably plays a subtle, moderatory role in processes controlling mood, social behavior, and sleep patterns. In certain pathological conditions such as dysthymia, generalized anxiety and negative symptoms in schizophrenia, 5-HT_2 receptor stimulation is probably increased and the down regulatory response may be deficient. In such cases, therapeutic effects of 5-HT_2 antagonists are to be expected, by acutely blocking the receptors and by causing receptor down regulation upon prolonged treatment.

SUMMARY

Stimulation of 5-HT_2 receptors in mammals by agonists causes detrimental neurological, psychological, and circulatory effects. 5-HT_2 antagonists block the elicited effects, but by themselves, they do not cause any apparent behavioral, neurological or subjective effects. However, 5-HT_2 antagonists increase slow wave sleep and have a therapeutic action on impaired circulation, dysthymia, and negative symptoms in schizophrenia. Chronic treatment of rodents with various 5-HT_2 antagonists was reported to cause an anomalous desensitization and 5-HT_2 receptor down regulation.

In this study we further investigated the 5-HT_2 receptor regulation *in vivo* and *in vitro* by agonist and antagonist treatment. Treatment of rats with the 5-HT_2 agonist, 1-(2,5-dimethoxy-4-methylphenyl)-2-aminopropane (DOM) (2.5 mg/kg s.c., every 8 h), rapidly caused desensitization of the head twitch response (−20% and −80% after 2 and 4 injections) and a decrease in the number of frontal cortical 5-HT_2 receptors labeled with [^{3}H]ketanserin (−24% and −41%, 24 h after 2 and 4 injections). The receptor resynthesis/degradation revealed half-times of 5 days initially to 3 days in the later drug-free period. Administration of the antagonist ketanserin (2.5 mg/kg, s.c., every 8 h) 15 min before the agonist, antagonized the acute behavioral effect but did not prevent the 5-HT_2 receptor down regulation after 4 treatments. In contrast, ketanserin by itself, given 4 times, caused a reduction in the B_{max}-value of [^{3}H]ketanserin binding by 19% and given 10 times it caused a reduction in B_{max}-values by 28% and 31% of [^{3}H]ketanserin and [^{3}H]DOB binding in the frontal cortex. Hence 5-HT_2 receptors labeled by an antagonist and an agonist ligand were similarly decreased. In vascular smooth muscle cells in culture kept for at least 24 h in a serotonin-free medium before treatment, the 5-HT_2 receptor mediated 5-HT–induced inositol phosphate formation, was rapidly desensitized by agonist treatment: −20% after 15 min and −80% after 1 h incubation of the cells with 10^{-5} M 5-HT or DOM. After 2 h and 24 h treatment resensitization occurred with half-times of 5 h and 12 h, respectively. Pretreatment of the cells for 15 min or 24 h with 10^{-7} M of the antagonists setoperone or ketanserin, followed by extensive washing, caused a reduction in the 5-HT–induced inositol phosphate formation by about 50% with setoperone and by 30% with ketanserin. Effects of 15 min and 24 h drug pretreatment were similar. The effects could be due to insufficient wash out of the drugs or to acute effects on the receptor processing.

Consideration of the whole of previous and recent observations, prompted the hypothesis that 5-HT_2 receptors probably receive a very low stimulation under physiological conditions. Acute blockade does not disrupt the normal situation, explaining the lack of directly observable effects of 5-HT_2 antagonists. Because of the low stimulation the receptors could exist in a supersensitive state; chronic blockade or denervation does not produce further supersensitivity. The rapid and profound desensitization and receptor down regulation is an effective defensive mechanism against the detrimental effects of agonist stimulation. The mechanism of 5-HT_2 receptor down regulation by antagonists, which probably differs from the agonist-induced down regulation, is as yet not understood.

REFERENCES

1. LEYSEN, J. E. 1985. Characterization of serotonin receptor binding sites. *In* Neuropharmacology of Serotonin. A. R. Green, Ed. Oxford: Oxford Medical Publications, pp. 79–109.
2. LEYSEN, J. E., C. J. E. NIEMEGEERS, J. M. VAN NUETEN & P. M. LADURON. 1982.

[^{3}H]ketanserin (R 41 468), a selective ^{3}H-ligand for serotonin$_2$ receptor binding sites. Binding properties, brain distribution, and functional role. Mol. Pharmacol. **21**: 301–314.

3. Titeler, M., K. Herrick & R. A. Lyon. 1985. [^{3}H]DOB: A specific agonist radioligand for 5-HT$_2$ serotonin receptors. Eur. J. Pharmacol. **117**: 145–146.
4. Leysen, J. E., D. de Chaffoy de Courcelles, F. De Clerck, C. J. E. Niemegeers & J. M. Van Nueten. 1984. Serotonin-S$_2$ receptor binding sites and functional correlates. Neuropharmacology **23**: 1493–1501.
5. Conn, P. J. & E. Sanders-Bush. 1987. Central serotonin receptors: Effector systems, physiological roles and regulation. Psychopharmacology **92**: 267–277.
6. Awouters, F., C. J. E. Niemegeers, A. H. P. Megens, T. F. Meert & P. A. J. Janssen. 1988. Pharmacological profile of ritanserin: A very specific central serotonin S$_2$-antagonist. Drug Dev. Res. **15**: 61–73.
7. Glennon, R. A., M. Titeler & J. D. McKenney. 1984. Evidence of 5-HT$_2$ involvement in the mechanism of action of hallucinogic agents. Life Sci. **35**: 2505–2511.
8. Dugovic, C. & A. Wauquier. 1987. 5-HT$_2$ receptors could be primarily involved in the regulation of slow-wave sleep in the rat. Eur. J. Pharmacol. **137**: 145–146.
9. Idzikowski, C., F. J. Mills & R. Glennard. 1986. 5-Hydroxytryptamine-2 antagonist increases human slow wave sleep. Brain Res. **378**: 164–168.
10. Van Nueten, J. M., P. A. J. Janssen, J. Symoens, W. J. Janssens, J. Heykants, F. De Clerck, J. E. Leysen, H. Van Cauteren & P. M. VanHoutte. 1988. Ketanserin. *In* New Cardiovascular Drugs. A. Scriabine, Ed. New York: Raven Press, pp. 1–56.
11. Janssen, P. A. J. 1987. The development of new antipsychotic drugs: Towards a new strategy in the management of chronic psychoses. J. Drug Ther. Res. **12**: 324–328.
12. Reyntjens, A., Y. G. Gelders, M.-L. J. A. Hoppenbrouwers & G. Vanden Bussche. 1986. Thymostenic effects of ritanserin (R 55 667), a centrally acting serotonin-S$_2$ receptor blocker. Drug Dev. Res. **8**: 205–211.
13. Leysen, J. E., P. Van Gompel, W. Gommeren, R. Woestenborghs & P. A. J. Janssen. 1986. Down regulation of serotonin-S$_2$ receptor sites in rat brain by chronic treatment with serotonin-S$_2$ antagonists: Ritanserin and setoperone. Psychopharmacology **88**: 434–444.
14. Peroutka, S. J. & S. H. Snyder. 1980. Regulation of serotonin$_2$ (5-HT$_2$) receptors labeled with [^{3}H]spiroperidol by chronic treatment with the antidepressant amitriptyline. J. Pharmacol. Exp. Ther. **215**: 583.
15. Leysen, J. E., P. F. M. Janssen & C. J. E. Niemegeers. 1989. Rapid desensitization and down regulation of 5-HT$_2$ receptors by DOM treatment. Eur. J. Pharmacol. **163**: 145–149.
16. Pauwels, P. J., P. Van Gompel, D. de Chaffoy de Courcelles & J. E. Leysen. 1988. Agonist-induced desensitization of 5-HT$_2$ receptors on cultured calf aortic smooth muscle cells. Neurochem. Int. **12**: 409–414.
17. Bone, E. A., P. Fretten, S. Palmer, C. J. Kirk & R. H. Michell. 1984. Rapid accumulation of inositol phosphates in isolated rat superior cervical sympathetic ganglia exposed to V$_1$-vasopressin and muscarinic cholinergic stimuli. Biochem. J. **221**: 803–811.
18. Buckholtz, N. S., D. Zhou & D. X. Freedman. 1988. Serotonin$_2$ agonist administration down-regulates rat brain serotonin$_2$ receptors. Life Sci. **42**: 2439–2445.
19. McKenna, D. J., A. J. Nazarali, A. Himeno & J. M. Saavedra. 1989. Chronic treatment with (±)-DOI, a psychotomimetic 5-HT$_2$ agonist, downregulates 5-HT$_2$ receptors in rat brain. Neuropsychopharmacology **2**: 81–87.
20. Cholden, L. S., A. Kurland & C. Savage. 1955. Clinical reactions and tolerance to LSD in chronic schizophrenia. J. Nerv. Ment. Dis. **122**: 211–221.
21. Freedman, D. X., G. K. Aghajanian, E. M. Ornitz & B. S. Rosner. 1958. Patterns of tolerance to lysergic acid diethylamide and mescaline in rats. Science **127**: 1173–1174.
22. Trulson, M. E. & B. L. Jacobs. 1977. Usefulness of an animal behavioral model in studying the duration of action of LSD and the onset and duration of tolerance to LSD in the cat. Brain Res. **132**: 315–326.
23. Leysen, J. E., W. Gommeren & P. Van Gompel. 1988. The dissociation rate of unlabelled drugs from receptor sites: A poorly investigated, yet important aspect in receptor research. Studies on the serotonin-S$_2$ receptor. J. Recept. Res. **8**: 547–559.

DISCUSSION OF THE PAPER

MICHEL HAMON (*INSERM, Paris*): You showed that in fact the effect of the agonists/antagonists are not additive *in vivo* on the down regulation process, but you didn't mention this type of experiments in case of the smooth muscle cell culture. Did you check whether the effects of agonists and antagonists are additive in these cells?

LEYSEN: We did a preliminary experiment with antagonists in the smooth muscle cell culture. However, since you cannot wash cells as thoroughly as membranes, we were not entirely sure that we could completely wash out the antagonists. Some antagonists did produce an apparent desensitization. Otherwise, the agonists by themselves cause such a profound desensitization that you wouldn't be able to see an additive effect of the $5HT_2$ antagonist. In a preliminary experiment there was no reversal of the effect of the agonists.

SAUL MAAYANI(*Mt. Sinai School of Medicine, New York*): I like to confirm your finding yet on another tissue. The isolated guinea-pig trachea. We have this afternoon a poster to show that the half-time of the desensitization is 3 minutes. It gets back to a normal response in 3 hours. So what you found is completely confirmed by kinetic studies in isolated tissue.

LEYSEN: What I did not mention in my talk is that a very rapid desensitization of 5HT receptor–mediated responses was also seen with LSD in studies reported in the 1950s. The desensitization of the hallucinogenic response to LSD in humans is particularly worth mentioning.

MERRY SHERMAN (*Rutgers University, New Jersey*): Have you addressed the molecular mechanism of the desensitization in terms of demonstrating that *de novo* receptor resynthesis was essential in culture or demonstrating the lack of a simple modification of the protein such as phosphorylation, dephosphorylation.

LEYSEN: In the smooth muscle culture we assumed that there is no marked receptor degradation since the resensitization was very fast. We did not do yet the experiments which you suggested.

DANIEL HOYER (*Sandoz, Basel*): Since probes are available for $5HT_2$ receptors did you measure what happens to messengers when you desensitize?

LEYSEN: No, not yet. Didn't you do it?

HOYER: Not yet.

ELAINE SANDERS-BUSH (*Vanderbilt School of Medicine, Nashville, TN*): Could you review for us again the evidence that you have suggesting that the action of the agonist and the antagonist on down regulation *in vivo*, are by different mechanisms?

LEYSEN: When we looked at the effect of the agonist alone, a combination of the agonist and the antagonist, and the antagonist alone, then the down regulation produced by the combined administration of an agonist and an antagonist was just in between the effect of the drugs given separately. The agonist, by itself, produced the biggest decrease in receptor number (−40%). The antagonist, by itself, produced about a 20% decrease, which is half the effect of the agonist, a combination of the two produced a 30% decrease. So what I think, is that the antagonist can in fact partially prevent the down regulation by the agonist, but this is masked by the down regulation which the antagonist produces by itself, probably via a distinct mechanism.

D. L. NELSON (*Session Chair*): Josée I wonder if you'd be willing to speculate about whether this phenomena has any relationship to the model proposed by Kauman for the regulation of $serotonin_2$ receptors in the vasculature tissue where he has proposed that an allosteric site would be associated with the $serotonin_2$ receptor in certain blood vessels. I may get this wrong, but as I recall there's a situation where methysergide produces a non-competitive inhibition of the action to serotonin. You then add ritanserin, which is a serotonin antagonist, as well as methysergide and under these conditions methysergide now becomes a competitive antagonist to the actions of serotonin. On the basis of this he's proposed that there's an allosteric site associated with the $serotonin_2$ receptor complex.

LEYSEN: In binding studies we tried, by kinetic means, to look at competition between agonists and antagonists. When the agonist and the antagonist show a rapid dissociation from the receptor, then the mutual inhibition appears as competitive. This is not the case with ritanserin because that is such a slowly dissociating agent that you never get equilibrium. The inhibition by ritanserin appears as non-competitive, and I'm afraid that in the studies of Kauman the slow dissociation of ritanserin plays part in the results which he gets.

DAVID T. WONG (*Lilly Research Laboratories, Indianapolis, IN*): Have you thought about the idea that $5HT_{2A}$ and $5HT_{2B}$ receptors may exist in rat brain?

LEYSEN: According to the hypothesis by Peroutka *et al.* $5HT_{2A}$ receptor subtypes would be labeled by the radioactive 5HT agonists e.g. ^{3}H-DOB, whereas the antagonist ^{3}H-ketanserin would label a $5HT_{2A}$ and in addition $5HT_{2B}$-sites. In our *in vivo* study in rats on agonist-induced down regulation, we found the same per cent decrease of the sites labeled by ^{3}H-DOB and of the sites labeled by ^{3}H-ketanserin in the frontal cortex. This can be taken as an indication, that at least in rat brain, the two ligands label the same receptor subtype. I call it the $5HT_2$ receptor, our findings do not provide indications for further subclassification of this receptor.

5-HT_3 Receptors

M. B. TYERS

Department of Neuropharmacology
Glaxo Group Research Limited
Ware, Hertfordshire, SG12 ODP, England

No review of 5-HT_3 receptors would be complete without reference to the early experiments by Rocha e Silva *et al.*,[1] and Gaddum and Picarelli,[2] in which cocaine was first shown to have antagonist activity against serotonin-induced (neuronally mediated) contractions of guinea-pig isolated ileum. In these experiments cocaine at low concentrations specifically antagonized serotonin without affecting contractions induced by pilocarpine or histamine. After the publication of these now classical papers, interest in the so-called M-receptor virtually ceased, with fewer than five research reports published each year between 1960 and 1983. The current explosive increase in research activities relating to 5-HT_3 receptors and functions is largely attributable to the identification of selective 5-HT_3 receptor agonists and antagonists. In addition to the papers by Rocha e Silva on cocaine, can be added the experiments conducted by Fozard *et al.*,[3,4] who characterized the receptor mediating the neuronally mediated 5-HT-induced tachycardia response in the rabbit isolated heart as belonging to the same M-type described for the guinea-pig ileum. In addition, Fozard and Mobarok Ali[3] first described the M-receptor antagonist activity of metoclopramide. Since then, a number of research groups have designed more selective and potent antagonists (FIG. 1). These antagonists were identified in a number of isolated functional receptor preparations, but primarily, in the rabbit heart and isolated neuronal preparations such as the rabbit[5] and rat superior cervical ganglion[6] and vagus nerve[7] preparations. In these neuronal tissues serotonin causes a rapid membrane depolarization via activation of the same receptor as in the rabbit heart mediating tachycardia through noradrenaline release. The receptors mediating these actions of serotonin have been renamed as 5-HT_3[8] receptors and may be defined as mediating those responses of serotonin which are mimicked by the 5-HT_3 selective agonist 2-methyl-5-HT[9] and antagonized by MDL72222 and ICS 205-930 in nanomolar concentrations.

In addition to these characteristics, 5-HT_3 receptors have several distinguishable properties (TABLE 1). Receptors of the 5-HT_3-type are exclusively located on neurons of the peripheral and central nervous systems. Agonist actions induce a rapid membrane depolarization which desensitizes upon repeated exposure.[4,7] The rate of desensitization is dependent upon the concentration of the agonist[10] and appears to vary between tissues. For example, desensitization to the depolarizing response in the vagus nerve or N1E-115 cells occurs within minutes while in the dorsal root ganglion there is no indication of the development of desensitization.

Studies in NG108-15 cell lines[10] and in guinea-pig submucous plexus neurons[11] reveal that depolarization is associated with an inward sodium current which can be blocked by tetraethylammonium. Such studies have suggested that the 5-HT_3 receptor is a member of the ligand-gated ion channels and Derkach *et al.*[12] have recorded two channels in the guinea-pig submucous plexus preparation with conductancies of 9 and 15 pS which open following activation of 5-HT_3 receptors. These effects can be blocked by ICS 205-930 or GR38032 at concentrations typical of their affinities for 5-HT_3 receptors in this species. Whether these two channels represent two different receptors,

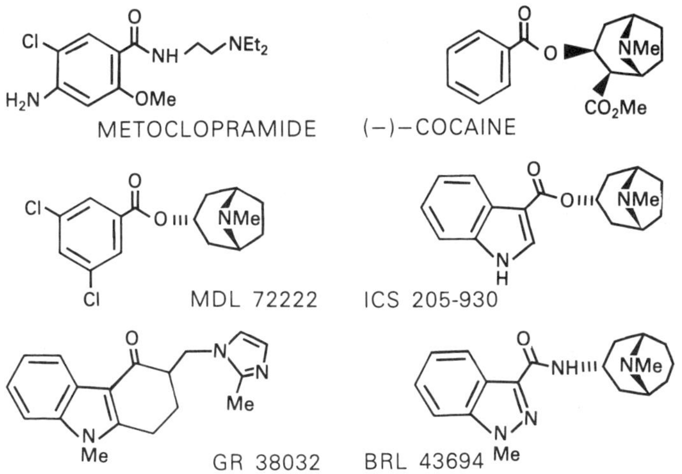

FIGURE 1. M-receptor antagonists.

or that they are both coupled to the same receptor type awaits more quantitative antagonist studies.

Many of the functions of 5-HT$_3$ receptors are concerned with the regulation of neurotransmitter release. In the guinea-pig ileum 5-HT$_3$ receptors mediate a contraction via at least two cholinergically mediated mechanisms although a contractile effect via substance P release is also probable[13,14]; the degree of involvement of the substance P mechanism appears to vary between strains of guinea-pig. Serotonin causes a biphasic dose-response curve in the guinea-pig longitudinal muscle myenteric plexus preparation (but not in the whole ileum): a high affinity portion of the dose response occurs at concentrations from 1×10^{-8} to 5×10^{-7}M and a lower affinity portion from 1×10^{-6}M. Atropine inhibits both responses through antagonism of muscarinic receptors on the smooth muscle; however, total inhibition by atropine is controversial.[14] In the totally atropine-sensitive preparations, it is only the lower affinity portion of the 5-HT$_3$ dose-response curve which is antagonized by 5-HT$_3$ antagonists—the high affinity portion is unaffected by 5-HT$_3$ antagonists and probably represents a different receptor sub-type similar to that reported by Dumuis *et al.*[15] for stimulation of adenylate cyclase in mouse embryo colliculi neuronal cell lines. The distinguishing characteristic is blockade by ICS 205-930 at micromolar concentrations. This poten-

TABLE 1. Characteristics of 5-HT$_3$ Receptors

- Located on neuronal membranes in peripheral and central nervous systems
- Membrane depolarization
- Rapidly desensitize
- Ligand-gate ion channels
- Serotonin active in micromolar concentrations
- Agonists: serotonin, 2-methyl-5-HT, phenylbiguanide
- Antagonists: MDL72222, ICS 205-930, ondansetron (GR38032)
- Binding ligands: ^{3}H-GR65630, ^{3}H-ICS 205-930, ^{3}H-zacopride

tially novel receptor has been proposed as the 5-HT_4 receptor but remains to be fully characterized before this nomenclature can become acceptable. Gaddum and Picarelli's M-receptor[2] was defined in the presence of dibenzyline which may or may not antagonize 5-HT_4 receptors; if not, the M-receptor can be assumed to include both the high and low affinity receptor types since Rocha e Silva *et al.*[1] had shown that cocaine completely antagonizes the serotonin contraction. However, in the guinea-pig *whole* ileum preparation, in contrast to the myenteric plexus longitudinal muscle preparation, serotonin only produces the low affinity portion of the dose response curve and hence it is more likely that the high affinity site was not observed by Rocha e Silva or Gaddum and Picarelli and, hence, the M-receptor is most probably the same as the guinea-pig 5-HT_3 receptor.

The suggestion for 5-HT_3 receptor sub-types[9] has not received general acceptance.[8] The differences observed in the affinities of antagonists for responses in different tissues have alternative explanations. For example, some of the distinguishing affinities, such as those obtained in the guinea-pig ileum and rabbit heart, are mediated indirectly through 5-HT_3 stimulated release of another neurotransmitter. The data obtained from such experiments can lead to misleading interpretations.[16] Furthermore, two of the antagonists used to classify these receptors were non-competitive in nature and hence pK_B values could be related to the pre-incubation time. Examination of values obtained from several laboratories support the concept that the affinities of antagonists vary considerably between tissues but closer examination reveals that these differences are more likely to be due to species differences than tissue differences (TABLE 2). Thus, in the guinea-pig ileum affinities appear always to be lower than in the rat vagus nerve; but guinea-pig ileum affinities are very similar to those in the guinea-pig vagus nerve.[17] Similarly, values for the rabbit heart are generally higher than for rat vagus nerve but values for rabbit vagus nerve are similar to those for rabbit heart. Hence the more reasonable conclusion on current data is that 5-HT_3 receptors differ *between* species but within a species there is no evidence for the occurrence of sub-types. Similar species' differences are found for binding affinities of antagonists.

In an exposed blister base in man, 5-HT causes a pain response mediated via 5-HT_3 receptors.[9] 5-HT_3 receptors are also located in the substantia gelatinosa of the spinal cord and these are markedly reduced in rats treated neonatally with capsaicin.[18,19] These results show that 5-HT_3 receptors occur presynaptically on both peripheral and central primary afferent terminals. Their functional role in the substantia gelatinosa is unknown although it is known that capsaicin sensitive afferent fibers conduct specific nociceptive information.[20]

The presence of 5-HT_3 receptors in the central nervous system was first demonstrated by Kilpatrick *et al.*[21] using the selective 5-HT_3 antagonist ligand [^{3}H]GR65630. Binding affinities for the antagonists tested correlated with affinity values determined in the rat vagus nerve. These studies showed relatively high binding in cortical and limbic areas in the rat. In rats, ferrets and man highest density binding to 5-HT_3 sites was found in the hindbrain locus of the chemoreceptor trigger zone.[21,22] The distribution of 5-HT_3 receptors is similar between species but varies in density.[24,25] Several other ligands have also been shown to label 5-HT_3 receptors in the central nervous system and in cell lines.[26,31] One species in which no 5-HT_3 binding has yet been demonstrated is the guinea-pig. The reason for this may be the lack of high affinity ligands for the guinea-pig 5-HT_3 receptors although [^{3}H]GR67330 ($pA_2 = 9.0$ in guinea-pig ileum) still failed to show any specific binding in guinea-pig brain.[31]

5-HT_3 receptors modulate neurotransmitter release in peripheral tissues, such as the guinea-pig ileum and rabbit heart, and several studies have shown this also to be the case in the central nervous system for dopamine[32,33] and acetylcholine.[34] The latter study showed that serotonin has two actions on K^+ stimulated acetylcholine release in

TABLE 2. Antagonist Affinities for 5-HT_3 Receptor Antagonists in Isolated Preparations from the Guinea-pig, Rat, and Rabbit

	Guinea-Pig			Rat			Rabbit	
Compound	Ileum pK_B	Vagus Nerve pK_B	Colon pK_B	Vagus Nerve pA_2	Brain Binding pK	Inh. ^{3}H ACh Release pK_B	Vagus Nerve pK_B	Heart pK_B
Metoclopramide	5.5[59]	5.4[17]	5.7[60]	6.6[59]	6.4[21]	5.8[34]	7.7[59]	7.21[4]
MDL72222	6.2[60]	6.4[17]	6.7[63]	7.9[59]	7.3[21]		7.9[3,54]	9.3
ICS 205-930	7.9[9]	7.6[54,61]	—	8.8[61]	8.5[21]		10.2	10.6
SDZ 206-830	9.1[9]	8.6[a]	—	9.3[9]	8.9[a]		13.1[60]	10.1[60]
SDZ 206-792	7.7[9]	—	—	—	—		8.9[60]	9.8[60]
GR38032	7.3[59]	7.0[54,61]	7.1[63]	8.6[59,9]	8.3[21]	9.0[34]	9.4[60]	—
BRL 43694	8.1[62]	7.9[17]	—	9.8[a]	9.21[21]		—	10.7[62]
Zacopride	8.5[54]	7.5[54,61]	—	9.1[9]	9.7[a]	9.4[34]	10.1[54]	—

[a]Burridge, Butler & Kilpatrick, unpublished.

the rat entorhinal cortex. Inhibition of release can be seen with serotonin or 2-methyl-5-HT but only in the presence of ritanserin; the selective 5-HT_3 agonist phenylbiguanide inhibits release in the absence of ritanserin. Serotonin and 2-methyl-5-HT can also *enhance* acetylcholine release but only in the presence of a 5-HT_3 antagonist. The inhibited release is due to activation of 5-HT_3 receptors since zacopride, GR38032 and metoclopramide antagonise the inhibition with pK_B values similar to their affinities for 5-HT_3 receptors in other receptor models in the rat (TABLE 2). The serotonin enhanced acetylcholine release may be due to an action on 5-HT_{1C} receptors for which both 2-methyl-5-HT and ritanserin have affinity at the concentrations used[35]; but further dose ranging studies are needed to establish a firm conclusion. The apparently surprising result here is that for the first time, 5-HT_3 receptors appear to be having an inhibitory effect—in all other known responses stimulation of 5-HT_3 receptors induces excitation. However, in other experiments, tetrodotoxin (TTX) a sodium channel blocker (i.e. blocks propagated action potentials) completely inhibited the 2-methyl-5-HT induced inhibition of release without affecting [^{3}H]acetylcholine release.[34] These data suggest that 5-HT_3 receptor-mediated inhibition of the release of acetylcholine is via excitation of an inhibitory interneuron. The inhibitory neurotransmitter is unknown, although there are several candidates.

The behavioral correlates of 5-HT_3 modulated acetylcholine release may be related to cognitive function. 5-HT_3 antagonists enhance performance in an object discrimination reversal learning task in marmosets[36,37] and reverse scopolamine-induced cognitive deficits in the food reinforced alternation task in rats.[38] These data suggest that cholinergic activity, at least in the entorhinal cortex, may be under chronic serotonin inhibitory influence so that 5-HT_3 antagonists cause an increase in release of acetylcholine *in vivo*.

In other CNS models, Hagan *et al.*[32] showed that behavioral hyperactivity and increased dopamine release in the nucleus accumbens following injection of di-Me-C_7 into the ventral tegmental area could be inhibited by 5-HT_3 antagonists, such as GR38032. This action has been confirmed using microdialysis *in vivo*.[39] However, unlike in the striatum[33] there is no evidence for a 5-HT_3 receptor modulation of dopamine release from nucleus accumbens slices *in vitro*.[40] It is likely, therefore, that the 5-HT_3 receptor modulation of mesolimbic hyperactivity is indirect probably via an action in the dorsal raphé nucleus or in the central nucleus of the amygdala.[41]

Inhibition of mesolimbic hyperactivity with 5-HT_3 antagonists has also been shown in rats and marmosets. In these studies, dopamine is infused continuously into the nucleus accumbens for 14 days; the resulting hyperactivity can be inhibited by 5-HT_3 antagonists.[42] Such studies have suggested that 5-HT_3 receptor antagonists may be useful in the treatment of psychotic disorders. However, the mesolimbic dopaminergic system also appears to mediate the rewarding and motor activating properties of substances of abuse.[43] Several studies have now indicated that serotonin has an important role to play in the maintenance of psychological dependency. The first studies to report on the effects of 5-HT_3 antagonists in rats treated with benzodiazepines, alcohol, nicotine, cocaine and amphetamine were by Oakley *et al.*[44,45] and Costall *et al.*[46] The behavioral consequences of withdrawing from chronic treatment with these substances of abuse can be completely suppressed with 5-HT_3 antagonists. Similarly, ethanol-preferring marmosets[45] or rats[46a] have significantly reduced consumption of alcohol during treatment with GR38032. Place-preference conditioning induced by morphine, nicotine[47,48] and amphetamine[49] is inhibited by 5-HT_3 antagonists. However, the effects on conditioned place-preference to amphetamine are controversial; Cooper *et al.*[49] find that this effect can be prevented by GR38032, while in the studies by Carboni *et al.*[48] amphetamine place-preference conditioning was not affected by ICS 205-930 or MDL72222. These studies have implicated the use of 5-HT_3 antagonists in the treatment of patients withdrawn from various substances of abuse. The various effects of 5-HT$_3$ antagonists in behavioral models in animals, suggesting their clinical utility as anxiolytics, antipsychotics, and in the treatment of drug abuse have been reviewed by Costall, Naylor and Tyers.[50]

The antiemetic properties of 5-HT$_3$ receptor antagonists against cancer chemotherapy- and radiotherapy-induced nausea and vomiting is now well established in both animals and man.[51–55] Studies to investigate the mechanism of action suggest two possible sites. Cisplatin, one of the most emetogenic drugs, causes a release of serotonin in the upper small intestine in ferrets.[56] The source is almost certainly the enterochromaffin cells (ECL) in the intestinal mucosa. The ECL cells are closely innervated by vagal afferents which probably form the rostral arm of the vomiting reflex, since vagotomy abolishes vomiting.[57] Following activation of 5-HT$_3$ receptors on the vagus nerve, the vomiting reflex projects to the nucleus tractus solitarius (NTS) and the area postrema (AP). The NTS and, to a lesser extent, the AP is rich in 5-HT$_3$ receptors.[24] 5-HT$_3$ antagonists injected discretely into this region of the hindbrain markedly inhibit emesis in ferrets, suggesting an additional central site of action.[23] Using the ligand [^{3}H]BRL 43694, Pratt and Bowery[58] found that vagotomy causes a loss of 5-HT$_3$ receptors in both areas, suggesting a presynaptic location of 5-HT$_3$ receptors on vagal afferent fibers.

In conclusion, the roles of 5-HT$_3$ receptors are numerous and widely distributed in the peripheral and central nervous systems. These functions have been identified following the development of highly selective 5-HT$_3$ antagonists. Selective 5-HT$_3$ angonists have also been identified, but their value in *in vivo* experiments is limited by lack of penetration into the CNS and the marked cardiovascular effects mediated by activation of the Bezold Jarisch reflex.

In addition to the antiemetic properties of 5-HT$_3$ antagonists, a number of potentially important uses in the treatment of psychiatric disorders have been proposed. Following ten years of revival for basic research on 5-HT$_3$ receptors, the next decade will show whether the research rationales and predictions from laboratory studies are supported by significant advances in psychiatric therapy.

REFERENCES

1. Rocha e Silva, M., J. R. Valle & Z. P. Picarelli. 1953. A pharmacological analysis of the mode of action of serotonin (5-HT) upon the guinea-pig ileum. Br. J. Pharmacol. **8:** 378–388.
2. Gaddum, J. H. & Z. P. Picarelli. 1957. Two kinds of tryptamine receptor. Br. J. Pharmacol. **12:** 323–328.
3. Fozard, J. R. & A. T. M. Mobarek Ali. 1978. Blockade of neuronal tryptamine receptors by metoclopramide. Eur. J. Pharmacol. **49:** 109–112.
4. Fozard, J. R., A. T. M. Mobarok Ali & G. Newgrosh. 1979. Blockade of serotonin receptors on autonomic neurones by (−) cocaine and some related compounds. Eur. J. Pharmacol. **59:** 195–210.
5. Wallis, D. I. & B. Woodward. 1975. Membrane potential changes induced by 5-HT in the rabbit superior cervical ganglion. Br. J. Pharmacol. **55:** 199–212.
6. Ireland, S. J., D. W. Straughan & M. B. Tyers. 1987. Influence of 5-HT uptake on the apparent 5-HT antagonist potency of metoclopramide in the rat isolated superior cervical ganglion. Br. J. Pharmacol. **90:** 151–160.
7. Ireland, S. J. & M. B. Tyers. 1987. Pharmacological characterisation of 5-HT-induced depolarisation of the rat isolated vagus nerve. Br. J. Pharmacol. **90:** 229–238.
8. Bradley, P. B., G. Engel, W. Feniuk, J. R. Fozard, P. P. A. Humphrey, D. N. Middlemiss, E. J. Mylecharane, B. P. Richardson & P. R. Saxena. 1986. Proposals for the classification and nomenclature of functional receptors for 5-HT. Neuropharmacology **25:** 563–576.
9. Richardson, B. P., G. Engel, P. Donatsch & P. A. Stadler. 1985. Identification of

serotonin M-receptor subtypes and their specific blockade by a new class of drugs. Nature **316:** 126–131.
10. NEIJT, H. E., H. P. M. VIJVERBERG & J. VAN DER BERCKEN. 1986. The dopamine response in mouse neuroblastoma cells is mediated by serotonin 5-HT_3 receptors. Eur. J. Pharmacol. **127:** 271–274.
11. SUPRENANT, A. & J. CRIST. 1988. Electrophysiological characterisation of functionally distinct 5-HT receptors on guinea-pig submucous plexus. Neuroscience **24:** 283–295.
12. DERKACH, V., A. SUPRENANT & R. A. NORTH. 1989. 5-HT_3 receptors are membrane ion channels. Nature **339:** 706–709.
13. FOX, A. J. & I. K. M. MORTON. 1989. Activities of some 5-HT analogues in 5-HT_3 receptor mediated release of [^{3}H]-Ach in guinea-pig ileum. Br. J. Pharmacol. **96:** 53P.
14. BUCHEIT, K-H., G. ENGEL, E. MUTSCHLER & B. RICHARDSON. 1985. Study of the contractile effect of 5-HT in the isolated longitudinal muscle strip from guinea-pig ileum. Naunyn-Schmied. Arch. Pharmacol. **329:** 36–41.
15. DUMUIS, A., R. BOUHELAL, M. SEBBEU & J. BODKAERT. 1988. A 5-HT receptor in the central nervous system, positively coupled to adenylate cyclase, is antagonised by ICS 205-930. Eur. J. Pharmacol. **146:** 187–188.
16. KENAKIN, T. P. 1987. *In* Pharmacological Analysis of Drug-Receptor Interaction. T. P. Kenakin, Ed. New York: Raven Press, pp. 241–243.
17. BURRIDGE, J., A. BUTLER & G. J. KILPATRICK. 1989. 5-HT_3 receptors mediate depolarisation of the guinea-pig isolated vagus nerve. Br. J. Pharmacol. **96:** 269P.
18. WAEBER, C. K., K. DIXON, D. HOYER & J. M. PALACIOS. 1988. Localisation by autoradiography of neuronal 5-HT_3 receptors in the mouse CNS. Eur. J. Pharmacol. **151:** 351–352.
19. HAMON, M., M. C. GALISSOT, F. MENARD, H. GOZLAN, S. BOURGOIN & D. VERGÉ. 1989. 5-HT_3 binding sites are on capsaicin-sensitive fibres in the rat spinal cord. Eur. J. Pharmacol. **164:** 315–322.
20. HAYES, A. G. & M. B. TYERS. 1980. Effects of capsaicin on nociceptive heat, pressure and chemical thresholds and on substance P levels in the rat. Brain. Res. **189:** 561–564.
21. KILPATRICK, G. J., B. J. JONES & M. B. TYERS. 1987. Identification and distribution of 5-HT_3 receptors in the rat brain using radioligand binding. Nature **330:** 746–748.
22. KILPATRICK, G. J., B. J. JONES & M. B. TYERS. 1989. Binding of the 5-HT_3 ligand [^{3}H]GR65630, to rat area postrema, vagus nerve and the brains of several species. Eur. J. Pharmacol. **159:** 157–164.
23. HIGGINS, G. H., G. J. KILPATRICK, K. T. BUNCE, B. J. JONES & M. B. TYERS. 1989. 5-HT_3 receptor antagonists injected into the area postrema inhibit cisplatin-induced emesis in the ferret. Br. J. Pharmacol. **97:** 247–255.
24. KILPATRICK, G. J., B. J. JONES & M. B. TYERS. 1989. The distribution of specific [^{3}H]GR65630 binding to the brains of several species. Characterisation of binding to rat area postrema and vagus nerve. Eur. J. Pharmacol. **159:** 157–164.
25. PEROUTKA, S. J. 1988. Species variations in 5-HT_3 recognition sites labeled by ^{3}H-quipazine in the central nervous system. Naunyn-Schmiedebergs. Arch. Pharmacol. **338:** 472–475.
26. BARNES, N. M., B. COSTAL & R. J. NAYLOR. 1988.^{3}H-Zacopride: A ligand for the identification of 5-HT_3 recogition sites. J. Pharm. Pharmacol. **40:** 668.
27. WATLING, K. T., S. ASPLEY, C. J. SWAIN & J. SAUNDERS. 1988. [^{3}H]Quarternised ICS205-930 labels 5-HT_3 receptor binding sites in rat brain. Eur. J. Pharmacol. **149:** 397.
28. HOYER, D. & H. C. NEIJT. 1987. Identification of 5-HT_3 recognition sites by radioligand binding in NG108-15 neuroblastomaglioma cells. Eur. J. Pharmacol. **143:** 291.
29. HOYER, D. & H. C. NEIJT. 1988. Identification of serotonin 5-HT_3 recognition sites in membranes of NIE-115 neuroblastoma cells by radioligand binding. Mol. Pharmacol. **33:** 303.
30. PEROUTKA, S. J. & A. HAMIK. 1988. [^{3}H]-Quipazine labels 5-HT_3 recognition sites in rat cortical membranes. Eur. J. Pharmacol. **148:** 297.
31. KILPATRICK, G. J., B. J. JONES & M. B. TYERS. 1989. [^{3}H]- GR67330, a very high affinity ligand for central 5-HT_3 receptors. Br. J. Pharmacol. **96:** 8P.
32. HAGAN, R. M., A. BUTLER, J. M. HILL, C. C. JORDAN, S. J. IRELAND & M. B. TYERS. 1987. Effect of the 5-HT_3 receptors antagonist GR38032F, on the responses to injection of a neurokinin agonist into the ventral tegmental area of the rat brain. Eur. J. Pharmacol. **138:** 303.

33. BLANDINA, P., J. GOLDFARB & J. P. GREEN. 1988. Activation of a 5-HT$_3$ receptor releases dopamine from striatal slice. Eur. J. Pharmacol. **155:** 349.
34. BARNES, J. M., N. M. BARNES, B. COSTALL, R. H. NAYLOR & M. B. TYERS. 1989. 5-HT$_3$ receptors mediate inhibition of acetylcholine release in cortical tissue. Nature **338:** 762–763.
35. ENGEL, G., M. GÖTHERT, D. HOYER, E. SCHLICKER & K. HILLENBRAND. 1986. Identifity of inhibitory presynaptic 5-HT autoreceptors in the rat brain cortex with 5-HT$_{1B}$ binding sites. Naunyn-Schmied. Arch. Pharmacol. **332:** 1–7.
36. BARNES, J. M., B. COSTALL, M. E. KELLY, R. J. NAYLOR, E. S. ONAIVI, D. M. TOMKINS & M. B. TYERS. 1989. GR38032F, improves performance in a mouse habituation test and inhibits cholinergic-linked deficits. Br. J. Pharmacol. **98:** 693P.
37. COSTALL, B., A. M. DOMENEY, P. A. GERRARD, D. N. C. JONES, R. J. NAYLOR & M. B. TYERS. 1989. Effects of GR38032F on performance of common marmosets (callithrix Jacchus) in an object discrimination reversal task. Br. J. Pharmacol. **98:**637P.
38. COSTALL, B., J. COUGHLAN, M. E. KELLY, R. J. NAYLOR & M. B. TYERS. 1989. Scopolamine-induced deficits in a T-maze reinforced alternation task are attenuated by 5-HT$_3$ receptor antagonists. Br. J. Pharmacol. **98:**636P.
39. CARBONI, E., E. ACQUAS, R. FRAU & G. DICHIARA. 1989. Differential inhibitory effects of a 5-HT$_3$ antagonist on drug induced stimulation of dopamine release. Eur. J. Pharmacol. **164:** 515–519.
40. BARNES, J. M., N. M. BARNES, B. COSTALL & R. J. NAYLOR. 1989. 5-Carboxamidotryptamine modulation of potassium-stimulated [^{3}H]-dopamine release in the rat nucleus accumbens. Br. J. Pharmacol. **97:** 378P.
41. COSTALL, B., M. E. KELLY, R. J. NAYLOR, E. S. ONAIVI & M. B. TYERS. 1989. Neuroanatomical sites of action of 5-HT$_3$ receptor agonist and antagonists for alteration of aversive behaviour in the mouse. Br. J. Pharmacol. **96:** 325–332.
42. COSTALL, B., A. M. DOMENEY, R. J. NAYLOR & M. B. TYERS. 1987. Effects of the 5-HT$_3$ receptor antagonist GR38032F on raised dopaminergic activity in the mesolimbic system of the rat and marmoset brain. Br. J. Pharmacol. **92:** 881–890.
43. WISE, R. A. & M. A. BOGARTH. 1982. Action of drugs of abuse on brain reward systems: an update with specific attention to opiates. Pharmacol. Biochem. Behav. **17:** 239.
44. OAKLEY, N. R., B. J. JONES & M. B. TYERS. 1988. Tolerance and withdrawal studies with diazepam and GR38032 in the rat. Br. J. Pharmacol. **96:** 764P.
45. OAKLEY, N. R., B. J. JONES, M. B. TYERS, B. COSTALL & A. M. DOMENEY. 1988. The effect of GR38032F on alcohol consumption in the marmoset. Br. J. Pharmacol. **96:** 870P.
46. COSTALL, B., A. M. DOMENEY, B. J. JONES, M. E. KELLY, P. A. GERRARD, R. J. NAYLOR & M. B. TYERS. Influence of GR38032F on the behavioural consequences of ceasing sub-chronic treatment with drugs of abuse. Br. J. Pharmacol. **96:** 905P.
46a. SELLERS, E. M., H. L. KAPLIN, M. O. LAWRIN & B. BRANDS. 1990. The 5-HT$_3$ receptor antagonist, GR38032F, decreases ethanol drinking by rats. Eur. J. Pharmacol. In press.
47. CARBONI, E., E. ACQUAS, P. LEONE, L. PEREZZANI & G. DICHIARA. 1988. 5-HT$_3$ receptor antagonist blocks morphine and nicotine-induced place preference conditioning. Eur. J. Pharmacol. **151:** 159.
48. CARBONI, E., E. ACQUAS, P. LEONE & G. DICHIARA. 1989. 5-HT$_3$ receptors antagonists block morphine and nicotine- but not amphetamine-induced reward. Psychopharmacology **97:** 175.
49. COOPER, S. J., G. VAN DER HOEK, B. J. JONES & M. B. TYERS. 1990. Antagonism of d-amphetamine induced place preference conditioning by the 5-HT$_3$ receptor antagonist GR38032F. Psychopharmacology. In press.
50. COSTALL, B., R. J. NAYLOR & M. B. TYERS. 1988. Recent advances in the neuropharmacology of 5-HT$_3$ agonists and antagonists. Rev. Neurosciences **2:** 41–65.
51. COSTALL, B., A. M. DOMENEY, R. J. NAYLOR & F. D. TATTERSALL. 1986. 5-HT M-receptor antagonism to prevent cisplatin-induced emesis. Neuropharmacology **25:** 959–961.
52. STABLES, R., P. L. R. ANDREWS, H. E. BAILEY, B. COSTALL, S. J. GUNNING, J. HAWTHORN, R. J. NAYLOR & M. B. TYERS. 1987. Antiemetic properties of the 5-HT$_3$ receptor antagonist, GR38032F. Cancer Treatment Rev. **14:** 333–336.

53. MINER, W. D. & G. J. SANGER. 1986. Inhibition of cisplatin-induced vomiting by selective 5-HT M-receptor antagonism. Br. J. Pharmacol. **88:** 497–499.
54. SMITH, W. L., C. B. JACKSON, A. G. PROAKIS, C. A. LEONARD, H. R. MUNSON & R. S. ALPIN. 1986. Zacopride (AHR-11190B): A unique and potent inhibitor of cancer chemotherapy induced emesis in dogs. Proc. Am. Soc. Clin. Oncology **5:** 260.
55. CUNNINGHAM, D., J. HAWTHORN, A. POPLE, J. C. GAZET, H. T. FORD, T. CHALLONER & R. C. COOMBERS. 1987. Prevention of emesis in patients receiving cytotoxic drugs by GR38032F, a selective 5-HT_3 receptor antagonist. Lancet **8:** 1461–63.
56. GUNNING, S., R. M. HAGAN & M. B. TYERS. 1987. Cisplatin induces biochemical and histological changes in the small intestine of the ferret. Br. J. Pharmacol. **90:** 135P.
57. ANDREWS, P. L. R., J. HAWTHORN & G. J. SANGER. 1987. The effect of abdominal visceral nerve lesions and a novel 5-HT-receptor antagonist on cytotoxic and radiation induced emesis in the ferret. J. Physiol. **382:** 47P.
58. PRATT, G. D. & N. G. BOWERY. 1989. The 5-HT_3 receptor ligand, ^{3}H-BRL43694 binds to presynaptic sites in the nucleus tractus solitarius of the rat. Br. J. Pharmacol. **97:** 414P.
59. BUTLER, A., J. M. HILL, S. J. IRELAND, C. C. JORDAN & M. B. TYERS. 1988. Pharmacological properties of GR38032F, a novel antagonist at 5-HT_3 receptors. Br. J. Pharmacol. **94:** 397–412.
60. DONATSCH, P., G. ENGEL, B. P. RICHARDSON & P. A. STADLER. 1984. ICS205-930: a highly selective and potent antagonist at peripheral neuronal 5-HT receptors. Br. J. Pharmacol. **81:** 34P.
61. LATTIMER, N., K. F. RHODES & V. L. SAVILLE. 1989. Possible differences in 5-HT_3-receptors in rat and the guinea pig. Br. J. Pharmacol. **96:** 270P.
62. SANGER, G. J. & D. R. NELSON. 1989. Selective and functional 5-HT_3 receptor antagonism by BRL43694 (granisetron). Eur. J. Pharmacol. **159:** 113–124.
63. GROSSMAN, C. J., K. T. BUNCE & P. P. A. HUMPHREY. 1989. Investigation of the 5-HT receptors in guinea pig descending colon. Br. J. Pharmacol. **97:** 451P.

DISCUSSION OF THE PAPER

MANFRED GÖTHERT, (*University of Bonn*): Did you also find ondansetron-induced inhibition in other brain regions than the one which you showed? And have you investigated also this effect in synaptosomal preparations?

TYERS: Discrete injections of 5-HT_3 antagonists and agonists into various brain regions can be shown to have a number of different effects. In the raphe nucleus and central amygdala 5-HT_3 antagonists have a disinhibitory effect while 5-HT_3 agonists can produce behavioral suppression. The neurochemical changes induced by these ligands have not been studied widely in brain slice preparations. However, there is some evidence that 5-HT_3 receptors control dopamine release in striatal slices.

We have also looked at the nucleus accumbens for effects on 5-HT induced dopamine release, but it doesn't appear that 5-HT_3 receptors have a local action. In *in vivo* experiments, 5-HT_3 antagonists can prevent the increase in dopamine release in the nucleus accumbens induced by stimulation of the ventral tegmental area, but this may not be mediated through an action in the nucleus accumbens.

GÖTHERT: Referring also to your ^{3}H-ACh release data, I have difficulty in interpreting your tetrodotoxin experiments because you were depolarizing with potassium which should also depolarize the nerve ending of the postulated interneuron so tetrodotoxin should effect the release of ACh.

TYERS: Tetrodotoxin blocks Na^+ permeability and axonal conduction of electrical activity without affecting the nerve terminals. Therefore, K^+-stimulated release of ACh should not be affected, but excitation of the interneuron would be inhibited.

R. MURPHY (*New York University*): Do you have sufficient resolution in your autoradiography, to look within the area postrema or other regions, like the NTS, where there's good evidence for topography of different groups of vagal afferents from different abdominal vagal branches?

TYERS: We haven't been able to do any microdistribution studies.

DAVID CLARKE (*Syntex, Palo Alto, CA*): With regard to the increase in release of acetylcholine with 2-methyl 5-HT, which is blocked by ritanserin, you used rather a high concentration of ritanserin, 1 micromolar, which is way above the affinity of ritanserin for 5-HT_2 or 5-HT_{1C} receptors. My question is quite simple, have you used lower concentrations?

TYERS: No. The evidence which I have presented with regard to the potentiating effect of 2-methyl-5-HT is preliminary and much further reserach has to be completed before this effect can be assigned to either 5-HT_2 or 5-HT_{1C} receptors, or neither.

ISTVAN TÖRK (*The University of New South Wales, Australia*): You mentioned that in the spinal cord the 5-HT_3 receptors are located on primary afferents. Is it possible that the same applies in the NTS where 5-HT_3 receptors could be located on the central terminals of the vagal afferents. And then the explanation that you provided regarding the vomiting reflex, may not be entirely correct because you inferred that the primary afferents act on the 5-HT_3 receptors within the area.

TYERS: There is some evidence that lesioning the vagus nerve causes a loss of 5-HT_3 receptors in the NTS. This finding is a little controversial because other groups have not found this. There is also some controversy as to whether the density of 5-HT_3 binding sites is higher in the area postrema than the NTS.

Another way of answering your question is that if 2-methyl-5-HT is injected directly into the area postrema, it produces some mild "nausea" but no vomiting. So I don't think that 5-HT is the primary neurotransmitter responsible for vomiting. But the fact that 5-HT_3 antagonists given into the AP can block the effects of cisplatin suggest the presence of 5-HT is necessary for the vomiting reflex to occur. So I think that there may be a dual system of neurotransmitters, acetylcholine from the vagus, and 5-HT from interneurons, and these together are necessary to propogate the vomiting reflex.

DANIEL HOYER (*Sandoz, Basel*): I would agree with most of what you have been saying except one thing. You say that 5-HT_3 antagonists as a class are extremely selective for 5-HT_3 receptors. That is true in most cases, but in some species, as you have shown, certain compounds, such as ICS 205-930, show affinities that are quite a bit lower than in the rat for instance. We know that some of these compounds, such as zacopride and BRL 24924, act as agonists at the "5-HT_4" receptors at about 0.1 μM. Also, ICS 205-930 acts as an antagonist at about the same concentration. So in some species, particularly the guinea pig, we have lost selectivity. Have you looked at human tissue?

TYERS: The affinities of 5-HT_3 antagonists for 5-HT_3 receptor binding to human cortex appears to correlate much better with the rat than other species.

HOYER: But binding is binding, what about functional responses?

TYERS: We've also investigated K^+ stimulated acetylcholine release in human entorhinal cortex where the affinity for GR 30832 in antagonizing the 5-HT induced inhibition of release is very similar to values obtained in rat tissues.

D. L. NELSON (*Session Chair*): Is there a difference between the binding in the rat vs the binding in the guinea pig?

TYERS: Binding to 5-HT_3 receptors in the guinea pig brain has yet to be shown.

NELSON: So that's consistent with the low potency of antagonists for functional 5-HT_3 receptor responses in guinea pig tissues.

ROBERTO INVERNIZZI (*Milan, Italy*): Could you explain your finding that 5-HT_3

antagonists inhibit the increased dopamine turnover in the nucleus accumbens by a local action on 5-HT_3 receptors in the VTA?

TYERS: I think that's possible.

INVERNIZZI: Are there any binding data?

TYERS: There is a low density of 5-HT_3 receptors in VTA, but as yet we have not investigated local effects of 5-HT_3 agonists or antagonists injected directly into this area.

INVERNIZZI: There are some data by Imperato showing that dopamine release in the nucleus accumbens is stimulated by morphine, given parenterally, and that this release can by regulated by MDL 72222 injected into the VTA.

B. RICHARDSON (*Sandoz, Basel*): I'm a bit confused about your data on the guinea pig ileum because I think that you showed that in the presence of atropine, 2-methyl 5-HT can still induce a contraction which could be antagonized with 5-HT_3 antagonists. Later, however, you showed a slide which indicated that 2-methyl 5-HT released acetylcholine and that this was also shifted in parallel by 5-HT_3 antagonists. So could you perhaps explain.

TYERS: In our hands, the full 5-HT mediated response in the guinea pig ileum longitudinal muscle preparation is blocked by atropine, showing that the response is entirely cholonergic. 5-HT_3 receptor antagonists cause a shift only of the upper, low affinity part of the biphasic dose response curve to 5-HT. The experiments with 2-methyl-5-HT were not carried out in the presence of atropine.

J. BLUNDELL (*University of Leeds, England*): Does repeated administration of 5-HT_3 antagonists lead to changes in the sensitivity of dopamine receptors?

TYERS: The answer to that really comes from some work where dopamine has been infused continuously into the nucleus accumbens over a period of about two weeks. In these experiments, chronic dosing with 5-HT_3 antagonists inhibits the induced hyperactivity response throughout that period. When the dosing with the 5-HT_3 antagonists is discontinued the animals behavior remains exactly as it did during the treatment. In contrast, neuroleptic drugs, such as haloperidol, also inhibit the hyperactivity throughout the two week period, but when withdrawn the animals become hyperactive and supersensitive to dopamine agonists, like *N*-propylapomorphine, and to further doses of dopamine. So I think the answer is no, there is no supersensitivity of dopamine receptors following chronic dosing with 5-HT_3 antagonists.

BLUNDELL: There are two aspects of your talk that suggest that 5-HT_3 receptors might be involved in appetite. One is the action in the mesolimbic system and the other the concentration of 5-HT_3 receptors in the area postrema and nucleus of the solitary tract. Do you have any evidence that 5-HT_3 receptors are involved in appetite control?

TYERS: 5-HT_3 antagonists do not appear to affect daily food intake. However, there are a number of studies that we have been conducting on the mesolimbic reward system. 5-HT_3 antagonists appear to be very potent at inhibiting the behavioral changes that occur following chronic treatment and withdrawal of drugs of abuse. Similarly, rats and marmosets that are "alcohol preferring" under the influence of a 5-HT_3 antagonist will no longer prefer alcohol. It is possible that bingeing, for, say highly palatable food, or sucrose may also be affected by 5-HT_3 antagonists, but this is something that has still to be determined. It may well involve the same reward pathways.

J. PALACIOS (*Sandoz, Basel*): Do you have any clinical evidence for antipsychotic effects of 5-HT_3 antagonists in man?

TYERS: No. Ondanseton (GR 38032) is in clinical trials in schizophrenia. But it takes several hundred patients and several years to be sure that a new drug has such activity.

D. GIETZEN (*University of California, Davis*): I would just like to say that we have some evidence with regard to conditioned taste aversions and 5-HT_3 antagonists. They seem to inhibit the acquisition of those aversions which may be consistent with your data on vomiting and nausea. Also with regard to appetite, and Dr. Blundell's comment, ICS205-930 and MDL 72222 will also inhibit anorexia induced by imbalanced amino acid diets.

Molecular Studies of 5-HT Receptors[a]

JEAN C. SHIH AND KEVIN CHEN

Division of Biological Sciences
School of Pharmacy
University of Southern California
Los Angeles, California 90033

INTRODUCTION

5-Hydroxytryptamine (5-HT, serotonin) appears to play a central role in many physiological processes including sleep,[1] appetite,[2] thermoregulation,[3] pain perception,[4] hormone secretion,[5] and sexual behavior.[6] Accordingly, knowledge of the structure of the 5-HT receptor would have far reaching implications ranging from basic understanding of the functioning of the receptor to the design of drugs which would be useful in the treatment of pathological states associated with these functions.

The early work aimed at understanding the 5-HT receptor, which utilized radioligand binding studies, resulted in the definition of at least six different receptor subtypes (5-HT_{1A}, 1B, 1C, 1D, 5-HT_2, and 5-HT_3).[6–13] The 5-HT_1 family of receptors exhibits a nanomolar affinity for ^{3}H-5-HT while on the other hand the 5-HT_2 receptors have low affinity for ^{3}H-5-HT but a high affinity for the 5-HT antagonists such as ^{3}H-ketanserin, -mesutergine, -metergoline, -*d*-LSD, etc. The receptors in the 5-HT_1 family are further defined on the basis of the differential binding to various radioligands, 5-HT_{1A} is labeled by ^{3}H-OH-DPAT[14,15] and ^{3}H-PAPP[6] while 5-HT_{1B} has a high affinity for ^{125}I-labeled cyanopindolol[15] and a relatively low affinity for 8-OH-DPAT. 8-OH-DPAT and busprione demonstrate micromolar affinities for the 5-HT_{1D} receptor while (−) pindolol is approximately two orders of magnitude less potent at these sites than at the 5-HT_{1B} receptors.[16] 5-HT_3 receptors were identified in the rat brain with ^{3}H-GR 65630.[17]

Following the identification of the multiple receptor subtypes, the next logical step would be to determine the structure of the various receptors which would then help shed light on the differences in radioligand binding. Two characteristics of these receptors make that next step very difficult to achieve by the conventional methods. First, the receptors are membrane bound. Once the receptors are solubilized they exhibit very low activity, if any at all. Secondly, these receptors exist in extremely low concentrations. As a result, traditional receptor purification and sequencing schemes are difficult to apply to the study of the 5-HT receptors, although our group expended tremendous effort on successfully solubilizing [20] and identifying the 5-HT_{1A} receptor.[38] In addition, we have pursued two further avenues of investigation which are discussed.

PHOTOAFFINITY LABELING OF THE 5-HT_{1A} RECEPTOR

Photoaffinity labeling technique has been shown to be useful for isolating membrane bound receptor proteins. Aryl azides have enjoyed the widest application in such

[a]This work was supported by Grants R01 MH 37020, R37 MH 39085 (Merit Award) and Research Scientist Award K05 MH 00796 to JCS from NIMH, DHHS. The support from the Boyd and Elsie Welin professorship is also greatly appreciated.

studies. These compounds, when photoirradiated, form a reactive nitrene intermediate that covalently insert into the receptor proteins to which they bind with high affinity. When radiolabeled photoaffinity probes are available, it would allow us to identify and to isolate the radiolabeled inactive receptor proteins and thus overcome the difficulties encountered in the solubilization of the receptors. To this end, our group synthesized ^{3}H-PAPP {1-[2-(4-aminophenyl)ethyl]-4-(3-trifluoromethylphenyl)piperazine}[18] and ^{3}H-*p*-azido-PAPP {1-[2-(4-azidophenyl)ethyl]-4-(3-trifluoromethylphenyl)piperazine}[19] and ^{125}I-*p*-azido-PAPP.[41] The initial work showed that displacement of the ^{3}H-5-HT binding by PAPP was biphasic with an equilibrium dissociation constant of 2.9nM and 328nM respectively, similar to that of spiperone.[6] In the presence of 1μM spiperone, when all the 5-HT_{1A} sites were occupied, the high affinity binding was not observed.[6] These results suggest that PAPP binds to the 5-HT-1A sites. Further studies with ^{3}H-PAPP demonstrated that: 1)^{3}H-PAPP bound to a single population of sites with a Kd of 1.6nM and a B_{max} of 162fmol/mg protein; and 2) the density of ^{3}H-PAPP sites in the hippocampus, cortex, striatum and cerebellum was similar to the density of the 5-HT_{1A} receptor in these areas.[18]

^{3}H-*p*-azido-PAPP binding experiments were performed initially without irradiation in a manner similar to that of ^{3}H-PAPP. Under this condition, binding was found to be completely reversible and specific as determined by competitive binding studies with other compounds known to bind specifically to the 5-HT_{1A} receptor.[19] The ^{3}H-*p*-azido-PAPP is specifically and covalently bound to the 5-HT_{1A} receptor protein upon irradiation, as demonstrated by separation of the labeled polypeptide on SDS-PAGE.[19] The incorporation of radioactivity was blocked if the membranes were incubated with ^{3}H-*p*-azido-PAPP in the presence of 1μM 5-HT[19] and other 5-HT_{1A} specific compounds, suggesting that the incorporation may be specific to 5-HT_{1A}.

Following solubilization of the 5-HT_{1A} receptor,[20] we partially purified the receptor using a PAPP affinity column. In order to elute the 5-HT_{1A} receptor from the affinity column, 1μM ^{3}H-*p*-azido-PAPP was used and the subsequent eluate, following concentration by ultrafiltration, was irradiated, concentrated further, and applied to SDS-PAGE. The results showed that radioactive peaks were associated with proteins with molecular weights of 55kD, 40kD, and 38kD.[19] The relationship among these proteins is not clear at this time. Unfortunately, the attempt to purify the 5-HT_{1A} receptor using photoaffinity radioligands is limited by the high non-specific labeling of proteins by these compounds. Subsequent to this work, exciting advances in the techniques of molecular cloning have led to another means of exploring the structure of the 5-HT receptor subtypes.

CLONING OF THE 5-HT RECEPTOR

Recently, three 5-HT receptor subtypes (5-HT_{1C}, 5-HT_{1A} and 5-HT_2) have been cloned by different approaches. The first 5-HT receptor to be cloned was the 5-HT_{1C}. A partial cDNA clone encoding mouse 5-HT_{1C} receptor was cloned by a hybrid depletion screening strategy.[21,22] Subsequently a full-length 5-HT_{1C} receptor was cloned by a different group.[23] The 5-HT_{1A} receptor was first isolated by screening a human genomic library at low stringency with a full length β-2 adrenergic receptor clone.[24] The function of the isolated intronless gene remained unknown until it was transfected and the expressed protein had all the typical ligand binding characteristics of the 5-HT_{1A} receptor.[25]

In collaboration with Dr. Peter Seeburg we have used the homology screening approach to clone the 5-HT_2 receptor.[26] Using two oligonucleotides directed against the amino acids of the 5-HT_{1C} receptor as probes,[23] a rat brain cDNA encoding 5-HT_2

receptor was cloned. We utilized the knowledge that the 5-HT receptor subtypes have a reasonable level of homology of the transmembrane regions. Thus, two oligonucleotides directed against amino acid residues 88–104 (transmembrane region II) and 134–149 (transmembrane region III) of the 5-HT_{1C} receptor were used to probe a cloned cDNA library constructed in λgt10 from rat forebrain mRNA. One of the resulting cloned cDNAs specified an open reading frame of 449 amino acids preceded by 668 nucleotides of 5′-untranslated sequence. The protein that is predicted from this mRNA was found to have seven putative transmembrane regions which is typical of G-protein coupled receptors. The homology of the predicted protein with the 5-HT_{1C} receptor was 51%. The highest identity was seen in the transmembrane regions with the fifth transmembrane region being identical in both. The lowest sequence identity occurs in the N-Terminal 60 amino acid residues which reside in an intracellular loop between the transmembrane regions V and VI, as well as in the C-terminus. This same pattern of sequence homology is seen in the muscarinic acetylcholine receptor subtypes[27,28] and suggests that the cloned cDNA encodes a 5-HT receptor subtype. FIGURE 1 shows a sequence alignment of the predicted sequence of the cloned cDNA with the 5-HT_{1C} receptor and other G-protein coupled neurotransmitter receptors.

In order to determine whether or not this cloned cDNA encoded for a 5-HT receptor, it was necessary to determine the pharmacological profile. To accomplish this, we expressed this cDNA transiently in a mammalian cell line by transfection utilizing a recombinant DNA plasmid containing the entire coding region of the cDNA under control of a cytomegalovirus promotor.[28,29] Membranes were prepared from these cells and the binding of a variety of 5-HT receptor ligands was studied.[26] We found high-affinity sites (0.5 ± 1nM) for spiperone, a 5-HT_2 and dopamine-2 receptor antagonist, while no such sites were found on the untransfected cells. Furthermore, the affinity of spiperone for our receptor was three orders of magnitude higher than those values reported for the 5-HT_{1B}, 5-HT_{1C}, and 5-HT_{1D} receptors and two orders of magnitude higher than that for the 5-HT_{1A} subtype.[30,31] This data combined with the finding that ^{3}H-5-HT did not show any detectable binding to the expressed receptor was evidence that this receptor is the 5-HT_2 subtype as 5-HT_{1A}, 5-HT_{1B}, and 5-HT_{1D} subtypes should bind the ^{3}H-5-HT.[30]

To further define the pharmacological profile of the receptor, competitive ligand binding studies were performed. We found the 5-HT antagonists, ketanserin and mianserin, to have affinities of 1.0 ± 0.5mM and 1.8 ± 0.3mM respectively. These values agreed closely with the values expected for interaction with the 5-HT_2 receptor.[7,15] The K_i for ketanserin with our receptor was two orders of magnitude less than the K_i observed with the 5-HT_{1C} receptor. Finally, haloperidol, a dopamine-2 antagonist, exhibited a K_i value of 125nM, a value two orders of magnitude larger than the 1.4nM value expected for a dopamine-2 receptor[7] and only slightly higher than the value reported for cortical 5-HT_2 receptors.[31] The fact that the ketanserin and mianserin displayed such high affinities for our receptor coupled with the high K_i exhibited by haloperidol allowed us to rule out that we were dealing with the dopamine-2 receptor. Our studies also demonstrated that 8-OH-DPAT binds the receptor with low affinity, which is consistent with the 5-HT_2 receptor subtype. Lastly, eliminating the fact that the cDNA encodes a 5-HT_3 receptor, studies with MDL-72222, a 5-HT_3 receptor antagonist, revealed that this drug was ineffective in competition binding studies. Thus, the results of the ligand binding studies clearly demonstrate that the cDNA encodes a 5-HT_2 receptor.

5-HT_2 receptor activation by 5-HT in the cortex,[33] thoracic aorta,[34] and platelets[35] appears to stimulate phosphoinositol (PI) hydrolysis. Activation of the 5-HT_{1C} receptor is known to stimulate this second messenger system and may effect a change in several cellular processes including Ca^{2+} release from intracellular stores and protein kinase C activation. Utilizing the Ca^{2+} sensitive dye, FURA-2/AM,[36] which undergoes spectral changes in response to the concentration of free Ca^{2+}, we studied the 5-HT triggered changes in the levels of intracellular Ca^{2+} in cells expressing the 5-HT_2 receptor.[26] Intracellular Ca^{2+} in these cells was routinely raised at least threefold at 10^{-7} M 5-HT,

```
5HT2  M A R - - - - G S T I M T S T P E M L N T S E A S N W T I D A E N R T N L S C E - - G Y L P P T C L S I L M L Q E
5HT1C M V N - - - - L G N A V R S L L M M L I G L L V W Q F D I S I S P V A G I V T D T F N S S D G G R L F Q F P D G V
G-21  M D V L S P G Q G N N T T S P P A P F E T G G N - - - - - - - - - - - - - - - - - T T G I S D V T V S Y Q V I T S L
BAR   M G P - - - - P G N D S D F L L T T N G S M V - - - - - - - - - - - - - - - - - - P D M D V T E E R D E A W V V
M4    E A G L P L G T V T Q L G S V N I S Q E T G M F S S M D - - - - - - - - - - - - - - T S D D P L G G M T I W Q V V

                     I                                                 II
52  K N W S A L L T T V V I I L - T I A G N I L V I M A V S L E K K L Q N A T N Y F L M S L A I A D M L L G F L V M P V
54  Q N W P A L - S I V V I I I M T I G G N I L V I M A V S M E K K L H H A T N Y F L M S L A I A D M L V G L L V M P L
43  L L G T L I F C A V L - - - - - - - G N A C V V A A I A L E R S L Q N V A N Y L I G S L A V T D L M V S V L V L P M
35  G M A I L M - S V I V L A I V F - - G N V L V I T A I A K F E R L Q T V T N Y F I T S L A C A D L V M G L A V V P F
69  F I A F L T - G F L A L V T I I - - G N I L V I V A F K V N K Q L K T V N N Y F L L S L A C A D L I I G V I S M M L

  II                                III
109 S M L T I L Y G Y R W P L P S K L C A I W I Y L D V L F S T A S I M M L C A I S L D R Y V A I Q N P I M M S R F N S
111 S L L A I L Y D Y V W P L P R Y L C P V W I S L D V L F S T A S I M H L C A I S L D R Y V A I R N P I E M S R E N S
94  A A L Y Q V - L N K W T L G Q V T C D L F I A L D V L C C T S S I L H L C A I A L D R Y W A I T D P I D Y V N K R T
90  G A S M I L - M K M W N F G N F W C E F W T S I D V L C V T A S I E T L C V I A V D R Y I A I T S P F K Y Q S L L T
124 F T T Y I I - M N R W A L G N L A C D L W L S I D Y V A S N A S V M N L L V I S F D R Y F S I T R P L T Y R A K R T

                        IV
167 R T K A F L K I I A V W T I S V G I S M - P I P V F G L Q D - D S K V F K E G S - C L L - - - - - - A D D N - F V L
169 R T K A I M K I A I V W A I S I G V S V - P I P V I G L R D - E S K V F V N N T T C V L - - - - - - N D P N - F V L
151 P - R P R A L I S L T W L I G F L I S I P P M L - - G W R T P E D R S D P D A - - C - - - - - - - - - T I S K D M G Y
147 K N K A R M V I L M V W I V S G L T S F L P I Q M M W Y R A T H Q K A I D - - - - C Y M K E T C C D F F T N - Q A Y
181 T K R R G V M I G L A W V I S F V L W A - P A I L F W Q Y F V G K R T V P P G E - C F I Q F L S E - - - - - - P T I

              V
215 - I G S F V A - F F I P L T I M V I T Y F - L T I K S L Q K E A --(30)-- Q R S I M R E P G S Y A G R R T M Q S I S
218 - I G S F V A - F F I P L T I M V I T Y F - L T I Y V L R R Q T --(39)-- D Q K P R R K K K E K R P R G T M Q A I N
196 T I Y S T F G A F Y I P L L L M L V L Y G P I F R A A R F R I R --(94)-- E R K N E R N A E A K R K - - - - - M A L
200 A I A S S I V S F Y V P L V V M V F V Y S R - V F Q V A K R Q L --(25)-- H G L - R R S S K F C L - - - - - - - - -
231 T F G T A I A A F Y M P V T I M T I L Y W R I Y K E T E K R T K --(207)- T R S Q I T K R K R M - - - - - - - - S L

                           VI                                          VII
295 N - E Q K A C K V L G I V F F L F V V M W C P F F I T N I M A - V I C K E S C N E N - - V I G A L L N V F V W I G Y
307 N - E K K A S K V L G I V F F V F L I M W C P F F I T N I L S - V L C G K A C N Q K - - L M E K L L N V F V W I G Y
338 A R E R K T V K T L G I I M G T F I L C W L P F F I I V A L V L P F C E S S C - H M P T L L G A I I N - - - W L C V
267 - K E H K A L K T L G I I M G T F T L C W L P F F I V N I V H - V I Q D N L I - P K E - - V Y I L L N - - - W L G Y
483 I K E K K A A Q T L S A I L L A F I I T W T P Y N I M V L V N T - F C D S C I - P K T - - Y W N L G Y - - - W L C Y

      VII
349 L S S A V N P L V Y T L F H K T Y R S A F S R Y I Q C Q Y K E N R K P L Q L I L V N T I P A L A Y K S S Q L Q V G Q
361 V C S G I N P V I Y T L F H K I Y R R A F S K Y L R C D Y K P D K K P P V R Q I P R V A A T A L S G R E L N V N I Y
391 I N S L L N P V I Y A Y F N K D F Q N A F K K I I K C N F C R Q
317 V N S A F N P L I Y C R S - P D F R I A F Q E L L C L R R S S S K A Y G N G Y S S N S N G K T D Y M G E A S G C Q L
534 I N S T V N P V C Y A L C N K T F R T T F K M L L L C Q C D K R K R R K Q Q Y Q Q R Q S V I F H K R V P E Q A L

407 K K N S Q E D A E Q T V D D C S M V T L G K Q Q S E E N C T D M I E T V N E K V S C V              5HT2  RAT
419 T H T N E R V A R K A N D P E P G I E M Q V E N L E L P V N P S N V - V S E R I S S V              5HT1C RAT
                                                                                                      G-21  HUMAN
374 G Q E K E S E R L C E D P P G T E S F V M C Q G T V P S L S L D S Q G R N C S T N D S P L      BAR   HAMSTER
                                                                                                      M4    RAT
```

FIGURE 1. Alignment of the primary structure of the $5HT_2$ receptor with the amino acid sequences of several G-protein-coupled receptors: rat $5HT_{1C}$,[23] human G-21,[24] which is serotonin $5HT_{1A}$ subtype,[25] hamster β_2-adrenergic,[39] and rat muscarinic acetylcholine M4 subtype.[40] Identical amino acids in all receptors are boxed, putative transmembrane segments are overlined. The residue numbers for each receptor are given at the beginning of each line.

whereas no changes could be detected in untransfected cells.The effects of 5-HT were completely blocked by 10nM spiperone. We also demonstrated that in response to saturating levels of 5-HT transfected cells showed a 10-fold stimulation of inositol monophosphate, while the inositol bis- and trisphosphates increased 4-fold and 1.5-

fold, respectively. The activation of PI turnover strongly suggests that the cloned receptor is a 5-HT_2 subtype, in strong agreement with its pharmacological profile.

FUTURE DIRECTIONS

The cloning of the 5-HT receptors has resulted in avenues of questions which can be asked as well as providing us with new approaches with which to answer these questions. One of the first questions which the molecular biological approach will allow us to answer is that of the relationships among the 5-HT receptor subtypes. Indeed, the current classification has already been called into question as the sequences of the various receptors has been determined. 5-HT_{1C} and 5-HT_2 receptors exhibit a high level of homology (51%) and they both appear to be linked to PI hydrolysis. On the other hand, 5-HT_{1A} and 5-HT_{1C} show less homology (39%) and appear to be coupled to different second messenger systems. This may indicate the need to reclassify the receptors in future work to 5-HT_{2A} and 5-HT_{2B}. Further, we may discover more than the present six receptor subtypes as the molecular biological techniques are applied.

Molecular biological techniques will also allow us to determine the chromosomal location of the 5-HT receptors, the localization of mRNA in the various regions of the brain, the regulation of the 5-HT receptor gene expression, as well as the role of the 5-HT receptors in the various physiological processes and pathological states involving serotonin.

One recent research paper outlined how introduction of the 5-HT_{1C} receptor into the unnatural environment of a fibroblast resulted in the generation of transformed, malignant foci.[37] This raises a question regarding the physiological functions of these receptors in various tissues. In neurons the 5-HT_{1C} receptor may be involved in neurotransmission via the regulation of ion channel function, while in fibroblasts, the same receptor alters the growth properties of cells and result in malignant transformation.

Understanding how signals are transmitted in the nervous system will eventually help us define basic brain functions such as learning and memory in molecular terms. This information will not only expand and stimulate the basic science field of neurology, but will also greatly enrich the fields of psychiatry and psychology with potential insights into the understanding and treatment of mental diseases and other degenerative brain diseases which are related to these receptors.

REFERENCES

1. BLUNDELL J. E. 1984. Neuropharmacology **23:** 1537–1551.
2. TRICKLEBANK, M. D., D. N. MIDDLEMISS & J. NEILL. 1986. Neuropharmacology. **25:** 877–886.
3. ROBERTS, M. H. T. 1984. Neuropharmcology **25:** 1529–1536.
4. CLEMENS, J. A., B. A. SAWYER & B. CERIMELE. 1977. Endocrinology. **100:** 692–698.
5. GESSA, G. L. & A. TAGLIAMONTE. 1979. *In* Serotonin in Health and Disease: Clinical Applications. W. B. Essman, Ed. Vol. **5:** 51–68. New York: Medical and Scientific Books.
6. ASARCH, K. B., R. W. RANSOM & J. C. SHIH. 1985. Life Sci. **36:** 1265–1273.
7. LEYSEN, J. E., J. E. NEIMEGEERS, J. M. VAN NEUTEN & P. M. LADURON. 1982. Mol. Pharmacol. **21:** 301–314.
8. PEROUTKA, S. J. & S. H. SNYDER. 1979. Mol. Pharmacol. **16:** 681–699.
9. PEDIGO, N. W., H. I. YAMAMURA & D. L. NELSON. 1981. J. Neurochem. **26:** 220–226.
10. YAGALOFF, K. A. & P. R. HARTIG. 1985. J. Neurosci. **5:** 3178–3183.

11. Pazos, A., D. Hoyer & J. M. Palacios. 1984. Eur. J. Pharmacol. **106:** 539–546.
12. Hamon, M., S. Bourgoin, H. Gozlan, M. D. Hall, C. Goetz, F. Artaud & A. S. Horn. 1984. Eur. J. Pharmacol. **100:** 263–276.
13. Bradley, P. B., G. Engel, W. Feniuk, J. R. Fozard, P. P. A. Humphrey, D. N. Middlemiss, E. J. Mylecharane, B. P. Richardson & P. R. Saxena. 1986. Neuropharmacology **25:** 563–573.
14. Gozlan, H., S. El Mestikawy, L. Pichat, J. Glowiniski & M. Hamon. 1983. Nature (London) **305:** 140–142.
15. Hoyer, D., G. Engel & H. O. Kalkman. 1985. Eur. J. Pharmacol. **118:** 13–23.
16. Heuring, R. E. & S. J. Peroutka. 1987. J. Neurosci. **7:** 894–903.
17. Kilpatrick, G. J., B. J. Jones & M. B. Tyers. 1987. Nature (London) **330:** 746–748.
18. Ransom, R. W., K. B. Asarch & J. C. Shih. 1986. J. Neurochem. **46:** 68–75.
19. Ransom, R. W., K. B. Asarch & J. C. Shih. 1986. J. Neurochem. **47:** 1066–1072.
20. Asarch, K. B. & J. C. Shih. 1987. J. Neurochem. **48:** 1494–1501.
21. Lubbert, H., B. J. Hoffman, T. P. Snutch, T. Van Dyke, A. J. Levine, P. R. Hartig, H. A. Lester & N. Davidson. 1987. Proc. Natl. Acad. Sci. USA **84:** 4332–4336.
22. Lubbert, H., T. P. Snuthc, N. Pascal, H. A. Lester & D. J. Davidson. 1987. J. Neurosci. **4:** 1159–1165.
23. Julius, D., A. B. MacDermott, R. Axel & T. M. Jessel. 1988. Science **241:** 558–564.
24. Kobilka, B. K., T. Frielle, S. Collins, T. Yang-Feng, T. S. Kobilka, U. Franke, R. J. Lefkowitz & M. G. Caron. 1987. Nature (London) **329:** 75–78.
25. Fargin, A., J. R. Raymond, M. J. Lohse, B. K. Kobilka, M. G. Caron & R. J. Lefkowitz. 1988. Nature (London) **335:** 358–360.
26. Pritchett, D. B., A. W. J. Bach, M. Wozny, O. Taleb, R. D. Toso, J. C. Shih & P. H. Seeburg. 1988. EMBO Journal **7:** 4135–4140.
27. Bonner, T. I., N. J. Buckley, A. C. Young & M. R. Brann. 1987. Science **237:** 527–532.
28. Peralta, E. G., A. Ashkenazi, J. W. Winslow, J. Ramachandran & D. J. Capon. 1988. Nature **334:** 434–436.
29. Gorman, C., R. Padmanabram & B. Howard. 1983. Science **221:** 551–553.
30. Peroutka, S. J. 1986. J. Neurochem. **47:** 529–540.
31. Leysen, J. E. 1988. *In* Neuromethods, Neuropharmacology II. A. A. Boulten, G. B. Baker, and A. V. Juorio, Eds. Clifton, NJ: Humana Press. In press.
32. Leysen, J. E., F. Awguters, L. Kennis, P. M. Laduron, J. Vandenberk & P. Janssen. 1981. Life Sci. **28:** 1015–1018.
33. Kendall, D. A. & S. R. Nahorski. 1985. J. Pharmacol. Exp. Ther. **233:** 473–479.
34. Roth, B. L., T. Nakaki, D. M. Chuang & E. Costa. 1984. Neuropharmacology **23:** 1223–1236.
35. de Chaffoy de Courcelles, D., J. E. Leysen, F. De Clerck, H. Van Velle & P. A. J. Janssen. 1985. J. Biol. Chem. **260:** 7603–7608.
36. Grynkiewicz, G. M. Peonie & R. Y. Tsien. 1985. J. Biol. Chem. **260:** 3440–3450.
37. Julius, D., T. J. Livelli, T. M. Jessell & R. Axel. 1989. Science. **244:** 1057–1062.
38. Tacheuchi, Y., W. Yang & J. C. Shih. 1988. J. Neurochem. **51:** 1343–1349.
39. Dixon, R. A. F. *et al.* 1986. Nature **321:** 75–79.
40. Braun, T., P. R. Schofield, B. D. Shivers, D. B. Pritchett & P. H. Seeburg. 1987. Biochem. Biophys. Res. Commum. **149:** 125–132.
41. Chumpradit, S., H. F. Kung, J. Billings, Y.-Z. Guo, W. Yang & J. C. Shih. 1989. J. Med. Chem. **32:** 543–547.

5-HT Receptors Linked to Adenylyl Cyclase Activity in Mammalian Brain[a]

FRANK D. YOCCA[b,c] AND SAUL MAAYANI[c,d]

[b]*CNS Pharmacology*
Pharmaceutical Research Institute
Bristol-Myers Squibb Company
5 Research Parkway, Department 404
Wallingford, Connecticut 06492-7660
and
Departments of Anesthesiology[c] and Pharmacology[d]
Mt. Sinai School of Medicine
City University of New York
New York, New York 10029

Serotonin (5-HT) membrane receptors that either stimulate or inhibit adenylyl cyclase (E.C. 4.6.1.1) activity have been a focus of numerous studies for nearly three decades. Several reviews on this subject have been published recently.[1–3] The summary presented here places emphasis on recent studies reported through June, 1990, and is focused mainly on mammalian brain. To date, four distinct 5-HT receptors linked to adenylyl cyclase have been *positively* classified in mammalian brain by pharmacologic methods: 5-HT_{1A},[4] 5-HT_{1B},[5] 5-HT_{1D},[6] and the putative 5-HT_4 receptor.[7] Positive classification[3] designates that selective drugs are available to allow pharmacologic distinction of a particular 5-HT receptor. In addition to a general review of these four 5-HT receptor subtypes, the following topics involving unique aspects of particular 5-HT receptor subtypes will be discussed in depth: 1) the putative 5-HT_4 receptor coupled to adenylyl cyclase activity in rat embryonic colliculi and guinea pig hippocampal membranes; 2) the apparent lack of coupling of the 5-HT_{1A} receptor to adenylyl cyclase activity in rat and bovine dorsal raphe membranes and 3) 5-HT receptors linked to adenylyl cyclase which cannot be positively classified with current pharmacologic probes ("orphan" receptors).[3] The review of these topics should help identify and focus attention on some issues important for understanding the mechanisms governing 5-HT receptor/adenylyl cyclase interactions.

MODULATION OF ADENYLYL CYCLASE ACTIVITY BY 5-HT: THE ROLE OF G PROTEINS

Guanine nucleotide-binding proteins (G proteins) comprise a family of unique proteins which function as intermediates in the mediation of the effects of endogenous substances, such as 5-HT, on adenylyl cyclase[1,2] and on ion channels.[8–10] Characterization studies have revealed that virtually all G proteins are heterotrimers which consist of an α, β, and γ subunit.[11] The apparent specificity of G proteins for a given transduction mechanism is most likely conferred from the α subunit,[12] which appears to

[a]Dr. Saul Maayani's research was supported by USPHS Grant GM-34852.
[b]Send reprint requests to Frank D. Yocca at Bristol-Myers Squibb Pharmaceutical Research Institute.

play a crucial role in transducing the intracellular response from the stimulated membrane receptor to the second messenger. Whether a specific G protein (or specific α subunit) can selectively couple to a specific subtype of membrane receptor remains to be demonstrated. Among the evidence which supports a critical role of the α subunit in the transduction process, the following appear to be most crucial. First, guanosine triphosphate (GTP), which binds to the α subunit is exchanged for guanosine diphosphate (GDP), a reaction which is catalyzed by hormone binding to the membrane receptor. The G-protein-GTP complex is short-lived, because of the hydrolysis of GTP by the enzyme GTPase, which is intrinsic to the α subunit.[11] Additionally, G proteins which mediate the stimulation of adenylyl cyclase activity (G_s) are activated in an irreversible manner by ADP ribosylation of the α_s subunit in the presence of cholera toxin.[13] In a somewhat opposite fashion, the coupling of G proteins to membrane receptors which mediate inhibition of adenylyl cyclase activity (G_i) are impaired by ADP ribosylation of the α_i subunit in the presence of pertussis toxin.[14]

The techniques used to determine the function of G proteins have also been useful in demonstrating their essential role in the activation and inhibition of adenylyl cyclase activity by 5-HT in brain. For example, GTP is a basic requirement in cell-free systems for mediation of 5-HT receptor stimulation/inhibition of adenylyl cyclase activity.[1,12,13] Furthermore, pertussis toxin blocks the 5-HT-induced inhibition of adenylyl cyclase activity through the 5-HT_{1A} receptor in hippocampal membranes,[8,10] demonstrating the importance of the α subunit in transducing the intracellular message of 5-HT to adenylyl cyclase. Additionally, high affinity 5-HT agonist binding to adenylyl cyclase-linked receptors[15–18] can be modulated by guanyl nucleotides. The reduction in agonist binding is thought to be due to a decrease in agonist affinity, which results from a destabilization of the ternary complex (hormone-receptor-G protein) triggered by the GTP-induced activation of the α subunit. Since GTP is a basic requirement for 5-HT receptors linked to adenylyl cyclase, one can speculate that the conformational state of the receptor which is functionally relevant may be the "low affinity" state. This may explain apparent dichotomies between affinity constants determined by [^{3}H]agonist binding assays and EC_{50} values generated in functional assays. Using as a model the rat hippocampal 5-HT_{1A} receptor which couples in a negative fashion to adenylyl cyclase through a pertussis toxin-sensitive G_i protein, a ratio of affinity constants (K_i) derived from radioligand binding studies and EC_{50} values from inhibitory cyclase studies were compared. As shown in TABLE 1, this ratio is greater than 10 for almost all the agonists tested. Similar ratios have also been determined for the 5-HT_{1B} and 5-HT_{1D} receptors (no binding sites have been reported to date for 5-HT_4 receptors). Differences in assay conditions not withstanding, the results support the notion that binding assays and functional studies sample different agonist-related states of 5-HT receptors which couple to adenylyl cyclase through a G_i protein. It is possible that functional assays sample the "low" affinity state (dissociated G protein with GTP bound to the α subunit) while binding studies sample the "high" affinity state (intact G protein with GDP bound to the α subunit). However, because the relationship between the two states is unknown (e.g., conversion of "high" to "low" affinity by agonist in the presence of guanyl nucleotides), the apparent lack of correlation between drug-related parameters, such as efficacy and binding affinity cannot be simply accounted for at this time.

As an understanding of the mechanisms which govern the functioning of G proteins becomes clearer, so too will be an understanding of receptors which couple to adenylyl cyclase in either a positive or negative fashion. To this end, a number of α subunits have recently been cloned and sequenced[19–21] with monospecific and monoclonal antibodies produced against these proteins or peptide fragments.[22,23] These tools, coupled with recently cloned and sequenced 5-HT receptors (i.e., the 5-HT_{1A} receptor[15]) should provide the means for precise identification of specific G proteins which couple to specific subtypes of 5-HT receptors that initiate the eventual activation/inhibition of adenylyl cyclase activity in a variety of tissues.

CENTRAL 5-HT RECEPTORS THAT STIMULATE ADENYLYL CYCLASE ACTIVITY

The study of 5-HT receptors in mammalian brain which stimulate adenylyl cyclase activity has encompassed a decade of research. We have recently reviewed several aspects of this topic.[1–3] The review presented here summarizes the more salient points of these digests and includes an update on recent reports.

The dawn of investigation into the enhancement of brain adenylyl cyclase activity by 5-HT began with fervor in the late 1970s and has continued into the current decade. Significant contributions were made by three groups of investigators working independently. Initially, Enjalbert and co-workers (1978) reported on the stimulation of adenylyl cyclase by 5-HT in membranes from rat and guinea pig colliculi, spinal cord

TABLE 1. EC_{50}/K_i Ratio of $5\text{-}HT_{1A}$ Agonists[a]

Drug	EC_{50} or K_D (nM)[b]	K_i (nM)[c]	EC_{50}/K_i
5-CT	5.2	0.47	11
BMY 7378	10.0	1.24	8
8-OH-DPAT	24.0	1.45	17
Ipsapirone	100.0	3.55	28
WB-4101	100.0	4.48	22
Buspirone	170.0	15.40	11
Gepirone	1000.0	90.50	11

[a]Data provided by authors.
[b]Determined using forskolin-stimulated adenylyl cyclase assay in rat hippocampus.
[c]Competition studies performed utilizing 0.5 nM of [^{3}H]8-OH-DPAT.

and hypothalamus.[24,25] The EC_{50} value for the response to 5-HT was consistent (1 μM) across the brain regions and species tested. At the time these experiments were performed, receptor classification could not be thoroughly addressed due to the absence of selective agonists and antagonists. These studies did, however, demonstrate that the stimulation of adenylyl cyclase activity was occurring through a 5-HT receptor which resided postsynaptic to 5-HT containing neurons since electrolytic ablation of raphe nuclei appeared not to alter the adenylyl cyclase response to this indoleamine.

Subsequent to these reports, Nelson *et al.* (1980) attempted to correlate the EC_{50} values of 5-HT agonists which stimulated adenylyl cyclase activity in collicular membranes with their affinity values for $5\text{-}HT_1$ binding sites and demonstrated that no relationship between the two existed.[26,27] This study constituted the first attempt to characterize a central 5-HT receptor by correlating receptor binding and functional data. Furthermore, these authors also showed that there was a lack of correlation between the regional density of $5\text{-}HT_1$ binding sites and the magnitude of adenylyl cyclase activity stimulated by 5-HT. Additionally, ontogenic changes in $5\text{-}HT_1$ binding sites appeared to be conversely related to the magnitude of the 5-HT-mediated adenylyl cyclase response. More importantly, however, Nelson and co-workers made a conceptual contribution by developing a viable approach to the study of 5-HT receptors, that is, the search for receptor binding and functional assays which when correlated, would allow positive classification of multiple subtypes of central 5-HT receptors.

In the early 1980s Shenker *et al.* demonstrated that 5-HT stimulated adenylyl cyclase activity in a concentration-dependent fashion in membranes from adult guinea pig hippocampus.[28,29] Utilizing curve fitting techniques, these investigators demonstrated that the response measured in hippocampal homogenates from guinea-pig appeared to be mediated by two distinct populations of 5-HT receptors.[30] The receptor which displayed "high affinity" for 5-HT displayed pharmacology which corresponded with the 5-HT_{1A} binding site. Interestingly, the results obtained with rat hippocampal membranes indicated that enhancement of adenylyl cyclase activity also occurred through the stimulation of 5-HT_{1A} receptors,[30,31] although the response in this species was quite variable. Furthermore, it was suggested by the authors that the "low affinity" receptor which stimulated adenylyl cyclase activity in guinea pig hippocampal membranes (originally termed R_L) demonstrated a pharmacologic profile that was similar to that of the 5-HT receptor which stimulates adenylyl cyclase activity in rat embryonic colliculi.[30] This was an astute observation in retrospect because recent work from the laboratory of Joel Bockaert confirmed this hypothesis.[32]

5-HT_4 RECEPTORS COUPLED POSITIVELY TO ADENYLYL CYCLASE: THE R_L REVISITED

While the initial work of Shenker and co-workers played a major role in the classification of central 5-HT_{1A} receptors,[30] the putative "low affinity" receptor for which 5-HT demonstrated low potency that was also (EC_{50} = 400 nM), present in guinea pig hippocampal membranes could not, at that time, be classified owing to a dearth of selective 5-HT agonists and antagonists. Subsequently, Dumuis *et al.* (1988), utilizing neurons from mouse embryo colliculi which were maintained in primary culture, described a 5-HT receptor which coupled in a positive fashion to adenylyl cyclase which, like the receptor in guinea pig hippocampal membranes, displayed relatively low potency for 5-HT (EC_{50} = 110 nM).[33] Furthermore, while broadly acting 5-HT_1 and 5-HT_2 antagonists did not block the effects of 5-HT in these cells, ICS-205-930, a selective 5-HT_3 antagonist demonstrated competitive antagonism, but at concentrations (1 μM) which were orders of magnitude higher than those required to block 5-HT_3 receptors. Upon rigid pharmacologic characterization with a variety of 5-HT agonists and antagonists,[34] Dumuis and co-workers (1988) compared the pharmacology of the 5-HT receptor in mouse embryonic colliculi and guinea pig hippocampus[30] (R_L) and proposed that these receptors were similar and belonged to a new category of 5-HT receptors, suggesting the nomenclature 5-HT_4.[34]

To further prove that this "low affinity" 5-HT receptor was indeed a distinct subtype, Dumuis *et al.* (1988) made a seminal discovery that a series of gastroprokinetic benzamide derivatives, particularly renzapride (BRL-24924) and cisapride stimulated the activity of adenylyl cyclase through an ICS-205-930-sensitive receptor, suggesting that these agents were agonists at this "non-classical" 5-HT receptor.[35] Interestingly, these authors speculated that a similarity might exist between the nonclassical 5-HT receptor which couples to adenylyl cyclase in colliculi neurons and the nonclassical 5-HT receptor involved in gastric and ileum motility which is specifically stimulated by substituted benzamide derivatives. These findings suggested that 5-HT_4 receptors may exist peripherally and centrally. If true, this suggestion adds the 5-HT_4 receptor to the growing list of 5-HT receptor subtypes that exist in both brain and peripheral tissues (5-HT_{1A}, 5-HT_2, 5-HT_3). More importantly, however, Bockaert *et al.* (1990), utilizing the pharmacologic tools (agonist and antagonist benzamides) which characterized the 5-HT_4 receptor in mouse embryonic colliculi, demonstrated that the low affinity 5-HT receptor (R_L) which stimulated adenylyl cyclase activity and which was originally discovered by Shenker and co-workers demonstrated a pharmacologic profile similar to the murine 5-HT_4 receptor.[32] The retrospective pharmacologic analysis performed by Bockaert and co-workers was the direct result of having access to 5-HT agonists and antagonists which were not available to Shenker *et al.*[30] and further emphasizes the importance of pharmacologic tools, not only for classification of 5-HT receptors but also for investigations into the mechanisms which govern their actions.

INHIBITION OF ENHANCED ADENYLYL CYCLASE ACTIVITY BY 5-HT_1 RECEPTOR SUBTYPES

In 1985, an important discovery was made by De Vivo and Maayani who described in a preliminary report that low concentrations of 5-HT agonists inhibited forskolin-stimulated adenylyl cyclase activity in cell-free preparations from guinea pig and rat hippocampus.[36] The 5-HT-mediated inhibition of adenylyl cyclase was a GTP-dependent effect, suggesting a role for an inhibitory G protein (G_i) in this newly discovered transduction process. Subsequently, with the aid of 5-HT agonists and antagonists and by correlating EC_{50} values in the adenylyl cyclase assay with affinities of several agonists for 5-HT sensitive [^{3}H]8-hydroxy-2-(di-*n*-propylamino)tetralin (8-OH-DPAT) binding in rat hippocampal membranes, the receptor that mediated the response in both species was classified as the 5-HT_{1A} receptor.[4] Similarly, Bockaert and colleagues (1988) observed that 5-HT inhibited the synthesis of cyclic adenosine monophosphate (cAMP) that was stimulated by vasoactive intestinal peptide in cultured neurons from mouse hippocampus, striatum and cortex, ultimately classifying this receptor as the 5-HT_{1A} subtype.[37] Interestingly, these investigations occurred simultaneously with the studies of Shenker and colleagues, who had discovered a "high affinity" 5-HT receptor which mediated stimulation of adenylyl cyclase in guinea pig hippocampal membranes.[28,29]

The impetus for exploring a possible inhibition of adenylyl cyclase activity by 5-HT was derived from a variety of electrophysiological studies. For example, Clarke and co-workers (1987) demonstrated that 5-HT_{1A} receptor stimulation *decreased* the stimulus evoked population spike amplitude in the rat hippocampal slice through a pertussis toxin sensitive G_i protein.[8] Also, drugs which enhance cAMP production, or for that matter, mimic cAMP, produced an increase in the population spike in rat hippocampal slices,[38] and since 5-HT_{1A} receptor stimulation reduces population spike amplitude, one would assume that a reduction in cAMP would also reduce the electrophysiological response. Furthermore, Andrade and co-workers (1986) demonstrated that 5-HT_{1A} receptors activate a potassium conductance in rat hippocampal pyramidal cells that is not blocked by the addition of exogenous cAMP analogs or by intracellular injection of cAMP.[39] Activation of this potassium channel is mimicked by GTP analogs and, similar to 5-HT_{1A} receptor inhibition of adenylyl cyclase in rat hippocampus, is blocked by pretreatment with pertussis toxin.[8] Therefore, inhibition of adenylyl cyclase and activation of potassium channels may occur through the activation of a G_i protein coupled to the 5-HT_{1A} receptor. This similarity in biochemical and electrophysiological phenomena lends credence to the notion that inhibition, rather than stimulation, of adenylyl cyclase by 5-HT_{1A} receptors may be of physiological relevance. On the other hand, one cannot exclude the possibility that 5-HT_{1A} receptor-mediated enhancement of adenylyl cyclase activity is also physiologically important. In any event, the possibility that the 5-HT_{1A} receptor subtype may be linked not only to different effector mechanisms but also to different G proteins raises an important question for other 5-HT receptors and their effectors: if a given receptor can couple to different G proteins and effectors in different tissues, is it possible that different pharmacologic profiles for this promiscuous receptor will be generated? Furthermore, because of differential coupling in different tissues, will the end result of pharmacologic analysis be assignment of multiple subtypes to what may be the same receptor?

Other subtypes of 5-HT_1 receptors appear to couple to adenylyl cyclase in rodent brain also. Utilizing membrane preparations from rat substantia nigra, a region which contains a high density of 5-HT_{1B} binding sites,[40,41] Bouhelal *et al.* (1988) demonstrated that 5-HT inhibited forskolin-stimulated adenylyl cyclase activity in a concentration-dependent, saturable manner, generating an EC_{50} (33 nM) comparable to that seen with 5-HT_{1A} receptor mediated inhibitory adenylyl cyclase assays, producing approximately 20% maximal inhibition of enzyme activity.[5] Pharmacological evidence supporting the positive classification of this receptor as the 5-HT_{1B} subtype was limited to the determination of competitive inhibition produced by (±)propranolol, (K_I for adenylyl cyclase = 82 nM,[5] K_I for binding = 85 nM[42]). Other 5-HT_{1B} antagonists, such as (±)cyanopindolol, displayed potencies (30–120 nM)[5] in the adenylyl cyclase assay that did not compare favorably with their potent inhibition of 5-HT_{1B} binding sites in the rat cortex.[42] Although the rank order potencies of eight 5-HT agonists and antagonists demonstrated a non-5-HT_{1A} receptor profile, these results do not conclusively support the classification of the receptor which inhibits forskolin-stimulated adenylyl cyclase activity in rat substantia nigra as the 5-HT_{1B} subtype.

Interestingly, the 5-HT_{1B} subtype is either not present or is barely detectable in the central nervous system of those species in which the 5-HT_{1D} subtype occurs.[43–45] It has been hypothesized that in such species (e.g., calf, guinea pig, pig, human), the 5-HT_{1D} subtype subserves functions mediated by the 5-HT_{1B} receptor, such as controlling the release of 5-HT by its presence on 5-HT containing terminals.[46] This hypothesis is supported by the findings that the anatomical distribution of 5-HT_{1D} receptor binding sites in bovine brain[43,47] was similar to that of the 5-HT_{1B} sites in rat brain.[40] Also, the pharmacology of the two sites is similar and the 5-HT_{1D} receptor, like the 5-HT_{1B} receptor, was also negatively linked to adenylyl cyclase activity. With this in mind, Schoeffter and colleagues (1988) extensively characterized the 5-HT receptor that mediated inhibition of forskolin-stimulated adenylyl cyclase in membranes from calf substantia nigra.[44] While 5-HT did not stimulate adenylyl cyclase activity, a saturable, GTP- and concentration-dependent inhibition of forskolin-stimulated adenylyl cyclase activity was observed with 23 percent maximal inhibition and an EC_{50} of 24 nM generated. Uniquely, four to eight tested agonists (RU 24969 > 8-OH-DPAT > buspirone > ipsapirone), as well as seven drugs which are reported to act as 5-HT antagonists in other systems (metergoline > methysergide ≥ rauwolscine ≥ cyanopindolol ≥ yohimbine > 21–009 > corynanthine) were found to be partial agonists in this receptor system. This receptor was described as the 5-HT_{1D} receptor subtype because good correlation was shown to exist between K_D (pK_D) values of a variety of previously mentioned agonists and antagonists for high affinity binding sites labeled with [^{3}H]-5-HT in bovine striatum and pEC_{50} values for agonists and pK_B values for antagonists in the adenylyl cyclase assay. Again, similar to the classification of the 5-HT_{1B} receptor, this receptor was classified but with agents that were non-selective 5-HT agonists and antagonists. While this pseudo-classification can be successful in regions which possess high concentrations of binding sites, comparative pharmacologic analysis with low receptor density brain regions or peripheral 5-HT receptors becomes difficult.

5-HT_{1A} RECEPTORS NEGATIVELY LINKED TO ADENYLYL CYCLASE: EXCLUSIVE POSTSYNAPTIC RECEPTOR COUPLING?

Advances in research and classification of central 5-HT receptors which couple to adenylyl cyclase have occurred rapidly in recent years for several reasons. First, the pharmacologic tools necessary for classification are more defined. Secondly, assay systems were developed in which responses to 5-HT were maximally efficacious and allow utilization of partial agonists for classification purposes. One reason for this is that brain regions which contain the highest concentrations of the receptor in question can be identified through ligand-binding studies. In fact, assays which allow for the study of 5-HT_{1A}, 5-HT_{1B}, and 5-HT_{1D} receptors linked negatively to adenylyl cyclase utilize tissues from brain regions which contain the highest concentration of those receptors. While this approach has been useful for assay development, it raises an important question regarding the coupling of 5-HT receptors which exist in low concentrations in different brain regions: do specific subtypes of 5-HT receptors which reside in different brain regions also couple to adenylyl cyclase? To answer this question, we selected the 5-HT_{1A} receptor as a model. The 5-HT_{1A} receptor has been the most extensively studied subtype of central 5-HT receptor coupled in a negative fashion to adenylyl cyclase.[4,36,48] This receptor appears to reside postsynaptic to 5-HT containing neurons in the hippocampus and demonstrates little ability to be sensitized since chemically-induced lesions of 5-HT neurons do not alter the 5-carboxamidotryptamine-induced inhibition of forskolin-stimulated adenylyl cyclase activity in hippocampal membranes[49] (FIG. 1). The fact that this receptor resides on hippocampal cells was corroborated by Dumuis and co-workers who demonstrated that cyclic AMP production was inhibited by a variety of 5-HT_{1A} agonists in hippocampal neurons in primary culture.[37]

Another area that contains a high concentration of 5-HT_{1A} receptor binding sites is the dorsal raphe nucleus.[40] We attempted to determine if this receptor, which resides on 5-HT containing cell bodies and/or dendrites, was coupled to adenylyl cyclase in a negative fashion. Utilizing bovine dorsal raphe membranes, we showed that while adenylyl cyclase activity could be stimulated by forskolin in a concentration-dependent fashion, no inhibition of activity could be detected in the presence of high concentrations of 5-HT.[50] In this study we demonstrated that 1) the pharmacology of the recognition sites related to the 5-HT_{1A} receptor in bovine hippocampus and dorsal raphe is indistinguishable,[50] and 2) this phenomenon cannot be related to a lack of guanyl nucleotide inhibitory binding protein (G_i) in dorsal raphe, since GppNHp produced a 30–40% inhibition of forskolin-stimulated cyclase.[50] These results suggested that not all central 5-HT_{1A} receptors could couple in an inhibitory fashion to adenylyl cyclase. This finding may be of importance in offering a molecular explanation for understanding the differences in responses and modulation of pre- and postsynaptic 5-HT_{1A} receptors to selective agonists. For example, it has been well documented in electrophysiological studies that 5-HT_{1A} agonists display different response parameters (EC_{50} and E_{max}) in rat dorsal raphe and hippocampus.[51] Furthermore, Blier, and de Montigny (1987) demonstrated that continuous treatment of rats with gepirone produced an apparent desensitization of 5-HT_{1A} somatodendritic autoreceptors controlling 5-HT neuronal cell-firing, while postsynaptic 5-HT_{1A} receptors which mediate the firing of pyramidal neurons in the CA3 region of the hippocampus were normosensitive.[52] Utilizing the technique of partial receptor alkylation, Meller and co-workers (1990) demonstrated that a large receptor reserve exists for full agonists at somatodendritic 5-HT_{1A} autoreceptors which mediate 5-HT synthesis,[53] while Yocca and co-workers (1990) indicated that postsynaptic 5-HT_{1A} receptors linked negatively to adenylyl cyclase in rat hippocampus lack a receptor reserve.[54] These functional differences, which occur for receptors with similar recognition domains, suggest that regional response and modulatory anomalies may be due to differences in post-receptor signal transduction processes (different G_i protein, coupling or stoichiometric differences between receptor-G_i proteins). Furthermore, the concept of region-dependent drug efficacy which results from differential receptor/G-

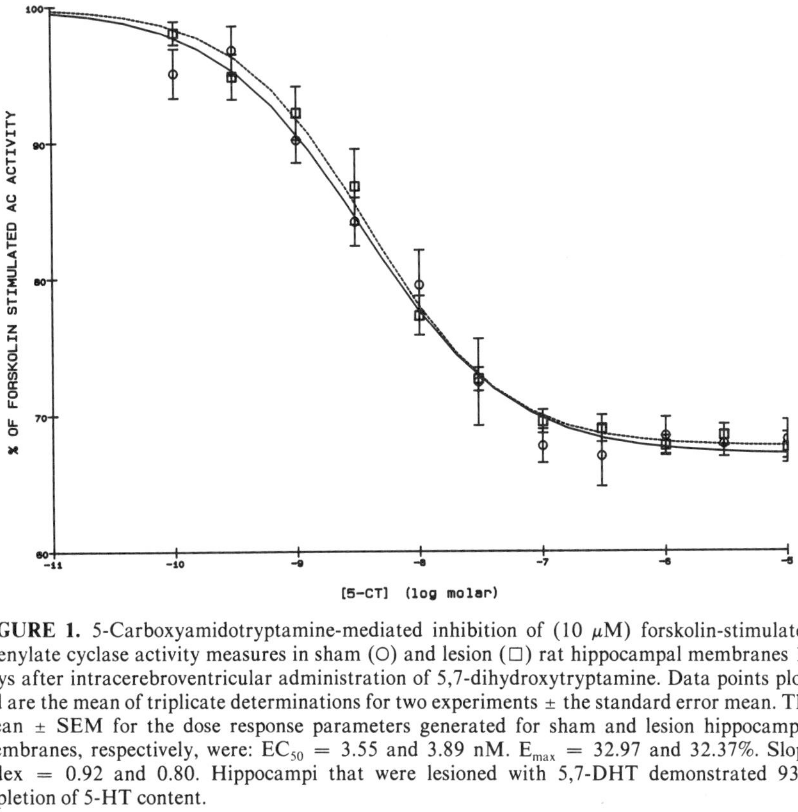

FIGURE 1. 5-Carboxyamidotryptamine-mediated inhibition of (10 μM) forskolin-stimulated adenylate cyclase activity measures in sham (○) and lesion (□) rat hippocampal membranes 14 days after intracerebroventricular administration of 5,7-dihydroxytryptamine. Data points plotted are the mean of triplicate determinations for two experiments ± the standard error mean. The mean ± SEM for the dose response parameters generated for sham and lesion hippocampal membranes, respectively, were: EC_{50} = 3.55 and 3.89 nM. E_{max} = 32.97 and 32.37%. Slope index = 0.92 and 0.80. Hippocampi that were lesioned with 5,7-DHT demonstrated 93% depletion of 5-HT content.

protein stoichiometry, different G protein distribution or different coupling may also explain the earlier observation made by Dumuis and co-workers who suggested that the spiperone-sensitive 5-HT_{1A} receptor that mediates inhibition of adenylyl cyclase in primary cultures of mouse embryo hippocampal cells is different from the "5-HT_{1A}-like" receptor in embryo cortical cells.[37] In general, regardless of the functional endpoint or tissue in question, these concepts must be considered before assigning the status of "orphan" to a receptor whose pharmacology is similar to yet different from positively classified 5-HT receptors.

UNCLASSIFIED ("ORPHAN") 5-HT RECEPTOR COUPLED TO ADENYLYL CYCLASE

There is an expanding group of 5-HT receptors which have been described in a variety of systems that cannot be positively classified as currently known receptor subtypes. One way to pharmacologically characterize these receptors is by describing, with the aid of selective drugs for other 5-HT receptor subtypes, what these receptors are not. These "orphan" receptors are frequently re-investigated when new experimental tools become available. One example is the 5-HT receptor that stimulates adenylyl cyclase activity in NCB-20 cells. This receptor was originally described more than a decade ago[55,56] and subsequently studied by Berry-Kravis and Dawson (1983).[59] Two recent reports revisited this receptor and described its pharmacology by applying different methodologies: measuring the increase in cAMP level at the cellular level by radioimmunoassay[57] and by conversion of [^{32}P]ATP to cAMP in membrane preparations.[58] Both groups reported on a twofold increase in the level of or formation of cAMP by saturating concentrations of 5-HT. Indolealkylamines proved to be the most active 5-HT agonists, with the exception of 5-carboxamidotryptamine, which exhibited a relatively high EC_{50} value (3.6 μM) and partial agonist activity (76% of the maximal response to 5-HT). The extensive structure-activity study of various 5-HT agonists reported by Conner and Mansour (1990) enabled them to propose that the indole ring modification appeared to modulate agonist affinity whereas *N*-methyl additions moderately reduced their efficacy.[58] For example, *N*-methylation of 5-HT decreased the EC_{50} more than twofold (320 vs. 130 nM, respectively) as well as the maximal response (71% vs. 5-HT). Interestingly, several ergoline derivatives acted as partial 5-HT agonists with EC_{50} values in the 16–400 nM range (LSD, BOL and methysergide). As expected, these drugs could also block the response to 5-HT with dissociation constant values similar to the EC_{50} values. The response could also be blocked in a competitive manner by methiothepin, mianserin and cyproheptadine. However, these agents do not display selectivity for any 5-HT receptor subtype and therefore do not aid in a positive classification of the 5-HT receptor in the NCB-20 cells. Among the reported negatives were the selective drugs 8-OH-DPAT, MDL 72222, ICS 205-930 and mesulergine. Despite over a decade of data accumulation, even with the use of contemporary 5-HT agonists and antagonists, this receptor still cannot be positively classified. This "orphan" receptor should be revisited again when new, selective drugs are discovered or when a new, perhaps multidisciplinary approach involving pharmacology and molecular biology can be applied to enable its definitive characterization.

ACKNOWLEDGMENTS

We are grateful to Drs. Duncan Taylor and Arlene Eison for their thoughtful review of the manuscript and to Jeanne Stockla for her assistance in producing the typescript.

REFERENCES

1. De Vivo, M. & S. Maayani. 1988. 5-HT receptors coupled to adenylate cyclase. *In* The Serotonin Receptors, E. Sanders-Bush, Ed. Clifton, NJ: Humana Press, pp. 141–179.
2. Maayani, S. & M. R. Sherman. 1990. Adenylate cyclase-linked 5-hydroxytryptamine receptors in the brain. *In* Serotonin: From Cell Biology to Pharmacology and Therapeutics. F. M. Maggi & N. Brunello, Eds. Italy: Kluwer, pp. 39–51.
3. Frazer, S., S. Maayani & B. B. Wolfe. 1990. Subtypes of receptors for serotonin. *In* Annual Review Pharmacology and Toxicology, Vol. 30: 307–348.

4. De Vivo, M. & S. Maayani. 1986. Characterization of the 5-hydroxytryptamine$_{1A}$ receptor-mediated inhibition of forskolin-stimulated adenylate cyclase activity in guinea pig and rat hippocampal membranes. J. Pharmacol. Exp. Ther. **238:** 248–253.
5. Bouhelal, R., L. Smounya & J. Bockaert. 1988. 5-HT$_{1B}$ receptors are negatively coupled with adenylate cyclase in rat substantia nigra. Eur. J. Pharmacol. **151:** 189–196.
6. Schoeffter, P., C. Waeber, J. M. Palacios & D. Hoyer. 1988. The 5-hydroxytryptamine 5-HT$_{1D}$ receptor subtype is negatively coupled to adenylate cyclase in calf substantia nigra. Naunyn-Schmiedebergs Arch. Pharmacol. **337:** 602–608.
7. Dumuis, A., R. Bouhelal, M. Sabben, R. Cory & J. Bockaert. 1988. A nonclassical 5-hydroxytryptamine receptor positively coupled with adenylate cyclase in the central nervous sytem. Mol. Pharmacol. **34:** 880–887.
8. Clarke, W. P., M. De Vivo, S. G. Beck, S. Maayani & J. Goldfarb. 1987. Serotonin decreases population spike amplitude in hippocampal cells through a pertussis toxin substrate. Brain Res. **410:** 357–361.
9. Innis, R. B., E. J. Nestler & G. K. Aghajanian. 1988. Evidence for G protein mediation of serotonin- and GABA$_{\beta}$-induced hyperpolarization of rat dorsal raphe neurons. Brain Res. **459:** 27–36.
10. Zgombick, J. M., S. G. Beck, C. D. Mahle, B. Craddock-Royal & S. Maayani. 1989. Pertussis toxin-sensitive guanine nucleotide-binding protein(s) couple adenosine A_1 and 5-hydroxytryptamine$_{1A}$ receptors to the same effector system in rat hippocampus: Biochemical and electrophysiological studies. Mol. Pharmacol. **35:** 484–494.
11. Gilman, A. C. 1987. G proteins: Transducers of receptor-generated signals. Ann. Rev. Biochem. **56:** 615–649.
12. Casey, P. J. & A. G. Gilman. 1988. G protein involvement in receptor-effector coupling. J. Biol. Chem. **263:** 2577–2580.
13. Cassel, D. & T. Pfeuffer. 1978. Mechanism of cholera toxin action: Covalent modification of the guanyl nucleotide-binding protein of the adenylate cyclase system. Proc. Natl. Acad. Sci. USA **75:** 2669–2673.
14. Katada, T. & M. Ui. 1982. Direct modification of the membrane adenylate cyclase system by islet-activating protein due to ADP-ribosylation of a membrane protein. Proc. Natl. Acad. Sci. USA **79:** 3129–3133.
15. Fargin, A., J. R. Raymond, M. J. Lohse, B. K. Kobilka, M. G. Caron & R. J. Lefkowitz. 1988. The genomic clone G-21 which resembles a β-adrenergic receptor sequence encodes the 5-HT$_{1A}$ receptor. Nature **335:** 358–360.
16. Hall, M. D., S. El Mestikawy, M. B. Emerit, L. Pichat, M. Hamon & H. Gozlan. 1985. [^{3}H]8-Hydroxy-2-(di-*n*-propylamino)tetralin binding to pre- and postsynaptic 5-hydroxytryptamine sites in various regions of the rat brain. J. Neurochem. **44:** 1685–1696.
17. Nelson, D. L., A. Harbet, E. Enjalbert, J. Bockaert & M. Hamon. 1980. Serotonin-sensitive adenylate cyclase and [^{3}H]serotonin binding sites in the CNS of rat—I. Kinetic parameters and pharmacological properties. Biochem. Pharmacol. **29:** 2445–2453.
18. Stratford, C. A., G. L. Tan, M. W. Hamblin & R. D. Ciaranello. 1988. Differential inactivation and G protein reconstitution of subtypes of [^{3}H]5-hydroxytryptamine binding sites in brain. Mol. Pharmacol. **34:** 527–536.
19. Bray, P., A. Carter, C. Simons, V. Guo, C. Puckett, J. Kamholz, A. Speigel & M. Nirenberg. 1986. Human cDNA clones for four species of $G\alpha_2$ signal transduction protein. Proc. Natl. Acad. Sci. USA **83:** 8893–8897.
20. Jones, D. T. & R. R. Reed. 1987. Molecular cloning of five GTP-binding protein cDNA species from rat olfactory neuroepithelium. J. Biol. Chem. **262:** 14241–14249.
21. Robishaw, J. D., M. D. Smigel & A. G. Gilman. 1986. Molecular basis for two forms of the G protein that stimulates adenylate cyclase. J. Biol. Chem. **261:** 9587–9590.
22. Goldsmith, P., P. Gierachik, G. Milligan, C. G. Unson, R. Vinitsky, H. L. Malech & A. M. Spiegel. 1987. Antibodies directed against synthetic peptides distinguish between GTP-binding proteins in neutrophils and brain. J. Biol. Chem. **262:** 14683–14688.
23. Milligan, G., P. Gierschik & A. M. Spiegel. 1987. The use of specific antibodies to identify and quantify guanine nucleotide-binding proteins. Biochem. Soc. Trans. **15:** 42–45.
24. Enjalbert, A., S. Bourgoin, M. Hamon, J. Adrien & J. Bockaert. 1978. Postsynaptic

serotonin-sensitive adenylate cyclase in the central nervous sytem (I). Mol. Pharmacol. **14:** 2–10.
25. Enjalbert, A., M. Hamon, S. Bourgoin & J. Bockaert. 1978. Postsynaptic serotonin-sensitive adenylate cyclase in the central nervous system (II). Mol. Pharmacol. **14:** 11–23.
26. Nelson, D. L., A. Herbert, A. Enjalbert, J. Bockaert & M. Hamon. 1980a. Serotonin-sensitive adenylate cyclase and [^{3}H]serotonin binding sites in the CNS of the rat—I. Biochem. Pharmacol. **29:** 2445–2453.
27. Nelson, D. L., A. Herbert, J. Adrien, J. Bockaert & M. Hamon. 1980b. Serotonin-sensitive adenylate cyclase and [^{3}H]serotonin binding sites in the CNS of the rat—II. Biochem. Pharmacol. **29:** 2455–2463.
28. Shenker, A., S. Maayani, H. Weinstein & J. P. Green. 1983. Enhanced serotonin-stimulated adenylate cyclase activity in membranes from adult guinea pig hippocampus. Life Sci. **32:** 2335–2352.
29. Shenker, A., S. Maayani, H. Weinstein & J. P. Green. 1985. Two 5-HT receptors linked to adenylate cyclase in guinea pig hippocampus are discriminated by 5-carboxamidotryptamine and spiperone. Eur. J. Pharmacol. **109:** 427–429.
30. Shenker, A., S. Maayani, H. Weinstein & J. P. Green. 1987. Pharmacologic characterization of two 5-hydroxytryptamine receptors coupled to adenylate cyclase in guinea pig hippocampal membranes. Mol. Pharmacol. **31:** 357–367.
31. Markstein, R., D. Hoyer & G. Engel. 1986. 5-HT$_{1A}$-receptors mediate stimulation of adenylate cyclase in rat hippocampus. Naunyn-Schmiedebergs Arch. Pharmacol. **333:** 335–341.
32. Bockaert, J., M. Sebben & A. Dumuis. 1990. Pharmacological characterization of 5-hydroxytryptamine$_4$ (5-HT$_4$) receptors positively coupled to adenylate cyclase in adult guinea pig hippocampal membranes effect of substituted benzamide derivatives. Mol. Pharm. **37:** 408–411.
33. Dumuis, A., R. Bouhelal, M. Sebben & J. Bockaert. 1988. A 5-HT receptor in the central nervous system, positively coupled with adenylyl cyclase is antagonized by ICS 205 930. Eur. J. Pharmacol. **146:** 186–188.
34. Dumuis, A., R. Bouhelal, M. Sebben, R. Cory & J. Bockaert. 1988. A nonclassical 5-hydroxytryptamine receptor positively coupled with adenylyl cyclase in the central nervous sytem. Mol. Pharmacol. **34:** 880–887.
35. Dumuis, A., M. Sebben & J. Bockaert. 1989. BRL 24924: A potent agonist at a non-classical 5-HT receptor positively coupled with adenylyl cyclase in colliculi neurons. Eur. J. Pharmacol. **162:** 381–384.
36. De Vivo, M. & S. Maayani. 1985. Inhibition of forskolin-stimulated adenylate cyclase activity by 5-HT receptor agonists. Eur. J. Pharmacol. **119:** 231–234.
37. Dumuis, A., M. Sebben & J. Bockaert. 1988. Pharmacology of 5-hydroxytryptamine-1A receptors which inhibit cAMP production in hippocampal and cortical neurons in primary culture. Mol. Pharmacol. **33:** 178–186.
38. Mueller, A. L., B. J. Hoffer & T. V. Dunwiddie. 1981. Noradrenergic responses in rat hippocampus: evidence for mediation by α and β receptors in the in vitro slice. Brain Res. **214:** 113–126.
39. Andrade, R., R. C. Malenka & R. A. Nicoll. 1986. A G protein couples serotonin and GABA$_\beta$ receptors to the same channels in hippocampus. Science **234:** 1261–1265.
40. Palacios, J. M. & M. M. Dietl. 1988. Autoradiographic studies of serotonin receptors. *In* The Serotonin Receptors. E. Sanders-Bush, Ed. Clifton, NJ: Humana Press, pp. 89–138
41. Vergé, D., G. Daval, M. Marcinkiewicz, A. Patey, S. El Mestikawy, H. Gozlan, & M. Hamon. 1986. Quantitative autoradiography of multiple 5-HT$_1$ receptor subtypes in the brain of control or 5,7-dihydroxytryptamine-treated rats. J. Neurosci. **6:** 3474–3482.
42. Engel, G., M. Göthert, D. Hoyer, E. Schlicker & K. Hillenbrand. 1986. Identity of inhibitory presynaptic 5-hydroxytryptamine (5-HT) autoreceptors in the rat brain cortex with 5-HT$_{1B}$ binding sites. Nauyn-Schmiedebergs Arch. Pharmacol. **332:** 1–7.
43. Heuring, R. E. & S. J. Peroutka. 1987. Characterization of a novel ^{3}H-5-hydroxytryptamine binding site subtype in bovine brain membranes. J. Neurosci. **7:** 894–903.
44. Hoyer, D., C. Waeber, A. Pazos, A. Probst & J. M. Palacios. 1988. Identification of a 5-HT$_1$ recognition site in human brain membranes different from 5-HT$_{1A}$, 5-HT$_{1B}$ and 5-HT$_{1C}$ sites. Neurosci. Lett. **85:** 357–362.

45. WAEBER, C., M. M. DIETL, D. HOYER, A. PROBST & J. M. PALACIOS. 1988. Visualization of a novel serotonin recognition site (5-HT_{1D}) in the human brain by autoradiography. Neurosci. Lett. **88:** 11–16.
46. HOYER, D., & D. N. MIDDLEMISS. 1989. Species differences in the pharmacology of terminal 5-HT autoreceptors in mammalian brain. Trends Pharmacol. Sci. **101:** 130–132.
47. WAEBER, C., P. SCHOEFFTER, J. M. PALACIOS & D. HOYER. 1988. Molecular pharmacology of 5-HT_{1D} recognition sites. Radioligand binding studies in human, pig and calf membranes. Naunyn-Schmiedebergs Arch. Pharmacol. **337:** 595–601.
48. DE VIVO, M. & S. MAAYANI. 1990. Stimulation and inhibition of adenylyl cyclase by distinct 5-hydroxytryptamine receptors. Biochem. Pharmacol. In press.
49. YOCCA, F. D., A. S. EISON, E. RYAN & J. TORRENTE. 1987. Effect of 5,7 dihydroxytryptamine-induced lesions on the 5-HT_{1A} receptor negatively coupled to adenylate cyclase in rat hippocampus. Soc. Neurosci. Abstr. **13:** 344.
50. YOCCA, F. D., J. M. ZGOMBICK, M. R. SHERMAN & S. MAAYANI. 1989. Differential modulation of adenylate cyclase (AC) activity by 5-HT_{1A} selective drugs in hippocampus and dorsal raphe (DR) of rat and cow. Conference on The Neuropharmacology of Serotonin, New York Academy of Sciences; Poster Abstract number 7.
51. SPROUSE, J. S. & G. K. AGHAJANIAN. 1988. Responses of hippocampal pyramidal cells to putative serotonin 5-HT_{1A} and 5-HT_{1B} agonists: a comparative study with dorsal raphe neurons. Neuropharmacology: **27:** 707–715.
52. BLIER, P. & C. DE MONTIGNY. 1987. Modification of 5-HT neuron properties by sustained administration of the 5-HT_{1A} agonist gepirone: Electrophysiological studies in rat brain. Synapse. **1:** 470–480.
53. MELLER, E., M. GOLDSTEIN & K. BOHMAKER. 1990. Receptor reserve for $5\text{-hydroxytryptamine}_{1A}$-mediated inhibition of serotonin synthesis: Possible relationship to anxiolytic properties of 5-HT_{1A} agonists. Mol. Pharmacol. **37:** 231–237.
54. YOCCA, F. D., L. IBEN & E. MELLER. 1990. Lack of receptor reserve (RR) at postsynaptic 5-HT_{1A} receptors coupled negatively to adenylyl cyclase. Soc. Neurosci. Abstr. In press.
55. MACDERMOT, J., H. HIGASHIDA, S. P. WILSON, H. MATSUZAWA, J. MINNA & M. NIRENBERG. 1979. Adenylate cyclase and acetylcholine release regulated by separate serotonin receptors of somatic cell hybrids. Proc. Natl. Acad. Sci. USA **76:** 1135–1179.
56. MACDERMOT, J. 1979. Guanosine 5′-triphosphate requirement for activation of adenylate cyclase by serotonin in a somatic cell hybrid. Life Sci. **25:** 241–246.
57. COSSERY, J. M., J. M. MIENVILLE, P. A. SHEEHY, A. M. MELLOW & D. M. CHUANG. 1990. Characterization of two distinct 5-HT receptors coupled to adenylate cyclase activation and ion current generation in NCB-20 cells. Neurosci. Lett **108:** 148–154.
58. CONNOR, D. A. & T. E. MANSOUR. 1990. Serotonin receptor-mediated activation of adenylate cyclase in the neuroblastoma NCB-20: A novel 5-hydroxytryptamine receptor. Mol. Pharmacol. **37:** 742–751.
59. BERRY-KRAVIS, E. & G. DAWSON. 1983. Characterization of an adenylate cyclase-linked serotonin (5-HT_1) receptor in neuroblastoma x brain explant hybrid cell line (NCB-20). J. Neurochem. **40:** 977–985.

DISCUSSION OF THE PAPER

DANIEL HOYER (*Sandoz, Basel*). Saul can you tell us what happens to the stimulation of the cyclase if you treat your animals with PTX? Do you see something?

MAAYANI: This is the obvious next experiment. I predict that if we do it, the stimulation will not be affected.

EFRAIN C. AZMITIA, (*New York University*) If you sensitize the system to one receptor and if they're linked, would you also get a sensitization to the second receptor?

MAAYANI: We haven't done it, but my model would suggest that it would.

Serotonin Receptors and Phosphatidylinositol Turnover[a]

ELAINE SANDERS-BUSH, MANAMI TSUTSUMI,[b]
AND KEVIN D. BURRIS

Department of Pharmacology
Vanderbilt University School of Medicine
Nashville, Tennessee 37232

The hydrolysis of membrane phosphoinositides is catalyzed by phospholipase C. This enzyme is analogous to adenylate cyclase in that it is localized in the plasma membrane and coupled to a cell surface receptor via a guanine nucleotide binding protein (G-protein). In some systems, the G-protein that links phospholipase C and the receptor is one of the G_i family, but, in most circumstances, the identity of the G-protein is unknown. Receptor-activated phospholipase C cleaves phosphatidylinositol-4,5-bisphosphate, leading to the formation of two products, inositol-1,4,5-trisphosphate (I-1,4,5-P_3) and diacylglycerol (DG), both of which function as second messengers (FIG. 1). I-1,4,5-P_3 releases calcium from intracellular stores; calcium then has multiple cellular effects, including activation of calcium/calmodulin-dependent protein kinases. Other inositol phosphate products have been identified and at least one, inositol-1,3,4,5-tetrakisphosphate (I-1,3,4,5-P_4) is also active as an intracellular mediator (see Majerus *et al.*[1] for a recent review). DG activates another protein kinase, protein kinase C. Protein kinase C regulates many cellular proteins, including components of the phosphoinositide hydrolysis and adenylate cyclase signaling cascade, ion channels and gene transcription factors. DG also releases arachidonic acid, a precursor of prostaglandins and leukotrienes. Thus, a myriad of cellular changes ensue with the activation of phosphatidylinositol turnover and the consequences of these changes are far-reaching.

Several receptors in the central nervous system are coupled to phosphatidylinositol turnover, including alpha$_1$-adrenergic receptors, cholinergic muscarinic sites, excitatory amino acid receptors and two 5HT sites, $5HT_2$ and $5HT_{1c}$ receptors. In the brain, it has been difficult to measure the initial products of the hydrolysis of phosphatidylinositol-4,5-bisphosphate and most studies, including our own, have utilized the method of Berridge *et al.*[2] that measures the accumulation of inositol-monophosphate (IP) in the presence of lithium as an indirect measure of the formation of I-1,4,5-P_3. The first 5HT receptor in the brain to be associated with the phosphatidylinositol turnover pathway was the $5HT_2$ site.[3] Since those original observations, many studies have confirmed that the $5HT_2$ receptor is coupled to phosphatidylinositol turnover in many tissues: the cerebral cortex of rats[4–7]; and mice[8]; astroglia from cerebral cortex and other brain areas[9]; cerebellar granule cells[10]; peripheral tissues such as vascular and other smooth muscle preparations[11–13] and platelets[14]; and various cell lines, C_6

[a]The original investigations of the authors were supported by research grants from the National Institute of Drug Abuse (DA 05181) and the National Institute of Mental Health (MH 34007) and by a National Research Service Award (GM 07628) from the National Institute of General Medical Sciences.

[b]Present address: Fishberg Center for Neurobiology, Mount Sinai School of Medicine, New York, NY 10029.

glioma cell,[15] A_7r_5 smooth muscle cells[16] and WRK1 cells derived from rat mammary tumors.[17] The $5HT_{1c}$ receptor was first characterized in radioligand binding studies and found to be highly localized in the choroid plexus.[18,19] This site was subsequently shown to be linked to phosphatidylinositol turnover in rat[5,20] and pig[21] choroid plexus. Antagonists potencies were the primary means of assigning 5HT-induced phosphatidylinositol turnover to a particular subtype of 5HT receptor. Studies utilizing receptor agonists are more problematic because their properties vary depending on the conditions and the sensitivity of the assay and uncontrollable factors such as receptor density and receptor reserve.

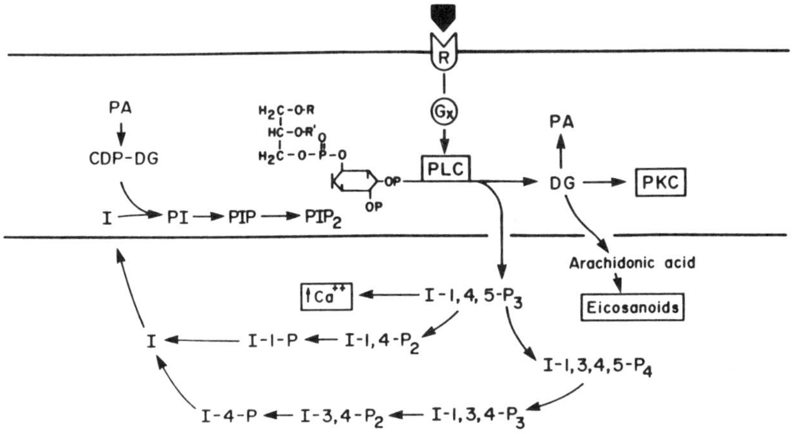

FIGURE 1. Schematic representation of receptor-initiated inositol lipid turnover. The interaction of an agonist with its receptor (R) leads to the activation of phospholipase C (PLC), a phosphodiesterase that cleaves the hydrolysis of membrane associated phosphoinositides, primarily phosphatidylinositol-4,5-bisphosphate (PIP_2). The receptor is linked to PLC via a G-protein (G_x), whose precise identity is unknown. Hydrolysis of PIP_2 generates two second messenger products, inositol-1,4,5-trisphosphate (I-1,4,5-P_3) and diacylglycerol (DG). I-1,4,5-P_3 releases calcium from intracellular stores and DG activates protein kinase C (PKC) and serves as a precursor for the eicosanoids prostaglandins and leukotrienes. DG is converted by DG kinase to phosphatidic acid (PA) which is then recycled into the inositol lipid pool. I-1,4,5-P_3 is inactivated by sequential dephosphorylation to form inositol (I), which also recycles into the inositol lipid pool. I-1,3,4,5-P_4 is formed by the action of a kinase and may act as an intracellular mediator as well. Other products of the hydrolysis of PIP_2 may occur and are not shown. Several reviews that give in-depth discussions of the inositol lipid pathway have appeared in the past few years and these should be consulted for more details (e.g., Berridge[49]).

AGONIST PROFILES

Extensive radioligand binding studies have provided estimates of the affinities of drugs for the various 5HT receptor subtypes. Such studies give accurate values for affinities, but can not unequivocally categorize the action of a drug, even in the most fundamental way, i.e., agonist versus antagonist. Thus, although a guanine nucleotide-induced shift of a drug's affinity has been useful for differentiating agonists from antagonists in some circumstances, this has been far less successful for the central 5HT receptors linked to phosphatidylinositol turnover. Furthermore, agonist affinities may not accurately reflect potency for eliciting a response, because of the existence of receptor reserve. Classical receptor theory states that the maximum response can be obtained with only a fraction of the receptors occupied; the remaining unoccupied sites are referred to as spare or reserve receptors. Thus, an agonist can have a greater potency at eliciting a response than what may be expected from its affinity at the receptor site measured by competition for a radioligand binding site. Another important property of an agonist, its efficacy, can not be discerned in binding studies. The term efficacy refers to the ability of an agonist to initiate a response when it binds to the receptor; a drug that gives the maximum possible response is referred to as a full agonist, while another drug that gives a response that is less than the maximum is referred to as a partial agonist. Stated another way: different drugs can give different responses for the same percent occupancy of a given receptor population. Efficacy reflects the efficiency of the coupling of the occupied receptor with the effector enzyme. Differences in the efficacies of drugs can be demonstrated experimentally by comparing the maximum responses of drugs. For example, in the rat choroid plexus, both 5HT (the presumed natural agonist) and trifluoromethylphenylpiperazine (TFMPP) elicit phosphatidylinositol turnover by interacting with the $5HT_{1c}$ receptor.[22] At concentrations that saturate the receptor, 5HT causes a 10-fold rise in radiolabeled IP formation, while the maximum response elicited by TFMPP is only one-half of that (FIG. 2). These substances are also different in their efficacies at the $5HT_2$ receptor in the rat cerebral cortex. 5HT causes a 2.5-fold increase in IP formation, but TFMPP-induced phosphatidylinositol turnover does not occur except at very high concentrations, i.e., >100 μM (FIG. 3). A similar finding was reported earlier.[6] The effect of high concentrations of TFMPP is not mediated by activation of $5HT_2$ receptors, since it is not blocked by ketanserin (FIG. 3). On the other hand, TFMPP blocks 5HT-elicited phosphatidylinositol turnover in cerebral cortex at a concentration of 25 μM.[22] These functional studies then confirm that TFMPP binds to the $5HT_2$ receptor as has been found in radioligand binding studies, but surprisingly TFMPP appears to be a pure

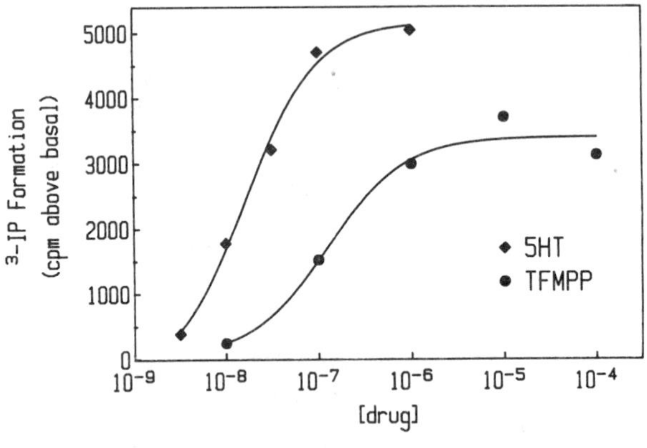

FIGURE 2. Comparison of the effects of 5HT and trifluoromethylphenylpiperazine (TFMPP) on phosphoinositide hydrolysis in choroid plexus. Phosphatidylinositol turnover was determined in intact choroid plexus of rats. ^{3}H-IP formation was measured in the presence of the indicated concentration of drug and was expressed as cpm above basal levels of ^{3}H-IP (901 cpm). The curves were fitted by the Graph-Pad computer program, using the equation for a sigmoid curve.

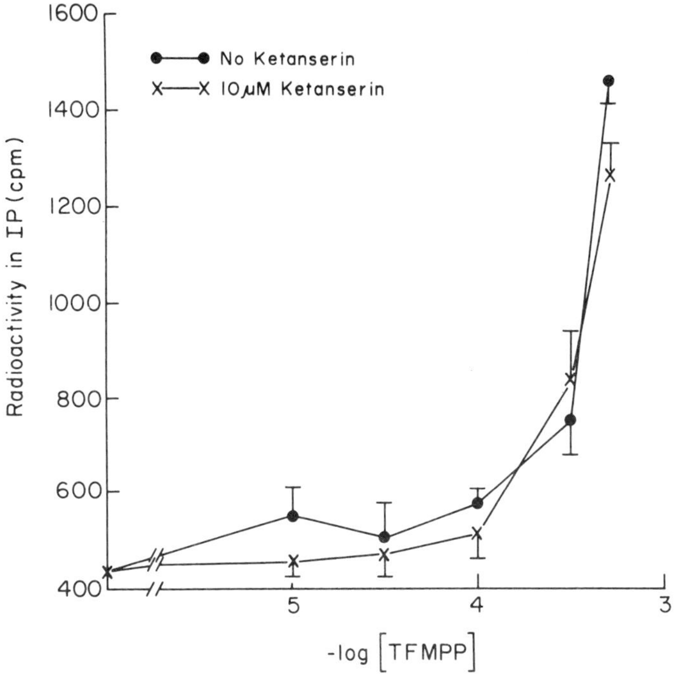

FIGURE 3. Trifluoromethylphenylpiperazine (TFMPP) does not activate $5HT_2$ receptors in cerebral cortex slices. ^{3}H-IP formation was measured in the presence of the indicated concentration of TFMPP with or without the addition of 10 μM ketanserin. The vertical bars represent SEM.

antagonist at the $5HT_2$ receptor. Studies such as these demonstrate the value of functional analyses of the transmembrane signaling pathway in addition to radioligand binding assays. As illustrated later, drug efficacy and spare receptors must be taken into account in studies of the regulation of 5HT receptors.

(+)Lysergic acid diethylamide (LSD) and the hallucinogenic amphetamines, 2,5-dimethoxy-4-methylamphetamine and 2,5-dimethoxy-4-bromoamphetamine act as agonists at both of the 5HT receptors linked to phosphatidylinositol turnover.[23] (+)LSD is a partial agonist at the $5HT_2$ receptor, eliciting a response that is only about 25% of that produced by 5HT (FIG. 4). Consistent with the interpretation that (+)LSD and 5HT activate phosphatidylinositol turnover by interacting with the same site, (+)LSD blocks the effect of 5HT (FIG. 4). However, the blockade is never complete because of the partial agonist activity of (+)LSD. Depending on the level of endogenous, full agonist (5HT) at a given time and place, partial agonists, such as (+)LSD, may have agonistic or antagonistic effects. This is consistent with the multiple effects described for (+)LSD in the literature. Furthermore, with such a low maximum response with the full agonist 5HT, the assay of phosphatidylinositol turnover must be optimized in order to get a detectable signal with a partial agonist. This may explain why Pierce and Peroutka[24] found that (+)LSD antagonized 5HT-induced phosphatidylinositol turnover in cerebral cortex, at concentrations that had no

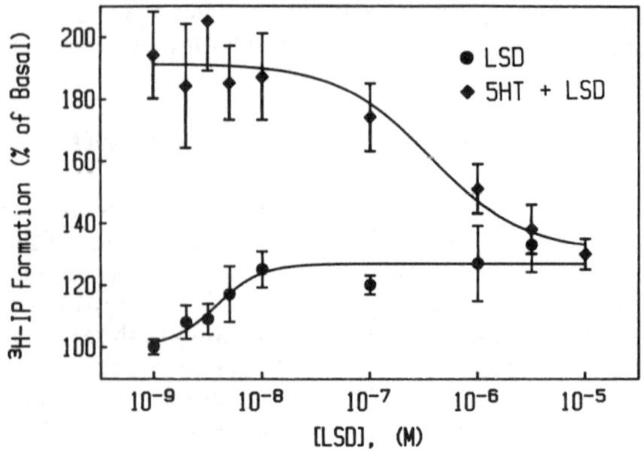

FIGURE 4. Partial agonist effect of (+)LSD in cerebral cortex. ^{3}H-IP formation was measured in cerebral cortex slices from rat brain. The indicated concentration of (+)LSD was added in the presence (♦) or absence (●) of 10 μM 5HT. ^{3}H-IP formation is expressed as percent of basal levels (624 cpm). Modified from Sanders-Bush *et al.*[23]

apparent agonist action. In the choroid plexus, (+)LSD elicits phosphatidylinositol turnover by activating the $5HT_{1c}$ receptor.[23] Once again, (+)LSD is a partial agonist, eliciting a phosphatidylinositol turnover signal that is about 50% of that elicited by 5HT (FIG. 5). Even so, (+)LSD gives almost a fivefold increase in IP formation in the choroid plexus. (−)LSD, on the other hand, does not activate the $5HT_{1c}$ receptor, even at a concentration of 100 μM (FIG. 5). (+)2,5-dimethoxy-4-methylamphetamine is

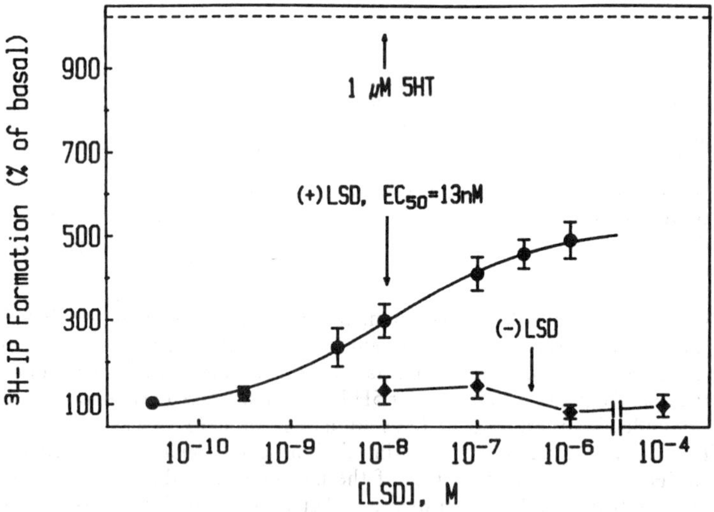

FIGURE 5. Effects of the optical isomers of LSD on phosphatidylinositol turnover in rat choroid plexus. ^{3}H-IP formation was measured in the presence of the indicated concentrations of (+) or (−) LSD. The data are expressed as % of basal (729 cpm) and are the mean of 4 determinations ± SEM. The maximum response to 5HT (1 μM) is shown for comparison (*dashed line*).

also a partial agonist at both the $5HT_{1c}$ and $5HT_2$ receptor,[23] while (+)2,5-dimethoxy-4-bromoamphetamine is essentially a full agonist at both sites (unpublished results).

COMPENSATORY CHANGES IN FUNCTION

Many studies show that the $5HT_2$ receptor binding site is not regulated in the predictable manner, i.e., with chronic inactivity, a compensatory supersensitivity is expected and vice versa with chronic activation, a compensatory subsensitivity. These studies have been thoroughly reviewed by Frazer *et al.*[25] The present discussion focuses on the studies of regulation of the transmembrane signaling pathway, phosphatidylinositol turnover. The most striking inconsistency is found after chronic reduction in 5HT neurotransmission either with lesions of 5HT pathways or chronic administration of $5HT_2$ antagonists (TABLE 1). Thus, in only one circumstance (denervation elicited by

TABLE 1. Regulation of $5HT_2$ Receptor–Initiated Phosphoinositide (PI) Hydrolysis

Treatment	PI Hydrolysis	Binding Site	References
Chronic Inactivity			
5,7 DHT	0[a]	0,↑	26,27
$5HT_2$ Antagonists	↓	↓	27,28
Chronic Activation			
5HT Uptake Inhibitors			
Selective	↓	0	31
Nonselective	↓	↓	7
$5HT_2$ Agonists	↓	↓,0[b]	28[d]
5HT *In vitro*	↓	?	32,47
Chronic Antidepressants			
ECS[c]	0	↑	8
Desipramine	↓	↓	8
Iprindole	↓	↓	7
Lithium	↓	?	48

[a]0 refers to no change; ↑ to an increase; ↓, a decrease.
[b]One of these drugs, TFMPP, may actually be an antagonist.
[c]Electroconvulsive shocks.
[d]Unpublished results.

the administration 5,7-dihydroxytryptamine in mice[8]) has a compensatory increase in receptor number been reported, but a functional supersensitivity did not accompany the change in $5HT_2$ receptor density. In rats, neither a change in $5HT_2$ receptor density (see for example Quik and Azmitia[26]) nor $5HT_2$ function[27] have been found. Even more difficult to understand are the findings of a loss of $5HT_2$ receptor sites and a subsensitive $5HT_2$-mediated phosphatidylinositol turnover response after chronic treatment with $5HT_2$ antagonists.[28] Studies of the mechanism of this effect suggest that these changes reflect an adaptive response rather than the direct effects of the 5HT antagonists.[28-30] One possible explanation for this paradoxical finding is that the antagonists are actually partial agonists. Agonists are known to elicit subsensitivity in the $5HT_2$ receptor, the expected compensatory change. However, extensive studies in our laboratory have failed to find evidence for partial agonist effects of those $5HT_2$

antagonists that cause down-regulation of the $5HT_2$ receptor. We are also not aware of any evidence of partial agonist properties in the literature, so this explanation seems unlikely. Another possibility that is difficult to rule out is that the effects of $5HT_2$ antagonists are unrelated to their effects at the $5HT_2$ receptor, but instead are secondary to some other unknown effects of the drugs. We are currently investigating the effects of $5HT_2$ antagonists on cells in culture in order to address this possibility. For the time being, the explanation for the down-regulation of $5HT_2$ receptors after chronic administration of $5HT_2$ antagonists remains a mystery. Thus, treatments that were predicted to elicit adaptive supersensitivity in the $5HT_2$ receptor system failed to do so. Indeed, so far, an adaptive supersensitivity of the $5HT_2$ receptor has not been demonstrated. On the other hand, the $5HT_2$ receptor adapts in the opposite direction with a subsensitive response after chronic activation by 5HT agonists *in vivo* or *in vitro* or by 5HT uptake inhibitors *in vivo* (see TABLE 1 for specific references). In addition to 5HT uptake inhibitors, other antidepressant treatments also elicit adaptive changes in $5HT_2$ receptors. Desipramine and the atypical antidepressant iprindole reduce the density of $5HT_2$ sites and desensitize the $5HT_2$-receptor phosphatidylinositol turnover signal. Chronic electroconvulsive shock treatment increases the density of $5HT_2$ sites, but a functional supersensitivity does not accompany this up-regulation of sites. A dissociation between $5HT_2$ binding and $5HT_2$ function is also found after other treatments, e.g., chronic administration of sertraline, a selective 5HT uptake inhibitor[31] or quipazine, a 5HT agonist (unpublished results). These treatments decrease $5HT_2$-receptor-mediated phosphatidylinositol turnover without decreasing $5HT_2$ receptor density. Thus, the $5HT_2$ receptor is apparently regulated at the level of the cell surface receptor site and also at a distal site, perhaps the G-protein-mediated coupling of the receptor site with phospholipase C. The heterogeneity of the brain makes it difficult, if not impossible, to determine the mechanism of the post-receptor regulatory process in brain slices. Such studies must await the development of less complex models of central $5HT_2$ receptors, e.g., cells in culture. Recent studies of cerebellar granule cells in primary culture show that $5HT_2$ receptors on these cells are rapidly desensitized after exposure to 5HT.[32] These cells may, therefore, be a suitable system for investigating the mechanism of regulation of the $5HT_2$ receptor.

The regulation of the $5HT_{1c}$ receptor has been less well characterized. Denervation supersensitivity has been reported for the $5HT_{1c}$ receptor linked to phosphatidylinositol turnover in the choroid plexus,[33] although this response is apparently not mediated by an increased number of $5HT_{1c}$ receptors.[34] Like the $5HT_2$ site, the $5HT_{1c}$ receptor is paradoxically down-regulated and desensitized by chronic antagonist treatment.[35] Furthermore, chronic administration of quipazine, a 5HT agonist, leads to a loss of $5HT_{1c}$ binding sites (FIG. 6). Studies of $5HT_{1c}$ receptor sensitivity after chronic quipazine administration show that phosphatidylinositol turnover is not significantly altered when the full agonist 5HT is used; however, a reduction is found for the partial $5HT_{1c}$ agonist TFMPP (FIG. 6). This difference is apparently explained by the existence of a $5HT_{1c}$ receptor reserve in the choroid plexus.[36] As pointed out earlier, 5HT gives a maximum phosphoinositide hydrolysis response when only about 50% of the $5HT_{1c}$ receptors are occupied, whereas TFMPP occupies all of the sites to elicit a maximum response. This may explain why chronic quipazine administration, which reduces the number of sites by 30%, lowers the TFMPP response to the same extent, but does not change the response to 5HT. Thus, the existence of a reserve for $5HT_{1c}$ receptors in the choroid plexus has functional significance in that it protects the system from perturbations in function even when there are alterations in the density of receptors. It remains to be determined if a reserve also exists for $5HT_2$ receptors in the cerebral cortex and if this phenomenon can explain some of the paradoxical findings for the regulation of the $5HT_2$ receptors.

PHYSIOLOGICAL FUNCTION OF 5HT RECEPTORS LINKED TO PHOSPHATIDYLINOSITOL TURNOVER

A major function of the choroid plexus is the regulation of cerebrospinal fluid (CSF) volume and composition. The choroid plexus controls CSF composition in three ways: (1) it acts as the blood/CSF barrier preventing the diffusion of polar substances from the blood to the CSF, (2) it controls the transcytosis of proteins from the CSF to the blood[37] and (3) it synthesizes and secretes a number of proteins, including the iron carrier transferrin, the copper carrier ceruloplasmin and transthyretin, which carries thyroid hormones.[38] This latter function is shared by other cells that form barriers between body fluids and may be important in the maintenance of the microenvironment of organs such as brain that are isolated from the general circulation. Given that

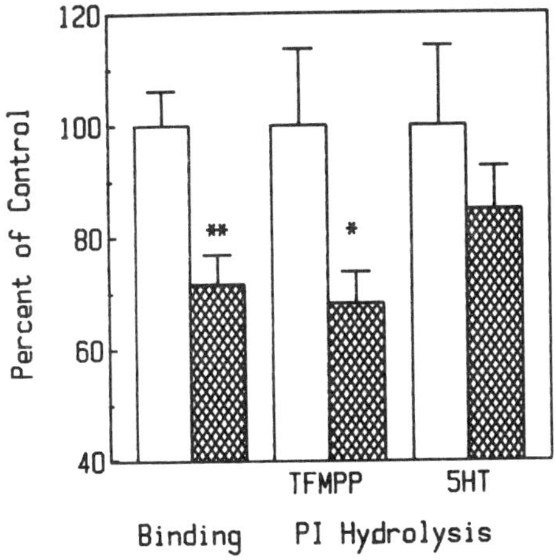

FIGURE 6. Down-regulation and desensitization of the $5HT_{1c}$ receptor after chronic agonist administration. Rats were injected i.p. with a loading dose of 20 mg/kg of quipazine followed by 7 injections of 10 mg/kg administered at 8 am and 4 pm on 4 consecutive days. One day after the last dose, ^{3}H-IP formation was determined in intact choroid plexus after incubating with 1 μM 5HT or 10 μM TFMPP. Saturation binding of ^{3}H-mianserin provided as estimate the density of $5HT_{1c}$ binding sites. The data are expressed as percent of control and are the mean of 4–8 determinations ± SEM. Stippled bars are quipazine-treated rats. $*p < 0.05$ $**p < 0.01$.

$5HT_{1c}$ sites are present at such high levels in the choroid plexus and that they apparently act as receptors to transmit a signal across the cell membrane via the phosphatidylinositol turnover pathway, we reasoned that these sites may play a role in the choroid plexus. 5HT has been shown to influence the formation of CSF in intact animals[39,40]; however, it is not clear that this reflects changes at the level of the choroid plexus or if the effects on CSF formation are secondary to changes in the vasculature or in the brain. In order to study the direct effects of 5HT on the function of the choroid plexus, we developed a model system of choroid plexus epithelial cells in primary culture. Although it is not possible to study the formation of CSF volume in these cells, we have found that the choroid plexus epithelial cells synthesize several proteins and secrete them into the cultured medium.[41] One of the major proteins is transferrin which was identified by gel electrophoresis and immunoprecipitation with transferrin antibody. Hybridization with transferrin cRNA confirmed that transferrin mRNA is present in these cells. We have utilized the synthesis of transferrin as a functional marker and have examined the effects of potential regulatory agents on this function. Transferrin levels in the cultured medium were measured by radioimmunoassay[42] and transferrin mRNA in the cells was quantitated by solution hybridization using a modification of the method of Newman *et al.*[43] with ^{32}P-labeled transferrin cRNA as a hybridization probe. These studies show that 5HT increases the level of transferrin in the cultured medium (FIG. 7) with an accompanying increase in the steady state level of transferrin mRNA (FIG. 8). We are presently exploring the mechanism of action of 5HT including possible changes in gene transcription and/or mRNA stability. Equally interesting is the question of the role of the $5HT_{1c}$ receptor in mediating the effect of 5HT on transferrin synthesis. There are no other known 5HT receptors in the choroid plexus[18,19]; therefore, it is likely that the effects of 5HT are mediated by the $5HT_{1c}$ site. The potency of 5HT to increase the synthesis of transferrin (EC_{50} = 17 nM) is consistent with its potency for activating the $5HT_{1c}$ receptor and eliciting phosphatidylinositol turnover. Furthermore, two other agents that activate $5HT_{1c}$ sites, MK212 and (+)LSD, also increase the levels of transferrin (FIG. 7). (−)LSD, on the other hand, is unable to increase transferrin levels (FIG. 7) or to activate the $5HT_{1c}$ receptor (FIG. 5). Thus, it is possible that activation of the $5HT_{1c}$ receptor has a growth hormone-like effect in the choroid plexus, promoting the expression of the gene for transferrin and perhaps other nutrient carrier proteins. Such a finding would be consistent with the recent evidence that the $5HT_{1c}$ receptor acts as a protooncogene when expressed in fibroblasts.[44] The increased synthesis of transferrin may reflect differentiation of the cells in culture. Thus, 5HT via the $5HT_{1c}$ receptor may promote differentiation of choroid plexus cells, as it has been shown to promote differentiation in other cell types (see e.g. Lauder and Krebs[45]). Interestingly, the beta-adrenergic agonist isoproterenol has the opposite effect of 5HT agonists, it decreases the level of transferrin in the cultured medium (FIG. 7). Isoproterenol activates adenylate cyclase in the choroid plexus[46] and it is possible that an increase in cyclic AMP mediates the effect of isoproterenol. Consistent with this interpretation, cyclic AMP analogues, such as 8-bromo-cyclic AMP (FIG. 7), also reduce transferrin levels. Thus it is tempting to speculate that the two major second messenger pathways, phosphatidylinositol turnover and adenylate cyclase, have opposing actions in the choroid plexus.

To date, there is no indication of the consequences of $5HT_2$ receptor-phosphatidyl-

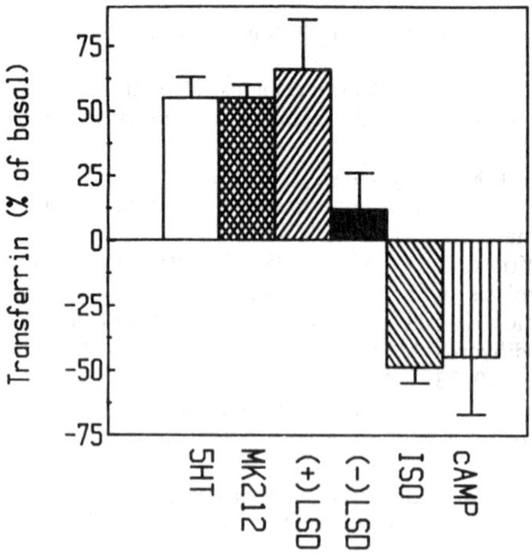

FIGURE 7. Transferrin levels in primary cultures enriched in choroid plexus epithelial cells. The cells were incubated with the regulatory agents for 4 days and transferrin levels in the conditioned medium were determined by radioimmunoassay. The data are expressed as % of basal levels and are the mean of at least 3 determinations ± SEM. The final concentration of regulatory agents was 10 μM, except cAMP (100 μM) and the LSD isomers (1 μM). Abbreviations: ISO, isoproterenol; cAMP, 8-bromo-cyclic AMP.

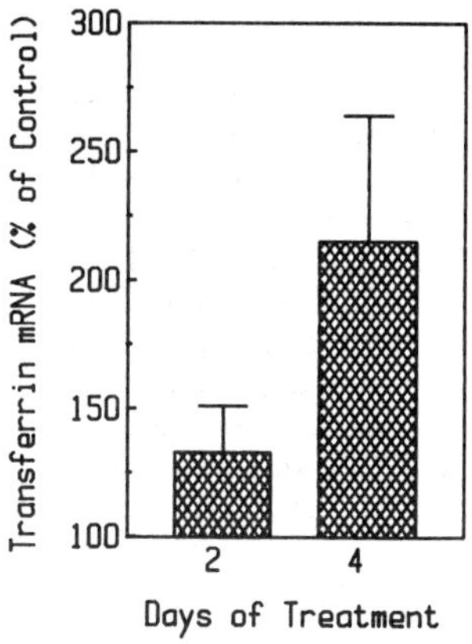

FIGURE 8. Effect of 5HT on the steady-state levels of transferrin mRNA. Epithelial cells from rat choroid plexus were treated with 10 μM 5HT for the indicated time. Total RNA was isolated from cells and hybridized in solution with ^{32}P-transferrin cRNA and excess yeast RNA. Values were normalized to a constitutive mRNA probe, 1B15. The data are expressed as percent of control (untreated cells) and are the mean of 7–8 determinations ± SEM.

inositol hydrolysis in the cerebral cortex or the consequences of activation of $5HT_{1c}$ receptors that are found in brain. Only time will tell if these receptors are involved in early signals of neurotransmission, or if, like in the choroid plexus, they are involved in more long-term regulation of brain functions, such as gene expression and/or the synthesis of proteins.

REFERENCES

1. Majerus, P. W., T. M. Connolly, V. S. Bansal, R. C. Inhorn, T. S. Ross & D. L. Lips. 1988. Inositol phosphates: synthesis and degradation. J. Biol. Chem. **263:** 3051–3054.
2. Berridge, M. J., C. P. Downes & M. R. Hanley. 1982. Lithium amplifies agonist-dependent phosphatidylinositol responses in brain and salivary glands. Biochem. J. **206:** 587–595.
3. Conn, P. J. & E. Sanders-Bush. 1984. Selective 5HT-2 antagonists inhibit serotonin stimulated phosphatidylinositol metabolism in cerebral cortex. Neuropharmacology **23:** 993–996.
4. Conn, P. J. & E. Sanders-Bush. 1985. Serotonin-stimulated phosphoinositide turnover: mediation by the S_2 binding site in rat cerebral cortex but not in subcortical regions. J. Pharmacol. Exp. Ther. **234:** 195–203.
5. Conn, P. J. & E. Sanders-Bush. 1986. Agonist-induced phosphoinositide hydrolysis in choroid plexus. J. Neurochem. **47:** 1754–1760.
6. Claustre, Y., L. Rouquier & B. Scatton. 1988. Pharmacological characterization of serotonin-stimulated phosphoinositide turnover in brain regions of the immature rat. J. Pharmacol. Exp. Ther. **244:** 1051–1056.
7. Kendall, D. A. & S. R. Nahorski. 1985. 5-hydroxytryptamine-stimulated inositol phospholipid hydrolysis in rat cerebral cortex slices: Pharmacological characterization and effects of antidepressants. J. Pharmacol. Exp. Ther. **233:** 473–479.
8. Godfrey, P. P., S. J. McClue & M. M. Young & D. J. Heal. 1988. 5-Hydroxytryptamine-stimulated inositol phospholipid hydrolysis in the mouse cortex has pharmacological characteristics compatible with mediation via $5\text{-}HT_2$ receptors but this response does not reflect altered $5HT_2$ function after 5,7-dihydroxytryptamine lesioning or repeated antidepressant treatments. J. Neurochem. **50:** 730–738.
9. Hansson, E., P. Simonsson & C. Alling. 1987. 5-Hydroxytryptamine stimulates the formation of inositol phosphate in astrocytes from different regions of the brain. Neuropharmacology **26:** 1377–1382.
10. Xu, J & D-M. Chuang. 1987. Serotonergic, adrenergic and histaminergic receptors coupled to phospholipase C in cultured cerebellar granule cells of rats. Biochem. Pharmacol. **36:** 2353–2358.

11. ROTH, B. L., T. NAKAKI, D. CHUANG & E. COSTA. 1986. 5-Hydroxytryptamine$_2$ receptors coupled to phospholipase C in rat aorta: Modulation of phosphoinositide turnover by phorbol ester. J. Pharmacol. Exp. Ther. **238:** 480–485.
12. CORY, R. N., P. BERTA, J. HAIECH & J. BOCKAERT. 1986. 5-HT$_2$ receptor-stimulated inositol phosphate formation in rat aortic myocytes. Eur. J. Pharmacol. **131:** 153–157.
13. COHEN, M. L. & L. A. WITTENAUER. 1987. Serotonin receptor activation of phosphoinositide turnover in uterine, fundal, vascular, and tracheal smooth muscle. J. Cardiovasc. Pharmacol. **10:** 176–181.
14. DE CHAFFOY DE COURCELLES, D., J. E. LEYSEN, F. DE CLERCK, H. VAN BELLE & P. A. J. JANSSEN. 1986. Evidence that phospholipid turnover is the signal transducing system coupled to serotonin-S$_2$ receptor sites. J. Biol. Chem. **260:** 7603–7608.
15. ANANTH, U. S., U. LELI & G. HAUSER. 1987. Stimulation of phosphoinositide hydrolysis by serotonin in C6 glioma cells. J. Neurochem. **48:** 253–262.
16. DOYLE, V. M., J. A. CREBA, U. R. RUEGG & D. HOYER. 1986. Serotonin increases the production of inositol phosphates and mobilizes calcium via the 5HT2 receptor in A7r5 smooth muscle cells. Naunyn-Schmniederberg's Arch. Pharmacol. **333:** 98–103.
17. CORY, R. N., B. ROUOT, G. GUILLON, F. SLADECZEK, M.-N. BALESTRE & J. BOCKAERT. 1987. The 5-hydroxytryptamine (5-HT2) receptor stimulates inositol phosphate formation in intact and broken WRK1 cells: Determination of occupancy-response relationships for 5-HT agonists. J. Pharmacol. Exp. Ther. **241:** 258–267.
18. YAGALOFF, K. A. & P. R. HARTIG. 1985. ^{125}I-Lysergic acid diethylamide binds to a novel serotonergic site on rat choroid plexus epithelial cells. J. Neurosci. **5:** 3178–3183.
19. PAZOS, A., D. HOYER & J. M. PALACIOS. 1984. The binding of serotonergic ligands to the porcine choroid plexus: Characterization of a new type of serotonin recognition site. Eur. J. Pharmacol. **106:** 539–546.
20. CONN, P. J., E. SANDERS-BUSH, B. J. HOFFMAN & P. R. HARTIG. 1986. A unique serotonin receptor in choroid plexus is linked to phosphatidylinositol turnover. Proc. Natl. Acad. Sci. USA **83:** 4086–4088.
21. HOYER, D. 1988. Molecular pharmacology and biology of 5-HT1c receptors. TIPS **9:** 89–94.
22. CONN, P. J. & E. SANDERS-BUSH. 1987. Relative efficacies of piperazines at the phosphoinositide hydrolysis-linked 5HT-2 and 5HT-1c receptors. J. Pharmacol. Exp. Ther. **242:** 552–557.
23. SANDERS-BUSH, E., K. D. BURRIS & K. KNOTH. 1988. Lysergic acid diethylamide and 2,5-dimethoxy-4-methylamphemaines are partial agonists at serotonin receptors linked to phosphoinositide hydrolysis. J. Pharmacol. Exp. Ther. **246:** 924–928.
24. PIERCE, P. A. & S. J. PEROUTKA. 1988. Antagonism of 5-hydroxytryptamine 2 receptor-mediated phosphatidylinositol turnover by d-lysergic acid diethylamide. J. Pharmacol. Exp. Ther. **247:** 918–925.
25. FRAZER, A., S. J. OFFORD & I. LUCKI. 1988. Regulation of serotonin receptors. *In* The Serotonin Receptors. E. Sanders-Bush, Ed., Clifton, NJ: Humana Press, pp. 319–365.
26. QUIK, M. & E. AZMITIA. 1983. Selective destruction of the serotonergic fibers of the fornix-fimbria and cingulum bundle increases 5HT-1 but not 5HT-2 receptors in rat midbrain. Eur. J. Pharmacol. **90:** 377–384.
27. CONN, P. J. & E. SANDERS-BUSH. 1986. Regulation of serotonin-stimulated phosphoinositide hydrolysis: Relation to the serotonin 5HT-2 binding site. J. Neurosci. **6:** 3669–3675.
28. BLACKSHEAR, M. A., L. L. MARTIN & E. SANDERS-BUSH. 1986. Adaptive changes in the 5HT-2 binding site after chronic administration of agonists and antagonists. Neuropharmacology **25:** 1267–1271.
29. LEYSEN, J. E., P. VAN GOMPEL, W. GOMMEREN, R. WOESTENBORGHS & P. A. J. JANSSEN. 1986. Down regulation of serotonin-S$_2$ receptor sites in rat brain by chronic treatment with the serotonin-S$_2$ antagonists: Ritanserin and setoperone. Psychopharmacology **88:** 434–444.
30. SANDERS-BUSH, E., M. BREEDING & M. ROZNOSKI. 1987. 5HT-2 binding sites after mianserin: Comparison of loss of sites and brain levels of drug. Eur. J. Pharmacol. **133:** 199–204.
31. SANDERS-BUSH, E., M. BREEDING, K. KNOTH & M. TSUTSUMI. 1989. Sertraline-induced

desensitization of the serotonin 5HT-2 receptor transmembrane signaling system. Psychopharmacology. **99:** 64–69.
32. Dillon-Carter, O. & D. Chuang, 1989. Homologous desensitization of muscarinic cholinergic, histaminergic, adrenergic, and serotonergic receptors coupled to phospholipase C in cerebellar granule cells. J. Neurochem. **52:** 598–603.
33. Conn, P. J., A. Janowsky & E. Sanders-Bush. 1986. Denervation supersensitivity of 5HT-1C receptors in rat choroid plexus. Brain Res. **400:** 396–398.
34. Palacios, J. M. & M. M. Dietl. 1988. Autoradiographic studies of serotonin receptors. *In* The Serotonin Receptors. E. Sanders-Bush, Ed. Clifton, NJ: Humana Press, pp. 89–140.
35. Sanders-Bush, E. & M. Breeding. 1990. Ritanserin and other putative selective 5HT-2 antagonists block serotonin 5HT-1c receptors in the choroid plexus. J. Pharmacol. Exp. Ther. **247:** 169–173.
36. Sanders-Bush, E. & M. Breeding. 1990. Serotonin $5HT_{1c}$ receptor reserve in choroid plexus masks receptor desensitization. J. Pharmacol. Exp. Ther. **252:** 984–988.
37. Balin, B. J. & R. D. Broadwell. 1988. Transcytosis of protein through the mammalian cerebral epithelium and endothelium. I. Choroid plexus and the blood-cerebrospinal fluid barrier. J. Neurocytol. **17:** 809–826.
38. Dickson, P. W., A. R. Aldred, P. D. Marley, T. Guo-Fen, G. J. Howlett & G. Schreiber. 1985. High prealbumin and transferrin mRNA levels in the choroid plexus of rat brain. Biochem. Biophys. Res. Commun. **127:** 890–895.
39. Lindvall-Axelsson, M., C. Mathew, C. Nilsson & Ch. Owman. 1988. Effect of 5-hydroxytryptamine on the rate of cerebrospinal fluid production in rabbit. Exp. Neurol. **99:** 362–368.
40. Maeda, K. 1983. Monoaminergic effect on cerebrospinal fluid production. Nihon Univ. J. Med. **25:** 155–174.
41. Tsutsumi, M., M. K. Skinner & E. Sanders-Bush. 1989. Transferrin gene expression and synthesis by cultured choroid plexus epithelial cells: Regulation by serotonin and cyclic 3′,5′-adenosine monophosphate. J. Biol. Chem. **264:** 9626–9631.
42. Skinner, M. K. & M. D. Griswold. 1982. Secretion of testicular transferrin by cultured sertoli cells is regulated by hormones and retinoids, Biol. Reprod. **27:** 211–221.
43. Newman, T. C., P. A. Dawson, L. L. Rudel & D. L. Williams. 1985. Quantitation of apolipoprotein E mRNA in the liver and peripheral tissues of nonhuman primates. J. Biol. Chem. **260:** 2452–2457.
44. Julius, D., T. J. Livelli, T. M. Jessell & R. Axel. 1989. Ectopic expression of the serotonin 1c receptor and the triggering of malignant transformation. Science **244:** 1057–1062.
45. Lauder, J. M. & H. Krebs. 1978. Serotonin as a differentiation signal in early neurogenesis. Dev. Neurosci. **1:** 15–30.
46. Nathanson, J. A. 1980. β-Adrenergic-sensitive adenylate cyclase in choroid plexus: Properties and cellular localization. Mol. Pharmacol. **18:** 199–209.
47. Pauwels, P. J., P. van Gompel, D. de Chaffoy de Courcelles & J. E. Leysen. 1988. Agonist-induced desensitization of 5HT2 receptors on cultured calf aortic smooth muscle cells. Neurochem. Int. **13:** 409–414.
48. Kendall, D. A. & S. R. Nahorski. 1987. Acute and chronic lithium treatments influence agonist and depolarization-stimulated inositol phospholipid hydrolysis in rat cerebral cortex. J. Pharmacol. Exp. Ther. **241:** 1023–1027.
49. Berridge, M. J. 1987. Inositol trisphosphate and diacylglycerol: Two interacting second messengers. Ann. Rev. Biochem. **56:** 159–193.

DISCUSSION OF THE PAPER

S. Maayani (*Mount Sinai School of Medicine, New York*): Is it possible, Elaine, that two receptors exist on the same cell? Let's say 5-HT_2 and one of the 5-HT_1 family that is linked to adenylate cyclase. Then when you put in the agonist you get, number one, coactivation of two receptors and second, interaction between two effectors. And maybe one of them, the 5-HT_2 receptor, densensitizes very quickly and the complexity that you see is part of this interaction between pathways.

SANDERS-BUSH: It's very possible. Of course when we do a study in the cerebral cortex slice we have no idea whether or not the receptors are on the same cells, etc. To answer those kinds of questions we need simpler model systems such as cultured cells.

G. AGHAJANIAN (*Yale University, New Haven, CT*): You're showing some data where there is dissociation between PI activation and receptor density changes. And work of others seems to indicate that PI activation doesn't necessarily go along with changes in 5-HT_2 receptor binding and it has been suggested that there might be two kinds of coupling. One set of 5-HT_2 receptors coupled to PI hydrolysis, and another set of 5-HT_2 receptors coupling to other types of transducer mechanisms. I was really struck in your talk by the TFMPP/MCPP data because we had some unpublished data with Kurt Rasmussen looking at 5-HT_2 or 5-HT_{1c} responses in motor neurons, and TFMPP and MCPP are wonderful agonists—very potent and induce a very strong depolarization response. In fact one could view this in terms of a negative feedback hypothesis in the following way—that if an agonist fails to activate PI that makes it all the better an agonist in terms of its physiological response. So your thought about trying to analyze more carefully what's going on with receptor binding, the status of the receptors, possibly two pools of receptors, versus what's going on with PI hydrolysis makes sense. I think that's a very good piece of advice in not assuming that all of the 5-HT_2 receptors are necessarily coupled through the PI pathway.

SANDERS-BUSH: That's right, and of course the other problem is that when we talk about cerebral cortex slices we're talking about a very large area and it's possible that receptor occupancy versus response varies from area to area as well as whether or not a receptor is coupled to PI hydrolysis from area to area. So those are all possibilities to consider in the future.

J. LEYSEN (*Janssen Research Foundation, Beerse, Belgium):* I would like to partly respond to what Dr. Aghajanian was suggesting and then I have one other question. I think that the discrepancy between the effect on the B-max and PI hydrolysis response may also be explained by the fact that you can have internalization of the receptor, or an uncoupling of receptor and transducing system, and not yet a receptor degradation. So, it does not necessarily have to be that you see a decrease in the maximum number of sites when you see a desensitization of the response. And I think it is perfectly possible that all of these 5-HT_2 receptors are coupled to PI responses, but just the fact that the receptors are not degraded yet, you do not see a loss of sites.

But then the question I have for you, I very much liked your work on transferrin, and I'd like to congratulate you for that. But my question is have you tried antagonists, and do you see antagonism of the effect of 5-HT on the induction of transferrin secretion?

SANDERS-BUSH: We have looked at the effect of antagonists, that's an obvious thing to do, and those studies have so far been inconclusive. I'm not sure why. One of the problems is that the effect of serotonin is a delayed effect. The cells have to be exposed to serotonin for greater than two days: four days is our common treatment paradigm, i.e. we expose the cells for four days to serotonin. And so we tried to expose the cells simultaneously for four days with an antagonist, a 5-HT_{1C} antagonist, for example, and we have not seen clear-cut antagonism of the effect of serotonin.

Animal Models of Serotonin Behavior

A. RICHARD GREEN[a]

Astra Neuroscience Research Unit
1 Wakefield Street
London WC1N 1PJ England

LISA I. BACKUS

MRC Unit and University Department of Clinical Pharmacology
Radcliffe Infirmary
Oxford OX2 6HE England

INTRODUCTION

The term 'animal model' in behavioral pharmacology can imply two somewhat different approaches.

The first is the study of physiological and psychological responses of experimental animals together with the assessment, by the use of selective drugs, of the involvement of a specific neurotransmitter or the sub-type of a neurotransmitter receptor. In this regard models can include, for example, those thought to allow prediction of efficacy of drugs in psychiatric states such as anxiety, depression and schizophrenia or pathological states such as nausea and vomiting or hypertension. While it is perhaps not unreasonable to assume that a modeled pathological state such as emesis or hypertension is closely analogous to the clinical situation, such an assumption would be most unwise in the cases of modeled psychiatric state since there is no evidence that such states can exist in experimental animals. What, in fact, can be seen in such models is that the behavior observed is altered by drugs already known to be of value clinically. The models have therefore been used to examine novel compounds to screen for their possible therapeutic potential. While predictive value in such models can be high, both false positives and false negatives can occur. Examples of this type of approach and evidence for its value are found elsewhere in this volume.

The other approach, and it is the one that will be examined in this chapter, is the use of a simple behavioral change as a marker of serotonin function. Monitoring the change and the way that psychoactive compounds alter the response allows predictions to be made as to what alterations are occurring in the function of the transmitter following drug administration. In this case the behavior is essentially a type of bioassay. What should be emphasized therefore is that the behavior is not a 'model' of a physiological or pathological condition. If, for example, the response being measured is altered by an antidepressant drug then this may give the scientist an insight as to how that antidepressant can change serotonin function. Interpretation may then be made as to whether such a change is involved in the therapeutic mechanism of action of the drug; but not that depression is in any way being modeled. The value of this approach is that it allows a rapid evaluation as to whether a drug is altering neurotransmitter function and it becomes powerful when combined with other techniques such as ligand binding.

[a]Author to whom correspondence should be addressed.

It is clear that animal models of serotonin behavior have not kept pace with the recent explosion of serotonin receptor sub-type identification, at least as established by ligand receptor binding. Functional tests remain vital since ligand binding cannot show whether the identified site is a receptor or merely an acceptor. However, to develop a functional model it is necessary to have both a clear behavioral change and the appropriate 'tools' to characterize it. These 'tools' are specific and selective agonists and antagonists. In the case of several of the serotonin putative receptor sub-types (and this is particularly true of some 5-HT_1 sub-types), it is obvious such drugs are lacking. Even with the 5-HT_{1A} site the most used, and therefore presumably the best, antagonists remain the β-adrenoceptor antagonists. While these may be relatively 5-HT_{1A} selective (as opposed to 5-HT_2 or 5-HT_3) they are undoubtedly not specific.

ANIMALS MODELS FOR THE 5-HT_{1A} RECEPTOR

Locomotor-Related Responses

One of the traditional 5-HT behavioral models has been the syndrome induced in rats by the administration of a monoamine oxidase inhibitor and L-tryptophan.[1-3] The syndrome can also be induced by a variety of serotonin agonists such as 5-methoxy-*N*,*N*-dimethyltryptamine (5-MeODMT),[4] quipazine[5] or the serotonin-releasing drug fenfluramine.[6] Basically the syndrome consists of reciprocal forepaw treading (which has also been called 'piano playing'), head weaving, hind-limb abduction which leads to a flat body posture, Straub tail, and piloerection.The syndrome has been reviewed in detail elsewhere.[1-3]

At the time many of these investigations were performed, little, if anything, was known of serotonin receptor sub-types. However in 1981 Arvidsson and colleagues[7] reported that the 5-HT_{1A} agonist 8-hydroxy-2-(di-*n*-propylamino) tetralin (8-OH-DPAT) produced the serotonin behavioral syndrome.

The behavior has been shown to be initiated by 8-OH-DPAT acting at post-synaptic receptors since a lesion of presynaptic serotonergic terminals by intracerebral injection of 5,7-dihydroxytryptamine (5,7-DHT) did not diminish the response but rather enhanced it,[8] suggesting a supersensitive receptor following denervation. The receptors are presumably of the 5-HT_{1A} sub-type given the selectivity of the drug for this receptor.[9] Other behavioral,[10,11] and ligand binding studies[12] support the proposal that the receptors are located post-synaptically.

The syndrome is antagonized stereoselectively by (-)-propranolol[8] and (-)-pindolol,[11] putative 5-HT_{1A} antagonists and studies by Tricklebank and colleagues[11] indicated the involvement of catecholamines in the response.

One interesting recent observation[13] has been that ritanserin, the 5-HT_2 antagonist, does not inhibit the behavioral syndrome induced by 8-OH-DPAT in rats but rather can enhance it in a dose dependent way, suggesting perhaps a functional 5-HT_{1A}/5-HT_2 receptor interaction (FIG. 1).

It is important to note that the syndrome is almost certainly hindbrain and spinally mediated.[14] This can prove a problem since ligand binding studies are seldom performed in these areas, and a failure to correlate binding and behavioral data may be due to region specific changes.

There have been a considerable number of studies on the behavioral syndrome when induced by administration of a monoamine oxidase inhibitor and L-tryptophan (see ref. 3 for review), but rather fewer when the syndrome has been induced by 8-OH-DPAT. However, recent examples of the use of the 8-OH-DPAT-induced syndrome include investigations on the responses of rats following repeated administra-

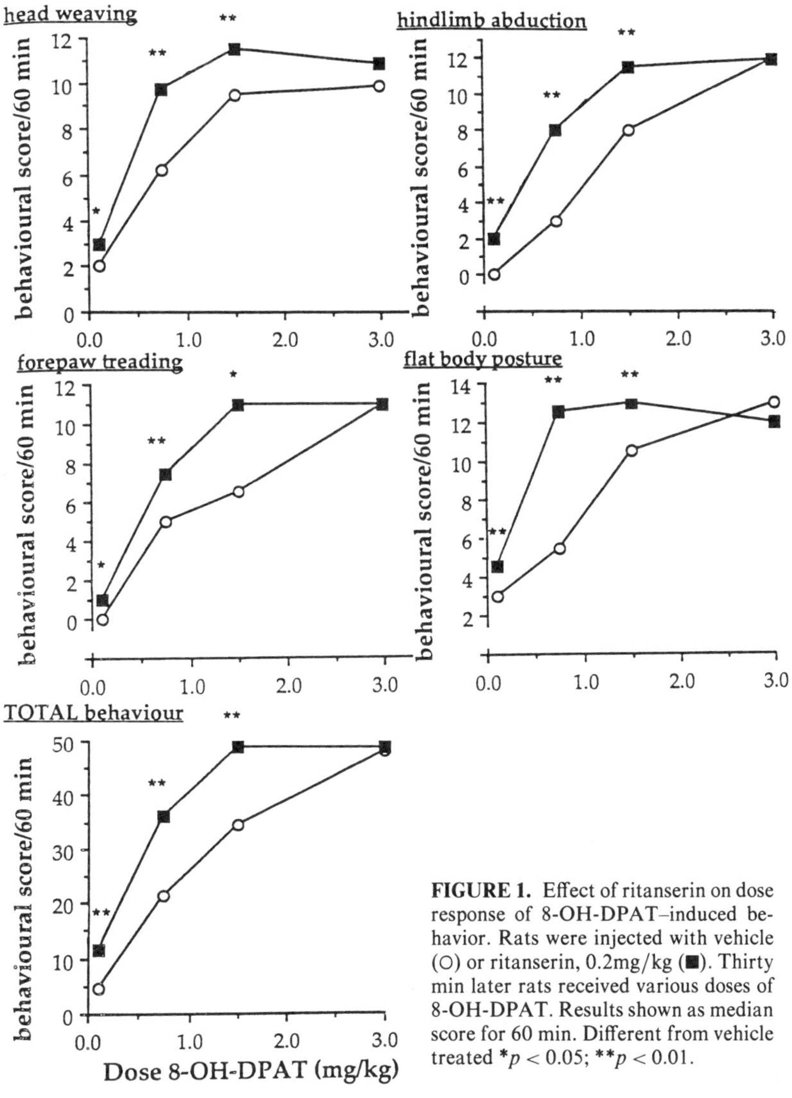

FIGURE 1. Effect of ritanserin on dose response of 8-OH-DPAT–induced behavior. Rats were injected with vehicle (○) or ritanserin, 0.2mg/kg (■). Thirty min later rats received various doses of 8-OH-DPAT. Results shown as median score for 60 min. Different from vehicle treated $^{*}p < 0.05$; $^{**}p < 0.01$.

tion of antidepressant drugs[15] and lithium.[16] Administration of the former leads to a diminution of the behavioral response[15] while treatment with the latter results in a marked enhancement of the response[16] thereby demonstrating the possible involvement of 5-HT_{1A} receptors in the mechanisms of action of antidepressant and antimanic drugs.

Another motor test recently described to be influenced by 8-OH-DPAT is that of supression of "exploratory" behavior. At low dose (0.1 $mgkg^{-1}$) 8-OH-DPAT decreases the normal locomotor behavior seen when rats are put into a novel environment, and this decrease was reversed by administration of (-)-pindolol.[17] When 8-OH-DPAT was injected into the dorsal raphe nucleus it also produced a dose-dependent decrease in locomotor behavior while injection into the median raphe produced an increase.[18]

It is certainly possible that the hypoactivity response to low dose of 8-OH-DPAT is related to another response recently reported by Renyi and colleagues.[19,20] They observed that rats normally leave their home cages within a few minutes of the cage top being removed. This response was inhibited by 8-OH-DPAT and the inhibitory effect of the drug antagonized by (-)-alprenolol but not pirenpirone.

Another motor response that is now being characterized is the rotational behavior induced in rats with unilateral lesions of the median forebrain bundle (MFB) or dorsal raphe nucleus (DRN).

In unilateral MFB lesioned animals 8-OH-DPAT has been found to induce contralateral turning (FIG. 2) and while both gepirone and ipsapirone were inactive when given alone they antagonized 8-OH-DPAT-induced rotation, consistent with the partial agonist characteristics of these compounds. Similar results have been reported by Blackburn *et al.*[21,22] in unilateral DRN-lesioned animals. This group of workers also

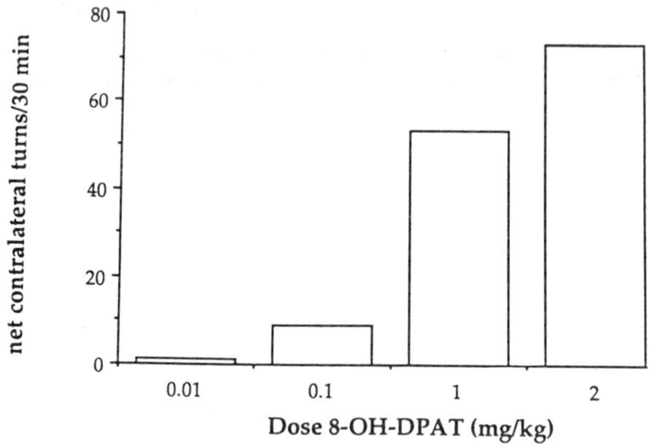

FIGURE 2. Dose response of 8-OH-DPAT—induced contralateral rotation in unilaterally MFB-lesioned rats. Rats were given various doses of 8-OH-DPAT and the rotations induced in the next 30 min recorded. Saline did not induce rotation.

noted that the 5-HT_2 agonists quipazine and MK 212 failed to induce rotation. Surprisingly (-)-pindolol had no effect on 8-OH-DPAT–induced rotation in unilateral DRN-lesioned rats. Furthermore, mixed agonists such as 5-MeODMT and RU 24969 both induced rotation in DRN-lesioned rats[20,21] but not in MFB-lesioned rats.

Finally it is worth noting that the severity of the turning response in the unilateral MFB-lesioned rats correlated closely with the percentage depletion of serotonin produced by the lesion (FIG. 3).

Hypothermic Response

The first reports that 8-OH-DPAT produced hypothermia were in rats.[23,24] However, the response was found to be complicated by the behavioral changes induced at higher doses, which may well be responsible for bell-shaped temperature curve (FIG. 4). In mice, by contrast, overt behavioral changes have not generally been reported.

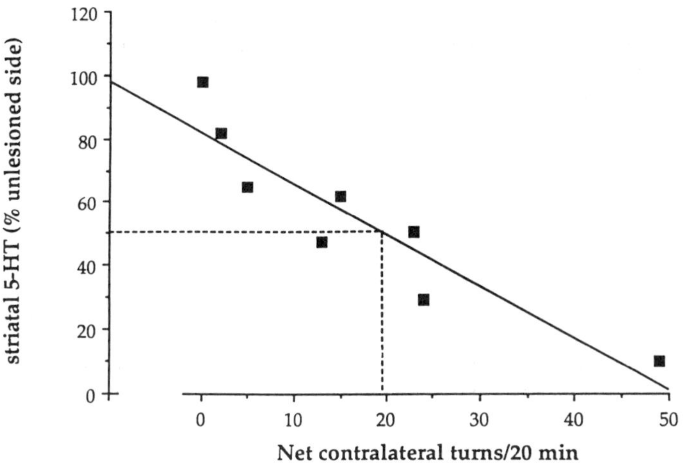

FIGURE 3. Relationship between number of contralateral turns and striatal serotonin depletion in unilaterally MFB-lesioned rats. Each point represents net contralateral turns in 20 min plotted against the % loss of serotonin in the lesioned versus the unlesioned side in the same animal. Correlation = 0.92.

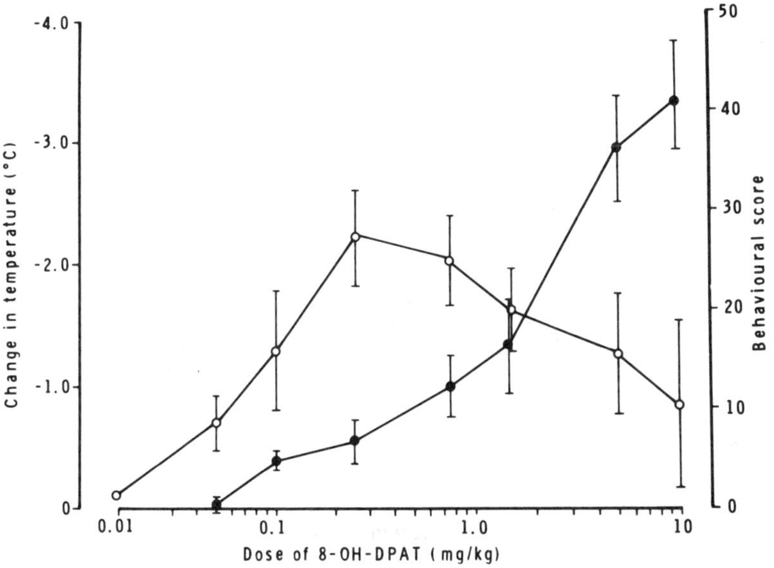

FIGURE 4. Dose/effect of 8-OH-DPAT in rats. Increasing doses of 8-OH-DPAT (s.c.) produced a hypothermic response (○) shown as mean I.S.D. of temperature decrease 30 min after injection. In the same animals the simultaneous motor behavioral syndrome (●) scored over 30 min is shown as mean ± SD). (Reproduced from Goodwin *et al.*[8] with permission of Springer-Verlag.)

Nevertheless a recent paper suggests that this absence of behavioral change in mice may be strain dependent, since it reported that serotonergic mediated behavioral changes such as head weaving and forepaw treading can occur in DBA mice.[25] This nevertheless appears to be a unique report of the behavior occurring in mice. While C57 black mice do not display behavioral changes following 8-OH-DPAT, they do show a clear dose-dependent hypothermia which is not bell-shaped (FIG. 5).[26]

The hypothermia probably results from activation of presynaptic 5-HT_{1A} receptors since lesioning of serotonergic terminals by injection of 5,7-DHT intracerebroventricularly essentially abolishes the temperature response (FIG. 5),[26] a finding recently confirmed.[27]

Although the hypothermic response of rats is antagonized by (-)-propranolol,[23] this is not true of mice.[26] However, the response in mice is antagonized by two drugs known to have 5-HT presynaptic receptor antagonist activity, namely methiothepin and quipazine.[26] The butyrophenone haloperidol is also an antagonist, perhaps not surprisingly given that the 5-HT_{1A} receptor is butyrophenone sensitive. The dopamine antagonist flupenthixol, in contrast, was without effect as were drugs acting on a variety of other receptor types (TABLE 1).

One interesting recent finding has been the effect of ritanserin on hypothermia. Low doses were found not to antagonize the response. However, what was observed was a small but significant enhancement of the 8-OH-DPAT response. Similar enhancement was observed after administration of two other 5-HT_2 antagonists, namely ICI 170,809 and ketanserin. Such findings might be explained using the framework proposed by Gudelsky *et al.*[28] that 5-HT_{1A} and 5-HT_2 receptors work in opposition to regulate body temperature.

Chronic 8-OH-DPAT administration attenuates the hypothermic response, suggesting receptor subsensitivity.[29] This interpretation was given weight by the observation that these animals showed a 'cross tolerance' to the hypothermic effect of gepirone.

The hypothermia model has been used to examine the effects of antidepressant

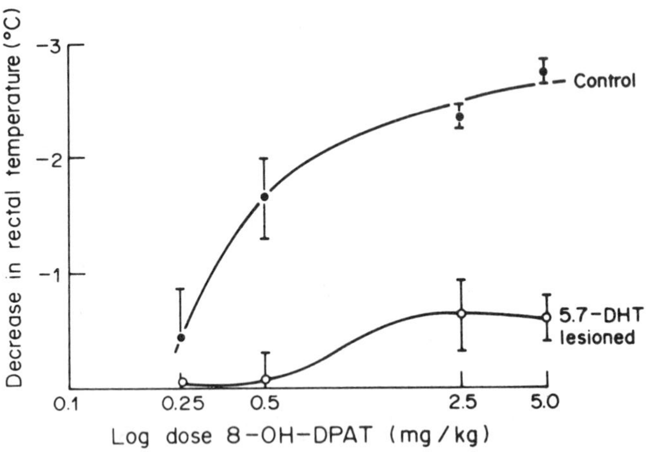

FIGURE 5. Decrease in rectal temperature produced 20 min after subcutaneous injection of different doses of 8-OH-DPAT, plotted as log-dose/response for groups of animals pretreated 6–8 days previously with intraventricular injection of either 50 μg 5,7-DHT creatinine sulphate (○) or vehicle (●). Results are shown as mean of at least 5 animals for each dose; vertical bars show SD (Reproduced from Goodwin *et al.*[26] with permission of Pergamon Press.)

TABLE 1. Effect of Various Drugs on the Hypothermic Response of Mice to 8-OH-DPAT

Drug Administration	Receptor Type Antagonists	Effect on 8-OH-DPAT–induced Hypothermia
Prazosin	α_1-adrenoceptor	None
Idazoxan	α_2-adrenoceptor	None
Pindolol	β_1-β_2-adrenoceptor	None
Metoprolol	β_1-adrenoceptor	None
ICI 118551	β_2-adrenoceptor	None
Ritanserin	5-HT_2	None
Cinanserin	5-HT	None
Methysergide	5-HT	None
Metergoline	5-HT	Enhances
Methiothepin	5-HT	Enhances
Quipazine	5-HT (presynaptic)	Attenuates
Flupenthixol	Dopamine	None
Haloperidol	Dopamine	Attenuates
RO 15-1788	Benzodiazepine	None
Propranolol	5-HT_{1A} and β-adrenoceptor	None

drugs and repeated electroconvulsive shock (ECS) on a presynaptic 5-HT_{1A} receptor–mediated action in both mice[30] and rats.[18] Repeated administration over 10–14 days of a wide range of antidepressant drugs and ECS resulted in an almost total attenuation of the hypothermic response to 8-OH-DPAT in both species. In the case of ECS administration to mice, this effect lasted for nearly a month after the last treatment (FIG. 6). These findings have now been confirmed by others[31] and again suggest the possible involvement of 5-HT_{1A} receptors in the therapeutic mechanism of action of antidepressant treatments.

5-HT_{1B} RECEPTOR MODELS

Administration of the suggested 5-HT_{1B} receptor agonist RU 24969 either systematically[32] or centrally[29] produces a pronounced locomotor response in mice and rats which is dose dependent. This response can be particularly antagonized by (-)-propranolol[32] and (-)-pindolol[33] but there are questions as to whether this effect is stereoselective in rats[33] if not in mice[23] and thus the results raise doubts about the role of 5-HT in the response. Reserpine pretreatment before RU 24969 removed the locomotor component and resulted in the appearance of the 5-HT behavioral syndrome[33] thereby illustrating, in line with ligand binding data, that RU 24969 probably acts at both 5-HT_{1A} and 5-HT_{1B} receptors.

5-HT_2 RECEPTOR–MEDIATED BEHAVIOR

The observation that several serotonin agonists produced a characteristic head twitch behavior in mice was first reported in 1963 by Corne *et al.*[34] The head twitch behavior can be elicited by administration of the serotonin precursor, 5-hydroxytryptophan (5-HTP) either with or without administration of a peripheral decarboxylase inhibitor such as cardidopa. If cardidopa is given a lower dose of 5-HTP can be injected to produce the same degree of response. The dose response of the head twitch is well documented (FIG. 7) as is the fact that a variety of serotonin agonists will also produce the response, including lysergic acid diethylamide (LSD), quipazine, and 5-MeODMT (see ref. 3).

The head-twitch response can be inhibited by low doses of ritanserin but not (-)-propranolol,[23] and there are data which support the contention that other neurotransmitters including GABA and noradrenaline are involved in the expression of the response (see ref. 3).

Lesioning of serotonin terminals by injection of 5,7-DHT intracerebroventricularly results in an enhanced response, and this enhancement correlates with the degree of serotonin depletion.[35]

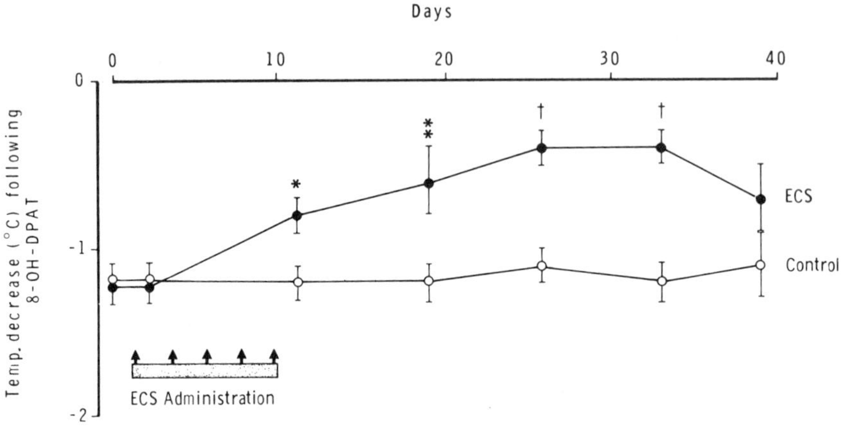

FIGURE 6. Effect of repeated ECS on the hypothermic response of mice to 8-OH-DPAT. The graph shows mean ± SEM temperature decrease in control (halothane anaesthesia, ○) and ECS (90 V, 50 Hz, 1 s through earclip electrodes to halothane-anesthetized mice, ●) 20 min after 8-OH-DPAT (0.5 mg/kg s.c.). The ECS-treated group shows a significantly smaller temperature decrease than the control group as shown: $*p < 0.05$; $**p < 0.005$; $\dagger p < 0.001$. There is no difference in the basal temperature between the two groups at any time-point shown. (Reproduced from Goodwin *et al.*[30] with permission of *Nature* (Macmillan Press).)

In the rat, administration of cardidopa and 5-HTP results in the appearance of a rather similar behavior in which the rats not only produce a head twitch but a shake down the body trunk as well.[36] This has been named descriptively the 'wet dog shake' and the pharmacology of the response suggests that it is probably essentially the same behavior as the head twitch in mice.[3] However there is one important difference. Head twitch behavior in mice is extremely rapid in onset with a peak response occurring within the 30 min following injection of 5-HTP. The wet dog shake response in rats in contrast is slower in onset and lasts over several hours.[36]

Lesions of central serotonergic neurones with 5,7-DHT resulted in the rats demonstrating enhanced wet dog shake behavior to serotonin agonists.[37]

The anatomical locus of head twitch behavior in mice appears to be the hind brain, since the response correlated positively with the brain stem but not whole brain serotonin content.[34] In the rat brain transection at the posterior but not anterior

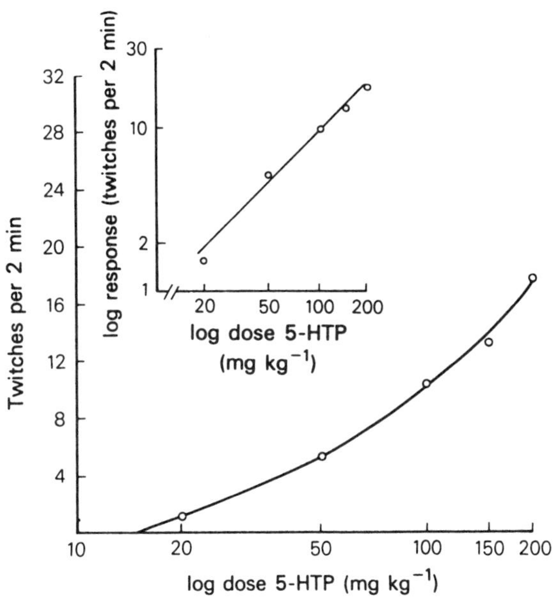

FIGURE 7. Effect of dose of 5-hydroxytryptophan on the behavioral response to carbidopa and 5-hydroxytryptophan. Mice were given carbidopa (25mg/kg) followed by various doses of 5-hydroxytryptophan. Main graph shows the log dose of 5-HTP versus total number of head twitches in 2 min, 30 min after the 5-HTP. Small graph shows the log dose-log response curve.

commisure abolished wet dog shakes[34] so in both species the lower diencephalon and brain stem appears to initiate the response.

Although 5-HT_2 receptor binding characteristics were performed on the frontal cortex and the behavior is apparently initiated in the hind brain a study of the consequences of repeated antidepressant drug and ECS administration on these two parameters in mice demonstrated good concordance (TABLE 2). Thus repeated administration of antidepressant decreased the density of 5-HT_2 receptor sites and also

TABLE 2. Effect of Antidepressant Drugs and Electroconvulsive Shock (ECS) on Serotonin$_2$ (5-HT_2)–Mediated Head Twitch Behavior and Receptor Number

	Behavior	5-HT_2 Receptor Binding	
	Head Twitches/2 min (n = 8)	B_{max} (n = 4)	K_D (n = 4)
Control	7.2	200 (20)	1.42 (0.20)
Zimelidine	2.4	151 (27)	1.30 (0.24)
Desipramine	4.3	145 (10)	1.50 (0.17)
Control	5.9	207 (28)	1.57 (0.23)
ECS	14.2	248 (25)	1.48 (0.42)

NOTE: K_D expressed in nM and B_{MAX} in fmol/mg protein, mean (± SD).

decreased the head twitch response.[38] Repeated ECS, in contrast, increased both the behavioral response and the 5-HT_2 receptor density (TABLE 2).[38]

Such apparent correlations are not always seen, however. Repeated administration for 4 days of 8-OH-DPAT to mice produced a clear enhancement of head twitch behavior but a decrease in 5-HT_2 receptor density in the frontal cortex.[29] Presumably this reflects region-specific differences in the effect of 8-OH-DPAT on 5-HT_2 receptor mechanisms.

SUMMARY

This chapter has outlined several major models being used to examine serotonin function. While these models are often used routinely (as has been shown in the outline on the effects of psychoactive drugs) our knowledge of the models themselves is still increasing. The studies reported here on the possible interactions between 5-HT_{1A} and 5-HT_2 receptors for example are novel. These models therefore are useful in increasing knowledge about neurotransmitter interactions, particularly when combined with selective lesioning techniques. However the main value of the models is likely to remain that of allowing rapid initial assessment of the serotonin antagonist or agonist properties of drugs or the consequences on serotonin function of administration of psychoactive drugs. Information obtained in this way can then be used in the design of more sophisticated experimentation.

REFERENCES

1. GRAHAME-SMITH, D. G. 1971. Studies in vivo on the relationship between brain tryptophan, brain 5-HT synthesis and hyperactivity in rats treated with a monoamine oxidase inhibitor and L-tryptophan. J. Neurochem. **18:** 1053–1066.
2. JACOBS, B. L. 1976. An animal behavior model for studying central serotonergic synapses. Life Sci. **19:** 777–786.
3. GREEN, A. R. & D. J. HEAL. 1985. The effects of drugs on serotonin-mediated behavioural models. *In* Neuropharmacology of Serotonin. A. R. Green, Ed. Oxford: Oxford University Press, pp. 326–365.
4. GRAHAME-SMITH, D. G. 1971. Inhibitory effects of chloropromazine on the syndrome of hyperactivity produced by L-tryptophan or 5-methoxy-N,N-dimethyltryptamine in rats treated with a monoamine oxidase inhibitor. Br. J. Pharmacol. **43:** 856–864.
5. GREEN, A. R., M. B. H. YOUDIM & D. G. GRAHAME-SMITH. 1976. Quipazine: Its effects on rat brain 5-hydroxytryptamine metabolism, monoamine oxidase activity and behaviour. Neuropharmacology **15:** 173–179.
6. GREEN, A. R. & P. H. KELLY. 1976. Evidence concerning the involvement of 5-hydroxytryptamine in locomotor activity produced by amphetamine or tranylcypromine plus L-dopa. Br. J. Pharmacol. **57:** 141–147.
7. ARVIDSSON, L. E., U. HACKSELL, J. L. G. NILSSON, S. HJORTH, A. CARLSSON, P. LINDBERG, D. SANCHEZ & H. WIKSTROM. 1981. 8-Hydroxy-2-(di-n-propylamino)tetralin, a new centrally acting 5-hydroxytryptamine receptor agonist. J. Med. Chem. **24:** 921–923.
8. GOODWIN, G. M., R. J. DE SOUZA, A. R. GREEN & D. J. HEAL. 1987. The pharmacology of the behavioural and hypothermic responses of rats to 8-hydroxy-2-(di-n-propylamino) tetralin (8-OH-DPAT). Psychopharmacology **91:** 506–511.
9. MIDDLEMISS, D. N. & J. R. FOZARD. 1983. 8-Hydroxy-2-(di-n-propylamino)tetralin discriminates between subtypes of 5-HT_1 recognition site. Eur. J. Pharmacol. **90:** 151–153.
10. HJORTH, S., A. CARLSSON, P. LINDBERG, D. SANCHEZ, H. WIKSTROM, L.-E ARVIDSSON, U. HACKSELL & J. L. G. NILSSON. 1982. 8-Hydroxy-2-(di-n-propylamino)tetralin, 8-OH-

DPAT, a potent and selective simplified ergot congener with central 5-HT-receptor stimulating activity. J. Neural Transm. **55:** 169–188.
11. TRICKLEBANK, M. D., C. FORLER & J. R. FOZARD. 1984. The involvement of subtypes of the 5-HT_1 receptor and of catecholaminergic systems in the behavioural response to 8-hydroxy-2-(di-n-propylamino)tetralin in the rat. Eur. J. Pharmacol. **106:** 271–282.
12. HALL, M. D., S. EL MESTIKAWY, M. B. EMERIT, L. PICHAT, M. HAMON & H. GOZLAN. 1985. [^{3}H]8-hydroxy-2-(di-n-propylamino)tetralin binding to pre- and postsynaptic 5-hydroxytryptamine sites in various regions of the rat brain. J. Neurochem. **44:** 1685–1696.
13. BACKUS, L. I., T. SHARP & D. G. GRAHAME-SMITH. 1989. Behavioural evidence for an interaction between 5-HT_2 and 5-HT_{1A} receptor function. Br. J. Pharmacol. **97:** 411P.
14. JACOBS, B. L. & H. KLEMFUSS. 1975. Brainstem and spinal cord mediation of serotonergic behavioural syndrome. Brain Res. **100:** 450–457.
15. GOODWIN, G. M., R. J. DE SOUZA & A. R. GREEN. 1987. Attenuation by electroconvulsive shock and antidepressant drugs of the 5-HT_{1A} receptor mediated hypothermia and serotonin syndrome produced by 8-OH-DPAT. Psychopharmacology **91:** 500–505.
16. GOODWIN, G. M., R. J. DE SOUZA, A. R. GREEN & A. J. WOOD. 1986. The enhancement by lithium of the 5-HT_{1A} mediated serotonin syndrome produced by 8-OH-DPAT in the rat: Evidence for a post synaptic mechanism. Psychopharmacology **90:** 488–493.
17. JOHANSSON, C. & S. AHLENIUS. 1989. Evidence for the involvement of 5-HT_{1A} receptors in the mediation of exploratory locomotor activity in the rat. J. Psychopharmacol. **3:** 32–35.
18. HILLEGAART, V., S. AHLENIUS & K. LARSSON. 1989. Effects of local application of 5-HT into the median and dorsal raphe nuclei on male rat sexual and motor behavior. Behav. Brain Res. **33:** 279–286.
19. RENYI, L., T. ARCHER, B. G. MINOR, B. TANDBERG, A. FREDRIKSSON & S. B. ROSS. 1986. The inhibition of the cage leaving response—a model for studies of serotoninergic neurotransmission in the rat. J. Neural Transm. **65:** 193–210.
20. RENYI, L. 1987. The involvement of 5-HT_1 and 5-HT_2 receptor in different components of the 5-HT syndrome in the rat. Neuroscience **22:** 57S.
21. BLACKBURN, T. P., J. D. KEMP, D. A. MARTIN & B. COX. 1984. Evidence that 5-HT agonist induced rational behaviour in the rat is mediated via 5-HT_1 receptors. Psychopharmacology **83:** 163–165.
22. BLACKBURN, T. P., D. A. MARTIN & B. COX. 1987. 8-OH-DPAT-induced rational behaviour is mediated via 5-HT_1 like receptors. *In* Brain 5-HT_{1A} Receptors. C. T. Dourish, S. Ahlenius and P. H. Hutson, Eds. Chichester: Ellis Horwood.
23. GOODWIN, G. M. & A. R. GREEN. 1985. A behavioural and biochemical study in mice and rats of putative selective agonists and antagonists for 5-HT_2 receptors. Br. J. Pharmacol. **84:** 734–753.
24. HJORTH, S. 1985. Hypothermia in the rat induced by the potent serotonergic agent 8-OH-DPAT. J. Neural Transm. **61:** 131–135.
25. YAMADA, J., Y. SUGIMOTO & K. HORISAKA. 1988. The behavioural effects of 8-hydroxy-2-(di-n-propylamino)tetralin (8-OH-DPAT) in mice. Eur. J. Pharmacol. **154:** 299–304.
26. GOODWIN, G. M., R. J. DE SOUZA & A. R. GREEN. 1985. The pharmacology of the hypothermic response in mice to 8-hydroxy-2-(di-n-propylamino)tetralin (8-OH-DPAT), a model for presynaptic 5-HT_1 function. Neuropharmacology **24:** 1187–1194.
27. HEAL, D. J., M. R. PROW, K. F. MARTIN & W. R. BUCKETT. 1989. Effects of lesioning central 5-HT and noradrenaline containing neurones on 8-OH-DPAT hypothermia, mydriasis and hypoactivity. Br. J. Pharmacol. **97:** 406P.
28. GUDELSKY, G. A., J. I. KOENIG & H. Y. MELTZER. 1986. Thermoregulatory responses to serotonin (5-HT) receptor stimulation in the rat: Evidence for opposing roles of 5-HT_2 and 5-HT_{1A} receptors. Neuropharmacology **25:** 1307–1313.
29. DE SOUZA, R. J., G. M. GOODWIN, A. R. GREEN & D. J. HEAL. 1986. Effects of chronic treatment with 5-HT_1 agonist (8-OH-DPAT and RU 24969) and antagonist (isapirone) drugs on the behavioural responses of mice to 5-HT_1 and 5-HT_2 agonists. Br. J. Pharmacol. **89:** 377–384.
30. GOODWIN, G. M., R. J. DE SOUZA & A. R. GREEN. 1985. Presynaptic serotonin receptor-mediated response in mice attenuated by antidepressant drugs and electroconvulsive shock. Nature **317:** 531–533.

31. MARTIN, L. F., I. PHILLIPS, D. J. HEAL & W. R. BUCKETT. 1989. Is down regulation of 5-HT_{1A} receptor function a common feature of the action of antidepressant drugs? Br. J. Pharmacol. **97:** 407P.
32. GREEN, A. R., A. P. GUY, & C. R. GARDNER. 1984. The behavioural effects of RU 24969, a 5-HT_1 receptor agonist in rodents and the effects on the behaviour of various antidepressant treatments. Neuropharmacology **23:** 655–661.
33. TRICKLEBANK, M. D., D. N. MIDDLEMISS & J. NEILL. 1986. Pharmacological analysis of the behavioural and thermo regulatory effects of the putative 5-HT_1 receptor agonist RU 24969 in the rat. Neuropharmacology **25:** 877–886.
34. CORNE, S. J., R. W. PICKERING & B. T. WARNER. 1963. A method for assessing the effects of drugs on the central actions of 5-hydroxytryptamine. Br. J. Pharmacol. **20:** 106–120.
35. HEAL, D. J., J. PHILPOT, S. G. MOLYNEUX & A. METZ. 1985. Intracerebroventricular administration of 5,7-dihydroxytryptamine to mice increase both head-twitch response and the number of cortical 5-HT_2 receptors. Neuropharmacology **24:** 1201–1204.
36. BEDARD, P. & C. J. PYCOCK. 1977. "Wet dog" shake behaviour in the rat: A possible quantitive model of central 5-hydroxytryptamine activity. Neuropharmacology **16:** 663–670.
37. BARBEAU, H. & P. BEDARD. 1981. Denervation supersensitivity to 5-hydroxytryptophan in rats following spinal transection and 5,7-dihydroxytryptamine injection. Neuropharmacology **20:** 611–616.
38. GOODWIN, G. M., A. R. GREEN & P. JOHNSON. 1984. 5-HT_2 receptor characteristics in frontal cortex and 5-HT_2 receptor mediated head-twitch behaviour following antidepressant treatment to mice. Br. J. Pharmacol. **83:** 235–242.

DISCUSSION OF THE PAPER

DAVID T. WONG (*Lilly Research Laboratories, Indianapolis, IN*): You didn't comment on the behavioral model used for 5-HT_3 receptor. Could you comment on this?

GREEN: Yes, I deliberately left out 5-HT_3 because it seems to me, at the moment, that mostly the 5-HT_3 antagonists are used to alter other systems. If you give 5-HT_3 agonists themselves you don't get much in the way of behavioral change. What you're doing with the drugs then is altering other systems, be it locomotor behavior, produced by amphetamine and apomorphine. We haven't got, nicely, a clear behavioral response from giving 5-HT_3 agonists alone. Anyway, I thought Dr. Tyers would be covering this area yesterday and it didn't seem worth me going over it again.

MANFRED GÖTHERT, (*University of Bonn*): I have a question concerning the terminology you are using. You said that the 8-OH-DPAT–induced hypothermia is due to a presynaptic site of action. You mean this in the sense that receptors are located at the serotonergic nerve terminals or just in the sense that the site of action is the serotonin neuron itself?

GREEN: Yes, I'm sorry about that. I meant merely that the responses are removed once you lesion, and I assume that it is somatic receptors that we're removing, that's all I meant by that, rather than, for example, in the behavior where after the lesion they're post-synaptic, they are downstream and therefore the response remains.

QUESTION : I want to go into species differences. The myoclonus in guinea pig has been proposed as a model for 1B receptors, and now we know that the guinea pig has no or very little 1B binding. Do you know if somebody has re-evaluated this behavior?

GREEN: Yes, I left out the myoclonias because I think it's such a mess at the moment. There are clear species differences in all these systems. For example we find that the temperature response in rats is very nicely stereo-specifically blocked by propranolol. It doesn't block the mouse hypothermia and I don't know why. And we're not the only people who have found that. So the pharmacology between the mouse and the rat alone in these systems is certainly different. There are other examples and it's a real problem. For example, again, in the circling we can not get gepirone to make the animal circle even though it produces hypothermia. But you will block the circling produced by 8-OH-DPAT. Gepirone seems to be acting more as an antagonist rather than agonist. So I can't give you a clear story on this other than to say I think the best thing if you are trying to define a functional response is to try and use several drugs and clarify it as far as possible.

M. FINK (*Stony Brook, NY*): Richard, you have spoken of the species specificity and yet you're using these models in animals to predict anti-depressant drug effects in man. What is the prediction rate from the animal data to man and secondly, you also made a general statement that the effects of ECS and the effects of these drugs are similar, that was in the first part of your talk, in the second you differentiated. Is it possible, or is it true, that the ECS mechanism is similar to the antidepressant drugs that you're working with?

GREEN: All right. I knew there would be an ECT question with you in the audience! I think prediction is fairly difficult because I don't think anyone is clear as to what biochemical or pharmacological features are required to indicate an antidepressant drug. So I think almost any company now would do several screens and they'll say, if a drug comes up positive in this, this and this test, then it's worth pursuing further. And I know a couple of companies now are using the hypothermia as one of the tests they take in line.

Regarding the ECT question—in terms of the 5-HT_1 hypothermia response, ECS and antidepressants look exactly the same. In terms of the 5-HT_2 response, the head twitch, they are different. Now, you know, it could be that part of the antidepressant features of ECT go along with the 5-HT_1 responses because it looks the same and maybe the reason that ECT is more efficacious in some patients when conventional drugs don't work is because of the differences you get, for example, in 5-HT_2. I think that's pushing hypotheses quite hard. We always feel more confident when we see a range of different treatments and drugs doing the same thing. We say, gosh maybe that's how they work. But I think we shouldn't avoid the fact that when a drug does something different that may be why that particular treatment is very good in a patient in which other treatments have failed. So it's just the different receptors where ECT can either be the same (on 5-HT_1 responses) or different (5-HT_2 receptors).

C. S. AULAKH (*NIMH, Bethesda, MD*): You mentioned that the head twitch response is 5-HT_2 mediated. Since we know now that the 5-HT_2 antagonists also antagonize 5-HT_{1C} effects, have you done any studies showing that 5-HT_{1C} antagonists do not block it?

GREEN: No, I think it would be interesting but again I think the problem there is having the specific drugs to be able to do this. I mean looking at the drugs for the 5-HT_{1C} receptor it's difficult saying any of them will be selective for that receptor and wouldn't be affecting other systems as well.

F. YOCCA (*Bristol-Meyers, Wallingford, CT*): You mentioned the work of Backus where they showed that by pretreating with ritanserin they could potentiate the syndrome induced by 8-hydroxy-DPAT. A recent paper by Arat and Hytell shows that by pretreatment with DOI they could also potentiate the syndrome produced by 8-hydroxy-DPAT. Can you speculate as to how a serotonin type 2 agonist and antagonist produce the same effect on a type 1 syndrome?

GREEN: I think probably not without more warning for that question, no.

Endocrinological Responses to 5-HT[a]

P. J. COWEN, I. M. ANDERSON AND S. E. GARTSIDE

MRC Clinical Pharmacology Unit and
University Department of Psychiatry
Littlemore Hospital
Oxford OX4 4XN England

INTRODUCTION

Serotonin (5-hydroxytryptamine; 5-HT) pathways have been implicated in the regulation of a number of anterior pituitary hormones particularly prolactin (PRL), corticotropin (ACTH) and growth hormone (GH).[1] Measurement of the endocrine responses following administration of drugs that increase brain 5-HT function provides one of the few practicable ways of assessing brain 5-HT function in humans. Recent studies in both animals and humans have focused on the role of different 5-HT receptor subtypes in endocrine regulation.

Studies in human 5-HT neuroendocrinology are limited by the availability of selective ligands that can be safely administered; however, data from animal investigations, in particular those using rodents, may not extrapolate to man. For human studies two approaches are available: first, assessment of the effect of selective 5-HT receptor antagonists (TABLE 1) on endocrine responses to drugs that increase brain 5-HT function by pre-synaptic means; second, assessment of the endocrine effects of selective 5-HT receptor agonists.

EFFECT OF 5-HT ANTAGONISTS ON ENDOCRINE RESPONSES TO 5-HT PRECURSORS

Endocrine Effects of L-Tryptophan (LTP)

In our laboratory we have employed the 5-HT precursor L-tryptophan (LTP) which is believed to increase brain 5-HT function through the enhancement of presynaptic 5-HT release. Intravenous LTP in doses of 5g or greater, produces a dose-related increase in plasma PRL and GH with minimal side-effects.[2,3] Because intravenous LTP increases brain 5-HT synthesis it has been presumed that the endocrine responses to LTP are mediated by increased brain 5-HT function.[2,3] However, some have doubted this interpretation, pointing out that LTP administration may decrease the central availability of tyrosine and thus reduce catecholamine function.[4] In addition it has been suggested that while LTP may increase brain 5-HT synthesis, it does not necessarily enhance 5-HT release from nerve terminals.[4] Recently, however, using *in vivo* brain dialysis, Sharp (personal communication) has shown that tryptophan treatment does enhance 5-HT release following electrical stimulation of the raphe.

In animal studies 5-HT uptake inhibitors enhance the endocrine response to 5-HT precursor loading.[1] Consistent with this we found that in humans pretreatment with a

[a]The studies of the authors were supported by the Medical Research Council IMA is an MRC training fellow and SEG holds an MRC studentship.

single dose of the 5-HT uptake inhibitor, clomipramine, markedly potentiated the PRL response to LTP.[5] Interestingly Price *et al.*[6] demonstrated a similar striking increase in the PRL response to LTP following one weeks' treatment with the selective 5-HT uptake inhibitor fluvoxamine. Both these studies suggest that the PRL response to LTP is mediated by 5-HT pathways.

Effect of 5-HT Antagonists on PRL Response to LTP

The PRL response to LTP was abolished by the non-selective 5-HT receptor antagonist, metergoline,[7] which is consistent with the mediation of this response by a post-synaptic 5-HT receptor (TABLE 1). In contrast, the selective $5\text{-HT}_{2/1C}$ receptor antagonist, ritanserin potentiated LTP-induced PRL release,[8] while pretreatment with the selective 5-HT_3 receptor antagonist, BRL 43694, produced no effect.[9]

Our interpretation of these findings is that the PRL response to LTP in humans is mediated by post-synaptic 5-HT_1 receptors. Since 5-HT_{1B} receptors do not occur in the human brain[10] it is possible that LTP-induced PRL release involves a post-synaptic 5-HT_{1A} receptor. A further possibility that stems from our data is that functional interactions may occur between 5-HT receptor subtypes, such that in some circumstances, $5\text{-HT}_{2/1C}$ receptor blockade may actually increase post-synaptic 5-HT_{1A} receptor responses. Interestingly, there is evidence that such interactions may occur in animal studies where ritanserin has been shown to enhance behavioral responses produced by the 5-HT_{1A} receptor agonist, 8-hydroxy-2-(di-*n*-propylamino)tetralin (8-OH-DPAT) (Backus, personal communication).

TABLE 1. Some 5-HT Receptor Antagonists

Drug	Receptor
Metergoline	$5\text{-HT}_{1A\text{-}D}$, 5-HT_2
Ritanserin	$5\text{-HT}_{2/1C}$
BRL 43694	5-HT_3

It seems worth noting that studies of 5-HT–mediated PRL release in rodents reveal a rather different picture. Thus while the PRL response to the 5-HT precursor, 5-HTP, in the rat is attenuated by metergoline, a marked attenuation is also seen following ritanserin pretreatment (Gartside, unpublished observations) (FIG. 1). This suggests that in the rat PRL responses to 5-HT precursor administration are mediated by $5\text{-HT}_{2/1C}$ receptors (see Meltzer *et al.*[11])

GH Response to LTP

While the GH response to LTP was enhanced by clomipramine[5] it was not altered by metergoline,[7] ritanserin,[8] or BRL 43694.[9] This casts some doubt on whether the GH response to LTP is indeed mediated by 5-HT receptors, though it is possible that for some reason metergoline is not a suitable drug with which to block GH responses. However, patients receiving chronic neuroleptic treatment did have a blunted GH increase to LTP,[3] so it is possible that dopaminergic pathways could be involved in this endocrine response.

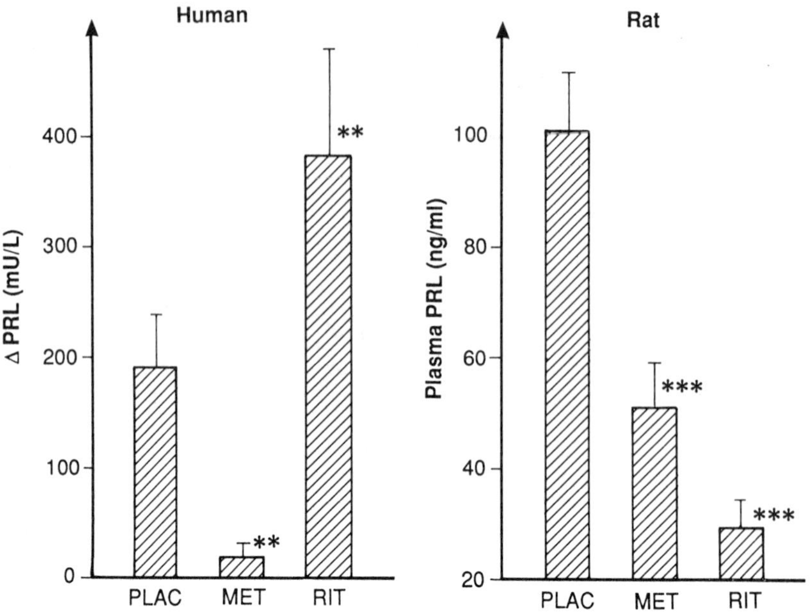

FIGURE 1. The effect of metergoline (MET) and ritanserin (RIT) on the PRL responses to LTP (5g) (humans) and 5-HTP (100mg/kg) (rat). ** significantly different from control, p = 0.01; ***, p = 0.001.

ENDOCRINE EFFECTS OF SELECTIVE 5-HT AGONISTS

m-Chlorophenylpiperazine (m-CPP)

The trazodone metabolite, *m*-chlorophenylpiperazine (*m*-CPP), has biochemical and behavioral properties consistent with direct agonist effects at post-synaptic 5-HT receptors in rodents (see Murphy *et al.*[12]). Administration of *m*-CPP to humans results in an increase in plasma PRL and ACTH and cortisol, and an elevation in oral temperature has also been noted.[13] Intravenous administration of *m*-CPP may also increase plasma GH levels,[14] but this is apparently not seen following oral administration.

Pretreatment of normal volunteers with metergoline abolished the PRL and ACTH and cortisol responses to *m*-CPP and also blunted the temperature increase.[13] These effects suggest that post-synaptic 5-HT receptors do mediate the neuroendocrine effects of *m*-CPP. However, determining the subtype of 5-HT receptor involved has proved problematic.

Ligand binding studies in both rodent and human brain suggest that *m*-CPP is rather a non-selective 5-HT receptor agonist,[15,16] though its highest affinity is apparently for the 5-HT_{1C} receptor subtype[15] (TABLE 2). In animal studies an increase in temperature is associated with 5-HT_2 receptor agonists,[17] so it is possible that the neuroendocrine and temperature effects of *m*-CPP in humans could be mediated through $5\text{-HT}_{2/1C}$ receptors. If this is the case then these effects should be abolished by pretreatment with ritanserin.

6-Chloro-2-(1-piperazinyl)-pyrazine (MK-212)

The neuroendocrine effects of 6-chloro-2-(1-piperazinyl)-pyrazine (MK-212) in humans are similar to those of *m*-CPP in that oral administration increases plasma PRL, ACTH and cortisol, without significantly altering plasma GH.[18] Furthermore, MK-212 increases core temperature in rodents. Ligand-binding studies in the rat suggest that MK-212 is rather a non-selective 5-HT receptor agonist,[15] but its highest affinity is for 5-HT_{1C} receptors (TABLE 2).

The increase in plasma cortisol in the rat which follows MK-212 administration is attenuated by $5\text{-HT}_{2/1C}$ antagonists such as ritanserin.[19] However, the increase in cortisol was not antagonized by spiperone[19] which is a potent 5-HT_2 receptor antagonist with less affinity for the 5-HT_{1C} receptor subtype.[15] Accordingly it is possible that some of the neuroendocrine effects of MK-212 are mediated by 5-HT_{1C} receptors. In contrast, the elevation in core temperature produced in the rat by MK-212 was blocked by both ketanserin and spiperone[17]; therefore hyperthermic responses to MK 212 may be caused by stimulation of post-synaptic 5-HT_2 receptors. In conclusion, while further studies with selective antagonists are needed, current evidence suggests that the neuroendocrine and temperature effects of MK-212 are mediated by $5\text{-HT}_{2/1C}$ receptors.

Gepirone

Gepirone is a pyrimidinylpiperazine derivative which in ligand-binding studies has a high affinity for 5-HT_{1A} receptors in rodent brain[20,21] (TABLE 2). Gepirone may be relatively selective for the 5-HT_{1A} receptor subtype and appears to have little effect on other neurotransmitters,[21] though a dopamine agonist action has recently been proposed.[22]

We studied the neuroendocrine effects of 10 mg and 20 mg of gepirone in normal male volunteers. We found that a dose of 20 mg, gepirone produced significant stimulation of the hypothalamic-pituitary-adrenal (HPA) axis, evidenced by increases in plasma ACTH, β-endorphin and cortisol (FIG. 2). In addition an elevation in plasma PRL was obtained, though at a dose of 10 mg there was evidence that PRL secretion may, in fact, have been decreased. Both the 10 and 20 mg doses of gepirone increased plasma GH, while gepirone, 20 mg, significantly lowered oral temperature (FIG. 3).

The stimulation of the HPA axis by gepirone is consistent with studies in rodents which suggest that a variety of 5-HT_{1A} agonists, including gepirone, increase plasma ACTH[23] and corticosterone[19] and that this response can be attenuated by the 5-HT_{1A} antagonist, pindolol.[19,23] It is also well-established that 5-HT_{1A} receptor agonists decrease core temperature in rodents,[24,25] though it is disputed whether the 5-HT_{1A} receptors involved are located pre- or post-synaptically.[24,25]

The effect of 5-HT_{1A} agonists, such as 8-OH-DPAT, on PRL secretion in rodents is

TABLE 2. Ligand binding[a] and Functional Effects of Some 5-HT Receptor Agonists

	5-HT_{1A}	5HT_{1C}	5-HT_2	PRL	ACTH	GH	TEMP
Gepirone	7.05	4.64	5.44	↑	↑	↑	↓
MCPP	6.49	7.68	6.70	↑	↑	0↑	↑
MK-212	5.32	6.16	4.76	↑	↑	0	↑

[a] pK_D, $-\log$ mol/l (from ref. 15)

unclear, though at best, a modest increase in plasma PRL is sometimes obtained.[26,27] However, Nash and Meltzer[22] have reported that gepirone treatment in the rat may actually reduce the secretion of PRL at pituitary level, perhaps by a direct dopaminergic mechanism.

While we did note small reduction in plasma PRL following gepirone 10 mg, the 20 mg dose was associated with a significant increase in plasma PRL, raising the possibility that in humans 5-HT_{1A} agonists do facilitate PRL release. In line with this suggestion is the finding that gepirone increases plasma PRL in non-human primates.[28]

In our study gepirone was most potent in increasing plasma GH, and in this property appears to differ from MK-212 and *m*-CPP. It is therefore possible that 5-HT_{1A} receptors may have a specific role in the facilitation of GH release but clearly more studies with selective antagonists are required to test this proposal.

Other 5-HT_{1A} Receptor Agonists

Buspirone is a 5-HT_{1A} receptor agonist which is closely related to gepirone.[21] Buspirone, however, has complex interactions with dopamine pathways which makes its effects on neuroendocrine function difficult to assess.[21] In normal volunteers, buspirone elevates plasma PRL, GH, and cortisol[29] but Meltzer *et al.*[30] have provided evidence that in the rat, at least, the effect on PRL may be mediated by dopamine

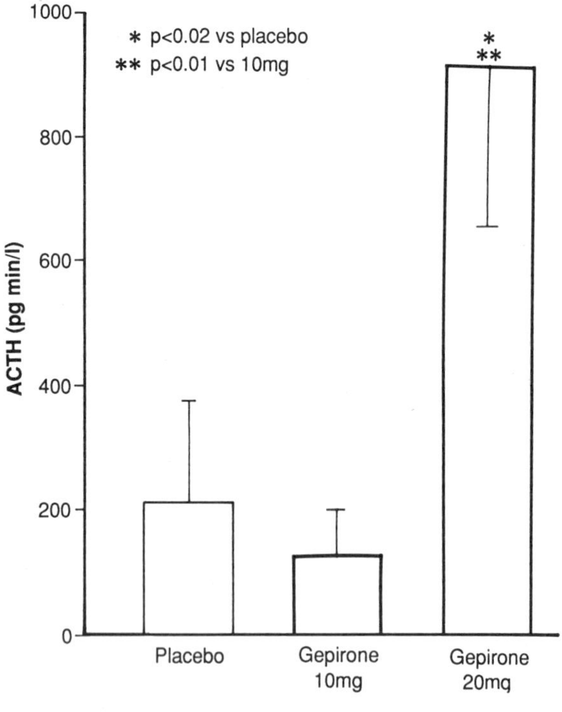

FIGURE 2. Mean (±SEM) area under the curve plasma ACTH following gepirone 10 mg and 20 mg in 12 healthy male volunteers.

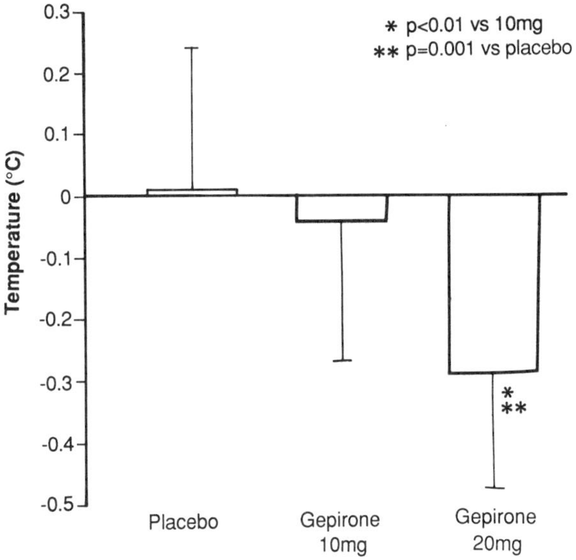

FIGURE 3. Mean (±SEM) maximum change in oral temperature following gepirone 10 mg and 20 mg in 11 healthy male volunteers.

receptor blockade at pituitary level. The increase in plasma GH is of interest in view of the similar effect of gepirone, but in a recent investigation we found that metergoline did not antagonize the GH response to buspirone (Anderson and Cowen, unpublished observations). It is therefore uncertain at present whether this GH response is mediated by 5-HT_{1A} receptors.

Ipsapirone is another selective 5-HT_{1A} receptor agonist, closely related to gepirone.[21] In a small study in human volunteers, Lesch *et al.*[31] found a significant increase in plasma cortisol but no change in plasma PRL or GH, though a trend to an increase was observed with the latter hormone. These findings confirm that an increase in HPA activity probably represents a common effect of 5-HT_{1A} receptor agonists. The increases in plasma PRL and GH seen with gepirone and buspirone are harder to interpret and it is likely that selective antagonists will be required to establish the role of 5-HT_{1A} receptors in these endocrine responses.

SUMMARY

There is increasing evidence that neuroendocrine and temperature responses in humans can be employed to study the functional sensitivity of different 5-HT receptor subtypes. The evidence suggests that the PRL response to LTP is mediated by 5-HT_1 receptors, perhaps the 5-HT_{1A} subtype, though further studies are needed to confirm this effect. It is uncertain whether the PRL responses to other 'presynaptic' challenges of 5-HT function, for example, fenfluramine, are mediated by the same post-synaptic 5-HT receptor subtype as that for LTP.

Conversely it seems likely that agonists which stimulate $5\text{-HT}_{2/1C}$ receptors increase both plasma PRL and ACTH in humans. There is also evidence that 5-HT_{1A}

receptors can increase ACTH secretion. This suggests that in humans as in animals both the 5-HT_{1A} and 5$HT_{2/1C}$ receptors can facilitate ACTH release (TABLE 2), though the significance of this dual control is not understood. It is also possible that both 5-HT_{1A} and 5-$HT_{2/1C}$ receptors stimulate PRL release (TABLE 2), but 5-HT_{1A} receptors may have a more prominent role in GH secretion (TABLE 2). In both human and animal studies 5-HT_{1A} and 5-HT_2 receptor agonists may produce opposite effects on body temperature (TABLE 2).

These recent developments in 5-HT neuroendocrinology have been of great interest, but much is still uncertain. Progress in this field will be considerably advanced by the availability of new selective 5-HT receptor ligands, particularly selective receptor antagonists.

REFERENCES

1. TUOMISTO, J. & P. MANNISTO. 1985. Neurotransmitter regulation of anterior pituitary hormones. Pharmacol. Rev. **37:** 249–332.
2. CHARNEY, D. S., G. R. HENINGER, J. F. RENHARD, D. E. STERNBERG & K. M. HAFSTEAD. 1982. Effect of intravenous L-tryptophan on prolactin and growth hormone and mood in healthy subjects. Psychopharmacology **77:** 217–222.
3. COWEN, P. J., H. GADHUI, B. GOSDEN & T. KOLAKOWSKA. 1985. Responses of prolactin and growth hormone to L-tryptophan infusion: Effects in normal subjects and healthy patients receiving neuroleptics. Psychopharmacology **86:** 164–169.
4. VAN PRAAG, H. M., C. LEMUS & R. KAHN. 1986. The pitfalls of serotonin precursors as challengers in hormonal probes of central serotonergic activity. Psychopharm. Bull. **22:** 565–570.
5. ANDERSON, I. M. & P. J. COWEN. 1986. Clomipramine enhances prolactin and growth hormone responses to L-tryptophan. Psychopharmacology **89:** 131–133.
6. PRICE, L. H., D. S. CHARNEY, P. L. DELGADO, G. M. ANDERSON & G. R. HENINGER. 1989. Effects of desipramine and fluvoxamine treatment on the prolactin response to L-tryptophan: Serotonergic function and the mechanism of action of antidepressant treatment. Arch. Gen. Psychiat. **46:** 625–631.
7. MCCANCE, S. L., P. J. COWEN, H. WALLER & D. G. GRAHAME-SMITH. 1987. The effect of metergoline on endocrine responses to L-tryptophan. J. Psychopharmacology **2:** 90–94.
8. CHARIG, E. M., I. M. ANDERSON, J. M. ROBINSON, D. J. NUTT & P. J. COWEN. L-tryptophan and prolactin release: Evidence for interaction between 5-HT_2 and 5-HT_2 receptors. Human Psychopharmacology **1:** 91–97.
9. ANDERSON, I. M., P. J. COWEN & D. G. GRAHAME-SMITH. 1988. The effect of BRL 43694 on the neuroendocrine responses to L-tryptophan infusion. J. Psychopharmacology **2:** 2.
10. HOYER, D., A. PAZOS, A. PROBST & J. M. PALACIOS. 1986. Serotonin receptors in the human brain. 1. Characterization and autoradiographic localization of 5-HT_{1A} recognition sites. Apparent absence of 5-HT_{1B} recognition sites. Brain Res. **376** 85–96.
11. MELTZER, H. Y., M. SIMONOVIC & G. A. GUDELSKY. 1983. Effects of pireneperone and ketanserin on rat prolactin secretion in vivo and in vitro. Eur. J. Pharmacol. **92:** 83–89.
12. MURPHY, D. L., E. A. MUELLER, J. L. HILL, T. J. TOLLIVER & F. M. JACOBSEN. 1989. Comparative, anxiogenic, neuroendocrine and other physiologic effects of m-chlorophenylpiperazine given intravenously or orally to healthy volunteers. Psychopharmacology **98:** 275–282.
13. MUELLER, E. A., D. L. MURPHY & T. SUNDERLAND. 1986. Further studies of the putative serotonin agonist, *m*-chlorophenylpiperazine: Evidence for a serotonin receptor mediated mechanism of action in humans. Psychopharmacology **89:** 388–391.
14. CHARNEY, D. S., S. W. WOODS, W. K. GOODMAN & G. R. HENINGER. 1987. Serotonin function in anxiety: II. Effects of the serotonin agonist m-CPP in panic disorder patients and healthy subjects. Psychopharmacology **92:** 14-24.
15. HOYER, D. 1988. Functional correlates of serotonin 5-HT_1 recognition sites. J. Receptor Research **8:** 59–81.

16. HAMIK, A. & S. J. PEROUTKA. 1989. l-(m-chlorophenyl)piperazine (mCPP) interactions with neurotransmitter receptors in the human brain. Biol. Psychiat. **25:** 569-575.
17. GUDELSKY, G. A., J. I. KOENIG & H. Y. MELTZER. 1986. Thermoregulatory responses to serotonin (5-HT) receptor stimulation in the rat. Neuropharmacology **25:** 1307–1313.
18. LOWY, M. T. & H. Y. MELTZER 1988. Stimulation of serum cortisol and prolactin secretion in humans by MK-212, a centrally active serotonin agonist. Biol. Psychiat. **23:** 818–828.
19. KOENIG J. I., G. A. GUDELSKY & H. Y. MELTZER. 1987. Stimulation of corticosterone and β-endorphin secretion in the rat by selective 5-HT receptor subtype activation. Eur. J. Pharmacol. **137:** 1–8.
20. PEROUTKA, S. J. 1985. Selective interaction of novel anxiolytics with 5-hydroxytryptamine$_{1A}$ receptors. Biol. Psychiat. **20:** 971–979.
21. TRABER J. & T. GLASER. 1987. 5-HT$_{1A}$ receptor anxiolytics. Trends Pharmac. Sci. **8:** 432–437.
22. NASH, J. F. & H. Y. MELTZER. 1989. The effect of gepirone and ipsapirone on the stimulated and unstimulated secretion of prolactin in the rat. J. Pharm. Exp. Ther. **249:** 236–241.
23. GILBERT, F., C. BRAZELL, M. D. TRICKLEBANK & S. M. STAHL. 1988. Activation of the 5-HT$_{1A}$ receptor subtype increases rat ACTH concentration. Eur. J. Pharmacol. **147:** 431–439.
24. GOODWIN, G. M., R. J. DESOUZA, A. R. GREEN & D. J. HEAL. 1987. The pharmacology of the behavioural and hypothermic responses of rats to 8-hydroxy-2-(di-n-propylamino)tetralin (8-OHDPAT). Psychopharmacology **91:** 506–511.
25. HUTSON, P. H., T. P. DONOHUE & G. CURZON. 1987. Hypothermia induced by the putative 5-HT$_{1A}$ agonists LY 65163 and 80HDPAT is not prevented by 5-HT depletion. Eur. J. Pharmacol. **143:** 221–228.
26. SIMONOVIC M., G. A. GUDELSKY & H. Y. MELTZER. 1984. Effect of 8-hydroxy-2-(di-n-propylamino)tetralin on rat prolactin secretion. J. Neural Transm. **59:** 143–149.
27. VAN DE KAR, L. D., S. A. LORENS, J. H. URBAN & C. L. BETHEA. 1989. Effect of selective serotonin (5-HT) agonists and 5-HT$_2$ antagonists on prolactin secretion. Neuropharmacology **28:** 299–305.
28. HENINGER, G. R., D. S. CHARNEY & A. SMITH. 1987. Comparison of the 5-HT agonists, tryptophan, mCPP, buspirone and gepirone on prolactin, growth hormone and cortisol release in Rhesus monkeys: Effects of lithium. Soc. Neurosci. Abst. **12:** 574.
29. MELTZER, H. Y. & J. F. NASH. 1988. Neuroendocrine studies of 5-HT and mood. *In* Neuroendocrinology of Mood (Current Topics in Neuroendocrinology, Vol. 8). D. Ganten & D. Pfaff, Eds. Berlin: Springer Verlag, pp. 181–210.
30. MELTZER, H. Y., M. SIMONOVIC, V. S. FANG & G. A. GUDELSKY. 1982. Effect of buspirone on rat plasma prolactin levels and striatal dopamine turnover. Psychopharmacology **78:** 49–53.
31. LESCH, K. P., R. RUPPRECHT, B. POTEN, U. MULLER, K. SOHNLE, J. FRITZE & H. M. SCHULTE. 1989. Endocrine responses to 5-HT-1A receptor activation by ipsapirone in humans. Biol. Psychiat **26:** 203–205.

DISCUSSION OF THE PAPER

A. SJOERDSMA (*Merrell Dow Research Institute, Cincinnati, OH*): You mentioned that you did observe drowsiness with your tryptophan loads. I'm curious whether your receptor blockers had any effect or do you have any notion whether they did because I've always felt the mechanism of action on sleep has been rather difficult to understand. Do you have any statement about that?

COWEN: No, we hoped that we'd be able to block the sleepiness-inducing effect of tryptophan, but we were never able to do that in a convincing way with receptor antagonists.

SJOERDSMA: What's your opinion on the mechanism of sleep production with tryptophan?

COWEN: I'm not even sure that it's mediated by 5-HT.

SJOERDSMA: I don't think it is either.

DUNCAN TAYLOR (*Bristol-Myers Co., Wallingford, CT*): Just two comments. Long-term administration of buspirone has been shown in patients with anxiety by Herb Meltzer (he's over there in the audience) not to affect levels of prolactin and that's the most relevant population to look at clinically (Cohn, Wilcox & Meltzer, Am. J. Med. **80**(Suppl. 3B):36–40, 1986.)

The second comment is that at high doses of buspirone, above what the FDA has given us clearance to use, you do see elevations of prolactin, but, interestingly, Neuhauser, Laakmann and colleagues (Neuhauser, Laakmann & Knossalla, Psychopharmacology **96**(Suppl.):393, 1988) have shown that those elevations can be blocked by serotonin antagonists such as methysergide, so that's consistent with the last point you brought up in you summary.

COWEN: Yes, we've actually blocked that response as well using metergoline, but as I said I think there are problems with metergoline and methysergide when you're trying to block prolactin responses. We plan to carry out a study using pindolol because I think that will be the best way of establishing the mechanism of buspirone-induced prolactin release.

RAY FULLER (*Lilly Research Laboratories, Indianapolis, IN*) Phil you had metergoline designated as a 1A and 1B antagonist, I think, in one of your early slides which I guess means you would expect it would block the effect of gepirone if they were mediated by 1A receptors. Have you done that study? I ask because in rats we find metergoline unable to block corticosteroid elevation mediated by 5-HT_{1A} agonists in general. We have not tried gepirone specifically.

COWEN: Yes, that's a problem, and in fact when one looks in the literature you don't tend to see people reporting the effect of non-selective antagonists on endocrine response to 5-HT_{1A} agonists; they go straight for drugs like pindolol. We have not studied the effect of metergoline on endocrine response to gepirone. I think, in fact, that the effects of pindolol might be easier to interpret.

G. HENINGER (*Connecticut Mental Health Center, New Haven*): We have used the rhesus monkey to try to probe this more clearly, relative to the human studies, and the summary of that is that in acute IV doses buspirone looks very much like haloperidol, and it's very different from gepirone.

And the second thing is that the 1A agonists, ipsapirone, gepirone, 8-hydroxy-DPAT, are not blocked by ritanserin. There wasn't a potentiation either in the rhesus monkeys. *m*-CPP is about 50% blocked by ritanserin and mescaline is 100% blocked by ritanserin. So that in the rhesus monkey to the degree that it's some model for the human situation the *m*-CPP looks like 50/50 while the 1A agonist looks like true 1A agonists in that regard.

COWEN: Thanks very much.

J. BLUNDELL (*University of Leeds, England*): Phil, what you've described today illustrates that pharmacological manipulation can adjust the prolactin response to tryptophan, but another aspect of your work has illustrated that dietary manipulation, that is subjects eating a thousand calorie a day diet, also alters the prolactin response to tryptophan, apparently increasing the sensitivity of the system. Now this is a dietary manipulation that many people consider to be very mild and must be being done by millions of people all over the western part of the world anyway. And this seems to suggest that the 5-HT system is very sensitive to mild dietary manipulation. Can you put your dietary work in the context of the pharmacological receptor profiles you've spoken about today?

COWEN: The effect of dieting is striking. A number of other groups are working in this area and it seems that when you go into quite a mild diet you lower plasma tryptophan and that seems to be enough to give you an enhanced prolactin response to tryptophan subsequently. There are data to show that in a number of animal models the same thing occurs. That is, if you give rats a low tryptophan diet there are enhanced endocrine responses to 5-HT precursors so it fits in quite well with the human model. What no one has done as far as I know is to probe the post-synaptic receptor to identify whether it is supersensitive, but I think that's the line we would want to take next.

BLUNDELL: The other aspect of that work is that the enhanced sensitivity of the 5-HT system was shown in women and yet I think most of the studies you reported today were done in male subjects, is that correct?

COWEN: Yes, that's right. We find that during a diet both females and males lower their plasma tryptophan to the same extent. The altered endocrine responses are seen only in females, which may have implications for eating disorders. But we learned that female primates seem to have rather higher rates of brain 5-HT turnover, so females could be more sensitive to decreased 5-HT synthesis where precursor availability is limited.

H. MELTZER (*Case Western Reserve University, Cleveland OH*): A couple of points. We just showed that in rats, gepirone had an inhibitory effect on prolactin secretion *in vivo* and *in vitro* through a direct dopamine agonist–mediated effect of the pituitary. Perhaps at the higher concentration, where you saw the increase in prolactin, this effect was mediated by an indirect agonist effect.

COWEN: Partial dopamine agonist effect. Yes it is possible.

MELTZER: the other point I wanted to make is I still think some caution is needed about the interpretation of the tryptophan experiment. I'm not at all convinced yet that until you show antagonism by specific blockers that this is a serotonin-mediated mechanism and that's for several reasons. First of all, for all the other serotonin agonists, the effect is much greater on cortisol than on prolactin and growth hormone. And you hardly see any response at all from L-tryptophan even with high intravenous doses. Secondly, as I'll show tomorrow, with 5-hydroxy-tryptophan mediated cortisol and prolactin we get a very strong blockade with pretreatment with even lower doses of ritanserin than you used in your experiments. I would say, until you get some more direct evidence, I would withhold final interpretation of that.

COWEN: Yes, I think that your views are shared by some other workers, and I agree that there are problems in interpretation. The crucial thing is for us to try to block the prolactin response to tryptophan with a selective 5-HT_{1A} blocker.

D. HOYER (*Sandoz Ltd., Basel*): We forgot what Ray Fuller just said about metergoline. A little note of caution. Metergoline is a fairly potent 5-HT_{1C} and 5-HT_2 receptor antagonist, at least in biochemical models for those receptors. Now on 5-$HT_{1A,1B}$ and 5-HT_{1D} receptors, let's not worry about 5-HT_{1B}, depending on species and tissue you're going to see partial agonism or you may see antagonism by this compound — plus the other activities of metergoline.

COWEN: We have tended to use metergoline as a 5-HT antagonist. However, there clearly are difficulties with this compound and the advent of more selective blockers would be a great help.

Neurophysiological and Neurochemical Studies of Brain Serotonergic Neurons in Behaving Animals[a]

BARRY L. JACOBS, CASIMIR A. FORNAL,
AND LYNN O. WILKINSON

Program in Neuroscience, Department of Psychology
Princeton University
Princeton, New Jersey 08544

INTRODUCTION

Brain serotonergic neurons lie at the heart of the neuropharmacology of central serotonin. Serotonergic neurotransmission may be altered by drugs acting at any of a variety of sites on the serotonergic neuron itself, or at its postsynaptic receptors. Thus, these neurons constitute the essential interface between such drugs and their functional effects. Basic comprehension of the variables that control and influence the activity of brain serotonergic neurons is therefore a prerequisite to fully understanding the action of drugs upon the central serotonergic system. In turn, this is critical for understanding the effect of these drugs upon physiology and behavior.

Therefore, because a number of other contributors to this volume have provided detailed descriptions of various drug effects on the brain serotonergic system, our discussion will focus on the variables that control, or correlate with, the electrophysiological activity of serotonergic neurons. Furthermore, since a separate chapter in this volume is devoted to the cellular neurophysiology of mammalian serotonergic neurons (see Aghajanian *et al.*, this volume), we will restrict our discussion to studies of the single unit activity of these neurons in behaving animals, exploring a wide spectrum of behavioral/environmental variables, as well as basic physiological manipulations.

We will limit our description of drug actions to two topics in which recordings of brain serotonergic neurons in behaving animals have had a particular impact: the mechanism of actions of the major hallucinogenic drugs, such as psilocybin, mescaline, and LSD; and, drug effects at the somatodendritic autoreceptor, because this controls the intrinsic activity of serotonergic neurons.

Finally, we will address the question of whether single unit activity accurately reflects the functional output of serotonergic neurons, or, whether under certain conditions, release of neurotransmitter can become uncoupled or partially dissociated from neuronal activity. This issue is approached through the measurement of changes in extracellular serotonin by means of *in vivo* brain microdialysis (several other chapters in this volume are also devoted specifically to this topic).

[a]The authors' work described in this chapter was supported by grants from the National Institute of Mental Health (MH23433) and U.S. Air Force (AFOSR 85-0034).

SINGLE UNIT ACTIVITY OF BRAIN SEROTONERGIC NEURONS IN FREELY MOVING CATS

To date, examination of the activity of serotonergic neurons in behaving animals has been limited to studies employing the domestic cat. The four largest clusters of serotonergic neurons in the cat brain (and that of most mammals) roughly correspond to the dorsal raphe nucleus (DRN), nucleus centralis superior (NCS), nucleus raphe magnus (NRM), and nucleus raphe pallidus (NRP). Our discussion will focus on studies of the largest cluster of serotonergic neurons, those in the DRN. We will also briefly mention studies of NRM serotonergic neurons because of their presumed involvement in nociception and/or analgesia. Detailed descriptions of the activity of serotonergic neurons in NCS and NRP can be found in our published papers.[1,2]

Sleep-Wake-Arousal Cycle/Circadian Rhythms

Our initial objective was to characterize the activity of serotonergic neurons in the DRN across the sleep-wake-arousal cycle.[3] This analysis would then provide a firm data base for use in future experiments. Serotonergic neurons were identified on-line by their characteristic slow and rhythmic activity. Further support for their neurochemical identity was provided by the fact that their spontaneous activity was strongly suppressed by low doses of the serotonergic agonist drugs 5-methoxy-*N*,*N*-dimethytryptamine (5-MeODMT) and LSD. Furthermore, histological examination revealed that neurons with these characteristics were localized precisely and exclusively in those brain areas where serotonergic neurons are found. Although the activity of DRN serotonergic neurons is slow and regular when the cat is in a quiet waking state, this activity shows dramatic changes across the sleep-wake-arousal cycle.[3,4] From a quiet waking rate of approximately 3 spikes/sec, the activity of these neurons will often phasically increase to 6 spikes/sec in response to an arousing stimulus. Reciprocally, the activity of these neurons shows a gradual decline as the cat becomes drowsy and enters slow wave sleep. A decrease in the regularity of firing accompanies this overall slowing of neuronal activity. Finally, the culmination of this state-dependent decrease in unit activity occurs as that cat enters REM sleep, when the activity of these neurons typically becomes completely quiescent. Thus, although serotonergic neurons display a slow, clock-like activity under quiet waking conditions, this activity can be drastically altered in conjunction with changes in sleep-wake-arousal state.

Because brain serotonin metabolism is strongly influenced by the light-dark cycle (true circadian rhythmicity has not been closely examined), we decided to examine whether the light-dark cycle would influence the activity of DRN serotonergic neurons.[5] When we held state constant, that is, studied the activity of a given neuron during REM sleep or quiet waking, at various times during the light-dark cycle, we found the activity to be completely stable. Thus, the activity of these cells is not influenced by the light-dark cycle, nor, by inference, the circadian cycle. Why then have a number of investigators reported that serotonin metabolism varies as a function of the light-dark cycle? There are two obvious, non-mutually exclusive, possibilities. First, since neurotransmitter metabolism occurs predominantly at the axon terminal, local conditions at these sites may modulate metabolism somewhat independently from a constant spike discharge rate being generated at the cell body (this issue is discussed below). Secondly, the variation in serotonin metabolism across the light-dark cycle may simply be secondary to the well-known distribution of sleep and waking across this cycle. We favor this latter explanation.

Sensory Stimuli/Painful Stimuli

Repetitive presentation (once every 2 sec) of phasic auditory (click) or visual stimuli (flash) both produce a similar effect on DRN serotonergic unit activity, of excitation followed by inhibition.[6] The excitation has a latency of approximately 40–50 msec and a duration of 60–80 msec. The inhibitory period which follows the excitation has a mean duration of approximately 250 msec. No evidence of habituation of these responses of DRN serotonergic neurons was seen, even after hundreds of stimulus presentations.[6,7] By contrast, serotonergic neurons in the NRM show only a modest excitation-inhibition response to these same auditory and visual stimuli.[8] Perhaps of greater interest, however, is our failure to observe any specific response of NRM serotonergic neurons to both phasic and tonic painful or noxious stimuli. There was also no change in neuronal activity in response to the systemic administration of morphine, in a dose that produces analgesia.[9] This is somewhat surprising in light of the large literature which involves the NRM in nociception and analgesia. The implications of these results are discussed below.

Environmental Stressors

We have also examined the activity of DRN serotonergic neurons during exposure of the cat to several different environmental challenges or stressors.

When loud white noise (100 dB) is presented for 15 minutes, the stimulus elicits strong sympathetic activation, as indicated by significant increases in tonic heart rate and plasma norepinephrine. It also evokes a stereotyped behavioral response in cats of crouching, with ears flattened. Despite this, the activity of a group of DRN neurons was not significantly different from that observed during an undisturbed active waking baseline.[10]

In a parallel experiment, cats were restrained for 15 minutes by being placed in a canvas bag. This also evoked a strong sympathetic activation. Struggling and vocalizations during the restraint provided additional behavioral evidence for the stressful nature of the stimulus. Despite this behavioral and physiologic activation, the activity of a group of DRN neurons was not significantly different from that observed during an undisturbed active waking baseline.[10]

In the final experiment in this series, a dog was brought into view of the cat for 5 minutes. This evoked the typical stereotyped feline defense reaction of arched back, facing broadside, piloerection, often growling and hissing, and physiologic indices of sympathetic activation. However, despite this behavioral and physiologic activation, the discharge rate of a group of DRN serotonergic neurons was unchanged from that observed during an appropriate control condition.[10]

Physiologic Challenges

In a series of studies paralleling those in the preceding section, we examined the response of DRN serotonergic neurons during perturbation of three physiologic systems: thermoregulatory, cardiovascular, and glucoregulatory.

We examined the response of DRN neurons to both increased ambient temperature and pyrogen-induced fever.[11] For environmental heating, the temperature in the experimental chamber was maintained at 43°C until each cat displayed continuous panting. The activity of DRN neurons remained unaffected during the interval when ambient temperature was increased from 25°C to 43°C. During this initial phase of heating, no appreciable behavioral or physiologic responses were seen. However, following prolonged heat exposure, cats displayed intense panting, relaxation of posture, and a progressive rise in body/brain temperature (range: 0.5–2.0°C). Once again, however, no change in DRN single unit activity was observed.

In studies of DRN neuronal activity during the febrile response, a synthetic pyrogen was administered systemically.[11] Body/brain temperature began to increase within 30 min, reached a peak at 1–2 hrs, and returned to pre-drug level by 6 hrs. The peak elevation of body temperature that was attained was typically 1.5°–2.5°C. Once again, no change in DRN single unit activity was observed during any phase of the pyrogen-induced febrile response.

The response of DRN neurons was also studied in relation to changes in blood pressure induced by peripherally acting drugs: phenylephrine, to induce increases in blood pressure; sodium nitroprusside, to phasically decrease blood pressure; and hydralazine, to produce prolonged hypotension. Over a range of 20–70 mm Hg increases and 10–50 mm Hg decreases in mean arterial pressure, there was no change in the discharge rate of DRN serotonergic neurons.[12] This is despite the fact that these blood pressure changes were of sufficient magnitude to produce dramatic reflexive changes in heart rate and plasma catecholamines.

In the final study in this series, we manipulated blood glucose level in both directions. The activity of serotonergic DRN neurons in behaving cats was not significantly altered by bolus injection of glucose (500 mg/kg, i.v.) that elevated blood glucose levels 3- to 5-fold.[13] Likewise, the activity of these neurons was not significantly affected by the administration of a dose of insulin (2–4 IU/kg, i.v.), which lowered blood glucose by 50%, or more, or following the rapid reversal of this hypoglycemia by subsequent glucose administration.

Summary

A large literature implicates brain serotonergic neurons in the control of physiological processes (such as nociception, cardiovascular function, and thermoregulation) and behavioral/environmental processes (such as feeding, aggression, and stress). However, the experiments described in this section imply that at least DRN (and possibly NRM) serotonergic neurons are not critically involved in these activities, since no effect on serotonergic single unit activity in behaving animals was seen, independent of changes along the sleep-wake-arousal continuum. This was true in spite of the fact that many of these studies employed stimuli chosen explicitly for the strength of their impact on the organism. The tonic activity of serotonergic neurons appears in *general* to vary in a stereotypical manner in association with behavioral state, and not *specifically* in association with any process. Therefore, the primary role of these neurons in physiology and behavior may be to coordinate the activity of target structures (or set the tone) in conjunction with the organism's level of behavioral arousal. This is a major task for the nervous system because, not only behavior, but the output of all physiological systems in the organism is strongly modulated across the sleep-wake-arousal cycle.[14,15]

A recent finding has caused some revision in our thinking about serotonergic neurons. We have discovered a subgroup of DRN neurons in cats that manifest a strong relation to a particular type of movement. When the animal licks, chews, bites, or grooms with its mouth, the activity of these neurons increases 2- to 5-fold—from 1–3 spikes/sec to 2–15 spikes/sec.[16] Interestingly, during other types of movements their activity decreases from that observed during quiet waking behavior. For example, during gross body movements the cells may fall silent for several seconds, while during eye movements, they may pause briefly or simply slow. Since this increase in serotonergic neuronal activity is seen exclusively in association with oral-buccal movements it raises a number of interesting questions. Is this relation peculiar to the cat—would rats or monkeys manifest a different type of movement-related activity? Will other movement-related subgroups be found in the cat? What are the efferent projections of these serotonergic neurons and from what sources do their afferents derive? Finally, why does the oral-buccal region have such a privileged relation with the serotonergic system? Along this line, it is interesting to note that serotonin has been directly implicated as a modulator of feeding in two different invertebrate species.[17,18]

Finally, it is of interest to compare the response of serotonergic neurons to environmental and physiologic challenges, with that of noradrenergic neurons. As we have seen, serotonergic neurons display a general lack of responsiveness to a variety of challenges or stressors. In contrast, when cats are exposed to the same challenges, noradrenergic neurons in the locus coeruleus are highly responsive.[19] This difference in response of these two brainstem neurochemical groups has important implications for theories regarding their roles in the control of physiology and behavior.

MEASUREMENT OF EXTRACELLULAR SEROTONIN IN FREELY MOVING CATS

The data in the previous section indicate that the electrophysiologic activity of serotonergic neurons is very difficult to perturb by any of a variety of strong behavioral/environmental or physiological challenges. However, it is possible that under certain conditions changes in serotonergic neurotransmission may become partially uncoupled from neuronal activity. For example, serotonin release may be globally modulated by alterations in the regulation of synthesis or breakdown, or locally modulated by regional influences on the serotonergic nerve terminal.

One technique which can be used to investigate the regulation of serotonin release in the behaving animal is *in vivo* dialysis. In this technique, a small probe with a permeable membrane is inserted into the brain structure of interest while artificial CSF is perfused through the probe. Neurochemicals passively diffuse across the membrane and are collected in the artificial CSF. Measurement of neurotransmitter and metabolite levels collected in this sample over some period of time are thought to reflect extracellular concentrations of these compounds. We measured extracellular serotonin in the anterior hypothalamus and caudate nucleus under several conditions identical to those in which we examined serotonergic single unit activity. Thus, we examined the relationship between extracellular serotonin and behavioral state, with the hypothesis that serotonin would increase during periods of arousal, and decrease during sleep. Under these conditions, we expected extracellular serotonin to parallel neuronal activity. Secondly, we examined the possibility of a dissociation between neuronal activity and serotonin release by measuring extracellular serotonin during a thermoregulatory challenge. Since we had found that neuronal activity was not specifically altered during pyrogen-induced fever, and the literature has implicated brain serotonin in the regulation of body temperature,[20,21] pyrogen-induced fever appeared to be a situation in which nerve terminal modulation might play a significant role in the determination of extracellular serotonin release.

Cats were implanted with gross electrodes and guide cannulae for the dialysis probes which were aimed at the anterior hypothalamus or caudate nucleus. Following a recovery period of at least 1 week, cats were placed in a recording booth identical to that used in our single unit studies, and a concentric dialysis probe (5 mm tip length[22]) was slowly lowered into the brain and cemented firmly in place with dental acrylic. A modified Ringers solution was perfused through the probe at a rate of 2 μl/min, and collection of dialysis samples began no sooner than 8 hours following probe implantation.

High pressure liquid chromatography with electrochemical detection was used for analysis of 5-HT levels as described previously.[23] Electrochemical detection was performed by a dual potentiostat electrochemical detector with two parallel working electrodes at 590 and 520 mV relative to an Ag/AgCl reference electrode. This dual electrode system was used to enhance our ability to detect co-elution of unknown substances with serotonin. Thus, for all samples, the ratio of serotonin peak heights at maximal and half maximal potentials was calculated (the current ratio) and compared to the current ratio in pure serotonin standards. This ratio was used to determine, before performing any experimental manipulation, whether the observed peak represented uncontaminated serotonin.

A significant contribution of neuronally released serotonin to extracellular serotonin has been demonstrated using this technique.[23] Thus, measured extracellular serotonin is decreased following addition of tetrodotoxin to the Ringers solution, or elimination of Ca^{++} with addition of EGTA to chelate extracellular Ca^{++} ions. Extracellular serotonin is increased following addition of the serotonin-releasing agent fenfluramine, the reuptake inhibitor fluoxetine, or increased K^+ concentrations in the Ringers. In addition, it has been demonstrated that a decrease in neuronal activity produced by systemic administration of the autoreceptor agonists 5-MeODMT[24] or 8-hydroxy-dipropylaminotetralin (8-OH-DPAT)[23] decreases levels of extracellular serotonin by approximately 50%. We investigated the decrease in extracellular serotonin produced by administration of 8-OH-DPAT in cat, and related effects of this drug on extracellular serotonin to the effects of this drug on serotonergic neuronal activity. In cat, extracellular serotonin was decreased by 8-OH-DPAT to a similar degree as in rat (44%), and the time course of this effect was parallel to the time course of the effects of this drug on serotonergic unit activity. These findings were very consistent with previous investigations of the neuronal source of extracellular serotonin using a fluorimetric detection technique,[24] and support the conclusion that extracellular serotonin as measured with electrochemical detection using the described criterion reflects neuronal release.

Our first experiments examined the relationship between extracellular serotonin and behavioral arousal. If neuronal activity contributes significantly to the control of serotonin release, extracellular serotonin should increase during periods of increased behavioral arousal, and decrease during sleep periods. We measured extracellular serotonin and simultaneously monitored behavioral state using a variety of polygraphic criteria. As predicted, extracellular serotonin increased and decreased in parallel with behavioral state changes.[25,26] Thus, extracellular serotonin appears to be controlled at least in part by fluctuations in the neuronal activity of serotonin containing neurons.

We next explored the hypothesis that the release of extracellular serotonin might be modulated independently of neuronal activity by local factors in the nerve terminal region. Thus, we examined release of serotonin in the anterior hypothalamus during fever induced by a synthetic pyrogen. As described above, serotonergic neuronal activity is not significantly changed during fever, and yet a substantial literature suggests that serotonin release in the anterior hypothalamus is increased by this manipulation.[27] Thus this manipulation seemed to be a likely candidate for significant contribution of nerve terminal modulation (i.e. one in which serotonin release might be dissociated from the neuronal activity of serotonergic neurons).

We administered synthetic pyrogen to six cats while brain temperature, behavioral state and extracellular serotonin were measured continuously. In no instance did we observe a consistent change in serotonin release that could not be attributed to a change in behavioral state during this manipulation.[26] A time series regression analysis demonstrated that extracellular serotonin was significantly related to behavioral state change consequent to the drug-induced febrile response, and not significantly affected by changes in body temperature per se.

PHARMACOLOGIC STUDIES

Historically, serotonin has been implicated in the mechanism of action of hallucinogenic drugs. This was based initially on the structural similarity of serotonin and the indole nucleus hallucinogens, such as psilocybin, and the ergot hallucinogens, such as LSD. It was also based on pharmacologic evidence that LSD blocked serotonin's actions in the periphery. In the past several years, the relationship between the three major types of hallucinogenic drugs (indole, ergot, and phenylethylamine) and serotonin has been elucidated. Evidence from pharmacologic, behavioral, electrophysiologic, and receptor binding studies now indicates that all three of these types of hallucinogenic drugs exert an agonist action at the 5-HT_2 receptor (see refs. 28 and 29 for a review of this evidence). Thus, the critical action of the major types of hallucinogenic drugs is mediated by a single common receptor. This is consistent with the fact that these drugs display cross-tolerance and elicit similar phenomenologic effects.

Somatodendritic autoreceptors (presumably of the 5-HT_{1A} variety) appear to play an important role in regulating the firing rate of brain serotonergic neurons. They are thought to do so by means of a local negative feedback mechanism, i.e. greater neuronal activity releases more serotonin which, in turn, decreases neuronal activity. We have examined the effects of systemic administration of drugs which are agonists or antagonists at this receptor.[30] Low doses of 8-OH-DPAT (5-20 μg/kg i.v.), an agonist at the 5-HT_{1A} receptor, completely suppress the activity of DRN serotonergic neurons in cats. Reciprocally, spiperone (1 mg/kg i.v.), an antagonist at the 5-HT_{1A} receptor produces an approximate doubling of the spontaneous activity of these neurons. (This implies that under normal conditions serotonin is exerting a tonic inhibitory action at the autoreceptor.) Spiperone pretreatment also blocked the ability of 8-OH-DPAT to depress DRN neuronal activity.

ACKNOWLEDGMENT

The authors thank Ms. Arlene Kronewitter for preparing the manuscript.

REFERENCES

1. RASMUSSEN, K., J. HEYM, & B. L. JACOBS. 1984. Activity of serotonin-containing neurons in nucleus centralis superior of freely moving cats. Exp. Neurol. **83:** 302–317.
2. HEYM, J., G. F. STEINFELS & B. L. JACOBS. 1982. Activity of serotonin-containing neurons in the nucleus raphe pallidus of freely moving cats. Brain Res. **251:** 259–276.
3. TRULSON, M. E. & B. L. JACOBS. 1979. Raphe unit activity in freely moving cats: Correlation with level of behavioral arousal. Brain Res. **163:** 135–150.
4. MCGINTY, D. J. & R. M. HARPER. 1976. Dorsal raphe neurons: Depression of firing during sleep in cats. Brain Res. **101:** 569–575.
5. TRULSON, M. E. & B. L. JACOBS. 1983. Raphe unit activity in freely moving cats: Lack of diurnal variation. Neurosci. Lett. **36:** 285–290.

6. HEYM J., M. E. TRULSON & B. L. JACOBS. 1982. Raphe unit activity in freely moving cats: Effects of phasic auditory and visual stimuli. Brain Res. **232:** 29–39.
7. RASMUSSEN, K., R. E. STRECKER & B. L. JACOBS. 1986. Single unit response of noradrenergic, serotonergic and dopaminergic neurons in freely moving cats to simple sensory stimuli. Brain Res. **369:** 336–340.
8. FORNAL, C., S. AUERBACH & B. L. JACOBS. 1985. Activity of serotonin-containing neurons in nucleus raphe magnus in freely moving cats. Exp. Neurol. **88:** 590–608.
9. AUERBACH, S., C. FORNAL & B. L. JACOBS. 1985. Response of serotonin-containing neurons in nucleus raphe magnus to morphine, noxious stimuli and PAG stimulation in freely moving cats. Exp. Neurol. **88:** 609–628.
10. WILKINSON, L. O. & B. L. JACOBS. 1988. Lack of response of serotonergic neurons in the dorsal raphe nucleus of freely moving cats to stressful stimuli. Exp. Neurol. **101:** 445–457.
11. FORNAL, C. A., W. J. LITTO, D. A. MORILAK & B. L. JACOBS. 1987. Single-unit responses of serotonergic dorsal raphe nucleus neurons to environmental heating and pyrogen administration in freely moving cats. Exp. Neurol. **98:** 388–403.
12. FORNAL, C. A. & B. L. JACOBS. 1988. Physiological and behavioral correlates of serotonergic single unit activity. *In* Neuronal Serotonin. N. N. Osborne & M. Hamon, Eds. New York: John Wiley, pp. 305–345.
13. FORNAL, C. A., W. J. LITTO, D. A. MORILAK & B. L. JACOBS. 1989. Single-unit responses of serotonergic neurons to glucose and insulin administration in behaving cats. Am. J. Physiol. **257:** R1345–R1353.
14. OREM, J. & C. D. BARNES, EDS. 1980. Physiology in Sleep. New York: Academic Press.
15. ZANCHETTI, A., G. BACCELLI & G. MANCIA. 1982. Cardiovascular regulation during sleep. Arch. Ital. Biol. **120:** 120–137.
16. RIBEIRO-DO-VALLE, L. E., C. A. FORNAL, W. J. LITTO & B. L. JACOBS. 1989. Serotonergic dorsal raphe unit activity related to feeding/grooming behaviors in cats. Soc. Neurosci. Abstr. **15:** 1283.
17. LENT, C. M. 1985. Serotonergic modulation of the feeding behavior of the medicinal leech. Brain Res. Bull. **14:** 643–655.
18. ROSEN, S. C., K. R. WEISS, R. S. GOLDSTEIN & I. KUPFERMANN. 1989. The role of a modulatory neuron in feeding and satiation in Aplysia: effects of lesioning of the serotonergic metacerebral cells. J. Neurosci. **9:** 1562–1578.
19. JACOBS, B. L. In press. Locus coeruleus neuronal activity in behaving animals. *In* The Pharmacology of Noradrenaline in the Central Nervous System. D. J. Heal & C. A. Marsden, Eds. Oxford: Oxford University Press.
20. MYERS, R. D. 1980. Hypothalamic control of thermoregulation. *In* Handbook of the Hypothalamus. P. J. Morgane & J. Panksepp, Eds., Vol. **3:** 83–210. New York. Marcel Dekker.
21. DASCOMBE, M. J. 1985. The pharmacology of fever. Prog. Neurobiol. **25:** 327–373.
22. HERNANDEZ, L., B. G. STANLEY & B. G. HOEBEL. 1986. A small, removable microdialysis probe. Life Sci. **39:** 2629–2637.
23. AUERBACH, S. A., M. J. MINZENBERG & L. O. WILKINSON. 1989. Extracellular serotonin and 5-HIAA in hypothalamus of the unanesthetized rat measured by in vivo dialysis coupled to HPLC-EC: Dialysate serotonin reflects neuronal release. Brain Res. **499:** 281–290.
24. KALEN, P., R. E. STRECKER, E. ROSENGREN & A. BJORKLUND. 1988. Endogenous release of neuronal serotonin and 5-HIAA in the caudateputamen of the rat as revealed by intracerebral dialysis coupled to high performance liquid chromatography with fluorimetric detection. J Neurochem. **51:** 1422–1435.
25. WILKINSON, L. O., K. F. MARTIN, S. B. AUERBACH, C. A. MARSDEN & B. L. JACOBS. 1988. Relationship between dialysate serotonin and raphe unit activity in freely moving cats. Br. J. Pharmacol. (suppl.) **872.**
26. WILKINSON, L. O. 1989. Relationship between the single unit activity of serotonin containing neurons and extracellular serotonin in the freely moving cat. Doctoral dissertation, Princeton University.
27. MYERS, R. D. 1977. New aspects of the role of hypothalamic calcium ions, 5-HT and PGE

during normal thermoregulation and pyrogen fever. *In* Drugs, Biogenic Amines and Body Temperature. K. E. Cooper, P. Lomax, and E. Schonbaum, Eds. Basel: Karger, pp. 51–53.
28. Jacobs, B. L. 1984. Postsynaptic serotonergic action of hallucinogens. *In* Hallucinogens: Neurochemical, Behavioral and Clinical Perspectives. B. L. Jacobs, Ed. New York: Raven Press, pp. 183–202.
29. Heym, J. & B. L. Jacobs. 1988. 5-HT_2 agonist activity as a common action of hallucinogens. *In* 5-HT Agonists as Psychoactive Drugs. R. H. Rech and G. A. Gudelsky, Eds. Ann Arbor, MI: NPP Books, pp. 95–106.
30. Fornal, C. A., W. J. Litto, L. E. Ribeiro-do-Valle & B. L. Jacobs. 1989. Role of somatodendritic autoreceptors in the regulation of serotonergic neuronal activity in behaving cats. Soc. Neurosci. Abstr. **15:** 553.

DISCUSSION OF THE PAPER

W. Clark (*Mount Sinai School of Medicine, New York*): Do you see changes depending on the state of the animal? If you give spiperone for example during REM sleep does the animal wake up?

Jacobs: The spiperone-induced changes are in fact state dependent. If the animal goes to sleep, for example, after we've given spiperone, you're very hard pressed to tell the difference between the activity of that neuron during sleep with spiperone and sleep without spiperone. Our explanation for that is that during sleep you have a very low spontaneous level of activity and you're not activating autoreceptors, and therefore you're not seeing the manifestation of this blockade of autoreceptors. That's not proven, that's our interpretation of those data.

Question: You mentioned that you have a subgroup of cells in the raphe that have very strong correlation with grooming and feeding. Is this a special subgroup of cells?

Jacobs: The problem is, as you know, because of the technique that we use it's very hard to ultimately identify the specific neurons that we're recording from. In all other ways they look like serotonergic neurons. The little anatomy that we've done indicates that they're pretty much localized to the same area of the dorsal raphe nucleus. At this point, we can't discriminate them in any other way except that they often manifest non-clock like types of activities during waking. I asked George Aghajanian yesterday if he thought there might be two groups of serotonergic neurons in rat dorsal raphe nucleus and he said "well. . . " .

Question: Could they be adrenergic cells?

Jacobs: You mean these oral activity related cells?

Comment: Yes.

Jacobs: No, not according to all the criteria that we now use. For example, they're exquisitely sensitive to 8-hydroxy-DPAT, and noradrenergic neurons, as far as we can tell, are completely unresponsive to 8-hydroxy-DPAT. Thus, by all criteria that we know, they look exactly like other raphe neurons.

Question: What proportion of the cells are they?

Jacobs: It's hard to say, they're not trivial. I'd say $1/5$, $1/4$, something of that order. But again it's hard to put numbers to that and I'd hesitate to say that more strongly than to say that's kind of a ball park figure.

Richard Katz (*CIBA Geigy*): I was struck by what appears to be a rather low responsivity of the raphe cells to morphine. Morphine in cats can be highly hallucinogenic. Is it not surprising that the raphe is so unresponsive to a potential hallucinogen?

Jacobs: Well let me say a couple of things. It's not a hallucinogen of the type that most of us have studied with respect to raphe neurons. The data are absolutely clear that neurons in the dorsal raphe nucleus and the nucleus raphe magnus, are completely unresponsive to morphine. I think you should be careful about calling morphine an hallucinogen in cats, I think most people would not.

KATZ: I don't know that one would see the same kind of twitches or hind limb effects one sees with LSD, but in unrestrained cats on morphine it really looks like they're chasing things that aren't there. I don't know what you'd call it, it's an odd response.

JACOBS: Our studies used doses that didn't produce that. This is a relatively low dose. Feline mania is seen at much, much higher doses than that.

H. AKBARI (*New York University*): I'm curious about the physiological and/or environmental challenge. Have you looked at the more caudal serotonin nuclei, or did you just concentrate on the dorsal nucleus?

JACOBS: We've done some of these studies in nucleus raphe magnus, and in general, we also see no change. It mimics what I showed you for the dorsal raphe nucleus. These cells are simply unresponsive to physiologic challenge. Dr. Robert McCall has just published a paper on nucleus raphe pallidus or magnus, and also reported no change with blood pressure.

M. BIBER (*Virginia Commonwealth University, Richmond*): Barry, there's a lot of evidence in rats that, in projection areas, you get very marked increases in 5-HT turnover in response to stresses. Do you think that there's a dissociation between firing rate and the biochemical changes that one sees?

JACOBS: I would question the turnover data. What I'd like to see people try now is *in vivo* dialysis to see what kinds of changes they get. A lot of it depends on the paradigm. It really depends on what you're calling your baseline and then what your manipulation is. For example if you leave a rat alone, especially during the light phase, they will go to sleep. Therefore, if you take a measure when an animal has been asleep and then stress the animal, my response to that is what you've simply done is taken a sleeping animal and compared it to a waking animal. If you do an appropriate control, and then you say you stress an animal and see a dramatic change in turnover, I'd be more impressed. I think you have to very careful about what the control condition is that you're comparing the stressor to.

We've also looked in the corpus striatum during this and we've done a bit with white noise, and again we simply don't see changes that can't be accounted for by simple state change in the animal.

BIBER: What do you think about the morphine data? I mean there's an enormous literature on enhanced 5-HT turnover with an analgesic.

JACOBS: Well, we haven't looked at that neurochemically. That's something that we or someone really ought to do. At the cellular level, the data are absolutely clear, and that is there is absolutely no change in neuronal activity in spite of the fact that you're producing analgesia. At the terminal sites, for example, in spinal cord, you may be getting dramatic changes in turnover in the dorsal horn, but I'd like to see that question studied directly.

BIBER: Maybe it suggests that there's a dissociation between firing and biochemical changes.

JACOBS: That's why we carried out these studies in the anterior hypothalamus with thermoregulatory challenge looking at the neurochemical change, and our data indicate that you simply don't find it.

V. PIERIBONE (*New York University*): Since the raphe magnus is a heterogeneous cyto-architectonic area, what kind of criteria do you use to insure that what you're recording from are actually serotonergic nerves.

JACOBS: Not to skip your question, I think the real question is not are these serotonergic neurons, but are there serotonergic neurons that we don't identify. I don't think there's any question that the neurons that I showed you were serotonergic because of the slow regular activity, long duration action potential, response to 8-hydroxy-DPAT, off in REM sleep, etc., etc. I think the interesting thing is are there serotonergic neurons, for example, that don't have those characteristics. That's a very difficult question to answer in the backward way.

PIERIBONE: Have you ever looked at dorsal raphe neurons and antidromically driven them from cortex to see if these are the same sort of neurons?

JACOBS: No we haven't.

M. MOLLIVER (*The Johns Hopkins School of Medicine, Baltimore, MD*): Barry, you pointed out that there were these oral facial active cells in the dorsal raphe nucleus that are extremely active during feeding behavior. And there were several points. I'd like to suggest a more theoretical point that it appears that there's a dual serotonergic innervation to most of the brainstem cranial nerve motor nuclei. And indeed our laboratory has shown that about 30% of the serotonin input to the motor nucleus of the trigeminal comes from the dorsal raphe nucleus, and about 70% comes from the medullary raphe group, particularly pallidus and obscurus.

George Aghajanian has reported several times that there's a potent excitatory effect of serotonin in the facial nucleus. And then there's a British group, the paper came out about two years ago, I've forgotten the authors, who reported that the medullary groups inhibit the phrenic nucleus that innervates the diaphragm whereas the dorsal raphe has a powerful excitatory influence on that. And all of these data taken together would suggest that there are two opposing influences upon the activity of motor nuclei by serotonin probably mediated by 5-HT_{1a} inhibitory and 5-HT_2 excitatory receptors. And we thought that this was of particular interest because it may be the first link to explain why, for example, in severe depression where the prominent feature is a mental change, there's a profound loss of facial expression. It may be that there's a disinhibition or loss of the excitation of the cranial nerve motor nuclei, and an unleashing of the inhibitory effects of the medullary groups. I wonder if you or George (Aghajanian) would care to comment on that hypothesis.

G. K. AGHAJANIAN (*Yale University, New Haven, CT*): As far as the inhibitory component that's been harder to demonstrate with serotonin on facial motoneurons. The really prominent effect is the slow depolarization facilatory action. Just to address a comment to Barry to look at his REM data in a very positive way as far as a functional change with serotonin. Within the context of seeing serotonin as facilitating motoneuronal activity, during the REM state one could see motoneurons as having been disfacilitated which would make a lot of sense physiologically, functionally. And that rebound from REM one could think of the motoneuron, at that point, as suddenly being refacilitated and one could see how this might have some adaptive significance.

JACOBS: I agree.

J. CHIASSON (*Institut de Recherches, Clinique de Montreal, Quebec, Canada*): You suggested that you were going to show data on the effect of insulin or hypoglycemia on his system, and I didn't see any. Is there anything that you would like to share with us on the effect of insulin on your model or on the 5-HT system.

JACOBS: I was trying to spare the audience and not show all the data, but in fact, the data are all exactly the same. In spite of insulin challenge or glucose administration, you essentially see no change at all in the activity of brain serotonergic neurons but brain noradrenergic neurons are rather responsive to those same manipulations.

B. HOEBEL (*Princeton University, Princeton, NJ*). In the context of this discussion, I'd like to make a comment relating to this question of describing serotonin's functions in terms of arousal vs something more specific. Barry and his group have shown beautifully that serotonin in the brain, particularly in the hypothalamus, is released as a function of arousal corresponding to the firing rate of the cell bodies. We've also been

measuring release, but in the rat, and find that serotonin is released in the hypothalamus when the animal eats. It is even released apparently before the animal eats if we use fluoxetine to boost basal levels and potentiate release so that we can see it better by microdialysis, and serotonin comes out even when the animal sees and smells food before it even eats it. And then when it eats more is released.

So the point I'd like to make is that yes, serotonin is released as a function of arousal but perhaps in various parts of the brain its function can be described much more specifically. In the case of the hypothalamus we determine its function by injecting serotonin to mimic what we see released naturally. And in both the medial and the lateral hypothalamus injections of serotonin suppress ongoing feeding suggesting that one, perhaps just one of its functions there is what is often called satiety. Even though it's released before the animal starts eating it does seem to play a role in suppressing feeding.

J. P. GREEN (*Session Chairman*): Barry do you have a comment? He shrugged his shoulders.

GARY MCCARVER (*UCLA*): I would like to make a comment on Barry Jacobs' beautiful paper which I enjoyed very much, and I appreciate the tremendous difficulties which methodologically they overcame. But he has faced and is facing still in the future tremendous challenges because I can foresee that he probably aims very much at analyzing all, or most, of the prominent representatives of agonist, antagonist and partial agonist for the different receptors subtypes.

Another challenge probably would be to extend these observations to other mammalian species, particularly primates and I'm sure that it's very much in his mind, which would then delineate whether these behavioral responses are really linked in different species to the same firing and inhibitory phenomena of neuronal discharges.

JACOBS: Probably the thing that I'd most like to see is studies like this done in another species, and especially with respect to the data that I showed, these oral-buccal related activities. It would be fascinating to see if they do or don't appear, for example, in rats or in primates, or in any species. It's a real drawback of the whole body of data that it's all derived from the cat.

Imaging the Serotonergic System by Positron Emission Tomography

J. JAMES FROST

Department of Radiology
The Johns Hopkins University
School of Medicine
Baltimore, Maryland 21205

INTRODUCTION

Positron emission tomography (PET) has been developed during the last two decades primarily as a method to quantify regional cerebral blood flow, regional cerebral utilization of oxygen and glucose as well as cerebral blood volume, tissue pH, amino acid utilization, and blood-brain permeability. Metabolic studies of the brain are clearly relevant to psychiatric and neurodegenerative disorders since metabolism of glucose and oxygen are closely related to neuronal activity. However, the study of neuroreceptors and neurotransmitters provides additional chemical information as they play a role in the control of neuronal activity. Measurement of neuroreceptors and their transmitters permits an enormous increase in chemical specificity as compared to metabolism and flow.

Receptor imaging was motivated by new developments in the neurosciences and the demonstration that many centrally active drugs act by stimulating or inhibiting chemical neurotransmission at specific receptor sites. Furthermore, measurements of neuroreceptors in human autopsy tissue from patients with a variety of neurologic and psychiatric disorders have demonstrated changes in the number or affinity of a variety of receptors. For example, schizophrenia is associated with an increased concentration of dopamine D_2 receptors in the caudate nucleus and is treated with a dopaminergic antagonist such as haloperidol.

PET is a sophisticated method of measuring the distribution of positron emitting isotopes in the brain or other body regions. The principle of radioactivity measurement is based on the simultaneous detection of two 511-keV gamma rays emitted 180° apart by a circular array of opposed detectors. The acceptance of only coincident gamma rays by detectors 180° opposed results in an improved spatial localization of the decaying isotope and a simpler correction for gamma ray attenuation than methods utilizing the detection of single gamma rays, i.e., single photon emission computed tomography (SPECT). The existence of positron emitting isotopes of several elements of biological interest makes PET even more suitable for imaging neurochemical systems. These isotopes include ^{11}C ($t_{1/2}$ = 20 min), ^{15}O ($t_{1/2}$ = 2 min), ^{13}N ($t_{1/2}$ = 10 min) and ^{18}F ($t_{1/2}$ = 110 min). For example, oxygen metabolism can be measured with $^{15}O-O_2$ and glucose utilization with ^{18}F-flurodeoxyglucose.

Receptor imaging is possible because of the high sensitivity of PET compared with other imaging modalities. Most neuroreceptors exist in nanomolar concentrations and thus very small quantities of radioactive ligands must be administered to avoid receptor saturation; as receptor saturation increases, the ratio of receptor-to-nonreceptor binding decreases. Ideally, receptor measurements should be performed at occupancies no greater than 10%, which means that subnanomolar concentrations of radioligand must

be detectable. Since millimolar concentrations are typically necessary for magnetic resonance imaging, this method would be inadequate for receptor measurements.

The advantages of PET over a postmortem neurochemical method such as receptor autoradiography suggest several uses for receptor imaging. First, in contrast to receptor autoradiography, PET permits measurements at several times during a patient's illness. Whereas receptor autoradiography is often limited to tissue representing the end stage of a disease, PET can be performed very early, possibly permitting the detection of abnormalities before structural changes are present. It is possible that alterations in specific neuroreceptors on specific neurons could become evident before regional metabolic abnormalities, as all neurons utilize glucose and oxygen. Thus, early diagnosis and identification of patients at risk are possible applications of receptor imaging. Identification of subcategories of clinically homogeneous diseases through neurochemical typing is another goal of receptor quantification. Examples of results from these studies are the elevation of μ-opiate receptors in epilepsy[1] and dopamine D_2 receptors in schizophrenia.[2]

Second, individualized drug therapy may be aided by receptor imaging. If an individual is found to have increased or decreased levels of a neuroreceptor, appropriate treatment would consist of a receptor antagonist or agonist, respectively. Furthermore, the occupancy of receptors from a therapeutic drug may be assessed with PET because of competitive inhibition of the tracer ligand and the drug. This approach could help achieve an optimal therapeutic response while minimizing side effects. For example, if total occupancy of dopamine receptors by haloperidol was reached in a schizophrenic patient without control of symptoms, higher doses would not be warranted because of the risk of tardive dyskinesia.

Third, receptor imaging in the future may also be used for anatomic localization of lesions for surgery, biopsy, radiation therapy, or other local approaches to diagnosis and therapy. The specificity of receptor radiotracers results in high signal-to-noise ratios and high image contrast. Commercially available PET scanners have resolution of 4–5 mm, permitting functional localization not possible a few years ago. For example, functional localization of areas of reduced glucose utilization in temporal lobe epilepsy has been successfully used as a guide to temporal lobe surgery for intractable epilepsy.[3-5] Recently, increases in μ-opiate receptors have been identified in temporal lobe epilepsy, making possible a neurochemical characterization of these patients and permitting new approaches to lesion identification for surgical purposes.[1] Another example is that of brain tumors, in which increased glucose utilization and uptake of methionine has been used to characterize tumors and to direct biopsies to regions of highest metabolism.[6,7]

Altered chemical neurotransmission may result from abnormalities in 1) neurotransmitter synthesis, storage or release; 2) neurotransmitter metabolism and reuptake; 3) neuroreceptor binding; or 4) coupling of neuroreceptors to second messengers. Although this discussion will be limited to recent developments in neuroreceptor imaging, the principles developed from these studies may be applied to imaging other components of neurochemical systems. For example, it now appears promising that serotonin uptake sites will be imaged by PET in the near future.

SEROTONIN RECEPTORS

Over the last decade there has been enormous growth in our knowledge of serotonin receptors and their roles in chemical neurotransmission and cellular function in the central and peripheral nervous systems. On the other hand, our understanding of serotonin receptors in human health and disease is much less well understood.

Accordingly, there is a critical need to perform direct measurements of regional serotonin receptor concentration in a number of neurological and psychiatric disorders. Animal and human studies have suggested a role for serotonin and serotonin receptors in sleep,[8] aggression,[9] pain transmission,[10] Alzheimer's disease,[11-14] normal aging,[15-17] depression,[18-20] and suicide.[12,21,22] The development of serotonin receptor ligands for PET or SPECT imaging has the potential to significantly advance our understanding of the pathophysiology these disorders and improve approaches to treatment.

There currently appear to be about seven different subtypes of serotonin receptors identified by *in vitro* ligand binding studies,[23-25] although this estimate is likely to increase or decrease according to the results presented at this meeting. In spite of the large number of serotonin receptor subtypes which could be imaged, the majority of imaging research has concentrated on the 5-HT_2 receptor. This is largely a result of the several high affinity ligands which exist for 5-HT_2 receptors which may be labeled with positron-emitting isotopes such as ^{11}C or ^{18}F. Other ligand properties which are desirable for *in vivo* imaging studies are a slow rate of metabolism, rapid transport across the blood-brain barrier, slow dissociation from receptors, and a low level of non-specific binding.

To summarize the subsequent discussion, it is clear that none of the ligands evaluated in man, to date, have the properties to permit optimal quantification of 5-HT_2 receptors in man. While there are many ligands which serve the purposes of *in vitro* receptor binding studies, there are many fewer which may be successfully used *in vivo*. In comparison to other receptor systems such as the D_2-dopamine receptor and opiate receptors were very good *in vivo* ligands exist, imaging studies of serotonin receptors could profit enormously by the development of new high affinity ligands.

The ligands which have been successfully used to image 5-HT_2 receptors, to date, have been derivatives of spiperone, ketanserin, and LSD (FIG. 1). These ligands and their analogues are typically first evaluated in mice and then in humans or baboons. A key parameter which is monitored closely is the ratio of total-to-nonspecific binding which is an indication of the signal-to-noise ratio for a specific ligand. This ratio is typically one at early times after injection and thereafter will increase to some maximal value at later times. The summary of such results from a variety of laboratories is presented in TABLE 1 where the maximum ratio of total-to-nonspecific binding (estimated as the ratio of frontal cortex-to-cerebellum binding) is given for mouse and human studies. Although the total-to-nonspecific binding ratio is not the only criteria by which ligand binding *in vivo* is judged, it is important that this ratio be as high as possible (preferably five or greater for human imaging studies).

Although several ligands display high total-to-nonspecific ratios in rodent brain, the corresponding values are considerably lower in human studies. Whether this is due to the lower concentration of serotonin receptors human brain, partial volume effects due to the limited spatial resolution of PET, or differences in metabolism or nonspecific binding remains to be determined. However, it is clear that the *in vivo* total-to-nonspecific binding ratios observed for the highly selective 5-HT_2 receptor ligands are considerably lower than those observed for other receptor systems. Future advances in development of drugs for serotonin receptors should improve this situation.

The first imaging of 5-HT_2 receptors in man was accomplished by the use of ^{11}C-*N*-methylspiperone (^{11}C-NMSP).[26] ^{11}C-NMSP was developed as a high affinity ligand to image dopamine D_2 receptors, but owing to its only fivefold lower affinity for 5-HT_2 receptors,[27] PET images demonstrate a moderate level of binding in the cerebral cortex. Whereas binding to dopamine D_2 receptors in the striatum displayed irreversible binding characteristics, the binding in the cortex was reversible, consistent with the lower affinity binding of ^{11}C-NMSP to 5-HT_2 receptors in cerebral cortex where dopamine D_2 receptors exist in very low concentration. Detailed *in vivo* studies in mice

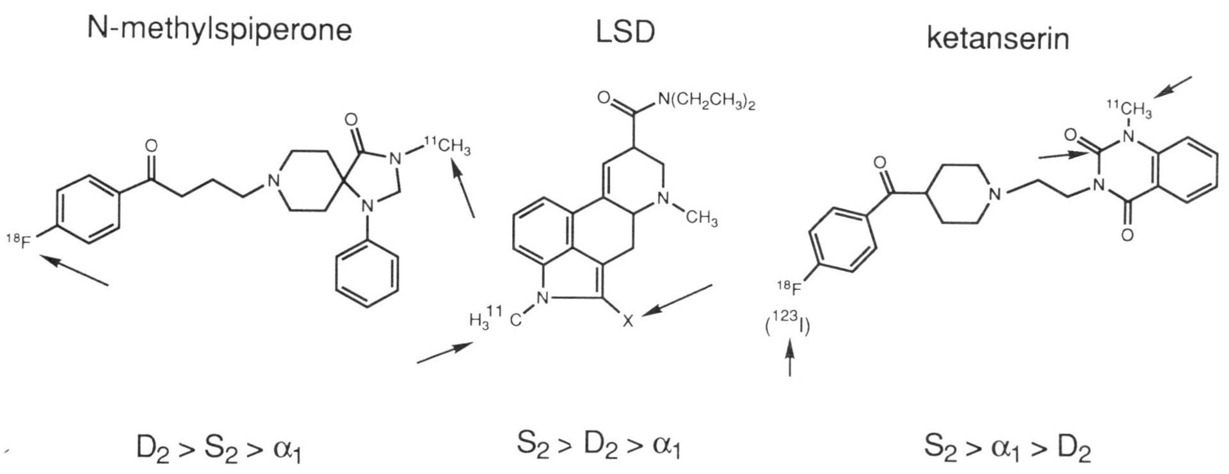

FIGURE 1. The available classes of ligands for imaging 5-HT_2 receptors are based on spiperone, LSD and ketanserin. Various structural analogues (TABLE 1) have been labeled with the positron emitting isotopes ^{11}C and ^{18}F as well as the single photon-emitting isotope ^{123}I. Whereas *N*-methylspiperone has greatest affinity for D_2 dopamine receptors, derivations of LSD and ketanserine have greatest affinity for 5-HT_2 receptors.

further demonstrated that at least 90% of the total binding in the frontal cortex represented labeling of 5-HT_2 receptors.[28] Although 11C-NMSP could be used to localize 5-HT_2 receptors in the cortex, its use was limited in other regions (e.g. striatum) where binding to dopamine D_2 receptors also occurs. A further limitation of ^{11}C-NMSP is the low total-to-nonspecific binding ratio observed in the cortex. Nonetheless, clinical results have been obtained using ^{11}C-NMSP in patients with stroke.[29] Additionally, an age-related decline of cortical 5-HT_2 receptors in male and female subjects has been demonstrated.[30]

Subsequent investigations of ligands for imaging 5-HT_2 receptors have concentrated on ligands with higher selectivity for 5-HT_2 receptors over dopamine D_2, α_1-adrenergic and H_1-histaminergic receptors. ^{11}C-ketanserin and ^{11}C-*N*-methylketanserin (TABLE 1) have subnanomolar affinities for 5-HT_2 receptors and excellent selectivity against dopamine D_2 receptors but ^{11}C-ketanserin suffers from only a fivefold selectivity for 5-HT_2 receptors over α_1-adrenergic or H_1-histaminergic receptors.[31] ^{11}C-*N*-methylketanserin has greater selectivity over α_1-adrenergic receptors.[32] Both ligands have been evaluated in man using PET[32,33] and, although

TABLE 1. *In Vivo* Binding to 5-HT_2 Receptors in Mouse and Human Brain

	Maximum Frontal Cortex/Cerebellum	
	Mouse	Human (young)
^{11}C-NMSP	6.0	1.8
^{11}C-Me ketanserin	3.1	2.1
^{11}C-ketanserin	2.1	1.2
^{11}C-Me Br LSD	5.4	2.3
^{18}F-setoperone	10	3.1
^{18}F-altanserin	11	2.6
^{125}I-I LSD	2.6	—
^{125}I-MeI LSD	3.8	—
^{125}I-EtI LSD	12	—
^{123}I-iodoketanserin	5	—

some degree of selective labeling of cortical regions was observed, (particularly with ^{11}C-*N*-methylketanserin), total-to-nonspecific binding ratios were relatively low.

The second series of compounds for imaging 5-HT_2 receptors is based on LSD (TABLE 1). These compounds tend to have greater selectivity against α-$_1$-adrenergic and H_1-histaminergic receptors as compared to ketanserin, but poorer selectivity against dopamine D_2 receptors.[34–39] Many of the LSD-based ligands display good *in vivo* binding properties in rodent brain and the most promising, ^{11}C-*N*-methyl-2-Br-LSD, (^{11}C-MBL) was used in human PET imaging studies.[39] Nonetheless, the average total-to-nonspecific binding ratio for several human subjects was low.

New ligands continue to be developed and tested in humans (TABLE 1). ^{18}F-setoperone has been evaluated in human subjects and displays a moderately high ratio of total-to-nonspecific binding, although there is still nonsignificant crossbinding with dopamine D_2 receptors.[40] ^{18}F-altanserin[41] has not yet been tested in humans but does display high total-to-nonspecific binding ratios in rodent studies (M. Guillame, unpublished results). The use of ^{18}F($t_{1/2}$ = 110 min) will permit longer imaging times and possibly the observation of larger total-to-nonspecific binding ratios as compared to ^{11}C-labeled ligands. Interestingly, new ligands are also being developed for SPECT. ^{123}I-ketanserin has been recently developed and evaluated in rodents and human subjects and the preliminary results are encouraging.[42]

SEROTONIN UPTAKE SITES

Although the majority of studies which have imaged components of chemical neurotransmission have dealt with postsynaptic neurotransmitter receptors, recent studies indicate that imaging of presynaptic neurotransmitter uptake sites will soon be possible.[43,44] This appears most promising for the serotonergic system since ^{3}H-paroxetine has been shown to selectively label serotonin reuptake sites *in vivo*.[43] The same applications outlined above for imaging serotonin receptors also apply to serotonin uptakes sites. In addition, the evaluation of neurotoxic damage to serotonergic neurons in the CNS of drug abusers who have taken MDMA would be an important application.[45]

^{3}H-paroxetine has been used to label serotonin uptake sites *in vivo* owing to its very high affinity for serotonin uptake sites (0.2nM) and its high selectivity against other uptake and receptor sites. *In vivo* ^{3}H-paroxetine binding corresponds to the regional distribution of serotonin uptake sites known from previous *in vitro* binding studies, is saturable, is stereospecific, and displays appropriate pharmacology. These studies were performed at three hours after injection since at earlier times the binding was predominately nonspecific. Therefore, the use of ^{11}C-paroxetine ($t_{1/2}$ = 20 min) would not be as desirable as paroxetine labeled with a longer lived isotope such as ^{18}F (requiring the study of a new compound as paroxetine itself does not contain fluorine). Another problem with paroxetine is the relatively low maximum ratio of total-to-nonspecific binding (3.3 in hypothalamus). Imaging studies would result in still lower ratios due to partial volume effects although this might be offset to some extent by the higher concentration of serotonin uptake sites in human brain.[46] These limitations notwithstanding, initial baboon imaging studies have been performed with an analogue of paroxetine, ^{11}C-*N*-methylparoxetine, and have provided encouraging initial results.[47] It is clear, though, that many of the current limitations for imaging serotonin receptors exist to a much greater degree for serotonin uptake sites. Future advances will depend on availability of uptake site ligands with higher affinity, lower nonspecific binding and the ability to be labeled with longer lived positron emitting isotopes.

CONCLUSION

Imaging of postsynaptic serotonin receptors and presynaptic uptake sites is an area of active investigation. Further advances in ligand development are still needed and approaches to quantitation using tracer kinetic modeling are at an embryonic stage. Imaging serotonin receptor subtypes other than the 5-HT_2 receptor is essentially an untouched area which should expand rapidly once selective, high affinity ligands become available. The basic science and clinical questions which could be addressed once the imaging methods are perfected are numerous and continue to provide the motivation to improve our ability to noninvasively image and quantify the serotonergic system.

REFERENCES

1. Frost, J. J., H. S. Mayberg, R. S. Fisher, *et al.* 1988. Muopiate receptors measured by positron emission tomography are increased in temporal lobe epilepsy. Ann. Neurol. **23:** 231–237.
2. Wong, D. F., H. N. Wagner, Jr., L. E. Tune, *et al.* 1989. Positron emission tomography reveals elevated D_2 dopamine receptors in drug-naive schizophrenics. Science **234:** 1558–1563.

3. ENGEL, J., JR., D. E. KUHL, M. E. PHELPS & J. C. MAZZIOTTA. 1982. Interictal cerebral glucose metabolism in partial epilepsy and its relation to EEG changes. Ann. Neurol. **12:** 510–517.
4. ENGEL, J., JR., D. E. KUHL, M. E. PHELPS, P. H. CRANDALL. 1982. Comparative localization of epileptic foci in partial epilepsy by PCT and EEG. Ann. Neurol. **12:** 529–537.
5. ENGEL, J., JR., R. F. ACKERMANN, D. E. KUHL & M. E. PHELPS. 1985. Brain imaging of glucose utilization in convulsive disorders. *In* Brain Imaging and Brain Function. L. Sokoloff, Ed. Association for Research in Nervous and Mental Disease, Vol. **63:** 163–184. New York: Raven Press.
6. DICHIRO, G. 1986. Positron emission tomography using [^{18}F] fluorodeoxyglucose in brain tumors: A powerful diagnostic and prognostic tool. Invest. Radiol. **22:** 360–371.
7. PATRONAS, N. J., G. DICHIRO, C. KUFTA, *et al.* 1985. Prediction of survival in glioma patients by means of PET. J. Neurosurg. **62:** 816–822.
8. WESEMANN, W., N. WEINER, M. ROTSCH & E. SCHULTZ. 1983. Serotonin binding in rat brain: Circadian rhythm and effect of sleep deprivation. J. Neural Transm. **18:** 287–294.
9. PUCILOWSKI, O. & W. KOSTOWSKI. 1983. Aggressive behavior and the central serotonergic systems. Behav. Brain Res. **9:** 33–48.
10. MESSING, R. B. & L. D. LYTLE. 1977. Serotonin-containing neurons: their possible role in pain and analgesia. Pain **4:** 1–21.
11. BOWEN, D. M, S. J. ALLEN, J. S. BENTON, M. J. GOODHARDT, E. A. HAAN, A. M. PALMER, N. R. SIMS, C. C. T. SMITH, J. A. SPILLANE, M. M. ESIRI, D. NEARY, J. S. SNOWDON, G. K. WILCOCK & A. N. DAVISON. 1983. Biochemical assessment of serotonergic and cholinergic dysfunction and cerebral in Alzheimer's disease. J. Neurochem. **41**(1):266–272.
12. CROW, T. J., A. J. CROSS, S. J. COOPER, J. F. W. DEAKIN, I. N. FERRIER, J. A. JOHNSON, M. H. JOSEPH, F. OWEN, M. POULTER, R. LOFTHOUSE, J. A. N. CORSELLIS, D. R. CHAMBERS, G. BLESSED, E. K. PERRY, R. H. PERRY & B. E. TOMLINSON. 1984. Neurotransmitter receptors and monoamine metabolities in the brains of patients with Alzheimer-type dementia and depression, and suicides. Neuropharmacology **23**(12B): 1561–1569.
13. MIDDLEMISS, D. N., A. M. PALMER, N. EDEL & D. M. BOWEN. 1986. Binding of the novel serotonin agonist 8-hydroxy-2-(Di-n-propylamino) tetralin in normal and Alzheimer brain. J. Neurochem. **46**(3): 993–996.
14. REYNOLDS, G. P., L. ARNOLD, M. N. ROSSOR, L. L. IVERSEN, C. Q. MOUNTJOY & M. ROTH. 1984. Reduced binding of [^{3}H]ketanserin to cortical 5-HT2 receptor in senile dementia of the Alzheimer type. Neurosci. Lett. **44:** 47–51.
15. MARCUSSON, J., L. ORELAND & B. WINBLAD. 1984. Effects of age on human brain serotonin (S1) binding sites. J. Neurochem. **43:** 1699–1705.
16. MARCUSSON, J., L. ORELAND & B. WINBLAD. 1984. Serotonin-2 binding sites in human frontal cortex and hippocampus. Brain Res. **311:** 51–56.
17. SHIH, J. C. & H. YOUNG. 1978. The alteration of serotonin binding sites in aged human brain. Life Sci. **23:** 1441–1448.
18. FERRIER, I. N., I. G. MCKEITH, A. J. CROSS, E. K. PERRY, J. M. CANDY & R. H. PERRY. 1986. Postmortem neurochemical studies in depression. Ann. NY. Acad. Sci. **487:** 128–142.
19. GROSSER, B. I., R. W. FULLER, F. SULSER, J. K. WAMSLEY & J. MENDELS. 1987. Serotonin in depression: A reevaluation. J. Clin. Psychiatry **48:** 3–33.
20. MENDLEWICZ, J., B. SHOPSIN & H. M. VAN PRAAF. 1984. Serotonin in affective disorders. Adv. Biol. Psychiatry **14:** 1–86.
21. STANLEY, M. & J. J. MANN. 1984. Increased serotonin-2 binding sites in the frontal cortex of suicide victims. Lancet **i:** 214–216.
22. OWEN, F., D. R. CHAMBERS, S. J. COOPER, T. J. CROW, J. A. JOHNSON, R. LOFTHOUSE & M. POULTER. 1986. Serotonergic mechanisms in brains of suicide victims. Brain Res. **362:** 185–188.
23. RICHARDSON, B. P. & G. ENGEL. 1986. The pharmacology and function of $5HT_3$ receptors. Trends Neurosci. **9:** 424–428.
24. MAWE, G. M., T. A. BRANCHEK & M. D. GERSHON. 1986. Peripheral neural serotonin

receptors: Identification and characterization with specific antagonists and agonists. Proc. Natl. Acad. Sci. USA **83:** 9799–9803.

25. HEURING, R. E. & S. J. PEROUTKA. 1987. Characterization of a novel ^{3}H-5-Hydroxytryptamine binding site subtype in bovine membranes. J. Neurosci. **7**(3): 894–903.
26. WAGNER, H. N. JR., H. D. BURNS, R. F. DANNALS, D. F. WONG, B. LANGSTROM, T. DUELFER, J. J. FROST, H. T. RAVERT, J. M. LINKS, S. B. ROSENBLOOM, S. E. LUKAS, A. V. KRAMER & M. J. KUHAR. 1983. Imaging dopamine receptors in the human brain by positron tomography. Science **221:** 1264–1266.
27. LYON, R. A., M. TITELER, J. J. FROST, P. J. WHITEHOUSE, D. F. WONG, H. N. WAGNER, JR., R. F. DANNALS, J. M. LINKS & M. J. KUHAR. 1986. [^{3}H]3-N-Methylspiperone labels D_2 dopamine receptors in basal ganglia and S2 serotonin receptors in cerebral cortex. J. Neurosci. **6:** 2941–2949.
28. FROST, J. J., A. C. SMITH, M. J. KUHAR, R. F. DANNALS & H. N. WAGNER, JR. 1987. In vivo binding of ^{3}H-N-Methylspiperone to dopamine and serotonin receptors. Life Sci. **40:** 987–995.
29. MAYBERG, H. S., R. G. ROBINSON, D. F. WONG, R. PARIKH, P. BOLDUC, T. PRICE, R. F. DANNALS, J. M. LINKS, A. A. WILSON, H. T. RAVERT & H. N. WAGNER, JR. 1988. PET imaging of cortical S2-serotonin receptors following stroke: Lateralized changes and relationships to depression. Am. J. Psychiatry **145**(8): 937–943.
30. WONG, D. F., H. N. WAGNER, JR., R. F. DANNALS, J. M. LINKS, J. J. FROST, H. T. RAVERT, A. A. WILSON, A. E. ROSENBAUM, A. GJEDDE, K. H. DOUGLASS, J. D. PETRONIS, M. F. FOLSTEIN, J. K. T. TOUNG, H. D. BURNS & M. J. KUHAR. 1984. Effects of age on dopamine and serotonin receptors measured by positron emission tomography in the living human brain. Science **226:** 1393–1396.
31. LEYSEN, J. E., C. J. E. NEIMEGEERS, J. M. VAN NUETEN & P. M. LADURON. 1982. [^{3}H]Ketanserin (R41 468), a selective ^{3}H-ligand for serotonin$_2$ receptor binding sites. Mol. Pharmacol. **21:** 301–314.
32. FROST, J. J., R. F. DANNALS, H. S. MAYBERG, J. M. LINKS, H. T. RAVERT, M. J. KUHAR & H. N. WAGNER, JR. 1987. Regional localization of serotonin-2 receptors in man using C-11-N-methylketanserin (NMKET) and PET. J. Nucl. Med. **28**(Suppl): 600.
33. BARON, J. C., Y. SAMSON, D. COMAR, C. CROUZEL, P. DENIKER & Y. AGID. 1985. Etude in vivo des recepterus serotoninergiques centraux chez l'homme par tomographie a positons. Rev. Neurol. **141:** 537–545.
34. FREEDMAN, D. X. & A. E. HALARIS. 1978. Monoamines and the biochemical mode of action of LSD at synapses. *In* Psychopharmacology: A Generation of Progress. M.A. Lipton, A. Dimascio, and K.F. Killam, Eds. New York: Raven Press.
35. ENGEL, G., E. MULLER-SCHWEINITZER & J. M. PALACIOS. 1984. 2-[125Iodo]LSD, a new ligand for the characterization and localization of 5-HT$_2$ receptors. Nauyn-Schmeidebergs Arch. Pharmacol. **325:** 328–336.
36. HARTIG, P. R., M. J. EVANS, A. M. KROHN, S. A. LEDER, P. C. SZE & D. A. STOFFERS. 1985. [125I]LSD binding to serotonin and dopamine receptors in bovine caudate membranes. Neurochem. Int. **7:** 699–707.
37. HOFFMAN, B. J., U. SCHEFFEL, J. R. LEVER, M. D. KARPA & P. R. HARTIG. 1987. N1-Methyl-2-^{125}I-Lysergic acid diethylamide, a preferred ligand for in vitro and in vivo characterization of serotonin receptors. J. Neurochem. **48:** 115–124.
38. HOFFMAN, B., M. KARPA, J. LEVER & P. HARTIG. 1985. N1-methyl-2[^{125}I]LSD ([^{125}I]MIL), a preferred ligand for serotonin 5-Ht$_2$ receptors. Eur. J. Pharmacol. **110:** 147–148.
39. WONG, D. F., J. R. LEVER, P. R. HARTIG, R. F. DANNALS, V. VILLEMAGNE, B. J. HOFFMAN, A. A. WILSON, H. T. RAVERT, J. M. LINKS, U. SCHEFFEL & H. N. WAGNER, JR. 1987. Localization of serotonin 5-HT$_2$ receptors in living human brain by positron emission tomography using N1-([^{11}C]-Methyl)-2-Br-LSD. Synapse **1:** 393–398.
40. BLIN, J., G. SETTE, M. FIORELLI, O. BLETHRY, J. L. ELGHOZI, C. CROUZEL & J. C. BARON. 1989. ^{18}F-Setoperone: A radioligand for serotonin-S$_2$ receptor studies using PET in the human brain. J. Cereb. Blood Flow Metab. **9**(suppl. 1): S113.
41. LEMAIRE, C., R. CANTINEAU, L. CHRISTIAENS & M. GUILLAUME. 1989. N.C.A. radiofluorination of altanserine a potential serotonine receptor-binding radiopharmaceutical for positron emission tomography. J. Labelled Comp & Radiopharm. **XXVI:**336.

42. MERTENS, J., C. C. BOSSUYT-PIRON, M. GUNS, A. BOSSUYT, J. LEYSEN & C. JANSSEN. 1989. High selective serotonin S_2 receptor mapping with SPECT in baboon brain. J. Nucl. Med. **30:** 741.
43. SCHEFFEL, U. & P. R. HARTIG. 1989. In vivo labeling of serotonin uptake sites with [^{3}H] paroxetine. J. Neurochem. **52:** 1605–1612.
44. WOLFE, J., M. S. KREIDER, C. GOODMAN & D. J. BRUNSWICK. 1987. Labeling in vivo of serotonin uptake sites in rat brain after administration of [^{3}H]cyano-imipramine. J. Pharmacol. Exp. Ther. **241:** 196–203.
45. RICAURTE, G. A., L. E. DELANNEY, I. IRWIN & J. W. LANGSTON. 1988. Toxic effects of MDMA on central serotonergic neurons in the primate: Importance of route and frequency of drug administration. Brain Res. **446:** 165–168.
46. LANGER, S. Z., F. JAVOY-AGID, R. RAISMAN, M. BRILEY & Y. AGID. 1981. Distribution of specific high affinity binding sites for [^{3}H]imipramine in human brain. J. Neurochem. **37:** 267–271.
47. VILLEMAGNE, V. Personal communication.

DISCUSSION OF THE PAPER

MANFRED GÖTHERT (*University of Bonn*): I would like to put a question concerning setoperone, the compound that was already introduced yesterday by Dr. Leysen. She already mentioned that this compound has a certain affinity, I think it's even in the low nanomolar range for the two receptors. And now I would like to learn the exact ratio of the affinity's for both 5-HT_2 and D_2 receptors.

FROST: I think that D2 binding is about fivefold lower than that of the S2 binding.

GÖTHERT: So in that case you could have considerable binding also to D2 receptors.

FROST: Yes, well that's what I showed. That's one of the limitations that you saw in the basal ganglia binding. There was significant displacement by sulperide.

GÖTHERT: Yesterday we also learned that this compound is less lipophilic than, for instance, ketanserin. And ketanserin is assumed to act mainly in the periphery. How can this compound enter the CNS if it is less lipophilic?

FROST: It may be less lipophilic but it is lipophilic enough. It certainly does enter the CNS.

JOSEE LEYSEN (*Janssen Research Foundation, Beerse, Belgium*): If you wish I have about 20 slides on the activity of setoperone so I can give a talk on the *in vitro* and *in vivo* binding data on setoperone. In fact the *in vitro* ratio of 5-HT_2 to D_2 is nearly 100 fold when you're measuring *in vitro* KI values. Yet in spite of the fact that setoperone is more hydrophilic than ketanserin, it very easily crosses the blood/brain barrier and it's one of the best tracks for *in vivo* binding.

FROST: I think the point is still that setoperone is not an ideal ligand because it does have significant cross-reactivity with D_2 receptors. So for example it would be very difficult to study 5-HT_2 receptors in the basal ganglia with setoperone. It might be fine for the cortex.

J. PALACIOS (*Sandoz, Ltd., Basel*): Yesterday it was mentioned that the D1 antagonist from Schering is also a serotonin 1_C ligand. And I understand that derivatives from this Schering antagonist have been used in positron emission tomography in humans. Do you know if they reveal any 1_C binding?

FROST: That's a good question. There is certainly avid uptake in the striata and there's also cortical binding. I think, in the past people have attributed that to D1 receptors in the cortex. But I guess now it remains to be shown whether that's in fact serotonin binding, if that's the point of your question.

PALACIOS: Yes. Also, can the choroid plexus be seen with your imaging system? Could you see the choroid plexus, if there is a high uptake?

FROST: We haven't actually done but a few of the D_1 studies at Hopkins, so I haven't really had a chance to look at those images in detail. Certainly the choroid plexus is of the size that it could be resolved by our PET system.

Peripheral Indices of Central Serotonin Function in Humans

DENNIS L. MURPHY[a]

Laboratory of Clinical Science
National Institute of Mental Health
Bethesda, Maryland 20892

INTRODUCTION

Serotonin (5-HT), its metabolites, and some of its metabolic processes which are accessible outside of brain or spinal cord tissue have been evaluated as potential peripheral indices of the functional status of central nervous system (CNS) 5-HT subsystems in humans. 5-HT and its major metabolite, 5-hydroxyindoleacetic acid (5-HIAA), have been quantitated in cerebrospinal fluid (CSF), other bodily fluids, and 5-HT-containing cells such as blood platelets.

The cellular processes by which 5-HT is concentrated, stored, released, and metabolized have also been studied in human platelets. Kinetic and other characteristics of these mechanisms, including the 5-HT transport system and the closely related imipramine binding site, the 5-HT storage vesicle, the 5-HT release reaction, and the degradative enzyme, monoamine oxidase (MAO), have been comparatively examined in human platelets vs. various brain preparations from humans, rodents, and other species.[1] Blood platelets also possess 5-HT_2 receptors which, when activated by 5-HT, produce a platelet shape change and partial platelet aggregation. These effects are mediated by the same signal transduction system used in brain, phosphoinositide hydrolysis.

In addition to direct measurements in the periphery of 5-HT and its metabolites, and in platelets of 5-HT uptake, release, metabolism, and receptor-mediated processes, some indirect approaches have been used which depend upon the peripheral administration of 5-HT-selective drugs or radiolabeled ligands for 5-HT binding sites to assess the functional status of brain 5-HT subsystems.[2,3] Neuroimaging of radiolabeled ligands (e.g., a ^{11}C-LSD derivative for 5-HT_2 receptor sites)[4] requires positron emission tomography (PET) or single photon emission tomography techniques; alternatively 5-HT-selective agonists are used to produce changes in ^{18}F-deoxyglucose metabolism which are examined by PET techniques[5]; these neuroimaging strategies are discussed in another chapter (Frost, J. J., this volume). More frequently, 5-HT-selective agonists or antagonists have been administered intravenously or orally to elicit CNS 5-HT-mediated changes which are reflected in behavioral or physiological alterations. For example, 5-HT receptors appear to mediate a number of hypothalamic-pituitary responses leading to elevated concentrations of plasma prolactin, adrenocorticotropin (ACTH), and cortisol. In addition, temperature, cardiovascular, sleep, behavioral, and other physiological changes appear to be modulated, in part, by brain 5-HT subsystems in humans.

Some of these peripherally measured responses to selective 5-HT agonists or antagonists have provided useful indices of apparent changes in the function of human

[a]Mailing address: Dr. Dennis L. Murphy, Chief, LCS, NIMH, NIH Clinical Center, 10-3D41, 9000 Rockville Pike, Bethesda, MD 20892; (301)-496-3421.

brain 5-HT subsystems which parallel responses produced by some drugs (e.g., the tricyclic antidepressants) in animal studies. Such responses have also been used to examine hypotheses regarding functional 5-HT neuronal changes suggested by postmortem studies or by other evidence from studies of physiological processes like aging or of neuropsychiatric disorders like dementia of the Alzheimer type or obsessive-compulsive disorder.

This chapter briefly catalogs some examples of the most well-studied peripheral indices of central 5-HT function in humans. It emphasizes, where possible, the indices for which there is evidence indicating parallels between the peripheral indices and direct measurements in human, nonhuman primate, or rodent brain. Some newer, promising approaches are also noted, as are some instances where dissimilarities exist between peripheral indices and human brain measurements or processes. The use of many of these indices is presented in several other chapters in this volume dealing with 5-HT function in various human physiological processes such as sleep, appetite, and circadian function and in neuropsychiatric disorders ranging from pain and anxiety to migraine and depression.

DIRECT MEASUREMENTS OF 5-HT AND ITS METABOLITES IN CSF, PLASMA, PLATELETS, AND URINE

Although there have been multiple reports of measurements of 5-HT content in human CSF, values obtained have ranged from 0.064 to 12.7 ng/ml, according to a recent review of 25 studies.[6] Using more specific assay techniques, two reports published in 1989 found only very low concentrations of 5-HT in lumbar CSF,[6,7] with one reporting a median value of 16 pg/ml obtained from over 100 subjects, many of whom had nondetectable values. Thus, with the possible exception of situations such as treatment with MAO inhibitors alone or in combination with tryptophan, which cause markedly increased CSF 5-HT concentrations in rodents and nonhuman primates,[2,8] measurement of human lumbar CSF content seems unlikely to provide a practicable index of brain 5-HT.

CSF 5-HIAA has been examined as a possible index of CSN 5-HT turnover in many clinical studies. Many separate metabolic steps, however, are involved in the formation, storage, and release of 5-HT and 5-HIAA. Dietary changes in carbohydrate intake or in the 5-HT precursor, L-tryptophan, or in neutral amino acids which compete with L-tryptophan for transport into brain, can change brain 5-HT and 5-HIAA concentrations. In addition, there is considerable evidence for functional and nonfunctionally related pools of 5-HT (and, correspondingly, of 5-HIAA), as exemplified in a recent study that found minimal association between 5-HT and 5-HIAA changes and alterations in the firing rates of 5-HT-containing dorsal raphe neurons.[9]

As expected from animal studies, 5-HIAA measured in human CSF has been shown to be sensitive to drugs affecting 5-HT synthesis (L-tryptophan, para-chlorophenylalanine), release (fenfluramine), uptake (tricyclics), and metabolism (MAO inhibitors).[10] Some examples validating substantially greater CSF 5-HIAA changes than other monoamine metabolite changes after treatment with 5-HT-selective agents are presented in TABLE 1. Such changes can be helpful in assessing drug effects, but at the same time exemplify that 5-HIAA or 5-HT differences found in some neuropsychiatric disorders may reflect the summed effects of many processes.[16,17] It is possible, for instance, that altered CSF 5-HIAA might reflect a gross abnormality in 5-HT synthesis, storage, transport, or a change in the total number of 5-HT neurons. Clearly, functional consequences cannot readily be inferred from CSF 5-HIAA values alone, and other studies are needed to identify what processes are involved in any 5-HIAA difference.

HUMAN BLOOD PLATELET 5-HT CONTENT, RELEASE, AND METABOLISM

A series of papers have reviewed the extent that platelets, which unlike other blood cells, possess specialized processes for the uptake, storage, and endocytotic release of 5-HT, can serve as models for 5-HT neurons in other tissues.[1,18,19] Only a few points of special interest that relate to the validity of measurements of platelet 5-HT content, transport, and metabolism as indices of brain serotonergic function can be summarized in this review.

Platelets do not synthesize 5-HT, but rather can rapidly accumulate 5-HT via a

TABLE 1. Differential Effects of Some 5-HT-Selective Uptake-Inhibiting Antidepressants and Some Nonselective Antidepressants on Monoamine Metabolites in CSF[a]

Antidepressant Drugs	Percent Change		
	5-HIAA	MHPG	5-HIAA/MHPG
5-HT Selective Agents			
Clomipramine	−48	−33	1.46
Zimelidine	−38	−20	1.95
Citalopram	−28	−11	2.55
Fluvoxamine	−46	(NS)	—
Means	−40	−21	1.99
Other Agents			
Nortriptyline	−20	−36	0.56
Imipramine	−35	−39	0.90
Amitriptyline	−36	−41	0.88
Clorgyline	−45	−87	0.52
Means	−34	−51	0.73

[a]The data in this table were collated from references[11-15]; changes in homovanillic acid were negligible in all these studies; MHPG, 3-methoxy-4-hydroxyphenylglycol.

sodium-dependent transport process which, because of effective vesicular and cytoplasmic storage processes, can achieve intracellular to extracellular concentration gradients of over 1000:1, with estimated concentrations of 5-HT within platelet storage vesicles of 0.6 M.[20,21] Platelet 5-HT content is a very sensitive index of the effects of 5-HT-selective uptake inhibitors, as indicated in TABLE 2, and there are instances in which the magnitude of clinical changes has correlated positively with the magnitude of this drug effect.[22]

Platelets can release 5-HT both via a calcium-dependent exocytotic mechanism accompanied by a proportionate disappearance of 5-HT storage vesicles, adenosine triphosphate, and calcium,[26] and also by a nonexocytotic process as in response to reserpine.[27,28]

5-HT is a poor substrate for the form of its major degradatory enzyme, MAO type B, which is present both within platelets[29] and within 5-HT cell bodies in the brain.[30,31] 5-HT accumulation within platelets is not affected by selective inhibition of MAO-B by deprenyl,[32] although increased 5-HT accumulation within platelets follows drug-

TABLE 2. Effects of Clomipramine vs. Fluoxetine on Platelet 5-HT Content

	Clomipramine[22,23]	Fluoxetine[24]
Platelet 5-HT Content	2–4% of Control	20% of Control
Drug Dose	140,278 mg/day	30 mg/day
Plasma Concentrations of Drug (ng/ml)	CMI: 81,176 ng/ml DCMI: 208,414 ng/ml	FLX: 80ng/ml NORFLX: 150 ng/ml
5-HT Uptake Inhibiting Potency (IC_{50}, nM)[25]	1.5 nM	6.9 nM

CMI: clomipramine; DCMI: desmethylclomipramine; FLX: fluoxetine; NORFLX: norfluoxetine.

induced MAO-A inhibition or a lack of MAO-A and MAO-B activity in individuals lacking the genes for these enzymes.[33]

An important consideration in evaluating smaller differences in platelet 5-HT content, MAO-B activity, 5-HT uptake, imipramine binding sites, and other platelet characteristics is that the circulating platelet pool is heterogeneous. There is considerable evidence that platelet release from the bone marrow may be stimulated by a number of events, including stress-related factors, and that newly released platelets are denser, larger, and contain up to twofold greater numbers of platelet storage vesicles, enzymes, and other constituents (TABLE 3). While there is evidence from monozygotic vs. dizygotic twin studies that some platelet characteristics are stable enough to detect genetic influences[37,38] and that some drug effects such as those summarized in TABLE 2 or gene deletion effects[33] are large enough to permit valid *in vivo* extrapolation from platelet measurements to brain and other tissues, considerable care is required in the interpretation of smaller differences such as those not infrequently observed in comparisons of neuropsychiatric patients vs. healthy controls.

5-HT UPTAKE AND THE IMIPRAMINE BINDING SITE IN HUMAN BLOOD PLATELETS

The 5-HT uptake process in human platelets appears nearly identical to that present in rodent and human brain 5-HT terminals; a few minor differences in affinity

TABLE 3. Heterogeneity in Blood Platelet Subpopulations: Density-Dependent Differences in Subcellular Constituents

Platelet Measures[34–36]	High Density Platelets	Lower Density Platelets
Density (g/ml)[34]	≤1.084	≤1.062***
Volume (fL)[35]	7.54	6.63**
5-HT Content (ng/mg protein)[34]	460	220***
MAO Activity (nmol/mg protein/hr)[34]	64	48**
5-HT uptake (mol/10^{-21} platelets/10s)[34]	1.61	0.98**
^{3}H-imipramine binding, B_{max} (fmol/mg protein)[36]	1010	734**

$^{**}P < 0.01$; $^{***}p < 0.001$ for mean differences between high vs. lower density platelet populations.

for tryptamine and several tryptolines vs. 5-HT for the platelet vs. brain transport process have been noted in a recent review.[39] Very high correlations ($r = 0.99$) between the inhibitory potencies of a series of drugs including 5-HT-selective and nonselective tricyclics and other drugs have been observed in many studies comparing 5-HT uptake in human platelets vs. rat and human brain synaptosomes.[1,39]

Similarly, high correlations have been found between inhibitory potencies for 5-HT uptake and the binding affinities of various radiolabeled tricyclics including ^{3}H-imipramine, ^{3}H-paroxetine, ^{3}H-cyanoimipramine, and other ligands.[1,39] The tricyclic binding site and the 5-HT transporter are closely related, and while the tricyclic site had for some time been regarded as a closely localized modulatory site, recent evidence from studies with ^{3}H-paroxetine and ^{3}H-cyanoimipramine indicate that these agents most likely label the 5-HT transporter itself, providing a substantial technical advantage for autoradiographic studies of this component of 5-HT neurons in brain.[39–42] In at least one study, chronic antidepressant treatment led to a reduced number of ^{3}H-imipramine binding sites in platelets, a finding similar to that observed in many similar studies of rodents treated chronically with tricyclic antidepressants.[43]

THE 5-HT$_2$ RECEPTOR IN HUMAN BLOOD PLATELETS

Multiple ligands, including some with very high selectivity for 5-HT$_2$ sites in human brains and other tissues (e.g., ^{3}H-ketanserin, ^{3}H-spiperone) bind with similar nanomolar affinities to a 5-HT$_2$ site on human platelets.[44,45] Very high correlations (e.g., $r = 0.97$) have been reported when inhibitory potencies of a series of drugs have been compared between human platelets and human or rodent cortex membrane preparations.[46,47] The 5-HT$_2$ binding site in platelets is a component of a functional receptor, as it mediates a platelet shape change response associated with partial platelet aggregation; moreover, 5-HT$_2$ binding site numbers and functional responses are fairly highly correlated in individual subjects.[44,48,49] Like the brain 5-HT$_2$ receptor system, the platelet 5-HT$_2$ receptor utilizes phosphoinositide hydrolysis as a second messenger system.[50] In one study, long-term tricyclic antidepressant treatment was associated with an increase in 5-HT$_2$ sites labeled with ^{125}I-LSD, unlike the reduced numbers of 5-HT$_2$ sites observed in over ten similar chronic tricyclic treatment studies carried out in rodents.[51]

Although several studies using ^{3}H-5-HT and other ligands for 5-HT$_1$ subtypes failed to identify any site besides the 5-HT$_2$ site,[52] one very recent report using ^{3}H-SCH 23390 described a novel site, tentatively labeled 5-HT$_{1E}$, which had 5-HT binding site characteristics.[53] ^{3}H-SCH 23390 was competitively displaced by nanomolar concentrations of 5-HT and RU24969. A site with similar characteristics was not identified in some initial studies of several brain regions.[53]

INDIRECT APPROACHES TO ASSESS CENTRAL 5-HT FUNCTION USING HORMONAL RESPONSES TO 5-HT PRECURSORS, SELECTIVE 5-HT AGONISTS, AND OTHER DRUGS THAT AFFECT 5-HT IN HUMANS

Several neuroendocrine responses to the 5-HT precursors, L-tryptophan and 5-hydroxytryptophan (5-HTP), or to drugs with selective serotonergic actions have been studied to evaluate functional components of the serotonergic systems in humans. Changes in plasma prolactin, cortisol, or ACTH concentrations have been used most frequently as end points which have been suggested to provide "probes" of central 5-HT function,[2,3,54–56] although this approach has been questioned.[57]

The 5-HT precursors, L-tryptophan and 5-HT (administered with a peripheral decarboxylase inhibitor), reliably increase plasma prolactin and growth hormone when administered intravenously; oral administration of either L-tryptophan or 5-HTP has generally not provided an effective stimulus.[58–61] Changes in cortisol and ACTH have also been evaluated in these studies but the results obtained have been highly variable.

A number of selective 5-HT agonists and 5-HT-releasing or uptake-inhibiting drugs reliably increase plasma prolactin, cortisol, and ACTH concentrations, while these agents less uniformly affect growth hormone responses (TABLE 4).

EFFECTS OF 5-HT ANTAGONISTS AND OTHER DRUGS ON HORMONAL RESPONSES TO 5-HT-SELECTIVE AGENTS

The 5-HT antagonists, metergoline, cyproheptidine, and methysergide, consistently block prolactin and cortisol responses to L-tryptophan, fenfluramine, and

TABLE 4. Effects of 5-HT-Selective Agonists and Other 5-HT-Related Drugs on Neuroendocrine Measures in Humans[a]

	Neuroendocrine Response: Plasma Hormone Changes		
Agent	Prolactin	Cortisol	Growth Hormone
m-Chlorophenyl-piperazine (oral)	Increased	Increased	No change
m-Chlorophenyl-piperazine (iv)	Increased	Increased	Increased
MK 212	Increased	Increased	No change
Quipazine	No change	Increased	No change
Buspirone	Increased		Increased
N,N-Dimethyltryptamine	Increased		Increased
Fenfluramine	Increased	Increased	No change
Clomipramine	Increased		Variable
Imipramine	Increased	Increased	Increased

[a]The data in this table have been collated from references 62–74.

m-chlorophenylpiperazine (*m*-CPP) (TABLE 5). It seems likely that, as has been demonstrated in rodents and nonhuman primates, these $5\text{-HT}_1/5\text{-HT}_2$ nonselective antagonists are acting at 5-HT_1 sites, as the selective 5-HT_2 antagonists, ketanserin or ritanserin, had either no blocking effects or resulted in enhanced prolactin responses to L-tryptophan.[78,81,82]

Prolactin, cortisol, and growth hormone responses to L-tryptophan, 5-HTP, or fenfluramine were enhanced by pretreatment with a number of drugs expected to increase the synaptic availability of 5-HT, including lithium and clomipramine.[84–88] In contrast, prolactin or cortisol responses to the direct receptor agonist, *m*-CPP, were either unchanged or decreased during chronic clomipramine treatment.[89] The prolactin response to L-tryptophan was also decreased after acute but not chronic diazepam administration.[90] Similarly, the enhanced prolactin response to L-tryptophan observed after single dose administration of ritanserin disappeared when ritanserin was given for two weeks.[82]

As there is abundant evidence implicating brain and peripheral 5-HT in appetite regulation and food intake, it is of interest that two studies in humans showed changes in response to serotonergic agents following dietary manipulations. In one study, consumption of a diet with markedly reduced tryptophan content enhanced prolactin responses to L-tryptophan.[91] Similarly, reduced food intake (a 1200 calorie/day diet) also led to increased prolactin and growth hormone responses to L-tryptophan in healthy volunteers.[92] An earlier study has demonstrated associations between CSF 5-HIAA concentrations and changes in plasma and CSF tryptophan.[93]

OTHER INDIRECT APPROACHES TO ASSESS CENTRAL 5-HT FUNCTION IN HUMANS

Other physiological and behavioral responses to 5-HT-selective agents have occasionally been studied in humans as potential end points to assess the functional status of brain serotonergic subsystems. However, in only a few instances have comparisons been made among different agents using the same paradigm. Likewise, there are only a few examples where antagonist blockade or changes in response to chronic treatment with agents known to modulate such responses in animals have been examined. A few such examples are deserving of comment, nonetheless.

TABLE 5. Effects of 5-HT Antagonists on Agonist-Induced Neuroendocrine Responses in Humans [a]

	Neuroendocrine Response: Plasma Hormone Changes		
Antagonist Agent	Prolactin	Cortisol	Growth Hormone
Metergoline			
L-Tryptophan	Decreased		No change
Fenfluramine	Decreased		
m-CPP	Decreased	Decreased	—
Cyproheptidine			
L-Tryptophan	Decreased		Decreased
5-HTP		Decreased	Decreased
Fenfluramine		Decreased	
Methysergide			
L-Tryptophan	Decreased		
Mianserin	No change		
Ritanserin			
L-Tryptophan	Enhanced		No change
Ketanserin			
L-Tryptophan	No change		

[a]The data in this table have been collated from references 58, 63, 66, and 75–83.

Body temperature is highly reliably elevated by the 5-HT agonist, *m*-chlorophenylpiperazine (*m*-CPP), in humans, monkeys, and rodents[70,74,94,95]; this response is blocked by the $5\text{-HT}_1/5\text{-HT}_2$ antagonist, metergoline, in all three species, as well as by other selective 5-HT antagonists in rodents.[77,94,95] Furthermore, chronic clomipramine administration in humans and rodents leads to a diminished ("down-regulated") hyperthermic response to *m*-CPP.[89,95] It is of note that highly selective agonists at other

5-HT sites, specifically 8-hydroxy-2(di-*n*-propylamino)tetralin (8-OH-DPAT) acting at 5-HT_{1A} sites in rodents, produce hypothermic rather than hyperthermic effects, which are blocked by different classes of antagonists,[96] indicating functional subsystem heterogeneity within brain serotonergic pathways. Further evidence of subsystem heterogeneity in humans comes from the observations that temperature and behavioral subsensitivity to repeat *m*-CPP challenges followed chronic clomipramine administration in one study; however, neuroendocrine (plasma prolactin and cortisol) increases were unchanged in magnitude by chronic clomipramine treatment.[89]

Sleep responses to selective serotonergic agents including L-tryptophan, fenfluramine, trazodone, fluoxetine, zimelidine, indalpine, ritanserin, and *m*-CPP have been evaluated and, in some cases, contrasted with agents with selective effects on norepinephrine and other neurotransmitter systems, as reviewed in another chapter (A. Wauquier & C. Dugovic, this volume). Sleep impairment with *m*-CPP found in healthy human volunteers[97] was also observed in rodents, who exhibited rapid eye movement (REM) suppression,[98] in contrast to the slow-wave sleep enhancing effects of the 5-HT_2 antagonist, ritanserin, again indicating 5-HT subsystem heterogeneity in humans which deserves further exploration.

Behavioral, cognitive, locomotor, cardiovascular, and other effects of selective 5-HT agents also represent very interesting measures which have only recently begun to be explored in detail in humans. Again, only brief mention can be made of the contrasting anxiogenic effects of some agents such as *m*-CPP[71,77,99] with the anxiolytic actions of drugs acting selectivity at other 5-HT receptors such as the partial agonists at 5-HT_{1A} sites, buspirone, gepirone, and ipsapirone,[100] and some of the 5-HT_2 antagonists such as ritanserin.[101] Diminution in anxiogenic responses to *m*-CPP with chronic clomipramine administration have been observed,[89] as well as evidence for a decline with age in the behavioral responses to *m*-CPP.[102] Although not a practicable end point for studies in humans, it is of note that attention was first drawn to the fact that the so-called "5-HT syndrome" of behaviors which has been extensively studied in rodents[103] may have a counterpart in humans who received clomipramine while residual effects of MAO-inhibitor treatment were still present.[104] Similar phenomena were reported when tryptophan was co-administered with MAO-inhibitors.[105–108]

REFERENCES

1. Da Prada, M., A. M. Cesura, J. M. Launay & J. G. Richard. 1988. Platelets as a model for neurons? Experientia **44:** 115–126.
2. Murphy, D. L., E. A. Mueller, N. A. Garrick & C. S. Aulakh. 1986. Use of serotonergic agents in the clinical assessment of central serotonin function. J. Clin. Psychiatry **47** (Suppl. 4): 9–15.
3. Murphy, D. L., A. M. Mellow, T. Sunderland, C. S. Aulakh, B. A. Lawlor & J. Zohar. 1989. Strategies for the study of serotonin in humans. *In* Serotonin in Major Psychiatric Disorders. E. F. Coccaro & D. L. Murphy, Eds. American Psychiatric Press. Washington, DC. In press.
4. Wong, D. F., J. R. Lever, P. R. Hartig, R. F. Dannals, V. Villemagne, B. J. Hoffman, A. A. Wilson, H. T. Ravert, J. M. Links, U. Scheffel & H. N. Wagner. 1987. Localization of serotonin 5-HT_2 receptors in living human brain by positron emission tomography using N1-([^{11}C]-methyl)-2-Br-LSD. Synapse **1:** 393–398.
5. Grome, J. J. & A. M. Harper. 1986. Local cerebral glucose utilisation following indoleamine- and piperazine-containing 5-hydroxytryptamine agonists. J. Neurochem. **46:** 117–124.
6. Anderson, G. M., I. N. Mefford, T. J. Tolliver, M. A. Riddle, D. M. Ocame, J. F. Leckman & D. J. Cohen. 1990. Serotonin in human lumbar cerebrospinal fluid: A reaassessment. Life Sci. **46:** 247–255.

7. TOLLIVER, T. J., S. J. HUANG, I. N. MEFFORD, N. A. GARRICK & D. L. MURPHY. 1988. Sensitive HPLC method for serotonin in cerebrospinal fluid. Soc. Neurosci. **14:** 212.
8. ANDERSON, G. M., K. L. TEFF & S. N. YOUNG. 1987. Serotonin in cisternal cerebrospinal fluid of the rat: Measurement and use as an index of functionally active serotonin. Life Sci. **40:** 2253–2260.
9. TRULSON, M. E. 1985. Dietary tryptophan does not alter the function of brain serotonin neurons. Life Sci. **37:** 1067–1072.
10. POST, R. M. & D. L. MURPHY. 1979. Research planning: Studies of cerebrospinal fluid metabolites and blood cell enzymes. *In* Handbook of Biological Psychiatry. H. M. van Praag, Ed. New York: Marcel Dekker, 51–84.
11. MAJOR, L. F., C. R. LAKE, S. LIPPER, P. LERNER & D. L. MURPHY. 1979. The central noradrenergic system and affective response to MAO inhibitors. Prog. Neuropsychopharmacol. **3:** 535–542.
12. BERTILSSON, L. & M. ASBERG. 1984. Amine metabolites in the cerebrospinal fluid as a measure of central neurotransmitter function: Methodological aspects. Adv. Biochem. Psychopharmacol. **39:** 27–34.
13. BOWDEN, C. L., S. H. KOWLOW, I. HANIN, J. W. MAAS, J. M. DAVIS & E. ROBINS. 1985. Effects of amitriptyline and imipramine on brain amine neurotransmitter metabolites in cerebrospinal fluid. Clin. Pharmacol. Ther. **37:** 316–324.
14. BJERKENSTEDT, L., L. FLYCKT, K. F. OVER & O. LINGJAERDE. 1985. Relationship between clinical effects, serum drug concentration and serotonin uptake inhibition in depressed patients treated with citalopram: A double-blind comparison of three dose levels. Eur. J. Pharmacol. **28:** 553–557.
15. MARTIN, P. R., B. ADINOFF, G. A. H. BONE, J. M. STAPLETON, M. J. ECKARDT & M. LINNOILA. 1987. Fluvoxamine treatment of alchoholic chronic organic brain syndromes. Clin. Pharmacol. Ther. **41:** 211 (abstract).
16. KUHN, D. M., W. A. WOLF & M. B. H. YOUDIM. 1986. Serotonin neurochemistry revisited: A new look at some old axioms. Neurochem. Int. **8:** 141–154.
17. MURPHY, D. L. 1986. Serotonin neurochemistry: A commentary on some of its quandaries. Neurochem. Int. **8:** 161–163.
18. PLETSCHER, A. & A. LAUBSCHER. 1980. Blood platelets as models for neurons: Use and limitations. J. Neural Transm. **16** (Suppl.): 7–16.
19. STAHL, S. M. 1985. Platelets as pharmacological models for the receptors and biochemistry of monoaminergic neurons. *In* The Platelets: Physiology and Pharmacology. G. L. Longenecker, Ed. New York: Academic Press, pp. 307–340.
20. COSTA, J. L., T. S. REESE & D. L. MURPHY. 1974. Serotonin storage in platelets: Estimation of storage-packet size. Science **183:** 537–538.
21. COSTA, J. L., H. STARK, B. SHAFER, L. CORASH, M. A. SMITH & D. L. MURPHY. 1978. Maximal packet size for serotonin in storage vesicles of intact human platelets. Life Sci. **23:** 2193–2197.
22. FLAMENT, M. F., J. L. RAPOPORT, D. L. MURPHY, C. J. BERG & C. R. LAKE. 1987. Biochemical changes during clomipramine treatment of childhood obsessive-compulsive disorder. Arch. Gen. Psychiatry **44:** 219–225.
23. ZOHAR, J., D. L. MURPHY, R. C. ZOHAR-KADOUCH, M. A. T. PATO, K. M. WOZNIAK & T. R. INSEL. 1989. Serotonin and obsessive-compulsive disorders. *In* Serotonin in Major Psychiatric Disorders. E. F. Coccaro & D. L. Murphy, Eds. American Psychiatric Press. Washington, DC. In press.
24. LEMBERGER, L., R. F. BERGSTROM, R. L. WOLEN, N. A. FARID, G. G. ENAS & G. R. ARONOFF. 1985. Fluoxetine: Clinical pharmacology and physiologic disposition. J. Clin. Psychiatry **46:** 14–19.
25. HYTTEL, J. & J. J. LARSEN. 1985. Serotonin-selective antidepressants. Acta Pharmacol. Toxicol. **56** (Suppl.1): 146–153.
26. COSTA, J. L., T. C. DETWILER, R. D. FEINMAN, D. L. MURPHY, C. S. PATLAK & K. D. PETTIGREW. 1977. Quantitative evaluation of the loss of human platelet dense bodies following stimulation by thrombin or A23187. J. Physiol. **264:** 297–306.

27. COSTA, J. L., S. A. SILBER & D. L. MURPHY. 1977. Effects of reserpine and imipramine on vesicular serotonin uptake and storage in intact human platelets. Life Sci. **21:** 181–188.
28. COSTA, J. L., K. D. PETTIGREW & D. L. MURPHY. 1979. Electron probe microanalysis of changes in dense-body phosphorus and calcium content following alterations in platelet serotonin levels. Biochem. Pharmacol. **28:** 23–26.
29. DONNELLY, C. H. & D. L. MURPHY. 1977. Substrate- and inhibitor-related characteristics of human platelet monoamine oxidase. Biochem. Pharmacol. **26:** 853–858.
30. WESTLUND, K. N., R. M. DENNEY, L. M. KOCHERSPERGER, R. M. ROSE & C. W. ABELL. 1985. Distinct monoamine oxidase A and B populations in primate brain. Science **230:** 181–183.
31. GARRICK, N. A. & D. L. MURPHY. 1982. Monoamine oxidase type A: Differences in selectivity towards l-norepinephrine compared to serotonin. Biochem. Pharmacol. **31:** 4061–4066.
32. COSTA, J. L., D. L. MURPHY & J. REVEILLE. 1977. Evaluation of the uptake of various amines into storage vesicles of intact human platelets. Br. J. Pharmacol. **61:** 223–228.
33. MURPHY, D. L., K. B. SIMS, F. KAROUM, A. DE LA CHAPELLE, R. NORIO, E.-M. SANKILA & X. O. BREAKEFIELD. 1990. Marked amine and amine metabolite changes in Norrie disease patients with an X chromosomal deletion affecting monoamine oxidase. J. Neurochem. **54:** 242–247.
34. CORASH, L., J. L. COSTA, B. SHAFER, J. A. DONLON & D. MURPHY. 1984. Heterogeneity of human whole blood platelet subpopulations: III Density-dependent differences in subcellular constituents. Blood **64:** 185–193.
35. CHAMBERLAIN, K. G., M. TONG, E. CHIU & D. G. PENINGTON. 1989. The relationship of human platelet density to platelet age: Platelet population labeling by monoamine oxidase inhibition. Blood **73:** 1218–1225.
36. ARORA, R. C. & H. Y. MELTZER. 1984. Imipramine binding in subpopulations of normal human blood platelets. Biol. Psychiatry **19:** 257–262.
37. NIES, A., D. S. ROBINSON, K. R. LAMBORN & R. P. LAMPERT. 1973. Genetic control of platelet and plasma monoamine oxidase activity. Arch. Gen. Psychiatry **28:** 834–838.
38. MURPHY, D. L. & J. L. COSTA. 1975. Utilization of cellular studies of neurotransmitter-related enzymes and transport processes in man for the investigation of biological factors in behavioral disorders. *In* The Psychobiology of Depression. J. Mendels, Ed. New York: Spectrum Publications, Inc. pp. 27–46.
39. GRAHAM, D. & S. Z. LANGER. 1988. The neuronal sodium-dependent serotonin transporter: Studies with [^{3}H]imipramine and [^{3}H]paroxetine. *In* Neuronal Serotonin. N. N. Osborne & M. Hamon, Eds. Chichester, England: John Wiley & Sons Ltd., pp. 367–391.
40. MARCUSSON, J. O., M. BERGSTROM, K. ERIKSSON & S. B. ROSS. 1988. Characterization of [^{3}H]paraxetine binding in rat brain. J. Neurochem. **50:** 1783–1790.
41. KOVACHICH, G. B., C. E. ARONSON, D. J. BRUNSWICK & A. FRAZER. 1988. Quantitative autoradiography of serotonin uptake sites in rat brain using [^{3}H]cyanoimipramine. Brain Res. **454:** 78–88.
42. CORTES, R., E. SORIANO, A. PAZOS, A. PROBST & J. M. PALACIOS. 1988. Autoradiography of antidepressant binding sites in the human brain: Localization using [^{3}H]imipramine and [^{3}H]paroxetine. Neuroscience **27:** 473–496.
43. POIRIER, M. F., H. LOO, C. BENKELFAT, D. SECHTER, E. ZARIFIAN, A. M. GALZIN, H. SCHOEMAKER, A. SEGONZAC & S. Z. LANGER. 1984. [^{3}H]imipramine binding and [^{3}H]5HT uptake in human blood platelets: Changes after one week chlorimipramine treatment. Eur. J. Pharmacol. **106:** 629–633.
44. LEYSEN, J. E., D. DE CHAFFOY DE COURCELLES, F. DE CLERCK, C. J. E. NIEMEGEERS & J. M. VAN NUETEN. 1984. Serotonin-S_2 receptor binding sites and functional correlates. Neuropharmacology **23:** 1493–1501.
45. VICTORZON, M., C. TAPPARELLI & E. MULLER-SCHWEINITZER. 1986. Comparison of the actions of serotoninergic agents of human saphenous veins and platelets. Eur. J. Pharmacol. **124:** 107–111.
46. DE CLERCK, F., B. XHONNEUX, J. LEYSEN & P. A. J. JANSSEN. 1984. Evidence for

functional 5-HT_2 receptor sites on human blood platelets. Biochem. Pharmacol. **33:** 2807–2811.

47. ELLIOT, J. M. & A. KENT. 1989. Comparison of [^{125}I]iodolysergic acid diethylamide binding in human frontal cortex and platelet tissue. J. Neurochem. **53:** 191–196.
48. MCBRIDE, P. A., J. J. MANN, M. J. POLLEY, A. J. WILEY & J. A. SWEENEY. 1987. Assessment of binding indices and physiological responsiveness of the 5-HT_2 receptor on human platelets. Life Sci. **40:** 1799–1809.
49. GEANEY, D. P., M. SCHACHTER, J. M. ELLIOT & D. G. GRAHAME-SMITH. 1984. Characterisation of [^{3}H]lysergic acid diethylamide binding to a 5-hydroxytryptamine receptor on human platelet membranes. Eur. J. Pharmacol. **97:** 87–93.
50. DE CHAFFOY DE COURCELLES, D., P. ROEVENS, J. WYNANTS & H. VAN BELLE. 1987. Serotonin-induced alterations in inositol phospholipid metabolism in human platelets. Biochim. Biophys. Acta **927:** 291–302.
51. COWEN, P. J., E. M. CHARIG, S. FRASER & J. M. ELLIOT. 1987. Platelet 5-HT receptor binding during depressive illness and tricyclic antidepressant treatment. J. Affective Dis. **13:** 45–50.
52. MCBRIDE, P. A., J. J. MANN, B. MCEWEN & A. BIEGON. 1983. Characterization of serotonin binding sites on human platelets. Life Sci. **33:** 2033–2041.
53. DE KEYSER, J., H. WALRAEVENS, A. CONVENTS, G. EBINGER & G. VAUQUELIN. 1989. [^{3}H]SCH 23390 labels a novel 5-hydroxytrptamine binding site in human blood platelet membranes. Eur. J. Pharmacol. **162:** 437–445.
54. MELTZER, H., B. WIITA, B. J. TRICOU, M. SIMONOVIC, V. FANG & G. MANOV. 1982. Effect of serotonin precursors and serotonin agonists on plasma hormone levels. *In* Serotonin in Biological Psychiatry. B. T. Ho, J. C. Schoolar & E. Usdin, Eds. New York: Raven Press, pp. 117–136.
55. TUOMISTO, J. & P. MANNISTO. 1985. Neurotransmitter regulation of anterior pituitary hormones. Pharmacol. Rev. **37:** 249–332.
56. MELLOW, A. M., C. S. AULAKH & D. L. MURPHY. 1988. Neuroendocrine effects of serotonergic agents: Applications to the study of central serotonin function in humans. *In* 5-HT Agonists as Psychoactive Drugs. R. H. Rech & G. A. Gudelsky, Eds. Ann Arbor, MI: pp. 259–282.
57. VAN PRAAG, H. M., C. LEMUS & R. KAHN. 1987. Hormonal probes of central serotonergic activity. Do They really exist? Biol. Psychiatry **22:** 86–98.
58. MACINDOE, J. H. & R. W. TURKINGTON. 1973. Stimulation of human prolactin secretion by intravenous infusion of L-tryptophan. J. Clin Invest. **52:** 1972–1978.
59. CHARNEY, D. S., G. R. HENINGER, J. F. REINHARD, D. E. STERNBERG & K. M. HAFSTEAD. 1982. The effect of intravenous L-tryptophan on prolactin and growth hormone and mood in healthy subjects. Psychopharmacology **77:** 217–222.
60. WINOKUR, A., N. D. LINDBERG, I. LUCKI, J. PHILLIPS & J. D. AMSETERDAM. 1986. Hormonal and behavioral effects associated with intravenous L-tryptophan administration. Psychopharmacology **88:** 213–219.
61. LANCRANJAN, I., A. WIRZ-JUSTICE, W. PUHRINGER & E. DEL POZO. 1977. Effect of L-5-hydroxytryptophan infusion on growth hornome and prolactin secretion in man. J. Clin. Endocrinol. Metab. **45:** 588–593.
62. PARATI, E. A., P. ZANARDI, D. COCCHI, T. CARACENI & E. E. MULLER. 1980. Neuroendocrine effects of quipazine in man in healthy state or with neurological disorders. J. Neural Transm. **47:** 293–297.
63. QUATTRONE, A., G. TEDESCHI, U. AGUGLIA, F. SCOPACASA, G. F. DIRENZO & L. ANNUNZIATO. 1983. Prolactin secretion in man: A useful tool to evaluate the activity of drugs on central 5-hydroxytryptaminergic neurons: Studies with fenfluramine. Br. J. Clin. Pharmacol. **16:** 471–475.
64. MELTZER, H. Y., T. FLEMMING & A. ROBERTSON. 1983. The effect of buspirone on prolactin and growth hormone secretion in man. Arch. Gen. Psychiatry **40:** 1099–1102.
65. SIEVER, L. J., D. L. MURPHY, S. SLATER, E. DE LA VEGA & S. LIPPER. 1984. Plasma prolactin changes following fenfluramine in depressed patients compared to controls: An evaluation of central serotonergic responsivity in depression. Life Sci. **34:** 1029–1039.

66. LEWIS, D. A. & B. M. SHERMAN. 1984. Serotonergic stimulation of adrenocorticotropin in secretion of man. J. Clin. Endocrinol. Metab. **58:** 458–462.
67. CASANUEVA, F. F., L. VILLANUEVA, A. PENALVA & J. CABEZAS-CERRATO. 1984. Depending on the stimulus, central serotoninergic activation by fenfluramine blocks or does not alter growth hormone secretion in man. Neuroendocrinology **38:** 302–308.
68. LAAKMANN, G., M. GUGATH, H. J. KUSS & K. ZYGAN. 1984. Comparison of growth hormone and prolactin stimulation induced by chlorimipramine and desimipramine in man in connection with chlorimipramine metabolism. Psychopharmacology **82:** 62–67.
69. PETRAGLIA, F., F. FACCHINETTI, E. MARTIGNONI, G. NAPPI, A VOLPE & A. R. GENAZZANI. 1984. Serotoninergic agonists increase plasma levels of β-endorphin and β-lipotropin in humans. J. Clin Endocrinol. Metab. **59:** 1138–1142.
70. MUELLER, E. A., D. L. MURPHY, T. SUNDERLAND & J. JONES. 1985. A new postsynaptic serotonin receptor agonist suitable for studies in humans. Psychopharmacol. Bull. **21:** 701–704.
71. CHARNEY, D. S., S. W. WOODS, W. K. GOODMAN & G. R. HENINGER. 1987. Serotonin function in anxiety: II. Effects of the serotonin agonist *m*-CPP in panic disorder patients and healthy subjects. Psychopharmacology **92:** 14–24.
72. NUTT, D., H. MIDDLETON & M. FRANKLIN. 1987. The neuroendocrine effects of oral imipramine. Psychoneuroendocrinology **12:** 367–375.
73. LOWY, M. T. & H. Y MELTZER. 1988. Stimulation of serum cortisol and prolactin secretion in humans by MK-212, a centrally active serotonin agonist. Biol. Psychiatry **23:** 818–828.
74. MURPHY, D. L., E. A. MUELLER, J. L. HILL, T. J. TOLLIVER & F. M. JACOBSEN. 1989. Comparative anxiogenic, neuroendocrine, and other physiologic effects of m-chlorophenylpiperazine given intravenously or orally to healthy volunteers. Psychopharmacology **98:** 275–282.
75. FRASER, W. M., H. S. TUCKER, S. R. GRUBB, J. P. WIGAND & W. G. BLACKARD. 1979. Effect of L-tryptophan on growth hormone and prolactin release in normal volunteers and patients with secretory pituitary tumors. Horm. Metab. Res. **11:** 149–155.
76. MCCANCE, S. L., P. J. COWEN & D. G. GRAHAME-SMITH. 1987. Methergoline attenuates the prolactin responses to L-tryptophan. Br. J. Clin. Pharmacol. **23:** 607–608.
77. MUELLER, E. A., D. L. MURPHY & T. SUNDERLAND. 1986. Further studies of the putative serotonin agonist m-chlorophenylpiperazine: Evidence for a serotonin receptor mediated mechanism of action in humans. Psychopharmacology **89:** 388–391.
78. CHARIG, E. M., I. M. ANDERSON, J. M. ROBINSON, D. J. NUTT & P. J. COWEN. 1986. L-tryptophan and prolactin release: Evidence for interaction between 5-HT_1 and 5-HT_2 receptors. Human Psychopharmacol. **1:** 93–97.
79. NAKAI, Y., H. IMURA, H. SAKURI, H. KURAHACHI & T. YOSHIMI. 1974. Effect of cyproheptadine on human growth hormone secretion. J. Clin. Endocrinol. Metab. **3:** 446–449.
80. COWEN, P. J., H. GADHVI, B. GOSDEN & T. KOLAKOWSKA. 1985. Responses of prolactin and growth hormone to L-tryptophan infusion: Effects in normal subjects and schizophrenic patients receiving neuroleptics. Psychopharmacology **86:** 164–169.
81. COWEN, P. J. & I. M. ANDERSÓN. 1986. 5-HT neuroendocrinology of changes during depressive illness and antidepressant drug treatment. *In* Advances in the Biology of Depression. J. F. W. Deakin & H. Freeman, Eds. London: Royal College of Psychiatrists, pp. 71–89.
82. IDZIKOWSKI, C., P. J. COWEN, D. NUTT & F. J. MILLS. 1987. The effects of chronic ritanserin treatment on sleep and the neuroendocrine response to L-tryptophan. Psychopharmacology **93:** 416–420.
83. COWEN, P. J. 1988. Prolactin response to tryptophan during mianserin treatment. Am. J. Psychiatry **145:** 740–741.
84. SLATER, S., E. DE LA VEGA, J. SKYLER & D. L. MURPHY. 1976. Plasma prolactin stimulation by fenfluramine and amphetamine. Psychopharmacol. Bull. **123:** 26–27.
85. MELTZER, H. Y., M. LOWY, A. ROBERTSON, P. GOODNICK & R. PERLINE. 1984. Effect of

5-hydroxytryptophan on serum cortisol levels in major affective disorders: III. Effect of antidepressants and lithium carbonate. Arch. Gen. Psychiatry **41:** 391–397.

86. CHARNEY, D. S., G. R. HENINGER & D. E. STERNBERG. 1984. Serotonin function and mechanism of action of antidepressant treatment. Arch. Gen. Psychiatry **41:** 359–365.
87. ANDERSON, I. M. & P. J. COWEN. 1986. Clomipramine enhances prolactin and growth hormone responses to L-tryptophan. Psychopharmacology **89:** 131–133.
88. GLUE, P. W., P. J. COWEN, D. J. NUTT, T. KOLAKOWSKA & D. G. GRAHAME-SMITH. 1986. The effect of lithium on 5-HT-mediated neuroendocrine responses and platelet 5-HT receptors. Psychopharmacology **90:** 398–402.
89. ZOHAR, J., T. R. INSEL, R. C. ZOHAR-KADOUCH, J. L. HILL & D. L. MURPHY. 1988. Serotonergic responsivity in obsessive-compulsive disorder: Effects of chronic clomipramine treatment. Arch. Gen. Psychiatry **45:** 167–172.
90. NUTT, D. J. & P. J. COWEN. 1987. Diazepam alters brain 5-HT function in man: Implications for the acute and chronic effects of benzodiazepines. Psychol. Med. **17:** 601–607.
91. DELGADO, P. L., D. S. CHARNEY, L. H. PRICE, H. LANDIS & G. R. HENINGER. 1989. Neuroendocrine and behavioral effects of dietary tryptophan restriction in healthy subjects. Life Sci. In press.
92. GOODWIN, G. M., C. G. FAIRBURN & P. J. COWEN. 1987. The effects of dieting and weight loss on neuroendocrine responses to tryptophan, clonidine, and apomorphine in volunteers: Important implications for neuroendocrine investigations in depression. Arch. Gen. Psychiatry **44:** 952–957.
93. PEREZ-CRUET, J., T. N. CHASE & D. L. MURPHY. 1974. Dietary regulation of brain tryptophan metabolism by plasma ratio of free tryptophan and neutral amino acids in humans. Nature **248:** 693–695.
94. ALOI, J. A., T. R. INSEL, E. A. MUELLER & D. L. MURPHY. 1984. Neuroendocrine and behavioral effects of m-chlorophenylpiperazine administration in rhesus monkeys. Life Sci. **34:** 1325–1331.
95. WOZNIAK, K. M., C. S. AULAKH, J. L. HILL & D. L. MURPHY. 1989. Hyperthermia induced by *m*-CPP in the rat and its modification by antidepressant treatments. Psychopharmacology **97:** 269–274.
96. GUDELSKY, G. A., J. I. KOENIG & H. Y. MELTZER. 1986. Thermoregulatory responses to serotonin (5-HT) receptor stimulation in the rat: Evidence for opposing roles of 5-HT_2 and 5-HT_{1A} receptors. Neuropharmacology **25:** 1307–1313.
97. LAWLOR, B. A., P. A. NEWHOUSE, T. BALKIN, S. E. MOLCHAN, A. M. MELLOW & D. L. MURPHY. 1989. A preliminary study of the effects of *m*-CPP on sleep architecture and behavior in healthy volunteers. Society of Biological Psychiatry, San Francisco, CA (abstract).
98. PASTEL, R. H. & J. D. FERNSTROM. 1987. Short-term effects of fluoxetine and trifluoromethylpiperazine on electroencephalographic sleep in the rat. Brain Res. **436:** 92–102.
99. MUELLER, E. A., D. L. MURPHY & T. SUNDERLAND. 1985. Neuroendocrine effects of m-chlorophenylpiperazine, a serotonin agonist, in humans. J. Clin. Endocrinol. Metab. **61:** 1179–1184.
100. TRABER, J. & T. GLASER. 1987. 5-HT_{1A} receptor-related anxiolytics. Trends Pharmacol. Sci. **8:** 432–437.
101. CEULEMANS, D. L. S., M.-L. J. A. HOPPENBROUWERS, Y. G. GELDERS & A. J. M. REYNTJENS. 1985. The influence of ritanserin, a serotonin antagonist, in anxiety disorders: A double-blind placebo-controlled study versus lorazepam. Pharmacopsychiatry **18:** 303–305.
102. LAWLOR, B. A., T. SUNDERLAND, J. L. HILL, A. M. MELLOW, S. E. MOLCHAN, E. A. MUELLER, F. M. JACOBSEN & D. L. MURPHY. 1989. Evidence for a decline with age in behavioral responsivity to the serotonin agonist, m-chlorophenylpiperazine, in healthy human subjects. Psychiatry Res. **29:** 1–10.
103. JACOBS, B. L. 1976. An animal behaviour model for studying central serotonergic synapses. Life Sci. **19:** 777–786.
104. INSEL, T. R., B. F. ROY, R. M. COHEN & D. L. MURPHY. 1982. Possible development of the serotonin syndrome in man. Am. J. Psychiatry **139:** 954–955.

105. PARE, C. M. B. 1963. Potentiation of monoamine-oxidase inhibitors by tryptophan. Lancet **ii:** 527–528.
106. BALOH, R. W., J. DIETZ & J. W. SPOONER. 1982. Myoclonus and ocular oscillations induced by L-tryptophan. Ann Neurol. **11:** 95–97.
107. THOMAS, J. M. & E. H. RUBIN. 1984. Case report of a toxic reaction from a combination of tryptophan and phenelzine. Am. J. Psychiatry **14:** 281–283.
108. PRICE, W. A., B. ZIMMER & P. KUCAS. 1986. Serotonin syndrome: A case report. J. Clin. Pharmacol. **26:** 77–78.

DISCUSSION OF THE PAPER

I. TÖRK (*The University of New South Wales, Kensintgon, Australia*): I just was interested to hear about the genetic case, the Norrie disease patients. What are the clinical symptoms of these patients? Would one expect an increased serotonin function?

MURPHY: First of all, these individuals were identified because they have a genetic disorder Norrie disease, which begins with blindness and variable mental retardation. But these individuals, with a submicroscopic X-chromosome deletion leading to the Norrie disease syndrome and also to loss of MAO-A and MAO-B, are even more severely affected, with marked mental retardation, autistic type syndrome, and autonomic dysfunction. These disorders may reflect serotonergic as well as noradrenergic and other neurotransmitter system changes, although, as I said, we are puzzled by the fact that 5-HIAA changes, measured only in urine so far, are negligible.

B. CAMPBELL (*Servier R&D, Ltd. London*): This is a comment rather than a question. One of the problems with humans is you can't take their brains out and measure binding characteristics as we can with rats. And that's the problem of doing peripheral measurements for anything. But you can get over it by something that you haven't mentioned, and that's using kinetic dynamic analysis of the actual drug levels in blood. If you make a few assumptions that after a period of time there is an uptake from the plasma into the brain, mathematically you can overcome the problem, that you're not actually measuring within the site in the brain. And if you measure the drug levels in the blood, and you compare that directly with some behavioral measurement in man, you in fact can compose the same sort of B-max dose response curves that you do actually within the brain. And it's quite possible.

For example we've done it with fenfluramine, where we've actually looked at sensitivity curves in a number of different subjects looking at food intake as an index of central activity. And you can quite easily show differences in sensitivity curves which you can then relate presumably to central serotonergic receptors because there's a big difference depending on the actual body size of the patient.

So taking that one step further to many other drugs, if you use plasma level measurements of the various drugs with and without inhibitors you in fact can get rid of a lot of the problems that you've been talking about earlier on.

MURPHY: Thank you. That's a very good point that needs to be made, and I know a number of groups are making drug plasma level measurements a necessary part of all their investigations.

HERBERT Y. MELTZER (*Case Western Reserve University, Cleveland, OH*): I wanted to comment on your conclusions about the discrepancies or the differences between the effect of chronic clomipramine on the endocrine, temperature, anxiety and other responses. I think the point really is that there's a whole diversity of effects on these different endpoints, and we really need to look at all of them. It's not that it is one of them that is giving us a false image of what's happening. I'm going to show tomorrow that with chronic fluoxetine, and we've seen this thing with sertraline, there's a highly significant enhancement of the cortisol and prolactin responses to MK212 as well as to 5-hydroxy tryptophan which is the exact opposite of what you would conclude from the Elaine Sanders-Bush data in terms of the down regulation of 5-HT_2 density and PI responses. I think the important point here is it's very, very difficult to generalize from either the animal models or even the human models. There's a very complex set of changes happening here both presynaptically and postsynaptically varying in different regions of the brain. And the real problem is to try to theorize on the basis of any of these things to what's happening in disease processes or in the mechanism of action of a drug because you can find virtually any range of parameters changing.

MURPHY: Yes, I concur. We are not stopping measuring neuroendocrine endpoints because I think every bit of information we can get about human responses is going to be valuable in trying to comparatively interpret or evaluate responses in animals and other situations. I did not mean to overemphasize neuroendocrine versus behavioral differences, or suggest that we or others should abandon the neuroendocrine approach.

Ontogeny of the Serotonergic System in the Rat: Serotonin as a Developmental Signal[a]

JEAN M. LAUDER[b]

Department of Cell Biology & Anatomy
University of North Carolina School of Medicine
Chapel Hill, North Carolina 27599-7090

INTRODUCTION

The development of the serotonergic system was first described by Olsen and Seiger[1-2] and later by Levitt and Moore[3] in the rat embryo using fluorescence histochemistry. These studies provided topographic information concerning the appearance and subsequent development of the different components of the raphe nuclei and their axonal projections. At the same time we initiated studies to date the onset of neuronal differentiation (time of origin; day of last cell division) of the raphe nuclei and other monoamine neurons in the rat embryo.[4] We related this information to the onset of neurotransmitter synthesis as determined by the fluorescence histochemical study of Olsen and Seiger.[1] From this correlation, it became clear that developing dorsal and medial raphe neurons contain serotonin (5-HT) soon after they begin to differentiate and rapidly extend axons towards the forebrain.[1]

The early expression of 5-HT in developing raphe neurons and their growing axons, prior to the onset of neurotransmission or synaptogenesis, led us to speculate that 5-HT might act as a "differentiation signal" for embryonic target cells. This hypothesis was tested in an analysis of the effects of inhibition of embryonic 5-HT synthesis by maternal administration of *p*-chlorophenylalanine (*p*CPA) on the onset of differentiation of neuronal populations known to receive a 5-HT innervation in the adult (presumptive target cells).[5] Results of this study indicated that *p*CPA treatment, which inhibited embryonic 5-HT synthesis[6] led to a delay in the onset of differentiation of presumptive target cells. However, little was known as to the true nature of the embryonic targets of developing 5-HT axons and whether they were specifically affected.

The advent of immunocytochemistry with specific 5-HT antisera[7,8] made possible the detailed mapping of the developing 5-HT system in the embryonic and fetal rat.[9-11] Such information also made it possible to examine the relationship of developing 5-HT neurons to embryonic target cells and to reassess the data from our previous *p*CPA study in light of this information. This analysis has demonstrated a strong correlation between the spatio-temporal distribution of developing 5-HT axons,[12] inhibition of 5-HT synthesis within these cells by *p*CPA,[13] and delayed onset of neuronal differentiation in specific regions of the embryonic rat brain.[12] In a more recent collaborative study, we have demonstrated the ability of maternal *p*CPA treatment to alter the postnatal expression of 5-HT_1 receptors in the rat brain, suggesting that such receptors

[a]This work was supported by National Institutes of Health grants NS 15706 and NS 21840.
[b]Address for correspondence: Department of Cell Biology & Anatomy, CB# 7090, University of North Carolina School of Medicine, Chapel Hill, NC 27599-7090.

are present prenatally and could mediate the effects of 5-HT on neuronal differentiation.[14] The results of these studies support the hypothesis that 5-HT acts as a "differentiation signal" for specific target cells during neuroembryogenesis by receptor-mediated processes.

The present paper describes the development of the serotonergic system in the developing rat brain, including formation of the raphe nuclear groups and their axonal projections, together with *in vitro* evidence that 5-HT acts as a differentiation signal for raphe neurons and target cells via neuronal and glial receptors.

DEVELOPMENT OF THE SEROTONERGIC SYSTEM

Cell Bodies

As determined using immunocytochemistry with specific 5-HT antisera in whole-mounts of embryonic rat brain,[15,16] expression of the serotonergic phenotype begins on embryonic day 12 (E12; crown-rump length 6 mm) just caudal to the mesencephalic flexure (FIG. 1) in the cells of the B7–9 group (rostral raphe nuclei), which are generated slightly earlier.[4] This complex of neurons, which includes the anlage of the dorsal and medial raphe nuclei (B7, B8), increases in size until E15 when generation of 5-HT neurons in this region is complete (FIG. 2).

Under the new terminology proposed by the 5-HT nomenclature committee, the B4–6 nuclei are no longer considered to be separate nuclear groups, although we originally included them in this cluster as part of the "B4–9 complex,"[11] which we will now re-name the "B7–9 complex."

The rostral raphe nuclei exist throughout most of their embryonic development as bilateral groups of cells connected across the midline by fibers which cross in the marginal zone of the floorplate (FIG. 4A–D). These bilateral groups fuse during the late fetal period to form midline nuclei (FIG. 4E, G, H). In addition, a variety of directional migrations occur to produce the individual nuclei derived from the B7–9 complex beginning at about E15.[9,11] For example, the cells destined to form the dorsal raphe nucleus (B7) already can be subdivided into two groups by E15 (FIG. 4A), one dorsal and one ventral to the medial longitudinal fasciculus (MLF) which runs through the middle of the B7–9 complex (see FIG. 3). The dorsal subgroup is distinguishable not only by its position, but by the large, thick processes which extend into the ventricular zone, sometimes as far as the lumen. With further development, the ventral subgroup appears to move ventromedially to fuse along the midline in a rostro-caudal fashion. The more dorsal portion, however, remains unfused and forms the bilateral "wings" of the dorsal raphe nucleus (FIG. 4E). Another subgroup which undergoes translocation around E15 develops into the B8 group or medial raphe nucleus (FIG. 4C, D, G, H). The cells making up this nucleus migrate ventrally away from the lumen within the medial portions of the B7–9 complex. The rostrally located B9 group, which is bilateral in adulthood, appears to be formed by a secondary lateral migration following an initial migration ventrally along the midline (FIG. 4F, G).[11]

The 5-HT neurons of the B1–3 groups (caudal raphe nuclei), which send descending projections through the myelencephalon to the spinal cord, are generated beginning on E11–12,[17] yet they do not express immunoreactivity for 5-HT until E14.[11,16] This is in contrast to the cells of the B7–9 complex which express 5-HT shortly after they are generated.[4,11] It has been suggested that the cells of the B3 group (raphe magnus) express transmitter more rapidly than those forming the B2 (raphe obscurus) and B1 (raphe pallidus) nuclei, since only they can be detected at E14 in whole-mounts.[15] However, we have found that all three groups are present at E14 in our own

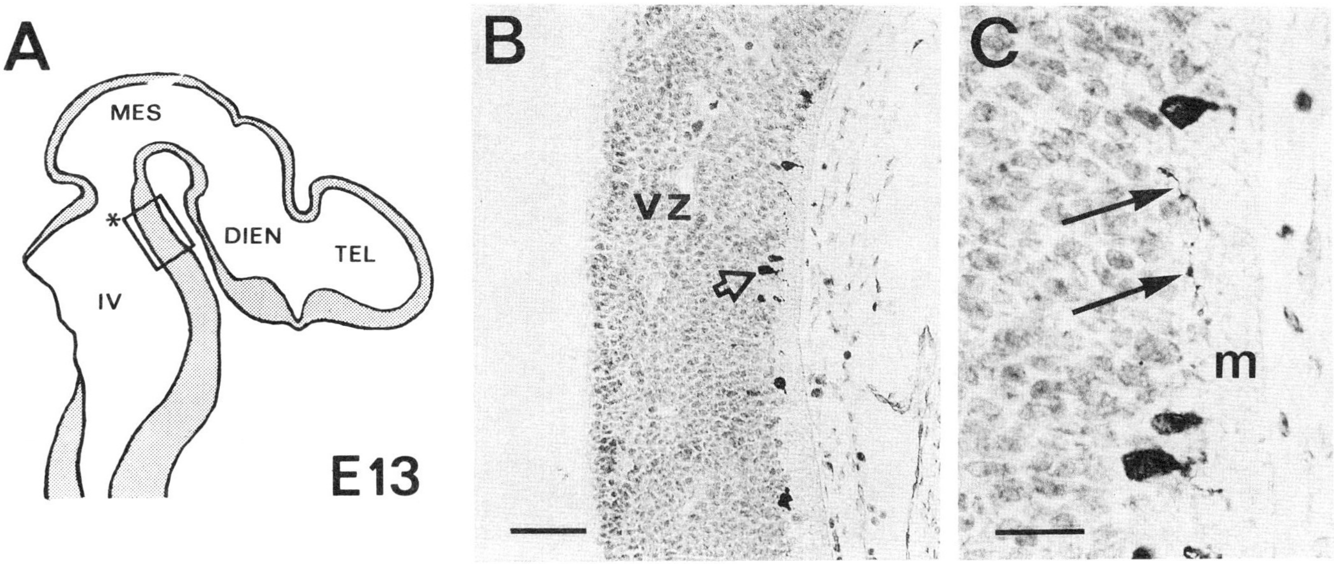

FIGURE 1. Appearance of 5-HT neurons in the E13 rat brain.[11] The first cells to express 5-HT immunoreactivity are located in the rhombencephalon just caudal to the mesencephalic flexure (box* in A). B and C illustrate the appearance of these cells from region of box. Rhombencephalic cells expressing 5-HT have been detected at E12 in whole-mounts.[16] **Abbreviations:** MES: mesencephalic flexure, DIEN: diencephalon, TEL: telencephalon, IV: fourth ventricle, VZ: ventricular zone, M: marginal zone. Bars: B = 50 μm, C = 20 μm.

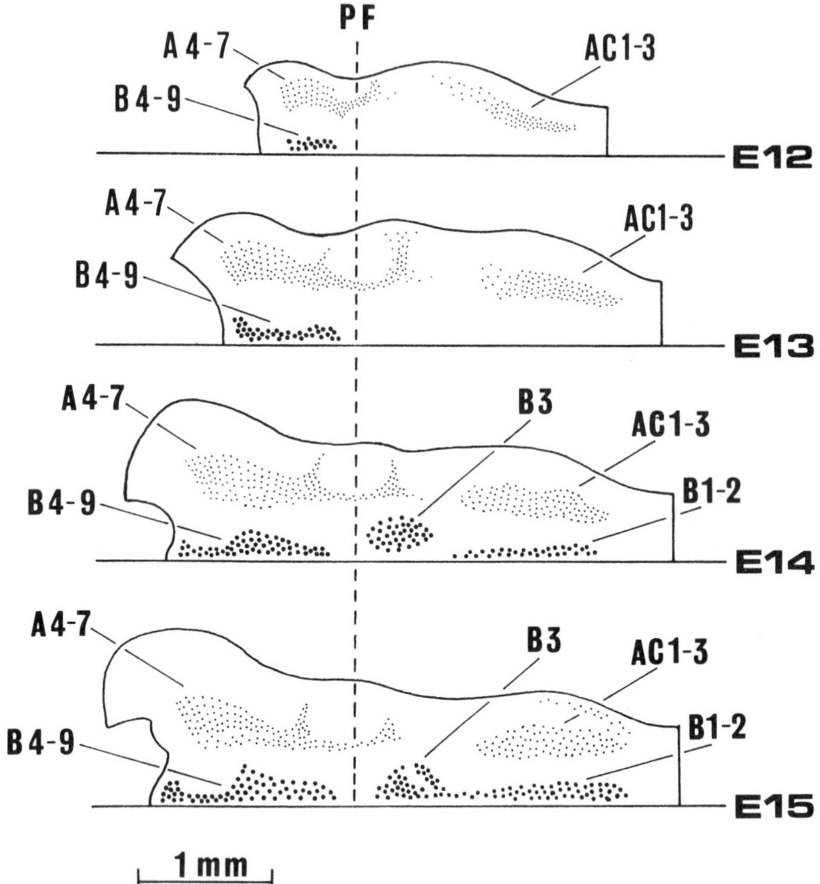

FIGURE 2. Schematic representation of the distribution of 5-HT and tyrosine hydroxylase (TH) immunoreactive neurons in whole-mounts of the E13–15 rat rhombencephalon.[16] Small dots indicate TH neurons, large dots represent 5-HT neurons. **Note:** according to the new nomenclature devised by the nomenclature committee, B4–6 are no longer considered to represent separate groups. Thus B4–9 should be changed to B7–9, as discussed in the text.

whole-mount study[16] (FIG. 2), although all of these cells undergo further migrations to their final positions during the next few days.[11]

Axonal Projections

The ascending projections of the raphe nuclei begin to be detectable by E12, as soon as 5-HT cell bodies are visible in the rostral rhombencephalon. This pathway is derived entirely from the B7–9 nuclei, and consists of a compact bundle of axons emanating from the rostral tip of this complex (see FIG. 3) which courses around the mesencephalic flexure within the marginal zone by E13–14 and into the diencephalon by E14–

15.[10,11,15] These fibers remain unbranched and fasciculated together within the confines of the median forebrain bundle (MFB) until after they have reached the caudal diencephalon, where they spread out in both dorso-ventral and medio-lateral directions and enter the intermediate zone (see FIG. 9 in ref. 11).

By E15, many 5-HT immunoreactive (IR) fibers course between the mammillary complex and the ventral thalamic area in association with the MFB. These fibers are usually thick, unbranched and often have growth cone-like endings. In addition, 5-HT axons are found ascending along pre-existing pathways such as the fasciculus retroflexus (shown at E17 in FIG. 5A). By late E15–early E16, the number of 5-HT IR fibers

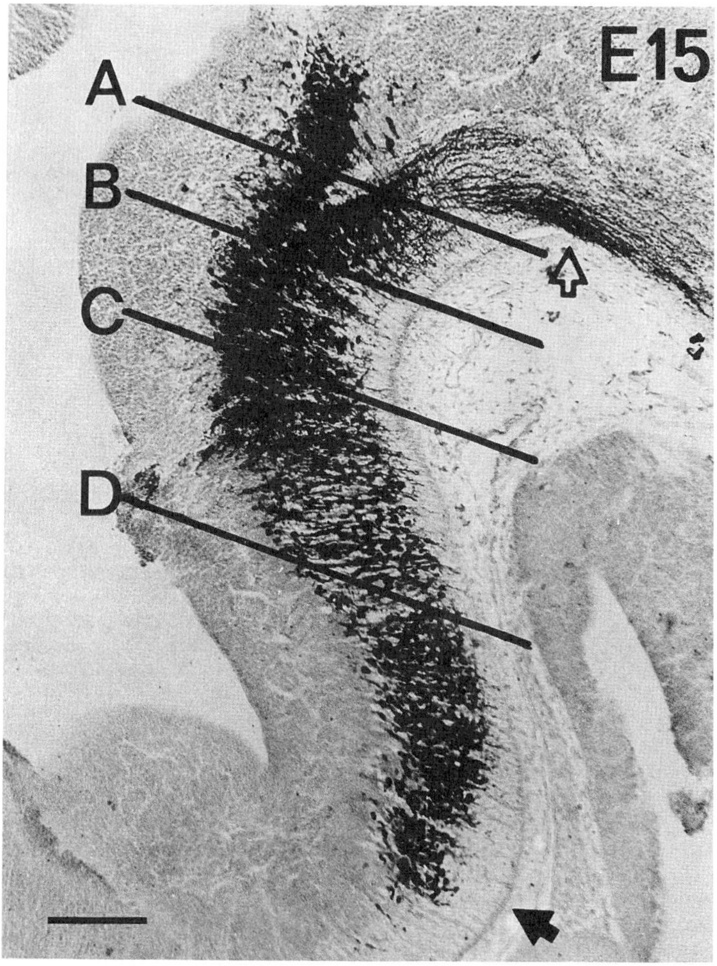

FIGURE 3. Sagittal section through the rostral raphe (B7–9) complex in the E15 rat brain.[11] Sections in the transverse plane at levels designated by lines A–D are shown in FIGURE 4,A–D. Note the rostrally projecting bundle of 5-HT immunoreactive axons coursing around the mesencephalic flexure (*arrow*). Bar = 150 μm.

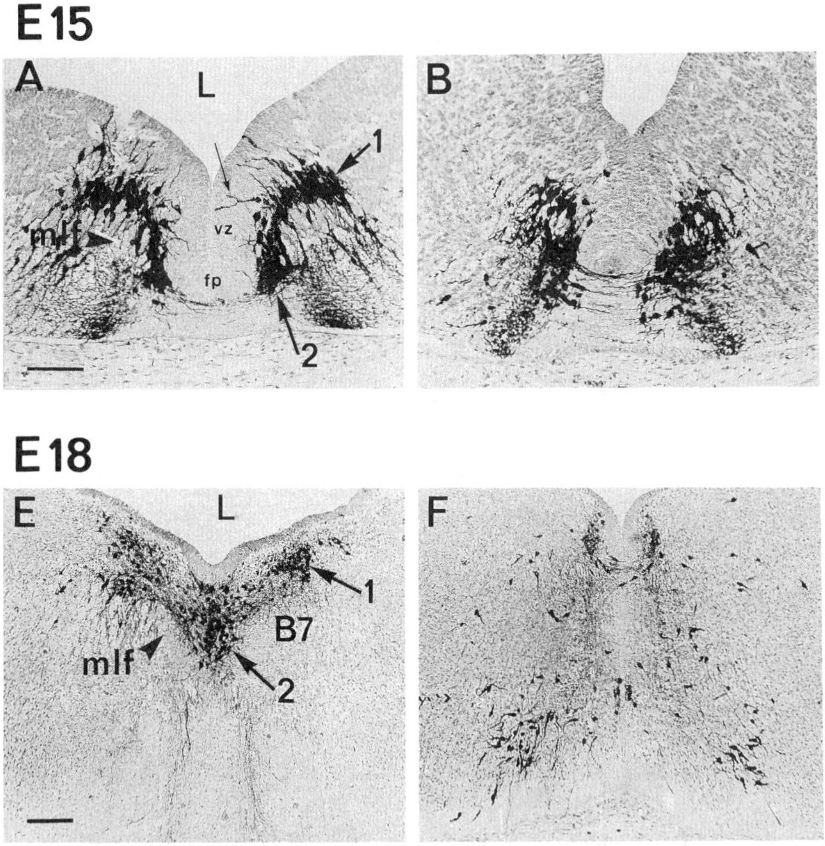

FIGURE 4. (**A–D**): Transverse sections through the E15 rostral raphe complex[11] at the levels indicated in FIGURE 3. (**E–H**): Transverse sections through the rostral raphe nuclei at E18 at levels comparable to those shown in **A–D.**[11] **Abbreviations:** L: lumen of neural tube, vz: ventricular zone, fp: floorplate; mlf: medial longitudinal fasciculus, B7: dorsal raphe nucleus, B8: medial raphe nucleus, 1,2: separate parts of the dorsal raphe nucleus which undergo different migrations (see text). Bars: E15 = 100 μm, E18 = 150 μm.

increases greatly in the lateral hypothalamus, reaching as far rostrally as the supraoptic region. Some fibers can also be seen deviating at right angles from the main ascending pathway to project ventrally towards the developing optic chiasm, where at E17 (FIG. 5B) they cross the midline just caudal to the chiasm in the supraoptic commissure.[10,11,15] Together with the crossing fibers observed at the level of the B7–9 complex, these fibers represent the only other example of contralateral 5-HT projections so far detected in the rat embryo.[11]

By E16, 5-HT IR fibers can be seen at the border of the diencephalon and telencephalon and by E17 this projection has reached the frontal pole of the telencephalon.[9–11,15] At mid-hypothalamic levels, some 5-HT fibers then turn laterally, pass ventral to the internal capsule, and enter the ganglionic eminence through the primordium of the globus pallidus (FIG. 5C). From the lateral hypothalamus, the

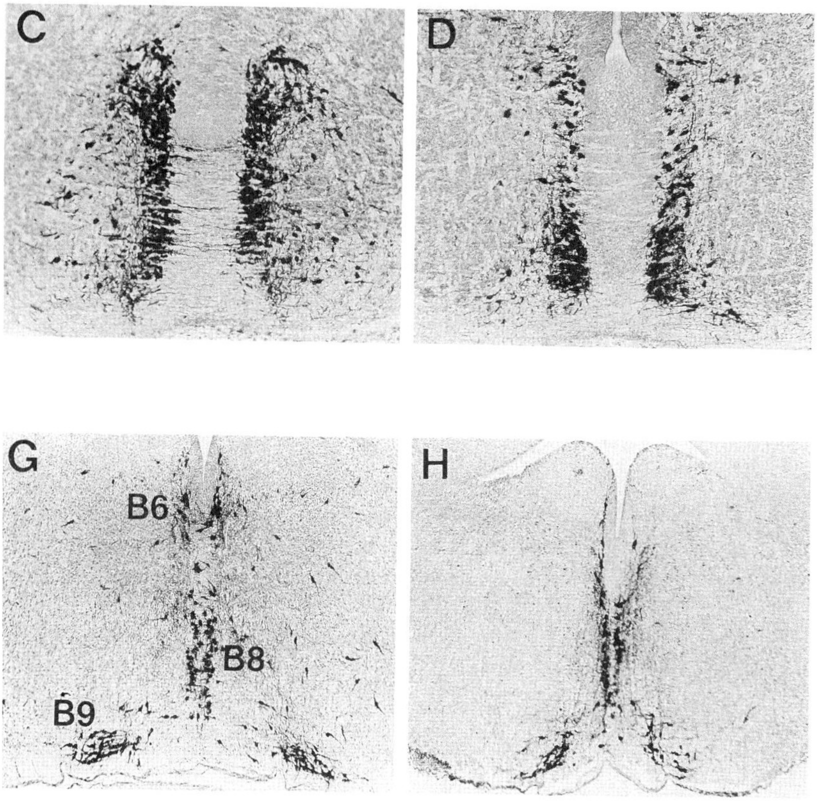

majority of 5-HT axons project medially to the septal region via the diagonal band of Broca (FIG. 5D). Other fibers enter the piriform cortex. Many fibers pass through the septal region to enter the medial aspect of the frontal pole (FIG. 5E) where they start to form a superficial and deep bundle on either side of the cortical plate.[10,11] In addition to the growth of the main bundle of ascending 5-HT fibers within and around the MFB, there is a large increase in the number of 5-HT IR fibers in other brain regions, such as the brainstem.[10,11]

According to the whole-mount study of Aitken and Törk,[15] the 5-HT fibers contained within the MFB appear to be segregated into medial and lateral components which project to different regions of the forebrain. The medial part of the MFB contains fibers destined for the frontal pole of the telencephalon, whereas the lateral part contains fibers which project medially into the hypothalamus and cross in the supraoptic commissure.

By E18, a dense plexus of 5-HT IR fibers exists in the septal area, many of which pass through this region to enter the neocortex medially. As at E17, these fibers form two bundles, above and below the cortical plate.[10,11] This bilaminar projection courses along the entire surface of the frontal pole of the cortex and extends laterally to include a small portion of the lateral cortical surface. Other 5-HT fibers enter the lateral cortex by passing ventral to the internal capsule after traversing ventral portions of the ganglionic eminence. A third group of 5-HT fibers, after entering the cortex ventrally,

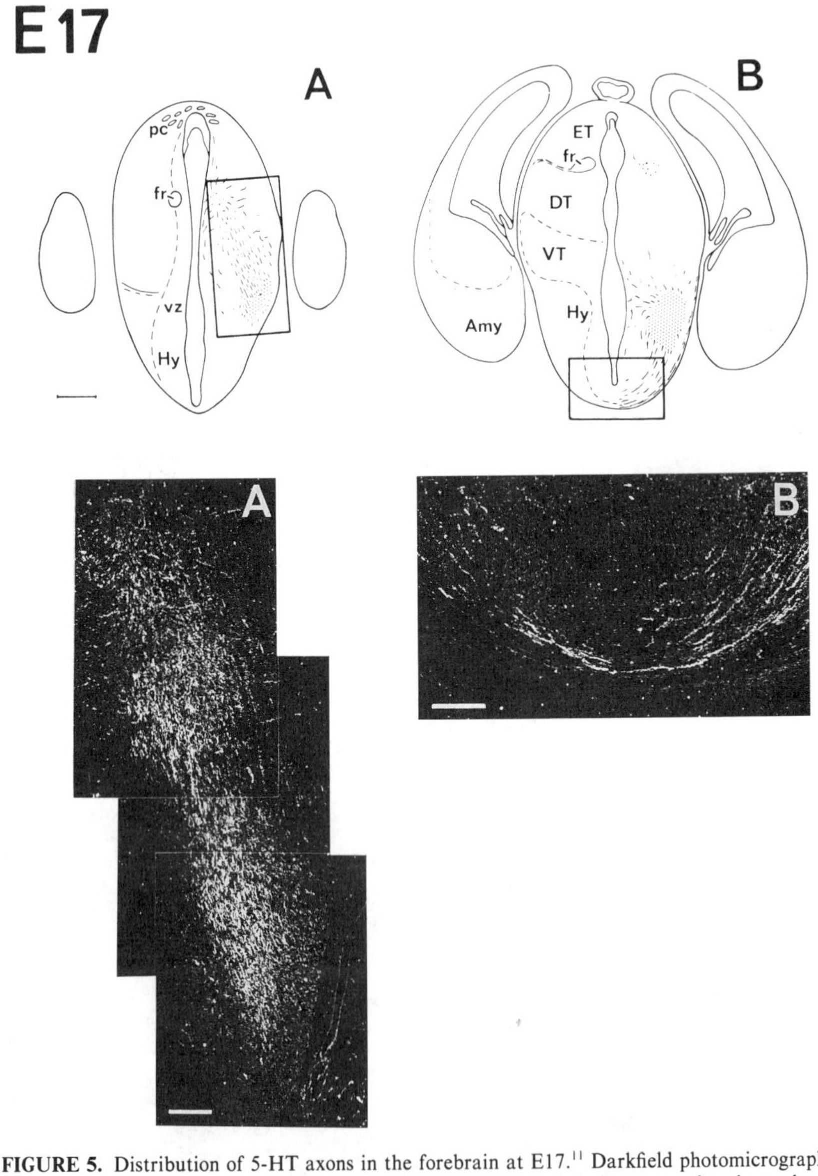

FIGURE 5. Distribution of 5-HT axons in the forebrain at E17.[11] Darkfield photomicrographs illustrate actual appearance of immunoreactive fibers located in boxes on drawings above. **Abbreviations:** Amy: amygdala, DT: dorsal thalamus, ET: epithalamus, GE: ganglionic eminence, Hy: hypothalamus, MC: mammillary complex; SE: septal nucleus, VT: ventral thalamus, ZI: zona incerta, db: diagonal band, fr: fasiculus retroflexus, ic: internal capsule, oc: optic commissure, pc: posterior commissure, III: third ventricle, LV: lateral ventricle. Bars: line drawings = 50 μm, micrographs = 150 μm.

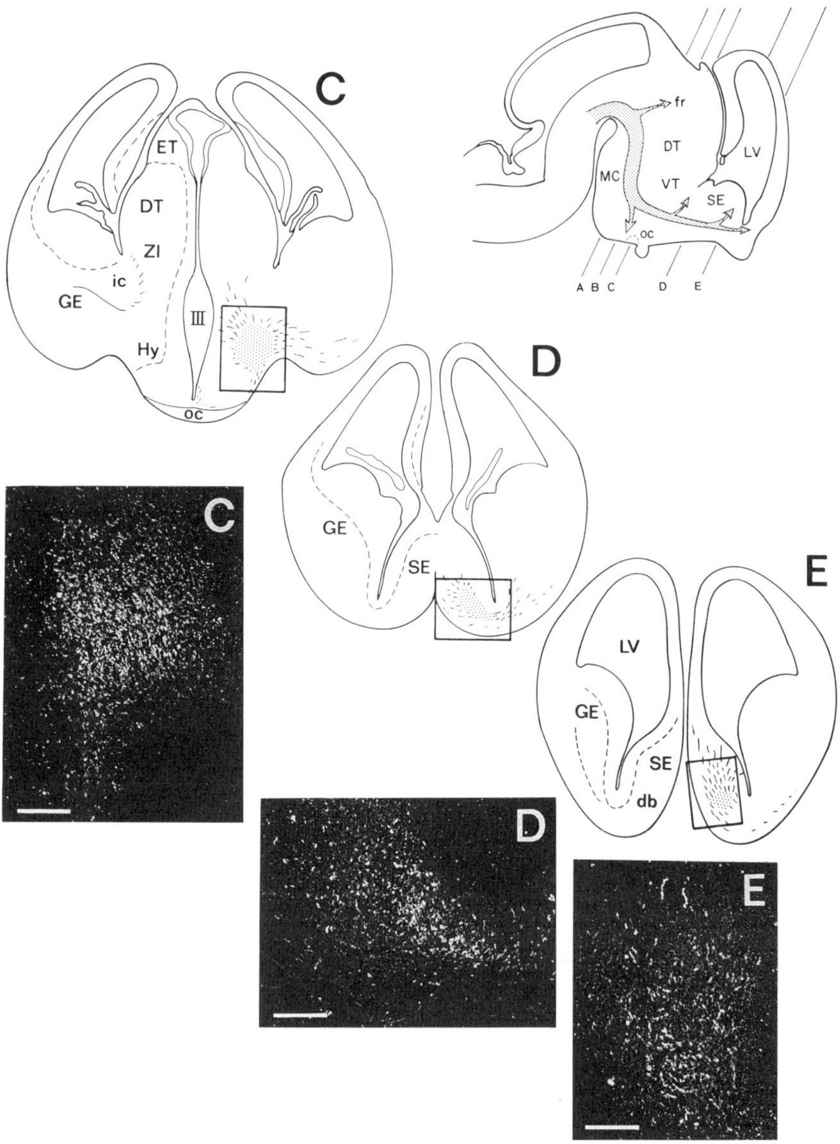

courses dorsally and rostro-caudally in a radial fashion along the medial and frontal aspects of the cortical poles. A distinct group of these fibers projects towards the hippocampal analge along the cingulate bundle. These fibers course through the anterior cingulate cortx, then enter the retrosplenial cortex near the hippocampal primordium. After entering the subiculum, they are located in two bands on either side of the developing pyramidal cell layer as they grow towards CA1. Some 5-HT fibers appear to reach the hippocampus via the fornix from the septum. All of these projections are to the dorsal hippocampus at this age, with ventral pathways forming later. In the diencephalon, small caliber 5-HT fibers appear in the mammillary complex for the first time, and a considerable number of fibers course medially from the MFB towards the third ventricle where some of them appear to make direct contact with the ventricular lining. Other fibers enter the stria terminalis, arch dorsally and caudally over the internal capsule and project towards the amygdala.[11] Serotonergic projections to the cerebellum develop during the postnatal period.[10]

Descending projections from the caudal raphe nuclei to the spinal cord have been reported in the cervical region by E14, coursing in the marginal zone along the ventral and lateral aspects of the cord (ventral and lateral funiculi).[11,19] At variance with these studies is the report by Bregman[20] that the 5-HT innervation of the spinal cord does not begin before E18. By E16–17, branches from these fibers penetrate the ventral gray matter from the marginal zone. This innervation is quite dense by E17, and descending fibers project as far as thoracic levels, innervating both the intermediolateral column (preganglionic sympathetic neurons) and the anterior horn.[2,11,18,19] Synapses begin to be made by E17–18, and following birth these fibers develop from a diffuse to a more restricted distribution, reaching their adult pattern by postnatal day (P) 21.[19] Innervation of the dorsal horn begins to develop only around E19 when fibers enter from the lateral funiculus, and by P7 these fibers have profusely innervated layers I and II of the dorsal horn, where they reach their adult configuration by P21.[19] Rajaofetra *et al.*[19] have proposed that the developing 5-HT innervation of the spinal cord may act as a differentiation or stabilization signal for developing neurons in the anterior and dorsal horns and the intermediolateral columns.

SEROTONIN AS A DEVELOPMENTAL SIGNAL

Neurotransmitters such as 5-HT, which act as specific intercellular signaling devices in the adult nervous system, may have evolved to this role from more general humoral functions in primitive organisms where they were used as both inter- and intracellular signaling devices.[21–24] These phylogenetically old functions are reiterated in the embryo[25] where these substances appear to influence a variety of morphogenetic events both inside and outside the nervous system (see reviews by Lauder[26] and Mattson[27]).

Serotonin and Cell Proliferation

The ability of maternally administered *p*CPA to inhibit 5-HT synthesis in embryonic raphe neurons[13] and to delay and prolong the onset of differentiation of specific neuronal populations[5] (see INTRODUCTION) implies that 5-HT could act as a modulator of cell proliferation within the neuroepithelium where these cells are being generated. Moreover, 5-HT fibers appear to contact proliferating cells in the ventricular zone adjacent to the developing raphe nuclear groups.[12]

On the basis of this evidence, we have conducted studies aimed at evaluating the possibility that 5-HT acts as a modulator of cell proliferation. Using dissociated cell cultures, we have extensively tested the ability of 5-HT added to the medium (with and without an MAO inhibitor) to affect the incorporation of ^{3}H-thymidine into proliferating neuroepithelial cells from E13–14 rat embryos or postnatal cortical astrocytes.

However, we have been unable to demonstrate a significant effect of 5-HT on the amount of ^{3}H-thymidine incorporation into such cultures. This may be related to the lack of appropriate receptors on these cells or to the absence of growth factors within the culture medium (10% NuSerum + BME). The latter possibility is raised since studies with fibroblasts have shown that 5-HT, which is mitogenic for smooth muscle cells,[28] can also stimulate proliferation of fibroblasts, but only in the presence of growth factors such as FGF, insulin or EGF, with which it acts synergistically.[29] Moreover, fibroblasts transfected with cDNA clones encoding the 5-HT_{1c} receptor become transformed into actively proliferating tumor cells, suggesting that this receptor may function as a proto-oncogene.[30] Thus, 5-HT might affect proliferation of neuroepithelial cells or glia under the right culture conditions, a possibility which is presently being explored. Effects of 5-HT on neuronal differentiation have been much easier to demonstrate, as discussed below.

Serotonin and Neuronal Differentiation

In cultures of identified neurons from the CNS of the snail *Helisoma,* 5-HT neurons and their target cells respond to direct application of 5-HT by an inhibition of growth cone activity and a retraction of affected neurites, as well as by alterations in synapse formation.[31,32] Moreover, if *Helisoma* embryos are treated with 5,7-dihydroxytryptamine, which transiently lowers 5-HT levels, and raised to adulthood, they exhibit alterations in the dendritic morphology of specific neurons which are normally innervated by 5-HT axons. Together, these studies suggest that 5-HT has a direct inhibitory effect on neurite outgrowth and synapse formation by receptive cells in the snail embryo.[33] However, in organotypic cultures of neonatal rat brain regions containing 5-HT target cells (cortex, hippocampus) addition of 5-HT to the medium stimulates neurite outgrowth, glial proliferation and synaptogenesis.[36] Such differences in the effects of direct application of 5-HT to individual neurons compared to complex, organotypic cultures where 5-HT is added to the medium may be explained in terms of activation of specific classes of 5-HT receptors on different cell types. For example, activation of receptors on glial cells in complex cultures might cause them to release growth factors important for neuronal differentiation and synaptogenesis,[34,37,38] whereas direct application of 5-HT to growing neurites might activate inhibitory autoreceptors,[34] as discussed below.

The importance of autoreceptors for growth regulation of developing 5-HT neurons has been demonstrated in cell culture studies by Whitaker-Azmitia and Azmitia[34] in which uptake of ^{3}H-5-HT was used as an index of neurite outgrowth by cultured 5-HT neurons, and by Jonakait *et al.*[35] who used the activity of tryptophan hydroxylase as an index of biochemical differentiation of these cells. In both cases, 5-HT inhibited differentiation of raphe neurons, and these effects could be mimicked by the 5-HT_1 agonist 5-methoxytryptamine (5-MT),[34] or inhibited by the 5-HT_1 antagonist methiothepin.[35] In co-cultures of dissociated embryonic raphe and hippocampus, 5-HT and 5-MT either inhibited or stimulated differentiation of 5-HT neurons depending on their concentration in the culture medium.[34] These inhibitory and stimulatory effects of 5-HT may be due to differential activation of autoreceptors on 5-HT neurons or postsynaptic-like receptors on glia or neuronal targets.[34,38]

The inhibitory role of autoreceptors in 5-HT neuronal differentiation may be related to the fact that the raphe nuclei begin their development as paired bilateral groups, connected across the midline by crossing fibers, as discussed above (see FIGS. 4A–C). It is possible that this "auto-innervation" serves to control the form or directionality of subsequent neurite outgrowth by these cells[34] or regulates other aspects of their development such as 5-HT synthesis (activity of tryptophan hydroxylase).[35]

To specifically investigate the direct and indirect effects of 5-HT on neuronal differentiation *in vitro* we have developed a dissociated cell culture system for specific populations of 5-HT and catecholamine neurons within the rhombencephalon and mesencephalon, based on maps of developing 5-HT and tyrosine hydroxylase (TH) immunoreactive neurons in the rat embryo[16] (see FIG. 2). We have also obtained cultures highly enriched in radial glia and astrocytes from these same embryonic brain regions, which may be important for stimulation of neuronal differentiation.[39,46] Interestingly, these glial cultures express immunoreactivity when stained with an anti-idiotypic antiserum[40,41] which appears to recognize 5-HT_{1c} receptors, since it stains fibroblasts transfected with cDNA clones encoding this receptor.[42] These results are consistent with the reports by Whitaker-Azmitia and Azmitia that astrocyte cultures derived from newborn rat brainstem express 5-HT_1 receptors[43] and can be stimulated to produce neuronotrophic factors by 5-HT_1 receptor ligands.[38]

In co-culture experiments[44] we have grown dissociated 5-HT neurons from the E14 rhombencephalon together with TH neurons from the mesencephalon (dopaminergic cells of the substantia nigra). In these experiments, we found that co-culture stimulated neurite outgrowth by both 5-HT and TH neurons (FIG. 6). One interpretation is that this effect may be related to the combined presence of glial elements from both brain regions in co-cultures, which may produce synergistic effects. We are presently addressing the issue of glial involvement in the differentiation-related effects of 5-HT, as discussed below. Whether the stimulatory effects of co-culture are related to the release of 5-HT from raphe neurons and its action on glial or neuronal receptors remains to be determined.

In recent studies (Liu and Lauder, in preparation), we have cultured 5-HT neurons from the rostal raphe nuclei of E14 rat embryos and TH neurons from either the rhombencephalon (FIG. 2) or mesencephalon (substantia nigra) using media supplemented with 5-HT (with or without nialamide, a non-specific MAO inhibitor). We have analyzed the effects of these treatments on a series of parameters related to the morphological differentiation of immunocytochemically identified 5-HT neurons (TABLE 1) or TH neurons using a computerized morphometric program.[46] The results of this study demonstrate that both 5-HT and nialamide stimulate the growth of 5-HT neurons, producing larger cell bodies (soma area), neuritic fields which cover a larger surface area (field area), and more numerous neurites with a higher degree of complexity (length and number of initial and terminal segments). Interestingly, nialamide appears to be more potent than 5-HT in this regard. Some results of this study suggest that 5-HT causes growth inhibition, but that the net effect is a stimulation of the differentiation of 5-HT neurons. However, this stimulatory effect is less robust than that caused by nialamide. Therefore, in addition to interacting with inhibitory autoreceptors, 5-HT may activate receptors on other cells such as radial glia or astrocytes (which are present in these cultures) leading to a net stimulation of neurite outgrowth by 5-HT neurons. Nialamide may either enhance the efficacy of 5-HT at such receptor sites or inhibit the effects of 5-HT on autoreceptors (or both).

The underlying reasons for the stimulatory effects of nialamide on 5-HT neuronal differentiation are unclear since Whitaker-Azmitia and Azmitia[34] have reported that pargyline (another MAO inhibitor) inhibits neurite outgrowth by 5-HT neurons and have suggested that it enhances the effects of 5-HT on autoreceptors. Moreover, Jonakait *et al.*[35] have reported that pargyline can mimic the inhibitory effects of 5-HT on the biochemical differentiation of cultured 5-HT neurons, and that this can be

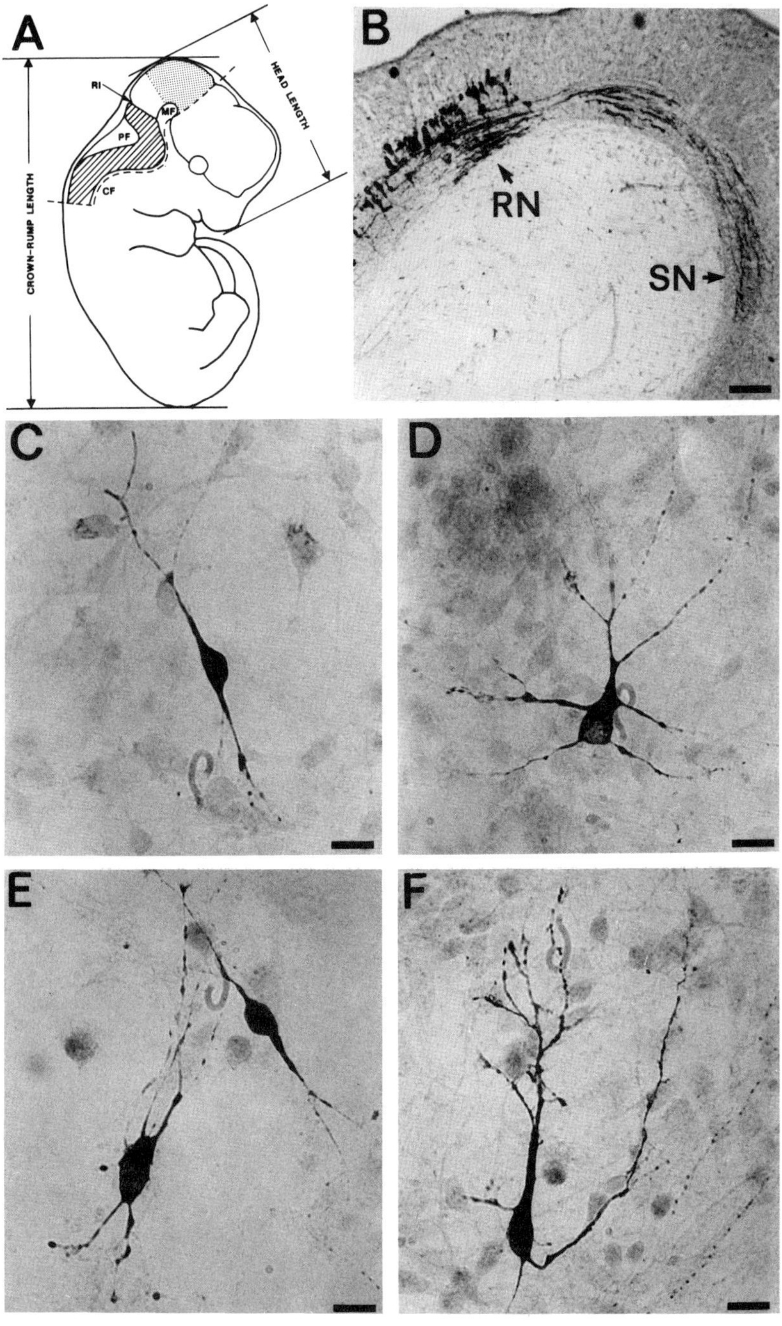

FIGURE 6. Effects of co-culture of rhombencephalon and mesencephalon (**D, F**) on the morphological differentiation of 5-HT (**C, D**) and TH (**E, F**) neurons[44] compared to single cultures (**C, E**). **A:** schematic drawing of an E13–14 rat embryo, sagittal view, illustrating regions dissected for cell cultures (*dotted area* = mesencephalon; *hatched area* = rhombencephalon). **B:** sagittal section from similar embryo double immuno-stained for 5-HT (RN: raphe nuclei; DAB reaction) and TH (SN: substantia nigra; 4-chloronapthol reaction). **Abbreviations:** MF: mesencephalic flexure, PF: pontine flexure, CF: cervical flexure. Bars: B = 75 μm, C–F = 15 μm.

TABLE 1. Effects of 5-HT and Nialamide on Morphological Differentiation of 5-HT Neurons from the Rostral Raphe Nuclei

Morphometric Measurement	Treatment Comparisons					
	5-HT + C vs. Control	5-HT + NC vs. Control	NC vs. Control	NC vs. 5-HT + C	NC vs. 5-HT + NC	5-HT + NC vs. 5-HT + C
Soma area	++	+++	++++	++	0	++
Soma form factor	0	0	0	0	0	+
Field area	+	++	+++	++++	+++	++
Field form factor	+	0	0	+	0	+
Number of neurites	+	+	++	+	+	0
Initial segment length	0	+	0	0	0	++
Initial segment bending	0	0	0	0	0	0
Number of terminal segments	++	+	++++	+++	+	0
Terminal segment length	–	0	0	0	0	++
Terminal segment bending	0	+	0	0	0	+

Significant differences between treatment groups at $p < .05$ (Fisher's least significant difference test): X vs. Y: (+) = X > Y; (−) = X < Y. Number of + or − indicates significant differences/4 expts. **Abbreviations:** (N) = nialamide; (C) = L-cysteine. (From Liu and Lauder, in preparation.)

blocked by the addition of the 5-HT_1 antagonist methiothepin. This suggests that pargyline increases the 5-HT activation of inhibitory autoreceptors on raphe neurons, whereas nialamide appears to have the opposite effect. Our own studies with pargyline, however, have shown it to be stimulatory, although less so than nialamide.

The effects of 5-HT on the morphological differentiation of cultured TH neurons from the rhombencephalon (noradrenergic cells) and mesencephalon (dopaminergic cells) appear to be more variable than on 5-HT neurons, and consist of a mixture of inhibitory and stimulatory effects (Liu and Lauder, in preparation). In general, both types of TH neuron are more strongly affected by 5-HT than by nialamide, in contrast to the effects of nialamide on 5-HT neurons. Morphometric parameters affected by 5-HT and/or nialamide relate mainly to soma size and field area for both types of TH neuron, but these effects seem to be more pronounced in cultures containing dopaminergic neurons. This is interesting since these cells appear to be targets of growing 5-HT axons from the rostral raphe nuclei by E13–14 (FIG. 6B). Therefore, these effects might reflect the influence of 5-HT on the differentiation of one of its earliest targets, the dopaminergic cells of the substantia nigra.

The possible involvement of glia in the effects of 5-HT on cultured 5-HT and TH neurons is presently being investigated in experiments where conditioned media[38,43] collected from embryonic radial glial/astrocyte cultures treated with 5-HT or related compounds are applied to rhombencephalic or mesencephalic cultures followed by immunocytochemistry and morphometric analysis of 5-HT or TH neurons. We are also examining the effects of 5-HT and related substances on differentiation of 5-HT and TH neurons co-cultured with radial glial/astrocytes.

SUMMARY AND CONCLUSIONS

The serotonergic system is an early forming component of the CNS circuitry, beginning its development on gestational days 11–12 in the rat. Owing to its early presence in the embryonic nervous system, 5-HT has been proposed to act as a developmental signal for receptive cells. *In vivo* and *in vitro* evidence that 5-HT can influence both biochemical and morphological differentiation of raphe neurons and receptive target cells suggests that this neurotransmitter may have an organizing function in the developing nervous system which involves effects on neurite outgrowth and other aspects of neuronal differentiation, including synaptogenesis. Such functions may be mediated by a variety of 5-HT receptors located on both neuronal and non-neuronal cells. The apparent function of 5-HT as a differentiation signal in the developing nervous system raises important issues regarding the use of psychoactive serotonergic drugs by pregnant women, since these drugs may act as neural teratogens in the unborn child.

ACKNOWLEDGMENTS

The author would like to thank Dr. Hadassah Tamir for the gift of the anti-idiotypic antiserum recognizing 5-HT receptors, and Mary Beth Wilkie and Dr. Jiangping Liu for performing cell culture experiments and morphometric analyses.

REFERENCES

1. OLSEN, L. & Å. SEIGER. 1972. Early prenatal ontogeny of central monoamine neurons in the rat: Fluorescence histochemical observations. Z. Anat. Entwickl.-Gesch. **137:** 301–316.
2. SEIGER, Å. & L. OLSEN. 1973. Late prenatal ontogeny of central monoamine neurons in the rat. Z. Anat. Entwickl.-Gesch. **140:** 281–318.
3. LEVITT, P. & R. Y. MOORE. 1978. Developmental organization of raphe serotonin neuron groups in the rat. Anat. Embryol. **154:** 241–251.
4. LAUDER, J. M. & F. E. BLOOM. 1974. Ontogeny of monoamine neurons in the locus coeruleus, raphe nuclei, and substantia nigra of the rat. I. Cell differentiation. J. Comp. Neurol. **155:** 469–481.
5. LAUDER, J. M. & H. KREBS. 1978. Serotonin as a differentiation signal in early embryogenesis. Dev. Neurosci. **1:** 15–30.
6. LAUDER, J. M., P. Y. SZE & H. KREBS. 1981. Maternal influences on tryptophan hydroxylase activity in embryonic rat brain. Dev. Neurosci. **4:** 291–295.
7. STEINBUSCH, H. W. M., A. A. J. VERHOFSTAD & H. W. J. JOOSTEN. 1978. Localization of serotonin in the central nervous system by immunohistochemistry: Description of a specific and sensitive technique and some applications. Neuroscience **3:** 811–819.
8. WALLACE, J. A., P. PETRUSZ & J. M. LAUDER. 1982. Serotonin immunocytochemistry in the adult and developing rat brain: Methodological and pharmacological considerations. Brain Res. Bull. **8:** 117–129.
9. LIDOV, H. G. W. & M. E. MOLLIVER. 1982a. Immunohistochemical study of the development of serotonergic neurons in the rat CNS. Brain Res. Bull. **9:** 559–604.
10. LIDOV, H. G. W. & M. E. MOLLIVER. 1982b. An immunohistochemical study of serotonin neuron development in the rat: Ascending pathways and terminal fields. Brain Res. Bull. **8:** 389–430.
11. WALLACE, J. A. & J. M. LAUDER. 1983. Development of the serotonergic systems in the rat embryo: An immunocytochemical study. Brain Res. Bull. **10:** 459–479.
12. LAUDER, J. M., J. A. WALLACE, H. KREBS, P. PETRUSZ & K. MCCARTHY. 1983. *In vivo* and *in vitro* development of serotonergic neurons. Brain Res. Bull. **9:** 605–625.
13. LAUDER, J. M., A. C. TOWLE, K. PATRICK, P. HENDERSON & H. KREBS. 1985. Decreased serotonin content of embryonic raphe neurons following maternal administration of p-chlorophenylalanine: A quantitative immunocytochemical study. Dev. Brain Res. **20:** 107–114.
14. WHITAKER-AZMITIA, P. M., J. M. LAUDER, A. SHEMMER & E. C. AZMITIA. 1987. Postnatal changes in serotonin$_1$ receptors following prenatal alterations in serotonin levels: Further evidence for functional fetal serotonin$_1$ receptors. Dev. Brain Res. **33:** 285–289.
15. AITKEN, A. R. & I. TORK. 1988. Early development of serotonin-containing neurons and pathways as seen in wholemount preparations of the fetal rat brain. J. Comp. Neurol. **274:** 32–47.
16. KÖNIG, N., M. B. WILKIE & J. M. LAUDER. 1988. Tyrosine hydroxylase and serotonin containing cells in embryonic rat rhombencephalon: A whole-mount immunocytochemical study. J. Neurosci. Res. **20:** 212–223.
17. ALTMAN, J. & S. A. BAYER. 1981. Development of the brainstem in the rat. V. Thymidine-radiographic study of the time of origin of neurons in the midbrain tegmentum. J. Comp. Neurol. **198:** 677–716.
18. NEWTON, B. W., A. B. BURKHART & R. W. HAMILL. 1989. Immunohistochemical distribution of serotonin in spinal autonomic nuclei: II. Early and late postnatal ontogeny in the rat. J. Comp. Neurol. **279:** 82–103.
19. RAJAOFETRA, N., F. SANDILLON, M. GEFFARD & A. PRIVAT. 1989. Pre- and post-natal ontogeny of serotonergic projections to the rat spinal cord. J. Neurosci. Res. **22:** 305–321.
20. BREGMAN, B. S. & P. J. REIER. 1986. Neural tissue transplants rescue axotomized rubrospinal cells of rat and goldfish. J. Comp. Neurol. **236:** 265–273.
21. BUZNIKOV, G. A. 1984. The action of neurotransmitters and related substances on early embryogenesis. Pharmac. Ther. **25:** 23–59.
22. MCMAHON, D. 1974. Chemical messengers in development: A hypothesis. Science **185:** 1012–1021.
23. ROTH, J., D. LE ROITH, J. SHILOACH, J. L. ROSENZWEIG, M. A. LESNIAK & J. HAVRAN-

KOVA. 1982. The evolutionary origins of hormones, neurotransmitters, biology. N. Engl. J. Med. **306:** 523–527.
24. TOMPKINS, G. 1975. The metabolic code. Science **189:** 760–763.
25. HARRIS, W. A. 1981. Neural activity and development. Ann. Rev. Physiol. **43:** 689–710.
26. LAUDER, J. M. 1988. Neurotransmitters as morphogens. Prog. Brain Res. **73:** 365–387.
27. MATTSON, M. P. 1988. Neurotransmitters in the regulation of neuronal cytoarchitecture. Brain Res. Rev. **13:** 179–212.
28. NEMECEK, G. M., S. R. COUGHLIN, D. A. HANDLEY, & M. A. MOSKOWITZ. 1986. Stimulation of aortic smooth muscle cell mitogenesis by serotonin. Proc. Natl. Acad. Sci. USA **83:** 674–678.
29. SEUWEN, K., I. MAGNALDO & J. POUYSSEGUR. 1988. Serotonin stimulates DNA synthesis in fibroblasts acting through 5-HT_{1b} receptors coupled to a G_i protein. Nature **335:** 254–256.
30. JULIUS, D., T. J. LIVELLI, T. M. JESSELL & R. AXEL. 1989. Ectopic expression of the serotonin 1c receptor and the triggering of malignant transformation. Science **244:** 1057–1062.
31. HAYDON, P. G., D. P. MCCOBB & S. B. KATER. 1984. Serotonin selectively inhibits growth cone motility and synaptogenesis of specific identified neurons. Science **226:** 561–564.
32. MCCOBB, D. P., P. G. HAYDON & S. B. KATER. 1988. Dopamine and serotonin inhibition of neurite elongation of different identified neurons. J. Neurosci. Res. **19:** 19–26.
33. GOLDBERG, J. I. & S. B. KATER. 1989. Expression and function of the neurotransmitter serotonin during development of *Helisoma* nervous system. Dev. Biol. **131:** 483–495.
34. WHITAKER-AZMITIA, P. M. & E. C. AZMITIA. 1986. Autoregulation of fetal serotonergic neuronal development: Role of high affinity serotonin receptors. Neurosci. Lett. **67:** 307–312.
35. JONAKAIT, G. M., S. SCHOTLAND & L. NI. 1988. Development of serotonin, substance P and thyrotrophin-releasing hormone in mouse medullary raphe grown in organotypic tissue culture: Developmental regulation by serotonin. Brain Res. **473:** 336–343.
36. CHUBAKOV, A. R., E. A. GROMOVA, G. V. KONOVALOV, E. F. SARKISOVA & E. I. CHUMASOV. 1986. The effects of serotonin on the morphofunctional development of rat cerebral neocortex in tissue culture. Brain Res. **369:** 285–297.
37. LAUDER, J. M. & K. D. MCCARTHY. 1986. Neuronal-glial interactions. *In* Astrocytes. S. Fedoroff & A. Vernadakis, Eds. Orlando, FL: Academic Press, pp. 295–314.
38. WHITAKER-AZMITIA, P. M. & E. C. AZMITIA. 1989. Stimulation of astroglial serotonin receptors produces culture media which regulates growth of serotonergic neurons. Brain Res. **497:** 80–85.
39. WILKIE, M. B. & J. M. LAUDER. 1988. Simple method for the culture of glial cells from embryonic rat brain: Implications for regional heterogeneity and the radial glial lineage. J. Neurosci. Res. **21:** 220–225.
40. KIRSCHGESSNER, A. L., M. D. GERSHON, K. P. LIU, P. Y. YU & H. TAMIR. 1989. Immunocytochemical localization of sites in rat brain that bind anti-idiotype antibodies prepared from antisera to serotonin. Unpublished abst. N.Y. Acad. Sci. Symposium Neuropharmacology of Serotonin.
41. LIU, K. P., S. H. HSIUNG, P. Y. YU & M. D. GERSHON. 1989. Characterization of anti-idiotype antibodies prepared from antisera to serotonin. Unpublished abst. N.Y. Acad. Sci. Symposium Neuropharmacology of Serotonin.
42. JULIUS, D., A. B. MACDERMOTT, R. AXEL & T. M. JESSELL. 1989. Molecular characterization of a functional cDNA encoding the serotonin 1c receptor. Science **241:** 558–564.
43. WHITAKER-AZMITIA, P. M. 1988. Astroglial serotonin receptors. *In* Glial Cell Receptors. H. Kimelberg, Ed. New York: Raven Press, pp. 107–120.
44. KÖNIG, N., V. HAN, E. LIETH & J. LAUDER. 1987. Effects of coculture on the morphology of identified raphe and substantia nigra neurons from the embryonic rat brain. J. Neurosci. Res. **17:** 349–360.
45. AZMITIA, E. C. & P. M. WHITAKER-AZMITIA. 1987. Target cell stimulation of dissociated serotonergic neurons in culture. Neuroscience **20:** 47–63.
46. LIETH, E., A. C. TOWLE & J. M. LAUDER. 1989. Neuronal-glial interactions: Quantitation of astrocytic influences on development of catecholamine neurons. Neurochem. Res. **14:** 979–985.

DISCUSSION OF THE PAPER

M. SEGAL (*The Weizmann Institute, Rehovot, Israel*): Jean, do you have any idea which receptors mediate the effects of pCPA on neuronal differentiation, since most 5-HT receptors are developed postnatally?

LAUDER: It may not be true that 5-HT receptors develop mainly postnatally given the results of a study I did with Pat Whitaker-Azmitia where we treated pregnant rats with pCPA on gestational days 12–17 and found that as neonates these animals had increased numbers of 5-HT binding sites in forebrain and brainstem. Conversely, treatment with the agonist 5-MT had the opposite effect, namely a decrease in the number of binding sites postnatally. In other words, functional receptors must have been present during the prenatal treatment period in order to get such responses to these drugs. Pat will discuss this in more detail in her presentation. In addition, a previous study by UZBEKOV, MURPHY and ROSE (Brain Res. **168:** 195, 1979) reported the presence of relatively high levels of 5-HT binding sites at birth, which fell, then rose again to adult levels during the postnatal period. These studies suggest that functional 5-HT receptors may be present during prenatal development in the rat.

SEGAL: Have you tried treating with antagonists?

LAUDER: No, not yet. We do have additional evidence that 5-HT receptors are present, however. We have stained cell cultures and tissue sections from embryonic rat and mouse brain with an anti-idiotypic antiserum raised against our 5-HT antiserum, kindly given to us by Hadassah Tamir. This antiserum stains a subpopulation of neuroepithelial cells which may be glial progenitor cells since in cultures from embryonic rat brain which are enriched in radial glia and astrocytes these cells also stain with anti-GFAP and the radial glial marker RC-1.

T. BRANCHEK (*Neurogenetic Corp., Paramus, NJ*): I wonder if you would comment on David Julius's findings regarding the mitogenic effects of 5-HT in fibroblasts.

LAUDER: Yes, that is a very interesting study. JULIUS *et al.* 1989 (Science **244:** 1057–1062) have shown that 5-HT acts synergistically with growth factors such as FGF, EGF and insulin to stimulate proliferation of fibroblasts. Serotonin by itself, however, has no effect. We intend to use this same strategy in our own cell culture studies where we have been attempting to show effects of 5-HT on proliferation of embryonic glia, astrocytes and neuroepithelial cells.

QUESTION: It would seem to me that if you upset the time of origin of selected groups of neurons in the developing brain that you should see some abnormalities. Do you have any evidence for this?

LAUDER: We do not see any gross malformations of the CNS such as neural tube defects. The animals also have normal crown-rump-lengths and birthweights. However, they are consistently born almost a day later so that gestation lasts 23 instead of 22 days. We have not analyzed these animals extensively on behavioral tests, but they do not seem to be significantly different from controls on a number of standard tests such as open field, etc. They are different in their response to a thermal gradient as neonates, however, moving much faster than controls towards the warm end.

Role of High Affinity Serotonin Receptors in Neuronal Growth[a]

PATRICIA M. WHITAKER-AZMITIA,[b] ANN V. SHEMER,[b] JEFFREY CARUSO,[b] LORY MOLINO,[b] AND EFRAIN C. AZMITIA[c]

[b]Department of Psychiatry
State University of New York
Stony Brook, New York 11794
and
[c]Department of Biology
Washington Square Center for Neural Science
New York University
New York, New York 10003

INTRODUCTION

In the mature central nervous system, serotonin has long been known to act as a neurotransmitter, however, there is a considerable literature showing that before serotonin functions as a neurotransmitter, it serves a vital role in regulating brain development. Several observations have led to this conclusion. Serotonin cells are one of the earliest cells to develop in the fetal brain. Moreover, as the terminals of serotonin neurons reach their target areas, these areas become differentiated just after they are innervated and animals which have been prenatally depleted of serotonin (such as with PCPA) show lack of proper differentiation.[1] Finally, the time during which development of the brain occurs, is also the time at which serotonin levels are highest, serotonin turnover is most rapid and serotonin receptors are highest in number.[2] Our own work has focused on the role of serotonin in regulating development of the serotonin neuron itself. However, these studies may also have implications for the role of serotonin as a developmental regulator for other neuronal types.

In approaching the problem of what might be the mechanism of action by which serotonin exerts its effect, it is first necessary to consider what components of serotonergic transmission are present in the fetal brain and then to determine if any of these are involved. TABLE 1 summarizes our current knowledge regarding what aspects of serotonin neurotransmission are present during fetal or early neonatal brain development.

We chose to focus on high affinity serotonin receptors for two reasons. First, high affinity serotonin receptors show an interesting developmental pattern in which there appears to be a peak in receptor number around the time of birth. This number then slowly declines for a short period of time before it increases again to adult levels. FIGURE 1 shows our results on serotonin receptor development in rat, but similar results have been found by others.[3,4] Interestingly, this pattern has also recently been described in primate brain (P. Rakic, personal communication). Other neurotransmit-

[a]This work is supported by grants from the National Institute for Neurological Disorders and Stroke and the National Institute for Child Health and Human Development.

TABLE 1. Serotonin Neurochemistry of Developing Rat Brain

Present	Absent
$5\text{-}HT_1$ Receptors and adenylate cyclase	Terminal Synapses
Tryptophan Hydroxylase	
MAO	
Serotonin Binding Protein	
Uptake	
Immature Ion Channels	
?	
$5\text{-}HT_2$ Receptors	
$5\text{-}HT_3$ Receptors	
PI Hy drolysis	

ter systems[5] show a similar pattern for development, and these receptors have also been suggested to play a role in development.

Secondly, prenatal treatment with drugs which affect the serotonin receptor can be shown to have a long-lasting effect on the binding parameters (particularly B_{max}) of receptors in the pup after birth.[6,7] This is interesting, because the receptors are not functioning in neurotransmission, per se, but they are nonetheless capable of responding to neurotransmitter.

For our studies, we chose to use the serotonin receptor agonist 5-methoxytryptamine (5-MT). While this agonist cannot be described as having high affinity for any one subtype of receptor over others, it does recognize virtually all serotonin receptors of both the $5\text{-}HT_1$ and $5\text{-}HT_2$ receptor families with affinities varied enough

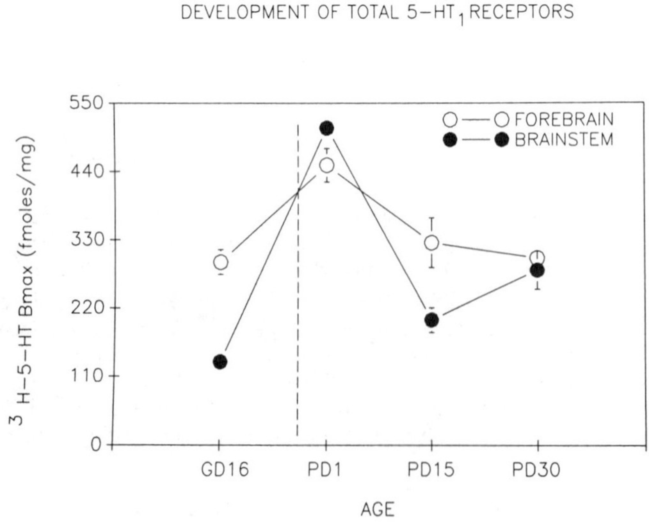

FIGURE 1. The total number of high affinity serotonin receptors in developing rat brain as measured by binding of radiolabeled ^{3}H-serotonin. Each point represents the mean of at least five determinations from saturation analysis using six to ten concentrations of radiolabel. Error bars indicate standard errors of the mean. GD = gestational day; PD = postnatal day; the dashed vertical line indicates birth.

so that multiple responses could be observable. We used 5-MT both in tissue culture models of the developing serotonin neuron and in pregnant animals. In the tissue culture studies, assessment of serotonin neuronal growth was by assay of specific uptake of ^{3}H-serotonin and by morphological analysis of immunocytochemically identified serotonin neurons. The whole animal studies were evaluated by assay of ^{3}H-serotonin uptake as well as by behavioral development in the rat pups.

In a final group of tissue culture experiments, we looked at the role which astroglial cells may play in the development of serotonin neurons. Astroglial cells make up one-half of the cells in the mature animal. They have active uptake systems for many neurotransmitters, a number of receptor types, MAO, and other catabolizing and synthesizing enzymes and in some cases appear to be capable of releasing neurotransmitter.[8] Thus, in the mature brain these cells should be considered as neuromodulating cells. In the immature brain, these cells also play an important role, in regulating brain development. This is accomplished by two major mechanisms—one is by providing a template for neurons to grow along[9] (these are often referred to as radial glial cells) and the other is by producing growth factors.[10]

We had previously found that astroglial cells contain a very high number of high affinity serotonin receptors in the immature state and that this number decreases with the maturational state of the cell.[11] FIGURE 2a is an autoradiographic study of high affinity serotonin receptors on a culture of astroglial cells containing both mature and immature cell types. FIGURE 2b shows the number of high affinity serotonin receptors found in each cell type. By exposing astroglial cultures to serotonin receptor agonists, and assaying the resultant glial-conditioned medium (GCM) for effects on neuronal growth, we were able to screen for factors which may be produced by astroglial cells in response to receptor stimulation and which could regulate serotonin neuronal development.

As a result of our studies, we propose that serotonin neuronal development is dependent on a balance between two types of serotonin receptors, one of which is on astroglial cells and which decreases with age.

METHODS

Tissue Culture Experiments

Neuronal Cultures

For preparation of neuronal cultures, rat embryos at 13–14 days of gestation were removed by Caesarean section from timed pregnant rats (Hilltop Breeding Laboratories). The brains were removed and placed in Eagle's minimum essential medium with 1% glucose. A raphe slice was dissected which contained the majority of the midbrain serotonergic cell bodies. Briefly, two scalpels with #11 blades were used to transect the fetal brain at the mesencephalic and pontine flexures. The cut was then made through the cerebral aqueduct which exposes the tegmentum. Two longitudinal cuts were made approximately 0.5 mm off the midline. The underlying meninges were removed and the strip transferred to Ca^{++}/Mg^{++}–free MEM which contained 1% glucose. The strips were transferred to 5–10 ml of Versene (EDTA) solution, minced, and gently agitated by repeated trituration using a pasteur pipette with a fire-polished tip. The cell suspension was spun at 500 × *g* for 5 minutes and the pellet resuspended in complete media (MEM with 1% glucose, non-essential amino acids, and 5% fetal calf serum), before being spun again. The final pellet was resuspended in complete media (approximately 0.5 cc/raphe slice). The complete media consisted of MEM with non-essential

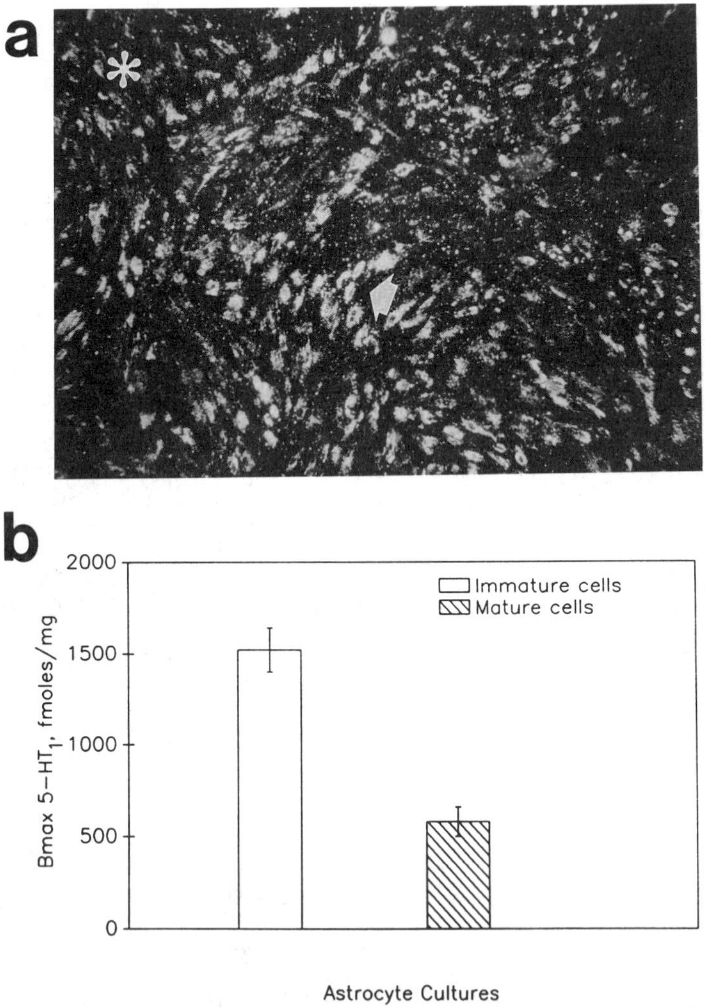

FIGURE 2. Number of high affinity serotonin receptors in mature and immature astroglial cell cultures derived from rat brain. **a:** Autoradiograph of ^{3}H-serotonin binding to primary cultures of rat astroglial cells, viewed under darkfield conditions. The culture shown contains both mature and immature cell types. The highest density of labeling is seen over the cells which are immature (i.e. few processes, large flat cell body), as indicated by the arrow. An area of the culture containing mature cells is indicated by the asterisk. **b:** Saturation analysis of ^{3}H-serotonin binding to homogenates of primary cultures of astroglial cells, showing the higher number of binding sites in the immature cultures.

amino acids (180 ml) fetal calf serum (20 ml, Gibco) and 1% glucose. The cells were plated onto 96-well Linbro plates (Falcon Labware) previously coated with poly-l-lysine (25 μg/ml, Sigma) at initial plating densities (IPD) ranging between 0.30 and 2.4×10^6 cells/cm^2 by adding 200 μl of the cell suspension to each well (area of 0.28 mm^2). Most experiments were performed at 1.2×10^6 cells/cm^2.

Neuronal growth was assessed by the specific fluoxetine-sensitive high affinity uptake of 50 nM ^{3}H-serotonin (a concentration of ^{3}H-serotonin not significantly accumulated by our astroglial cultures). We have previously found this method to be an accurate reflection of serotonergic neuronal membrane surface area.

Primary Astroglial Cultures

Cultures of brain astroglial cells were derived from newborn (1 to 3 day old) Sprague-Dawley rat pups. The pups were decapitated, the heads rinsed with 70% ethanol and the brains removed into minimal essential medium containing 12.5% horse serum and 4% fetal calf serum. Meninges were removed by adhesion onto sterile gauze and brain areas of interest dissected out by use of a dissecting microscope. The brain tissue was minced using two scalpel blades in 0.02% EDTA and then repeatedly triturated into a 21 gauge needle. The resultant cell suspension was centrifuged at 500 × *g*. for twenty minutes and resuspended in culture medium at a concentration of one brain per 5 cc. The cell suspension was then plated into 25 cm^2 flasks, or multiwell plates (Falcon Labware). After one day in culture, non-adhering cells were removed by agitating the flasks and changing the medium. Medium was subsequently changed twice weekly. For older cultures, cells were passed by incubating the culture with pancreatin and sub-culturing 1 to 10. Purity of the cultures was determined by staining the cells for the glial specific protein glial fibrillary acidic protein (GFA) and the fibroblast specific protein fibronectin. Generally, our cultures are 90–95% pure. Production of glial-conditioned medium (GCM) is as follows: Primary astroglial cultures prepared as described above, were rinsed twice with MEM containing 5 mcg/insulin and 0.5 mg/ml albumin but no serum. The cells were left to incubate at 37°C for twelve hours before replacing medium with fresh serum-free medium containing selective agents. After twenty-four hours, the medium was collected and stored frozen in aliquots at −70°C until used in neuronal cultures.

Whole Animal Studies

Pregnant Sprague Dawley rats were housed individually and maintained on a 12 hour light-dark cycle with food and water *ad libitum*. From D12 of gestation until the day of birth the rats were given 5-MT 1 mg/kg or 3 mg/kg or the equivalent volume of saline once a day i.p. Litters were divided and cross-fostered. Neonates were sacrificed on D1, D15, and D30 for biochemical analyses or tested for the following behaviors: spontaneous alternation (days 15, 16, and 30), passive avoidance (day 20–23) or open field activity (day 30).

Synaptosomal Uptake of ^{3}H-Serotonin

On postnatal D1, D15 and D30 brainstem and forebrain sections were removed from the neonate by blunt dissection and homogenized in 10 vol/weight ice cold 0.32 M sucrose with 10 strokes of a loosely fitting glass homogenizer and a teflon pestle. The homogenate was centrifuged (500 × *g*, 5 min) and the supernatant carefully removed and saved. The pellet P1 was resuspended in 2 ml of 0.32 M sucrose and centrifuged again at low speed 500 × *g*, 5 min to obtain a second supernatant. The two supernatants were combined and centrifuged at 10,000 × *g* for 15 min. The resulting crude synaptosomal pellet was resuspended in 10 vols of the original tissue weight of MEM with 1% glucose. Incubations for uptake were performed in triplicate in tissue culture multiwell plates (Linbro 96-well) with a total reaction volume of 200 microliters containing 20 microliters of the P2 suspension. The wells contained MEM as a buffer and 50 nM of ^{3}H-5HT (NEN 26.2 Ci/mole). Non-specific uptake was defined as that occurring in the presence of an excess of unlabeled 5HT (5×10^{-5} M, Sigma). The incubation solution was raised to 37°C and the reaction was started by introducing 20 microliters of the P2 suspension in triplicate. The reaction was terminated after 20 mins. by filtering the incubation medium through Whatman filters and washing for 15 sec. with 0.1 M phosphate buffered saline, with a Titertek cell harvester (Flow Laboratories). Protein levels in the P2 fraction were determined by the Lowry method. Kinetic analysis was performed using concentrations of ^{3}H-5HT from 5 nM to 100 nM. All determinations were in triplicate and uptake expressed as CPM/mg protein/20 min.

Behavioral Studies

Open field activity was measured in a Lehigh Valley Electronics photoactometer drug 85 centimeters in diameter equipped with 12 photobeams. Activity scores were recorded during 5 consecutive blocks of 2 minutes each. Passive avoidance was measured in a simple step through test, in which animals were allowed to cross from one side to the other of a two compartment box. Once the animal had crossed over a door was closed behind it and a shock of 0.3 amps was delivered to the feet for 10 seconds. Three days later the animals were returned to the original side of the box and latency, in seconds, to cross over to the side where they had been shocked was

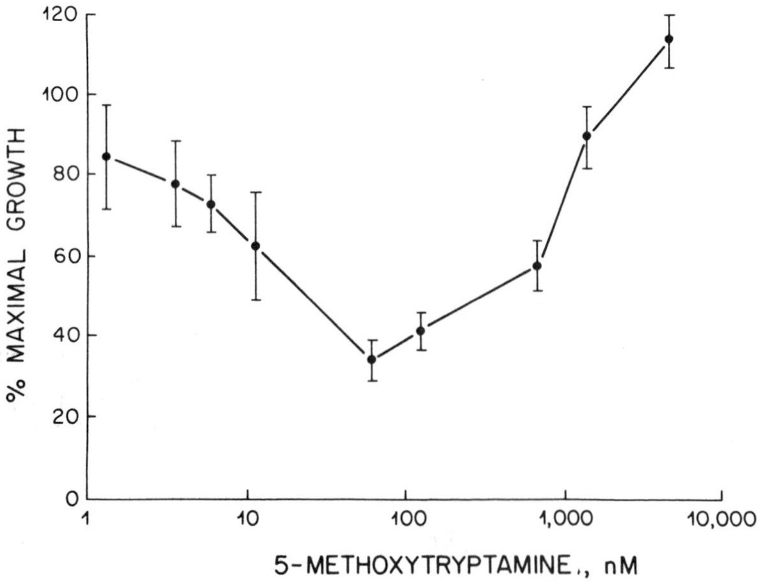

FIGURE 3. Effects of increasing concentrations of 5-methoxytryptamine (5-MT) on the growth of serotonergic neurons in tissue culture, expressed as percent of growth in control cultures. Each point represents the mean result from at least four cultures while the error bars indicate standard errors of the mean.

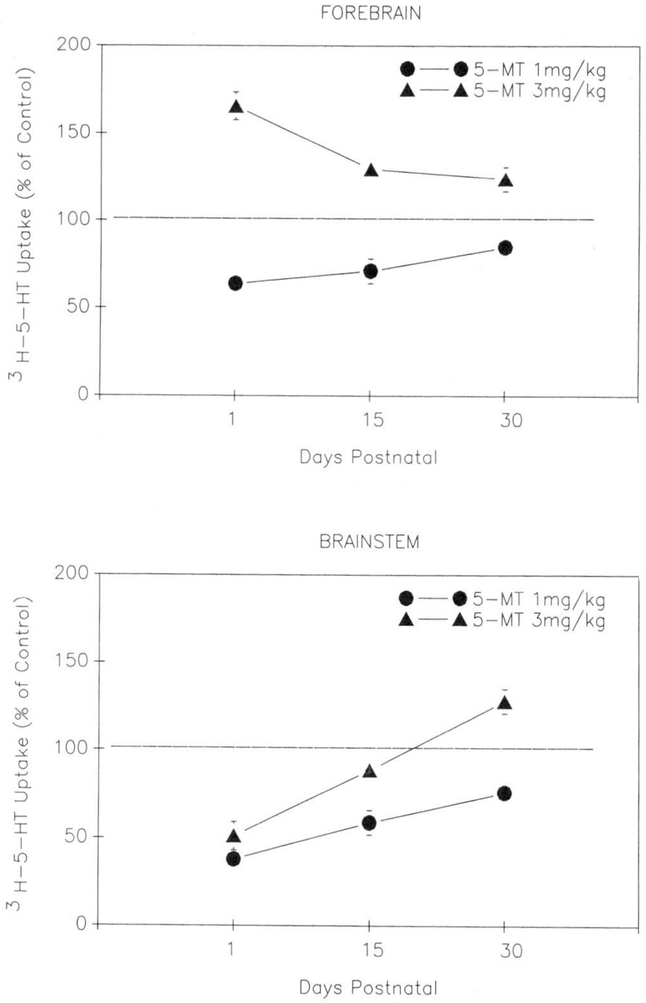

FIGURE 4. Development of serotonergic terminals after prenatal treatment with two different doses of 5-methoxytryptamine (5-MT). Terminal density is measured by specific uptake of 50 nM ^{3}H-serotonin into synaptosomes and expressed as percent of control. In forebrain, the pattern of development is similar to that seen in tissue culture, that is high doses of 5-MT stimulate outgrowth, while low doses inhibit outgrowth, up to 30 days postnatally. The brainstem pattern is more complex, with the increased outgrowth after high doses only seen after 30 days.

measured. Long latencies were considered a sign of memory and/or of behavioral inhibition.

Spontaneous alternation was measured in an apparatus consisting of a start box (4½ in by 4 in) with a door opening into an intermediate compartment (4¾ in by 5½ in) flanked on the right and left by two further chambers (4½ in by 9½ in each). Animals were placed in the start box, the door was raised and then from the intermediate compartment they could choose to go into the right or left chamber, where they were left for 30 seconds with the door closed behind them. They were then placed directly into the start box for a second trial. Animals were allowed eight trials. Making the opposite choice on a subsequent trial was scored as a successful spontaneous alternation. Thus, a total of seven correct responses were possible for each animal. Results were expressed as percentages.

RESULTS

In both tissue culture and whole animal studies, 5-MT had a biphasic effect on the development of serotonin terminals. At low doses (1–100 nM) in tissue culture, 5-MT was strongly inhibitory to growth (decreased to 37% of control). As the dose was increased, this inhibition was overcome, and in fact at high doses (>1,000 nM), a slight stimulation of growth was seen (FIG. 3). Similarily, in whole animals, the uptake of serotonin was differentially affected by the two doses of 5-MT tested. These results are given in FIGURE 4.

The results with the astroglial-conditioned medium (GCM) are given in FIGURES 5–7. These show that a growth-regulating factor is released by astrocytes through

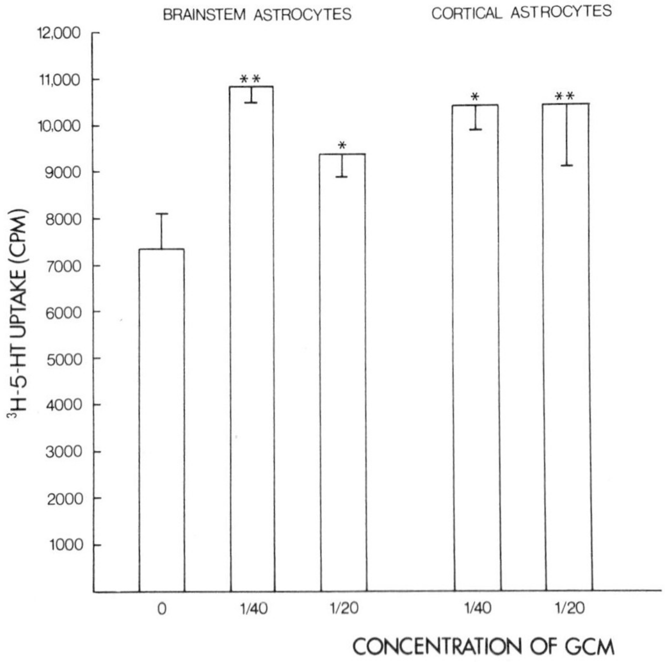

FIGURE 5. Effects of astroglial-conditioned media from different brain regions, on the development of serotonergic neurons in tissue culture. The astroglial cells were derived from the brainstem or cortex of newborn rat pups, and the medium conditioned by exposure of the cells to 500 nM serotonin.

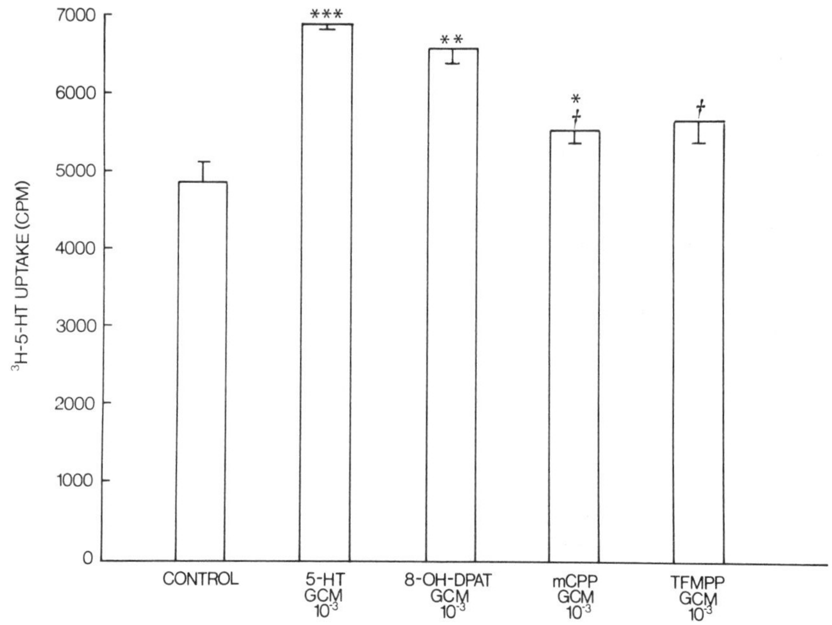

FIGURE 6. Effects of specific serotonin receptor agonists on the production of growth-promoting factors by cortical astroglial cells. Astroglial cultures were exposed for 24 hours to 500 nM serotonin, 500 nM 8-OH-DPAT (a 5-HT_{1a} agonist) or 500 nM mCPP or TFMPP (5-HT_{1b} agonists). The medium was collected, diluted 1/1000, and tested in cultures of serotonergic neurons.

stimulation of serotonin receptors in either brainstem or cortical cultures (FIG. 5). This factor is linked to stimulation of the 5-HT_{1a} receptor (FIG. 6) and requires time to be produced (FIG. 7) suggesting that this is a substance which requires protein synthesis and is not merely a breakdown product of serotonin, or a second messenger (such as cAMP) or a product of intermediate metabolism, such as glucose. In FIGURE 8, the effect of GCM on neuronal cultures of different plating densities is shown. At high plating densities the GCM is inhibitory, while at low plating densities it is mostly stimulatory. FIGURE 9 gives the morphology of the astroglial cultures used to raise the factor. This figure illustrates that as the astroglial cell is being stimulated to produce a growth-regulating factor, it is at the same time being matured itself. Moreover, we have found that mature cultures such as that pictured in FIGURE 9, b and c, do not produce significant amounts of the factor.

The behavioral results are summarized in FIGURES 10–12. The earliest behavior measured, spontaneous alternation, was impaired at 15 and 16 days post-natally, but had returned to normal levels by day 30 (FIG. 10). Similarily, passive avoidance deficits were measurable at day 20, but not at day 30 (FIG. 11). However, using pharmacological challenges, alterations in open field activity were observed at 30 days postnatal. In addition to altered response to 5-MT, the animals also apparently had dopamine deficits, as the response to apomorphine was also altered by prenatal 5-MT treatment (FIG. 12).

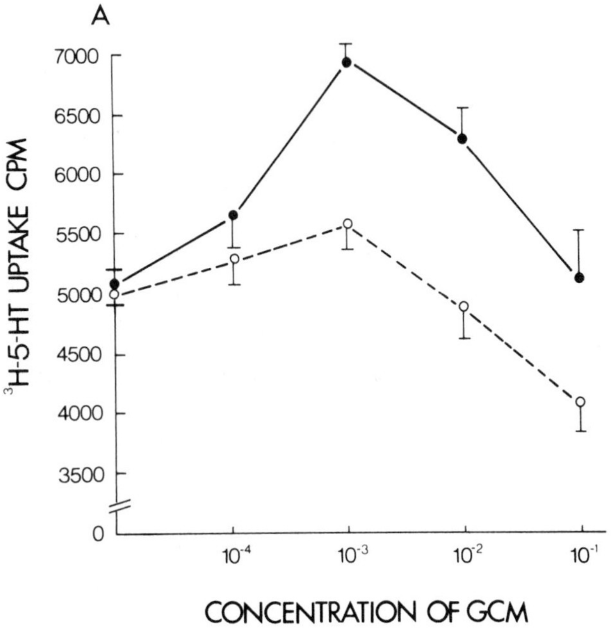

FIGURE 7. Dose-response curves for the effects of serotonin-GCM on growth of serotonergic neurons. The solid line represents data from GCM taken from astroglial cultures exposed for 24 hours to serotonin, while the dashed line represents data from astroglial cultures exposed to serotonin for only 4 hours.

DISCUSSION

Most of the work in the past on development of serotonin or other monoamine-systems in animals has focused only on development taking place after birth. This is an understandable approach considering the small amount of tissue available in fetal animals; however, a large proportion of development has already taken place by birth. Our two approaches to studying brain development—tissue culture and examination of offspring following pre-natal manipulations—should prove useful in increasing our knowledge of the processes taking place at earlier timepoints. In our studies, these two methodologies have given surprisingly similar results in terms of the role serotonin may play in regulating the development of its own neurons. On the basis of this work, we can propose that the development of serotonin neurons is dependent on the functional balance between two high affinity serotonin receptors, which are described below:

1. A receptor recognizing low concentrations of 5-MT which is localized on serotonin neurons and regulates release. Stimulation of this receptor results in disoriented outgrowth, where neurites encircle the cell body or become stunted. We have been unable to characterize this receptor in terms of the pharmacology known for serotonin receptors in the adult brain—that is we have not been able to reproduce the effects of 5-MT, in tissue culture, with known selective receptor agonists including 8-OH-DPAT (5-HT_{1a}), TFMPP (5-HT_{1b}), DOI (5-HT_2) or 2-methylserotonin (5-HT_3). However, the morphology of the cells is similar to that seen after treatment with serotonin-depleting drugs such as MDMA, PCA, or reserpine and for this reason we

have proposed a role for release/depletion in this inhibitory effect of 5-MT on neuronal growth.

2. 5-HT_{1a} receptors, localized primarily on astroglial cells. Stimulation of this receptor results in the astroglial cells producing a factor or factors which can regulate the development of serotonin neurons. When the 5-HT_{1a} receptor on astrocytes is stimulated, several consequences are observable: i) release of the growth-regulating factor(s); ii) maturation of the astroglial cells; and iii) decrease in the number of high affinity serotonin receptors on the astroglial cell. This decrease may lead to the biphasic developmental curves for whole brain serotonin receptors shown in FIGURE 1.

The astroglial-produced medium has several interesting properties. First, the medium has a biphasic effect on growth of serotonin neurons in that there appears to be a peak concentration for maximal stimulation of outgrowth above which the media becomes inhibitory. Secondly, the plating density of the neuronal cultures determines the effects of the medium, so that cells plated at high density are inhibited by the medium while cells plated at low density are stimulated. This is interesting in view of the fact that the serotonin system, like many systems, "overshoots" during development and actually makes many synaptic connections which will subsequently be removed.[12] It is possible that the high plating density represents this overproduction of terminals. Therefore the conditioned medium, and the astroglial 5-HT_{1a} receptors which regulate the production of the factors in the media, are involved in selection of correct synapses and removal of excess or incorrect ones.

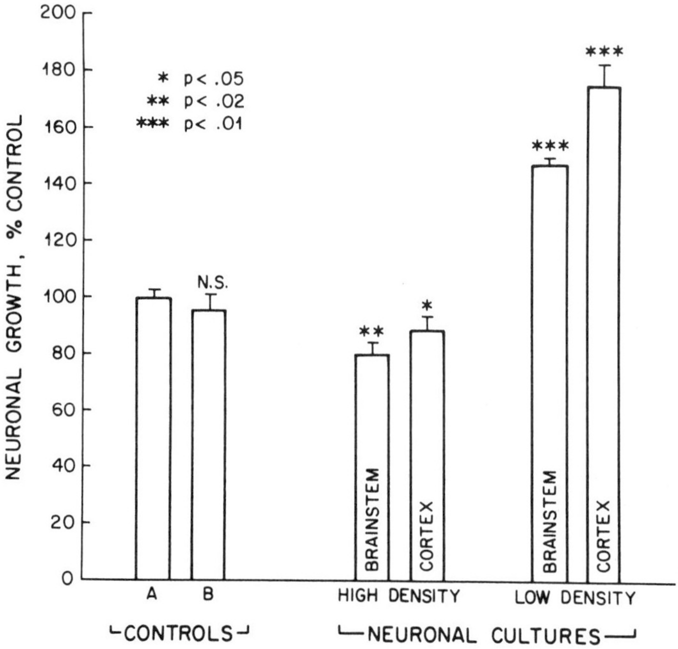

FIGURE 8. Effects of plating density of serotonergic neurons on the ability of serotonin-GCM to promote growth. Control A is medium from unexposed cultures while control B is medium containing 500 nM serotonin which was left at 37°C for 24 hours but not in the presence of cells. Both controls were done in cultures at a medium plating density (approximately 1.2×10^6 cells/cm^2). High plating density was 2.4×10^6 cells/cm^2 and low density was 0.3×10^6 cells/cm^2.

It is also interesting to observe that interfering with development of the serotonin system *in utero* can lead to behavioral changes in the adult animal. The behaviors which we have chosen are all considered to be related to an intact and correctly balanced serotonergic system. In this chapter we have reported changes occurring up to

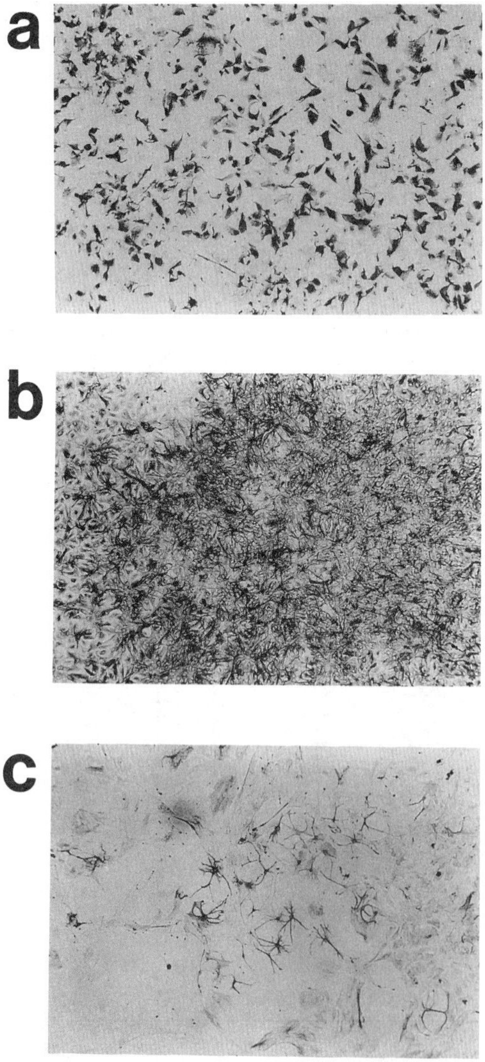

FIGURE 9. Morphology of astroglial cells used to raise the conditioned medium. **a.** Cultures after 24 hours exposure to media containing no agonists (200×). **b.** Cultures after 24 hours exposure to 500 nM serotonin (200×). **c.** Cultures after 24 hours exposure to 500 nM ipsaperone (400×). The addition of the serotonin agonist has caused the astroglial cells to take on a more mature morphology, that is with processes and smaller cell bodies.

30 days, although using other prenatal treatment paradigms we have found changes up to 90 days. It is interesting that we have found changes in behavior even after the neurochemical changes have disappeared. This may be because an adult animal's behavior is dependent upon acquiring certain behaviors at specific times during

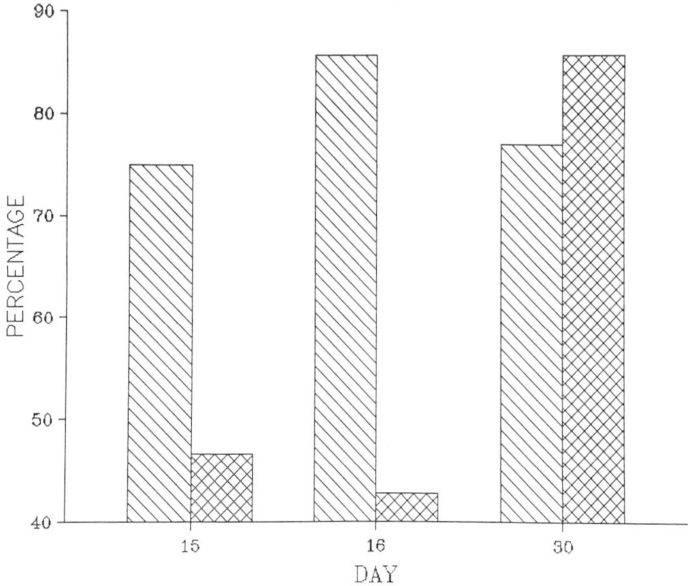

FIGURE 10. Spontaneous alternation in rat pups prenatally treated with 5-methoxytryptamine. Scores are expressed as a percentage of trials in which the animals successfully alternate choices. On days 15 and 16 postnatally the pretreated animals (*diagonal bars*) showed deficits compared to saline treated animals (*cross-hatched bars*). By day 30, the animals had returned to normal levels of alternation.

development. That is, if a particular behavior in a young animal is dependent on specific neurochemistry, and that neurochemistry is altered, then all behaviors acquired subsequently which build on the initial behavior will be altered.

The behavioral response to apomorphine after prenatal 5-MT treatment is suggestive that the dopamine and serotonin systems may in some way rely upon each other for

FIGURE 11. Passive avoidance in rat pups prenatally treated with 5-methoxytryptamine. Results are expressed as seconds before crossing over into the side where a shock had previously been received.

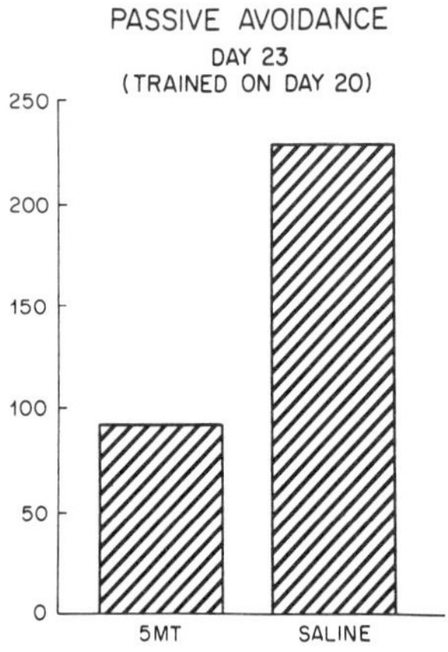

proper development. We are currently focusing on this area of research as the implications for understanding a variety of human mental disorders are obvious. Also, it appears that altering the serotonin system through the dopamine system produces more permanent changes.

The role of astroglial 5-HT_{1a} receptors in regulating development may have parallels in other neurotransmitting systems. For example, beta-adrenergic receptors on astroglial cells are thought to release nerve growth factor (NGF).[13] Indeed, it has

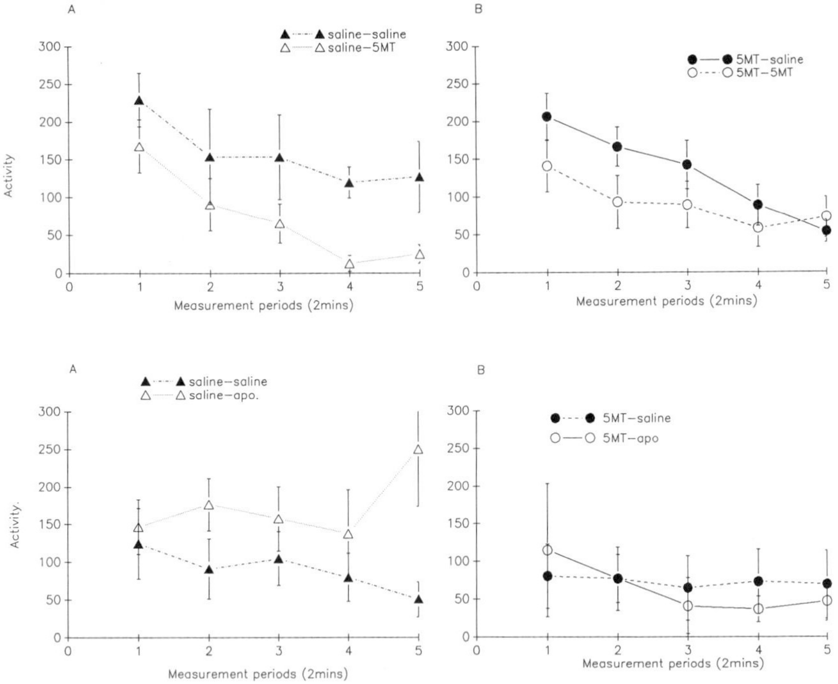

FIGURE 12. Open field activity in 30-day-old rat pups treated prenatally with 5-methoxytryptamine (5-MT) and the response to pharmacological challenges. The top panel illustrates response to acute 5-MT challenges and the bottom panel represents response to an apomorphine challenge. In both panels, **A** is data from animals prenatally treated with saline and **B** is from animals prenatally treated with 5-MT. Animals prenatally treated with saline showed decreased activity after a 5-MT challenge (*top panel,* **A**), but an increased activity after an apomorphine challenge (*bottom panel,* **A**). However, animals prenatally treated with 5-MT, showed blunted response to both 5-MT challenge (*top panel,* **B**) and apomorphine challenge (*bottom panel,* **B**).

been postulated that brain development in general is dependent on cAMP production through receptor stimulation.[14] If this is the case, that neurotransmitter receptors play an important role in regulation of brain development, there are several important consequences.

Firstly, one must consider the potential teratology of drugs used to treat mental illness, since these drugs largely have effects on neurotransmitter receptors. There is evidence, for example, that the use of benzodiazepines or phenytoin during pregnancy results in mental deficiencies in the offspring.[15] The concept of behavioral or mental teratology, as opposed to physical or physiological teratology, is only recently becoming accepted. However, the fact that psychotropic drugs may alter brain development may also put to good use, for example in the case of childhood developmental disorders such as autism. The human brain does not fully mature until the peripubertal period. Theoretically, therefore, if the neurochemical basis of a specific developmental disorder was known and the appropriate pharmacological tools were available, it may be possible to correct an abnormality.

A second ramification of receptors playing a role in development relates to interpretation of research showing neurochemical changes in mental diseases. When receptors or uptake systems are found to be different in a particular mental disease, are we really saying that the brain has not developed properly? The concept of psychiatric illnesses being neurodevelopmental disorders has gained popularity, particularly in schizophrenia research.[16]

The final consequences of receptor-mediated development is that psychotropic drugs may actually be neurontropic—which is to say that these drugs may not only alter neurochemistry but may also alter brain structure. This idea is more understandable when one considers that the mature brain is not static, but in fact shows quite dynamic characteristics, such as synaptic turnover. That is, new synapses are constantly being made while old ones are eliminated. If the same factors which regulate development are responsible for this as well, and it seems probable that they are, then altering receptor function would also alter synaptic profiles in the mature brain. This would also explain the long onset for therapeutic response seen with psychoactive drugs.

In summary we have described a new and exciting role for serotonin—the regulation of brain development. Considering that the factors which regulate development of the brain also play important roles in maintenance, plasticity, and aging, our work could have implications for the understanding of human brain function, dysfunction, and response to drugs throughout life.

REFERENCES

1. LAUDER, J. M. 1981. Hormonal and humoral influences on brain development. Psychoneuroendocrinology **8**(2): 121–155.
2. HAMON, M. & S. BOURGOIN. 1982. Endogenous levels of 5-HT, tryptophan and 5-hydroxyindoleacetic acid (5-HIAA) in the brain of developing rats. *In* Biology of Serotonergic Transmission. N. N. Osborne, Ed. Chichester, England: John Wiley & Sons Ltd., pp. 197–220.
3. UZBEKOV, M. G., S. MURPHY & S.P.R. ROSE. 1979. Ontogenesis of serotonin "receptors" in different regions of rat brain. Brain Res. **168:** 195–199.
4. DUVAL, G., D. VERGE, A. BECERRIL, H. GOZLAN, U. SPAMPINATO & M. HAMON. 1987. Transient expression of 5-HT_{1A} receptors binding sites in some areas of the rat CNS during postnatal development. Int. J. Dev. Neurosci. **5:** 171–180.
5. SEEMAN, P., N. H. BZOWEJ, H-C. GUAN, C. BERGERON, L. BECKER, G. P. REYNOLDS, E. D. BIRD, P. RIEDERER, K. JELLINGER, S. WATANABE, & W. W. TOURTELLOTTE. 1987. Human brain dopamine receptors in children and aging adults. Synapse **1:** 399–404.
6. WHITAKER-AZMITIA, P. M., J. M. LAUDER, A. SHEMER & E. C. AZMITIA. 1987. Prenatal plasticity of high affinity serotonin receptors. Further evidence for a role of receptors in development. Dev. Brain Res. **35:** 285.
7. SOBLOSKY, J. S. & I. JENG. 1985. Influences on 5-[^{3}H]hydroxytryptamine binding site development in chick embryo by serotonergic compounds. J. Neurochem. **44:** 544–551.
8. HERTZ, L. & J. S. RICHARDSON. 1984. Is neuropharmacology merely the pharmacology of neurons—or are astrocytes important too? TIPS: 272–276.

9. SILVER, J., S. E. LORENZ, D. WAHLSTEN & J. COUGHLIN. 1982. Axonal guidance during development of the great cerebral commissures: Descriptive and experimental studies, in vivo, on the role of preformed glial pathways. J. Comp. Neurol. **210**(1): 10–29.
10. ALEXANDER, A. T. & J. DE VELLIS, 1981. Glial-released proteins: III. Influence on neuronal morphological differentiation. Brain Res. **224:** 117–127.
11. WHITAKER-AZMITIA, P. M. & E. C. AZMITIA. 1986. ^{3}H-5-Hydroxytryptamine binding to brain astroglial cells: Differences between intact and homogenized preparations and mature and immature cultures. J. Neurochem. **46**(4): 1186.
12. D'AMATO, R. J., M. E. BLUE, B. L. LARGENT, D. R. LYNCH, D. J. LEDBETTER, M. E. MOLLIVER & S. H. SNYDER. 1987. Ontogeny of the serotonergic projection to rat neocortex: Transient expression of a dense innvervation to primary sensory areas. Proc. Natl. Acad. Sci. USA **84**(12): 4322–4326.
13. SCHWARTZ, J. P. & E. COSTA. 1977. Regulation of nerve growth factor content in C6 glioma cells by β-adrenergic receptor stimulation. Naunyn-Schmiedeberg's Arch. Pharmacol. **300:** 123–129.
14. MCMAHON, D. 1974. Chemical messengers in development: A hypothesis. Science **185:** 1012–1021.
15. HANSON, J. W. & D. W. SMITH. 1975. The fetal hydantoin syndrome. J. Pediatr. **87:** 285–290.
16. FEINBERG, I. 1983. Schizophrenia: Caused by a fault in programmed synaptic elimination during adolescence? J. Psychiatr. Res. **17**(4): 319–334.

DISCUSSION OF THE PAPER

S. J. BOYSON (*University of Colorado H. S. C., Denver):* Astrocytomas are very common in the adult brain and the more malignant grades are characterized by very bizarre looking juvenile type astrocytes. Have you done any receptor binding on this type of cell and is it possible that a serotonin 1A agonist might be a useful treatment for astrocytomas?

WHITAKER-AZMITIA: We have cultured a human glioma and the cells appear to have a very high number of serotonin receptors. Of course I can only speculate on whether or not 5-HT_{1a} receptor agonists would mature them enough that they would no longer be proliferative but it's possible.

PAUL WALKER (*Rutgers University*): Is it possible to have increases in serotonin uptake without having neurite outgrowth?

WHITAKER-AZMITIA: The amount of serotonin uptake by the cultures can be shown very clearly to be correlated with amount of neurite outgrowth. This means that the easiest way for us to quantify outgrowth is simply to do the uptake.

MICHEL HAMON (*INSERM, Paris*) : Do you have direct evidence for the presence of 5-HT_{1a} receptors on astroglial cells?

WHITAKER-AZMITIA: We've done competition studies using autoradiography in our cultures, and I would say we lose 90% of the binding in the presence of 5-HT_{1a} agonists.

The Hyperserotonemia of Autism[a]

GEORGE M. ANDERSON, WILLIAM C. HORNE,
DIPTENDU CHATTERJEE, AND DONALD J. COHEN

Child Study Center and
The Department of Laboratory Medicine
Yale University School of Medicine
New Haven, Connecticut 06510

INTRODUCTION

The first interest in a role for serotonin (5-HT) in mental disorders arose from a consideration of the powerful effects of serotonergic agents, such as lysergic acid diethylamide (LSD), on behavior.[1] Subsequent research has clearly demonstrated the importance of 5-HT for a range of central processes.[2,3] Reports of a critical role for 5-HT during embryogenesis[4] and in the developing nervous system[5–7] has served to further stimulate interest in the possible involvement of 5-HT in development disorders such as autism. In addition, reports of symptom amelioration after treatment with the serotonergic agent fenfluramine[8] and the reported presence of autoantibodies to the 5-HT_{1A} receptor in one autistic individual[9] have also linked 5-HT and autism (although it should be pointed out that neither finding has been fully replicated). However, more than these suggestive connections and findings, it has been the robust and well-replicated finding of increased whole blood or platelet 5-HT in autistic individuals that has most drawn attention to the role of 5-HT in autism. As seen in TABLE 1, a number of groups have replicated Schain and Freedman's initial report of 1961[10] of hyperserotonemia in autism. Group means observed for autistic individuals have ranged from 121% to 237% of those in control groups, with the median increase about 50%. An increase of this magnitude was observed in our relatively recent study which employed an HPLC-fluorometric method to determine whole blood 5-HT.[20] As shown in FIGURE 1, A and B, the increase was observed in unmedicated subjects, was seen across the age range examined, and was apparent whether 5-HT concentration was expressed as ng/ml or ng/10^9 platelets. Autistic subjects appear normal in terms of mean platelet counts[20,24] and platelet size.[24] We also observed a marked effect of neuroleptic and anticonvulsant medication. In medicated autistic subjects, mean levels were actually slightly lower than in normal subjects (see FIG. 1). The drug effect was also apparent when examined within individuals, significantly lower whole blood 5-HT levels being observed when subjects were medicated.[20,25] In addition, an increased individual variance across time was observed in medicated subjects.[25] This drug-induced decrease and temporal instability in 5-HT levels may have contributed to some of the variability in the previous reports (TABLE 1). As had been previously reported,[11,12] 5-HT levels in normal subjects tended to decrease with age during childhood. At least some of this decline can be attributed to decreasing platelet counts. It is also worth mentioning that, although most studies have measured total whole blood 5-HT, the several studies which assayed 5-HT in isolated platelets also observed increased group means in autistic

[a] We gratefully acknowledge the support of NIH (grant MH-30929), the Marks Foundation, and the Gettner Research Fund.

TABLE 1. Blood Serotonin in Autism

Study (Reference)	Sample	Finding (% increase)
Shain & Freedman, 1961 (10)	Whole blood	117%
Ritvo *et al.*, 1970 (11)	Whole blood	22%
Yuwiler *et al.*, 1971 (12)	Whole blood	49%
Campbell *et al.*, 1974 (13)	Platelet-rich plasma	65%
Yuwiler *et al.*, 1975 (14)	Whole blood	33%
Goldstein *et al.*, 1976 (15)	Whole blood	18%
Takahashi *et al.*, 1976 (16)	Platelet pellet	21%
Hanley *et al.*, 1977 (17)	Whole blood	137%
Hoshino *et al.*, 1979 (18)	Serum	25%
Hoshino *et al.*, 1984 (19)	Whole blood	40%
Anderson *et al.*, 1987 (20)	Whole blood	51%
Minderaa *et al.*, 1987 (21)	Whole blood	40%
Badcock *et al.*, 1987 (22)	Whole blood	20%
Launay *et al.*, 1988 (23)	Whole blood	~300%

subjects.[13,16] This is certainly expected, given that nearly all (>99%) of blood 5-HT is found sequestered in the platelet.

Although the basic finding of a group mean increase in autistic subjects is well-established, neither the relationship of blood 5-HT to the behavioral aspects of autism, nor the distribution of the measure in the autistic group are clear. Serotonin levels have not been consistently correlated with degree of mental retardation or other symptomatology. In a group of 21 unmedicated autistic individuals which we studied the distribution of blood 5-HT levels was apparently gaussian, suggesting that the group mean increase was not due to a subgroup of hyperserotonemic individuals.[20] The possibility of a bimodal distribution cannot be completely ruled out, however, given the size of the group examined.

Unfortunately, although the group mean increase is relatively robust, the large degree of overlap with the normal population has prevented the use of blood 5-HT levels in screening or diagnosis of autism. Numerous investigators have been intrigued, however, by the possibility of discovering the cause(s) of the elevation. It has been hoped that an understanding of the mechanism of hyperserotonemia would lead to the development of more specific and useful markers, elucidation of the central abnormality, and useful approaches to the treatment of the condition.

STUDIES OF CAUSATION

Two main possibilities have been considered, at least implicitly, as potential causes of the hyperserotonemia: i) increased exposure of the platelets to free 5-HT as a result of either decreased catabolism or increased synthesis of 5-HT, and ii) altered platelet handling of 5-HT.[12,17,20,26–31]

Increased 5-HT Exposure

The issue of decreased catabolism of 5-HT has been examined by measurements of platelet monoamine oxidase (MAO) activity, and by measuring monoamine substrates and metabolites. The studies of platelet MAO activity have consistently found no difference between autistic and normal subjects.[32-35] This approach is somewhat compromised, however, as platelets contain the B-form of MAO, while 5-HT is predominantly metabolized by MAO-A. In a different approach, we have attempted to assess *in-vivo* MAO activity by determining urine concentration ratios for several substrate/metabolite pairs in autistic and normal subjects. Ratios of dopamine (DA)/homovanillic acid,[36] norepinephrine (NE)/vanillylmandelic acid, NE/3-methoxy-4-hydroxyphenylglycol,[37] and 5-HT/5-hydroxyindoleacetic acid (5-HIAA)[38] in the autistic group were either similar to (DA and NE) or slightly lower than (5-HT) normal values. These results, together with the normal excretion rates generally seen for the

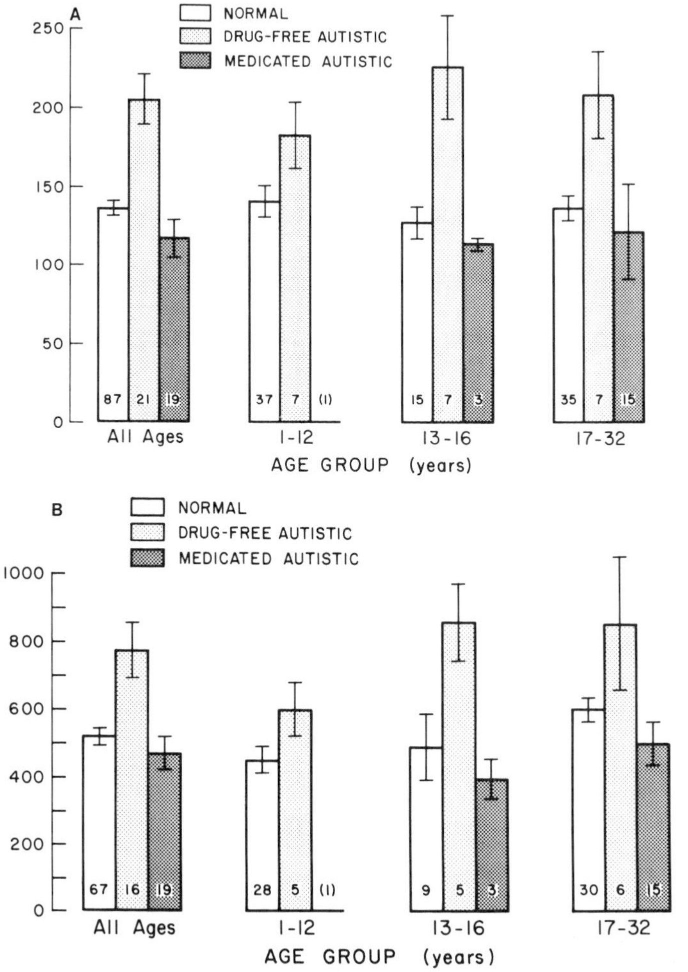

FIGURE 1. Whole blood levels of serotonin determined in autistic and normal subjects. **A:** whole blood serotonin–ng/ml (mean ± SEM); **B:** whole blood serotonin–ng/10^9 platelets (mean ± SEM).

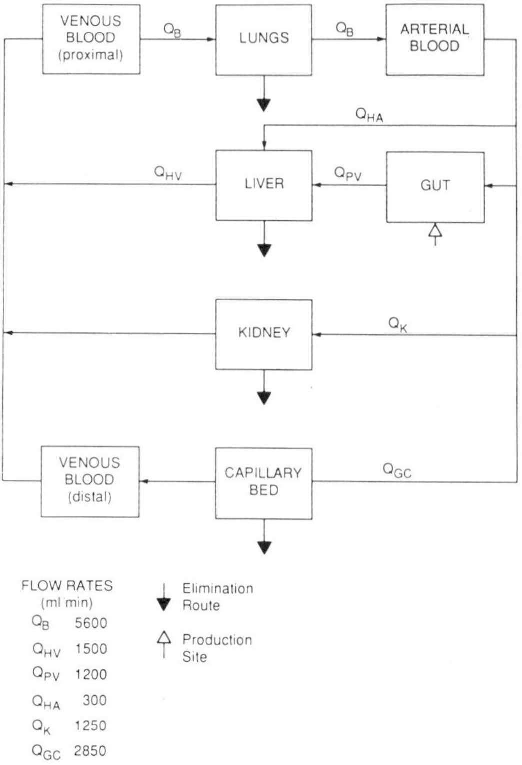

FIGURE 2. Physiological flow model for peripheral serotonin in man. Flow rates (Q values) are for a 170 lb man. A gut production rate of 3 μg/min was estimated based on excretion rates of serotonin and 5-hydroxyindoleacetic acid.

5-HT metabolite, 5-HIAA, strongly suggest that MAO functioning, and hence 5-HT catabolism, is unaltered in autism.

The normal excretion rates observed for 5-HIAA in autistic subjects in most,[10,21,39] but not all[17,40] studies, along with the evidence of normal catabolism and the observation that 5-HIAA is the end-point metabolite for over 95% of 5-HT[41] also strongly indicate that 5-HT synthesis is not increased in autism. Other findings related to this issue are the observation of normal whole blood levels of the 5-HT precursor tryptophan (TRP),[20] and the presence of normal plasma levels of Substance P[38] in autistic subjects.

While these data strongly suggest that 5-HT metabolism is unaltered in autism, they do not definitely indicate whether or not the circulating platelet is exposed to increased levels of 5-HT. This issue can be addressed most directly by determining plasma free levels of 5-HT. We have developed an HPLC-fluorometric method which permits the analysis of 5-HT in platelet poor plasma and plasma ultrafiltrate.[42] This in turn has allowed us to test a model developed for free 5-HT in blood.[43] In the model, shown in FIGURE 2, we have applied concepts of physiological-flow models used to predict levels of infused drugs. No new assumptions have been made concerning 5-HT physiology. Serotonin produced in the gut is known to be cleared by the liver, lungs, capillary endothelium and kidneys. When estimated values for the extraction efficiencies and production rate are substituted into equations describing the model, good agreement with measured plasma ultrafiltrate 5-HT levels was observed.[43] The measured mean value of 397 ± 80 pg/ml in normal subjects was 3–50 times lower than those previously reported for plasma 5-HT.[42] Although it is not possible to say with certainty that the free levels measured *in vitro* are those present *in vivo*, the good agreement between predicted and observed plasma (and urine) levels is encouraging both in terms of the model and with respect to the validity of the *in vitro* measurements.

When the method was used to determine platelet-poor plasma 5-HT levels in autistic subjects, the levels were similar to those seen in normal controls.[38,44] We have also measured urinary excretion rates for 5-HT in autistic subjects and found the group mean to be slightly lower than that seen for normal subjects (104.0 ± 39.4 versus 72.8 ± 19.8 μg/g creatinine, $p = 0.04$).[45] Jouve and colleagues have observed normal levels of urinary 5-HT in autistic children.[46] These more direct measures of plasma 5-HT levels, as well as the evidence regarding 5-HT metabolism, indicate that the platelet is not exposed to increased concentrations of 5-HT. This, in turn, strongly implies that the platelet's handling of 5-HT is altered in autism.

Platelet Research

We believe that the accumulated evidence warrants focusing research on the platelet. Aspects which bear close scrutiny are uptake, storage, and release of 5-HT. After reviewing what has been done in these areas, our ongoing and planned studies, as well as additional possibilities for study, will be discussed.

Despite work in a number of laboratories, no consensus has been reached concerning relative rates of 5-HT uptake by platelets from autistic and normal subjects. As seen in TABLE 2, rates in autistic individuals have been reported to be higher than, similar to, and lower than those in normal subjects. The studies have been more consistent in not finding differences in the affinities for uptake in the two groups. We[35] and others[54] have also examined the number (B_{max}) and affinity (K_D) of the platelet imipramine (IMI) binding site in autistic and normal individuals. In neither study were significant group differences seen in either measure of imipramine binding, binding now thought to occur predominantly directly at the plasma membrane 5-HT uptake site.[55,56] The question of altered efflux has been studied in some detail after an initial report[49] indicated that platelets from autistic subjects had a greater efflux of pre-loaded [^{3}H]-5-HT during *in vitro* incubation. In a multicenter study, the concensus was that autistic and normal groups had similar mean efflux rates.[57] Unfortunately, the techniques used have failed to distinquish between possible alternative mechanisms of

TABLE 2. Platelet Serotonin Uptake in Autism

Study	Reference	Finding
Sankar & Sankar, 1962	47	Decrease
Sankar, 1970	48	Decrease
Boullin *et al.*, 1971	49	Increase
Lucas *et al.*, 1971	50	No difference
Yuwiler *et al.*, 1975	14	No difference
Rotman *et al.*, 1980	51	Increase
Katsui *et al.*, 1986	52	Increase
Launay *et al.*, 1988	23	Decrease/no difference

efflux, e.g., passive diffusion, active secretion, and release as a result of damage to plasma and organelle membranes during *in vitro* manipulation. Also, the reuptake of extra-platelet 5-HT has not been inhibited, thus additionally complicating analysis of the results.

A study by Safai-Kutti *et al.*[58] addressed the question of *in vivo* platelet secretion by measuring plasma levels of beta-thromboglobulin (BTG) and platelet factor 4 (PF4), two platelet-specific secretory proteins found in platelet alpha-granules. Plasma levels of these two proteins are considered to be indicative of the amount of *in vivo* platelet activation and secretion. They also determined bleeding times, a measure of *in vivo* platelet function. The autistic group was observed to have substantially lower plasma levels of BTG and PF4, as well as increased bleeding times, suggesting that platelets of the autistic subjects were less responsive to physiological stimulation. McBride and colleagues[59] have examined the 5-HT_2 receptor-mediated augmentation of ADP-induced platelet aggregation and found lower 5-HT augmentation in the autistic group than in normals. In addition to the blunted augmentation response, the autistic group had lower numbers (B_{max}) of 5-HT_2 receptors detected by [^{125}I]-LSD binding. Interestingly, there was an inverse correlation between 5-HT augmentation and platelet 5-HT levels, with the less responsive platelets having higher 5-HT content.

These clinical studies raise several interesting issues related to platelet serotonin uptake, storage, and efflux, as well as platelet activation and responsiveness. Clarification in these areas would be expected to hasten our understanding of the underlying causes of the hyperserotonemia of autism and of the role of 5-HT in platelet physiology in general. One issue is that of compartmentalization and storage of 5-HT within the platelet. It is well established that most platelet 5-HT is stored within the dense granule,[60] and an average amount of about 0.2 fg 5-HT per vesicle has been estimated.[61] Studies by Crawford have indicated that the dense granule normally contains only about 30% of its total capacity.[62] However, careful research by Costa, Murphy and colleagues[63] appears to show that a three-compartment model for platelet 5-HT (extra-platelet, cytosolic and granular) may be oversimplified. At least two cytosolic 5-HT pools appear to exist, and some 5-HT may move directly from outside the platelet to the granule.

Another unresolved issue is that of the extent of uptake and release of 5-HT by platelets in the circulation. Although it would appear that the platelet continues to take up 5-HT throughout its 9–10 day lifespan,[64,65] it is unclear how much 5-HT leaves the platelet during this period, and whether platelet 5-HT content increases continuously or if an equilibrium loaded level is attained. Studies of the *in vivo* half-life of platelet radiolabeled 5-HT suggest that platelet 5-HT has a low turnover, but they are not definitive.[66,67] Resolution of this issue would have important implications for understanding the cause of autistic hyperserotonemia, for if it were shown that little 5-HT normally leaves the circulating platelet, then a decrease in this small efflux or release rate would not be expected to increase platelet 5-HT levels significantly. Such a result would strongly suggest that increases in platelet 5-HT seen in autism are caused by an increased uptake.

It should be pointed out that increased platelet 5-HT uptake could result from increased rates of uptake or from a longer period of uptake. That is to say, an increased platelet lifespan could lead to a greater accumulation of platelet 5-HT. The tentative finding of lower *in vivo* platelet activation in autistic subjects[58] may be interpreted as suggestive of increased platelet lifespan. Although the area is unclear, in converse situations of increased platelet activation, decreased platelet lifetimes have been observed, as well as decreased platelet 5-HT levels.[68] The lifespan question can be addressed by measuring the rate of recovery of platelet cyclooxygenase activity after administration of aspirin,[69,70] or by more precise methods involving *in vitro* radiolabeling platelets with ^{111}In or ^{51}Cr followed by reinfusion.[71]

Two separate components of platelet physiology, passive efflux of 5-HT down the concentration gradient established by the transport proteins and secretion of 5-HT from storage pools subsequent to platelet activation, must be studied in order to successfully characterize efflux of 5-HT from the platelet. Careful study of 5-HT efflux rates from unstimulated platelets of normal and autistic subjects, carried out in the presence of uptake inhibitors, should specifically address the first of these concerns. In addition, the extent of active 5-HT secretion must be determined, for while rates of unstimulated 5-HT efflux may be similar in autistic and normal subjects, the secretory response may be decreased in autism. Somewhat surprisingly, given the number of studies on platelet 5-HT uptake and efflux, there has been no previous investigation of platelet 5-HT secretion (the release reaction) *per se*.

The finding of McBride and colleagues[59] of an altered 5-HT-mediated aggregation response calls for expanded studies of platelet aggregation and shape change[72] in response to 5-HT and other weak agonists, such as ADP and epinephrine. Analysis of platelet responsiveness to a variety of agonists should help define the extent of possible receptor alterations. Studies of the three platelet responses should complement one another and should lead to a better characterization and further localization of the alteration(s) which might be present in platelets from autistic subjects. Specific aspects of platelet stimulus-response coupling may well become of particular interest depending on the nature of the findings.

In addition to reexamining the questions of uptake and efflux, and expanding the scope of investigation of platelet responsiveness, several other strategies also appear worthwhile. The first, which we have begun, is a broad-based study of platelet composition. This involves the study of platelet proteins, lipids, cations, and other amines and amino acids. The work on cations is directed toward revealing alterations in ion gradients which might affect the uptake of 5-HT at the plasma membrane. Analysis of platelet levels of amines and amino acids such as histamine, catecholamines, glutamate, and taurine address the issue of the degree of specificity of the alteration in platelet 5-HT levels. The protein research has involved the creation of platelet protein inventories using giant-gel (35 cm × 40 cm) two-dimensional gel electrophoresis. By mapping normal human platelet proteins and comparing the more than 1500 resolved proteins to those of platelets from autistic subjects, we hope to discover absent, altered, or novel proteins in the autistic platelet. Although this approach is very broadly focused, the ability of the giant-gels to resolve and detect a majority of the platelet proteins makes it attractive. Work on lipid composition of platelet membranes is also proceeding, with special attention being paid to ganglioside profiles in autistic subjects.

A second area of interest has been the characterization of the platelet 5-HT storage organelle, the dense granule. We have improved granule isolation procedures in order to better study granule composition (stored amines and amino acids, soluble and membrane proteins, and lipids) and function. Using granules isolated by the improved procedure, we have made the first determination of the tetrabenazine binding site in platelet dense granules[73] and are attempting to apply this method to granules isolated from platelets of autistic subjects. In addition, we hope to be able to assess the proton-ATPase responsible for the intragranular acidification, a process which plays a critical role in 5-HT transport into the granule.[74]

Finally, the general area of the regulation of platelet 5-HT uptake, storage, and efflux is of interest. It is not known how, or to what extent, these platelet processes are normally regulated. One can ask, for example, whether 5-HT uptake rates change if the platelet is exposed to more or less 5-HT, or if other compensatory mechanisms exist. Some intriguing data exist which suggest a possible link between the platelet 5-HT_2 receptor and platelet 5-HT levels. As previously mentioned, autistic subjects appear to have lower numbers of 5-HT_2 receptors and increased 5-HT levels.[59] Conversely, neuroleptic treatment increases 5-HT_2 receptor number[75,76] and, as we have discussed, lowers platelet 5-HT.[20] One initial approach to clarifying regulatory aspects might be to study the effect of a range of pharmacological agents, including subtype-specific 5-HT agonists and antagonists, on platelet 5-HT content. Agents found to elevate platelet 5-HT might prove particularly useful in tracking down the mechanisms of hyperserotonemia. While it seems prudent to first investigate the processes themselves in autistic and normal subjects, a better understanding of the factors controlling these processes (uptake, storage and efflux) and their degree of regulation might serve to direct the research.

SUMMARY

The planned and ongoing studies of platelet function and composition should allow us to better define the alteration which we presume to be present in platelets of autistic subjects. Although much of the research focuses on serotonergic aspects, the more general research should permit a better delineation of the extent of the alteration and will protect against a premature narrowing of the inquiry. The methodological development which has been a necessary aspect of the work should contribute to an improved understanding of platelet function and composition, as well as result in improved clinical tools for the assessment of platelet functioning in neuropsychiartic disorders and hematology. As an example, improvements in short-term *in vitro* storage conditions to stabilize aggregation and shape change responses over time were found to be necessary, and are probably critical to an optimal comparison of these phenomena across groups.[77]

The identification of the platelet alteration which is responsible for the hyperserotonemia of autism should prove useful in several ways. It would be expected that assessment of the altered function would provide a marker with less overlap with the normal population than the mutidetermined measure of blood 5-HT. Determination of the specific protein(s) involved in the altered platelet should lead directly to gene probes and chromosomal location. These, in turn, should prove useful for neonatal screening, subtyping, and more powerful genetic and family studies. Work of this sort might also allow early intervention and improved treatment. Finally, characterization of the physiological alteration would provide a basis for focusing studies of brain neurochemistry and should, as well, suggest modes of neuropharmacological intervention. The confidence that one can have in the basic finding of hyperserotonemia in autism and the potential benefits to be derived from its explication make further research in this area of great interest.

REFERENCES

1. WOOLLEY, D. W. 1962. The Biochemical Basis of Psychoses or the Serotonin Hypothesis about Mental Diseases. New York: John Wiley and Sons.
2. IVERSON, S. D. & L. L. IVERSON. 1981. Behavioral Pharmacology 2d edit. New York: Oxford University Press.
3. GREEN, A. R., ED. 1985. Neuropharmacology of Serotonin. New York: Oxford University Press.

4. Buznikov, G. A. 1984. Pharmacol. Ther. **25:** 23–59.
5. Zhou, E. C., S. Auerbach & E. C. Azmitia. 1987. J. Neurosci. Res. **17:** 235–246.
6. Whitaker-Azmitia, P. M., J. M. Lauder, A. Shemmer & E. C. Azmitia. 1987. Brain Res. **430:** 285–289.
7. Lauder, J. M. & H. Krebs. 1986. *In* Developmental Neuropsychobiology. W. T. Greenough and J. M. Juraska, Eds. Orlando, FL: Academic, Press, pp. 119–174.
8. Ritvo, E. R., B. J. Freeman, A. Yuwiler, E. Geller, P. Schroth, A. Yokota, A. Mason-Brothers, G. J. August, W. Klykylo, B. Leventhal, K. Lewis, L. Piggott, G. Realmuto, E. G. Stubb & R. Umansky. 1986. Psychopharmacol. Bull. **22:** 133–140.
9. Todd, R. D. & R. D. Ciaranello. 1985. Proc. Natl. Acad. Sci. USA **82:** 612–616.
10. Schain, R. J. & D. X. Freedman. 1961. J. Pediatr. **58:** 315–320.
11. Ritvo, E. R., A. Yuwiler, E. Geller, E. M. Ornitz, K. Saeger & S. Plotkin. 1970. Arch. Gen. Psychiatry **23:** 566–572.
12. Yuwiler, A., E. R. Ritvo, D. Bald, D. Kyper & A. Kopen. 1971. J. Autism Child. Schizophr. **1:** 421–435.
13. Campbell, M., E. Friedman, E. DeVito, L. Greenspan & P. J. Collins. 1974. J. Autism Child. Schizophr. **4:** 33–41.
14. Yuwiler, A., E. R. Ritvo, E. Geller, R. Glousman, G. Schneiderman & D. Matsuno. 1975. J. Autism Child. Schizophr. **5:** 83–98.
15. Goldstein, M., P. Mahanand, J. Lee & M. Coleman. 1976. *In* The Autistic Syndrome. M. Coleman, Ed. Amsterdam: North-Holland, pp. 57–63.
16. Takahashi, S., H. Kanai & Y. Miyamoto. 1976. J. Autism Child. Schizophr. **6:** 317–326.
17. Hanley, H. G., S. M. Stahl & D. X. Freedman. 1977. Arch. Gen. Psychiatry **34:** 521–531.
18. Hoshino, Y., H. Kumashiro & M. Kaneko. 1979. Fukushima J. Med. Sci. **26:** 79–91.
19. Hoshino, Y., T. Yamamoto, M. Kaneko, R. Tachibana, M. Watanabe, Y. Ono & H. Kumashiro. 1984. Neuropsychobiology **11:** 22–27.
20. Anderson, G. M., F. R. Volkmar, E. L. Hoder, P. McPhedran, R. B. Minderaa, J. G. Young, C. R. Hansen, Jr. & D. J. Cohen. 1987. J. Child Psychol. Psychiatr. **28:** 885–900.
21. Minderaa, R. B., G. M. Anderson, F. R. Volkmar, G. W. Akkerhuis & D. J. Cohen. 1987. Biol. Psychiatry **22:** 933–940.
22. Badcock, N. R., J. G. Spence & L. M. Stern. 1987. Ann. Clin. Biochem. **24:** 625–634.
23. Launay, J. M., P. Ferrari, M. Haimart, C. Bursztejn, F. Tabuteau, A. Braconnier, D. Pasques-Bondoux, C. Luong & C. Dreux. 1988. Neuropsychobiology **20:** 1–11.
24. Geller, E., A. Yuwiler, B. J. Freeman & E. Ritvo. 1988. J. Autism Dev. Disord. **18:** 119–126.
25. Minderaa, R. B., G. M. Anderson, F. R. Volkmar, G. W. Akkerhuis & D. J. Cohen. 1989. J. Autism Dev. Disord. **19:** 129–136.
26. Ritvo, E. R. 1977. J. Child Psychol. Psychiatr. **13:** 373–379.
27. Ritvo, E. R., K. Rabin, A. Yuwiler, B. J. Freeman & E. Geller. 1978. *In* Autism: A Reappraisal of Concepts and Treatment. M. Rutter & E. Schopler, Eds. New York: Plenum Press, pp. 163–183.
28. Young, J. G., M. E. Kavanagh, G. M. Anderson, B. A. Shaywitz & D. J. Cohen. 1982. J. Autism Dev. Disord. **12:** 147–165.
29. Anderson, G. M. & Y. Hoshino. 1987 *In* Handbook of Autism and Pervasive Developmental Disorders. D. J. Cohen & A. Donnellan, Eds. New York: John Wiley & Sons, Inc., pp. 166–191.
30. Yuwiler, A., E. Geller & E. Ritvo. 1985. *In* Handbook of Neurochemistry. E. Lajtha, Ed. New York: Plenum Press, pp. 671–691.
31. Anderson, G. M. 1987. Med. Biol. **65:** 67–74.
32. Boullin, D. J., H. N. Bhagavan, R. A. O'Brian & M. B. H. Youdin. 1976. *In* M. Coleman, Ed. The Autistic Syndromes. Amsterdam: North-Holland, pp. 100–110.
33. Cohen, D. J., J. G. Young & J. A. Roth. 1977. Arch. Gen. Psychiatry **34:** 534–537.
34. Lake, R., M. G. Ziegler & D. L. Murphy. 1977. Arch. Gen. Psychiatry **35:** 553–556.
35. Roth, J. A., J. G. Young & D. J. Cohen. 1976. Life Sci. **18:** 919–924.

36. MINDERAA, R. B., G. M. ANDERSON, F. B. VOLKMAR, G. W. AKKERHUIS & D. J. COHEN. 1989. J. Am. Acad. Child Psychiatry **28:** 190–194.
37. MINDERAA, R. B., G. M. ANDERSON, F. R. VOLKMAR, G. W. AKKERHUIS & D. J. COHEN. 1989. Unpublished data.
38. ANDERSON, G. M. 1989. Unpublished data.
39. SHAW, C. R., J. LUCAS & R. D. RABINOVITCH. 1959. Arch. Gen. Psychiatry **1:** 366–371.
40. SUTTON, H. E., J. H. READ & A. ARBOR. 1958. Am. J. Dis. Child. **96:** 23–28.
41. UDENFRIEND, S., E. TITUS, H. WEISSBACH & R. E. PETERSON. 1959. J. Biol. Chem. **219:** 335–344.
42. ANDERSON, G. M., F. C. FEIBEL & D. J. COHEN. 1987. Life Sci. **40:** 1063–1070.
43. ANDERSON, G. M., J. M. STEVENSON & D. J. COHEN. 1987. Life Sci. **41:** 1777–1785.
44. COOK, E. H., B. L. LEVENTHAL & D. X. FREEDMAN. 1988. Biol. Psychiatry **24:** 488–491.
45. ANDERSON, G. M., R. B. MINDERAA, S. C. CHOE, F. R. VOLKMAR & D. J. COHEN. 1989. J. Autism Dev. Disord. **19:** 349–351.
46. JOUVE, J., J. MARTINEAU, N. MARIOTTE, C. BARTHELEMY, J. P. MUH & G. LELORD. 1986. J. Chromatogr. **378:** 437–443.
47. SANKAR, D. V. S. & D. B. S. SANKAR. 1962. Fed. Proc. **21:** 248.
48. SANKAR, D. V. S. 1970. Acta Paedopsychiatr. **37:** 174–183.
49. BOULLIN, D. J., M. COLEMAN, R. A. O'BRIEN & B. RIMLAND. 1971. J. Austism Child. Schizophr. **1:** 63–71.
50. LUCAS, A. R., K. WARNER & J. S. GOTTLIEB. 1971. Biol. Psychiatry **38:** 39–48.
51. ROTMAN, A., R. CAPLAN & G. A. SZEKELEY. 1980. Psychopharmacology **67:** 245–248.
52. KATSUI, T., M. OKUDA, S. USUDA & T. KOIZUMI. 1986. J. Autism Dev. Disord. **16:** 69–76.
53. ANDERSON, G. M., R. B. MINDERAA, P-P. VAN BENTHEM, F. R. VOLKMAR & D. J. COHEN. 1984. Psychiatry Res. **11:** 133–141.
54. WEIZMAN, A., N. GONEN, S. TYANO, G. A. SZEKELY & M. REHAVI. 1987. Psychopharmacology **91:** 101–103.
55. HUMPHREYS, G. J., J. LEVIN & G. RUDNICK. 1988. Mol. Pharm. **33:** 657–663.
56. MARCUSSON, J. & K. ERIKSSON. 1988. J. Neurochem. **51:** 1477–1481.
57. BOULLIN, D., B. J. FREEMAN, E. GELLER, E. RITVO, E. RUTTER & A. YUWILER. 1982. J. Autism Dev. Disord. **12:** 97–98.
58. SAFAI-KUTTI, S., J. KUTTI & C. GILLBERG. 1985. Acta Paediatr. Scand. **74:** 799–800.
59. MCBRIDE, A. P., G. M. ANDERSON, M. E. HERTZIG, J. A. SWEENEY, J. KREAM, D. J. COHEN & J. J. MANN. 1989. Arch. Gen. Psychiatry **46:** 213–221.
60. DAPRADA, M., J. G. RICHARDS & R. KETTLER. 1981. *In* Platelets in Biology and Pathology. J. L. Gordon, Ed. Vol. **2:** 105–145. New York: Elsevier.
61. COSTA, J. L., T. S. REESE & D. L. MURPHY. 1973. Science. **183:** 537–538.
62. CRAWFORD, N. 1967. Clin. Chim. Acta. **18:** 297–307.
63. COSTA, J. L., K. L. KIRK, D. L. MURPHY & H. STARK. 1981. Br. J. Pharmacol. **72:** 449–459.
64. WHITE, J. G. 1971. Am. J. Pathol. **63:** 403–408.
65. MEZZANO, D., E. ARANDA, S. RODRIQUEZ, A. FORADORI & P. LIVA. 1984. Am. J. Hematol. **17:** 11–21.
66. HEYSSEL, R. M. 1961. J. Clin. Invest. **40:** 2134–2139.
67. ZUCKER, M. B., L. HELLMAN & B. ZUMOFF. 1964. J. Lab. Clin. Med. **63:** 137–146.
68. CLAGETT, G. P., M. RUSSO & H. HUFNAGEL. 1981. J. Lab. Clin. Med. **97:** 345–359.
69. KRAYTMAN, M., 1973. Blood **41:** 587–597.
70. STUART, M. J., S. MURPHY & F. A. OSKI. 1975. N. Engl. J. Med. **292:** 1310–1313.
71. KELTON, J. J. 1986. Transfusion **26:** 19–22.
72. LEYSEN, J. E., D. DE CHAFFOY DE COURCELLES, F. DECLERCK, C. J. E. NIEMEGEERS & J. M. VAN NUETEN. 1984. Neuropharmacology **23:** 1493–1501.
73. CHATTERJEE, D., G. M. ANDERSON, M. CHAKRABORTY & D. J. COHEN. 1990. Life Sci. **46:** 1755–1764.
74. RUDNICK, G. 1986. Ann. Rev. Physiol. **48:** 403–413.
75. BORN, G. V. R. 1988. Experientia **44:** 113–115.
76. GRAHAME-SMITH, D. G., D. P. GEANLY, M. SCHACTER & J. M. ELLIOT. 1988. Experientia **44:** 142–145.
77. HORNE, W. C., G. M. ANDERSON & D. J. COHEN. Unpublished data.

DISCUSSION OF THE PAPER

R. Y. MOORE (*Session Chairman*): Perhaps I can begin by asking you George has anyone looked at the relatives of individuals with autism to see whether there's any abnormality in blood serotonin content?

ANDERSON: There have been several family studies and there is some suggestion that there is some genetic contribution to the levels that are observed. But it's not clear at this point.

I. TÖRK (*University of New South Wales, Kensington, N.S.W., Australia*): I would like to ask whether anybody's looked at the brain serotonin content because I believe that all these changes that you were referring to regarding platelets are expression of some changes that occur in the brain as well regarding receptors. And I don't think that the serotonin changes in platelets themselves cause all this.

ANDERSON: As I said I think our main goal, at this point, is to understand the mechanism behind the change in platelet levels, and then once we have that we have a much better approach to the question of whether a similar alteration exists centrally. In terms of what's been measured in the brain no one has done the neurochemical post-mortem studies in autistic subjects and that's something that needs to be done. There are several brains at the McLean brain bank now and we hope, we're trying, to recruit more subjects for that type of study.

We have also looked at CSF 5-HIAA in specimens that were part of the multi-center study of fenfluramine, which was thought to perhaps be useful in the treatment of autism, and in those samples we didn't see any differences between CSF 5-HIAA in autistic and normal control subjects. Unfortunately, our recent studies on CSF serotonin, which we were also hoping to measure, have indicated that that would not be a useful measure because the levels are so low. In addition to these chemical measures, there are also neuroendocrine measures that have been made. One is the study of Ann McBride which I mentioned looking at the platelet 5-HT_2 receptor B_{max}. In that study, we also looked at the fenfluramine-stimulated prolactin release and found a blunted response there also.

There was an earlier study by a Japanese researcher Yoshihiko Hoshino who, with the 5-hydroxytryptophan, also showed a blunted prolactin response. So there's some suggestion there that there might be altered central 5-HT receptors.

B. KOSOFSKY (*Massachusetts General Hospital, Boston*): There is also evidence that in Down's syndrome there is an increased concentration of serotonin in the blood. Are you aware of any similarities or differences between those two syndromes?

ANDERSON: Well it's my understanding in Down's syndrome actually there's a reported decrease in platelet serotonin. I don's think anyone's actually measured free plasma serotonin in those subjects. But there are a number of reports of decreased platelet serotonin and that's been thought to reflect an alteration in the sodium/potassium ATP-ase at the membrane that results in lower ion gradients and leads to lower uptake rates of serotonin.

R. MAUREEN MURPHY (*University of Cincinnati, OH*): You mentioned that there were some sexual differences in serotonin levels, boys being higher. Do you notice that they present clinically different and also following any of the drugs that alter serotonin levels do you see any differences in their clinical presentation?

ANDERSON: The large groups of boys and girls that we looked at were normal children so I couldn't say from that study. We did look at autistic subjects, but we only had five unmedicated female autistics—which is what one would expect given the 4:1 male to female ratio in the disorder. So I can't really answer the question.

B. J. GUARDIOLA-LEMAITRE (*Paris, France*): Did you look at the glucuronide pathway in urine? I mean, do you look at 5-HT in urine, or conjugated 5-HT in urine, that can explain the difference between 5-HT in blood on 5-HIAA?

ANDERSON: No we did not, we looked at free 5-HT in urine, but we did not look at conjugated.

Plasticity of Fetal and Adult CNS Serotonergic Neurons: Role of Growth-Regulatory Factors[a]

EFRAIN C. AZMITIA

Department of Biology
New York University
New York, New York 10003

MAYA FRANKFURT

The Rockefeller University
New York, New York 10021

MARTHA DAVILA

Department of Pharmacology
University of California College of Medicine
Irvine, California 92717

PATRICIA M. WHITAKER-AZMITIA

Department of Psychiatry
State University of New York
Stony Brook, New York 11794

FENG C. ZHOU

Department of Anatomy
Indiana University School of Medicine
Indianapolis, Indiana 46223

INTRODUCTION

The ability of the serotonergic neurons to regulate brain function is dependent on the number and type of neurons which receive a 5-HT connection (Descarries *et al.*; this volume) and on the number and type of receptors which recognize 5-HT for directing the processing of the 5-HT response. But what determines the strength of the signal? The physiology of the serotonergic system is remarkably stable (Jacobs *et al.*, this volume). Therefore, the strength of the signal is probably independent of firing rate and dependent on the termination density.

In the preceding chapter on development, the regulatory factors which regulate survival and neurite outgrowth were discussed (Whitaker-Azmitia *et al.*, this volume). In this chapter we shall review a variety of biological processes which underlie brain plasticity of the serotonergic neurons. In addition, we shall discuss serotonergic growth-regulatory factors studied with tissue culture of fetal 5-HT neurons. Finally, in

[a]The research described in this report was supported by NSF grant (BNS-8812892) to ECA and NIMCDS grant (NS23027) to FCZ.

the last section we shall illustrate the interactions between both adult and fetal neurons following transplantation.

The purpose of this review will be to illustrate the dynamics of the mammalian brain serotonergic neurons in a variety of circumstances (regeneration and collateral sprouting) and to understand the molecular and cellular basis of that plasticity.

TISSUE CULTURE

Introduction

The first cultured serotonergic CNS neurons were explants of midbrain raphe taken from newborn rats.[38] These postnatal serotonergic neurons survive for up to 16 days in culture. The neurons synthesize and metabolize 5-HT. Detectable levels of ^{3}H-5-HIAA are measured in the medium when radioactive tryptophan is added.

In a dissociated preparation of embryonic (14–15 dg) rat brainstem raphe, the 5-HT neurons were visualized by immunocytochemistry and by specific reuptake of ^{3}H-5-HT.[75] Within 24 hours of plating, 3.3% of the surviving neurons are immunoreactive (IR) to the 5-HT antibody. The percent of surviving neurons IR for 5-HT increases over time up to 36% of the surviving neurons by 28 days in culture. It is interesting to note that a large number of GABA-ergic neurons are observed in this raphe culture system (up to 40% of the neurons in 14-day-old cultures) but GABA is not co-localized with 5-HT.

Dissociated mesencephalic raphe cells from fetal rat (14–18 dg) are grown in microculture in 96 well plates.[5,10] The maturation of serotonergic cells is qualitatively studied using immunocytochemistry and quantitatively by measuring the retention of [^{3}H]5-hydroxytryptamine (6×10^{-8}M). These neurons form varicose fibers and growth cones within hours and survive for up to three weeks in culture. Each serotonergic neuron concentrates approximately 1 fmol of serotonin after 20 minutes of incubation.

ACCELERATED MATURATION

Serotonergic neurons develop more rapidly in culture than most other neurotransmitter systems. Process outgrowth occurs within hours in culture,[9] and Ca^{++}-dependent uptake and release are functionally mature as early as 2–3 days in culture.[9,59] Noradrenergic neurons from 18 dg fetuses show an immature phenotype and do not show substantial uptake for at least three weeks, while specific uptake into dopaminergic neurons occurs after six days in culture.[26] Aggregate cultures of dopaminergic neurons after four days in culture[45] and explant cultures of locus coeruleus neurons after 8–14 days in culture[62,25] contain fluorescent cells with fine fibers. Dissociated mesencephalic cultures do not have fluorescence cell bodies or axonal varicosities for up to three weeks in culture.[56,57] These studies illustrate the variation in morphogenesis of identified transmitter systems in culture.

Target Cell Interaction

Serotonergic cell growth in culture can be stimulated by a variety of normal target and non-target cells. Hippocampal cells induce the greatest increase of [^{3}H]5-HT uptake capacity (FIG. 1). Immunocytochemical staining shows that co-culturing with hippocampal cells does not significantly change the survival of 5-HT-IR cells between 48 and 96 hours. Thus, the increase in accumulation represents a maturation of the uptake and storage capacity of the existing cells.

Other target areas, normally innervated by mesencephalic raphe cells (cortex, caudate and olfactory bulb) stimulate the [^{3}H]5-HT uptake capacity. In addition,

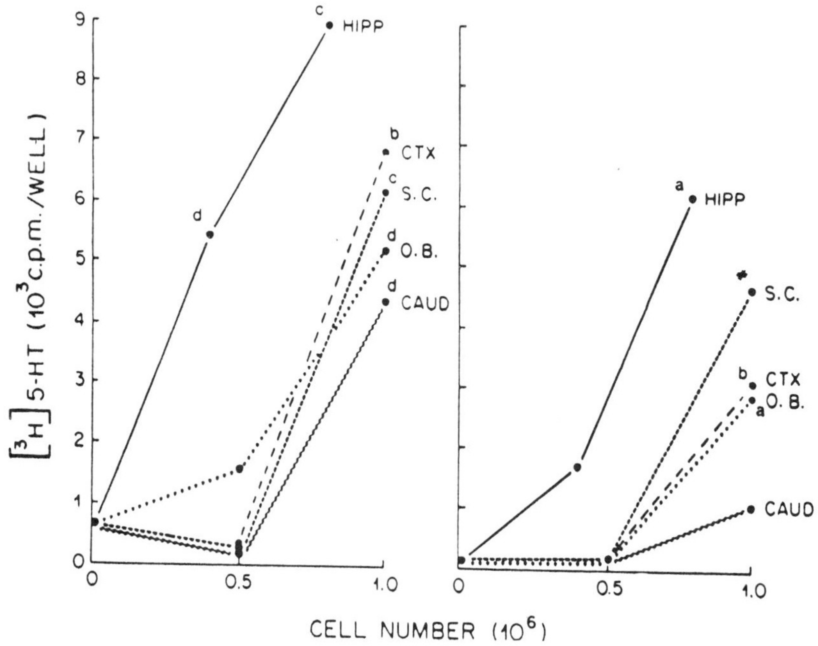

FIGURE 1. The effects of various target cells grown on collagen substrate on the uptake and storage of [^{3}H]5-HT (*ordinate*) by mesencephalic raphe at different initial plating densities (*abscissa*). The initial plating density of the raphe cells in 0.8×10^6(M) on the left graph and 0.4×10^6(M) on the right graph. The most effective target area was the hippocampus which caused a large increase in co-cultures with either low or high densities of raphe cells. Increasing the hippocampus density from 0.4 to 0.8×10^6 cells/cm^2 caused a further increase in [^{3}H]5-HT accumulation. All other target areas were effective at initial plating densities of 1×10^6 cells/cm^2 but not at half this amount. Statistical comparison with raphe alone cultures. Student's t-test: **a**, 0.001; **b**, 0.01; **c**, 0.02; and **d**, 0.05. # = n = 1. HIPP, hippocampus; CAUD, caudate; CTX, cortex; O.B., olfactory bulb; S.C., spinal cord. Each point represents at least four separate cultures. (Reprinted from Azmitia & Whitaker-Azmitia.[9])

serotonergic neurons from the superior mesencephalic group are stimulated by spinal cord cells which they do not normally innervate. The degree of stimulation observed is not directly related to the innervation density reached in the adult brain since hippocampal serotonin levels and fiber density is relatively low.[46,60,66] However, the stimulation by target tissue is related to 5-HT receptor density since hippocampal

5-HT receptors are among the highest in the CNS.[13,52] Pharmacological studies are consistent with this latter explanation.[70,73]

In summary, cultured serotonergic neurons respond to a wide variety of target neurons within a wide developmental window. The combination of diverse target cells and extended time frames could explain the expansive distribution of serotonergic fibers in the vertebrate brain. The characteristics of the target cell stimulation described for 5-HT neurons are unique. These findings hold out the possibility that a widespread maturational factor exists for CNS serotonergic neurons.

SEROTONERGIC GROWTH REGULATORY FACTORS (SGRF)

In this section, we shall focus on a wide range of factors which have significant effects on the survival, attachment, biochemical maturation, and neurite extension of fetal serotonergic neurons in tissue culture. Our selection of factors is broadened to include ions, intermediate metabolites, anti-oxidants, neuropeptides, gangliosides, neurotransmitters and proteins. A detailed treatment of this topic has been published.[10] Furthermore, the results of prior depletion of serotonin on general forebrain maturation has been extensively reviewed.[44] In this section we shall limit our discussion to calcium, free hydroxyl radicals and neurotransmitters.

Calcium and Free Hydroxyl Radicals

Calcium is a divalent cation (Ca^{++}) with a myriad of biologic effects such as inhibition of phosphorylation, uncoupling of mitochondria electron transport, activation of phospholipase and proteases, cell adhesion, and process outgrowth.[34,76,74] In cultures of mouse cortical neurons, low Ca^{++} results in decreased process outgrowth.[74] Conversely, high Ca^{++} levels intracellularly result in neurite retraction and eventual neuronal death.[18,16] Raphe cultures which grow in low Ca^{++} medium have fewer 5-HT-IR cells with smaller somas and fewer processes after three days in culture.[10] Conversely, high levels of Ca^{++} (2–4mM) result in neuronal death. If a specific L-type Ca^{++} channel antagonist (nimodipine, 10^{-9}M) is added to the cultures at the time of plating a substantial potentiation in serotonergic survival and process outgrowth is seen over normal conditions,[12] This indicates that regulating L-type Ca^{++} channels can produce supranormal growth. *In vivo* studies support this concept of L-type Ca^{++} channel stimulation of serotonergic growth. The treatment of pregnant rats with nimodipine results in accelerated outgrowth of serotonergic fibers to forebrain structures.[54] In addition, use of nimodipine in conjunction with neural transplantation results in accelerated growth and survival of the donor fetal neurons.[27,32]

Free hydroxyl radicals are a major cause of neuronal death.[39,41] The production of H_2O_2 by monoamine oxidase contributes to the generation of free radicals inside neurons.[17] Various neurotoxins, such as 1-methyl-4-phenyl-1,2,3,6-tetrahydropyridine (MPTP) and 6-hydroxydopamine (6-OH-DA), act by producing an excess of free hydroxy radicals,[51,67] and neurotransmitters themselves can be converted to potent oxidizing neurotoxins under certain conditions.[37,64] In these studies, the addition of free radical scavengers (e.g. pyruvate, vitamin E) or antioxidant enzymes (e.g. supraoxide dismutase, catalase or glutathione peroxidase) prevent the buildup of oxidants and the corresponding cell death.[49] Serotonergic neurons in culture are suppressed by increased levels of H_2O_2 and enhanced by addition of catalase and glutathione peroxidase (unpublished observation). These results indicate that serotonergic neurons are adversely affected by conditions which generate hydroxy radicals or stimulated by conditions where free-radical buildup is prevented.

Neurotransmitters

Serotonergic neuronal maturation is influenced by its own neurotransmitter during development. Pargyline (a non-specific monoamine oxidase inhibitor) alone and with serotonin (10^{-8}M to 10^{-6}M) inhibit growth of serotonergic neurons. High affinity serotonin receptors in fetal brainstem and in forebrain tissue regulate the direction and extent of growth. These receptors are present in the fetal brain.[72] 5-Methoxytryptamine (5-MT), a serotonin agonist with selectivity for the 5-HT autoreceptors,[49] inhibits growth at nanomolar concentrations. 5-MT produces stunted processes, increases autoinnervation, and leads to death of 5-HT-IR neurons. These inhibitory effects on the growth of cultured 5-HT neurons are reversed at higher concentrations of 5-MT (micromolar) where an actual stimulation is found.

In vivo confirmation of these results uses pregnant rats treated with 5-MT. The pups show suppressed outgrowth of 5-HT fibers towards hippocampus (two weeks to one month) after treatment with low doses of 5-MT (1 mg/kg).[65] The mechanism for these 5-HT receptor mediate stimulatory or inhibitory effects may involve glial cells (see Whitaker-Azmitia *et al.*, this volume).

The autoregulation of fetal serotonergic neuronal development provides a unique way for the CNS 5-HT system to innervate the brain. This mechanism implies that a serotonergic neuron will grow until it reaches a brain region where it produces excess 5-HT due to the activation of the synthesis and/or release of 5-HT. The 5-HT would interact with the terminal autoreceptors and stop the serotonergic cells own growth. Since serotonergic fibers contain terminal autoreceptors,[36] such a mechanism is feasible as a feedback brake on neurite extension.

The excitatory amino acid (EAA) glutamate can produce a Ca^{++}-related cell death in neurons containing an NMDA (*N*-methyl-D-aspartate) receptor.[16,43] The serotonergic raphe neurons have NMDA receptors in association with the excitatory input derived from the lateral habenula. Addition of 0.5 mM glutamate to serotonergic cultures raised with low glucose produces evidence of neuropathology.[12] The addition of MK-801, a specific NMDA antagonist, prevents the decrease in the 5-HT uptake capacity seen in cultured serotonergic neurons exposed to high glutamate. Interestingly, nimodipine, the specific L-type Ca^{++} channel antagonist, also attenuate the effects of glutamate on serotonergic maturation. These results indicate the excess Ca^{++} entry via a variety of channels are additive in producing neuronal toxicity.

The role of neuropeptides as neuronal growth regulating factors has been reviewed.[10,22] Neuropeptides comprise the largest group of putative neurotransmitters (neuroregulators) in the brain and serve as growth factors which are multifunctional.

In dissociated cultures of mesencephalic neurons, low doses of ACTH (1–10 ng/ml) stimulate serotonergic development.[8] The ACTH family of peptides ($ACTH_{4-10}$, $ACTH_{1-24}$ and $ACTH_{1-39}$ and Organon 2766) are all effective in culture. However, if hippocampal fetal neurons (target cells) are present, exogenous ACTH is no longer stimulatory on serotonergic maturation. Interestingly, injections of the ACTH peptides to pregnant rats produce pups which had high levels of hippocampal 5-HT uptake capacity up to 21 days of age (FIG. 2).[23] At longer times, there is no difference between treated and control offsprings. This overriding of the trophic peptide effect by target neurons supports the idea that positive neuropeptide growth factors are general signals during development or after damage but have little direct action on neuronal maintenance in the adult.

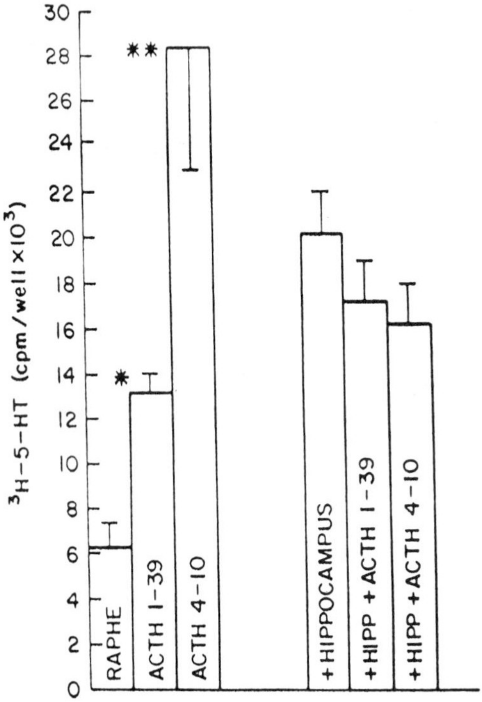

FIGURE 2. The effects of ACTH 1–39 and ACTH 4–10. Peptides were administered daily for 5 days to cultures either of mesencephalic raphe cells (raphe: three columns on left) or of raphe and hippocampal cells (three columns on right). The columns represent the average and SEM of the high affinity uptake of [^{3}H]5-HT in three separate cultures. *$p < 0.05$; **$p < 0.1$. (Reprinted from Azmitia & De Kloet.[8])

Enkephalin is found to be inhibitory to cultured fetal 5-HT neurons at low concentrations (<1 ngm/ml) (FIG. 3).[22] Interestingly, the addition of fetal hippocampal or spinal cord neurons to the raphe cultures enhance the inhibitory properties of this opioid. This result with a negative neuropeptide growth factor, indicates that these peptides have the ability to counteract the trophic effect of the target and halt or repel the growing fiber. There is evidence that the opioid action may be due to a closing of a Ca^{++} channel.[20]

REGENERATION

The ability of injured neurons to send their axons back into the same area they vacated following axotomy is termed "regeneration." The early studies of serotonin regeneration in spinal cord and brainstem were previously reviewed.[4,14] The regrowth of 5-HT fibers after 5,7-DHT lesions in the hypothalamus, where the regrowing fibers not only re-establish their normal synaptic contacts, but can also reestablish function is now discussed.

The unilateral microinjection of 5,7-DHT into the MFB of adult male rats results in destruction of the serotonergic fibers ascending in this pathway. Within one week, the hypothalamus is largely devoid of 5-HT-IR fibers,[28] and the levels of 5-HT and 5-HIAA are greatly reduced (e.g. 90% decrease in dorsomedial nucleus).[30] Behaviorally, estogen-primed ovariectomized rats show a facilitated lordosis following bilateral

5,7-DHT injections.[48] Between 14–19 days after the neurotoxin injection, the 5-HT-IR axons in the caudal hypothalamus have smooth bulbous tips which begin to emit fine wispy processes. 5-HT-IR fibers extend forward into the previously denervated hypothalamic nuclei by one month with the levels of 5-HT increasing to 50% of control levels. The results at 8 weeks after lesion show that 5-HT-IR fibers have hyperinnervated the hypothalamus and the levels of 5-HT and 5-HIAA have returned to control

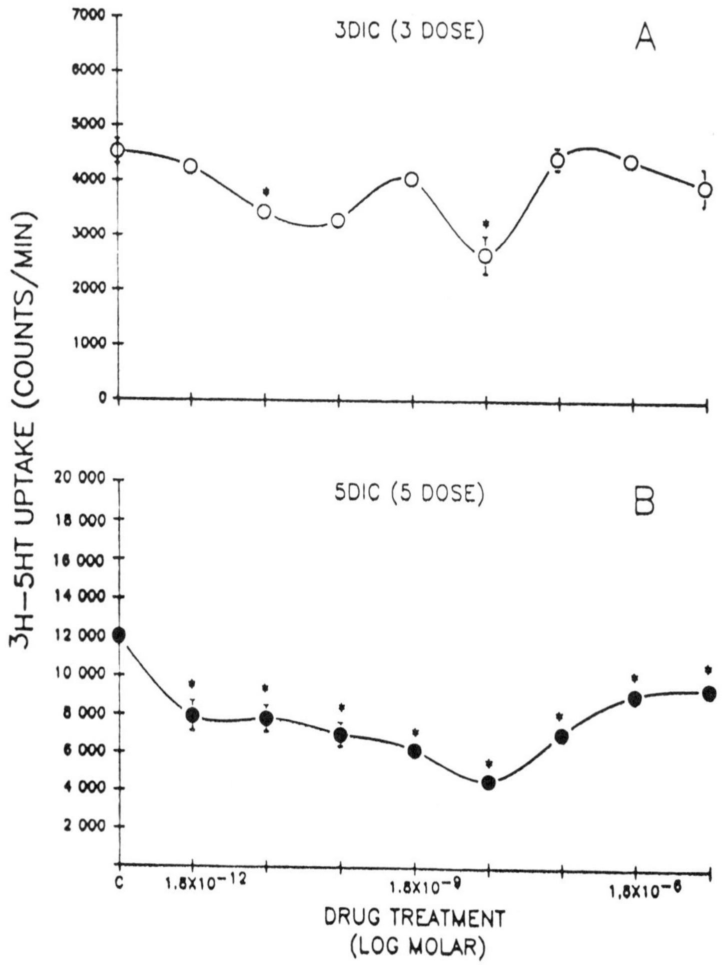

FIGURE 3. Effects of daily application of Leu-enkephalin on fetal brainstem raphe cells. The opioid was added at the time of plating and once a day thereafter at concentrations between 18 mM and 1.8pM in 1:10 serial dilutions for 3 (3 DOSE) or 5 (5 DOSE) days. High affinity uptake of [^{3}H]5-HT was measured after 3(A) or 5(B) days in culture (DIC). Each point is the mean ± SEM (n = 4). Within-group variations were calculated by one-way ANOVA (A,F = 5.2447, $p < 0.00051$; B,F = 9.3435, $p < 0.00002$) and statistical significance was determined by Newman-Keuls test (*$p < 0.05$). (Reprinted from Davila-Garcia & Azmitia.[22])

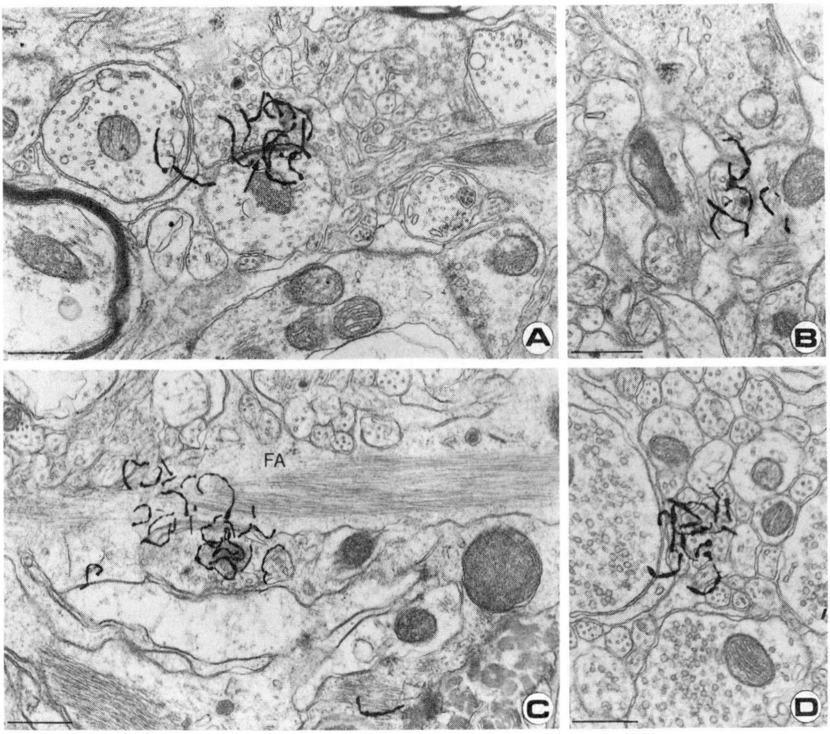

FIGURE 4. Electron microscopic radioautographs from 5,7-DHT- and sham-treated animals 30 or 50 days post-lesion. **A:** [^{3}H]5-HT-labeled terminal or intervaricose axon segment (50 days, sham, ipsilateral) almost entirely surrounded by glia; **B:** [^{3}H]5-HT-labeled terminal (30 days, 5,7-DHT, contralateral) partly embedded in a fibrous astrocyte (FA). **C:** Small [^{3}H]5-HT-labeled terminal or intervaricose axon segment (50 days, sham, ipsilateral) almost entirely surrounded by glia. **D:** [^{3}H]5-HT-labeled terminal or intervaricose axon segment (50 days, sham, ipsilateral) adjacent to a fasicle of small, unmyelinated axons. Scale bars: 0.5mm. (Reprinted from Frankfurt & Beaudet.[29])

levels in many areas. The facilitated lordosis response to estrogen returns to "normal" at 9 weeks, presumably due to the ingrowth of the regenerating 5-HT fibers.

This series of studies indicates that damaged serotonergic fibers can survive an injury and with time send back their axons to reinnervate the denervated region. The hyperinnervation observed by immunocytochemistry is confirmed by light and electron microscopic radioautography.[29] Fifty days after 5,7-DHT neurotoxin injection, there is no significant difference in the types of appositions or synaptic frequency between [^{3}H] 5-HT–labeled axons and the post-synaptic neurons, as compared to sham injected (FIG. 4). Thus the regenerating serotonergic axons grow back to some of the same areas (the dorsomedial hypothalamus and zona incerta), reestablish synaptic and neurotransmitter function and restore the behavior to normal levels. These results provide the morphological substrate for functional recovery and indicate a target directed regrowth signal must exist.

Successful regeneration in the adult CNS requires months while growing fetal neurons can innervate the hypothalamus within a few days. This temporal discrepancy might be due to the lack of a proper substrate in the adult for the growing axonal sprouts to attach. The substrate molecule laminin is not detected in the adult neuropil.[40,47] However, injections of laminin accelerates adult regeneration.[32] The fibers perform properly when they arrive at the target site as indicated by the ability to restore morphological, biochemical, and functional connections. The problem is getting to the target, and laminin may be an important chemical substrate to direct the growing sprouts.

SEROTONERGIC COLLATERAL SPROUTING

Homotypic collateral sprouting is the ability of an undamaged neuron to expand its terminal innervation density in response to removal of fibers containing the same neurotransmitter. This plastic response was first shown in the CNS by 5,7-DHT selective destruction of the serotonergic hippocampal afferents traveling in the cingulum bundle (CB) and monitoring the induced expansion of the undamaged serotonergic fibers traveling in the fornix-fimbria (FF) of adult rats (FIG. 5).[4] Evidence for a hyperinnervation after 90 days is noted. These unilateral CB-lesioned animals also show an asymmetrical turning response to a 5-HT agonist which return to normal after 42 days.

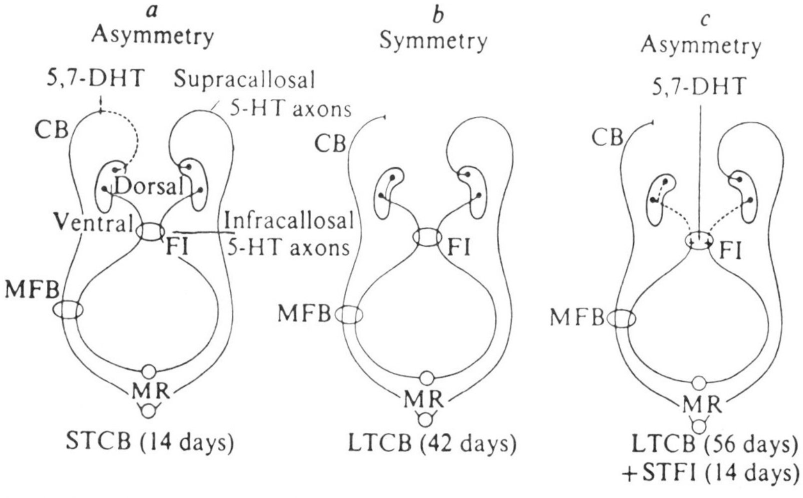

FIGURE 5. Schematic representation of the 5-HT innervation of the hippocampus by supracallosal fibers traveling in the cingulum bundle (CB) and by infracallosal fibers traveling in the fimbria (FI). These fibers originate from different neurons in the medial raphe nucleus and ascend to the forebrain in the medial forebrain bundle (MFB). **a:** The degenerating 5-HT axons (*dashed line*) after unilateral short-term cingulum bundle (STCB) microinjection of 5,7-DHT, and the resultant anatomical asymmetry produced. **b:** The collateral sprouting of the infracallosal 5-HT axons into the dorsal hippocampus which restores symmetry (after long-term cingulum bundle (LTCB) injection of 5,7-DHT). **c:** The degenerating 5-HT axons (*dashed line*) produced by a midline microinjection of 5,7-DHT into the fibria (STFI) of LTCB rats and the resultant asymmetry produced. The midline microinjection into the fimbria of normal rats (not shown) produced no such asymmetry. (Reprinted from Azmitia *et al.*[4])

Further evidence in support of this phenomenon involve HRP-retrograde transport studies.[78] With the latter procedure a significant increase is seen in HRP-labeled neurons in the MRN at 21 and 42 days compared to three days after CB lesion, with the number of cells at 42 days actually being supranormal (121% of sham-injected).

5-HT-IR fibers are distributed in a laminar pattern in the hippocampus.[80] These fibers have large varicosities and are densely distributed in the infragranular layer of dentate gyrus, in the stratum lacunosum-moleculare of the cornu Ammonis and in the area fasciola cinerea (FC) (FIG. 6). The density of the 5-HT-IR fibers in the dorsal hippocampus is greatly decreased but maintains a similar laminar pattern three days after lesioning by 5,7-DHT into the CB. An apparently normal density and distribution pattern of the 5-HT-IR fiber is seen by 42 days post-lesion. The FC in the hippocampus is among the first regions reinnervated by 5-HT-IR fibers with very dense and large varicosities. The restitution of 5-HT-IR fibers in the dorsal hippocampus after the 5,7-DHT lesion in the CB is accompanied by a marked increase in the number and intensity of 5-HT-IR fibers in the FF. No evidence of a regeneration of 5-HT-IR fibers is seen distal to the site of lesion in the CB.

Collateral sprouting of cholinergic and noradrenergic fibers occurs after partial deafferentation of the hippocampus.[24,31,33] The phenomenon of homotypic collateral sprouting is slower than reactive synaptogenesis (RS) where the sprouting begins within days and is essentially complete by two weeks.[19] Several other differences between serotonergic collateral sprouting (SCS) and RS have been noted.[7] One is that the chemical specificity of the lesioned fiber is critical for SCS but not for RS. In RS the proximity of the fiber to the site of lesion is critical whereas in SCS the collateral sprouts can travel long distances (up to a centimeter). Further evidence that SCS is a separate process from RS is their differential response to adrenal steroids. In RS of hippocampal afferents after a lesion of the entorhinal cortex, the removal of circulating adrenal steroids enhances the appearance of new sprouts,[61] while in SCS the adrenal steroids are necessary for sprouting.[79]

When the 5-HT fibers are damaged, there is a period of time before the new collateral sprouts are formed.[7,81] This appears to be due to the formation of a soluble trophic factor by the target cells. The hippocampus can be removed at various times after 5,7-DHT lesions and a soluble extract studied in culture. No significant enhancement of trophic activity is seen until two weeks post lesion.[7] This is in marked contrast to the production of NGF which increases within a few days and peaks in the hippocampus at two weeks post-lesion.[21] Furthermore, when the 5-HT fibers begin to grow there is evidence that they can overshoot their normal levels and hyperinnervate the target neurons. The molecule responsible for these changes may be the glial specific protein S-100 (see ref. 12a).

TRANSPLANTATION STUDIES

Introduction

The first transplantation of serotonergic neurons in to the brain was made into a cavity in the retrosplenial cortex of the rat.[14] The transplantation of serotonergic neurons into the spinal cord were first performed by Nygren *et al.*[55] Only 25% of the transplants were successful. The serotonergic neurons survived for up to three months, and fluorescent fibers could be traced 10mm cranially or caudally from the cell bodies.

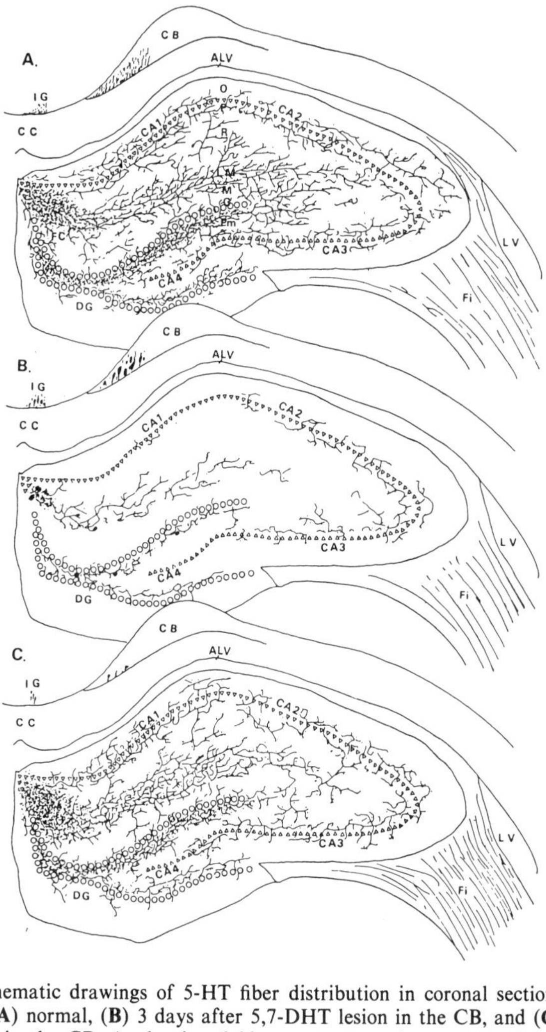

FIGURE 6. Schematic drawings of 5-HT fiber distribution in coronal sections of the dorsal hippocampi of (**A**) normal, (**B**) 3 days after 5,7-DHT lesion in the CB, and (**C**) 42 days after 5,7-DHT lesion in the CB. **A:** the dorsal hippocampus receives a 5-HT innervation via the supracallosal pathway (medial CB-IG) and the infracallosal pathway (FF). These afferent 5-HT fibers mainly terminate in stratum lacunosum-moleculare of CA, infragranular layer of dentate gyrus and the fasciola cinerium (FC). Less dense 5-HT fibers are distributed in stratum oriens and stratum radiatum of the CA areas. The density of the 5-HT fibers in CA areas appears in a sequence of CA1, CA4 and CA3. **B:** 3 days after 5,7-DHT lesion in the CB, degenerated fibers are observed in the CB, IG, CA areas and dentate gyrus. Density of 5-HT-IR fibers are significantly decreased in CB, IG, infragranular layer of dentate, stratum lacunosum-moleculare, of CA and in the FC. The 5-HT-IR fibers in the FF appear normal. **C:** 42 days after lesion, the 5-HT-IR fibers are restored to a normal density and pattern, except, for a greater than normal fiber density in FC. On the lesion side 5-HT fibers regenerate from the degenerated stumps, but do not extend far from the lesion site. A reactive growth of compensatory 5-HT fibers in observed in the FF, which results in an abnormally high density of 5-HT-IR fibers in the ipsilateral FF. **Abbreviations:** CC, corpus callosum; O, stratum oriens; P, stratum pyramidalis; r, stratum radiatum; LM, stratum lacunosum-moleculare; LV, lateral ventricle; m, stratum moleculare (Reprinted from Zhou & Azmitia.[80])

Descriptive Studies and Hyperinnervation

The first microinjections of fetal serotonergic neurons was made into the hippocampus of adult and aged animals.[5] 5-HT-IR fibers hyperinnervate both Cornu Ammonis and the dentate gyrus. This ability to supersede the normal levels of innervation is unique to serotonin. If fetal hippocampus is transplanted, the adult host serotonergic fibers grow into the donor target tissue. The studies demonstrate: (1) the innervation density of the brain by serotonergic neurons can exceed normal levels when fetal neurons are supplied; and (2) fetal hippocampus can induce an expansive response from the adult 5-HT neurons.

The hyperinnervation of target areas by the fetal serotonergic neurons is confirmed by anatomical and biochemical measures of $[^3H]$-5-HT uptake, and 5-HT and 5-HIAA levels. If transplants are made into unlesioned brains the high-affinity uptake of $[^3H]$ 5-HT increases by 53% over normal levels, the 5-HT levels increase by 50% and 5-HT-IR cells and fiber density of 5-HT-IR fibers are supranormal after one month (FIG. 7).[3,81] A prior 5,7-DHT lesion reduces both the uptake of $[^3H]$5-HT and the 5-HT levels to 36% and 30% of normal values, respectively. If the transplants are made into these animals, a dramatic increase of high-affinity uptake (345% over control and 946% over 5,7-DHT primed hippocampus) and 5-HT levels (225% over control and 850% over 5,7-DHT primed hippocampus) is found. Similar results are seen two weeks after intracisternal injection of 5,7-DHT.[68] The levels are 205% and 234% above normal levels and 3383% and 1504% above the 5,7-DHT treated levels for hippocampal 5-HT and 5-HIAA, respectively. These workers find evidence for a change in rotational behavior after ipsilateral transplantation, which is significantly correlated to the asymmetry in 5-HT levels.

Other workers note a hyperinnervation by fetal serotonergic neurons. Holets and Cotman[42] transplanted fetal raphe neurons (16–18dg) into entorhinal cortex of 6-day-old neonatal rat recipients which received a fimbria and entorhinal cortex ablation three days earlier. One week after transplant, some 5-HT-IR fibers are seen in host cortex but none in hippocampus. Between 14–21 days, some 5-HT-IR fibers are seen in CA_2-CA_3 and some in infragranular dentate gyrus. After three weeks, a hyperinnervation is noted in the infra- and supragranular layers of the dentate gyrus, especially in the hilus area.

In ultrastructural studies of the raphe transplants, fifteen percent of 5-HT-IR fibers in the hippocampus form asymmetrical contacts in the normal rat. This percentage is not changed after fetal raphe transplants, even though the number of fibers is greatly increased.[2]

Functional Recovery

The most complete study of functional recovery by transplanted fetal serotonergic neurons is presented by Luine *et al.*[48] It is shown that neurotoxin-induced destruction of serotonergic terminals in the medial hypothalamus facilitates lordosis, decreases 5-HT-IR fiber density, and reduces 5-HT and 5-HIAA levels. Transplantation of the serotonergic neurons restores the 5-HT and 5-HIAA levels 50 days after transplantation, and completely restores the change in lordosis to control levels or below. 5-HT-IR fibers are also seen at normal or hyperinnervated levels in the medial hypothalamus. The studies demonstrate that not only motor behavior, but complex hormonal-dependent sex-linked behaviors require the presence of serotonin, whether derived from the normal midbrain raphe neurons or from transplanted fetal neurons.

Several other complex behaviors can function when fetal are substituted for adult

5-HT neurons. The transplantation of fetal serotonergic neurons into the fourth ventricle may correct sleep disturbances.[35,50] 5,7-DHT microinjection into the fornix-fimbria results in a facilitation of complex maze learning by reducing the mean number of errors and trials to criterion in rats.[1] The transplantation of these animals with fetal serotonergic neurons produces deficits in the maze-learning ability.[58] Finally, it has

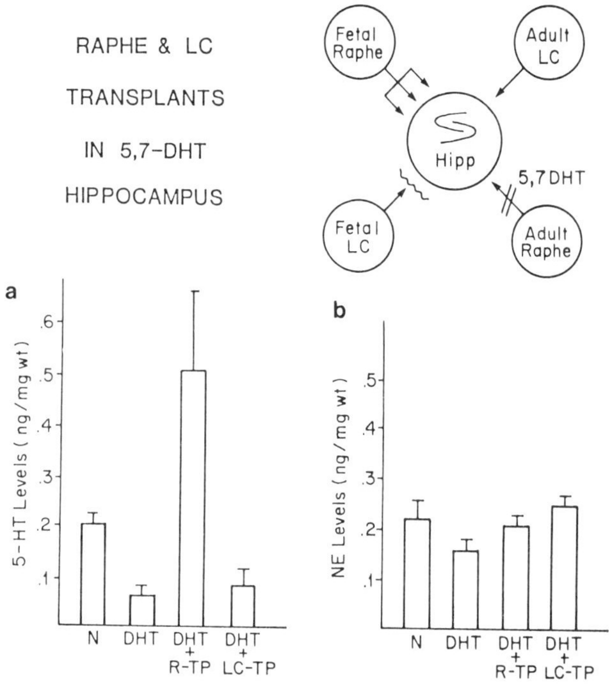

FIGURE 7. The 5-HT and NE levels in the control hippocampus (N), hippocampus with 5,7-DHT lesions, and after raphe or locus coeruleus transplantation in the hippocampus with lesion. When the hippocampus was deprived of 5-HT the transplanted fetal raphe was greatly stimulated, but the growth of the transplanted fetal NE-LC neurons was not facilitated (*see drawing*). Panel **a** on left shows 5-HT levels. The transplanted raphe neurons (R-TP) had a greater increase in level of 5-HT when placed in the hippocampus with lesion than when placed in control hippocampus. The locus coeruleus transplants (LC-TP) did not increase 5-HT level in the hippocampus with 5,7-DHT lesion. Panel **b** on right shows NE levels. There was no difference in NE levels between control hippocampus and hippocampus with 5,7-DHT lesion, nor LC-TP or R-TP in the hippocampus with 5,7-DHT lesions. The NE levels were not higher in the LC-TP in 5,7-DHT lesioned hippocampus as compared with LC-TP in control hippocampus. (Reprinted from Zhou *et al.*[82])

recently been shown that cognitive functions associated with the cortex and hippocampus are dependent on both the serotonergic and cholinergic system.[53,69] Rats injected with 5,7-DHT or administered atropine (a cholinergic antagonist) show only slight dysfunction in spatial memory, but when both treatments are given to the same rats there is a severe loss of spatial memory.[63] If these 5,7-DHT lesioned animals receive a serotonergic fetal neuronal transplant into the dorsal hippocampus, then the atropine injection is no longer effective at disrupting memory. Thus, in cortically based cognitive behaviors, fetal serotonergic neurons provide the input for normal responses. The only requirement is a trophic one—the growth of the fetal 5-HT neurons into the proper target zone. Once the 5-HT fibers are in place, the brain appears to function normally once again.

Host Trophic Factor

Partial removal of 5-HT fibers from the adult hippocampus by injections of 5,7-DHT can induce homotypic collateral sprouting of the undamaged 5-HT fiber.[4,7,79,80] These results indicate the presence of an endogenous trophic factor for 5-HT fibers which can induce adult 5-HT sprouting. Can fetal 5-HT neurons be similarly affected?

This question of a lesion-induced neuronotrophic response was studied in greater detail. Transplants of raphe neurons into an intact adult hippocampus shows increases in high-affinity uptake of [^{3}H]5-HT, 5-HT levels and 5-HT-IR fiber density above normal but not as high as that seen if the transplants are made into brains previously injected with 5,7-DHT.[3,82] Although the extent of fiber outgrowth of the 5-HT-IR neurons is increased in a lesioned primed hippocampus, the number of surviving 5-HT-IR cells is not changed statistically. On the other hand, fetal 5-HT neurons grafted into the cerebellum (which normally receives a scant 5-HT innervation) do not survive well. The co-administration of the hippocampal soluble extract in the transplanted cerebellum is able to prevent 5-HT neuronal loss, stimulate 5-HT fiber growth and increase the 5-HT levels.[85] The neuronotrophic effect of the 5,7-DHT lesion is attributed to a soluble factor since an extract from a lesioned-primed hippocampus could stimulate the development of transplanted 5-HT neurons when transplanted to an intact hippocampus.[82] The hyperinnervation of the transplanted neurons do not appear to displace the normal afferent inputs to the hippocampus from the midbrain raphe neurons.[83] Thus, fetal and adult afferents can coexist in an adult target area and establish a much higher degree of innervation. It is important to note that the enhanced development of fetal serotonergic neurons after lesions with 5,7-DHT, do *not* enhance the development of fetal locus ceruleus neurons.[81] In fact, the noradrenergic neurons *grow worse* into the 5-HT denervated than into the normal hippocampus (see FIG. 7). This is important evidence that the damaged adult brain can produce specific neuronotrophic factors which differentiate between serotonin- and norepinephrine-producing neurons.

One of the problems with fetal axonal growth into adult tissue may be the lack of a suitable substrate for the axons to attach to when growing. To test whether laminin has a similar role in the adult brain, sequentially injected laminin and transplanted fetal neurons are made into various brain regions to determine if the fetal neurons will preferentially grow along a laminin injection tract (FIG. 8). Serotonergic neurons are transplanted to the motor cortex, neostriatum or hippocampus of adult animals.[84] The tract used for microinjection of cell suspension is then immediately filled with suspension medium or laminin. In other animals, laminin or control solution is injected in a separate needle tract displaced 0.3–1 mm from the transplant injection tract. Straight and thick 5-HT-IR fibers are predominant within the laminin-treated tracts, or are directed toward the laminin-treated parallel tracts when it is positioned less than 0.5 mm from the transplant site. The density of 5-HT-IR fibers in the injection tracts in all three brain areas is much higher for laminin than that seen with control medium.

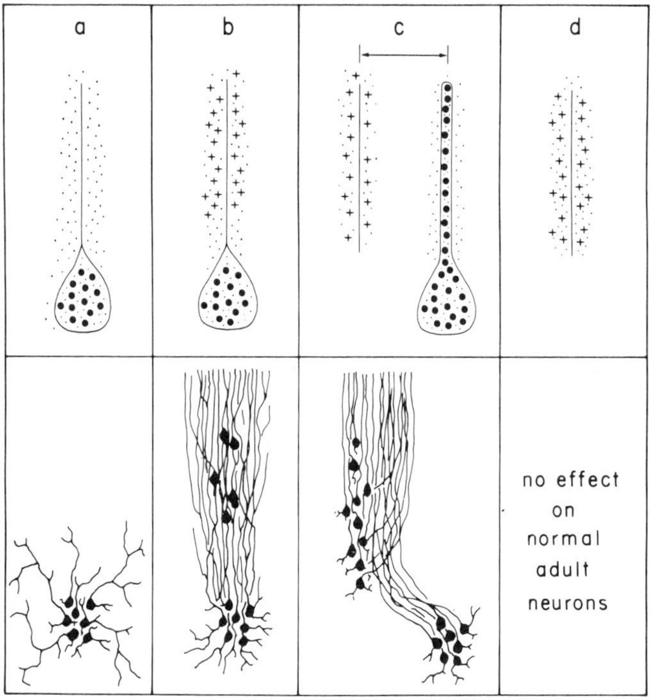

FIGURE 8. Schematic drawings show the effect of laminin on developing and adult neurons: upper panels, experimental designs; lower panels, results. (a) VEHICLE CONTROL: *Upper Panel*—Dissociated brainstem raphe, locus ceruleus and nigra cells (solid circles) were transplanted in an adult brain region, and Hank's solution vehicle (dots) were added dorsal to the graft site through a micropipet immediately after grafting. *Lower Panel*—the fibers of the grafted neurons grew randomly with no recognizable pattern. Few neurons, if any were observed in the graft-tract. (b) LAMININ GROUP: *Upper Panel*—laminin (crosses) was injected dorsal to the transplant. *Lower Panel*—the fibers of transplanted neurons grew preferentially along the laminin-treated tract. A number of neurons migrated up the tract. (c) PARALLEL INJECTION GROUP: *Upper Panel*—raphe and locus ceruleus cells were transplanted in a brain region and a thin layer of cells was deposited in the pipette tract during retracting. Laminin solution was added in a tract 0.3-1mm lateral and parallel to the grafted tract immediately after transplantation. *Lower Panel*—fibers of transplanted neurons grew toward, and a number of neurons migrated to the laminin-treated tract, when they were placed less than 0.5 mm away. (d) LAMININ CONTROL: *Upper Panel*—laminin solution was injected in a brain region where no graft was made. *Lower Panel*—the undamaged adult 5-HT, NE and DA neurons did not respond to laminin treatment. *Solid circle,* transplanted cells; *dots,* Hank's medium (vehicle); *crosses,* laminin. (Reprinted from Zhou & Azmitia.[83])

Thin axonal fibers of fetal 5-HT neurons are observed surrounding the laminin-treated tracts, but not in vehicle-treated tracts. Finally, laminin injection to the hippocampus, motor cortex or neostriatum of the adult brain do not stimulate sprouting of undamaged adult 5-HT fibers. These results suggest that purified laminin can facilitate and guide process outgrowth of 5-HT neurons during an early developmental stage, but does not induce sprouting of these same fiber types in the adult brain.

CONCLUSION

The results we have reviewed indicate that the serotonergic CNS system is extremely plastic and can respond to wide variety of neuronotrophic signals. The plastic responses that have been documented in the adult brain include regeneration of

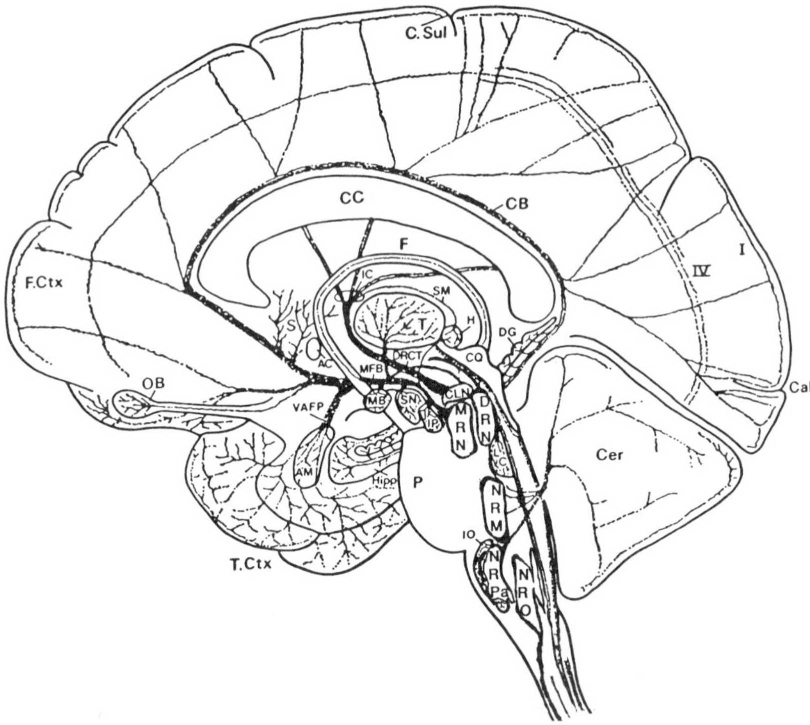

FIGURE 9. Summary diagram of the primate serotonergic system. The main nuclei are shaded. The fiber pathways are shown as broken lines. *Abbreviations:* AC, anterior commissure; Am, amygdaloid nuclei; CC, corpus callosum; Cer, cerebellum; CLN, caudal linear nucleus; CQ, corpus quadrigimini; C.Sul, central sulcus; DG, dentate gyrus; DRCT, dorsal raphe cortical tract; DRN, dorsal raphe nucleus; F, fornix; F.Ctx, frontal cortex; Hipp, hippocampus; IC, internal capsule; IO, inferior olivary nuclei; IP, interpeduncular nucleus; LC, locus coeruleus; MB, mammillary body; MFB, medial forebrain bundle; MRN, median raphe nucleus; NRM, nucleus raphe magnus; NRO, nucleus raphe obscurus; NRPa, nucleus raphe pallidus; OB, olfactory bulb; P, pons; S, septum; SM, stria medullaris; SN, substantia nigra; T, thalamus; T.Ctx, temporal cortex; VAFP, ventro amygdalofugal pathway. (Reprinted, with modification, from Azmitia & Gannon.[4a])

the damaged 5-HT fibers back to their original contacts in the hypothalamus. In this case *morphological, biochemical,* and *functional* restoration of the lost functions were noted. In the process of homotypic collateral sprouting a similar response in these three parameters after sprouting of undamaged 5-HT fibers in the hippocampus was seen.

Finally, neuronal transplantation of fetal neurons in the adult hypothalamus and hippocampus can also lead to restoration of lost functions.

There are several important rules to be extracted from this collection of work. (1) The adult neural circuitry can be rewired by use of trophic and toxic factors or by the introduction of fetal neurons. These changes are functional and relatively permanent. (2) Adult 5-HT neurons are capable of further growth provided the appropriate trophic signals are activated. (3) Fetal and adult neurons generally respond similarly to same trophic and toxic factors, although the fetal neurons are more sensitive. This fact allows the result obtained from controlled tissue culture studies to be applied to the neurons in the whole animal's brain. Understanding the interactions of the trophic and toxic molecules on the developing 5-HT CNS might help explain the principles governing the building of one of the most expansive neural systems in the mammalian brain (FIG. 9).

The role of specific receptor subtypes in regulating the synthesis, release and function of neuronotrophic molecules is just now being investigated. The direct action of 5-HT in regulating process outgrowth and in neuropathology is evident from the reports from a variety of laboratories, many of which are included in this symposium. Harvesting the beneficial results of neuropharmacological agents in promoting neuronal survival and maintenance should produce an environment for more efficient and prolonged use of the finite number of neurons present in the adult brain.

ACKNOWLEDGMENTS

Thanks to Xiao-Ping Hou, Tajrena Alexi, and Susannah Tapley for technical and editorial support.

REFERENCES

1. ALTMAN, H. J., M. P. NORMILE, T. GALLOWAY, A. RAMIREZ & E. C. AZMITIA. 1988. Facilitation of complex maze learning in rats following selective 5-HT deafferentation of the dorsal hippocampus. Soc. Neurosci. Abstr., **14:** 725.
2. ANDERSSON, K. J., V. R. HOLETS, P. C. MAZUR, R. LASHER & C. W. COTMAN. 1986. Immunocytochemical localization of serotonin in the rat dentate gyrus following raphe transplants. Brain Res. **369:** 21–28.
3. AUERBACH, S., F. ZHOU, B. L. JACOBS & E. C. AZMITIA. 1985. Serotonin turnover in raphe transplanted into rat hippocampus. Neurosci. Lett. **61:** 147–152.
4. AZMITIA, E. C., A. M. BUCHAN & J. H. WILLIAMS. 1978. Structural and functional restoration by collateral sprouting of hippocampal 5-HT axons. Nature **274:** 374–377.
4a. AZMITIA, E. C. & P. J. GANNON. 1986. Anatomy of the serotonergic system in the primate and sub-primate brain. *In* Myoclonus. S. Fahn, C. D. Marsden, and M. Van Woert, Eds. Adv. Neurol. **43:** 407–468.
5. AZMITIA, E. C., M. J. PERLOW, M. J. BRENNAN & J. M. LAUDER. 1981. Fetal raphe and hippocampal transplants into adult and aged C57BL/6N mice: A preliminary immunocytochemical study. Brain Res. Bull. **7:** 703–710.
6. AZMITIA, E. C., M. I. DAVILA, P. J. LAMA, R. B. MURPHY, R. PESASEYED & P. M. WHITAKER-AZMITIA. 1986. Serotonergic growth factor from adult hippocampus supernatant stimulated the maturation of serotonergic neurons in dissociated mesencephalic cultures. Soc. Neurosci. Abstr. **12:** 576.
7. AZMITIA, E. C. & F. C. ZHOU. 1986. Chemically induced homotypic collateral sprouting of hippocampal serotonergic afferents. *In* Process of Recovery from Neural Trauma. G. M. Gilad, A. Gorio, and G. W. Kreutzberg, Eds. Exp. Brain Res. (Suppl. 13): 129–141.
8. AZMITIA, E. C. & E. R. DE KLOET. 1987. ACTH neuropeptide stimulation of serotonergic

neuronal maturation in tissue culture: modulation by hippocampal cells. *In* Progress in Brain Research, Vol. 72. E. R. de Kloet, V. M. Wiegant, and D. de Wied, Eds. Amsterdam: Elsevier, pp. 311–317.

9. Azmitia, E. C. & P. M. Whitaker-Azmitia. 1987. Target cell stimulation of serotonergic neurons in primary dissociated tissue culture. Neuroscience **20:** 47–63.
10. Azmitia, E. C., P. M. Whitaker-Azmitia & R. Bartus. 1988. Use of tissue culture models to study neuronal regulatory trophic and toxic factors in the aged brain. Neurobiol. Aging **9:** 743–758.
11. Azmitia, E. C. 1990. Microcultures of dissociated primary CNS neurons. *In* Methods in Neuroscience, Vol. 2. P. M. Conn, Ed. Orlando, FL: Academic Press. In press.
12. Azmitia, E. C. 1990. Nimodipine attenuates NMDA- and MDMA-induced toxicity of cultured fetal serotonergic neurons: Evidence for a generic model of calcium toxicity. *In* Nimodipine and Central Nervous System Functions—New Vistas. W. H. Gispen, and J. Traber, Eds. Berlin: Springer-Verlag. In press.

12a. Azmitia, E. C., K. Dolan & P. M. Whitaker-Azmitia. 1990. S100β but not NGF, EGF, insulin or calmodulin is a CNS serotonergic growth factor. Brain Res. **576:** 354–356.

13. Biegon, A., T. C. Rainbow & B. S. McEwen. 1982. Quantitative autoradiography of serotonin receptors in rat brain. Brain Res. **242:** 197–204.
14. Bjorklund, A., U. Stenevi & N. A. Svendgaard. 1976. Growth of transplanted monoaminergic neurons into the adult hippocampus along the perforant path. Nature **262:** 787–790.
15. Bjorklund, A., L. Wiklund & L. Descarries. 1981. Regeneration and plasticity of central serotonergic neurons: a review. J. Physiol. (Paris) **77:** 247–255.
16. Choi, D. W. 1985. Glutamate neurotoxicity in cortical cell culture is calcium dependent. Neurosci. Lett. **58:** 293–297.
17. Cohen, G. 1983. Catalase, glutathione peroxidase superoxide dismutase, and cytochrome P-450. *In* Handbook of Neurochemistry, Vol. 4, Enzymes in the Nervous System. A. Lajtha, Ed. New York: Plenum Press, pp. 315–328.
18. Cohan, C. S., J. A. Connor & S. B. Kater. 1987. Electrically and chemically mediated increases in intracellular calcium in neuronal growth cones. J. Neurosci. **7:** 3588–3599.
19. Cotman, C. W. & G. S. Lynch. 1976. Reactive synaptogenesis in the adult nervous system: the effects of partial deafferentation on new synapse formation. *In* Neuronal Recognition. S. H. Barondes, Ed. New York: Plenum Press, pp. 69–108.
20. Craine, S. M., K. F. Shen & A. Chalazonitis. 1988. Opioids excite rather than inhibit sensory neurons after chronic opioid exposure of spinal cord-ganglion cultures. Brain Res. **455:** 99–109.
21. Crutcher, K. A. & F. Collins. 1986. Entorhinal lesions result in increased nerve growth factor-like growth-promoting activity in medium conditioned by hippocampal slices. Brain Res. **399:** 383–389.
22. Davila-Garcia, M. I. & E. C. Azmitia. 1989. Effects of acute and chronic administration of leu-enkephalin on cultured serotonergic neurons. Dev. Brain Res. **49:** 97–103.
23. Davila-Garcia, M. I. & E. C. Azmitia. 1990. Neuropeptides as positive or negative neuronal growth regulatory factors: Effects of ACTH and leu-enkephalin on cultured serotonergic neurons. *In* Molecular Aspects of Development and Aging of the Nervous System. J. M. Lauder, A. Privat, E. Giacobini, P. Timiras, and A. Vernidakis, Eds. New York: Plenum Press. In press.
24. Dravid, A. R. & E. B. Van Deusen. 1984. Recovery of enzyme markers for cholinergic terminals in septo-temporal regions of the hippocampus following selective fimbrial lesions in adult rats. Brain Res. **324:** 199–228.
25. Dreyfus, C. F., M. D. Gershon & S. M. Crain. 1979. Innervation of hippocampal explants by central catecholaminergic neurons in co-cultured fetal mouse brainstem explants. Brain Res. **161:** 431–445.
26. Engele, J., C. Pilgrim, M. Kirsch & I. Reisert. 1989. Different developmental potentials of diencephalic and mesencephalic dopaminergic neurons in vitro. Brain Res. **493:** 98–109.
27. Finger, S. & S. B. Dunnett. 1989. Nimodipine enhances growth and vascularization of neural grafts. Exp. Neurol. **104:** 1–9.
28. Frankfurt, M. & E. C. Azmitia. 1983. The effect of intracerebral injections of 5,7-

dihydroxytryptamine and 6-hydroxydopamine on the serotonin immunoreactive cell bodies and fibers in the adult rat hypothalamus. Brain Res. **261:** 91–99.
29. FRANKFURT, M. & A. BEAUDET. 1988. Reinnervation of dopamine neurons by regenerating serotonin axons in the rat medial zona incerta. A combined radioautographic and immunocytochemical ultrastructural study. Exp. Brain Res. **72:** 473–480.
30. FRANKFURT, M., K. J. RENNER, E. C. AZMITIA & V. LUINE. 1985. Intra-hypothalamic 5,7-dihydroxytryptamine: temporal analysis of effects on 5-HT content in brain nuclei and on facilitated lordosis behavior. Brain Res. **340:** 127–133.
30a. FRANKFURT, M. & A. BEAUDET. 1987. Ultrastructural organization of regenerated serotonin axons in the dorsomedial hypothalamus of the adult rat. J. Neurocytol. **16:** 799–809.
31. GAGE, F. H., A. BJORKLUND & U. STENEVI. 1983. Reinnervation of the partially deafferented hippocampus by compensatory collateral sprouting from spared cholinergic and noradrenergic afferents. Brain Res. **268:** 27–37.
32. GARCIA-HERNANDEZ, F., F. C. ZHOU & E. C. AZMITIA. 1989. Laminin injections guide and promote fiber growth of transplanted and regenerating neurons in adult but not in aged brain. Soc. Neurosci. Abstr. **15:** 1370.
33. GASSER, U. E. & A. R. DRAVID. 1986. Recovery of rat hippocampal monoaminergic and cholinergic markers following dorsal input lesions. Neurosci. Abstr. **12:** 1281.
34. GIBSON, G. E. & C. PETERSON. 1987. Calcium and the aging nervous system. Neurobiol. Aging **8:** 329–343.
35. GLIN, L., G. GANDOLFO, A. MCRAE-DEGUEURCE & C. GOTTESMANN. 1986. Neonatal administration of 5,7-dihydroxytryptamine induces in the adult rat a decrease in paradoxical sleep restored by an intracisternal injection of fetal raphe cells. Rev. Electroencephal. Neurophysiol. Clin. **16:** 273–282.
36. GÖTHERT, M. & E. SCHLICKER. 1983. Autoreceptor-mediated inhibition of ^{3}H-5-hydroxytryptamine release from rat brain cortex slices by analogues of 5-hydroxytryptamine. Life Sci. **32:** 1183–1192.
37. GRAHAM, D. G., S. M. TIFFANY, W. R. BELL, JR. & W. F. GUTKNECTH. Autoxidation versus covalent binding of quinones as the mechanism of toxicity of dopamine, 6-hydroxydopamine, and related compounds toward C1300 neuroblastoma cells in vitro. Mol. Pharmacol. **14:** 644–653, 1978.
38. HALGREN, E. & S. VARON. 1972. Serotonin turnover in cultured raphe nuclei from newborn rat: In vitro development and drug effects. Brain Res. **48:** 438–442.
39. HALLIWELL, B. & J. M. GUTTERIDGE. 1986. Oxygen-free radicals and iron in relation to biology and medicine: Some problems and concepts. Arch. Biochem. Biophys. **246:** 501–514.
40. HAMMARBACK, J. A., S. L. PALM, L. T. FURCHT & P. C. LETOUNEAU. 1985. Guidance of neurite outgrowth by pathways of substratum-absorbed laminin. J. Neurosci. Res. **13:** 213–220.
41. HARMON, D. 1981. The aging process. Proc. Natl. Acad. Sci. USA **78:** 7124–7128.
42. HOLETS, V. R. & C. W. COTMAN. 1984. Postnatal development of the serotonin innervation of the hippocampus and dentate gyrus following raphe implants. J. Comp. Neurol. **226:** 457–476.
43. KOH, J. Y., S. PETERS & D. W. CHOI. 1986. Neurons containing NADPH-Diaphorase are selectively resistant to quinolinate toxicity. Science **234:** 73–76.
44. LAUDER, J. M. 1983. Hormonal and humoral influences on brain development. Psychoneuroendocrinology **8:** 121–155.
45. LEVITT, P., R. Y. MOORE & B. B. GARBER. 1976. Selective cell association of catecholamine neurons in brain aggregates in vitro. Brain Res. **111:** 311–320.
46. LIDOV, H. G. W., R. GRZANNA & M. E. MOLLIVER. 1980. The serotonin innervation of the cerebral cortex in the rat—an immunohistochemical analysis. Neuroscience **5:** 207–227.
47. LIESI, P. 1985. Do neurons in the vertebrate CNS migrate on laminin? EMBO J. **4:** 1163–1170.
48. LUINE, V. N., K. J. RENNER, M. FRANKFURT & E. C. AZMITIA. 1984. Facilitated sexual behavior reversed and serotonin restored by raphe nuclei transplanted into denervated hypothalamus. Science xx: 1436–1439.
49. MARTIN, E. M., S. D. SKAPER & S. VARON. 1987. Catalase protection of neuronal survival

in vitro is not directed to the accumulation of peroxidase in the culture medium. Int. J. Dev. Neurosci. **5:** 1–10.

50. McRae-Degueurce, A., G. Gandolfo, L. Glin & C. Gottesmann. 1988. Fetal serotonergic transplants in the fourth ventricle of adult rodents reverse sleep disturbances induced by neonatal administration of 5,7-dihydroxytryptamine. Neurosci. Lett. **89:** 108–113.
51. Mytilineou, C. & G. Cohen. 1987. Tissue culture model for studying MPTP toxicity to dopamine neurons. Adv. Neurol. **45:** 145–148.
52. Nelson, D. L., A. Herbert, S. Bourgoin, J. Glowinski & M. Hamon. 1978. Characteristics of central 5-HT receptors and their adaptive changes following intracerebral 5,7-dihydroxytryptamine administration in the rat. Mol. Pharmacol. **14:** 983–995.
53. Nilsson, O. G., R. E. Strecker, A. Daszuta & A. Bjorklund. 1988. Combined cholinergic and serotonergic denervation of the forebrain produces severe deficits in a spatial learning task in the rat. Brain Res. **453:** 235–246.
54. Nyakas, C., E. Gaspar & P. G. M. Luiten. 19XX. Effects of nimodipine on hypoxia induced learning and memory deficits. *In* Nimodipine and Central Nervous Functions—New Vistas. W. H. Gispen and J. Traber, Eds. Berlin: Springer-Verlag. In press.
55. Nygren, L. G., L. Olson & A. Seiger. 1977. Monoaminergic reinnervation of the transected spinal cord by homologous fetal brain grafts. Brain Res. **129:** 227–235.
56. Prochiantz, A., U. Di Porzio, A. Kato, B. Bergen & J. Glowinski. 1979. In vitro maturation of mesencephalic dopaminergic neurons from mouse embryos is enhanced in presence of their striated target cells. Proc. Natl. Acad. Sci. USA **76:** 5387–5391.
57. Prochiantz, A., M. C. Daguet, A. Herbert & J. Glowinski. 1981. Specific stimulation of in vitro maturation of mesencephalic dopaminergic neurons by striatal membranes. Nature **293:** 570–572.
58. Ramirez, T. M., H. J. Altman, H. J. Normile, D. M. Kuhn & E. C. Azmitia. 1989. The effects of serotonergic intrahippocampal neonatal grafts on learning and memory in rats trained in the stone 14-unit t-maze. Soc. Neurosci. Abstr. **15:** 465.
59. Reisert, I., J. Engele, G. Pilgrim & M. Kirsch. 1989. Different development potentials of diencephalic and mesencephalic dopaminergic neurons in vitro. Brain Res. **483:** 98–109.
60. Saavedra, J. M., M. Brownstein & M. Palkovitz. 1974. Serotonin distribution in the limbic system of the rat. Brain Res. **79:** 438–441.
61. Scheff, S. W. & C. W. Cotman. 1982. Chronic glucocorticoid therapy alters axon sprouting in the hippocampal dentate gyrus. Exp. Neurol. **76:** 644–654.
62. Schlumpf, M., W. J. Shoemaker & F. E. Bloom. 1977. Explant culture of catecholamine-containing neurons from rat brain: Biochemical histofluorescence and electron microscopic studies. Proc. Natl. Acad. Sci. USA **74:** 4471–4475, 1977.
63. Segal, M. & G. Richter-Levin. 1990. Raphe cells grafted into the hippocampus can ameliorate spatial memory deficits in rats with combined serotonergic/cholinergic deficiencies. Brain Res. In press.
64. Seiden, L. S. & G. Vosmer. 1984. Formation of 6-hydroxydopamine in caudate nucleus of rat brain after a single large dose of methamphetamine. Pharmacol. Biochem. Behav. **21:** 29–31.
65. Shemmer, A., P. M. Whitaker-Azmitia & E. C. Azmitia. 1988. Effects of prenatal 5-methoxytryptamine and parachlorophenylalanine on serotonergic uptake and behavior in the neonatal rat. Pharmacol. Biochem. Behav. **30:** 847–851.
66. Steinbusch, H. W. M. 1981. Distribution of serotonin-immunoreactivity in the central nervous system of the rat—cell bodies and terminals. Neuroscience **4:** 557–618.
67. Tiffany-Castiglioni, E., R. P. Saneto, P. H. Proctor & J. R. Perez-Polo. 1982. Participation of active oxygen species in 6-hydroxydopamine toxicity to a human neuroblastoma cell line. Biochem. Pharmacol. **31:** 181–188.
68. Tsubokawa, T., Y. Katayama, S. Miyazaki, H. Ogawa, M. Iwasaki, S. Shibanoki & K. Ishikawa. 1988. Supranormal levels of serotonin and its metabolite after raphe cell transplantation in serotonin-enervated rat hippocampus. Brain Res. Bull. **20:** 303–306.
69. Vanderwolf, C. H. & G. B. Baker. 1986. Evidence that serotonin mediates non-cholinergic neocortical low voltage fast activity, non-cholinergic hippocampal rhythmical slow activity and contributes to intelligent behavior. Brain Res. **374:** 342–356.

70. WHITAKER-AZMITIA, P. M. & E. C. AZMITIA. 1984. A pharmacological model for studying the development of serotonergic neurons in tissue culture: The role of high-affinity serotonin receptors. *In* Developmental Neuroscience: Physiological, Pharmacological and Clinical Aspects. F. Caciagli, E. Giacobini, and R. Paoletti, Ed. Amsterdam: Elsevier, pp. 251–257.
71. WHITAKER-AZMITIA, P. M. & E. C. AZMITIA. 1986. Autoregulation of fetal serotonergic neuronal development: Role of high-affinity serotonin receptor. Neurosci. Lett. **67:** 307–312.
72. WHITAKER-AZMITIA, P. M. & E. C. AZMITIA. 1986. ^{3}H-5-hydrotryptamine binding to brain astroglial cells: differences between intact and homogenized preparations and mature and immature cultures. J. Neurochem. **46:** 1186–1189.
73. WHITAKER-AZMITIA, P. M. & E. C. AZMITIA. 1989. Stimulation of astroglial serotonin receptors produces culture media which regulates growth of serotonergic neurons. Brain Res. **497:** 80.
74. WONG, A. & D. E. JONES. 1983. The effects of calcium on the surface characteristics of dissociated fetal mouse cerebral cells: Morphological aspects. Dev. Brain Res. **6:** 259–268.
75. YAMAMOTO, M., H. W. M. STEINBUSCH & T. M. JESSELL. 1981. Differential properties of identified serotonin neurons in dissociated cultures of embryonic rat brain stem. J. Cell Biol. **91:** 142–152.
76. YOUNG, W. 1987. The post-injury responses in trauma and ischemia: Secondary injury or protective mechanism? Cen. Nerv. Syst. Trauma **4:** 27–42.
77. ZHOU, F. C. & E. C. AZMITIA. 1983. Differentiation of two median raphe-hippocampal 5-HT pathways, a quantitation HRP study. Brain Res. Bull. **10:** 445–451.
78. ZHOU, F. C. & E. C. AZMITIA. 1984. Induced homotypic collateral sprouting of serotonergic fibers in hippocampus. Brain Res. **308:** 53–62.
79. ZHOU, F. C. & E. C. AZMITIA. 1985. Effect of adrenalectomy and corticosterone on the sprouting of serotonergic fibers in hippocampus. Neurosci. Lett. **54:** 111–116.
80. ZHOU, F. C. & E. C. AZMITIA. 1986. Induced homotypic collateral sprouting of serotonergic fibers in the hippocampus. II. An immunocytochemistry study. Brain Res. **373:** 337–348.
81. ZHOU, F. C., S. B. AUERBACH & E. C. AZMITIA. 1987. Stimulation of serotonergic neuronal maturation after fetal mesencephalic raphe transplantation into the 5,7-DHT-lesioned hippocampus of the adult rat. Ann. N. Y. Acad. Sci. **495:** 138–152.
82. ZHOU, F. C., S. B. AUERBACH & E. C. AZMITIA. 1987. Denervation of serotonergic fibers in the hippocampus induced a trophic factor which enhances the maturation of transplanted serotonergic neurons but not norepinephrine neurons. J. Neurosci. Res. **17:** 235–246.
83. ZHOU, F. C. & E. C. AZMITIA. 1988. Laminin facilitates and guides fiber growth of transplanted neurons in adult brain. J. Chem. Neuroanatomy **1:** 133–146.
84. ZHOU, F. C., S. B. AUERBACH & E. C. AZMITIA. 1988. Transplanted raphe and hippocampal fetal neurons do not displace afferent inputs to the dorsal hippocampus from serotonergic neurons in the median raphe nucleus of the rat. Brain Res. **450:** 51–59.
85. ZHOU, F. C. & E. C. AZMITIA. 1990. Neurotrophic factor for serotonergic neurons prevents degeneration of grafted raphe neurons in the cerebellum. Brain Res.

DISCUSSION OF THE PAPER

H. G. BAUMGARTEN (*Free University, Berlin*): I've got a question concerning the first part of your talk. You were injecting 5,7-DHT and did you protect the dopaminergic and the noradrenergic neurons? It's well known that the compound is not absolutely specific. These catecholaminergic neurons might also be involved in serotonergic sprouting as a trophic signal.

AZMITIA: Yes. 5,6-DHT is selective for serotonergic neurons and has little effect on dopamine. In all of our studies with 5,7-DHT we pretreated with desipramine, a catecholaminergic uptake blocker to protect those neurons. We have measured norepinephrine and acetyl choline systems after these hippocampal 5,7-DHT microinjections in desipramine pretreated animals and found no significant reductions. Thus catecholaminergic neurons are *not* involved as a 5-HT trophic signal.

BAUMGARTEN: Only with desipramine, not with a more specific dopamine uptake blocker?

AZMITIA: In studies with hypothalamic injections, dopamine levels are reduced if we use desipramine. Therefore, for those studies we use normifensen. The microinjections are always made through a glass pipette to limit the non-specific damage. We can largely confine the extent of damage to serotonin neurons.

ELIZABETH ABERCROMBIE (*University of Pittsburgh, PA*): There's another type of plasticity. We've observed a biochemical type of plasticity in noradrenergic and dopaminergic systems following partial lesions of those systems with 6-hydroxydopamine. We see increases in tyrosine hydroxylase indicating increased synthesis and increased release from residual neurons that can restore levels of transmitter. And I'm wondering is there any evidence for that sort of biochemical plasticity in the serotonin system following partial lesion.

AZMITIA: We have seen similar changes in serotonin turnover after a partial lesion both in hypothalamus (Luine *et al.,* Brain Res. **264:** 344, 1983) and in hippocampus (Auerbach *et al.,* Neurosci. Lett. **61:** 147, 1985). These observations of rapid plasticity are fairly new. When I was a graduate student about 2 decades ago, we looked at serotonin metabolism by measuring changes in tryptophan hydroxylase activity. We saw that if we had adrenalectomized the animals or subjected them to severe stress over a 3 and 4 day period, activity of the enzyme was changed. At that time, most of our colleagues were surprised since the prevalent view was that the brain biochemistry just doesn't change that quickly. Nowadays it's known that rapid changes can occur. There are posters this evening on changes in tryptophan hydroxylase and metabolic activities in the neuron which happen within minutes. (See poster paper by Boadle-Biber and co-workers [poster abstract #47 from this conference].) A number of neurotransmitter systems show these rapid biochemical shifts, but these kinds of changes, for the most part, are short in duration, lasting only a few hours after these types of incidents. But what we find here, in terms of collateral sprouting, regeneration, transplantation, and drug-induced neuropathology are structural changes in these systems which last from months to a year. So, the concept that we have of the serotonin system is one of a very plastic reorganizing neural system. Hopefully, by using the neuropharmacology of serotonin, we can learn to regulate these types of growth. We can manipulate, actually re-engineer, the rate these fibers can grow and actually set their local termination densities.

QUESTION: Do you know whether the extract that you inject into the cerebellum before a raphe transplant promotes a growth because it contains some growth factor or because it antagonizes some inhibitory effect within the cerebellum? In other words if you took an extract from cerebellum would you inhibit growth in other tissues where you would have growth?

AZMITIA: That's an excellent idea. We've emphasized the fact that the brain contains positive signals. But we, and others, have shown a large number of factors which can be inhibitory—free hydroxyl radicals, opioids, excess calcium, high pH, etc. All these conditions can stop these neural systems from growing, and it's possible that a number of these "trophic effects" we see are actually cancellation of inhibitory effects. And no we haven't looked at cerebellar extracts in raphe transplants into hippocampus but I'll tell Dr. Zhou that those studies should be done.

QUESTION: Did you look at the serotonin levels in the hippocampus to see whether they would digress after cocaine?

AZMITIA: No we didn't. That cocaine study where we showed the depletion in the hippocampus in the young animals was tried after doing a large number of studies in culture. Under culture condition we have shown neurons to be adversely affected by cocaine. We saw evidence of neuronal death by counting cell numbers, reduced neurite outgrowth by morphometric analysis and large decreases in the neuronal uptake capacity.

COMMENT: In our studies of older animals (about 90 days), we didn't see any effects of cocaine on dopamine, 5-HT, or serotonin.

AZMITIA: Right, Dr. De Souza was telling me last night that they've injected adult animals for weeks with cocaine at very high doses without seeing changes. Our animals were injected quite early on and we did see pronounced changes, again suggesting that neurons before they reach their target territory are very sensitive to a number of changes.

QUESTION: Dr. Fuller mentioned in a talk earlier yesterday that some uptake sites were on glia cells. Do you think there's a chance that the *in vivo* regeneration and sprouting that you're looking at might be partially due to glial proliferation?

AZMITIA: No, there is no chance our result merely reflects an increase in glial cells in the target area. We've mentioned 5-HT uptake and level changes in the talk today. We've also done immunocytochemistry, track tracing both anterograde and retrograde, receptor changes, electrophysiology, and behavioral studies. So although there is a remote possibility the uptake study could be due to glial proliferation (we're using 50 nanomolar of tritiated 5-HT to measure uptake and this is not taken up normally by glial cells), all the other evidence is against such an explanation.

In collateral sprouting, a normal adult neuron can sprout. Adult brain cells are not at full capacity. Given new territory by transplanting target (e.g. hippocampus) or by partial lesions they can expand to establish a new homeostatic balance, and promote neuronal expansion. It is possible that a glial-derived growth factor may stimulate the 5-HT growth.

J. LEYSEN (*Janssen Research Foundation, Beerse, Belgium*): In your cultures of raphe and cortex, did you ever look at trophic factors which are required for the development of the postsynaptic neurons? Did you see changes in the neurons which contain the postsynaptic serotonin receptors?

AZMITIA: No, we haven't. Dr. Jean Lauder has evidence that 5-HT can regulate the maturation of target neurons, but receptors on the target cells were not measured.

LEYSEN: We have done some neuronal cultures and tried to measure 5HT-2 receptors and we were never able to find them.

AZMITIA: One suggestion might be to try co-cultures between serotonergic raphe neuron and their target cells in cortex or hippocampus. It's possible that 5-HT fibers can induce their own receptor.

Serotonergic Control of Cerebral Activity and Behavior: Models of Dementia

C. H. VANDERWOLF,[a] G. B. BAKER,[b] AND C. DICKSON[a]

[a] *Department of Psychology*
University of Western Ontario
London, Ontario, Canada N6A 5C2
and
[b] *Department of Psychiatry*
University of Alberta
Edmonton, Alberta, Canada T6G 2B7

CEREBRAL ACTIVITY

Recent evidence indicates that the ascending cholinergic projections from the basal forebrain and the ascending serotonergic projections from the brain stem exert a powerful control over the generalized electrical activity of the hippocampal formation and neocortex.[1] These inputs have a very general effect, altering the pattern of unitary activity and the development of evoked potentials as well as spontaneous slow wave activity. However, a number of the basic effects can be conveniently studied using spontaneous slow wave recording techniques in freely moving animals.[2]

One of the results of cholinergic and serotonergic inputs to the hippocampal formation in normal rats is the generation of rhythmical slow activity (RSA) which, during waking, occurs in close correlation with spontaneous motor activities such as walking, changing posture, turning or raising the head, and manipulating objects with the forelimbs (Type 1 behavior). During other behaviors such as standing or lying immobile, face-washing, chewing, gnawing, licking, or shivering (Type 2 behavior) hippocampal activity is usually quite irregular (large amplitude irregular activity, LIA) but RSA can be elicited by a variety of experimental procedures. Such RSA appears to be dependent on cholinergic inputs to the hippocampus while the RSA occurring during Type 1 behavior is dependent on both cholinergic and serotonergic inputs.

Neocortical activity also appears to be under the control of ascending cholinergic and serotonergic inputs although the generalized patterns of electrical activity present in the neocortex differ in several respects from the patterns occurring in the hippocampal formation. During Type 1 behavior, low voltage fast activity (LVFA) is always present as a result of combined cholinergic and serotonergic inputs. During Type 2 behavior, LVFA is often present but large irregular slow waves or rhythmical spindles may be present instead, especially during behavioral immobility. Any LVFA that occurs during Type 2 behavior appears to be dependent on cholinergic inputs to the neocortex while the LVFA occurring during Type 1 behavior is dependent on both cholinergic and serotonergic inputs. Thus, the control of neocortical LVFA appears to parallel closely the control of hippocampal RSA.

The effects of serotonergic control of cortical activity can be observed in the absence of the masking effect of concurrent cholinergic activity following the administration of centrally-acting anti-muscarinic drugs. The hippocampal RSA and neocortical LVFA that persist after such treatment can be referred to as atropine-resistant RSA and atropine-resistant LVFA even though other anti-muscarinic drugs (ditran, quinuclidinyl benzilate, scopolamine) have the same effect as atropine. Atropine-resistant RSA and atropine-resistant LVFA can be referred to jointly as atropine-resistant cerebral activation (ARCA).

FIGURES 1–4 illustrate some of the effects of cholinergic and serotonergic blockade on hippocampal and neocortical slow wave activity. The records in FIGURE 1, obtained from a control rat that had received intra-cerebral injections of Locke's solution, are very similar to those seen in intact normal rats or rats that have received systemic injections of a drug vehicle. FIGURE 2 shows that *p*-chlorophenylalanine (PCPA, an inhibitor of the synthesis of serotonin[3]) acting alone does not have a marked effect on spontaneous hippocampal activity. RSA continues to occur during Type 1 behavior in the PCPA-treated rat as a result, presumably, of the action of cholinergic inputs to the hippocampus. When these inputs are blocked by a large dose of scopolamine all RSA disappears, presumably because the PCPA had previously eliminated the normally present atropine-(or scopolamine-) resistant RSA.[4] Intra-cerebral injections of 5,7-dihydroxytryptamine (5,7-DHT, a selective serotonergic neurotoxin[5–8]) or systemic injections of reserpine (a depletor of serotonin and other monoamines[9]) have similar effects.[10,11] However, FIGURE 2 does not show another of the effects of PCPA which is to increase the occurrence of atropine-sensitive RSA during Type 2 behavior.[4] This may suggest that one of the functions of ascending serotonergic projections is to regulate the activity of the cholinergic septo-hippocampal pathway.[11]

FIGURES 3 and 4 show how atropine-(or scopolamine-)resistant LVFA is normally correlated with the occurrence of Type 1 behavior and how such LVFA is lost in rats treated with PCPA or intracerebral injections of 5,7-DHT. Reserpine has similar effects.[12] These findings, together with a wealth of other data,[1] have suggested the diagrammatic summary shown in FIGURE 5.

The fact that atropine-resistant RSA and LVFA correlate with the occurrence of Type 1 behavior is consistent with the hypothesis that the neuronal activity that generates these waveforms also plays a role in the causation of Type 1 behavior. However, since correlation does not necessarily indicate causality, much additional evidence is necessary before such a hypothesis can be accepted. One means of assessing the significance of serotonergic control of cerebral activation would be to destroy serotonergic neurons or block their action and determine the effect of this on behavior. Before discussing such work it is necessary to evaluate various methods of blocking serotonergic activity in the brain. We have carried out a series of studies on the effects of various anti-serotonergic procedures on atropine-resistant cortical activation, making use of conventional histological procedures (Nissl stain) plus high pressure liquid chromatography (HPLC) with electrochemical detection as means of assessing the effects on the brain. In preparation for HPLC, the rats were decapitated, the brain rapidly removed and chilled in ice-cold saline for 2 min prior to dissection and freezing over solid carbon dioxide. Further details are given by Baker *et al.*[13] and Vanderwolf *et al.*[11]

TABLE 1 shows that a single dose of PCPA (500 mg/kg, i.p.) results in an approximately 90° decline in whole-brain levels of serotonin and 5-hydroxyindoleacetic acid (5-HIAA) 3 days later. Three daily doses of PCPA at 500 mg/kg have a somewhat greater effect, especially on 5-HIAA. The difference in effect of a single dose of PCPA, as compared to 3 doses, appears to have functional consequences. A single dose of PCPA of 500 mg/kg has no obvious effect on atropine-resistant RSA and LVFA[14] while 3 daily doses of 500 mg/kg produce marked impairment or loss of these waveforms.[4,15] Three doses of PCPA also have a greater effect behavior than a single dose (see below). However, PCPA has clear non-specific actions as well, producing moderate but significant reductions in brain levels of tryptophan, noradrenalin,

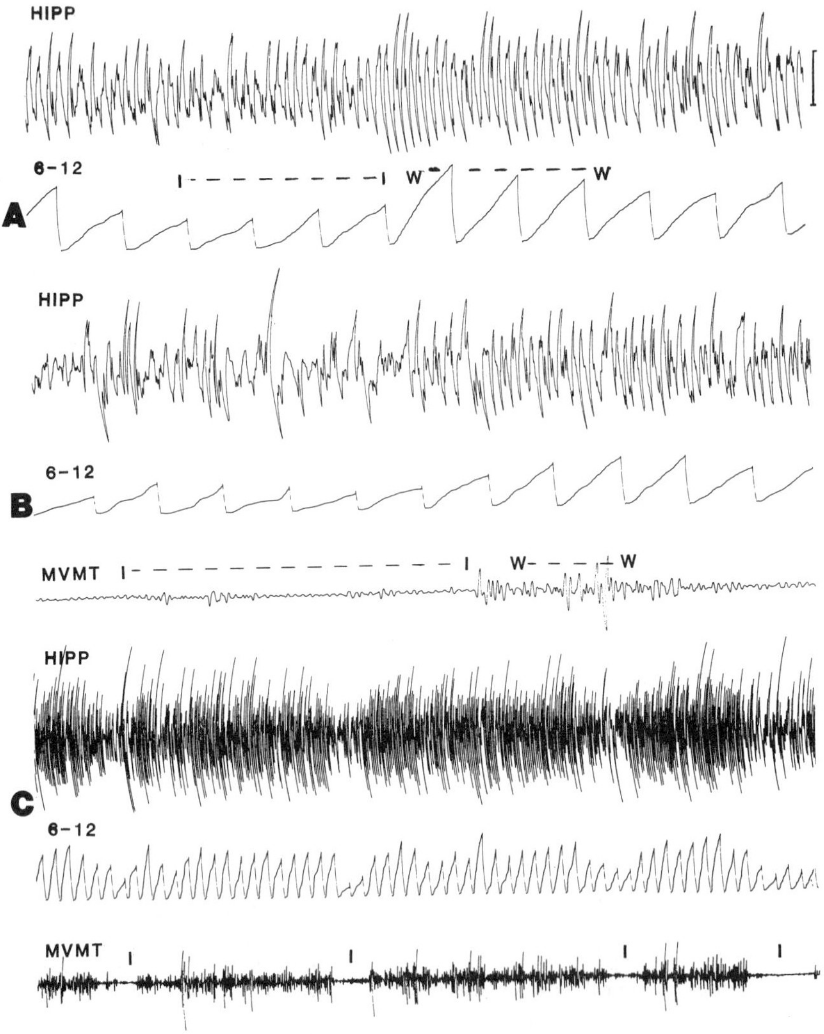

FIGURE 1. Hippocampal slow wave activity in a freely moving rat that had received 4 intra-brain stem injections of Locke's solution (1.0 μl in each of: the dorsal raphe, the median raphe, and bilaterally in the B9 region dorsal to the medial lemniscus). HIPP, hippocampus; 6–12 Hz, hippocampal activity after passage through a 6–12 Hz band-pass filter, rectification, and integration over 1 sec intervals; MVMT, output from a magnet-and-coil type of movement sensor (omitted in trace **A** to conserve space). **A:** no drug; **B:** after treatment with scopolamine HBr (5 mg/kg, s.c.); **C:** lower speed record after scopolamine. I-----I, period of immobility; W-----W, period of walking. In both **A** and **B** note that well developed rhythmical slow activity (RSA) results in increased 6–12 Hz activity during walking while more irregular activity and lower levels of 6–12 Hz activity occur during immobility. Activity during immobility is more rhythmical in the normal state than after scopolamine owing, presumably, to blockade of the cholinergic input to the hippocampus by this drug. Trace **C** illustrates the correlation between scopolamine-resistant RSA and motor activity (head movement, stepping, walking) over a longer period. Cal: 1 mV. Time is indicated in sec by the 6–12 Hz integrator trace.

dopamine, and homovanillic acid. Previous research has also demonstrated moderate non-specific effects of PCPA.[16]

TABLES 2–4 shows the effect of intra-brain stem injections of 5,7-DHT creatinine sulfate (25 μg/μl, given in 0.5–1.0 μl volumes at each of 2–4 sites in the brain stem, i.e., dorsal raphe and median raphe nuclei, and bilaterally in the B9 region dorsal to the medial lemniscus). Data from the combined raphe nuclei plus B9 group have been taken from Vanderwolf *et al.*[11] All these rats were treated with desipramine (25 mg/kg, i.p.) prior to surgery to protect noradrenergic neurons and killed 2–4 months

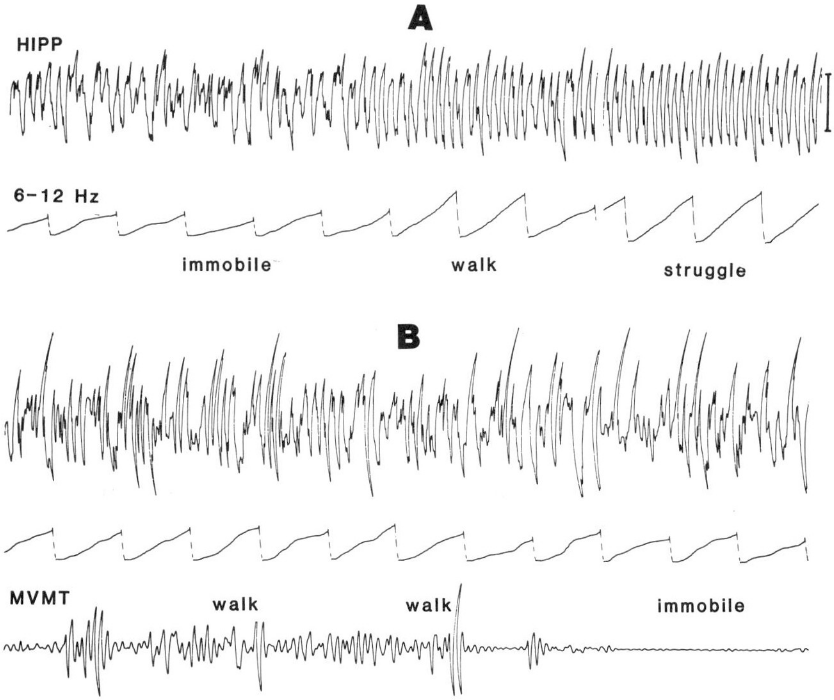

FIGURE 2. Hippocampal slow wave activity in a freely moving rat that had received *p*-chlorophenylalanine (500 mg/kg/day, i.p.) for 3 days prior to recording, Abbreviations and calibration as in FIGURE 1. **A:** note irregular activity and correspondingly low levels of 6–12 Hz activity during immobility but RSA and higher levels of 6–12 Hz activity during spontaneous walking and during struggling elicited by manually pushing the rat forward. **B:** after treatment with scopolamine (5 mg/kg, s.c.) RSA is almost completely suppressed (some traces remain); hippocampal activity is similar during walking and during immobility.

after the brain stem injections. Tables 2–4 show that the injections of 5,7-DHT in either: (1) the median and dorsal raphe nuclei, or (2) bilaterally in the B9 region produced 80–90% depletions of serotonin and 5-HIAA throughout the forebrain. This is to be expected, since previous anatomical data indicate that both regions contribute to ascending serotonergic pathways.[17,18] When injections were made in the raphe and B9 regions in combination, the depletions of serotonin and 5-HIAA were increased to 93–97 percent in most of the forebrain but did not increase systematically in the hippocampus (TABLE 3).

In another study[11] rats receiving injections of 5,7-DHT (300 μg in 10 μl) in one lateral ventricle sustained only mild depletions of serotonin in the neocortex (52.5% of control) and hippocampal formation (45.2% of control). Very large intraventricular injections of 5,7-DHT (600 μg in 40 μl of solution) produced depletions of serotonin and 5-HIAA that were almost equivalent to those achieved by 4 injections of much smaller amounts of the drug directly into the brain stem.

These neurochemical data were correlated with the electrophysiological data. Injections of 5,7-DHT directly into 4 sites in the brain stem produced not only a more consistent depletion of serotonin but also had a more severe effect on atropine-resistant RSA and LVFA than the 2 site intra-brain stem injection procedures. If ARCA was absent at all recording sites (in these experiments there were electrode pairs in frontal and parietal neocortex and in the dorsal hippocampus) it was rated as totally lost. However, if: (1) ARCA occurred occasionally in correlation with walking at a given

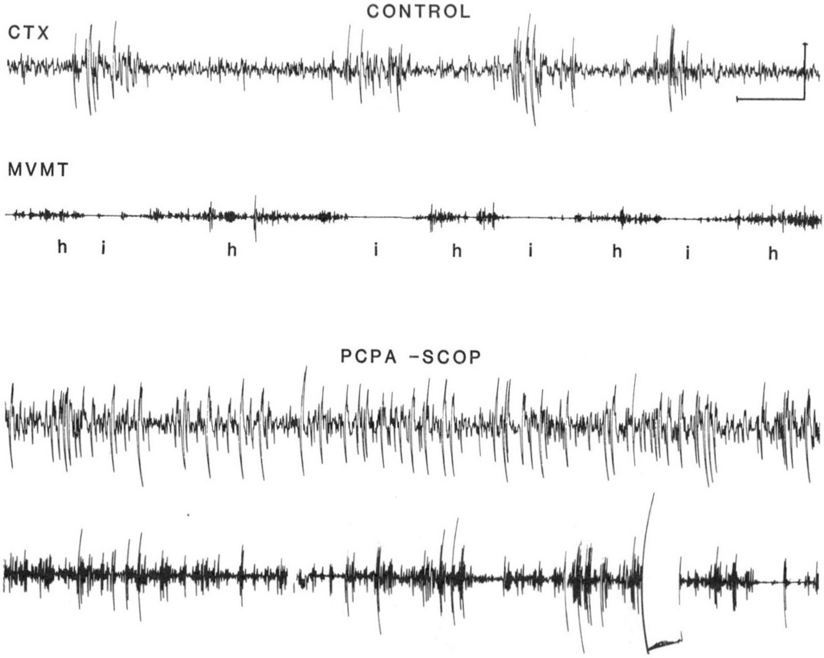

FIGURE 3. Effects of *p*-chlorophenylalanine on scopolamine-resistant neocortical low voltage fast activity. CTX, neocortical slow wave activity from a surface-to-depth electrode pair; MVMT, output of magnet-and-coil type of movement sensor. CONTROL, rat treated with drug vehicle (3 daily doses, i.p.) plus scopolamine HBr (5 mg/kg, s.c., given on the day after the last dose of drug vehicle shortly before the record was taken); h, spontaneous head movements; i, behavioral immobility. PCPA-SCOP, rat treated with *p*-chlorophenylalanine (500 mg/kg, i.p. on each of 3 successive days) plus scopolamine HBR (5 mg/kg, s.c., given on the day after the last dose of PCPA shortly before the record was taken). Note that the control rat displays large irregular slow waves during immobility and low voltage fast activity (LVFA) during head movement. The PCPA-SCOP rat displays continuous large irregular slow waves despite continuous head movement, stepping, and walking. Virtually all LVFA is abolished. Cal: 1 mV, 5 s.

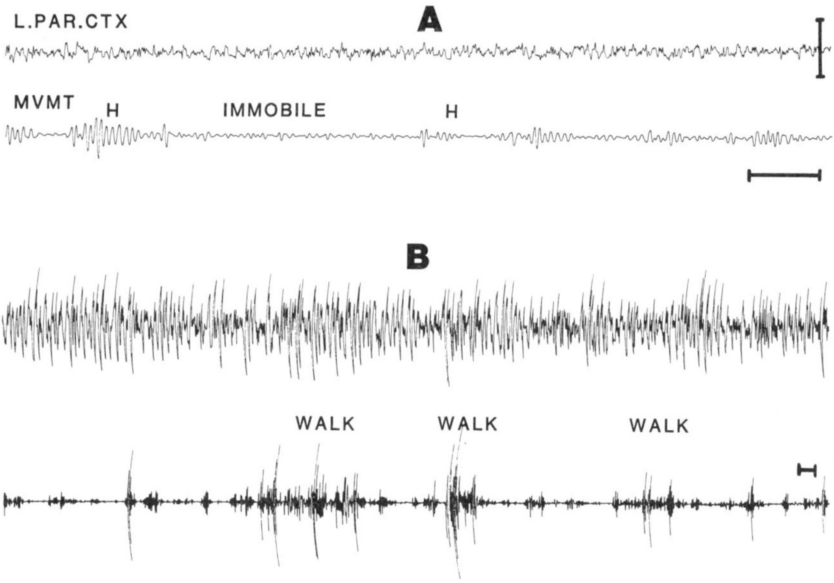

FIGURE 4. Neocortical slow wave activity in a rat 37 days after 4 intra-brain stem injections of Locke's solution, containing 5,7-dihydroxytryptamine creatinine sulfate (5,7-DHT, Sigma, 25 μg/μl, 0.5 μl/site) in each of: the dorsal raphe, the median raphe, and bilaterally in the B9 region. Noradrenergic neurons were protected by desmethylimipramine HCl (25 mg/kg, i.p.). L.PAR.CTX., left parietal neocortex, MVMT, output from movement sensor; H, head movement. **A:** predrug record showing apparently normal LVFA, presumably dependent on cholinergic inputs to the neocortex; **B:** after scopolamine (5 mg/kg, s.c.). Note that LVFA has disappeared although rat walks about spontaneously Cal: 1mV, 1 sec.

electrode site, or (2) was absent at one electrode site but at least occasionally present at another site at the same time, then ARCA was said to be partially lost. The 4-site injection procedure produced partial or total loss of ARCA in 18 of a series of 22 rats. In comparison, injections into the dorsal raphe and median raphe nuclei alone never produced more than a partial loss of ARCA and produced that in only 7 of a group of 16 rats.[11] Bilateral injections into the B9 region produced partial losses of ARCA in 3 rats in a group of 5. Intraventricular injections of 5,7-DHT, in quantities ranging from 120–600 μg in 8–40 μl of solution, produced partial losses of ARCA in only 14 of a series of 32 rats.

These data indicate that injections in the dorsal and median raphe nuclei and bilaterally in the B9 region have an additive effect on both depletion of serotonin and 5-HIAA and on ARCA. As a result, combined injection at all 4 of these sites is an especially effective means of depleting forebrain serotonin and eliminating ARCA. An additional advantage of the 4 site intra-brain stem injection procedure is that it has little effect on forebrain levels of dopamine and homovanillic acid and only moderate effects on noradrenalin. However, a serious disadvantage of this procedure, with the concentrations of 5,7-DHT that we have used, is that definite lesions (cavities) may be created in the injection zone in the brain stem in some rats. In a recent series of 18 rats, serial Nissl stained sections were prepared of the brain stem in rats that had received

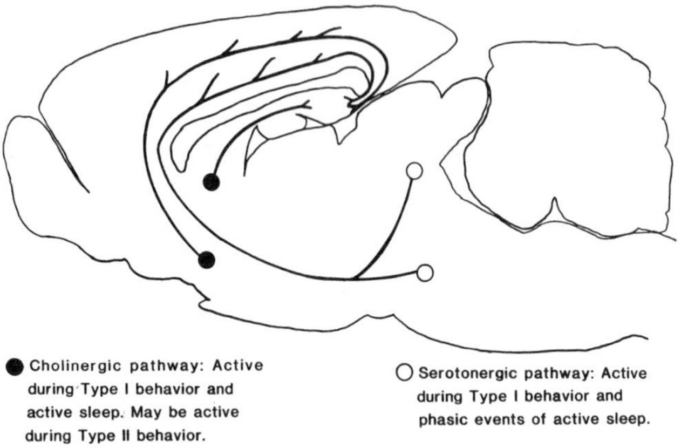

FIGURE 5. Diagram illustrating part of the proposed pathways that determine the occurrence of low voltage fast activity in the neocortex and rhythmical slow activity in the hippocampus. Filled circles represent cholinergic projections which are presumed to be active during Type 1 behavior and active sleep, and may be active during Type 2 behavior. Open circles represent serotonergic projections which are presumed to be active during Type 1 behavior and the phasic events of active sleep. (From Vanderwolf (1988)[1] by permission of Academic Press.)

0.5 μl injections of a solution of 5,7-DHT (25 μg/μl, dissolved in Locke's solution) 3–4 months earlier. In 10 of these 18 cases, the 5,7-DHT injections had created definite cavities, usually centered in the median raphe, and ranging in the size from a pin head to a 2-mm-wide hole. Although these lesions destroyed tissue rich in serotonergic neurons such as the median raphe nucleus, the nucleus linearis oralis, and the B9 region, there is no doubt that non-serotonergic neurons, especially in the nucleus pontis oralis, were destroyed as well. Long fiber pathways such as the medial lemniscus or the medial longitudinal fasciculus were usually not damaged by the injection procedure.

In conclusion, the 5,7-DHT intracerebral injection procedure and the procedure of

TABLE 1. Whole-brain Levels (given as percent of control) of Some Amines, Amine Metabolites, and Tryptophan in Rats Treated with One (PCPA-1) or Three (PCPA-3) Daily Doses (i.p.) of *p*-Chlorophenylalanine

	NA	DA	HVA	5-HT	5-HIAA	TRP
PCPA-1 (N = 5)	73.5 ± 4.5[a]	77.9 ± 3.0[a]	76.8 ± 2.9	10.6 ± 1.3[a]	10.5 ± 1.7[a]	68.5 ± 4.9[a]
PCPA-3 (N = 4)	51.3 ± 2.6[a]	69.0 ± 3.6[a]	65.9 ± 9.6[a,b]	8.0 ± 0.4[a]	6.0 ± 0.3[a,b]	52.6 ± 1.4[a,b]
Control levels (ng/g of tissue)	524.3 ± 14.2	920.7 ± 22.6	51.3 ± 1.9	467.7 ± 14.5	260.0 ± 8.3	4695.8 ± 177.3

Values in all Tables are means ± standard errors of the means. *a*, differs from control, $p < .01$; *b*, differs from PCPA-1 group, $p < .02$. NA, noradrenaline; DA, dopamine; HVA, homovanillic acid; 5-HT, 5-hydroxytryptamine; 5,HIAA, 5-hydroxyindoleacetic acid; TRP, tryptophan. Control levels refer to absolute values in 5 vehicle-injected rats. In all cases rats were decapitated on the 4th day after the first (or only) injection of PCPA or drug vehicle (0.5 percent solution of gum arabic in saline). Data in TABLES 1–4 were analyzed with the help of a parametric analysis of variance and the non-parametric Mann-Whitney U test.

TABLE 2. Neocortical Levels (given as a percent of control) of Some Amines, Amine Metabolites, and Tryptophan in Rats Injected with 5,7-Dihydroxytryptamine in Various Brain Regions

	NA	DA	HVA	5-HT	5-HIAA	TRP
Dorsal + median raphe nuclei (N = 5)	91.6 ± 8.1	102.9 ± 21.7	71.0 ± 3.0	11.4 ± 2.0[a]	10.7 ± 1.8[a]	96.3 ± 9.9
B9 region (N = 3)	80.4 ± 21.3	132.0 ± 54.7	94.9 ± 48.3	12.7 ± 0.6[a]	11.9 ± 0.9[a]	112.7 ± 7.5
Combined raphe + B9 (N = 21)	54.1 ± 5.0[a]	230.5 ± 51.4	90.1 ± 0.6	7.0 ± 0.7[a]	3.1 ± 1.0[a]	120.6 ± 6.0
Control levels (ng/g)	238.3–363.2	32.1–47.0	17.8–32.8	327.6–367.2	191.6–259.7	2504.8–3424.0

[a]Differs from corresponding vehicle-injected control group, $p < .02$.
Abbreviations as in TABLE 1. Control levels in TABLES 2–4 refer to the range of mean absolute values in 3 vehicle-injected groups (6–7 rats/group) prepared at the same time and with the same injection procedure as the corresponding experimental group.

systemic administration of PCPA both severely depleted the forebrain of serotonin but both had non-specific side effects as well. However, since the side effects of the two procedures are quite different, any effect that is common to them is likely to be due to interference with serotonergic functions.

The data obtained from both of these procedures indicate that forebrain depletion of serotonin and 5-HIAA must be greater than about 90% if reasonably consistent losses of ARCA are to be achieved. However, Vanderwolf *et al.*[11] found that 4 rats in a group of 22 that had received 4 intra-brain stem injections of 5,7-DHT retained apparently normal ARCA despite 93 percent depletions of serotonin in the neocortex and hippocampus. This may suggest that even minute amounts of serotonin may sometimes permit relatively normal function.

BEHAVIOR

During the procedure of recording hippocampal and neocortical electrical activity from rats treated with PCPA (500 mg/kg/day × 3) plus scopolamine (5 mg/kg) it

TABLE 3. Hippocampal Levels (given as percent of control) of Some Amines, Amine Metabolites, and Tryptophan in Rats Injected with 5,7-Dihydroxytryptamine in Various Brain Regions

	NA	DA	HVA	5-HT	5-HIAA	TRP
Dorsal + median raphe nuclei (N = 5)	98.6 ± 9.4	146.5 ± 22.9	74.3 ± 53.7	8.6 ± 1.1[a]	7.5 ± 1.6[a]	99.1 ± 7.0
B9 region (N = 3)	71.6 ± 16.0	93.0 ± 18.0	35.8 ± 28.4	7.7 ± 0.6[a]	12.3 ± 0.9[a]	111.2 ± 3.3
Combined raphe + B9 (N = 21)	65.4 ± 5.8[a]	76.0 ± 24.5	—	6.0 ± 0.4[a]	9.3 ± 0.5[a]	121.5 ± 10.1
Control levels (ng/g)	237.3–309.7	41.9–76.0	13.6–57.3	291.9–377.0	150.5–268.7	2786.1–3813.3

[a]Differs from corresponding control group, $p < .02$ or less.
Abbreviations as in TABLE 1.

TABLE 4. Levels of Some Amines, Amine Metabolites, and Tryptophan (given as percent of control) in the Forebrain, Exclusive of the Neocortex and Hippocampal Formation, in Rats Injected with 5,7-Dihydroxytryptamine in Various Brain Regions

	NA	DA	HVA	5-HT	5-HIAA	TRP
Dorsal + median raphe nuclei (N = 5)	91.0 ± 4.1	93.8 ± 5.0	74.7 ± 7.8^a	15.5 ± 4.3^a	14.0 ± 4.2^a	96.0 ± 10.8
B9 region (N = 3)	90.8 ± 10.3	97.0 ± 7.4	84.7 ± 4.3	19.1 ± 3.4^a	16.5 ± 2.6^a	106.7 ± 7.7
Combined raphe + B9 (N = 21)	79.4 ± 4.2^a	91.7 ± 3.2	92.5 ± 3.7	3.9 ± 0.3^a	3.5 ± 1.0^a	108.0 ± 4.7
Control levels (ng/g, N = 6–7)	421.1–577.2	2306.2–2620.3	147.3–182.9	521.6–557.2	214.3–450.8	2731.2–3735.6

^{a}Differs from corresponding control group, $p < .02$.
Abbreviations as in TABLE 1.

became apparent that the animals' behavior was grossly abnormal. Records were taken as the rats moved about freely on a movement sensor platform measuring 35 × 35 cm and elevated about 65 cm above a table. Normal rats, or those treated with scopolamine alone, rarely jump or fall off the platform but the rats treated with PCPA plus scopolamine would walk off the edge repeatedly without the slightest hesitation. Constant attention and physical restraint were necessary to keep them from falling. The rats were easy to handle, showing no unusual tendency to bite, and usually walked about and reared without obvious ataxia.

More systematic behavioral tests confirmed the presence of a severe disorganization of behavior in rats treated with a large dose of PCPA plus scopolamine. A quick, simple test which was useful in this work is a modification of swim-to-platform test devised by Morris.[19] The apparatus consisted of a glass aquarium (43 × 90 cm, and 45 cm deep) filled with clean water (20°C) to a depth of 25 cm. An overturned rat cage provided a wire mesh platform (21.5 × 18.5 cm), raised about 1 cm above the water surface, and located in the center of the aquarium. Rats were dropped in one corner of the aquarium and the time taken (to a maximum of 60 sec) to climb up on the platform was recorded. Ten such trials were given with an inter-trial interval of 10–15 sec. Trials on which more than 10 sec were required to climb up on the platform were considered errors. Normal rats or rats treated with a drug vehicle or scopolamine methyl bromide (a muscarinic receptor blocker which, owing to its quaternary ammonium structure, does not penetrate the brain readily) made only 1–2 such errors on 10 trials but rats treated with 3 doses of PCPA or a single dose of scopolamine performed more poorly, making 2–3 times as many errors as the controls. A single dose of PCPA had no detectable effect (TABLE 5). When PCPA and scopolamine were combined the rats made errors on virtually every trial, swimming repeatedly around the tank. Rats treated with this drug combination swam as well and as fast as normal and climbed up promptly if they encountered the platform head on. However, if they merely touched the platform with a limb or one side of the body, they often failed to turn toward it (as normal rats did) and often pushed themselves vigorously away from it.

The detrimental effects of PCPA and scopolamine were attenuated by previous training if the drugs were given separately but when the two drugs were given in combination, severe behavioral impairments were present in trained as well as in untrained rats. This is shown in an experiment in which groups of rats were given 10 training trials (acquisition) on the swim-to-platform test followed by 10 retention trials 7 days later. On the 3 days preceding the retention tests different groups were given daily injections of PCPA (500 mg/kg, i.p.), an equivalent volume of drug vehicle, or no treatment. About 15 min prior to the retention test, the rats received either scopolamine (5 mg/kg, s.c.) or an equivalent volume of drug vehicle, yielding a total of 6 groups: 1) PCPA + scopolamine; 2) PCPA + vehicle; 3) vehicle + scopolamine; 4) vehicle + vehicle; 5) no treatment + scopolamine; and 6) no treatment + vehicle. FIGURE 6 shows that PCPA and scopolamine, acting alone, produced only a minor deficit in retention performance, or no deficit at all. However, when the 2 drugs acted in combination, the rats performed very poorly, swimming repeatedly around the tank on every trial. Similar effects have been reported by Richter-Levin and Segal[20] who found that small doses of atropine (20 mg/kg, i.p.) and PCPA (a single dose of 300 mg/kg) that had no effect on swim-to-platform performance when administered separately, produced a severe deficit when administered together.

Potentiating effects of combined cholinergic and serotonergic blockade can also be demonstrated in the swim-to-platform test using lesion techniques. Nilsson *et al.*[21] showed that the impairment produced by a surgical lesion of the septal nuclei (affecting cholinergic input to the hippocampus) was greatly increased in rats that had also received an intraventricular injection of 5,7-DHT even though rats that had been treated with 5,7-DHT alone performed almost as well as normal.

The potentiating effects of combined cholinergic and serotonergic blockade can also be demonstrated using a shock avoidance test.[22] Avoidance training was given in a 30.5 × 30.5 × 30.5 cm wooden box with a floor consisting of a grid of round steel bars. After a preliminary period of exploration, and tests to determine if the rats would jump out prior to training (none did so), the rats were given a series of 15 trials on each of which they were dropped in the box and the time taken to jump out was recorded. Shock was applied after a delay of 10 sec and continued until the rat jumped out (maximum shock duration of 30 sec trial). Control rats made about 12 avoidance responses (jumping out before shock was applied), but rats treated with scopolamine made only about 5 avoidance responses in 15 trials. PCPA alone had very little effect on avoidance performance but when it was combined with scopolamine, avoidance behavior disappeared completely. Most of the rats treated with the drug combination were not even able to escape shock by jumping out after it was applied. The escape

TABLE 5. Number of Errors (Trials out of 10 on Which Swimming Time Exceeded 10 Sec) Made in 10 Trials by Rats Swimming to a Visible Platform after Various Drug Treatments

	N	Errors
1. Untreated controls	11	1.5 ± 0.4
2. Vehicle-injected controls	6	1.4 ± 0.2
3. Scopolamine methyl bromide (5 mg/kg, s.c.)	8	1.4 ± 0.2
4. PCPA (500 mg/kg, i.p.)	8	1.4 ± 0.3
5. PCPA (500 mg/kg/day × 3, i.p.)	10	2.6 ± 0.7[a]
6. Scopolamine hydrobromide (5 mg/kg, s.c.)	6	4.7 ± 1.3[a]
7. PCPA (500 mg/kg/day × 3, i.p.) plus scopolamine hydrobromide (5 mg/kg, s.c.)	6	9.0 ± 0.8[a,b]

PCPA, *p*-chlorophenylalanine. *a*, Differs from control groups, $p < .05$; *b*, differs from scopolamine or *p*-chlorophenylalanine groups, $p < .05$. Mann-Whitney tests. Except for Groups 3 and 4, data are taken from Vanderwolf (1987).[22]

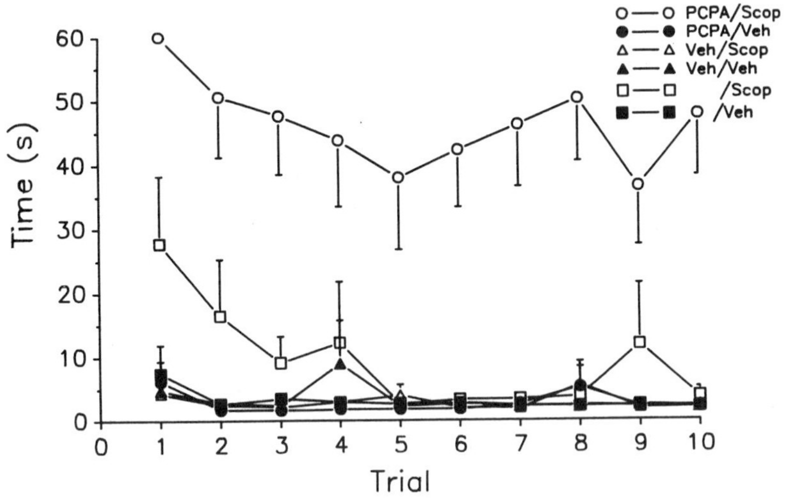

FIGURE 6. Effect of *p*-chlorophenylalanine and scopolamine on retention of swim-to-platform behavior. Rats received 10 training trials (not shown) and 10 retention trials 7 days later. On the 3 days preceding retention testing the rats received *p*-chlorophenylalanine (PCPA, 500 mg/kg/day, i.p.), drug vehicle (Veh) or no treatment; 15 min before retention testing they received scopolamine (5 mg/kg, s.c.) or saline (Veh). Symbols identifying the 6 groups so formed are shown in the upper right (6 rats/group). Analysis of variance reveals significant group differences ($p < .00001$), and a significant effect of trials ($p < .00006$) but no significant interaction. Mann-Whitney U tests based on errors (trials in which more than 10 s were taken to reach the platform) show that the PCPA/Scop group differs from each of the other groups ($p < .02$ or less, in each case) but that the latter do not differ among themselves.

deficit could be attenuated by experience. If rats were given avoidance training while under the influence of PCPA alone and then given scopolamine followed by additional testing, good escape behavior was maintained but avoidance behavior virtually disappeared.[22]

TABLE 6 shows the results of experiments on the effects of 4 intra-brain stem injections of 5,7-DHT (as described above) on shock avoidance behavior requiring a jump response. The rats were tested 2–4 months after the injection of the neurotoxin. The 5,7-DHT-treated rats acquired the avoidance response more slowly than control rats that had received 4 intra-brain stem injections of Locke's solution. However, after a saline injection and additional training (retention), performance improved to nearly normal levels in the 5,7-DHT-treated group. In rats that received scopolamine after acquisition, additional testing (retention) revealed a complete disappearance of avoidance behavior in the 5,7-DHT-treated rats and an intermediate level of avoidance behavior in the rats that had been treated with intracerebral Locke's solution. Escape behavior was retained in most cases in the 5,7-DHT plus scopolamine treated rats even though avoidance behavior disappeared.

Since systemic injections of PCPA and intracerebral injections of 5,7-DHT affect cerebral levels of norepinephrine as well as serotonin, it is important to consider the possible role of the former neurotransmitter in the control of cerebral activity and behavior. Selective neurotoxin lesions of the ascending noradrenergic projections have modest effects, or no effect at all, on rats' performance in various behavioral tests[23–26] and potentiate slightly, or not at all, the disruptive effects of scopolamine.[27,28] In correlation with this, elimination of ascending noradrenergic influences by means of pharmacological treatments, surgical lesions and treatment with selective neurotoxins has been shown to have no effect on atropine-resistant hippocampal RSA and atropine-resistant neocortical LVFA.[14,29–31] The virtual elimination of dopamine also has no direct effect on ARCA but, nonetheless, results in a disruption of higher level control of behavior[32] as well as akinesia and catalepsy, as is well known.[33] It appears that the behavioral effects of dopaminergic blockade are due to factors that are not directly involved in the generation of ARCA.

As already mentioned, reserpine provides an additional means of eliminating serotonergic influences in the brain and of partially or totally eliminating ARCA. Reserpine, in a dose of 10 mg/kg, consistently produces a total elimination of atropine-resistant LVFA. However, reserpine also depletes the brain of catecholamines, resulting in severe akinesia and catalepsy. Spontaneous motor activity can be restored by L-dopa or catecholaminergic agonists such as amphetamine,[34,35] but these compounds do not fully restore higher level control of behavior. Thus, avoidance response is only partly restored[36] and the restored behavior is abnormally sensitive to the disruptive effect of antimuscarinic drugs. Vanderwolf *et al.*[37] found that *d*-amphetamine would restore hypothalamic self-stimulation behavior in rats that were well trained and then given reserpine (10 mg/kg). However, if the reserpinized rats were also given atropine or scopolamine in addition to amphetamine all self-stimulation behavior ceased even though the rats were very active. Combinations of atropine and amphetamine had little effect on self-stimulation in non-reserpinized rats.

These results are consistent with the results of experiments on the effect of reserpine on neocortical electrical activity. L-dopa and other catecholaminergic agonists do not restore atropine-resistant LVFA in reserpinized rats although 5-hydroxytryptophan and various serotonergic agonists were able to do so.[1] It is probable that the behavioral impairments that persist in reserpinized rats after the administration of catecholaminergic agonists are due primarily to persisting low levels of cerebral serotonin.

TABLE 6. Shock Avoidance Performance in Rats Following Treatment with Intracerebral Injections of 5,7-Dihydroxytryptamine and Scopolamine

Initial Injection	N	Acquisition	Second Injection	Retention
1. Locke's solution in brainstem	6	12.3 ± 3.3	saline	14.8 ± 0.2
2. Locke's solution in brainstem	8	10.1 ± 1.4	scopolamine	5.6 ± 1.7[b]
3. 5,7-DHT in brainstem	8	6.5 ± 1.5[a]	saline	13.5 ± 0.7
4. 5,7-DHT in brainstem	8	8.1 ± 0.8[a]	scopolamine	0.1 ± 0[c]

Scores indicate successful avoidances in 15 trials during both acquisition and retention. Retention was tested 15–30 min after acquisition and the injection of saline or scopolamine hydrobromide (5 mg/kg, s.c.). 5,7-DHT, 5,7-dihydroxytryptamine creatinine sulfate (Sigma), 25 $\mu g/\mu l$, given in 0.5–1.0 μl volumes at each of 4 sites in the brainstem; controls received equivalent injections of Locke's solution. *a*, differs from either Group 1 or 2 during Acquisition, $p < .01$; *b*, differs from Group 1 during Retention, $p < .01$; *c*, differs from each of the other groups during retention, $p < .01$; Mann-Whitney tests. All rats given saline after acquisition showed improved performance during retention. All of the 5,7-DHT-treated rats, and 6 of the 8 Locke's solution-treated rats that were given scopolamine after acquisition performed worse during retention than during acquisition.

Interference with serotonergic input to the forebrain appears to have a very general effect on behavior. In addition to the points already mentioned, published reports have described disturbances in: feeding behavior[38–40]; response to noxious stimulation[41,42]; social and sexual behavior[43–46]; predatory aggression[47,48]; frequency of spontaneous locomotion[49,50]; 2-way shock avoidance behavior[51,52]; anxiety[53]; and performance in a maze test.[54]

Several general conclusions can be drawn from the behavioral data. (1) Central serotonergic blockade produces impairments in a broad spectrum of behaviors. (2) The effects produced are, in many cases, rather modest. (3) When central serotonergic blockade is combined with central cholinergic blockade, the behavioral deficits produced are much greater than is produced by cholinergic blockade alone or by serotonergic blockade alone.

INTERPRETATION AND POSSIBLE CLINICAL APPLICATION

In the years since the landmark paper by Moruzzi and Magoun[55,56] it has become conventional to state that the reticular formation is responsible for the maintenance of consciousness, attention, and the waking state and that these functions are achieved through an ascending effect on the thalamus which, in turn, activates the neocortex. Large lesions of the medial brain stem are often said to produce unconsciousness or coma as a result of interruption of this ascending system.

Recent data indicate the necessity for extensive revision of these classical concepts.[1,57] Activation of both the neocortex and hippocampal formation is dependent on activity in ascending cholinergic and serotonergic pathways rather than thalamocortical pathways. Cerebral activation patterns correlate primarily with concurrent behavior rather than with consciousness or attention. Selective and total blockade of cerebral activation produces a dementia-like syndrome of behavioral disorganization rather than the sleep or coma that would be expected on the basis of classical concepts. In retrospect, it seems likely that the loss of postural tonus and spontaneous motor activity following large lesions of the brain stem are due to destruction of descending motor pathways.[58] Loss of cerebral activation following combined cholinergic and serotonergic blockade does not produce such motoric depression because descending brain stem pathways remain intact.

The evidence that serotonin plays an essential role in ARCA may suggest an explanation for: (1) the wide spectrum of behavioral deficits that results from central serotonergic blockade, and (2) the potentiating effect of combining serotonergic blockade with cholinergic blockade. With respect to the first point, ARCA consistently accompanies Type 1 behavior, a class that includes walking, moving the head about, and manipulating objects with the limbs. These motor activities play a very general role in virtually all behavioral performances such as feeding, sexual activity, escape from danger, or behavioral thermoregulation. If ascending serotonergic pathways are part of the central mechanism that controls Type 1 behavior, one would expect a generalized deficit in behavior when these pathways are blocked. This expectation is borne out by observation. The fact that paralysis, i.e., loss of movement, is not one of the consequences of central serotonergic blockade may seem puzzling by this account. However, given the general nature of cerebral control of movement, it is what should be expected. Removal of the entire cerebral cortex does not by any means abolish spontaneous behavior in most mammals. Decorticate rats walk and run about, face-wash, etc., but their behavior is aimless and erratic.[59,60] This suggests that brain stem, cerebellar, and spinal circuits are sufficient to co-ordinate muscle activity to form recognizable movement patterns but that cortical circuits normally control the occurrence of the lower level circuits to produce intelligent adaptive behavior. If the cerebral cortex is intact but cholinergic *or* serotonergic inputs to it are blocked, cerebral activation is impaired and moderate deficits in the cerebral control of behavior will be observed. If *both* cholinergic and serotonergic inputs are blocked simultaneously, so that all cerebral activation is lost, cortical circuits presumably cannot function normally at any time and a severe disorganization of behavior is the result. This may be the main reason that combined cholinergic and serotonergic blockade produces much more severe behavioral deficits than blockade of either system in isolation. The precise nature of cholinergic and serotonergic effects on cerebral neuronal activity is discussed in a recent review.[1]

Data on the role of cholinergic and serotonergic systems in animals may be relevant to certain neurological and psychiatric conditions in man. There is currently an abundance of evidence available to show that Alzheimer's disease is associated with a loss of both cholinergic and serotonergic projections to the cerebral cortex.[61–64] In the light of the animal work these losses should be sufficient to produce both the severe behavioral deficits and the large amplitude irregular electroencephalogram that are characteristic of this disorder.[65,66] The animal experiments also suggest that attempts to restore cholinergic and serotonergic inputs together might be much more effective in ameliorating the behavioral disturbances in Alzheimer's disease than attempts to restore either input alone. Animal experiments could provide additional data relevant to this question and the effects of combined cholinergic and serotonergic enhancement could easily be tested in demented patients.

REFERENCES

1. VANDERWOLF, C. H. 1988. Cerebral activity and behavior: Control by central cholinergic and serotonergic systems. Int. Rev. Neurobiol. **30:** 225–340.
2. VANDERWOLF, C. H. & L.-W. S. LEUNG. 1985. Brain electrical activity in relation to behavior. *In* Neuromethods, Volume I, General Neurochemical Techniques. A. A. Boulton & G. B. Baker Eds. Clifton, NJ: Humana Press, pp. 305–341.
3. KOE, K. B. & A. WEISSMAN. 1966. *p*-Chlorophenylalanine: A specific depletor of brain serotonin. J. Pharmacol. Exp. Ther. **194:** 499–516.
4. VANDERWOLF, C. H. & G. B. BAKER. 1986. Evidence that serotonin mediates noncholinergic neocortical low voltage fast activity, non-cholinergic hippocampal rhythmical slow activity and contributes to intelligent behavior. Brain Res. **374:** 342–356.
5. BAUMGARTEN, H. G., A. BJÖRKLUND, L. LACHENMAYER & A. NOBIN. 1973. Evaluation of the effects of 5,7-dihydroxytryptamine on brain serotonin and catecholamine neurons in the rat CNS. Acta Physiol. Scand. Suppl. **391:** 1–22.
6. BAUMGARTEN, H. G., H. P. KLEMM, L. LACHENMAYER, A. BJÖRKLUND, W. LOVENBERG & H. G. SCHLOSSLERGER. 1978. Mode and mechanism of action of neurotoxic indoleamines: A review and a progress report. Ann. N.Y. Acad. Sci. **305:** 3–24.
7. BAUMGARTEN, H. G. & L. LACHENMAYER. 1972. 5,7-Dihydroxytryptamine: Improvement in chemical lesioning of indoleamine neurons in the mammalian brain. Z. Zellforsch. Mik. Anat. **135:** 399–414.
8. BJÖRKLUND, A., H. G. BAUMGARTEN & A. RENSCH. 1975. 5,7-Dihydroxytryptamine: Improvement of its selectivity for serotonin neurons in the CNS by pretreatment with desipramine. J. Neurochem. **24:** 833–835.
9. SHORE, P. A. & A. GIACHETTI. 1978. Reserpine: Basic and clinical pharmacology. *In* Handbook of Psychopharmacology, Neuroleptics and schizophrenia. L. L. Iversen, S. D. Iverson & S. H. Snyder, Eds. Vol. **10:** 197–219. New York: Plenum Press.
10. LEUNG, L.-W. S. 1984. Pharmacology of theta phase shift in the hippocampal CAI region of freely moving rats. Electroenceph. Clin. Neurophysiol. **58:** 457–466.
11. VANDERWOLF, C. H., L.-W. S. LEUNG & G. B. BAKER. The role of serotonin in the control of cerebral activity: studies with intracerebral 5,7-dihydroxytryptamine. 1989. Brain Res. **504:** 181–191.

12. VANDERWOLF, C.H. & B. A. PAPPAS. 1980. Reserpine abolishes movement-correlated atropine-resistant neocortical low voltage fast activity. Brain Res. **202:** 79–94.
13. BAKER, G. B., R. T. COURTTS & T. S. RAO. 1987. Neuropharmacological and neurochemical properties of N-(2-cyanoethyl)-2-phenylethylamine, a prodrug of 2-phenylethylamine. Br. J. Pharmacol. **92:** 243–255.
14. VANDERWOLF, C. H., R. KRAMIS & T. E. ROBINSON. 1978. Hippocampal electrical activity during waking behavior and sleep: Analyses using centrally acting drugs. Functions of the Septo-hippocampal System. Ciba Foundation Symposium. **58:** 199–226. Amsterdam: Elsevier, Excerpta Medica, North-Holland.
15. VANDERWOLF, C. H., G. C. HARVEY & L.-W. S. LEUNG. 1987. Transcallosal evoked potentials in relation to behavior in the rat: Effects of atropine, p-chlorophenylalanine, reserpine, scopolamine and trifluoperazine. Behav. Brain Res. **25:** 31–48.
16. READER, T. A., R. BRIERE, L. GRONDIN & A. FERRON, 1986. Effects of p-chlorophenylalanine on cortical monoamines and on the activity of noradrenergic neurons. Neurochem. Res. **11:** 1025–1035.
17. FUXE, K. & G. JONSSON. 1974. Further mapping of central 5-hydroxytryptamine neurons: studies with the neurotoxic dihydroxyptryptamines. *In* Serotonin: New Vistas, Histochemistry and Pharmacology. E. Costa, G. L. Gessa & M. Sandler, Eds. Advances in Biochemical Psychopharmacology **10:** 1–12.
18. O'HEARN, E. & M. E. MOLLIVER. 1984. Organization of raphe-cortical projections in rat: A quantitative retrograde study. Brain Res. Bull. **13:** 709–726.
19. MORRIS, R. G. M. 1981. Spatial localization does not require the presence of local cues. Learn. Mot. **12:** 239–260.
20. RICHTER-LEVIN, G. & M. SEGAL. 1989. Spatial performance is severely impaired in rats with combined reduction of serotonergic and cholinergic transmission. Brain Res. **477:** 404–407.
21. NILSSON, O. G., R. E. STRECKER, A. DASZUTA & A. BJÖRKLUND. 1988. Combined cholinergic and serotonergic denervation of the forebrain produces severe deficits in spatial learning in the rat. Brain Res. **453:** 235–246.
22. VANDERWOLF, C. H. 1987. Near-total loss of "learning" and "memory" as a result of combined cholinergic and serotonergic blockade in the rat. Behav. Brain Res. **23:** 43–57.
23. ARCHER, T., G. JONSSON & S. B. ROSS. 1985. Active and passive avoidance following the administration of systemic DSP-4, xylamine or *p*-chloroamphetamine. Behav. Neurol. Biol. **43:** 238–249.
24. HAGAN, J. J., J. E. ALPERT, R. G. M. MORRIS & S. D. IVERSEN. 1983. The effects of central catecholamine depletions on spatial learning in rats. Behav. Brain Res. **9:** 83–104.
25. PISA, M. & H. C. FIBIGER. 1983. Intact selective attention in rats with lesions of the dorsal noradrenergic bundle. Behav. Neurosci. **97:** 519–529.
26. PISA, M. & H. C. FIBIGER. 1983. Evidence against a role of the rat's dorsal noradrenergic bundle in selective attention and place memory. Brain Res. **272:** 319–329.
27. DECKER, M. W. & M. GALLAGHER. 1987. Scopolamine-disruption of radial arm maze performance: Modification by noradrenalin depletion. Brain Res. **417:** 59–69.
28. DECKER, M. W. & J. L. MCGAUGH. 1989. Effects of concurrent manipulations of cholinergic and noradrenergic function on learning and retention in mice. Brain Res. **477:** 29–37.
29. KOLB, B. & I. Q. WHISHAW. 1977. Effects of brain lesions and atropine on hippocampal and neocortical electroencephalograms in the rat. Exp. Neurol. **56:** 1–22.
30. ROBINSON, T. E., C. H. VANDERWOLF & B. A. PAPPAS. 1977. Are the dorsal noradrenergic bundle projections from the locus coeruleus important for neocortical or hippocampal activation? Brain Res. **8:** 75–98.
31. WHISHAW, I. Q., T. E. ROBINSON, T. SCHALLERT, M. DERYCK & V. D. RAMIREZ. 1978. Electrical activity of the hippocampus and neocortex in rats depleted of brain dopamine and norepinephrine: Relations to behavior and effects of atropine. Exp. Neurol. **62:** 748–767.
32. WHISHAW, I. Q. & S. B. DUNNETT. 1985. Dopamine depletion, stimulation or blockade in the rat disrupts spatial navigation and locomotion dependent upon beacon or distal cues. Behav. Brain Res. **18:** 11–29.

33. HORNYKIEWICZ, O. 1966. Dopamine (3-hydroxytyramine) and brain function. Pharmacol. Rev. **18:** 925–964.
34. CARLSSON, A., M. LINDQVIST & T. MAGNUSSON. 1957. 3,4 Dihydroxyphenylalanine and 5-hydroxytryptophan as reserpine antagonists. Nature (London) **180:** 1200.
35. RECH, R. H. 1964. Antagonism of reserpine behavioral depression by d-amphetamine. J. Pharmacol. Exp. Ther. **146:** 369–376.
36. AHLENIUS, S. 1974. Effects of L-dopa on conditioned avoidance responding after behavioural supression by α-methyltyrosine or reserpine in mice. Neuropharmacology **13:** 729–739.
37. VANDERWOLF, C. H., M. GUTMAN & G. B. BAKER. 1984. Hypothalamic self-stimulation: The role of dopamine and possible relations to neocortical slow wave activity. Behav. Brain Res. **12:** 9–19.
38. BREISCH, S. T., F. P. ZEMLAN & B. G. HOEBEL. 1976. Hyperphagia and obesity following serotonin depletion by intraventricular p-chlorophenylalanine. Science **192:** 382–385.
39. CLINESCHMIDT, B. V. 1973, 5,6-Dihydroxytryptamine: Suppression of the anorexigenic action of fenfluramine. Eur. J. Pharmacol. **24:** 405–409.
40. SALLER, C. F. & E. M. STRICKER. 1976. Hyperphagia and increased growth in rats after intraventricular injection of 5,7-dihydroxytryptamine. Science **192:** 385–387.
41. MESSING, R. B. & L. D. LYTLE. 1977. Serotonin-containing neurons: Their possible role in pain and analgesia. Pain **4:** 1–21.
42. TENEN, S. S. 1968. Antagonism of the analgesic effect of morphine and other drugs by p-chlorophenylalanine, a serotonin depletor. Psychopharmacology (Berlin) **12:** 278–285.
43. FILE, S. E. & J. F. W. DEAKIN. 1980. Chemical lesions of both dorsal and median raphe nuclei and changes in social and aggressive behavior in rats. Pharmacol. Biochem. Behav. **12:** 855–859.
44. SHEARD, M. H. 1969. The effect of parachlorophenylalanine on behavior in rats: Relation to brain serotonin and 5-hydroxyindoleacetic acid. Brain Res. **15:** 524–528.
45. SHILLITO, E. E. 1969. The effect of parachlorophenylalanine on social interaction of male rats. Br. J. Pharmacol. **36:** 193–194.
46. TAGLIAMONTE, A., P. TAGLIAMONTE, G. GESSA & B. BRODIE. 1969. Compulsive sexual activity induced by parachlorophenylalanine in normal and pinealectomized male rats. Science **166:** 1433–1435.
47. DICHIARA, G., R. CAMBA & P. F. SPANO. 1971. Evidence for inhibition by brain serotonin of mouse killing behaviour in rats. Nature **233:** 272–273.
48. EICHELMAN, B. S. JR. & N. B. THOA. 1973. The aggressive monoamines. Biol. Psychiat. **6:** 143–164.
49. FIBIGER, H. C. & B. A. CAMPBELL. 1971. The effect of parachlorophenylalanine on spontaneous locomotor activity in the rat. Neuropharmacology **10:** 25–32.
50. GERSON, S. C. & R. J. BALDESSARINI. 1980. Motor effects of serotonin in the central nervous system. Life Sci. **27:** 1435–1451.
51. KÖHLER, C., S. B. ROSS, B. SREBRO & S.-O. ÖGREN. 1978. Long-term biochemical and behavioral effects of p-chloroamphetamine in the rat. Ann. N.Y. Acad. Sci. **305:** 645–663.
52. ÖGREN, S. O. 1985. Evidence for a role of brain serotonergic neurotransmission in avoidance learning. Acta Physiol. Scand. **125** (Suppl. 554)**:** 7–71.
53. GARDNER, C. R. 1985. Pharmacological studies on the role of serotonin in animal models of anxiety. *In* Neuropharmacology of Serotonin A. R. Green, Ed. Oxford: Oxford University Press, pp 281–325.
54. PANG, K., A. CHAVEZ & G. ROSE. 1985. Effects of central depletion of norepinephrine or serotonin upon memory tasks using the radial maze. Neurosci. Abst. **11:** 534.
55. MORUZZI, G. & H. W. MAGOUN. 1949. Brainstem reticular formation and activation of the EEG. Electroenceph. Clin. Neurophysiol. **1:** 455–473.
56. MORUZZI, G. 1972. The sleep-waking cycle. Ergeb. Physiol. **64:** 1–165.
57. VANDERWOLF, C. H. & T. E. ROBINSON. 1981. Reticulo-cortical activity and behavior: A critique of the arousal theory and a new synthesis. Behav. Brain Sci. **4:** 459–514.
58. KUYPERS, H. G. J. M. 1964. The descending pathways to the spinal cord, anatomy and function. *In* Progress in Brain Research. J. C. Eccles & J. P. Schade, Eds. Vol. **11:** 178–202.

59. VANDERWOLF, C. H., B. KOLB & R. K. COOLEY. 1978. The behavior of the rat following removal of the neocortex and hippocampal formation. J. Comp. Physiol. Psychol. **92:** 156–175.
60. WHISHAW, I. Q., B. KOLB & R. J. SUTHERLAND. 1983. The analysis of behavior in the laboratory rat. *In* Behavioral Approaches to Brain Research. T. E. Robinson, Ed. New York: Oxford University Press, pp. 141–211.
61. BOWEN, D. M. & A. N. DAVISON. 1986. Biochemical studies of nerve cells and energy metabolism in Alzheimer's disease. Br. Med. Bull. **42**(1)**:** 75–80.
62. COLLERTON, D. 1986. Cholinergic function and intellectual decline in Alzheimer's disease. Neuroscience **19:** 1–28.
63. ROSSOR, M. & L. L. IVERSEN. 1986. Non-cholinergic neurotransmitter abnormalities in Alzheimer's disease. Br. Med. Bull. **42**(1)**:** 70–74.
64. WHITFORD, G. M. 1986. Alzheimer's disease and serotonin: A review. Neuropsychobiology **15:** 133–142.
65. FENTON, G. W. 1986. Electrophysiology of Alzheimer's disease. Br. Med. Bull. **42**(1)**:** 29–33.
66. MAHENDRA, B. 1984. Dementia: A Survey of the Syndrome of Dementia. Lancaster, England: MTP Press Ltd.

DISCUSSION OF THE PAPER

H. J. ALTMAN (*Lafayette Clinic, Detroit MI*): I wondered if you might consider the possibility that part of the behavioral effects that you see is somewhat similar to what Cools would describe as a failure to shift behavioral set. You described the animals not being able to find the platform unless they literally bumped into it, and then they might escape, sort of the same description that Cools would have used in a swim test that he used.

VANDERWOLF: Well I don't have any argument with giving any name to it you like. The fact is simply that they do have a huge behavioral deficit, whatever one wants to call it.

M. SEGAL (*Weizmann Institute, Rehovot, Israel*): I like your results very much. The fact that you show a moment-to-moment variation in the EEG dependence on serotonin is very nice. I do have a problem with the results presented earlier today by Barry Jacobs showing that it's a pretty boring nucleus in the firing rates and the fact that the cells cease to fire during REM sleep when you have a lot of activity there.

I also have some problem with Descarries's results yesterday showing that it's not actually producing any synapses really anywhere, it's just leaking out. How would you control activity on a moment to moment basis if the cells don't do this.

VANDERWOLF: Well the last question is the easiest to answer because I have no idea how to answer it. I don't know that. It puzzles me, and it may be that the serotonin is simply one of several factors, of at least two, that are necessary to produce the electrocortical patterns. For example, another way of blocking these patterns, which is probably not dependent on serotonin, is with some psychomimetic drugs like cyclazocine and SKF 10,047. These are phencyclidine receptor ligands which produce psychosis in people, and they also will block scopolamine-resistant electrocortical patterns. So it may be that there's some interaction between serotonin and amino acids since the phencyclidine receptor binders are believed to block excitatory amino acids.

Now with respect to your first question on the behavior of units in the dorsal raphe and median raphe, I haven't done experiments in this field so I have to be rather hesitant in pronouncing on them, but I wonder whether there might not be more differentiation of unit types than has been reported so far. During REM sleep there is a rather remarkable effect in that RSA is present for minutes at a time. Now you never see that under normal circumstances in a waking animal unless it runs in a treadmill. Ordinarily, you don't see anything like that. Most of the RSA that occurs during REM sleep is of the atropine-sensitive type, but during the muscular twitches there are short bursts of atropine-resistant activity. So there seems to be a release of the cholinergic-dependent RSA during REM sleep and I wonder whether that might not be a consequence of the shutting down of one kind of serotonergic neuron since one also sees this release of scopolamine sensitive RSA in animals that have lost their serotonin.

Now what the function of that is I have no idea but it might account for these phenomena that way.

GEORGE K. AGHAJANIAN (*Yale University School of Medicine, New Haven, CT*): Well, the effects of scopolamine or atropine, or the anti-muscarinic drugs, have been studied very well in people and they do produce a delirium. Parachloraphenylalanine hasn't been studied that intensively in people. As I recollect the few studies that have been done showed that one of the effects of parachloraphenylalanine, and this wasn't really being looked for, was confusion. So that would fit with what you were presenting.

But the question I wanted to ask was about the noradrenergic system. Have you tried to couple lesions of the noradrenergic system with the muscurinic antagonism?

VANDERWOLF: Oh, yes, I've been working in this field for a long time, but without realizing that I was dealing with serotonin all the time. It took me a long time to come to this. I first found, in 1973 actually, that reserpine would eliminate these atropine-resistant phenomena. A combination of reserpine with scopolamine makes all electrocortical activation disappear. My first thought was obviously this has to be due to a loss of dopamine or norepinephrine. So we spent a long time eliminating dopamine and norepinephrine in every conceivable way, and it . . .

AGHAJANIAN: Like dorsal bundle lesions . . .

VANDERWOLF: All that sort of stuff, and α-methyltyrosine and 6-hydroxy dopamine in the ventricles and in neonatal rats and all these treatments were ineffective.

R. Y. MOORE (*Session Chairman*): It's disappointing, George, but you'll have to take it.

AGHAJANIAN: I really like serotonin most of all.

QUESTION: Have you looked at any 5-HT receptor antagonists in potentiating the behavorial deficit? Which receptors do you think are involved?

VANDERWOLF: I have no idea. This is something that I've been muddling along with for some years and most of the receptor antagonists are not active. I can give you a list of those we tested if I can remember them all. We've tried methysergide, metergoline, cyproheptadine, mianserin, pizotifen, and maybe one or two others, in a huge range of doses and they're all completely inactive. Now the only one that shows some signs of activity is methiothepin, but chlorpromazine has some activity in big doses. But those are the only two antagonist drugs that I know of that will affect atropine-resistant electrocortical activation.

B. SOMMER (*State University of New York, Stony Brook*): And 5-HT receptor agonists, have you tried any of those like 8-hydroxy-DPAT or . . .

VANDERWOLF: No, we haven't done that. A master's student, Barbara Robertson, has a paper in preparation, in which we describe experiments with tranylcypromine and tryptophan, like Grahame-Smith, that kind of thing, and we also used parachloroamphetamine. Basically these drugs don't really modify the relation of the electrical activity to behavior. As in a normal animal, whenever it walks then you get these atropine-resistant patterns. And if the drugs make it walk more than normal then the atropine-resistant patterns occur more often. But the patterns remain correlated with behavior under the action of these drugs.

Aging and Regenerative Capacity of the Rat Serotonergic System

A Morphological, Neurochemical and Behavioral Analysis after Transplantation of Fetal Raphe Cells [a]

H. W. M. STEINBUSCH,[b] M. G. P. A. VAN LUIJTELAAR,
H. DIJKSTRA, A. NIJSSEN,[c] AND J. A. D. M. TONNAER[d]

Department of Pharmacology
Faculty of Medicine
Free University
Amsterdam, The Netherlands

[c]*Department of Psychonomy*
Faculty of Psychology
University of Amsterdam
Amsterdam, The Netherlands

[d]*Department of CNS Pharmacology*
Organon
Oss, The Netherlands

INTRODUCTION

One of the phenomena of pathological aging is the progressive deterioration of the brain. Until recently aging of neurotransmitter systems was studied primarily by behavioral[1] and clinical[2,3] observations and biochemical techniques[4,5] in animals and humans. Cell counts of neurons in certain brain nuclei, as for example, the substantia nigra, gave additional information about alterations in the aging as well as the pathological brain.[6–9] Normal aging has been characterized neurochemically among others by a decline of cholinergic, dopaminergic and noradrenergic transmission. It is well known that the dopaminergic system is involved in Parkinson's disease, and that the acetylcholine and serotonin transmitter systems play a role in Alzheimer's disease.[9–12] Only recently changes in serotonergic parameters have been distinguished, not only in these neuronal diseases,[13] but also in normal aging.[14] Biochemical studies of the human brain reveal no change with aging in 5-HT concentrations in frontal cortex, hippocampus or putamen, however in the hippocampus a positive correlation with age has been found for its metabolite 5-HIAA. This latter effect points towards an enhanced 5-HT turnover.[15] Consequently, decreases in [^{3}H]-5-HT binding capacity in frontal cortex and hippocampus and declines in affinity in the caudate-putamen complex have been reported.[16,40] In patients with Alzheimer's disease the concentrations 5-HT and 5-HIAA,

[a]This work was supported by Grant 900-552-072 from the Netherlands Foundation for Medical Research (MEDIGON).

[b]Address for correspondence: Dr. H. W. M. Steinbusch, Department of Pharmacology, Faculty of Medicine, Vrije Universiteit, van der Boechorststraat 7, 1081 BT Amsterdam, The Netherlands (tel: 31- (20)- 5482722; fax: 31- (20)- 462228).

the 5-HT uptake capacity, and the K^+-stimulated release of 5-HT were found to be reduced below control levels in frontal cortex, which is suggestive for a serotonergic denervation. Patients with Parkinson's disease showed also a reduction of hippocampal and striatal 5-HT and 5-HIAA levels and decreased 5-HT levels in the frontal cortex.[17,45–47] Controversial data have been published concerning the serotonergic system in aged animals. While the 5-HT uptake capacity in mouse cortex, hippocampus and basal ganglia[18] or rat striatum[5] does not change with aging, either no change[19] or decreases in 5-HT levels[20] and declines in the activity of the 5-HT deaminating enzyme MAO-A have been found in rats.[34,36] In this respect biochemical markers in aged brain structures should be considered as an integral reflection of changes in innervation and compensatory processes occurring in the remaining fibers. However, information on morphological changes in the 5-HT systems in the aged brain or pathological subjects is remarkably missing in the literature.

Brain cell transplantation has been used to study neuronal specificity and plasticity both in developmental and regenerative processes, as well as in aging and degenerative processes. To examine the involvement of neurochemically defined neuronal systems in behavior, interruption of central neuronal connections are induced by chemical lesions. Subsequently, implantation of neuronal cells with the capacity of adaptive outgrowth may functionally restore or replace neuronal communication. This technique is a valuable tool to create animal models for age-related neurodegenerative diseases. Using this model, potential therapeutic effects of implants can be estimated both in lesioned young adult or in intact aged rats. It has recently been demonstrated that lesion-induced impairments of serotonergic transmission in rats can be partially reversed biochemically by grafts of fetal serotonergic neurons.[32] Raphe cell suspensions have been implanted into several sites in the central nervous system, such as spinal cord, hippocampus, striatum, and hypothalamus.[33,35,41,44,49] These studies have demonstrated that grafted serotonergic neurons can survive for prolonged periods in the host brain, develop normal physiological properties, establish synaptic graft-host connections with neuronal elements, and improve at least certain types of lesion-induced functional deficits.[33,42] Hippocampal dysfunctioning, caused by the depletion of its serotonergic system, might induce disturbances in spatial memory similar to age-related impairments seen in humans and rats. In this respect, reinstatement of serotonergic functioning by a serotonergic transplant into the hippocampus could be expected to restore orientation. Recent reports suggest that neuronal transplants can survive also in the aged rat brain.[21]

The purpose of the present study was to add morphological, neurochemical and behavioral data to the knowledge of the aging process of the serotonergic system in the rat brain. We report that fetal mesencephalic serotonergic neurons can be grafted to the neostriatum of aged rat brain. Since we have shown previously that the neurotoxic-induced denervation of the serotonergic system in the rat forebrain can serve as model for the aging of the serotonergic system, we further studied the morphological, neurochemical, and behavioral effects of fetal raphe transplants on the induced degenerative processes in the hippocampus of the young adult rat brain.

Behavioral observations were made in two sets of experiments. In the first experiment we compared the behavior of young adult versus aged rats; in the second experiment we investigated the effects of lesioning and subsequent grafting in young adult rats. In this second experiment behavioral tests were conducted not before 4 weeks after transplantation in order to allow maximal outgrowth of the implanted neurons. The following behaviors were tested consecutively: sexual behavior, territorial aggression and scent marking, dominance aggression, sleep, and behavior in the spatial memory task. Several of these behaviors tested are not related directly to hippocampal functioning, but rather to hypothalamic or extra-hypothalamic functions related to 5-HT, such as sexual behavior and aggression. We nevertheless studied these behaviors to test our assumption that denervation was restricted to the hippocampus since we used low doses of 5,7-DHT.

MATERIALS AND METHODS

The study was conducted using a large number of young (3 months; 160–180 g b. wt.) and aged (28 months or older; >400 g b. wt.) male rats at the start of the study.

Immunohistochemistry

For serotonin-immunohistochemistry the animals were anesthetized and perfused with 500 ml of ice-cold 4% (w/v) paraformaldehyde, 0.05% (v/v) glutaraldehyde and 0.2% picric acid in 0.1 M sodium phosphate buffer (pH 7.4) for about 20 min. Brains were quickly removed, immersed in fixative without glutaraldehyde of 4°C for 2 hrs and cut on a Vibratome microtome. Free floating 50 μm thick sections were collected in Tris-buffered saline and incubated with rabbit anti-5-HT serum for 40 hrs at 4°C at a dilution of 1:10,000. Details about the specificity and sensitivity of antibodies to serotonin and the immunocytochemical procedure have been published elsewhere.[22,23]

Lesioning and Transplantation Surgery

Young host recipients were, prior to transplantation surgery, denervated by injecting 10 μg 5,7-DHT into both lateral ventricles. Two hours prior to the 5,7-DHT lesion the rats received an injection with desmethylimipramine (25 mg/kg body weight, i.p.) Aged rats were not denervated, due to the high mortality rate after such a lesion. Rats were used for grafting at least 12 days after the lesion.

Grafts were taken from E 15 fetuses of the same inbred strain with a crown-rump length of 14 mm. Serotonergic cells were selected from the ventral surface between the mesencephalic and pontine flexure.[24] The tissue was disrupted by repeated pipetting. The resulting suspension was kept at room temperature for a maximum of 4 hours. Two μl of a cell suspension (containing approximately 80,000 viable cells) was injected unilaterally as a column in the neostriatum of the young and aged rat or as a single 2 μl injection in the hippocampus of the young adult rat during a period of 4 minutes. Some young rats received for behavioral observations a bilateral injection into the hippocampus.

Behavioral Observations

In two sets of experiments behavioral observations were conducted. In the first experiment we compared the behavior of young adult versus aged rats; in the second experiment we investigated the effects of lesioning and subsequent grafting in young adult rats.

In the first, aging experiment, the males were subjected to a spatial orientation task in the Morris water maze paradigm[37] to estimate memory functioning. In addition their social behavior was observed in our colony cage. Besides observable changes in individual behavioral of the experimental animal itself "hidden" changes may exist, which only are revealed by the social responses of group mates. First separate colonies were formed of young and old males; subsequently both groups were mixed. Initiator behavior and recipient in social encounters were recorded.

In the second, transplantation experiment several behavioral functions were tested. Sexual behavior was tested 4 weeks after implantation during the last part of the dark period. In this test copulatory behavior towards a sterilized female (ligated oviducts) in natural behavioral estrous was recorded for 15 minutes, or until the first ejaculation. Mount and intromission latency (the time lapse between introduction of the female and the first occurrence of mount or intromission), mount frequency, ejaculation latency Mount and intromission latency (the time lapse between introduction of the female and the first occurrence of mount or intromission), mount frequency, ejaculation latency (time between first intromission and ejaculation) intromission percentage (the proportion of mount with intromission) and mount rate were scored.

Behaviorally defined sleeping was studied 12 to 13 weeks after transplantation. It was sampled on videotape during one light period, which lasted 12 hours. It was defined as: any posture, in which the animal was lying down the abdomen and head resting on the substratum. The first half hour was recorded to estimate sleep latencies, the last half hour was sampled to detect anticipatory activity in the males preceding the dark period. Between these two samples, six additional, equally spaced 15-minute samples were taken to detect sleeping patterns.

Aggression was tested 16 to 18 weeks after implantation either against intruders in the home cage (intruder attack) or within a colony (dominant attack). Intruder attack was tested in three consecutive trials, performed at intervals of one week. A male intruder was released in the home cage of the animal to be tested (the resident). During 15 minutes, bouts of "offensive behavior" by the resident (sideways attack, frontal attack, push over, and keep down), and the frequencies of both "clinch and bite" and homosexual "mounts" were recorded. To determine the dominance status of males in colonies the frequencies of attack, respectively avoidance were recorded for all possible dyadical male combinations.

Spatial memory was also tested in a large colony cage (2 × 3 m) 23 to 29 weeks after transplantation. In this cage the animals were submitted to a spatial orientation task in which the acquisition and the reversal of correct orientation was investigated. Correct orientation to visuo-spatial cues was positively reinforced by food reward, using a restricted feeding schedule. Errors, however, brought the subject in contact with an aggressive dominant colony male which caused social avoidance (negative reinforcement). To validate the results from this social orientation task they were referred to those obtained in the Morris Water maze task during week 33 to 38. In this task the rats' capacity was tested to learn and recall the spatial location of a small platform, which is hidden in a circular black pool (2.1 m diameter).

Combined Morphological and Neurochemical Studies

Eleven months after implantation combined morphological and neurochemical experiments were started for which only the young adult rats were used. The experimental design of these experiments has been described in detail elsewhere.[25] The major steps will be briefly summarized. Each experiment consisted of one animal out of a total of four groups (control, sham-lesioned, lesioned and lesioned+transplanted). Rats were decapitated one at a time, the brains were removed, and the hippocampal lobes were dissected on ice. They were cut into 300 μm slices with a tissue chopper and immediately divided into 5 levels (FIG. 1). From these levels the third slice was immediately fixed and further processed for serotonin-immunocytochemistry as described above. The first two slices were transferred to Krebs-Ringer bicarbonate medium and used for uptake and release studies. The uptake and release studies were performed on slices which were incubated with combined radiolabeled serotonin and choline and superfused as described previously.[26] The last two slices (per level) were immediately frozen on dry ice, stored at −70°C until HPLC analysis followed as described elsewhere.[25] Some old rats were used only for HPLC measurements of 5-HT and 5-HIAA levels.

RESULTS AND DISCUSSION

Morphological and Behavioral Differences of the Serotonergic System in the Young Adult and Aged Rat

The comparison of the serotonergic system in the young adult and aged rat brain revealed some morphological differences. The overall distribution of 5-HT-immunoreactive fibers in the aged rat forebrain resembled the innervation patterns observed in young animals.[27,28] However, in the hippocampus and to a lesser degree in the frontoparietal cortex and the neostriatum, we observed semi-quantitatively a decrease in fiber density (FIG. 2, B, C & D). This decline indicated a regional loss of serotonergic fibers. In addition, we noted predominantly in the frontoparietal cortex and the neostriatum fibers with an aberrant morphology besides of many "normal" varicose fibers, resembling those appearing in the brain of young adult rats (FIG. 4, B, C & D). These aberrrant fibers occurred as saucer-like swollen varicosities, and aggregates with a crumpled and folded structure which made them look like small fiber-knots. We have recently demonstrated that these aberrant fibers are actually degenerating fibers since their morphological appearance resembles the one seen after a 5,7-DHT–induced denervation.[29,30]

An important question to be raised is at what age a rat has to be considered as being aged or senescent, in comparison to the aged human brain. Behavioral, physiological and morphological criteria, together with survivor curves specific to the strain of rats in use, have to be weighed to justify age classifications. In our laboratory, rats of 27 months or older showed neuromorphological degenerative symptoms. They reach the limit of 50 percent mortality at an age of 28 months. We measured the 5-HT and 5-HIAA levels in the four forebrain areas and the raphe area. HPLC analysis revealed that 5-HT and 5-HIAA levels in the frontoparietal cortex, hippocampus and raphe area were increased in the aged rat, while the 5-HT level in the caudate-putamen complex was not different from the young adult rat. The ratio 5-HIAA/5-HT appeared increased in the frontoparietal cortex, sensoric part, the caudate-putamen and the raphe area, while this ration in the frontoparietal cortex, motoric part and the hippocampus was not altered in the aged rat (TABLE 1). These data indicate that aging of the serotonergic system is a local and mainly morphologically defined degeneration process. It is known that serotonergic cell bodies give rise to projections throughout the entire brain. Retrograde tracing studies have shown that mesencephalic raphe cells send collaterals to various brain regions.[28] Therefore we assume that the local degeneration of the serotonergic system might indicate the appearance of an extraneurotoxic compound, possibly specific for the serotonergic system, since for instance in the caudate-putamen or several cortical areas no degeneration of the dopaminergic system occurred.[14] Alternatively, a diminished production of neurotrophic factors at high age,

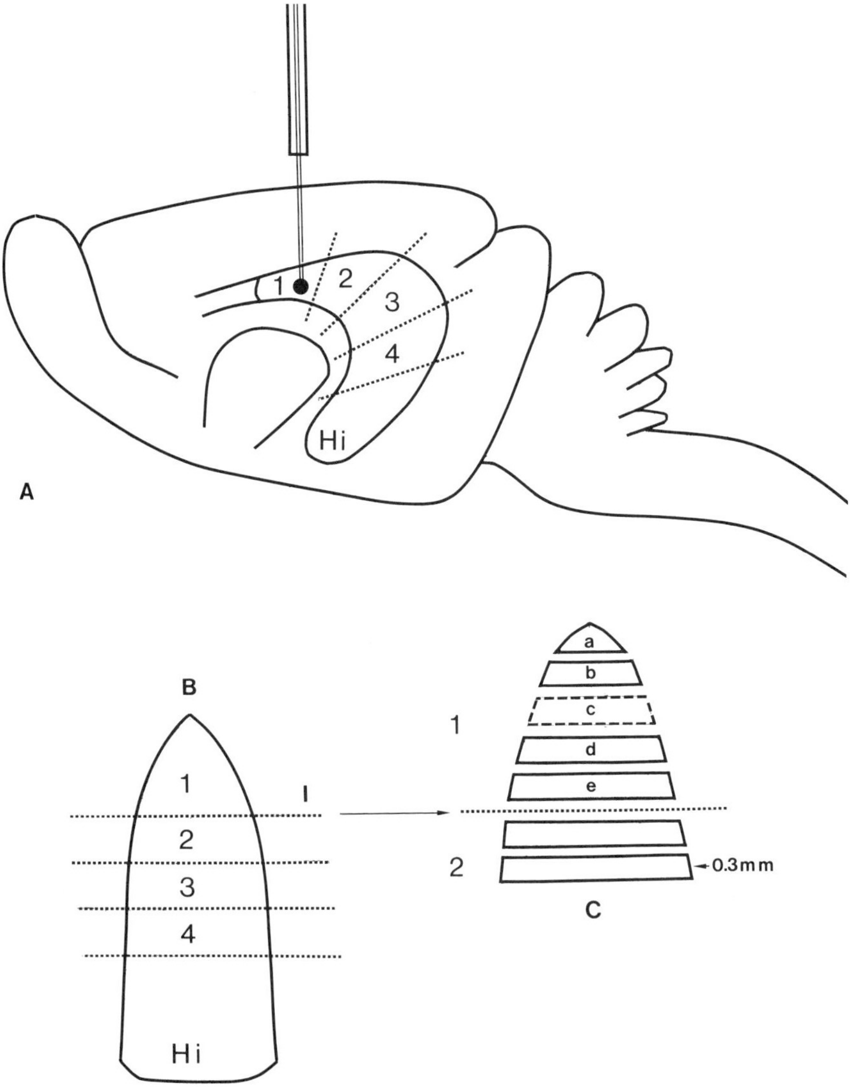

FIGURE 1. A: Schematic sagittal drawing through the rat brain, showing the hippocampal formation (Hi) and a raphe cell suspension graft deposit in area T1. The dashed lines indicate the approximate extent of the areas numbered T1–T4 used for the combined morphological and neurochemical assays as further indicated in **B.** Horizontal view of the hippocampal formation indicating the four levels to be used for the combined morphological and neurochemical analysis as indicated in **C.** Schematic drawing of level T1 indicating the five tissue pieces of 300 μm collected from one larger area. Sections a and b are used for uptake and release, section c for immunohistochemical observations and sections d and e for HPLC evaluation.

critical for the support and maintenance of innervating serotonergic neurons, may be hypothesized to result in a retrograde degeneration of particular serotonergic afferents. In comparison to young males spatial memory was impaired in aged males, as revealed by their performance in the Morris water maze task.

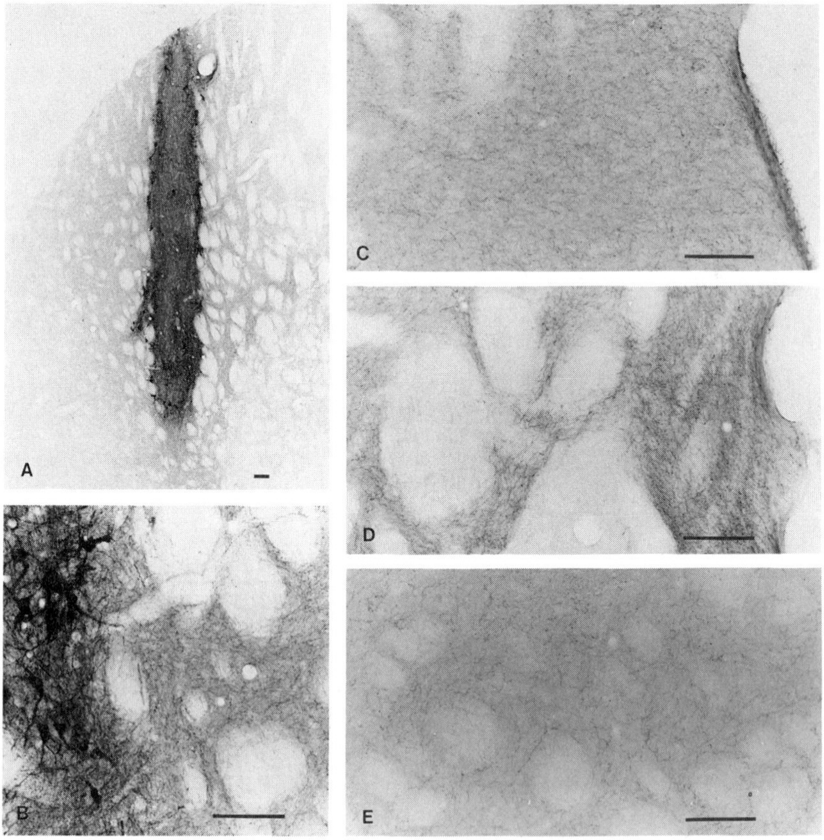

FIGURE 2. Photomicrographs of serotonin-immunoreactive fibers and cell bodies in the unilateral denervated young adult rat. **A:** The rat has received a mesencephalic raphe cell suspension into the caudate-putamen. **B:** Higher magnification of A. **E:** The serotonergic cells survive and send axons in the host tissue. The newly formed serotonergic fibers give only rise to a scattered serotonergic reinnervation, which is considerably less than the one in the contralateral non-denervated caudate-putamen. **C, D:** endogenous or host serotonin-immunoreactive fibers in the dorsomedial (**C**), or ventromedial contralateral part of the host caudate-putamen. Bars indicate 50 μm.

In our colony cage we compared groups of aged male rats to a group of young adult male rats. Whereas in the aged rats social interactions were virtually absent, nonsocial encounters were rare in a young control colony. Normally these interactions are not very aggressive in the absence of females. After pooling both colonies the nonsocial

TABLE 1. HPLC Data of 5-HT/5-HIAA Levels in the Aged Rat Forebrain[a]

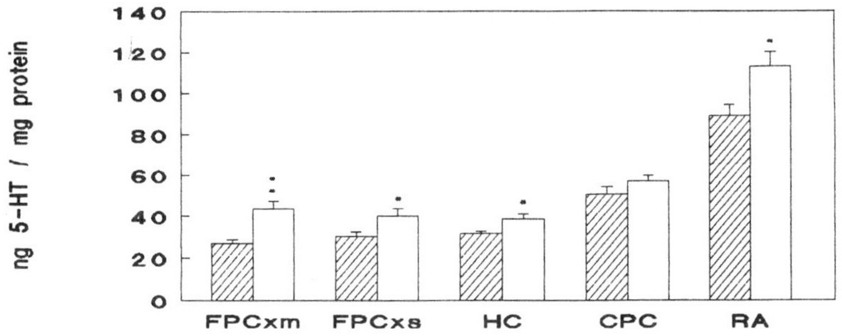

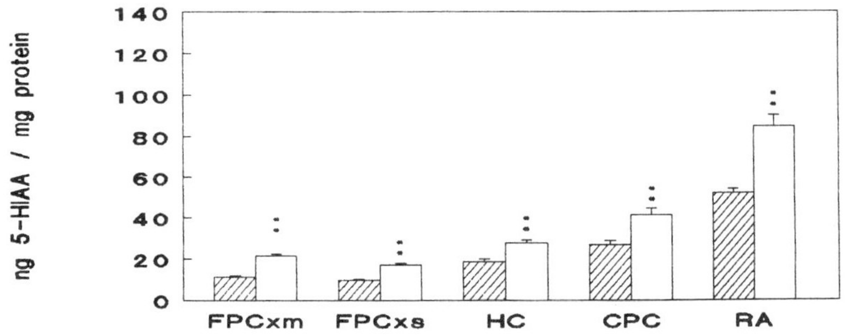

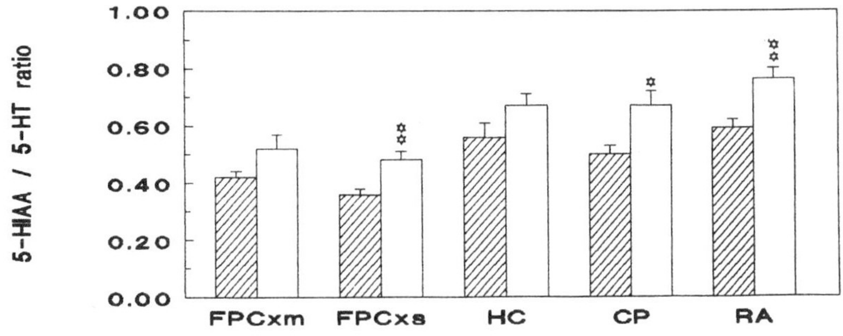

[a]FPCxm: frontoparietal cortex, motoric part; FPCxs: frontoparietal cortex, sensoric part; HC: hippocampal formation; CPC: caudate-putamen complex; RA: raphe area.

aged males unexpectedly evoked very aggressive responses in young control males. The adoption of a "defensive upright" posture, presumably facilitated by impaired sensory-motor functioning, apparently reinforced rather than inhibited attack tendencies in the young rat.

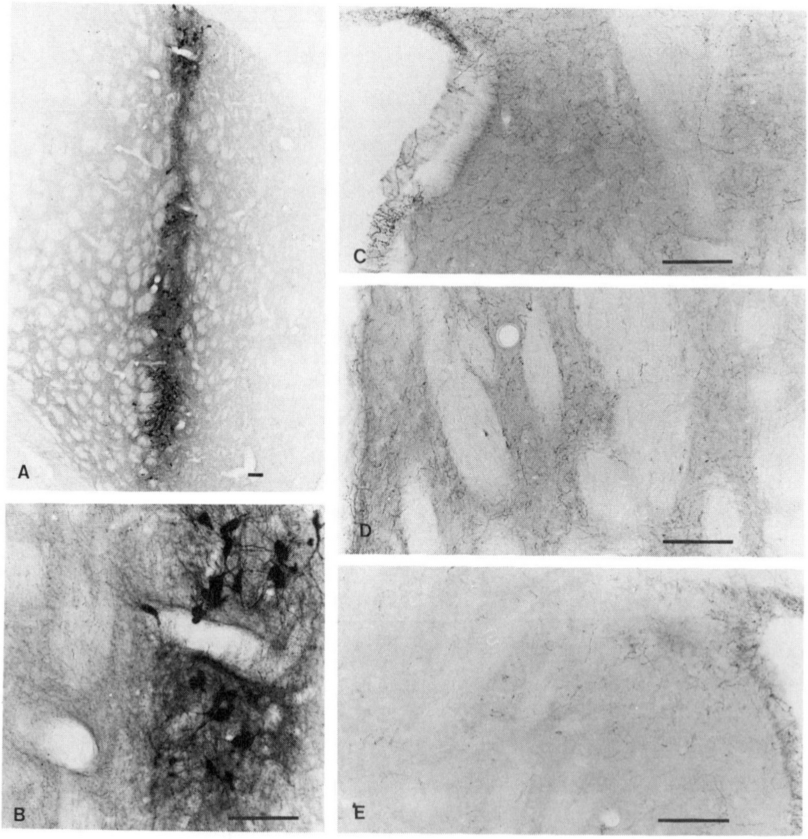

FIGURE 3. Photomicrographs of serotonin-immunoreactive fibers and cell bodies in a non-denervated young adult rat. **A:** The rat has received a mesencephalic raphe cell suspension into the caudate-putamen. **B:** Higher magnification of **A. E:** The serotonergic cells survive and send axons into the host tissue. The newly formed serotonergic fibers give rise to a serotonergic reinnervation, which is less than the one seen in the contralateral caudate-putamen. **C, D:** endogenous or host serotonin-immunoreactive fivers in the dorsomedial (**C**), or ventromedial contralateral part of the host caudate-putamen. Bars indicate 50 μm.

Morphological Observations of Fetal Raphe Cells Transplanted in Neostriatum of the Young Adult and Aged Rat

To investigate whether increased levels of neurotoxic compounds or decreased levels of trophic activity can be influence on the regenerative capacity of serotonergic

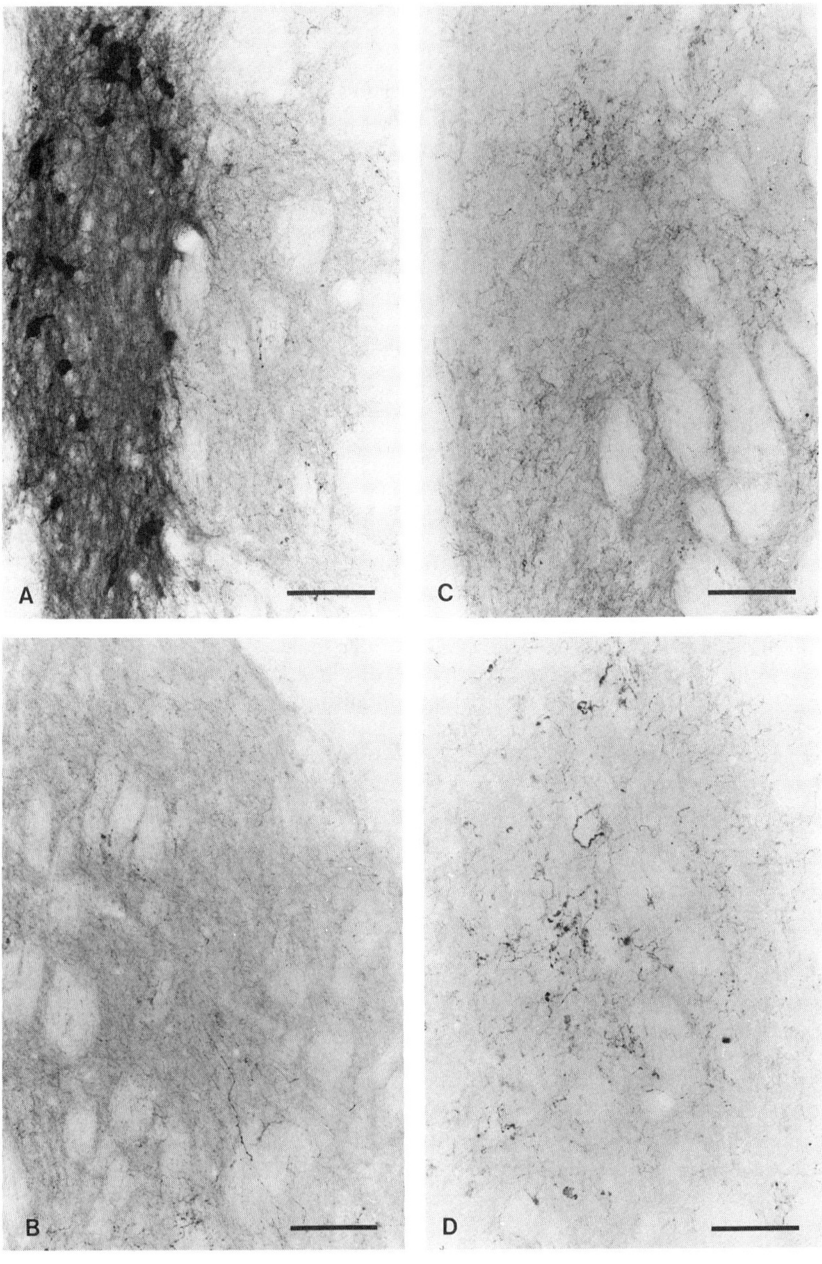

FIGURE 4. Photomicrographs of serotonin-immunoreactive fibers and cell bodies in the aged rat. **A:** The non-denervated aged rat has received a mesencephalic raphe cell suspension into the caudate-putamen. The serotonergic cells survive and send axons into the host tissue. **B, C, D:** endogenous or host serotonin-immunoreactive fibers in the medial (**C**), lateral (**B**) or ventral part of the host caudate-putamen. Note the large number of aberrant fibers. Bars indicate 50 μm.

cells in the aged rat brain, we implanted fetal raphe cells either into the denervated neostriatum of the young adult or the non-denervated, normal neostriatum of the aged rat (FIGS. 2, A & B; 3, A & B; and 4 A). The grafted cells survived well in both groups of rats with the same type of outgrowth into the host neostriatum (FIGS. 2, B; 3, B; and 4, A). These results do not point towards a possible release of endogenous neurotoxic-factors in the aged rat brain. However, they support the notion that implantation might release endogenous neurotropic compounds in the transplant itself or in the surrounding host tissue that stimulates outgrowth in the host tissue.

Behavioral Observations of Transplantation of Fetal Raphe Cells in the Denervated Hippocampus of the Young Rat

Spatial memory performance or reference memory, particularly linked with hippocampal functioning, was tested using a social avoidance and food reward paradigm on a one trial a day basis. It turned out that both in the acquisition phase as well as in the reversal training, the performance of the lesioned males and lesioned + transplanted males was similar to controls. This implies that after disturbance of the serotonergic input to the hippocampus spatial orientation and memory was still intact. In lesioned + transplanted males the choice latencies appeared to be significantly decreased. Perhaps for some reason their avoidance tendency during the decision phase was decreased, resulting in a sooner decision. This test lead to the same conclusion: neither lesion nor transplant affected orientation. Recently, Nilsson *et al.*[44] and Richter-Levin and Segal[43] showed that even intraventricular administration of 150 μg 5,7-DHT did not seriously deteriorate spatial orientation in the Morris water maze. Learning and memory parameters in experimental aging research are based on performance in passive and active avoidance tests. To test memory with respect to spatial orientation, the radial maze and the Morris water maze proved to be useful tools.[37,38]

Depletion of the serotonergic forebrain system induced minor changes in the pattern of sexual and aggressive behavior, which were "normalized" after transplantation of fetal raphe cells into the hippocampus. It has been mentioned already that both aggressive as well as sexual behavior are phenomena which are not directly associated with hippocampal performance, but rather hypothalamic functioning. However, aggressive behavior together with scent marking, which was increased, is seen as territorial behavior and is in this respect associated with hippocampal functioning. Genital grooming after mounts without intromission during mating was decreased in a number of lesioned males. Further, they showed more territorial scent marking and were more aggressive towards male intruders.

Regarding sleeping pattern it was noted that interruptions of sleeping periods occurred in the form of short lasting jerks and shivers (lasting one second or less) as well as longer-lasting changes in position of the posture. It turned out that lesion nor transplantation affected the latency or duration of sleeping behavior or the occurrence of interruptions.

FIGURE 5. Darkfield illumination of a transversal section through the dorsal hippocampus, showing a raphe cell suspension and its needle track. The rat was previously denervated and a mesencephalic raphe cell suspension was implanted. The cell suspension shows after a survival time of 6 weeks a hyperinnervation with serotonin-immunoreactive fibers. The remaining host hippocampus is reinnervated although their levels are slightly above control values. Bar indicates 50 μm.

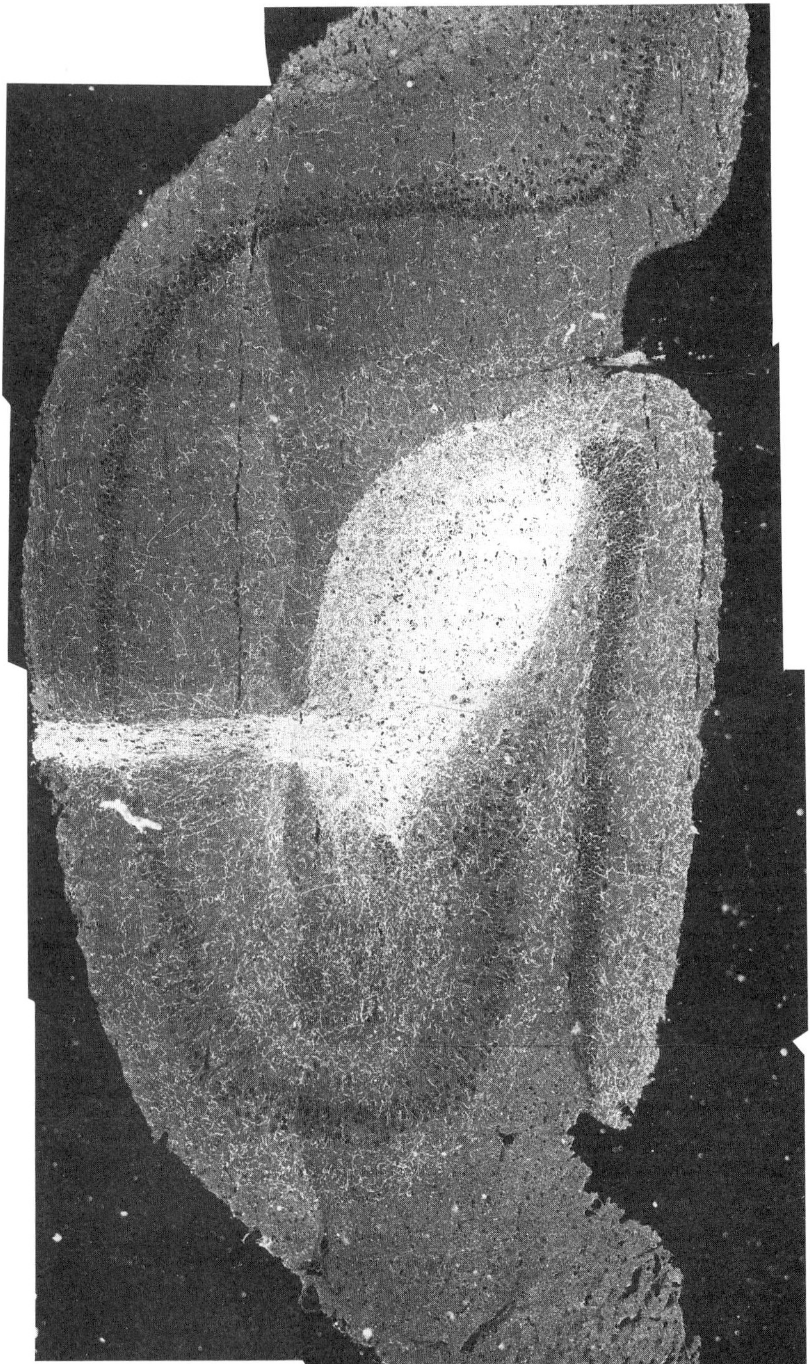

Morphological and Neurochemical Observations of Fetal Raphe Cells Transplanted in the Denervated Hippocampus of the Young Rat

Rats which were previously subjected to behavioral analysis were further morphologically and neurochemically characterized according to a new experimental setup.[25] This had the major advantage of permitting the study simultaneously of both morphological and neurochemical parameters within the same animal. In addition, such a system could be also applied to a survey in which the contralateral non-implanted side serves as an internal control.

The morphological data showed that after grafting, the implant spread out in the hippocampal fissure (FIG. 5). The fetal cells migrated towards the borders of the implant and reinnervated the surrounding host hippocampus. The implant itself was hyperinnervated (FIG. 5), as compared to the normal serotonergic innervation of the hippocampus (FIG. 6). Following the outgrowth over a distance of 1,500, 3,000, 4,500, 6,000 μm, respectively, levels T1 to T5, the density of innervation gradually decreased. Even in the most distal part of the hippocampus the fiber density exceeded the density observed in the lesioned side. These data confirm earlier observations by Amitia *et al.*,[21] Auerbach *et al.*,[31] and Daszuta *et al.*[32]

The release experiments showed that the mean uptake of radiolabeled serotonin in the sham-lesioned group did not differ from the control group. Slices from the lesioned group showed a statistically significant decrease in uptake. The most rostral level of the implanted hippocampal lobes (T1), containing the grafted cells, showed a fivefold increase in uptake in comparison to the lesioned countermates. This rise in uptake capacity decreased caudally to the transplantation side. Uptake in levels T2 and T3 was significantly higher than in the lesioned countermates, but did not exceed the control uptake capacity. In all groups, K^+-stimulation yielded a release of radiolabeled serotonin in surplus of the spontaneous efflux during standard superfusion conditions. In both the lesioned as well as the transplanted group of K^+-stimulated release of [^{3}H]-5-HT was enhanced above control levels. Sham-lesioning had no effect on this parameter.

The HPLC measurements showed that in the sham-lesioned rats no altered 5-HT levels were found when compared to control. 5,7-DHT–induced denervation caused a reduction of the 5-HT levels to half of control values. Implantation of fetal mesencephalic raphe cells raised 5-HT levels to hypernormal values. When the overall measurements were split into the rostral (T1) and the middle (T3) levels, the effect of the outgrowth of the fetal mesencephalic cells became more evident. At the most rostral levels, containing the implanted cells or the major outgrowth, 5-HT levels were doubled when compared to control and four times as high as in the lesioned hippocampi. 5-HT levels at the more distal level (T3) were equal to those in the control animals, but still twice as high as in the lesioned animals. The same pattern of changes was observed for 5-HIAA levels: sham-lesioning did not affect 5-HIAA levels, while treatment with 5,7-DHT reduced the 5-HIAA levels to half of control values. The overall levels of 5-HIAA in the transplanted group were slightly above control levels. Within the implanted hippocampal lobes, T1 showed the major increase in 5-HIAA content, while the levels in T3 were comparable to control. When compared to the lesioned hippocampi, level T1 showed a threefold increase and level T3 a twofold increase. Considering the tremendous increase in 5-HT and 5-HIAA levels at the

FIGURE 6. Darkfield illumination of serotonin-immunoreactive fibers in a transversal section through the dorsal hippocampus. Note that there is a dense distribution of serotonergic fibers in the molecular cell laye, while the pyramidal cell layer receives only a few scattered fibers. Bar indicates 50 μm.

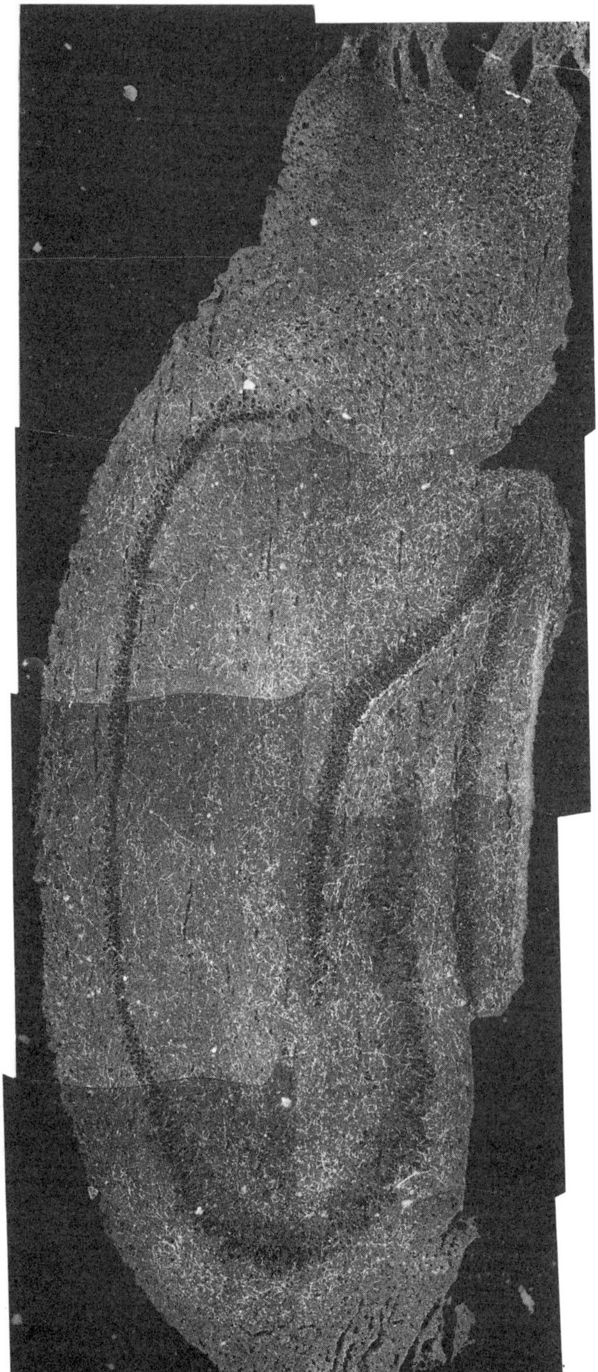

hippocampal level that contained the implanted fetal cells (T1) it is important to realize that the 5-HT and 5-HIAA content of the fetal cells contribute to these levels. The high levels do not reflect only the high fiber density, but also the presence of the serotonergic cells. The same counts for the uptake capacity. Despite the observed parallelism between changes in innervation density and neurochemical data, the extent of the alterations were unequal. Lesioned hippocampi, stained for serotonin-immunoreactivity, showed hardly any remaining serotonergic fibers, whereas neurochemical data obtained from both HPLC analysis and release experiments are indicative of a reduction to "only" half of control values. Concerning the uptake measurement this apparent discrepancy could be explained by considering the fact that an excess of radiolabeled serotonin was added to the slices. Therefore low affinity uptake sites could contribute to the final amount of uptake that was measured.

Previous studies showed a pronounced depletion in circumventricular areas after icv injection of 150 μg, 5,7-DHT.[39] This high concentration of 5,7-DHT was avoided since it was our goal to study only behavioral effects of a serotonin-depleted hippocampus. It was confirmed that low concentrations (20 μg) of 5,7-DHT resulted in a $>95\%$ decrease of the density in serotonergic innervation in the hippocampus. The discrepancy between morphological and neurochemical data can be explained by assuming a detection limit in the immunocytochemical procedure procedure. Although specificity and sensitivity of the antibody used in this study were shown to be high, the in vitro tests can never be compared totally with tissue staining.[22,23] Therefore it might be possible that in the lesioned hippocampal lobes fibers from the serotonergic system still remain, though undetectable by the antiserum used.

After 5,7-DHT treatment changes in turnover levels are reported, indicative of a compensation by the remaining fibers for the loss of innervation density. Our results indicated no change in turnover due to the 5,7-DHT treatment. The enhanced stimulated release that was observed in the denervated hippocampal lobes, however, indicates an enhanced reactivity of the surviving fibers upon K^+-stimulation. Interestingly an enhanced reactivity of serotonergic fibers were also reported in aged rats.[34]

Our data confirm the changes in 5-HT and 5-HIAA levels with the distance from the grafted cells that were previously described by Daszuta *et al.*[32] However, a major difference is that we are able to make a selection between successful and nonsuccessful operations, using the strict criterion that we only used those rats in which we had a successful transplant with the hippocampus itself. Transplants situated at the outer boundaries of the hippocampus, or even partly within the lateral ventricle or between the corpus callosum and the hippocampus were not considered to be successful. They were not excluded from our study. This criterion leads to the removal of approximately 60% of our transplanted rats.

CONCLUSION

In this study we showed the occurrence of aberrant and a regional loss of serotonergic fibers in the forebrain of the aged rat, indicating that the 5-HT system in aged rats is degenerating. The increased 5-HIAA/5-HT ratio is indicative of a compensatory mechanism or a decreased clearance of 5-HIAA in the aged rat. Morphological and neurochemical parameters concerning the effects of 5,7-DHT treatment and subsequent grafting of fetal raphe cells in the hippocampus demonstrated the functionality of newly formed serotonergic fibers. Depletion of the serotonergic forebrain system induces changes in sexual and aggressive behavior, which do not occur if fetal raphe cells are transplanted into the hippocampus. However, despite evident serotonin denervation no changes are seen with respect to sleep, exploration and, in particular, spatial orientation.

SUMMARY

Morphological dissimilarities between the brains of young (3 months) and aged (28 months and older) rats were demonstrated using serotonin-immunocytochemistry. A degeneration of the serotonergic system, noted as a decreased innervation and the appearance of enlarged or swollen varicosities, was observed particularly in the frontoparietal cortex, and the neostriatum of the aged rat brain. No direct relationship between this aberrant morphology and decrease in density of serotonin-innervation was found as we demonstrated a decline in fiber density without the appearance of aberrant serotonergic fibers in the hippocampus. HPLC analysis revealed that serotonin (5-HT) and 5-hydroxyindolacetic acid (5-HIAA) levels in the frontoparietal cortex, hippocampus and raphe area were increased in the aged rat, while the 5-HT level in the caudate-putamen complex was not different from the young adult rat. The ratio 5-HIAA/5-HT, indicative of 5-HT turnover, appeared increased in the frontoparietal cortex, sensoric part, the caudate-putamen and the raphe area, while this ration in the frontoparietal cortex, motoric part and the hippocampus was not altered in the aged rat. Behavioral screening revealed a decrease spatial performance of aged males in a Morris Water-Maze task.[37]

To investigate whether the age of the host recipient was of influence on the regenerative capacity, a fetal raphe cell suspension of embryonic day E 15 was implanted in the caudate-putamen of young adult as well as aged rats. Neither differences in survival of the serotonergic cells nor in fiber outgrowth between both groups appeared five weeks after transplantation. Subsequently, transplantation of raphe cells in the hippocampus of young adult rats, after lesioning the hippocampal serotonergic innervation with 5,7-DHT, was performed to compare behavioral, morphological and neurochemical effects of the implants. It appeared that 11 months after transplantation the serotonergic innervation of the previously denervated hippocampus was greatly restored. There was a striking resemblance between the immunohistochemical and neurochemical data with respect to the increase in the amount of newly formed serotonergic fibers, the increase in uptake of [^{3}H]-5-HT and in 5-HT and 5-HIAA levels. Also the behavior of lesioned and lesioned + transplanted males was rather similar to controls. In the behavioral tests we were mainly interested in hippocampal functioning, therefore orientation was of our prime interest. The other behavioral tests were only to confirm that the possible changes were linked to hypothalamic or extra-hypothalamic functions. No significant change was detected with respect to exploration, sleep, sexual activity, intruder attack and spatial memory performance. However, lesioned males showed less genital grooming during sexual behavior, and more scent marking during intruder attack. Lesioned males which received a transplant behaved similar to controls in this respect.

ACKNOWLEDGMENTS

We want to thank John G. J. M. Bol for excellent technical assistance, Henk Nordsiek for processing the photomicrographs and Margriet A. Hiehle for assistance in behavioral experiments.

REFERENCES

1. Collier, T. J., D. M. Gash & J. R. Sladek, Jr. 1987. Norepinephrine deficiency and behavioral senescence in aged rats. Transplanted locus coeruleus neurons as an experimental replacement therapy. Ann. N.Y. Acad. Sci. **495:** 396–403.
2. McGeer, P. L., E. G. McGeer & J. S. Suzuki. 1976. Aging and extrapyramidal function. Arch. Neurol. **34:** 33–35.
3. Beck, C. H. M. 1978. Functional implications of changes in the senescent brain. A review. Can. J. Neurol. Sci. **5:** 417–424.
4. Algeri, S., G. Calderini, G. Toffano & F. Ponzio. 1983. Neurotransmitter alterations in aging rats. *In* Aging of the Brain. D. Samuel, F. Algeri, S. Gershon, V. E. Grimm and G. Toffano, Eds. New York: Raven Press, pp. 227–224.
5. Allen S. J., J. S. Benton, M. J. Goodhardt, E. A. Haan, N. R. Sims, C. C. T. Smith, J. A. Spillane, D. M. Bowen & A. N. Davison. 1983. Biochemical evidence of selective nerve cell changes in the normal ageing human and rat brain. J. Neurochem. **41:** 256–265.
6. McGeer, E. G. & P. L. McGeer. 1980. Aging and neurotransmitter systems. *In* Ergot Compounds and Brain Function: Neuroendocrine and Neuropsychiatric Aspects. M. Goldstein, A. Liberman, D. B. Calne and M. O. Thorner, Eds. Adv. Biochem. Psychopharmacol. **23:** 305–314. New York: Raven Press.
7. Mann, D. M. A., P. O. Yates & B. Marcyniuk. 1984. Alzheimer's presenile dementia, senile dementia of Alzheimer's type and Down's syndrome in middle age form an age related continuum of pathological changes. Neuropathol. Appl. Neurobiol. **10:** 185–207.
8. Marcyniuk, B., D. M. A. Mann & P. O. Yates. 1986. Loss of nerve cells from the locus coeruleus in Alzheimer's disease is topographically arranged. Neurosci. Lett. **64:** 247–252.
9. Herregodts, P., P. Cras, M. Bruyland, Y. Michotte, J. J. Martin & G. Ebinger. 1987. Degeneration of the serotonergic system in Alzheimer's disease. Clin. Neurol. Neurosurg. **89:** 20–21.
10. Adolfson, R., C.-G. Gottfries, B.-E. Roos & B. Winblad. 1979. Changes in the brain catecholamines in patients with dementia of the Alzheimer type. Br. J. Psychiatry **135:** 216–223.
11. Hardy, J., R. Adolfsson, I. Alafuzoff, G. Bucht, J. Marcusson, P. Nyberg, E. Perdahl, P. Wester & B. Winblad. 1985. Transmitter deficits in Alzheimer's disease. Neurochem. Int. **7:** 545–563.
12. Cross, A. J., T. J. Crow, J. A. Johnson, M. H. Joseph, E. K. Perry, R. H. Perry, G. Blessed & B. E. Tomlinson. 1983. Monoamine metabolism in senile dementia of Alzheimer type. J. Neurol. Sci. **60:** 383–392.
13. Bowen, D. M., S. J. Allen, J. S. Benton, M. J. Goodhardt, E. A. Haan, A. M. Palmer, C. C. T. Smith, J. A. Spillane, M. M. Esiri, P. Neary, J. S. Snowdon, G. R. Wilcock & A. N. Davidson. 1983. Biochemical assessment of serotonergic and cholinergic dysfunction and cerebral atrophy in Alzheimer's disease. J. Neurochem. **41:** 266–272.
14. Steinbusch, H. W. M., J. G. Wolters, M. G. P. A. van Luijtelaar & J. A. D. M. Tonnaer. 1988. The serotonergic and dopaminergic system in the aged rat brain as studied by immunocytochemistry with antibodies to serotonin and dopamine. Preliminary observations. *In* New Trends in Aging Research. G. C. Pepeu, B. Tomlinson and C. M. Wischik, Eds. **15:** 193–203. Fidia Research Series, Liviana Press, Springer Verlag.
15. Bucht, G., R. Adolfsson, C. G. Gottfries, B. E. Roos & B. Winblad. 1981. Distribution of 5-hydroxytryptamine and 5-hydroxyindoleacetic acid in human brain in relation to age, drug influencc, agonal status and circadian variation. J. Neural Transm. **51:** 185–203.
16. Marcusson, J. O., L. Oreland & B. Winblad. 1984. Effect of age on human brain serotonin (S-1) binding sites. J. Neurochem. **43:** 1699–1705.
17. Scatton, B., F. Javoy-Agid, L. Rouquier, B. Dubois & Y. Agid. 1983. Reduction of cortical dopamine, noradrenaline, serotonin and their metabolites in Parkinson's disease. Brain Res. **275:** 321–328.
18. Azmitia, E. C., D. Quartermain & M. J. Brennan. 1979. Hippocampal [^{3}H] 5-hydroxy-

triptamin uptake and behavioral changes in the young and aged mice. Exp. Brain Res. **36:** R4.

19. TIMIRAS, P. S., D. B. HUDSON, & R. E. SEGALL. 1984. Life-time serotonin: Regional effects of age and precursor availability. Neurobiol. Aging **5:** 235–342.
20. MEEK, J. L., L. BERTILSSON, D. L. CHENEY, G. ZSILLO & E. COSTA. 1977. Aging induced changes in acetylcholine and serotonin content of discrete brain nuclei. J. Gerontol. **32:** 129–131.
21. AZMITIA, E. C., M. J. PERLOW, M. J. BRENNAN & J. M. LAUDER. 1981. Fetal raphe and hippocampal transplants into adult and aged C57BL/6N mice: A preliminary study. Brain Res. Bull **7:** 703–710.
22. STEINBUSCH, H. W. M., J. DE VENTE & J. SCHIPPER. 1986. Immunocytochemistry of monoamines in the central nervous system. *In* Neurohistochemistry Today, Neurology and Neurobiology. P. Panula, H. Päivärintä & S. Soinila, Eds. **16:** 75–106. New York: Alan R. Liss, Inc.
23. STEINBUSCH, H. W. M. & F. J. H. TILDERS. 1987. Immunohistochemical techniques for light-microscopical localization of dopamine, noradrenaline, adrenaline, serotonin and histamine in the central nervous system. *In* Monoaminergic Neurons: Light Microscopy and Ultrastructure. IBRO Handbook Series: Methods in the Neurosciences. H. W. M. Steinbusch, Ed. **10:** 125–166. Chicester, UK: Wiley.
24. SEIGER, A. 1985. Preparation of immature central nervous system regions for transplantations. *In* Neural Grafting in the Mammalian CNS. A. Björklund and U. Stenevi, Eds. Amsterdam: Elsevier, pp. 71–77.
25. VAN LUIJTELAAR, M. G. P. A., J. A. D. M. TONNAER, J. J. HAGAN, A. L. FRANKHUIJZEN, H. H. DIJKSTRA & H. W. M. STEINBUSCH. 1990. Morphological, neurochemical and behavioral aspects of serotonergic grafts in the denervated rat hippocampus. Neuroscience. Submitted.
26. MULDER, A. H., A. L. FRANKHUYZEN, J. C. STOOF, J. WEMER & A. N. M. SCHOFFELMEER. 1984. Catecholamine receptors, opiate receptors, and presynaptic modulation of neurotransmitter release in the brain. *In* Catecholamines: Neuropharmacology and Central Nervous System—Theoretical Aspects. New York: Alan R. Liss, pp. 47–58.
27. STEINBUSCH, H. W. M. 1981. Distribution of serotonin-immunoreactivity in the central nervous system of the rat. Cell bodies and terminals. Neuroscience **6:** 557–618.
28. STEINBUSCH, H. W. M. & R. NIEUWENHUYS. 1983. The raphe nuclei of the rat brainstem: a cytoarchitectonic and immunohistochemical study. *In* Chemical Neuroanatomy. P. Emson, Ed. New York: Raven Press, pp. 131–207.
29. VAN LUIJTELAAR, M. G. P. A., H. W. M. STEINBUSCH & J. A. D. M. TONNAER. 1988. Aberrant morphology of serotonergic fibers in the forebrain of the aged rat. Neurosci. Lett. **95:** 93–96.
30. VAN LUIJTELAAR, M. G. P. A., H. W. M. STEINBUSCH & J. A. D. M. TONNAER. 1990. Similarities between aberrant serotonergic fibers in the aged and 5,7-DHT denervated the young adult rat brain. Exp. Brain Res. **78:** 81–89.
31. AUERBACH, S., F. ZHOU, B. L. JACOBS & E. AZMITIA. 1985. Serotonin turnover in raphe neurons transplanted in the rat hippocampus. Neurosci. Lett. **61:** 147–152.
32. DASZUTA, A., R. E. STRECKER, P. BRUNDIN & A. BJÖRKLUND. 1988. Serotonin neurons grafted to the adult hippocampus. I. Time course of growth as studied by immunohistochemistry and biochemistry. Brain Res. **458:** 1–19.
33. LUINE, V., K. RENNER, M. FRANKFURT & E. AZMITIA. 1985. Raphe transplants into hypothalamus reverse facilitation of sexual behaviour in 5,7-dihydroxytryptamine-treated female rats: Immunocytochemical, neurochemical and behavioral studies. *In* Neural Grafting in the Mammalian CNS. A. Björklund and U. Stenevi, Eds. Amsterdam: Elsevier, pp. 655–662.
34. SIMPKINS, J. W., G. P. MUELLER, H. H. HUANG & J. MEITES. 1979. Evidence for depressed catecholamine and enhanced serotonin metabolism in aging male rats: Possible relation to gonadotropin secretion. Endocrinology **100:** 1672–1678.
35. TSUBOKAWA, T., Y. KATAYAMA, S. MIYAZAKI, H. OGAWA, M. IWASAKI, S. SHIBANOKI & K. ISHIKAWA. 1988. Supranormal levels of serotonin and its metabolite after raphe cell transplantation in serotonin-denervated rat hippocampus. Brain Res. Bull. **20:** 303–306.

36. STROLIN-BENEDETTI, M. & P. E. KEANE. 1980. Differential changes in monoamine oxidase A and B activity in the aging rat brain, J. Neurochem. **35:** 1026–1032.
37. GAGE, F. H., A. BJÖRKLUND, U. STEVENI, S. B. DUNNETT & P. A. T. KELLY. 1984. Intrahippocampal grafts ameliorate learning impairments in aged rats. Science **225:** 553–536.
38. GASH, D. M., T. J. COLLIER & J. R. SLADEK, JR. 1985. Neural transplantations: A review of recent developments and potential applications to the aged brain. Neurobiol. Aging **6:** 131–150.
39. BJORKLUND, A., H. G. BAUMGARTEN & A. RENSCH. 1974. 5,7-Dihydroxytryptamine: improvement of its selectivity for serotonin neurons in the central nervous system by pretreatment with desipramine. J. Neurochem. **24:** 883–835.
40. MORGAN, D. G. 1987. The dopamine and serotonin systems during aging in human and rodent brain. Prog. Neuropsychopharmacol. Biol. Psychiatry **11:** 153–157.
41. FOSTER, G. A., M. SCHULTZBERG, A. BJORKLUND, F. H. GAGE & T. HOKFELT. 1985. Fate of embryonic mesencephalic and medullary raphe neurons transplanted to the striatum, hippocampus or spinal cord of the adult rat: Analysis of 5-hydroxytryptamine-, substance P- and tyrotropin-releasing hormine-immunoreactivity. *In* Neural Grafting in the Mammalian CNS. A. Björklund and U. Stenevi, Eds. Amsterdam: Elsevier, pp. 179–189.
42. SEGAL, M. & E. C. AZMITIA. 1986. Fetal raphe neurons grafted into the hippocampus develop normal adult physiological properties. Brain Res. **364:** 162–166.
43. RICHTER-LEVIN, G. & M. SEGAL. 1989. Raphe cells grafted into the hippocampus can ameliorate spatial memory deficits in rats with combined serotonergic/cholinergic deficiencies. Brain Res. **477:** 184–187.
44. NILSSON, O. G., R. E. STRECKER, A. DASZUTA & A. BJORKLUND. 1988. Combined cholinergic and serotonergic denervation of the forebrain produces severe deficits in spatial learning in the rat. Brain Res. **453:** 235–246.
45. ARAI, H., K. KOSOKA & R. IIZUKA. 1984. Changes of biogenic amines and their metabolites in postmortem brains from patients with Alzheimer-type dementia. J. Neurochem. **43:** 388–393.
46. BAKER, G. B. & G. P. REYNOLDS. 1989. Biogenic amines and their metabolites in Alzheimer's disease: Noradrenaline, 5-hydroxytryptamine and 5-hydroxyindole-3-acetic acid depleted in hippocampus but not in substantia innominata. Neurosci. Lett. **100:** 335–339.
47. PALMER, A. M., P. T. FRANCIS, J. S. BENTON, N. R. SIMS, D. M. A. MANN, D. NEARY, J. S. SNOWDEN & D. M. BOWEN. 1987. Presynaptic serotonergic dysfunction in patients with Alzheimer's disease. J. Neurochem. **488:** 15.
48. YAMAMOTO, T. & A. HIRANO. 1985. Nucleus raphe dorsalis in Alzheimer's disease: Neurofibrillary tangles and loss of large neurons. Ann. Neurol. **17:** 573–577.
49. ANDERSON, K. J., V. R. HOLETS, P. C. MAZUR, R. S. LASHER & C. W. COTMAN. 1986. Immunocytochemical localization of serotonin in the rat dentate gyrus following raphe transplants. Brain Res. **369:** 21–28.

DISCUSSION OF THE PAPER

MANFRED GÖTHERT (*University of Bonn*): I was particularly interested in the functional studies you did in the transplanted serotonergic neurons in which you showed that the nerve fibers can take up and release the tritated neurotransmitter. Did you try to modulate the release? In particular, are there some presynaptic receptors also present on those newly sprouted serotonergic nerve terminals?

STEINBUSCH: We have not yet studied the modulation of the presynaptic serotonergic receptor. This might be possible in a collaborative study with your department.

GÖTHERT: As you know, Dr. Mulder has been working with alpha-2 adrenoreceptors situated on serotonergic neurons. This would also be a very interesting question whether the serotonergic nerve fibers are actually endowed with alpha-2 adrenoreceptors.

STEINBUSCH: One of my colleagues, Dr. Frankhuyzen is currently studying this particular aspect.

E. SCHLICKER (*University of Bonn*): Just an appendix to Dr. Göthert's statement. We have examined serotonin release in young adults and in aged rats of 2.5 years, very comparable to the conditions in your recent paper in *Neuroscience Letters,* and we have also found that serotonin release in the aged rat brain is diminished by about 35%. The paper will appear in some weeks in Neuropharmacology.

STEINBUSCH: Thank you.

VINCENT A. PIERIBONE (*New York University*): Since you first reported this we had the opportunity to look at serotonin systems in the aged primate. We had two rather old monkeys and compared them to two young monkeys and I just wanted to make the observation that we saw the same type of swollen varicosities and aberrant fibers that you reported here in the rat cortex. In the striatum we have seen similar things, and even more caudally also in the dorsal raphe and in the colliculus area we saw the same type of clustered fibers and swollen varicose fibers.

STEINBUSCH: Thank you.

MARK MOLLIVER (*Johns Hopkins University, Baltimore, MD*): Harry, that was a very elegant talk and I want to compliment you on that. Would you clarify for me, did I understand that you did not see the aged fibers in cortex prior to 12 months of age, and then do they increase in frequency after that?

STEINBUSCH: We see the aberrant fibers, like you call them, starting from 12 months old, mainly in forebrain regions.

MOLLIVER: Yes, hereafter they increase in frequency?

STEINBUSCH: Then they increase and we see an additional decrease of density of the normal fibers. But again, the appearance of the abberant fibers and their increase is always regional specific.

MOLLIVER: In cortex you said it was primarily in the frontal cortex.

STEINBUSCH: It's primarily in the frontoparietal cortex.

MOLLIVER: What about parietal and occipital?

STEINBUSCH: We have not seen this type of degenerating fibers.

MOLLIVER: And is there any lamina preference?

STEINBUSCH: There is not a defined delineation of the aberrant fibers, although there is a preference for the upper laminae 1, 2 and 3.

E. VAN BOCKSRAELE (*Hahnemann University, Philadelphia*). Dr. Azmitia showed a very impressive slide in his talk when he used laminin and he got a tremendous outgrowth of his transplanted neurons. I was wondering if you tried that and, if you didn't, why you think that you have such outgrowth without laminin.

STEINBUSCH: Well, this is a good question. We don't use any laminin at all. We have seen this papers and we're thinking about using it, but since we see a very rich innervation without laminin, we don't see the necessity to use it. In our opinion, it is important to mix the fetal raphe cell suspension with the various trophic factors to enhance the outgrowth. Although we get the same type serotonergic outgrowth, the speed of outgrowth is quite different. I would like to forward this question also to Dr. Azmitia or to Dr. Zhou.

Have you compared the grafting of those cells with those cells previously cultured in a tissue chamber?

FENG C. ZHOU (*Indiana University, Indianapolis*): The advantage of laminin is that you can guide the fibers towards a given direction. The transplant, at least when you transplant the serotonergic neurons into the hippocampus, depends on structures. It's easier to grow from the dorsal to ventral hippocampus, but they don't cross midline.

Laminin has the advantage to guide fibers to a certain direction, and of course can also facilitate the growth over longer distances.

ROBERT T. MOORE (*Session Chairman*): Harry, there was one thing that puzzled me. When you give the 5,7-DHT, you make the interpretation that those are degenerating fibers. Since the response of the fibers to the toxin is probably going to be to attempt to regenerate, I'm not clear how you can make the distinction that the aberrant fibers are necessarily degenerating as opposed to regenerating.

STEINBUSCH: We assume that since 5,7-DHT has been noted as a neurotoxic compound, that the first reaction of the brain to this neurotoxic compound will be a degeneration, a retraction, rather than some kind of a regeneration effect. So I think, and I don't know what kind comments Dr. Azmitia would like to make on this point, that the enlargement and swelling of the varicose fibers after the 5,7-DHT treatment is a degeneration effect.

AZMITIA: That is exactly what we observed in our studies of degenerating fibers after 5,7-DHT both in the hypothalamus and in the cortex. After several weeks, fine wispy regenerating sprouts can be seen extending from the swollen varicosities.

Serotonin in Alzheimer-type Dementia and Other Dementing Illnesses

ALAN J. CROSS

Astra Neuroscience Research Unit
1, Wakefield Street,
London, WC1N 1PJ, England

INTRODUCTION

Alzheimer-type dementia (ATD) is the most prevalent dementing illness in the elderly. The major clinical features of ATD include a progressive cognitive decline involving both memory impairments and focal cortical deficits such as aphasia and agnosia. In addition affective changes are also common, the most prominent being depression, agitation and aggression.

Neurochemical studies have demonstrated an extensive loss of cholinergic markers (such as choline acetyltransferase, ChAT) throughout the cortex and associated regions of brains from ATD patients.[1-3] The loss of cortical ChAT activity correlates with both the pathological severity of the disease and the degree of cognitive impairment.[4,5] Although there can be little argument that the cholinergic deficit is of central importance to the cognitive changes in ATD, it is now clear that several other neurotransmitter systems are also involved. In particular it has become apparent that the serotonergic system is involved at an early stage of the disease[6,7]; however, the relationship between the changes in serotonergic mechanisms and the features of the disease are not clearly understood. Many of the symptoms associated with ATD are present in other neurodegenerative diseases, both cortical and subcortical. In particular Parkinsons disease (PD) is frequently associated with depression and memory impairment, and dementia is a symptom of Huntington's disease.[8] By studying serotonergic mechanisms in these diseases in comparison with ATD it may be possible to determine which features of the disease are likely to be associated with serotonergic deficits.

SEROTONIN NEURONS IN ATD

A variety of neurochemical markers of serotonin containing nerve terminals have been studied in brains of ATD patients. A consistent reduction of these markers in the cortex of ATD patients is apparent (TABLE 1). It is notable that in the temporal cortex serotonin uptake sites are lost to a similar extent as serotonin and 5-HIAA concentrations. Thus it is reasonable to assume that these changes represent a loss of serotonergic terminals rather than a functional imbalance of these terminals. Neuropathological studies have shown that the large neurons of the dorsal raphe nucleus frequently contain neurofibrillary tangles in ATD.[19,20] More recent studies have also shown a substantial loss of neurons in the raphe nuclei in ATD.[21-23] The relationship between the loss of serotonergic terminals in forebrain regions and the losses and abnormalities of serotonergic neurons in the brain stem have not been studied directly. The distribu-

TABLE 1. Serotonergic Markers in Alzheimer-Type Dementia[a]

Neurochemical Marker	Brain Region	% Reduction in ATD	References
Serotonin concentration	Temporal (biopsy)	69	Palmer[9]
	Cortex (autopsy)	48	Francis[10]
5-HIAA concentration	Temporal (biopsy)	56	Palmer[9]
	Cortex (autopsy)	18	Francis[10]
	Cortex (autopsy)	54	Cross[7]
[^{3}H]-5-HT uptake	Temporal (biopsy)	61	Palmer[9]
	Cortex		
[^{3}H]-paroxetine binding		48	Cross & Slater[11]
[^{3}H]-Citalopram binding		44	D'Amato[12]

[a]These studies relate to temporal cortex and are included as representative examples (see also references 13–18).

tion of the reduction in 5-HIAA concentrations is shown in FIGURE 1. It is clear that the temporal cortex and hippocampus are the most severely depleted brain regions, with other brain regions (and particularly other cortical regions) being markedly less affected. Similar observations have been made for the distribution of serotonin in ATD brain, in this case the temporal and parietal cortex being the most severely affected regions.[24]

If the loss of cell bodies in the dorsal raphe nucleus were the primary change in serotonin neurons in ATD, then the loss of serotonergic terminals might be expected to be evenly distributed throughout the projection areas of this system. The serotonergic cells of the dorsal raphe nucleus are known to be highly collateralized. In the rat the dorsal raphe nucleus projects to both the caudate putamen and the temporal-parietal

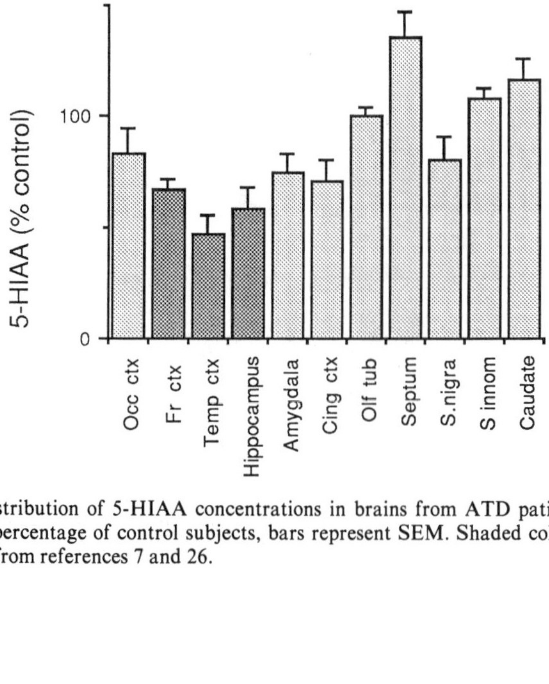

FIGURE 1. Distribution of 5-HIAA concentrations in brains from ATD patients. Values are expressed as a percentage of control subjects, bars represent SEM. Shaded columns, $p < 0.05$. Data are taken from references 7 and 26.

cortex.[25] However, it is clear that in ATD the loss of serotonergic terminals is restricted mainly to temporal-parietal cortex and hippocampus. On this basis it would seem highly unlikely that the primary change in serotonergic neurons occurs at the level of the cell body, but rather that these neurons degenerate in a retrograde manner.[26] The histological studies of Mann and colleagues[27] show that the cells of the raphe nucleus present the same morphological changes as those of the basal forebrain cholinergic neurons and the noradrenergic locus ceruleus cells. Moreover in Downs syndrome patients (all of whom develop the pathological changes of ATD at about 30–35 years of age) the appearance of senile plaques and neurofibrillary tangles in the cortex and hippocampus occurs at an earlier age than the pathological changes in the dorsal raphe nucleus.[28] It has previously been proposed that the cholinergic[29,30] and to a lesser extent noradrenergic[31] and adrenergic[32] neurons also undergo retrograde degeneration in ATD. Taken together these data strongly imply that in ATD serotonergic neurons also undergo retrograde degeneration in response to a primary cortical change. It is

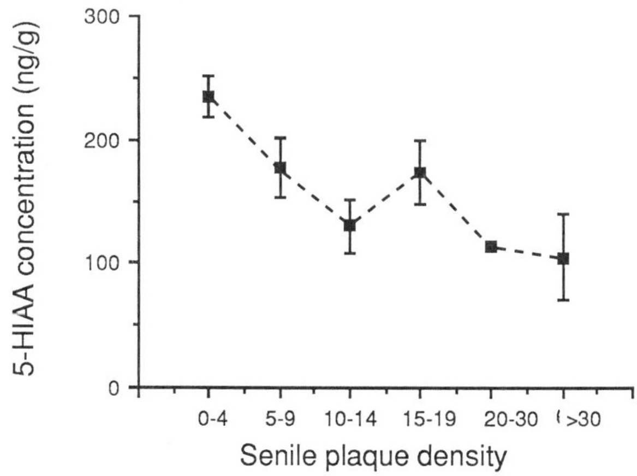

FIGURE 2. Relationship between 5-HIAA concentrations and senile plaque density in control and ATD patients. For all cases, $r = 0.45, p < 0.05$, for ATD patients alone, $r = 0.03$.

interesting to note that in PD the loss of serotonin-containing nerve terminals occurs extensively throughout the cerebral cortex and basal ganglia,[33] consistent with anterograde degeneration.

The loss of high affinity serotonin uptake in biopsy tissue from temporal cortex was observed in ATD patients within two years of the onset of symptoms.[9,10] Given that the average duration of illness from the appearance of symptoms is at least seven years,[35,36] the serotonergic deficit would appear to be an early change in the disease process.

The reduction in 5-HIAA in ATD does not correlate with the density of senile plaques (FIG. 2); however, Bowen and colleagues have shown that the degree of loss of 5-HIAA correlates with the presence of neurofibrillary tangles in temporal cortex. Neurofibrillary tangles are present mainly in the pyramidal cells of the cortex,[37] which probably use glutamate as a transmitter. Thus it has been suggested that the serotonergic degeneration in ATD may be mediated by glutamate.[9]

SEROTONIN RECEPTORS IN ATD

Initial studies of serotonin receptor changes in ATD have been confirmed in many subsequent reports.[26,38–42] The reduction in serotonin receptors is selective, many other neurotransmitter receptor binding sites being unaltered (FIG. 3). 5-HT_1 receptors labeled with [^{3}H]-5-HT are reduced in temporal cortex and some limbic structures (FIG. 4a). Further analysis has shown that both 5-HT_{1A} and "non-5-HT_{1A}" receptors are reduced in ATD.[40] The loss of 5-HT_1 receptors in ATD may be limited to younger (presenile) patients.[26] As these patients tend to have a more severe form of the disease, the reduction of 5-HT_1 receptors may be a later change in the disease.

5-HT_2 receptors (labeled with [^{3}H]-ketanserin) are reduced more extensively than 5-HT_1 receptors (FIG. 4b). The reduction in 5-HT_2 receptors occurs in both young and old patients and may therefore be a relatively early change in the disease.[26] In the rat, destruction of ascending cholinergic neurons by injection of ibotenic acid into the basal forebrain results in a significant reduction in 5-HT_1 receptors in the denervated cortex.[43] In contrast 5-HT_2 receptors are unaffected. Therefore it is likely that a proportion of 5-HT_1 receptors are present on cholinergic terminals. In confirmation of this it has been shown that 5-HT_1 receptors can modulate the release of acetylcholine *in vitro*.[44] Indirect evidence that this interaction may also occur in human brain has come from the study of patients with Parkinsons disease. Patients with PD plus concomitant dementia (but without the cortical pathology of ATD) show a loss of both ChAT activity and 5-HT_1 receptors in parietal cortex.[39] Non-demented PD patients have control values of ChAT activity and no reduction of 5-HT_1 receptors. Given that these patients show little pathological change in the cortex it would be reasonable to conclude that 5-HT_1 receptors are present on cholinergic terminals in human cortex. 5-HT_2 receptors are not reduced in these patients.[39]

Autoradiographic studies have extended these findings, and provided information

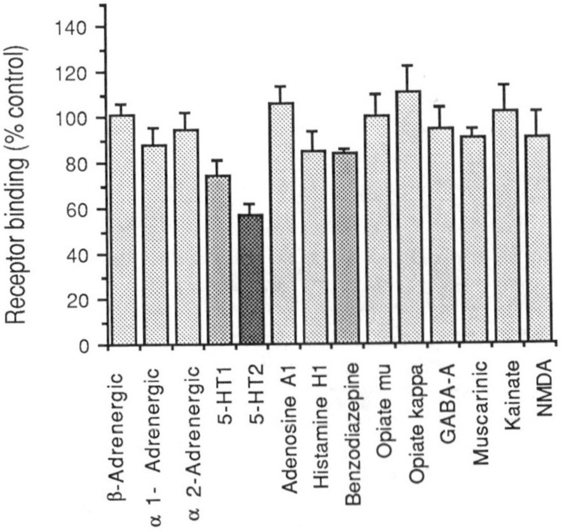

FIGURE 3. Neurotransmitter receptor binding sites in temporal cortex in ATD. Values are expressed as a percentage of an appropriate control group, bars represent SEM. Shaded columns, $p < 0.05$. Data are taken from references 26 and 49.

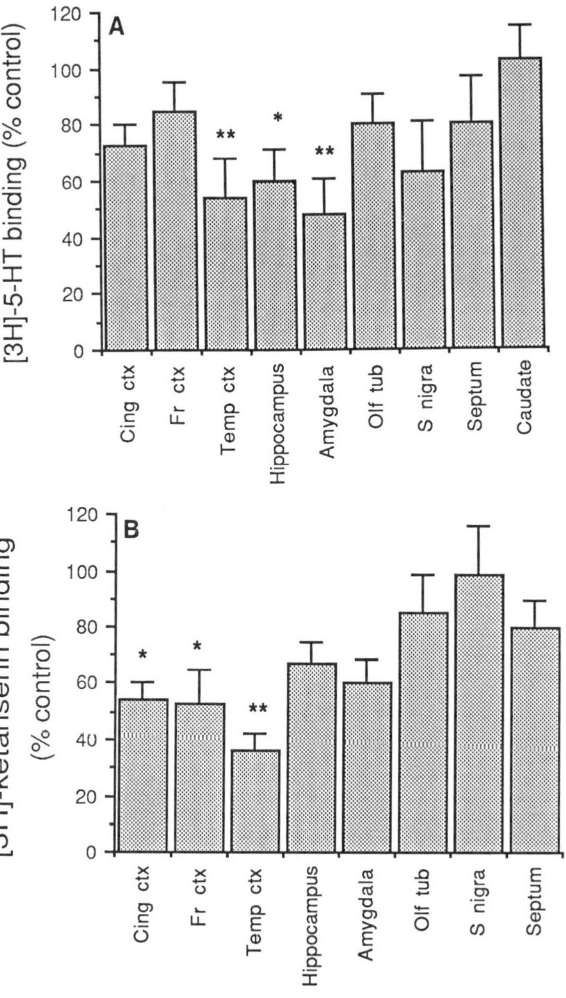

FIGURE 4. (**A**) 5-HT_1 receptor binding in ATD patients expressed as a percentage of control subjects, bars represent SEM. $*p < 0.05$, $**p < 0.01$. (**B**) 5-HT_2 receptor binding in ATD patients using [3H]-ketanserin as ligand; details as above.

on the distribuion of serotonin receptors through the depth of the cortex in ATD.[45] In temporal cortex [^{3}H]-DPAT binding is concentrated in the superficial layers of the cortical grey matter, and this dense layer of 5-HT_{1A} receptors is reduced in ATD (FIG. 5). In contrast [^{3}H]-ketanserin binding to 5-HT_2 receptors is concentrated in the middle layers of the cerebral cortex. In ATD these bands of high 5-HT_2 receptor binding are almost totally lost (FIG. 6). The dense band of 5-HT_2 receptors at 40% of cortical depth which is lost in ATD probably corresponds to layer III. A second band of 5-HT_2 receptors at 80% of cortical depth is also markedly depleted. This band may correspond to layer V. Both of these cortical layers contain pyramidal cells which are particularly susceptible to neurofibrillary tangle formation in ATD.[37] In the entorhinal

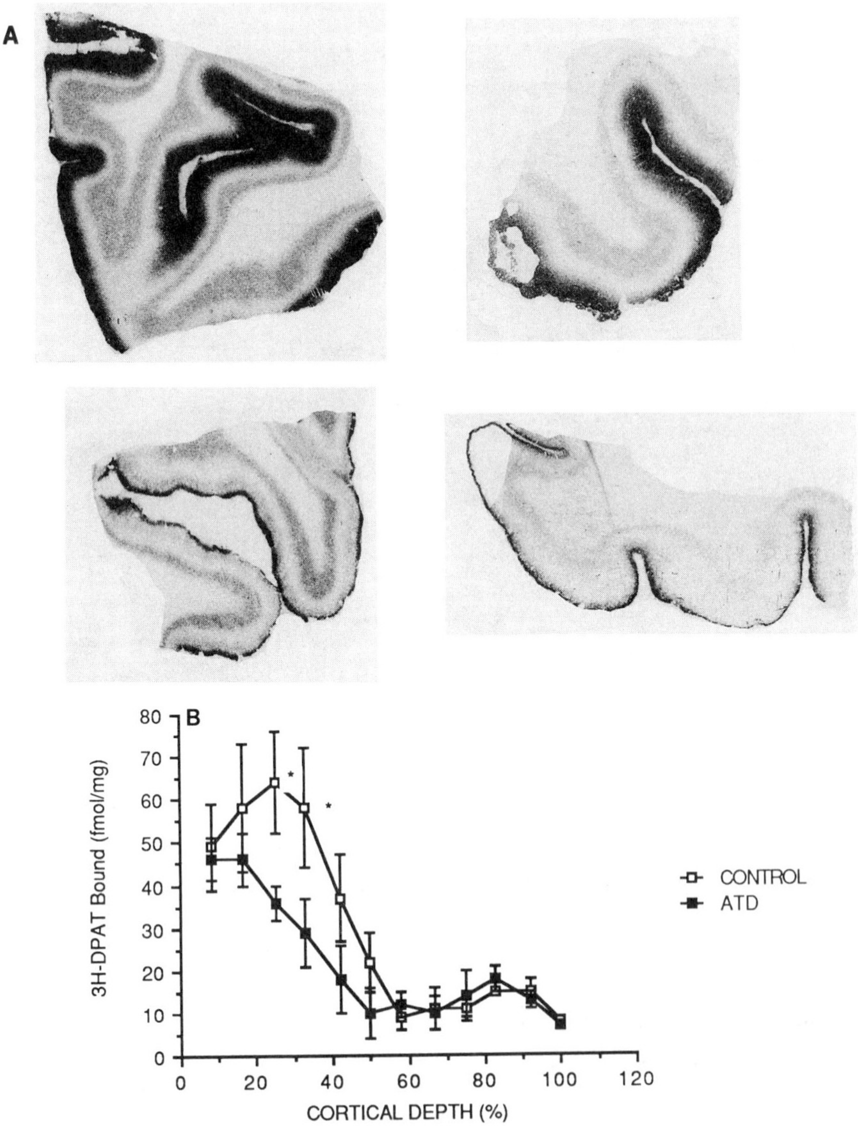

FIGURE 5. (**A**) Autoradiographs of [^{3}H]-DPAT binding to temporal cortex of 2 control subjects (*upper*) and 2 ATD patients (*lower*). (**B**) Semi-quantitative analysis of [^{3}H]-DPAT autoradiographs from control subjects and ATD patients (n = 5 for both groups). *$p < 0.05$.

cortex the density of [^{3}H]-ketanserin binding is inversely related to the density of neurofibrillary tangles, again suggesting a relationship between 5-HT_2 receptors and tangle-bearing pyramidal neurons.[39] Other evidence suggests that 5-HT_2 receptors maybe associated with intrinsic cortical neurons. In experimental animals excitotoxin lesions of the neocortex result in marked losses of 5-HT_2 receptors.[46] In ATD patients the reduction of [^{3}H]-ketanserin binding parallels the reduction in somatostatin immunoreactivity in temporal and frontal cortex.[26] Immunohistochemical studies have shown that samatostatin-like immunoreactivity is localized to intrinsic cortical neurons having a characteristic bipolar or multipolar morphology.[47,48]

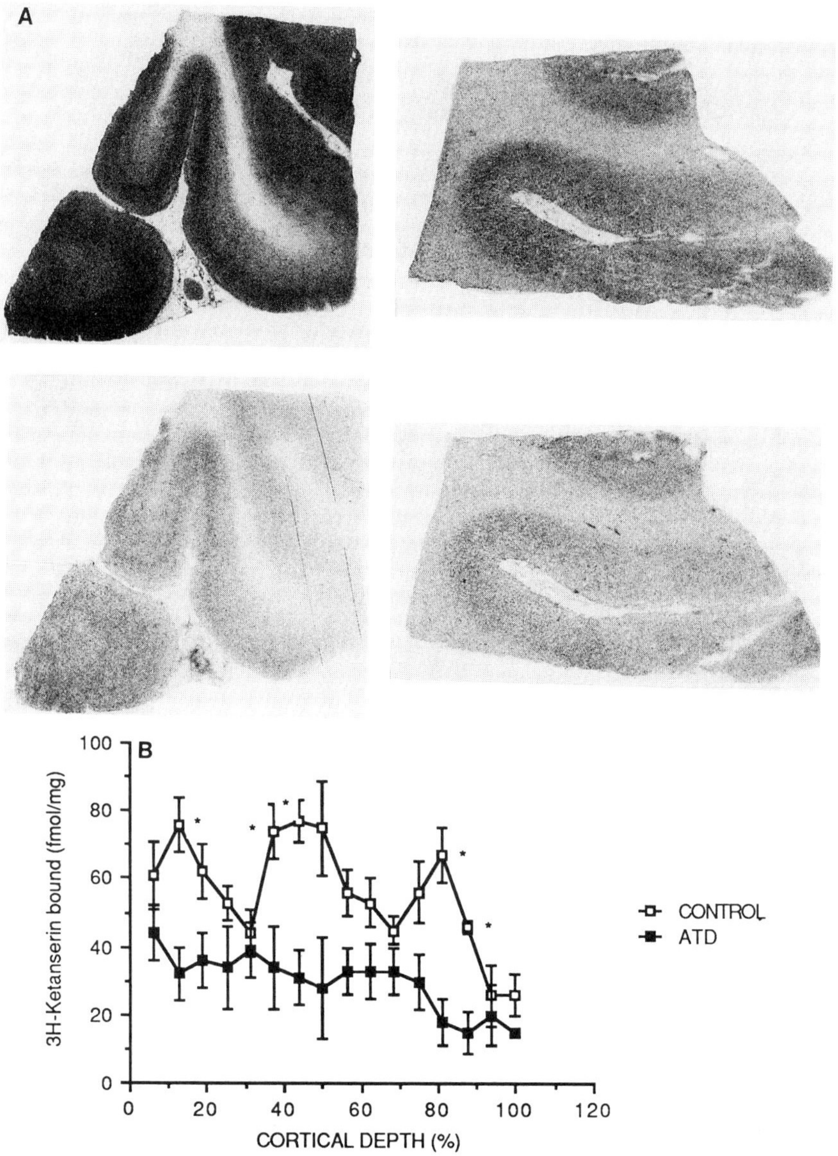

FIGURE 6. (**A**) Autoradiographs of [^{3}H]-ketanserin binding to a control (*left*) and ATD (*right*) temporal cortex. (**B**) Semi-quantitative analysis of [^{3}H]-ketanserin autoradiographs from control and ATD patients. *$p < 0.05$.

TABLE 2. Relationship between 5-HIAA Concentration and Mental Test Score[a]

Brain Region	All Patients	ATD Patients Only
Frontal cortex	0.25 (13)	0.46 (11)
Temporal cortex	—	−0.13 (9)
Hippocampus	0.54 (12)	0.01 (9)

[a]Values are correlation coefficients derived from linear regression analysis, number of samples is given in parentheses. The non-ATD patients were chronic unipolar depressives.

CLINICAL ASPECTS OF THE SEROTONERGIC CHANGES

Although 5-HIAA concentrations in cortex correlate with cognitive function in a group containing both controls and ATD patients, this relationship is not statistically significant in ATD patients alone (TABLE 2). A similar observation has been reported by Palmer and colleagues[9] in a study of post-mortem samples. It was also reported that in biopsy samples no significant correlation was apparent between cognitive function and high affinity [^{3}H]-5-HT uptake.[9] In patients with Parkinsons disease, sub-divided for the presence and absence of dementia, forebrain serotonin concentrations were reduced to a similar extent in both groups.[33] Thus it would seem reasonable to assume that in ATD and PD the serotonergic deficit does not contribute significantly to the overall cognitive decline. It remains to be determined whether the serotonergic deficit relates to specific components of the memory deficits. This conclusion is strengthened by the results of clinical studies with serotonergic drugs in ATD. In the main these studies have used selective 5-HT reuptake inhibitors, and have failed to demonstrate any improvement in cognitive function (TABLE 3). In contrast some of the noncognitive[35] features of ATD consistently show an improvement. In particular aggression and agitation seem to be consistently ameliorated. It is interesting to note that in the study of Palmer *et al.*[9] the only clinical correlate of brain 5-HIAA concentrations in ATD was an association of low 5-HIAA with aggressive behavior. Thus it may well be the case that the serotonergic denervation in ATD relates more closely to the noncognitive changes associated with the disease.

Whilst the relationship between the serotonergic deficit and non-cognitive changes in ATD is mainly a matter of conjecture, correlates of the loss of 5-HT_2 receptors are even more tenuous. The reduction in 5-HT_2 receptors in ATD is a consistent and selective change; it would be surprising if this change does not contribute to the clinical picture of ATD. The loss of 5-HT_2 receptors does not relate to mental test scores in ATD patients.[49] 5-HT_2 receptors are not reduced in demented PD patients, although such patients have a cholinergic deficit. It is tempting to speculate that changes in 5-HT_2 receptors contribute to the differences observed in cognitive change in ATD patients compared to PD patients. Prominent amongst those differences are in the focal cortical changes (such as aphasia, agnosia, etc.) associated with ATD but not PD.[8]

TABLE 3. Serotonin Drugs on Alzheimer-type Dementia

Cognitive Effects		
Dehlin *et al.*[50]	Alaproclate	No effect
Cutler *et al.*[51]	Zimeldine	No effect
Gottfires *et al.*[52]	Citalopram	No effect
Non-Cognitive effects		
Gottfries *et al.*[52]	Citalopram	Improved sleep agitation, aggression
Wilcock *et al.*[53]	Trazodone/tryptophan	Improved aggression, agitation

In conclusion, it is now clear that changes in serotonergic systems are intimately involved in ATD and other neurodegenerative disorders. It is likely that some of these changes are associated with the non-cognitive features of ATD. In many cases the non-cognitive features of ATD are the main reason for hospitalization and cause the greatest distress to caregivers. It would seem essential that the large number of novel serotonergic drugs currently under investigation be examined for efficacy in ATD.

REFERENCES

1. PERRY, E. K., P. H. GIBSON, G. BLESSED, R. H. PERRY & B. E. TOMLINSON. 1977. Neurotransmitter enzyme abnormalities in senile dementia. J. Neurol. Sci. **34:** 247–265.
2. BOWEN, D. M., C. B. SMITH, P. WHITE & A. N. DAVISON. 1976. Neurotransmitter-related enzymes and indices of hypoxia in senile demential and other abiotrophies. Brain **99:** 459–496.
3. DAVIES, P. & A. J. MALONEY. 1976. Selective loss of central cholinergic neurons in Alzheimer's disease. Lancet **ii:** 1043.
4. PERRY, E. K., B. E. TOMLINSON, G. BLESSED, K. BEZMANN, P. H. GIBSON & R. H. PERRY. 1978. Correlation of cholinergic abnormalities with senile plaques and mental test scores in senile dementia. Br. Med. J. **II:** 1457–1459.
5. NEARY, D., J. S. SNOWDEN, D. M. A. MANN, D. M. BOWEN, N. R. SIMS, B. NORTHERN & P. O. YATES. 1986. Alzheimer's disease: A correlative study. J. Neurol. Neurosurg. Psychiatry **49:** 229–237.
6. BOWEN, D. M., P. WHITE, J. A. SPILLANE, M. J. GOODHARDT, G. CURZON, P. IWANGOFF, W. MEIER-RUGE & A. N. DAVISON. 1979. Accelerated ageing or selective neuronal loss as an important cause of dementia? Lancet **i:** 11–14.
7. CROSS, A. J., T. J. CROW, J. A. JOHNSON, E. K. PERRY, R. H. PERRY, G. BLESSED & B. E. TOMLINSON. 1984. Studies on neurotransmitter receptor systems in neocortex and hippocampus in senile dementia of the Alzheimer gype. J Neurol. Sci. **64:** 109–111.
8. BROWN, R. G. & C. D. MARSDEN. 1988. "Subcortical dementia" The neuropsychological evidence. Neuroscience **25:** 363–387.
9. PALMER, A. M., P. T. FRANCIS, J. S. BENTON, N. R. SIMS, D. M. A. MANN, D. NEARY, J. S. SNOWDON & D. M. BOWEN. 1987. Presynaptic serotonergic dysfunction in patients with Alzheimers disease. J. Neurochem. **48:** 8–15.
10. FRANCIS, P. T., A. M. PALMER, N. R. SIMS, D. M. BOWEN, A. N. DAVISON, M. M. ESIRI, D. NEARY, J.S. SNOWDEN & G. K. WILCOCK. 1985. Neurochemical studies of early-onset Alzheimers disease: possible influence on treatment. New Engl. J. Med. **313:** 7–11.
11. CROSS, A. J. & P. SLATER. Unpublished observations.
12. D'AMATO, R. J., R. M. ZWEIG, P. J. WHITEHOUSE, G. C. WENK, H. S. SINGER, R. MAYEUX, D. L. PRICE, & S. H. SNYDER. 1987. Aminergic systems in Alzheimers disease and Parkinsons disease. Ann. Neurol. **22:** 229–236.
13. REINIKAINEN, K. J., L. PALJARVI, M. HUUSKONEN, H. SOININEN, M. LAAKSO & P. J. RIEKKENEN. 1988. A post-mortem study of noradrenergic serotonergic and GABAergic neurones in Alzheimers disease. J. Neurol. Sci. **84:** 101–116.
14. EBINGER, G., M. BRUYLAND, J. J. MARTIN, P. HERRESGOOTS, P. CRAS, Y. R. MICHOTTI & L. GOMME. 1987. Distribution of biogenic amines and their catabolites in brains from patients with Alzheimers disease. J. Neurol. Sci. **77:** 267–287.
15. ARAI, H, K. KOSAKA & R. IIZUKA. 1984. Changes in biogenic amines and their metabolites in postmortem brains from patients with Alzheimer-type dementia. J. Neurochem. **43:** 388–397.
16. MARCUSSON, J. O., I. ALAFUZOFF, I. T. BACKSTROM, E. ERICSON, C. G. GOTTFRIES & B. WINBLAD. 1987. 5-hydroxytriptamine-sensitive [^{3}H]imipramine binding of protein nature in human brain. II Effects of natural ageing and dementia disorder. Brain Res. **425:** 137–145.
17. SPARKS, D. L., S. T. DEKOSKY & W. R. MARKESBERY. 1988. Alzheimers disease. Aminergic-cholinergic alterations in hypothalamus. Arch. Neurol. **45:** 994–999.

18. GOTTFRIES, C. G., R. ADOLFSSON, S. M. AQUILONIUS, A. CARLSSON, S.-A, ECKERNAS, L. NORDBERG, L. ORELAND, L. SVENNERHOLM, A. WIBERG & B. WINBLAD. 1983. Biochemical changes in dementia disorders of Alzheimer type (AD/SDAT). Neurobiol. Aging **4:** 261–271.
19. ISHII, T. 1966. Distribution of Alzheimer neurofibrillary tangles in the brain stem and hypothalamus in senile dementia. Acta. Neuropathol. **6:** 181–187.
20. HIRANO, A. & H. M. ZIMMERMAN. 1962. Alzheimer's neurofibrillary tangles, a topographical study. Arch. Neurol. **7:** 227–242.
21. CURCIO, C. A. & T. KEMPER. 1984. Nucleus raphe dorsalis in dementia of the Alzheimer type: neurofibrillary changes and neuronal packaging density. J. Neuropathol. Exp. Neurol. **43:** 359–368.
22. MANN, D. M. A. & P. O. YATES. 1983. Serotonergic nerve cells in Alzheimer's disease. J. Neurol. Neurosurg. Psychiatry **46:** 96.
23. YAMAMOTO, T. & A. HIRANO. 1985. Nucleus raphe dorsalis in Alzheimers disease: Neurofibrillary tangles and loss of large neurones. Ann. Neurol. **17:** 573–577.
24. PALMER, A. M., G. C. STRATMANN, A. W. PROCTOR & D. M. BOWEN. 1988. Possible neurotransmitter basis of behavioural changes in Alzheimers disease. Ann. Neurol. **23:** 616–620.
25. AZMITIA, E. F. & M. SEGAL. 1978. An autoradiographic analysis of the differential ascending projections of the dorsal and median raphe made in the rat. J. Comp. Neurol. **176:** 641–668.
26. CROSS, A. J., T. J. CROW, I. N. FERRIER, J. A. JOHNSON, S. R. BLOOM & J. A. N. CORSELLIS. 1984. Serotonin receptor changes in dementia of the Alzheimer type. J. Neurochem. **43:** 1574–1581.
27. MANN, D. M. A. 1985. The neuropathology of Alzheimers disease: a review with pathogenetic, acetiological and therapeutic considerations. Arch. Ageing Develop. **31:** 213–255.
28. MANN, D. M. A., P. O. YATES & B. MARCYNIUK. 1984. Alzheimers presenile dementia, senile dementia of the Alzheimer-type and Downs syndrome in middle age form an age related continuum of pathological changes. Neuropathol. Appl. Neurobiol. **10:** 185–207.
29. PEARSON, R. C. A., M. V. SOFRONIEW, A. C. CUELLO, T. P. S. POWELL, F. ECKENSTEIN, M. M. ESIRI & G. K. WILCOCK. 1983. Persistence of cholinergic neurones in the basal nucleus in the brain with senile dementia of the Alzheimer type demonstrated by immunohistochemical staining for choline acetyltransferase. Brain Res. **289:** 370–374.
30. PERRY, R. H., J. M. CANDY, E. K. PERRY, D. IRVINE, G. BLESSED, A. F. C. FAIRBAIRN & B. E. TOMLINSON. 1982. Extensive loss of cholineacetyltransferase activity is not reflected by neuronal loss in the nucleus of Maynert in Alzheimers disease. Neurosci. Lett. **33:** 311–315.
31. CROSS, A. J., T. J. CROW, E. K. PERRY, R. H. PERRY, G. BLESSED & B. E. TOMLINSON. 1981. Reduced dopamine-β-hydroxylase activity in Alzheimers disease. Br. Med. J. **I:** 93–94.
32. BURKE, W. J., M. D. CHUNG, J. S. HUANG, S. S. HUANG, J. H. HARING, R. STRONG, G. L. MARSHALL & J. H. JOH. 1988. Evidence for retrograde degeneration of epinephrine neurons in Alzheimers disease. Ann. Neurol. **24:** 532–536.
33. SCATTON B., F. JAVOY-AGID, L. ROUQUIER, B. DUBOIS & Y. AGID. 1983. Reduction of cortical dopamine, noradrenaline, serotonin and their metabolites in Parkinsons disease. Brain Res. **275:** 321–328.
34. RAISMAN, R., R. CASH & Y AGID. 1986. Parkinsons disease: Decreased density of [^{3}H]-imipramine and [^{3}H]-paroxetine binding sites in putamen. Neurology **36:** 556–560.
35. BLESSED, G. 1980. Clinical aspects of the senile dementias. *In* Biochemistry of Dementia. P. J. Roberts, Ed. London: John Wiley & Sons, pp. 1–14.
36. BERG, L., J. P. MILLER, M. STORANDT, J. DUCHEK, J. C. MORRIS, E. H. RUBIN, W. J. BURKE & L. A. COBEN. 1988. Mild senile dementia of the Alzheimer-type : 2. Longitudinal assessment. Ann. Neurol. **23:** 477–484.
37. PEARSON, R. C. A., M. M. ESIRI, R. W. HIORNS, G. K. WILCOCK & T. P. S. POWELL. 1985. Anatomical correlates of the distribution of the pathological changes in the neocortex in Alzheimer's disease. Proc. Natl. Acad. Sci. USA **82:** 4531–4534.
38. REYNOLDS, G. P., L. ARNOLD, M. N. ROSSOR, L. L. IVERSEN, C. Q. MOUNTJOY & M.

ROTH. 1984. Reduced binding of [^{3}H]ketanserin to cortical 5-HT_2 receptors in senile dementia of the Alzheimer type. Neurosci. Lett. **44:** 47–51.
39. PERRY, E. K., R. H. PERRY, J. M. CANDY, A. F. FAIRBAIRN, D. J. DICK, G. BLESSED & B. E. TOMLINSON. 1984. Cortical serotonin S2 receptor binding abnormalities in patients with Alzheimers disease: comparisons with Parkinsons disease, Neurosci. Lett. **51:** 353–357.
40. MIDDLEMISS, D. N., A. M. PALMER, N. EDEL & D. M. BOWEN. 1986. Binding of the novel serotonin agonists 8-hydroxy-2-(dipropylamino) tetralin in normal and Alzheimer brain. J. Neurochem. **46:** 993–996.
41. QUIRION, R., J. C. MARTEL, Y. ROPITAILLE, P. ETIENNE, P. WOOD, N. P. V. NAIR & S. GAUTHIER. 1986. Neurotransmitter and receptor deficits in senile dementia of the Alzheimer-type. Can. J. Neurol. Sci. **13:** 503–510.
42. SPARKES, D. L. 1989. Ageing and Alzheimers disease altered serotonergic binding. Arch. Neurol. **46:** 138–140.
43. CROSS, A. J. & J. F. W. DEAKIN. 1985. Cortical serotonin receptor subtypes after lesion of ascending cholinergic neurones. Neurosci. Lett. **60:** 261–265.
44. MAURA, G. & M. RAITERI. 1986. Cholinergic terminals in rat hippocampus possess 5-HT_{1B} receptors mediating inhibition of acetylcholine release. Eur. J. Pharmacol. **129:** 333–337.
45. CROSS, A. J. & P. SLATER. 1988. An autoradiographic analysis of serotonin receptors in human temporal cortex changes in Alzheimer-type dementia. Neurochem. Int. **13:** 89–96.
46. LEYSEN, J. E. P. VAN GOMPEL, M. VERWIMP & C. J. E. NIEMEGEERS. 1983. Role and localisation of serotonin S″ receptor-binding sites: effects of neuronal lesions. *In* CNS Receptors—from Molecular Pharmacology to Behaviour. P. Mandel and F. V. De Feudis, Eds., New York: Raven Press, pp. 373–383.
47. VINCENT, S. R., O. JOHANSSON, T. HOKFELT, B. HEYERSON, C. SACHS, R. P. ELDE, L. TERENIUS & J. KINMEL. 1982. Neuropeptide coexistence in human cortical neurones. Nature **298:** 65–67.
48. SORENSEN, K. V. 1982. Somatostatin: Localisation and distribution in the cortex and the subcortical white matter of the human brain. Neuroscience **7:** 1227–1232.
49. CROSS, A. J., T. J. CROW, I. N. FERRIER, & J. A. JOHNSON. 1986. The selectivity of the reduction of serotonin S2 receptors in Alzheimer type dementia. Neurobiol. Aging, **7:** 3–7.
50. DEHLIN, O., B. HENDENRUD, P. JANSSON & J. NORGARD. 1985. A double-blind comparison of alaproclate and placebo in the treatment of patients with senile dementia. Acta Psychiatr. Scand. **71:** 190–196.
51. CUTLER, N. R., J. HAXBY, A. D. KAY, P. K. NARANG, L. J. LESKO, J. L. COSTA, M. NINOS, A. M. LINNOILA, W. Z. POTTER, J. W. REINFREW & A. M. MOORE. 1985. Evalution of zimeldine in Alzheimers disease, cognitive and biochemical measures. Arch. Neurol. **42:** 744–748.
52. GOTTFRIES, C. G. 1985. Normal aging, Alzheimers disease and senile dementia. Aspects on etiology, pathogenesis, diagnosis and treatment. Belgium: Editors de L' Universite de Belgium, pp. 11–18.
53. WILCOCK, G. K., J. STEVENSON & A. PERKINS. 1987. Trazodone/tryptophan for aggressive behaviour. Lancet **i:** 929–930.

DISCUSSION OF THE PAPER

QUESTION: How do you distinguish between the transporter of the granules, which avidly binds ketanserin and 5-HT_2 particularly in lesions in areas where you find diminution of the synapse and the diminishing of serotonin?

CROSS: We have done a number of studies to characterize [^{3}H]-ketanserin binding in human brain and as far as our studies go they're quite consistent with binding to a 5-HT_2 receptor. In addition we use LSD to define our non-specific binding, not ketanserin, and I don't think LSD has that much affinity for the transporter. We've done similar studies using [^{3}H]-LSD rather than [^{3}H]-ketanserin. So I think we can be fairly certain that it's the 5-HT$_2$ receptor that we're looking at.

RICHARD WURTMAN (*MIT, Cambridge, MA*): The difference you observe between the early onset and late onset patients is quite interesting. I wonder do you have any thought as to which population of neurons that contains the receptors is more vulnerable in the early onset patients than in the late onset patients?

CROSS: That's an interesting question. The evidence is not good but there is evidence that pyramidal neurons are lost fairly early in the disease, and there is some evidence that somatostatin losses only occur in the really severely affected patients. So I would, if I had to, put my money on pyramidal cells. I know that you don't necessarily agree with the anatomy and so on but that would be my affirmation.

WURTMAN: So again you think the pyramidal neurons are the ones that are more likely to disappear in the early onset patients. It's interesting.

HERBERT Y. MELTZER (*Case Western Reserve University, Cleveland, OH*): Given the frequent use of neuroleptics and antidepressants in Alzheimer's patients have you explicitly ruled that out as a cause of the decrease in 5-HT$_2$ receptors?

CROSS: No. I don't think we can do that, not definitely. I showed you data on Huntington's patients who have very similar profiles of neuroleptic intake to the demented patients as far as we can determine, and we don't see any change in 5-HT$_2$ receptors. We also have a group of chronic unipolar depressed patients. Now these are institutionalized patients; they're treatment resistant so they've been taking really high doses of many psychoactive drugs. Again we find no loss of 5-HT$_2$ receptors in that group of patients. Now they're a pretty odd group of patients, but that seems to me the only way that we can control it. By going back and looking at drug histories, we can identify patients who we don't think have had neuroleptics but we can't exclude that they weren't swapping drugs in the institution.

MELTZER: Another reason why you might want to go back and take a look at their drug histories is that it might support your suggestion that the serotonin receptor deficit is more related to behavorial agitation. That suggestion was based primarily, from what I could see, on the drug studies that had been done. Did you happen to take a look at that because what you might want to take a look at is deficits in 5-HIAA or 5-HT levels and the use of neuroleptics to control the behavorial agitation. Did you happen to notice whether patients were getting more Haldol or more neuroleptics to control the behavorial agitation?

CROSS: No, I don't have any data on that. Clinical notes in Britain may be somewhat different to clinical notes in the States. But there is one study from David Bowen's lab which looked at precisely that. A history of agitation and aggression. And that subgroup of Alzheimer's patients had the lowest 5-HIAA concentrations of all the Alzheimer's patients. But that's the only study that I know of that's looked at that. But clearly this is something that is worth looking at in more detail, I think.

JOSEE LEYSEN (*Janssen Research Foundation, Beerse, Belgium*). In your ibotenic acid lesions in the animals you were looking for an animal correlate. We published a very old study (1982) where we did kainic acid lesions in the cortex and we had to inject a large amount of kainic acid in order to see any changes. So with 4 to 5 micrograms at several locations in the cortex we did not see any change in CAT activity but we saw profound decrease in the GAD activity, glutamic acid dicarboxylase. And this was accompanied by the same amount of decrease in 5-HT$_2$ receptor binding, and when we did very profound lesions we went up to 70% decrease. So I wonder whether it's possible to look in the Alzheimer's diseased brains at GAD which may be related to 5-HT$_2$ receptor changes.

CROSS: I'm aware of your study and I'm sorry I didn't refer to it. I normally do. As far as your studies with kainic acid go then I think they complement quite nicely the study with ibotenic acid in the nucleus basalis in that they suggest that 5-HT_2 receptors are present on intrinsic cortical neurons. In our patients we do have data on GABA concentrations, we don't look at GAD because there are a number of problems with that. And we have found, in some studies, that we do get a correlation between 5-HT_2 receptor binding and GABA concentrations, but it's not as consistent as the data we have on somatostatin. But there is clearly something there.

CORNELIUS VANDERWOLF (*University of Western Ontario, London*): I think that Bowen found some decreases in glutamate levels in Alzheimers patients as well, and I wondered do you have any data on whether that precedes or follows the loss of cholinergic and serotonergic markers since the glutamate might be associated particularly with the pyramidal neurons.

CROSS: We've spent a lot of time trying to look at glutamate uptake or some other excitatory amino acids. So I think, although we can't show it neurochemically, the loss of glutamate neurons is there and probably is an early change.

Serotonin Regulation of Circadian Rhythmicity[a]

L. P. MORIN,[b,c,f] K. M. MICHELS,[d] L. SMALE,[b]
AND R. Y. MOORE[d,e]

Departments of [b]Psychiatry, [c]Psychology,
[d]Neurology, [e] Neurobiology and Behavior
State University of New York, Stony Brook
Stony Brook, New York 11794

THE SUPRACHIASMATIC NUCLEUS AND ENTRAINING PATHWAYS

The suprachiasmatic nuclei (SCN) of the hypothalamus generate electrophysiological and metabolic cycles which repeat about every 24 hours.[1-3] Ordinarily, this rhythm is synchronized to the environmental photoperiod that also has a period of about 24 hours. The phase angle of entrainment of the endogenous rhythms to the photoperiod remains relatively constant. The circadian clock of the SCN is coupled to the locomotor system and hierarchically controls a circadian rhythm of running activity. Under a typical 14:10 light:dark cycle (LD), a hamster will often run 7–10 km during the dark phase, starting its activity 15–20 min after the lights go off. Thus, under LD conditions, the entrained locomotor rhythm has a period of 24 h and a fixed phase relationship to the photoperiod as well as a relatively constant rhythm amplitude. True circadian rhythmicity is demonstrated when, in the absence of a time-giving stimulus (e.g., under constant light or dark), the measured event oscillates in a "freerunning" pattern. The freerunning rhythm is self-sustained and repeats approximately every 24 h according to the dictates of an endogenous clock in the SCN.

Elimination of photic input to the brain results in a freerunning rhythm; destruction[4] of the SCN causes loss of circadian locomotor rhythmicity.[5] Given that there are no extraretinal photoreceptors in adult mammals, entrainment of the hamster locomotor rhythm could be accomplished through a retinohypothalamic tract (RHT) that conveys photic information directly to the SCN[6,7] or through an indirect route consisting of primary optic tract afferents to the lateral geniculate complex and a secondary visual input to the SCN through the geniculohypothalamic tract (GHT).[6,8,9] Recent research has shown that the RHT is necessary for circadian rhythm entrainment.[10] The GHT is not necessary for entrainment, but does modulate circadian period under constant light, the phase angle of entrainment, as well as the rate at which rhythm reentrainment is accomplished after a phase shift of the light:dark cycle.[11-13]

FUNCTION OF THE GENICULOHYPOTHALAMIC TRACT

The GHT originates in the intergeniculate leaflet (IGL) of the thalamus.[6,8,9,14] The IGL is identifiable from the location of certain terminal fields from retinal ganglion

[a]Supported by NIH Grants NS-22168 and NS-16304.

[f]Address Correspondence to: Dr. Lawrence P. Morin, Department of Psychiat:y, Health Science Center, SUNY, Stony Brook, NY 11794.

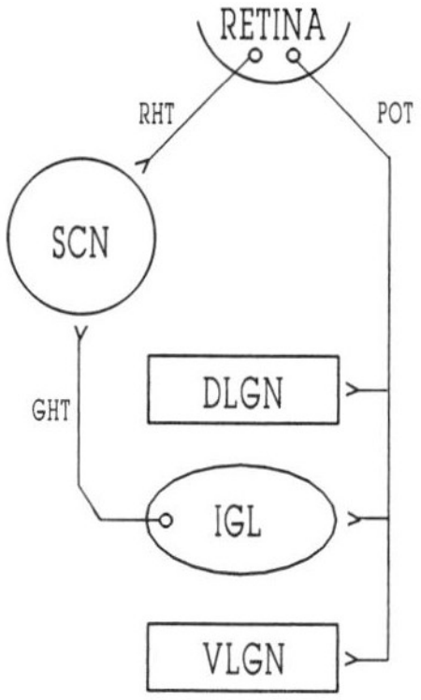

FIGURE 1. Schematic of the visual pathways involved in the entrainment of circadian rhythms to photic stimuli. *Abbreviations:* DLGN—dorsolateral geniculate nucleus; GHT—geniculohypothalamic tract; IGL—intergeniculate leaflet; POT—primary optic tract; RHT—retinohypothalamic tract; SCN—suprachiasmatic nucleus; VLGN—ventrolateral geniculate nucleus.

cells projecting through the optic tract.[7,8] However, it is most dramatically visualized following immunohistochemical evaluation for neuropeptide Y (NPY) and leu-enkephalin (ENK).[7] In the hamster, cell bodies containing NPY and ENK are found throughout the rostro-caudal extent of the IGL. These cells project to the SCN of the hamster; destruction of the IGL eliminates the NPY and ENK innervation normally found in the SCN.[9,12] A schematic diagram of the visual pathways is shown in FIGURE 1.

NPY is a behaviorally active substance in the hamster.[15,16] When infused intrahypothalamically at different times of the circadian day, the peptide induces systematic changes in rhythm phase: the magnitude and direction of change are dependent upon the circadian rhythm time at which the peptide is applied (FIG. 2). A "phase response curve" (PRC) for NPY shows maximal phase advances occurring about 6 h prior to normal activity onset with maximal phase delays occurring in response to NPY given 4–6 h after activity onset.[16] This type of PRC differs from that for light stimulation. Light has no effect on phase when administered during the "dead zone," 1–11 h prior to

FIGURE 2. Comparison of the predominant types of circadian rhythm phase response curves (PRCs). Activity onset is defined as circadian time 12 hours. Note that the light type PRC contains a "dead zone" during which a light stimulus elicits no phase change and that the phase advance portion of the curve occurs during the middle to late subjective night. In contrast, the NPY type PRC has no dead zone and maximum phase advances occur in response to a stimulus applied at circadian time 6 hours. (Light type PRC after Takahashi *et al.*[17]; NPY type PRC after Albers and Ferris[16]).

activity onset, but induces phase delays during the initial (0–5) hours after activity onset followed by phase advances during the later (5–10) hours after activity onset.[17]

TRIAZOLAM AND CIRCADIAN FUNCTION

The NPY-type of PRC is mimicked by electrical[18] or chemical[12] stimulation of the lateral geniculate region. This has led to the suggestion that the IGL can modify circadian clock phase through its NPY innervation of the SCN.[12] The widely used hypnotic benzodiazepine, triazolam, also produces a NPY-type PRC,[19] suggesting that triazolam may affect circadian rhythm phase through the IGL and GHT. This hypothesis was tested in hamsters with bilateral neurotoxic lesions of the IGL. Triazolam given 6 h prior to activity onset caused a 90-min phase advance in control animals, but no change in animals with complete IGL destruction (FIG. 3). At CT 16, triazolam induced a 45-min phase delay, whereas animals with IGL lesions showed no significant change.[20]

This experiment indicates that the IGL is necessary for expression of triazolam effects on circadian phase. However, it is unlikely that the leaflet is the site of the benzodiazepine effect on rhythm phase. Neither immunohistochemistry nor autoradiography reveal the presence of $GABA_A$/benzodiazepine receptors in the hamster IGL.[21] Receptor concentration is modest in the SCN and high in the dorsal raphe nucleus. Because of its high benzodiazepine receptor concentration,[21] the known, dense serotonergic input to the SCN[22–24] and the ability of intra-raphe benzodiazepine to enhance rat locomotion,[25] the raphe became the logical candidate for the site at which triazolam acts to affect circadian rhythm phase. Immunohistochemical analysis of the hamster brain confirms the extensive serotonergic innervation of the SCN.[6,9] In addition, the IGL and ventral LGN receive a well defined serotonergic projection.[7,26] Both the IGL and the SCN are also identified by substance P immunoreactivity.[9,62] However, the substance P appears to be derived from distinctly separate sources in the two nuclei.[9] A distinct, discrete subnucleus of cells accounts for most, if not all, the substance P in the SCN in the hamster. In contrast, the substance P in the IGL is found entirely in fibers and terminals and is probably derived from cells of origin in the dorsal raphe nucleus.[9] It is not yet certain whether the substance P in the IGL is co-localized on 5-HT containing cells of the raphe. Regardless, both substance P and 5-HT originating in the raphe might influence circadian rhythm phase through the IGL or, for 5-HT, through a dense, direct projection to the SCN.

SEROTONIN AND CIRCADIAN FUNCTION

A serotonergic contribution to circadian rhythm regulation has been postulated precisely because the SCN receives such a large input from the midbrain raphe. However, the available data from rats do not permit a resolution of the role of serotonin. Neither Block and Zucker,[27] using electrolytic raphe lesions, nor Honma *et al.*,[24] using chemical lesions of the serotonin system, observed any lasting disruption of, or change in, circadian locomotor periodicity. Block and Zucker[27] note that some of their animals appeared to have increased locomotor activity (but cf. ref. 29). In contrast, Levine *et al.*[30] found greatly disrupted running rhythmicity in rats with raphe lesions. Such lesions have also been reported to result in the loss of the rat corticosterone rhythm[31–33] or to alter that rhythm's phase angle of entrainment.[28] It is not clear that this effect is specifically caused by disruption of circadian clock mechanisms that generate the rhythm.

More recently, several laboratories have begun to investigate the role of the serotonergic system as a regulator of hamster circadian rhythmicity. Clorgyline, a monoamine oxidase (type A) inhibitor, given chronically to entrained animals delays the rate of reentrainment,[34] lengthens the circadian period[35] and increases the magni-

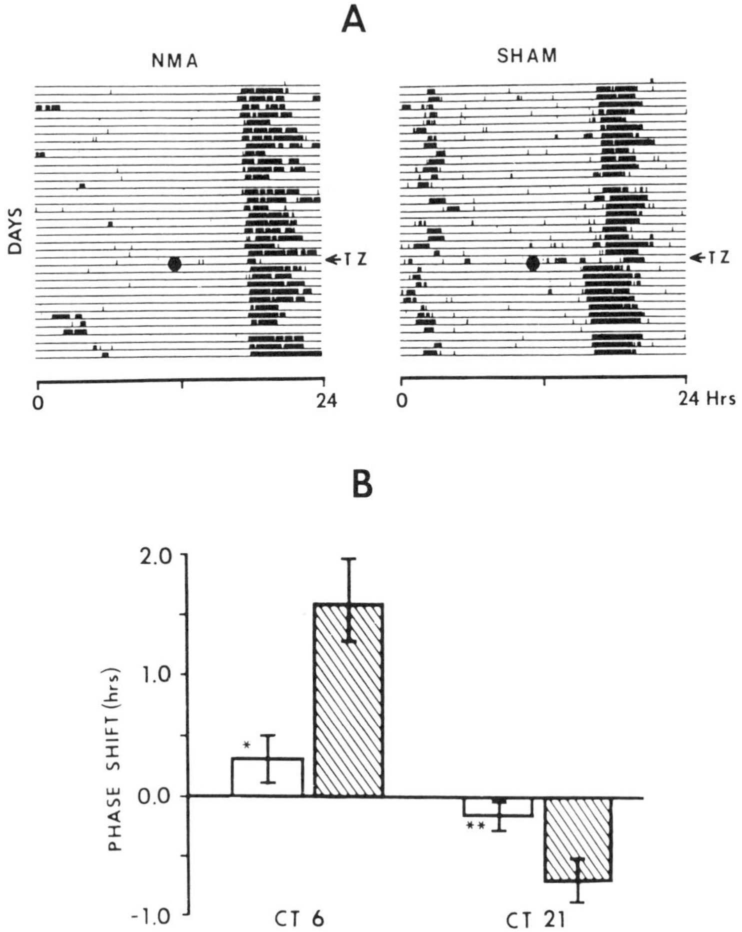

FIGURE 3. (**A**) Effect on circadian rhythm phase of triazolam (TZ) given i.p. at circadian time 6 h to blind, freerunning male hamsters that had received complete bilateral, NMA-induced lesions of the lateral geniculate region or sham lesions. Note the large phase advance by the unlesioned animal and the absence of any advance by the lesioned animal. (**B**) Summary of the mean phase changes by NMA-lesioned or sham-lesioned animals given triazolam at circadian time 6 h or 21 h (mean ± standard errors). The only experimental animals that responded to triazolam had incomplete lesions of the IGL (after Johnson *et al.*[20]).

tude of light pulse induced phase delays.[36] Onset of the daily activity phase may[34] or may not[36] be delayed by clorgyline and the amount of daily activity within the dark phase may[36] or may not[34] be redistributed from early in the night to later in the night. In addition, clorgyline may provoke dissociation of the circadian locomotor rhythm into several components.[35,36] The major interpretive difficulty with the foregoing concerns the mechanism through which clorgyline or other monoamine oxidase (type A) inhibitors work to alter circadian clock function. This issue has been discussed by Duncan *et al.*[36] from an anatomical and neuropharmacological perspective. There is no compelling evidence that the actions of monoamine oxidase (type A) inhibitors on the hamster circadian system are mediated exclusively, or even partly, through the expected elevation of serotonin. Despite this, Duncan *et al.*[36] suggested that clorgyline's actions may be mediated by the IGL and have recently shown that the ability of clorgyline to provoke circadian rhythm dissociation can be abolished by IGL lesions.[37] At this time, it appears that both triazolam and clorgyline require an intact IGL in order to exert their respective effects on the rhythm system, although there is no clear site or mechanism of action of either drug.

In a recent study, 5,7-dihydroxytryptamine (5,7-DHT) was applied directly to the SCN. This selective neurotoxin eliminated 5-HT immunoreactivity in the nucleus, but did not alter either activity onset or offset.[38] The only reported effect of the 5,7-DHT lesions of the SCN is to prevent the delays in activity offset seen consequent to administration of a monoamine oxidase (type A) inhibitor.[38]

The above discussion emphasizes that little is certain with respect to raphe or serotonergic control of circadian rhythms in either rats or hamsters. Therefore, we designed an experiment to evaluate the role of the serotonergic system in adult hamster circadian rhythm regulation. In addition, we tested whether serotonergic system damage would alter the ability of triazolam to accelerate reentrainment to a shifted light-dark cycle.

Male hamsters housed under L:D 14:10 received intraventricular 5,7-DHT (75 μl in 0.5% ascorbic acid) or vehicle. The 5,7-DHT treated animals showed an immediate, lasting change in the pattern of wheelrunning. This was characterized by 1) advanced activity onsets (beginning an average 0.71 ± 0.07 h before lights off compared to 0.18 ± 0.04 h after lights off for controls) and 2) delayed activity offsets (9.8 ± 0.2 h after lights off for 5,7-DHT treated animals compared to 8.5 ± 0.3 h for controls; see FIG. 4). The earlier onset and later offset yielded a large difference in total activity duration between the groups. Lesioned animals had an activity phase that lasted 10.5 ± 0.3 h compared to 8.4 ± 0.3 h for controls (FIG. 4). When the activity patterns were stable after the lesions, the LD 14:10 cycle was advanced such that lights off occurred 8 h earlier. Half of the lesioned and half the control animals each received 2 mg triazolam i.p. in dimethylsulfoxide 6 h before activity onset on the day of the phase shift. The remaining animals received vehicle injections. Animals were then allowed to reentrain to the shifted LD cycle before being placed in constant darkness (DD) (FIG. 4). The earlier activity onsets and later offsets of lesioned animals persisted through the 8 h advance of the photoperiod and under the DD conditions. Despite causing changes in the activity phase, lesions of the serotonergic system did not appear to modify the ability of triazolam to facilitate the rate of accomplishing a phase shift. Triazolam-treated animals reentrained at the same rate, whether they had been 5,7-DHT lesioned or not. Triazolam also hastened the rate of reentrainment, as expected.[39]

It is clear, however, that the removal of serotonergic innervation to the SCN and IGL does not abolish hamster circadian rhythmicity. The innervation is, therefore, neither part of the clock mechanism nor a necessary mediator of clock output and the serotonin system does not appear to be necessary for clock regulation in the adult hamster. Rate of reentrainment to the shifted photoperiod is not altered by the 5,7-DHT lesions nor is the freerunning circadian period measured in constant darkness.

Although these data clearly demonstrate that entrained circadian periodicity is maintained despite near-total destruction of the forebrain serotonin system, there are two tantalizing results that require further exploration. First, despite the close similar-

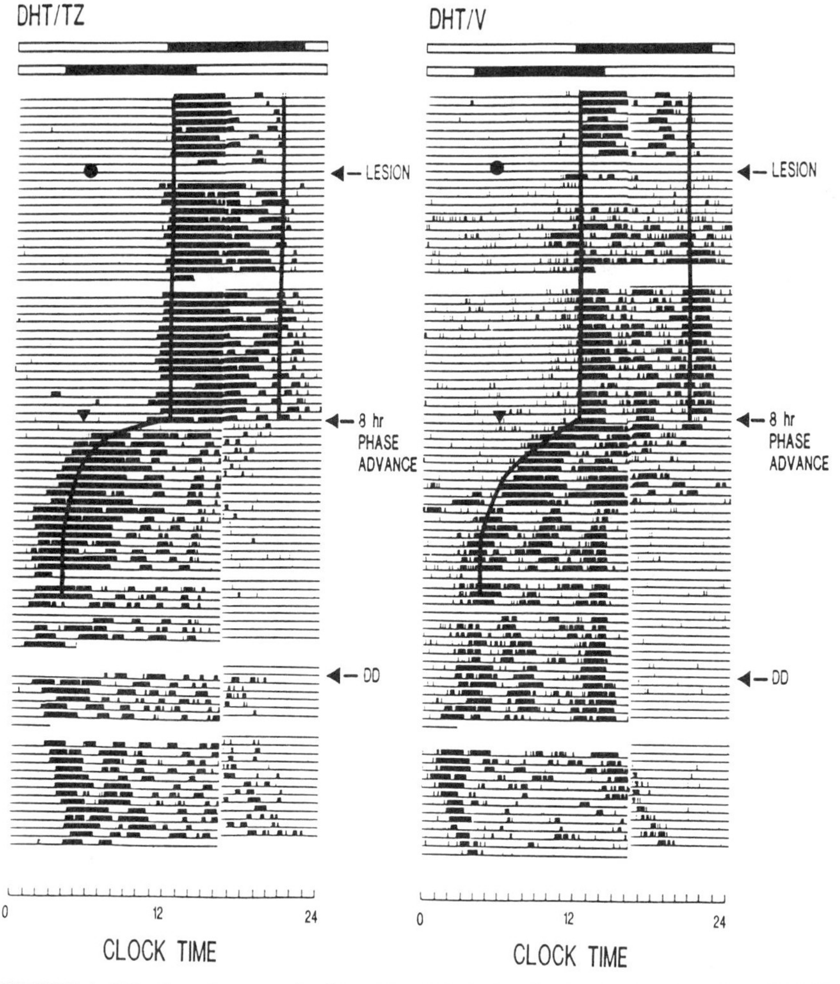

FIGURE 4. Wheelrunning records of two 5,7-DHT–lesioned animals given triazolam (TZ) 6 h before activity onset on the day of an 8 h advance of the LD cycle (*left*) or given vehicle at that time (*right*). The solid circle indicates the time of the lesion; the triangle indicates the time of the injection. After reentrainment to the shifted light cycle, animals were placed in constant darkness (DD). The parallel vertical lines bracket the average nocturnal activity phase of sham lesioned animals. Note that after the 5,7-DHT lesion, the activity phase of each animal expands to an earlier onset and a later offset. The subsequent curving line shows the average activity onset times of sham lesioned animals given TZ (*left*) or vehicle (*right*) during reentrainment to a light cycle shifted 8 h. 5,7-DHT lesioned animals did not differ from controls in rate of reentrainment following either TZ or vehicle. The horizontal white and black bars at the top of each record indicate the initial light-dark cycle and the shifted cycle.

ity in the mean freerunning circadian periods under constant darkness for the control and 5,7-DHT groups, the range was much greater for the group with lesions. The mean period for controls was 23.97 h with a range of 23.86 to 24.03 and for the 5,7-DHT group it was 23.99 h with a range of 23.80 to 24.26. Second, the 5,7-DHT treated animals were much more likely to have rhythm anomalies during reentrainment after the phase shift or during constant darkness. In fact, 4 animals became more or less arrhythmic in constant darkness. This result is most unusual. Hamsters occasionally become arrhythmic in constant light,[40] but almost never in darkness.[4] Both the greater variability in circadian period and the rhythm anomalies are consistent with a loss of cohesiveness in the expression of the circadian clock. Indeed, a role for serotonin as a regulator of circadian rhythm coherency has previously been suggested.[30] A clear demonstration that the serotonergic system mediates rhythm cohesiveness requires further study using environmental conditions that are much more likely to precipitate loss of normal circadian rhythmicity.[30,34]

CONCLUSIONS

Photic entrainment of the hamster locomotor circadian rhythm is accomplished through a retinohypothalamic projection to the SCN. Entrainment occurs when light-induced rhythm phase advances are canceled by equivalent light-induced phase delays. The light-type PRC shows that phase advances are stimulated by light during the last half of the subjective night, whereas phase delays occur in response to light during first half of the subjective night. In contrast, the phase advance region of the NPY-type PRC occurs during the mid-subjective day and the phase delay region during the mid-subjective night. The short-acting hypnotic, triazolam, produces an NPY-type PRC. The effects of triazolam on rhythm phase are blocked by IGL lesions indicating a role for the IGL in phasic triazolam action. However, autoradiography and immunohistochemistry reveal that neither the IGL nor SCN are likely to be sites at which triazolam acts to alter rhythm phase. The midbrain dorsal raphe nucleus has a high concentration of benzodiazepine receptors and communicates, via its serotonergic efferent system, with both the IGL and SCN. Recent work demonstrates that following depletion of forebrain serotonin hamsters continue to show normal phase shifts in response to triazolam. It is therefore unlikely that the serotonin system mediates the phasic action of triazolam. However, the serotonin system clearly regulates the distribution of wheelrunning within the nocturnal activity phase and may modulate cohesiveness of the circadian locomotor rhythm.

ACKNOWLEDGMENT

We thank D. Caselles for typing the manuscript.

REFERENCES

1. INOUYE, S. T. & H. KAWAMURA, 1979. Persistence of circadian rhythmicity in a mammalian hypothalamic "island" containing the suprachiasmatic nucleus. Proc. Natl. Acad. Sci. USA **76:** 5962–5966.
2. SHIBATA, S. & R. Y. MOORE. 1988. Electrical and metabolic activity of suprachiasmatic nucleus neurons in hamster hypothalamic slices. Brain Res. **438:** 374–378.
3. NEWMAN, G. C. & F. E. HOSPOD. 1986. Rhythm of suprachiasmatic nucleus 2-deoxyglucose uptake *in vitro.* Brain Res. **1:** 245–250.
4. MORIN, L. P. & L. A. CUMMINGS. 1981. Effect of surgical or photoperiodic castration, testosterone replacement or pinealectomy on male hamster running rhythmicity. Physiol. Behav. **26:** 825–838.

5. RUSAK, B. 1977. The role of the suprachiasmatic nuclei in the generation of circadian rhythms in the golden hamster, Mesocricetus auratus. J. Comp. Physiol. **118:** 145–164.
6. MOORE, R. Y. & J. P. CARD. 1985. Visual pathways and the entrainment of circadian rhythms. Ann. N.Y. Acad. Sci. **453:** 123–133.
7. JOHNSON, R. F., L. P. MORIN & R. Y. MOORE. 1988. Retinohypothalamic projections in the hamster and rat demonstrated using cholera toxin. Brain Res. **462:** 301–312.
8. HARRINGTON, M. E., D. W. NANCE & B. RUSAK. 1985. Neuropeptide Y immunoreactivity in the geniculosuprachiasmatic tract. Brain Res. Bull. **15:** 465–472.
9. MORIN, L. P., L. SMALE, K. M. MICHELS & R. Y. MOORE. 1988. NPY, GFAP, GAD, 5-HT and enkephalin in the intergeniculate leaflet and suprachiasmatic nuclei of the hamster. Soc. Neurosci. **14:** 51.
10. JOHNSON, R. F., R. Y. MOORE & L. P. MORIN. 1988. Loss of entrainment and anatomical plasticity after lesions of the hamster retinohypothalamic tract. Brain Res. **460:** 297–313.
11. HARRINGTON, M. E. & B. RUSAK. 1986. Lesions of thalamic intergeniculate leaflet alter hamster circadian rhythms. J. Biol. Rhythms, **1:** 309–325.
12. JOHNSON, R. F., R. Y. MOORE & L. P. MORIN. 1989. Lateral geniculate lesions alter activity rhythms in the hamster. Brain Res. Bull. **22:** 411–422.
13. PICKARD, G. E., M. R. RALPH & M. MENAKER. 1987. The intergeniculate leaflet partially mediates effects of light on circadian rhythms. J. Bio. Rhythms **2:** 35–56.
14. RIBAK, C. E. & A. PETERS. 1975. An autoradiographic study of the projections from the lateral geniculate body of the rat. Brain Res. **92:** 341–368.
15. ALBERS, H. E., C. F. FERRIS, S. E. LEEMAN & B. D. GOLDMAN. 1984. Microinjection of neuropeptides into the suprachiasmatic region of the hypothalamus phase shifts circadian activity rhythms of Syrian hamsters. Science **222:** 833–835.
16. ALBERS, H. E. & C. F. FERRIS. 1984. Neuropeptide Y: Role in light-dark entrainment of hamster circadian rhythms. Neurosci. Lett. **50:** 163–168.
17. TAKAHASHI, J. S., P. J. DECOURSEY, L. BAUMAN & M. MENAKER. 1984. Spectral sensitivity of a novel photoreceptive system mediating entrainment of mammalian circadian rhythms. Nature **308:** 186–188.
18. RUSAK, B., J. H. MEIJER & M. E. HARRINGTON. 1989. Hamster circadian rhythms are phase shifted by electrical stimulation of the geniculohypothalamic tract. Brain Res. In press.
19. TUREK, F. W. & S. LOSEE-OLSON. 1986. A benzodiazepine used in the treatment of insomnia phase-shifts the mammalian circadian clock. Nature **321:** 167–168.
20. JOHNSON, R. F., L. SMALE, R. Y. MOORE & L. P. MORIN. 1988. Lateral geniculate lesions block circadian phase-shift responses to a benzodiazepine. Proc. Natl. Acad. Sci. USA **85:** 5301–5304.
21. MICHELS, K. M., L. SMALE, L. P. MORIN & R. Y. MOORE. 1988. GABA-A/Benzodiazepine receptors in the rat and hamster circadian system. Soc. Neurosci **14:** 345.
22. AZMITIA, E. C. & M. SEGAL. 1978. An autoradiographic analysis of the differential ascending projections of the dorsal and median raphe nuclei in the rat. J. Comp. Neurol. **179:** 641–668.
23. MOORE, R. Y., A. E. HALARIS & B. E. JONES. 1978. Serotonin neurons of the midbrain raphe: Ascending projections. J. Comp. Neurol. **180:** 417–438.
24. VAN DE KAR, L. D. & S. A. LORENS. 1979. Differential serotonergic innervation of individual hypothalamic nuclei and other forebrain regions by the dorsal and median midbrain raphe nuclei. Brain Res. **162:** 45–54.
25. SAINATI, S. M. & S. A. LORENS. 1983. Intra-raphe benzodiazepines enhance rat locomotor activity: Interactions with GABA. Pharmacol. Biochem. Behav. **18:** 407–414.
26. MANTYH, P. W. & J. A. KEMP. 1983. The distribution of putative neurotransmitters in the lateral geniculate nucleus of the rat. Brain Res. **288:** 344–348.
27. BLOCK, M. & I. ZUCKER. 1976. Circadian rhythms of rat locomotor activity after lesions of the midbrain raphe nuclei. J. Comp. Physiol. **109:** 235–247.
28. HONMA, K., K. WATANABE & T. HIROSHIGE. 1979. Effects of phenylalanina and 5,6-Dihydroxytryptamine on the free-running rhythms of locomotor activity and plasma corticosterone in the rat exposed to continuous light. Brain Res. **169:** 531–544.
29. KAM, L. M. & G. P. MOBERG. 1977. Effect of raphe lesions on the circadian pattern of wheel

running in the rat. Physiol. Behav. **18:** 213–217.
30. LEVINE, J. D., A. M. ROSENWASSER, J. A. YANOVSKI & N. T. ADLER. 1986. Circadian activity rhythms in rats with midbrain raphe lesions. Brain Res. **384:** 240–249.
31. SZAFARCZYK, A., G. IXART, F. MALAVAL, J. NOUGUIER-SOULE & I. ASSENMACHER. 1979. Effects of lesions of the suprachiasmatic nuclei and of p-chlorophenylalanine on the circadian rhythms of adrenocorticotrophic hormone and corticosterone in the plasma, and on locomotor activity in rats. J. Endocrinol. **83:** 1–16.
32. WILLIAMS, J. H., V. M. MIALL-ALLEN, M. KLINOWSKI & E. C. AZMITIA. 1983. Effects of microinjections of 5,7-dihydroxytryptamine in the suprachiasmatic nuclei of the rat on serotonin reuptake and the circadian variation of corticosterone levels. Neuroendocrinology **36:** 431–435.
33. DUNN, J. D., D. C. JOHNSON & A. J. CASTRO. 1983. The effect of raphe transection on circadian patterns of serum corticosterone. Neuroendocrinol. Lett. **5:** 233–238.
34. TAMARKIN, L., C. CRAIG, N. GARRICK & T. WEHR. 1983. Effect of clorgyline (a MAO type A inhibitor) on locomotor activity in the syrian hamster. Am. J. Physiol. **245:** R215–R221.
35. WIRZ-JUSTICE, A. & I. C. CAMPBELL. 1982. Antidepressant drugs can slow or dissociate circadian rhythms. Experientia **38:** 1301–1309.
36. DUNCAN, W. C., L. TAMARKIN, P. G. SOKOLOVE & T. A. WEHR. 1988. Chronic clorgyline treatment of Syrian hamsters: An analysis of effects on the circadian pacemaker. J. Biol. Rhythms **3:** 305–322.
37. DUNCAN, W., E. SUTIN, W. OREM, B. GAO, P. G. SOKOLOVE & T. A. WEHR. 1989. Prevention of light-induced dissociation wheelrunning activity by IGL lesions of chronic clorgyline treated Syrian hamsters. Soc Neurosci. **15.** In press.
38. KRUSE, J. S. 1988. Destruction of suprachiasmatic nuclei serotonergic fibers alters response of hamster activity rhythms to a monoamine oxidase inhibitor. Ann. Rev. Chronopharm. **5:** 131–134.
39. VAN REETH, O. & F. W. TUREK. 1987. Adaptation of circadian rhythmicity of shift in light-dark cycle accelerated by a benzodiazepine. Am. J. Physiol. **253:** R204–R207.
40. MORIN, L. P. & L. A. CUMMINGS. 1982. Splitting of wheelrunning rhythms by castrated or steroid-treated male and female hamsters. Physiol. Behav. **29:** 665–675.

DISCUSSION OF THE PAPER

QUESTION: Would you like to comment on the role of the pineal in the behavior and the possible role of 5-HT and the interaction of that system?

MORIN: The pineal plays no direct role in the regulation of circadian rhythms in mammals. It plays an indirect role insofar as the pineal regulates gonadal activity under conditions of short photoperiods. The testes of hamsters regress and become atrophic and do not secrete any androgens. Under those circumstances, the lack of availability of behaviorally facilitating hormones, the animals' level of locomotion goes way down. But it has no direct effect on the rate of reintrainment, on the period of the rhythm; only on the amplitude of the rhythm.

Effects of Steroid Hormones on the Serotonergic System

ANAT BIEGON

Department of Neurobiology
Weizmann Institute of Science
Rehovot, Israel

INTRODUCTION

Serotonin is implicated in a variety of physiological and behavioral functions, some of which, e.g., ovulation,[1] sexual behavior,[2] and the response to stress[3] are tightly linked to steroid hormones in experimental animals. Thus serotonin is a good candidate for the role of a mediator of various central actions of gonadal and adrenocortical steroids. This being the case, one would expect to encounter steroid hormone modulation of at least some aspects of serotonergic activity. The evidence reviewed in the following pages does indeed suggest that almost every component of the serotonin neurotransmission chain is modulated by some steroid hormone in some brain region. Theoretically, various mechanisms could be responsible for steroid hormone effects on the serotonergic system. Steroids penetrate the blood brain barrier easily and interact with intracellular receptors concentrated in specific brain regions.[4–6] The hormone-receptor complexes may then interact directly with the genome to produce changes in the expression of specific messages, possibly including neurotransmissions related molecules such as snythetic and metabolic enzymes, postsynaptic receptors, etc. Steroids may also alter serotonergic transmission indirectly by altering neural activity in a related system and causing adaptive changes, or directly by acting on neuronal membranes to change the binding properties of neural receptors. Again, we have evidence that all of the above mechanisms may be operational in the brain at some point in the animal's life cycle. The relevance of the animal findings to human beings and clinical situations can be indirectly evaluated by studies of sex differences and hormonal cycle linked changes in serotonergic parameters in the human brain post mortem or in peripheral models of central serotonergic activity, and a short review of this kind of studies concludes the present paper.

EFFECTS OF STEROID HORMONES ON PRESYNAPTIC PROCESSES

Both gonadal and adrenal steroids appear to affect various presynaptic processes, such as synthesis, release, uptake and "turnover" of the transmitter, often estimated by the ratio between the level of serotonin and 5-hydroxy indoleacetic acid (5-HIAA), the major metabolite.

Gonadal Steroids

Physiological changes in gonadal hormone secretion during the rat estrous cycle are associated with a significant increase in hypothalamic 5-HT uptake, which occurs on proestrous day.[7] This is probably not a simple direct effect of the hormones on the uptake mechanism, since in the hypothalamic slice preparation, *in vitro* treatment with estrogen and progesterone inhibits, rather than increases, 5-HT uptake.[8] *In vivo* injection of estradiol and progesterone results in time-, dose- and region-specific changes in 5-HT uptake.[8–10]

Changes in brain 5-HT levels and turnover were also noted during the estrous cycle.[11] In this case, too, injections of estrogen, progesterone or both result in complex, dose-, time- and region-dependent effects.[12–16] In the dorsal raphe nucleus, estrogen is reported to increase 5-HT levels and synthesis,[13,14] while many other brain regions are unaffected.[12] Increased hypothalamic 5-HIAA but not 5-HT has been observed following estradiol treatment[15] while progesterone causes a decrease in hypothalamic 5-HT accumulation in the estrogen primed rat.[16]

Sex differences in serotonin levels, especially in hypothalamic nuclei, have also been reported.[17] Although sex differences could reflect testicular as well as ovarian hormone actions, studies of androgen effects on serotonergic presynaptic activity are relatively scarce. In male rats, testosterone injections beginning at weaning lead to decreased 5-HT and 5-HIAA levels by age 45 days, while gonadectomy at weaning was associated with increased 5-HT levels at this age.[18] In the medial preoptic nucleus, the sexual dimorphism in serotonin fibre distribution appears to be dependent on perinatal androgen, since females exposed to prenantal testosterone showed complete masculinization of the 5-HT fibers distribution.[19]

Corticosteroids

The activity of brain tryptophan hydroxylase, the rate-limiting enzyme in serotonin synthesis, is affected by adrenalectomy[20] and glucocorticoids[21] in the rat. Adrenalectomy also decreases 5-HT uptake into hypothalamic but not cortical synaptosomes. Corticosterone replacement reverses the effect of adrenalectomy on hypothalamic 5-HT uptake, while increasing the uptake in sham-operated animals.[2] Under physiological conditions, plasma corticosterone was found to be inversely correlated with 5-HT levels in the amygdala and directly correlated with hypothalamic 5-HT content.[23] Effects of adrenalectomy and corticosteroid replacement on 5-HT turnover appear to be region dependent: Serotonin turnover is decreased in the hippocampus following adrenalectomy, and corticosterone replacement restores turnover to control values.[24] In the hypothalamus, adrenalectomy decreases 5-HT/5-HIAA levels in some nuclei while the supraoptic nucleus exhibits a large increase. These effects, too, are at least partially reversible by corticosterone treatment.[25]

EFFECTS OF STEROID HORMONES ON POSTSYNAPTIC ACTIVITY

The major subtypes of serotonin receptors in the brain, 5-HT_1 and 5-HT_2, are largely postsynaptic, and local modulation of their affinity and/or density is yet another means through which steroids can affect serotonergic activity. Serotonin-dependent behaviors supply a crude but irreplaceable tool for evaluating the "final outcome" of different regulatory inputs on the system.

Gonadal Steroids

The estrous cycle of the rat is accompanied by regionally selective changes in serotonin receptors of the 5-HT_1 type.[26] A significant decrease in B_{max} occurs on proestrous and estrous as compared to diestrous, in estrogen concentrating brain regions (preoptic-hypothalamic area) but not in the cortex and hippocampus. *In vivo* treatment with estradiol exerts a biphasic effect on the density of $5\text{-}HT_1$ receptors in the female rat brain.[27] An acute reduction in receptor density throughout the brain is followed 48 hours later by a selective increase in those brain regions known to contain estrogen receptors—hypothalamus, preoptic area and amygdala. A quantitative autoradiographic study[28] of the effects of estrogen on $5\text{-}HT_1$ receptors, facilitating the examination of small, discrete nuclei, has revealed an effect in the lateral septum as well.[29] The acute reduction in receptor density can be mimicked by estradiol *in vitro*.[27] Chronic (14 days) exposure to estradiol or progesterone induces a down regulation in $5\text{-}HT_1$ receptors, while $5\text{-}HT_2$ receptor number goes up, in female rat cortex.[30] Progesterone given to estradiol-treated rats reverses the estrogen-induced increase in $5\text{-}HT_2$ receptors but has no effect on the estrogen-induced reduction in $5\text{-}HT_1$ receptors. In addition, there is a sex difference in the response of $5\text{-}HT_1$ receptors to gonadal steroids.[31]

The "serotonin behavioral syndrome"[32] is also dependent on circulating gonadal hormones. Initially, a large sex difference had been observed in the sensitivity of male and female rats to the combination of pargyline and tryptophan used to induce the syndrome. Females responded at significantly lower doses of tryptophan than males. In addition, females exhibited a biphasic effect on body temperature (initial hypothermia followed by hyperthermia) which was not seen in the males at all—males either had hyperthermia or no response.[33] Gonadectomy and hormone replacement studies on normal and TFM (testosterone feminized males, lacking androgen receptors) rats have demonstrated that testosterone, acting on androgen receptors, is responsible for this sex difference.[34] Sex differences and gonadal hormone regulation of a serotonin dependent behavior have been reported also in Guinea pigs, where 5-hydroxytryptophan induced myoclonus is enhanced by estradiol in both sexes.[35] Progesterone given to estrogen primed animals enhanced myoclonus in males but reversed the facilitatory effect of estradiol in females. As in the rat, this is a qualitative rather than quantitative sex difference in the serotonergic response to steroids.

Corticosteroids

Adrenalectomy and corticosterone replacement have selective effects on serotonin receptors: $5\text{-}HT_1$ receptor binding in the hippocamnpus goes up after adrenalectomy. Corticosterone replacement reverses the effect of adrenalectomy and down regulates the receptors.[36]

Another study has demonstrated that $5\text{-}HT_1$ receptors in dorsal raphe and substantia nigra were also increased following adrenalectomy, in addition to the hippocampal increase in binding. Unaffected regions include cerebral cortex, central grey, ventral hippocampus and median raphe. Corticosterone but not aldosterone replacement reversed the effects of adrenalectomy of $5\text{-}HT_1$ binding in the rat brain.[37] Chronic (4 days) corticosterone treatment reduces serotonin dependent behavior in rats, elicited by both *p*-chloramphetamine (a serotonin releaser) and 5-MeODMT, an agonist, implying the involvement of postsynaptic receptor down regulation.[38]

STUDIES IN HUMAN SUBJECTS

Studies of steroid modulation of serotonergic activity in the human central nervous system are difficult due to the lack of specific peripheral indices of central serotonergic activity. Is is not at all clear to what degree peripheral 5-HT and 5-HIAA levels reflect changes in brain serotonin metabolism, and probing brain metabolism directly *in vivo* is just becoming possible now with the advent of positron emission tomography. Other possibilities are post-mortem studies and studies on peripheral models of serotonergic neurons, such as the platelet.[39]

Gonadal Steroids

Blood MAO activity and 5HT levels fluctuate with the human menstrual cycle,[40] and these changes can be correlated with cyclic mood changes. Platelet 5-HT_2 receptors, which have been shown to be upregulated in depression,[41,42] also show dependence on circulating ovarian steroids across the menstrual cycle, with a minimum around midcycle and a maximal binding during the premenstrual period (TABLE 1). In manic depressive patients treated with lithium, withdrawal of the drug causes a significant increase in platelet 5-HT_2 receptors in women but not in men,[43] yet again suggesting a sex difference in the regulation of serotonergic activity.

TABLE 1. Effect of the Menstrual Cycle on 5-HT_2 Receptor Binding on Platelets[a]

Phase of cycle	N	^{3}H-Ket binding, fm/mg P.
Beginning (1–13)	14	34.6 ± 5.1
Midcycle (13–17)	7	18.8 ± 2.5*
End (18–25)	11	42.1 ± 6.4

*Significantly different from midcycle by one way ANOVA (p = 0.043) followed by Duncan's post-hoc test ($p < 0.05$).

Paired Comparisons on 7 Individuals			
Phase of cycle:	Beginning	Midcycle	End
^{3}H-Ket binding	33.9 ± 7.6	18.8 ± 2.5*,**	52.2 ± 11.5

*$p = 0.044$, paired t-test, compared to end; **$p = 0.053$, paired t-test, compared to beginning.
[a]Blood donors were medical students aged 18–20, not taking any medication including birth control pills. Blood was collected in the morning at 3 or 4 weekly intervals from each individual, and assigned to cycle phase according to the subject's report on the date of the last period. ^{3}H-Ketanserine, 3.0 nM, was used to assess binding.[41]

CONCLUSIONS

The data reviewed above indicate very clearly that steroid hormones have significant modulatory influence on all aspects of the serotonergic system of mammalian brains. Thus, both gonadal and adrenal steroids have region-specific effects on serotonin synthesis, turnover, uptake, release, receptors and behavior. This is not a general phenomenon, since although steroids have been shown to influence some parameters of other neurotransmitter systems, such as muscarinic and noradrenergic receptors,[30,36] many of the studies quoted above examined other neurotransmitter markers along with serotonergic markers and the latter are usually the ones most influenced by steroids.

One of the most striking features of the regulation of serotonergic activity by steroid hormones is the marked regional specificity of the observed effects. This highly localized modulation of pre- and post-synaptic processes may in fact be the mechanism making it possible for a single neurotransmitter to subserve many different and sometimes even contradictory functions—e.g., serotonin has both facilitatory and inhibitory effects on ovulation and female sexual behavior and this is probably achieved by modulation of different receptor subtypes in different locations in the brain.

REFERENCES

1. MEYER, D. C. 1978. Endocrinology **103:** 1067–1074.
2. MEYERSON, B. J., H. CORRER & M. ELIASSON. 1974. *In* Advances in biochemical psychopharmacology. E. Costa, G. L. Gessa and M. Sandler, Eds. Vol. **2:** 229–242. New York: Raven Press.
3. BORANIC, M., D. PERICIC, M. POLJAK-BLAZI, V. SVERKO & T. MAROTTI. 1987. Ann. NY. Acad. Sci. **496:** 485–491.
4. STUMPF, W. E. 1980. *In* The endocrine functions of the brain. M. Motta, Ed. New York: Raven Press, pp. 43–49.
5. PFAFF, D. & M. KEINER. 1973. J. Comp. Neurol. **151:** 121–158.
6. GERLACH, J. L. & B. S. MCEWEN. 1972. Science **175:** 1133–1136.
7. MEYER, D. C. & W. B. QUAY. 1976. Endocrinology **98:** 1160–1164.
8. ENDERSBY, C. A. & C. A. WILSON. 1974. Brain Res. **73:** 321–331.
9. WIRZ-JUSTICE, A., E. HACKMAN & M. LICHTSTEINER. 1974. J. Neurochem. **22:** 187–189.
10. KENDALL, D. A. & S. R. TONGE. 1977. Br. J. Pharmacol. **60:** 310 p.
11. FLUDDER, J. M. & S. R. TONGE. 1975. J. Pharm. Pharmacol. **27:** 1–39.
12. CROWLEY, W. R., T. L. O'DONOHUE, E. A. MUTH & D. M. JACOBOWITZ. 1979. Brain Res. Bull. **4:** 571–574.
13. CONE, R. I., G. A. DAVIS & R. W. GOY. 1981. Brain Res. Bull. **7:** 639–644.
14. DIPAOLO, T., M. DAIGLE, V. PICARD & N. BARDEN. 1983. Exp. Brain Res. **51:** 73–76.
15. RENNER, K. & V. LUINE. 1986. Brain Res. **366:** 4–71.
16. RENNER, K. J., L. C. KREY & V. N. LUINE. 1987. Brain Res. Bull. **19:** 195–202.
17. RENNER, K. J., A. BIEGON & V. N. LUINE. 1985. Exp. Brain Res. **58:** 198–201.
18. MARTINEZ-CONDE, E., M. L. LERET & S. DIAZ. 1985. Comp. Biochem. Physiol. **80:** 411–414.
19. SIMERLY, R. B., L. W. SWANSON & R. A. GORSKI. 1985. Brain Res. **340:** 91–98.
20. AZMITIA, E. C. & B. S. MCEWEN. 1974. Brain Res. **78:** 291–302.
21. SZE, P. Y., L. NECKERS & A. C. TOWLEL. 1976. J. Neurochem. **26:** 169–173.
22. LEE, P. H. & M. Y. CHAN. 1985. Neuropharmacology **24:** 1043–1050.
23. ALDEGUNDE, M., E. ARNAEZ, I. MIGUEZ & P. FERNANDEZ. 1984. Int. J. Neurosci. **24:** 233–238.
24. DE KLOET, E. R., G. L. KOVACS, G. SZABO, G. TELEGDY, B. BOHUS & D. H. G. VERSTEEG. 1982. Brain Res. **239:** 659–663.
25. JHANWAR-UNIYAL M., K. RENNER, M. BAILO, V. N. LUINE & S. F. LEIBOWITZ. 1987. Neurosci. Lett. **79:** 145–150.
26. BIEGON, A., H. BERCOVITZ & D. SAMUEL. 1980. Brain Res. **187:** 221–225.
27. BIEGON, A. & B. S. MCEWEN. 1982. J. Neuroscience **2:** 199–205.
28. BIEGON, A., T. C. RAINBOW & B. S. MCEWEN. 1982. Brain Res. **242:** 197–204.
29. BIEGON, A., C. T. FISCHETTE, T. C. RAINBOW & B. S. MCEWEN. 1982. Neuroendocrinology **35:** 287–291.
30. BIEGON, A., A. RECHES, L. SNYDER & B. S. MCEWEN. 1983. Life Sci. **32:** 2015–2022.
31. FISCHETTE, C. T., A. BIEGON & B. S. MCEWEN. 1983. Science **222:** 333–335.
32. JACOBS, B. L. 1976. Life Sci. **9:** 777–786.
33. BIEGON, A., M. SEGAL & D. SAMUEL. 1979. Psychopharm. **61:** 77–80.
34. FISCHETTE, C. T., A. BIEGON & B. S. MCEWEN. 1984. Life Sci. **35:** 1197–1206.
35. O'CONNOR, L. H. & H. H. FEDER. 1985. Brain Res. **330:** 121–125.
36. BIEGON, A., T. C. RAINBOW & B. S. MCEWEN. 1985. Brain Res. **332:** 309–314.
37. DEKLOET E. R., H. SYBESMA & H. M. REUL. 1986. Neuroendocrinology **42:** 513–521.
38. DICKINSON, S. L., G. A. KENNET & G. CURZON. 1985. Brain Res. **345:** 10–18.
39. SHEDDON, J. M. 1973. *In* Progress in Neurobiology. G. A. Kerkut, and J. W. Phillis, Eds. Oxford: Pergamon Press, p. 153.

40. WARREN, D. E., W. H. TEDFORD & W. E. FLYNN. 1979. Medical Hypotheses **5:** 359–364.
41. BIEGON, A., A. WEIZMAN, L. KARP, A. RAM, S. TIANO & M. WOLF. 1987. Life Sci. **41:** 2485–2492.
42. ARORA, R. C. & H. Y. MELTZER. 1989. Life Sci. **44:** 725–734.
43. CARNI, N., S. SCHWARTS & A. BIEGON. 1987. British Lithium Congress Abstracts **2:** 5.

DISCUSSION OF THE PAPER

JEAN C. SHIH (*USC, Los Angeles, CA*): It's a really nice paper. In the benzodiazepine receptor system there is a well-characterized allosteric regulation site, and I was wondering whether you think it's a similar situation here? In other words you think the steroids allosterically regulate 5-HT receptors.

BIEGON: We don't have evidence for allosteric regulation although as I showed, there are two types, if you will, of regulation, the acute and the delayed type. From studies done on the muscarinic receptor system it appears that the modulation of a receptor number by steroids, the delayed one, is strictly an estrogen receptor mediated affair dependent on protein synthesis, since Tom Rainbow and colleagues have shown that the increase in muscarinic receptors upon an injection of estradiol can be blocked by anisomycin, which is a protein synthesis inhibitor. So I think that we are talking about a direct effect of steroids, via steroid receptors, possibly on the synthesis of the receptor molecule, protein, itself, or something very close to it like a precursor.

QUESTION: So you're thinking that gene expression is involved?

BIEGON: Yes.

QUESTION: Is there an effect of serotonin on steroids and their receptors?

BIEGON: Oh greatly so. If I had another two days or 20 minutes I would tell you about the reverse side, but it's true that serotonin affects steroids, ovarian, testicular and the adreno-cortico steroids. I said hardly anything about corticosterone but there are very strong reciprocal relationships between the serotonergic system and the corticosteroids, absolutely.

JOSEE LEYSEN (*Janssen Research Foundation, Beevse, Belgium*): I would like to know how you measured the ketanserin binding on the human platelet.

BIEGON: We have started with the Leysen paper which says there are no 5-HT_2 receptors on human platelets, but there were quite a few on cat platelets. I thought it may be that the human platelets have fewer 5-HT_2 receptors than the cat and that they'd be more sensitive to very strong manipulations of the membrane preparation. So what I did is a relatively gentle preparation of membranes from the platelets and then the binding procedure itself was done with ketanserin using mianserin as a displacer. So I think the only difference between my procedure and yours is a gentler preparation. We didn't do freezing and thawing of the membrane preparation. Just preparing platelets, lysing them, spinning down the membranes and then using them for straight ketanserin binding assay. Ketanserin does tend to bind to filter papers, and since the number of receptors on the platelets in the humans is very low, this could increase the noise in the assay, so what we also do is soak the filters in BSA, 1%, and dry them prior to the filtration. These two steps are actually necessary to see a good signal with ketanserin on a human platelet.

LEYSEN: Did you check whether there was displacement possible by tetrabenezine?

BIEGON: Not with tetrabenezine but with phentolamine. None.

LEYSEN: No? Thank you.

THOMAS R. INSEL (*NIMH, Bethesda, MD*): I wonder if you, or anyone else, has looked at the relationship of gonadal steroids to the coupling of receptors to second messengers, especially for the 5-HT system? And I had a second question as well which has to do with whether the changes that you showed, in the rat anyway, were mostly with estrogen or DHT administration. Have you looked at just physiologic changes across the estrous cycle or across pregnancy and did you see the same kinds of relationships there?

BIEGON: Well the answer to your first question is no we haven't looked into any second messenger systems, although other people may have done. As far as the estrous cycle, I think I started with it. I mean the first slides I showed you were changes in 5-HT_1 receptors across the cycle which actually then was followed by the hormonal dissection of the mechanism. I haven't done it across pregnancy. There are lots and lots of papers showing changes in serotonin turnover during pregnancy and estrous cycle, etc. I haven't seen anybody do receptors during pregnancy, I haven't done it myself.

ISTVAN TÖRK (*University of New South Wales, Kensington, Australia*): If the increase of serotonin receptors in the platelets is due to changed genetic expression then it could only be affecting the new platelets that are being released from the bone marrow because the circulating platelets do not have the apparatus to do that. So, in fact, the increase might be even higher and it's diluted by the existing platelets that you are sampling.

BIEGON: I agree.

TÖRK: So that means that if there is any change in the brain then it could be, and if you're trying to gauge that on the basis of platelets increase, it might be actually really more.

BIEGON: I'm very careful about trying to generalize from the platelet to the brain because we see instances where the serotonergic markers in the platelets react the same way as serotonin receptors in the brain, but we also see instances where they go exactly the opposite way. For example people have shown that imipramine treatment reduces imipramine binding in the brain and in the platelets, but if you're looking at imipramine binding with aging then binding in the human brain post mortem increases with age, but in the platelets it decreases with age. So we see examples of platelet receptors or markers behaving like the ones in the brain, and we see instances where they go exactly the opposite way. So I would be cautious about speculating on the direction of serotonin receptor changes in the brain based on the platelet work. I still think it's an interesting observation that they had this link to changing steroid levels and they may be, perhaps, used as a peripheral marker for emotional lability or depression regardless of what exactly happens in the brain.

My suspicion is that in the brain changes will be occurring which are region specific. Some of them may be like the changes in the platelets and some of them may go a different way.

HERMANN VAN PRAAG (*Session Chairman*): Let me be permitted to ask the last question. You noted a similarity between the effect of estrogens on beta adrenergic receptors, serotonin receptors, and the effect of certain antidepressant devices. Would you very briefly speculate about the possibility of the reverse, a deficiency of estrogen as a possible depressant mechanism.

BIEGON: Willingly. All you have to do is look at the life cycle of the woman and look at the periods in women's life cycle which are associated with depressive episodes and emotional lability. All of these periods are periods of low estrogen secretion. For example premenstrually where the premenstrual syndrome occurs in those women who are sensitive to it, estrogen levels are the lowest throughout the cycle. Post partum, this is another period in women's life cycle which is associated with an increased incidence of depression and sometimes psychosis, and this is an all time low in terms of estrogen levels. Estrogen levels and progesterone levels increase tremendously during pregnancy, reach a peak just before birth, and then plummet down so this is another period where estrogen is low and emotional lability and depression are high. Then of course the last is the menopausal period where, again, estrogen levels go down and this period in a woman's life is associated with increased emotional lability. The occurrence of depression and hospitalization for depression in women are increased premenstrually, post partum and menopausally.

Serotonin Receptor Subtypes and Sexual Behavior[a]

BORIS B. GORZALKA, SCOTT D. MENDELSON, AND NEIL V. WATSON

Department of Psychology
The University of British Columbia
Vancouver, British Columbia, Canada V6T 1Y7

INTRODUCTION

In the early 1960s Meyerson set forth the hypothesis that serotonergic activity inhibits sexual receptivity in the female rat.[1-3] During the latter part of the 1960s, various investigators proposed that serotonergic activity also inhibits sexual behavior in the male rat.[4-7] While much of the data gathered since then appears consistent with serotonergic inhibition of lordosis[8] and male sexual behavior,[9] many inconsistencies have been found in the literature.[10,11] These inconsistent results may reflect, at least in part, differential roles of subtypes of 5-HT receptors in the control of sexual behavior.

The following is a review and re-evaluation of the effects of 5-HT antagonists and agonists on female and male sexual behavior in the rat. Although none of the drugs discussed bind exclusively to 5-HT receptors, in most cases their effects appear to be mediated primarily by serotonergic systems. Therefore, our focus is to consider the effects of these drugs only in terms of probable action at specific subtypes of 5-HT receptors.

FEMALE SEXUAL BEHAVIOR

The primary indicator of sexual receptivity in the female rat is the display of lordosis behavior. Lordosis is the downward flexion of the back and lifting of the rump and tail displayed by a receptive female in response to the mounting and pelvic thrusting of a male. The proportion of lordosis responses elicited by a male is a function of the extent of the female rat's receptivity. In addition to exhibiting lordosis responses, female rats in heat frequently display stereotyped behavior indicative of proceptivity. Proceptive behavior is not dependent on mounting by a male. On the contrary, proceptive behavior may facilitate the elicitation of mounting. Because there are currently insufficient data to permit useful conclusions about serotonergic regulation of proceptivity, the present review of female sexual behavior is restricted to a consideration of receptivity.

Female Sexual Behavior: 5-HT Antagonists

While having affinity for 5-HT_1 receptors, the classical 5-HT antagonists show varying degrees of preference for 5-HT_2 receptors.[12] Central administration of the

[a]This research was supported by a Natural Sciences and Engineering Research Council of Canada grant to B. B. Gorzalka.

classical antagonist methysergide into limbic sites facilitates lordosis in estrogen-primed, ovariectomized females.[13-16] Inhibition has been observed following injection of methysergide into the preoptic area.[17]

Consistent with the results of most studies employing central administration sites, peripheral administration of methysergide also facilitates lordosis in estrogen-primed, ovariectomized females.[14-16,18-23] However, the maximal facilitatory effects of peripherally administered methysergide have been reported to occur 2 to 6 h after treatment.[13,19] Pharmacokinetic data indicate that the maximal antiserotonergic effects of intraperitoneally administered methysergide can occur 30 min to 1 h after its peripheral administration. We have observed that depending on the interval between administration and behavioral testing, methysergide can either facilitate or inhibit lordosis, or have no effect.[22] However, at the times when it is most effective as a 5-HT antagonist, peripherally administered methysergide is inhibitory. It is conceivable that the lordosis-facilitating effect of methysergide after 2 h is due to the action of a metabolite.

Cinanserin is a classical 5-HT antagonist that is somewhat selective for 5-HT_2 receptors.[24] When administered into the medial preoptic or posterior areas of the hypothalamus, cinanserin facilitates lordosis in estrogen-primed females.[13,14] However, when administered peripherally, neither 25 mg/kg[25] nor 5 mg/kg cinanserin facilitated lordosis.[26] Hunter *et al.*[21] observed that 10 mg/kg cinanserin may produce either a strong inhibition or a slight facilitation of lordosis behavior depending on whether the animals were initially receptive or not. However, we have found that when animals are subdivided into "receptive" and "nonreceptive" groups prior to statistical analysis, even saline treatment can produce an apparent inhibition or facilitation of receptivity, most likely as a regression toward the mean.[27]

Metergoline, like methysergide, is somewhat nonselective in its binding to the various 5-HT receptor subtypes.[28] Peripheral administration of 0.05 mg/kg metergoline facilitates lordosis.[29] However, doses greater than 0.5 mg/kg metergoline inhibit lordosis.[21,26,29]

Other classical antagonists that have been evaluated for their effects upon lordosis behavior are methiothepin, mianserin, cyproheptadine, and pizotefin. Methiothepin is somewhat nonselective in its serotonergic binding,[28] whereas mianserin, cyproheptadine and pizotefin bind preferentially to 5-HT_2 receptors.[24] Following systemic administration, methiothepin,[30] cyproheptadine,[26,30] mianserin,[21,26] and pizotefin[30] all inhibit lordosis.

More highly selective 5-HT_2 antagonists have become available recently. The 5-HT_2-selective antagonist pirenperone inhibits lordosis when administered peripherally[11,26] or centrally.[31] Moreover, pirenperone enhances the inhibitory effect of the serotonin precursor 5-hydroxytryptophan (5-HTP) on lordosis.[32] In at least three studies, ketanserin, a 5-HT_2 antagonist related in structure to pirenperone, has been found to inhibit lordosis.[11,21,30] LY53857 and ritanserin are also potent and highly selective 5-HT_2 antagonists. Moreover, unlike most 5-HT_2 antagonists, LY53857 and ritanserin are relatively inactive at α_1-adrenergic and dopaminergic receptors.[33,34] We have found that both LY53857 and ritanserin inhibit lordosis.[31] However, the 5-HT_2 selective antagonist altanserin is ineffective at doses up to 0.2 mg/kg.[26]

Recently, selective 5-HT_3 antagonists have been evaluated for effects on lordosis. Observations in our laboratory reveal that high doses of ICS 205930 facilitate lordosis in estrogen-primed females,[35] while other data have revealed facilitation with more moderate doses of GL 38032.[36] Although we have found little or no effect of MDL 72222 on lordosis behavior,[35] others have reported a facilitatory effect.[36] However, pending the evaluation of other 5-HT_3 selective drugs, this suggestion, and the proposal of an inhibitory function for 5-HT_3 activity,[35,36] must be regarded as speculative.

Female Sexual Behavior: 5-HT Agonists

Like the pharmacological agents that mimic 5-HT, 5-HT itself can either inhibit or facilitate lordosis, depending on which area of the brain receives the exogenous 5-HT and, perhaps, on which subtypes of receptors are activated. For example, the administration of 5-HT into the preoptic and hypothalamic areas inhibits[15,17] whereas infusion into the lateral ventricles facilitates lordosis.[37]

LSD was the first 5-HT agonist to be evaluated for effects on lordosis. *In vitro* binding data indicate that the binding of LSD to the various subtypes of 5-HT receptors is relatively nonselective. It binds with roughly equally high affinity to 5-HT_2, 5-HT_{1A}, and 5-HT_{1C} receptors, and with slightly lower affinity to 5-HT_{1B} sites.[38] Peripheral administration of LSD inhibits lordosis in rats treated with estrogen and progesterone.[39-43] This has been taken as evidence of serotonergic inhibition of lordosis. However, this interpretation ignores the fact that LSD may act as either an agonist or an antagonist, depending, apparently, on the subtype of 5-HT receptor activated. The ability of LSD to reduce the rate of firing of neurons in the dorsal raphe,[44] and to inhibit the release of serotonin from terminals,[45] suggests that LSD may act as an agonist at 5-HT_{1A} and 5-HT_{1B} receptors, respectively. However, LSD has also been found to block the excitatory and, most likely, 5-HT_2–mediated[12] effects of serotonin in the cortex[46] and reticular formation.[47] Thus, inhibition of lordosis by LSD could be due to either stimulation of 5-HT_1 activity or antagonism of 5-HT_2 activity.

LSD in very low doses facilitates lordosis in estrogen-treated females.[25,48] This has been attributed to inhibition of serotonergic activity through action upon autoreceptors in the dorsal raphe. However, the same low doses of LSD (5–10 μg/kg) enhance the ability of serotonin to facilitate the glutamate-induced excitation of motor neurons in the rat facial nucleus.[49] This effect of serotonin appears to be mediated by 5-HT_2 receptors.[50] Thus the facilitatory effect of low doses of LSD could be due to either stimulation of 5-HT_2 activity or presynaptic inhibition of 5-HT_1 activity.

The hallucinogenic phenylalkylamines 2,5-dimethoxy-4-methylamphetamine (DOM), 2,4,5,-trimethoxyamphetamine (TMA), 2,5,-dimethoxy-4-methylphenylethylamine (DOMPE), and mescaline have also been found to facilitate lordosis at low doses, and to inhibit lordosis at higher doses.[51] The facilitatory effects of the phenylalkylamines, like the effects of LSD, have been attributed to reductions in serotonergic activity.[51] However, neither mescaline[52] nor DOM[50] has significant effects on serotonergic autoreceptors in the dorsal raphe. Like LSD, mescaline[49] has been found to enhance, and DOM[50] to mimic the 5-HT_2 receptor-mediated excitatory effect of serotonin on neurons of the facial motor nucleus. Furthermore, the affinities of DOM and TMA for 5-HT_2 sites have been found to be 30-fold higher,[53] and the affinity of mescaline 12-fold higher,[24] than their affinities for 5-HT_1 sites. Thus, the range of doses in which these drugs produced facilitation and inhibition of lordosis[51] appears to reflect the drugs' relative affinities for 5-HT_2 and 5-HT_1 sites. Additional support for a stimulatory role of 5-HT_2 activity comes from evidence that the phenylalkylamine 2,5-dimethoxy-4-iodoamphetamine (DOI), a more selective 5-HT_2 agonist, facilitates lordosis behavior.[36]

Low doses of the hallucinogenic tryptamine derivatives *N,N*,dimethyltryptamine (DMT), 5-methoxydimethyltryptamine (5-MeODMT), and psilocybin also facilitate lordosis whereas high doses are inhibitory.[51] Tryptamine derivatives have been thought to bind with highest affinity to 5-HT_1 receptors.[12] Within this class of receptors, 5-MeODMT and DMT show a marked selectivity for 5-HT_{1A} sites.[54] Thus the facilitatory and inhibitory effects of the *N*-methylated tryptamines could be due to activity at somato-dendritic autoreceptors and postsynaptic 5-HT_{1A} receptors, respectively. However, at doses comparable to those that facilitate lordosis, DMT mimics the neural excitatory effect of serotonin in the facial motor nucleus.[49] Thus, activation of 5-HT_2 receptors provides an alternative explanation for the facilitation of lordosis.

Although the piperazine derivative quipazine binds to both 5-HT_1 and 5-HT_2 receptors,[28] it appears to act more as an agonist at 5-HT_2 receptors.[33,55] Quipazine was first reported to inhibit lordosis in females primed with estrogen and progesterone.[20] However, in later studies comparable doses of quipazine were found to be ineffective in females treated with estrogen and progesterone[11,56] and to facilitate lordosis in females treated with estrogen alone.[21,31]

Quipazine attenuates the lordosis-inhibiting effects of the 5-HT_2 antagonists pirenperone, ketanserin, methysergide, cyproheptadine, and pizotefin.[11,30] In view of the much higher affinity of antagonists for 5-HT_2 receptors, the suggestion that quipazine could compete effectively with these drugs for 5-HT_2 binding sites is surprising. However, recent data indicate that a subpopulation of 5-HT_2 binding sites possesses a conformational state with high affinity for 5-HT_2 agonists.[57] Quipazine, DMT, 5-MeODMT, and phenylalkylamines similar to DOM have been found to bind with high affinity to the agonist binding state of the 5-HT_2 receptor.[57] At moderately high doses, quipazine could be expected to displace 5-HT_2 antagonists from these sites. Finally, it is worth noting that quipazine's action as an antagonist at 5-HT_3 receptor sites[58] may partially contribute to the facilitatory effect of the drug on lordosis.

The piperazine MK 212 facilitates lordosis, and this effect has been attributed to stimulation of 5-HT_2 receptors.[37] However, binding studies show a very low affinity of MK 212 for 5-HT_2 and 5-HT_1 receptors.[38] It would be of interest to consider the affinity of MK 212 for the agonist binding state of the 5-HT_2 receptor or for the 5-HT_3 receptor.

Recently, drugs with very high selectivity for 5-HT_{1A} receptors have become available. The highly selective 5-HT_{1A} agonist 8-OH-DPAT, and the slightly less selective partial agonists buspirone, ipsapirone, and gepirone inhibit lordosis in females primed either with estrogen, or with estrogen and progesterone.[59–61] The β-blocker (−) pindolol, which also antagonizes activity at 5-HT_{1A} and 5-HT_{1B} receptors, reverses the 8-OH-DPAT-induced inhibition of lordosis in females primed with estrogen, but fails to reverse the inhibition in females primed with estrogen and progesterone.[32] Neither the selective β-blocker betaxolol, the 5-HT_2-selective antagonist pirenperone nor the non-selective 5-HT antagonist methiothepin reverses the inhibition induced by 8-OH-DPAT.[32] Extremely small doses of the 5-HT_{1A} agonist lisuride[54] also inhibit lordosis in females primed with estrogen and progesterone.[26] At lower doses, ipsapirone and gepirone facilitate lordosis in estrogen-primed females.[61] Because both drugs reduce the activity of serotonergic neurons in the dorsal raphe,[62,63] the lordosis-facilitating effects of these drugs may be due to activity at somato-dendritic 5-HT_{1A} autoreceptors.

1-(3-Trifluoromethylphenyl)piperazine (TFMPP) and m-chlorophenyl piperazine (MCPP) appear to act as agonists at 5-HT_{1B} receptors.[64,65] Low, but not high, doses of peripherally administered TFMPP or MCPP facilitate lordosis in estrogen-primed females.[66] We have very recently observed that 0.03 mg/kg of the putative 5-HT_{1B} agonist CGS 12066B[67] facilitates lordosis in estrogen-primed females.[31] Together, these data suggest that 5-HT_{1B} autoreceptors mediate lordosis-facilitating effects of 5-HT. In apparent contradiction to this possibility, the 5-HT_{1B} agonist RU 24969 has been reported to inhibit lordosis.[21] We have observed either facilitation or inhibition of lordosis by RU 24969 in estrogen-primed females depending on the dose of drug and the presence or absence of progesterone.[31] Although RU 24969 is often characterized as a selective 5-HT_{1B} agonist, it is worth noting that the drug binds with nearly equal high affinity to 5-HT_{1A} receptors.[68]

MALE SEXUAL BEHAVIOR

In the male rat, copulatory behavior typically consists of repeated mounts with pelvic thrusting, many of which result in penile intromission. The copulatory bout culminates in an ejaculation, followed by a brief refractory period.

The behavioral events and their temporal patterning permit the measurement of a number of parameters. These include: the total number of mounts (M), intromissions (I), and ejaculations (E); the latency from introduction of a receptive female to the first mount (ML), or intromission (IL); the latency from the first intromission to ejaculation (EL); and the post-ejaculatory interval (PEI) between ejaculation and the first intromission of the next copulatory bout. These measures can be used to derive mount frequencies (MF), intromission frequencies (IF), inter-copulatory intervals (ICI) and copulatory efficiency (CE = I/I + M). In addition, drug treatment effects on male sexual behavior are sometimes quantified in terms of percentages of animals mounting (%M), intromitting (%I) or ejaculating (%E) Inhibition or facilitation of male sexual behavior may thus refer to changes in any of these parameters. Serotonergic mediation of penile reflexes, *ex copula*, has also been examined.[69,70] However, these reflexes appear to rely in large part on spinal mechanisms, and it is not clear to what extent they are related to either *in copula* behavior or central serotonergic mechanisms. Although intrinsically interesting, these studies are excluded from the present review.

Male Sexual Behavior: 5-HT Antagonists

Reports on the effects of the non-selective 5-HT antagonist methysergide suggest that, in general, the drug is either ineffective or stimulatory. Methysergide has been reported to produce a slight facilitation of homosexual mounting behavior in male rats at a dose of 1 mg/kg,[71] and to reverse the quipazine-induced inhibition of PCPA-induced homosexual mounting at 5 mg/kg.[72] However, others have reported that 10 mg/kg methysergide did not facilitate homosexual mounting.[73] In heterosexual tests, chronic administration of 2.5 mg/kg methysergide has been reported to decrease IL, EL and PEI and increase %E.[74] In other heterosexual tests, acute administration of large doses (10–20 mg/kg) was found to produce reductions in EL and I.[75] However, in another study 1.0 mg/kg methysergide was found to increase M and I, but to have no effect on ML, IL or EL.[76] In a fourth report, administration of 4–18 mg/kg methysergide had no effect on male copulatory parameters, other than a slight increase in E at the lower doses.[77] In contrast to the data from female rats, in these studies there was no obvious relationship between the effects reported and the length of the interval between drug administration and testing.

Systemic administration of metergoline has also yielded mixed effects on male sexual behavior. Earlier reports suggested either no effect or a slight facilitation,[78,79] at doses ranging from 1 to 10 mg/kg. However, a potent *inhibitory* influence of metergoline was recently reported, expressed as decreases in %M, %I and %E. Curiously, these effects were observed only when metergoline (3 mg/kg) was initially dissolved in a few drops of glacial acetic acid.[80] When prepared with other solvents, metergoline had no effect.

Methiothepin, at doses of 0.125 to 1 mg/kg, has been found to have little effect on male sexual behavior, although at higher doses the percentage of animals achieving ejaculation is reduced.[79] In contrast, cyproheptadine at a higher dose of 2.5 mg/kg has been found to facilitate male sexual behavior as demonstrated by an increase in %E, and decreases in EL and IL.[74]

Pirenperone[10,79] (.075 – 1.0 mg/kg), and ketanserin[10] (5 mg/kg) have both been found to inhibit male sexual behavior, although, paradoxically, pirenperone also reversed the inhibition induced by 5-HTP.[79] These data suggested that 5-HT_2 activity might serve a stimulatory function with respect to male sexual behavior. More recently, however, Foreman *et al.*[81–83] have reported facilitatory effects of the 5-HT_2-specific antagonists LY 53857, LY 237733 and LY 281067. It is noteworthy that these drugs appear to lack the alpha-adrenergic activity seen with the other antagonists. Indeed, administration of the α-adrenergic antagonist prazosin has been reported to block the facilitation seen with LY 53857.[83]

Lastly, there is preliminary evidence that 5-HT_3 receptors may mediate some of the inhibitory effects of 5-HT on male sexual behavior. Although ICS 205930 (0.05–5 mg/kg) alone has no effect on sexual behavior in male rats, high doses appear to attenuate certain sexually-inhibitory effects of morphine.[84]

Male Sexual Behavior: 5-HT Agonists

As was the case with female receptivity, LSD was the first 5-HT agonist tested for effects on male sexual behavior. Although there has been some suggestion of a modest facilitation (increased IF, decreased IL, PEI and inter-intromission interval) at low doses of LSD (i.e. 10 and 20 μg/kg),[85] others have reported no effect,[86] or a decrease in copulatory activity.[87,88] Higher doses of LSD (i.e. 0.5 and 1.0 mg/kg) have been reported to inhibit certain parameters of male sexual behavior[85] and may reduce the number of animals achieving ejaculation.[86]

Decreases in %E have been reported for 3 and 6 mg/kg quipazine.[10,86] Inhibition of homosexual mounting has also been reported for comparable doses.[72,89] However, when administered in combination with the 5-HT_2 antagonist pirenperone, quipazine (3 mg/kg) appears to be facilitatory in that it attenuates pirenperone-induced inhibition.[10] These seemingly paradoxical effects of quipazine may simply reflect the drug's relative nonselectivity for 5-HT receptor subtypes. While thought to act primarily as an agonist at 5-HT_2 receptors, quipazine binds with nearly equal high affinity to 5-HT_{1B} receptors, and with slightly lesser affinity to the 5-HT_{1A} site.[28] Thus, the apparent sexually-facilitatory effects of quipazine may be mediated by stimulation of 5-HT_2 receptors, while its inhibitory effects are controlled by another subtype of receptor. This possibility is consistent with the report that centrally active doses of the 5-HT_2 agonist DOM (0.25–1.0 mg/kg) have no inhibitory effects on male sexual behavior.[86] It must be noted, however, that the related compound DOI potently inhibits sexual behavior in male rats.[82]

In marked contrast to the female pattern, activity at 5-HT_{1A} receptors appears to facilitate male copulatory behavior. Doses of the selective 5-HT_{1A} agonist 8-OH-DPAT ranging from 0.025 to 4 mg/kg are uniformly facilitatory, as expressed by decreased I and EL[60,79,86,90–93] and decreased M.[79,92–94] Moreover, 8-OH-DPAT also appears to induce copulation, including the ejaculatory pattern, in castrates and in animals that have undergone penile deafferentation.[92] Facilitation has also been reported to occur after various doses of the structurally related 5-HT agonist 8-methoxy-2-(di-*n*-propylamino)tetralin (8-OMe-DPAT).[86]

Building on previous evidence of a generally inhibitory influence of 5-HT on male sexual behavior, Ahlenius and Larsson[95] have suggested that the facilitatory effects of 8-OH-DPAT are the result of pre-synaptic inhibition of serotonergic activity, or the drug's action as a partial agonist at some 5-HT receptors. The finding that the 8-OH-DPAT-induced reduction of I prior to ejaculation persisted after administration of 5-HTP,[91] suggests presynaptic mediation of the facilitatory effect. However, 8-OH-

DPAT facilitates male sexual behavior at doses that generate reciprocal forepaw-treading and flat body posture, thought to be indicative of post-synaptic activity.[96] The 8-OH-DPAT-induced facilitation of male sexual behavior is also at least partially attenuated by (−)-pindolol and (−)-alprenolol, which exhibit 5-HT_{1A} and β-adrenergic antagonist properties.[95] Finally, we note that 8-OH-DPAT facilitates male sexual behavior at doses that are as much as 30 times greater than doses that virtually eliminate lordosis behavior in female rats.[60] Ostensibly, the lordosis-inhibiting effects of 8-OH-DPAT are mediated postsynaptically. Taken together, the data lead us to suggest that the facilitatory effects of 8-OH-DPAT on male rat copulatory behavior are attributable to activation of certain post-synaptic 5-HT_{1A} receptors.

RDS-127,[80,97–99] which potently displaces [^{3}H]DPAT from 5-HT_{1A} recognition sites,[99,100] and the potent though less selective 5-HT_{1A} agonist lisuride,[90,91,101–104] have also been found to facilitate male sexual behavior (primarily by reducing I and EL). In addition, both drugs have been reported to restore the full copulatory pattern in castrates.[97,101] At relatively high doses (10 mg/kg), the 5-HT_{1A} partial agonist ipsapirone also facilitates sexual behavior in the male rat (decreases in M, I, and EL).[93] Further support for a stimulatory role of 5-HT_{1A} activity comes from the observation that high doses of the weak partial agonist BMY 7378 facilitate male sexual behavior whereas low doses are apparently inhibitory.[31]

Intraperitoneal administration of 0.25−1.0 mg/kg 5-MeODMT has no effect on sexual behavior in castrates maintained on estradiol and/or dihydrotestosterone.[105] However, in intact males, 1.0 mg/kg 5-MeODMT facilitates sexual behavior, in terms of decreased I prior to ejaculation.[104] These effects of 5-MeODMT are consistent with a facilitatory role of 5-HT_{1A} receptors in male sexual behavior.[100] However, the effects of 5-MeODMT might be mediated at least partially by 5-HT_2 receptors.[57]

The 5-HT_{1B} agonists TFMPP and MCPP inhibit sexual behavior in male rats.[66] This contrasts with the facilitation of lordosis obtained with identical doses in female rats. The mixed 5-HT_{1A}/5-HT_{1B} agonist RU 24969 also inhibits male sexual behavior.[31]

CONCLUSIONS

From a review of the effects of 5-HT antagonists and agonists it becomes apparent that serotonin can either inhibit or facilitate the expression of sexual behavior in the rat, depending on which subtypes of 5-HT receptors are activated. In addition, the effects on sexual behavior of activation of certain subtypes of 5-HT receptors differ between males and females.

In the female rat, the lordosis-inhibiting effects of serotonin appear to be mediated primarily by postsynaptic 5-HT_{1A} receptors. On the basis of preliminary evidence, we further suggest that 5-HT_3 receptors may mediate some of the lordosis-inhibiting effects of serotonin.

The lordosis-facilitating effects of serotonin appear to be mediated primarily by 5-HT_2 receptors. Moreover, it is tempting to suggest that the subpopulation of 5-HT_2 receptors possessing a high affinity agonist binding state may play a particularly important role in mediating these effects. Of course, we must note that drugs selective for 5-HT_2 receptors tend also to bind with high affinity to 5-HT_{1C} receptors. We therefore cannot eliminate the possibility that activity at certain populations of 5-HT_{1C} receptors enhances lordosis.

Evidence suggests that activity at somato-dendritic 5-HT_{1A} autoreceptors may also facilitate lordosis behavior. Ostensibly, this would be due to reductions in the activity of central lordosis-inhibiting serotonergic pathways. Activity at 5-HT_{1B} receptors similarly appears to facilitate lordosis, although it is unclear whether the effect is mediated pre- or postsynaptically.[106]

In contrast to its effects on lordosis behavior, activation of 5-HT_{1A} receptors appears to facilitate sexual behavior in the male. Although we suggest postsynaptic mediation of this effect, we recognize the possibility that presynaptic activity might contribute to the facilitatory effects of 5-HT_{1A} agonists, through reduction of sexually inhibiting serotonergic activity.

Unlike the effects seen in females, activation of 5-HT_{1B} receptors appears to strongly inhibit male sexual behavior. Although highly speculative at present, preliminary evidence suggests that 5-HT_3 receptor activity may also contribute to the inhibitory aspects of serotonergic activity.

The effects of activity at 5-HT_2 receptors on the expression of sexual behavior in the male rat are somewhat uncertain at this time. Inconsistencies in the literature on the effects of 5-HT_2 agonists and antagonists remain to be resolved, but the most recent data with the more selective 5-HT_2 antagonists suggest an inhibitory role for 5-HT_2 activity.

REFERENCES

1. MEYERSON, B. J. 1964. Arch. Int. Pharmacodyn. **150:** 4–33.
2. MEYERSON, B. J. 1964. Psychopharmacologia. **6:** 210–218.
3. MEYERSON, B. J. 1964. Acta Physiol. Scand. **63**(Suppl. 241): 3–32.
4. SHEARD, M. H. 1969. Brain Res. **15:** 524–528.
5. SHILLITO, E. E. 1969. Br. J. Pharmacol. **36:** 193–194.
6. TAGLIAMONTE, A., P. TAGLIAMONTE, G. L. GESSA & B. B. BRODIE. 1969. Science **166:** 1433–1435.
7. MALMNAS, C. O. & B. J. MEYERSON. 1971. Nature **232:** 398–399.
8. MEYERSON, B. J., C. O. MALMNAS & B. J. EVERITT. 1985. *In* Handbook of Behavioral Neurobiology. N. Adler, D. Pfaff and R. W. Goy, Eds. Vol. **7:** 495–536. New York: Plenum Press.
9. BITRAN, D. & E. M. HULL. 1987. Neurosci. Biobehav. Rev. **11:** 365–389.
10. MENDELSON, S. D. & B. B. GORZALKA. 1985. Pharmacol. Biochem. Behav. **22:** 565–571.
11. MENDELSON, S. D. & B. B. GORZALKA. 1985. Pharmacol. Biochem. Behav. **22:** 1025–1033.
12. PEROUTKA, S. J., R. M. LEBOVITZ & S. H. SNYDER. 1981. Science **212:** 827–829.
13. ZEMLAN, F. P., I. L. WARD, W. R. CROWLEY & D. L. MARGULES. 1973. Science **179:** 1010–1011.
14. WARD, I. L., W. R. CROWLEY, F. P. ZEMLAN & D. L. MARGULES. 1975. J. Comp. Physiol. Psych. **88:** 53–61.
15. FOREMAN, M. M. & R. L. MOSS. 1978. Horm. Behav. **10:** 97–106.
16. FRANCK, J. E. & I. L. WARD. 1981. Neuroendocrinology **32:** 50–56.
17. CLEMENS, L. G. 1978. *In* Sex and Behavior. T. E. McGill, D. A. Dewsbury and B. D. Sachs, Eds. New York: Plenum Press, pp. 243–266.
18. HENRIK, E. & A. A. GERALL. 1976. J. Comp. Physiol. Psych. **90:** 590–600.
19. DAVIS, G. A. & R. L. KOHL. 1978. Pharmacol. Biochem. Behav. **9:** 487–491.
20. RODRIGUEZ-SIERRA, J. F. & G. A. DAVIS. 1979. Neuropharmacology **18:** 335–339.
21. HUNTER, A. J., D. R. HOLE & C. A. WILSON. 1985. Pharmacol. Biochem. Behav. **22:** 5–13.
22. MENDELSON, S. D. & B. B. GORZALKA. 1986. Neuropharmacology **25:** 749–755.
23. ULIBARRI, C. & P. YAHR. 1987. Physiol. Behav. **39:** 767–774.
24. LEYSEN, J. E. & J. P. TOLLENAERE. 1982. Ann. Rep. Med. Chem. **17:** 1–10.
25. EVERITT, B. J., K. FUXE, T. HOKFELT & G. JONSSON. 1975. J. Comp. Physiol. Psychol. **89:** 556–572.
26. SIETNIEKS, A. 1985. J. Neural Trans. **61:** 81–94.
27. RAIBLE, L. H. & B. B. GORZALKA. 1986. Peptides **7:** 581–586.
28. HOYER, D., G. ENGEL & H. O. KALKMAN. 1985. Eur. J. Pharmacol. **118:** 13–23.

29. FUXE, K., B. J. EVERITT, L. AGNATI, B. FREDHOLM & G. JONSSON. 1976. *In* Schizophrenia Today. D. Kemali, G. Bartholini and D. Richter, Eds. Oxford: Plenum Press, pp. 135–157.
30. MENDELSON, S. D. & B. B. GORZALKA. 1986. Life Sci. **38:** 33–39.
31. MENDELSON, S. D. & B. B. GORZALKA. Unpublished data.
32. AHLENIUS, S., K. LARSSON & A. FERNANDEZ-GUASTI. 1989. Psychopharmacology **98:** 440–444.
33. COHEN, M. L., R. W. FULLER & K. D. KURZ. 1983. J. Pharmacol. Exp. Ther. **227:** 327–332.
34. JANSSEN, P. A. J. 1985. J. Cardiovasc. Pharmacol. **7:** S2–S11.
35. MENDELSON, S. D. & B. B. GORZALKA. 1989. *In* Behavioural Pharmacology of 5-HT. P. Bevan, A. Cools and T. Archer, Eds. Hillsdale, NJ: Lawrence Erlbaum, pp. 55–71.
36. JAMES, M. D., S. M. LANE, D. R. HOLE & C. A. WILSON. 1989. *In* Behavioural Pharmacology of 5-HT. P. Bevan, A. Cools and T. Archer, Eds. Hillsdale, NJ: Lawrence Erlbaum, pp. 73–77.
37. WILSON, C. A. & A. J. HUNTER. 1985. Brain Res. **333:** 223–229.
38. ENGEL, G., M. GOTHERT, D. HOYER, E. SCHLICKER & K. HILLENBRAND. 1986. Nauyn-Schmiedebergs Arch. Pharmacol. **332:** 1–7.
39. ELIASSON, M., A. MICHANEK & B. J. MEYERSON. 1972. Acta Pharmacol. Toxicol. **31**(Suppl. 1): 22.
40. MEYERSON, B. J., H. CARRER & M. ELIASSON. 1974. *In* Advances in Biochemical Psychopharmacology. E. Costa, G. L. Gessa and M. Sandler, Eds. Vol. **II**: 229–242. New York: Raven Press.
41. ELIASSON, M. & B. J. MEYERSON. 1976. Neuropharmacology **16:** 37–44.
42. ELIASSON, M. & B. J. MEYERSON. 1977. Psychopharmacology **49:** 301–306.
43. SIETNIEKS, A. & B. J. MEYERSON. 1980. Eur. J. Pharmacol. **63:** 57–64.
44. HAIGLER, H. J. & G. K. AGHAJANIAN. 1974. J. Pharmacol. Exp. Ther. **188:** 688–699.
45. MIDDLEMISS, D. N. 1982. *In* Presynaptic Receptors: Mechanism and Function. J. De Belleroche, Ed. Chichester: Ellis Horwood, pp. 46–74.
46. ROBERTS, M. H. T. & D. W. STRAUGHAN. 1967. J. Physiol. **193:** 269–294.
47. BOAKES, R. J., P. B. BRADLEY, I. BRIGGS & A. DRAY. 1970. Br. J. Pharmacol. **40:** 202–218.
48. SIETNIEKS, A. & B. J. MEYERSON. 1983. Neuroendocrinology **36:** 462–467.
49. MCCALL, R. B. & G. K. AGHAJANIAN. 1980. Life Sci. **26:** 1149–1156.
50. PENINGTON, J. J. & R. J. REIFFENSTEIN. 1986. Can. J. Physiol. Pharmacol. **64:** 1302–1309.
51. EVERITT, B. J. & K. FUXE. 1977. Neurosci. Lett. **4:** 215–220.
52. HAIGLER, H. J. & G. K. AGHAJANIAN. 1973. Eur. J. Pharmacol. **21:** 53–60.
53. SHANNON, M., G. BATTAGLIA, R. A. GLENNON & M. TITELER. 1984. Eur. J. Pharmacol. **102:** 23–29.
54. PEROUTKA, S. J. 1986. Brain Res. **344:** 167–171.
55. GLENNON, R. A., R. YOUNG & J. A. ROSENCRANS. 1983. Eur. J. Pharmacol. **91:** 189–196.
56. ARENDASH, G. W. & R. A. GORSKI. 1983. Pharmacol. Biochem. Behav. **19:** 351–357.
57. LYON, R. A., K. H. DAVIS & M. TITELER. 1987. Mol. Pharmacol. **31:** 194–199.
58. PEROUTKA, S. J. & A. HAMIK. 1988. Eur. J. Pharmacol. **148:** 297–299.
59. AHLENIUS, S., A. FERNANDEZ-GUASTI, S. HJORTH & K. LARSSON. 1986. Eur. J. Pharmacol. **124:** 361–363.
60. MENDELSON, S. D. & B. B. GORZALKA. 1986. Physiol. Behav. **37:** 345–351.
61. MENDELSON, S. D. & B. B. GORZALKA. 1986. Eur. J. Pharmacol. **132:** 323–326.
62. DOURISH, C. T., P. H. HUTSON & G. CURZON. 1986. Trends Pharmacol. Sci. **7:** 212–214.
63. EISON, A. S., M. S. EISON, M. STANLEY & L. A. RIBLET. 1986. Pharmacol. Biochem. Behav. **24:** 701–708.
64. CUNNINGHAM, K. A. & J. B. APPEL. 1986. J. Pharmacol. Exp. Ther. **237:** 369–377.
65. GLENNON, R. A., J. D. MCKENNY & R. YOUNG. 1984. Life Sci. **35:** 1475–1480.
66. MENDELSON, S. D. & B. B. GORZALKA. Neuropharmacology. In press.
67. NEALE, R. F., S. L. FALLON, W. C. BOYAR, J. W. F. WASLEY, L. L. MARTIN, G. A. STONE, B. S. GLAESER, C. M. SINTON & M. WILLIAMS. 1987. Eur. J. Pharmacol. **136:** 1–9.

68. TRICKLEBANK, M. D., D. N. MIDDLEMISS & J. NEILL. 1986. Neuropharmacology **25:** 877–886.
69. RENYI, L. 1985. Neuropharmacology **24:** 697–704.
70. BERENDSEN, H. H. G. & C. L. E. BROEKKAMP. 1987. Eur. J. Pharmacol. **135:** 279–287.
71. BENKERT, O. & T. EVERSMANN. 1972. Experientia **28:** 532–533.
72. GRABOWSKA, M. 1975. *In* Sexual Behavior: Pharmacology and Biochemistry. M. Sandler and G. Gessa, Eds. New York: Raven, pp. 59–62.
73. BAGGIO, G., & F. FERRARI. 1983. Psychopharmacology **80:** 38–42.
74. ABRAHAM, E. M., P. M. VIESCA, A. V. PLAZA & B. MARIN. 1988. Behav. Brain Res. **30:** 251–258.
75. LARSSON, K., & S. AHLENIUS. 1986 *In* Psychopharmacology of Sexual Disorders. M. Segal, Ed. London: John Libbey, pp. 15–32.
76. LIEBLICH, I., P. SHAVIV & E. COHEN. 1985. Horm. Behav. **10:** 278–291.
77. SOULAIRAC, M.-L. & A. SOULAIRAC. 1975 *In* Sexual Behavior: Pharmacology and Biochemistry. M. Sandler and G. Gessa, Eds. New York: Raven Press, pp. 99–116.
78. AHLENIUS, S., K. LARSSON & L. SVENSSON. 1980. Psychopharmacology **68:** 217–220.
79. AHLENIUS, S. & K. LARSSON. 1984. Eur. J. Pharmacol. **99:** 279–286.
80. CLARK, J. T. & E. R. SMITH. 1986. Physiol. Behav. **37:** 47–52.
81. FOREMAN, M. M., R. L. LOVE, J. L. HALL, G. P. MARZONI & W. L. GARBRECHT. 1987. Soc. Neurosci. Abstr. **13**(1): 407.
82. FOREMAN, M. M., R. L. LOVE & J. L. HALL. 1988. Soc. Neurosci. Abstr. **14**(1): 374.
83. FOREMAN, M. M., J. L. HALL, & R. L. LOVE. 1989. Life Sci. **45:** 1263–1270.
84. MENDELSON, S. D. & J. G. PFAUS. Unpublished data.
85. BIGNAMI, G. 1966. Psychopharmacologia (Berl.) **10:** 44–58.
86. AHLENIUS, S., K. LARSSON, L. SVENSSON, S. HJORTH, A. CARLSSON, P. LINDBERG, H. WIKSTROM, D. SANCHEZ, L.-E. ARVIDSSON, U. HACKSELL & J. L. G. NILSSON. 1981. Pharmacol. Biochem. Behav. **15:** 785–792.
87. MALMNAS, C.-O. 1973. Acta Physiol. Scand. **395**(Suppl.): 1–128.
88. MALMNAS, C.-O. 1974. *In* Advances in Biochemical Psychopharmacology. E. Costa, G. L. Gessa and M. Sandler, Eds. Vol. **II:** 243–248. New York: Raven Press.
89. DA PRADA, M., E. P. BONETTI & H. H. KELLER. 1977. Neurosci. Lett. **6:** 349–353.
90. AHLENIUS, S. & K. LARSSON. 1984. Psychopharmacology **83:** 330–334.
91. AHLENIUS, S., & K. LARSSON. 1985. Eur. J. Pharmacol. **110:** 379–381.
92. DAHLOF, L.-G., S. AHLENIUS & K. LARSSON. 1988. Physiol. Behav. **43:** 841–843.
93. GLASER, T., W. U. DOMPERT, T. SCHUURMAN, D. G. SPENCER JR. & J. TRABER. 1987. *In* Brain 5-HT_{1A} Receptors: Behavioural and Neurochemical Pharmacology. C. T. Dourish, S. Ahlenius and P. H. Hutson, Eds. Chichester: Ellis Harwood, pp. 106–119.
94. MORALI, G. & K. LARSSON. 1984. Pharmacol. Biochem. Behav. **20:** 185–187.
95. AHLENIUS, S. & K. LARSSON. 1987. *In* Brain 5-HT_{1A} Receptors: Behavioural and Neurochemical Pharmacology. C. T. Dourish, S. Ahlenius and P. H. Hutson, Eds. Chichester: Ellis Harwood, pp. 185–198.
96. TRICKLEBANK, M. D. 1984. Eur. J. Pharmacol. **106:** 271–282.
97. CLARK, J. T., E. R. SMITH, M. L. STEFANICK, S. P. ARNERIC, J. P. LONG & J. DAVIDSON. 1982. Physiol. Behav. **29:** 1–6.
98. CLARK, J. T., M. L. STEFANICK, E. R. SMITH & J. M. DAVIDSON. 1983. Pharmacol. Biochem. Behav. **19:** 781–786.
99. CLARK, J. T., S. J. PEROUTKA, R. D. CIARENELLO, E. R. SMITH & J. M. DAVIDSON. 1985. Behav. Brain Res. **18:** 251–260.
100. KWONG, L. L., E. L. SMITH, J. M. DAVIDSON & S. J. PEROUTKA. 1986. Behav. Neurosci. **100:** 664–668.
101. AHLENIUS, S., K. LARSSON & L. SVENSSON. 1980. Eur. J. Pharmacol. **64:** 47–51.
102. HANSEN, S. 1982. Neurosci. Lett. **33:** 329–332.
103. HLINAK, Z., J. MADLAFOUSEK & I. KREJCI. 1983. Psychopharmacology **79:** 231–235.
104. FERNANDEZ-GUASTI, A., S. HANSEN, T. ARCHER & G. JONSSON. 1986. Brain Res. **377:** 112–118.
105. BAUM, M. J. & M. S. STARR. 1980. Pharmacol. Biochem. Behav. **13:** 57–67.
106. KENNETT, G. A., C. T. DOURISH & G. CURZON. 1987. Eur. J. Pharmacol. **141:** 429–435.

DISCUSSION OF THE PAPER

MANFRED GÖTHERT (*University of Bonn*): A brief comment and a question. The comment concerns quipazine and MK212. Both compounds are very dirty drugs. They not only are agonists at the 5-HT_2 receptor, but they also act as antagonists at the 5-HT_{1B} receptor. The group of Bauman and Vitemeyer, and my own group, have shown this. I think the partial reversal you get with quipazine may be related to the antagonism at the 5-HT_{1B} receptor.

And my question is about one available clinical compound which lowers blood pressure, and which exerts its action mainly by inactivation of 5-HT_{1A} receptors. And this is *urapadil*. I'm fully aware of the fact that this compound blocks α_1-adrenoceptors, but I think it becomes more and more clear that it's mainly acting via 5-HT_{1A} receptors. My question now is do you know whether some alterations in sexual behavior have been observed with this compound?

GORZALKA: In response to your first comment we, of course, realize about the numerous effects of quipazine and MK-212. Many of those effects are mentioned in our paper. No, I don't know whether that drug has been tested for effects on sexual behavior. I suspect is hasn't. You might be interested to know that buspirone, a partial 5-HT_{1A} agonist, has now been patented for treatment of male sexual dysfunction.

QUESTION: Dr. Gorzalka, did you try to antagonize the MK212 effect in the rat with any drugs, and is there any evidence about the role of receptor subtypes in the sexuality of men and women.

GORZALKA: No, we've not attempted yet to antagonize the effect of MK 212 in the rat. The human evidence on receptor subtypes has not yet been published. The earlier published evidence in humans did not employ drugs selective for receptor subtypes.

H. W. M. STEINBUSCH (*Free University, Amsterdam*): Did you ever check whether your male rat was a dominant rat or a subordinate rat in a colony cage?

GORZALKA: Our impression is that it makes no difference at all whether the male is dominant or submissive to other male rats in terms of his interactions with the female rat. Dominance may, however, be relevant to the sexual behavior of the hamster. The female hamster is larger than and more powerful than the male hamster. The only other species of mammal where the female is clearly more aggressive than the male is the brown-spotted hyena. To the best of my knowledge, serotonergic drugs have not been administered to this species.

FRANK D. YOCCA (*Bristol-Myers, Wallingford, CT*): You mentioned your work with BMY 7378 and you said that it was an antagonist. We find it to be a weak partial agonist. Do you see any agonist activity at all with this compound?

GORZALKA: In terms of effects on ejaculation latencies and number of intromissions, BMY 7378 seemed to act as an agonist at high doses. We have also found it to be an antagonist in some of our studies. Our evidence is consistent with the drug being a weak partial agonist.

J. BLUNDELL (*University of Leeds, England*): I'd like to ask about the interpretation of these changes in the copulatory sequence where you claim that 5-HT_{1A} agonists facilitate in the male. Now this involves a decrease in ejaculatory latency and a decrease in the number of intromissions. It appears that the drugs quicken the sequence but that may be biologically inappropriate because in terms of survival of offspring the full copulatory sequence is required and has a clear biological value. So the application of these drugs to facilitate sexual behavior seems to me to be debatable. What you call facilitatory involves a quickening or shortening of the sequence which may be biologically inappropriate.

GORZALKA: I agree with your point which is that an increase in sexual potency could lead to a decrease in fertility. Furthermore, I'd like to point out that the standard definition of facilitation of sexual behavior in the male rat is precisely a reduction in the ejaculation latency and a reduction in the number of intromissions. This does indeed seem counter-intuitive relative to human sexuality. I think most men would view a reduction in their ejaculation latency as not being especially desirable. Their partners might feel the same way. However I also want to point out that when you do get this reduction in ejaculation latency in the rat, if you were to extend your test to an hour or longer, you would find a dramatic increase in the number of ejaculations per hour. Some humans might view that as desirable.

BLUNDELL: It seems to me very dangerous to speculate from these rat data to human copulatory sequences.

GORZALKA: I agree completely and that is why we haven't speculated from rats to humans in our paper.

ISTVAN TÖRK (*University of New South Wales, Kensington, Australia*): Propranolol, which is a β-blocker used for treatment of hypertension, was shown early in the conference to also have an affinity to 5-HT_{1A} receptors. It is also known that this drug causes a reduction of libido and sexual activity in males. Do you think that this side effect of propranolol is actually caused by its affinity to the 5-HT_{1A} receptor?

GORZALKA: We published a paper in which we looked at the effect of a variety of different β-adrenergic drugs on sexual activity in the rat and we believe that the effect of propranolol is due neither to β-adrenergic nor to serotonergic activity (Mendelson, S. D. & B. B. Gorzalka. 1988. Pharmacol. Biochem. Behav. **29:** 717–723).

Serotonin and Sleep-Wakefulness

A. WAUQUIER[a] AND C. DUGOVIC

Department of Neuropsychopharmacology
Janssen Research Foundation
B-2340 Beerse, Belgium

INTRODUCTION

Sleep is the result of an integrated action of the central nervous system. During sleep (S) and wakefulness (W) there are many different and even apparently opposite changes in the organism. Some neurons, as those from cortical regions, display a very high activity during W and paradoxical sleep (PS), others such as in the basal forebrain are very active during slow wave sleep (SWS); those located in the brain stem, are very active during W, but those from the substantia nigra remain active throughout the S-W cycle. Such diversity of local activity and the conjunctive differential activity occurring simultaneously during the different stages of S-W, instigated Koella[1] to discuss S-W patterns in terms of "local vigilances," by which he understands that these are the consequence of reactivity of local neuronal networks, which are responsible for the output of these systems. Because sleep is such a complex behavior, it is not astonishing that different neurotransmitters are involved in the organization and regulation of sleep.[2] Possibly they are involved in different aspects (or local vigilances) of this complex behavior. 5-Hydroxytryptamine (5-HT) has been attracting much attention in sleep research since it was considered *the* transmitter involved in the induction and maintenance of sleep.

The emphasis in the present paper is on reviewing pharmacological experiments on 5-HT and sleep, and providing an interpretation of the classical and new concepts on serotonergic mechanisms in the control of the S-W cycle.

SPONTANEOUS VARIATIONS OF 5-HT

In hamsters, Matussek and Patschke[3] found a high level of serotonin during sleep and a low level during motor activity. This has been repeatedly shown later.[4] In the encéphale isolé cat, Puizillout *et al.*[5] found a decreased release of 5-HT in striatum and associated cortical areas when raphe nuclei activity diminished. The data of Hery *et al.*[6] and Cespuglio *et al.*[7] suggested that in rats 5-HT release was greater during the dark period, which is the active phase in nocturnal animals. In rats, Tsuchiya *et al.*[8] found an increase in 5-HT levels during sleep following sleep deprivation. On the basis of a previous finding[9] that a 24-h sleep deprivation resulted in an enhancement of SWS and PS, Borbély *et al.*,[10] using a particular sleep deprivation technique, found an increase of 5-HT, reflecting a sleep-related reduction of transmitter release. Furthermore, there was a similar time course of the increase of 5-HT and the enhancement of SWS and PS.

[a] *Address for correspondence:* Prof. Dr. A. Wauquier, Medical College of Ohio, Department of Neurology, 3000 Arlington Avenue, Toledo, OH 43699-0008.

Thus, there are some suggestions that there is a relation between the endogenous variations of 5-HT and sleep. However, the sleep deprivation studies still suffer from the fact that they do not necessarily imply a relation between normal variations in 5-HT concentrations and S-W patterns, in part because sleep deprivation also affects mechanisms not related to sleep.

PHARMACOLOGICAL INCREASE OF 5-HT LEVELS

Serotonin Application

In spite of the fact that 5-HT cannot cross the blood-brain barrier, several authors used a systemic route of administration. An i.v. injection of 5–30 μg/kg of 5-HT[11] and an intracarotid injection or an injection in the vertebral artery of 0.2–5 μg/kg of 5-HT in cats induced a high voltage low frequency (HVLF) EEG which could be preceded by a period of low voltage high frequency (LVHF) activity.[12] These effects were due partly to an action on the carotid sinus[13] and partly to an action at the level of the area postrema, a structure localized outside the blood brain barrier.[12] Doses of 5×10^{-9} M of 5-HT into the 4th ventricle of immobilized cats produced EEG signs of sleep, whereas 5.5×10^{-9} M of 5-HT applied in the 4th ventricle of awake cats produced sleep (mostly SWS) for 8 hours, preceded by a short period (30 min) of arousal.[1,12] Koella[1] suggested that the effects obtained after intravasal 5-HT application are possibly due to an effect on receptors outside the blood brain barrier.

L-Tryptophan

Exogenously administered tryptophan raises brain 5-HT levels, the effect being dependent on circulating plasma levels.[13] Administration of L-tryptophan was shown to produce sleep in human volunteers and insomniac patients.[14] In contrast, no major sleep stage variation was found in rats at doses up to 600 mg/kg i.p.[15-17] The authors only observed a reduction in SWS and/or PS latencies. Also, in cats no hypnotic effect of L-tryptophan was obtained.[18,19] Only at the dose of 300 mg/kg i.p. did the compound produce drowsiness and prolong the SWS latency.[18] The results were interpreted as indicating that this serotonin precursor had a "general de-activating effect" on sleep without necessarily having sleep-inducing or sleep-maintaining function.[18]

5-Hydroxytryptophan (5-HTP)

In various species (cats, rabbits and dogs), 5-HTP was reported to produce slow waves in the EEG[20-22] resulting from increased levels of 5-HT. Whether such an EEG effect truly represents sleep has been a matter of dispute for many years. A series of other studies further investigated the effects of 30–50 mg/kg 5-HTP given i.p., i.v. or orally in cats. The authors described a SWS effect but stated that the EEG features did not totally resemble SWS[23]: sleep spindles[18,23] disappeared and there was a waking state with periods of drowsiness.[24] In all experiments there was a suppression of PS which could be followed by a rebound effect. Two effects are thus apparent: 5-HTP causes a HVLF EEG, by some indicated as sleep, by others as a sign of increased drowsiness; PS is suppressed.

5-HT Uptake Inhibitors

5-HT uptake inhibitors enhance the stimulation of 5-HT receptors indirectly by increasing the concentration of 5-HT at the synaptic cleft, and thus potentiate serotonergic transmission. Actions on S-W patterns of drugs blocking 5-HT uptake in the brain have been extensively studied, and the main reported finding was an inhibition of PS. Recent data in healthy man indicated that compounds such as zimeldine, indalpine and fluoxetine not only reduced rapid eye movement sleep (REMS) duration but also the total sleep time whereas W and stage 1 were increased.[25,26] In animal studies, fluoxetine, paroxetine, zimeldine, alapocrate and citalopram produced a dose-related suppression of PS,[27–31] but had differential actions on SWS. Fluoxetine was reported to have only small SWS effects in rats,[27] but with zimeldine SWS was found to be reduced in rats[28] or increased in both rats and cats.[29] Citalopram was also found to increase DSWS in cats.[31] Thus, enhancing 5-HT transmission by 5-HT uptake blockers generally results in PS suppression whereas the other sleep stages could be differently altered.

DEPLETION OF BRAIN SEROTONIN

Lesions

The involvement of serotonergic systems originating in raphe nuclei, in S-W has also been studied using lesioning techniques. Total coagulation of the raphe system generated behavioral W and cortical low voltage high frequency activity; small amounts of SWS (10%) did occur, but no PS.[23] The magnitude of the insomnia was related to the extent of the lesion of 5-HT containing cell bodies and consequently to the decrease of cerebral 5-HT. There seemed to be a topographical organization involved since rostral structures (raphe dorsalis and centralis) were suggested to be implicated in SWS and intermediate structures (raphe pontis and magnus) in triggering PS.[32] Conversely, a lesion of the noradrenergic dorsal bundle led to an increase in SWS and PS which was related to an increase in the 5-HT turnover.[33] This also suggested an interrelationship between 5-HT and catecholaminergic (CA) neurons.

Many of the above findings on the role of 5-HT in sleep using lesioning techniques were disputed mainly because the recordings were too brief following the lesioning, such that an eventual functional recovery could not be detected. Furthermore, the lesions might have been less specific than suggested. Mouret and Coindet[34] studied the effects of lesions of the locus coeruleus and raphe nuclei on S-W in rats. Most lesions of the raphe nuclei only transiently affected sleep. Nevertheless, lesions of the lateral part of the central gray and the rostral end of the nucleus raphe dorsalis substantially changed SWS, whereas lesions around the dopaminergic A_{10} region led to PS alterations. Further, rostral raphe structures led to selective SWS suppression.[35]

Neurotoxin: 5,6- or 5,7-Dihydroxytryptamine (5,6 DHT and 5,7 DHT)

Experiments by Ross *et al.*[36] addressed the problem of the relationship between 5-HT and sleep using a selective lesioning of the serotonergic system through an i.cv. administration of 200 μg of 5,7 DHT, which depleted 5-HT in forebrain by 78%. At the time of maximum serotonin depletion, no statistically significant difference was noted on SWS, only a transient insomnia was produced. There was an increased PS 6–14 days after treatment with 5,7 DHT. Froment *et al.*[37] found a reduction of SWS following i.cv. injection of 5,6 DHT in conjunction with a reduced serotonin content.

An injection of 0.5 μg of 5,6 DHT in the rostral raphe of cats only caused a short-lasting increase of PS over 24 h.[32] However, an i.cv. injection of 1000 μg of 5,6 DHT led to a significant decrease of stage 2 SWS and PS. Further, there was a significant correlation with the decrease in 5-HT and 5-HIAA. These results were interpreted as being consistent with the hypothesis that serotonergic neurons play a major role in the occurrence of SWS and PS, and are at variance with the study of Ross *et al.*[36] The latter authors refer to Dement *et al.*[38] where it is claimed that the insomnia produced by parachlorophenylalanine (PCPA) may result from hyperresponsivity to sensory events.

Parachlorophenylalanine (PCPA)

In the late 1960s various studies described insomnia induced by PCPA, a compound known to inhibit tryptophan hydroxylase: in rat,[39] cat,[40] rabbit,[41] and monkey.[42] This has later been confirmed in rat,[43] cat,[44] rabbit,[45] and primates.[46]

According to Koella *et al.*[47] an i.p. injection of 5–200 mg/kg of PCPA in cats produced a long-lasting reduction of sleep, exceeding two weeks at the highest dose. Following an injection of 400 mg/kg in cats, the sleep reduction has a slow onset of activity (>24 h) and gradually recovers after 80 h.[40] In rats, an i.p. injection of 600 mg/kg of PCPA given in two doses at 2-h interval, EEG HVLF activity occurred, reaching a maximum value between the 3rd and 4th day, returning to normal after 6–7 days.[39]

Much emphasis was given to the insomnia. However, Borbély *et al.*[43] stated that this phase was preceded by an initial sedative effect. During the first 24 h following PCPA treatment (300 mg/kg i.p.) waking time was significantly reduced and PS was decreased by about 30% within the light period. The day after PCPA administration, especially during the light phase, sleep was depressed followed by a rebound effect. Two days later, there was a complete recovery to control values. There is thus agreement with those studies reporting a sleep reduction 1 to 2 days after PCPA administration, and a disagreement with those showing no PS suppression. The rebound has also not been observed by all authors.

There is an apparent relation between the decrease of sleep and the decrease of 5-HT,[39] but EEG and behavioral recovery is seen in spite of the fact that the levels of 5-HT remained reduced up to 8 days after drug administration.[48] Different interpretations in terms of synaptic adaptation processes or receptor supersensitivity were formulated by Borbély *et al.*,[43] but none appears sufficiently adequate to explain the rapid recovery of the sleep states.

Restoration after 5-HT Depletion

In view of the antagonistic actions of 5-HTP and PCPA, a variety of studies demonstrated that PCPA-induced insomnia was reversible by 5-HTP. Normal levels of sleep can be restored by an injection of 5-HTP (10–50 μg/kg) during the insomnia state induced by 400 mg/kg of PCPA in cats.[40] Jouvet[32] reported that a single i.v. or i.p. injection of 2–5 mg/kg of 5-HTP given at the maximum of the insomnia both restored the normal quantity and quality of sleep patterns for 6–8 h. Similar effects were reported in rabbits[41] and cats.[49] The doses of both substances are of importance; a "balanced" dosing results in normal amounts of sleep.[50] However, it is difficult to explain the long latency of PS induction and the very long "hypnogenic effect" on the basis of a neurotransmitter action.[51]

Interestingly, Dement *et al.*[38] interpreted the restoration of sleep by 5-HTP in PCPA-treated animals not as a restoration of transmitter but as an antagonism of the hyperresponsitivity towards external and internal sensory events ensuing after PCPA treatment. There is no further experimental research on this hypothesis.

PHARMACOLOGICAL MANIPULATIONS OF 5-HT RECEPTORS

LSD

It is known that LSD increases the levels of brain 5-HT[52] and lowers the firing rate of raphe cell neurons.[53] The effects of LSD on the S-W cycle have mostly been investigated in the cat. In general, at low dosage PS was firstly reduced and W was increased, while at high dosage SWS was also decreased.[54–57] Similarly, subcutaneous injection of 4–16 μg/kg LSD to Beagle dogs caused a deficit of PS and an enhancement of W (Wauquier *et al.*, unpublished results). Analysis of EEG spectra showed a decrease in the EEG amplitude of cats,[55,56] whereas in midpontine decerebrated dogs Pickworth *et al.*[58] found an increase in the amplitude of the low frequency waves. In certain experiments EEG frequency was also altered.[56,58]

Serotonin Agonists

Direct stimulation of 5-HT receptors has been investigated earlier using quipazine. In cats, systemic administration of the compound (0.3–3 mg/kg i.v.) induced behavioral alterations (sham-rage reactions and catatonic stereotyped postures) as well as EEG synchronization, whereas its local application to the pontine raphe accelerated sleep onset and prolonged total sleep time.[59] Rats treated with 1–10 mg/kg quipazine i.p. exhibited a dose-related suppression of both SWS and PS.[60]

The effects of new agents which selectively stimulate a given 5-HT receptor subtype have been recently tested on the S-W cycle of rats. The 5-HT_1 agonists, whether 5-HT_{1A} (buspirone and 8-OH-DPAT)[61–65] or 5-HT_{1B} agonists (RU24969)[62,64] increased W and reduced both SWS and PS following i.p. or s.c. administration using various dose-ranges. The 5-HT_2 agonists DOM and DOB (0.16–2.5 mg/kg i.p.) produced similar sleep suppressing effects in adult, neonate and juvenile rats.[64,65] Thus, in rats a sleep deficit seems to be a common response to the stimulation of different 5-HT receptor subtypes. With regard to clinical studies, buspirone has been shown to have no sedative or hypnotic effect in anxious patients.[66]

Serotonin Antagonists

In animal studies, most of early developed serotonin antagonists have been reported to reduce sleep. Cyproheptadine, metergoline, methiothepine and methysergide produced a deficit in both SWS and PS in cats,[19,44,57] and had different sleep suppressing effects in rats.[67–69] Conversely, in humans methysergide reduced REM sleep and stage 4 but increased stage 3[70] and FQ27-096 and FU29-245 (reported to display serotonin receptor blocking properties) increased SWS.[71,72] Because of the non-specificity of these compounds (mixed serotonin/dopamine activities and mixed antagonistic/

agonistic properties) the interpretation of the data remained unclear. However, it is interesting to note that in both cats[59] and rats[60] the stimulating action of quipazine was prevented by pretreatment with a 5-HT antagonist, cinanserin and metergoline, respectively.

The development of a new class of serotonin antagonists acting specifically at 5-HT_2 receptor subtypes, ritanserin being the prototype of a centrally active compound,[73] provided the possibility to study the significance of these receptor subtypes on S-W regulation. In human volunteers,[74,75] insomniac patients (Oswald, unpublished observations) and in dysthymic patients,[76] ritanserin specifically increased deep (D) SWS. In rats, ritanserin (0.63–2.5 mg/kg i.p.) given at the beginning of the light period induced an increase in DSWS duration at the expense of W, light (L) SWS and PS,[64,77] and enhanced EEG slow-wave activity.[78] Similar effects have been observed in neonate and juvenile rats treated with 0.1–1 mg/kg ritanserin.[65] More recently, two chemically different 5-HT_2 receptor antagonists ICI 169,369 and ICI 170,809 have been tested in rats at oral doses of 1–40 mg/kg during the light period. Both drugs suppressed PS and only 20 mg/kg of ICI 170,809 produced a delayed enhancement of SWS duration.[79]

Interestingly, ritanserin was able to reverse the DSWS deficit induced by the 5-HT_2 receptor agonist DOM in a dose-dependent manner, but not the S-W alterations produced either by the 5-HT_{1A} agonist 8-OH-DPAT or the 5-HT_{1B} agonist RU24969.[64] The hypothesis emerged that the 5-HT_2 receptors play an active role in the regulation of DSWS. Furthermore, 5-HT_1 receptor subtypes are unlikely to interact with 5-HT_2 receptors in the S-W regulation mediated through 5-HT_2 receptors. In neonate and juvenile rats treated with the 5-HT_2 agonist DOB, ritanserin induced a dose-related return to control values of all the S-W states.[65] This suggests a functional role of 5-HT_2 receptors in both sleep states (quiet sleep or SWS and active sleep or PS) during ontogenesis.

ROLE OF 5-HT MECHANISMS IN SLEEP-WAKEFULNESS

The classical concept of the role of serotonin as a neurotransmitter involved in the induction and maintenance of sleep has been challenged on the basis of a number of pharmacological and neurophysiological findings. In contrast to expectations, the activity of the dorsal raphe nucleus neurons and of the nucleus raphe pallidus showed a gradual decline with sleep and firing rate of these neurons was at its lowest level during PS.[80] Similarly, there was a 5-HT release during waking and a decrease during SWS and PS.[5,81] In addition, following PCPA treatment the 5-HTP restoration of sleep does not necessarily indicate that 5-HTP is necessary, since PS can be restored by cerebrospinal fluid transfer from PS-deprived cats.[82] Furthermore, there was a long latency to PS and a very long hypnogenic effect. These considerations led to new concepts on the role(s) of 5-HT in S-W regulation.

Jouvet[83] advanced the theory that 5-HT acts both as a transmitter and a neurohormone. In his view 5-HT might be released as a neurotransmitter during waking and then act as a neurohormone resulting in the liberation of hypnogenic factors responsible for SWS and PS. SWS and PS would then be caused by two different 5-HT-dependent factors and SWS could be regulated by receptor-mediated mechanisms.

A number of arguments suggest that 5-HT has a 'de-activating' influence on wakefulness (W), rather than being directly responsible for sleep induction. Bonvallet and Block[84] and Bonvallet and Allen[85] already suggested a negative feedback system between the nucleus tractus solitarius and the reticular formation. Koella[86,87] pursued this idea and suggested that 5-HT exerts an anti-waking effect.

Koella[88] proposed that activation of the solitary tract nucleus by 5-HT through its action on the area postrema has an inhibitory effect on the arousal activity of the reticular formation. The fact that in cat serotonin precursors increased drowsiness rather than SWS[18] was also taken as evidence for a de-waking effect. These effects were interpreted as demonstrating an indirect effect. Thus, "serotonin, in addition to its (putative) direct sleep-promoting effect, also acts as an 'antiwaking' agent."[89] Moreover, 5-HT may antagonize CA involved in arousal and thus 5-HT is a transmitter which had an inhibitory effect on W.

As far as PS is concerned according to Jouvet[83] PS would be influenced by a gate-controlled executive mechanism. Thus, the expression of PS might be 5-HT-dependent. The antagonism of PS by 5-HT antagonists would involve two mechanisms. The first affects the series of events leading to the synthesis and release of factors responsible for PS. 5-HT is a part of this sequence. A second mechanism actively inhibits systems such as the rostral raphe or locus coeruleus (origin of the CA system).

Further, Koella[1] supports the idea of Jouvet that there is an indolamine independent but non-indolamine PS factor, which may play a role as PS facilitation instrument, and thus as a trigger factor for the shift towards PS. However, a specific role for 5-HT in triggering PS[90,91] is, to some extent challenged by the fact that 5-HTP, 5-HT uptake inhibitors, 5-HT agonists and antagonists, all reduce PS. These data would rather favor the idea that 5-HT should function within normal limits or operate above a critical level for the induction and maintenance of sleep.[92]

Finally, in the last decade multiple 5-HT receptor subtypes have been identified (5-HT_{1A}, 5-HT_{1B}, 5-HT_{1C}, 5-HT_{1D}, 5-HT_2, 5-HT_3) and consequently numerous drugs which selectively stimulate or block a given receptor site are becoming available.[93,94] This led to the hypothesis that each receptor subtype might be differently involved in sleep regulation. Recent studies using specific 5-HT agonists/antagonists suggested that 5-HT_2 receptors are involved in the maintenance of DSWS. However, the functional role of 5-HT_1 sites in S-W regulation remains to be defined since specific 5-HT_1 antagonists are not yet available. In addition, the precise site of action for most of these new synthetized drugs is not known. It would be of interest to find out whether they act on the hypnogenic structures suggested by Jouvet.[95] However, it has to be considered that sleep is not localized in a center, but depends on the integrated action of a system.

CONCLUSIONS

Serotonergic mechanisms are vigilance (or reactivity) suppressing instruments, according to the terminology of Koella.[1] They are necessary for the induction of SWS and in addition have an anti-waking effect. Moreover, 5-HT may antagonize catecholamines involved in arousal mechanisms. Further, the expression of PS is 5-HT-dependent. Apart from its transmitter role, serotonin could act as neurohormone, albeit indirectly, thereby being responsible for the liberation of hypnogenic factors. This might be in conjunction with or in addition to an increasing inhibitory feedback, interacting with 5-HT receptors at the level of the area postrema. The possible differential role of the various 5-HT receptor subtypes, is not yet clarified, but it is suggested that 5-HT_2 receptors have an inhibitory role on SWS.

REFERENCES

1. KOELLA, W. P. 1985. Serotonin and sleep. *In* Sleep: Neurotransmitters and Neuromodulators. A. Wauquier, J. M. Gaillard, J. M. Monti and M. Radulovacki, Eds. New York: Raven Press, pp. 185–196.
2. WAUQUIER, A., J. M. GAILLARD, J. M. MONTI & M. RADULOVACKI. 1985. Sleep, Neurotransmitters and Neuromodulators. New York: Raven Press.

3. Matussek, N. & U. Patschke. 1963. Beziehungen des Schlaf- und Wachrhythmus zum Noradrenalin- und Serotoningehalt im Zentralnervensystem von Hamstern. Mol. Exp. **11:** 81–87.
4. Kovacecic, R. & M. Radulovacki. 1976. Monoamine changes in the brain of cats during slow wave sleep. Science **193:** 1025–1027.
5. Puizillout, J. J., G. Gaudin-Chazal, A. Daszuta, N. Seyfritz & J. P. Ternaux. 1979. Release of endogenous serotonin from "encéphale isolé" cats. II. Correlations with raphe neuronal activity and sleep and wakefulness. J. Physiol. **75:** 531–537.
6. Hery, F., E. Rouer & J. Glowinski. 1972. Daily variations of serotonin metabolism in the rat brain. Brain Res. **43:** 445–465.
7. Cespuglio, R., F. Riou, H. Faradji, J. L. Ponchon, M. Buda, F. Gonon, M. Jouvet & J. F. Pujol. 1979. The release of monoamines during the sleep waking cycle as studied with the voltametric technique. Neurosci. Lett. (Suppl.) **3:** S322.
8. Tsuchiya, K., M. Toru & T. Kobayashi. 1969. Sleep deprivation: Changes of monoamines and acetylcholine in rat brain. Life Sci. **8:** 867–873.
9. Borbely, A. A. & H. U. Neuhaus. 1979. Sleep-deprivation: Effects on sleep and EEG in the rat. J. Comp. Physiol. **133:** 71–87.
10. Borbely, A. A., P. Steigrad & I. Tobler. 1980. Effect of sleep deprivation on brain serotonin in the rat. Behav. Brain Res. **1:** 205–210.
11. Rothballer, A. B. 1959. Effects of catecholamines on the central nervous system. Pharmacol. Rev. **11:** 494–547.
12. Koella, W. P. & J. Czicman. 1966. Mechanism of the EEG-synchronizing action of serotonin. Am. J. Physiol. **211:** 926–934.
13. Fernstrom, J. R. & R. J. Wurtmann. 1971. Brain serotonin content: Physiological dependence on plasma tryptophane levels. Science **173:** 149–152.
14. Schlossberger, H. G., W. Kochen, B. Linzen & H. Steinhart. 1984. Progress in Tryptophan and Serotonin Research. Berlin, New York: Walter de Gruyter & Co.
15. Hartmann, E. & R. Chung. 1972. Sleep-inducing effects of l-tryptophane. J. Pharm. Pharmacol. **24:** 252–253.
16. Hill, S. Y. & R. B. Reyes. 1978. Effects of L-tryptophan and ethanol on sleep parameters in the rat. Psychopharmacology **58:** 229–233.
17. Wojcik, W. J., C. Fornal & M. Radulovacki. 1980. Effect of tryptophan on sleep in the rat. Neuropharmacology **19:** 163–167.
18. Ursin, R. 1976. The effects of 5-hydroxytryptophan and L-tryptophan on wakefulness and sleep patterns in the cat. Brain Res. **106:** 105–115.
19. Radulovacki, M. 1982. L-Tryptophan's effects on brain chemistry and sleep in cats and rats: A review. Neurosci. Biobehav. Rev. **6:** 421–427.
20. Bogdanski, D. F., H. Weisbach & S. Udenfriend. 1957. The distribution of serotonin, 5-hydroxytryptophan decarboxylase and monoamine oxidase in brain. J. Neurochem. **1:** 227–228.
21. Costa, E., G. R. Pscheidt, W. G. Van Meter & H. E. Himwich. 1960. Brain concentrations of biogenic amines and EEG patterns of rabbits. J. Pharmacol. Exp. Ther. **130:** 81–88.
22. Monnier, M. & R. Tissot. 1958. Correlated effects on behavioural and electrical brain activity evoked by stimulation of the reticular system, thalamus and rhinencephalon in the conscious animal. *In* Neurological Basis of Behavior. G. T. W. Wolstenholme and M. O'Connor, Eds. London: Churchill.
23. Jouvet, M. 1967. Neurophysiology of the states of sleep. Physiol. Rev. **47:** 117–177.
24. Buckingham, R. L. & M. Radulovacki. 1976. The effects of reserpine, L-dopa and 5-hydroxytryptophan on 5-hydroxyindole acetic and homovanilic acids in cerebrospinal fluid, behaviour and EEG in cats. Neuropharmacology **15:** 389–392.
25. Nicholson, A. N. & P. A. Pascoe. 1986. 5-Hydroxytryptamine and noradrenaline uptake inhibition: Studies on sleep in man. Neuropharmacology **25:** 1079–1083.
26. Nicholson, A. N. & P. A. Pascoe. 1988. Studies on the modulation of the sleep-wakefulness continuum in man by fluoxetine, a 5-HT uptake inhibitor. Neuropharmacology **27:** 597–602.
27. Pastel, R. H., & J. D. Fernstrom. 1987. Short-term effects of fluoxetine and trifluoro-

methylphenylpiperazine on electroencephalographic sleep in the rat. Brain Res. **436:** 92–102.

28. KLEINLOGEL, H. & H. R. BURKI. 1987. Effects of the selective 5-hydroxytryptamine uptake inhibitors paroxetine and zimeldine on EEG sleep and waking stages in the rat. Neuropsychobiology **17:** 206–212.
29. SOMMERFELT, L., E. R. HAUGE & R. URSIN. 1987. Similar effect on REM sleep but differential effect on slow wave sleep of the two 5-HT uptake inhibitors zimeldine and alaproclate in cats and rats. J. Neural Transm. **68:** 127–144.
30. SOMMERFELT, L. & R. URSIN. 1987. The effects of zimeldine and alaproclate combined with a small dose of 5-HTP on waking and sleep stages in cats. Behav. Brain Res. **24:** 1–10.
31. HILAKIVI, I., T. KOVALA, A. LEPPÄVUORI & A. SHVALOFF. 1987. Effects of serotonin and noradrenaline uptake blockers on wakefulness and sleep in cats. Pharmacol. Toxicol. **60:** 161–166.
32. JOUVET, M. 1974. The regulation of the sleep-waking cycle by monoaminergic neurons in the cat. *In* Brain and Sleep. H. M. Van Praag and H. Meinardi, Eds. Amsterdam: De Ewen Bohn, B. V., pp. 22–38.
33. PETITJEAN, F. & M. JOUVET. 1970. Hypersomnie et augmentation de l'acide 5-hydroxyindolacétique cérébral par lésion isthmique chez le chat. C. R. Soc. Biol. **164:** 2288–2293.
34. MOURET, J. & J. COINDET. 1980. Polygraphic evidence against a critical role of the raphe nuclei in sleep in the rat. Brain Res. **186:** 273–287.
35. PETITJEAN, F., C. BUDA, M. JANIN, M. SAKAI & M. JOUVET. 1978. Patterns of sleep alterations following selective raphe nuclei lesions. Sleep Res. **7:** 40.
36. ROSS, C. A., M. E. TRULSON & B. L. JACOBS. 1976. Depletion of brain serotonin following intraventricular 5,7-dihydroxytryptamine falls to disrupt sleep in the rat. Brain Res. **114:** 517–523.
37. FROMENT, J. L., F. PETITJEAN, N. BERTRAND, C. COINTRY & M. JOUVET. 1974. Effets de l'injection intracérébrale de 5,6-dihydroxy-tryptamine sur les monoamines cérébrales et les états de sommeil du chat. Brain Res. **67:** 405–417.
38. DEMENT, W. C., S. HENRIKSEN, B. L. JAWLES & M. MITLER. 1973. Biogenic amines, phasic events, and behavior. *In* Pharmacology and the Future of Man. F. E. Bloom and J. H. Acheson, Eds. New York: Karger, pp. 74–89.
39. MOURET, J., J. L. FROMENT, P. BOBILLIER & M. JOUVET. 1967. Etude neuropharmacologique et biochemique des insomnies provoquées par la *p*-chlorophénylalanine. J. Physiol. (Paris) **59:** 463–464.
40. DELORME, F., J. L. FROMENT & M. JOUVET. 1966. Suppression du sommeil par la *p*-chlorométhamphétamine et la parachlorophénylalanine. C. R. Soc. Biol. (Paris) **160:** 2347–2351.
41. FLORIO, V., A. S. DE CAROLIS & V. G. LONGO. 1968. Observations on the effect of DL-parachlorophenylalanine on the electroencephalogram. Physiol. Behav. **3:** 861–863.
42. WEITZMAN, E. D., M. M. RAPPORT, P. MCGREGOR & T. JAWBY. 1968. Sleep patterns of the monkey and brain serotonin concentration: Effect of *p*-chlorophenylalanine. Science **160:** 1361–1363.
43. BORBELY, A. A., H. Y. NEUHAUS & I. TOBLER. 1981. Effect of *p*-chloro-phenylalanine and tryptophan on sleep, EEG and motor activity in the rat. Behav. Brain Res. **2:** 1–22.
44. SALLANON, M., C. BUDA, M. JANIN & M. JOUVET. 1982. 5HT antagonists suppress sleep and delay its restoration after 5-HTP in *p*-chlorophenylalanine- pretreated cats. Eur. J. Pharmacol. **82:** 29–35.
45. GAILLARD, J. M., B. HERKERT & R. TISSOT. 1974. Reversal of the reserpine electroencephalographic synchronization in the rabbit by parachlorophenylalanine. Neuropharmacology **13:** 789–793.
46. BERT, J. & F. BALZAMO. 1974. Differential effects of parachlorophenylalanine on sleep in two primates, *Papio papio* and *Papio homadryas*. Electroencephal. Clin. Neurophysiol. **37:** 161–166.
47. KOELLA, W. P., A. FELDSTEIN & J. S. CZICMAN. 1968. The effect of parachlorophenylalanine on the sleep of cats. Electroencephal. Clin. Neurophysiol. **25:** 481–490.
48. STEIGRAD, P., I. TOBLER, P. G. WASER & A. A. BORBELY. 1978. Effect of *p*-chlorophenyla-

lanine on cerebral serotonin binding, serotonin concentration and motor activity in the rat. Arch. Pharmacol. **305:** 143–148.
49. Petitjean, F., M. Touret & C. Buda. 1981. Monoamines et sommeil: identification de neurones contenant des indolamines dans une situation pharmacologique particulière (PCPA, 5-HTP) en relation avec le sommeil. J. Physiol. (Paris) **77:** 237–240.
50. Aprison, N. H. & J. N. Hingtgen. 1972. Serotonin and behavior: A brief summary. Fed. Proc. **31:** 121–129.
51. Sallanon, M., C. Buda, M. Janin & M. Jouvet. 1985. Implication of serotonin in sleep mechanisms: Induction or facilitation? *In* Sleep: Neurotransmitters and Neuromodulators. A. Wauquier, J. M. Gaillard, J. M. Monti and M. Radulovacki, Eds. New York: Raven Press, pp. 135–140.
52. Freedman, D. X. 1961. Effects of LSD-25 on brain serotonin. J. Pharmacol. Exp. Ther. **134:** 160–166.
53. Aghajanian, J. K., W. E. Foote & M. N. Sheard. 1968. Lysergic acid diethylamide: Sensitive neuronal units in the midbrain raphé. Science **161:** 706–708.
54. Hobson, J. P. 1964. The effect of LSD on the sleep cycle of the cat. Electroencephal. Clin. Neurophysiol. **17:** 52–56.
55. Stern, W. C. & P. J. Morgane. 1972. Serotonin and EEG spiking activity in the lateral geniculate nucleus. Proceedings, 80th Annual Convention, APA, pp. 837–838.
56. Kay, D. C. & W. R. Martin. 1978. LSD and tryptamine effects on sleep-wakefulness and electrocorticogram patterns in intact cats. Psychopharmacology **58:** 223–228.
57. Polc, P., J. Schneeberger & W. Haefely. 1979. Effects of several centrally active drugs on the sleep-wakefulness cycle of cats. Neuropharmacology **18:** 259–267.
58. Pickworth, W. B., C. G. Sharpe & W. R. Martin. 1977. Transcallosally evoked potentials and the EEG in the decerebrate dog: Actions of tryptaminergic, dopaminergic and adrenergic agonists. Electroencephal. Clin. Neurophysiol. **42:** 809–816.
59. Rodriguez, R., J. A. Rojas-Ramirez & R. R. Drucker-Colin. 1973. Serotonin-like actions of quipazine on the central nervous system. Eur. J. Pharmacol. **24:** 164–171.
60. Fornal, C. & M. Radulovacki. 1981. Sleep suppressant action of quipazine: Relation to central serotonergic stimulation. Pharmacol. Biochem. Behav. **15:** 937–944.
61. Lerman, J. A., K. I. Kaitin, W. C. Dement & S. J. Peroutka. 1986. The effects of buspirone on sleep in the rat. Neurosci. Lett. **72:** 64–68.
62. Dzoljic, M. R., P. R. Saxena & O. E. Ukponmwan. 1987. 5-HT_1 receptor agonists enhance alertness. Sleep Res. **16:** 88.
63. Depoortere, H. 1988. Effects of 5-HT_{1A}-agonists on the sleep-wakefulness cycle in rats. *In* Sleep 86. W. P. Koella, F. Obal, H. Schulz and P. Visser, Eds. Stuttgart, New York: Gustav Fischer Verlag, pp. 346–348.
64. Dugovic, C., A. Wauquier, J. Leysen, R. Marrannes & P. A. J. Janssen. 1989. Functional role of 5-HT_2 receptors in the regulation of sleep and wakefulness in the rat. Psychopharmacology **97:** 436–442.
65. Davenne, D., C. Dugovic, B. Franc & J. Adrien. 1989. Ontogeny of slow wave sleep. *In* Slow wave sleep: Physiological, pathophysiological and functional aspects. A. Wauquier, C. Dugovic and M. Radulovacki, Eds. New York: Raven Press, pp. 21–30.
66. De Roeck, J., R. Cluydts, C. Schotte, D. Rouckhout & P. Cosyns. 1989. Explorative single-blind study on the sedative and hypnotic effects of buspirone in anxiety patients. Acta Psychiatr. Scand. **89:** 129–135.
67. Fuxe, K. & K. Kiianmaa. 1978. 5-Hydroxytryptamine neurons and the sleep wakefulness cycle. Effects of methergoline and zimeldine. Neurosci. Lett. **8:** 55–58.
68. Adrien, J., R. Laguzzi, S. Bourgoin & M. Hamon. 1980. Le sommeil du rat à lésion précoce du raphé: pharmacologie due système sérotoninergique. Waking Sleeping **4:** 119–129.
69. Laguzzi, R. 1981. Effects of intraventricular injection of methiotepin and metergoline on the sleep stage in the rat. C. R. Soc. Biol. (Paris) **175:** 160–163.
70. Mendelson, W. B., L. S. Jacobs, J. D. Reichman, E. Othmer, P. E. Cryer, B. Trivedi & W. H. Daughaday. 1975. Methysergide: Suppression of sleep related prolactin excretion and enhancement of sleep related growth hormone secretion. J. Clin. Invest. **56:** 690–697.

71. SPIEGEL, R. 1981. Increased slow-wave sleep in man after several serotonin antagonists. *In* Sleep 1980. Basel: Karger, pp. 275–278.
72. OSWALD, I., K. ADAM & R. SPIEGEL. 1982. Human EEG slow-wave sleep increased by a serotonin antagonist. Electroencephalogr. Clin. Neurophysiol. **54:** 583–586.
73. LEYSEN, J. E., W. GOMMEREN, P. VAN GOMPEL, J. WIJNANTS, P. F. M. JANSSEN & P. M. LADURON. 1985. Receptor binding properties *in vitro* and *in vivo* of ritanserin: A very potent and long acting serotonin-S_2 antagonist. Mol. Pharmacol. **27:** 600–611.
74. DECLERCK, A. C., A. WAUQUIER, P. H. M. VAN DER HAM-VELTMAN & Y. GELDERS. 1987. Increase in slow wave sleep in human treated with the serotonin-5-HT_2 antagonist ritanserin (The first exploratory polygraphic study). Curr. Ther. Res. **41:** 427–432.
75. IDZIKOWSKI, C., F. J. MILLS & R. GLENNARD. 1986. 5-Hydroxytryptamine$_2$ antagonist increases human slow wave sleep. Brain Res. **378:** 164–168.
76. PAIVA, T., F. ARRIAGA, A. WAUQUIER, E. LARA, R. LARGO & J. N. LEITAO. 1988. Effects of ritanserin on sleep disturbances of dysthymic patients. Psychopharmacology **96:** 395–399.
77. DUGOVIC, C. & A. WAUQUIER. 1987. 5-HT_2 receptors could be primarily involved in the regulation of slow-wave sleep in the rat. Eur. J. Pharmacol. **137:** 145–146.
78. BORBELY, A. A., L. TRACHSEL & I. TOBLER. 1988. Effects of ritanserin on sleep stages and sleep EEG in the rat. Eur. J. Pharmacol. **156:** 275–278.
79. TORTELLA. F. C., E. ECHEVARRIA, R. H. PASTEL, B. COX & T. P. BLACKBURN. 1989. Suppressant effects of selective 5-HT_2 antagonists on rapid eye movement sleep in rats. Brain Res. **485:** 294–300.
80. JACOBS, B. L. 1985. Overview of the activity of brain monoaminergic neurons across the sleep-wake cycle. *In* Sleep: Neurotransmitters and Neuromodulators. A. Wauquier, J. M. Gaillard, J. M. Monti and M. Radulovacki, Eds. New York: Raven Press, pp. 1–14.
81. CESPUGLIO, R., H. FARADJI, M. E. GOMEZ & M. JOUVET. 1981. Single unit recordings in the nuclei raphe dorsalis and magnus during the sleep-waking cycle of semi-chronic prepared cats. Neurosci. Lett. **24:** 133–138.
82. SALLANON, M., C. BUDA, M. JANIN & M. JOUVET. 1982. Restoration of paradoxical sleep by cerebrospinal fluid transfer to PCPA pretreated insomnia cats. Brain. Res. **251:** 137–147.
83. JOUVET, M. 1983. Hypnogenic indolamine-dependent factors and paradoxical sleep rebound. *In* Sleep 1982. W. P. Koella, Ed. Proc. Eur. Sleep Res. Basel: Karger, pp. 2–18.
84. BONVALLET, M. & V. BLOCK. 1961. Bulbar control of cortical arousal. Science **133:** 1133–1134.
85. BONVALLET, M. & M. B. ALLEN. 1963. Prolonged spontaneous and evoked reticular activation following discrete bulbar lesions. Electroencephal. Clin. Neurophysiol. **15:** 969–988.
86. KOELLA, W. P. 1970a. Serotonin oder Somnotonin. Schweiz. Med. Wochenschr. **100:** 357–364.
87. KOELLA, W. P. 1970b. Serotonin oder Sommotonin. Schweiz. Med. Wochenschr. **100:** 424–430.
88. KOELLA, W. P. 1977. Beta-blockers and sleep. *In* Beta Blockers and the Central Nervous System. P. Kielholz, Ed. Bern: Hans-Huber, pp. 174–183.
89. KOELLA, W. P. 1982. A modern neurobiological concept of vigilance. Experientia **38:** 1426–1437.
90. JOUVET, M. 1969. Biogenic amines and the states of sleep. Science **163:** 32–41.
91. JOUVET, M. 1974. Monoaminergic regulation of the sleep-waking cycle in the cat. *In* The Neurosciences, 3rd Study Program. F. O. Schmitt and F. G. Worden, Eds. Cambridge: The MIT Press, pp. 499–508.
92. KOELLA, W. P. 1981. Neurotransmitters and sleep. *In* Psychopharmacology of Sleep. D. Wheatley, Ed. New York: Raven Press, pp. 19–52.
93. BRADLEY, P. B., G. ENGEL, W. FENIUK, J. R. FOZARD, P. P. A. HUMPHREY, D. N. MIDDLEMISS, E. J. MYLECHARANE, B. P. RICHARSON & P. R. SAXENA. 1986. Proposals for the classification and nomenclature of functional receptors for 5-hydroxytryptamine. Neuropharmacology **25:** 563–576.

94. CONN, P. J. & E. SANDERS-BUSH. 1987. Central serotonin receptors: effector systems, physiological roles and regulation. Psychopharmacology **92:** 267–277.
95. JOUVET, M., M. DENOYER, K. KITAHAMA & M. SALLANON. 1989. Slow wave sleep and indolamines: A hypothalamic target. *In* Slow Wave Sleep: Physiological, Pathophysiological and Functional Aspects. A. Wauquier, C. Dugovic and M. Radulovacki, Eds. New York: Raven Press, pp. 91–108.

DISCUSSION OF THE PAPER

MANFRED GÖTHERT (*University of Bonn*): Did you observe any tolerance to the effect of ritanserin?

WAUQUIER: I can refer to a human study which was done by Dr. Izikowsky in the U.K. where he gave a chronic administration of ritanserin and found that the slow sleep increasing effects were persistent. There's also a study by Adam and Oswald from Edinburgh where they treated insomniac patients with ritanserin for 28 days and found after those 28 days the same slow-sleep–increasing effects as at the beginning. These experiments suggest there's no tolerance.

CORNELIUS VANDERWOLF (*University of Western Ontario, London*): I don't have a question really but I have a comment. I recently reviewed most of the work on serotonin and sleep, the early work, and it seems to me the experiments that you discussed at the beginning, of Jouvet, are in fact not very compelling. If small doses of 5-HTP are given without a peripheral decarboxylase inhibitor in the way that Jouvet did almost the entire amount of the compound is decarboxylated peripherally so you get serotonin being formed in the peripheral tissues and having multiple effects there as we all know. And in fact when 5-HTP is given with a decarboxylase inhibitor there are about half a dozen different published studies showing what you get is not sleep but in fact you get a serotonin syndrome, if the dose is big enough to have an effect. So I don't think the classical experiments are particularly compelling at all, and in fact I think the evidence is much stronger in favor of the idea that what the central serotonergic projections do is to produce an activation of the neocortex which has relation to waking behavior rather than the generation of sleep.

The other thing I would like to comment on is sleep is a sort of residual activity that we do when there's nothing else happening, and we all know perfectly well that if we have any kind of visceral upset or if we're worried about anything or are too hot or too cold or something we can't sleep. And I think that when you give drugs and you affect the amount of sleep that an animal engages in, you have to take into account the possibility that the effect on sleep may be secondary to changes that are occurring elsewhere in the animal. It's not an easy thing to demonstrate a primary effect of a drug or a lesion, a brain lesion, on sleep itself.

Well all I can ask is whether you have any comments on these remarks I suppose.

WAUQUIER: As far as the first comment is concerned I agree. I told the story based on what Jouvet and others have been discussing many years ago and the hypothesis he formulated, have been dominating the field for two decades. So what I tried to do was to describe the historical development up to the present day. By no means I have been giving all the critical comments which could be given on these experiments.

As far as the second comment is concerned, of course one has to be cautious to derive conclusions as to whether a drug has, or the changes one is observing is a direct drug-related effect rather than being an indirect effect. But, therefore, you do experiments to find that out. There's substantial evidence that indeed, for example

$5\text{-}HT_2$ antagonists such as ritanserin are increasing slow wave sleep. This is observed in cats, rats, dogs, and in man, volunteers as well as in patients. So there is quite a consistent pattern which is emerging from those experiments.

If the question is whether the drug is producing physiological sleep then we are entering a different issue. There is one recent experiment published by Borbely's group where he attempted to address that question studying the frequency pattern induced by ritanserin in comparison with sleep-deprived animals and then studying the subsequent sleep. So sleep pressure increased and he was studying the physiologically enhanced sleep. Also, ritanserin enhanced sleep. Now, what he demonstrated was that in terms of frequency pattern, ritanserin was doing the same thing as the sleep deprivation, yet there was an additional peak in the spectrum, in the theta band, which was evident after ritanserin and which was not seen after sleep deprivation. And this led him to suggest that we have to be cautious in interpreting the slow-sleep–increasing effects of ritanserin.

QUESTION: Have you considered the role of the release of corticotropin-releasing factor and its modulation by $5\text{-}HT_2$ ligands as the mechanism for the resynchronization by ritanserin in view of the fact that cortisol, and cortisol analogs, also induce resynchronization after jetlag.

WAUQUIER: Short answer: no. We didn't consider this at this moment. What I describe here are very recent experiments. We recently showed that melatonin had a role in the increase of slow wave sleep when the compound was given before the sleep periods. This might be important, for it is suggested that ritanserin may have a role in resynchronization of sleep.

HERMANN VAN PRAAG (*Session Chairman*): In the rats you said ritanserin's effect on slow wave sleep is light/dark dependent. That's not the case in man I think, or is it?

WAUQUIER: Well in man there is not such a black/white picture as you see in rats, but there's certainly a difference when the compound is given before the evening, say half an hour before going to bed, versus when the compound is given in the morning. The compound might be of interest in jetlag, for example. In pathological conditions or when sleep is disturbed, such light-dark dependency (i.e., also circadian effects) might be more evident.

5-HT$_{1A}$ Receptors and 5-HT$_{1A}$-Mediated Responses: Effect of Treatments That Modify Serotonergic Neurotransmission

ALAN FRAZER AND JULIE G. HENSLER

Department of Psychiatry
University of Pennsylvania School of Medicine and
Neuropsychopharmacology Unit (151E)
Veterans Affairs Medical Center
Philadelphia, Pennsylvania 19104

Classically, a decrease in exposure of a tissue to its endogenous transmitter produces a supersensitive or exaggerated response to exogenous agonist. This increase in sensitivity or responsiveness may be accounted for by an increase in the density of postsynaptic receptors to the transmitter (homologous up-regulation). Conversely, an increase in exposure of a tissue to agonists will, over time, result in a decreased responsiveness or sensitivity to that agonist, owing to a decrease in receptor density or "down-regulation."[1–4] Central noradrenergic and dopaminergic systems undergo compensatory or regulatory processes similar to that originally described for nicotinic cholinergic responses following denervation of skeletal muscle.

The density of β-adrenoceptors (BARs) changes upon repeated under- or over-exposure to either norepinephrine (NE) or other β-adrenoceptor agonists. Destruction of central noradrenergic neurons by the neurotoxin 6-hydroxydopamine (6-OHDA) increases the density of BARs in many areas of brain[5,6] as well as responses linked to these receptors.[7] By contrast, repeated exposure *in vitro* of cells containing BARs to β-adrenoceptor agonists causes decreased responsiveness with a subsequent decrease in density or "down-regulation" of the receptors.[8,9] Repeated treatment of rats with β-adrenoceptor agonists also causes desensitization of cardiac and lung responses to catecholamines and a down regulation of the BARs in these tissues.[10,11] Similarly, we have shown that repeated treatment of rats with the lipophilic partial β_2 agonist clenbuterol[12] reduces the density of β_2-adrenoceptors throughout the central nervous system (CNS).[13] More recently, we found that intraventricular administration of the full β-adrenoceptor agonist, isoproterenol, decreased the density of BARs, particularly the β_2 subtype, in many areas of brain.[14] Down regulation of BARs also occurs after repeated administration of treatments that enhance noradrenergic transmission, in particular antidepressants.[15,16] The effect of antidepressants seems to be confined to the β_1 subtype,[14,17] possibly because this subtype has about a 10-fold greater affinity for NE than the β_2 subtype does.[18]

Central α_1-adrenoceptors can also be "up-" or "down-regulated." For example, following either intraventricular injection of 6-OHDA or 7 days of repeated administration of reserpine, the density of α_1-receptors in the frontal cortex of the rat is increased by 50%.[6,19] Similarly, lesioning the dorsal noradrenergic bundle with 6-OHDA produces a substantial increase in ^{3}H-WB 4101 labeled binding sites in the frontal cortex.[20] Whereas a loss of neurotransmitter usually follows a lesion of the neuronal system, microinjection of 6-OHDA into the dorsal tegmental bundle produces an increase in cerebellar levels of NE. This increase in the content of NE is associated with a 25% decrease in α_1-adrenoceptor density in the cerebellum.[6] In general, then,

α_1-adrenoceptors do seem to regulate in response to changes in exposure to agonist in a manner similar to that found with nicotinic cholinergic and β-adrenergic receptors.

Although few studies have evaluated the regulation of dopamine receptor subtypes, from the results obtained to date it may be inferred that both D_1 and D_2 receptors undergo the same regulatory processes as nicotinic cholinergic receptors. For example, Hess *et al.*[21] and McGonigle *et al.*[22] reported an increase in the number of striatal D_1 receptors following repeated treatment with SCH 23390, a selective D_1 antagonist. This increase in receptor number occurred in conjunction with enhancement of functional indices of D_1 receptor activation. Adenylyl cyclase formation as well as stereotypy and increased locomotion induced by the D_1 agonist SKF 38398 were augmented following cessation of repeated administration of the selective D_1 antagonist.[21] With regard to the D_2 receptor subtype, destruction of central dopaminergic pathways produces supersensitive behavioral responses accompanied by increases in the density of D_2 receptors in the corpus striatum[23] as does chronic blockade of these receptors by antipsychotic drugs.[24,25]

The above phenomena are not seen consistently for serotonergic receptors, even though serotonergic responses seem to exhibit compensatory changes in sensitivity.[26] There is a disparity between the ability of drugs that alter serotonergic neurotransmission to change the sensitivity of behavioral and electrophysiological responsiveness to serotonin (5-HT) and the ability of these same drugs to change 5-HT (5HT) receptors. In the remainder of this chapter, we will focus primarily on our results that show that even when treatments alter responses that are linked to the $serotonin_{1A}$ ($5HT_{1A}$) subtype, such treatments do not modify $5HT_{1A}$ receptors in brain.

EFFECTS OF SEROTONERGIC LESIONS WITH 5,7-DIHYDROXYTRYPTAMINE (5,7-DHT)

There are reports indicating either behavioral or electrophysiological supersensitivity to serotonergic agonists following the lesioning of central serotonergic pathways. In particular, responses that appear to be mediated by $5HT_{1A}$ receptors exhibit supersensitivity. For example, a behavioral syndrome termed the serotonin syndrome can be induced in rats by serotonergic agonists.[27–29] This behavioral syndrome is linked to activation of $5HT_1$ receptors[30] with specific components being linked to the $5HT_{1A}$ subtype.[31] The lesion induced by 5,7-DHT results in behavioral supersensitivity to the direct acting agonist 5-methoxy-*N,N*-di-methyltryptamine (5-MeODMT).[32] The time course for the development of supersensitivity corresponds to the time course of presynaptic terminal degeneration after 5,7-DHT. However, supersensitivity did not develop upon treatment with the inhibitor of serotonin synthesis, parachlorophenylalanine (PCPA), even though PCPA produced similar depletion of 5HT in the spinal cord as that caused by 5,7-DHT.[32] From such a result, it may be inferred that it is the loss of afferent innervation (or perhaps nerve impulse activity) that is responsible for the development of supersensitivity rather than the loss of serotonin *per se*. Very recently, Lucki[33] reported that the serotonin syndrome in rats caused by the $5HT_{1A}$ selective agonist 8-hydroxy-2-(di-*n*-propylamino)-tetralin (DPAT) became supersensitive following lesioning with 5,7-DHT. However, the lesion-induced shift of the dose-response curve to DPAT was modest.

In addition to this behavioral effect, certain electrophysiological responses to serotonergic agonists show supersensitivity after 5,7-DHT-induced lesions. In 1979, Wang *et al.*[34] used a *cerveau isolé* preparation to record the responsiveness of cells in either the amygdala or ventral lateral geniculate nucleus (VLGN) of rats that had received 5,7-DHT into the midline of the ventromedial tegmentum. Within one day of lesioning, cells in the amygdala or VLGN exhibited a supersensitive response to 5HT and the effect became maximal after 7 days. At early times, no supersensitivity was seen to LSD administered iontophoretically. However, if more than 7 days had elapsed since the lesion was produced, then cells became supersensitive to LSD. Since LSD is not taken up by serotonergic neurons,[35,36] this latter result was interpreted as being due to the development of post-junctional supersensitivity. Development of post-junctional supersensitivity was not limited to serotonin agonists as cells in these areas of brain also showed supersensitive responses to NE and to γ-aminobutyric acid (GABA). In contrast to the supersensitivity that developed in response to the lesion, no supersensitivity developed in rats treated with either PCPA for 17–20 days or with reserpine for 3 days, either alone or in combination.[34] Results very similar to these were obtained subsequently by Ferron *et al.*[37] who recorded from spontaneously firing cells in the frontoparietal cortex of rats maintained under anesthesia. Two weeks after intraventricular administration of 5,7-DHT, the duration of inhibition of firing caused by iontophoretic 5HT was markedly enhanced in about 75% of the cells. As found previously, treatment of rats with PCPA did not affect responses elicited by 5HT. However, the response of frontoparietal cortical cells to either DA, NE or the α-adrenoceptor agonist, phenylephrine, was markedly shortened in either lesioned or PCPA-treated rats.[37] The effect of lesioning central serotonergic neurons on single cell responsiveness may be specific for certain brain areas. No supersensitivity in response to 5HT was measured in hippocampal pyramidal cells in the CA_3 region after intraventricular 5,7-DHT, even though the lesion reduced the content of 5HT by about 90%.[38]

The type of receptors for 5HT mediating the electrophysiological responses in the amygdala, ventral lateral geniculate nucleus, frontoparietal cortex and CA_3 region of the hippocampus has not been determined. However, activation of $5HT_{1A}$ receptors by serotonergic agonists can produce electrophysiological changes in both serotonergic cell body[39–42] and terminal field areas, particularly the CA_1 region of the hippocampus.[43–48] Given these lesion-induced changes in responses, we have looked at the effect of 5,7-DHT treatment on $5HT_{1A}$ receptors.

Rats were first pre-treated with desmethylimipramine[49] and then administered 5,7-DHT (200 μg free base) into the lateral ventricles and decapitated 18 days later. $5HT_{1A}$ receptors were measured using the technique of quantitative autoradiography with ^{3}H-DPAT (1nM).[50] Non-specific binding was defined using 5HT (2μM) and in most areas of brain represented 12–15% of total binding. Seventeen areas in brain were quantified, such as the dentate gyrus, CA_1 and CA_2/CA_3 regions of the hippocampus; ventromedial and dorsomedial hypothalamic nuclei; basolateral, lateral and medial amygdaloid nuclei; entorhinal cortex; layers 1–3, 4 and 5–6 of the frontal cortex; as well as the dorsal and median raphe nuclei. The lesion did not cause any significant increase in the binding of ^{3}H-DPAT in any of these areas. This was not because the lesion did not destroy serotonergic neurons. As shown in FIGURE 1, in terminal field areas, there were significant and marked decreases in the binding of ^{3}H-cyano-imipramine (^{3}H-CN-IMI) to serotonin uptake sites.[51] 5,7-DHT treatment also produces marked decreases in binding of ^{3}H-CN-IMI in the dorsal raphe nucleus.[52] Although no increases in the binding of ^{3}H-DPAT to $5HT_{1A}$ receptors were observed, significant decreases were found in several areas of brain (FIG. 2). The greatest decrease was found in the dorsal raphe nucleus. In addition, the binding of ^{3}H-DPAT in both the pyramidal and molecular layers of the CA_2/CA_3 regions of the hippocampus of lesioned rats was reduced by 30–35%.

The decrease in ^{3}H-DPAT binding in the dorsal raphe nucleus in 5,7-DHT-treated rats has been found by others.[50] Binding of ^{3}H-DPAT in this area most probably reflects binding to $5HT_{1A}$ receptors located on serotonergic cell bodies and their

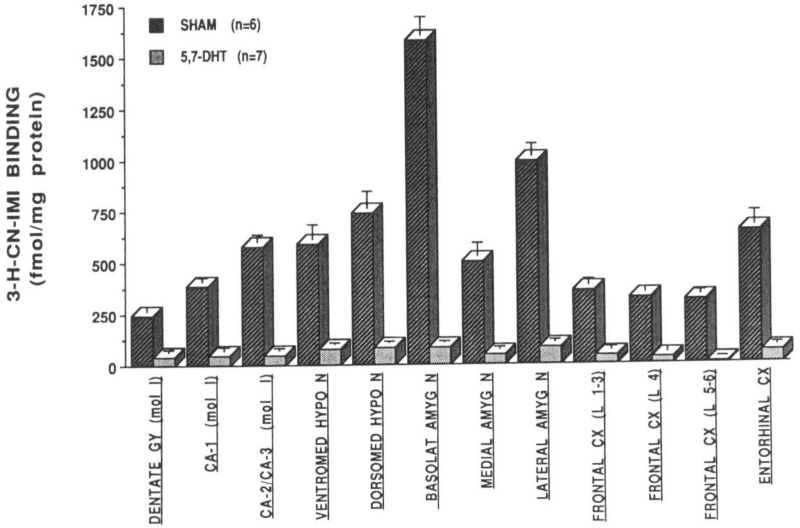

FIGURE 1. Effect of treatment with 5,7-dihydroxytryptamine on the binding of 3H-cyanoimipramine (^{3}H-CN-IMI) to serotonin uptake sites in serotonergic terminal field areas of rat brain. Shown are the X ± SEM of specific binding (fmol/mg protein) of ^{3}H-CN-IMI (0.3nM). Non-specific binding was defined using 2 μM sertraline. A significant effect of the lesion was seen in all brain regions (Student's *t*-test: $p < 0.001$).

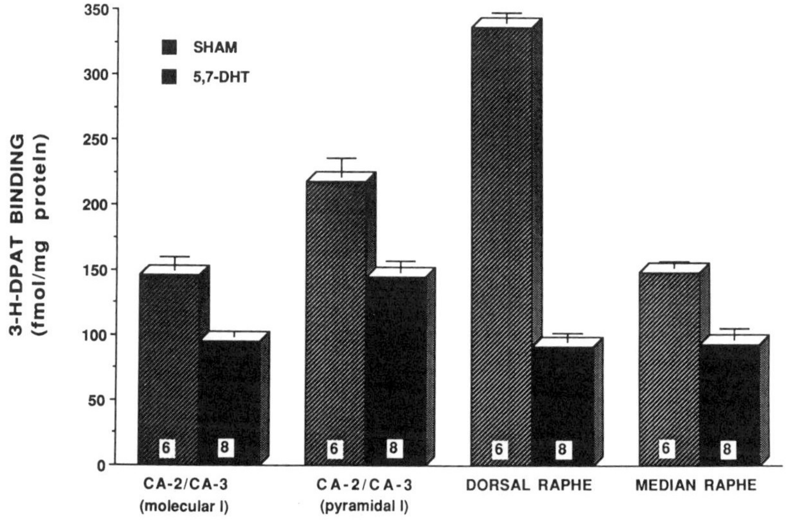

FIGURE 2. Areas of rat brain where the treatment with 5,7-dihydroxytryptamine decreased the binding of ^{3}H-DPAT to serotonin$_{1A}$ receptors. Shown are X ± SEM of specific binding (fmol/mg protein) of ^{3}H-DPAT (1nM). Because multiple comparisons were to be made, the level of statistical significance was set *a priori* at $p < 0.01$ (Student's *t*-test).

dendrites, although serotonergic axon collaterals cannot be excluded (Hensler & Frazer, unpublished observations). In contrast to our results, Verge *et al.*[50] did not report any lesion-induced decrease in any terminal field area, including the CA_2/CA_3 region. However, Fischette *et al.*[53] measured $5HT_1$ receptors by quantitative autoradiography using ^{3}H-5HT and did find a decrease in binding in the CA_3 region of the hippocampus (as we have) as well as in the dentate gyrus.

One explanation for the reduction in ^{3}H-DPAT binding in the CA_2/CA_3 region of the hippocampus of 5,7-DHT-treated rats is that some binding sites in this region are on serotonergic terminals that have been destroyed by the lesion. If so, then perhaps $5HT_{1A}$ receptors function as autoreceptors in this region. Evidence has been provided that terminal field 5HT autoreceptors in the rat brain may be the $5HT_{1B}$ subtype.[54,55] Such evidence has been obtained primarily on frontal cortical or cerebellar tissue. Middlemiss[56] did report that DPAT was ineffective in inhibiting the potassium-evoked release of ^{3}H-5HT from hippocampal slices as well as frontal cortical or striatal slices. It should be emphasized that even if $5HT_{1A}$ receptors are on serotonergic terminals in the CA_2/CA_3 region, they may not function as autoreceptors but rather exert some other type of modulatory role on serotonergic neurons.

Thus, we are unable to provide any evidence for "up-regulation" of $5HT_{1A}$ receptors in any area of brain as measured by the binding of ^{3}H-DPAT even though such lesions can cause supersensitive responsiveness to serotonergic agonists. Although some investigators have reported increases in ^{3}H-5HT binding to homogenates of certain brain regions after lesioning rats with 5,7-DHT,[57,58] most investigators have not found that lesioning serotonergic nerves causes any increase in $5HT_1$ receptors in homogenate preparations.[59–64]

From the perspective of denervation-induced supersensitivity, it may be inferred from our data that $5HT_{1A}$ receptors do not "up-regulate" in response to denervation in the manner seen for nicotinic cholinergic, β-adrenergic and dopaminergic receptors. A clue that the regulatory process for serotonergic receptors may be more complex than that of these other receptors is the fact that depletion of 5HT with PCPA does not mimic the effect of lesions on serotonergic responsiveness.[32,37] It would appear that a profitable area for future research aimed at understanding regulation of serotonergic responsiveness would be the study of post-receptor phenomena.

EFFECTS OF ANTIDEPRESSANTS

Many neurotransmitter receptors that up-regulate in response to a persistent decrease in exposure to an agonist also down-regulate when exposure to agonist increases over time.[6,8,65,66] There are no studies of how $5HT_1$ receptors respond to repeated exposure to serotonergic agonists. Rather, most investigators have attempted to increase serotonergic neurotransmission *in vivo* and then examine whether compensatory changes in serotonergic responsiveness or receptors occur. Antidepressants have been used extensively in studies such as this, given their ability to block the re-uptake of 5HT or inhibit the catabolism of 5HT by monoamine oxidase (MAO).

Many investigators have found that repeated treatment of mice or rats with antidepressants decreases serotonergic responsiveness.[26] We have been interested particularly in responses mediated by $5HT_{1A}$ receptors. In an early study, we used the serotonin behavioral syndrome as a measure of a $5HT_1$-mediated response.[30] Subsequent work has indicated that some components of this syndrome are caused by activation of the $5HT_{1A}$ subtype.[31] We found that repeated administration of MAOI inhibitors, but not tricyclic antidepressants or fluoxetine, decreases the ability of serotonergic agonists to elicit the serotonin behavioral syndrome.[67] More recently, we have begun to use hypothermia produced in rats by DPAT as a response linked to activation of $5HT_{1A}$ receptors. DPAT produces a dose-related hypothermia in rats that can be studied in the absence of a behavioral syndrome.[68–72] The hypothermic response to DPAT has been reported by some to be attenuated following depletion of 5HT by repeated administration of PCPA or by injection of 5,7-DHT into the third ventricle; because of this, DPAT-induced hypothermia has been proposed to be mediated by presynaptic $5HT_{1A}$ receptors.[71] By contrast, others have found DPAT-induced hypothermia to be *enhanced* following pretreatment with PCPA[68,73] or 5,7-DHT-induced

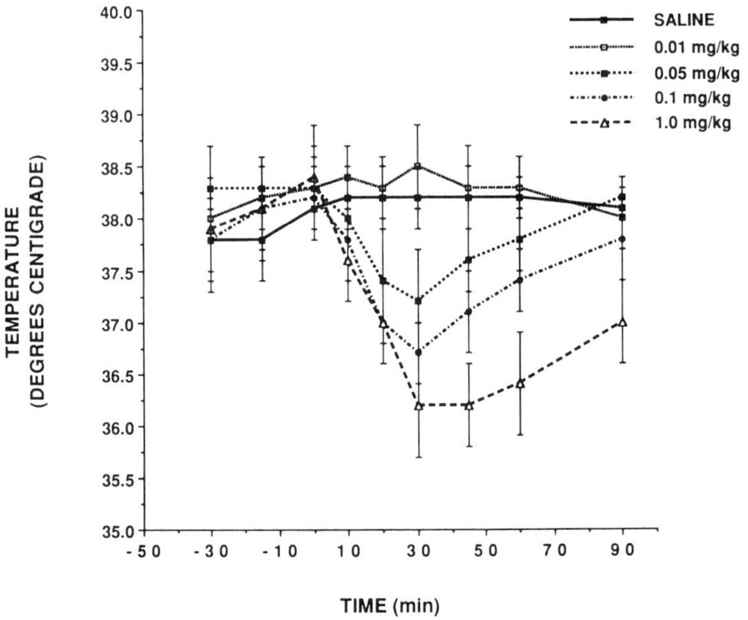

FIGURE 3. Hypothermic response in rats to acute injection of DPAT. Shown are X ± SEM. Body measurements were taken by means of rectal probe and digital thermometer. The probe, lubricated with petroleum jelly was inserted 5cm into the rectum, with the rat unrestrained. Animals were habituated to the probe and observation room prior to the experiment. Baseline temperature measurements were taken 30 min, 15 min, and immediately prior to subcutaneous injection of DPAT or saline. Temperature measurements were recorded 10, 20, 30, 45, 60 and 90 min after injection of DPAT (0.01–1mg/kg) or saline.

lesion of serotonergic neurons,[73] and have concluded that the action of DPAT is not dependent on presynaptic 5HT stores but rather is probably mediated by postsynaptic $5HT_{1A}$ receptors.

Initially, a time course and dose response relationship of the hypothermia induced by subcutaneous injection of DPAT in the rat was measured (see FIG. 3). Maximal decreases in body temperature were observed after 30 min at all doses of DPAT tested, whereas administration of saline caused no change in body temperature. A dose of DPAT of 0.05 mg/kg produced a modest, but reproducible and statistically significant ($p < 0.001$, paired *t*-test), decrease in body temperature after 30 min. This dose and time were selected in subsequent studies investigating the effect of repeated administration of antidepressant drugs on DPAT-induced hypothermia, as it was considered that the lower the dose of DPAT used the more likely it was that its hypothermic effect was mediated through selective effects on the $5HT_{1A}$ receptor. Antidepressants selected for study were inhibitors of type A and type B monoamine oxidase (tranylcypromine and phenelzine), a selective inhibitor of type A monoamine oxidase (clorgyline) and a selective inhibitor of the uptake of 5HT (sertraline). The dose, route of administration and duration of administration of these drugs are shown in TABLE 1. DPAT was administered approximately 24 hours after the final injection of drug or saline. The results of this study are shown in TABLE 2. All of these treatments attenuated significantly the hypothermia induced by DPAT. Other investigators have found also that repeated administration of antidepressant treatments to rats, such as electroconvulsive shock (ECS), desmethylimipramine, zimelidine, tranylcypromine and clorgyline, reduces the hypothermic effect of DPAT.[72,74] Thus, it appears that this effect of DPAT, presumably mediated by $5HT_{1A}$ receptors,[70] is attenuated by repeated administration of different types of antidepressant treatments.

In addition to hypothermia and the serotonin behavioral syndrome, the effect of antidepressants on other responses that may be linked to $5HT_1$ receptors have been studied. For example, neuroendocrine responses can be elicited by DPAT and other $5HT_{1A}$ selective agonists. The DPAT-induced rise of adrenocorticotropin hormone (ACTH) and corticosterone[75] in plasma may be mediated by $5HT_{1A}$ receptors[76] but it is

TABLE 1. Antidepressant Treatments

Drug	Dose, Schedule	Duration
Citalopram	20 mg/kg, q.d., i.p.	14 days
Clorgyline	1 mg/kg, q.d., s.c.	21 days
Phenelzine	5 mg/kg, q.d., i.p.	21 days
Tranylcypromine	5 mg/kg, q.d., i.p.	21 days
Sertraline	5 mg/kg, bid., i.p.	21 days

not clear that this subtype mediates the increase in prolactin caused by DPAT.[77] Given this, it is interesting that Aulakh *et al.*,[78] reported that repeated treatment of rats with imipramine, clorimipramine or clorgyline had no effect on the rise of ACTH caused by DPAT. By contrast, the DPAT-induced rise in prolactin was attenuated by chronic treatment with clorgyline.[78] In contrast to the results with hypothermia and the serotonin behavioral syndrome, these results do not provide any evidence that repeated administration of antidepressants cause subsensitive neuroendocrine responsiveness linked to $5HT_{1A}$ receptors.

There are data, however, to indicate that repeated administration of antidepressants causes subsensitive responsiveness of somatodendritic autoreceptors for 5HT in the dorsal raphe nucleus. The autoreceptors in this region appear to be the $5HT_{1A}$ subtype.[39–42] In a series of investigations by deMontigny and his associates[79–81] it was found that repeated treatment of rats with inhibitors of serotonin uptake, namely zimelidine, indalpine and citalopram, caused a diminished response of raphe cells to iontophoretically applied 5HT or LSD. Inhibitors of MAO, such as phenelzine and clorgyline, produced the same effect as the inhibitors of 5HT uptake.[82] It was concluded that long-term treatment with any of these drugs causes autoreceptor subsensitivity. Thus, it seems that an autoreceptor-mediated electrophysiological response does show subsensitivity to agents that might be expected to increase the exposure of the raphe somatodendritic autoreceptor to 5HT. As this response may be

TABLE 2. Effect of Treatment with Antidepressants on DPAT-Induced Hypothermia in Rats

Treatment	n	Decrease in Body Temperature (Degrees Centigrade)[a]
Saline	10	1.5 ± 0.1
Tranylcypromine	5	0.4 ± 0.2[b]
Phenelzine	5	0.6 ± 0.2[b]
Clorgyline	5	0.7 ± 0.1[b]
Sertraline	5	0.8 ± 0.1[b]

[a]Shown are X ± SEM.
[b]Treatments altered significantly (one-way ANOVA, F = 10.01) the hypothermic response to acute injection of DPAT (0.05 mg/kg, s.c.); $p < 0.001$ by post-hoc Dunnett's test.

mediated by $5HT_{1A}$ receptors, an important question would be what these treatments do to $5HT_{1A}$ receptors in the dorsal raphe nucleus?

We have addressed this issue by measuring the binding of ^{3}H-DPAT (1nM) in the dorsal raphe nucleus using quantitative autoradiography. Rats were treated either with saline, or inhibitors of MAO (phenelzine or clorgyline) or inhibitors of the uptake of 5HT, (citalopram or sertraline) (see TABLE 1 for dosage schedules). As is shown in TABLE 3, none of these treatments changed significantly the binding of ^{3}H-DPAT in the dorsal raphe nucleus. Thus, even though these treatments appear to cause somatodentritic autoreceptor subsensitivity when measured electrophysiologically, we do not find any evidence that this effect is accompanied by (or due to) down-regulation of $5HT_{1A}$ receptors, as measured by the binding of ^{3}H-DPAT.

Even though certain antidepressants cause subsensitive electrophysiological responses of dorsal raphe cells to 5HT, some post-synaptic electrophysiological responses seem to be *enhanced* by certain of these treatments. The inhibitory responsiveness caused by 5HT are enhanced by 14 days of treatment with drugs like imipramine, desmethylimipramine, clorimipramine, amitriptyline, iprindole and mianserin or with ECS in the ventral lateral geniculate nucleus and pyramidal cells in the CA_3 region of the hippocampus.[80,83] However, not all serotonergic terminal areas demonstrate enhanced sensitivity to iontophoretic application of 5HT after treatment with tricyclic antidepressants. Olpe and Schellenberg[84] found no change in the sensitivity of rostral and cingulate cortical neurons to 5HT after four weeks of treatment with either clorimipramine or desmethylimipramine. In this study, treatment with the MAO inhibitor clorgyline reduced the sensitivity of these neurons to 5HT.[84] These investigators subsequently used similar treatments and studied serotonergic responsiveness *in vitro* by recording from pyramidal cells of the hippocampal slice preparation.[85] As they found before, treatment with the clorimipramine did not change responsiveness to 5HT whereas treatment with clorgyline reduced responsiveness significantly. Blier *et al.*[86]

TABLE 3. Specific Binding of ^{3}H-DPAT in the Dorsal Raphe Nucleus of the Rat Following Chronic Treatment with Antidepressants

Areas	Saline (n = 10)	Phenelzine (n = 4)	Clorgyline (n = 4)	Citalopram (n = 5)	Sertraline (n = 5)
Dorsal Raphe N	372 ± 12[a]	376 ± 26	371 ± 17	396 ± 15	386 ± 10

[a]Shown are X ± SEM (fmol/mg protein). None of the treatments changed significantly the specific binding of ^{3}H-DPAT (1nM) in the dorsal raphe nucleus (one-way ANOVA).

found also that repeated treatment of rats with clorgyline reduced the responsiveness of hippocampal pyramidal cells to iontophoretic 5HT.

The types of 5HT receptors mediating the electrophysiological effects of 5HT in these areas of brain is unknown. Not much work has been done using DPAT administered iontophoretically probably because it appears to be much less effective than 5HT on cells in serotonergic terminal fields in contrast to its full agonist activity on cells of the dorsal raphe nucleus.[46] Nevertheless, $5HT_{1A}$ receptors are present in high density in the hippocampus[50] (see TABLE 4) and do appear to mediate certain electrophysiological effects of 5HT in the hippocampus.[44-48] Accordingly, we have examined what certain antidepressant treatments do to $5HT_{1A}$ receptors in terminal field areas of serotonergic innervation with particular emphasis on hippocampal, cortical and hypothalamic areas. As shown in TABLE 4, none of the antidepressant treatments (clorgyline, phenelzine, citalopram or sertraline) caused any significant effect on ^{3}H-DPAT binding in the brain areas examined. Thus, even though such treatments can change electrophysiological responsiveness to 5HT in terminal fields, as well as causing a reduction in certain responses elicited by DPAT (e.g., see TABLE 2), we have found little evidence that this altered responsivity is associated with changes in $5HT_{1A}$ receptors in terminal field areas.

Since antidepressant-induced changes in responses that might be linked to $5HT_{1A}$ receptors do not seem to be able to be accounted for by changes in the density of these receptors (in contrast to the situation with β-adrenergic-mediated effects),[16] what other factors could contribute to the altered responsiveness? One possibility is that changes occur in cellular processes distal to receptor activation. There is good evidence that $5HT_{1A}$ receptors are linked through G proteins to adenylyl cyclase; in particular, 5HT inhibits forskolin-stimulated adenylyl cyclase *via* $5HT_{1A}$ receptors in the hippocampus.[87] Consequently, we examined whether treatment of rats with phenelzine (selected because its repeated administration blocks both DPAT-induced hypothermia and the serotonin behavioral syndrome[33]) would shift the dose-response curve for the inhibition of forskolin-stimulated adenylyl cyclase by 5HT in hippocampal homogenates. Adenylyl cyclase was measured essentially as described by deVivo and Maayani[88] with the conversion of ^{32}P-ATP to ^{32}P-cyclic AMP measured as described by Salomon.[89] As is shown in FIGURE 4, serotonin caused a dose-dependent inhibition of forskolin-

TABLE 4. Effect of Antidepressant Drug Treatments on the Specific Binding of ^{3}H-DPAT in Rat Brain

	Saline (n = 10)	Phenelzine (= 4)	Sertraline (n = 5)	Citalopram (n = 5)	Clorgyline (n = 4)
Hippocampus					
Ca_1 (Pyramidal 1.)	352 ± 14[a]	401 ± 36	330 ± 25	365 ± 28	338 ± 28
CA_1 (Molecular 1.)	397 ± 17	433 ± 44	367 ± 26	426 ± 33	373 ± 35
$CA_{2/3}$ (Pyramidal 1.)	167 ± 14	187 ± 26	204 ± 15	225 ± 11	169 ± 19
$Ca_{2/3}$ (Molecular 1.)	118 ± 9	131 ± 15	147 ± 8	160 ± 15	130 ± 18
Frontal Cortex					
Layer 5-6	161 ± 10	172 ± 11	185 ± 10	195 ± 12	153 ± 9
Hypothalamus					
Ventromedial N.	217 ± 11	181 ± 16	221 ± 13	234 ± 10	187 ± 16
Dorsomedial N.	124 ± 9	104 ± 11	130 ± 17	132 ± 17	126 ± 18
Amygdala					
Medial N.	160 ± 13	160 ± 21	180 ± 12	156 ± 16	146 ± 11

[a] X ± SEM (fmol/mg protein). None of the treatments changed significantly the binding of ^{3}H-DPAT (1nM) in these brain areas (one-way ANOVA).

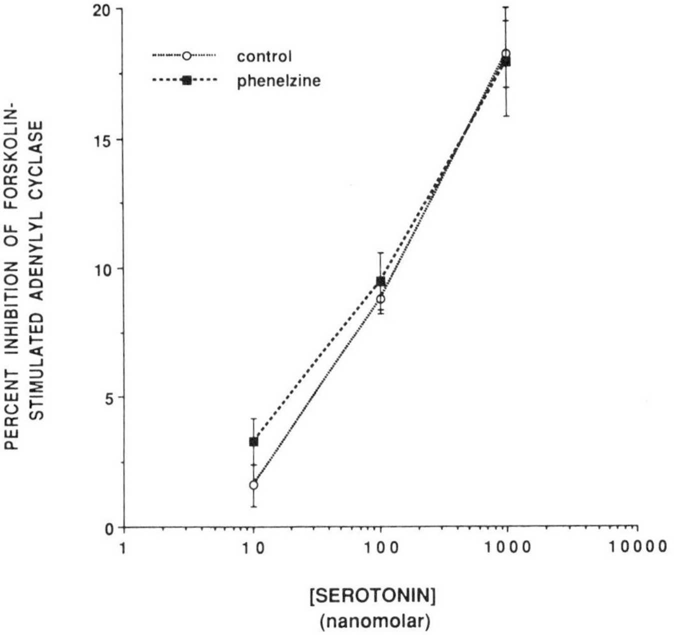

FIGURE 4. The effect of treatment with phenelzine on 5HT-induced inhibition of adenylyl cyclase. Rats were treated with saline or phenelzine (10mg/kg, i.p., once daily) for 10 days and decapitated 24 hours after the final injection. Shown are X ± SEM of the percent inhibition caused by 5HT of forskolin-stimulated adenylyl cyclase. In control hippocampal homogenates, the value for forskolin-stimulated (330μM) adenylyl cyclase was 383 ± 24 pmoles/min/mg protein whereas it was 393 ± 28 pmoles/min/mg protein in homogenates obtained from phenelzine-treated rats. These values represent approximately a fivefold increase in enzyme activity over basal values.

stimulated adenylyl cyclase in hippocampal homogenates. Treatment of rats with phenelzine caused no change in the dose response curve to 5HT (FIG. 4). Thus, a second messenger response linked to $5HT_{1A}$ receptors does not appear to be affected by repeated treatment of rats with phenelzine. Although we found no effect of treatment with phenelzine, Newman and Lerer[90] reported recently that repeated treatment of rats with either electroconvulsive shock or desmethylimipramine did reduce significantly the extent of inhibition caused by 5HT of forskolin-stimulated adenylyl cyclase. Clearly, more studies of this parameter are indicated.

In conclusion, $5HT_{1A}$ receptors do not appear to undergo compensatory changes in response to treatments which are expected to change the exposure of the receptor to endogenous transmitter. Although the sensitivity of $5HT_{1A}$-mediated responses to exogenous agonist is altered following such treatments, the density of $5HT_{1A}$ receptors, as measured by the binding of ^{3}H-DPAT does not exhibit concomitant changes. This contrasts with what has been demonstrated for dopaminergic, noradrenergic and cholinergic neurotransmitter systems using radioligand techniques. This may not be surprising as there is evidence that alterations in serotonergic responsivity may not simply be due to changes in the exposure of receptors to 5HT. For example, administration of PCPA does not mimic the effect of 5,7-DHT on the serotonin behavioral syndrome[32] or certain electrophysiological responses.[34,37] Were the 5,7-DHT-induced enhancement of responsivity simply due to the loss of 5HT, it would be expected that PCPA would mimic the effect. Another possible factor to explain the absence of an effect on ^{3}H-DPAT binding, particularly in the dorsal raphe nucleus, is that there may be other types of serotonergic receptors that function as the somatodendritic autoreceptor in addition to the $5HT_{1A}$ subtype. Sprouse and Aghajanian[41] found, for example, that (-) propranolol was very effective in inhibiting the effects of DPAT or ipsapirone on the firing of raphe cells, but not that of 5HT itself. Finally, it may be that changes in the sensitivity of $5HT_{1A}$-mediated responsiveness are due to changes in cellular responses that follow cell surface receptor activation and membrane associated events. Nevertheless, the key question still remains as to how changes in serotonergic responsivity seem to occur by treatments that modify serotonergic neurotransmission. The answer to this question may be relevant for understanding better how changes in responsiveness occur in other neuronal systems as well.

REFERENCES

1. CANNON, W. B. & A. ROSENBLUETH. 1949. The supersensitivity of denervated structures. New York: Macmillian Press.
2. AXELSSON, J. & S. THESLEFF. 1958. The densensitizing effect of acetycholine on the mammalian end plate. Acta Physiol. Scand. **43:** 15–26.
3. HARTZELL, H. C. & D. M. FAMBROUGH. 1972. Acetylcholine receptors: Distribution and extrajunctional density in rat diaphram after denervation correlated with actylcholine sensitivity. J. Gen. Physiol. **60:** 248–262.
4. BERG, D. K., R. B. KELLY, P. B. SARGENT, P. WILLIAMSON & Z. W. HALL. 1972. Binding of α-bungarotoxin to acetylcholine receptors in mammalian muscle. Proc. Natl. Acad. Sci. USA **69:** 147–151.
5. SPORN, J. R., B. B. WOLFE, T. K. HARDEN & P. B. MOLINOFF. 1977. Supersensitivity in rat cerebral cortex: Pre- and post-synaptic effects of 6-hydroxydopamine at noradrenergic synapses. Mol. Pharmacol. **13:** 1170–1180.
6. U'PRICHARD, D. C., T. D. REISINE, S. T. MASON, H. C. FIBIGER & H. I. YAMAMURA. 1980. Modulation of rat brain α and β-adrenergic receptor populations by lesion of the dorsal noradrenergic bundle. Brain Res. **187:** 143–154.
7. WOLFE, B. B., T. K. HARDEN, J. R. SPORN & P. B. MOLINOFF. 1978. Presynaptic modulation of β-adrenergic receptors in rat cerebral cortex after treatment with antidepressants. J. Pharmacol. Exp. Therap. **207:** 446–457.
8. SU, Y. F., T. K. HARDEN & J. P. PERKIN. 1980. Catecholamine-specific desensitization of adenylate cyclase. Evidence for a multistep process. J. Biol. Chem. **255:** 7410–7419.
9. HARDEN, T. K. 1983. Agonist-induced desensitization of β-adrenergic receptor-linked adenylate cyclase. Pharmacol. Rev. **35:** 5–32.
10. ABRASS, I. B., J. L. DAVIS & P. J. SCARPACE. 1982. Isoproterenol responsiveness and myocardial β-adrenergic receptors in young and old rats. J. Gerentol. **37:** 156–160.
11. CHANG, H. Y., R. M. KLEIN & G. KUNOS. 1982. Selective desensitization of cardiac β-adrenoceptors by prolonged *in vivo* infusion of catecholamines in rats. J. Pharmacol. Exp. Therap. **221:** 784–789.
12. ORDWAY, G. A., J. M. O'DONNELL & A. FRAZER. 1987. Effects of clenbuterol on central β_1 and β_2 adrenergic receptors of the rat. J. Pharmacol. Exp. Therap. **241:** 187–195.
13. VOS, P., P. A. DAVENPORT, R. P. ARTYMYSHYN, A. FRAZER & B. B. WOLFE. 1987. Selective regulation of β_2 adrenergic receptors by the chronic administration of the lipophilic β-adrenergic receptor agonist clenbuterol: An autoradiographic study. J. Pharmacol. Exp. Therap. **242:** 707–712.
14. ORDWAY, G. A., C. GAMBARANA & A. FRAZER. 1988. Quantitative autoradiography of central β adrenoceptor subtypes: Comparison of the effects of chronic treatment with desmethylimipramine or centrally administered 1-isoproterenol. J. Pharmacol. Exp. Therap. **247:** 379–389.

15. BANERJEE, S. P., L. S. KUNG & S. K. CHANDA. 1977. Development of β-adrenergic receptor subsensitivity by antidepressants. Nature (Lond) **268:** 455–456.
16. SUGRUE, M. F. 1983. Chronic antidepressant therapy and associated changes in central monoaminergic receptor functioning. Pharmacol. Therap. **21:** 1–33.
17. MINNEMAN, K. P., M. D. DIBNER, B. B. WOLFE & P. B. MOLINOFF. 1979. β_1- and β_2-adrenergic receptors in rat cerebral cortex are independently regulated. Science **204:** 866–869.
18. MINNEMAN, K. P., L. R. HEGSTRAND & P. B. MOLINOFF. 1979. The pharmacological specificity of β_1 and β_2 adrenergic receptors in rat heart and lung *in vitro.* Mol. Pharmacol. **16:** 21–33.
19. BYLUND, D. B. & J. R. MARTINEZ. 1980. α_2-Adrenergic receptors appear in rat salivary glands after reserpine treatment. Nature **285:** 229–230.
20. U'PRICHARD, D. C., W. D. BECHTEL, B. M. ROUOT & S. H. SNYDER. 1979. Multiple apparent α-noradrenergic receptor binding sites in rat brain: Effect of 6-hydroxydopamine. Mol. Pharmacol. **16:** 47–60.
21. HESS, E. J., L. J. ALBERS, H. LE & I. CREESE. 1986. Effects of chronic SCH 23390 treatment on the biochemical and behavioral properties of D_1 and D_2 dopamine receptors: Potentiated behavioral responses to a D_2 dopamine agonist after selective D_1 dopamine receptor upregulation. J. Pharmacol. Exp. Therap. **238:** 846–854.
22. MCGONIGLE, P., S. J. BOYSON, S. REUTER & P. B. MOLINOFF. 1986. Effects of chronic treatment with selective and nonselective antagonists on the subtypes of dopamine receptor. Synapse **3:** 74–82.
23. CREESE, I., D. R. BURT & S. H. SNYDER. 1977. Dopamine receptor binding enhancement accompanies lesion-induced behavioral supersensitivity. Science **197:** 596–598.
24. MULLER, P. & P. SEEMAN. 1978. Brain neurotransmitter receptors after long-term haloperidol: Dopamine, acetylcholine, serotonin, α-noradrenergic and naloxone receptors. Life Sci. **21:** 1751–1758.
25. OWEN, F., A. J. CROSS, J. L. WADDINGTON, M. POULTER, S. J. GAMBLE & T. J. CROW. 1980. Dopamine-mediated behavior and [^{3}H]spiperone binding to striatal membranes in rats after nine months of haloperidol administration. Life Sci. **26:** 55–59.
26. FRAZER, A., S. J. OFFORD & I. LUCKI. 1988. Regulation of serotonin receptors and responsiveness in the brain. *In* The Serotonin Receptors. E. Sanders-Bush, Ed. Clifton, NJ: Humana Press, pp. 319–362.
27. GRAHAME-SMITH, D. G. 1971. Studies *in vivo* on the relationship between brain tryptophan, brain 5HT synthesis and hyperactivity in rats treated with a monoamine oxidase inhibitor and 1-tryptophan. J. Neurochem. **18:** 1053–1066.
28. MODIGH, K. & T. H. SVENSSON. 1972. On the role of central nervous system catecholamines and 5-hydroxytryptamine in the nialamine-induced behavioral syndrome. Br. J. Pharmacol. **46:** 32–45.
29. TRULSON, M. E. & B. L. JACOBS. 1976. Behavioral evidence for the rapid release of CNS serotonin by PCA and fenfluramine. Eur. J. Pharmacol. **36:** 149–154.
30. LUCKI, I., M. S. NOBLER & A. FRAZER. 1984. Differential actions of serotonin antagonists on two behavioral models of serotonin receptor activation in the rat. J. Pharmacol. Exp. Therap. **228:** 133–139.
31. TRICKLEBANK, M. D., C. FORLER & J. R. FOZARD. 1985. The involvement of subtypes of the $5HT_1$ receptor and of catecholaminergic systems in the behavioral responses to 8-hydroxy-2-(di-*N*-propylamino) tetralin in the rat. Eur. J. Pharmacol. **106:** 271–282.
32. TRULSON, M. E., E. E. EUBANKS & B. L. JACOBS. 1976. Behavioral evidence for supersensitivity following destruction of central serotonergic nerve terminals by 5,7-dihydroxytryptamine. J. Pharmacol. Exp. Therap. **198:** 23–32.
33. LUCKI, I. 1989. Behavioral responses associated with serotonin receptors. *In* Neurobehavioral Pharmacology, Vol. 7. P. Thompson, B. P. Dews, and J. E. Barrett, Eds. Hillsdale, NJ: Lawrence Erlbaum Assoc.
34. WANG, R. Y., C. DEMONTIGNY, B. I. GOLD, R. H. ROTH & G. K. AGHAJANIAN. 1979. Denervation supersensitivity to serotonin in rat forebrain: Single cell studies. Brain Res. **178:** 479–497.

35. Ross, S. B. & A. L. Renyi. 1967. Accumulation of tritiated 5-hydroxytryptamine in brain slices. Life Sci. **6:** 1407–1415.
36. Ross, S. B. & A. L. Renyi. 1969. Inhibition of the uptake of tritiated 5-hydroxytryptamine in brain tissue. Eur. J. Pharmacol. **7:** 270–277.
37. Ferron, A., L. Descarries & T. Reader. 1982. Altered neuronal responsiveness to biogenic amines in rat cerebral cortex after serotonin denervation or depletion. Brain Res. **231:** 93–108.
38. deMontigny, C., R. Y. Wang, T. A. Reader & G. K. Aghajanian. 1980. Monoaminergic denervation of the rat hippocampus: Microiontophoretic studies on pre- and postsynaptic supersensitivity to norepinephrine and serotonin. Brain Res. **200:** 363–376.
39. deMontigny, C., P. Blier & Y. Chaput. 1984. Electrophysiologically-identified serotonin receptors in the rat CNS. Neuropharmacology **23:** 1511–1520.
40. VanderMaelen, C. P., G. K. Matheson, R. C. Wilderman & L. A. Patterson. 1986. Inhibition of serotonergic dorsal raphe neurons by systemic and iontophoretic administration of buspirone, a non-benzodiazepine anxiolytic drug. Eur. J. Pharmacol. **129:** 123–130.
41. Sprouse, J. S. & G. K. Aghajanian. 1986. (-)-Propranolol blocks the inhibition of serotonergic dorsal raphe cell firing by $5HT_{1A}$ selective agonists. Eur. J. Pharmacol. **128:** 295–298.
42. Sprouse, J. S. & G. K. Aghajanian. 1987. Electrophysiological responses of serotonergic dorsal raphe neurons to $5HT_{1A}$ and $5HT_{1B}$ agonists. Synapse **1:** 3–9.
43. Beck, S. G., W. P. Clarke & J. Goldfarb. 1985. Spiperone differentiates multiple 5-hydroxy-tryptamine responses in rat hippocampal slices *in vitro*. Eur. J. Pharmacol. **116:** 195–197.
44. Andrade, R. & R. A. Nicoll. 1987. Pharmacologically distinct actions of serotonin on single pyramidal neurons of the rat hippocampus recorded *in vivo*. J. Physiol. **394:** 99–124.
45. Colino, A. & J. V. Halliwell. 1987. Differential modulation of three separate K-conductances in hippocampal CA_1 neurons by serotonin. Nature **328:** 73–77.
46. Sprouse, J. S. & G. K. Aghajanian. 1988. Responses of hippocampal pyramidal cells to putative serotonin $5HT_{1A}$ and $5HT_{1B}$ agonists: A comparative study with dorsal raphe neurons. Neuropharmacology **27:** 707–715.
47. Wu, P. H., N. Gurevich & P. L. Carlen. 1988. $Serotonin_{1A}$ receptor activation in hippocampal CA_1 neurons by 8-hydroxy-2-(di-n-propylamino)tetralin, 5-methoxytryptamine and 5-hydroxytryptamine. Neurosci. Lett. **86:** 72–76.
48. Ropert, N. 1988. Inhibitory action of serotonin in CA_1 hippocampal neurons *in vitro*. J. Neurosci. **26:** 69–81.
49. Bjorklund, A., H. G. Baumgartin & A. Rensch. 1975. 5,7-Dihydroxytryptamine: Improvement of its selectivity for serotonin neurons in the CNS by pretreatment with desipramine. J. Neurochem. **24:** 833–835.
50. Verge, D., G. Davel, M. Marcinkiewicz, A. Patey, S. El Mestikawy, H. Gozlan & M. Hamon. 1986. Quantitative autoradiography of multiple $5HT_1$ receptor subtypes in the brain of control or 5,7-dihydroxytryptamine treated rats. J. Neurosci. **6:** 3474–3482.
51. Kovachich, G. B., C. E. Aronson, D. J. Brunswick & A. Frazer. 1988. Quantitative autoradiography of serotonin uptake sites in rat brain using ^{3}H-cyanoimipramine. Brain Res. **454:** 78–88.
52. Hensler, J., G. Kovachich, C. Aronson & A. Frazer. 1988. Heterogeneous distribution of ^{3}H-DPAT and ^{3}H-cyanoimipramine (CN-IMI) binding sites in dorsal raphe nucleus (DRN) of the rat. Soc. Neurosci. Abstr. **14:** 778.
53. Fischette, C. T., B. Nock & K. Renner. 1987. Effects of 5,7-dihydroxytryptamine on $serotonin_1$ and $serotonin_2$ receptors throughout the rat central nervous system using quantitative autoradiography. Brain Res. **421:** 263–279.
54. Raiteri, M., G. Maura, G. Bonanno & A. Pittaluga. 1986. Differential pharmacology and function of two $5HT_1$ receptors modulating transmitter release in rat cerebellum. J. Pharmacol. Exp. Therap. **237:** 644–648.
55. Engel, G., M. Göthert, D. Hoyer, E. Schlicker & K. Hillenbrand. 1986. Identity of inhibitory presynaptic 5-hydroxytryptamine (5HT) autoreceptors in the rat brain cortex with $5HT_{1B}$ binding sites. Naunyn-Schmiedebergs Arch. Pharmacol. **332:** 1–7.

56. MIDDLEMISS, D. N. 1984. 8-Hydroxy-2-(di-n-propylamino)tetralin is devoid of activity at the 5-hydroxytryptamine autoreceptor in rat brain. Implications for the proposed link between the autoreceptor and the [^{3}H]5HT recognition site. Naunyn-Schmiedebergs Arch. Pharmacol. **327:** 18–22.
57. NELSON, D. L., H. S. BOURGOIN, J. GLOWINSKI & M. HAMON. 1978. Characteristics of central 5HT receptors and their adaptive changes following intracerebral 5,7-dihydroxytryptamine administration in the rat. Mol. Pharmacol. **14:** 983–995.
58. SAVAGE, D. D., J. MENDELS & A. FRAZER. 1980. Decrease in [^{3}H]serotonin binding in rat brain produced by the repeated administration of either monoamine oxidase inhibitors or centrally acting serotonin agonists. Neuropharmacology **19:** 1063–1070.
59. BENNETT, J. P. & S. H. SNYDER. 1976. Serotonin and lysergic acid diethylamide binding in rat brain membranes: Relationship to postsynaptic receptors. Mol. Pharmacol. **12:** 373–389.
60. BLACKSHEAR, M. A., L. R. STERANKA & E. SANDERS-BUSH. 1981. Multiple serotonin receptors: Regional distribution and effect of raphe lesions. Eur. J. Pharmacol. **76:** 325–334.
61. FILLION, G. M. B., J. C. ROUSSELLE, M. P. FILLION, D. M. BEAUDOIN, M. R. GOINY, J. M. DENIAU & J. J. JACOB. 1978. High-affinity binding of ^{3}H-5-hydroxytryptamine to brain synaptosomal membranes: Comparison with ^{3}H-lysergic acid diethylamide binding. Mol. Pharmacol. **14:** 50–59.
62. SEGAWA, T., T. MIZUTA & Y. NOMURA. 1979. Modifications of central 5-hydroxytryptamine binding sites in synaptic membranes from rat brain after long-term administration of tricyclic antidepressants. Eur. J. Pharmacol. **58:** 75–83.
63. HALL, M. D., S. EL MESTIKAWY, M. B. EMERIT, L. PICHAT, M. HAMON & H. GOZLAN. 1985. [^{3}H]8-Hydroxy-2-(di-n-propylamino)tetralin binding to pre- and post-synaptic 5-hydroxy-tryptamine sites in various regions of the rat brain. J. Neurochem. **44:** 1685–1696.
64. PRANZATELLI, M. R., G. RUBIN & S. R. SNODGRASS. 1986. Serotonin-lesion myoclonus syndromes. I. Neurochemical profile and S_1 receptor binding. Brain Res. **364:** 57–66.
65. NOBLE, M. D., T. H. BROWN & J. H. PEACOCK. 1978. Regulation of acetylcholine receptor levels by a cholinergic agonist in mouse muscle cell cultures. Proc. Natl. Acad. Sci. USA **75:** 3488–3492.
66. MITRIUS, J. C. & D. C. U'PRICHARD. 1985. Regulation of α_2 adrenoceptors by nucleotides, ions and agonists: Comparison in cells of neural and nonneural origin. Adv. Cyclic Nucleotide Protein Phosphorylation Res. **19:** 57–73.
67. LUCKI, I. & A. FRAZER. 1982. Prevention of the serotonin syndrome in rats by repeated administration of monoamine oxidase inhibitors but not tricyclic antidepressants. Psychopharmacology **77:** 205–211.
68. HJORTH, S. 1985. Hypothermia in the rat induced by the potent serotonergic agent 8-OH-DPAT. J. Neural Transm. **61:** 131.
69. GOODWIN, G. M. & A. R. GREEN. 1985. A behavioral and biochemical study in mice and rats of putative agonists and antagonists for $5HT_1$ and $5HT_2$ receptors. Br. J. Pharmacol. **84:** 743–753.
70. GUDELSKY, G. A., J. I. KOENIG & H. Y. MELTZER. 1986. Thermoregulatory responses to serotonin (5HT) receptor stimulation in the rat: Evidence for opposing roles of $5HT_2$ and $5HT_{1A}$ receptors. Neuropharmacology **12:** 1307–1313.
71. GOODWIN, G. M., R. J. DE SOUZA, A. R. GREEN & D. J. HEAL. 1987. The pharmacology of the behavioral and hypothermic responses of rats to 8-hydroxy-2-(di-n-propylamino)tetralin (8-OH-DPAT). Psychopharmacology **91:** 506–511.
72. WOZNIAK, K. M., C. S. AULAKH, J. L. HILL & D. L. MURPHY. 1988. The effect of 8-OH-DPAT on temperature in the rat and its modification by chronic antidepressant treatments. Pharm. Biochem. Behav. **3:** 451–456.
73. HUTSON, P. H., T. P. DONOHOE & G. CURZON. 1987. Hypothermia induced by the putative $5HT_{1A}$ agonist LY16516 and 8-OH-DPAT is not prevented by 5HT depletion. Eur. J. Pharmacol. **143:** 221–228.
74. GOODWIN, G. M., R. J. DE SOUZA & A. R. GREEN. 1987. Attenuation by electroconvulsive shock and antidepressant drugs of the $5HT_{1A}$ receptor-mediated hypothermia and

serotonin syndrome produced by 8-OH-DPAT in the rat. Psychopharmacology **91:** 500–505.
75. KOENIG, J. I., G. A. GUDELSKY & H. Y. MELTZER. 1987. Stimulation of corticosterone and β-endorphin secretion in the rat by selective 5HT receptor subtype activation. Eur. J. Pharmacol. **137:** 1–8.
76. GILBERT, F., C. BRAZELL, M. D. TRICKLEBANK & S. M. STAHL. 1988. Activation of the $5HT_{1A}$ receptor subtype increases rat plasma ACTH concentration. Eur. J. Pharmacol. **147:** 431–439.
77. AULAKH, C. S., K. M. WOZNIAK, M. HAAS, J. L. HILL, J. ZOHAR & D. L. MURPHY. 1988. Food intake, neuroendocrine and temperature effects of 8-OH-DPAT in the rat. Eur. J. Pharmacol. **146:** 253–259.
78. AULAKH, C. S., K. M. WOZNIAK, J. L. HILL & D. L. MURPHY. 1988. Differential effects of long-term antidepressant treatments on 8-OH-DPAT-induced increases in plasma prolactin and corticosterone in rats. Eur. J. Pharmacol. **156:** 395–400.
79. BLIER, P. & C. DEMONTIGNY. 1983. Electrophysiological investigations on the effect of repeated zimelidine administration on serotonergic neurotransmission in the rat. J. Neurosci. **3:** 1270–1278.
80. BLIER, P., C. DEMONTIGNY & J. TARDIF. 1984. Effect of two antidepressant drugs, mianserin and indalopine, on the serotonergic system: Single cell studies in the rat. Psychopharmacology **84:** 242–249.
81. CHAPUT, Y., C. DEMONTIGNY & P. BLIER. 1986. Effects of a selective 5HT reuptake blocker, citalopram, on the sensitivity of 5HT autoreceptors: Electrophysiological studies in the rat brain. Naunyn-Schmiedebergs Arch. Pharmacol. **333:** 342–348.
82. BLIER, P. & C. DEMONTIGNY. 1985. Serotonergic but not noradrenergic neurons in rat CNS adapt to long-term treatment with monoamine oxidase inhibitors. Neuroscience **16:** 949–955.
83. DEMONTIGNY, C. & G. K. AGHAJANIAN. 1978. Tricyclic antidepressants: Long-term treatment increases responsivity of forebrain neurons to serotonin. Science **202:** 1303–1306.
84. OLPE, H. R. & A. SCHELLENBERG. 1981. The sensitivity of cortical neurons to serotonin: Effect of chronic treatment with antidepressants, serotonin uptake inhibitors and monoamine-oxidase-blocking drugs. J. Neural Transm. **51:** 233–244.
85. OLPE, H. R., A. SCHELLENBERG & R. S. G. JONES. 1984. The sensitivity of hippocampal pyramidal neurons to serotonin *in vitro:* Effect of prolonged treatment with clorgyline or clomipramine. J. Neural Transm. **60:** 265–271.
86. BLIER, P., C. DEMONTIGNY & A. J. AZZARO. 1986. Modification of serotonergic and noradrenergic neurotransmission by repeated administration of monoamine oxidase inhibitors: Electrophysiological studies in the rat central nervous system. J. Pharmacol. Exp. Therap. **237:** 987–994.
87. DEVIVO, M. & S. MAAYANI. 1988. 5HT receptors coupled to adenylate cyclase. *In* The Serotonin Receptors. E. Sanders-Bush, Ed. Clifton, NJ: Humana Press, pp. 141–179.
88. DEVIVO, M. & S. MAAYANI. 1986. Characterization of the $5\text{-hydroxytrypamine}_{1A}$ receptor-mediated inhibition of forskolin-stimulated adenylate cyclase activity in guinea pig and rat hippocampal membranes. J. Pharmacol. Exp. Therap. **238:** 248–253.
89. SALOMON, Y. 1979. Adenylate cyclase assay. Adv. Cyclic Nucleotide Res. **10:** 35–55.
90. NEWMAN, M. E. & B. LERER. 1988. Chronic electroconvulsive shock and desimipramine reduce the degree of inhibition by 5HT and carbachol of forskolin-stimulated adenylate cyclase in rat hippocampal membranes. Eur. J. Pharmacol. **148:** 257–260.

DISCUSSION OF THE PAPER

S. MAAYANI (*Mt. Sinai School of Medicine, New York*): Sometimes we see that partial agonists are more sensitive to minute changes in functionality. Did you check agonists other than 5-HT in the cyclase assay?

FRAZER: That's a good suggestion. We have not done that and we should do that.

GEORGE AGHAJANIAN (*Yale University, New Haven, CT*): The adenylate cyclase studies that you did would have been in the hippocampus, is that correct?

FRAZER: Yes.

AGHAJANIAN: I'd like to refer to one of the posters, Frank Yocca and Saul Maayani were involved in, where there is a lack of coupling of the 1A receptor to adenylate cyclase in the raphe. You cited the work of deMontigny showing a desensitization of the response to serotonin at the somatodendritic receptor, which is a 1A-mediated response. The question becomes, if there is not down regulation of the 1A receptor in the raphe, could the desensitization be due to a step distal to the receptor? But how do you examine that, if the receptor is not coupled to the cyclase? Well one way to do that would be to see if there's any alteration in the sensitivity to GTP analogs as far as the binding is concerned. I was wondering if you were doing that or contemplating doing that.

FRAZER: I think that's a very good suggestion also. George's point is that we have set these experiments up to optimize the effects of serotonin on the cyclase using maximally effective concentrations of GTP. When you set your experiment up that way to get maximal coupling, we didn't see any effect of treatment with phenelzine. The question would become, for example, in the phenelzine-treated animals if there would be a difference in the GTP sensitivity needed to induce the cyclase response. Those are experiments that we're planning to do but we have no data on it yet. I think it's a very good suggestion.

GERALD CURZON (*Institute of Neurology, London*): A brief comment on the question. The comment is that we found that a large single dose of a 1A agonist will downregulate presynaptic 1A sites by many criteria. The question is I'm not too sure of the extent to which you agree and disagree with deMontigny because my understanding of what deMontigny says is that most antidepressants by some mechanism or other increase post-synaptic 5-HT function. Some of them do it by down regulating the somatodendritic sites, some of them do it by affecting release, some of them do it by affecting post synaptic receptor sensitivity. And I don't think he was making a single generalization about the somatodendritic sites.

FRAZER: I do not think that our data are in contradiction to the data that Dr. deMontigny has presented. All I'm saying is that the changes in responsivity which he has found to occur in the dorsal raphe nucleus, for example, after antidepressant treatment cannot be accounted for by a change in receptor number. We have seen also changes in responsivity to a direct-acting serotonergic agonist on body temperature after antidepressants. Changes in responsivity do occur. However, other mechanisms will have to be invoked to explain the changes in responsiveness that he, as well as we and many others in this audience, have seen. That would be my point.

FRANK D. YOCCA (*Bristol-Myers, Wallingford, CT*): We have done similar studies to what you have using the 5-HT1_A–mediated adenyl cyclase assay and looked at a number of different antidepressants to see if they had any change on the ability of an agonist to inhibit forskolin stimulation; our data agree with yours. We did not see any changes!

FRAZER: Also, in reference to George's comment, your work, and ours, was done at maximally effective concentrations of GTP, was it not?

YOCCA: Yes, if you are referring to the cyclase studies. In reference to the poster, however, we demonstrated that the inhibition of 8-OH-DPAT binding in cow raphe was more sensitive to the effects of low GppNHp concentrations. George learned quickly from the poster I see. Almost too quickly. But yes, it's a good point!

Serotonin and Suicidal Behavior[a]

J. JOHN MANN,[b] VICTORIA ARANGO,
AND MARK D. UNDERWOOD

Laboratories of Neuropharmacology
Western Psychiatric Institute and Clinic
Department of Psychiatry
University of Pittsburgh
Pittsburgh, Pennsylvania 15213

Alterations in the serotonin system have been found in both suicide completers and in suicide attempters.[1] Several questions are raised by these findings and will be addressed in this paper.

1. Do these changes represent an alteration in net serotonin transmission and, if so, does this functional change represent an increase or decrease?
2. Is the change in the serotonin system in the brain apparent in all serotonin terminal fields in the brain or is there a degree of regional selectivity?
3. Given that altered serotonergic indices have also been observed in association with physical violence, what determines whether the violence is externally or internally (suicide) directed?

REGULATION OF SEROTONIN TRANSMISSION IN THE CENTRAL NERVOUS SYSTEM

In order to provide some background that will permit interpretations of the results of studies of brain biochemistry and neuroendocrine challenge tests we will begin by a brief review of the regulation of serotonin receptor populations in response to alterations in transmitter levels.

The most probable pathophysiological alteration in the serotonin system associated with suicidal behavior involves a reduction in net transmission. This effect could be generated by two different mechanisms. The first involves a reduction in transmitter availability and would potentially be associated with compensatory receptor changes such as up-regulation and/or supersensitivity. The second possible mechanism involves a decrease in signal transduction due to a primary, or direct receptor, defect which may be genetic such as has been described in pseudohypoparathyroidism type 1a.[2] Alternatively, this primary defect may be acquired, as has been described in some forms of hypertension.[3] In this latter situation transmitter levels may be normal or even increased.

Regulations of 5-HT$_2$ Receptors

The details of serotonin receptor regulation are described elsewhere in this volume. Only highlights are presented here, focusing mainly on the effects of increased or

[a]This work was partly supported by PHS Grant MH40210 awarded to J.J. Mann.
[b]Author to whom correspondence should be addressed.

decreased transmitter levels and on the effects of receptor blockade. The 5-HT_2 family of serotonin (5-HT) receptors includes not only the 5-HT_2 subtype, but also, based on a common second messenger system (phosphoinositide hydrolysis) are related to the 5-HT_{1C} class of receptors. 5-HT_2 receptors are heterogeneously distributed and can be found in a variety of locations in brain including cerebral cortex, striatum, olfactory bulb, and the nucleus accumbens. Outside the brain 5-HT_2 receptors are found in platelets, blood vessels, the trachea, and the ileum. 5-HT_{1C} receptors are most dense in choroid plexus, but are also found in ventral caudate-putamen and the intralaminar centromedian and centrolateral nuclei of the thalamus.

The central effects of in vivo administration of 5-HT_2 agonists include twitches and seizures in animals and in man are most notably associated with hallucinations and psychotic symptoms. The most pronounced peripheral effect of administration of 5-HT_2 agonists in animals and man is impairment of blood flow in both the systemic and cerebral vascular beds. Antagonists, by contrast, have little or no behavioral or autonomic effect, but result in enhanced slow wave sleep.

Down-regulation and Desensitization of 5-HT_2 Receptors

Regulation of 5-HT_2 receptors *in vivo* is complex. They have been demonstrated to be down-regulated by a variety of antidepressants, including MAO inhibitors[4] and 5-HT reuptake inhibitors such as the tricyclics (e.g. imipramine),[5] agonists (e.g. 1-(2,5-dimethoxy-4-methyl-phenyl)2-amino propane, DOM, Leysen *et al.*, this volume) antagonists (e.g. mianserin)[6,7] and non-specific neuroleptics (e.g. chlorpromazine).[8] Chronic treatments are most effective in eliciting down-regulation, but acute administration of certain psychotropic drugs has also been shown to be effective. 5-HT_2 bindings sites are decreased 48 h following a single dose of the tetracyclic antidepressant mianserin[6] or following 3 days of treatment with chlorpromazine.[8] Both of these drugs are 5-HT_2 antagonists with nanomolar affinity for the receptor. Although 5-HT_2 receptors are rapidly degraded they are only slowly resynthesized over the following 4–10 days.[6,8] Thus, paradoxically, 5-HT_2 antagonists have been demonstrated to decrease 5-HT_2 binding sites and this effect is only slowly reversible. The transduction mechanism of 5-HT_2 and 5-HT_{1C} receptor stimulation appears to be mediated by phosphoinositide hydrolysis (PI) and the subsequent release of second messengers.[7] In addition, the 5-HT_2 mediated PI response has been studied in vivo and demonstrates a correspondence with change in receptor density.[7] Chronic blockade of 5-HT_2 receptors with mianserin resulted in a paradoxical reduction in 5-HT_2 binding sites and a proportionally similar decrease in PI hydrolysis.[7] Other treatments without effect on 5-HT_2 receptor binding similarly elicited no change in PI response to 5-HT. Thus, PI hydrolysis in response to 5-HT stimulation in cortical tissue is mediated by 5-HT_2 or 5-HT_{1C} receptors and appears to represent the signal transduction mechanism for these receptor populations.

Up-regulation of 5-HT_2 Receptors

The complexity of regulation is further demonstrated by the contrasting effects of pharmacological antidepressants and electroconvulsive treatment. The density of 5-HT_2 receptors is decreased by a variety of antidepressant drugs and is increased by electroconvulsive shock (ECS) treatment. Chronic administration of amitriptyline, iprindole or tranylcypromine decreases ^{3}H-spiperone binding in homogenates of rat cortex[9] while chronic ECS treatment increases the binding to this receptor.[9,10] A single

ECS does not increase ^{3}H-spiperone binding but it is increased after as few as 3 treatments.[9]

The effect of ECS on up-regulating 5-HT_2 receptors seems dependent on intact 5-HT axons since it does not occur in rats lesioned with 5,7-dihydroxytryptamine (5,7-DHT).[10] Interestingly, Stockmeier and Kellar (1986)[10] did not observe any change in the presynaptic markers high-affinity uptake of ^{3}H-5-HT or ^{3}H-imipramine binding following repeated ECS treatment. Whether the down-regulation of 5-HT_2 receptors by antidepressants is dependent on intact serotonergic neurons is unclear since conflicting results have been reported.[5,11]

The effects of lesioning serotonergic neurons in the CNS on 5-HT_2 binding is unclear. Stockmeier and Kellar (1986) directly infused 5,7-DHT into the dorsal and median raphe nuclei to produce serotonergic axon lesions, as evidenced by a 90% reduction of high-affinity uptake of ^{3}H-5-HT, but observed no change in ^{3}H-ketanserin binding 4 weeks after the lesion. In contrast, depletion of brain 5-HT by chronic treatment with the tryptophan hydroxylase inhibitor *p*-chlorophenylalanine did result in up-regulation of 5-HT_2 receptor binding in the study of Roth *et al.* (1987)[12] but not significantly so in the study of Stockmeier and Keller (1986).[10] Reserpine or tetrabenazine treatment, which depletes brain of both serotonin and catecholamines, did increase 5-HT_2 receptor binding,[10,13] suggesting a possible interaction between serotonergic, catecholaminergic and perhaps other neurotransmitters in the regulation of these receptors.

Further support for the involvement of transmitter interactions in 5-HT_2 receptor regulation comes from the data of Wenk and Engisch (1986)[14] who examined the effects of ibotenic acid lesions of the basal forebrain cholinergic system on ^{3}H-ketanserin binding. Four months, but not one week, following basal forebrain lesion of the cholinergic cell bodies the number of ^{3}H-ketanserin binding sites and B_{max} was significantly increased in rat frontal cortex. The effect was specific for frontal cortex, as binding was unchanged in the hippocampus, which also displayed decreased choline acetyltransferase activity. Interestingly, ^{3}H-quinuclidinyl benzilate (^{3}H-QNB) binding was unchanged in cortex or hippocampus, suggesting that cholinergic receptors in cortex do not up-regulate or that changes in postsynaptic binding are offset by a loss of presynaptic binding.

Regulation of 5-HT_1 Receptor Subtypes

The 5-HT_1 family of receptors is made up of the 5-HT_{1A} and the 5-HT_{1B}/5-HT_{1D} receptor subtypes. The 5-HT_{1B} and 5-HT_{1D} subtypes differ in both binding specificity as well as species differences in that the 5-HT_{1B} but not the 5-HT_{1D} receptor subtype is found in rat and mouse and only the 5-HT_{1D} is found in human brain. Transmitter action in the 5-HT_1 family of receptors (with the exception of the 5-HT_{1C} receptor which is linked to phosphoinositide hydrolysis) is mediated by adenylate cyclase. These receptors are thought to function as autoreceptors or postsynaptic receptors depending on brain region. The 5-HT_{1A} autoreceptors are on terminals and the 5-HT_{1D} are somatodendritic and inhibit serotonin release.

Regulation of 5-HT_1 receptors appears to be complicated. Drug-induced enhanced exposure of 5-HT_1 receptors to serotonin does not have consistent effects. Monoamine oxidase inhibitors but not serotonin reuptake blockers reduce the number of 5-HT_1 (and 5-HT_2) binding sites.[4,5] Wirz-Justice *et al.* (1978)[15] reported that clomipramine, a 5-HT reuptake inhibitor, had no effect on 5-HT_1 binding. Chronic desipramine, a selective norepinephrine reuptake inhibitor, has been reported to decrease 5-HT_1 binding in some studies,[16,17] but not in others.[18,19] Electroconvulsive treatment does not alter 5-HT_1 binding but does increase 5-HT_2 binding.[18] 5-HT_1 binding is reported to be increased in hippocampus in response to denervation,[20,21] but the up-regulation is not of the 5-HT_{1A} receptor subtype (Frazer & Hensler, this volume). The effect of denervation on 5-HT_{1B}/5-HT_{1D} receptors is unclear. Studies are required that distinguish subtypes of 5-HT_1 receptors and assess regional brain differences.

In summary, 5-HT_2 agonists appear to be clinically detrimental and result in adverse neurological and psychotic conditions and produce impaired blood flow. The 5-HT_2 receptors can be rapidly desensitized and downregulated, but are only slowly resynthesized over the next 3–6 days. 5-HT_2 antagonists have few detectable behavioral effects in healthy controls or animal studies but may have valuable psychotropic properties. These observations would suggest that serotonergic neurons acting at 5-HT_2 receptors under normal conditions have a low tonic activity and their acute blockade is not disruptive to normal behavior. The effects of denervation are at present unclear, but appear to be under complex regulatory control and perhaps involves interactions with other neurotransmitter systems. 5-HT_1 receptors function via adenylate cyclase and, at least in certain brain regions, serve as autoreceptors. Differences in how these different classes of receptors are regulated and the second messenger system to which they couple indicates that they are pharmacologically, physiologically and anatomically distinct. The involvement of specific serotonin receptor subtypes may vary significantly in different aspects of pathological behavior and psychiatric disorders.

Correlation between Completed Suicide and the Serotonin System in the Brain

The earliest and most numerous studies of the brain of suicide victims have measured levels of serotonin (5-HT) or its major metabolite, 5-hydroxyindoleacetic acid (5-HIAA). Five out of seven studies that we reviewed[22] found lower levels of 5-HT or 5-HIAA in the brain stem. In contrast, only one out of six studies of frontal cortex found a lower level of either 5-HT or 5-HIAA. Data from other brain regions are too few to permit a conclusion. Thus, it appears that although lower levels of 5-HT and 5-HIAA may be present in brain stem, this does not hold true for frontal cortex. The reason for this regional difference is not known. Brain stem levels reflect changes at the level of the serotonin cell bodies whereas cortical measures reflect some aspect of nerve terminal activity.

At a functional level it is not clear how to interpret these data. Much of the serotonin being assayed is probably in a longer term storage pool and not available for immediate release. The level of 5-HIAA is a better index of turnover and therefore neuronal activity. However, both indices are imperfect measures of transmission. For example, administration of serotonin reuptake inhibitors increases net serotonin transmission but 5-HIAA levels fall. Nevertheless, a finding of low brain stem 5-HT and 5-HIAA most likely means reduced serotonergic activity.

Measurement of serotonin receptors offers another approach for assessing the brain serotonin system in post mortem tissue. There are at least six studies that have assayed the imipramine binding sites which are associated with the serotonin transporter on the serotonin nerve terminals. One study found a decrease in prefrontal cortex[23] and four did not.[24–27] One study found a decrease in the hypothalamus[28] and another found decreases in postcentral cortical gyrus, insular cortex and claustrum.[26] One study found increased ^{3}H-imipramine binding in prefrontal cortex[25] and another found increases in pyramidal and molecular layers of the CA hippocampal fields and the hilus of the dentate gyrus.[26] Finally there is one report of a laterality effect involving differences in the proportion of ^{3}H-imipramine binding in the left versus the right side of the brain in suicide victims compared to controls.[29] Thus, overall no consistent findings have been made. One problem may be lack of specificity of imipramine as a ligand. This ligand binds with high affinity to other receptors including α_1-adrenergic, cholinergic muscarinic, and histamine receptors. Use of desipramine to assess specific binding does not add any selectivity. Thus, changes in binding may also be due to alterations in these other receptor populations. A second cause of variance in the results may be regional specificity of the biochemical changes which may be critical if anatomical landmarks are unreliable or not consistently utilized in different studies. A third factor may be laterality effects as reported by Arató and colleagues.[29]

Another series of studies have assayed the postsynaptic 5-HT_2 receptor. Most of these studies have focused on prefrontal cortex. Three out of four studies[24,27,30,31] have reported an increase in the number of 5-HT_2 receptors in the brain of suicide victims. One study has also reported an increase in 5-HT_2 binding in the prefrontal cortex of depressed patients dying of causes other than suicide.[32] This finding raises the possibility that an increase in 5-HT_2 binding may be partly or entirely a correlate of a depressive illness. Given that some studies report an up-regulation of 5-HT_2 receptors in response to depletion of serotonin,[10,12,13] it may be inferred that decreased serotonin release is one possible cause of an increase in 5-HT_2 receptors in the prefrontal cortex of suicide victims.

More recently we have investigated other cortical regions and found that, in contrast to the prefrontal cortex, 5-HT_2 receptors are not increased in temporal pole.[33] Thus, there may be some degree of regional specificity of the correlation between suicidal behavior and 5-HT_2 receptors. If so, it may be possible to identify the brain regions most involved in suicidal behavior as being those areas where the increase in 5-HT_2 receptor number is greatest relative to a control group.

Studies of the 5-HT_1 receptor subpopulations in suicide are more preliminary than those of the serotonin transporter site or the 5-HT_2 receptor. An early study by Mann *et al.* (1986)[31] which did not discriminate between the subclasses of 5-HT_1 receptors did not observe a statistically significant increase in prefrontal cortex. More recently Meltzer (personal communication) has found an increase in 5-HT_{1A} receptors in non-violent suicides. Studies of the 5-HT_{1A}, 5-HT_{1C} and 5-HT_{1D} receptor populations would be of great interest. However, it should be noted that in certain brain regions these are presynaptic autoreceptors and in the other brain regions they are postsynaptic. Such regional differences in the pre- versus postsynaptic locations of 5-HT_{1A} or 5-HT_{1D} receptors emphasizes the necessity of studying selected brain regions. Studies of 5-HT_{1C} receptors are also necessary since these receptors are not confined to the choroid plexus. The mechanism of regulation of these 5-HT_1 receptor subtypes and their physiological roles remain to be more fully defined. Such information will be necessary before alterations in 5-HT_1 binding can be interpreted.

STUDIES OF SEROTONERGIC INDICES IN SUICIDE ATTEMPTERS

There is debate as to the relationship between attempted suicide and completed suicide. Netertheless, important data have been collected in suicide attempters compared to patient controls who have not attempted suicide but have comparable psychiatric diagnosis. A detailed review of these studies is beyond the scope of this paper but the main findings will be summarized.

Cerebrospinal fluid (CSF) levels of 5-HIAA are lower in suicide attempters who have affective disorders, personality disorders or schizophrenia.[1] These findings are strongest in unipolar depression and apparently absent in bipolar disorder. There is some evidence that the type of suicidal behavior may correlate with the level of CSF 5-HIAA. For example, several studies report correlations of CSF 5-HIAA with violent suicidal methods.[34–36] Since, the selection of suicide method is heavily dependent on availability and modeling, it is probable that the association is actually with the seriousness of the intent or medical lethality of the method. If the type of suicide attempt is an important variable, then this may explain some of the differences in the results reported in the literature that are currently attributed to diagnostic subgroups. For example, the failure to find lower CSF 5-HIAA in all studies of schizophrenic suicide attempters may be due to differences in the proportion of highly lethal suicide attempts in the study population.

Another approach has been the use of neuroendocrine challenges using agents that are either direct or indirect serotonin agonists. Clearly this strategy is limited to assessing hypothalamic serotonin pathways. Nevertheless the results have been consistent with the serotonin hypothesis of suicidal behavior. Two studies have found a correlation between suicide attempts and blunting of the prolactin response to the indirect serotonin agonist fenfluramine[37] (Mann, De Meo, McBride *et al.*, unpublished data). Both study populations included patients with an affective disorder. In addition Coccaro and colleagues included some patients without an affective disorder. The association of blunted prolactin responsivity with a history of suicide attempts was found both in patients with an affective disorder as well as in patients who had personality disorder and did not have an affective disorder. Thus, the neuroendocrine data suggest that this found in suicidal behavior is not confined to patients with an affective disorder.

Meltzer *et al.* (1984)[38] reported that the cortisol response to the serotonin precursor 5-hydroxytryptophan was greater in psychiatric patients with a past history of a suicide attempt. They interpreted this finding as being a reflection of increased brain 5-HT_2 receptor number. Support for this position came from their follow-up study with the direct 5-HT_2 receptor agonist, MK-212 in which they found a positive correlation between prolactin response to MK-212 and the prolactin response to 5-hydroxytryptophan. These data suggest that 5-hydroxytryptophan may act as a direct agonist on 5-HT_2 sites. More recently, it has been observed that MK-212 mediates ACTH release via 5-HT_{1C} and not 5-HT_2 receptors.[39] Thus, the results of Meltzer *et al.* (1984)[38] may reflect involvement of 5-HT_{1C} receptors in addition to any changes in 5-HT_2 receptors. With the recognition of multiple serotonin receptor subtypes it will be necessary to use more selective serotonin agonists in future studies.

Another approach that has been used to study the serotonin system in suicidal patients has been to assess serotonin-related proteins in the platelet. The proteins studied include: 1) monoamine oxidase (MAO) B; 2) the serotonin transporter; and 3) serotonin content. No clear relationship has been found between suicidal behavior and either platelet MAO B[40] or with brain MAO B enzyme kinetics in suicide victims.[41] In contrast, low platelet MAO B activity is found in bipolar disorder[42] and probands with relatives who have suicided or made a suicide attempt.[43]

The serotonin transporter has been studied both by measurement of 5-HT uptake kinetics as well as by assaying binding kinetics of ligands for the so-called imipramine binding site.

Platelet serotonin uptake kinetics (V_{max} and K_m) do not appear to differ between suicide attempters and non-attempters.[40] The number of imipramine binding sites have not been found to correlate with suicide attempter status or type of suicide attempt although a trend has been reported for a small positive correlation between K_D and the suicide item score on the Hamilton Depression Scale.[40]

Platelet serotonin content does not appear to distinguish between suicide attempters and non-attempters[40] (McBride, Anderson, Mann *et al.*, unpublished data).

SUMMARY

Studies of the brain of suicide victims indicate there is a decrease in brain stem levels of 5-HT and/or 5-HIAA. There also appears to be a region-specific increase in 5-HT_2 receptors, which are post synaptic and may therefore be increased in number secondary to decreased serotonin levels. Lack of information about how 5-HT_2 and other serotonin receptor populations are regulated hamper our ability to explain the mechanisms underlying these findings. If the initial reports of a decrease in the number of imipramine binding sites prove to be correct then this finding would be further evidence for an effect involving the serotonin neurons, seen in this case at the level of the terminals.

The relationship between suicide attempters and completers remains to be worked out. However, studies of suicide attempters, particularly those making more lethal attempts, appear to confirm the findings made in the brain of suicide completers. Neuroendocrine and CSF studies indicate the presence of serotonin subresponsivity and lower levels of CSF 5-HIAA. Thus, the overall direction of change is towards a weaker serotonin signal which rather than being due to a primary receptor defect (a possibility that cannot be ruled out but for which there is no current evidence), appears to be due to reduced levels of serotonin release. The causes of this effect represent a research challenge.

It is clear that reduced levels of serotonin alone cannot explain the timing and type of suicidal behavior. Future studies must address the role of other neurotransmitters which may explain why some aggression is directed outward (towards other people) and in other cases the aggression is directed toward the self (suicidal behavior). Defining the role of serotonin and other involved transmitter systems is a necessary step before a comprehensive pharmacological treatment plan can be designed and tested.

REFERENCES

1. MANN, J. J. & M. STANLEY, Eds. 1986. Psychobiology of Suicidal Behavior. Ann. N.Y. Acad. Sci. **487.**
2. VAN DOP, C. & H. R. BOURNE. 1983. Pseudohypoparathyroidism. Ann. Rev. Med. **34:** 259–266.
3. FELDMAN, R. D., W. J. LAWTON & W. L. MCARDLE. 1987. Low sodium diet corrects the defect in lymphocyte β-adrenergic responsiveness in hypertensive subjects. J. Clin. Invest. **79:** 290–294.
4. SAVAGE, D. D., J. MENDELS & A. FRAZER. 1980. Monoamine oxidase inhibitors and serotonin uptake inhibitors: differential effects on ^{3}H-serotonin binding sites in rat brain. J. Pharmacol. Exp. Ther. **212:** 259–263.
5. DUMBRILLE-ROSS, A. & S. W. TANG. 1983. Noradrenergic and serotonergic input necessary for imipramine-induced changes in beta but not S_2 receptor densities. Psychiatry Res. **9:** 207–215.
6. BLACKSHEAR, M. A. & E. SANDERS-BUSH. 1982. Serotonin receptor sensitivity after acute and chronic treatment with mianserin. J. Pharmacol. Exp. Ther. **221:** 303–308.
7. CONN, P. J. & E. SANDERS-BUSH. 1986. Regulation of serotonin-stimulated phosphoinositide hydrolysis: Relation to the serotonin 5-HT_2 binding site. J. Neurosci. **6:** 3669–3675.
8. ANDREE, T. H., M. MIKUNI, C. Y. TONG, J. I. KOENIG & H. Y. MELTZER. 1986. Differential effects of subchronic treatment with various neuroleptic agents on serotonin 2 receptors in rat cerebral cortex. J. Neurochem. **46:** 191–197.
9. KELLAR, K. J., C. S. CASCIO, J. A. BUTLER & R. N. NURTZKE. 1981. Differential effects of electroconvulsive shock and antidepressant drugs on serotonin-2 receptors in rat brain. Eur. J. Pharmacol. **69:** 515–518.

10. STOCKMEIER, C. A. & K. J. KELLAR. 1986. In vivo regulation of serotonin-2 receptor in rat brain. Life Sci. **38:** 117–127.
11. GANDOLFI, O., M. L. BARBACCIA & E. COSTA. 1985. Differential effects of serotonin antagonists on ^{3}H-mianserin and ^{3}H-ketanserin recognition sites. Life Sci. **36:** 713–721.
12. ROTH, B. L., S. MCLEAN, X.-Z. ZHU & D.-M. CHUANG. 1987. Characterization of two [^{3}H]ketanserin recognition sites in rat striatum. J. Neurochem. **49:** 1833–1838.
13. TAKAHASHI, R., T. TATEISHI, H. YOSHIDA & I. HIRONAKA. 1982. Effects of chronic treatment with antidepressant drugs on serotonergic receptor binding activity in normal and tetrabenazine depression rat. *In* New Vistas in Depression. S. Langer, R. Takahashi, T. Segawa & M. Briley, Eds. New York: Pergamon Press, pp. 29–36.
14. WENK, G. L. & K. L. ENGISCH. 1986. [^{3}H]Ketanserin (serotonin type 2) binding increases in rat cortex following basal forebrain lesions with ibotenic acid. J. Neurochem. **47:** 845–850.
15. WIRZ-JUSTICE, A., K. KRAUCHI, M. LICHTSTEINER & H. FEER. 1978. Is it possible to modify serotonin receptor sensitivity? Life Sci. **23:** 1249–1254.
16. SEGAWA, T., T. MIZUTA & Y. NOMURA. 1979. Modifications of central 5-hydroxytryptamine binding sites in synaptic membranes from rat brain after long-term administration of tricyclic antidepressants. Eur. J. Pharmacol. **58:** 75–83.
17. MAGGI, A., D. C. U'PRICHARD & S. J. ENNA. 1980. Differential effects of antidepressant treatment on brain monoamine receptors. Eur. J. Pharmacol. **61:** 91–98.
18. BERGSTROM, D. A. & K. J. KELLAR. 1979. Effect of electroconvulsive shock on monoaminergic receptor binding sites in rat brain. Nature **278:** 464–466.
19. PEROUTKA, S. J. & S. H. SNYDER. 1980. Long-term antidepressant treatment decreases spiroperidol-labeled serotonin receptor binding. Science **210:** 88–90.
20. NELSON, D. L., A. HERBERT, S. BOUGOIN, J. GLOWINSK & M. HAMON. 1978. Characteristics of central 5-HT receptors and their adaptive changes following injections of 5,7-dihydroxy tryptamine administration in the rat. Mol. Pharmacol. **14:** 983–995.
21. SEEMAN, P., K. WESTMAN, D. COSCINA & J. J. WARSH. 1980. Serotonin receptors in hippocampus and frontal cortex. Eur. J. Pharmacol. **66:** 179–191.
22. MANN, J. J., V. ARANGO, P. M. MARZUK, S. THECCANAT & D. J. REIS. 1989. Evidence of the serotonin hypothesis of suicide: A review of postmortem studies. Br. J. Psychiatry. **155** (suppl. 8): 7–14.
23. STANLEY, M., J. VIRGILIO & S. GERSHON. 1982. Tritiated imipramine binding sites are decreased in the frontal cortex of suicides. Science **216:** 1337–1339.
24. CROW, T. J., A. J. CROSS, S. J. COOPER, J. F. W. DEAKIN, I. N. FERRIER, J. A. JOHNSON, M. H. JOSEPH, F. OWEN, M. POULTER, R. LOFTHOUSE, J. A. N. CORSELLIS, D. R. CHAMBERS, G. BLESSED, E. K. PERRY, R. H. PERRY & B. E. TOMLINSON. 1984. Neurotransmitter receptors and monoamine metabolites in the brains of patients with Alzheimer-type dementia and depression, and suicides. Neuropharmacology **23:** 1561–1569.
25. MEYERSON, L. R., L. P. WENNOGLE, M. S. ABEL, J. COUPET, A. S. LIPPA, C. E. RAUH & B. BEER. 1982. Human brain receptor alterations in suicide victims. Pharmacol. Toxicol. **17:** 159–163.
26. GROSS-ISSEROFF, R., M. ISRAELI & A. BIEGON. 1989. Autoradiographic analysis of tritiated imipramine binding in the human brain post mortem: Effects of suicide. Arch. Gen. Psychiatry **46:** 237–241.
27. MELTZER, H. Y., J. F. NASH, T. OHMORI, L. KREGEL & R. ARORA. 1987. Neuroendocrine and biochemical studies of serotonin and dopamine in depression and suicide. International Conference on New Directions in Affective Disorders, **S60:**
28. PAUL, S. M., M. REHAVI, P. SKOLNICK & F. K. GOODWIN. 1984. High affinity binding of antidepressants to a biogenic amine transport site in human brain and platelet; studies in depression. *In* Neurobiology of Mood Disorders. R. M. Post & C. J. Bellinger, Eds. Baltimore: Williams and Wilkins, pp. 845–853.
29. ARATÓ M., K. TEKES, M. PALKOVITS, E. DEMETER & A. FALUS. 1987. Serotonergic split brain and suicide. Psychiatry Res. **21:** 355–356.
30. STANLEY, M. & J. J. MANN. 1983. Increased serotonin-2 binding sites in frontal cortex of suicide victims. The Lancet **i:** 214–216.

31. MANN, J. J., M. STANLEY, P. A. MCBRIDE & B. S. MCEWEN. 1986. Increased $serotonin_2$ and β-adrenergic receptor binding in the frontal cortices of suicide victims. Arch. Gen. Psychiatry **43:** 954–959.
32. FERRIER, I. N., I. G. MCKEITH, A. J. CROSS, E. K. PERRY, J. M. CANDY & R. H. PERRY. 1986. Postmortem neurochemical studies in depression. Ann. N.Y. Acad. Sci. **487:** 128–142.
33. ARANGO, V., P. ERNSBERGER, H. TIERNEY, M. STANLEY, D. J. REIS & J. J. MANN. 1987. Autoradiographic demonstration of increased 5-HT_2 receptors in frontal cortex of suicide victims. Soc. Neurosci. Abstr. **13:** 216.
34. ÅSBERG, M., L. TRÄSKMAN & P. THORÉN. 1976. 5-HIAA in the cerebrospinal fluid. A biochemical suicide predictor?. Arch. Gen. Psychiatry **33:** 1193–1197.
35. ÅGREN, H. 1980. Symptom patterns in unipolar and bipolar depression correlating with monoamine metabolites in the cerebrospinal fluid. II. Suicide. Psychiatry Res. **3:** 225–236.
36. BANKI, C. M. & M. ARATÓ. 1983. Amine metabolites and neuroendocrine responses related to depression and suicide. J. Affect. Disord. **5:** 223–232.
37. COCCARO, E. F., L. J. SIEVER, H. M. KLAR, G. MAURER, K. COCHRANE, T. B. COOPER, R. C. MOHS & K. L. DAVIS. 1989. Serotonergic studies in patients with affective and personality disorders. Arch. Gen. Psychiatry **46:** 587–599.
38. MELTZER, H. Y., R. PERLINE, B. J. TRICOU, M. LOWY & A. ROBERTSON. 1984. Effect of 5-hydroxytryptophan on serum cortisol levels in major affective disorders. II. Relation to suicide, psychosis and depressive symptoms. Arch. Gen. Psychiatry **41:** 379–387.
39. KING, B. H., C. BRAZELL, C. T. DOURISH & D. N. MIDDLEMISS. 1989. MK-212 increases rat plasma ACTH concentration by activation of the 5-HT_{1C} receptor subtype. Neurosci. Lett. **105:** 174–176.
40. MELTZER, H. Y. & R. C. ARORA. 1986. Platelet markers of suicidality. Ann. N.Y. Acad. Sci. **487:** 271–280.
41. MANN, J. J. & M. STANLEY. 1984. Postmortem monoamine oxidase enzyme kinetics in the frontal cortex of suicide victims and controls. Acta Psychiatr. Scand. **69:** 135–139.
42. MANN, J. J. 1979. Altered platelet monoamine oxidase activity in affective disorders. Psychol. Med, **9:** 729–736.
43. BUCHSBAUM, M. S., R. J. HAIER & D. L. MURPHY. 1977. Suicide attempts, platelet monoamine oxidase and the average evoked response. Acta Psychiatr. Scand. **56:** 69–79.

DISCUSSION OF THE PAPER

QUESTION: Just a technical question. How do you define the high affinity β-adrenergic receptor, what is that?

MANN: A biphasic competition curve can be demonstrated if you use pindolol with isoproterenol as the competition drug. The high affinity component of binding can be abolished by GTP analogs. By selecting the appropriate concentration of isoproterenol you can either detect high plus low affinity specific binding or you can detect a component of binding that is almost entirely high affinity.

QUESTION: Do the changes in high and low affinity β-adrenergic binding have any relationship to the β_1 and β_2 subtypes of these receptors?

MANN: We haven't done that experiment yet, but it's a good question.

MARK MOLLIVER, (*Johns Hopkins, Baltimore, MD*): I was very interested in your distribution of 5-HT_2 receptors in frontal and temporal cortex. My question is what part of temporal cortex were you looking at because MaryAnn Wilson in my laboratory has found that distinctly different patterns of serotonin enervation in different portions of temporal cortex in the primate.

MANN: Well we'll be very interested to hear the results of her work, but we used area 38, which is easily identifiable as temporal pole.

JOHN SCHMIDTS (*NIH, Bethesda, MD*): I know historically speaking corticosteroids and their metabolites were first investigated as possible biological markers for suicidal activity. Can you comment on the most recent thoughts regarding the interplay between corticosteroids and serotonin?

MANN: The interplay. Well perhaps what you should really ask is what is the story with corticosteroids and suicidal behavior, and the answer to that is that it seems pretty difficult, at least certainly for us, to demonstrate any specific relationship that is independent of the affective disorder. There are serotonergic mechanisms involved in the regulation of the HPA axis and we have looked at the cortisol response, for example, to fenfluramine. We find the changes are in the same general direction as the prolactin response, but not as clear. Part of the reason is that we are looking for an effect super-imposed on a significant diurnal variation, which decreases the signal to noise ratio.

JOSE PALACIOS (*Sandoz, Basel, Switzerland*): The ligand that you used, iodo-LSD, could also label serotonin 1-C receptors. Did you examine in detail if part of your binding can be linked to serotonin 1-C and is this altered in suicide?

MANN: That's a good question. We haven't analyzed all our autoradiograms. One of the reasons we chose LSD was that while at the same time it is a nonselective sort of ligand it did give us the opportunity to save tissue and measure several subclasses of receptors.

Role of Serotonin in Depression[a]

HERBERT Y. MELTZER

Department of Psychiatry
Case Western Reserve University School of Medicine
Cleveland, Ohio 44106

There is extensive evidence that serotonin (5-HT) may play an important role in making persons vulnerable to depression or mania, in the precipitation of an episode of depression or mania, or in the mechanism of action of drugs which have an antidepressant, antimanic or mood stabilizing effect.[1,2] Some of the evidence for these conclusions will be critically reviewed and some new data from our laboratory will be presented.

EFFECT OF SEROTONIN ON MOOD, AFFECT AND SOMATIC FUNCTIONS

Through its wide distribution over the neuraxis, serotonergic neurons arising from cell bodies in the mid-brain raphe play a key role in the modulation of excessive stimuli of a wide variety and in the organization and coordination of appropriate responses. It has been suggested that the overall function of the serotonergic system is to enable the organism to ward off feelings of fear, helplessness, and depression.[3] This concept of the role of 5-HT is consistent with the evidence that 5-HT has a significant, but certainly not exclusive, role in the core behaviors that are disturbed in the affective disorders: e.g., mood, sleep, sexual activity, appetite, circadian and seasonal rhythms, neuroendocrine function, body temperature, motor activity, and cognitive function.[2]

Among the most persuasive evidence for a role of 5-HT in mood disorders is the effect on mood itself of lowering brain 5-HT concentration in depressed patients[4,5] and normals.[6,7] Decreasing the availability of tryptophan (TRP), the precursor of 5-HT, by restricting TRP intake can induce a mild depression in normals[5,6] or a more severe depression in recovering depressed patients.[7] Interestingly, when the same TRP-lowering amino acid mixture was given to unmedicated depressed patients, it was reported to have a mood elevating effect.[7]

Meta-chlorophenylpiperazine (mCPP), a 5-HT agonist with moderate affinity for 5-HT_2, 5-HT_{1C} and a somewhat weaker affinity for 5-HT_{1D} sites,[8] has been reported to produce elevated mood in normal volunteers.[9] There is, however, little evidence that other 5-HT agonists, e.g. buspirone (a 5-HT_{1A} agonist),[10] or MK-212,[11] a 5-HT agonist with effects on various subtypes of 5-HT_1 and 5-HT_2 receptors, can produce elevated mood in normal controls or depressed patients (Meltzer, unpublished data). Charney *et al.*[12] reported a mood elevating effect of i.v. L-tryptophan in normals. However, Winokur *et al.*[13] observed only sedative effects of an i.v. tryptophan load in normal subjects.

Sleep disturbances are another major feature of depression which could relate to an abnormality of the serotonergic system. There is a large body of evidence which indicates that 5-HT is essential to sleep.[14] Short rapid eye movement (REM) latency is frequently found in depressed patients, although it is also present in other psychiatric

[a]Supported in part by NIMH grant number MH-41684, MH-41683, MH-41594, grants from the Cleveland Foundation, The Prentiss Foundation, and the Sawyer Foundation. Dr. Meltzer is the recipient of a Research Scientist Award grant number MH-47808.

disorders as well. Pharmacological studies suggest that short REM latency may be due to decreased serotonergic activity,[15,16] although cholinergic mechanisms may also be involved.[17]

Numerous disturbances in circadian rhythms, e.g. the hypothalamic-pituitary-adrenal axis,[18] the sleep-wake cycle[19] and body temperature[20] have been observed in depressed patients[21] and could be related to disturbances in the regulatory mechanisms concentrated in the Zeitgeber of the suprachiasmatic nucleus, an area with a dense serotonergic innervation.[22] Circannual rhythms in serotonergic function could play a role in seasonal affective disorder (SAD), a form of depression characterized by recurrent fall and winter episodes marked by hypersomnia, overeating and carbohydrate craving. There is some evidence for a disturbance of 5-HT function in SAD.[23] In addition, 5-HT plays an important role in motor function,[24,25] anxiety,[26] memory,[27] and aggression,[28] all of which may be disturbed in affective disorders.

These considerations of the role of 5-HT in functions that are disturbed in depression suggests that an abnormality of the serotonergic system could contribute to some aspects of depression. However, other neurotransmitters and modulators, especially norepinephrine, acetylcholine and dopamine, certainly affect these behaviors as well.[2] Abnormality of 5-HT may be the primary disturbance in some of these behaviors or may only play a contributory role in others. There is also no reason to exclude the possibility that the neural basis for disturbances of particular functions differ in various subtypes of depression, e.g. unipolar vs. bipolar, or even within a given subtype of depression. Any particular symptom of depression may be the expression of diverse biological deficits or failures of adaptation.[2]

BIOCHEMICAL ABNORMALITIES OF THE SEROTONERGIC SYSTEM IN MAJOR DEPRESSION

Numerous biochemical abnormalities in various aspects of serotonergic neurotransmission have been reported in major depression.[1,2] Only some of the more important studies can be considered here.

CSF 5-Hydroxyindoleacetic Acid

The basal concentration in cerebrospinal fluid (CSF) of 5-hydroxyindoleacetic acid (5-HIAA), the major metabolite of 5-HT, and the concentration of 5-HIAA in CSF after blockade of its reabsorption by pretreatment with probenecid, have been studied as an index of central serotonergic activity in depressed patients. Lumbar CSF 5-HIAA concentration has been reported to correlate with brain 5-HIAA in man.[29] Most studies report no differences in CSF 5-HIAA levels in depressed patients, with[30-32] or without[33,34] probenecid treatment. However, Agren[35] and Asberg *et al.*[36] both found decreased CSF 5-HIAA levels in hospitalized, drug-free unipolar and bipolar depressed patients. Three studies have reported positive correlations between CSF 5-HIAA levels and various depressive symptoms[31,35,37] but others have found no relationship or the opposite.[35,38] Antidepressant treatment with drugs such as imipramine or amitriptyline, which inhibit 5-HT uptake *decrease* CSF 5-HIAA levels, but the magnitude of the decrease does not predict clinical response.[39] This has been suggested to be due to feedback effects caused by increased post-synaptic receptor stimulation. We have found that the log CSF 5-HIAA levels in 25 unmedicated depressed patients were not significantly correlated ($\rho = 0.034, p =$ NS) with severity of depression as measured by the Depression subscale of the Schedule for Affective Disorders and Schizophrenia[40] (Meltzer *et al.*, in preparation).

Plasma Tryptophan

Brain 5-HT concentration may be related to the ratio of total or free tryptophan (TRP) to large neutral amino acids (LNAA) in plasma.[41] Since tryptophan (TRP) concentration is the rate limiting factor in the synthesis of brain 5-HT and tryptophan competes with LNAA for entry into the brain,[42] there has been extensive investigation of plasma TRP and LNAA concentrations in unmedicated depressed patients. Basal plasma TRP levels do not appear to be decreased in depressed patients[42,43] but Cowen *et al.*[44] did report such a decrease. We have found no significant difference in plasma TRP levels in 38 unmedicated depressed patients compared to 25 normal controls (36.8 SE 1.3 vs 37.3 ± 1.6μmol/l). Free, rather than total, TRP concentration in plasma has been suggested to be the critical determinant of brain TRP and hence 5-HT concentration. Plasma free TRP may also be decreased in depressed patients.[45] However, we have not found any difference in plasma free TRP or the free TRP/total TRP ratio in unmedicated major depression (Meltzer and Schrieber, unpublished data). There is conflicting evidence concerning plasma TRP/LNAA ratio in unmedicated depressed patients. Some studies report a decrease in the plasma TRP/LNAA ratio in depressed patients,[44–48] while others do not.[49] However, the plasma ratio TRP/LNAA has been reported to predict clinical response to treatment with L-TRP[43] or amitriptyline[45] but not nortryptiline.[49] These findings are intriguing but requires independent replication before much can be made of them. The ratio of plasma TRP/valine + leucine as well as plasma TRP was reported to be significantly lower in depressed patients who are dexamethasone non-suppressors and predicted post-dexamethasone serum cortisol.[48] However, Joseph *et al.*[47] did not find any differences in the TRP/LNAA ratio between dexamethasone suppressors and non-suppressors.

We have found no difference in the concentration of plasma 5-HIAA levels between drug-free depressed patients and normal controls (Meltzer, unpublished results). However, there was a significant positive correlation between plasma TRP and 5-HIAA concentrations in normal controls (Spearman $\rho = 0.36$, $N = 27$, $p = 0.06$) which was not present in patients with major depression ($\rho = 0.14$, $N = 45$, $p = 0.36$). The origin of plasma 5-HIAA is unknown. It is presumably mainly peripheral and may be related to dietary factors. The lack of correlation between plasma TRP and 5-HIAA in depressed patients may be related to abnormalities in the regulation of plasma TRP concentrations in major depression. There were significantly higher plasma TRP levels in the male depressed patients (42.7 ± 1.5, N = 27) vs. the female depressed (30.9 ± 2.2, N = 11) ($p = 0.0001$) and a similar trend in the normal controls (40.3 ± 2.0, N = 14, vs. 34.2 ± 2.6, N = 8). It is conceivable that the lower plasma TRP levels may be related to the much higher rate of depression in females than males.

Plasma TRP levels following ingestion of large oral doses of TRP have been reported to be lower in depressed patients in two studies,[50,51] but not in two others.[52,52] We have previously reported that TRP concentrations following an i.v. TRP load were significantly lower in major depression than normal controls and that this persisted after treatment.[51] However, Hoes *et al.*[52] found no difference in plasma TRP levels following a 5 gram oral TRP load between depressed patients and controls during the absorption, distribution and elimination phases. If TRP were rapidly cleared from plasma in some but not all cases of major depression, this could contribute to the absence of a correlation between plasma TRP and 5-HIAA levels we observed in depressed patients.

Specific groups of depressed patients have been reported to have increased red cell transport of TRP which was associated with decreased plasma TRP levels.[54] These investigations also noted increased removal of TRP from plasma into brain of depressed patients in a study of arteriovenous differences.[55] It is possible that more rapid transport of TRP from the plasma into non-central nervous system compartments, especially muscle and liver, may contribute to a serotonergic deficit in depression.[51]

In all, the hypothesis that an abnormality in the availability of TRP, perhaps because of excessive metabolism of TRP in the periphery, leading to a decrease in the TRP/LNAA ratio as a possible cause of decreased brain 5-HT in some depressed patients, remains an attractive hypothesis worthy of study. If confirmed, it would suggest possible therapeutic approaches designed to enhance the availability of TRP in those individuals.

An abnormality in the availability of TRP could explain why treatments which affect TRP concentration can affect mood in major depression. Shopsin *et al.*[4] reported that *p*-chlorophenylalanine (PCPA), which inhibits 5-HT synthesis, rapidly induced a recurrence of depression in antidepressant-treated, recovered depressed patients. Recently, Heninger *et al.*[5] reported that administration of a large dose of an amino acid mixture devoid of TRP to depressed patients receiving a TRP-free diet produced a 90% decrease in plasma TRP levels and an increase in depressive symptoms, in 15/21 antidepressant-treated recovered depressed patients. However, they also found that this same procedure improved depressive symptoms in 14 of 22 untreated depressed patients. This latter finding is quite difficult to reconcile with the hypothesis that too little 5-HT is critical to the maintenance of depression. However, the ability of PCPA to precipitate relapse and the effect of the low TRP diet plus large doses of LNAA on mood in recovering depressed patients does suggest that 5-HT synthesis is critical to the action of antidepressant treatment.

Administration of TRP by itself to depressed patients sometimes but rarely has an antidepressant effect. TRP administration is sometimes able to facilitate the antidepressant effect of MAO inhibitors.[56] These results might suggest that antidepressant treatments increase the requirement for TRP while in the untreated state, TRP concentrations are not rate limiting.

There have been numerous reports that about 50% of major depressed patients who fail to respond to antidepressants alone will respond when given lithium carbonate (see Meltzer and Lowy[2] for review). Usually the response is elicited within 3–14 days. Even cases of psychotic depression have responded and some patients cycle up into a hypomanic condition.[57] This effect may be due to the ability of lithium to enhance pre- and post-synaptic serotonergic mechanisms.[58,59] There have been a series of neuroendocrine studies in depressed patients or rodents which have assessed the effect of subchronic lithium treatment on serotonergic activity. We reported that the cortisol response to 5-hydroxytryptophan (5-HTP) in a small number of depressed patients[60] and the prolactin (PRL) response in rodents[61] was enhanced by subchronic lithium carbonate administration. However, we have not found that the cortisol or PRL responses to 5-HTP were enhanced after 4–6 weeks lithium carbonate treatment in a subsequent group of depressed patients (Meltzer *et al.*, in preparation). Glue *et al.*[62] reported that lithium administration increased the PRL, but not the growth hormone (GH) response, to i.v. TRP in eight normals. Price *et al.*[63] reported that lithium treatment for less than one week significantly increased the PRL response to i.v. TRP whereas 3–4 weeks treatment with lithium had no effect on the PRL response. Neither brief or subacute lithium treatment had any effect on the slight tendency of i.v. TRP to worsen mood in these patients. Since the PRL response to L-TRP is enhanced by ritanserin,[64] a 5-HT_2 antagonist, but the PRL response to 5-HTP is blocked by ritanserin (Meltzer *et al.*, unpublished results), it is possible that the effect of L-TRP is not mediated by a serotonergic mechanism. In sum, chronic lithium administration does not appear to lead to enhancement of the effects of 5-HT_2 receptor stimulation, at least in man, but further study is required.

BLOOD PLATELET SEROTONIN MECHANISM IN DEPRESSION

There has been extensive utilization of platelets to study selected serotonergic mechanisms in major depression because the processes of 5-HT uptake and ^{3}H-imipramine binding in platelets are comparable to those in brain 5-HT terminals.[65] There is controversial evidence for a decrease in the number of platelet 5-HT uptake and ^{3}H-imipramine binding sites in major depression.[2] A decrease in the number of brain 5-HT uptake sites would be expected to increase the availability of 5-HT at nerve terminals. This is counter to the hypothesis of too little 5-HT as a cause of depression. Therefore, we suggested that any decrease of 5-HT uptake sites in brain may be part of a compensatory mechanism to enhance the availability of 5-HT.[66] The decrease in platelet ^{3}H-imipramine binding and an analogous decrease in brain 5-HT terminals would tend to decrease the availability of 5-HT and hence is more consistent with the hyperserotonergic hypothesis of mood disorders. ^{3}H-Paroxetine has been suggested to be a better candidate for labeling the 5-HT transporter than ^{3}H-imipramine.[67] A recent report found no differences in the number of ^{3}H-paroxetine binding sites in the blood platelets of 23 patients with major depression who have been withdrawn from antidepressants 7–14 days previously compared to 23 normal controls.[68] The discrepancies between various platelet studies are probably related to differences in subtypes and severity of depression, lingering effects of previous drug treatments, duration of illness, circadian and seasonal variations in platelet 5-HT uptake and ^{3}H-imipramine binding, as well as differences in the methods of platelet isolation and determination of the uptake and binding parameters.

There have been several recent studies of apparent 5-HT_2 binding sites and receptors in blood platelets of depressed patients. The platelet receptor appears to mediate platelet aggregation.[69] McBride *et al.*[70] and Cowen *et al.*,[71] used ^{125}iodo-(I) LSD as ligand while Biegon *et al.*[72] and Arora and Meltzer,[73] used ^{3}H-LSD. The first two groups found no difference in ^{125}I-LSD binding between the two groups. However, Biegon *et al.*[72] and Arora and Meltzer[73] found significant increases. The increase noted by Arora and Meltzer[73] was found only in females. McBride *et al.*[70] also reported a significant increase in the number of platelet 5-HT_2 binding sites in depressed patients who had made serious suicide attempts. Arora and Meltzer[73] found no relation between the number of platelet ^{3}H-LSD binding sites and severity of depression or suicide rating. They also found no correlation between the number of ^{3}H-LSD sites and the number of platelet 5-HT uptake or ^{3}H-imipramine binding sites. Since there is evidence for increased numbers of 5-HT_2 sites in the frontal cortex of suicide victims,[74,75] many of whom may be presumed to be depressed, the results of Biegon *et al.*[72] and Arora and Meltzer[73] take on increased interest. However, a number of potential artifacts must be controlled for, especially drug treatment. Cowen *et al.*[71] have reported that tricyclic antidepressants increase the number of ^{3}H-I^{125}-LSD binding sites in blood platelets. We have also obtained similar results with ^{3}H-LSD (Arora and Meltzer, unpublished data). However, tricyclic antidepressants as well as fluoxetine and iprindole decrease the number of 5-HT_2 binding sites in rat brain.[76] This discrepancy points out the limitation of the platelet as a model.

NEUROENDOCRINE CHALLENGE STUDIES

The secretion of various hormones, e.g. cortisol, PRL and growth hormone, are promoted by 5-HT in both man and rodents.[77] There has been a voluminous literature exploring the utility of using the effects of serotonergic agents on plasma concentrations of these hormones to study 5-HT mechanisms in depression. We have reviewed these elsewhere,[2,77] so only some highlights will be reported here. The cortisol response to D,L,5-HTP, 200 mg, p.o. was reported to be increased in patients with major depression.[60] Maes *et al.*[48] found an enhanced cortisol response to 200 mg of L-5-HTP, the active form of 5-HTP, in female major depressed patients compared to females with minor depression. However, we have found no difference in the cortisol or PRL response to L-5-HTP, 200 mg in male or female depressed patients compared to normal controls (Meltzer *et al.*, unpublished data). This may be due to differences in dose as well as patient population. It is possible that the D,L-5-HTP in our first study selectively interfered with the uptake or metabolism of L-5-HTP in depressed patients. We found no PRL response to D,L-5-HTP, 200 mg, whereas there is a slight but significant PRL response to L-5-HTP, 200 mg, in both depressed patients and normal controls. We have also found no difference in the cortisol the PRL responses to MK-212, 20 mg, in major depression and controls (unpublished results). MK-212 is believed to stimulate the HPA axis via a 5-HT_2 or 5-HT_{1C} mechanism, or both.[78] These results suggest the 5-HT_2 or 5-HT_{1C} receptors which stimulate PRL and cortisol secretion in major depression are not upregulated.

It is noteworthy that the plasma cortisol responses to 5-HTP and MK-212 in normal volunteers are markedly greater than the PRL response when both are expressed as a function of basal levels. Two other presumably serotonergic challenge agents which have shown a blunted PRL response in depression, i.v. TRP[79] and oral fenfluramine,[80] both produce much greater increases in PRL than cortisol secretion in normal subjects. The PRL response to i.v. TRP in depressed patients is enhanced by 28 days treatment with desipramine or amitriptyline.[81] Cowen *et al.*[82] also found an increase in the PRL response to TRP following 16 days treatment with desipramine in 9 normal volunteers, and this enhancement was positively correlated with the increase in the number of platelet ^{3}H-LSD binding sites. Kahn *et al.*[84] found no differences in the cortisol responses to oral mCPP, a 5-HT_{1C} and 5-HT_2 agonist, in major depression compared to normal controls. The PRL response to tryptophan was reported to be enhanced by ritanserin, a 5-HT_2 antagonist,[84] whereas we have found that the cortisol response to 5-HTP is blocked by ritanserin.[85] The PRL response to 5-HTP also appeared to be blocked by ritanserin.[85] Since the cortisol response to serotonergic agents appears to be greater than the PRL response, these results suggest that L-TRP may not be a valid probe of the serotonergic system.

These results are difficult to interpret until it is clarified which, if any, of these agents, are stimulating cortisol or PRL secretion by 5-HT mechanisms. It is possible that effects of 5-HTP, TRP, MK-212 and mCPP on various 5-HT receptor subtypes or other neurotransmitters, e.g. dopamine or norepinephrine, may explain the discrepancies between these findings. At the current time, there are no clear conclusions to be drawn about the status of pre- or post-synaptic serotonergic mechanisms in depression on the basis of these studies.

SEROTONIN UPTAKE INHIBITORS

The recent development of drugs which are specific 5-HT uptake blockers, e.g. zimelidine, fluoxetine, citalopram, fluvoxamine, and sertaline, provides a new set of

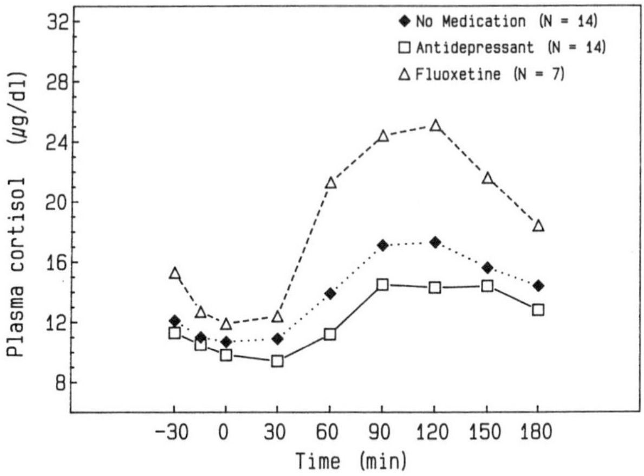

FIGURE 1. Potentiation of cortisol response to L-5-HTP by fluoxetine. Patients with OCD or MD were given L-5-HTP 100 mg p.o. first unmedicated and then after fluoxetine treatment or treatment with a tricyclic antidepressant (desipramine, nortriptyline, etc.) or maprotiline (N = 1). All subjects were studied after an overnight fast and placement of an indwelling venous catheter beginning at 9 A.M. At 9:30 the first blood samples were obtained and the 5-HTP was given 30 min later. Cortisol was determined by radioimmunoassay. The study was repeated 3–6 weeks after treatment with antidepressants and 4–12 weeks (median 7) weeks after fluoxetine. Only the pretreatment cortisol response for the anti-depressant group is given. Fluoxetine significantly potentiated the response to 5-HTP in both the OCD and MD patients. The antidepressants did not.

tools to study the role of 5-HT in depression. Although these drugs have variable effects on catecholaminergic mechanisms,[2] they share in common the ability to promote serotonergic activation by a more potent effect on 5-HT uptake than imipramine or desipramine. Thus, we have found that administration for 1–4 weeks of fluoxetine and sertraline potentiates the effect of L-5-HTP on PRL and cortisol secretion but amitriptyline, imipramine etc. do not (FIG. 1). Moreover, fluoxetine, but not tricyclic antidepressants, potentiates the response to MK-212 (FIG. 2). In view of the lack of evidence that 5-HT uptake blockers are no more effective or more rapidly effective than tricyclic antidepressants for the treatment of major depression,[2,86–90] whereas they are more effective in the treatment of obsessive-compulsive disorder[91] and perhaps panic attacks.[92] These results suggest that the further enhancement of serotonergic activity by specific 5-HT uptake blockers compared to tricyclic antidepressants is not of great value in the treatment of major depression.

Serotonin and Suicide

The role of 5-HT in suicide will be discussed in detail elsewhere in this symposium. Our finding of increased 5-HT_2 receptor binding in homogenates of frontal cortex in violent suicide completers (e.g. gunshot wounds or hanging, N = 21) as compared to non-violent suicide completers (e.g. carbon monoxide poisoning or drug overdosage, N = 11) or matched controls (N = 37) has been previously reported.[73] Contrary to this, Owen *et al.*[93,94] and Crow *et al.*[95] did not find any difference in cortical 5-HT_2 binding between suicides and age-matched controls. These investigators, however, have studied binding at only one concentration of ligand whereas Stanley and Mann,[96] Mann *et al.*,[72] and Arora and Meltzer[73] performed full saturation studies. In a recent study, Cheetham *et al.*[97] reported that 5-HT_2 receptor binding was not altered in frontal cortex of 19 suicide completers with definite evidence of depression as compared to 19 matched controls. However, only five of the antidepressant-free suicide completers died by violent means (i.e., hanging or self-inflicted wounds), and in these subjects there was a trend toward increased binding in the frontal cortex and amygdala. Taken together, these studies indicate there may be a strong correlation between increased numbers of 5-HT_2 receptors in frontal cortex and suicide via violent means.

Binding to 5-HT_1 receptors has also been evaluated in homogenates of frontal cortex from suicides and controls. In contrast to the findings with 5-HT_2 receptors, there was no difference in $[^3H]$5-HT binding to "total" 5-HT receptors between suicides and normals.[72,93–95] However, in a recent study by Matsubara and Meltzer (in preparation), the binding of $[^3H]$8-OH-DPAT to 5-HT_{1A} receptors in homogenates of frontal cortex was significantly increased (higher B_{max}) in non-violent suicides (N = 12) compared to matched controls (N = 40) and violent suicides (N = 11) (TABLE 1). The increase in 5-HT_{1A} receptor binding in nonviolent suicides was not correlated with the increase in 5-HT_2 receptor binding in violent suicides. These results suggest that up-regulation of 5-HT_{1A} and 5-HT_2 receptors in suicide occur by separate mechanisms.

In human cortex, greater than 60 percent of $[^3H]$5-HT binding is to a putative 5-HT_{1D} binding site distinct from the 5-HT_{1A} or 5-HT_{1C} binding sites.[98] Inasmuch as other studies show no change in $[^3H]$5-HT binding in suicide brains, but an increase in $[^3H]$8-OH-DPAT binding to 5-HT_{1A} receptors in victims of nonviolent suicide, measurement of $[^3H]$5-HT binding to 5-HT_{1D} receptors in suicides might be decreased. If

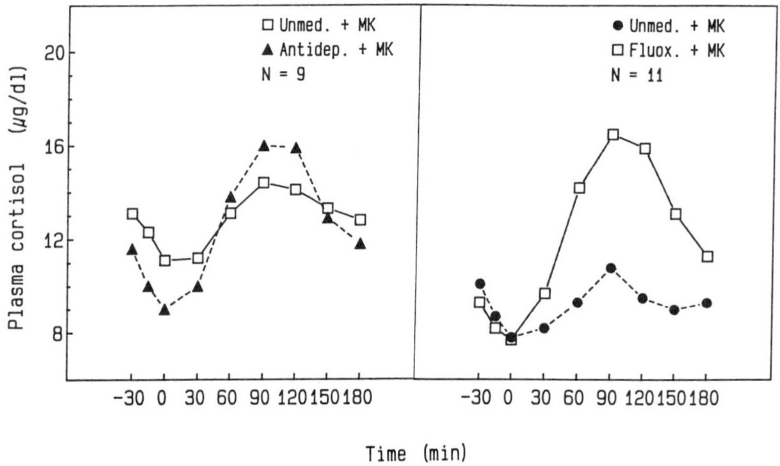

FIGURE 2. Potentiation of cortisol response to MK-212 by fluoxetine. Patients with OCD or MD were given MK-212, 20 mg, and studied as described in legend to FIGURE 1. The *panel at the left* included 7 patients with MD and 2 with OCD. Three-six weeks after beginning treatment with a tricyclic antidepressant or maprotiline (N = 1), they were again studied in the same fashion. No significant difference was noted. The *group on the right* consisted of 5 patients with MD and 6 with OCD. They were studied before and 4–12 (median 7) weeks after fluoxetine. The cortisol response was markedly potentiated.

the proposed studies of [^{3}H]8-OH-DPAT binding in suicides substantiate the previous observation, a pilot study of 5-HT_{1D} receptor binding in suicides might be warranted.

CONCLUSIONS

We have reviewed some of the evidence which supports the role of 5-HT in depression e.g. 1) that 5-HT plays an important role in some of the behaviors that are abnormal in depression, e.g. mood, sleep, appetite; 2) that there are abnormalities in tryptophan metabolism in major depression; 3) that presumptive decreases in brain 5-HT concentrations can precipitate depression in recovered depressed patients; 4) that platelet 5-HT_2 receptors are increased in depression; and 5) that treatments which are effective in depression promote serotonergic activity. The lack of a decrease in CSF 5-HIAA concentrations, of a clear cut abnormality in platelet 5-HT uptake and the lack of superiority of 5-HT uptake blockers as antidepressants is not consistent with a primary role of 5-HT in depression. Serotonin may be more important as a vulnerability factor in major depression as originally suggested by Prange *et al.*[99] than as a proximate cause of depression. Further studies of post-mortem material with more sophisticated techniques may be helpful to better define the range of serotonergic abnormalities in major depression. Interactions of 5-HT with other neurotransmitters, especially norepinephrine and dopamine, and interactions between 5-HT receptor subtypes may also be important factors in major depression.

TABLE 1. B_{max} and K_d Values of [^{3}H]8-OH-DPAT Binding in Controls and Suicides

Subjects	B_{max} (pmol/g)	K_d (nM)
non-suicide controls (39)	4.49 ± 0.80	0.69 ± 0.23
non-homicides (29)	4.47 ± 0.76	0.68 ± 0.23
homicides (10)	4.56 ± 0.94	0.70 ± 0.25
suicides (23)	4.95 ± 1.40	0.71 ± 0.22
nonviolent (12)	5.48 ± 1.52	0.80 ± 0.25
violent (11)	4.38 ± 1.03	0.62 ± 0.14

Data are expressed as mean ±SD obtained from saturation experiment using six different concentrations of [^{3}H]8-OH-DPAT (0.1 – 4.0 nM).

[**Note added in proof:** Deakin *et al.* recently confirmed the finding of decreased plasma tryptophan levels following I.V. tryptophan and a diminished GH response in patients with major depression. Deakin, J. F. W., I. Pennell, A. J. Upadhyaya & R. Lofthouse. 1990. A neuroendocrine study of 5HT function in depression for biological mechanisms of endogenous and psychosocial causation. Psychopharmacology **101**(1): 85–92.]

REFERENCES

1. MURPHY, D. L., I. CAMPBELL & J. COSTA. 1978. Current status of the indoleamine hypothesis of the affective disorders. *In* Psychopharmacology: A generation of progress. M. A. Lipton, A. DiMascio, and K. F. Killam. New York: Raven Press, pp. 1235–1248.
2. MELTZER, H. Y. & M. T. LOWY. 1987. The serotonin hypothesis of depression. Psychophar-

macology: The Third Generation of Progress. H.Y. Meltzer, Ed. New York: Raven Press, pp. 513–526.
3. VOGT, M. 1982. Some functional aspects of central serotonergic neurones. *In* Biology of Serotonergic Neurotransmission. N. N. Osborn, Ed., Chichester, Eng.: John Wiley & Sons, pp. 299–316.
4. SHOPSIN, B., S. GERSHON & G. SELZER. 1976. The neuropsychopharmacology of mania. *In* Drug Treatment of Mental Disorders. L. L. Simpson, Ed. New York: Raven Press, pp. 175–193.
5. DELGADO, P. L., P. S. CHARNEY, L. H. PRICE, G. K. AGHAJANIAN, H. L. ANDES & G. R. HENINGER. 1990. Serotonin function and the mechanism of antidepressant action. Reversal of anti-depressant induced remission by rapid depletion of plasma tryptophan. Arch. Gen. Psychiatry **47:** 411–418.
6. YOUNG, S. N., S. E. SMITH, R. O. PIHL & F. R. ERVIN. 1985. Tryptophan depletion causes a rapid lowering of mood in normal subjects. Psychopharmacology **87:** 173–177.
7. SMITH, W. K. 1987. The stress analogy. Schiz. Bull. **13:** 215–225.
8. HAMIK, A. & S. J. PEROUTKA. 1989. 1-(*m*-Chlorophenyl)piperazine (mCPP) interactions with neurotransmitter receptors in the human brain. Biol. Psychiatry **25:** 569–575.
9. MUELLER, E. A., D. L. MURPHY & T. SUNDERLAND. 1985. Neuroendocrine effects of M-chlorophenylpiperazine, a serotonin agonist, in humans. J. Clin. Endocrinal. Metab. **61:** 1179–1184.
10. PEROUTKA, S. J. 1985. Selective interaction of novel anxiolytics with 5-hydroxytryptamine$_{1A}$ receptors. Biol. Psychiatry **20:** 971–979.
11. MARTIN, G. E., R. J. ELGIN, JR., J. R. MATHIASEN, C. B. DAVIS, J. M. KESSLICK, W. J. BALDY, R. P. SHANK, D. L. DISTEFANO, C. L. FEDDE & M. K. SCOTT. 1989. Activity of aromatic substituted phenylpiperazines lacking affinity for dopamine binding sites in a preclinical test of antipsychotic anxiety. J. Med. Chem. **32:** 1052–1056.
12. CHARNEY, D. S., G. R. HENINGER, J. F. REINHARD, D. E. STERNBERG & K. M. HOFSTEAD. 1982. The effect of intravenous L-tryptophan on prolactin and growth hormone and mood in healthy subjects. Psychopharmacology **77:**217–222.
13. WINOKUR, A., N. D. LINDBERG, I. LUCKI & J. PHILLIPS & J. D. AMSTERDAM. 1986. Hormonal and behavioral effects associated with intravenous L-tryptophan administration. Psychopharmacology **88:** 213–219.
14. KOELLA, W. P. 1988. Serotonin and sleep. *In* Neuronal Serotonin. N. N. Osborn & M. Hamon, Eds. Chichester, Eng.: John Wiley & Sons, Ltd., pp. 153–170.
15. FORNAL, C. & M. RADULAVACKI. 1983. Sleep suppressant action of fenfluramine in rats. I. Relation to postsynaptic serotonergic stimulation. J. Pharmacol. Exp. Therap. **225:** 667–674.
16. REYES, R. B., S. Y. HILL & D.J. KUPFER. 1983. Effects of acute doses of zimelidine on REM sleep in rats. Psychopharmacology **80:** 214–216.
17. SITARAM, N., J. I. NURNBERGER, E. J. GERSHON & J. C. GALLIN. 1982. Cholinergic regulation of mood and REM sleep: A potential model and marker for vulnerability to depression. Am. J. Psychiatry **139:** 571–576.
18. SACHAR, E. J., J. HELLMAN, H. P. ROFFWARG, F. S. HALPERN, D. K. FUKUSHIMA & T. F. GALLAGHER. 1973. Disrupted 24-hour patterns of cortisol secretion in psychotic depression. Arch. Gen. Psychiatry **28:** 19–24.
19. WEHR, T. A. & F. K. GOODWIN. 1981. Biological rhythms and psychiatry. *In* American Handbook of Psychiatry, Vol. 7, 2d ed. S. Arieti & H. K. H. Brodie, Eds. New York: Basic Books, pp. 46–74.
20. AVERY, D. H., G. WILDSCHIODTZ & O. J. RAFAELSEN. 1982. Nocturnal temperature in affective disorder. J. Affective Disord. **4:** 61–71.
21. SACK, D. A., N. E. ROSENTHAL, B. L. PARRY & T. A. WEHR. 1987. Biological rhythms in psychiatry. *In* Psychopharmacology: Third Generation of Progress. H. Y. Meltzer, Ed. New York: Raven Press, pp. 669–685.
22. MOORE, R. Y. 1981. The anatomy of central serotonin neuron systems in the rat brain. *In* Serotonin, Neurotransmission and Behavior. B. L. Jacobs and A. Gelperin, Eds. Cambridge, MA/London: MIT Press, pp. 35–71.
23. JACOBSON, F. M., D. A. SACK, T. A. WEHR, S. ROGERS & N. E. ROSENTHAL. 1987.

Neuroendocrine response to 5-hydroxytryptophan in seasonal affective disorder. Arch. Gen. Psychiatry **44:** 1086–1091.
24. Modigh, K., J. Balldin, S. Eden, A.-K. Granerus & J. Wahlinder. 1971. Electroconvulsive therapy and receptor sensitivity. Acta Psych. Scand. **63**(Supp. 290): 91–99.
25. Cross, A. J. 1988. Serotonin in neurodegenerative disorders. *In* Neuronal Sertonin. N. N. Osborn and M. Hamon, Ed. Chichester, Eng. John Wiley & Sons, Ltd., pp. 231–253.
26. Iversen, S. D. 1984. 5-HT and anxiety. Neuropharmacology **23**(12B): 1553–1560.
27. Altman, H. J., D. A. Nordy & S. O. Ogren. 1984. Role of serotonin in memory: Facilitation by alaproclate and zimledine. Psychopharmacology **84:** 496–502.
28. Valzelli, L. 1981. Psychopharmacology of aggression: An overview. Int. Pharmacopsychiatry **16:** 39–48.
29. Stanley, M., L. Traskman-Bendz & K. Dorovini-Zis. 1985. Correlations between aminergic metabolites simultaneously obtained from human CSF and brain. Life Sci. **37:** 1279–1286.
30. Banki, C. M., M. Vojnik & G. Molnar. 1981. Cerebrospinal fluid amine metabolites, tryptophan and clinical parameters in depression. J. Affective Disord. **3:** 81–89.
31. Roy, A., D. Pickar, M. Linnoila, A. R. Doran, P. Ninan & S. M. Paul. 1985. Cerebrospinal fluid monoamine and monoamine metabolite concentrations in melancholia. Psychiatry Res. **15:** 281–292.
32. Vestergaard, P., T. Sorenson, E. Hoppe, O. J. Rafaelsen, C. M. Yates & N. Nicholaauo. 1978. Biogenic amine metabolites in cerebrospinal fluid of patients with affective disorders. Acta Psychiatr. Scand. **63:** 179–190.
33. Berger, P. A., K. F. Faull, J. Kilkowski, P. J. Anderson, H. Kraemer, K. L. Davis & J. D. Barchas. 1980. CSF monoamine metabolites in depression and schizophrenia. Am. J. Psychiatry **137:** 174–180.
34. Korf, J., W. van den Burg & H. van den Hofdaker. 1983. Acid metabolites and precursor amino acids of 5-hydroxytryptamine and dopamine in affective and other psychiatric disorders. Psychiat. Clin. **16:** 1–16.
35. Agren, H. 1980. Symptom patterns in unipolar and bipolar depression correlating with monoamine metabolites in the cerebrospinal fluid: 1. General patterns. Psychiatry Res. **3:** 211–223.
36. Asberg, M., L. Bertilsson, B. Martensson, G.-P., Scalla-Tomba, P. Thoren & L. Traskman-Bendz. 1984. CSF monoamine metabolites in melancholia. Acta Psychiatr. Scand. **69:** 201–219.
37. Leckman, J. F., D. S. Charney, C. R. Nelson, G. R. Heninger & M. D. Bowers, Jr. 1981. CSF tryptophan, 5-HIAA and HVA in 132 psychiatric patients categorized by diagnosis and clinical state. *In* Recent Advances in Neuropsychopharmacology. B. Angrist, G. D. Burrow, M. Lader, O. Lingjaerde, G. Sedvall & D. Wheatley, Eds. Oxford: Pergamon Press, pp. 289–297.
38. Davis, K. L., L. E. Hollister, A. A. Mathé, B. M. Davis, A. B. Rothpearl, K. F. Faull, J. Y. Hsieh, J. D. Barchas & P. A. Berger. 1981. Neuroendocrine and neurochemical measurements in depression. Am. J. Psychiatry **138:** 1555–1562.
39. Bowden, C. L., S. H. Koslow, I. Hanin, J. W. Maas, J.M. Davis & E. Robins. 1985. Effects of amitriptiline and imipramine on brain amine neurotransmitter metabolites in cerebrospinal fluid. Clin. Pharmacol. Therap. **37:** 316–324.
40. Endicott, J. & R. Spitzer. 1978. A diagnostic interview: A schedule for affective disorders and schizophrenia. Arch. Gen. Psychiatry **35:** 837–844.
41. Olendorf, W. H. 1971. Brain uptake of radiolabeled amino acids, amines, and hexoses after arterial injection. Am. J. Physiol. 221:1629–1639.
42. Moller, S. E., L. Kirk & K. H. Fremming. 1976. Plasma amino acids as an index for subgroups in manic depressive psychosis: Correlation to effect of tryptophan. Psychopharmacology **49:** 205–213.
43. Moller, S. E., L. Kirk & P. Honoré. 1980. Relationship between plasma ratio of tryptophan to competing amino acids and the response to L-tryptophan treatment in endogenous depressed patients. J. Affective Disord. **2:** 47–59.
44. Cowen, P. J., E. M. Charig, S. Fraser & J. M. Elliott. 1989. Platelet 5-HT receptor

binding during depressive illness and tricyclic antidepressant treatment. J. Affective Dis. **13:** 45–50.
45. Moller, S. E., P. Honoré & O. B. Larsen. 1983. Tryptophan and tyrosine ratios to neutral amino acids in endogenous depression: relation to antidepressant response to amitriptyline and lithium plus L-tryptophan. J. Affect. Dis. **5:** 67–79.
46. De Meyer, M. K., P. A. Shea, H. C. Hendrie & N. N. Yoshimura. 1981. Plasma tryptophan and five other amino acids in depressed and normal subjects. Arch. Gen. Psychiatry **38:** 642–646.
47. Joseph, M., T. Brewerton, V. Reus & B. Stebbins. 1984. Plasma L-tryptophan/neutral amino acid ratio and deamethasone suppression in depression. Psychiatry Res. **11:** 185–192.
48. Maes, M., M. De Ruyter, R. Claes, G. Bosma & E. Suy. 1987. The cortisol responses to 5-hydroxytryptophan, orally, in depressive inpatients. J. Affective Dis. **13:** 23–30.
49. Moller, S. E., K. Odum, L. Kirk, M. Bjerre, F. Fog-Moller & A. Knudsen. 1985. Plasma tyrosine/neutral amino acid ratio correlated with clinical response to nortriptyline in endogenously depressed patients. J. Affect. Dis. **9:** 223–229.
50. Smith, D. F. & L. S. Stromgren. 1981. Influence of unilateral ECT on tryptophan metabolism in endogenous depression. Pharmacopsychiatry **14:** 135–138.
51. Koyama, T. & H. Y. Meltzer. 1986. A biochemical and neuroendocrine study of the serotonergic system in depression. *In* New Results in Depression. H. Hippius *et al.*, Eds. Berlin: Springer-Verlag, pp. 169–188.
52. Hoes, M. J. A. J. M., T. Loeffen & T. B. Vree. 1981. Kinetics of L-tryptophan in depressive patients: A possible correlation between the plasma concentrations of L-tryptophan and some psychiatric rating scales. Psychopharmacology **75:** 350–353.
53. Moller, S. E., L. Kirk & P. Honoré. 1982. Tryptophan tolerance and metabolism in endogenous depression. Psychopharmacology **76:** 79–83.
54. Bovier, P., J. Widmer, J.-M. Gaillard & R. Tissot. 1988. Evolution of red blood cell membrane transport and plasma level of L-tyrosine and L-tryptophan in depressed treated patients according to clinical improvement. Neuropsychobiology **19:** 125–134.
55. Gaillard, J. M. & R. Tissot. 1979. Blood-brain movements of tryptophan and tyrosine in manic-depressive illness and schizophrenia. J. Neural Trans., Suppl. 15: 189–196.
56. Glassman, A. M. & S. R. Platman. 1969. Potentiation of a monoamine oxidase inhibitor by tryptophan. J. Psychiatr. Res. **7:** 83–88.
57. Louie, A. K. & H. Y. Meltzer. 1984. Lithium potentiation of antidepressant treatment. Journal of Clinical Psychopharmacology **4**(6): 316–322.
58. De Montigny, C., G. Cournoyer, R. Morissette, R. Langlois & G. Caille. 1983. Lithium carbonate addition in tricyclic antidepressant-resistant unipolar depression: Correlations with the neurobiologic actions of tricyclic antidepressant drugs and lithium ion on the serotonin system. Arch. Gen. Psychiatry. **40:** 1327–1334.
59. Meltzer, H. Y., M. T. Lowy, A. Robertson, P. Goodnick & R. Perline. 1984. Effect of 5-hydroxytryptophan on serum cortisol levels in major affective disorders. III. Effect of antidepressants and lithium carbonate. Arch. Gen. Psychiatry **41:** 391–402.
60. Meltzer, H. Y., B. Umberhoman-Wiita, A. Robertson, B. J. Tricou, M. T. Lowy & R. Perline. 1984. Effect of 5-hydroxytryptophan on serum cortisol levels in the major affective disorders and normal controls. I. Enhanced response in depression and mania. Arch. Gen. Psychiatry **41:** 366–374.
61. Meltzer, H. Y., M. Simonovic, R. D. Sturgeon & V. S. Fang. 1980. Effect of antidepressants, lithium and electroconvulsive treatment on rat serum prolactin levels. Acta Psychiat. Scand. **63:** 100–121.
62. Glue, P. W., P. J. Cowen, D. J. Nutt, T. Kolakowska & D. G. Grahame-Smith. 1986. The effect of lithium on 5-HT-mediated neuroendocrine responses and platelet 5-HT receptors. Psychopharmacology **90:** 398–402.
63. Price, L. H., D. S. Charney, P. L. Delgado & G. R. Heninger. 1984. Lithium treatment and serotonergic function: Neuroendocrine and behavioral responses to intravenous tryptophan in affective disorder. Arch. Gen. Psychiatry **46:** 13–19.
64. Idzikowski, C. & P. J. Cowen. 1987. The effects of chronic ritanserin treatment on sleep and the neuroendocrine response to L-tryptophan. Psychopharmacology **93:** 416–420.

65. Stahl, S. M., P. J. Woo, I. N. Mifford, P. A. Berger & R. D. Ciaranello. 1977. Hyperserotonemia and platelet serotonin uptake and release in schizophrenia and affective disorders. Am. J. Psychiatry **140:** 26–30.
66. Meltzer, H. Y., R. C. Arora, R. Baber & B. J. Tricou. 1987. Serotonin uptake in blood platelets of psychiatric patients. Arch. Gen. Psychiatry **30:** 1322–1326.
67. Mellerup, E. T. & P. Plenge. 1989. High-affinity binding of [^{3}H]paroxetine and [^{3}H]imipramine to rat neuronal membranes. Psychopharmacology **89:** 436–439.
68. D'Haenen, H., M. DeWaile & J. E. Leysen. 1988. Platelet ^{3}H-paroxetine binding in depressed patients. Psychiatry Res. **26:** 11–17.
69. De Clerck, F., B. Xhonneux, J. Leysen & P. A. J. Janssen. 1987. Evidence for functional 5-HT$_2$ receptor sites on human blood platelets. Biochem. Pharmacology **33:** 45–50.
70. McBride, P. A., J. J. Mann, B. McEwen & A. Biegon. 1987. Characterization of serotonin binding sites on human platelets. Life Sci. **33:** 2033–2041.
71. Cowen, P. J., E. M. Charig, S. Fraser & J. M. Elliott. 1987. Platelet 5-HT receptor binding during depressive illness and tricyclic antidepressant treatment. J. Aff. Dis. **13:** 45–50.
72. Biegon, A., A. Weizman, L. Korp, A. Ram, S. Tiano & M. Wolff. 1987. Serotonin 5-HT$_2$ receptor binding on blood platelets—A peripheral marker for depression. Life Sci. **41:** 2485–2492.
73. Arora, R. C. & H. Y. Meltzer. 1989. Serotonergic measures in suicide brain - III: Increased 5-HT$_2$ binding sites in frontal cortex suicide victims. Am. J. Psychiatry **146:** 730–736.
74. Mann, J. J., M. Stanley, A. McBride & B. S. McEwen. 1986. Increased serotonin$_2$ and β-adrenergic receptor binding in the frontal cortices of suicide victims. Arch. Gen. Psychiatry **43:** 954–959.
75. Arora, R. C. & H. Y. Meltzer. 1989. Serotonergic measures in the brains of suicide victims: 5-HT$_2$ binding sites in the frontal cortex of suicide victims and control subjects. Am. J. Psychiatry **146:** 730–736.
76. Stolz, J. F., C. A. Marsden & D. N. Middlemiss. 1983. Effect of chronic antidepressant treatment and subsequent withdrawal of [^{3}H]-5-hydroxytryptamine and [^{3}H]-spiperone binding in rat frontal cortex and serotonin receptor mediated behavior. Psychopharmacology **80:** 150–155.
77. Meltzer, H. Y. & J. F. Nash, Jr. 1988. Serotonin and mood: Neuroendocrine aspects. *In* Current Topics in Neuroendocrinology, Vol. 8. D. Granten and D. Pfaff, Eds. Heidelberg: Springer-Verlag, pp. 183–210.
78. Koenig, J. I., G. A. Gudelsky & H. Y. Meltzer. 1987. Stimulation of corticosterone and β-endorphin secretion by selective 5-HT receptor subtype activation. Eur. J. Pharmacol. **137:** 1–8.
79. Heninger, G. R., D. S. Charney & D. E. Steinberg. 1984. Serotonergic function in depression: Prolactin response to intravenous tryptophan in depressed patients and healthy subjects. Arch. Gen. Psychiatry **41:** 398–402.
80. Siever, L. J., D. L. Murphy, S. Slater, E. de la Vega & S. Lipper. 1984. Plasma prolactin changes following fenfluramine in depressed patients compared to controls: An evaluation of central serotonergic responsivity in depression. Life Sci. **34:** 1029–1039.
81. Charney, D. S., G. R. Heninger & D. E. Sternberg. 1984. Serotonin function and mechanism of action of antidepressant treatment: Effect of amitriptyline and desipramine. Arch. Gen. Psychiatry **41:** 359–365.
82. Cowen, P. J., D. P. Geaney, M. Schächter, A. R. Green & J. M. Elliott. 1986. Desipramine treatment in normal subjects: Effects on neuroendocrine responses to tryptophan and on platelet serotonin (5-HT)-related receptors. Arch. Gen. Psychiatry **43:** 61–67.
83. Lee, M., J. F. Nash, M. Barnes & H. Y. Meltzer. 1990. Inhibitory effect of ritanserin, a serotonin$_2$ antagonist, on 5-hydroxytryptophan-mediated cortisol and prolactin secretion in humans. Psychopharmacology. In press.
84. Kahn, R. S. & H. G. M. Westenberg. 1985. L-5-Hydroxytryptophan in the treatment of anxiety disorder. J. Aff. Dis. **8:** 197–200.

85. CHARIG, E. M., I. M. ANDERSON, J. M. ROBINSON, Q. J. NUTT & J. COWEN. 1986. L-tryptophan and prolactin release: Evidence for interaction between 5-HT and 5-HT_2 receptors. Hum. Psychopharmacol. **1:** 93–97.
86. LINGJAERDE, O., O. BRATFOS, T. BRATLID & J. O. HAND. 1983. A double-blind comparison of zimelidine and desipramine in endogenous depression. Acta Psychiatr. Scand. **68:** 22–30.
87. DAHL, L.-E., L. LUNDIN, LEFÉVRE, P. HONRÉ & S. J. DENCKER. 1982. Antidepressant effect of femoxetine and desipramine and relationship to the concentration of amine metabolites in cerebrospinal fluid. Acta Psychiatr. Scand. **66:** 9–17.
88. KNUDSEN, P., N. BJORNDAL, T. JOHNSEN & K. PFEIFFER PETERSEN. 1984. Zimeldine versus nomifensine: A double blind study of depressed patients. Neuropsychobiology **11:** 236–242.
89. NYSTRÖM, C. & T. HÄLLSTROM. 1985. Double-blind comparison between a serotonin and a noradrenaline reuptake blocker in the treatment of depressed outpatients. Acta Psychiatr. Scand. **72:** 6–15.
90. NOLEN, W. A., J. J. VAN DE PUTTE, W. A. DIJKEN, J. S. KAMP, B. A. BLAUSJAAR, H. J. KRAMER & J. HOFFMANS. 1988. Treatment strategy in depression: 1. Non-tricyclic and selective reuptake inhibitors in resistant depression: A double-blind partial crossover study on the effect of oxaprotiline and fluvoamine. Acta Psychiatr. Scand. **78:** 668–675.
91. ZOHAR, J. & T. R. INSEL. 1987. Obsessive-compulsive disorder: Psychobiological approaches to diagnosis, treatment and pathophysiology. Biol. Psychiatry **22:** 667–687.
92. EVANS, L., J. KENARDY, P. SCHNEIDER & H. HOEY. 1986. Effect of a selective serotonin uptake inhibitor in agoraphobia with panic attacks: A double blind comparison of zimelidine, imipramine and placebo. Acta Psychiatr. Scand. **73:** 49–53.
93. OWEN, F., A. J. CROSS, T. J. CROW, J. F. W. DEAKIN, I. N. FERRIER, R. LOFTHOUSE & M. POULTER. 1983. Brain 5-HT_2 receptors and suicide. Lancet **2:** 1256.
94. OWEN, F. J., D. R. CHAMBERS, S. J. COOPER, T. J. CROW, J. A. JOHNSON, R. LOFTHOUSE & M. POULTER. 1986. Serotonergic mechanisms in brains of suicide victims. Brain Res. **362:** 185–188.
95. CROW, T. J., A. J. CROSS, S. J. COOPER, J. F. W. COOPER, I. N. FERRIER, J. A. JOHNSON, F. OWEN, M. POULTER, R. LOFTHOUSE, J. A. N. CORSELLIS, D. R. CHAMBERS, G. BLESSED, E. K. PERRY, R. H. PERRY & B. E. TOMLINSON. 1984. Neurotransmitter receptors and monoamine metabolites in the brains of patients with Alzheimer-type dementia and depression, and suicides. Neuropharmacology **23:** 1561–1569.
96. STANLEY, M. & J. MANN. 1983. Increased serotonin-2 binding sites in frontal cortex of suicide victims. Lancet **1:** 214–216.
97. CHEETHAM, S. C., M. R. CROMPTON, C. L. E. KATONA & R. W. HORTON. 1988. Brain 5-HT_2 receptor binding sites in depressed suicide victims. Brain Res. **443:** 272–280.
98. WAEBER, C., P. SCHOEFFTER, J. M. PALACIOS & D. HOYER. 1988. Molecular pharmacology of 5-HT_{1D} recognition sites: radioligand binding studies in human, pig and calf brain membranes. Naunyn-Schmiedeberg's Arch. Pharmacology **337:** 595–601.
99. PRANGE, A. J., JR., I. C. WILSON, C. W. LYNN, L. B. ALLTOP & R. A. STIKELEATHER. 1974. L-Tryptophan in mania: Contribution to a permissive hypothesis of affective disorders. Arch. Gen. Psychiatry **30:** 56–62.

DISCUSSION OF THE PAPER

MANFRED GÖTHERT (*University of Bonn*): Langer has put forward a very interesting hypothesis concerning the genesis of depression. His opinion is that depression may be due to an increased function of presynaptic autoreceptors which induces a decreased release. And there is also some evidence favoring this because, for instance, two groups have shown that prolonged treatment with antidepressant drugs, with tricyclic antidepressants, decreases the function of the autoreceptor (MAURA, G. & M. RAITER. 1984. Functional evidence that chronic drugs induce adaptive changes of central autoreceptors regulating serotonin release. Eur. J. Pharmacol. **97:** 309–313). In this connection I would like to ask you whether you know whether there is any report in the literature that indicates that there is a change in 5-HT_{1D} binding sites in depressed or suicidal patients, because in humans the autoreceptor probably is of the 5-HT_{1D} subtype.

MELTZER: I don't know of any reports on the 5-HT_{1D}. We are, in fact, looking at that in brains from suicides at the moment. I think the possibility of a dysregulation of the autoreceptor or supersensitivity, leading to decreased release of serotonin is a plausable hypothesis. However, most investigators have not found that chronic treatment with antidepressants produces 5-HT autoreceptor subsensitivity. This has been reviewed by Willner. (Willner, P. 1985. Antidepressants and Serotonergic neurotransmission. An integrative review. Psychopharmacology **85:** 387–404). One of the points I made is that we really ought to consider the heterogeneity of depression. We don't have to have one mechanism that explains all types of depression. We could have a situation in which for some depressed patients a presynaptic mechanism may be the critical one, whereas for others a postsynaptic one may be most important. In other patients, there may be varying mixtures of both these types of defects. There may also be problems in compensatory mechanisms. If that kind of heterogeneity does exist, it's going to be terribly difficult to sort it out. It's not impossible, but we're going to need to look in a very conservative way for integrated pieces of evidence that suggest which of these is going on in a given group or given individual.

QUESTION: You mentioned that you found an increased 5-HT_{1A} receptor binding in brains from suicides. Could you clarify what ligand you used?

MELTZER: We used 8-hydroxy-DPAT and I'm glad you asked me to clarify this because what was interesting about this increased binding, wasn't present in the group as a whole, but only in the non-violent suicides. And that contrasts with the finding for the 5-HT_2 receptors which were elevated only in violent suicides by John Mann's studies. Given all the suggestive evidence that John Mann and I reviewed which indicate that 5-HT is related to suicide, the biological studies of depression need to covary and adjust for the presence of suicidality; not just for the failed attempts, but for suicidal ideation as well.

GEORGE R. HENINGER (*Yale University, New Haven, CT*): I might add just one piece of new data. We'll hear later about OCD and serotinergic functioning, and in the tryptophan-depletion studies that we've done we've given that to a number of OCD patients, have not found a worsening of OCD symptoms even though these patients, in some recent studies by Dr. Goodman, have been shown to be responsive to fluvoxamine, but not to desipramine. So there's some treatment specificity on the serotonergic system and OCD but they're not showing a relapse like the depressed patients with this lower tryptophan in plasma.

MELTZER: I think that's really a fascinating observation and I wanted to make the point that fluoxetine in our hands, and chlorimipramine, are so much more effective in OCD than the typical antidepressants. And, as you saw from our data with the hormone challenges, these drugs differ from the typical tricyclics in potentiating both the 5-HTP and the M-212 response. It would suggest to me that evidence for serotonin being involved in OCD is stronger than it is for depression. So what you just said kind of muddies the water a bit, but points out perhaps that the tryptophan is not as important in OCD as it is for major depression.

COMMENT: I'll just bring it out that probably the theme of the meeting is heterogeneity so even within the disease syndromes as we see them there's heterogeneity in terms of depletion vs. treatment. They diverge that way.

Nosological Tunnel Vision in Biological Psychiatry

A Plea for a Functional Psychopathology

H. M. VAN PRAAG, G. M. ASNIS, R. S. KAHN,
S. L. BROWN, M. KORN, J. M. HARKAVY FRIEDMAN,
AND S. WETZLER

Department of Psychiatry
Albert Einstein College of Medicine/Montefiore Medical Center
Bronx, New York 10467

FUNCTIONAL PSYCHOPATHOLOGY: A LOOK BACKWARD

From the very beginning, biological psychiatry has oriented itself towards nosology, a way of looking at behavioral disorders as distinct, clearly separable entities, each with their own symptomatology, causation, outcome and treatment. Finding a marker and, eventually, the cause of, for instance, schizophrenia, depression or panic disorder has been and still is the major goal of biological psychiatric research. The introduction of the DSM III in 1980 forcefully reinforced biological psychiatry's nosological orientation.

For many years now we have advocated and pursued another route, marked by what we have called a functional approach to psychopathology (van Praag and Leijnse 1965[1]; van Praag *et al.* 1975,[2] 1987a[3]). Most parsimoniously formulated, this concept holds three tenets.

1. The basic units of classification in psychopathology are not syndromes or nosological entities, but psychological dysfunctions, such as disturbances in perception, cognition, memory, information processing, and many others. *They* are the elementary constituents of which psychiatric syndromes are made.

2. Nosological classification of a given psychiatric disorder is the first step in diagnosis. The next step in the diagnostic process is dissection of the syndrome into its component parts; i.e., the psychological dysfunctions and their subsequent measurement.

Analogies of two-tier diagnosing are provided by medicine. Myocardial infarction, for example, is a nosological diagnosis which will be routinely complemented by an analysis of its functional sequelae, in terms of, for instance, heart beat frequency and rhythm, conductance and output. Treatment is predicated on the functional analysis, not on the nosological diagnosis.

3. In biological psychiatric research, correlations between biological and psychological dysfunctions should be sought.

Functional psychopathology is dimensional in orientation, viewing a given psychiatric disorder as a conglomerate of psychological dysfunctions, most of them nosologically non-specific and occurring in different severities and in different combinations in the various psychiatric syndromes. The nosological approach is categorical, viewing psychiatric disorders as discrete entities, each with its own causation, symptomatology and course.

The term "functional" is not identical with "symptomatological." Psychiatric symptoms are the behavioral expression of a psychological dysfunction, not the dysfunction itself. A visual hallucination for example is a symptom, a particular perceptual disturbance the underlying dysfunction. Within the functional framework, the latter is the target of biological research.

We have pursued the functional approach in psychiatry for many years and believe this route to be productive. We will illustrate this point of view reviewing data derived from serotonin (5-hydroxytryptamine; 5-HT) research, but the same reasoning holds if the focus would have been on the catecholamines (van Praag *et al.* 1990[4]) (FIG. 1).

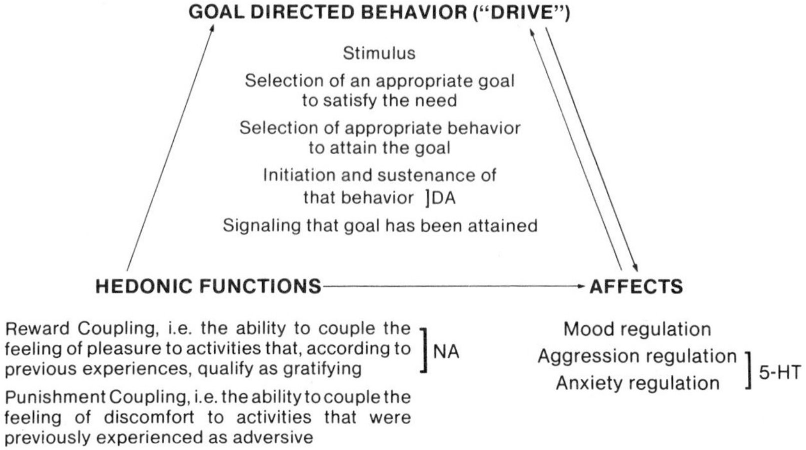

FIGURE 1. Depiction of the major domains of psychological dysfunction in major depression (melancholic type) and the hypothesized MA-ergic dysfunctions underlying these psychological dysfunctions (van Praag *et al.* 1990[44]).

THE ROLE OF SEROTONIN IN ABNORMAL BEHAVIOR

"5-HT Depression"

Disturbances in central 5-HT metabolism were first reported in depression. They were inferred to exist based on the finding of lowered post-probenecid minus baseline levels of 5-hydroxyindole acetic acid (5-HIAA, the major degradation product of 5-HT) in cerebrospinal fluid (CSF) of depressed individuals, indicating diminished 5-HT metabolism in the CNS (van Praag *et al.* 1970[5]).

Subsequently, disturbances in other systems, indicative of central 5-HT-ergic functioning were observed. First, disturbances in platelet 5-HT, such as deviations in 5-HT content and uptake and in the density of imipramine receptors were reported. Blood platelets have been considered as a model of central 5-HT-ergic nerve endings. Furthermore, disturbances in the ratio of the plasma concentration of tryptophan and the so-called competing amino acids were found. This ratio determines the influx of tryptophan into the CNS, and CNS availability of tryptophan, in its turn, is an important determinant of 5-HT synthesis rate. Moreover, several hormonal challenge tests indicated disturbances in 5-HT-ergic functioning. Finally, psychopharmacological experiments with indirect 5-HT agonists such as the 5-HT precursor 5-hydroxytryptophan (5-HTP), the selective 5-HT uptake inhibitors, and 5-HT antagonists, such as parachlorophenylalanine (PCPA), a 5-HT synthesis inhibitor, suggested that manipulation of central 5-HT can effect mood changes (for a review see Brown *et al.* 1990[6]).

The data most clearly and directly related to central 5-HT metabolism are derived from CSF studies, although these studies are the most cumbersome to conduct. The interrelation between the various 5-HT indicators is insufficiently known. (For a review on 5-HT and depression see e.g. van Praag 1982[7]; Goodwin and Post, 1983[8]; Brown *et al.* 1990[6]).

Initially, depressed patients with and without demonstrable disturbances in central 5-HT seemed indistinguishable in psychopathological terms. In 1971, interpreting the available data at that time, we introduced the concept of the biochemical heterogeneity of depression (van Praag and Korf, 1971[9]). Some forms of depression, we postulated, are linked to disturbances in 5-HT functions; others are not or are to a lesser extent only. The same syndrome, then, could be the ultimate outcome of various pathophysiological processes. As an analogy we mentioned the anemia syndrome. Anemia patients present by and large the same symptoms, but the pathogenesis of the syndrome is heterogeneous, and with that the treatment.

Subsequent studies, however, made the postulate of a separate category of "5-HT depressions" untenable (van Praag and Lemus, 1986[10]). They indicated that increasing 5-HT availability *alone* is not a sufficient antidepressant measure. This argues against the concept of a depression-type that is generated predominantly by central 5-HT disturbances, leading to decreased 5-HT availability.

Whither "5-HT Depression"

First we demonstrated in a double-blind, placebo-controlled, comparative study that 5-HTP is an active antidepressant in hospitalized patients with rather severe forms of melancholic depression, while tryptophan turned out to be no better than placebo (van Praag 1984[11]). Both 5-HT precursors increase central 5-HT metabolism to an equal extent, as reflected in the probenecid-induced accumulation of 5-HIAA in CSF. They appeared to differ, however, in their effect on catecholamine (CA) metabolites in CSF (reflecting intracerebral metabolism of the mother amines). 5-HTP increases the metabolism of DA as well as of NA, while tryptophan does not (van Praag 1984[11]). In high doses (>5g i.v.) tryptophan even lowers DA and NA metabolism, possibly by interfering with the influx of tyrosine, the precursor substance of the CAs, in the CNS (van Praag *et al.* 1987b[12]).

5-HTP's ability to stimulate CA metabolism is thought to be due to the presence of aromatic amino acid decarboxylase in CA-ergic nerve cells. Hence, 5-HTP is converted to 5-HT not only in 5-HT-ergic but also in CA-ergic neurons. In the latter, 5-HT functions as a false transmitter, as a consequence of which the synthesis rate of CA increases. The net functional effect of the two opposing processes; i.e., false transmitter formation leading to decreased function and increased CA production leading to augmented function, is probably heightened CA-ergic activity.

If 5-HTP's therapeutic superiority to tryptophan is related to its combined effects on 5-HT and CA systems, one would expect tryptophan's antidepressant efficacy to be raised above the significance level by combining it with a compound capable of increasing CA availability selectively. We showed in humans that tyrosine is presumably such a compound. It increases CA metabolism, while 5-HT metabolism is not measurably changed. Combining tryptophan with tyrosine indeed led to significant antidepressant activity (van Praag 1986a[13]).

A second observation supports the hypothesis of 5-HTP deriving its therapeutic potential from its dual influence of 5-HT-ergic and CA-ergic systems. In about 25% of patients initially treated successfully with 5-HTP, the response subsided in the second month of treatment (van Praag 1984[11]). This phenomenon is paralleled by normalization of CA metabolism (again, as reflected in the CSF concentration of CA metabolites) while the metabolism of 5-HT remained increased. We hypothesized that clinical relapse and normalization of CA metabolism were related. If so, re-activating CA metabolism should restore clinical remission and, indeed, it did. Adding tyrosine to the 5-HTP regime in the group of relapsers once again led to subsidence of the depression and to a rise of CA metabolism (van Praag and Lemus, 1986[10]).

These data suggest that *combined* augmentation of 5-HT and CA availability in the CNS provides the best conditions for antidepressant activity, and they argue as such against the existence of a separate diagnostic entity called "5-HT depression." 5-HTP's behavioral effects in monkeys have also been demonstrated to be a function of its influence on 5-HT and CA systems (Raleigh 1987[14]).

If 5-HT disturbances are not linked to a particular syndromal subtype of depression, could it be that they relate to particular psychopathological dimensions that may occur in depression, but might be lacking as well, explaining the occurrence of 5-HT disturbances in some, but not other depressions (van Praag *et al.* 1987c[3]).

This hypothesis seems plausible in the light of recent observations, the dimensions in question being disregulation of aggression and heightened anxiety. The relevant data will now be briefly summarized.

5-HT and Disturbed Aggression Regulation

In 1976 Asberg *et al.*[15] confirmed our findings of a subgroup of depressed individuals with low CSF 5-HIAA, and added the important observation that in that subgroup suicide attempters seemed to accumulate. This finding was confirmed by many, though not all investigators. Subsequent studies, revealed moreover that low CSF 5-HIAA was not only observed in depressed suicide attempters, but likewise in non-depressed, non-psychotic (e.g. personality disturbed) suicide attempters and in schizophrenic patients who had made suicide attempts because "voices" had ordered them to do so, but who were not suffering from depression, as defined by the DSM III (review van Praag 1986b[16]).

Lowered concentration of CSF 5-HIAA has also been reported in individuals with increased outward directed aggression, but belonging to different diagnostic categories (Brown *et al.* 1979[17], 1982[18]; Bioulac *et al.* 1980[19]; Linnoila *et al.* 1983[20]; Lidberg *et al.* 1984[21], 1985[22]; Virkkunen *et al.* 1987[23]). The biological study of aggression in humans is known to be fraught with methodological difficulties such as the usually short duration of aggressive outbursts and the fact that these explosions are more often than not intertwined with a host of factors, social or toxicological (drugs, alcohol), for which it is hard to control. The fact that all seven studies published so far on the subject have reported low CSF 5-HIAA, seems to indicate that the finding is a robust one remaining erect in spite of considerable background noise.

Dysregulation of aggression can be a major feature of depression, as evidenced by the greatly increased incidence of both suicidal behavior and signs of outward directed aggressiveness seen in depressed individuals (Plutchik *et al.* 1989[24]). Depressed patients with low CSF 5-HIAA show not only an increased rate of suicide attempts but also an increased frequency of signs of outward directed aggression (van Praag *et al.*, 1986[25]).

The facts, then, seem sufficiently robust to carry the hypothesis that the signs of diminished 5-HT metabolism in the CNS, as originally observed in depression, are related to dysregulated aggression, irrespective of the direction it takes and irrespective of the nosological context in which it occurs.

5-HT and Heightened Anxiety

Anxiety is a second psychopathological dimension occurring in depression, but being by no means specific for that disorder, and also seems to be 5-HT related. Several pieces of evidence point in that direction.

5-HT potentiating drugs, in particular 5-HTP, increasing 5-HT synthesis, and the 5-HT uptake inhibitors such as clomipramine, trazadone, fluvoxamine and fluoxetine were reported to have therapeutic effects in panic disorder (Evans and Moore 1981[26]; Koczkas *et al.* 1981[27]; Evans *et al.* 1986[28]; Kahn *et al.* 1984[29], 1987[30]; Den Boer *et al.* 1987[31]). Even more interesting, the effect of those indirect 5-HT agonists seems to be biphasic, at least in some patients (Kahn and Westenberg 1985[32]). An initial phase of increased anxiety, lasting 2–3 weeks is followed by a phase of amelioration in which anxiety and panic attacks diminish. We explained this biphasic effect by postulating a) hypersensitivity of post-synaptic 5-HT receptors and b) relatedness of this phenomenon to the anxiety attacks. Under these circumstances, increasing 5-HT availability would first lead to additional stimulation of an already hyperactive 5-HT system and hence, to clinical deterioration. Down regulation of the receptor system due to increased 5-HT availability would eventually lead to amelioration of the anxiety disorder (Kahn *et al.* 1988[33]).

Recently, additional evidence favoring this hypothesis was acquired. Panic disorder patients were challenged with *m*-chlorophenylpiperazine (MCPP), a relatively selective 5-HT1 receptor agonist. Normals and patients suffering from major depression served as controls. The major findings were twofold. Patients with panic disorder showed a substantial increase in anxiety during the test, in comparison to placebo. In the control groups no significant response occurred, on either MCPP or placebo (Kahn *et al.* 1988a[34]). Moreover, the cortisol response to MCPP was augmented in panic disorder patients; again, as compared to a) the response to placebo and b) the MCPP response in the two control groups (Kahn *et al.* 1988[35]).

Another piece of confirmatory evidence is found in the development of drugs that diminish central 5-HT activity, and possess anxiolytic potential. Buspirone is an established anxiolytic. It is a substance with a high affinity for presynaptically located 5-HT1A receptors, and probably decreases 5-HT-ergic activity (Taylor *et al.* 1985[36]). Ritanserin is a selective antagonist of postsynaptically located 5-HT2 receptors, of which anxiolytic actions have been reported (Ceulemans *et al.* 1985[37]). Both drugs seem to be effective in generalized anxiety disorder; their effect in panic disorder has not yet been well studied. Recently, Jones *et al.* (1988[38]) reported animal data indicating that GR38032F, a selective antagonist of 5-HT3 receptors, appears to be a potential anxiolytic agent. In humans, the various 5-HT receptor systems cannot yet be studied separately and independently.

If, indeed, in (certain) states of increased anxiety (certain) postsynaptic 5-HT receptors are hypersensitive, one would expect a diminution of central 5-HT synthesis and a lowering of CSF 5-HIAA as a consequence. No data are available on CSF 5-HIAA in anxiety disordered patients. In depressed patients, a negative correlation between CSF 5-HIAA and anxiety levels has indeed been observed (Banki 1977[39]; Rydin *et al.* 1982[40]; van Praag unpublished[41]).

The present data do not permit extending the 5-HT hypersensitivity hypothesis beyond panic disorder, but some preliminary data indicate 5-HT receptor hypersensitivity may be related to increased anxiety in general rather than to panic disorder alone. First, we found a correlation between cortisol response to MCPP and anxiety level as measured one week prior to the test in both panic disorder and major depression, suggesting a relationship between the general state of anxiety and 5-HT receptor hypensensitivity. The relation between CSF 5-HIAA and anxiety in depression, mentioned above, points in the same direction, and so does the finding that suppressors of 5-HT activity like ritanserin and buspirone are effective in generalized anxiety disorder.

A DIMENSIONAL 5-HT HYPOTHESIS

Signs of decreased central 5-HT metabolism, then, seem to correlate not with a particular syndromal depression type but with particular affective disturbances, i.e. increased (auto) aggression and augmented anxiety.

If the psychopathological dimensions of increased aggression and heightened anxiety are both related to disordered 5-HT metabolism, one would expect these dimensions to intercorrelate, and indeed they do. In a recent series of studies on predictors of suicide and violence in groups of psychiatric patients with a diversity of diagnoses, we found trait anxiety to be strongly correlated with both suicide risk and violence risk (Apter *et al.* 1990[42]).

The data favoring a dimensional 5-HT hypothesis, linking 5-HT disturbances to disordered aggression- and anxiety-regulation, across diagnoses, seem fairly convincing.

FUNCTIONAL PSYCHOPATHOLOGY: A LOOK FORWARD

5-HT disturbances found in psychiatric disorders look chaotic if viewed from a nosological vantage point, being demonstrable in so many different diagnostic entities. Interpreting them from the functional viewpoint brings order to the apparent chaos. Thus, 5-HT research clearly demonstrates the usefulness of the functional approach in psychopathology. The implications of this approach, however, reach beyond the boundaries of biological psychiatry in several important ways.

First, they reach into the realm of clinical psychopharmacology. Several psychotropic drugs have turned out to be diagnostically non-specific. Tricyclic antidepressants are efficacious in certain forms of depression *and* in certain anxiety states. Clomipramine is an effective antidepressant *and* a useful treatment in obsessive-compulsive disorder. Carbamazepine is an established antiepileptic agent *as well as* an asset in the prophylaxis of episodic mood disorders. These observations are puzzling from a nosological but not from a functional point of view. One could conceive of a particular drug influencing a particular functional system in the brain, and thus, a particular set of psychological dysfunctions, irrespective of diagnosis. The greater the biochemical specificity of a drug, one could argue, the greater the chance it will be nosologically non-specific, but effective against certain (clusters of) psychological dysfunctions, across diagnoses.

Once correlations between psychological and biological dysfunctions have been established, the next step inevitably will be the search for drugs that can selectively correct the biological dysfunctions. Such drugs would be expected to ameliorate the corresponding psychological dysfunctions, irrespective of nosological diagnosis. Within a nosological frame of reference treatment is directed towards a particular disorder, within a functional framework towards the underlying or resulting dysfunctions. In the former approach treatment is preferably restricted to one drug; in the latter approach combinations of drugs will be used, contingent on the existing spectrum of dysfunctions. Functional psychopathology will lead to functional psychopharmacology and functional psychopharmacology will inevitably tend to use drugs in combination. Elsewhere, we have given an (hypothetical) example of trans-nosological indications of a group of psychotropic drugs, i.e., the selective 5-HT reuptake inhibitors, the first group that seems to possess a high degree of selectivity, at least as far as monoamines (MA) are concerned (van Praag *et al.*, 1987c[43]). One could object that neuronal systems do not act in isolation, but interact with numerous others. That does not mean, however, that selectivity is illusory and that limiting a drug's primary actions is meaningless. Fluoxetine, for example, is a selective inhibitor of 5-HT uptake with little effect on noradrenaline (NA) reuptake. Its impact on NA-ergic functioning is not negligible, but certainly much less than that of a drug that primarily inhibits the uptake of both MA, such as amitriptyline. Indeed, perfect selectivity is unattainable, but that goal can certainly be advanced.

Goal-directed, dysfunction-oriented multi-pharmacy is not a chimera, but a strategy legitimately employed in other medical disciplines. For example, once a myocardial infarction has been diagnosed and the cardiac condition functionally analyzed, drugs will be administered to regulate the discrete disturbances in, for instance, rhythm, conduction, frequency, and output. These drugs are often prescribed simultaneously. Thus, this represents goal-directed, dysfunction-oriented multi-pharmacy. It is in this direction that we expect to see clinical psychopharmacology develop. It is *the* scientific approach to drug treatment, at least as long as we cannot address the cause(s) of the disease directly.

A *second* field that will benefit from a functionally oriented psychopathology is that of measurement of abnormal human behavior. So far, the presence of a psychiatric syndrome or symptom is expressed as a rough estimate, that is as mildly, moderately or severely present, or anywhere in between. Many psychological functions and dysfunctions, however, are measurable in real quantitative terms and with considerable precision and reliability. Functional psychopathology will lay the groundwork for a scientifically based psychopathology.

Finally, the field of the psychotherapies would benefit from the functional approach, because it would allow, much more precisely than is possible today, exposure and delineation of mental domains that are dysfunctional and those that are still relatively normal. This would permit a rather precise definition of treatment goals and a more rational choice of treatment method. Treatment planning would thus be furthered and provided with a rationale.

Thus, functional psychopathology is not only a productive framework for biological research in psychiatry, but could, in addition, be evolving into a forceful catalyst for psychiatry's ongoing scientific orientation.

We introduced the functional/dimensional approach in psychopathology as a complement of and not as an alternative to the nosological/categorical approach. We still fully subscribe to this standpoint, as much as we do to the view that the exclusive adherence to nosology has not served biological psychiatry well. In fact, it has been a factor stunting its growth.

SUMMARY

Classical nosology has been the major cornerstone of biological psychiatric research; finding biological markers and eventually causes of disease entities has been the major goal. Another approach, one we have designated as "functional," seems possible, attempting to correlate biological variables with psychological dysfunctions, the latter being considered to be the basic units of classification in psychopathology. We have pursued this route for many years, and based on the resulting findings we formulated the following hypothesis. Signs of diminished dopamine, serotonine, and noradrenaline metabolism, as have been found in psychiatric disorders, are not disorder-specific, but rather are related to psychopathological dimensions; i.e., hypoactivity/inertia; increased aggression/anxiety and anhedonia, independent of the nosological framework in which these dysfunctions occur (van Praag *et al.* 1990[44]). In this paper only the 5-HT data have been discussed. Implications of the functional approach for psychiatry are discussed, including a shift from nosological to functional application of psychotropic drugs. Functional psychopharmacology will be dysfunction-oriented and therefore inevitably geared towards utilizing drug combinations. This prospect is hailed as progress, both practically and scientifically.

REFERENCES

1. VAN PRAAG, H. M. & B. LEIJNSE. 1965. Neubewertung des syndroms. Skizze einer funktionellen Pathologie. Psychiatr. Neurol. Neurochir. **68:** 50–66.
2. VAN PRAAG, H. M., J. KORF, J. P. W. F. LAKKE & T. SCHUT. 1975. Dopamine metabolism in depression, psychoses, and parkinson's diesase: The problem of the specificity of biological variables in behaviour disorders. Psychol. Med. **5:** 138–146.
3. VAN PRAAG, H. M., R. KAHN, G. M. ASNIS, S. WETZLER, S. BROWN, A. BLEICH & M. KORN. 1987a. Denosologization of biological psychiatry on the specificity of 5-HT disturbances in psychiatric disorders. J. Affect. Dis. **13:** 1–8.
4. VAN PRAAG, H. M., G. M. ASNIS, S. L. BROWN, M. KORN, J. M. HARKAVY FRIEDMAN & S. WETZLER. 1990. Beyond serotonin. A multi-aminergic perspective on abnormal behavior. *In* Serotonin in Psychiatric Disorders. S. L. Brown and H. M. van Praag, Eds. New York: Brunner/ Mazel. In press.
5. VAN PRAAG, H. M., J. KORF & J. PUITE. 1970. 5-hydroxyindoleacetic acid levels in the cerebrospinal fluid of depressive patients treated with probenecid. Nature **225:** 1259–1260.
6. BROWN, S. L., A. BLEICH & H. M. VAN PRAAG. 1990. The monoamine hypothesis of depression: The case for serotonin. *In* The Role of Serotonin in the Psychiatric Disorders. S. L. Brown & H. M. van Praag, Eds. New York: Brunner/Mazel. In press.
7. VAN PRAAG, H. M. 1982. Neurotransmitters and CNS disease: Depression. Lancet **ii:** 1259–1264.
8. GOODWIN, F. V. & R. M. POST. 1983. 5 hydroxytryptamine and depression: A model for the interaction of normal variances and pathology. Br. J. Clin. Pharmacol. **15:** 393–3405.
9. VAN PRAAG, H. M. & J. KORF. 1971. Endogenous depressions with and without disturbances in the 5-hydroxytryptamine metabolism: A biochemical classification? Psychopharmacology **19:** 148–152.
10. VAN PRAAG, H. M. & C. LEMUS. 1986. Monoamine precursors in the treatment of psychiatric disorders. *In* Nutrition and the Brain. R. J. Wurtman and J. J. Wurtman, Eds. New York: Raven Press, pp. 89–138.
11. VAN PRAAG, H. M. 1984. In search of the mode of action of antidepressants: 5-HT-tryosine mixtures in depressions. Neuropharamacology **22:** 433–440.
12. VAN PRAAG, H. M., C. LEMUS & R. KAHN. 1987b. Hormonal probes of central serotonergic activity. Do they really exist? Bio. Psych. **22:** 86–98.
13. VAN PRAAG, H. M. 1986a. Serotonin precursors with and without tyrosine in the treatment

of depression. *In* Biological Psychiatry. C. Shagrass, R. Josias, W. Bridger, K. Weiss, D. Stoff and J. Simpson, Eds. New York: Elsevier Science Publishers Inc.
14. RALEIGH, M. J. 1987. Differential behavioral effects of tryptophan and 5-hydroxytryptophan in vervet monkeys: Influence of catecholaminergic systems. Psychopharmacology **93:** 44–50.
15. ASBERG, M., L. TRAKSMAN & P. THOREN. 1976. 5-HIAA in the cerebrospinal fluid: A biochemical suicide predictor? Arch. Gen. Psych. **33:** 1193–1197.
16. VAN PRAAG, H. M. 1986b. Biological suicide research. Outcome and limitations. Biol. Psych. **21:** 1305–1323.
17. BROWN, G. L., F. K. GOODWIN, J. C. BALLENGER, P. F. GOYER & L. F. MAJOR. 1979. Aggression in humans correlates with cerebrospinal fluid metabolites. Psych. Res. **1:** 131–139.
18. BROWN, G. L., M. E. EBERT, P. F. GOYER, D. C. JIMERSON, W. J. KLEIN, W. E. BUNNEY & F. K. GOODWIN. 1982. Aggression, suicide and serotonin: Relationships to CSF amine metabolites. Am. J. Psych. **139:** 741–746.
19. BIOULAC, B., M. BENEZICH, B. RENAUD, B. NOEL & D. ROCHE. 1980. Serotonergic functions in the 47, XYZ syndrome. Biol. Psych. **15:** 917–23.
20. LINNOILA, M., M. VIRKHUNEN, M. SCHEININ, A. NUUTILA, R. RIMON & F. K. GOODWIN. 1983. Low cerebrospinal fluid 5-hydroxyindoleacetic acid concentration differentiates impulsive from nonimpulsive violent behavior. Life Sci. **33:** 2609–2614.
21. LIDBERG, L., M. ASBERG & U. B. SUNDQUIST-STENSMAN. 1984. 5-Hydroxyindoleacetic acid in attempted suicides who have killed their children. Lancet **ii:** 928.
22. LIDBERG, L., J. R. TUCK, M. ASBERG, G. P. SCALIA-TOMBA & L. BERTILSSON. 1985. Homicide, suicide and CSF 5-HIAA. Acta Psychiatr. Scand. **71:** 230–236.
23. VIRKKUNEN, M., A. NUUTILA, F. K. GOODWIN & M. LINNOILA. 1987. Cerebrospinal fluid monoamine metabolite levels in male arsonists. Arch. Gen. Psych. **44:** 241–247.
24. PLUTCHIK, R., H. M. VAN PRAAG & H. R. CONTE. 1989. Correlates of suicide and violence risk, III. A 2-stage model of countervailing forces. Psych. Res. **28:** 215–225.
25. VAN PRAAG, H. M., R. PLUTCHIK & H. CONTE. 1986. The serotonin-hypothesis of (auto) aggression. Critical appraisal of the evidence. Ann. N.Y. Acad. Sci. **487:** 150–167.
26. EVANS, L., & G. MOORE. 1981. The treatment of phobic anxiety by zimeldine. Acta Psychiatr. Scand. **63**(290): 342–345.
27. KOCZKAS, S., G. HOLMBERG & L. WEDIN. 1981. A pilot study of the effect of the 5-HT uptake inhibitor, zimelidine, on phobic anxiety. Acta Psychiatr. Scand. **63**(290): 328–341.
28. EVANS, L., J. KENARDY, P. SCHNEIDER & H. HOEY. 1986. Effect of a selective serotonin uptake inhibitor in agoraphobia with panic attacks. Acta Psychiatr. Scand. **73:** 49–53.
29. KAHN, R. S., H. G. M. WESTENBERG & J. JOLLES. 1984. Zimelidine treatment of obsessive-compulsive disorder. Acta Psychiatr. Scand. **69:** 259–61.
30. KAHN, R. S., H. G. M. WESTENBERG, W. M. A. VERHOEVEN, C. C. GISPEN-DE WIED & W. D. J. KAMERBEEK. 1987. Effect of a serotonin precursor and uptake inhibitor in anxiety disorders: A double-blind comparison of 5-hydroxytryptophan, clomipramine and placebo. Inter. Clin. Psychopharmacol. **2:** 33–45.
31. DEN BOER, J. A., H. G. M. WESTENBERG, W. D. J. KAMERBEEK, W. M. A. VERHOEVEN & R. S. KAHN. 1987. Effect of serotonin uptake inhibitors in anxiety disorders: a double-blind comparison of clomipramine and fluvoxamine. Inter. Clin. Psychopharm. **2:** 21–32.
32. KAHN, R. S. & H. G. M. WESTENBERG. 1985. *l*-5-Hydroxytryptophan in the treatment of anxiety disorders. J. Affect. Dis. **8:** 197–200.
33. KAHN, R. S., H. M. VAN PRAAG, S. WETZLER, G. M. ASNIS & G. BARR. 1988. Serotonin and anxiety revisited. Biol. Psych. **23:** 189–208.
34. KAHN, R. S., S. WETZLER, H. M. VAN PRAAG & G. M. ASNIS. 1988. Behavioral indications for receptor hypersensitivity in panic disorder. Psych. Res. **25:** 101–104.
35. KAHN, R. S., S. WETZLER, H. M. VAN PRAAG & G. M. ASNIS. 1988. Neuroendocrine evidence for serotonin receptor hypersensitivity in panic disorder. Psychopharmacology **96:** 360–364.
36. TAYLOR, D. P., M. S. EISON, L. A. RIBLET & C. P. VANDERMAELEN. 1985. Pharmacological and clinical effects of buspirone. Pharmacol. Biochem. Behav. **23:** 687–694.

37. CEULEMANS, D. L. S., M. I. J. A. HOPPENBROUWERS, Y. GELDERS & A. J. M. REYNTJENS. 1985. The influence of ritanserin, a serotonin antagonist, in anxiety disorders: A double-blind placebo-controlled study versus lorazepam. Pharmakopsychiatria **18:** 303–305.
38. JONES, B. J., B. COSTALL, A. M. DOMENEY, M. E. KELLY, R. J. NAYLOR, N. R. OAKLEY & M. B. TYERS. 1988. The potential anxiolytic activity of GR38032F, a 5-HT3-receptor antagonist. Br. J. Pharmacol. **93:** 985–993.
39. BANKI, C. M. 1977. Correlation between CSF metabolites and psychomotor activity in affective disorders. J. Neurochem. **28:** 255–57.
40. RYDIN, E., D. SCHALLING, & M. ASBERG. 1982. Rorschach ratings in depressed and suicidal patients with low CSF 5-HIAA. Psych. Res. **7:** 229–243.
41. VAN PRAAG, H. M. Diagnosing depression. Looking backward into the future. Develop. in Psychiat.
42. APTER, A., H. M. VAN PRAAG, R. PLUTCHIK, S. SEVY & S. L. BROWN. 1990. The relationships between a serotonergically linked series of psychopathological dimensions. Psychiatry Res. **32:** 191–199.
43. VAN PRAAG, H. M., R. KAHN, G. M. ASNIS, C. Z. LEMUS & S. L. BROWN. 1987c. Therapeutic indications for serotonin potentiating compounds. A hypothesis. Bio. Psych. **22:** 205–212.
44. VAN PRAAG, H. M., G. M. ASNIS, R. S. KAHN, S. L. BROWN, M. KORN, J. M. HARKAVY FRIEDMAN & S. WETZLER. 1990. Monoamines and abnormal behavior. A multi-aminergic perspective. Br. J. Psych. In press.

Serotonin and Nociception

BRIAN P. RICHARDSON

Preclinical Research Department
Sandoz Pharma Ltd.
CH-4002 Basel, Switzerland

A key rôle of serotonin (5-hydroxytryptamine, 5-HT) in regulating the transmission of nociceptive information at various levels of the peripheral and central nervous systems is undisputed. In fact a recent review of research in this area quoted more than 400 papers on this subject, most of which have appeared in the last 5 years.[1] Moreover neuronal systems containing serotonin seem not only to be involved in regulation of segmental nociceptive input at a spinal and supraspinal level, but also in mediating the nociceptive action of morphine and that resulting from electrical stimulation of certain brain stem nuclei.[2,3]

Despite a general acceptance of the above statements, much of the pharmacology of nociception is nevertheless contraindicatory and this almost certainly relates to differences in the species studied, the type of anesthesia employed, the mode of drug application and the responses measured. In addition, there is a general lack of appreciation that the purportedly selective drug tools which have been used in the past, in fact have actions at more than one of the seven 5-HT receptor subtypes which are now known to exist.[4] However, with the advent of truly selective compounds it is finally becoming possible to dissect these responses pharmacologically and thereby identify the particular subtype of 5-HT receptor involved in eliciting them.[5] The main purpose of the current review is to define the particular subtypes of 5-HT receptor involved in regulating the forward transmission of nociceptive input at various hierachial levels of the peripheral and central nervous systems and thereby indicate approaches to the development of novel analgesic compounds.

THE PERIPHERY

More than three decades ago it was shown that serotonin induces pain when applied to a blister base produced on the forearm of human volunteers.[6] Although the mono- and di-*N*-methylated analogues could mimic the action of serotonin, the antagonists available at that time did not block its painful effects in this system. During the late 1970s work in the author's laboratories led to the development of selective agonists and antagonists which could respectively mimic and block the painful action of serotonin on the human blister base.[7] Hence it could be shown that 2-CH_3-5-HT, a selective agonist at 5-HT_3 receptors, was equipotent with 5-HT in inducing painful responses, whereas S-(+)-α-CH_3-5-HT, a selective 5-HT_2 receptor agonist, was more than 100 times less potent in this regard (FIG. 1A). Studies with antagonists showed that methysergide, a 5-HT_1 and 5-HT_2 receptor antagonist, did not block the painful effect of serotonin, whilst the selective 5-HT_3 receptor antagonist, ICS 205-930, was an extremely potent inhibitor (FIG. 1B). It could thus be demonstrated that serotonin induces pain in man through activation of 5-HT_3 receptors located on the cutaneous terminals of primary nociceptive afferent nerves. In addition to producing pain *per se,* serotonin also potentiates the painful effects of other algetic agents such as bradykinin. This was first demonstrated by Sicuteri *et al.* who infused these substances into the dorsal hand vein of human volunteers.[8] Again this potentiating action of serotonin could not be blocked by the then available antagonists such as methysergide. Studies on the human blister base subsequently showed essentially the same synergistic phenomenon between 5-HT and bradykinin, which could be inhibited by prior local treatment of the blister base

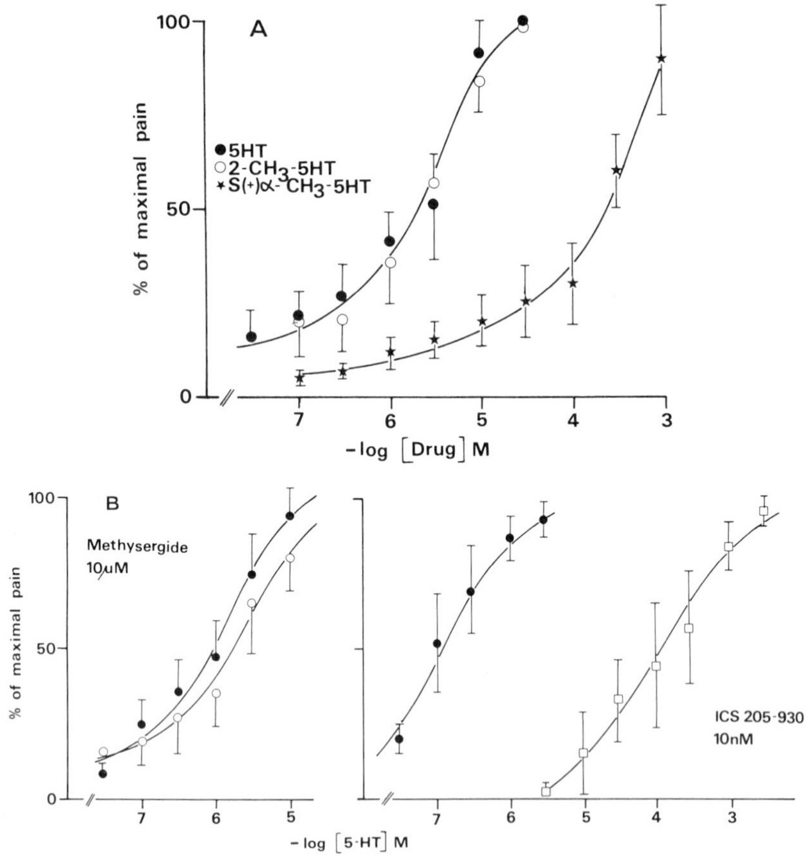

FIGURE 1. Painful effects of serotonin and methylated analogues when applied to a blister base on the human forearm. **A.** Effects of 5-HT (●, n = 5), 2-methyl-5-HT (○, n = 3) and S(+)-α-methyl-5-HT (*, n = 3) in causing pain. **B.** Concentration response curves for 5-HT in the absence (●, n = 4) or presence of 10^{-5}M methysergide (○, n = 3; left panel) or 10^{-8}M ICS 205-930 (□, n = 4; right panel). Antagonists were applied to the blister base for 20 min before applying 5-HT in their continued presence. Vertical bars indicate SEM in each case. (Reproduced with permission from *Nature*.[7])

with the 5-HT_3 receptor antagonist ICS 205-930.[7] Thus not only the pain-producing effects of serotonin are mediated by 5-HT_3 receptors, but also its ability to potentiate the algetic effects of bradykinin.

Concomitant with the serotonin-induced pain response in the blister base model, we noticed the appearance of a marked cutaneous flare reaction, which was also blocked by ICS 205-930.[7] More detailed analysis of this phenomenon by Orwin and Fozard confirmed that this response was also 5-HT_3 receptor-mediated since it was blocked by MDL 72222, another selective 5-HT_3 receptor antagonist.[9] Thus the activation of primary afferent nociceptive nerve endings by serotonin presumably not only depolarizes these cells and releases substance P in the dorsal horn of the spinal cord, but also elicits an axonal reflex with the release of substance P in the skin, this giving rise to a flare response. Such a mechanism provides the basis for a sterile inflammatory response, referred to as neurogenic inflammation. As discussed in detail elsewhere, such a cascade of events may very well explain some aspects of the vascular pain encountered in migraine.[10,11]

It is very likely that the pain potentiating action of serotonin is more relevant than its ability to produce pain *per se,* since the former is both considerable and protracted whilst the latter is mild, transient and exhibits tachyphylaxis. Electrophysiological correlates of the pain produced by applying serotonin to primary afferent fibers have been established *in vitro* by Neto who showed that serotonin can depolarize nerve fibers in the rabbit vagus.[12] Subsequent studies in my laboratory showed that serotonin selectively depolarizes a subset of vagal afferent C fibers and that this effect is mimicked by the selective 5-HT_3 receptor agonist 2-CH_3-5-HT, but not by the selective 5-HT_2 agonist S-(+)-α-CH_3-5-HT.[7] The depolarizing action of serotonin is competitively antagonized by 5-HT_3 receptor antagonists, but not by the 5-HT_1 and 5-HT_2 receptor antagonist methysergide.[7] This shows that the depolarizing action of serotonin on 'C' fibers in this preparation is mediated by 5-HT_3 receptors. Similar results have been obtained using the rat vagus nerve, where it has also been shown that phenylbiguanide is a 5-HT_3 receptor agonist.[13] Consistent with this, phenylbiguanide produces pain in the human blister base.[14]

Studies performed on the rabbit nodose ganglion, which houses the cell bodies of vagal afferent 'C' fibers, have shown a depolarizing action of serotonin similar to that produced in the cervical vagus nerve.[15] Again ICS 205-930 and MDL 72222 are potent antagonists of this action of serotonin, confirming the involvement of 5-HT_3 receptors in this response.[16] Intracellular microelectrode recordings show that the rapid depolarization produced by serotonin in this preparation results from the opening of Na^+ channels in the nerve membrane which are insensitive to tetrodotoxin.[17] This is consistent with the recent suggestion that 5-HT_3 receptors may actually be part of a cation channel complex analogous to nicotinic and glutamate receptors.[17]

The electrophysiological mechanisms underlying the ability of serotonin to potentiate the nociceptive action of bradykinin are not fully understood. Nevertheless both bradykinin and serotonin receptors do occur on the same cutaneous nociceptive afferent nerve fibers in cats, although little evidence for potentiation of each other's ability to increase discharge rates in this preparation have been observed.[19] Interestingly, serotonin and bradykinin depolarize primary afferent 'C' fibers in the rabbit nodose ganglion by different ionic mechanisms—serotonin opening Na^+ channels as previously mentioned and bradykinin by reducing K^+ permeability. Bradykinin depresses the after hyperpolarization caused by serotonin in these neurons, probably by virtue of its ability to block K^+ conductance. This effect is long lasting and may explain synergism between serotonin and bradykinin in producing pain.[20]

Although exogenous serotonin is clearly algetic when applied locally or injected, the question of the involvement of endogenously released serotonin in naturally occurring pain is obviously more relevant. There is much circumstantial evidence to suggest a key rôle for serotonin in producing pain in migraine[21] and it is probable that in myocardial infarction and venous thrombosis, where massive release from platelets occurs, that it is involved too. Recent studies have shown that the 5-HT_3 receptor antagonist ICS 205-930 is extremely effective in inhibiting carrageenan-induced hyperalgesia in rats, although the accompanying edema remains uninfluenced.[22] Thus serotonin may also be involved in some forms of inflammatory pain where its actions are again mediated through 5-HT_3 receptors.

SPINAL CORD

Several investigators have demonstrated that intrathecal serotonin induces nociception in a variety of different animal species.[23-27] This effect is potentiated by serotonin reuptake blocking drugs and inhibitors of monoamine oxidase, both of which increase the availability of the locally applied serotonin.[24] Pharmacological studies of this phenomenon have generated conflicting results, and it is still not clear which subtype of serotonin receptor is involved. Early studies performed by Wang showed that intrathecal serotonin blocked the tail flick response to noxious heat[23] and shortly after Yaksh and Wilson showed that methysergide could reverse this effect when given systemically.[24] However, since there is a supraspinal serotoninergic pathway originating in the brain stem that is activated through a 5-HT_2 receptor mechanism and which tonically inhibits nociceptive input at the spinal level (see below), it is not clear whether methysergide was really acting at the spinal level in these studies, or whether it inhibited control mechanisms at a higher level. Subsequent experiments designed to clarify this point confirmed that methysergide, 2-bromolysergic acid diethylamide and ketanserin could all antagonize the antinociceptive action of serotonin when locally applied to the cord.[25] However, the doses of ketanserin required to produce this effect were high relative to its potency as an inhibitor of 5-HT_2 receptors and metergoline, which is an extremely potent antagonist at this receptor subtype, was hardly effective at all. It thus seems likely, that the ability of serotonin to increase tail flick latency under these circumstances is mediated by a 5-HT_1-like receptor rather than a 5-HT_2 receptor. Interestingly none of the antagonists examined influenced the ability of 5-HT to increase nociceptive threshold in the hot plate test performed in the same study. This is probably because the behavioral response measured involves supraspinally organized neuronal pathways, using 5-HT_1 receptor subtypes which are not influenced by the drugs employed.

To confuse the situation further, a recent study has shown that the ability of intrathecal serotonin to increase nociceptive thresholds in both the tail flick and hot plate tests in rats can be potently antagonized by the 5-HT_3 receptor antagonist ICS 205-930, also given intrathecally.[28] Moreover, at higher doses, ICS 205-930 alone could induce a state of hyperalgesia suggesting that descending serotoninergic pathways are tonically active. The doses of ICS 205-930 used in these studies were indeed very low and clearly point to an involvement of 5-HT_3 receptors in the action of serotonin at the spinal level.

In contrast to the results obtained in intact animals, intrathecal serotonin decreases the nociceptive threshold for a series of different reflexes in spinal rats and expands the receptive field.[29] This suggests that the descending bulbospinal serotoninergic pathway normally facilitates local spinal nociceptive reflexes whilst inhibiting ascending sensory input. Unfortunately quipazine was used as a central serotoninergic agonist in these studies and this compound has recently been shown to have effects on several different subtypes of serotonin receptor, acting as an agonist or antagonist at 5-HT_1 and 5-HT_2 receptors and an antagonist at 5-HT_3 receptors.[13,30,31] This complicates the interpretation of the results, especially in light of the findings with ICS 205-930 in intact animals mentioned previously. Nevertheless, an increase in 5-HT_1 binding sites in the spinal cord of rats that underwent spinal transection 7 days previously, coupled with a parallel and progressive increase in their sensitivity to the central serotoninergic agonist

5-methoxy-*N'N*-dimethyltryptamine, suggests that the reported facilitation of spinal nociceptive reflexes was mediated by $5\text{-}HT_1$ receptors under these circumstances. More recently evidence supporting this interpretation has been obtained using selective serotoninergic agonists which have now become available. Thus 8-OH-*N'N*-dipropyl-2-aminotetraline, (8-OH-DPAT), which is a selective agonist at $5\text{-}HT_{1A}$ receptors, decreased tail flick latency, even in intact rats.[32] In contrast, compounds like 1-(2,5-dimethoxy-4-iodophenyl)-2-aminopropane (DOI), 6-chloro-2-(1-piperazinyl)-pyrazine (MK 212) and quipazine, which were claimed by the authors to be $5\text{-}HT_2$ receptor selective agonists, clearly increased the tail flick latency in a dose dependent manner. However, all three compounds have appreciable activity at some $5\text{-}HT_1$ receptor subtypes, so that the conclusion that the observed antinociceptive effects of intrathecal serotonin result from stimulation of spinal $5\text{-}HT_2$ receptors is clearly not definitive.[30,33,34]

From the foregoing discussion, it is apparent that the effects of serotonin applied locally to the cord are complex and that facilitatory effects on local nociceptive spinal reflexes contrast with its inhibitory action on ascending sensory input. It is probable that a subtype of $5\text{-}HT_1$ receptor mediates the former and that $5\text{-}HT_1$, $5\text{-}HT_3$ and, possibly, even $5\text{-}HT_2$ receptors are involved in the latter. In this regard it should be borne in mind that serotonin applied through an intrathecal catheter may have more generalized effects than that released endogenously from descending bulbospinal neurons or intrinsic segmental spinal neurons. A parsimonious explanation for the observed inhibitory action of $5\text{-}HT_1$, $5\text{-}HT_2$ and $5\text{-}HT_3$ receptor antagonists on the antinociceptive action of intrathecal serotonin is that each receptor subtype exists in the spinal cord, but on different neurons in the nociceptive pathway. Thus $5\text{-}HT_3$ receptors have been shown to exist on the terminals of primary afferent fibers in the dorsal horn,[35] and it has been suggested that their activation could lead to presynaptic inhibition due to depolarization blockade, thereby inhibiting, incoming nociceptive signals from the periphery.[36] $5\text{-}HT_1$ receptors probably mediate the neuronal inhibition observed when serotonin is applied by iontophoresis to secondary nociceptive or convergent neurons which ascend in the spinothalmic tract (see below). Thus activation at this level would also inhibit the forward transmission of nociceptive input. Some dorsal horn interneurons in the spinal cord are excited by iontophoresis of serotonin, and this effect is totally inhibited by the selective $5\text{-}HT_2$ antagonist cinanserin.[37] If these interneurons release an inhibitory neurotransmitter onto the convergent neuron, then again nociceptive transmission would be interrupted. Thus it is possible to conceive of a situation where $5\text{-}HT_1$, $5\text{-}HT_2$ or $5\text{-}HT_3$ receptor antagonists each could block the antinociceptive action of intrathecal serotonin. Whether this is really the case can only be established by further experimentation with more selective agonist and antagonist drugs.

Electrophysiological equivalents of these behavioral end points for nociceptive input have also been comprehensively documented. Thus iontophoretic application of 5-HT to dorsal horn neurons has been shown to result in neuronal inhibition in several species.[38–44] Although serotonin produces either inhibition or excitation of dorsal horn interneurons in laminae I and II,[38] it inhibits a majority of convergent neurons located deeper in laminae IV and V.[39,41,42] Thus the depressant action of iontophoretically applied serotonin on the firing rate of secondary nociceptive neurons may be the electrophysiological counterpart of the antinociceptive obtained when it is applied intrathecally. The depressant action of serotonin on the firing of dorsal horn interneurons seems to be mediated by a $5\text{-}HT_1$ receptor since methysergide, but not cinanserin, can inhibit this effect.[42,43] It therefore seems likely that serotonin inhibits the transmission of nociceptive information in the dorsal horn through activation of $5\text{-}HT_1$ receptors.

BRAIN STEM

Reynolds was the first to show that electrical stimulation of midbrain central gray matter in rats produces powerful analgesia, sufficient to perform a laparotomy without use of chemical anesthetics.[45] Later it was demonstrated that electrical stimulation of several discrete brain stem areas could produce a similar effect and, in particular, that activation of the nucleus raphe magnus (NRM) in cats, which houses the cell bodies of serotoninergic neurons projecting to the dorsolateral funiculus of the spinal cord, was very effective in this regard.[46,47] Stimulation of the descending bulbospinal pathway has subsequently been shown by many workers to result in inhibition of secondary nociceptive neurons in the dorsal horn of the spinal cord, an effect that can be reversed, at least partly, by the intrathecal application of the 5-HT_1 and 5-HT_2 receptor antagonist methysergide.[48] Hammond and coworkers showed that electrical stimulation of NRM indeed causes release of serotonin from the spinal cord.[49] Evidence has also been provided to show that this descending inhibitory pathway is normally tonically active, since the intravenous administration of the 5-HT_2 receptor antagonist cinanserin significantly increased the discharge rate of convergent dorsal horn neurons activated by transcutaneous electrical stimulation of peripheral C fibers in rats.[50] Since cinanserin has a high degree of selectivity for 5-HT_2 receptors, and is ineffective at blocking the inhibitory action of serotonin on secondary dorsal horn nociceptive neurons,[43] it appears probable that its action to increase the discharge rate of these cells occurs at the level of the brain stem. This suggests that a serotoninergic tract in the brain stem is involved in regulating the activity of the descending inhibitory pathway. Direct evidence for this possibility has been obtained by Llewelyn *et al.* who showed that microinjection of serotonin into the NRM of rats can increase tail flick latency, and that this is blocked when cinanserin is coadministered with the serotonin.[51] Moreover, analgesia induced by electrical stimulation of the periaqueductal gray matter (PAG) in rats can be inhibited by injection of methysergide into the NRM.[52] Since electrical activation of PAG also results in the release of serotonin from the spinal cord,[53] it appears that there is a serotoninergic pathway from the PAG to the NRM. Activation of this pathway leads to the release of serotonin in the NRM, activation of descending bulbospinal serotoninergic neurons carried in the dorsolateral funiculus and release of serotonin in the dorsal horn of the spinal cord. This results in inhibition of segmental nociceptive input. In support of this is the observation that serotoninergic nerve fibers originating from PAG do innervate the NRM in rats.[54] Single cell recordings from the cell bodies of nerves in the NRM, show that the iontophoretic application of serotonin excites a majority of these neurons by a 5-HT_2 receptor–mediated mechanism.[51,55] This is consistent with the observation that the 5-HT_2 receptor antagonist cinanserin decreases tonic descending inhibitory control of nociceptive input (see above).

Much evidence exists to suggest the involvement of this serotoninergic inhibitory pathway in mediating the analgesic action of morphine. In conscious rats with chronically implanted brain cannulae, the injection of morphine into NRM induces antinociception,[56,57] and this can again be antagonized by the concomitant administration of cinanserin.[57] This suggests that morphine produces its analgesic effects through the local release of serotonin under these circumstances. Direct support for this interpretation was provided by Pycock *et al.* who showed that both morphine and the stable enkephalin analogue, FK 33-824, release serotonin from superfused slices of PAG and NRM.[58] Moreover injection of morphine directly into the NRM also releases serotonin in the spinal cord[59–61] and produces antinociception, as does electrical stimulation of this nucleus (see above). Both these effects of morphine can be inhibited by the systemic administration of naloxone.[59,60] It thus appears that at least part of the antinociceptive action of morphine is mediated through activation of serotoninergic neurons in the brain stem. This results in the release of serotonin in the NRM with subsequent activation of 5-HT_2 receptors on the cell bodies of neurons which project to the dorsolateral funiculus of the spinal cord to inhibit nociceptive inputs at a segmental level.

CONCLUSIONS

This review shows that serotonin regulates nociceptive input at various levels of the nervous system. When applied to the nerve terminals of primary afferents in the periphery serotonin has algetic properties *per se* and considerably potentiates the pain-producing action of other substances such as bradykinin. Although the mechanism of this potentiation is not fully understood at the cellular level, both this and the algetic action of serotonin alone are mediated by 5-HT_3 receptors. It is thus possible that the selective 5-HT_3 receptor antagonists which have recently become available for clinical use will prove of therapeutic value in conditions such as migraine, inflammatory states, and myocardial infarction where pain subsequent to serotonin release occurs. Indeed there is already some clinical and experimental data to support this contention.[7,22,62]

In the spinal cord local application of serotonin to the cell bodies of secondary nociceptive afferents in the dorsal horn can selectively inhibit both the depolarizing action of substance P on these cells, and the increased firing rate recorded when primary afferent fibers in their receptive field are activated by nociceptive stimuli. The serotonin receptor subtype involved in mediating this action has not been definitively characterized, but is probably a 5-HT_1-like receptor. Since a descending inhibitory serotoninergic pathway carried in the dorsolateral funiculus is believed to synapse onto such secondary nociceptive dorsal horn neurons and inhibit the forward transmission of nociceptive information by the spinothalamic tract at the segmental level, this inhibitory serotonin receptor may well have physiological relevance for pain gating. In agreement with this is the observation that intrathecal serotonin increases the pain threshold in several species. The serotonin receptor involved in mediating this effect has also not been definitively characterized, although both 5-HT_1 and more recently 5-HT_3 receptors have been implicated. In some studies it was also claimed that the 5-HT_2 receptor antagonist ketanserin blocks the nociceptive action of serotonin. Further studies using more selective agonists and antagonists are required before a definitive statement regarding the subtype(s) of serotonin receptors involved can be made. Nevertheless the results obtained so far indicate that selective agonists of the receptor involved could inhibit nociceptive input at the spinal level and thereby provide another new type of analgesic agent.

In the brain stem the local application of serotonin to cells of the NRM, which is believed to be the origin of the serotoninergic pathway mentioned above, causes excitation through a 5-HT_2 receptor mechanism. Since there is evidence that the NRM receives serotoninergic input from the ventromedial and ventrolateral parts of the periaqueductal gray, and that electrical stimulation of these regions leads to activation of NRM through a 5-HT_2 receptor mediated mechanism, it seems likely that the descending spinal inhibitory serotoninergic pathway is normally activated via this mechanism. Consistent with this is the observation that activation of the descending serotoninergic system at any level (i.e. by electrical stimulation of PAG or NRM, microinjection of serotonin into NRM, iontophoretic application of serotonin to the dorsal horn etc.) leads to inhibition of the forward transmission of nociceptive input. It thus seems likely that selective 5-HT_2 receptor agonists could provide novel types of centrally acting analgesic agents. However such agents would obviously have to be targeted to the NRM by some means to avoid a whole array of systemic and central side effects related to generalized 5-HT_2 receptor stimulation.

Thus, in summary, serotonin produces algetic or analgetic effects, depending on the point in the nervous system at which it is released. In the periphery it depolarizes afferent fibers and causes pain as well as potentiating the algetic action of other substances. Conversely in the spinal cord and brain stem it plays an inhibitory rôle in pain transmission and increases pain gating. The actions of 5-HT in the periphery, spinal cord and brain stem are each mediated by different subtypes of 5-HT receptor, raising the possibility of selective pharmacological intervention at various levels of pain transmission.

REFERENCES

1. LE BARS, D. 1988. Serotonin and Pain. *In* Neuronal Serotonin. N. N. Osborne & M. Hamon, Eds. Chichester, Eng.: John Wiley and Sons Ltd., pp. 171–229.
2. FIELDS, H. L. & J. M. BESSON, Eds. 1988. Progress in Brain Research Vol. 77. Amsterdam: Elsevier Science Publishers B. V.
3. ROBERTS, M. H. T. 1989. Drug Design and Delivery. **4:** 77–83.
4. RICHARDSON, B. P. & D. HOYER. 1988. Selective agonists at 5-hydroxytryptamine receptor subtypes. *In* Serotonin: From Cell Biology to Pharmacology and Therapeutics. R. Paoletti & P. M. Vanhoutte, Eds. Dordrecht, The Netherlands; Kluwer Academic Publishers.
5. HUMPHREY, P. P. A. & B. P. RICHARDSON. 1989. 5-HT receptor classification. *In* Serotonin. E. Mylecharane, J. Angus, I. De La Lande & P. P. A. Humphrey. Eds. Basingstoke, Eng.: Macmillan Press.
6. KIELE, C. A. & D. ARMSTRONG. 1964. *In* Substances producing Pain and Itch. London: Arnold.
7. RICHARDSON, B. P., G. ENGEL, P. DONATSCH & P. A. STADLER. 1985. Nature **316:** 126–131.
8. SICUTERI, F., M. FANICIULLACCI, G. FRANCHI & P. L. DEL BIANCO. 1965. Life Sci. **4:** 309–316.
9. ORWIN, J. M. & J. R. FOZARD. 1986. Eur. J. Clin. Pharmacol. **30:** 209–212.
10. MOSKOWITZ, M. 1984. Ann. Neurol. **16:** 157–168.
11. FOZARD, J. R. 1982. Adv. Neurol. **33:** 295–307.
12. NETO, F. R. 1978. Eur. J. Pharmacol. **49:** 351–356.
13. IRELAND, S. J. & M. B. TYERS. 1987. Br. J. Pharmacol. **90:** 229–238.
14. FASTIER, F. N., M. A. MCDOWALL & H. WAAL. 1959. Br. J. Pharmac. Chemother. **14:** 527–535.
15. WALLIS, D. I., C. E. STANSFIELD & H. L. NASH. 1982. Neuropharmacology. **21:** 31–40.
16. ROUND, A. & D. I. WALLIS. 1986. Br. J. Pharmacol. **88:** 485–494.
17. HIGASHI, H. & S. NISHI. 1982. J. Physiol. **323:** 543–567.
18. D. HOYER. 1989. 5-Hydroxytryptamine receptors and effector coupling mechanisms in peripheral tissues. *In* Peripheral Actions of 5-HT. J. R. Fozard, Ed. Oxford: Oxford University Press, pp. 72–99.
19. BECK, P. W. & H. O. HANDWERKER. 1974. Pflügers Arch. **347:** 209–222.
20. HIGASHI, H., N. UEDA & S. NISHI. 1982. Brain Res. Bull. **8:** 23–32.
21. FOZARD, J. R. 1985. 5-hydroxytryptamine in the pathophysiology of migraine. *In* Proc. 5th International Symposium on Vascular Neuroeffector Mechanisms. J. A. Bevan, Ed. Amsterdam: Elsevier, pp. 321–328.
22. ESCHALIER, A., V. KAYSER & G. GUILBAUD. 1989. Pain **36:** 249–255.
23. WANG, J. K. 1977. Anaesthesiology **47:** 269–271.
24. YAKSH, T. L. & P. R. WILSON. 1979. J. Pharmacol. Exp. Ther. **208:** 446–453.
25. SCHMAUSS, C., D. L. HAMMOND, J. W. OCHI & T. L. YAKSH. 1983. Eur. J. Pharmacol. **90:** 349–357.

26. ARCHER, T., B. G. MINOR & C. POST. 1985. Brain Res. **333:** 55–61.
27. MINOR, B. G., C. POST & T. ARCHER. 1985. Neurosci. Lett. **54:** 39–44.
28. GLAUM, S. R., H. K. PROUDFIT & E. G. ANDERSON. 1989. Neurosci. Lett. **95:** 313–317.
29. ZELMAN, F. P., L. M. KOW & D. F. PFAFF. 1983. J. Pharmacol. Exp. Ther. **226:** 447–485.
30. SCHOEFFTER, P. & D. HOYER. 1989. Naunyn Schmiedeberg's Arch. Pharmacol. **339:** 675–683.
31. COHEN, M. L. & R. W. FULLER. 1983. Life Sci. **32:** 711–718.
32. SOLOMON, R. E. & G. F. GEBHART. 1988. J. Pharmacol. Exp. Ther. **245:** 905–912.
33. HOYER, D., C. WAEBER, P. SCHOEFFTER, J. M. PALACIOS & A. DRAVID. 1989. Naunyn Schmiedeberg's Arch. Pharmacol. **339:** 252–258.
34. ENGEL, G., M. GÖTHERT, D. HOYER, E. SCHLICHER & K. HILLENBRAND. 1986. Naunyn Schmiedeberg's Arch. Pharmacol. **332:** 1–7.
35. GOZLAN, H., M. C. GALLISSOT, S. BOURGOIN, D. VERGÉ, F. MÉNARD & M. HAMON. 1989. Proceedings of International Symposium on Serotonin: From Cell Biology to Pharmacology and Therapeutics. Florence, Italy, April 1989. Abstract Book p. 32.
36. ANDERSON, E. G. & S. TODOROVIC. 1989. Proceedings of International Symposium of Serotonin: From Cell Biology to Pharmacology and Therapeutics. Florence, Italy. April 1989. Abstract Book p. 33.
37. DAVIES, J. E. & M. H. T. ROBERTS. 1981. Brain Res. **217:** 399–404.
38. RANDIĆ, M. & H. H. YU. 1976. Brain Res. **111:** 197–203.
39. HEADLEY, P. M., A. W. DUGGAN & B. T. GRIERSMITH. 1978. Brain Res. **145:** 185–189.
40. BELCHER, G., R. W. RYALL & R. SCHAFFNER. 1978. Brain Res. **151:** 307–321.
41. JORDAN, L. M., D. R. KENSHALO, R. F. MARTIN, L. H. HABER & W. D. WILLIS. 1979. Brain Res. **164:** 342–346.
42. GRIERSMITH, B. T. & A. W. DUGGAN. 1980. Brain Res. **187:** 231–236.
43. DAVIES, J. E. & M. H. T. ROBERTS. 1981. Brain Res. **217:** 399–404.
44. JEFTINIJA, S., G. RASPANTINI, M. RANDIĆ, T. L. YAKSH, V. L. W. GO & A. A. LARSON. 1986. Brain Res. **368:** 107–115.
45. REYNOLDS, D. V. 1968. Science **164:** 444–445.
46. FIELDS, H. L. & A. I. BASBAUM. 1978. Ann. Rev. Physiol. **40:** 193–221.
47. OLIVERAS, J. L., G. GUILBAUD & J. M. BESSON. 1979. Brain Res. **164:** 317–322.
48. BARBARO, N. M., D. L. HAMMOND & H. L. FIELDS. 1985. Brain Res. **343:** 223–229.
49. HAMMOND, D. L., G. M. TYCE & T. L. YAKSH. 1985. J. Physiol. **359:** 151–162.
50. RIVOT, J. P., B. CALVINO & J. M. BESSON. 1987. Brain Res. **403:** 142–146.
51. LLEWELYN, M. B., J. AZAMI & M. H. T. ROBERTS. 1983. Brain. Res. **258:** 59–68.
52. ROBERTS, M. H. T. 1984. Neuropharmacology. **23:** 1529–1536.
53. YAKSH, T. L. & G. M. TYCE. 1980. Brain Res. **192:** 133–146.
54. BEITZ, A. J. 1982. J. Neurosci. **2:** 829–842.
55. DAVIES, M., L. S. WILKINSON & M. H. T. ROBERTS. 1988. Br. J. Pharmacol. **94:** 483–491.
56. DICKENSON, A. H., J. L. OLIVERAS & J. M. BESSON. 1979. Brain Res. **170:** 95–111.
57. LLEWELYN, M. B., J. AZAMI & M. H. T. ROBERTS. 1984. Brain Res. **306:** 165–170.
58. PYCOCK, C. J., S. BURNS & R. MORRIS. 1981. Neurosci. Lett. **22:** 313–317.
59. SHIOMI, H., H. MURAKAMI & H. TAKAGI. 1978. Eur. J. Pharmacol. **52:** 335–344.
60. YAKSH, T. L. & G. M. TYCE. 1979. Brain Res. **171:** 176–181.
61. VASKO, M. R. & M. VOGT. 1982. Neurosci. **7:** 1215–1225.
62. LOISY, C., S. BEORCHIA, V. CENTONZE, J. R. FOZARD, P. J. SCHECHTER & G. P. TELL. 1985. Cephalalgia **5:** 79–82.

DISCUSSION OF THE PAPER

QUESTION: The last point about the 5-HT_2 agonists in the CNS, was that to imply they might be analgesic or pain causing?

RICHARDSON: Well at least in rats if you apply 5-HT_2 agonists to the nucleus raphe magnus you can induce analgesia. And the analgesic effects of morphine applied to the nucleus raphe magnus are blocked by 5-HT_2 antagonists.

QUESTION: And you would suggest a drug development program based on 5-HT_2 agonists for analgesia?

RICHARDSON: Well you would, of course, have to somehow target it because you've got a lot of potentially unwanted effects of 5-HT_2 agonists such as hallucinogenic actions and effects on the vasculature, etc.—so it wouldn't be an easy task.

P. HUMPHREY (*Glaxo Group Research, Ltd., Ware, Hertfordshire, Eng.*): Brian you've showed us some very nice pharmacology in the rat, but I wonder how predictive it is for pain in man. I mean for instance in the carrageenan test, you told us that the 5-HT which causes the pain came from mast cells and I think rodents are the only species in which the mast cells contain vast amounts of 5-HT. So again I take a point made by the last speaker that perhaps rats aren't very predictive of man.

RICHARDSON: Well I think that's right. Of course, we do not know what might happen from a clinical point of view in inflammatory pain in man where platelet 5-HT could play a role, but based on the information that we have in rats, it would seem that the release of 5-HT locally from mast cells in the skin follows the same time course as the inhibitory effect of ICS 205-930 on the carrageenan-induced hyperalgesia in rats. After about 60 minutes the cutaneous 5-HT has disappeared and doesn't participate in the pain response to carrageenan. If that's the case, then I would agree with you. I would therefore doubt that 5-HT_3 receptor antagonists have a clinically useful analgesic effect, but perhaps they should be examined specifically in inflammatory pain.

What I can say is that we, at Sandoz, have looked clinically at ICS 205-930 as an analgesic in man in post-surgical pain and it's clearly not active. The question is whether in particular situations, such as migraine or inflammation, where 5-HT is released, 5-HT_3 receptor antagonists may be effective even if they're not general analgesic compounds.

Serotonin and Appetite

G. CURZON

Institute of Neurology
Queen Square
London WC1N 3BG England

INTRODUCTION

Neurochemical research on relationships between tryptophan (TRP), 5-hydroxytryptamine (5-HT) and feeding has centered on how feeding affects brain TRP concentration and on the synthesis and availability of 5-HT. Pharmacologists have focused on the control of appetite by 5-HTergic drugs. These findings lead to questions such as whether effects of feeding on TRP and on 5-HT function are involved in the normal control of appetite and in disorders such as anorexia nervosa. Research in this area has been stimulated by recent evidence for numerous 5-HT receptor subtypes. This chapter will mainly concern how drugs with selectivity towards these receptors affect feeding in the rat. As a drug used clinically to decrease appetite (fenfluramine, Ponderax®) acts by releasing 5-HT,[1] it is probable that other 5-HT-ergic drugs can influence appetite in the rat and do not decrease food intake merely, for example, due to general malaise. Before describing research on this topic the effects of feeding on brain TRP and of TRP administration on feeding will be briefly indicated.

EFFECTS OF FEEDING ON BRAIN TRP AND 5-HT

Rat brain tryptophan increases on giving a large high carbohydrate–low protein meal, because of a fall of plasma concentrations of amino acids which compete with TRP transport to the brain.[2] A similar increase occurs on overnight fasting because TRP is freed from binding to plasma albumin so that it becomes more available to the brain.[3-5] While these mechanisms may *normally* have no more than slight effects under normal dietary conditions,[6] it should be remembered that this conclusion is based on data from relatively few subjects studied over short periods. As values obviously range more widely within larger groups or over longer periods of study, subgroups could exist with brain TRP concentrations that are highly sensitive to dietary influences. Such TRP changes can influence 5-HT synthesis, but there is still no direct convincing evidence that giving TRP, even at pharmacological dosage, increases 5-HT at receptor sites. Nevertheless, TRP and 5-HT-ergic drugs affect feeding and it is particularly interesting to note that feeding has recently been reported to increase extracellular 5-HT in the lateral hypothalamus of the rat.[7] These findings suggest that feeding may lead to central TRP and 5-HT changes which normally mediate its termination and that inappropriate relationships between feeding or food deprivation and changes of brain indole chemistry and 5-HT function could be involved in disorders of appetite such as anorexia nervosa.

EFFECT OF L-TRP ON FEEDING

As 5-HT-ergic drugs affect feeding[8] and brain 5-HT synthesis is increased by L-TRP[9] the effects of TRP on feeding have received much attention. Early work on rats failed to reveal changes[10] perhaps because the amino acid was given after food deprivation as intake was decreased under other conditions, i.e. on the first day of exposure to a diet to which L-TRP was added.[11] Similar results were obtained on humans.[12] The association of anorexia in cancer patients with high plasma free TRP and CSF TRP concentrations[13] may therefore be of causal significance.

L-TRP can also affect food choice as acute[14] or chronic[11] administration to rats within meals decreased subsequent carbohydrate intake and increased protein intake. Similarly, 1g L-TRP given to humans in a high protein meal decreased carbohydrate intake in a later, free choice meal.[15] Conversely, if plasma TRP was decreased by giving a TRP deficient amino acid mixture then protein intake in a later, free choice meal was decreased.[16]

These findings are *consistent* with an important role for 5-HT in feeding. However the influence of other pathways of TRP metabolism on feeding remains largely unexplored. Nevertheless evidence, summarized below, on the effects of 5-HT receptor subtype agonists on feeding strengthens interpretation in terms of 5-HT.

EFFECTS OF 5-HT AGONISTS ON FEEDING

5-HT$_{1A}$ Agonists

The putative 5-HT$_{1A}$ agonists 8-hydroxy-2-(di-*n*-propylamino)tetralin (8-OH-DPAT)[17,18] buspirone, ipsapirone,[19] gepirone[20] and LY165163[21] increase food intake in free feeding rats. Typical data is shown in TABLE 1. As the anorexic effects of non-selective, indirect 5-HT agonists (e.g. fenfluramine and 5-hydroxytryptophan) and the hyperphagic effects of the 5-HT synthesis inhibitor *p*-chlorophenylalanine (*p*CPA) suggest that 5-HT inhibits food intake,[1,8] 8-OH-DPAT was proposed to cause hyperphagia by activating presynaptic 5-HT$_{1A}$ receptors so that 5-HT release from terminals is decreased.[19] This mechanism is supported[21,22] by the absence of 8-OH-DPAT- and LY165163-induced hyperphagia after depletion of 5-HT by pCPA. Presynaptic 5-HT$_{1A}$ receptors do not occur on 5-HT$_{1A}$ terminals[23] but are found at high

TABLE 1. Effects of 5-HT$_{1A}$Agonists on Food Intake in Freely Feeding Rats[a]

Treatment (S.C.)	Food Intake (g ± SEM)
	Two hours
Saline	0.8 ± 0.3
8-OH-DPAT 0.125 mg/Kg	2.2 ± 0.4
8-OH-DPAT 0.250 mg/Kg	2.7 ± 0.4
Saline	0.4 ± 0.1
LY165163 0.5 mg/Kg	1.5 ± 0.5
LY165163 1mg/Kg	2.3 ± 0.3
	Four hours
Saline	1.6 ± 0.4
Gepirone 4 mg/Kg	3.4 ± 0.4
Gepirone 8 mg/Kg	4.4 ± 0.3
Saline	1.5 ± 0.3
Buspirone 4 mg/Kg	3.2 ± 0.5
Ipsapirone 5 mg/Kg	2.2 ± 0.5

[a]Effects of drugs were all $p < 0.05$

TABLE 2. Effects of Antagonists on the Hyperphagic Response to 8-OH-DPAT (1mg/Kg s.c.) in Normally Fed Rats[a]

Treatment	Blockade	Interpretation
Metergoline 5mg/Kg	Yes	5-HT receptor
Ketanserin 2.5mg/Kg	No	Not 5-Ht$_2$, α_1
ICS-205-930 1mg/Kg	No	Not 5-HT$_3$
(−) Pindolol 4mg/Kg	Yes	5-HT$_{1A}$, 5-HT$_{1B}$, or β
(+) Pindolol 4mg/Kg	No	Not β
Spiperone 0.05mg/Kg	Yes	5-HT$_{1A}$ or DA
Haloperidol 0.1mg/Kg	No	Not DA or α_1
Idazoxan 3mg/Kg	No	Not α_2

[a]Data are from ref. 28.

density on 5-HT cell bodies in the raphe nuclei[24]; injecting 8-OH-DPAT therein increased feeding.[25]

Although 8-OH-DPAT has high affinity for 5-HT$_{1A}$ sites it also has lower affinity for 5-HT$_{1B}$, 5-HT$_{1C}$, 5-HT$_{1D}$, and 5-HT$_2$ sites[26] and also for α_2-adrenergic[27] sites. However, antagonist experiments[28] point strongly to the hyperphagic effect being mediated at 5-HT$_{1A}$ sites (TABLE 2).

It is conceivable that 8-OH-DPAT affects feeding via a general activation of motor and consumatory behavior due to a widespread decrease of 5-HT release at terminals. It is known that raphe lesions can enhance dopamine-dependent behavior such as locomotion and gnawing.[29] Therefore 8-OH-DPAT might increase feeding secondarily to a release of these behavioral components. Indeed it has been suggested[30,31] that it elicits eating merely as a consequence of gnawing. However, animals given a choice between food pellets and wood blocks chose to eat the pellets[17] and both 8-OH-DPAT and gepirone dose-dependently increased consumption of a liquid diet.[32]

TRP, which might increase 5-HT at postsynaptic receptors, decreases carbohydrate intake.[11,14–16] Therefore it is of interest that 8-OH-DPAT, which *decreases* 5-HT at these sites[33] had the opposite effect, increasing carbohydrate intake by rats given a carbohydrate/protein choice (Sarna, Whitton, Curzon and Leathwood, unpublished). These findings are consistent with the hypothesis that the choice depends on 5-HT function[34] but it seems unlikely that this mechanism operates in the normal animal as a result of normal patterns of food intake[6,35] even though it may occur on drug treatment. Also, it has been speculated that the resultant changes at 5-HT receptors may not influence discrimination between carbohydrate and protein as such, but instead some unrelated sensory difference between the characteristics of the diets.[36,37]

5-HT$_{1B}$ and 5-HT$_{1C}$ Agonists

If decreased 5-HT at postsynaptic sites is responsible for the hyperphagic effects of 5-HT$_{1A}$ agonists then two questions arise: which 5-HT receptor subtypes are involved and where do they occur? Postsynaptic 5-HT$_{1B}$ receptors are possibly implicated as the 5-HT$_{1B}$ agonist RU 24969 causes hypophagia.[38,39] Also 1-(3-chlorophenyl) piperazine (mCPP) and 1-[3-(trifluoromethyl) phenyl] piperazine (TFMPP) which have substantial affinity[40,41] for 5-HT$_{1B}$ 5-HT$_{1C}$ and 5-HT$_2$ sites caused hypophagia in normally fed and food-deprived rats.[38,42] Results on the above three drugs are given in TABLE 3.

RU 24969 also causes hyperactivity[43,44] whilst mCPP and TFMPP cause hypoactivity which is mediated by central 5-HT$_{1C}$ receptors.[45] It therefore seemed possible

TABLE 3. Effects of 5-HT_{1B}/5-HT_{1C} Agonists on Food Intake in Freely Feeding Rats[a]

Treatment (i.p.)	Food intake (g/4h ± SEM)
Saline	1.14 ± 0.18
RU 24969 1mg/Kg	0.55 ± 0.12
RU 24969 2mg/Kg	0.30 ± 0.10
RU 24969 5mg/Kg	0.07 ± 0.16
Saline	1.21 ± 0.16
mCPP 1mg/Kg	0.31 ± 0.17
mCPP 5mg/Kg	0.15 ± 0.03
TFMPP 1mg/Kg	0.52 ± 0.07
TFMPP 5mg/Kg	0.15 ± 0.03

[a]Data are from ref. 38. Effects of drugs were all $p < 0.05$.

that mCPP- and TFMPP-induced hypophagia also involved stimulation of 5-HT_{1C} receptors. Indeed, evidence[45] suggested this was so for the hypophagia seen when mCPP and TFMPP-treated rats were placed for 20 min in a novel cage. Responses to RU 24969, mCPP and TFMPP were therefore investigated under conditions more appropriate to the study of hypophagia, i.e., using animals previously deprived of food and measuring intake over longer periods.[46]

When this was done, the effects of antagonists (TABLES 4 and 5) indicated that mCPP (and probably also TFMPP) caused hypophagia by stimulating 5-HT_{1C} receptors although functional 5-HT_{1B} receptors were also needed for the response to occur. However, RU 24969 acted solely by direct stimulation of 5-HT_{1B} receptors. Since completing this work, it was reported that both mCPP and TFMPP bind somewhat more strongly to 5-HT_{1C} receptors than to 5-HT_{1B} sites *in vitro.*[26]

Affinities of the antagonists for non-5-HT receptors are unable to explain their effects on hypophagias due to RU 24969 or mCPP. Thus although mianserin (but not metergoline or mesulergine) has high affinity for histamine H1 receptors,[47,48] reduced histaminergic activity reduces food intake.[49] Mediation by α_2 receptors for which mCPP and mianserin have some affinity[50,51] is also unlikely, since the specific α_2 antagonist idazoxan did not oppose mCPP-induced hypophagia. The α_1 antagonist phentolamine is also ineffective.[42] Lastly, mCPP has little affinity for dopamine receptors.[52] Two anomalous findings must be mentioned. Firstly, cyproheptadine (10

TABLE 4. Effects of Antagonists (S.C.) on the Hypophagic Response to RU 24969 (5mg/Kg i.p.)[a]

Treatment	Blockade	Interpretation
Metergoline 5mg/Kg	Yes	5-HT receptor
Ketanserin 0.2 mg/Kg	No	Not 5-HT_2, α_1
ICS 205930 1mg/Kg	No	Not 5-HT_3
Mianserin 5mg/Kg	No	Not 5-HT_{1C}, 5-HT_2, H_1
Mesulergine 0.2mg/Kg	No	Not 5-HT_{1C}, 5-HT_2, DA
1-NP 2mg/Kg	No	Not 5-HT_{1C}, 5-HT_2
(±) Cyanopindolol 8mg/Kg	Yes	5-HT_{1A}, 5-HT_{1B}, β
(−) Pindolol 2mg/Kg	Yes	5-HT_{1A}, 5-HT_{1B}, β
(+) Pindolol 2mg/Kg	No	Not β
Spiperone 0.10mg/Kg	No	Not 5-HT_{1A}, 5-HT_2, DA

[a]Data are mostly on food-deprived rats (ref. 46); some on freely feeding rats (ref 39). 1-NP = 1-Naphthyl piperazine.

mg/Kg, not shown), unlike other 5-HT_{1C} antagonists, did not block the hypophagic effect of either drug, although it antagonized mCPP-induced hypolocomotion.[45] Secondly, unlike other 5-HT_{1A} and 5-HT_{1B} antagonists, (−) pindolol (2 mg/Kg, not shown) did not block mCPP-induced hypophagia even though it prevented hypophagia due to RU 24969.

Notwithstanding the above anomalies, antagonist experiments as a whole strongly suggest that mCPP and TFMPP cause hypophagia by action at 5-HT_{1C} receptors. These results,[46] together with our previous data on the hypolocomotor effect of mCPP[45] and our subsequent findings on its anxiogenic effect[53] were of some potential importance as they were the first evidence for behavioral consequences of activating 5-HT_{1C} receptors. We have therefore paid particular attention to the fact that 5-HT_{1C} sites and 5-HT_2 sites have much in common. Thus, most antagonists with high affinity for 5-HT_{1C} sites also bind strongly to 5-HT_2 sites[26] and activation of both sites induces phosphoinositide hydrolysis.[54] Furthermore, activation of 5-HT_2 sites has been suggested to mediate fenfluramine-induced hypophagia, as it was blocked by ketanserin.[55] This drug and two other 5-HT_2 antagonists, LY 53857 and 1-naphthylpiperazine (1-NP), also blocked hypophagia due to 1-(2,5-dimethoxy-4-iodophenyl)-2-aminopropane (DOI).[56] These considerations may seem to cast doubt on our evidence that activation of 5-HT_{1C} sites causes hypophagia. On the other hand since 1-NP is also a potent 5-HT_{1C} antagonist[26,54] and DOI and LY 53857 have high affinity for the 5-HT_{1C} site,[26] the evidence for the involvement of 5-HT_2 sites relies heavily upon the selectivity of ketanserin. These problems were investigated (Kennett and Curzon, unpublished) by determining the *in vivo* antagonist potencies of nine antagonists to both 5-HT_2 and 5-HT_{1C} sites (ketanserin, ritanserin, mianserin, 1-NP, LY 53857, methysergide, altanserin, metergoline and (±) propranolol) against 5-HT_2-mediated head twitch behavior[57] and mCPP-induced hypophagia. The ratios of the ID_{50} values obtained correlated highly significantly ($p < 0.001$) with the ratios of the affinities for 5-HT_{1C} and 5-HT_{1C} sites as calculated from published data.[26] This strengthens our evidence that mCPP-induced hypophagia is mediated by 5-HT_{1C} rather than by 5-HT_2 receptors. Relating the *in vivo* effects of drugs to their *in vitro* binding affinities eliminates uncertainties due to differences of metabolism of different drugs, as such differences would comparably affect ability to inhibit both hypophagia and head twitch.

TABLE 5. Effects of Antagonists (S.C.) on the Hypophagic Response to mCPP (5mg/Kg i.p.)[a]

Treatment	Blockade	Interpretation
Metergoline 5mg/Kg	Yes	5-HT receptor
Ketanserin 0.2 mg/Kg	No	Not 5-HT_2, α_1
Ritanserin 0.6mg/Kg	No	Not 5-HT_2
ICS 205930 1mg/Kg	No	Not 5-HT_3
Mianserin 5mg/Kg	Yes	5-HT_{1C}, 5-HT_2, H_1
Mesulergine 0.2 mg/Kg	Yes	5-HT_{1C}, 5-HT_2, DA
1-NP 2mg/Kg	Yes	5-HT_{1C}, 5-HT_2,
(±) Cyanopindolol 8mg/Kg	Yes	5-HT_{1A}, 5-HT_{1B}, β
(−) Propanolol 16mg/Kg	Yes	5-HT_{1A}, 5-HT_{1B}, β
Idazoxan 1mg/Kg	No	Not α_2

[a]Data are on food-deprived rats (ref 46). 1-NP = 1-naphthyl piperazine.

Other recent data are also consistent with 5-HT_{1C} rather than 5-HT_2 sites mediating hypophagia.[58] Thus, a dose of ritanserin causing 50% occupation of 5-HT_2

sites had no effect on D-fenfluramine–induced hypophagia while 4× the dose inhibited it about 50%. This suggests mediation by 5-HT_{1C} sites as it can be calculated from the affinity constants[26] that the above increase of ritanserin dosage would be sufficient to half saturate 5-HT_{1C} receptors but would have little effect on either 5-HT_{1A} or 5-HT_{1B} sites.

Are RU 24969 or mCPP and TFMPP-induced hypophagias secondary to other behavioral effects? Hypophagia due to RU 24969 is unlikely to be secondary to its hyperlocomotor effect, as haloperidol[38,43,44] and (+) pindolol[44] block this but not the hypophagia.[38] Conversely, metergoline blocks the hypophagia[38,46] but potentiates the hyperlocomotion.[43,44] Also, infusing RU 24969 into the hypothalamus caused hypophagia only.[39] Evidence also points to mCPP (and TFMPP) hypophagia and hypoactivity being separately induced since (±) cyanopindolol or (−) propanolol did not alter the activity, yet prevented the hypophagia,[45,46] while TFMPP caused hypophagia but not hypoactivity on hypothalamic infusion.[39] Lastly, while the 5-HT_{1C} antagonists mianserin, 1-NP, cyproheptadine, and mesulergine opposed both mCPP-induced hypophagia and hypoactivity they increased food intake, but not activity, when given alone.[45,46] These increases were seen in normally fed but not in food-deprived rats. Therefore, the hyperphagia may either be rate dependent or due to effects on appetite rather than satiety. This may explain contradictory reports on the effect of cyproheptadine on rat food intake.[59,60]

It is conceivable that 5-HT_{1B} and 5-HT_{1C} agonists cause hypophagia merely due to drug-induced malaise. This was investigated[61] using the antiemetic drug trimethobenzamide (TMB). This is a rational procedure, even though rats cannot vomit, as TMB prevents both the putative malaise-induced hypophagic response to cholecystokinin[62] and lithium chloride-induced taste aversion.[63] As it prevented the hypophagic response of rats to acetyl salicylate (a known emetic in man and dogs) but did not affect hypophagia due to RU 24969, mCPP and TFMPP, the latter effects probably do not reflect a malaise-dependent mechanism.

As infusing 5-HT or norfenfluramine into the paraventricular nucleus (PVN) of the hypothalamus causes hypophagia,[64] it seemed likely that RU 24969, mCPP and TFMPP had a similar effect at this site; as briefly mentioned above, this was shown. Thus 0.5, 1.0 and 2.0 μg of both RU 24969 and TFMPP caused dose dependent hypophagia on PVN infusion.[39] Similar infusion of 8-OH-DPAT was without effect.

SEX DIFFERENCES IN 5-HT DEPENDENT HYPOPHAGIA

The revelation of sex differences in the hypophagic effects of stress have led us to similar studies on hypophagic 5-HT agonists. In the former work, a single 2 h restraint stress caused hypophagia over the next 24 h. This was not merely a result of the gastric lesions which can result from restraint as the histamine-H_2 antagonist ranitidine decreased their number but not the hypophagia.[65] On repeating the restraint daily for 5 days,[66,67] food intake returned to normal in male but not in female rats. It has also been reported that chronic fenfluramine infusion is more hypophagic in female rats than in males.[68] Subsequently, it was shown that the hypophagic effects of RU 24969 and mCPP were greater in female than in male rats if they had been deprived of food for 24 h but not in freely feeding animals.[69] These results suggest the existence of a 5-HT dependent hypophagic mechanism which is more active in food-deprived female rats than in males. This could conceivably have a role in the greater incidence of anorexia nervosa in women.

SUMMARY

Feeding or food withdrawal can affect the supply of tryptophan to the brain and hence (in some circumstances) 5-HT synthesis therein. Also fenfluramine which releases 5-HT to postsynaptic receptors suppresses appetite, and there are reports that tryptophan can have a similar effect. Furthermore, feeding is reported to release hypothalamic 5-HT. Therefore 5-HT could have a role in the normal termination of feeding and perhaps also in disorders of appetite. The recognition of various 5-HT receptor subtypes has stimulated research in this area. We have now investigated the involvement of the subtypes in the pharmacological control of feeding.

Thus, 5-HT_{1A} agonists (8-OHDPAT, buspirone, gepirone, etc.) stimulate intake in freely feeding rats, probably by activating autoreceptors on the cell bodies of 5-HT neurons so that 5-HT release at terminals is decreased. The hyperphagia is not explicable by increased activity or gnawing and is strikingly manifest against carbohydrate in carbohydrate vs. protein choice experiments.

Feeding in previously food-deprived rats is decreased by the 5-HT agonists RU 24969, 1-(3-chlorophenyl) piperazine (mCPP) and 1-(3-trifluoromethyl) phenyl) piperazine (TFMPP). Effects of antagonists on these properties suggest that RU 24969-induced hypophagia depends on 5-HT_{1B} receptors only, while mCPP and TFMPP induce hypophagia at 5-HT_{1C} sites, though this effect also requires 5-HT_{1B} receptors for its expression. Responsible sites occur in the paraventricular nucleus of the hypothalamus as infusing either RU 24969 or TFMPP therein causes hypophagia. On systemic injection, the hypophagic drugs are particularly active in female rats, an effect of conceivable relevance to human anorexic illness.

REFERENCES

1. ROWLAND, N. E. & J. CARLTON. 1986. Neurobiology of an anorectic drug: Fenfluramine. Prog. Neurobiol. **27:** 13–62.
2. FERNSTROM, J. D. & R. J. WURTMAN. 1972. Brain serotonin content physiological regulation by plasma neutral amino acids. Science **178:** 414–416.
3. KNOTT, P. J. & G. CURZON. 1972. Free tryptophan in plasma and brain tryptophan metabolism. Nature **239:** 452–453.
4. GILLMAN, P. K., J. R. BARTLETT, P. K. BRIDGES, A. HUNT, A. J. PATEL, B. D. KANTAMANENI & G. CURZON. 1981. Indolic substances in plasma, cerebrospinal fluid and frontal cortex of human subjects infused with saline or tryptophan. J. Neurochem. **37:** 410–417.
5. SARNA, G. S., B. D. KANTAMANENI & G. CURZON. 1984. Variables influencing the effect of a meal on brain tryptophan. J. Neurochem. **44:** 1575–1580.
6. CURZON, G. 1988. Feeding, stress, exercise and the supply of tryptophan to the brain. *In* Amino Acid Availability and Brain Function in Health and Disease. G. Huether, Ed. Berlin: Springer, pp. 39–59.
7. SCHWARTZ, D. H., S. MCCLANE, L. HERNANDEZ & B. G. HOEBEL. 1989. Feeding increases extracellular serotonin in the lateral hypothalamus of the rat as measured by microdialysis. Brain Res. **479:** 349–354.
8. BLUNDELL, J. E. 1977. Is there a role for serotonin (5-hydroxytryptamine) in feeding? Int. J. Obesity **1:** 15–42.
9. FERNSTROM, J. D., & R. J. WURTMAN. 1971. Brain serotonin content: physiological dependence on plasma tryptophan levels. Science **173:** 149–152.
10. WEINBERGER, S. B., S. KNAPP & A. J. MANDELL. 1978. Failure of tryptophan load-induced increases in brain serotonin to alter food intake in the rat. Life Sci. **22:** 1595–1602.
11. WHITE, P. J., K. A. CYBULSKI, R. PRIMUS, D. F. JOHNSON, G. H. COLLIER & G. L. WAGNER. 1988. Changes in macronutrient selection as a function of dietary tryptophan. Physiol. Behav. **43:** 73–77.

12. HRBOTICKY, N., L. A. LEITER & G. A. ANDERSON. 1985. Effects of L-tryptophan on short term food intake in lean men. Nutrition Res. **5:** 595–607.
13. CANGIANO, C., A. CASCINO, F. CECI, M. MUSCARITOLI, F. ROSSI FANELLI, E. T. MENICHETTI & M. MULIERI. 1987. Plasma and CSF tryptophan in patients with cancer anorexia. *In* Progress in Tryptophan and Serotonin Research. D. A. Bender, M. H. Joseph, W. Kochen, and H. Steinhart, Eds. Berlin: De Gruyter, pp. 269–272.
14. LI, E. T. S. & G. H. ANDERSON. 1984. 5-Hydroxytryptamine: a modulator of food composition but not quantity? Life Sci. **34:** 2453–2460.
15. BLUNDELL, J. E. & A. J. HILL. 1987. Influence of tryptophan on appetite and food selection in man. *In* Amino Acids in Health and Disease: New Perspectives. S. Kaufman, Ed. New York: Alan R. Liss, pp. 403–419.
16. YOUNG, S. N., S. V. TOURJMAN, K. L. TEFF, R. O. PIHL & H. G. ANDERSON. 1988. The effect of lowering tryptophan on food selection in normal males. Pharmacol. Biochem. Behav. **31:** 149–152.
17. DOURISH, C. T., P. H. HUTSON & G. CURZON. 1985. Low doses of the putative serotonin agonist 8-hydroxy-2-(di-n-propylamino) tetralin (8-OH-DPAT) elicit feeding in the rat. Psychopharmacology **86:** 197–204.
18. BENDOTTI, C. & R. SAMANIN. 1987. The role of putative 5-HT_{1A} and 5-HT_{1B} receptors in the control of feeding in rats. Life Sci. **41:** 635–642.
19. DOURISH, C. T., P. H. HUTSON, G. A. KENNETT & G. CURZON. 1986. 8-OH-DPAT induced hyperphagia: Its neural basis and possible therapeutic relevance. Appetite **7**(Suppl.): 127–140.
20. GILBERT, F. & C. T. DOURISH. 1987. Effects of the novel anxiolytics gepirone, buspirone and ipsapirone on free feeding and on feeding induced by 8-OH-DPAT. Psychopharmacology **93:** 349–352.
21. HUTSON, P. H., T. P. DONOHOE & G. CURZON. 1987. Neurochemical and behavioural evidence for an agonist action of 1-[2-(4-aminophenyl)-4-(3-trifluoromethylphenyl) piperazine at central 5-HT receptors. Eur. J. Pharmacol. **138:** 215–223.
22. DOURISH, C. T., P. H. HUTSON & G. CURZON. 1986. Parachlorophenylalanine prevents feeding induced by the serotonin agonist 8-hydroxy-2-(di-n-propylamino) tetralin (8-OH-DPAT). Psychopharmacology **89:** 467–471.
23. MIDDLEMISS, D. N. 1984. 8-Hydroxy-2-(di-n-propylamino) tetralin is devoid of activity at the 5-hydroxytryptamine autoreceptor in rat brain. Implications for the proposed link between the autoreceptor and the [^{3}H]5-HT recognition site. Naunyn-Schmiedeberg's Arch. Pharmacol. **327:** 18–22.
24. VERGE, D., G. DUVAL, A. PATEY, S. GOZLAN, S. EL MESTIKAWY & M. HAMON. 1985. Presynaptic 5-HT autoreceptors on serotonergic cell bodies and/or dendrites but not terminals are of the 5-HT_{1A} subtype. Eur. J. Pharmacol. **113:** 463–464.
25. HUTSON, P. H., C. T. DOURISH & G. CURZON. 1986. Neurochemical and behavioural evidence for mediation of the hyperphagic action of 8-OH-DPAT by 5-HT cell body autoreceptors. Eur. J. Pharmacol. **129:** 347–352.
26. HOYER, D. 1988. Functional correlates of serotonin 5-HT_1 recognition sites. J. Receptor Res. **8:** 59–81.
27. MIDDLEMISS, D. N. 1987. Lack of effect of the putative 5-HT_{1A} receptor agonist, 8-OH-DPAT on 5-HT release in vitro. *In* Brain 5-HT_{1A} Receptors: Behavioural and Neurochemical Pharmacology. C. T. Dourish, S. Ahlenius, and P. H. Hutson, Eds. Chichester, Eng.: Ellis Horwood, pp. 82–93.
28. HUTSON, P. H., C. T. DOURISH & G. CURZON. 1988. Evidence that the hyperphagic response to 8-OH-DPAT is mediated by 5-HT_{1A} receptors. Eur. J. Pharmacol. **150:** 361–366.
29. COSTALL, B. & R. J. NAYLOR. 1974. Stereotyped and circling behaviour induced by dopaminergic agonists after lesions of the midbrain raphe nuclei. Eur. J. Pharmacol. **29:** 206–222.
30. MONTGOMERY, A. M. J., P. WILLNER & R. MUSCAT. 1988. Behavioural specificity of 8-OH-DPAT induced feeding. Psychopharmacology **94:** 110–114.
31. CHAOULOFF, F., B. SERRURRIER, D. MERINO, D. LAUDE & J. L. ELGHOZI. 1988. Feeding

responses to a high dose of 8-OH-DPAT in young and adult rats: Influence of food texture. Eur. J. Pharmacol. **151:** 267–273.

32. DOURISH, C. T., M. L. CLARK & S. D. IVERSEN. 1988. 8-OH-DPAT elicits feeding and not chewing: Evidence from liquid diet studies and a diet choice test. Psychopharmacology **95:** 185–188.
33. HUTSON, P. H., G. S. SARNA, M. T. O'CONNELL & G. CURZON. 1989. Hippocampal 5-HT synthesis and release in vivo is decreased by infusion of 8-OH-DPAT into the nucleus raphe dorsalis. Neurosci. Lett. **100:** 276–280.
34. WURTMAN, J. J. & R. J. WURTMAN. 1979. Drugs that enhance central serotonergic transmission diminish elective carbohydrate consumption by rats. Life Sci. **24:** 895–904.
35. PETERS, J. C. & A. E. HARPER. 1987. A skeptical view of the role of central serotonin in the selection and intake of protein. Appetite **8:** 206–210.
36. BOOTH, D. A. 1987. Central dietary "feedback onto nutrient selection": Not even a scientific hypothesis. Appetite **8:** 195–201.
37. FERNSTROM, J. D. 1987. Food-induced changes in brain serotonin syntheses: Is there a relationship to appetite for specific macronutrients. Appetite **8:** 163–182.
38. KENNETT, G. A., C. T. DOURISH & G. CURZON. 1987. 5-HT_{1B} agonists induce anorexia at a postsynaptic site. Eur. J. Pharmacol. **141:** 429–435.
39. HUTSON, P. H., T. P. DONOHOE & G. CURZON. 1988. Infusion of the 5-hydroxytryptamine agonists RU24969 and TFMPP into the paraventricular nucleus of the hypothalamus causes hypophagia. Psychopharmacology **95:** 550–552.
40. SILLS, M. A., B. B. WOLFE & A. FRAZER. 1984. Determination of selective and non-selective compounds for the 5-HT_{1A} and 5-HT_{1B} receptor subtypes in rat frontal cortex. J. Pharmacol. Exp. Ther. **231:** 480–487.
41. ASARCH, K. E., R. W. RANSOM & J. S. SHIH. 1985. 5-HT_{1A} and 5-HT_{1B} heterogeneity. Life Sci. **36:** 1265–1273.
42. SAMANIN, R., T. MENNINI, A. FERRARIS, C. BENDOTTI, F. BORSINI & S. GARATTINI. 1979. m-Chlorophenylpiperazine: A central agonist causing powerful anorexia in rats. Naunyn-Schmiedeberg's Arch. Pharmacol. **308:** 159–163.
43. GREEN, A. R., A. P. GUY & C. R. GARDNER. 1984. The behavioural effects of RU24969, a suggested 5-HT receptor agonist in rodents and the effect on the behaviour of treatment with antidepressants. Neuropharmacology **23:** 655–661.
44. TRICKLEBANK, M. D., D. M. MIDDLEMISS & J. NEILL. 1986. Pharmacological analysis of the behavioural and thermoregulatory effects of the putative 5-HT_1 receptor agonist RU24969 in the rat. Neuropharmacology **25:** 877–886.
45. KENNETT, G. A. & G. CURZON. 1988. Evidence that mCPP may have behavioural effects mediated by 5-HT_{1C} receptors. Br. J. Pharmacol. **94:** 137–147.
46. KENNETT, G. A. & G. CURZON. 1988. Evidence that hypophagia induced by mCPP and TFMPP requires 5-HT_{1C} and 5-HT_{1B} receptors: Hypophagia induced by RU 24969 only requires 5-HT_{1B} receptors. Psychopharmacology **96:** 93–100.
47. LEYSEN, J. E., F. AWOUTERS, L. KENNIS, P. M. LADURON, J. VANDENBERK & P. A. J. JANSSEN. 1981. Receptor binding profile of R41 468, a novel antagonist of 5-HT_2 receptors. Life Sci. **28:** 1015–1022.
48. CLOSSE, A. 1983. [^{3}H] mesulergine, a selective ligand for serotonin-2 receptors. Life Sci. **32:** 2485–2495.
49. MENON, M. K., W. G. CLARK & D. AURES. 1971. Effect of Thiazol-4-xyl-methoxyamine, a new inhibitor of histamine synthesis on brain histamine, monoamine levels and behaviour. Life Sci. **10:** 1097–1109.
50. CLINESCHMIDT, B. V., L. M. FLATAKER, E. FAISON & R. HOLMES. 1979. An in vivo model for investigating α_1 and α_2 receptors in the CNS: Studies with mianserin. Arch. Int. Pharmacodyn. **242:** 59–76.
51. SMITH, T. M. & R. F. SUCKOW. 1985. Trazodone and m-chlorophenylpiperazine: Concentration in brain and receptor activity in regions of the brain associated with anxiety. Neuropharmacology **24:** 1067–1071.
52. INVERNIZZI, R., S. COTECCHIA, A. DE BLASI, T. MENNINI, R. PATACCINI & R. SAMANIN. 1981. Effects of m-chloro-phenylpiperazine on receptor binding and brain metabolism of monoamines in rats. Neurochem. Int. **3:** 239–244.

53. KENNETT, G. A., P. WHITTON, K. SHAH & G. CURZON. 1989. Anxiogenic-like effects of mCPP and TFMPP in animal models are opposed by 5-HT_{1C} receptor antagonists. Eur. J. Pharmacol. **164:** 445–454.
54. CONN, P. J. & E. SANDERS-BUSH. 1987. Relative efficacies of piperazines at the phosphoinositide hydrolysis-linked serotonergic (5-HT_2 and 5-HT_{1C}) receptors. J. Pharmacol. Exp. Ther. **242:** 552–557.
55. HEWSON, G., G. E. LEIGHTON, R. G. HILL & J. HUGHES. 1988. Ketanserin antagonises the anorectic effect of DL-fenfluramine in the rat. Eur. J. Pharmacol. **145:** 227–230.
56. SCHECHTER, L. E. & K. J. SIMANSKY. 1988. 1-(2,5-Dimethoxy-4-iodophenyl)-2-aminopropane (DOI) exerts an anorexic action that is blocked by 5-HT_2 antagonists in rats. Psychopharmacology **94:** 342–346.
57. YAP, C. Y. & D. A. TAYLOR. 1983. Involvement of 5-HT_2 receptors in the wet-dog shake behaviour induced by 5-hydroxytryptophan in the rat. Neuropharmacology **22:** 801–804.
58. SAMANIN, R., T. MENNINI, C. BENDOTTI, D. BARONE, S. CACCIA & S. GARATTINI. 1989. Evidence that central 5-HT_2 receptors do not play an important role in the anorectic activity of D-fenfluramine in the rat. Neuropharmacology **28:** 465–469.
59. BAXTER, M. J., A. A. MILLER & F. E. SOROKO. 1970. The effect of cyproheptadine on food consumption in the fasted rat. Br. J. Pharmacol. **39:** 229–230.
60. GHOSH, M. N. & S. PARVATHY. 1973. The effect of cyproheptadine on water and food intake and on body weight in the fasted adult and weanling rat. Br. J. Pharmacol. **48:** 328–329.
61. KENNETT, G. A. & G. CURZON. 1988. The antiemetic drug trimethobenzamide prevents hypophagia due to acetyl salicylate but not to 5-HT_{1B} or 5-HT_{1C} agonists. Psychopharmacology **96:** 101–103.
62. MOORE, B. D., & J. R. DEUTSCH. 1985. An antiemetic is antidotal to the satiety effect of cholecystokinin. Nature **315:** 321–322.
63. COIL, J. D., W. G. HANKINS, A. J. JENDEN & J. GARCIA. 1978. The attenuation of a specific cue to consequence association by antiemetic agents. Psychopharmacology **56:** 21–25.
64. SHOR-POSNER, G., J. A. GRINKER, C. MARMESON, O. BROWN & S. F. LEIBOWITZ. 1986. Hypothalamic serotonin in the control of meal patterns and macronutrient selection. Brain Res. Bull. **17:** 663–671.
65. DONOHOE, T. P., G. A. KENNETT & G. CURZON. 1987. Immobilisation stress-induced anorexia is not due to gastric ulceration. Life Sci. **40:** 467–472.
66. KENNETT, G. A., F. CHAOULOFF, M. MARCOU & G. CURZON. 1986. Female rats are more vulnerable than males in an animal model of depression: The possible role of serotonin. Brain Res. **382:** 416–421.
67. HALEEM, D. J., G. KENNETT & G. CURZON. 1988. Adaptation of female rats to stress: Shift to male pattern by inhibition of corticosterone synthesis. Brain Res. **458:** 339–347.
68. ROWLAND, N. E. 1986. Effect of continuous infusions of dexfenfluramine on food intake, body weight and brain amines in rats. Life Sci. **39:** 2581–2586.
69. HALEEM, D. J. 1988. Serotonergic functions in rat brain: Sex-related differences and responses to stress. Ph.D. Thesis, University of London.

DISCUSSION OF THE PAPER

R. MURPHY (*New York University, New York City*): A question with regard to your hypophagia. I gathered from your slides that some of the animals are food deprived and some of the animals are free feeding. I'm just curious if you've looked systematically at food-deprived animals vs. the non-deprived animals in terms of the effect.

CURZON: Yes. Hypophagia was seen in freely feeding or food-deprived rats. But ED_{50} values are greater in the food-deprived ones.

S. LIEBOWITZ (*Rockefeller University, New York City*): I agree with that comment. In freely feeding animals we get lower threshold doses in terms of the effectiveness of the drugs. I also wanted to ask if a circadian rhythm in the effects of serotonin could be modulating nutrient intake.

CURZON: That's a very important point. However, we have usually done our experiments in the afternoon. It's a real gap in our work that we haven't looked at different times of day. I think there's been at least one paper in the literature where DPAT was shown not to be hyperphagic when given during the night, which is when the animals normally eat most. There seems to be a ceiling level of food intake and DPAT can't bring that up any higher.

LIEBOWITZ: I was particularly intrigued by your last slide in terms of showing, I think it was a surge in the hypophagic effects of MCPP after puberty. Is that correct?

CURZON: Yes. The surge was very clear. The sex difference was maintained with increasing age. We wondered whether it was merely due to a sex difference in metabolism. This now seems unlikely because the hypolocomotor effect of MCPP is identical in both male and female rats.

P. HARTIG (*Neurogenetic Corporation, Paramus, NJ*): Just a quick technical question. I think that we often overlook that a dirty old drug, spiperone, is a fairly good differentiator between 5-HT_2 and $_{1c}$. It's got about 2½ log units difference higher affinity for 5-HT_2. Did you try that in your series?

CURZON: No. The other drugs we've looked at as well as the four mentioned in my talk are LY-53857, methysergide, altanserine, metergoline and propranolol. ID_{50} data on all nine antagonists strongly point to mCPP causing hypophagia by acting on 5-HT_{1c} receptors.

Serotonin in Human Eating Disorders

DAVID C. JIMERSON

Department of Psychiatry
Beth Israel Hospital
Boston, Massachusetts 02215
and
Harvard Medical School
Boston, Massachusetts

MICHAEL D. LESEM

Harris County Psychiatric Center and
Department of Psychiatry
University of Texas Medical School
Houston, Texas 77225

ARLENE P. HEGG

Suicide and Suicidal Behavior Program
Mood, Anxiety and Personality Disorders Research Branch
National Institute of Mental Health
Rockville, Maryland 20857

TIMOTHY D. BREWERTON

Department of Psychiatry and Behavioral Sciences
Medical University of South Carolina
Charleston, South Carolina 29425

INTRODUCTION

During the decade of the 1980s there has been increasing clinical and research focus on the eating disorders of anorexia nervosa and bulimia nervosa. Both syndromes are characterized by abnormal eating patterns coupled with psychological symptoms involving preoccupation with body shape and weight (TABLE 1). These eating disorder syndromes are estimated to affect two to five per cent of young women, the population at highest risk for these disorders, as well as a smaller number of young men.[1-3] Anorexic patients often encounter serious medical complications as a result of severe cachexia. Many patients with anorexia nervosa experience a prolonged course with recurrent relapse to low weight episodes, with poor long term outcome in approximately one third of patients, and a mortality rate as high as 18 per cent in long-term follow-up studies.[4] While antidepressants may diminish depressive and other psychological symptoms in anorexic patients,[5-7] current pharmacological approaches are of limited benefit as adjunctive treatment in weight restoration programs for low weight patients. The severity of symptoms in bulimia nervosa is apparent from test meal studies yielding estimates of mean caloric intake during binge episodes ranging from 1500 to 4400 kcal.[8-10] While short term outcome studies have demonstrated significant therapeutic benefit of pharmacotherapy and of psychotherapy for the majority of patients with bulimia nervosa, a substantial number of patients appear to have persistent symptomatology on follow-up.[11,12]

Indirect clinical evidence suggests that biological factors may contribute to the development and/or persistence of eating disorders. Twin studies have shown a high concordance rate for anorexia nervosa in monozygotic twin pairs in comparison to the rate for dizygotic twins.[13] Among patients with bulimia nervosa and anorexia nervosa, more than half have a history of major depression,[14,15] which in turn may reflect a neurobiological component. Patients with bulimia as well as those with anorexia nervosa have significantly increased frequency of eating disorders and affective illness in first degree relatives, consistent with the possibility of genetic predisposition to these disorders.[15–17] As reviewed below, antidepressant drugs significantly reduce symptoms of bingeing and vomiting in bulimic patients.

While the current manuscript focuses on the clinical syndromes of anorexia nervosa and bulimia nervosa, it should be noted that serotonin dyregulation may also contribute to symptoms in some patients with obesity. Moreover, serotonin active drugs may have a role in the pharmacological treatment of obesity.[18]

TABLE 1. Characteristic Symptoms for Eating Disorder Syndromes[a]

Bulimia Nervosa
• Recurrent episodes of binge eating, averaging at least two binge episodes per week.
• Regular efforts to prevent weight gain, such as intensive dieting, exercise or self-induced vomiting.
• Persistent overconcern with body shape and weight.
Anorexia Nervosa
• Refusal to maintain body weight of at least 85% of normal.
• Intense fear of feeling fat associated with weight gain.
• In females, amenorrhea lasting at least three months.

[a]Adapted from Diagnostic and Statistical Manual of Mental Disorders, Third Edition, Revised[94] (See reference for formal diagnostic criteria.)

SEROTONIN AS A MODULATOR OF EATING BEHAVIOR

Preclinical pharmacological studies of meal size and patterns of feeding and associated behaviors suggest that endogenously released hypothalamic serotonin and exogenously administered serotonin agonists contribute to post-prandial satiety. Activation of serotonergic pathways does not alter the latency for the initiation of feeding, but decreases the rate and duration of food intake.[19] Conversely, administration of serotonin antagonists or interruption of central serotonin pathways increases meal size. These effects are observed with localized manipulations of serotonin in the paraventricular nucleus of the medial hypothalamus, where serotonin may act to antagonize feeding stimulated through α_2-adrenoceptors.[20] In addition to decreasing the quantity of food consumed, central serotonin appears to affect selection of dietary composition, having a predominant effect in decreasing the consumption of carbohydrate in comparison to lesser effects on consumption of protein and fat.[20,21] Human volunteers show decreased food intake but modest effects on dietary composition following acute administration of serotonin agonists (e.g., L-tryptophan, fenfluramine).[22,23]

Development of relatively selective pharmacological probes has provided some information on serotonin receptor subtypes involved in regulation of food intake. Serotonin agonists active at serotonin-1_B and serotonin-1_C receptors decrease food intake in laboratory rodents,[24,25] although serotonin-1_B receptors do not appear to be present in human brain.[26] Other serotonin-1 receptor subtypes may also be involved in the suppression of eating.[27] Serotonin-1_A agonists increase feeding behavior in rodents, most likely reflecting inhibitory effects of stimulation of pre-synaptic autoreceptors.[28]

RESEARCH PROBES FOR CLINICAL STUDIES

Indirect probes have been developed to assess pre-synaptic neuronal activity and post-synaptic receptor sensitivity in clinical studies. Concentration of the major serotonin metabolite 5-hydroxyindole acetic acid (5-HIAA) in lumbar cerebrospinal fluid (CSF) reflects release and metabolism of the neurotransmitter, providing an index of presynaptic activity in serotonergic pathways. CSF studies are limited, however, by lack of anatomical specificity (e.g., contributions from the spinal cord to metabolite levels) and by the difficulty in repeating studies to monitor longitudinal changes in clinical state and treatment conditions.[29]

Measurement of amino acid concentrations in blood provides indirect information on the availability of the serotonin precursor L-tryptophan in the central nervous system, which in turn influences the rate of central serotonin synthesis.[30] Since tryptophan competes with other large neutral amino acids (LNAA) in blood for transport into the brain, the relative concentration of tryptophan in comparison to that of the other neutral amino acids determines its rate of entry into the CNS. The tryptophan/LNAA ratio (and hence the availability of tryptophan in brain) is increased by consumption of meals relatively rich in carbohydrate, and decreased by consumption of meals relatively rich in protein.[30,31] Dietary manipulations designed to decrease central tryptophan availability have been reported to induce depressive symptoms in healthy human volunteers, consistent with a decrease in central serotonin synthesis.[32] Similar dietary manipulations augmented the increase in plasma prolactin concentration produced by acute pharmacological challenge with intravenous L-tryptophan.[33]

Assessment of central serotonin function in clinical studies has come to rely more heavily on behavioral and physiological responses to the administration of relatively selective pharmacological agonists (e.g., L-tryptophan, 5-hydroxytryptophan, *m*-chlorophenylpiperazine (*m*-CPP) and fenfluramine) and antagonists (e.g., metergoline).[34,35] Serotonergic drugs act in the raphe nuclei or in the medial basal hypothalamus to stimulate prolactin secretion.[36,37] Secretion of cortisol is also stimulated by serotonergic agonists, although this is often a relatively weak response and may involve non-serotonergic pathways.[35,36,38] There has been considerable recent interest in studies with the serotonin-1 receptor agonist *m*-CPP, which stimulates prolactin secretion relatively selectively through a serotonergic mechanism.[39,40]

SEROTONIN FUNCTION IN ANOREXIA NERVOSA

Clinical Symptom Profile

Along with preoccupation with weight and distorted perception of body image, patients with anorexia nervosa show altered appetite and satiety responses in test meal studies.[41,42] Early symptoms experienced by patients with anorexia nervosa include efforts to control food intake and body weight, which often arise in the context of psychosocial (e.g., family) stresses. It is not known whether abnormal satiety responses exist prior to the onset of dieting and weight loss. Pre-existing impairment in satiety responses could contribute to preoccupation with food intake and concerns about body weight. Conversely, abnormal sensitivity in postprandial satiety, perhaps increased by stresses and hormonal changes associated with puberty, might contribute to conscious restriction of food intake as a patient searches for mastery of psychological conflicts. This latter model would correspond to eating patterns predicted from animal models of excess hypothalamic serotonin function.[19]

Approximately half of patients with low weight episodes of anorexia nervosa also develop bulimic symptoms,[2,43,44] with recurrent binge eating suggestive of impaired central serotonin function. Development of bulimic symptoms in anorexic patients is not necessarily a result of weight loss, since the onset of bulimic symptoms may predate the onset of anorexia.[45] The possibility of differential serotonergic regulation in non-bulimic ("restrictor") and bulimic subgroups of anorexic patients is consistent with previous pharmacological data showing that the serotonin antagonist cyproheptadine slowed weight gain in hospitalized bulimic anorexic patients, while slightly accelerating weight gain in non-bulimic patients.[46] Development of major depression in the majority of patients with anorexia nervosa is also suggestive of dysregulation of central serotonin.[47] Depressive episodes may be more common in bulimic than in restrictor anorexic patients,[48,49] a pattern that would be consistent with other evidence for decreased serotonin function in the former subgroup. Evidence that restrictor anorexics have fewer impulsive and more obsessional symptoms than bulimic anorexic patients,[50] could also be consistent with differences in central serotonin activity in these patient subgroups.

It is unclear whether central serotonergic changes in anorexia nervosa (and bulimia nervosa) suggested by the neurobiological studies reviewed below are pre-existing factors predisposing to the onset of the disorders. Dieting behavior with mild to moderate restriction in caloric intake, resulting from preoccupation with body weight and shape, is an early symptom of anorexia nervosa as well as of bulimia nervosa.[2,51,52] A prospective study of the effects of a low calorie diet with resultant mild weight loss in healthy female volunteers, for example, demonstrated increased prolactin response to acute pharmacological challenge with the serotonin precursor L-tryptophan,[53] as well as increased serotonin binding sites on blood platelets.[54] Alteration in dietary composition (e.g., low carbohydrate diet), as well as reduction in caloric intake, could result in secondary changes in hypothalamic serotonin function by the time patients seek medical treatment and enter research studies. Research strategies to help clarify this question include prospective longitudinal studies of patients, family studies, and prospective studies of high risk groups.

Precursor and Metabolite Studies

Initial studies of serotonin function in anorexia nervosa focused primarily on measurement of concentrations of the amino acid precursor L-tryptophan and the major serotonin metabolite 5-HIAA in plasma, urine and CSF. Studies of plasma tryptophan concentrations yielded inconsistent findings (TABLE 2). These results suggest that nutritional status at the time of study may contribute to reported differences between patients and controls. Schweiger and colleagues studied blood amino acid responses following standardized test meals in hospitalized low weight anorexic patients.[55] They observed that at several time points following protein-rich and carbohydrate-rich meals, anorexic patients had lower values than controls for the tryptophan/LNAA ratio.

An initial case report[56] and a subsequent controlled study by Kaye and colleagues[57] showed CSF 5-HIAA levels to be substantially reduced in low weight anorexic patients (TABLE 3). Metabolite values were not significantly different for restrictor and bulimic subgroups of anorexic patients studied at low weight.[58] Gerner and coworkers observed

TABLE 2. Plasma Tryptophan Levels in Anorexia Nervosa

	Free TRP		Total TRP		TRP/LNAA
	Low Wt	Wt Recov	Low Wt	Wt Recov	Ratio
Coppen *et al.*, 1976[95]	↓	↓	↓	N	—
Johnston *et al.*, 1984[96]	—	—	↓	—	↓
Kaye *et al.*, 1984[57]	N	N	N	N	N
Gerner *et al.*, 1984[59]	—	—	N*	—	—

N indicates values similar to controls. *Tryptophan levels measured in CSF. TRP = tryptophan; LNAA = large neutral amino acids.

no difference in CSF 5-HIAA levels between controls and anorexic patients, who in this study had begun a weight restoration program.[59] Another study reported significantly low urinary excretion of 5-HIAA in low weight anorexic patients, with an increase following treatment.[60] Interpretation of these latter findings is limited, however, by uncertainty as to the source of the serotonin metabolite in urine.

Study of anorexic patients following recovery to a normal weight range reduces the influence of malnutrition on neurobiological measures. Moreover, given the frequent clinical difficulty in preventing relapse to low weight after an anorexic patient is refed during hospitalization, studies of weight-recovered patients provide the possibility of identifying possible neurobiological vulnerability factors for relapse. In studies at the National Institute of Mental Health in Bethesda, CSF 5-HIAA levels returned to a normal range when patients were restudied several weeks following weight restoration.[57] It is known that following weight restoration in anorexic patients, some physiological alterations that may be a consequence of malnutrition and weight loss per se (e.g., amenorrhea) can take a number of months to resolve. Preliminary evidence suggests that some patients who have been able to maintain a stable weight for at least six months may have CSF 5-HIAA levels higher than healthy controls.[57,61] When anorexic patients studied after weight restoration were subgrouped according to the presence or absence of bulimic symptoms, patients with bulimic symptoms had significantly lower CSF accumulations of 5-HIAA (measured following probenecid pretreatment) than did non-bulimic anorexics.[58] If confirmed in larger studies, these findings suggest that an underlying increase in central serotonin function, particularly in non-bulimic anorexic patients, may play a role in relapse to low weight episodes.

Measurement of serotonin uptake and ^{3}H-imipramine binding in blood platelets provide additional clinical probes which have been used as indirect models for regulatory activities at central serotonin synapses. Studies in anorexic patients have shown reduced ^{3}H-imipramine binding but normal serotonin uptake (TABLE 4).

TABLE 3. Studies of CSF 5-HIAA in Anorexia Nervosa

Author	No. of Subjects	CSF 5-HIAA (% of control)	
		Low Weight	Weight-Recovered
Gillberg, 1983[56]	2	35[a]	—
Kaye *et al.*, 1984[57]	8	78	117[b]
Gerner *et al.*, 1984[59]	33	88	—
Kaye *et al.*, 1987[97]	14	65[c]	85[b]

[a]Recalculated using literature control values.[59]
[b]Significantly increased from low weight value.
[c]Patients vs. controls, $p < 0.01$.

Pharmacological Challenge Studies

Results of pharmacological challenge studies conducted at NIMH have provided additional evidence for altered central serotonin function in anorexia nervosa.[62-64] Patients met DSM III diagnostic criteria for anorexia nervosa, and were hospitalized for longitudinal research studies and a behaviorally oriented treatment program. Patients were free of significant physical illness other than problems related to malnutrition, were medication free, and were abstinent from bingeing and vomiting while participating in the research and treatment programs. Pharmacological challenge studies in low weight patients followed a period of medical evaluation and nutritional stabilization on a low calorie diet lasting three weeks. Following the phase of low weight studies, patients participated in a refeeding program based on progressively increasing caloric intake. After patients achieved their goal weight (approximately 85% age- and height-adjusted average body weight), caloric intake was gradually reduced to allow patients to stabilize at this body weight. Repeat pharmacological challenge studies were conducted after three weeks' stabilization at goal weight.

In low weight patients, plasma prolactin responses were significantly blunted in response to challenge with *m*-CPP (0.5 mg/kg p.o.) ($p < 0.005$), and in response to

TABLE 4. Serotonin Measures in Blood Platelets in Eating Disorders

	Platelet ^{3}H-Imipramine Binding (Bmax)	Platelet Serotonin Uptake (Vmax)
Anorexia Nervosa		
Weizman *et al.*, 1986[98]	↓	N
Zemishlany *et al.*, 1987[99]	—	N
Bulimia Nervosa		
Marazziti *et al.*, 1988[71]	↓	—
Goldbloom *et al.*, 1988[72]	—	↑

N indicates values similar to controls.

L-tryptophan administration (100 mg/kg i.v.) ($p < 0.05$), in comparison to responses in healthy controls. Since the majority of the patients had a history of bulimic symptomatology, it was not possible to make a meaningful comparison according to bulimic or non-bulimic subtype. The marked blunting of serotonergic responses observed in low weight patients is likely to be in part a consequence of malnutrition, as indicated by the partial normalization with weight gain. One specific factor may be the hypercortisolism of low weight anorexia, since long-term administration of cortisol to rats resulted in blunted prolactin responses to *m*-CPP.[65]

SEROTONIN FUNCTION IN BULIMIA NERVOSA

Clinical Symptom Profile

The pattern of large binge meals observed in bulimic patients is consistent with preclinical models of impaired function in hypothalamic serotonin pathways mediating satiety responses. Bulimic patients manifest abnormal patterns of food consumption and altered satiety ratings in studies utilizing test meals.[42,66,67] Decreased central serotonin could also contribute to associated symptom patterns observed in bulimic patients, including episodes of major depression and impulsive behavior patterns.[64] As noted earlier, interpretation of psychobiological studies in patients with bulimia needs to take into account the possible effects of recent and long-term changes in dietary pattern and body weight. Study of patients in a symptomatic phase is more likely to reveal biological changes which contribute to perpetuation or progression of symptoms, while studies during a phase of symptom remission may implicate biological factors contributing to initial symptom onset or subsequent relapse.

Precursor and Metabolite Studies

Because of the relatively recent identification of the syndrome of bulimia nervosa in patients at normal weight, there is limited data available from neurotransmitter studies in this patient group. Evaluation of plasma amino acid responses to a standardized protein rich test meal showed lower tryptophan/LNAA ratios in bulimic patients in comparison to controls, such that altered precursor availability might lead to reduced central serotonin synthesis.[68] In a study of naturalistic binge episodes, Kaye and coworkers reported that the duration of binge episodes was longer in bulimic patients with smaller increases in plasma tryptophan/LNAA ratio and with smaller plasma prolactin responses to the binge meal.[69] The authors suggested that increased duration of binge eating might be associated with impairment of the serotonin-mediated satiety response.

CSF data have been preliminarily analyzed for a group of normal weight patients who met DSM III criteria for bulimia and were hospitalized on a clinical research unit at NIMH in Bethesda.[70] Patients were free of significant physical illness, were medication free, followed a nutritionally balanced diet, maintained relatively stable body weight, and were abstinent from bingeing and vomiting while participating in the research and treatment programs. Healthy volunteer controls were admitted overnight to the research unit prior to neurobiological studies. CSF concentrations of the serotonin metabolite 5-HIAA and of the dopamine metabolite homovanillic acid (HVA) were similar in patients and controls. When patients were stratified according to binge frequency, however, the patients with a history of more frequent bingeing had significantly reduced CSF 5-HIAA and HVA levels on the admission study in comparison to values for the patients with lower binge frequency or for the controls.[70] Differences in 5-HIAA levels between the two subgroups of bulimic patients did not appear to be related to differences in body weight or history of depression. These results suggested that increased binge frequency in bulimic patients was associated with relatively persistently reduced central serotonin and dopamine turnover.

An initial report of platelet ^{3}H-imipramine binding in non-depressed bulimic patients showed significantly reduced density of binding sites for patients in comparison to results for controls.[71] Values for Kd were not altered in the patient group. In contrast, a preliminary report on serotonin uptake in blood platelets showed significantly increased values in patients in comparison to controls.[72]

Pharmacological Challenge Studies

Double-blind, placebo-controlled neuroendocrine challenge studies were conducted after a three week stabilization period in normal weight bulimic patients admitted to a clinical research unit at NIMH. In comparison to responses in healthy controls, patients demonstrated significantly ($p < 0.002$) blunted plasma prolactin responses to the serotonin receptor agonist *m*-chlorophenylpiperazine (*m*-CPP) administered at a dose of 0.5 mg/kg p.o.[64,73] Blunted prolactin responses in the patients did not appear to be explained by differences in age, weight, history of depression, or reduced plasma concentrations of *m*-CPP in comparison to values for the controls. Moreover, evidence for greater prolactin response in bulimic patients than in controls following challenge with thyrotropin releasing hormone (TRH),[74] argues against non-specific impairment of secretion by the pituitary lactotroph. Administration of the serotonin precursor L-tryptophan (100 mg/kg i.v.) produced increments in plasma prolactin levels that did not differ significantly between bulimic patients and controls. It was of interest that the tryptophan-induced prolactin response was significantly blunted in the subset of bulimic patients with major depression, similar to the blunted prolactin response to L-tryptophan previously reported in depression.[75] Plasma cortisol responses to *m*-CPP and to L-tryptophan were not significantly different in patients and controls.

Pharmacological Treatment Studies

Single dose administration of the serotonin agonist fenfluramine was shown to reduce the size of a test meal in bulimic patients.[76] Initial studies on longer term administration of fenfluramine have yielded modestly positive results.[77,78] A preliminary therapeutic trial with L-tryptophan (at a relatively low dose) in bulimic patients did not demonstrate significant reduction in binge frequency.[79] In double-blind controlled trials, antidepressant medications have produced significant decreases in binge frequency in a majority of bulimic patients enrolled in the studies. These studies have included tricyclic and other heterocyclic antidepressants,[80–87] as well as monoamine oxidase inhibitors.[88,89] Efficacy of antidepressant medications in bulimic patients may be related to drug-induced increase in efficiency of hypothalamic serotonergic neurotransmission.[90,91] Initial data suggest that the selective serotonin reuptake blocker fluoxetine is effective in bulimia.[92] The specificity of these treatment effects requires additional clarification, since psychotherapeutic interventions are also helpful in a majority of bulimic patients.[84,93]

SUMMARY

The onset and progression of symptomatology in patients with anorexia nervosa and bulimia nervosa is complex. It is unlikely that dysregulation of a single neurotransmitter system would be sufficient to explain the pathophysiology of these disorders. The studies reviewed above provide preliminary evidence that decreased central serotonin function may contribute to the onset or persistence of binge eating episodes in patients with bulimia nervosa, including low weight anorexic patients with bulimic symptoms. Future clinical studies will benefit from the availability of selective serotonin receptor agonists and antagonists. Longitudinal studies through progressive phases of treatment and clinical remission will be important to clarify the contribution of dietary and body weight changes to results of neurotransmitter studies with eating disorder patients.

REFERENCES

1. CRISP, A. H., R. L. PALMER & R. S. KALUCY. 1976. How common is anorexia nervosa? Br. J. Psychiatry **128:** 549–554.

2. JOHNSON, C., C. LEWIS & J. HAGMAN. 1984. The syndrome of bulimia: Review and synthesis. Psychiatr. Clin. North Am. **7:** 247–273.
3. SCHOTTE, D. E. & A. J. STUNKARD. 1987. Bulimia vs bulimic behaviors on a college campus. JAMA **258:** 1213–1215.
4. THEANDER, S. 1985. Outcome and prognosis in anorexia nervosa and bulimia: Some results of previous investigations, compared with those of a Swedish long-term study. J. Psychiatr. Res. **19:** 493–508.
5. GWIRTSMAN, H. E., W. KAYE, M. WEINTRAUB & D. C. JIMERSON. 1984. Pharmacologic treatment of eating disorders. Psychiatr. Clin. North Am. **7:** 863–878.
6. BIEDERMAN, J., D. B. HERZOG, T. M. RIVINUS, G. P. HARPER, R. A. FERBER, J. F. ROSENBAUM, J. S. HARMATZ, R. TONDORF, P. J. ORSULAK & J. J. SCHILDKRAUT. 1985. Amitriptyline in the treatment of anorexia nervosa: A double-blind, placebo-controlled study. J. Clin. Psychopharmacol. **5:** 10–16.
7. CRISP, A. H., J. H. LACEY & M. CRUTCHFIELD. 1987. Clomipramine and 'drive' in people with anorexia nervosa: An in-patient study. Br. J. Psychiatry **150:** 355–358.
8. MITCHELL, J. E. & D. C. LAINE. 1985. Monitored binge-eating behavior in patients with bulimia. Int. J. Eating Disorders **4:** 177–183.
9. KAYE, W. H., H. E. GWIRTSMAN, D. T. GEORGE, S. R. WEISS & D. C. JIMERSON. 1986. Relationship of mood alterations to bingeing behavior in bulimia. Br. J. Psychiatry **149:** 479–485.
10. KISSILEFF, H. R., B. T. WALSH, J. G. KRAL & S. M. CASSIDY. 1986. Laboratory studies of eating behavior in women with bulimia. Physiol. Behav. **38:** 563–570.
11. SWIFT, W. J., M. RITHOLZ, N. H. KALIN & N. KASLOW. 1987. A follow-up study of thirty hospitalized bulimics. Psychosom. Med. **49:** 45–55.
12. HERZOG, D. B., M. B. KELLER, P. W. LAVORI & I. L. OTT. 1987. Short-term prospective study of recovery in bulimia nervosa. Psychiatry Res. **23:** 45–55.
13. NOWLIN, N. S. 1983. Anorexia nervosa in twins: Case report and review. J. Clin. Psychiatry **44:** 101–105.
14. HUDSON, J. I., H. G. POPE, JR., D. YURGELUN-TODD, J. M. JONAS & F. R. FRANKENBURG. 1987. A controlled study of lifetime prevalence of affective and other psychiatric disorders in bulimic outpatients. Am. J. Psychiatry **144:** 1283–1287.
15. GERSHON, E. S., J. L. SCHREIBER, J. R. HAMOVIT, E. D. DIBBLE, W. KAYE, J. I. NURNBERGER, JR., A. E. ANDERSEN & M. EBERT. 1984. Clinical findings in patients with anorexia nervosa and affective illness in their relatives. Am. J. Psychiatry **141:** 1419–1422.
16. HUDSON, J. I., H. G. POPE, JR., J. M. JONAS, D. YURGELUN-TODD & F. R. FRANKENBURG. 1987. A controlled family history study of bulimia. Psychol. Med. **17:** 883–890.
17. KASSETT, J. A., E. S. GERSHON, M. E. MAXWELL, J. J. GUROFF, D. M. KAZUBA, A. L. SMITH, H. A. BRANDT & D. C. JIMERSON. 1989. Psychiatric disorders in the relatives of probands with bulimia nervosa. Am. J. Psychiatry. **146:** 1468–1471.
18. BRAY, G. A. & D. S. GRAY. 1988. Treatment of obesity: An overview. Diabetes Metab. Rev. **4:** 653–679.
19. BLUNDELL, J. E. 1986. Serotonin manipulations and the structure of feeding behaviour. Appetite **7**(Suppl.): 39–56.
20. LEIBOWITZ, S. F. & G. SHOR-POSNER. 1986. Brain serotonin and eating behavior. Appetite **7**(Suppl. 1): 1–14.
21. WURTMAN, J. J. & R. J. WURTMAN. 1979. Drugs that enhance central serotoninergic transmission diminish elective carbohydrate consumption by rats. Life Sci. **24:** 895–904.
22. SILVERSTONE, T. & E. GOODALL. 1986. Serotoninergic mechanisms in human feeding: The pharmacological evidence. Appetite **7**(Suppl.): 85–97.
23. BLUNDELL, J. E. & A. J. HILL. 1987. Serotoninergic modulation of the pattern of eating and the profile of hunger-satiety in humans. Int. J. Obesity **11**(Suppl. 3): 141–155.
24. SAMANIN, R., T. MENNINI, A. FERRARIS, C. BENDOTTI, F. BORSININ & S. GARATTINI. 1979. m-Chlorophenylpiperazine: A central serotonin agonist causing powerful anorexia in rats. Naunyn-Schmiedebergs Arch. Pharmacol. **308:** 159–163.
25. KENNETT, G. A. & G. CURZON. 1988. Evidence that hypophagia induced by mCPP and

TFMPP requires 5-HT_{1C} and 5-HT_{1B} receptors; Hypophagia induced by RU 24969 only requires 5-HT_{1B} receptors. Psychopharmacology **96:** 93–100.
26. HOYER, D., A. PAZOS, A. PROBST & J. M. PALACIOS. 1986. Serotonin receptors in the human brain. I. Characterization and autoradiographic localization of 5-HT_{1A} recognition sites. Apparent absence of 5-HT_{1B} recognition sites. Brain Res. **376:** 85–96.
27. WONG, D. T. & R. W. FULLER. 1987. Serotonergic mechanisms in feeding. Int. J. Obes. **11**(Suppl. 3): 125–133.
28. DOURISH, C. T., P. H. HUTSON & G. CURZON. 1986. Para-chlorophenylalanine prevents feeding induced by the serotonin agonist 8-hydroxy-2-(di-n-propylamino) tetralin (8OH-DPAT). Psychopharmacology **89:** 467–471.
29. JIMERSON, D. C. & W. BERRETTINI. 1985. Cerebrospinal fluid amine metabolite studies in depression: Research update. *In* Pathochemical Markers in Major Psychoses. H. Beckmann and P. Riederer, Eds. Berlin: Springer-Verlag, pp. 129–143.
30. CURZON, G. 1985. Effect of food intake on brain transmitter amine precursors and amine synthesis. *In* Psychopharmacology and Food. M. Sandler and T. Silverstone, Eds. Oxford: Oxford University Press, pp. 59–70.
31. WURTMAN, R. J. & J. J. WURTMAN. 1984. Nutritional control of central neurotransmitters. *In* The Psychobiology of Anorexia Nervosa. K. M. Pirke and D. Ploog, Eds. Berlin: Springer-Verlag, pp. 4–11.
32. YOUNG, S. N., S. E. SMITH, R. O. PIHL & F. R. ERVIN. 1985. Tryptophan depletion causes a rapid lowering of mood in normal males. Psychopharmacology **87:** 173–177.
33. DELGADO, P. L., D. S. CHARNEY, L. H. PRICE, G. ANDERSON, H. LANDIS & G. R. HENINGER. 1987. Dietary tryptophan restriction produces an upregulation of the neuroendocrine response to infused tryptophan in healthy human subjects. Soc. Neurosci. Abstracts **13**(Part 2): 227. (Abstract)
34. MURPHY, D. L., E. A. MUELLER, N. A. GARRICK & C. S. AULAKH. 1986. Use of serotonergic agents in the clinical assessment of central serotonin function. J. Clin. Psychiatry **47**(Suppl. 4): 9–15.
35. VAN PRAAG, H. M., C. LEMUS & R. KAHN. 1987. Hormonal probes of central serotonergic activity: Do they really exist? Biol. Psychiatry **22:** 86–98.
36. VAN DE KAR, L. D., M. KARTESZI, C. L. BETHEA & W. F. GANONG. 1985. Serotonergic stimulation of prolactin and corticosterone secretion is mediated by different pathways from the mediobasal hypothalamus. Neuroendocrinology **41:** 380–384.
37. WILLOUGHBY, J. O., M. F. MENADUE & H. J. LIEBELT. 1988. Activation of 5-HT 1 serotonin receptors in medial basal hypothalamus stimulates prolactin secretion in the unanaesthetized rat. Neuroendocrinology **47:** 83–87.
38. VAN DE KAR, L. D., J. H. URBAN, K. D. RICHARDSON & C. L. BETHEA. 1985. Pharmacological studies on the serotoninergic and nonserotonin-mediated stimulation of prolactin and corticosterone secretion by fenfluramine. Neuroendocrinology **41:** 283–288.
39. MUELLER, E. A., D. L. MURPHY & T. SUNDERLAND. 1986. Further studies of the putative serotonin agonist, m-chlorophenylpiperazine: Evidence for a serotonin receptor mediated mechanism of action in humans. Psychopharmacology **89:** 388–391.
40. HAMIK, A. & S. J. PEROUTKA. 1989. 1-(m-Chlorophenyl)piperazine (mCPP) interactions with neurotransmitter receptors in the human brain. Biol. Psychiatry **25:** 569–575.
41. GARFINKEL, P. E. 1974. Perception of hunger and satiety in anorexia nervosa. Psychol. Med. **4:** 309–315.
42. HALMI, K. A. 1988. Taste, hunger, and satiety perceptions in anorexia nervosa and bulimia. *In* Eating Behavior in Eating Disorders. B. T. Walsh, Ed. Washington DC: American Psychiatric Press, pp. 113–124.
43. RUSSELL, G. 1979. Bulimia nervosa: An ominous variant of anorexia nervosa. Psychol. Med. **9:** 429–448.
44. CASPER, R. C., E. D. ECKERT, K. A. HALMI, S. C. GOLDBERG & J. M. DAVIS. 1980. Bulimia: Its incidence and clinical significance in patients with anorexia nervosa. Arch. Gen. Psychiatry **37:** 1030–1035.
45. KASSETT, J. A., H. E. GWIRTSMAN, W. H. KAYE, H. A. BRANDT & D. C. JIMERSON. 1988. Pattern of onset of bulimic symptoms in anorexia. Am. J. Psychiatry **145:** 1287–1288.

46. HALMI, K. A., E. ECKERT, T. J. LADU & J. COHEN. 1986. Anorexia nervosa: Treatment efficacy of cyproheptadine and amitriptyline. Arch. Gen. Psychiatry **43:** 177–181.
47. MELTZER, H. Y. 1987. The serotonin hypothesis of depression. *In* Psychopharmacology: The Third Generation of Progress. H. Y. Meltzer, Ed. New York: Raven Press, pp. 513–526.
48. PIRAN, N., S. KENNEDY, P. E. GARFINKEL & M. OWENS. 1985. Affective disturbance in eating disorders. J. Nerv. Ment. Dis. **173:** 395–400.
49. LAESSLE, R. G., S. KITTL, M. M. FICHTER, H.-U. WITTCHEN & K. M. PIRKE. 1987. Major affective disorder in anorexia nervosa and bulimia: A descriptive diagnostic study. Br. J. Psychiatry **151:** 785–789.
50. GARNER, D. M., P. E. GARFINKEL & M. O'SHAUGHNESSY. 1985. The validity of the distinction between bulimia with and without anorexia nervosa. Am. J. Psychiatry **142:** 581–587.
51. POLIVY, J. & C. P. HERMAN. 1985. Dieting and bingeing: A causal analysis. Am. Psychol. **40:** 193–201.
52. MITCHELL, J. E., D. HATSUKAMI, R. L. PYLE & E. D. ECKERT. 1986. The bulimia syndrome: Course of the illness and associated problems. Compr. Psychiatry **27:** 165–170.
53. GOODWIN, G. M., C. G. FAIRBURN & P. J. COWEN. 1988. The effects of dieting and weight loss on neuroendocrine responses to tryptophan, clonidine, and apomorphine in volunteers: Important implications for neuroendocrine investigations in depression. Arch. Gen. Psychiatry **44:** 952–957.
54. GOODWIN, G. M., S. FRASER, K. STUMP, C. G. FAIRBURN, J. M. ELLIOTT & P. J. COWEN. 1988. Dieting and weight loss in volunteers increases the number of α_2-adrenoceptors and 5-HT receptors on blood platelets without effect on [^{3}H]imipramine binding. J. Affective Disord. **12:** 267–274.
55. SCHWEIGER, U., M. WARNHOFF, J. PAHL & K. M. PIRKE. 1986. Effects of carbohydrate and protein meals on plasma large neutral amino acids, glucose, and insulin plasma levels of anorectic patients. Metabolism **35:** 938–943.
56. GILLBERG, C. 1983. Low dopamine and serotonin levels in anorexia nervosa [letter]. Am. J. Psychiatry **140:** 948–949.
57. KAYE, W. H., M. H. EBERT, M. RALEIGH & C. R. LAKE. 1984. Abnormalities in CNS monoamine metabolism in anorexia nervosa. Arch. Gen. Psychiatry **41:** 350–355.
58. KAYE, W. H., M. H. EBERT, H. E. GWIRTSMAN & S. R. WEISS. 1984. Differences in brain serotonergic metabolism between nonbulimic and bulimic patients with anorexia nervosa. Am. J. Psychiatry **141:** 1598–1601.
59. GERNER, R. H., D. J. COHEN, L. FAIRBANKS, G. M. ANDERSON, J. G. YOUNG, M. SCHEININ, M. LINNOILA, B. A. SHAYWITZ & T. A. HARE. 1984. CSF neurochemistry of women with anorexia nervosa and normal women. Arch. Gen. Psychiatry **41:** 1441–1444.
60. RIEDERER, P., K. TOIFL & P. KRUZIK. 1982. Excretion of biogenic amine metabolites in anorexia nervosa. Clin. Chim. Acta **123:** 27–32.
61. KAYE, W. H., H. E. GWIRTSMAN & M. H. EBERT. 1989. Serotonin: A trait disturbance in anorexia nervosa? Sci. Proceeds. Am. Psychiatr. Assoc.: New Res. Abst. **142:** p. 205.
62. BREWERTON, T. D., E. A. MUELLER, H. A. BRANDT, M. D. LESEM, D. L. MURPHY & D. C. JIMERSON. 1987. Evidence for serotonin dysregulation in anorexia. Sci. Proceeds. Am. Psychiatr. Assoc.: New Res. Abst. **140:** p. 123.
63. JIMERSON, D. C., H. A. BRANDT & T. D. BREWERTON. 1988. Evidence for altered serotonin function in bulimia and anorexia nervosa: Behavioral implications. *In* Psychobiology of Bulimia Nervosa. K. M. Pirke, W. Vandereycken and D. Ploog, Eds. Berlin: Springer-Verlag, pp. 83–89.
64. BREWERTON, T. D., H. A. BRANDT, M. D. LESEM, D. L. MURPHY & D. C. JIMERSON. 1990. Serotonin in eating disorders. *In* Serotonin in Major Psychiatric Disorders. E. F. Coccaro and D. L. Murphy, Eds. Washington DC: American Psychiatric Press. In press.
65. BAGDY, G., A. E. CALOGERO, C. S. AULAKH, K. SZEMEREDI & D. L. MURPHY. 1989. Long-term cortisol treatment impairs behavioral and neuroendocrine responses to 5-HT_1 agonists in the rat. Neuroendocrinology **50:** 241–247.

66. CHIODO, J. & P. R. LATIMER. 1986. Hunger perceptions and satiety responses among normal-weight bulimics and normals to a high-calorie, carbohydrate-rich food. Psychol. Med. **16:** 343–349.

67. HETHERINGTON, M. & B. J. ROLLS. 1988. Sensory-specific satiety and food intake in eating disorders. *In* Eating Behavior in Eating Disorders. B. T. Walsh, Ed. Washington DC: American Psychiatric Press, pp. 141–160.

68. BROOCKS, A., M. M. FICHTER & K. M. PIRKE. 1988. Effects of test meals on insulin, glucose, plasma large neutral amino acids and norepinephrine in patients with bulimia nervosa. Proc. Third Int. Conf. on Eating Disorders: Abst. 242.

69. KAYE, W. H., H. E. GWIRTSMAN, T. D. BREWERTON, D. T. GEORGE & R. J. WURTMAN. 1988. Bingeing behavior and plasma amino acids: A possible involvement of brain serotonin in bulimia nervosa. Psychiatry Res. **23:** 31–43.

70. JIMERSON, D. C., M. D. LESEM, W. H. KAYE & T. D. BREWERTON. 1988. Symptom severity and neurotransmitter studies in bulimia. Psychopharmacology **96:** S124 (Abst).

71. MARAZZITI, D., E. MACCHI, A. ROTONDO, G. F. PLACIDI & G. B. CASSANO. 1988. Involvement of serotonin system in bulimia. Life Sci. **43:** 2123–2126.

72. GOLDBLOOM, D. S., L. HICKS & P. E. GARFINKEL. 1988. Platelet serotonin uptake in bulimia nervosa. Proceeds. Am. Psychiatr. Assoc.: New Res. Abst. **141:** p. 137.

73. BREWERTON, T. D., E. A. MUELLER, H. A. BRANDT, M. D. LESEM, D. C. JIMERSON & D. L. MURPHY. 1987. Neuroendocrine studies with 5-HT agents in bulimia. Proceeds. Am. Psychiatr. Assoc. **140:** 85–86.

74. LEVY, A. B., K. N. DIXON & W. B. MALARKEY. 1988. Pituitary response to TRH in bulimia. Biol. Psychiatry **23:** 476–484.

75. HENINGER, G. R., D. S. CHARNEY & D. E. STERNBERG. 1984. Serotonergic function in depression. Arch. Gen. Psychiatry **41:** 398–402.

76. ROBINSON, P. H., S. A. CHECKLEY & G. F. M. RUSSELL. 1985. Suppression of eating by fenfluramine in patients with bulimia nervosa. Br. J. Psychiatry **146:** 169–176.

77. BLOUIN, A. G., J. H. BLOUIN, E. L. PEREZ, T. BUSHNIK, C. ZURO & E. MULDER. 1988. Treatment of bulimia with fenfluramine and desipramine. J. Clin. Psychopharmacol. **8:** 261–269.

78. RUSSELL, G. F. M., S. A. CHECKLEY, J. FELDMAN & I. EISLER. 1988. A controlled trial of d-fenfluramine in bulimia nervosa. Clin. Neuropharmacol. **11**(Suppl. 1): S146–S159.

79. KRAHN, D. & J. MITCHELL. 1985. Use of L-tryptophan in treating bulimia. Am. J. Psychiatry **142:** 1130.

80. SABINE, E. J., A. YONACE, A. J. FARRINGTON, K. H. BARRATT & A. WAKELING. 1983. Bulimia nervosa: A placebo controlled double-blind therapeutic trial of mianserin. Br. J. Clin. Pharmacol. **15:** 195S–202S.

81. POPE, H. G., JR., J. I. HUDSON, J. M. JONAS & D. YURGELUN-TODD. 1983. Bulimia treated with imipramine: A placebo-controlled double-blind study. Am. J. Psychiatry **140:** 554–558.

82. MITCHELL, J. E. & R. GROAT. 1984. A placebo-controlled, double-blind trial of amitriptyline in bulimia. J. Clin. Psychopharmacol. **4:** 186–193.

83. HUGHES, P. L., L. A. WELLS, C. J. CUNNINGHAM & D. M. ILSTRUP. 1986. Treating bulimia with desipramine. Arch. Gen. Psychiatry **43:** 182–186.

84. AGRAS, W. S., B. DORIAN, B. G. KIRKLEY, B. ARNOW & J. BACHMAN. 1987. Imipramine in the treatment of bulimia: A double-blind controlled study. Int. J. Eating Disorders **6:** 29–38.

85. HORNE, R. L., J. M. FERGUSON, H. G. POPE, J. I. HUDSON, C. G. LINEBERRY, J. ASCHER & A. CATO. 1988. Treatment of bulimia with bupropion: A multicenter controlled trial. J. Clin. Psychiatry **49:** 262–266.

86. BARLOW, J., J. BLOUIN, A. BLOUIN & E. PEREZ. 1988. Treatment of bulimia with desipramine: A double-blind crossover study. Can. J. Psychiatry **33:** 129–133.

87. POPE, H. G., JR., P. E. KECK, JR., S. L. MCELROY & J. I. HUDSON. 1989. A placebo-controlled study of trazodone in bulimia nervosa. J. Clin. Psychopharmacol. **9:** 254–259.

88. WALSH, B. T., M. GLADIS, S. P. ROOSE, J. W. STEWART, F. STETNER & A. H. GLASSMAN.

1988. Phenelzine vs placebo in 50 patients with bulimia. Arch. Gen. Psychiatry **45:** 471–475.
89. Kennedy, S. H., N. Piran, J. J. Warsh, P. Prendergast, E. Mainprize, C. Whynot & P. E. Garfinkel. 1988. A trial of isocarboxazid in the treatment of bulimia nervosa. J. Clin. Psychopharmacol. **8:** 391–396.
90. Willner, P. 1985. Antidepressants and serotonergic neurotransmission: An integrative review. Psychopharmacology **85:** 387–404.
91. Heninger, G. R. & D. S. Charney. 1987. Mechanism of action of antidepressant treatments: implications for the etiology and treatment of depressive disorders. *In* Psychopharmacology: The Third Generation of Progress. H. Y. Meltzer, Ed. New York: Raven Press, pp. 535–544.
92. Freeman, C. P. L. & M. Hampson. 1987. Fluoxetine as a treatment for bulimia nervosa. Int. J. Obesity **11**(Suppl. 3): 171–177.
93. Laessle, R. G., C. Zoettl & K. M. Pirke. 1987. Metaanalysis of treatment studies for bulimia. Int. J. Eating Disorders **6:** 647–653.
94. American Psychiatric Assoc. 1987. Diagnostic and Statistical Manual of Mental Disorders, Third Edition—Revised. Washington DC: American Psychiatric Press.
95. Coppen, A. J., R. K. Gupta, E. G. Eccleston, K. M. Wood, A. Wakeling & V. F. A. DeSousa. 1976. Plasma-tryptophan in anorexia nervosa [letter]. Lancet **1:** 961.
96. Johnston, J. L., L. A. Leiter, G. N. Burrow, P. E. Garfinkel & G. H. Anderson. 1984. Excretion of urinary catecholamine metabolites in anorexia nervosa: Effect of body composition and energy intake. Am. J. Clin. Nutr. **40:** 1001–1006.
97. Kaye, W. H., H. E. Gwirtsman, D. T. George, D. C. Jimerson & M. H. Ebert. 1987. CSF 5-HIAA concentrations in anorexia nervosa: Reduced values in underweight subjects normalize after weight gain. Biol. Psychiatry **23:** 102–105.
98. Weizman, R., M. Carmi, S. Tyano, A. Apter & M. Rehavi. 1986. High affinity [^{3}H]imipramine binding and serotonin uptake to platelets of adolescent females suffering from anorexia nervosa. Life Sci. **38:** 1235–1242.
99. Zemishlany, Z., I. Modai, A. Apter, Z. Jerushalmy, E. Samuel & S. Tyano. 1987. Serotonin (5-HT) uptake by blood platelets in anorexia nervosa. Acta Psychiatr. Scand. **75:** 127–130.

DISCUSSION OF THE PAPER

R. Murphy (*New York University, New York City*): In the bulimic group, where you saw the *m*-CPP effect in terms of blunted prolactin response, did you look at any measures of hypothalamic-pituitary-adrenal function by some other means?

Jimerson: We did look at the cortisol response to the serotonergic challenges in these patients and this was not different from controls. This may fit in with some previous data that suggests that there is differential regulation of prolactin and cortisol secretion in response to serotonergic challenges. We did not utilize any neuropharmacological challenges to assess dopaminergic responses in these patients.

Serotonin Agents in Anxiety

DUNCAN P. TAYLOR

CNS Biology
Pharmaceutical Research and Development Division
Bristol-Myers Company
Wallingford, Connecticut 06492

The anticipation of a known danger or event may temporarily generate a feeling of fear or apprehension in a normal person. However, when anxiety, as manifested by the exhibition of at least three of the symptoms, motor tension, autonomic hyperactivity, apprehensive expectation, and vigilance and scanning, persists continuously for at least one month then a diagnosis of generalized anxiety disorder (GAD) may be made.[1] Other kinds of anxiety disorders include panic disorder, phobic disorder, obsessive-compulsive disorder, and post-traumatic stress disorder.[1] These disorders affect from four to eight percent of the U.S. population and have been treated with a variety of agents, most of which possess a sedative component to their action: alcohol, opiates, belladonna, barbiturates, propanediol carbamates, and benzodiazepines. This series represents a gradual improvement in safety with respect to danger in overdose when taken alone, but even the benzodiazepines have not been as safe as hoped. For instance, it has been found that benzodiazepine users are significantly more likely than nonusers to experience an accident or injury requiring medical attention, that users have a greater number of accident-related hospital admissions, and that they incur a greater number of accident-related in-patient days.[2] Moreover, the National Institute on Drug Abuse has found that abuse of prescriptions for benzodiazepines has doubled among adults, rising from 3.1 percent of benzodiazepine prescriptions in 1980 to 7.1 percent in 1985.[3] Furthermore, the problems of dependence upon benzodiazepines following long-term therapy, even at low doses, are well-known.[4] The concerns elicited by these findings have resulted in a continuing search by the pharmaceutical industry for newer agents which would be more selective, have fewer or diminished side effects, and present a profile consistent with safety during long-term use.

The first novel anxiolytic to emerge from this search which bears no obvious structural or biochemical homology with the benzodiazepines is buspirone (BuSpar®; see FIG. 1). As buspirone has come to be recognized as safe and effective in the treatment of GAD, there has been an explosion in the numbers of publications concerning it, over 800 to date (up from 500 in 1987[5]). In this paper I shall review the clinical studies which led to this recognition, cover some of the neurobiological results which implicate serotonin in its mechanism of action, ponder the evidence that serotonin plays a role in the pharmacotherapy of anxiety, and consider clinical and preclinical data obtained with other new potential serotonergic anxiolytics.

GAD AND BUSPIRONE: CLINICAL SAFETY AND EFFICACY

The efficacy of buspirone in the treatment of GAD was established in several double-blind three-arm trials with both active drug and placebo controls. In these trials the efficacy of buspirone was equivalent to that of the active benzodiazepine used (either diazepam[6–10,12] or clobazam[11]) and both drug treatment groups were significantly better than the placebo group when assessed with the Hamilton Anxiety Rating

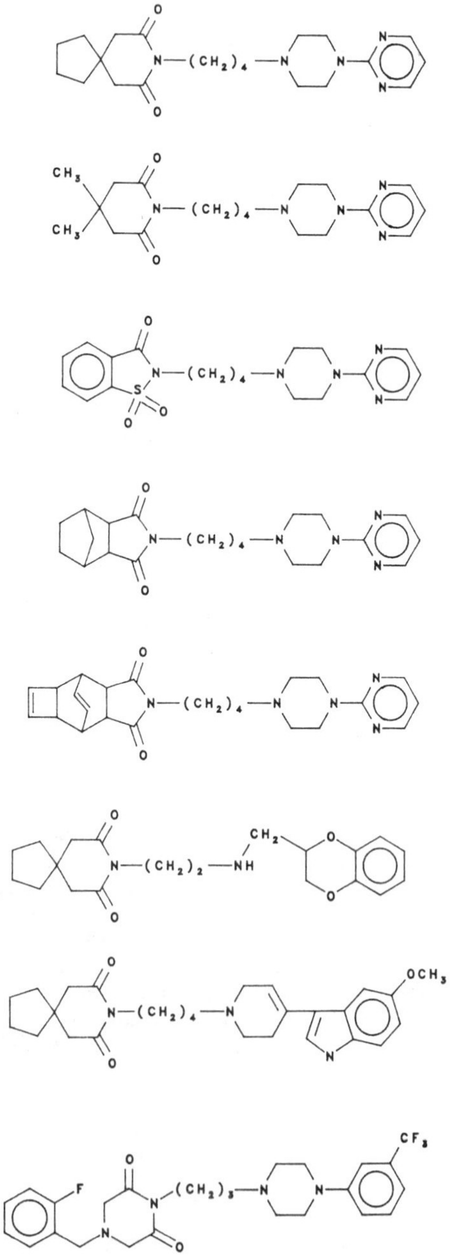

FIGURE 1. Serotonergic anxiolytics and potential anxiolytics.

Scale (HAM-A), a psychiatric instrument for evaluating the presence and extent of the symptoms of anxiety (see FIG. 2). In these studies, final mean daily doses of buspirone and benzodiazepine were 20–25 mg and 19–21 mg, respectively. Although both drugs produced a clinically beneficial effect on anxiety, patients receiving buspirone suffered less from feelings of sedation, drowsiness, tiredness, and fatigue than those receiving a benzodiazepine in all studies (see FIG. 2). Additional safety studies were carried out in a number of double-blind two-arm trials with an active drug control. Again, in the presence of equivalent effects on HAM-A scores, the incidence of sedation was considerably less in the buspirone-treated groups than in the groups receiving a benzodiazepine (clorazepate, see FIG. 3). Other safety studies included

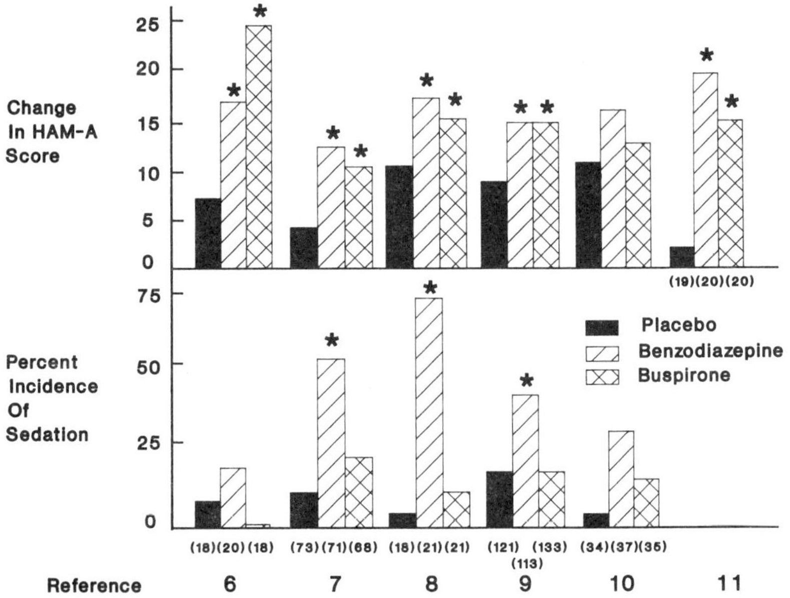

FIGURE 2. Effect of placebo or drug therapy on anxiety scores and incidence of sedation in three-arm trials. Anxiety scores are mean change from baseline on the Hamilton Anxiety Rating Scale (HAM-A) at study endpoint. Numbers in parenthesis represent numbers of patients in each treatment group at endpoint. Asterisks denote significant differences ($p < 0.05$) from placebo-treated groups.

comparisons of side effects produced by buspirone compared to a variety of standard benzodiazepine anxiolytics in clinical use. The incidences of drowsiness and fatigue produced by all of these benzodiazepines are significantly greater ($p < 0.05$) than those for buspirone or placebo, which do not differ from each other (see FIG. 4).[15] When specifically investigated in either subjects or patients it was found that buspirone did not impair daytime wakefulness, and its effects on sleep structure were minimal and clinically inconsequential.[16,17] Buspirone was as effective in 513 elderly patients (older than 65 years) as it was in 5,039 younger patients, with mean changes of 12.6 and 14.4 points on the HAM-A, respectively.[18] Although it might be supposed that the elderly would be more susceptible in both kind and degree, no significant differences in the profiles of side effects or their incidences were reported.[18] As it is known that

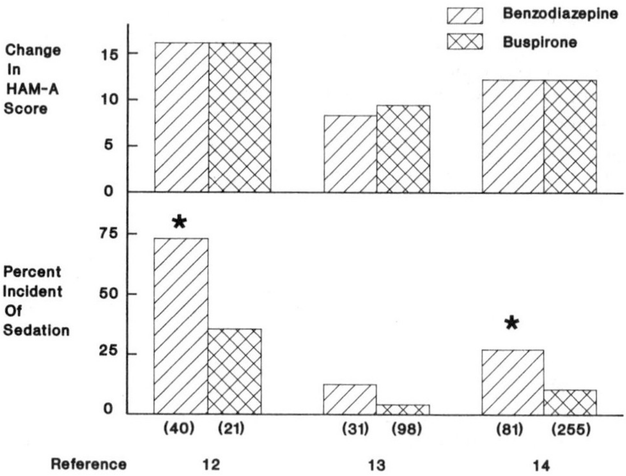

FIGURE 3. Effect of drug therapy on anxiety scores and incidence of sedation in two-arm trials. (See legend to FIGURE 2 for further details.) Asterisks denote significant differences ($p < 0.05$) between treatment groups.

benzodiazepines can impair memory function by causing short-term anterograde amnesia, a specific investigation of the effects of buspirone compared to diazepam and placebo was carried out in patients with GAD.[19] As expected, diazepam significantly increased the amount of information lost during a 20 minute delay before testing when compared to placebo or its own baseline, whereas buspirone had no significant effect

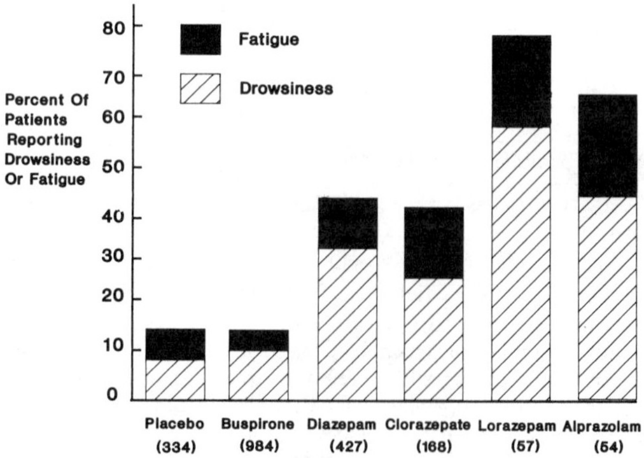

FIGURE 4. Percent of patients reporting incidence of drowsiness and/or fatigue during anxiolytic trials of drugs. Numbers in parentheses indicate number of patients in each treatment group. Incidence of both fatigue and drowsiness for all benzodiazepine-treated groups was significantly greater ($p < 0.05$) than for buspirone. There were no significant differences between buspirone and placebo for the incidences of either drowsiness or fatigue. (Data are from reference 15.)

(see FIG. 5). The ability of benzodiazepines to produce muscle relaxation and impair psychomotor performance alone or in combination with alcohol is well-known. Numerous studies have shown the buspirone does not share the muscle-relaxant properties of the benzodiazepines, or their psychomotor impairment alone, and it does not possess the ability to potentiate impairing effects of alcohol.[20]

In an open-field study involving a total of 5,414 patients, side effects with the greatest incidence were gastric complaints (6.6%), dizziness (6.0%), headache (4.9%), and sleeplessness (3.8%).[15] The findings with respect to gastric complaints, dizziness, and headache confirm accumulated safety data obtained during three-arm double-blind studies in 496 GAD patients.[21] The reports of sleeplessness and insomnia may reflect a difference in patients' or physicians' expectations as the benzodiazepines have such marked sedative/hypnotic properties. For example, when insomnia, a separate factor on the HAM-A scale, was broken out and examined both buspirone and

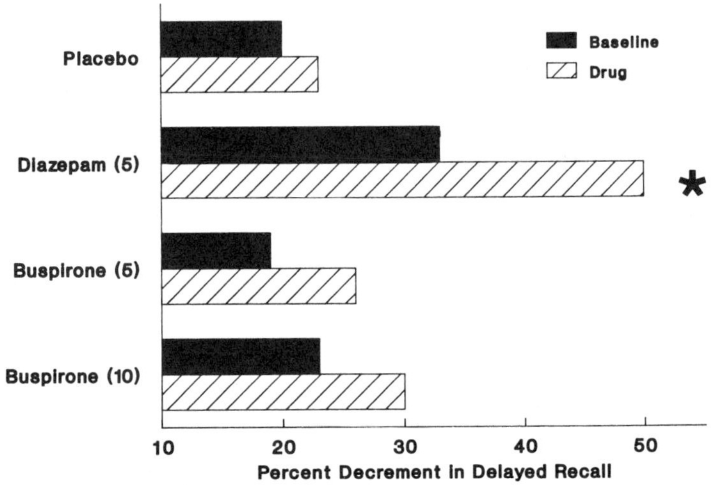

FIGURE 5. Percent decrement in delayed recall of word lists by patients following drug treatment. Asterisk indicates a significant difference ($p < 0.01$) between drug and baseline. (Adapted from reference 19.)

clobazam produced a decrease compared to placebo (−1.0 and −1.2, respectively) in one study,[22] and in another study buspirone and diazepam produced a similar, significant decrease compared to placebo (−1.3 and −1.1, respectively).[6] Furthermore, upon abrupt replacement on placebo from six months of drug treatment, significantly more patients who had previously received clorazepate complained of insomnia than those who had received buspirone.[23]

These observations bring up a significant consideration in anxiolytic therapy which will point out another significant difference between buspirone and the benzodiazepines, and that is the safety and efficacy of buspirone in long-term use and its lack of production of tolerance or dependence. In open-label trials in GAD patients, the use of buspirone was associated with a steady improvement in HAM-A scores.[24] Moreover, no upward titration of drug dose was apparent as mean daily doses appeared to decrease if anything after the first month.[25] Following abrupt withdrawal by placebo substitution, patients receiving diazepam or clorazepate showed significant increases in anxiety rating scores, whereas patients receiving buspirone did not.[12,26,27]

Finally, buspirone is clinically distinguished from the benzodiazepines in terms of its liability for abuse. For instance, sedative/hypnotic abusers sought to use diazepam more frequently than either placebo or buspirone. The latter two showed little likelihood for continued use and were no different from each other.[28] Similarly, buspirone failed to exhibit any potential for abuse in alcohol-dependent anxious patients.[29]

These observations and others led us to describe the clinical profile of buspirone as "anxioselective" when compared to the benzodiazepines,[30] that is, although both are clinically effective in the treatment of GAD, buspirone lacks the ancillary properties of the benzodiazepines which cause them to be used as sedative/hypnotics, muscle relaxants, and anticonvulsants. Moreover, buspirone does not produce psychomotor impairment, potentiate the deleterious effects of alcohol, produce amnesia, produce dependence or tolerance after long-term treatment, lacks the ability to evoke a withdrawal syndrome, and lacks the liability for abuse in sedative/hypnotic abusers or alcoholics.

BUSPIRONE'S MECHANISM OF ACTION

I have recently reviewed the mechanism of action of buspirone[5] and would like to summarize that discussion here. Buspirone appears to act primarily as a $serotonin_{1A}$ ($5\text{-}HT_{1A}$) partial agonist based on biochemical and physiological assessments. Buspirone binds *in vitro* to $5\text{-}HT_{1A}$ sites labeled by specific radioligands for these sites.[5] In addition $[^3H]$buspirone labels sites in the limbic system of rodents and man with a distribution which is strikingly similar to that seen with $[^3H]$5-HT.[31] Buspirone exhibits the functional effects of a partial $5\text{-}HT_{1A}$ agonist both biochemically in its effects on the serotonin-sensitive adenylate cyclase[32] and electrophysiologically in its effects on pyramidal cells in hippocampal slices.[33] Further evidence for a functional response to the binding of buspirone to $5\text{-}HT_{1A}$ sites comes from electrophysiological studies of the firing of serotonergic cells in the dorsal raphe nucleus. Here buspirone acts like a serotonin agonist, attenuating or abolishing the activity of spontaneously firing cells.[34] The neurochemical consequence of this is found in the decrease in hippocampal levels of the serotonin metabolite 5-hydroxy-indoleacetic acid.[35] That buspirone's clinical actions are mediated by its serotonergic actions is suggested by the fact that the elevations of prolactin produced by high doses of buspirone can be blocked by pretreatment with the serotonin antagonist methysergide.[36] Finally, buspirone is active in a variety of animal models of anxiety, one such being the attenuation of punishment-suppressed drinking (conflict test).[37] When the serotonergic systems of animals are lesioned with selective neurotoxins, the anticonflict effect of buspirone was lost.[37] In addition, buspirone produces some, but not all, signs of the serotonin syndrome (hindlimb abduction, flattened body posture, Straub tail) and antagonizes the other signs evoked by the full $5\text{-}HT_{1A}$ agonist 8-hydroxy-DPAT (reciprocal forepaw treading, head weaving, tremor).[38] Further support for the action of buspirone as a partial $5\text{-}HT_{1A}$ agonist comes from drug discrimination experiments where buspirone has been shown to generalize to the cue produced by 8-hydroxy-DPAT,[39] but buspirone itself cannot be used for training in such paradigms.[40]

SEROTONIN IN ANXIETY

That serotonin might play a role in the pharmacotherapy of anxiety is a concept almost 20 years old, supported by anatomical, neurochemical, electrophysiological, and behavioral data. Administration of parachlorophenylalanine, an inhibitor of serotonin synthesis, resulted in the attenuation of punishment-suppressed behavior, an animal model of anxiety.[41] In addition, the direct administration of serotonin antagonists could result in a similar release from suppression.[42] It is not surprising that serotonin could have such an effect since the serotonergic neurons which originate in the raphe nuclei innervate the cortex, hippocampus, and amygdala, limbic regions long associated with emotional behavior. Indeed, direct intra-raphe administration of serotonin released responding in a conditional emotional response paradigm.[43] That the clinically effective benzodiazepines might produce their anxiolytic effects by actions upon serotonergic effects was suggested by the observations that these drugs inhibited the firing of serotonergic neurons in the dorsal raphe nucleus, decreased turnover of serotonin in the brain (that is, they decreased levels of 5-hydroxyindoleacetic acid), and reversed the proconflict effects (enhanced suppression of behavior) of direct serotonergic stimulation of the dorsal raphe nucleus.[42] In addition, the efficacy of diazepam in the attenuation of punishment-suppressed responding was abolished when serotonergic neurons were destroyed by lesioning with selective neurotoxins.[37] Moreover, administration of benzodiazepines reduced serotonergic neuronal activity assessed electrophysiologically in rats[44] and cats.[45] Recently it was demonstrated that the utilization of [^{3}H]2-deoxyglucose, a marker for neuronal energy consumption, was decreased in the serotonin cell body–containing dorsal and median raphe nuclei following administration of the benzodiazepine alprazolam or buspirone.[46] Although the benzodiazepines have a well-recognized direct effect upon the γ-aminobutyric acid (GABA) chloride ionophore complex, a number of recent reviews support the candidacy of serotonin as a mediator of the anxiolytic effects of benzodiazepines downstream from the GABA receptor.[5,47,49]

A variety of new chemical entities are under development as anxiolytics, and what is currently known of their mechanism of action reflects the strong belief of the pharmaceutical industry that serotonin plays a role in the relief of anxiety. Many of these compounds resemble buspirone chemically (the generic term "azapirones" has been coined to describe them, see FIG. 1) and pharmacologically in that they may be 5-HT_{1A} partial agonists (see TABLE 1). Reviews of preclinical data on buspirone, gepirone, and ipsapirone are available.[5,48] Tandospirone (SM-3997, metanopirone) has been shown to decrease serotonin synthesis and share drug discrimination properties with buspirone, gepirone, ipsapirone and 8-hydroxy-DPAT.[50] MDL 73005EF bound to 5-HT_{1A} sites *in vitro* with high affinity, was active in conflict tests, and generalized to 8-hydroxy-DPAT in a drug discrimination paradigm.[51,52] Two other analogs with activity in some animal models of anxiety have been reported in the patent literature from Roussel UCLAF[53] and Adir et Cie[54] (see FIG. 1). A number of structurally distinct 5-HT_{1A} agonists and partial agonists (their pharmacological similarities to the azapirones include 5-HT_{1A} binding and anticonflict activity) are also in preclinical development at this time including flesinoxan (Duphar),[55] CGS 18102A (Ciba-Geigy),[56] isamoltane (Ciba-Geigy),[57] CM 57493 (Sanofi),[58] eltoprazine (Duphar),[59] LY 178210 (Lilly),[60] and NDO-008 (Astra).[61] In addition to chemical and pharmacological similarity to buspirone, preliminary evidence of efficacy in the treatment of GAD has been obtained with gepirone, ipsapirone, and tandospirone. Pilot studies of gepirone in phase II clinical trials revealed significant effects upon HAM-A scores (see FIG. 6).[62,63] In phase II trials of like design, anxiolytic efficacy with ipsapirone was also seen.[64,65] Similar results with tandospirone have been claimed,[66] although no data from these studies are yet available in the published literature. One interesting additional result of these studies was the observation that gepirone and ipsapirone appeared to have antidepressant properties. Perhaps this is not surprising as buspirone was reported early on to have similar effects in GAD patients.[6,7] In fact, both buspirone and gepirone have been shown to provide antidepressant benefit in patients with a primary diagnosis

TABLE 1. Biological Activity of Azapirones[a]

Test	Buspirone	Gepirone	Ipsapirone	Tandospirone	Wy-47,846	MDL 73005EF
Binds to 5-HT_{1A} sites	+	+	+	+	+	+
Binds to benzodiazepine sites	–	–	–	–	–	–
Active in conflict procedures	+	+	+	+	?	+
Shares 5-HT_{1A} drug discrimination properties	+	+	+	+	?	+
Decreases 5-HT neuronal activity	+	+	+	+	+	?
Decreases 5-HT synthesis	+	+	+	+	?	?
Autoradiographic localization overlaps 5-HT	+	+	+	+	?	?
Mimics 5-HT action on adenylate cyclase	+	+	+	?	?	?
Alters glucose metabolism in limbic system	+	+	+	?	?	?
Reduces aggression	+	+	+	?	?	?
Clinically active as an anxiolytic	+	+	+	+	?	?

[a] +, Activity present; –, activity absent; ?, activity not tested. See text for references.

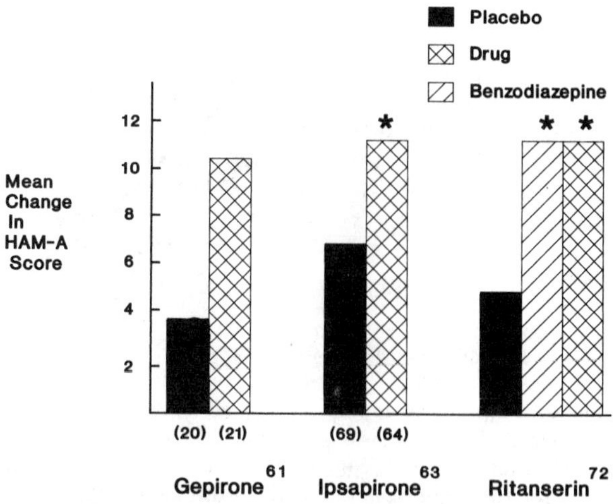

FIGURE 6. Effect of placebo or drug therapy on anxiety scores in clinical trials of other potential anxiolytics. (See legend to FIGURE 2 for further details.)

of major depression.[11,67] From anecdotal clinical case reports a variety of other utilities have been suggested for buspirone including other anxiety disorders, such as panic and obsessive-compulsive disorders, anorexia nervosa, agitation due to dementia or head trauma, sexual dysfunction, premenstrual syndrome, substance abuse, treatment of drug-induced extrapyramidal motor disorders, schizophrenia, childhood autism, dyspnea, control of blood pressure, and motion sickness.[68]

The antidepressant effects of buspirone, gepirone, and ipsapirone, as well as the ability of buspirone,[69] gepirone,[70] and 8-hydroxy-DPAT[71] to down-regulate 5-HT_2 receptors is reminiscent of the effects of 5-HT_2 antagonists, and might suggest that this latter class of agents may have utility not only in depression but anxiety as well. In fact, two such agents are in clinical development as anxiolytics, ritanserin and ICI 169,369.[61] Ritanserin is a 5-HT_2 antagonist which is active in at least one animal model of anxiety.[72] Significantly, ritanserin has demonstrated clinical efficacy in the reduction of HAM-A scores in a double-blind active drug and placebo-controlled trial (see FIG. 6).[73,74] ICI 169,369 is a potent 5-HT_2 antagonist.[75] No clinical details concerning ICI 169,369 have yet been published.

Another group of serotonin antagonists may also possess utility in the treatment of anxiety. The 5-HT_3 antagonists, such as ondansetron (GR 38032F), zacopride, and ICS 205-930, have been reported to be active in some animal models of anxiety,[76,77] and clinical trials are in progress.

SUMMARY

The side effects and unwanted or unnecessary ancillary pharmacological properties of benzodiazepine anxiolytic drugs resulted in a continuing search for new agents with improved profiles of activity. Buspirone was the first novel drug to emerge from this search in almost thirty years. Investigations into its mechanism of action revealed a key role for serotonin in the pharmacotherapy of anxiety. A variety of serotonergic agents are now in preclinical and clinical development as anxiolytics, including 5-HT_{1A} partial agonists, 5-HT_2 antagonists, and 5-HT_3 antagonists. In addition to the new drugs which will be developed as a consequence of these investigations, their clinical efficacy will prompt the development of new animal models of psychopathology, leading us ever closer to a full understanding of the neurobiological substrates of anxiety. In addition, deployment of these new agents in the armamentaria of the clinician and the basic scientist will lead to new insights into the treatment of other disorders and the biochemical mechanisms by which the effects of these drugs are obtained.

ACKNOWLEDGEMENTS

I am indebted to members of the Bristol-Myers CNS Clinical Research department, particularly D. S. Robinson, R. E. Gammans, and D. R. Alms for preprints, advice, and guidance. I am deeply appreciative of the quotidian discussions with A. S. Eison, M. S. Eison, S. L. Moon, C. P. VanderMaelen, and F. D. Yocca from Bristol-Myers CNS Biology which contributed materially to the organization and resources required for this manuscript. Finally, I am thankful to J. Stockla for her indispensable assistance in producing the typescript and artwork and to M. Haas for chemical structures.

REFERENCES

1. AMERICAN PSYCHIATRIC ASSOCIATION, ED. 1980. Diagnostic and Statistical Manual of Mental Disorders. 3d edit. American Psychiatric Association. Washington, D.C.
2. OSTER, G., M. W. RUSSELL, D. M. MUSE, S. F. ADAMS & J. IMBIMBO. 1987. J. Clin. Psychiatry **48** (12, suppl.): 17–21.
3. ANONYMOUS. 1988. Lancet **ii:** 1153.
4. WOODS, J. H., J. L. KATZ & G. WINGER. 1988. J. Am. Med. Assoc. **260:** 3476–3480.
5. TAYLOR, D. P. 1988. FASEB J. **2:** 2445–2452.
6. GOLDBERG, H. L. & R. J. FINNERTY. 1979. Am. J. Psychiatry **136:** 1184–1187.
7. RICKELS, K., K. WEISMAN, N. NORSTAD, M. SINGER, D. STOLTZ & J. DANTON. 1982. J. Clin. Psychiatry **43** (12, sect. 2): 81–86.
8. PECKNOLD, J. C., P. FAMILAMIRI, H. CHANG, R. WILSON, J. ALARCIA & D. J. MCCLURE. 1985. Prog. Neuro-psychopharmacol. Biol. Psychiatry **9:** 639–642.
9. COHN, J. B. & K. RICKELS. 1989. Curr. Med. Res. Opin. **11:** 304–320.
10. WHEATLEY, D. 1982. J. Clin. Psychiatry **43** (12, sect. 2): 92–94.
11. ROBINSON, D. S., R. C. SHROTRIYA, D. R. ALMS, M. MESSINA & J. ANDARY. 1989. Psychopharmacol. Bull. **25**: 21–26.
12. RICKELS, K., E. SCHWEIZER, I. CSANALOSI, W. G. CASE & H. CHUNG. 1988. Arch. Gen. Psychiatry **45:** 444–450.
13. GOLDBERG, H. L. & R. FINNERTY. 1982. J. Clin. Psychiatry **43** (12, sect. 2): 87–91.
14. COHN, J. B., C. L. BOWDEN, J. G. FISHER & J. J. RODOS. 1986. Am. J. Med. **80** (Suppl. 3B): 10–16.
15. NEWTON, R. E., J. D. MARUNYCZ, M. T. ALDERDICE & M. J. NAPOLIELLO. 1986. Am. J. Med. **80** (Suppl. 3B): 17–21.
16. SEIDEL, W. F., S. A. COHEN, N. G. BLIWISE & W. C. DEMENT. 1985. Psychopharmacology **87:** 371–373.
17. DE ROECK, J., R. CLUYDTS, C. SCHOTTE, D. ROUCKHOUT & P. COSYNS. 1989. Acta Psychiat. Scand. **79:** 129–135.
18. ROBINSON, D. S., M. J. NAPOLIELLO & J. SCHENK. 1988. Clin. Ther. **10:** 740–746.
19. LUCKI, E., K. RICKELS, M. A. GIESECKE & A. GELLER. 1987. Br. J. Clin. Pharmacol. **23:** 207–211.
20. TUNNICLIFF, G. & A. S. EISON. 1990. *In* Buspirone: Mechanisms and Clinical Aspects. G. Tunnicliff, A. S. Eison and D. P. Taylor, Eds. San Diego, CA: Academic Press. In press.
21. ROBINSON, D. S. 1987. Int. Med. Spec. (Suppl.): 35–39.
22. BOEHM, C., M. PLACCHI, F. STALLONE, R. GAMMANS, D. ALMS, R. C. SHROTRIYA & D. S. ROBINSON. 1990. J. Clin. Psychopharmacology. In press.
23. RICKELS, K. 1985. *In* Anxiety Disorders: An International Update. New York: Academy Press Information Services, pp. 25–30.
24. FEIGHNER, J. P. 1987. J. Clin. Psychiatry **48** (12, suppl.): 3–6.
25. ROBINSON, D. S. 1985. *In* Anxiety Disorders: An International Update. New York: Academy Press Information Services, pp. 93–99.
26. MURPHY, S. M., R. T. OWEN & P. J. TYRER. 1984. Lancet **ii:** 1389.
27. FONTAINE, R., P. BEAUDRY, L. BEAUCLAIR & G. CHOUINARD. 1987. Prog. Neuro-psychopharmacol. Biol. Psychiatry **11:** 189–197.
28. COLE, J. O., M. H. ORZACK, B. BEAKE, M. BIRD & Y. BAR-TAL. 1982. J. Clin. Psychiatry **43** (12, sect. 2): 69–74.
29. GRIFFITH, J. D., D. R. JASINSKI, G. P. CASTEN & G. R. MCKINNEY. 1986. Am. J. Med. **80** (Suppl. 3B): 30–35.
30. TAYLOR, D. P., D. K. HYSLOP & L. A. RIBLET. 1980. Soc. Neurosci. Abstr. **6:** 791.
31. MOON, S. L. & D. P. TAYLOR. 1985. Soc. Neurosci. Abstr. **11:** 114.
32. YOCCA, F. D., D. K. HYSLOP, D. P. TAYLOR & S. MAAYANI. 1986. Fed. Proc. **45:** 436.
33. ANDRADE, R. & R. A. NICOLL. 1987. Naunyn-Schmiedebergs Arch. Pharmacol. **336:** 5–10.
34. VANDERMAELEN, C. P., D. P. TAYLOR, G. GEHLBACH & M. S. EISON. 1986. Psychopharmacol. Bull. **22:** 807–812.
35. MENNINI, T., M. GOBBI, F. PONZIO & S. GARATTINI. 1986. Arch. Int. Pharmacodyn. Ther. **279:** 40–49.

36. NEUHAUSER, H., G. LAAKMAN & A. KNOSSALLA. 1988. Psychopharmacology **96** (Suppl.): 393.
37. EISON, A. S., M. S. EISON, M. STANLEY & L. A. RIBLET. 1986. Pharmacol. Biochem. Behav. **24:** 701–707.
38. SMITH, L. M. & S. J. PEROUTKA. 1986. Pharmacol. Biochem. Behav. **24:** 1513–1519.
39. CUNNINGHAM, K. A., P. M. CALLAHAN & J. B. APPEL. 1987. Eur. J. Pharmacol. **138:** 29–36.
40. HENDRY, J. S., R. L. BALSTER & J. A. ROSECRANS. 1983. Pharmacol. Biochem. Behav. **19:** 97–101.
41. GELLER, I. & K. BLUM. 1970. Eur. J. Pharmacol. **9:** 319–324.
42. STEIN, L., C. D. WISE & B. D. BERGER. 1973. *In* The Benzodiazepines. S. Garattini, E. Mussini and L. O. Russell, Eds. New York: Raven Press, pp. 229–326.
43. THIEBOT, M.-H., M. HAMON & P. SOUBRIE. 1984. *In* Psychopharmacology of the Limbic System. E. Zarifian and M. R. Trimble, Eds. New York: Wiley, pp. 164–173.
44. LAURENT, J. P., M. MANGOLD, U. HUMBEL & W. HAEFELY. 1983. Neuropharmacology **22:** 501–511.
45. TRULSON, M. E., D. W. PREUSSLER, G. A. HOWELL & C. J. FREDERICKSON. 1982. Neuropharmacology **21:** 1045–1050.
46. PIERCEY, M. F. & W. E. HOFFMAN. 1989. Abstr. Int. Symp. Serotonin from Cell Biol. to Pharmacol. Ther. p. 90.
47. IVERSEN, S. D. 1984. Neuropharmacology **23:** 1553–1560.
48. TRABER, J. & T. GLASER. 1987. Trends Pharmacol. Sci. **8:** 432–437.
49. GARDNER, C. A. 1988. Gen. Pharmacol. **19:** 347–356.
50. SEYMOUR, P., S. MCLEAN, E. MENA & J. HEYM. 1989. Abstr. 1st Princeton Drug Symp. p. 33.
51. HIBERT, M., A. K. MIR, G. MAGHIOROS, P. MOSER, D. N. MIDDLEMISS, M. D. TRICKLEBANK & J. R. FOZARD. 1988. Br. J. Pharmacol. **93** (Suppl.): 2P.
52. MOSER, P., M. HIBERT, D. N. MIDDLEMISS, A. K. MIR, M. D. TRICKLEBANK & J. R. FOZARD. 1988. Br. J. Pharmacol. **93** (Suppl.): 3P.
53. GASC, J. C., L. NEDELEC, D. HUMBERT & A. M. BOAVENTURA. 1988. West German Patent No. DE 3,742,124.
54. LAVIELLE, G. & J. C. POIGNANT. 1988. French Patent No. FR 2,601,364.
55. MOS, J. & B. OLIVIER. 1989. Abstr. 1st Princeton Drug Symp. p. 37.
56. SILLS, M. A., D. A. BENNETT, R. A. LOVELL, J. M. LIEBMAN, P. L. WOOD, B. S. GLAESER, M. WILLIAMS & A. S. HUTCHISON. 1989. Abstr. 1st Princeton Drug Symp. p. 44.
57. WALDMEIER, P. C., M. WILLIAMS, P. A. BAUMANN, S. BISCHOFF, M. A. SILLS & R. F. NEALE. 1988. Naunyn-Schmiedebergs Arch. Pharmacol. **337:** 609–620.
58. ADRIEN, J., L. LANFUMEY, H. GOZLAN, C. M. FATTACCINI & M. HAMON. J. Pharmacol. Exp. Ther. **248:** 1222–1230.
59. OLIVIER, B. & J. MOS. 1989. Abstr. 1st Princeton Drug Symp.: S10.
60. LEANDER, J. D. & M. E. FLAUGH. 1989. Abstr. 1st Princeton Drug Symp. p. 34.
61. ANONYMOUS. 1987. Scrip (No. 1265, December 11): 24–25.
62. COTT, J. M., N. M. KURTZ, D. S. ROBINSON, S. P. LANCASTER & J. E. COPP. 1988. Psychopharmacol. Bull. **24:** 164–167.
63. CSANALOSI, I., E. SCHWEIZER, W. G. CASE & K. RICKELS. J. Clin. Psychopharmacol. **7:** 31–33.
64. KUEMMEL, B., M. BENEKE, G. KROL, G. SCHOELLHAMMER & H. SPECHTMEYER. 1988. Psychopharmacology **96** (Suppl.): 353.
65. BENEKE, M., B. KUEMMEL, I. S. ROED & H. SPECHTMEYER. 1988. Psychopharmacology **96** (Suppl.): 353.
66. FEIGHNER, J. P. & W. F. BOYER. 1989. Psychopathology **22** (Suppl. 1): 21–26.
67. AMSTERDAM, J. D., N. BERWISH, L. POTTER & K. RICKELS. 1987. Curr. Ther. Res. **41:** 185–193.
68. EISON, A. S., F. D. YOCCA & D. P. TAYLOR. 1990. *In* Buspirone: Mechanisms and Clinical Aspects. G. Tunnicliff, A. S. Eison and D. P. Taylor, Eds. Academic Press. San Diego, CA. In press.

69. TAYLOR, D. P., J. A. BECKER, M. CRANE, P. K. MYSLOP, L. A. RIBLET & D. L. TEMPLE, JR. 1983. Soc. Neurosci. Abstr. **9:** 435.
70. EISON, A. S. & F. D. YOCCA. 1985. Eur. J. Pharmacol. **111:** 389–392.
71. DESOUZA, R. J., G. M. GOODWIN & D. J. HEAL. 1986. Br. J. Pharmacol. **89:** 377–384.
72. COLPAERT, F. C., T. F. MEERT, C. J. E. NIEMGEERS & P. A. J. JANSSEN. 1985. Psychopharmacology **86:** 45–54.
73. CEULEMANS, D. L. S., M. L. J. A. HOPPENBROUWERS, Y. G. GELDERS & A. J. M. REYNTJENS. 1985. Pharmacopsychiatry **18:** 303–305.
74. JANSEN, A. A. I. & E. A. PANGALILIA-RATU LANGI. 1988. Psychopharmacology **96** (Suppl.): 350.
75. BLACKBURN, T. P., B. COX, A. J. GUILDFORD, D. J. LE COUNT, D. N. MIDDLEMISS, R. J. PEARCE & C. W. THORNBER. 1987. J. Med. Chem. **30:** 2252–2259.
76. COSTALL, B., A. M. DOMENEY, P. A. GERRARD, M. E. KELLY & R. J. NAYLOR. 1988. J. Pharm. Pharmacol. **40:** 302–305.
77. COSTALL, B., M. E. KELLY, R. J. NAYLOR, E. S. ONAIRI & M. B. TYERS. 1989. Br. J. Pharmacol. **96:** 325–332.

DISCUSSION OF THE PAPER

DAVID T. WONG (*Lilly Research Laboratories, Indianapolis, IN*): We have done some receptor binding studies with alcohol-preferring animals and we reported that the preferring animals have higher 5-HT_{1A} receptors in both hippocampus and in cortex (Wong, Threlkeld, Lumeng & Li. Life Sci. **46:** 231–235, 1990). Also the work of Meyer has demonstrated that buspirone suppresses alcohol intake in some animals. Do you have any human data suggesting that it may suppress alcohol intake?

TAYLOR: This man works for Lilly I want to point out to you. I did not plant him. The potential utility of buspirone in the treatment of alcoholism was first suggested by Meyer (Meyer, J. Stud. Alcohol **47:** 269–273, 1986). Some success in reducing the craving for alcohol in alcoholics has been noted after treatment with buspirone by Kranzler (Kranzler, ACNP Abstracts, p. 87, 1987; J. Clin. Psychopharmacol. **9:** 153, 1989) and Bruno (Psychopathology **22**(Suppl.): 49–59, 1989). This is perhaps not too surprising as animal studies have revealed that buspirone can attenuate drug-induced alcohol consumption in rats (Privette, Hornsby & Myers, Alcohol **5:** 147–152, 1988) and volitional alcohol intake in monkeys (Collins & Myers, Alcohol **4:** 49–56, 1987). In contrast to the benzodiazepines, buspirone appears to have little abuse potential in alcohol-dependent patients (Griffith, Jasinski, Casten & McKinney, Am. J. Med. **80**(Suppl. 3B): 30–35, 1986). The potential utility of buspirone in the treatment of alcoholism has been patented (Shrotriya & Casten, U.S. Patent No. 4,777,173, 1988).

DEBRA H. BENNETT (*Ciba-Geigy, Summit, NJ 07901*): Duncan, do you find in the clinical tests for anxiety with BuSpar that somebody who has been previously maintained on a benzodiazepine has more difficulty responding to buspirone than someone who has been drug-naïve previously?

TAYLOR: It depends on when they stop using the benzodiazepine. There was a publication from Karl Rickels' laboratory that looked at former users and reported that former users were more likely to show lowered efficacy with buspirone (Schweizer, Rickels & Lucki, New Engl. J. Med **314:** 719–720, 1986). When, in fact, those data were restratified, it was found that the patients who had taken benzodiazepines within 1 month of the initiation of their trial on buspirone were in fact at risk for adverse effects which we think reflect withdrawal from the benzodiazepine. Those who had been off the drug for greater than 1 month showed the same therapeutic benefit from buspirone as those who were naive to benzodiazepines (Gammans *et al.*, in preparation).

MANFRED GÖTHERT (*University of Bonn*). Could you please comment on the potential role of dopamine receptors in the action of buspirone because it has a certain affinity for those receptors as well.

TAYLOR: I don't think that that's a serious problem. The difference in affinity is about 10-fold. At one point I published a review (Taylor, Riblet, Stanton, Eison, Eison & Temple, Pharmacol. Biochem. Behav. **17** (Suppl. 1): 25–35, 1982). which looked at the use of dopamine agonists, in fact, in the clinical treatment of anxiety, and biochemical evidence with buspirone suggests that it may have some agonist properties as well at very, very high doses. But the efficacy of gepirone and ipsapirone as pure 5-HT_{1A}, or more selective 5-HT_{1A} partial agonists, suggests to me, at least, that buspirone is achieving its anxiolytic efficacy through the serotonergic axis rather than a dopaminergic axis.

Serotonin Function and Human Anxiety Disorders[a]

DENNIS S. CHARNEY,[b,c] SCOTT W. WOODS,[d]
JOHN H. KRYSTAL,[c] AND GEORGE R. HENINGER[d]

[c]*Psychiatry Service*
West Haven VA Medical Center
950 Campbell Avenue
West Haven, Connecticut 06512
and
[d]*Ribicoff Research Facilities*
Clinical Neuroscience Research Unit
Connecticut Mental Health Center
New Haven, Connecticut 06519

In recent years, there has been an increase in the understanding of the neurobiological etiology of anxiety disorders and the mechanism of action of antianxiety drugs. There is preclinical and clinical evidence suggesting the involvement of norepinephrine, benzodiazepine, serotonin and dopamine neuronal systems and corticotropin releasing factor in the development and expression of anxiety or fear.

Interest in the role that dysfunction of serotonin neurons may play in anxiety disorders has increased recently because of the development of anxiolytic drugs with specific actions on 5-HT_{1A}, 5-HT_2, and 5-HT_3 serotonin receptor subtypes.

This chapter will focus on the preclinical evidence suggesting that alterations in serotonin function may relate to the development of certain types of anxiety and the therapeutic effectiveness of specific antianxiety drugs. The recent clinical investigations which have assessed serotonin function in two human anxiety disorders, generalized anxiety disorder and panic disorder,[1] before and during antianxiety and antipanic drug treatment will be discussed in this context.

NEUROANATOMICAL SUBSTRATES FOR ANXIETY OR FEAR AND THEIR RELATION TO THE SEROTONIN HYPOTHESIS OF ANXIETY

A comprehensive discussion of the neural substrates of anxiety or fear is beyond the scope of this chapter. However, briefly outlined below is neuroanatomical, neurophysiological and neurochemical evidence supporting a role for the amygdala, hippocampus, locus coeruleus, and cortex in the development of anxiety or fear and how it may relate to serotonin hypotheses of anxiety.[2]

From a neuroanatomical perspective, the *amygdala* is particularly well suited to be an important factor in the mediation of anxiety or fear. The amygdala receives afferents from cortical and thalamic extereoceptive systems, as well as from subcortical visceral afferent pathways. Projections from the amygdala go to autonomic pathways

[a]We thank Evelyn Testa for her expert secretarial assistance.
[b]To whom correspondence should be sent.

(lateral hypothalamus), neurohumoral pathways (paraventricular—supraoptic region, medial basal hypothalamus) and skeletal motor systems (ventral tegmental area→caudate, putamen, substantia nigra).[3–5] The amygdala may, therefore, serve as a general homeostatic function evaluating the significance of input from extereoceptive and intereoceptive sources and initiating behavioral and visceral responses accordingly. Amygdalectomy has been shown to reduce the response to threatening stimuli in monkeys, increase punished responding in rats, and block the conditioned fear induced by the potentiated startle response.[6–12] The presence of 5-HT_{1A}, receptors on the amygdala suggest these receptors may relate to the anxiety-fear regulating functions of the amygdala.[13] Therefore, the amygdala may be one of the brain sites involved in the anxiolytic mechanism of action of 5-HT_{1A} agonists, such as buspirone and gepirone.

The *hippocampus* has been hypothesized to play a key position in anxiety development. The septohippocampal system has connections with limbic structures and cortical sensory areas. Lesions increase punished responses like anxiolytic drugs.[14] However, lesions of this area have had equivocal effects on classically conditioned fear and do not consistently affect the anxiolytic properties of antianxiety drugs.[2,14] High densities of 5-HT_{1A} receptors are located in the hippocampus suggesting that the hippocampus may also be important in the antianxiety effects of 5-HT_{1A} receptor agonist anxiolytics.[13]

There is strong support, which has been reviewed in detail elsewhere, for the role of the *locus coeruleus–norepinephrine* system in the development of anxiety.[15,16] The locus coeruleus sends projections to other areas involved in fear or anxiety like the thalamus, cerebral cortex, amygdala, hippocampus, and hypothalamus. In addition, the locus coeruleus receives afferents which suggest a critical role in responses both to the external and internal environment.[17] For example, alterations in blood pressure, body temperature, distension of internal organs such as the bladder, colon and stomach, all activate the locus coeruleus.[18] Other evidence supporting the role for the locus coeruleus in anxiety or fear include the observation that drugs which activate the locus coeruleus are anxiogenic and drugs which decrease its function are anxiolytic.[16–19] In monkeys, stimulation of the locus coeruleus produces fearlike behavior and lesions result in behavior consistent with reduced fear responses.[19] In freely moving cats threatening stimuli results in an activation of the locus coeruleus.[20,21] The activity of the locus coeruleus is highly regulated with benzodiazepine, serotonin, and opiate receptors having inhibitory effects and VIP, CRF, substance P, and acetylcholine resulting in activation.[15,16] The ability of serotonin to decrease locus coeruleus firing appears to be mediated by 5-HT_2 receptors and suggests interactions between noradrenergic and serotonergic systems may be relevant to the expression and treatment of anxiety or fear.[22–24] It is possible that the antipanic properties of 5-HT reuptake inhibitors may relate to the ability of increased synaptic 5-HT to reduce noradrenergic neuronal activity.

It is probable that the sensory processing areas of the *neocortex* linked with the limbic forebrain are particularly important in the interpretation of sensory stimuli from emotionally significant events such as those involving threat.[2] Electrical stimulation of serotonin raphe nuclei produce well circumscribed increases in cortical glucose use in frontal motor, frontal sensorimotor, and frontoparietal somatosensory cortices. It has been suggested that this effect may relate to the serotonergic innervation found during the ontongeny of the somatosensory cortex.[25] The findings of high cortex concentrations of 5-HT_2 receptors suggest this receptor may be involved in the contribution of the neocortex to anxiety or fear. The recent reports of anxiolytic efficacy for the 5-HT_2 receptor antagonist, ritanserin, is relevant to this hypothesis.[26]

ANIMAL BEHAVIORAL MODELS OF ANXIETY AND FEAR: RELATIONSHIP TO SEROTONIN FUNCTION

Conflict Paradigms

The standard conflict paradigm consists of training laboratory animals in operant tasks with food reward and then introducing short signal periods during which responding for food is rewarded but also punished by a mild electric shock. Animals stop responding during these short periods exhibiting a so-called punishment-induced suppression. The punishment-induced blockade of ongoing behavior has been shown to be a useful method of identifying anxiolytic agents. Indeed, antianxiety drugs, such as the benzodiazepines and 5-HT_{1A} agonists, are the only group of agents consistently able to increase punished responding. In addition, there is a good correlation between the minimally effective antipunishment drug dose in animals and the average daily doses used in the treatment of anxiety. A variety of experimental paradigms have resulted in findings consistent with the hypothesis that depressed serotonin neural transmission is associated with attenuation of punishment-induced inhibition.[23,27–30] In contrast, the administration of serotonin agonists or electrical stimulation of the raphe nuclei do not consistently alter punished responding.[27]

Novelty-Induced Inhibition

Novel stimuli may increase the general level of fear. Often, but not invariably, one consequence of novelty is an inhibition of behavior. This includes inhibition of food intake in new surroundings, reduction of ambulation or exploratory behavior, and suppression of social interactions between a pair of rats. Antianxiety drugs generally release these forms of behavioral inhibition thus increasing eating or drinking, exploratory behavior and social interactions in unfamiliar situations.[14] Destruction of serotonergic neurons, like benzodiazepines, preferentially increase social contacts in animals placed in unfamiliar situations. Serotonin innervation of the lateral septum appears to be critically involved in this effect.[31] Very few reports are available to suggest that depression of serotonin transmission increases locomotor exploratory behavior in a novel situation.[27,31,32]

Nonrewarding-Induced Inhibition

Analyzing animal behavior under nonrewarding conditions may shed further light on whether serotonin neurons play an important role in the control of anxiety. This assumption derives from the concept that in animals, behavioral suppression under nonreward conditions is one possible adaptive response for avoiding frustration. Nonreward inhibition is highly sensitive to benzodiazepines. These drugs have been shown to increase the number of nonreward responses in the early trials of a nonrewarded session when the level of reward expectation is high. The effects of decreasing serotonin transmission has been inconsistent in regard to bolstering nonrewarding-induced inhibition. From these data is has been suggested one can extract the idea that a crucial dimension for the involvement of serotonin neurons under conditions of nonreward is the animal's level of control over reward delivery or its level of reward expectancy.[27]

Fear Potentiated Startle Reflex

The fear-potentiated acoustic startle reflex has shown the potential to be a useful model of conditioned fear and a method to identify anxiolytic drugs. The acoustic startle response, characterized by a distinctive rapidly occurring pattern of whole-body muscular responses, is elicited by a sudden auditory stimulus. The magnitude of the startle response in rats can be increased by eliciting startle in the presence of a cue (e.g., a light) that was previously paired with a footshock. Thus, after an initial training session, presentation of the light is a conditioned stimulus that presumably evokes a central fear state, which in turn produces an elevation of the acoustically elicited startle response (i.e., the fear-potentiated startle effect). Consistent with a fear interpretation, anxiolytic compounds such as sodium amytal,[33] diazepam,[34,35] or morphine[36] decrease the magnitude of fear-potentiated startle. In addition, fear-potentiated startle is also blocked by lesions of the central nucleus of the amygdala, a structure that has been implicated in fear in a variety of test situations.[9]

Buspirone and gepirone, but not their common metabolite, 1-PP, produced a dose-dependent reduction of fear-potentiated startle.[37] To evaluate the role of various neurotransmitter systems in mediating buspirone's blockade of the fear-potentiated startle effect, lesions of the dorsal and median raphe nuclei or IP injections of the serotonin antagonists cinanserin or cyproheptadine did not alter fear-potentiated startle, nor did these treatments prevent buspirone from blocking fear-potentiated startle. The 5-HT_{1A} agonist 8-OH-DPAT did not block fear-potentiated startle even at doses that produced a marked "5-HT syndrome." Another 5-HT_{1A} agonist, ipsapirone, blocked potentiated startle only at a very high dose. *p*-Chlorophenylalanine and *p*-chloroamphetamine did not alter fear-potentiated startle. Taken together, the data do not support the hypothesis that buspirone's anxiolytic effects are mediated by actions at 5-HT_{1A} receptors and more generally indicate that serotonergic neurons do not play an important role in fear-potentiated startle.[38]

TABLE 1. Animal Models of Anxiety: Relationship to Serotonergic Function[a]

Model	Involvement of Serotonergic Neurons
Conflict Paradigms	+ +[b]
Novelty-induced behavioral inhibition:	
Hypophagia	+
Exploratory Behavior	−
Nonreward-induced behavioral inhibition	+ −
Fear-potentiated acoustic startle	−

[a]See text for details.
[b]+ + Strong; + Moderate; + − Inconsistent; − Minimal.

The reason for the apparent lack of involvement of serotonin in potentiated startle relative to these other paradigms is currently not known. One plausible explanation has been provided by Soubrie,[27] who suggested that serotonin may not be involved specifically in anxiety, but more generally in the inhibition of inappropriate responses. Thus, in animal models where anxiety results in the animal's withholding responses to avoid punishment (i.e., not licking a water spout; not jumping off a platform onto an electrified grid), decreases in serotonergic function might generally diminish the ability of the animal to withhold that response. In the startle paradigm the potentiation of an elicited reflex response serves as the index of fear rather than the withholding of a response (TABLE 1).

THE EFFECT OF ANTIPANIC AND ANTIGENERALIZED ANXIETY DRUGS ON BRAIN SEROTONIN FUNCTION

Non-Benzodiazepine Antipanic Drugs

Tricyclic drugs (e.g., imipramine), serotonin reuptake inhibitors (e.g., fluvoxamine), and monoamine oxidase inhibitors (e.g., phenelzine) and documented antipanic medications. Neurophysiological studies have consistently indicated that these drugs increase serotonin neurotransmission with chronic administration. This appears to occur via at least two separate mechanisms. Specific serotonin reuptake inhibitors, such as fluoxetine or fluvoxamine, and monoamine oxidase inhibitors produce a decreased sensitivity of the serotonin autoreceptor, whereas a spectrum of other drug treatments including tricyclic norepinephrine reuptake inhibitors such as desipramine and imipramine produce postsynaptic serotonin supersensitivity by a yet undetermined mechanism. This increase in serotonin neurotransmission has not been associated with an increase in the density of 5-HT_1 receptors. In contrast, 5-HT_2 receptors are reduced by long-term administration of most antipanic and antidepressant treatments.[39–42]

The neurophysiological and binding studies of antipanic and antidepressant drug action must be considered in the context of the serotonin receptor subtypes which have been evaluated and the functional interaction between 5-HT_1 and 5-HT_2 receptors. The 5-HT_1 receptor subtypes can be differentiated physiologically in the dorsal raphe nucleus where 5-HT_{1A} but not 5-HT_{1B} ligands mimic the inhibitory action of serotonin.[43] It is this receptor, the 5-HT_{1A} receptor, that is subsensitive after long-term treatment with 5-HT reuptake inhibitors and monoamine oxidase inhibitors.

In contrast to the dorsal raphe, where the 5-HT_{1A} ligands are full agonists, in the hippocampus, these compounds are partial agonists or antagonists.[44] Neurophysiological investigations indicate hippocampal serotonin receptors become supersensitive after prolonged treatment with nonbenzodiazepine antipanic drugs. The enhancement of postsynaptic sites have also been shown to occur in the lateral geniculate, the amygdala, the facial motor nucleus, the superchiasmatic nucleus, and the somatosensory cerebral cortex.[42]

Neurophysiological studies have demonstrated that in brain areas where 5-HT_1 and 5-HT_2 receptors exist, the 5-HT_2 receptor reduces inhibitory and enhances excitatory effects produced by activation of the 5-HT_1 receptor. Of particular relevance is the finding in the prefrontal cortex that 5-HT_2 receptor activation works in opposition to the depressant effects of serotonin on cell firing.[45] Therefore, in prefrontal cortex, the antipanic drug induced decrease in 5-HT_2 receptors may be associated with a potentiation of the depressant effects of serotonin on cell firing.

Behavioral paradigms of serotonin function have been less helpful in assessing effects of different drugs on serotonin transmission in the brain. Most of the behavioral paradigms measure the function of 5-HT_2 receptors which are distinct from the 5-HT_1 receptors studied neurophysiologically. In addition, behaviors assessed are primarily mediated in the brain stem and the spinal cord, and not the forebrain, which is most relevant to antipanic and antidepressant properties and where many of the neurophysiological studies have been conducted.[46]

Anti-Generalized–Anxiety Drugs

Benzodiazepine drugs and 5-HT_{1A} receptor agonists, such as buspirone and gepirone, reduce serotonergic neurotransmission when given acutely. This is based upon electrophysiological studies in which these drugs decrease the firing of serotonergic cells in the dorsal raphe nucleus and decrease 5-HT turnover in brain regions such as the hippocampus.[43] In addition, it has recently been demonstrated that the utilization of (^{3}H)2-deoxyglucose was decreased in the dorsal and median raphe nuclei following alprazolam or buspirone.[47] There is evidence that tolerance may develop to the 5-HT reducing effects of benzodiazepines with chronic treatment.[48–51]

There has been one study assessing the effect on serotonin function of longer term 5-HT_{1A} receptor agonist treatment.[42] Gepirone acutely reduces the firing rate of dorsal raphe serotonin neurons followed by a gradual return to normal with 14 days of treatment. The responsiveness of dorsal raphe neurons to LSD, 5-HT, 8-OH-DPAT and gepirone, but not to GABA, is decreased after 14 days of gepirone treatment. A desensitization of the somatodendritic autoreceptor can thus account for the gradual recovery of the firing rate serotonin neurons. This may provide an explanation for the delayed onset of action of drugs such as buspirone and gepirone in generalized anxiety disorder. From this data, and from the fact that postsynaptic serotonin receptor function is unchanged by long-term gepirone administration, it has been hypothesized that serotonin neurotransmission is increased. This may result from normal activity of serotonin neurons and normal release of serotonin combined with tonic activation by gepirone and presumably other 5-HT_{1A} agonist drugs or normosensitive postsynaptic serotonin receptors[42] (TABLE 2).

TABLE 2. The Net Effect of Anti-Generalized–Anxiety and Antipanic Drugs on Serotonin Function[a]

	Acute[b]	Chronic
Benzodiazepines	↓	0
5HT1A Agonists (Buspirone)	↓	?↑
Tricyclics	0	↑
MAOIs	0	↑
Specific 5HT Reuptake Inhibitors	↑	↑

[a]See text for details.

[b]Acute (few days or less) and chronic (at least few weeks) refer to duration of drug administration; ↓Decreased; 0 No major effect; ?↑ Increased in one study, but replication required; ↑ Increased.

SUMMARY OF PRECLINICAL STUDIES

The neuroanatomical distribution of the serotonin neuronal system is consistent with an important role in the regulation of anxiety or fear behavior. The presence of specific serotonin receptors in brain regions such as the amygdala, hippocampus, locus coeruleus, and neocortex suggest specific sites for serotonin regulation of anxiety. In particular, the high density of 5-HT_{1A} receptors in the amygdala and hippocampus indicate these areas may be important in the antianxiety actions of the 5-HT_{1A} agonists such as buspirone and gepirone.[52] The high concentration of 5-HT_2 receptors in the neocortex raise the possibility that effects in this region may account for the putative anxiolytic properties of 5-HT_2 receptor antagonists such as ritanserin.[53] The functional interactions between the serotonin and norepinephrine systems may be relevant to the antipanic actions of potent and selective serotonin reuptake inhibitors.

The preclinical studies of behavioral models of anxiety and fear have provided an inconsistent picture of the relationship between serotonin function and different forms of anxiety. The inconsistency is due, in part to the differences that emerge between the effects of anxiolytic drugs and serotonin depletion in the different models and the fact that the presence of anxiogenic factors is neither sufficient or necessary to involve serotonin neurons. The probability that the different animal models are reflective of different types of anxiety or fear and that serotonin neurons are involved in some, but not all, forms of anxiety contributes to the confusion.

The neurobiological investigations of the mechanism of action of antipanic and non-benzodiazepine antigeneralized anxiety drugs suggest that drugs whose anxiolytic effects relate to actions on serotonin function (tricyclics, MAOIs, specific serotonin reuptake inhibitors, perhaps 5-HT_{1A} agonists) may be increasing net serotonin function, an hypothesis tantamount to a reversal of the original hypothesis that the ability of benzodiazepine drugs to reduce serotonin activity was therapeutically important (TABLE 2). The therapeutic mechanism of action of benzodiazepine drugs which are effective in the treatment of both panic disorder and generalized anxiety disorder, is probably related to actions on benzodiazepine-GABA receptors.

CLINICAL INVESTIGATIONS

Studies of Serotonin Function in Patients with Anxiety Disorders

There have been relatively few clinical investigations of serotonin function in patients with anxiety disorders (TABLE 3). In one study designed to evaluate serotonin function the effects of intravenous tryptophan on prolactin secretion were determined in panic disorder patients and compared to healthy subjects.[54] The ability of tryptophan to increase prolactin levels, which has been shown to reflect serotonin function, was not different between the patients and healthy subjects.

Another method used to assess the possible role of serotonin function in the development of panic anxiety, has been the measurement of behavioral and biochemical responses to the mixed serotonin receptor agonist, *m*-chlorophenylpiperazine (MCPP). MCPP has, at least, moderate affinity for both 5-HT_1 and 5-HT_2 receptors making it difficult to ascribe MCPP effects to a specific serotonin receptor subtype.[55] Studies in rhesus monkeys and healthy human subjects have shown that the 5-HT_2 and 5-HT_{1C} receptor antagonist, ritanserin, attenuates the neuroendocrine and behavioral effects of MCPP.[56,57]

In a recently reported investigation, intravenous MCPP (0.1 mg/kg over 20 minutes) had anxiogenic effects in both the healthy subjects and panic disorder patients. Panic attacks meeting DSM-III criteria, occurred following MCPP in approximately 45% of the patients, and 30% of the healthy subjects (nonsignificant).[58] Other ratings of anxiety did not distinguish the two groups. MCPP resulted in significant but similar increases in cortisol, prolactin and growth hormone in the healthy subjects and patients. The result of this investigation suggested that serotonin

TABLE 3. Clinical Investigations of Serotonergic Function in Panic Disorder Patients[a]

Paradigm	Dose	Method of Administration	Finding
1. Tryptophan-induced prolactin response	100 mg/kg	Intravenous	Prolactin response similar in patients and healthy subjects.
2. MCPP-induced changes in behavior and neuroendocrine function	0.1 mg/kg	Intravenous	Similar increases in cortisol, prolactin, growth hormone, and anxiety in patients and healthy subjects.
	0.25 mg/kg	Oral	Greater increases in cortisol and anxiety in patients.
3. 5-HTP-induced changes in behavior and neuroendocrine function	60 mg	Intravenous	Similar increases in cortisol and β-endorphin in patients and healthy subjects. Behaviorally, healthy subjects seemed more sensitive to 5-HTP than patients. Anxiety was not elicited in either group.
4. Fenfluramine induced changes in behavior and neuroendocrine function	60 mg	Oral	Greater increases in cortisol, prolactin and anxiety in patients.
5. Platelet imipramine binding	—	—	Similar in patients and healthy subjects in 4 of 5 studies.
6. Platelet serotonin uptake and peripheral serotonin levels	—	—	One study found increased B_{max} in patients; one study reported decreased 5-HT in patients; another, no difference in platelet 5-HT

[a]See text for details.

neuronal dysfunction may not be of etiologic significance in most panic disorder patients. However, the observed anxiogenic properties of MCPP suggested that additional studies of the role of serotonin systems in the pathophysiology of human anxiety disorders were indicated.

In a subsequent investigation, low oral doses of MCPP (0.25 mg/kg) were reported to increase anxiety and produce panic attacks in panic disorder patients with or without major depression. In normal controls and in patients with major depression but without panic disorder, the drug did not influence anxiety levels. The release of cortisol after MCPP was also augmented in panic disorder patients. This has led to a hypothesis that hypersensitivity of postsynaptic 5-HT receptors may exist in some panic disorder patients.[59]

Serotonin function in panic disorder patients has most recently been investigated by determining the behavioral and neuroendocrine responses to intravenous 5-hydroxy tryptophan (5-HTP), a serotonin precursor, and fenfluramine, an indirect serotonin agonist. 5-HTP (60 mg over 30 min) induced similar increases in cortisol and β-endorphin in panic disorder patients and healthy subjects and did not induce panic anxiety. These data support the intravenous MCPP findings that the sensitivity of serotonin neurons may not be aberrant in most panic disorder patients.[60] However, it has been reported that fenfluramine produces greater anxiogenic responses and increased stimulation of prolactin and cortisol secretion in panic disorder patients compared to healthy subjects and depressed patients.[61] Thus, considered together, the challenge paradigms in panic disorder patients using serotonin agonists are yielding inconsistent results regarding the functional sensitivity of the serotonin system in this disorder.

Radioactive imipramine has been shown to label a saturable high affinity binding site in brain and platelet. Several lines of evidence suggest that this binding site is associated with, although not identical to, the serotonin reuptake site. Most studies indicate depressed patients have reduced platelet imipramine binding.[62] However, in 4 of 5 studies platelet imipramine binding is normal in panic disorder patients.[63–67]

One study has measured platelet serotonin uptake in patients with panic attacks, and found higher V_{max} values in the patient group than in the controls while the affinity constant K_m was not significantly different.[68] The authors suggested that there may be a specific abnormality of platelet serotonin uptake in patients with panic attacks and the increased platelet reuptake of serotonin was consistent with a report of lower concentrations of this neurotransmitter in the plasma of panic disorder patients.[69] It should be noted, however, that the great bulk of plasma serotonin is in platelets and this serotonin is believed to originate mainly from chromaffin cells in the gut walls. In another study, platelet serotonin levels were found to be normal in patients with panic disorder compared to healthy subjects.[70]

Unfortunately, serotonin function has not been evaluated in patients with generalized anxiety disorder. Since the 5-HT_{1A} agonists are effective for GAD[71] and not PD,[72] it is perhaps more likely that serotonin dysfunction is of etiologic relevance in GAD. Studies of serotonin function in GAD by our research group are in progress

Efficacy of Serotonin-Specific Drugs in Anxiety Disorders

The most convincing clinical observations implicating serotonin systems in the pathophysiology of anxiety disorders remains the demonstrated anxiolytic properties of serotonin-specific drugs.

Serotonin Reuptake Inhibitors

There is emerging evidence that serotonin reuptake inhibitor drugs have efficacy in the treatment of panic disorder. Both zimelidine and chlorimipramine have been shown to be effective antipanic agents in placebo controlled investigations.[73,74] In addition, fluvoxamine was demonstrated to be effective in clinical studies comparing it to chlorimipramine and maprotoline.[75] Fluoxetine has been reported to have efficacy in some panic disorder patients in an open study.[76]

5-HT_{1A} *Receptor Agonists*

The 5-HT_{1A} agonists, buspirone and gepirone, are effective agents for Generalized Anxiety Disorder after several weeks of administration.[77,78] However, recent work indicates that buspirone is not an effective antipanic agent.[72]

5-HT_2 *Receptor Antagonists*

The role for serotonin in anxiolytic efficacy is supported by work indicating ritanserin, a 5-HT_2 receptor antagonist, has antianxiety properties.[26,53] However, ritanserin also appears ineffective in panic disorder.[75] This suggests 5-HT_{1A} and 5-HT_2 receptors may not be primarily involved in the pathogenesis of panic disorder.

5-HT_3 *Receptor Antagonists*

Very recent reports have attributed antianxiety properties to selective 5-HT_3 antagonists.[79] If these clinical findings are confirmed and extended, the 5-HT_3 receptor antagonists may represent a new class of antianxiety drugs.

Clinical Studies of the Mechanism of Serotonin-Specific Drugs in Anxiety Disorders

There are few clinical studies assessing the effects of antianxiety drugs on serotonin function. Both chlorimipramine and fluvoxamine have been demonstrated to enhance serotonin function based upon their ability to enhance the prolactin response to intravenous tryptophan.[80,81]

It is possible that some of the therapeutic effects of the 5-HT reuptake inhibitors are due to actions on other transmitter systems besides serotonin. These compounds appear to decrease noradrenergic function because serotonin is inhibitory to the firing of noradrenergic neurons such as those of the locus coeruleus. For example, fluvoxamine has been shown to lower CSF MHPG levels, a norepinephrine metabolite. In addition, some of the 5-HT inhibitors have shown in laboratory animals to downregulate beta adrenergic postsynaptic receptors.[82]

Long-term alprazolam and diazepam treatment has no effect on the prolactin response to tryptophan, suggesting no net increase or decrease in serotonin function.[53,54] In healthy subjects, a single diazepam dose attenuates the prolactin rise produced by tryptophan suggesting a reduction in serotonin function occurs following acute but not chronic benzodiazepine treatment.[51] It has been speculated that benzodiazepine actions which become tolerant, such as sedation, are related to actions on serotonin neurons rather than antianxiety effects. There have been no reported clinical studies of the effects of acute or chronic administration of 5-HT_{1A} agonists, 5-HT_2, and 5-HT_3 receptor antagonists on serotonin function.

CONCLUDING COMMENTS

This chapter has reviewed evidence suggesting possible roles for serotonin neurotransmission in the development and treatment of anxiety states. While preclinical neuroanatomical, neurophysiological, and behavioral studies support the concept, the specific nature of the involvement of serotonin in anxiety states remains to be defined. This may be related to several factors, including the existence of serotonin receptor subtypes with different functional properties, behavioral models of anxiety which reflect different types of anxiety or fear, and the paucity of studies comparing the acute and chronic effects of anxiolytic drugs.

At present, clinical investigations of the pathophysiology of anxiety disorders permit only preliminary conclusions. The demonstration that MCPP elicits anxiety is provocative and supportive of the hypothesis that serotonin dysfunction may be anxiogenic. However, there are not replicated reports of an abnormality in serotonin function in either panic disorder or generalized anxiety disorder. The demonstration that serotonin reuptake inhibitor drugs are antipanic is not consistent with the hypothesis of hypersensitive serotonin receptor function in panic disorder because chronic administration of these drugs (e.g., when anxiolytic action occurs) appear to enhance serotonin neurotransmission. The preliminary observation that the 5-HT_2 antagonist ritanserin is not antipanic also fails to support this hypothesis.

The net effect of 5-HT_{1A} agonists on serotonin function, such as buspirone and gepirone, should be studied in humans. It needs to be determined whether the ability of these drugs to acutely reduce serotonin firing is translated into reduced or increased serotonin neurotransmission with chronic treatment when therapeutic actions are commencing. Since these are drugs effective for generalized anxiety disorder but not panic disorder patients, it raises the possiblity that serotonin dysfunction, particularly involving 5-HT1A receptors, may occur in generalized anxiety disorder. Furthermore, the preliminary observations suggesting the efficacy of 5-HT_2 and 5-HT_3 receptor antagonists for generalized anxiety disorder indicate that the function of these receptors needs evaluation in a spectrum of anxiety disorder patients.

REFERENCES

1. DIAGNOSTIC AND STATISTICAL MANUAL OF MENTAL DISORDERS, 3RD EDITION, REVISED. 1987. American Psychiatric Association Press, Washington, D.C.
2. LEDOUX, J. E. 1987. Emotion. *In* Handbook of Physiology—The Nervous System V. V. B. Mountcastle, F. Plum and S. R. Geiger, Eds. Bethesda, MD: Am. Physiol. Soc.
3. JONES, E. G. & T. P. S. POWELL. 1970. An experimental study of converging sensory pathways within the cerebral cortex of the monkey. Brain **93:** 793–820.
4. MESULAM, M. M., G. VANHOESEN, D. N. PANDYA & N. GESCHWIND. 1977. Limbic and sensory connections of the IPL in the rhesus monkey. Brain Res. **136:** 293–414.
5. SWANSON, L. W. 1983. The hippocampus and the concept of the limbic system. *In* Neurobiology of the Hippocampus. W. Seifert, Ed. London: Academic Press.
6. BLANCHARD, D. C. & R. J. BLANCHARD. 1972. Innate and conditioned reactions to threat in rats with amygdaloid lesions. J. Comp. Physiol. Psychol. **81:** 281–290.
7. COHEN, D. H. 1989. The functional neuroanatomy of a conditioned response. *In* Neural Mechanisms of Goal-Directed Behavior and Learning. R. F. Thompson, L. H. Hicks and B. Shvyrkov, Eds. New York: Academic Press.

8. DOWNER, J. D. C. 1961. Changes in visual gnostic function and emotional behavior following unilateral temporal lobe damage in the "split-brain" monkey. Nature (Lond.) **191:** 50–51.
9. HITCHCOCK, J. M. & M. DAVIS. 1986. Lesions of the amygdala, but not of the cerebellum or red nucleus block conditioned fear as measured with the potentiated startle paradigm. Behav. Neurosci. **100:** 11–22.
10. IWATA, J., J. E. LEDOUX, M. P. MEELEY, S. ARNERIC & D. J. REIS. 1986. Intrinsic neurons in the amygdaloid field projected to by the medial geniculate body mediate emotional responses conditioned to acoustic stimuli. Brain Res. **383:** 195–214.
11. KAPP, B. S., R. C. FRYSINGER, M. GALLAGHER & J. HASELTON. 1979. Amygdala central nucleus lesions: Effects on heart rate conditioning in the rabbit. Physiol. Behav. **23:** 1109–1117.
12. KAPP, B. S., J. P. PASCOE & M. A. BIXLER. 1984. The amygdala: A neuroanatomical systems approach to its contributions to aversive conditioning. *In* The Neuropsychology of Memory. N. Butters and L. R. Squire, Eds. New York: Guilford.
13. PAZOS, A., A. PROBST & J. M. PALACIOS. 1987. Serotonin receptors in the human brain. III. Autoradiographic mapping of serotonin-1 receptors. Neuroscience **21:** 97–122.
14. GRAY, J. A. 1982. *Precis of* The neuropsychology of anxiety: An enquiry into the functions of the septo-hippocampal system. Behav. Brain Sci. **5:** 469–534.
15. CHARNEY, D. S., G. R. HENINGER & A. BRIER. 1984. Noradrenergic function in panic anxiety: Effects of yohimbine in healthy subjects and patients with agoraphobia and panic disorder. Arch Gen. Psychiatry **41:** 751–763.
16. CHARNEY, D. S., S. W. WOODS, W. K. GOODMAN & G. R. HENINGER. 1987. Neurobiological mechanisms of panic anxiety: Biochemical and behavioral correlates of yohimbine-induced panic attacks. Am. J. Psychiatry **144:** 1030–1036.
17. MOORE, R. Y. & F. E. BLOOM. 1979. Central catecholamine neuron systems: Anatomy and physiology of the norepinephrine and epinephrine systems. Ann. Rev. Neurosci. **2:** 113–168.
18. SVENNSSON, T. H. 1987. Peripheral, autonomic regulation of locus coeruleus noradrenergic neurons in brain: Putative implications for psychiatry and pharmacology. Psychopharmacology **92:** 1–7.
19. REDMOND, D. E., JR. 1979. New and old evidence for the involvement of a brain norepinephrine system in anxiety. *In* The Phenomenology and Treatment of Anxiety. W. E. Fann, Ed. New York: Spectrum Press.
20. RASMUSSEN, K., D. A. MORILAK & B. L. JACOBS. 1986. Single unit activity of locus coeruleus neurons in the freely moving cat. I. During naturalistic behaviors and in response to simple and complex stimuli. Brain Res. **371:** 324–334.
21. RASMUSSEN, K. & B. L. JACOBS. 1986. Single unit activity of locus coeruleus neurons in the freely moving cat. II. Conditioning and pharmacologic studies. Brain Res. **371:** 335–344.
22. RASMUSSEN, K. & G. K. AGHAJANIAN. 1986. Effect of hallucinogens on spontaneous and sensory-evoked locus coeruleus unit activity in the rat: Reversal by selective 5-HT_2 antagonists. Brain Res. **385:** 395–400.
23. RASMUSSEN, K., R. A. GLENNON & G. K. AGHAJANIAN. 1986. Pehenethylamine hallucinogens in the locus coeruleus; potency of action correlates with rank order of 5-HT_2 binding affinity. Eur. J. Pharmacol. **132:** 79–89.
24. GOREA, E. & J. ADRIEN. 1988. Serotonergic regulation of noradrenergic coeruleuan neurons: Electrophysiological evidence for the involvement of 5-HT_2 receptors. Eur. J. Pharmacol. **154:** 285–291.
25. CUDENNEC, A., D. DUVERGER, A. SERRANO, B. SCATTON & E. T. MACKENZIE. 1988. Influence of ascending serotonergic pathways on glucose use in the conscious rat brain. II. Effects of electrical stimulation of the rostral raphe nuclei. Brain Res. **444:** 227–246.
26. REYNTJENS, A., Y. GELDERS, M-L. J. A. HOPPENBRAVERS & G. VANDEN BUSSCHE. 1986. Thymostenic effects of ritanserin, a centrally acting Serotonin S-2 blocker. Drug Dev. Res. **8:** 205–211.
27. SOUBRIE, P. 1986. Reconciling the role of central serotonin neurons in human and animal behavior. Behav. Brain Sci. **92:** 319–364.
28. THIEBOT, M. H., M. HAMON & P. SOUBRIE. 1982. Attenuation of induced-anxiety in rats by

chlordiazepoxide: Role of raphe dorsalis benzodiazepine binding sites and serotonergic neurons. Neuroscience **7:** 2287–2294.

29. Thiebot, M. H., M. Hamon & P. Soubrie. 1983. The involvement of nigral serotonin innervation in the control of punishment-induced behavioral inhibition in rats. Pharmacol. Biochem. Behav. **19:** 225–229.
30. Wise, C. D., B. D. Berger & L. Stein. 1972. Benzodiazepines: Anxiety reducing activity by reduction of serotonin turnover in the brain. Science **177:** 180–183.
31. Clarke, A. & S. E. File. 1982. Selective neurotoxin lesions of the lateral septum: Changes in social and aggressive behaviors. Pharmacol. Biochem. Behav. **17:** 623–628.
32. Beninger, R. J. 1984. Effects of metergoline and quipazine on locomotor activity of rats in novel and familiar environments. Pharmacol. Biochem. Behav. **20:** 701–705.
33. Chi, C. I. 1965. The effect of amobarbital sodium on conditioned fear as measured by the potentiated startle response in rats. Psychopharmacologia **7:** 115–122.
34. Davis, M. 1979. Diazepam and flurazepam: Effects on conditioned fear as measured with the potentiated startle paradigm. Psychopharmacology **62:** 1–7.
35. Berg, W. K. & M. Davis. 1984. Diazepam blocks fear-enhanced startle elicited electrically from the brainstem. Physiol Behav. **32:** 333–336.
36. Davis, M. 1979. Morphine and naloxone: Effects on conditioned fear as measured with the potentiated startle paradigm. Eur. J. Pharmacol. **54:** 341–347.
37. Kehne, J. H., J. V. Cassella & M. Davis. 1988. Anxiolytic effects of buspirone and gepirone in the fear-potentiated startle paradigm. Psychopharmacology **94:** 8–13.
38. Davis, M., J. V. Cassella & J. H. 1988. Serotonin does not mediate anxiolytic effects of buspirone in the fear-potentiated startle paradigm: Comparison with 8-OH-DPAT and ipsapirone. Psychopharmacology **94:** 14–20.
39. Blier, P., C. deMontigny & Y. Chaput. 1987. Modifications of the serotonin system by antidepressant treatments: Implications for the therapeutic response in major depression. J. Clin. Psychopharmacol. **7:** 24S–35S.
40. Charney, D. S., D. B. Menkes & G. R. Heninger. 1981. Receptor sensitivity and the mechanism of action of antidepressant treatment. Implications for the etiology and therapy of depression. Arch. Gen. Psychiatry **38:** 1160–1180.
41. Heninger, G. R. & D. S. Charney. 1987. Mechanism of action of antidepressant treatments: Implications for the etiology and treatment of depressive disorders. *In* Psychopharmacology, A Generation of Progress. H. Y. Meltzer, Ed. New York: Raven Press.
42. Blier, P. & C. DeMontigny. 1987. Modification of 5-HT neuron properties by sustained administration of the 5-HT_{1A} agonist gepirone: Electrophysiological studies in the rat brain. Synapse **1:** 470–480.
43. Sprouse J. S. & G. K. Aghajanian. 1987. Electrophysiological responses of serotoninergic dorsal raphe neurons to 5-HT_{1A} and 5-HT_{1B} agonists. Synapse **1:** 3–9.
44. Sprouse, J. S. & G. K. Aghajanian. 1988. Responses of hippocampal pyramidal cells to putative serotonin 5-HT_{1A} and 5-HT_{1B} agonists: A comparative study with dorsal raphe neurons. Neuropharmacology **27**(7): 707–715.
45. Lakoski, J. M. & G. Aghajanian. 1985. Effects of ketanserin on neuronal responses to serotonin in the prefrontal cortex, lateral geniculate and dorsal raphe nucleus. Neuropharmacology **24:** 265–273.
46. Charney, D. S., L. H. Price & G. R. Heninger. The receptor sensitivity hypothesis of antidepressant action: An evaluation and recommendations for future investigations. In press.
47. Piercey, M. K. & W. E. Hoffman. 1989. The effects of buspirone and alprazolam on (^{3}H)2-deoxyglucose metabolism in raphe nuclei. Presented at International Symposium, Serotonin: From Cell Biology to Pharmacological Therapy.
48. Collinge, J., C. J. Pycock & P. V. Taberner. 1983. Studies on the interaction between cerebral 5-hydroxytryptamine and gamma-aminobutyric acid in the mode of action of diazepam in the rat. Br. J. Pharmacology **79:** 637–643.
49. Gonsalves, S. F. & D. W. Gallager. 1986. Tolerance to antipentylenetetrazol effects following chronic diazepam. Eur. J. Pharmacol. **121:** 181–184.
50. Lister, R. G. & S. E. File. 1983. Changes in regional concentrations in the rat brain of

5-hydroxytryptamine and 5-hydroxyindoleacetic acid during the development of tolerance to the sedative action of chlordiazepoxide. J. Phar. Pharmacol. **35:** 601–603.
51. NUTT, D. J. & P. J. COWEN. 1987. Diazepam alters brain 5-HT function in man: Implications for the acute and chronic effects of benzodiazepines. Psychol. Med. **17:** 601–607.
52. TRABER, J. & T. GLASER. 1987. 5-HT_{1A} receptor-related anxiolytics. TIPS **8:** 432–437.
53. CEULEMANS, D. L. S., M-L. J. A. HOPPENBROUWERS, Y. G. GELDERS & A. J. M. REYNTJENS. 1985. The influence of ritanserin, a serotonin antagonist, in anxiety disorders:A double-blind placebo-controlled study versus lorazepam. Pharmacopsychiatry **18:** 303–305.
54. CHARNEY, D. S. & G. R. HENINGER. 1986. Serotonin function in panic disorders: The effect of intravenous tryptophan in healthy subjects and patients with panic disorder before and during alprazolam treatment. Arch. Gen. Psychiatry **43:** 1059–1065.
55. KENNETT, G. A., P. WHITTON, K. SHAH & G. CURZON. 1989. Anxiogenic-like effects of MCPP and TFMPP in animal models are opposed by 5-HT_{1C} receptor antagonists. Eur. J. Pharmacol. **164:** 445–454.
56. HENINGER, G.R., D. S. CHARNEY & A. SMITH. 1987. Effects of serotonin receptor agonists and antagonists on neuroendocrine function in rhesus monkeys. Soc. Neurosci. Abstr. **224**(9): 801.
57. SEIBYL, J. P., D. S. CHARNEY, J.H. KRYSTAL, L. H. PRICE & G. R. HENINGER. 1989. 5-HT function in the biochemical response to MCPP. Presented at the American Psychiatric Association, 42nd Annual Meeting, San Francisco, CA.
58. CHARNEY, D. S., S. W. WOODS, W. K. GOODMAN & G. R. HENINGER. 1987. Serotonin function in anxiety. II. Effects of the serotonin agonist MCPP in panic disorder patients and healthy subjects. Psychopharmacology **92:** 14–24.
59. KAHN, R., G. ASNIS, S. WETZLER & H. VANPRAAG. 1988. Neuroendocrine evidence for a serotonin receptor supersensitivity in panic disorder. Psychopharmacology **96:** 360–364.
60. WESTENBERG, H. G. M. & J. A. DENBOER. 1989. Serotonin function in panic disorder: Effect of 1-5-hydroxytryptophan in patients and controls. Psychopharmacology **98:** 283–285.
61. TARGUM, S. D. & L. E. MARSHALL. 1989. Fenfluramine provocation of anxiety in patients with panic disorder. Psychiatry Res. **28:** 295–306.
62. LANGER, S. Z., A-M. GALZIN, M-F. POIRIER, H. LOO, D. SECHTER, *et al.* 1987. Association of ^{3}H-paroxetine binding with 5-HT transporter in brain and platelets: Relevance to studies in depression. J. Receptor. Res. **7:** 499–521.
63. INNIS, R. B., D. S. CHARNEY & G. R. HENINGER. 1987. Differential ^{3}H-Imipramine platelet binding in patients with panic disorder and depression. Psychiatry Res. **21:** 33–41.
64. LEWIS, D. A., R. NOYES, W. CORYELL & J. CLANCY. 1985. Tritiated imipramine binding to platelets is decreased in patients with agoraphobia. Psychiatry Res. **16:** 1–9.
65. NUTT, D. J. & S. FRASER. 1987. Platelet binding studies in panic disorder. J. Aff. Dis. **12:** 7–11.
66. SCHNEIDER, L. S., D. MUNJACK, J. A. SEVERSON & R. PALMER. 1987. Platelet [^{3}H]-Imipramine binding in generalized anxiety disorder, panic disorder, and agoraphobia with panic attacks. Biol. Psychiatry **22:** 59–66.
67. UHDE, T. W., W. H. BERRETTINI, P. P. ROY-BYRNE, J-P. BOULENGER & R. M. POST. 1987. Platelet [^{3}H]-Imipramine binding in patients with panic disorder. Biol. Psychiatry **22:** 52–58.
68. NORMAN, T. R., F. K. JUDD, M. GREGORY, R. H. JAMES, N. M. KIMBER, I. M. MCINTYRE & G. D. BURROWS. 1986. Platelet serotonin uptake in panic disorder. J. Aff. Dis. **11:** 69–72.
69. SCHNEIDER, P., L. EVANS, L. ROSS-LEE, B. WILTSHIRE, M. EADIE, J. KENARDY & H. HOEY. 1987. Plasma biogenic amine levels in agoraphobia with panic attacks. Pharmacopsychiatry **20:** 102–104.
70. BALON, R., R. POHL, V. YERAGANI, J. RAINEY & G. F. OXENKRUG. 1987. Platelet serotonin levels in panic disorder. Acta Psychiatr. Scand. **75:** 315–317.
71. TAYLOR, D. P. 1988. Buspirone, a novel approach to the treatment of anxiety. FASEB J. **2:** 2445–2452.

72. SHEEHAN, D. V., A. B. RAJ, K. H. SHEEHAN & S. SOTO. The relative efficacy of buspirone, imipramine, and placebo: A preliminary report. Pharmacol. Biochem. Behav. **29:** 815–817.
73. EVANS, L., J. KENARDY, P. SCHNEIDER & H. HOEY. 1986. Effect of a selective serotonin uptake inhibitor in agoraphobia with panic attacks. Acta Psychiatr. Scand. **73:** 49–53.
74. WESTENBERG, H. G. M. & J. A. DENBOER. 1989. Selective monoamine uptake inhibitors and a serotonin antagonist in the treatment of panic disorder. Psychopharmacol. Bull. **25**(1): 119–123.
75. DENBOER, J. A. & H. G. M. WESTENBERG. 1988. Effect of a serotonin and noradrenaline uptake inhibitor in panic disorder: A double-blind comparative study with fluvoxamine and maprotiline. Int. Clin. Psychopharmacol. **3:** 59–74.
76. GORMAN, J. M., M. R. LIEBOWITZ, A. J. FYER, D. GOETZ, R. B. CAMPEAS, M. R. FYER, S. O. DAVIES & D. F. KLEIN. 1987. An open trial of fluoxetine in the treatment of panic attacks. J. Clin. Psychopharmacol. **7:** 329–332.
77. CSANALOSI, I., E. SCHWEIZER, W. G. CASE & K. RICKELS. 1987. Gepirone in Anxiety: A pilot study. J. Clin. Psychopharmacol. **7:** 31–33.
78. GOA, K. L. & A. WARD. Buspirone. A preliminary review of its pharmacological properties and therapeutic efficacy as an anxiolytic. Drugs **32:** 114–129.
79. JONES, B. J., B. COSTALL, A. M. DOMENY, M. E. KELLY, R. J. NAYLOR, *et al.* 1988. The potential anxiolytic activity of GR38032F, a 5-HT_3-receptor antagonist Br. J. Pharmacol. **93**(4): 985–993.
80. ANDERSON, I. M. & P. J. COWEN. 1986. Clomipramine enhances prolactin and growth hormone responses to L-tryptophan. Psychopharmacology **89:** 131–133.
81. PRICE, L. H., D. S. CHARNEY, P. L. DELGADO, G. M. ANDERSON & G. R. HENINGER. 1989. Effects of desipramine and fluoxamine treatment on the prolactin response to L-Tryptophan: A test of the serotonergic function enhancement hypothesis of antidepressant action. Arch. Gen. Psychiatry **46:** 625–631.
82. MANIER, D. H., D. D. GILLESPIE & F. SULSER. 1989. Dual aminergic regulation of central beta adrenoceptors: Effect of "Atypical" antidepressants and 5-hydroxytryptophan. Neuropsychopharmacology **2**(2): 89–95.

DISCUSSION OF THE PAPER

COMMENT: You have mentioned that yohimbine can elicit panic sometimes and you have related that to the noradrenergic system. I think yohimbine is not an ideal choice of ligands for noradrenaline since the drug has also reasonable potency as a dopamine receptor antagonist and also reasonable activity, agonistic activity, at 5-HT_{1d} receptors.

CHARNEY: Yes, we're aware of that but it looks like in the doses that we use that we don't pick up those effects. For example you would expect the dopamine antagonists to increase prolactin levels and yohimbine, in the doses that we're using, does not. As clinical researchers you have to use the tools that are available.

D. W. GIETZEN (*University of California, Davis*): It seems that with your amino acid drink you're causing a plasma amino acid imbalance in your patients. Have you tried other amino acids?

CHARNEY: We plan to, to look at the specificity of this response.

GYORGY BAGDAY (*NJMH, Bethesda, MD 20892*): In animal studies both yohimbine and IV MCPP cause marked increases in plasma corticosterone which might be connected with the anxiety. In the case of MCPP we use several antagonists, and it turns out that probably it is mediated by 5-HT_{1C} receptors.

CHARNEY: Yes, with MCPP we have looked at its effect on plasma MHPG and it does not increase plasma MHPG like the yohimbine does. Recently we've completed a study looking at ritanserin's effect on MCPP's anxiogenic and neuroendocrine actions and we found that it antagonizes it by about 50–60%.

MAX FINK (*Session Chairman*): Dennis you have made a point, I think, that there is a distinct difference between two disorders, one called GAD, generalized anxiety disorder, and the other panic disorder, in terms of their effects, their relation to the serotonin system. Clinically how easy is it to separate these two disorders?

CHARNEY: There's some overlap in that most panic disorder patients have symptoms of generalized anxiety. But you can relatively easily define a pure group, or relatively pure group of GAD patients that do not have panic disorder.

Serotonin in Obsessions, Compulsions, and the Control of Aggressive Impulses

THOMAS R. INSEL,[a,c] JOSEPH ZOHAR,[b]
CHAWKI BENKELFAT,[a] AND DENNIS L. MURPHY[a]

[a]*Laboratory of Clinical Science*
National Institute of Mental Health
Bethesda, Maryland 20892

[b]*Beer-Sheva Mental Health Center*
Beer-Sheva, Israel 84170

Serotonin has been implicated in a number of psychiatric syndromes, but it is obsessive-compulsive disorder (OCD) which currently provides the most compelling evidence of a link between serotonin and psychopathology. The precise nature of this link remains unclear—in spite of strong evidence implicating serotonin in the treatment of OCD, we still lack a clear model of serotonergic dysfunction to explain the pathophysiology of this syndrome. This paper will review the current literature regarding serotonin and OCD with a perspective relating serotonin not only to this syndrome but to disorders of impulse control. To this end, we will review clinical studies implicating serotonin in the modulation of aggressive and impulsive behavior.

THE SELECTIVITY OF ANTI-OBSESSIONAL AGENTS

From a neuropharmacologic perspective, the most remarkable aspect of OCD is its high selectivity for drug response.[1,2] Most drugs that reduce anxiety, depression, or psychosis are not effective against obsessions. While individual OCD patients may respond to antidepressants such as the tricyclic imipramine or the monoamine oxidase inhibitor tranylcypromine, thus far the only medications which appear consistently effective in studies of OCD are clomipramine,[3–5] fluvoxamine,[6,7] and fluoxetine.[8,9] These three compounds are all antidepressants, and yet they reduce obsessional symptoms in nondepressed OCD patients. For this reason, they have been described as anti-obsessional drugs, in addition to being antidepressants. Just as in the treatment of depression, therapeutic effects only emerge after chronic (>5 weeks) administration of these drugs. What is surprising is that so many other excellent antidepressants, even those which are structurally related to these anti-obsessional drugs, appear ineffective for the symptoms of OCD. The salient difference may be that each of the anti-obsessional drugs is a potent inhibitor of serotonin reuptake. Antidepressant drugs without effects on serotonin reuptake, such as nortriptyline[3] and desipramine,[2] and drugs with weaker effects on serotonin reuptake, such as imipramine[10,11] and amitriptyline,[12] have all been demonstrated to be either less effective than clomipramine or essentially ineffective in the treatment of OCD (FIG. 1).

Changes in serotonergic function have also been associated with clinical improvement during treatment with anti-obsessional drugs. In some studies,[4,13] but not

[c]Address all correspondence to: Thomas R. Insel, M.D., NIHAC/NIMH, P.O. Box 289, Poolesville, MD 20837.

others,[3,5] the improvement in obsessional symptoms correlates with the plasma level of clomipramine and not its less selective metabolite desmethylclomipramine. With clomipramine treatment, the decrease in obsessional symptoms correlates ($r = 0.76$) with the decrease in the concentration of the serotonin metabolite 5-HIAA in cerebrospinal fluid—implicating the inhibition of uptake in the anti-obsessional effect.[14] A similar result obtains with serotonin content in blood platelets. During clomipramine treatment, the decrease in platelet serotonin content correlates highly ($r = 0.87$) with the decrease in obsessional symptoms.[15] It should be noted, however, that these measures—decreases in metabolite concentration and decreases in platelet content—

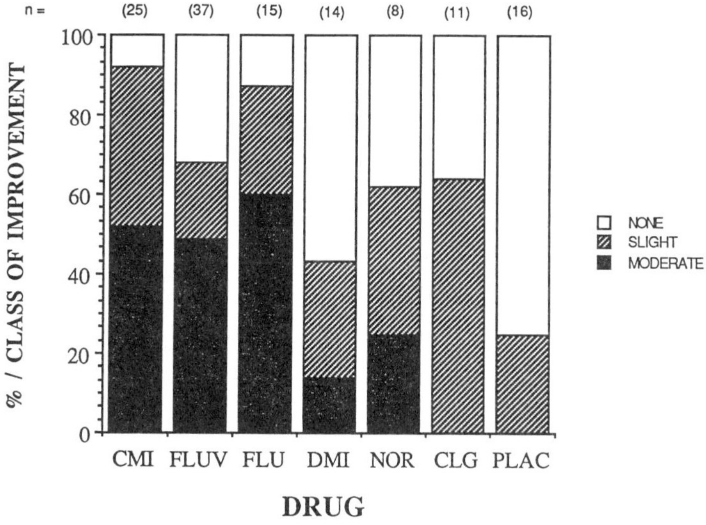

FIGURE 1. Summary of various treatment studies in OCD patients shows percentage with moderate (>40%), slight (5–40%), or no improvement at end points compared to baseline. Potent serotonin reuptake blockers show about 50% of patients in the "moderate" improvement category: clomipramine (CMI) data from ref. 1, fluvoxamine (FLUV) data from refs. 6 and 7, fluoxetine (FLU) data from open studies—refs. 8 and 9. Clinical response to antidepressants without potent serotonin uptake blockade effects appears considerably less: desipramine (DMI) and clorgyline (CLG) data from ref. 1, 4, nortriptyline (NOR) data from ref. 3. Placebo (PLAC) is consistently ineffective for reducing OCD symptoms.[1] Data from controlled trials with imipramine[10,11] and amitriptyline[12] are not included in this figure, but appear consistent with the observation that only potent serotonin reuptake blockers are anti-obsessional as these drugs are less effective than clomipramine[10,12] or not more effective than placebo[11] for the relief of OCD symptoms.

reflect acute drug effects which might be expected to emerge early, weeks prior to the decrease in obsessional symptoms.

Further evidence for a serotonergic mechanism underlying the anti-obsessional response comes from a recent study with metergoline, a 5-HT receptor antagonist.[16] Metergoline (4 mg) was administered orally to OCD patients who had received clomipramine for 4 months and manifested on average a 40% reduction in OC symptoms. The study was designed as a placebo-controlled crossover with 4 days of placebo and 4 days of metergoline treatment in addition to the daily clomipramine administration. Compared to the 4-day placebo period, OCD patients became progressively more anxious and obsessional during the metergoline phase (FIG. 2). In a subsequent study with a different cohort of clomipramine-treated patients, a single 12 mg dose of metergoline was associated with a similar worsening of OC symptoms (FIG. 2). Although an identical treatment regimen has yet to be investigated in untreated OCD patients, a single dose of 4 mg of metergoline does not appear to increase OC symptoms in untreated patients.[17] Thus, the effects during clomipramine administration may reflect a reversal of serotonergic changes associated with treatment and not an intrinsic "pro-obsessional" effect of metergoline. It seems likely that metergoline

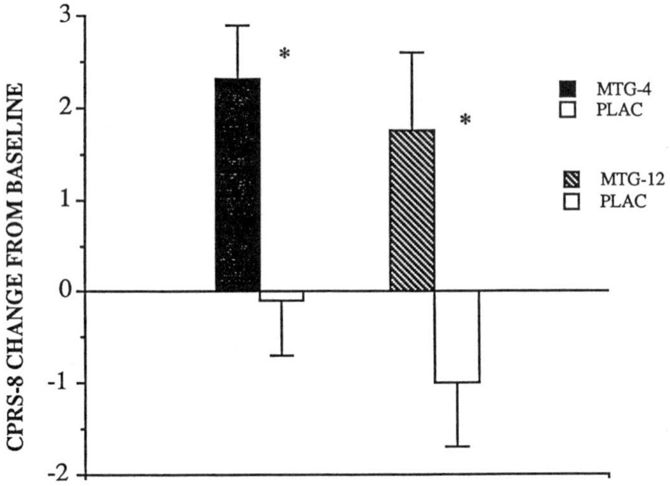

FIGURE 2. Response to the serotonin receptor antagonist metergoline in OCD patients receiving chronic clomipramine treatment. In the first study, 10 patients who were on average 40% improved on clomipramine received either placebo or metergoline (4 mg) orally for 4 days in a double-blind, crossover design. On the fourth day, changes in the OCD subscale of the Comprehensive Psychiatric Rating Scale (CPRS-8) indicated an increase in symptoms during metergoline compared to placebo treatment (* = $p < .05$). These data are modified from ref. 16. In a subsequent study, 5 OCD patients showing similar improvement on CMI received a higher oral dose of metergoline (12 mg) or placebo using the same experimental design with only a single day of treatment. Results again demonstrate a significant increase in CPRS-8 rated symptoms following metergoline compared to placebo (* = $p < .05$). Both studies indicate that serotonin receptor blockade may partially reverse the clinical response to clomipramine in OCD patients.

did, in fact, result in physiologic serotonin antagonism, as the drug lowered plasma prolactin concentrations in clomipramine-treated patients.

In summary, the link between serotonin and anti-obsessional effects has been suggested by the selectivity of the serotonin-reuptake inhibitors, the correlations between clinical response and changes in serotonergic markers, and the recent evidence of a partial relapse of symptoms following metergoline administration in OCD patients who had improved with clomipramine treatment. Several aspects of this emerging story still need to be addressed. Inhibition of serotonin reuptake occurs within minutes of the first dose of drug and yet anti-obsessional effects emerge only after several weeks of treatment. Is clinical response related to some late, compensatory response to reuptake inhibition and, if so, is this change ultimately an increase or a decrease in serotonergic neurotransmission? Recent evidence has suggested that some serotonin uptake inhibitors may not be anti-obsessional—a mixed set of results has been noted with zimelidine.[18–21] In one study,[20] patients receiving zimelidine had a decrease in CSF 5-HIAA and yet failed to improve clinically. Clearly drugs differ in their regional selectivity; they may also have some interfering nonserotonergic effects or lack effects on some essential serotonergic system. Perhaps blockade of serotonin reuptake is necessary but not sufficient for anti-obsessional efficacy. Finally, anti-obsessional effects have been reported recently for the 5-HT_{1A} agonist buspirone (ref. 22, although also see ref. 23). Ultimately, we may learn that it is not reuptake blockade per se, but indirect effects of reuptake blockade on serotonergic function which are key to anti-obsessional drug action.

EVIDENCE FOR SEROTONIN INVOLVEMENT IN THE PATHOPHYSIOLOGY OF OCD

The evidence from treatment studies of OCD has naturally led to a search for some abnormality in serotonin functioning in untreated OCD patients. This search has broadly followed three paths: studies of serotonin function in blood elements, measurement of CSF concentrations of the serotonin metabolite 5-hydroxyindoleacetic acid (5-HIAA), and challenge of untreated OCD patients with serotonergic agents. Although the data are inconsistent, several interesting observations are worth noting.

In terms of studies in blood elements, one early investigation[24] reported a decrease in serotonin concentration in whole blood from OCD patients, but this has not been replicated by a recent study of platelet serotonin content in OCD children.[15] Platelets, which are the major reservoir of serotonin in blood, show normal uptake of serotonin in OCD patients.[20,25] Serotonin uptake sites on platelets, labeled with ^{3}H-imipramine, appear reduced by nearly 50% in one study of 18 OCD children,[25] but not in another study of untreated OCD adults.[20]

The serotonin metabolite in CSF, 5-HIAA, may serve as an indicator of central serotonin turnover. CSF 5-HIAA was 30% higher in one small cohort ($n = 8$) of OCD patients compared to 23 controls.[20] In a larger study, Thoren and co-workers did not find a significant difference, although CSF 5-HIAA concentrations were 20% higher in OCD patients compared to healthy controls.[14] No differences in other CSF monoamine metabolites have been noted in OCD patients,[20] and the apparent 5-HIAA concentration differences cannot be explained by group differences in height, weight, diet, activity level, or season of sampling.[20]

To investigate further the possible role of 5-HT in the mediation of obsessional symptoms, our group began a series of studies examining the behavioral responses to direct activation of the serotonergic system with the selective 5-HT receptor agonist, *m*-chlorophenylpiperazine (*m*-CPP), in untreated patients with OCD and healthy controls.[17] Traditional drugs that affect serotonergic metabolism and action are not entirely specific for the serotonergic system (e.g., L-tryptophan or L-hydroxytryptophan)[26,27] or are associated with severe adverse effects, such as dimethyltryptamine,[28] which has hallucinogenic properties. In animal studies, *m*-CPP produces typical changes in 5-HT-mediated behaviors such as decreased food consumption[29,30]; and decreased locomotion[31]; neuroendocrine responses such as increased prolactin, cortisol, and corticotropin levels[32,33]; and physiologic reactions such as hyperthermia,[33] which are expected of a postsynaptic 5-HT receptor agonist. All these changes are reversed by the administration of the 5-HT receptor antagonist, metergoline, and are responsive to presumed alterations in 5-HT receptor sensitivity produced by lesions and drugs.[29,32,33]

Using *m*-CPP as a probe of the functional status of the brain serotonergic system, we investigated whether patients with OCD would differ from healthy controls in their behavioral responses to *m*-CPP, and whether *m*-CPP would affect OCD symptoms. Responses to single oral doses (0.5 mg/kg) of *m*-CPP in the patients with OCD were compared with responses to placebo under double-blind, random-assignment conditions on separate days.[17] Relative to healthy controls, the untreated patients with OCD became markedly more anxious and obsessional (FIG. 3A). The peak behavioral response was generally observed in the first three hours following *m*-CPP administration and the duration of the worsening ranged from several hours to 48 hours. Moreover, following *m*-CPP but not placebo, approximately 50% of the patients spontaneously described emergence of new obsessions or recurrence of obsessions that had not been present for several months. They also described themselves as being more

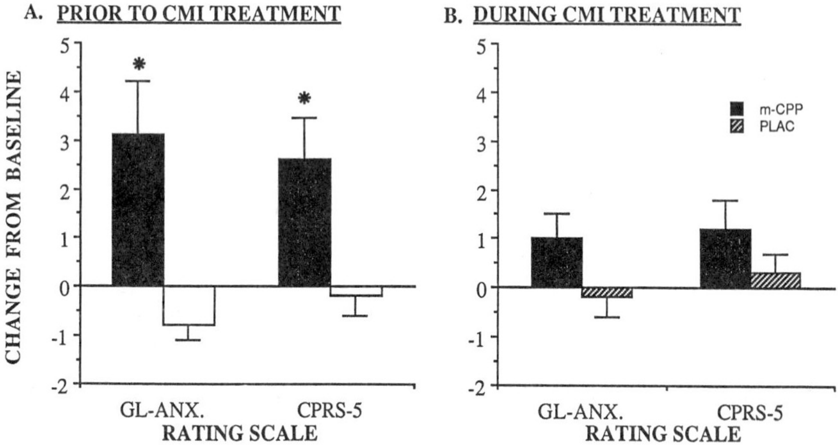

FIGURE 3. Response to the serotonin agonist *m*-chlorophenylpiperazine (*m*-CPP) and placebo administered as a single oral dose under double-blind conditions). **A**) Untreated OCD patients (n = 12) manifest a significant increase in anxiety (Global-anxiety scale) and OC symptoms (CPRS-5) following 0.5 mg/Kg of *m*-CPP administration relative to placebo (* signifies $p < .05$). **B**) A subgroup of these same patients (n = 9) receiving chronic clomipramine treatment were re-challenged with *m*-CPP and placebo with only slight, non-significant increases in these same measures. These results suggest that clomipramine treatment confers a sub-sensitivity in the serotonin receptor mediating *m*-CPP's effects.

depressed and dysphoric. The response to placebo as well as the side effects in both groups were mild and did not differ significantly between the patients and the controls.[2,17]

As oral administration of *m*-CPP has also been associated with anxiogenic effects in patients with panic disorders,[34] it was not immediately clear whether *m*-CPP was increasing OC symptoms selectively or whether the drug may cause a nonspecific exacerbation of diverse psychopathologic responses. Certainly, *m*-CPP affected several symptoms other than obsessions in the patients with OCD; thus, the drug might primarily increase anxiety or dysphoria, with OC symptoms increasing only as a secondary phenomenon. This possibility would be supported if patients with OCD were hypersensitive to other anxiogenic compounds. However, several recent studies with anxiogenic agents such as lactate,[35] yohimbine,[36] and caffeine (Zohar *et al.*, unpublished data) failed to increase either anxiety or OC symptoms in patients with OCD. It seems therefore that patients with OCD are not especially prone to respond to provocative agents that exacerbate symptoms in patients with non-OCD anxiety disorders. Also, OCD patients do not exhibit increased OC symptoms or anxiety in response to serotonin precursors such as L-tryptophan[37] or to serotonin-releasing agents such as fenfluramine.[38,39] These agents, while sharing some neuroendocrine and other physiologic effects with *m*-CPP, also act at many other serotonin terminal fields and receptors that have been identified in animal brain studies. Some of these receptors (e.g., the 5-HT_{1A} receptor) mediate a number of responses that are opposite to those elicited by *m*-CPP in animals. For example, agents acting at 5-HT_{1A} receptors in rodent brain elicit hypothermia, hypotension, increased food intake, and "anxiolytic" responses in some animal models of anxiety; in contrast, *m*-CPP has hyperthermic, hypertensive, anorectic, and "anxiogenic" actions in rodents (reviewed in ref. 40). Thus, the serotonin receptor agonist, *m*-CPP, is, so far, the only agent that has been associated with exacerbation of OC symptoms in a placebo-controlled study.[17] However, both the regional and pharmacologic specificity of *m*-CPP in human brain need to be elucidated before we can formulate a mechanism for its apparent "pro-obsessional" effects.

Aside from the questions of *m*-CPP's specificity, recently there has been a conflict in the literature regarding the drug's behavioral effects. Although a second study examining the effects of *m*-CPP in OCD patients also found that these patients were behaviorally hyperresponsive to this agonist,[39] a third study reported that *m*-CPP given intravenously to OCD patients and controls failed to elicit an increase in OC symptoms.[37] Both OCD patients and controls in this latter study developed marked anxiety and many other psychologic and somatic symptoms, results similar to observations from another study comparing i.v. and oral *m*-CPP in healthy volunteers.[40] A more recent study also found that *m*-CPP given intravenously to OCD patients and controls failed to elicit OC symptoms in the patients (Pigott *et al.*, submitted). Observations of the subjects and reports from the patients indicated that the abrupt, severe provocation of anxiety and many somatic symptoms of short duration appeared to be such an overwhelming experience that most patients were distracted from any obsessional thoughts or compulsive rituals they had experienced prior to i.v. *m*-CPP administration or during placebo administration (Pigott *et al.*, submitted). Thus, our present interpretation is that i.v.-administered *m*-CPP provides a considerably different experience from that following orally administered *m*-CPP, during which OC symptoms developed progressively over several hours, a time period which parallels quite closely the duration of the neuroendocrine and temperature responses and also the time course of plasma concentrations of *m*-CPP.[2,40,41]

Earlier we discussed the emerging evidence that the anti-obsessional effects of clomipramine may be mediated via serotonergic mechanisms. The apparent hypersensitivity to *m*-CPP in OCD patients provides a novel approach for assessing changes in serotonin receptor responsiveness during clomipramine treatment. To investigate possible changes in brain serotonergic responsivity, *m*-CPP and placebo were given under double-blind conditions to nine patients with OCD before and after treatment with clomipramine.[42] Unlike our previous observations of a marked transient increase in obsessional symptoms and anxiety following *m*-CCP administration in untreated patients, readministration of *m*-CPP after four months of treatment with clomipramine did not significantly increase obsessional symptoms and anxiety (FIG. 3B). The ability of clomipramine to reduce OC symptoms in patients with OCD coupled with the development of relative hyporesponsivity to *m*-CPP during clomipramine treatment strongly suggests that the therapeutic effects of clomipramine in patients with OCD are associated with the development of subsensitivity in a serotonergic receptor field involved in OC symptoms and related behaviors. It will be of great interest to determine whether evidence of a similar serotonergic receptor down-regulation occurs during the treatment of OCD patients with the more highly serotonin-selective uptake inhibitors such as fluoxetine, fluvoxamine, sertraline, and paroxetine, and also during treatment with some structurally different serotonergic agents such as buspirone and gepirone that act at 5-HT_{1A} sites.

SEROTONIN IN THE CONTROL OF AGGRESSIVE IMPULSES

From a phenomenologic perspective, the OCD patient is characterized by guilt, relentlessly worrying about aggressive impulses yet largely inhibited in the domain of action. Sociopaths who manifest virtually no guilt after committing an impulsive, violent crime may be considered the opposite extreme from OCD on this dimension of the control of aggressive impulses. There is an intriguing pre-clinical literature on serotonin and aggression which Soubrié has recently reviewed and re-interpreted as indicating that decreased serotonin neurotransmission is associated with increased performance of behaviors that are usually suppressed (i.e., impulsiveness).[43] Consistent with this concept, a recent developmental study in rhesus monkeys reported a positive correlation between CSF 5-HIAA and "behavioral inhibition."[44]

In spite of the formidable challenge of measuring "impulsiveness" in clinical investigations, data from several different studies suggest a relationship to one indirect measure of serotonin function, CSF 5-HIAA. This relationship was first described in depressed patients for whom low CSF 5-HIAA was associated with violent suicide attempts.[45] Several subsequent studies corroborating this finding are reviewed elsewhere in this volume (see the chapter by Mann *et al.* on serotonin and suicidal behavior). Worth noting in the current context is that the association between low CSF 5-HIAA and suicide spans several diagnostic categories and reflects a correlation with violent, usually unpredictable (i.e., impulsive) acts and not with suicidal ideation.[46] Comparable results with non-depressed, non-suicidal cohorts are summarized in TABLE 1.[47–52] Again, in subjects of various diagnostic groups (including healthy controls), individuals with low concentrations of CSF 5-HIAA appear to manifest more violence, more impulsiveness, or more hostility. Consistent with this model, OCD patients, in at least one study thus far, have increased concentrations of CSF 5-HIAA.[20]

The relationship of CSF 5-HIAA to the psychic concept of impulsiveness is tricky from both ends. Not only is impulsiveness difficult to validate and quantify, but the interpretation of CSF metabolite concentrations is unclear. Although CSF concentrations of 5-HIAA might be expected to reflect content in the spinal cord and periventricular regions, experimental data suggest a close correlation with concentrations of 5-HIAA in frontal cortex.[53] There are a number of "state" factors that have been shown to influence CSF 5-HIAA concentrations including age, height, season, diet, and activity level not to mention drug (especially alcohol) use.[54] Nevertheless, both clinical and pre-clinical studies demonstrate that CSF 5-HIAA may also be used as a "trait" measure. In serial sampling studies, metabolite concentrations are relatively stable over time.[55] Concentrations of CSF 5-HIAA are also more concordant in monozygotic than dizygotic twins.[56]

Further support for the CSF studies comes from a recent report using a different approach to assess central serotonergic function. Caccaro and co-workers[57] have used the plasma prolactin response to fenfluramine (60 mg, orally) to assess serotonergic responsiveness in patients with various personality disorders ($n = 25$), major affective disorder ($n = 20$), and healthy controls ($n = 18$). The plasma prolactin response was reduced in both patient groups relative to controls. Of note, the response to fenfluramine in the personality disorder patients was negatively correlated with measures of aggression (r = −.68 for Buss-Durkee "irritability" and prolactin response). In both patient groups, prolactin responses to fenfluramine were reduced in individuals with a history of suicide relative to those without a history of suicide. These results with a neuroendocrine marker for serotonergic responsiveness are broadly consistent with the hypothesis that reduced central serotonergic function is associated with impulsiveness, aggression, or irritability.

There is one additional, although highly speculative, link between the CSF 5-HIAA literature, aggression, and OCD. Frontal lobe injury has been associated traditionally with emotional lability including outbursts of anger. Curiously, frontal lobe lesions in OCD whether surgically[58] or self-administered[59] may reduce OC symptoms. Imaging studies have demonstrated increased metabolic activity in the frontal lobes of OCD patients,[60,61] leading to the notion that this disorder involves hyperfrontality and associated increases in worry, fixed beliefs, and emotional constriction.[62] In a Dutch study, patients who had sustained a fronto-temporal injury had significantly lower CSF 5-HIAA concentrations relative to those with diffuse brain injuries or non-injured controls.[63] Potentially, frontal lobe serotonergic terminals are the link between impulse control and CSF 5-HIAA.

TABLE 1. Studies of CSF 5-HIAA and Aggression/Impulsiveness in Non-depressed Cohorts

Study	Dx Group	n	CSF 5-HIAA Finding
Brown *et al.*, 1979	Personality disorders (non-depressed)	26	r = −.78 with index measure of aggression; for MHPG r = .64; 11/26 had suicidal history
Brown *et al.*, 1982	Borderline personality (non-depressed)	12	r = −.53 with aggressive history; no relationship for MHPG
Linnoila *et al.*, 1983	Criminal offenders	36	CSF 5-HIAA for impulsive < non-impulsive; 17 with suicide attempts < non-suicide; recidivist < non-recidivist
	(impulsive)	27	
	(non-impulsive)	9	
Lidberg *et al.*, 1985	Criminals	16	overall criminals = controls; suicides < controls; lowest in 5 criminals who murdered sexual partner
	Suicides	22	
	Controls	39	
Virkkunen *et al.*, 1987	Arsonists	20	arsonists < controls
	Controls	10	
Roy *et al.*, 1988	Controls	17	r = −.53 with Foulds acting out hostility scores

CONCLUSION

In this chapter, we have reviewed the accumulating evidence implicating serotonin in the treatment and the pathophysiology of OCD. Serotonin uptake blockers are thus far the only compounds that have been shown to be anti-obsessional, but whether their effects reflect an increase or a decrease in central serotonergic function remains unclear. The role of serotonin in the pathophysiology of OCD remains speculative, but is strongly suggested by the increased behavioral response to *m*-CPP. Finally, we considered the possibility that serotonin has a general role in the modulation of aggression or impulsivity, so that the results from studies of OCD could be understood in a larger context of investigations of several diagnostic categories varying on a dimension of impulsiveness. Although more studies are needed, the current data report high concentrations of CSF 5-HIAA in OCD patients (with low impulsiveness) and low concentrations of CSF 5-HIAA in violent criminals, arsons, and suicidal patients (with high impulsiveness). This notion linking CSF 5-HIAA to impulsiveness is largely consistent with Soubrié's hypothesis that central serotonin function is inversely correlated with the release of suppressed behavior.

REFERENCES

1. Insel, T. R. & J. Zohar. 1987. Psychopharmacologic approaches to obsessive-compulsive disorder. *In* Psychopharmacology: The Third Generation of Progress. H. Meltzer, Ed. New York: Raven Press, pp. 1205–1209.
2. Zohar, J. & T. R. Insel. 1987. Obsessive-compulsive disorder: Psychobiological approaches to diagnosis, treatment, and pathophysiology. Biol. Psychiatry **22:** 667–687.
3. Thoren, P., M. Asberg, B. Cronholm, L. Jornestedt & L. Traskman. 1980a. Clomipramine treatment of obsessive-compulsive disorder. I. A controlled clincial trial. Arch. Gen. Psychiatry **37:** 1281–1285.
4. Insel, T. R., D. L. Murphy, R. M. Cohen, I. Alterman, C. Kilts & M. Linnoila. 1983. Obsessive-compulsive disorder: A double-blind trial of clomipramine and clorgyline. Arch. Gen. Psychiatry **40:** 605–612.
5. Flament, M. F., J. L. Rapoport, C. J. Berg, W. Sceery, C. Kilts, B. Mellstrom & M. Linnoila. 1985. Clomipramine treatment of childhood obsessive-compulsive disorder: A double-blind controlled study. Arch. Gen. Psychiatry **42:** 977–983.
6. Perse, T. L., J. H. Greist, J. W. Jefferson, R. Rosenfeld & R. Dar. 1987. Fluvoxamine treatment of obsessive-compulsive disorder. Am. J. Psychiatry **144:** 1543–1548.
7. Goodman, W. K., L. H. Price, S. A. Rasmussen, P. L. Delgado, G. R. Heninger & D. S. Charney. 1989. Efficacy of fluvoxamine in obsessive-compulsive disorder: A double-blind comparison with placebo. Arch. Gen. Psychiatry **46:** 36–44.
8. Fontaine, R. & G. Chouinard. 1985. Fluoxetine in the treatment of obsessive-compulsive disorder. Prog. Neuropsychopharmacol. Biol. Psychiatry **9:** 605–608.
9. Turner, S., R. Jacob, D. C. Beidel & J. Himmelhoch. 1985. J. Clin. Psychopharm. **5:** 207–212.
10. Volavka, J., F. Neziroglu & J. A. Yaryaru-Tobias. 1985. Clomipramine and imipramine in obsessive-compulsive disorder. Psychiatry Res. **14:** 83–91.
11. Foa, E. B., G. Steketee, M. J. Kozak & D. Dugger. 1987. Effects of imipramine on depression and obsessive-compulsive symptoms. Psychiatry Res. **21:** 123–126.
12. Ananth, J., J. C. Pecknold, N. van den Steen & F. Engelsmann. 1981. Double-blind comparative study of clomipramine and amitriptyline in obsessive neurosis. Prog. Neuropsychopharmacol. Biol. Psychiatry **5:** 257–262.
13. Stern, R. S., I. M. Marks, J. Wright & D. K. Luscombe. 1980. Clomipramine: Plasma levels, side effects, and outcome in obsessive-compulsive neurosis. Postgrad. Med. J. **56:** 134–139.
14. Thoren, P., M. Asberg, L. Bertilsson, B. Mellstrom, F. Sjoqvist & L. Traskman. 1980b. Clomipramine treatment of obsessive-compulsive disorder. II. Biochemical aspects. Arch. Gen. Psychiatry **37:** 1289–1294.
15. Flament, M. F., J. L. Rapoport, D. L. Murphy, C. J. Berg & C. R. Lake. 1987. Biochemical changes during clomipramine treatment of childhood obsessive-compulsive disorder. Arch. Gen. Psychiatry **44:** 219–225.

16. Benkelfat, C., D. L. Murphy, J. Zohar, J. L. Hill, G. Grover & T. R. Insel. 1989. Clomipramine in obsessive-compulsive disorder: Further evidence for a serotonergic mechanism of action. Arch. Gen. Psychiatry **46:** 23–28.
17. Zohar, J., E. A. Mueller, T. R. Insel, R. C. Zohar-Kadouch & D. L. Murphy. 1987. Serotonergic responsivity in obsessive-compulsive disorder: Comparison of patients and healthy controls. Arch. Gen. Psychiatry **44:** 946–951.
18. Prasad, A. 1985. A double blind study of imipramine versus zimelidine in treatment of obsessive compulsive neurosis. Pharmacol. Biochem. Behav. **22:** 347–348.
19. Kahn, R., H. G. M. Westerberg & J. Jolles. 1984. Zimelidine treatment of obsessive-compulsive disorder. Acta Psychiatr. Scand. **69:** 259–261.
20. Insel, T. R., E. A. Mueller, I. Alterman, M. Linnoila & D. L. Murphy. 1985. Obsessive-compulsive disorder and serotonin: Is there a connection? Biol. Psychiatry **20:** 1174–1189.
21. Fontaine, R., G. Chouinard & L. Iny. 1985. An open clinical trial of zimelidine in the treatment of obsessive-compulsive disorder. Curr. Ther. Res. **37:** 326–332.
22. Pato, M. T., T. A. Pigott, J. L. Hill, G. Grover, S. E. Bernstein & D. L. Murphy. 1989. Clomipramine versus buspirone in OCD: A controlled trial. Presented at the 1989 annual meeting of the American Psychiatric Assn., San Francisco, CA, New Research abstract 14.
23. Jenike, M. A. & L. Baer. 1988. Buspirone in obsessive-compulsive disorder: An open trial. Am. J. Psychiatry **145:** 1285–1286.
24. Yaryura-Tobias, J. A. & H. N. Bhagavan. 1977. L-Tryptophan in obsessive-compulsive disorders. Am. J. Psychiatry **134:** 1298–1299.
25. Weizman, A., M. Carmi, H. Hermesh, A. Shahar, A. Apter, S. Tyano & M. Rehavi. 1986. High-affinity imipramine binding and serotonin uptake in platelets of eight adolescent and ten adult obsessive-compulsive patients. Am. J. Psychiatry **143:** 335–339.
26. Fernstrom, S. & R. J. Wurtman. 1974. Nutrition and the brain. Sci. Am. **230:** 85–96.
27. Saavedra, J. & J. Axelrod. 1974. Brain tryptamine and the effect of drugs. Adv. Biochem. Psychopharmacol. **10:** 135–139.
28. Meltzer, H., B. Wiita, B. J. Tricou, M. Simonovic, V. Fang & G. Manov. 1982. Effect of serotonin precursors and serotonin agonists on plasma hormone levels. *In* Serotonin in Biological Psychiatry. B. T. Ho, J. C. Schoolar & E. Usdin, Eds. New York: Raven Press.
29. Samamin, R., T. Mennini & A. Ferraris. 1979. M-chlorophenylpiperazine: A central serotonin agonist causing powerful anorexia in rats. Naunyn Schmiedebergs Arch. Pharmacol. **308:** 159–163.
30. Mennini, T., E. Poggasi, S. Caccia, C. Bendotti, F. Borsini & R. Samanin. 1980. Adaptive changes in central serotonin receptors after long-term treatment in rats. *In* Neurotransmitters and Their Receptors. U. Z. Littauer, Y. Dudai & I. Silman, Eds. New York: John Wiley & Sons, Inc.
31. Vetulani, J. M. Sansone, B. Bednarczyk & J. Hano. 1982. Different effects of 3-chlorophenylpiperazine on locomotor activity and acquisition of conditioned avoidance response in different strains of mice. Naunyn Schmiedebergs Arch. Pharmacol. **319:** 271–274.
32. Maj, J. & A. Lewandowska. 1980. Central serotoninmimetic action of phenylpiperazine. Pol. J. Pharmacol. Pharm. **32:** 495–504.
33. Aloi, J. A., T. R. Insel, E. A. Mueller & D. L. Murphy. 1984. Neuroendocrine and behavioral effects of m-chlorophenylpiperazine administration in rhesus monkeys. Life Sci. **34:** 1325–1331.
34. Kahn, R. S., S. Wetzler, H. M. Van Praag, G. M. Asnis & T. Strauman. 1988. Behavioral indications for serotonin receptor hypersensitivity in panic disorder. Psychiatry Res. **25:** 101–104.
35. Gorman, J. M., M. R. Leibowitz, A. J. Fyer, S. O. Davies & D. F. Klein. 1985. Lactate infusions in obsessive-compulsive disorder. Am. J. Psychiatry **142:** 864–866.
36. Rasmussen, S. A., W. K. Goodman, S. W. Woods, G. R. Heninger & D. S. Charney. 1987. Effects of yohimbine in obsessive-compulsive disorder. Psychopharmacology **93:** 308–313.
37. Charney, D. S., W. K. Goodman, L. H. Price, S. W. Woods, S. A. Rasmussen & G. P.

HENINGER. 1988. Serotonin function in obsessive-compulsive disorder: A comparison of the effects of tryptophan and m-chlorophenylpiperazine in patients and healthy subjects. Arch. Gen. Psychiatry **45:** 177–185.

38. INSEL, T. R., E. A. MUELLER, J. C. GILLIN, L. J. SIEVER & D. L. MURPHY. 1984. Biological markers in obsessive-compulsive and affective disorders. J. Psychiatr. Res. **18:** 407–423.
39. HOLLANDER, E., M. FAY, B. COHEN, R. CAMPEAS, J. M. GORMAN & M. R. LIEBOWITZ. 1988. Serotonergic and noradrenergic sensitivity in obsessive-compulsive disorder: Behavioral findings. Am. J. Psychiatry **145:** 1015–1017.
40. MURPHY, D. L., E. A. MUELLER, J. L. HILL, T. J. TOLLIVER & F. M. JACOBSEN. 1989. Comparative anxiogenic, neuroendocrine, and other physiologic effects of m-chlorophenylpiperazine given intravenously or orally to healthy volunteers. Psychopharmacology. **98:** 275–282.
41. MUELLER, E. A., D. L. MURPHY & T. SUNDERLAND. 1985. Neuroendocrine effects of m-chlorophenylpiperazine, a serotonin agonist, in humans. J. Clin. Endocrinol. Metab. **61:** 1179–1184.
42. ZOHAR, J., T. R. INSEL, R. C. ZOHAR-KADOUCH, J. L. HILL & D. L. MURPHY. 1988. Serotonergic responsivity in obsessive-compulsive disorder: Effects of chronic clomipramine treatment. Arch. Gen. Psychiatry **45:** 167–172.
43. SOUBRIÉ, P. 1986. Reconciling the role of central serotonin neurons in human and animal behavior. Behav. Brain Sci. **9:** 319–364.
44. HIGLEY, J. D., S. J. SUOMI & M. LINNOILA. 1990. Serotonin influences on timidity in the nonhuman primate. *In* Serotonin in Major Psychiatric Disorders. E. Coccaro and D. L. Murphy, Eds. Washington, D.C.: Am. Psych. Press. In press.
45. ASBERG, M., L. TRASKMAN & P. THOREN. 1976. 5-HIAA in the cerebrospinal fluid: A biochemical suicide predictor? Arch. Gen. Psychiatry **33:** 1193–1197.
46. VAN PRAAG, H. M. 1983. CSF 5-HIAA and suicide in non-depressed schizophrenics. Lancet **ii:** 977–978.
47. BROWN, G., F. GOODWIN, J. BALLENGER, P. GOYER & L. MAJOR. 1979. Aggression in humans correlates with cerebrospinal fluid metabolites. Psych. Res. **1:** 131–135.
48. BROWN, G., M. EBERT, P. GOYER, D. JIMERSON, W. KLEIN, W. BUNNEY & F. GOODWIN. 1982. Aggression, suicide, and serotonin: Relationship to CSF amine metabolites. Am. J. Psychiatry **139:** 741–746.
49. LINNOILA, M., M. VIKKUNEN, M. SCHEININ, A. NUUTILA, R. RIMON & F. K. GOODWIN. 1983. Low cerebrospinal fluid 5-HIAA concentration differentiates impulsive from non-impulsive violent behavior. Life Sci. **33:** 2609–2614.
50. LIDBERG, L., J. TUCK, M. ASBERG, G. SCALIA-TOMBA & L. BERTILSSON. 1985. Homicide, suicide, and CSF 5-HIAA. Acta Psychiatr. Scand. **71:** 230–236.
51. VIRKKUNEN, M., A. NUUTILA, F. GOODWIN & M. LINNOILA. 1987. Cerebrospinal fluid monoamine metabolite levels in male arsonists. Arch. Gen. Psychiatry **44:** 241–247.
52. ROY, A., B. ADINOFF & M. LINNOILA. 1988. Acting out hostility in normal volunteers: Negative correlation with CSF 5-HIAA levels. Psych. Res. **24:** 187–194.
53. STANLEY, M., L. TRASKMAN-BENDZ & K. DOROVIN-ZIS. 1985. Life Sci. **37:** 1279–1986.
54. JIMERSON, D. C. & W. BERRETTINI. 1985. Cerebrospinal fluid metabolite studies in depression: Research update. *In* Pathochemical Markers in Major Psychoses. H. Beckmann and P. Riederer, Eds. Berlin: Springer Verlag.
55. TRASKMAN-BENDZ, L., M. ASBERG, L. BERTILSSON & P. THOREN. 1984. Acta Psychiatr. Scand. **69:** 333–342.
56. SEDVALL, G., L. ISELIUS, H. NYBACK, L. ORELAND, G. OXENSTIERNA, S. B. ROSS & F. A. WIESEL. 1984. Genetics studies of CSF monoamine metabolites. Adv. Biochem. Psychopharmacol. **39:** 79–85.
57. COCCARO, E. F., L. J. SIEVER, H. KLAR, G. MAURER, K. RUBINSTEIN, T. B. COOPER, R. C. MOHS & K. L. DAVIS. 1989. Serotonergic studies in affective and personality disorder patients: Correlates with behavioral aggression and impulsivity. Arch. Gen. Psychiatry. **46:** 587–596.
58. KELLEY, D. 1980. Anxiety and Emotions. Springfield, IL: Charles C. Thomas.
59. SOLYOM, L., I. M. TURNBULL & M. WILENSKY. 1987. A case of self-inflicted leucotomy. Br. J. Psychiatry **151:** 855–857.

60. Baxter, L. R., M. E. Phelps, J. C. Mazziotta, B. H. Guze, J. M. Schwartz & C. E. Selin. 1987. Local cerebral glucose metabolic rates in obsessive-compulsive disorder: A comparison with rates in unipolar depression and normal controls. Arch. Gen. Psychiatry **44:** 211–218.
61. Nordahl, T. E., C. Benkelfat, W. Semple, M. Gross, A. C. King & R. M. Cohen. 1989. Cerebral glucose metabolic rates in obsessive compulsive disorder. Neuropsychopharmacology. **2:** 23–28.
62. Insel, T. R. & J. T. Winslow. 1990. Neurobiology of OCD. *In* Obsessive-Compulsive Disorders: Theory and Management, 2d edit. M. Jenike, L. Baer and M. Mimichiello, Eds. Littleton, MA: Yearbook Medical Publishers. pp. 118–131.
63. van Woerkom, T. C. A. M., A. W. Teelken & J. M. Minderhoud. 1977. Difference in neurotransmitter metabolism in frontotemporal lobe contusion and diffuse cerbral contusion. Lancet **1:** 812–813.

DISCUSSION OF THE PAPER

George Ricaurte (*The Johns Hopkins Hospital, Baltimore, MD*): Tom I just had one question and that is with regard to the CSF data that you presented, the interpretation that you gave it is perhaps that these patients with OCD have increased serotonin metabolism and turnover. And that perhaps their syndrome, if you will, is related to functional hyperactivity of the serotonergic system. I'm a little puzzled as to how that would fit in with the responsivity of this group of patients to agents that essentially would do, accomplish the same thing, i.e., the serotonin reuptake blockers, clomipramine and so forth, that would in essence end up augmenting serotonergic neurotransmission to even a greater extent.

Insel: The simple answer to that question is that I think there's a great deal of confusion about just what serotonin reuptake blockers do in any given terminal field. The general dogma is that these drugs increase serotonergic neurotransmission. For instance, there is good evidence of hyper-responsivity in terms of iontophoretic application of 5-HT in hippocampal terminal fields following chronic administration of serotonin reuptake blockers to rats. So that in some regions there is this increase in responsivity and their presumably increased serotonergic neurotransmission. In other regions, such as in neocortex, there's a decrease in responsivity, or at least a decrease in receptor number. So the ultimate effect of a drug such as clomipramine may be complex and regionally specific. This question is relevant to our *m*-CPP data, which appears to affect a terminal field that may be particularly implicated in this disorder, as the drug increases the symptoms in some way. The fact that the patients challenged with *m*-CPP no longer show the increase in symptoms following clomipramine administration would be consistent with the serotonin uptake inhibitor decreasing sensitivity at this site. Unfortunately, it is not clear in human brain where *m*-CPP acts or even how many receptor subtypes may be involved.

S. J. Boyson (*University of Colorado Health Science Center, Denver*): As you know Tourette's Disorder clusters with obsessive compulsive disorder. I wanted to know what you think about the pharmacology of Tourette's based on this. What I understand from psychiatrists is that fluoxetine is useful for OCD but not for the motor tics and vocal tics of Tourette's.

Insel: There are some very interesting recent genetic studies, from Pauls *et al.*, demonstrating that at least a subgroup of Tourette's patients are very likely to have first degree relatives with OCD. However, I know of no controlled studies demonstrating efficacy of serotonin uptake inhibitors, such as fluoxetine for the tics of Tourette's syndrome. Currently, it appears that dopamine receptor blockade is more important for treating the tics and motor symptoms of Tourette's. Possibly the cognitive aspects of the two syndromes are going to have a more similar pharmacology, but without additional studies of serotonin uptake blockers in Tourette's, this remains only speculative.

Serotonin and Migraine

PATRICK P. A. HUMPHREY, WASYL FENIUK, MARION J. PERREN, ISABEL J. M. BERESFORD, AND MALCOLM SKINGLE

Pharmacology Division
Glaxo Group Research Limited
Ware, Hertfordshire, SG12 ODP, England

ERIC T. WHALLEY

Department of Physiological Sciences
University of Manchester
Manchester, M13 9PT, England

WHAT IS MIGRAINE?

Symptomatology

It is said that the first clear description of the symptoms of migraine were left for posterity by Aretaeus of Cappadocia during the second century A.D. (see ref. 1). It is, therefore, all the more remarkable that even today we still do not understand the underlying pathogenesis. Furthermore, there still remains some difficulty in accurately diagnosing the disease and consistently differentiating an episode of migraine from an 'ordinary' headache. However, the "typical" migraine attack is very distinctive, being characterized by an often severe, unilateral headache and accompanied by nausea and vomiting. Estimates of the incidence of the disease suggest that 5–15% of the population of the Western world suffer from migraine.[2]

In some individuals prodromal symptoms and signs, such as mood or bowel disturbances, may give a premonition of an impending attack.[3] In what used to be called classical migraine, the headache phase is preceded by an aura. The aura symptomatology involves one or more specific neurological deficits, often visual such as an enlarging or moving scotoma with a scintillating edge, so often described in artistic form.[4] Other possible neurological symptoms include sensory or, as in the case of familial hemiplegic migraine, motor disturbance.[5] The proceeding headache is usually long lasting and in many individuals severe enough to require bedrest. During this headache phase other frequently associated symptoms include vomiting, anorexia, diarrhea or constipation, as well as photophobia and phonophobia in some individuals.[6] The classification of migraine, in relation to other types of headache, and the various sub-classifications of the symptoms has recently been updated by the International Headache Society.[7]

Pathogenesis

For about half a century, preceding the beginning of this decade, it was generally considered that migraine was a vascular disease associated primarily with a derangement of blood supply to the head. The view emanated largely from the work of Graham and Wolff,[8] who showed that ergotamine treatment of migrainous headaches was associated with a reduction of the amplitude of the pulsations of the superficial temporal artery. Wolff's "vascular hypothesis" of migraine proposed that the headache was caused by a period of extracranial vasodilatation while the preceding neurological symptoms (if present) were thought to result from a focal intracranial vasoconstriction (see ref. 9). The precipitating factors were considered to be varied, but humoral hypotheses were invariably implicated. However, it was never satisfactorily explained how the headache could be localized to one (sometimes varying) side of the head and why physiologically induced cephalic vasodilation did not induce an attack even in migraineurs (see ref. 5). This led to the opposite view that migraine was entirely a central nervous system disorder and that all the symptoms of the disease, including the head pain, could be explained by a derangement of brain function.[1,10]

SEROTONIN INVOLVEMENT

Numerous papers and books have dealt with the circumstantial evidence for an involvement of serotonin in migraine (e.g., refs. 5 and 11). The evidence will only briefly be summarized here (see TABLE 1). During the headache phase the 5-hydroxyindole acetic acid metabolite of serotonin, 5-hydroxyindoleacetic acid, is excreted in increased amounts in the urine.[12] This is associated with decreased levels of serotonin in the blood platelets and correlates with the known ability of amine-depleting agents, such as reserpine, to cause migraine in migraineurs.[14] Furthermore, a serotonin-releasing factor has been described by Anthony and colleagues which is present in the plasma of migraineurs only during the headache phase.[16] Its chemical identity has yet to be determined. These data suggest that migraine is a low plasma serotonin syndrome, though it has been inferred that an extra-platelet locus of depletion is more important—possibly the brain (see ref. 17)?

However, in contradiction to the view that migraine is caused by a depletion of central serotonin stores, it has been shown that intravenous serotonin will clinically ameliorate an attack[15] and as has already been argued exogenous serotonin is too basic a substance to be acting centrally "as a transmitter replacement." Rather, serotonin is likely to be acting as a 'vasoconstrictor', in much the same way as has been reported for noradrenaline, to "normalize" dilated and distended cranial vessels.[5,18] It should be noted parenthetically, however, that the many effects of intravenous serotonin or noradrenaline make their clinical use unacceptable.

It remains to be seen whether the apparent involvement of serotonin in the pathology of migraine is epiphenomenal or whether it has a primary involvement in the etiology of the disease. Regardless of such considerations, the efficacy of the newly identified selective serotonin mimetic, sumatriptan, shows great promise (see below) and adds further fuel to this interesting debate.

TABLE 1. Evidence for an Involvement of Serotonin in Migraine

Increased excretion of 5-hydroxyindoleacetic acid in urine	Sicuteri *et al.*, 1961[12]
Platelet serotonin falls 40% at onset of attack	Curran *et al.*, 1965[13]
Attacks can be triggered in migraineurs by amine releasing agents such as reserpine	Anthony *et al.*, 1967[14]
Intravenous serotonin alleviated the symptoms of migraine	Kimball *et al.*, 1960[15] Anthony *et al.*, 1967[14]

TABLE 2. Binding Affinities of Clinically Available Drugs for 5-HT_2 Sites in Rat Prefrontal Cortex

Drug	– Log K*i*
Pizotifen	8.85
Cyproheptadine	8.78
Spiperone	8.49
Methysergide	8.36
Astemizole	7.88
Amitriptyline	8.03
Imipramine	7.00
Phentolamine	6.73
Propranolol	6.23

Mean K_i values from experiments with [^{3}H]ketanserin and [^{3}H]spiperone described by Leysen *et al.*[20]

APPROACHES TO ANTI-MIGRAINE THERAPY

Numerous authors have pondered on the pharmacology of presently available anti-migraine drugs,[1,3,19] but it is difficult to fully rationalize their relatively modest clinical effects. Many of them interact with serotonin receptors in some way or the other but, until more selective agents are available, which have been properly examined clinically, the true significance remains merely tantalizingly elusive and fans the scientific controversy discussed in the previous section.

Prophylactics

The traditional prophylactic agents used in the treatment of migraine, such as pizotifen, cyproheptadine, methysergide and amitriptyline are all potent 5-HT_2 receptor antagonists (see TABLE 2) but it is by no means clear whether this is their mechanism of action, as they all have other effects.[21] Interestingly a variety of other clinically available drugs also share an affinity for 5-HT_2 receptors, like the H_1 receptor antagonist, astemizole, for example (TABLE 2). It would be interesting to know whether such agents would also be useful prophylactics for the treatment of migraine. Notwithstanding such considerations it is clear that only the careful clinical evaluation of a highly specific and selective 5-HT_2 receptor antagonist will resolve the question of whether 5-HT_2 receptor blockade per se is important mechanistically.

As regards the other prophylactics commonly used in the treatment of migraine, such as the β-adrenoceptor antagonists and the calcium channel blockers, no interaction with serotonin receptors appears to be important. Thus, propranolol has little or no affinity for 5-HT_2 receptors and although it does bind to 5-HT_1 brain binding sites so do other β-adrenoceptor antagonists even more avidly, which are not clinically effective.[19,22,23] Furthermore, it is evident that the calcium channel blockers, with claimed efficacy as migraine prophylactics, possess only weak or modest affinity for serotonin receptors.[19,24–26]

Ergot Alkaloids

Since the clinical studies of Wolff,[9] ergotamine has been extensively used in the treatment of acute migraine. It is often effective even in very severe cases of migraine, providing it is given promptly and administered parenterally to ensure adequate bioavailability. However, the pharmacology of the ergots is extensive[27] and explains the various side effects they have. The ergots interact with many different types of amine receptors and, for example, it would seem that agonism at dopamine D_2 receptors explains the nausea and emesis often experienced with ergotamine and dihydroergotamine in clinical practice. This and other side effects, such as those on the uterus and the peripheral vasculature, have led to the largely restricted use of these agents to intractable migraineurs.

5-HT$_3$ Receptor Antagonists

It has been argued that serotonin released from serotoninergic neurons innervating cerebral blood vessels causes the pain of migraine by activating 5-HT_3 receptors on sensory afferent fibers.[3,19] This, at first, attractive hypothesis gained credence with

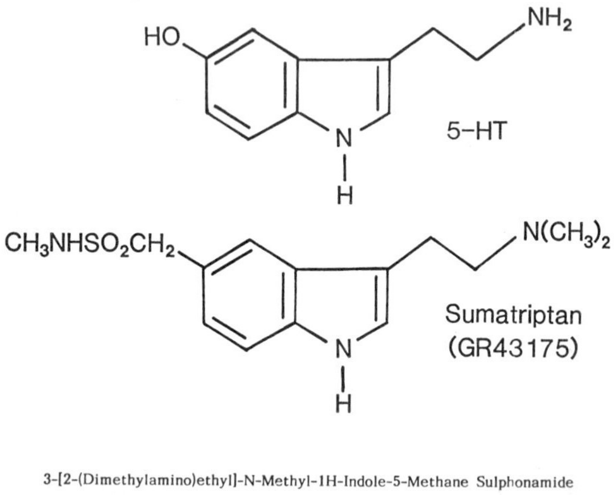

FIGURE 1. Chemical structure of 5-hydroxytryptamine (5-HT; serotonin) and sumatriptan.

early clinical data on the prototype 5-HT_3 receptor antagonist, MDL 72222, which showed it to have some efficacy in the acute treatment of migraine.[28] However, the more potent and selective 5-HT_3 receptor antagonist, ICS-205 930, has not been shown to be effective in the treatment of acute migraine.[29] Indeed the Sandoz group even tested MDL 72222 themselves and were unable to repeat the earlier encouraging findings.[30] With the advent of a number of other 5-HT_3 receptor–blocking drugs for clinical use it remains to be determined whether or not any of them will be efficacious in the treatment of migraine.

Sumatriptan

Sumatriptan is a newly identified serotonin receptor agonist which looks very promising in the acute treatment of migraine.[31,32] Its pharmacology has been described in some detail.[32–37] It is chemically related to serotonin (FIG. 1) but its pharmacology is much more restricted, being a highly selective agonist for a 5-HT_1-like receptor sub-type which is localized as far as the vasculature is concerned mainly to certain cranial blood vessels. Sumatriptan, unlike ergotamine and other ergots, has no action on other amine receptors.[37,38] The 5-HT_1-like receptor type found on cranial blood vessels also occurs on some nerves, such as sympathetic nerves innervating some peripheral vessels, where it inhibits noradrenaline release from the sympathetic nerve terminals.[33,39] Recent evidence suggests that a similar receptor may function as the autoreceptor on serotoninergic nerve terminals in mammalian brain.[40]

Sumatriptan undoubtedly has a selective cranial vasoconstrictor activity and in the absence of any demonstrable analgesic activity in animals it is the opinion of the authors that this is the mechanism of its anti-migraine action.[32,35] Others have suggested it may have a central component of action despite the low lipophilicity and high basicity of the compound (like serotonin). Regardless of the relative merits of these arguments, studies, both in the laboratory and in the clinic, should not only lead to a better understanding of the mechanism of action of sumatriptan but also hopefully a greater understanding of the pathophysiological mechanisms involved in the disease itself.

STUDIES ON THE POSSIBLE MECHANISM OF ACTION OF SUMATRIPTAN

The results from new studies on the mechanism of action of sumatriptan outlined below compound the view that its anti-migraine action results from a selective cranial vasoconstriction, though the precise vascular locus remains to be determined. Sumatriptan does not appear to have an effect on transmission in sensory afferent nerve pathways, nor to be a substance P antagonist, nor does it modify the release of substance P from trigeminal nerve endings in the rabbit iris or release of acetylchloline from the vagus nerve. It does, however, have a highly specific action in the meninges in rats, where it inhibits the leakage of plasma protein caused by stimulation of the trigeminal nerve. Furthermore studies in the human isolated dura mater clearly demonstrate the ability of sumatriptan to constrict the meningeal circulation. It may well be that it is this vasoconstrictor action of sumatriptan which explains its ability to reduce extravasation in the dura mater following trigeminal nerve stimulation and that this, together with the additional reduction in transmural pressure and consequent inhibition of painful sensory stimuli, is the underlying mechanism of its anti-migraine action.

Poor Brain Penetration by Sumatriptan

As mentioned above there is a current debate as to whether the real locus of action of sumatriptan is in the central nervous system. This consideration is particularly germane in view of recent findings that the sumatriptan-sensitive 5-HT_1-like receptor in cranial blood vessels is similar to the 5-HT_{1D} binding site in the human brain.[38,41,42] We have therefore carried out experiments to look at the degree of brain penetration by sumatriptan. Using a variety of radiolabeled ligands we have shown that in the conscious mouse, sumatriptan is poorly taken up into the brain following intravenous injection, even less well than inulin, a large molecular weight polysaccharide (TABLE 3).

TABLE 3. Penetration of Systemically Administered Compounds into the Central Nervous System of the Mouse

Treatment	Isotope	Total nMol/Mouse Injected	Equivalent Dose (mg/kg) i.v.	% of Total Radioactivity in Brain
Sumatriptan	^{14}C	0.04	0.166	0.006
		4	16.6	0.006
Inulin	^{3}H	0.04	2	0.07
		4	200	0.09
Spiperone	^{3}H	0.04	0.158	0.97
		0.4	1.58	1.19
Ketanserin	^{3}H	0.4	1.37	0.50
		4	13.7	2.93

Labeled drug (10μCi per mouse) was mixed with varying amounts of cold drug to produce the doses shown and injected intravenously (i.v.). Two minutes later the mouse was killed and the brain removed, quickly washed and then the radioactivity counted. Values shown are the mean percentage radioactivity of total dose from $n = 5$ or more animals.

Lack of Effect of Sumatriptan on Neurogenically (Trigeminal Nerve) Mediated Contractions of the Rabbit Isolated Iris Sphincter Muscle

Electrical field stimulation of the rabbit isolated iris sphincter muscle produces cholinergic and substance P–mediated contractile responses.[43] The substance P-component of the response has been shown to be due to the activation of nerves which originate from the trigeminal ganglion.[44] Since activation of the trigeminal system has been implicated in the pathophysiology of migraine, we have investigated the influence of sumatriptan on neurogenically mediated contractions of the rabbit iris sphincter muscle. These studies were carried out in the continual presence of atropine (1 μM) so that only the influence of sumatriptan on the substance P-component of the response could be investigated. Sumatriptan in concentrations as high as 10 μM did not modify contractile responses of the rabbit isolated iris sphincter muscle preparation to electrical field stimulation. However, these responses were markedly attenuated by tetrodotoxin demonstrating that they were neurogenically mediated (FIG. 2). These results suggest that sumatriptan does not inhibit the release of substance P from the neurons mediating contraction of the rabbit iris sphincter muscle and that sumatriptan is not a substance P antagonist.

Effect of Sumatriptan on Sensory Reflexes in Vivo

It has been suggested that migraine headaches are due to stimulation of sensory nerves in the walls and perivascular areas of the meningeal circulation.[45] Indeed it has been argued that 5-HT, which is present in the cerebral microcirculation,[46] is likely to activate these sensory nerve endings and to sensitize them to other nociceptive stimuli such as kinins.[47]

In keeping with its known neuronal depolarizing action on afferent nerves, the intravenous administration of 5-HT or the selective 5-HT$_3$ agonist, 2-methyl 5-HT,[48] to anaesthetized animals induces a vagally mediated reflex bradycardia as a consequence of the activation of sensory nerve endings located in the wall of the right ventricle.[49] It has been proposed that the effectiveness of intravenous metoclopramide in the acute treatment of migraine is due to its ability to block those 5-HT receptors responsible for the excitation of sensory afferent neurons.[47] Most certainly, metoclopramide and other 5-HT_3 receptor–blocking drugs have the ability to inhibit 5-HT– and 2-methyl 5-HT–induced bradycardia in anaesthetized animals (but see ***5-HT$_3$ Receptor Antagonists*** above).[47]

We have investigated the action of sumatriptan on 2-methyl 5-HT–induced bradycardia in anaesthetized cats. Sumatriptan (1 mg/kg i.v.) in a dose in excess of those used clinically in the treatment of migraine had no effect on 2-methyl 5-HT–induced bradycardia. In marked contrast, the 5-HT_3 receptor blocking drug, ondansetron (GR38032),[50] abolished the bradycardia (FIG. 3).

These studies demonstrate that sumatriptan does not in any way inhibit sensory afferent neurotransmission in the experimental situation described. Thus it does not act centrally to inhibit efferent vagal activity; furthermore sumatriptan has neither atropinic actions nor does it prevent the release of acetylcholine from vagal nerve endings. The studies also further confirm that sumatriptan does not have 5-HT_3 receptor–blocking activity.[33,38]

A Localized Intracranial Vasoconstrictor Action of Sumatriptan

The ability of sumatriptan to produce constriction of certain cerebral blood vessels from a range of species including man is now well documented.[51,52] Our recent studies have focused on the effect of sumatriptan on the meningeal circulation. In these studies human dura mater was perfused with Krebs solution via a branch of the meningeal

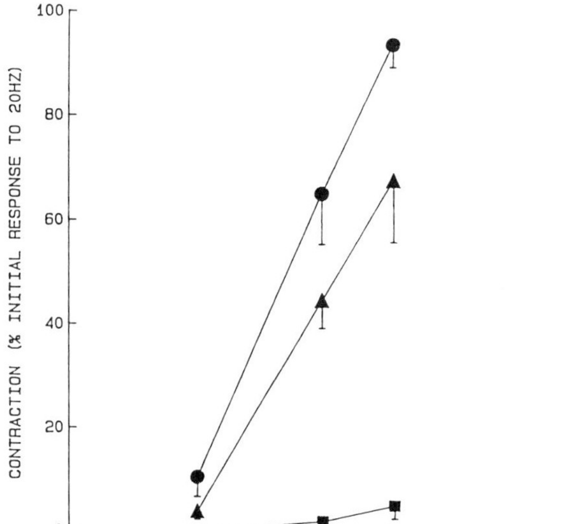

FIGURE 2. Effect of 10 μM sumatriptan (GR43175) (▲) on the substance P component of the contractile response to electrical field stimulation of rabbit isolated iris sphincter muscle (3–20 Hz for ten seconds every ten minutes at supramaximal voltage (16 V), ●). Note that responses in the presence of sumatriptan were little changed but were abolished by 1 μM tetrodotoxin (■). Values are the mean ± SE of 4 observations.

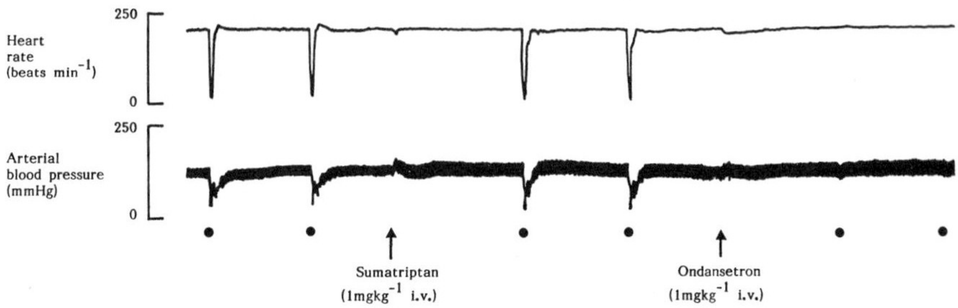

FIGURE 3. Effect of sumatriptan (1 mg/kg i.v.) and ondansetron (1 mg/kg i.v.) on the Bezold-Jarisch reflex response (fall in heart rate) produced by 2-methyl 5-HT (10 μg/kg^{-1} i.v., ●) in the chloralose-anaesthetized cat. Note the lack of effect of sumatriptan, whilst responses were abolished by ondansetron, a potent 5-HT_3 receptor blocker.[50]

artery and changes in perfusion pressure monitored. Sumatriptan caused a dose-dependent increase in perfusion pressure in the dura (FIG. 4), suggesting that sumatriptan-sensitive 5-HT_1-like receptors mediate vasoconstriction in these vessels. The finding that sumatriptan is highly effective in alleviating migraine headache would be consistent with the view that migraine headache is associated with a dilation of blood vessels in the meningeal circulation.[45,53]

We have also carried out a preliminary study in investigating the influence of sumatriptan on histamine-induced increases in intracranial pressure in anaesthetized dogs. In these studies, sumatriptan in doses up to 1.1 mg/kg i.v., caused marked increases in carotid arterial vascular resistance, but did not modify histamine-induced

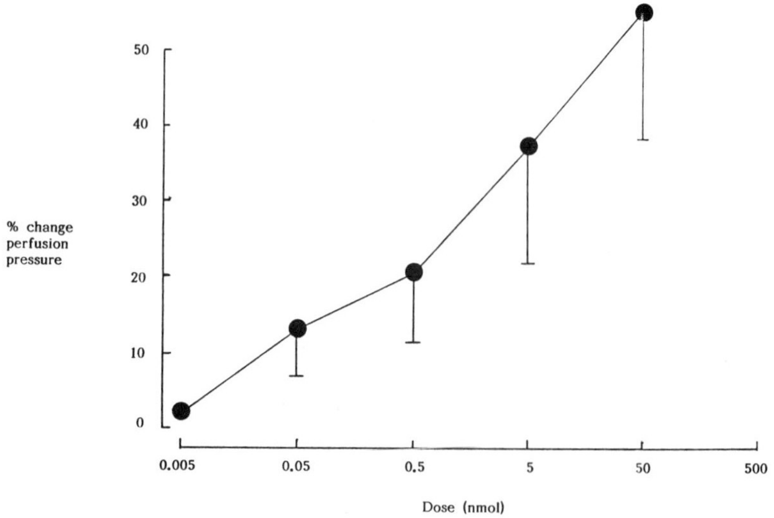

FIGURE 4. The dura mater was isolated from human cadavers 2–8 hours after death and the vasculature perfused with Krebs at a constant flow rate of 1 ml/min. Increasing doses of sumatriptan (50 μl volume) were injected as a bolus into the perfusion circuit and changes in perfusion pressure measured. Values shown are the mean ±SE from five experiments.

increases in intracranial pressure (FIG. 5). Although 5-HT_1-like receptors mediating vasoconstriction are clearly present in the cerebral vasculature of the dog,[51] their distribution does not appear to be sufficiently widespread or accessible to reduce the increase in intracranial pressure following the administration of histamine. Whether or not sumatriptan has the ability to reduce the increases in intracranial pressure produced by agents such as bradykinin or substance P is as yet unknown. The lack of any reduction in cerebral blood flow with sumatriptan in animals or man[34,54] further

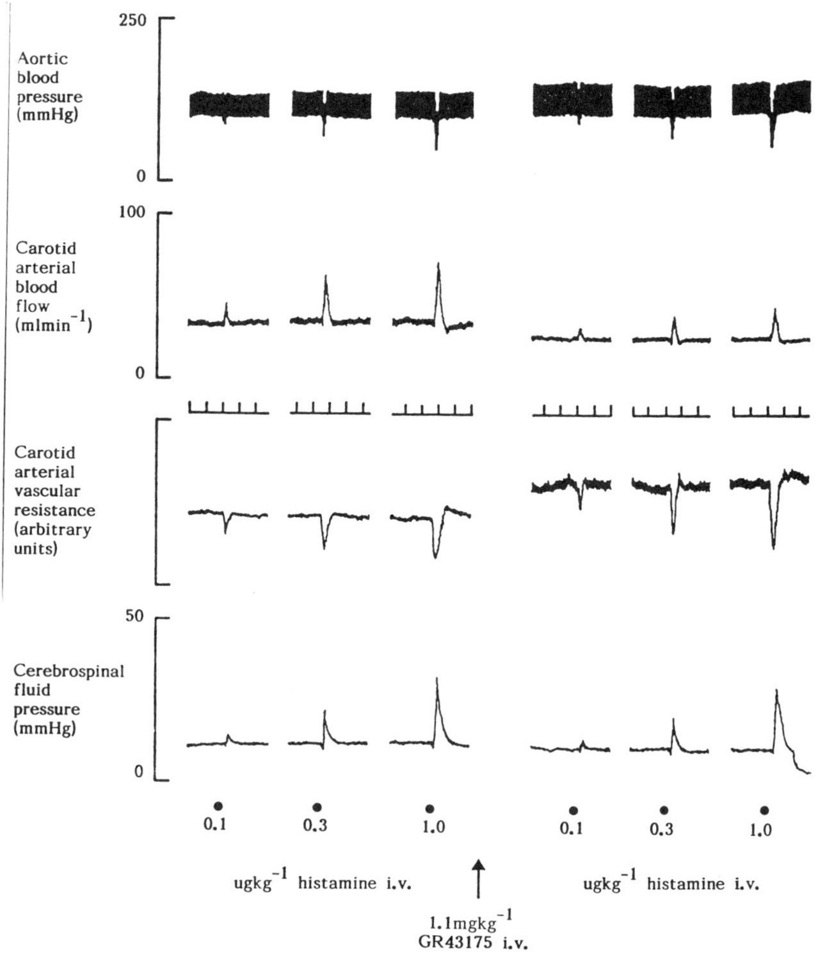

FIGURE 5. Effect of sumatriptan (GR43175; 1.1 mg/kg i.v.) on increases in cerebrospinal fluid (CSF) pressure and arterial vasodilator responses elicited by intravenously administered histamine (0.1–1 μg/kg) in the barbitone-anaesthetized dog. Sumatriptan had little or no effect on the changes produced by histamine on blood pressure though the increase in carotid flow produced by histamine was blunted. Note that sumatriptan itself produced a decrease in carotid flow and an increase in carotid arterial vascular resistance with little change in arterial blood pressure, demonstrating its selective vasoconstrictor action in the carotid circulation.

emphasizes its remarkably selective vasoconstrictor action, which is not only largely localized to the cranial vasculature but also *within* the cranial vasculature.

Effects of Sumatriptan on Neurogenically Mediated Plasma Protein Extravasation

The potential importance of the trigeminal nerve in the genesis of migraine has already been emphasized. Ergot alkaloids which despite side effects have long been the mainstay of acute migraine therapy inhibit neurogenic extravasation of plasma protein into the dura mater.[55] Recent studies in Moskowitz's laboratory[56] have shown that sumatriptan (30–100 μg/kg i.v.) inhibits plasma protein extravasation into the dura mater induced by electrical stimulation of the trigeminal nerve or following intravenous capsaicin (which stimulates unmyelinated sensory nerve fibres). Sumatriptan did not antagonize the extravasation of plasma protein induced by substance P or neurokinin A though it did antagonize that induced by bradykinin. It should be emphasized that the doses of sumatriptan employed were similar to those used clinically in the acute treatment of migraine.[57] Although Moskowitz's results could be interpreted to indicate that the effects of sumatriptan in these studies are neuronal, the absence of measurements of the influence of sumatriptan on the caliber of meningeal vessels does not exclude the strong possibility that its inhibitory action results from vasoconstriction per se. Indeed we ourselves have also carried out studies examining the influence of sumatriptan on the extravasation of plasma protein into the guinea-pig hind paw following stimulation of the saphenous and fibularis nerves (TABLE 4). In these studies where we presume sumatriptan to be devoid of vasoconstrictor activity, it did not have any inhibitory activity in doses as high as 0.5 mg/kg i.v. These studies further demonstrate the very selective and restricted action of sumatriptan.

TABLE 4. Lack of Effect of Sumatriptan against Neuronally Stimulated Extravasation in Guinea-pig Hind Paw

	Evans Blue μg/paw		
	Right Paw (Unstimulated)	Left Paw (Stimulated)	% Increase in Plasma Protein
Control	7.6 ± 1.9	14.8 ± 3.3	227 ± 61
Sumatriptan	5.5 ± 0.9	10.8 ± 1.5	231 ± 64

Extravasation of plasma proteins (Evans Blue μg/paw) into the guinea-pig left hind paw following electrical stimulation of the saphenous and fibularis nerves at the level of the thigh (20V, 10Hz, 5ms, 5min) in control and sumatriptan (0.5 mg/kg i.v.) treated guinea-pigs. The increase in extravasation was compared with that seen in the unstimulated right hind paw. Values are means ±SEM from six control experiments and five experiments with sumatriptan.

SUMMARY

Migraine has long been considered as a "vascular headache" but clearly neurological mechanisms are involved. The pathophysiology appears to somehow involve serotonin, both peripherally and centrally, but its involvement may be just epiphenomenal. Adding to the enigma it is apparent that many of the presently available drugs for the treatment of migraine interact in one way or another with serotonin receptors. However, they tend to have a number of other unrelated actions and they are only of limited clinical value. Interestingly a promising new drug for the treatment of the acute attack, sumatriptan, has a very selective action as an agonist at a specific 5-HT_1-like receptor sub-type, mediating vasoconstriction, which is localized on cranial blood vessels. Its action may, or may not, be independent of any involvement of serotonin in the genesis of migraine. Hopefully though, current attempts to determine sumatriptan's mechanism of action will shed further light on the pathology of migraine itself and the putative involvement of serotonin.

REFERENCES

1. Raskin, N. H., Ed. 1988. Headache. 2d edit. New York: Churchill Livingston.
2. Linet, M. S. & W. F. Stewart. 1987. The epidemiology of migraine headache. *In* Migraine—Clinical, therapeutic, conceptual and research aspects. J. N. Blau, Ed. London: Chapman and Hall, pp. 451–477.
3. Blau, J. N. 1987. Migraine—Clinical, therapeutic, conceptual and research aspects. London: Chapman and Hall.
4. Lashley, K. S. 1941. Arch. Neurol Psychiatry (Chicago) **46:** 331–339.
5. Lance, J. W. 1973. The mechanism and management of headache. 2nd edit. London: Butterworths.
6. Lance, J. W. & M. Anthony. 1966. Arch. Neurol. **15:** 356–361.
7. Olesen, J., A. Bes, R. Kunkel, J. W. Lance, G. Nappi, V. Pfaffenrath, F. C. Rose, B. S. Schoenberg, D. Soyka, P. Tfelt-Hansen, K. M. A. Welch & M. Wilkinson. 1988. Cephalalgia **8** (Suppl. 7): 1–96.
8. Graham, J. R. & H. G. Wolff. 1938. Arch. Neurol. Psychiatry (Chicago) **39:** 737–763.
9. Wolff, H. G. 1963. Headache and other head pain. London: Oxford University Press.
10. Raskin, N. H. 1981. Ann. Rev. Pharmac. Toxicol. **21:** 463–478.
11. Lance, J. W., G. A. Lambert, P. J. Goadsby & A. S. Zagami. 1989. Cephalalgia **8** (Suppl. 9): 7–13.
12. Sicuteri, F., A. Testi & B. Anselmi. 1961. Int. Arch. Allergy **19:** 55–58.
13. Curran, D. A., H. Hinterberger & J. W. Lance. 1965. Brain **88:** 997–1007.
14. Anthony, M., H. Hinterberger & J. W. Lance. 1967. Arch. Neurol. **16:** 544–552.
15. Kimball, R. W., A. P. Friedman & E. Valeejo. 1960. Neurology **10:** 107–111.
16. Anthony, M., H. Hinterberger & J. W. Lance. 1969. Res. Clin. Study Headache **2:** 29–59.
17. Fozard, J. R. 1982. Prog. Pharmacol. **4/4:** 134–146.
18. Ostfeld, A. M. & H. G. Wolff. 1955. Arch. Neurol. Psychiatry (Chicago) **74:** 131.
19. Peatfield, R. C., J. R. Fozard & F. C. Rose. 1986. Handbook of Clin. Neurology **4:** 173–216.
20. Leysen, J. E., C. J. E. Niemegeers, J. M. van Nueten & P. M. Laduron. 1982. Mol. Pharmacol. **21:** 301–314.
21. Fozard, J. R. 1975. J. Pharm. Pharamacol. **27:** 297–321.
22. Engel, G., M. Göthert, D. Hoyer, E. Schlicker & K. Hillenbrand. 1986.Naunyn Schmiedebergs Arch. Pharmacol. **332:** 107.
23. Herrick-Davis, K. & M. Titeler. 1988. J. Neurochem. **50:** 1624–1631.
24. Affolter, H., W. P. Burkard & A. Pletscher. 1985. Eur. J. Pharmacol. **108:** 157–162.
25. Auguet, M., S. Delaflotte, F. Clostre & F. V. Defeudis. 1986. Gen. Pharmacol. **17**(2): 133–135.
26. Peroutka, S. J. 1988. Ann. Neurol. **23:** 500–504.
27. Berde, B. & H. O. Schild. 1978. Ergot alkaloids and related compounds. Handbook of Experimental Pharmacology Vol. 49. New York: Springer-Verlag.
28. Loisy, C., S. Beorchia, V. Centonze, J. R. Fozard, P. J. Schechter & G. P. Tell. 1986. Cephalalgia **5:** 79–82.
29. Van Heuven-Nolsen, D. 1988. Trends in Pharmacol. Sci. **9:** 423–425.

30. LATASTE, X. 1988. Unpublished observations reported at International Symposium: Cardiovascular Pharmacology of 5-Hydroxytryptamine, October 1988, Amsterdam, The Netherlands.
31. DOENICKE, A., J. BRAND & V. L. PERRIN. 1988. Lancet **i:** 1309–1311.
32. HUMPHREY, P. P. A., W. FENIUK, M. J. PERREN, H. E. CONNOR & A. W. OXFORD. 1989. Cephalalgia **9** (Suppl. 9): 23–33.
33. HUMPHREY, P. P. A., W. FENIUK, M. J. PERREN, H. E. CONNOR, A. W. OXFORD, I. H. COATES & D. BUTINA. 1988. Br. J. Pharmacol. **94:** 1123–1132.
34. PERREN, M. J., W. FENIUK & P. P. A. HUMPHREY. 1989. Cephalalgia **9** (Suppl. 9): 41–46.
35. HUMPHREY, P. P. A., W. FENIUK, M. J. PERREN, A. W. OXFORD & R. T. BRITTAIN. 1989. Drugs of the Future **14**(1): 35–39.
36. FENIUK, W., P. P. A. HUMPHREY & M. J. PERREN. 1989. Br. J. Pharmacol. **96:** 83–90.
37. FENIUK, W., P. P. A. HUMPHREY & M. J. PERREN. 1989. Cephalalgia **9** (Suppl. 9): 35–39.
38. PEROUTKA, S. J. & B. G. MCCARTHY. 1989. Eur. J. Pharmacol. **163:** 133–136.
39. FENIUK, W., P. P. A. HUMPHREY & A. D. WATTS. 1981. J. Auton. Pharmacol. **1:** 127–132.
40. MIDDLEMISS, D. N., M. E. BREMER & S. M. SMITH. 1988. Eur. J. Pharmacol. **157:** 101–107.
41. HOYER, D. & P. SCHOEFFTER. 1988. Eur. J. Pharmacol. **147:** 145–147.
42. SUMNER, M. J. & P. P. A. HUMPHREY. 1989. Br. J. Pharmacol. **98:** 29–31.
43. FUJIWARA, M., H. HAYASHI, I. MURAMATSU & N. UEDA. 1984. J. Physiol. **350:** 580–597.
44. UEDA, N., I. MURAMATSU, H. HAYASHI & M. FUJIWARA. 1982. Life Sci. **31:** 369–375.
45. BLAU, J. N. 1978. Lancet **i;** 1136–1139.
46. REINHARD, J. F., J. E. LEIBMAN, A. J. SCHLOSBERG & M. A. MOSKOWITZ. 1979. Science **206:** 85–87.
47. FOZARD, J. R. 1982. Basic mechanisms of antimigraine drugs. *In* International Headache Congress. M. Critchley *et al.,* Eds. New York: Raven Press, pp. 295–308.
48. BRADLEY, P. B., G. ENGEL, W. FENIUK, J. R. FOZARD, P. P. A. HUMPHREY, D. N. MIDDLEMISS, E. J. MYLECHARANE, B. P. RICHARDSON & P. R. SAXENA. 1986. Neuropharmacology **25**(6): 563–576.
49. PAINTAL, A. S. 1973. Physiol. Rev. **53:** 159–227.
50. BUTLER, A., J. M. HILL, S. J. IRELAND, C. C. JORDAN & M. B. TYERS. 1988. Br. J. Pharmacol. **94:** 397–412.
51. CONNOR, H. E., W. FENIUK & P. P. A. HUMPHREY. 1989. Br. J. Pharmacol. **96:** 379–387.
52. PARSONS, A. A., E. T. WHALLEY, W. FENIUK, H. E. CONNOR & P. P. A. HUMPHREY. 1989. Br. J. Pharmacol. **96:** 434–449.
53. MOSKOWITZ, M. A. 1984. Ann. Neurol. **16:** 157–168.
54. FRIBERG, L. 1989. Cephalalgia **9** (Suppl. 10): 359–360.
55. SAITO, K., S. MARKOWITZ & M. A. MOSKOWITZ. 1988. Ann. Neurol. **24:** 732–737.
56. BUZZI, M. G. & M. A. MOSKOWITZ. 1989. Cephalalgia **9** (Suppl. 10): 27–28.
57. PERRIN, V. L., M. FÄRKKILÄ, J. GOASGUEN, A. DOENICKE, J. BRAND & P. TFELT-HANSEN. 1989. Cephalalgia **9** (Suppl. 9): 63–72.

DISCUSSION OF THE PAPER

COMMENT: There are some unique 5-HT receptors described before by you and Feniuk in the dog saphenous vein and then Peroutka, and then Nelson in the dog basilar artery.

HUMPHREY: Yes, we first identified the unique 5-HT_1-like receptor I've been talking about generically, in the dog saphenous vein. In order not to complicate my talk about cranial blood vessels and migraine, I did not mention it today. But the dog saphenous vein appears to contain the same receptor as occurs in the cranial blood vessels I discussed. The pharmacology seems identical, and in fact we recently published a paper on the human basilar artery as well and the 5-HT receptor there seems to be just the same as the one in the dog saphenous vein, too.

QUESTION: Vanhoutte tells us that some 5-HT receptors are not directly on the smooth muscle, but on endothelial cells. Did you check to what extent you activate selective vasoconstriction by maybe releasing something locally in those blood vessels attached.

HUMPHREY: Yes, we have looked in some blood vessels by taking off the endothelium, but we found it did not modify the effect at all. However, Hoyer's latest findings show that sumatriptan does seem to activate the endothelium in the pig coronary artery where it produces a small relaxant effect and that's unusual. It's the first time an endothelial-dependent effect of the drug has been described.

D. HOYER (*Sandoz, Ltd., Basel, Switzerland*): In the peripheral tissues you've been using, or even the vascular tissues, you should probably try to do some correlations to see whether what you are seeing corresponds to the 5-HT_{1D} receptor. I would guess from the order of potency of some of the agonists that you've been using, that it is 5-HT_{1D}-like, but have you been doing more work on this?

HUMPHREY: Not really, no. I mean I think one of the problems we've pointed out in one of our papers is that the dog saphenous vein 5-HT receptor is not blocked by metergoline, which has, as you know, high affinity for the 5-HT_{1D} site. So on that I have to rest my case. But I suppose definitive evidence will come with a good selective antagonist either for the dog saphenous vein receptor or the 5-HT_{1D} receptor.

MILT TEITLER (*Albany Medical College, NY*): I was under the impression there was clinical data on the drug and it was rather dramatic. Do you have clinical data?

HUMPHREY: Yes, there's lots of good clinical data but as a laboratory pharmacologist I chose not to show it today. But in a few months time, the results of some large placebo controlled clinical trials will be reported at the IVth International Headache Congress.

JOSEE LEYSEN (*Janssen Research Foundation, Beerse, Belgium*): Have you done any chronic studies with the compound and is there any indication of desensitization of these receptors?

HUMPHREY: I think that's a very interesting point. I think we have to do more work on this, but certainly in animals we've seen no evidence of desensitization but it is obviously a point we must bear in mind with sumatriptan being an agonist. Clinically, too, there's been no evidence to date of desensitization.

DAVID CLARKE (*Syntex, Palo Alto, CA*): Sumatriptan inhibits the release of transmitter from certain nerves. Is there any evidence that it might inhibit release of substance P or CGRP from sensory afferents supplying the cerebral vasculature.

HUMPHREY: I think that's a very good point. Indeed, Michael Moskowitz thinks that's how sumatriptan is working in the diagram I showed and so he's trying to measure substance P and CGRP release in rats. But they are difficult experiments to perform at the level of the meningeal vasculature.

QUESTION: Are there any positive data on that yet?

HUMPHREY: No. I mean I still believe that it's going to be a vasoconstrictor but we've got to do the experiments.

MANFRED GÖTHERT (*University of Bonn*): I would like to come back to your last scheme that you were showing concerning the efferent branch of the innervation of the blood vessels in the brain. And you showed us that the cell bodies are located in the raphe nuclei and I guess this is mainly based upon the work of Burnstock. But now a lot of evidence is available that the serotoninergic nerves that are visible there are in fact the noradrenergic nerves which have taken up serotonin. And I think much work has been done on this by Owman's group. Could you please comment on that. I think it wouldn't spoil your hypothesis.

HUMPHREY: Of course, you're right. I think there has been a lot of recent evidence suggesting that the serotoninergic innervation is not as dense as was first thought. I think many of the so called serotoninergic nerves probably are noradrenergic nerves that have taken up sérotonin as you suggest. Nevertheless I think there is a school of thought that says there are a few genuine serotoninergic nerves, but really it's irrelevant to my hypothesis. I could have put any efferent arm in my scheme; it's the afferent neuronal path via the trigeminal nerve I think that's important.

MDMA: Historical Perspectives

JAMES W. GIBB, MICHEL JOHNSON, DONNA STONE, AND GLEN R. HANSON

Department of Pharmacology and Toxicology
University of Utah
Salt Lake City, Utah 84112

I understand my role is to set the stage for this session of the symposium by reviewing the history of the toxicity of the substituted amphetamines and focus primarily on 3,4-methylenedioxymethamphetamine (MDMA). In discussing a topic dealing with the historical perspectives of any subject one runs a significant risk. In this kind of an assignment you refer heavily to the work of others as well as to your own. In spite of a concerted effort to give proper credit to the respective investigators, it is almost impossible to acknowledge the contribution of everyone. At the outset, then, I apologize to anyone whose work may not be mentioned or whom I may inadvertently exclude, misrepresent or underrepresent. Those present will, no doubt, provide the appropriate amendments; to those who are not in attendance today, I again apologize.

The first indication of toxicity associated with amphetamine congeners was observed by Professor Pletscher at Hoffmann-LaRoche in Basel, Switzerland in 1963 when he reported that *p*-chloral-*n*-methylamphetamine (RO 4-6861) caused a prolonged decrease in brain concentrations of 5-hydroxytryptamine (5-HT) and 5-hydroxyindoleacetic acid (5-HIAA) without significantly altering norepinephrine[1]; later he observed no change in dopamine concentrations.[2] This compound is now known as *p*-chloroamphetamine or PCA.

Fuller *et al.*[3] subsequently studied the structure-activity relationship of related compounds and found that the most toxic analog of amphetamine, as judged by depletion of brain 5-HT concentrations, was PCA. Throughout the 1970s, Fuller and his collaborators conducted extensive studies on the SAR and comparative metabolism of the substituted amphetamines.[4–6]

In 1970, Sanders-Bush and her colleagues[7] observed that, concurrent with the decrease in indoleamine concentrations, PCA inhibited the conversion of ^{14}C-labeled tryptophan to 5-HT. This was the first evidence that tryptophan hydroxylase (TPH) was markedly decreased by a substituted amphetamine compound. The decrease in TPH activity persisted for four months after a single dose of PCA.[8] These investigators also reported a persisting impairment of 5-HT uptake in animals treated with PCA.

Harvey *et al.*[9] later provided anatomical evidence that PCA was toxic to the serotonergic system; as early as one day after administration of the drug they observed histological alterations which persisted for 30 days after a single dose of PCA (20 mg/kg). PCA, then, appears to be a reasonable model for amphetamine congeners that impair the serotonergic system.

We entered the field of neurotoxicity of amphetamines in 1971.[10,11] A significant increase in the abuse of amphetamine and methamphetamine (METH) was causing considerable alarm. The question was raised as to what the effects of METH in the high doses used by drug abusers might have on dopaminergic parameters in the brain. To answer this question we attempted to simulate in rats the conditions in humans who abuse amphetamines. Five large doses of METH (15 mg/kg, at 6-h intervals) were administered to rats. Various brain areas were analyzed for concentrations of dopamine (DA) and its metabolites, dihydroxyphenylacetic acid (DOPAC) and homovanillic acid (HVA); the activity of the rate-limiting enzyme, tyrosine hydroxylase (TH), was also measured. In the neostriatum, each of these dopaminergic parameters was markedly decreased after METH.

In an effort to explain this fascinating response an attempt was made to block the effect.[12] At that time the drug of choice in the treatment of amphetamine toxicity in emergency rooms was chlorpromazine (CPZ). CPZ was administered concurrently with the METH. METH caused its characteristic decrease in TH activity, but when CPZ was administered with METH, the METH-induced decrease in TH activity was prevented in a dose-dependent manner.

Since CPZ blocks DA receptors, this blockade by CPZ of the METH effect suggested that DA was somehow involved in this decrease in serotonergic parameters. The next step, therefore, was to interrupt DA synthesis by giving an inhibitor of TH, α-methyl-*p*-tyrosine (AMT), along with the METH.[13] AMT essentially prevented the METH-induced decrease in striatal activity. When DA synthesis was reinstated by circumventing the rate-limiting TH step by administering L-DOPA and a peripheral decarboxylase inhibitor, a decrease in TH activity was again observed.

Seiden and his co-workers[14,15] also observed a decrease in DA concentrations after toxic doses of METH. Later they found that brain DA uptake and 5HT content were severely compromised after METH[16] and that the METH-induced decrease in DA uptake could be prevented by AMT.[17] Ricaurte *et al.*[18] subsequently reported histological evidence of tissue disruption using the Fink-Heimer staining technique.

In 1980, Hotchkiss and Gibb (FIG. 1; ref. 19) reported that METH not only impaired the dopaminergic system but decreased serotonergic parameters. When METH was administered the decrease in neostriatal 5HT and 5HIAA concentrations was even more pronounced than the decrease of neurotransmitter in the dopaminergic systems; moreover, TPH activity was also dramatically decreased.

We[20] demonstrated a role for DA in the serotonergic response (FIG. 2). When AMT was administered with METH, neostriatal TPH was not altered by METH. However, when DA synthesis was reinstated, by administering L-DOPA and the decarboxylase inhibitor, the decrease in TPH activity by METH was again realized.

In 1985, Ricaurte *et al.*[21] reported that the methylenedioxy congener of amphetamine, 3,4-methylenedioxyamphetamine (MDA), caused dramatic and long-lasting decreases in concentrations of 5-HT and 5-HIAA in the neostriatum and hippocampus in a dose-dependent manner; serotonin uptake sites were also impaired. Moreover, they

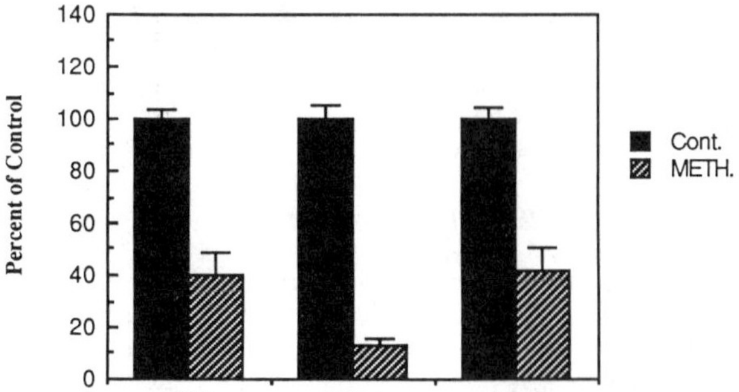

FIGURE 1. Effects of multiple doses of METH on striatal serotonergic system. Rats were injected with METH (15 mg/kg, s.c., every 6 h for 5 doses). Animals were sacrificed 18 h after the last dose.

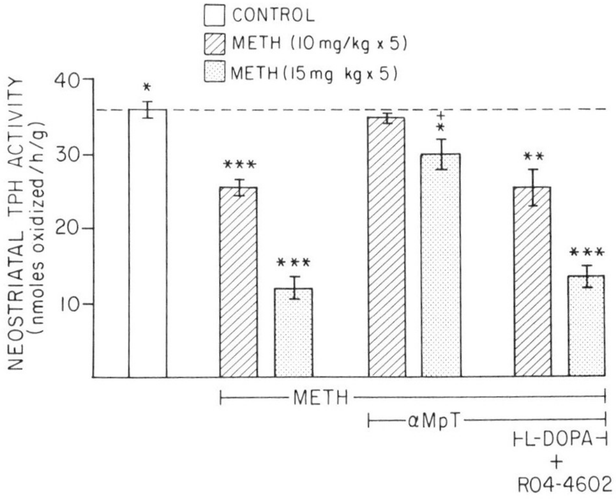

FIGURE 2. Effect of L-dopa and R04-4062 on the ability of αMpT to antagonize the depression of neostriatal TPH activity by multiple injections of METH. Each value represents the mean ± SEM for six or more animals. $*p < .05$; $**p < .01$; $***p < .001$ vs. control; $\dagger p < .001$ vs. METH alone by two-tailed Student's t-test. (After Schmidt *et al.*[2] Courtesy of *J. Pharmacol. Exp. Ther.*)

reported morphological evidence of tissue damage using the Fink-Heimer technique. Since that time Molliver and his co-workers[22,23] have also reported marked serotonergic deficits using an elegant histological technique. Interesting changes in receptors have been reported by Errol DeSouza and his colleagues.[24]

We investigated the effect of MDMA or MDA (10 mg/kg) on TH and TPH activity in the neostriatum (FIG. 3; refs. 25, 26). Animals were sacrificed 3 h after a single dose of either drug. TPH activity was decreased to less than 50% of control. Concentrations of 5-HT and 5-HIAA were decreased significantly after a single acute dose of MDMA (FIG. 4); the decreases in serotonergic parameters were more pronounced and persisted for longer periods of time after multiple doses of the drug. MDMA, then, presents a similar profile to that observed previously for PCA. Although dopamine content was increased initially with MDMA, it was essentially normal 24 h after administering the drug. This response differs from the effects of METH where concentrations of dopamine and its metabolites are decreased for long periods of time.

In view of increasingly widespread abuse of MDMA, the possible irreversible or prolonged effects of MDMA on the serotonergic system was investigated. The time-course of changes in serotonergic parameters after a single dose of MDMA (10 mg/kg) are depicted in FIGURE 5.[27] TPH activity was decreased within the first h after administration of MDMA and continued to decline over the 3–6-h period. 5HT concentrations began to decrease early on and were suppressed significantly by one h in all regions. 5-HIAA remained normal or slightly above normal at the early times and did not decrease below normal until approximately 3 h after the drug was administered. When the response was examined over the total 2-week period, as is depicted in the right side of the figure, TPH activity remained depressed throughout the 2-week period in all regions examined except the hypothalamus. The concentrations of 5-HT and 5-HIAA returned toward normal between 6 and 24 h, but decreased again at 2 weeks in both the neostriatum and hippocampus.

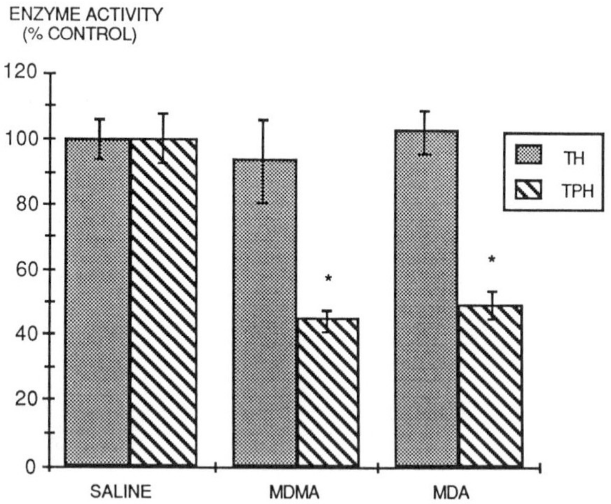

FIGURE 3. Effect of acute drug treatments on neostriatal tyrosine hydroxylase and tryptophan hydroxylase activities. Rats were killed 3 h after a single 10 mg/kg injection of drug. Results are presented as the means ± SEM (n = 6) and expressed as percent control. Control values for TH and TPH activities were 2178.8 and 38.2 nmol/g tissue per h, respectively. *$p < 0.001$ versus control, by the two-tailed Student's t-test. (After Stone *et al.*[25] Courtesy of *Eur. J. Pharmacol.*)

In contrast to FIGURE 5, in FIGURE 6 multiple doses of MDMA were administered; 5 doses of MDMA (5 or 10 mg/kg, every 6 h) were administered and no more drug was administered after that time. Animals were then sacrificed at varying times after the fifth dose and neostriatal serotonergic parameters were measured. At the higher dose, TPH activity and concentrations of 5-HT and 5-HIAA decreased to levels that were 20–40% of control at the time treatment ceased. Even 110 days after the last dose of

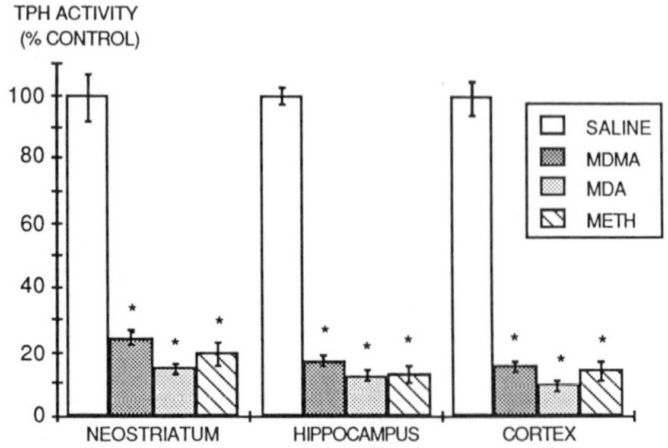

FIGURE 4. Effect of subacute drug treatments on tryptophan hydroxylase activity in the neostriatum, hippocampus and cortex. Treatment consisted of s.c. injection of MDA (10 mg/kg), MDMA (10 mg/kg) or METH (15 mg/kg) every 6 h for 5 doses. Rats were killed 18 h after the last dose. Combined results from 2 experiments ($n \geq 12$) are presented as the means ± SEM, expressed as percent control. Control (saline) values (in nmol $^{14}CO_2$ liberated/g tissue per h) were: neostriata, 45.0; hippocampus, 50.6; cortex, 75.4. *$p < 0.001$ versus control, by the two-tailed Student's t-test. (After Stone *et al.*[25] Courtesy of *Eur. J. Pharmacol.*)

MDMA, all three serotonergic parameters were only approximately 50–60% of control. These results are disturbing in view of the fact that MDMA is being abused significantly on a large number of college campuses. The doses employed in this study are only 2–4 times the acute human dose on a mg/kg basis; however, species variation may influence the response.

The short-term and long-term effects of the amphetamines can be illustrated in FIGURES 5 and 6; when a single low dose of MDMA is given there is the characteristic decrease in serotonergic parameters followed by a gradual return to normal. However,

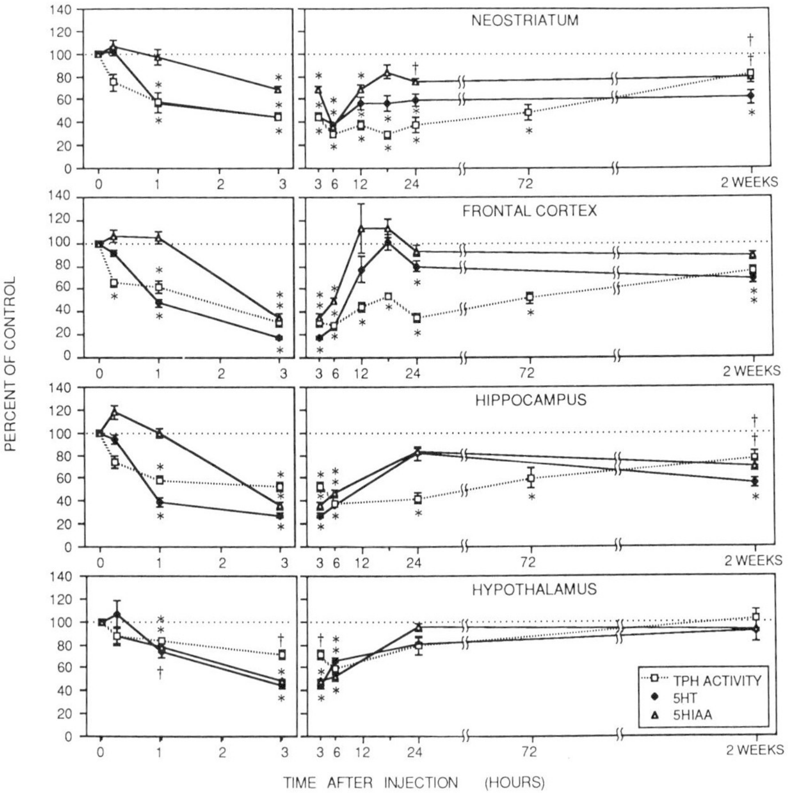

FIGURE 5. Time course of the regional serotonergic effects of acute administration of MDMA. A single dose of MDMA (10 mg/kg) or saline (control) was injected subcutaneously; rats were killed at specified times thereafter. Each point represents the mean ± SEM from 4–6 rats, expressed as a percentage of the corresponding control. Immediate effects (up to 3 h after injection) are represented in the left panel; the right panel diagrams longer-term regional responses (from 3 h–2 weeks after injection). One h control values ± SEM for neostriata (n), frontal cortex (fc), hippocampus (h) and hypothalamus (ht) were as follows: activity of tryptophan hydroxylase (TPH) (in nmol/g tissue/hr); n = 39.9 ± 3.6, fc = 80.1 ± 4.9, h = 63.0 ± 2.6, ht = 249.3 ± 7.0; concentrations of 5-HT and 5-HIAA, respectively (in μg/g tissue): n = 0.533 ± 0.019 and 0.557 ± 0.023, fc = 0.518 ± 0.018 and 0.218 ± 0.013, h = 0.359 ± 0.040 and 0.338 ± 0.009, ht = 0.905 ± 0.064 and 0.396 ± 0.018. Control values at other times did not vary significantly from those listed above. †$p < 0.05$, *$p < 0.005$ vs. corresponding control by the two-tailed Student's t-test. (After Stone *et al.*[27] Courtesy of *Neuropharmacology*.)

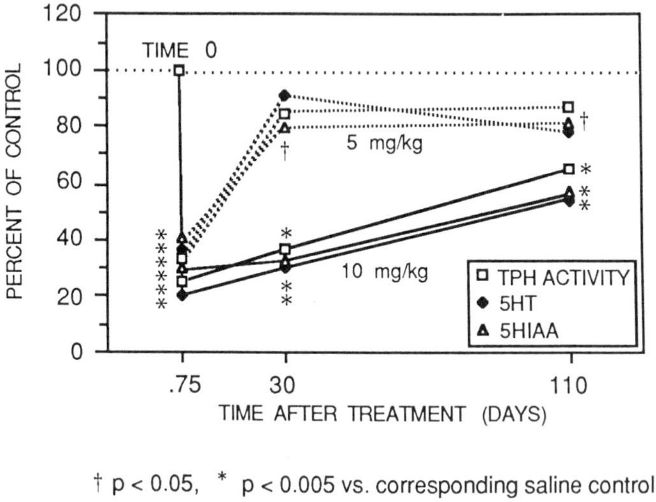

FIGURE 6. Long-term recovery of serotonergic parameters in the neostriatum after multiple doses of MDMA. Experimental details are described in FIGURE 3. Each point represents the mean value (error bars have been omitted for the sake of clarity) from 6–10 rats, expressed as a percentage of the corresponding control. Doses of 5 mg/kg (dotted lines) and 10 mg/kg (solid lines) are represented. Control enzyme activities are listed in FIGURE 3. Control concentrations, of 5-HT and 5-HIAA respectively, (in μg/g tissue) were: 0.469 ± 0.024 and 0.362 ± 0.023 at 18 h; 0.365 ± 0.061 and 0.381 ± 0.034 at 30 days; and 0.527 ± 0.047 and 0.453 ± 0.032 at 110 days. †$p < 0.05$ (5-HIAA only), *$p < 0.005$ vs. corresponding control, by the two-tailed Student's t-test. (After Stone *et al.*[27] Courtesy of *Neuropharmacology*.)

when a very large dose, or if repeated doses of the drug are administered, the serotonergic parameters are decreased and remain depressed for extended periods of time. Whether these are two separate phenomena remains a subject of debate.

Over the past 9 years, we have explored the possible role of DA in the neurochemical effects of amphetamine and its analogs on the serotonergic system. Does dopamine play a role in the neurochemical changes observed after MDMA? In an effort to decrease DA concentrations one dose of AMT was administered 90 minutes before a single dose of MDMA and the animals were sacrificed three days later.[28] TPH activity was not different from control; changes in 5-HT concentrations were attenuated, but AMT had no effect on the MDMA-induced decrease in 5-HIAA concentrations.

We next determined whether depletion of DA with reserpine would also prevent the MDMA-induced decrease in striatal TPH activity.[28] Animals were pretreated with reserpine (5 mg/kg, i.p.) 12 h before a single dose of MDMA (20 mg/kg, s.c.) and the rats were sacrificed three days later. Similar to AMT, reserpine pretreatment completely blocked the persistent decrease of striatal TPH activity induced by MDMA.

In an additional effort to demonstrate the role of DA in the response induced by MDMA, the dopaminergic input to the neostriatum was selectively removed prior to treating with MDMA. Lesions of the substantia nigra were performed by injecting 6-hydroxydopamine bilaterally into the nigra. After a 7 to 10–day recovery period, a single dose of MDMA (5 or 10 mg/kg, s.c.) was administered and the rats were sacrificed 3 h later (FIG. 7). In the neostriatum where dopaminergic innervation had been interrupted, MDMA had no effect. However, the frontal cortex was not protected by the lesion and the protection in the hippocampus was relatively less. This suggests that DA, or another catecholamine, is essential for the MDMA effects to be realized. In the same study GBR 12909, a specific DA uptake blocker, attenuated the MDMA-induced effects on TPH activity and concentrations of 5-HT and 5-HIAA. From these observations, it appears that DA is involved in the neurotoxic response caused by MDMA.

Over the past six years we have been interested in the interaction of the biogenic amines and the neuropeptides. Only one study to illustrate some of our findings is presented (FIG. 8). We now believe that DA release elevates neurotensin concentrations in various brain regions. Schmidt *et al.*[29] have demonstrated that MDMA and METH release large amounts of DA; thus, these drugs serve as excellent tools for studying the neuropeptides. Neurotensin concentrations in the striatum, substantia nigra and nucleus accumbens were elevated to 200–400% of control by MDMA, MDA

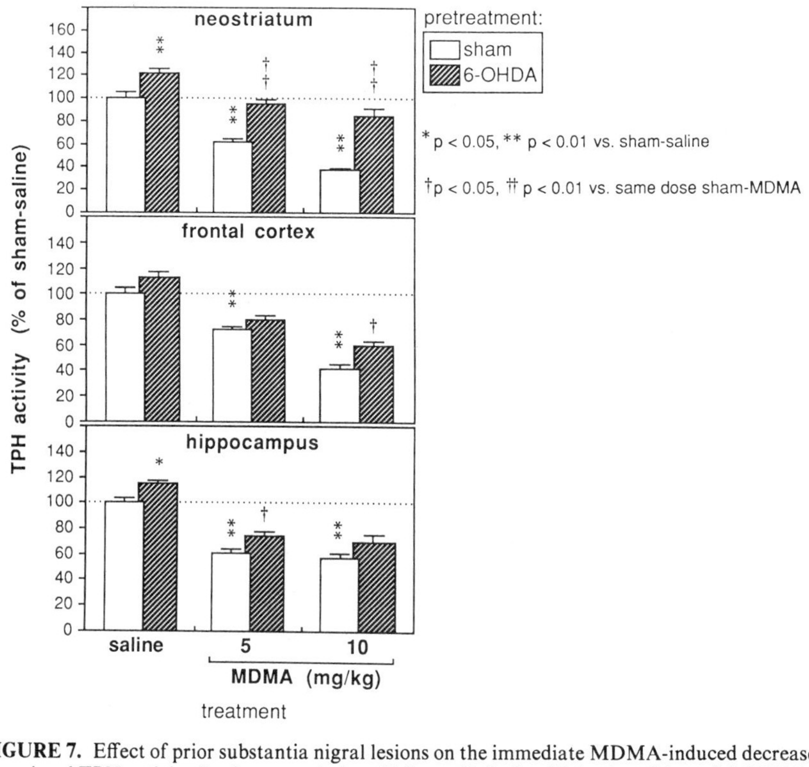

FIGURE 7. Effect of prior substantia nigral lesions on the immediate MDMA-induced decreases in regional TPH activity. Lesions were induced bilaterally by local injection of 4 μg of 6-OHDA-8 μl 0.1% ascorbate saline-side. Control rats received sham lesions of ascorbate vehicle alone. After a 7- to 10-day recovery period, acute MDMA (5 or 10 mg/kg) was administered s.c. and rats were sacrificed 3 h later. Results are the means ± SEM, expressed as a percentage of sham-saline (n = 22 for sham saline group, n = 14 for 6-OHDA-saline group, n = 6–12 for MDMA-treated groups). Control TPH activities (in nanomoles per gram of tissue per hour) were: striatum, 42.2 ± 2.3; frontal cortex, 77.3 ± 3.7; hippocampus, 52.2 ± 1.8. *$p < .05$; **$p < .01$ vs. sham-saline, †$p < 0.5$, ‡$p < .01$ vs. corresponding sham-MDMA. (After Stone *et al.*[28] Courtesy of *J. Pharmacol. Exp. Ther.*)

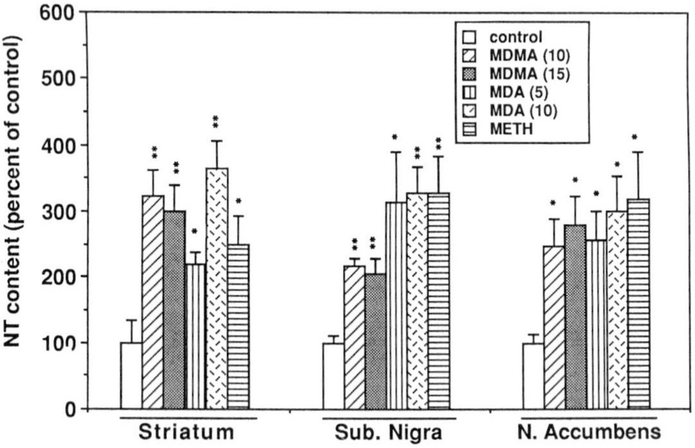

FIGURE 8. Effects of MDMA, MDA, and methamphetamine on neurotensin levels. Rats received multiple administrations of 2 different doses of MDMA (10 or 15 mg/kg/dose) or MDA (5 or 10 mg/kg/dose) or a single dose of METH (15 mg/kg/dose). *$p < 0.01$ and **$p < 0.001$ compared to respective controls. (After Hanson *et al.*[30] Courtesy of Stephen Peroutka.)

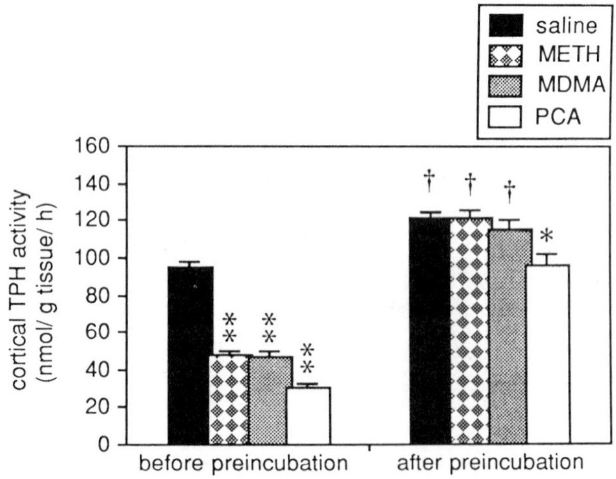

FIGURE 9. Comparison of cortical TPH activity, before and after reactivation, from rats treated with MDMA, methamphetamine (METH), or *p*-chloroamphetamine (PCA). Rats were administered a single subcutaneous injection of MDMA (10 mg/kg), METH (10 mg/kg), PCA (5 mg/kg), or saline, and sacrificed 3 h later. Cortical TPH activity (nmol/g tissue/h) was assessed following the addition of reactivation mixture, both before and after a 24-h preincubation under N_2. Results are the means ± SEM (n = 6). *$p < 0.05$, **$p < 0.001$ vs. time-matched saline. †denotes a significant difference ($p < 0.01$) between % after preincubation and % before preincubation [% = (TPH activity from drug-treated rats/TPH activity from time-matched saline-treated rats) × 100]. (After Stone *et al.*[35] Courtesy of *J. Neurochem.*)

or METH.[30] Similar alterations of other neuropeptides, including dynorphin A,[31] substance P,[32] and substance K[33] have been observed. The implications of these observations in psychiatric disorders, drug abuse, and neurological disorders are being actively explored.

Hopefully, I have provided an overview of the history of neurotoxicity and neurochemical effects associated with administration of large doses of the substituted amphetamines, especially MDMA. In my opinion, the most challenging and fascinating area of research in this field deals with the mechanism by which MDMA and related compounds cause their effects. We feel that dopamine and/or one of its metabolites is probably involved; this is not in conflict with the suggestion of Seiden and his co-workers[34] that 6-hydroxydopamine or some comparable metabolite may be implicated.

Recently we observed that when TPH, derived from MDMA-treated animals, was exposed to reducing conditions, there was complete reversal of the early effects of MDMA and related compounds on TPH activity (FIG. 9; ref. 35). This observation suggests that oxidative stress or related processes may be involved in the phenomenon. Recently, Sonsalla and her co-workers[36] have demonstrated that MK-801, an NMDA antagonist, protects against the METH-induced changes in dopaminergic parameters. Interestingly, however, we have seen protection of METH, but no attenuation of the MDMA-induced changes in serotonergic parameters.[37] This may be another avenue to explore in order to elucidate the mechanism(s) which are involved in the neurotoxicity associated with these compounds.

REFERENCES

1. PLETSCHER, A., W. P. BURKARD, H. BRUDERER & K. F. GEY. 1963. Decrease of cerebral 5-hydroxytryptamine and 5-hydroxyindolacetic acid by arylalkylamine. Life Sci. **11:** 828–833.
2. PLETSCHER, A., G. BARTHOLINI, H. BRUDERER, W. P. BURKARD & K. F. GEY. 1964. Chlorinated arylalkylamines affecting the cerebral metabolism of 5-hydroxytryptamine. J. Pharmacol. Exp. Ther. **145:** 344–350.
3. FULLER, R. W., C. W. HINES & J. MILLS. 1965. Lowering of brain serotonin level by chloroamphetamines. Biochem. Pharmacol. **14:** 483–488.
4. FULLER, R. W., H. D. SNODDY, B. W. ROUSH & B. B. MOLLOY. 1973. Further structure-activity studies on the lowering of brain 5-hydroxyindoles by 4-chloroamphetamines. Neuropharmacol. **12:** 33–42.
5. FULLER, R. W. & K. W. PERRY. 1977. Further studies on the effects of N-cyclopropyl-p-chloroamphetamine in rat brain. Neuropharmacol. **16:** 495–497.
6. FULLER, R. W., H. D. SNODDY, A. M. SNODDY, S. K. HEMRICK, D. T. WONG & B. B. MOLLOY. 1980. p-Iodoamphetamine as a serotonin depletor in rats. J. Pharmacol. Exp. Ther. **212:**115–119.
7. SANDERS-BUSH, E. & F. SULSER. 1970. p-Chloroamphetamine: In vivo investigations on the mechanism of action of the selective depletion of cerebral serotonin. J. Pharmacol. Exp. Ther. **175:** 419–426.
8. SANDERS-BUSH, E., J. A. BUSHING & F. SULSER. 1975. Long-term effects of p-chloroamphetamine and related drugs of central serotonergic mechanisms. J. Pharmacol. Exp. Ther. **192:** 33–40.
9. HARVEY, J. A., S. E. MCMASTER & L. M. YUNGER. 1975. p-Chloroamphetamine: selective neurotoxic action in brain. Science **187:** 841–843.
10. KODA, L. Y. & J. W. GIBB. 1971. The effect of repeated large doses of methamphetamine on adrenal and brain tyrosine hydroxylase. The Pharmacologist **13:** 253.
11. KODA, L. Y. & J. W. GIBB. 1973. Adrenal and striatal tyrosine hydroxylase activity after methamphetamine. J. Pharmacol. Exp. Ther. **185:** 42–48.

12. BUENING, M. K. & J. W. GIBB. 1974. Influence of methamphetamine and neuroleptic drugs on tyrosine hydroxylase activity. Eur. J. Pharmacol. **26:** 30–34.
13. GIBB, J. W. & F. J. KOGAN. 1979. Influence of dopamine synthesis on methamphetamine-induced changes in striatal and adrenal tyrosine hydroxylase. Naunyn-Schmiedebergs Arch. Pharmacol. **310:** 185–187.
14. SEIDEN, L. S., M. W. FISCHMAN & C. R. SCHUSTER. 1975/76. Long-term methamphetamine induced changes in brain catecholamines in tolerant rhesus monkeys. Drug Alcohol Depend. **1:** 215–219.
15. WAGNER, G. C., L. S. SEIDEN & C. R. SCHUSTER. 1979. Methamphetamine induced changes in brain catecholamines in rats and guinea pigs. Drug Alcohol Depend. **4:** 135–138.
16. RICAURTE, G. A., C. R. SCHUSTER & L. S. SEIDEN. 1986. Long-term effects of repeated methylamphetamine administration on dopamine and serotonin neurons in the rat brain: A regional study. Brain. Res. **193:** 153–163.
17. COMMINS, D. L. & L. S. SEIDEN. 1986. α-Methyltyrosine blocks methamphetamine-induced degeneration in the rat somatosensory cortex. Brain Res. **365:** 15–20.
18. RICAURTE, G. A., R. W. GUILLERY, L. S. SEIDEN, C. R. SCHUSTER & R. Y. MOORE. 1982. Dopamine nerve terminal degeneration produced by high doses of methylamphetamine in the rat brain. Brain Res. **235:** 93–103.
19. HOTCHKISS, A. & J. W. GIBB. 1980. Long-term effects of multiple doses of methamphetamine on tryptophan hydroxylase and tyrosine hydroxylase activity in rat brain. J. Pharmacol. Exp. Ther. **214:** 257–262.
20. SCHMIDT, C. J., J. K. RITTER, P. K. SONSALLA, G. R. HANSON & J. W. GIBB. 1985. Role of dopamine in the neurotoxic effects of methamphetamine. J. Pharmacol. Exp. Ther. **233:** 539–544.
21. RICAURTE, G., G. BRYAN, L. STRAUSS, L. SEIDEN & C. SCHUSTER. 1985. Hallucinogenic amphetamine selectively destroys brain serotonin nerve terminals. Science **229:** 986–988.
22. MULLEN, C., L. A. MAMOUNAS, E. O'HEARN & M. E. MOLLIVER. 1987. Dual serotonergic projections to forebrain in the rat: Two classes of axon terminals exhibit differential vulnerability to psychotropic drugs p-chloroamphetamine (PCA) and 3,4-methylenedioxyamphetamine (MDA). Soc. Neurosci. Abst. **13:** 907.
23. O'HEARN, E., G. BATTAGLIA, E. B. DESOUZA, M. J. KUHAR & M. E. MOLLIVER. 1988. Methylenedioxyamphetamine (MDA) and methylenedioxymethamphetamine (MDMA) cause selective ablation of serotonergic axon terminals in forebrain: Immunocytochemical evidence for neurotoxicity. J. Neurosci. **8:** 2788–2803.
24. BATTAGLIA, G., S. Y. YEH, E. O'HEARN, M. E. MOLLIVER, M. J. KUHAR & E. B. DESOUZA. 1987. 3,4-Methylenedioxymethamphetamine and 3,4-methylenedioxyamphetamine destroy serotonin terminals in rat brain: Quantification of neurodegeneration by measurement of [^{3}H] paroxetine-labeled serotonin uptake sites. J. Pharmacol. Exp. Ther. **243:** 911–916.
25. STONE, D. M., D. C. STAHL, G. R. HANSON & J. W. GIBB. 1986. The effects of 3,4-methylenedioxymethamphetamine (MDMA) and 3,4-methylenedioxyamphetamine (MDA) on monoaminergic systems in the rat brain. Eur. J. Pharmacol. **128:** 41–48.
26. STONE, D. M., M. JOHNSON, G. R. HANSON & J. W. GIBB. 1987. A comparison of the neurotoxic potential of methylenedioxyamphetamine (MDA) and its N-methylated and N-ethylated derivatives. Eur. J. Pharmacol. **134:** 245–248.
27. STONE, D. M., K. M. MERCHANT, G. R. HANSON & J. W. GIBB. 1987. Immediate and long-term effects of 3,4-methylenedioxymethamphetamine (MDMA) on serotonin pathways in brain of rat. Neuropharmacology **26:** 1677–1683.
28. STONE, D. M., M. JOHNSON, G. R. HANSON & J. W. GIBB. 1988. Role of endogenous dopamine in the central serotonergic deficits induced by 3,4-methylenedioxymethamphetamine. J. Pharmacol. Exp. Ther. **247:** 79–87.
29. SCHMIDT, C. J., J. A. LEVIN & W. LOVENBERG. 1987. In vitro and in vivo neurochemical effects of methylenedioxymethamphetamine on striatal monoaminergic systems in the rat brain. Biochem. Pharmacol. **36:** 747–755.
30. HANSON, G. R., K. M. MERCHANT, M. JOHNSON, A. A. LETTER, L. BUSH & J. W. GIBB. 1990. Effect of MDMA-like drugs on CNS neuropeptide systems. *In* Ecstasy; the

Clinical, Pharmacological and Neurotoxicological Effects of the Drug MDMA. Stephen Peroutka, Ed.
31. HANSON, G. R., K. M. MERCHANT, A. A. LETTER, L. BUSH & J. W. GIBB. 1988. Characterization of methamphetamine effects on the striatal-nigral dynorphin system. Eur. J. Pharmacol. **155:** 11–18.
32. RITTER, J. K., C. J. SCHMIDT, J. W. GIBB & G. R. HANSON. 1984. Increases of substance P-like immunoreactivity within striatal-nigral structures after subacute methamphetamine treatment. J. Pharmacol. Exp. Ther. **229:** 487–492.
33. HANSON, G. R., A. A. LETTER, K. MERCHANT & J. W. GIBB. 1988. Comparison of responses by striatonigral substance P and neurotensin A systems to methamphetamine treatment. Peptides **7:** 983–987.
34. SEIDEN, L. S. & G. VOSMER. 1984. Formation of 6-hydroxydopamine in caudate nucleus of the rat brain after a single large dose of methylamphetamine. Pharmacol. Biochem. Behav. **21:** 29–31.
35. STONE, D. M., G. R. HANSON & J. W. GIBB. 1990. In vitro reactivation of rat cortical tryptophan hydroxylase following in vivo inactivation by methylenedioxymethamphetamine. In press.
36. SONSALLA, P. K., W. J. NICKLAS & R. E. HEIKKILA. 1989. Role of excitatory amino acids in methamphetamine-induced nigrostriatal dopaminergic toxicity. Science **243:** 398–400.
37. JOHNSON, M., G. R. HANSON & J. W. GIBB. 1990. Effect of MK-801 on the decrease in tryptophan hydroxylase is decreased by methamphetamine and its methylenedioxy analog. Eur. J. Pharmacol. In press.

DISCUSSION OF THE PAPER

RICHARD J. WURTMAN (*MIT, Cambridge, MA*): The finding that alpha-methyltyrosine could block the effect, as you say, suggests that endogenous dopamine is mediating the response; it makes one wary about the millions of patients over the world who take L-dopa and the thousands who take it with deprenyl, an MAO inhibitor. I wonder if you or anyone else has done studies on widely used drugs, as opposed to misused drugs, like dopa plus an MAO inhibitor, or bromocriptine to see whether megadoses given chronically cause adverse effects.

GIBB: We performed some studies some years ago, where we tried L-dopa along with an MAO inhibitor and we were not able to duplicate the effects induced by methamphetamine. So I'm not sure just what the significance of that is. I think there are probably other components in the response to amphetamine and its substituted amines which we have not yet identified.

WURTMAN: Well I think now you have a much broader choice of dopaminergic agonists to test your hypothesis that dopamine is mediating it.

E. AZMITIA (*New York University, New York City*): I have two quick questions. The first deals with the recovery of function that you notice in the study with the two doses of MDMA where you injected 5 or 10 mg/kg; the group receiving 5 mg returned to control levels I think at 30 days whereas the 10 mg/kg group was recovering, but yet by 110 days it had not returned to normal. How do you explain the fact that although both of them seem to be decreasing to the same level, the recovery of function was much greater with the lower dose?

GIBB: I would assume that it is a dose-response phenomenon but your point is well taken. Why does the higher dose cause a persisting effect whereas the lower dose seems to be recovering quite nicely? I don't have a more complete answer at this time.

AZMITIA: The other point is just more of an observation with the dopamine link; the data with the α-methyltyrosine is very intriguing. But how would the MDMA be producing the toxicity in hippocampus where dopamine levels are really so low. It is understandable how it might be working in the striatum but what about the hippocampus? We have evidence also in raphe cultures of similar responses, where there is very little dopamine as well and these things seem to be toxic. Could it be that the dopamine attenuates the toxicity when it's around but that MDMA can also produce a toxicity directly on the serotonergic cells?

GIBB: That's a problem that we have been struggling with and we do not have an answer yet. It could be working via another catecholamine or it is possible, like you suggest, that there is a direct effect on serotonergic cells.

R. MURPHY (*New York University, New York City*): I was just curious about your reducing conditions; number one, if you could just specify a little bit more clearly what they were. And secondly if you think, for example endogenous ascorbate, plays a role; there is some evidence it is co-localized in dopamine systems. Could it be important in this regard?

GIBB: We have looked at that. Our reducing conditions are essentially exposing the TPH from the MDMA-treated animal to dithiothreatol and iron; the optimal conditions occur under nitrogen. With regard to your second question, we and others have examined ascorbate; the jury is still out as to how that will play out.

Structure-Activity Relationships of MDMA and Related Compounds: A New Class of Psychoactive Drugs?[a]

DAVID E. NICHOLS AND ROBERT OBERLENDER

Department of Medicinal Chemistry and Pharmacognosy
School of Pharmacy and Pharmacal Sciences
Purdue University
West Lafayette, Indiana 47907

INTRODUCTION

The popularity and illicit use of 3,4-methylenedioxymethamphetamine (MDMA; "Ecstasy") have been widely reported in the lay press over the past four years. As with any drug problem, scientific resources were quickly brought to bear on the study of MDMA, to examine such issues as toxicity, mechanism of action, reinforcing qualities, and dependence liability. Numerous studies have now been published dealing with all these areas.

However, it seems that most of the research on MDMA has been directed toward study of its neurotoxicity, and the mechanism of this toxic effect.[1] While interesting from a scientific viewpoint, and important because of the potential health risk to humans, interest in the neuronal damage resulting from high doses of MDMA has overshadowed the importance of studies on the basic psychopharmacology of behaviorally relevant doses. If one is to believe anecdotal user's accounts, the effect of MDMA is substantially different from that produced by hallucinogenic drugs such as LSD or DOM. One can make the definition of "hallucinogenic" so broad that it includes virtually all psychoactive phenethylamine derivatives. However, minor structural variations in the ring-substituted phenethylamines may have effects on their psychopharmacology ranging from profound, to very subtle. Profound changes are easy to detect, but the failure to recognize the subtle differences may result in the loss of new drug leads. Thus, it would seem prudent to define drug classes as narrowly as possible.

There seems little doubt that MDMA and similar substances, and the hallucinogenic amphetamines, alter serotonergic function in the central nervous system. However, they appear to do so in different ways. The fact that serotonin pathways are now recognized to be involved in modulation of various mood states, probably means that a variety of serotonergic and antiserotonergic drugs will ultimately be discovered that have specific effects on mental functioning.

It has been known for a long time that 3,4-methylenedioxyamphetamine (MDA) produces a strong need, in users of the drug, to be with and talk to other people.[2] MDMA seems to have a similar type of action, and it has been suggested that MDMA may facilitate development of the therapeutic alliance.[3] A recent survey of recreational users produced the finding that MDMA induces a sense of "closeness" with other individuals that is distinct from the psychoactive effects produced by hallucinogens or stimulants.[4] There are, presently, a number of psychopathologies that are extremely

[a]Supported in part by U.S. Public Health Service Grant 5 RO1 DA 04578 from the National Institute on Drug Abuse.

difficult to treat, for which a drug that would facilitate communication between patient and therapist might find great utility. Two examples are borderline personality disorder, and antisocial personality disorder. There is no reason to believe that psychotherapeutic agents cannot be developed to facilitate treatment of these, or other psychiatric disorders, but no one has apparently attempted to do this.

Thus, one must carefully compare the pharmacology of MDMA with that of other drugs, to identify features of MDMA's activity that may distinguish it from prototypes in existing drug classes. We believe that there is now enough known about MDMA to support the hypothesis that it, and related substances, are members of a new pharmacological class. We have called these drugs entactogens.[5] Furthermore, these types of substances may become useful as tools for developing a better understanding of the neurochemistry of emotional states.

The studies to be described in this presentation were directed toward further testing of the hypothesis that MDMA is a member of a novel drug class. Whether or not MDMA or pharmacologically similar substances will find medical utility is an altogether separate question, one which was not addressed.

To begin with, there is substantial evidence to support the idea that MDMA and pharmacologically related substances are not "hallucinogenic amphetamines." With-

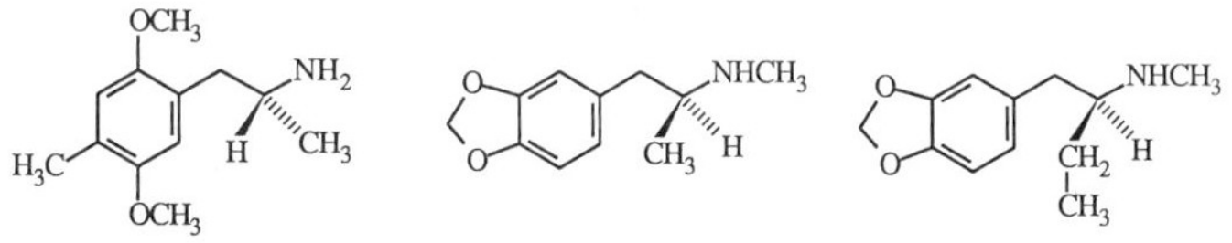

R-DOM *S*-MDMA *S*-MBDB

FIGURE 1. Structures of the biologically more potent enantiomers of the hallucinogen DOM, and of the entactogens MDMA and MBDB.

out reiterating the detailed arguments, which have already been published,[5–8] the following represent reasoning, based on structure-activity relationships, that leads to the conclusion that the entactogens are not acting at the same site(s) as the hallucinogenic amphetamines.

1. *N*-methylation of hallucinogenic amphetamines generally attenuates hallucinogenic activity from 3- to 10-fold.[9] *N*-methylation of MDA to give MDMA only has a slight effect on activity.
2. For the hallucinogenic amphetamines, it is the enantiomer with the *R*-(−) absolute configuration that is more potent. For MDMA it is the isomer with the reversed *S*-(+) configuration.[10] (See FIG. 1.)
3. Homologation of the α-methyl of MDMA to an α-ethyl gives a compound that retains MDMA-like activity and stereoselectivity.[5] A similar transformation in hallucinogenic amphetamines leads to a complete loss of hallucinogenic effect.[11]
4. Hallucinogenic amphetamines have high affinity for the agonist state of the 5-HT_2 receptor, with K_I values typically from 1–30 nM.[12] The K_I value for MDA at this site is in the 300–400 nM range,[12,13] and the K_I values for (+)-MDA and (+)-MDMA at the antagonist-labeled 5-HT_2 site are 13,000–16,000 nM.[14] It is important to note that at this receptor subtype, the *R*-enantiomer of substituted amphetamine derivatives has highest affinity, in contrast to the reversal of stereoselectivity observed for MDMA or MBDB (see point 2 above).

5. MDMA and its α-ethyl homologue are potent releasers and inhibitors of reuptake of monoamines, particularly serotonin, *in vitro* and *in vivo*.[6,7,15] Hallucinogenic amphetamines lack this effect.

These are some of the obvious structural and pharmacological features that distinguish MDMA from the hallucinogens. There is no strong evidence to suggest that MDMA is properly classified as an hallucinogen, even though high doses of MDMA can apparently provoke hallucinatory phenomena in some individuals.[16]

The question of whether MDMA should be classified as a "CNS stimulant" has been the subject of somewhat more controversy. While numerous studies have suggested a role for dopamine involvement in the action of MDMA, these have typically employed very large doses. Although at lower doses, and especially at later times, a stimulant component of MDMA's discriminative stimulus properties is detectable,[8,17] if one uses MBDB as the prototype drug for this class, it is clear that dopaminergic mechanisms do not play a significant role.[6–8,15]

Continuing efforts in our laboratory have therefore been directed toward an exploration of the animal behavioral effects of MDMA and related entactogens, in attempts to elucidate the mechanism of action. The possible importance of serotonin release to this mechanism was first postulated in 1982,[18] and numerous studies have now been published which seem to confirm this. Further, the most salient feature of the action of MDMA is its ability to induce the release of endogenous neuronal stores of serotonin (5-HT). Intuitively, it also makes some sense that this could be the primary mechanism of action for a new drug class, since currently there are no psychotropic drugs where 5-HT release has been identified as the mechanism of action.

Therefore, experiments have been largely directed toward studying the serotonergic component of the action of entactogens, and the importance of other monoaminergic systems to the mechanism of action of MDMA and related molecules. For behavioral studies, we have employed the two-lever drug discrimination paradigm. The training drug used in these studies was *S*-MBDB, the α-ethyl homologue of MDMA.[8] As previously reported,[8,15] MBDB has a less complex pharmacology than MDMA, whose pharmacological effects are probably mediated by actions in more than one monoamine system.

METHODS

Drug Discrimination Studies

Male Sprague-Dawley rats (Harlan Laboratories, Indianapolis, IN) weighing 200–240 g at the start of the experiments were used. None of the rats had previously received drugs or behavioral training. They were housed individually with access to water ad lib, but were given a rationed amount of feed (Lab Blox) after experimental sessions to maintain them at approximately 85% of free-feeding weight. Animals were maintained at a temperature of 22–24°C, with lights on at 6:00 and off at 20:00.

Six standard operant chambers (Coulbourn Instruments) consisted of modular test cages enclosed within sound-attenuated ventilated cubicles with white noise masking. Electronic pellet dispensers and data recording were controlled by solid state logic, interfaced through a Coulbourn Instruments Dynaport to an IBM PC running software developed in our laboratory, all located in an adjacent room.

A two-lever procedure with a fixed ratio (FR) 50 schedule of food reinforcement

(Bioserv, 45 mg dustless pellets) was used. Rats were randomly assigned to a training drug, with half in each group trained on drug-L, saline-R, and the other half trained on drug-R, saline-L. Discrimination training sessions lasted 15 min and were conducted at the same time each day for each group, 5 to 6 days per week. Levers were cleaned between subjects with 10% ethanol to avoid olfactory cues.[19]

After initial shaping, drug or saline was administered, by intraperitoneal injection, 30 min prior to sessions, randomly ordered, with neither treatment given for more than two consecutive sessions. In the studies reported here, *S*-fenfluramine hydrochloride (1.5 mg/kg) and *S*-MBDB hydrochloride (1.75 mg/kg) were used as the training drugs. The rats were required to respond on the stimulus-appropriate lever to receive food pellet rewards and the schedule of reinforcement was gradually increased from FR 1 to FR 50. This phase of the training continued until each rat had achieved an accuracy of at least 85% (100 × number of correct presses/number of total presses before first reinforcement) for eight of ten consecutive sessions. Then, every third or fourth session was used as a test period, as long as the rat maintained an accuracy of at least 85% on the preceding maintenance sessions (drug and saline). Test solutions were also administered IP, 30 min prior to the session. Test sessions were run under conditions of extinction, with the rats removed from the operant box immediately after 50 presses were completed on one lever. If 50 presses on one lever were not completed within 5 min, the session was ended and scored as a disruption. The data for a test session were discarded if the 85% criterion was not met on the following maintenance sessions.[20] Treatments were randomized at the beginning of the study for each trained group.

Data were scored in a quantal fashion. The lever on which the rat first emitted 50 responses was scored as the "selected lever." The percentage of rats selecting the drug lever for each dose of a test compound was then used to determine ED_{50} values and 95% confidence intervals, using the method of Litchfield and Wilcoxon.[21] In addition, the percentage of animals disrupted (%D) was calculated and used as a measure of the relative ability of a test drug to disrupt bar-pressing behavior. This value represents the percentage of rats scored as disrupted, out of the total number of rats tested at all doses.

Except for *S*-amphetamine sulfate, obtained from NIDA, and 6-chloro-2-aminotetralin, which was a gift from Eli Lilly, drugs were prepared in our laboratory. *S*- and *R*-MBDB·HCl, *S* and *R*-fenfluramine hydrochloride, and enantiomeric test drugs were prepared using an asymmetric synthesis method.[22]

RESULTS

TABLE 1 presents the results of substitution tests in *S*-MBDB-trained rats, with racemates and enantiomers of methylenedioxy-substituted compounds. The *S*-enantiomers are more potent than their antipodes in each series. We have already reported that MBDB produces complete substitution in MDMA-trained rats.[8] Therefore, the data suggest that the discriminative stimulus properties for these compounds substantially overlap.

Two additional cyclic analogues of MDA and MDMA also gave full substitution. The 5,6-methylenedioxy-2-aminoindan analog,[23] MDAI (see FIG. 2), and its *N*-methyl derivative displayed potencies comparable to racemic MDA and its *N*-methyl derivative MDMA, respectively.

In TABLE 2 are listed compounds that produced full substitution, but which lack the methylenedioxy substituent. The most potent of all the compounds tested was *para*-chloroamphetamine (PCA). Furthermore, there were relatively few disruptions with PCA at any dose, up to that which produced full substitution. Next in order of potency

TABLE 1. Methylenedioxy-substituted Drugs That Fully Substitute for *S*-MBDB

Drug	ED_{50} (95% C.I.) (μM/kg)	Potency Relative to *S*-MBDB	%D[a]
S-MDA	1.44 (0.87–2.39)	226	20
S-MDMA	1.67 (1.00–2.80)	195	13
MDAI	2.04 (1.26–3.29)	159	9
(±)-MDA	2.07 (1.31–3.26)	157	25
(±)-MBDB	2.91 (2.16–3.93)	112	2
N-Me-MDAI	3.01 (1.16–5.59)	108	14
R-MDA	3.05 (1.77–5.49)	107	29
R-MDMA	3.08 (1.73–5.49)	106	9
S-MBDB	3.25 (2.10–5.03)	100	0
(±)-MDMA	3.38 (2.31–4.95)	96	16
R-MBDB	6.48 (4.29–9.49)	50	11

[a]Percentage of the total rats tested at all doses, that failed to emit 50 lever presses within 5 min; see METHODS in text.

are meta-trifluoromethylphenylpiperazine (TFMPP) and the 5-HT_{1B}-selective agonist RU24969.[24] However, even though these two compounds fully substituted, a large number (ca. 50%) of the animals tested were disrupted, suggesting that TFMPP and RU24969 may have only partially overlapping profiles relative to *S*-MBDB.[25] A cyclic analogue of PCA, 6-chloro-2-aminotetralin (PCAT)[26], also substituted in *S*-MBDB-trained rats.

Fenfluramine also mimics *S*-MBDB, and as noted later, *S*-MBDB, MDMA, and PCA all completely substituted for *S*-fenfluramine. These results parallel those of Schechter and co-workers, who observed complete substitution of MDMA for fenfluramine, and norfenfluramine for MDMA.[17,27]

Given in TABLE 3 are results of substitution tests for compounds which failed to substitute completely for the *S*-MBDB cue. Among the compounds which fell into this category are the hallucinogens LSD, DOM and mescaline, and the CNS stimulants *S*-amphetamine, *S*-methamphetamine and cocaine. Interestingly, although the 5-HT releasing agents PCA and fenfluramine produced complete substitution, the selective 5-HT uptake inhibitor fluoxetine did not. Of particular note are the lack of substitution for *S*-MBDB by the 5-HT_{1A} selective agents 8-OH-DPAT[28] and buspirone.[29]

Thus far, most of our efforts have been directed toward stimulus generalization experiments, but a limited number of attempts have been made to block the *S*-MBDB cue with certain antagonists. The most effective pretreatment was found to be the nonselective serotonin antagonist metergoline. When rats were administered this compound 90 minutes prior to testing, followed by the training dose of *S*-MBDB at the regular time, the percentage of rats selecting the drug lever was reduced from 100% to 50% at a dose of 0.25 mg/kg, and to 29% at 0.5 mg/kg. Other antagonists that were tested include: ketanserin (44% at 2 mg/kg, 60 min), fluoxetine (63% at 5 mg/kg, 90 min), and haloperidol (86% at 0.25 mg/kg).

Listed in TABLE 4 are results of substitution tests in rats trained to discriminate 1.5 mg/kg of *S*-fenfluramine from saline. PCA, MDMA and *S*-fenfluramine were nearly

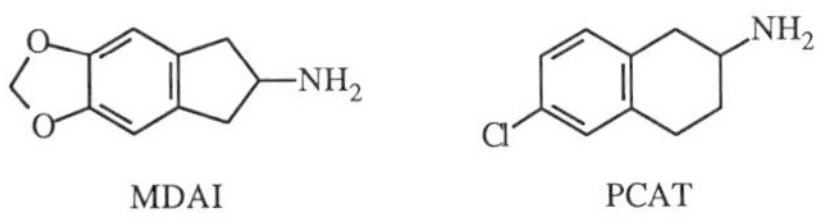

FIGURE 2. Structures of the non-neurotoxic analogues of MDMA and PCA.

TABLE 2. Non-Methylenedioxy–Substituted Drugs That Fully Substitute for *S*-MBDB

Drug	ED_{50} (95% C.I.) (μM/kg)	Potency Relative to *S*-MBDB	%D
PCA	0.82 (0.50–1.36)	396	16
TFMPP	0.92 (0.50–1.70)	353	48
RU24969	1.05 (0.69–1.60)	310	45
PCAT	1.40 (0.84–2.33)	232	7
Fenfluramine	1.98 (1.36–2.88)	164	11

equipotent in these animals. Surprisingly, there was about a twofold potency difference between MDMA and *S*-MBDB, whereas in the *S*-MBDB-trained rats these two were nearly equipotent. A maximum of 50% of the rats tested selected the *S*-fenfluramine lever when LSD (0.02–0.17 mg/kg) was tested and no more than 13% selected the drug lever after *S*-amphetamine (0.5–1.85 mg/kg). In previous studies, employing racemic fenfluramine as the training drug, PCA produced complete substitution[30,31] while LSD and *S*-amphetamine did not.[31]

DISCUSSION

As discussed previously,[23] the substitution data obtained in tests of *S*-MBDB-trained rats suggests that entactogens have similar behavioral profiles, which differ from those of hallucinogens and stimulants. Additional support for this can be found in recently published data from drug discrimination studies with MDMA-trained rats,[17] since the CNS stimulants amphetamine and cathinone, as well the hallucinogen DOI, all failed to produce complete substitution. One can also find evidence for the difference between the actions of entactogens, hallucinogens and stimulants, in the data obtained from studies of fenfluramine. This indirect 5-HT agonist seems to produce effects in rats that are perceived to be distinguishable from those of both stimulants and hallucinogens, but not entactogens.

The symmetrical substitution of MDMA and fenfluramine[17,27] and *S*-MBDB and fenfluramine suggests that the psychopharmacology of fenfluramine may be similar to that of MDMA and MBDB. However, although fenfluramine has been reported to produce hallucinatory effects at very high dosages,[32] dose ranges used to produce

TABLE 3. Drugs Which Did Not Substitute for *S*-MBDB

Drug	Maximum Percentage of Rats Selecting the *S*-MBDB Lever	Dose (μM/kg)	Dose Range Tested
Cocaine	63	29.4	7.36–36.8
8-OH-DPAT	50	0.18	0.09–0.70
Fluoxetine	50	14.5	7.23–43.4
S-Amphetamine	50	3.26	2.17–6.52
LSD	50	0.19	0.05–0.37
DOM	50	2.04	0.51–5.09
Mescaline	43	40.4	10.1–50.5
S-Methamphetamine	38	1.35	0.67–2.70
Buspirone	13	2.37	1.18–5.92

anorexia do not produce euphoria or effects resembling MDMA. Indeed, the side effects accompanying treatment with fenfluramine more often include dysphoria and sedation, actions quite unlike those of MDMA or MBDB.

One possible explanation for this is the apparent neuroleptic activity possessed by *R*-fenfluramine.[33–36] Thus, racemic fenfluramine might lack MDMA-like psychopharmacology, since a neuroleptic action by the *R*-enantiomer would block dopaminergic activity. By contrast, MDMA possesses a significant action on dopaminergic pathways.[6–8,17,37,38] However, as we have previously noted, this dopaminergic component to the action of MDMA is not essential to the primary discriminative cue.[8] This is based on the fact that *S*-MBDB has pharmacology similar to MDMA, but lacks significant effects on dopaminergic systems either *in vitro*[6,7] or *in vivo*.[15] In addition, we have recently shown that the α-ethyl homologue of methamphetamine has about a tenfold decrease in potency compared with methamphetamine, when tested in (+)-amphetamine-trained rats.[23] This further illustrates that the addition of the α-ethyl group to phenethylamines attenuates their ability to interact with dopaminergic neurons.

An additional question must now be asked. Is it possible that *S*-fenfluramine might have psychopharmacology similar to MDMA or MBDB? To our knowledge, no clinical studies have been conducted with high doses of *S*-fenfluramine, so this cannot at present be answered. However, the animal studies reported here suggest that *S*-fenfluramine and *S*-MBDB might well be similar in humans at appropriate dose

TABLE 4. Drugs Which Completely Substituted for *S*-Fenfluramine

Drug	ED_{50} (95% C.I.) (μM/kg)	Potency Relative to *S*-FEN	%D
PCA	1.84 (1.12–3.04)	137	15
(±)-MDMA	2.47 (1.14–5.35)	102	7
S-Fenfluramine	2.52 (1.98–3.20)	100	0
S-MBDB	6.53 (4.25–10.03)	39	9

ranges. Further, our results also suggest that *para*-chloroamphetamine might have clinical effects resembling MDMA and MBDB. Although *N*-methyl-PCA was studied as an antidepressant,[39,40] there are no reports of clinical studies that contain sufficient detail to determine this.

It might be concluded therefore, that there is a category of substances which can be grouped together based on similarity of animal pharmacology, that includes MDMA, MBDB, PCA, and *S*-fenfluramine. These are all potent releasers of serotonin from neurons. While fenfluramine has generally been studied in the context of its anorexigenic effects, and PCA was investigated as an antidepressant, MDMA has been difficult to classify. Part of the reason for this may have been that the pharmacological effects of MDMA involve both serotonergic and dopaminergic pathways,[6–8,17,37] However, MBDB appears to possess pharmacology similar to MDMA, but seems to be more selective for serotonergic systems,[6,23] lacking significant effects on dopamine pathways. Thus, we hypothesized that MDMA and MBDB may represent a unique category of psychoactive agent. Several other researchers have either reached a similar conclusion, or have indicated that it is not possible at the present time, to classify MDMA into a known category of drug.[41–44]

The complete substitution of the selective $5\text{-}HT_{1B}$ agonists RU-24969[24] and TFMPP[45] for *S*-MBDB (TABLE 2), may suggest a critical involvement of the $5\text{-}HT_{1B}$ receptor subtype. The lack of similarity between *S*-MBDB and 8-OH-DPAT and buspirone, as well as the reported inability of 8-OH-DPAT to substitute for MDMA,[17] suggests that 5-HT_{1A} receptors are not involved in the effects of entactogens. The superiority of metergoline over ketanserin, in blocking the *S*-MBDB cue, further indicates that the cue is probably not mediated via 5-HT_2 receptor interactions. This conclusion is also supported by the work of Glennon *et al.*,[43] where MDMA failed to substitute in DOM-trained animals, and by our earlier studies, where MDMA and MBDB failed to substitute in LSD-trained animals.[5]

However, the nature of the involvement, if any, of 5-HT_{1B} receptors is far from clear. The large number of disruptions occurring in rats tested with either RU-24969 or TFMPP suggests that these drugs are probably not identical in the nature of their cue to *S*-MBDB.[25] Schechter and coworkers found that although TFMPP completely substituted for MDMA,[27] MDMA did not completely substitute for TFMPP.[46] In fact, TFMPP also has significant affinity for the 5-HT_{1A} receptor[47] and is as potent a releaser of serotonin as is fenfluramine.[48] RU 24969 has also been shown to have significant interactions at non-5-HT_{1A} sites.[49] Therefore, additional studies are needed before a clear picture of the effects of these drugs on serotonergic systems can begin to emerge.

Finally, just as it was necessary to develop MBDB as a new structural analog of MDMA that would be useful in simplifying the pharmacological study of these kinds of agents, it was felt to be essential to eliminate one additional component of the pharmacology of MDMA and MBDB that complicates the interpretation of mechanism of action: the neurotoxicity. Schmidt and his co-workers, in a series of papers,[50–53] have shown that the acute pharmacology of MDMA is independent of the long-term neurotoxicity. An early, reversible phase of serotonin depletion was observed following MDMA treatment, with a later, irreversible phase of serotonin depletion that could be blocked by administration of the serotonin uptake inhibitors citalopram or fluoxetine.[50–52] Battaglia *et al.*[54] have also found that citalopram prevented the neurotoxicity induced by MDMA. The inability of fluoxetine to completely antagonize the *S*-MBDB cue in drug discrimination suggests that, while the long-term neurotoxic effects are blocked by fluoxetine, the acute psychopharmacology is not. Further, Battaglia *et al.*[54] also demonstrated that repeated injections were required to produce long-term serotonergic deficits.

In order to develop an MDMA analogue that might lack the neurotoxic component of action, it was noted that preliminary studies of the cyclic analogue of PCA, 6-chloro-2-aminotetralin (PCAT), suggested that, unlike PCA, this analogue did not produce long-term serotonin depletion.[26] As shown in TABLE 2, PCAT, as well as the indan analogues of MDA and MDMA, fully substituted in *S*-MBDB-trained animals. Furthermore, the indan analogue MDAI does not produce serotonin neurotoxicity.[23] Very recently, studies have been completed which also show that PCAT administration likewise does not lead to serotonergic deficits (M.P. Johnson and D.E. Nichols, unpublished). One might anticipate that these two analogues, or certain related structures, might well represent molecules that possess the unique acute pharmacology of MDMA and MBDB, presumably a serotonergic action, but which lack either dopaminergic or neurotoxic side effects.

In summary, the data continue to support the hypothesis that MDMA, and MBDB, possess novel psychopharmacology. Their effects appear to be mediated by serotonin release and/or blockade of reuptake, although it has been shown that MDMA has a secondary indirect component of release of endogenous dopamine. With respect to the serotonin release mechanism, there is some evidence that serotonin receptors of the 5-HT_{1B} subtype may be involved, although the lack of very specific 5-HT_{1B} agonists and antagonists prevents a firm conclusion on this point. Furthermore, structures have now been developed which lack the neurotoxicity characteristic of

MDMA, MBDB, PCA and fenfluramine. These molecules will undoubtedly be valuable in further elucidating the mechanism of action of this drug class, as well as in understanding the basis for the neurotoxicity of MDMA, MBDB, and related molecules.

REFERENCES

1. O'Hearn, E., G. Battaglia, E. B. DeSouza, M. J. Kuhar & M. E. Molliver. 1988. Methylenedioxyamphetamine (MDA) and methylenedioxymethamphetamine (MDMA) cause selective ablation of serotonergic axon terminals in forebrain: Immunocytochemical evidence for neurotoxicity. J. Neurosci. **8:** 2788–2803.
2. Jackson, B. & A. Reed, Jr. 1970. Another abusable amphetamine. JAMA **211:** 830.
3. Trager, J. 1988. A place for ecstasy in psychotherapy? Med. Tribune, Dec. **29:** 12–14.
4. Peroutka, S. J., H. Newman & H. Harris. 1988. Subjective effects of 3,4-methylenedioxymethamphetamine in recreational users. Neuropsychopharmacology **1:** 273–277.
5. Nichols, D. E., A. J. Hoffman, R. A. Oberlender, P. Jacob, III & A. T. Shulgin. 1986. Derivatives of 1-(1,3-benzodioxol-5-yl)-2-butanamine: representatives of a novel therapeutic class. J. Med. Chem. **29:** 2009–2015.
6. Johnson, M. P., A. J. Hoffman & D. E. Nichols. 1986. Effects of the enantiomers of MDA, MDMA, and related analogs on [^{3}H]serotonin and [^{3}H]dopamine release from superfused rat brain slices. Eur. J. Pharmacol. **132:** 269–276.
7. Steele, T. D., D. E. Nichols & G. K. W. Yim. 1987. Stereochemical effects of 3,4-methylenedioxymethamphetamine (MDMA) and related amphetamine derivatives on inhibition of uptake of [^{3}H]-monoamines into synaptosomes from different regions of rat brain. Biochem. Pharmacol. **36:** 2297–2303.
8. Oberlender, R. & D. E. Nichols. 1988. Drug discrimination studies with MDMA and amphetamine. Psychopharmacology **95:** 71–76.
9. Shulgin, A. T. 1978. Psychotomimetic drugs: structure-activity relationships. *In* Handbook of Psychopharmacology. S. D. Iversen and S. H. Snyder, Eds. Vol. **11:** 243–333. New York: Plenum.
10. Anderson, G. M., III, G. Braun, U. Braun, D. E. Nichols & A. T. Shulgin. 1978. Absolute configuration and psychotomimetic activity. *In* QuaSAR Quantitative Structure Activity Relationships of Analgesics, Narcotic Antagonists, and Hallucinogens. National Institute on Drug Abuse Research Monograph **22:** 27–32. G. Barnett, M. Trsic & R. E. Willette, Eds. DHEW Pub. No. (ADM) 78-729. U.S. Washington, DC: U.S. Government Printing Office.
11. Standridge, R. T., H. G. Howell, J. A. Gylys, R. A. Partyka & A. T. Shulgin. 1976. Phenylalkylamines with potential psychotherapeutic utility. 1. 2-Amino-1-(2,5-dimethoxy-4-methylphenyl)butane. J. Med. Chem. **19:** 1400–1404.
12. Titeler, M., R. A. Lyon & R. A. Glennon. 1988. Radioligand binding evidence implicates the brain 5-HT_2 receptor as a site of action for LSD and phenylisopropylamine hallucinogens. Psychopharmacology **94:** 213–216.
13. Nichols, D. E., R. Oberlender, K. Burris, A. J. Hoffman & M. P. Johnson. 1989. Studies of dioxole ring substituted 3,4-methylenedioxyamphetamine (MDA) analogues. Pharmacol. Biochem. Behav. **34:** 571–576.
14. Lyon, R. A., R. A. Glennon & M. Titeler. 1986. 3,4-Methylenedioxymethamphetamine (MDMA): stereoselective interactions at brain 5-HT_1 and 5-HT_2 receptors. Psychopharmacology **88:** 525–526.
15. Johnson, M. P. & D. E. Nichols. 1989. Neurotoxic effects of the alpha-ethyl homologue of MDMA following subacute administration. Pharmacol. Biochem. Behav. **33:** 105–108.
16. Siegel, R. 1986. MDMA: Nonmedical use and intoxication. J. Psychoact. Drugs **18:** 349–354.
17. Schechter, M. D. 1988. Serotonergic-dopaminergic mediation of 3,4-methylenedioxymethamphetamine (MDMA, "Ecstasy"). Pharmacol. Biochem. Behav. **31:** 817–824.
18. Nichols, D. E., D. H. Lloyd, A. J. Hoffman, M. B. Nichols & G. K. W. Yim. 1982.

Effects of certain hallucinogenic amphetamine analogs on the release of [^{3}H]serotonin from rat brain synaptosomes. J. Med. Chem. **25:** 530–535.

19. EXTANCE, K. & A. J. GOUDIE. 1981. Inter-animal olfactory cues in operant drug discrimination procedures in rats. Psychopharmacology **73:** 363–371.
20. COLPAERT, F. C., C. J. E. NIEMEGEERS & P. A. J. JANSSEN. 1982. A drug discrimination analysis of lysergic acid diethylamide (LSD): In vivo agonist and antagonist effects of purported 5-hydroxytryptamine antagonists and of pirenpirone, a LSD-antagonist. J. Pharmacol. Exp. Ther. **221:** 206–214.
21. LITCHFIELD, J. T., JR. & F. WILCOXON. 1949. A simplified method of evaluating dose-effect experiments. J. Pharmacol. Exp. Ther. **96:** 99–113.
22. BARFKNECHT, C. F., & D. NICHOLS. 1976. Asymmetric synthesis of phenylisopropylamines. U.S. Patent 4,000,197.
23. NICHOLS, D. E. & R. OBERLENDER. 1989. Structure-activity relationships of MDMA-like substances. National Institute on Drug Abuse Research Monograph Series. Washington, DC: U.S. Government Printing Office. **94:** 1–29.
24. MIDDLEMISS, D. N. 1984. The putative 5-HT$_1$ receptor agonist, RU 24969, inhibits the efflux of 5-hydroxytryptamine from rat frontal cortex slices by stimulation of the 5-HT autoreceptor. J. Pharm. Pharmacol. **37:** 434–437.
25. STOLERMAN, I. P. & G. D. D'MELLO. 1981. Role of training conditions in discrimination of central nervous system stimulants by rats. Psychopharmacology **73:** 295–303.
26. FULLER, R. W., D. T. WONG, H. D. SNODDY & F. P. BYMASTER. 1977. Comparison of the effects of 6-chloro-2-aminotetralin and of Org 6582, a related chloramphetamine analog, on brain serotonin metabolism in rats. Biochem. Pharmacol. **26:** 1333–1337.
27. SCHECHTER, M. D. 1986. Discriminative profile of MDMA. Pharmacol. Biochem. Behav. **24:** 1533–1537.
28. HAMON, M., S. BOURGOIN, H. GOZLAN, M. D. HALL, C. GOETZ, F. ARTAUD & A. S. HORN. 1984. Biochemical evidence for the serotonergic agonist properties of 8-OH-DPAT in the rat brain. Eur. J. Pharmacol. **100:** 263–276.
29. TAYLOR, D. P. 1988. Buspirone, a new approach to the treatment of anxiety. FASEB J. **2:** 2445–2452.
30. MCELROY, J. F. & R. S. FELDMAN. 1984. Discriminative stimulus properties of fenfluramine: Evidence for serotonergic involvement. Psychopharmacology **83:** 172–178.
31. WHITE, F. J. & J. B. APPEL. 1981. A Neuropharmacological analysis of the discriminative stimulus properties of fenfluramine. Psychopharmacology **73:** 110–115.
32. GRIFFITH, J. D., J. G. NUTT & D. R. JASINSKI. 1976. A comparison of fenfluramine and amphetamine in man. Clin. Pharmacol. Ther. **18:** 563–570.
33. BETTINI, E., C. ANGELO, R. SPINELLI & R. SAMANIN. 1987. Neuroleptic-like effects of the *l*-isomer of fenfluramine on striatal dopamine release in freely moving rats. Biochem. Pharmacol. **36:** 2387–2391.
34. BENDOTTI, C., F. BORSINI, M. G. ZANINI, R. SAMANIN & S. GARATINNI. 1980. Effect of fenfluramine and norfenfluramine stereoisomers on stimulant effects of d-amphetamine and apomorphine in the rat. Pharmacol. Res. Commun. **12:** 567–574.
35. SEBBAN, C., B. TESOLIN, A. SHVALOFF & B. LEMAITRE-GUARDIOLA. 1989. *l*-Fenfluramine and haloperidol in rats: a qEEG comparison. Psychopharmacology **98:** 131–138.
36. INVERNIZZI, R., C. BERETTERA, S. GARATTINI & R. SAMANIN. 1986. *d*- and *l*-isomers of fenfluramine differ markedly in their interaction with brain serotonin and catecholamines in the rat. Eur. J. Pharmacol. **120:** 9–15.
37. STONE, D. M., M. JOHNSON, G. R. HANSON, & J. W. GIBB. 1988. Role of endogenous dopamine in the central serotonergic deficits induced by 3,4-methylenedioxymethamphetamine. J. Pharmacol. Exp. Ther. **247:** 79–87.
38. HUBNER, C. B., M. BIRD, S. RASSNICK, & C. KORNETSKY. 1988. The threshold lowering effects of MDMA (Ecstasy) on brain-stimulation reward. Psychopharmacology **95:** 49–51. 42.
39. VAN PRAAG, H. M., J. KORF, F. VAN WOUDENBERG & T. P. KITS. 1968. Influencing the human indoleamine metabolism by means of a chlorinated amphetamine derivative with antidepressive action (p-chloro-N-methylamphetamine). Psychopharmacologia (Berl.) **13:** 145–160.
40. VAN PRAAG, H. M., T. P. KITS, T. SCHUT & P. DIJKSTRA. 1969. An attempt at indirect

evaluation of the noradrenaline hypothesis. Results of a pilot study of the antidepressive qualities of p-chloro-N-methylamphetamine. Behavioral Neuropsychiatry **1:** 17–24.
41. GOLD, L. H. & G. F. KOOB. 1988. Methysergide potentiates the hyperactivity produced by MDMA in rats. Pharmacol. Biochem. Behav. **29:** 645–648.
42. GOLD, L. H., G. F. KOOB & M. A. GEYER. 1988. Stimulant and hallucinogenic behavioral profiles of 3,4-methylenedioxymethamphetamine and N-ethyl-3,4-methylenedioxyamphetamine in rats. J. Pharmacol. Exp. Ther. **247:** 547–555.
43. GLENNON, R. A., M. YOUSIF & G. PATRICK. 1988. Stimulus properties of 1-(3,4-methylenedioxyphenyl)-2-aminopropane (MDA) analogs. Pharmacol. Biochem. Behav. **29:** 443–449.
44. NASH, J. F., H. Y. MELTZER & G. A. GUDELSKY. 1988. Elevation of serum prolactin and corticosterone concentrations in the rat after the administration of 3,4-methylenedioxymethamphetamine. J. Pharmacol. Exp. Ther. **245:** 873–879.
45. SILLS, M. A., B. B. WOLFE & A. FRAZER. 1984. Determination of selective and nonselective compounds for the 5-HT_{1A} and 5-HT_{1B} receptor subtypes in rat frontal cortex. J. Pharmacol. Exp. Ther. **231:** 480–487.
46. SCHECHTER, M. D. 1988. Use of TFMPP stimulus properties as a model of 5-HT_{1B} receptor activation. Pharmacol. Biochem. Behav. **31:** 53–57.
47. NEALE, R. F., S. L. FALLON, W. C. BOYAR, J. W. F. WASLEY, L. L. MARTIN, G. A. STONE, B. S. GLAESER, C. M. SINTON & M. WILLIAMS. 1987. Biochemical and pharmacological characterization of CGS 12066B, a selective serotonin-1B agonist. Eur. J. Pharmacol. **136:** 1–9.
48. PETTIBONE, D. J. & M. WILLIAMS. 1984. Serotonin-releasing effects of substituted piperazines *in vitro*. Biochem. Pharmacol. **33:** 1531–1533.
49. HEURING, R. E., J. R. SCHLEGEL & S. J. PEROUTKA. 1984. Species variation in RU 24969 interactions with non-5-HT_{1A} binding sites. Eur. J. Pharmacol. **122:** 279–282.
50. SCHMIDT, C. J., J. A. LEVIN & W. LOVENBERG. 1987. In vitro and in vivo neurochemical effects of methylenedioxymethamphetamine on striatal monoaminergic systems in the rat brain. Biochem. Pharmacol. **36:** 747–755.
51. SCHMIDT, C. J. & V. L. TAYLOR. 1987. Acute effects of methylenedioxymethamphetamine (MDMA) on 5-HT synthesis in the rat brain. Pharmacologist **29:** Abs 224.
52. SCHMIDT, C. J. 1987. Neurotoxicity of the psychedelic amphetamine, methylenedioxymethamphetamine. J. Pharmacol. Exp. Ther. **240:** 1–7.
53. SCHMIDT, C. J. & V. L. TAYLOR. 1988. Direct central effects of acute methylenedioxymethamphetamine on serotonergic neurons. Eur. J. Pharmacol. **156:** 121–131.
54. BATTAGLIA, G., S. Y. YEH & E. B. DESOUZA. 1988. MDMA-induced neurotoxicity: parameters of degeneration and recovery of brain serotonin neurons. Pharmacol. Biochem. Behav. **29:** 269–274.

DISCUSSION OF THE PAPER

DEREK MIDDLEMISS (*Merck Sharpe & Dohme, Harlow, Essex, Eng.*): One way to nail down whether it is actually a 5-HT_{1B} receptor would be to see if it's antagonized by the beta blocker, and to look at the stereospecificity of the antagonism.

NICHOLS: Yes, we haven't done that, but we have those studies planned.

GEORGE RICAURTE (*The Johns Hopkins Hospital, Baltimore, MD*): David, I just wanted to comment that in principle I agree with what you're saying, that perhaps the actions or psychoactive effects of MDMA may be distinct from some of the other substituted amphetamines and I base that on just a subjective account of human subjects who have experimented with the drug. And I appreciate your efforts to try to develop a non-toxic analogue. You're using the drug discrimination paradigm to help you distinguish a drug that retains the MDMA-like activity that you're seeing but is free of the toxicity. My question has to do with the drug substitution paradigm. It seems to me you're assuming that the *S*-MBDB is preserving MDMA-like activity and the first question is what would be the evidence for that?

And secondly I would like to know what you think the relationship is between the cue that the animals are responding to in the drug discrimination paradigm and the interesting subjective effects that humans report. You've alluded to the fact that you think serotonin release is involved, but I think there are certainly a number of people here who have vast experience with fenfluramine, a drug that also releases serotonin. And in general I think the subjective effects reported after fenfluramine and MDMA, both of which release serotonin, are truly vastly different. And to that extent it makes me question the extent to which the drug discrimination paradigm is going to give you information regarding the interesting subjective effects of MDMA.

NICHOLS: Well you've hit on one of the key issues. The fact that we did observe complete substitution with *d*-fenfluramine is perplexing and a number of people, as you know, have commented on that. *d*-Fenfluramine is able to antagonize the effects of amphetamine. Clinical studies, and some animal studies, have shown with the co-administration of amphetamine and *d*-fenfluramine, you can antagonize the CNS stimulant effect of amphetamine. Whereas clearly MDMA is still able to release dopamine in the presence of that serotonergic component.

I'm not really sure what the answer to the question is. If we have to rely on animal models, my own feeling is that drug discrimination is the most powerful behavioral model. As you know John Griffith and Don Jasinsky years ago at Lexington showed that high doses of fenfluramine can induce hallucinatory episodes in humans, but I'm not aware of any clinical pharmacology that's been published for high doses of *d*-fenfluramine. I suspect that high doses would be psychoactive. How similar they would be to MDMA I don't know. But one of the reasons that we obtained both *d*- and *l*-fenfluramine, and trained rats to discriminate *d*-fenfluramine, was to try to see if we could distinguish what the differences may be between *d*-fenfluramine and a compound like MBDB.

Also the other thing to point out with respect to MBDB vs MDMA, is that the pharmacology in the animal models is very similar. I think at the NIDA conference last year we pointed out the quantitative EEG studies where the EEG fingerprints of MDMA and MBDB look very similar. But in the studies that we published with Dr. Shulgin where MBDB was tested in humans, the compound did not produce the euphoria that you see with MDMA. So they're not identical. There is, I think, an importance of the dopamine releasing component for one aspect of it, but in terms of the serotonin release I think it's difficult to answer that question in the absence of clinical studies. But if you have something, a model better than drug discrimination, I'd be willing to try to do it. We've wracked our brains on this point and it is, I think, a key issue; fenfluramine vs MDMA serotonin release, and the disparity between the clinical effects of those drugs.

RICHARD J. WURTMAN (*MIT, Cambridge, MA*): Similarly, it's interesting that with the 15 mg per kg dose of fluoxetine you got about 50% substitution. S. Garattini, who unfortunately is no longer here, finds that fluoxetine in high doses is a releaser of serotonin as well as an reuptake blocker. And I wonder two things. Firstly have you checked higher doses of fluoxetine in proportion to the dose usually used. And secondly what about some of the other supposedly selective serotonin uptake blockers.

NICHOLS: The only uptake blocker we've looked at is fluoxetine. We have gone up to 10 mg per kilo in attempts to antagonize the cue, but we get lots of disruptions, so you can't really say you're getting antagonism.

As I said, we've given fluoxetine alone in these animals to see whether fluoxetine will substitute. The substitution tests at higher doses give even more disruption, so we can't conclude anything there. You can attenuate MBDB discrimination by pretreatment with fluoxetine, but not completely.

R. MURPHY (*New York University, New York City*): I was curious about the pharmacology of the aminoindane a little bit more. If I understand you correctly there is some serotonergic action in the sense it's releasing serotonin so it's not purely dopaminergic. But can you comment more on what the receptor profile of the compound is like, or do you have any information on it?

NICHOLS: We are just now starting biochemical pharmacology on that compound. For all we know, it may not even release serotonin. It may well hit some post synaptic site. And ideally it would be great if it was a post synaptic agonist. That would pin down the mechanism of action. I should mention we have tested that compound in animals trained to discriminate dextro amphetamine from saline, and with virtually all other compounds that we've looked at, including fenfluramine, some of the animals will select the amphetamine appropriate lever. With that aminoindane analogue not a single animal, out of a very large number, selected the amphetamine lever. So that's some evidence that it doesn't have an effect on dopamine systems.

B. CAMPBELL (*Servier R & D, Ltd., London*): A comment and a question. Fenfluramine even at very high doses doesn't produce hallucinations. The question is with the MDMA and like compounds with the methylenedioxy group on the ring we know that they're broken down very extensively to hydroxylated and methoxy compounds which are very much like catecholamines. Have you tried these metabolites because these could, in fact, be the active components?

NICHOLS: Well the catechols wouldn't cross the blood brain barrier so we wouldn't be able to administer them systemically.

CAMPBELL: Yes, but within the brain they might be broken down once MDMA is there.

NICHOLS: In terms of producing the cue?

CAMPBELL: Yes.

NICHOLS: I suppose that's a possibility. We have got compounds now that don't have the methylenedioxy that are very potent in substituting for MDMA. So my own feeling is that it's not a catechol metabolite that's producing the behavioral pharmacology effect.

STANLEY YEH (*NIDA, Baltimore, MD*): The catechol metabolites of MDA, such as α-methyldopamine, α-methyl-norepinephrine, α-methylepinephrine, or 4-hydroxy-3 methoxy derivatives, all these metabolites of MDA and MDMA have been tested *in vivo*. These metabolites appear to not cross the blood brain barrier, and not stimulate locomotor activity. They do not release serotonin or show any damage to the serotonin neuron.

J. BLUNDELL (*University of Leeds, Eng.*): I'm thinking about the mechanism underlying the full substitution. You've alluded to the activation of the 5-HT_{1B} receptor. In view of the data you reported here and all of the studies you've done with other compounds, is it your view that activation of the $_{1B}$ receptor constitutes a sufficient or a necessary condition, or neither, for a full substitution effect?

NICHOLS: I think it's difficult to really firmly conclude. We suspect that 5-HT_{1B} receptors may be involved but without, for example, studies with $_{1A}$ and $_{1B}$ antagonists, and some more selective $_{1B}$ agonists, I don't think we can firmly conclude that. My impression is that the $_{1B}$ interaction is important, but I couldn't say any more than that at this time.

RAY FULLER (*Lilly Research Laboratories, Indianapolis, IN*): Just one comment with regard to TFMPP. Although it, I think, can be a 5-HT_{1B} agonist, without any doubt, Dr. Pettibone and colleague's work has shown that it also is a very potent serotonin releaser *in vitro*. Whereas we've never seen any evidence that it releases serotonin *in vivo*, I think that it should be borne in mind as a possible explanation for this effect which it shares in common with other 5-HT releasing drugs.

Receptor Pharmacology of MDMA and Related Hallucinogens[a]

MILT TEITLER[b] AND SIGRUN LEONHARDT

Department of Pharmacology and Toxicology
Albany Medical College
Albany, New York 12208

NATHAN M. APPEL AND ERROL B. DE SOUZA

NIDA Addiction Research Center
Baltimore, Maryland 21224

RICHARD A. GLENNON

Department of Medicinal Chemistry
Virginia Commonwealth University
Richmond, Virginia 23298

INTRODUCTION

Research on the mechanism-of-action underlying the hallucinogenic and/or psychoactive properties of lysergic acid diethylamide (LSD), mescaline, the phenylisopropylamine (PIA) hallucinogens, methylenedioxymethamphetamine (MDMA) ("Ecstasy"), and methylenedioxyamphetamine (MDA) has recently focused on the interactions of these substances with brain serotonin (5HT) receptors. Evidence pointing to the involvement of serotonergic systems was provided by the electrophysiological studies of Aghajanian *et al.*[1-3] However, the identification of a pharmacologically defined receptor as the specific site-of-action of LSD accounting for its potent mind-altering effects has been elusive. Studies reviewed in this chapter have revealed evidence indicating that stimulation of brain $5HT_2$ receptors is the central mechanism producing the psychoactive effects of the hallucinogenic substances.

MATERIALS AND METHODS

Tissue preparation and radioligand binding assays procedures were as described previously with minor modifications.[4,5] Following decapitation, brain regions of male Sprague-Dawley rats or human cortical tissues (obtained at autopsy) were placed in 0.9% ice-cold saline. Freshly dissected tissue was homogenized (Polytron setting 6 for 20 seconds) in 30 volumes of ice-cold buffer containing 50 mM Tris-HCL (pH 7.4 at 37°C), 0.5 mM Na_2EDTA and 10 mM $MgSO_4$ and centrifuged at 30,000 × *g* for 15 minutes. The supernatant was discarded and the pellet resuspended and pre-incubated for 15 min at 37°C. The homogenate membranes were washed twice by centrifugation and resuspension. The final suspension buffer contained 10 μM pargyline and 0.1% ascorbate and was added last to the assay incubation medium. Following incubation with membranes and radioligand at 37°C for 15 minutes, samples were rapidly filtered

[a]This work was supported by PHS Grants MH40716 (M.T.) and DA01642 (R.A.G.).

[b]To whom all correspondence should be sent; formerly spelled Titeler.

over glass fiber filters and washed with 10 ml of ice-cold 50 mM Tris-HCl buffer. Individual filters were inserted into vials and equilibrated with 5 ml of scintillation fluid for six hours before counting at 50% efficiency. Results were analyzed using an updated version of the program EBDA[6] and statistically analyzed using RS1 (BBN Software). The values reported are the K_i values determined from experimentally determined IC_{50} values, adjusted for the presence of radioligand.

RESULTS AND DISCUSSION

Hallucinogen Interactions with ^{3}H-Ketanserin–Labeled 5HT$_2$ Receptors

The phenylisopropylamines (PIA) are a group of phenalkylamine derivatives with powerful hallucinogenic "LSD-like" effects.[7] In order to investigate the possibility that the effects of these drugs might involve the brain $5HT_2$ receptor, a series of psychoactive phenylisopropylamines were tested for their affinities in binding to the ^{3}H-ketanserin–labeled rat brain $5HT_2$ receptor.[8] As shown in FIGURE 1 and TABLE 1, the PIA derivatives displayed varying affinities for the radiolabeled $5HT_2$ receptors. The K_I values obtained in these *in vitro* radiolabeling studies were compared with the potencies of the PIAs in producing psychoactive (LSD-like) effects in humans and in cueing animals trained to discriminate the administration of the potent hallucinogen DOM.[9] As shown in FIGURE 2, the affinities of the PIAs for the radiolabeled $5HT_2$ receptor correlated strongly with both the human psychoactive properties of these drugs as well as their potencies in the rat drug discrimination model. These results were interpreted as indicating that a likely mechanism-of-action of the PIA hallucinogens involves their interaction with brain $5HT_2$ receptors. Two observations indicated that the PIAs were acting as agonists, i.e. stimulating brain $5HT_2$ receptors. In the radiolabeling studies, the Hill coefficients of the competition curves produced by the PIAs were significantly less than unity (FIG. 1; TABLE 1). Furthermore, GppNHp (a non-hydrolyzable, biologically active analog of GTP) shifted the competition curve produced by the potent hallucinogen 2,5-dimethoxy(4-iodo)phenylisopropylamine (DOI) (FIG. 3). Antagonist competition curves for ^{3}H-ketanserin–labeled $5HT_2$ receptors have been shown to produce Hill coefficients close to unity.[5] The shallow competition curves and the shifting of these curves by guanyl nucleotides have been shown to be highly predictive of agonist interactions with radiolabeled receptors. In drug discrimination studies the ability of PIAs to cue DOM-trained rats was potently blocked by the potent $5HT_2$ antagonists ketanserin and pirenpirone, at dosages selective for $5HT_2$ receptors.[10] These results indicated that the behavioral effects of the PIAs are probably mediated by brain $5HT_2$ receptors and that the PIAs are probably agonists, as the antagonists reversed the effects of the PIAs.

Direct Labeling of Brain 5HT$_2$ Receptors with the Radioactive Phenylisopropylamine Hallucinogens ^{3}H-DOB (2,5-dimethoxy(4-bromo)phenylisopropylamine and ^{125}I-DOI

The studies described above were supportive of the hypothesis that stimulation of brain $5HT_2$ receptors is necessary and sufficient for producing the psychoactive effects of the PIAs and LSD and that the drugs DOB and DOI were $5HT_2$ receptor agonists. In order to directly demonstrate the interaction of a PIA with the brain $5HT_2$ receptor, ^{3}H-DOB was prepared and incubated with rat frontal cortical homogenates.[4] The

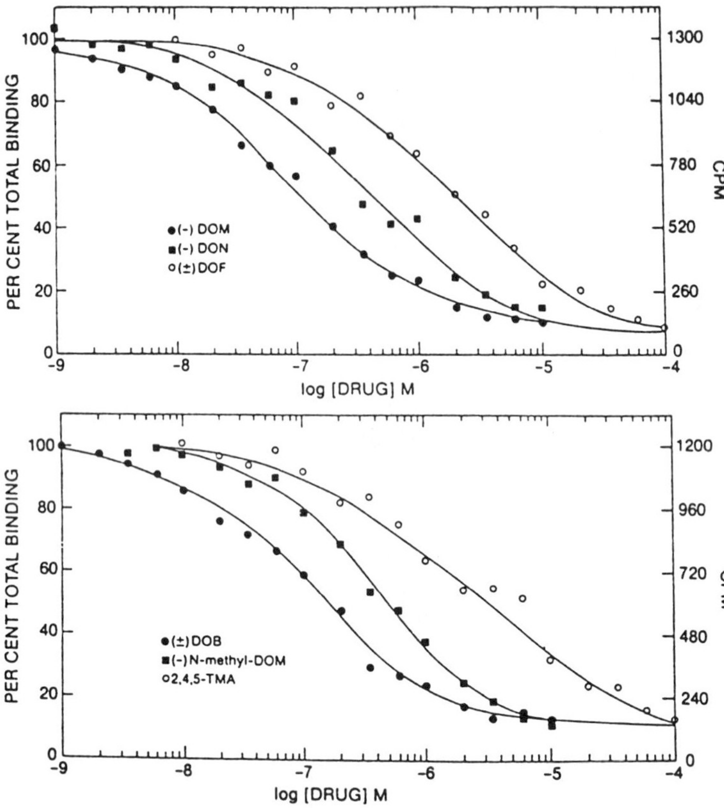

FIGURE 1. Competition curves of six PIA derivatives for the binding of ^{3}H-ketanserin to rat frontal cortex membranes. Membranes were prepared as described under METHODS and incubated with 0.4nM ^{3}H-ketanserin and increasing concentrations of the derivatives presented. The points represent experimental data determined from the means of three experiments, each point determined in triplicate. The lines drawn represent the best-fit of the data according to a model for two binding sites.

binding of this radioligand to homogenates of rat frontal cortex was saturable (K_D = 0.4nM; B_{max} = 17 fmol/mg) and the distribution throughout major rat brain regions closely matched the distribution of ^{3}H-ketanserin binding.[4] The affinities of a series of antagonists to compete for the ^{3}H-DOB-labeled sites closely correlated with the affinities of these drugs to compete for ^{3}H-ketanserin-labeled sites (FIG. 4). The similarity in the pharmacological profile of ^{3}H-ketnaserin and ^{3}H-DOB-labeled $5HT_2$ receptors in conjunction with the lower number of ^{3}H-DOB binding sites (17 fmol/mg versus approximately 200 fmol/mg for ^{3}H-ketanserin-labeled sites) indicated that ^{3}H-DOB was labeling a subset of $5HT_2$ receptors. Previous studies of ^{3}H-ketanserin-labeled $5HT_2$ receptors had revealed that two sub-populations of receptors exits: the majority of the ^{3}H-ketanserin-labeled sites appear to have low affinity for agonists and are not guanyl-nucleotide sensitive, while a minority of the sites display high affinity for agonists and are inhibited by guanyl nucleotides (FIG. 5; ref. 5). Antagonists appear to have similar affinities for both states of the receptor. This pattern is commonly observed for receptors coupled to GTP-binding proteins; the receptor-GTP-binding protein complex displays high affinity for agonists and is dissociated upon the binding of GTP (or an active non-hydrolyzable analog) to the G-protein. Therefore it was hypothesized that the sites labeled by ^{3}H-DOB represented the receptor-GTP-binding protein complex of the $5HT_2$ receptor. Consistent with this hypothesis was the observation that the potencies of the GTP analogs in inhibiting ^{3}H-DOB binding match their potencies in binding to GTP-binding proteins, and that adenyl derivatives are inactive (FIG. 6).

The affinities of the hallucinogenic PIAs were determined using ^{3}H-DOB competition assays. It was noted that the PIAs displayed higher affinities for the ^{3}H-DOB labeled $5HT_2$ receptor than for the ^{3}H-ketanserin-labeled $5HT_2$ receptor: this is the general pattern observed for agonists interacting with radiolabeled $5HT_2$ receptors.[4,5] There was a very strong correlation between the potencies of the PIAs as hallucinogens in man and their affinities for the ^{3}H-DOB-labeled agonist high affinity state of the $5HT_2$ receptor (FIG. 7). These results were consistent with the hypothesis that the PIAs act through stimulation of brain $5HT_2$ receptors. Although the binding of ^{3}H-DOB to an agonist high affinity state of the $5HT_2$ receptor provided important insights into the molecular interactions of hallucinogens with brain $5HT_2$ receptors, there are technical limitations in using ^{3}H-DOB. The original radioligand was produced at specific activities ranging from 16 to 40 Ci/mmol. The low specific radioactivity combined with the low density of binding sites for the agonist high affinity state of the $5HT_2$ receptor

TABLE 1. K_I Values and Hill Coefficient Determined by Competition Experiments for 0.4nM ^{3}H-Ketanserin–labeled $5HT_2$ Receptors in Rat Frontal Cortical Homogenates.[8]

	5-HT$_2$ Binding	
Agent	$K_{I(nM)}$	Hill Coefficient
R(−)-DOI	9.9 (±1.0)	0.72 (±0.03)
(±)-DOI	18.9 (±2.6)	0.73 (±0.04)
R(−)-DOM	60 (±5)	0.71 (±0.02)
(±)-DOM	100 (±17)	0.71 (±0.04)
(±)-DOB	63 (±2)	0.80 (±0.01)
α-demethyl DOM	110 (±8)	0.77 (±0.03)
R(−)-DON	210 (±35)	0.75 (±0.05)
(±)-DON	300 (±40)	0.79 (±0.05)
R(−)-N-methyl DOM	260 (±20)	0.91 (±0.03)
(±)-N-methyl DOM	415 (±40)	0.83(±0.04)
(±)-DOF	1110 (±150)	0.76 (±0.04)
(±)-2,4,5-TMA	1650 (±280)	0.68 (±0.04)
(±)-4-OEt 2,5-DMA	2220 (±300)	0.77 (±0.04)
(±)-4-Me PIA	3360 (±350)	0.89 (±0.04)
(±)-2,5-DMA	5200 (±360)	0.85 (±0.03)
(±)-PMA	33600 (±4480)	0.87 (±0.05)
(±)-3,4-DMA	43300 (±3300)	0.66 (±0.03)
R(−)-DOI + Gpp(NH)p(10^{-4} M)	35 (±2)	0.90 (±0.03)

K_L/K_H are the ratios between the affinities determined using ^{3}H-ketanserin (which primarily labels the agonist low affinity state) and ^{125}I-DOI (which primarily labels the agonist high affinity state).

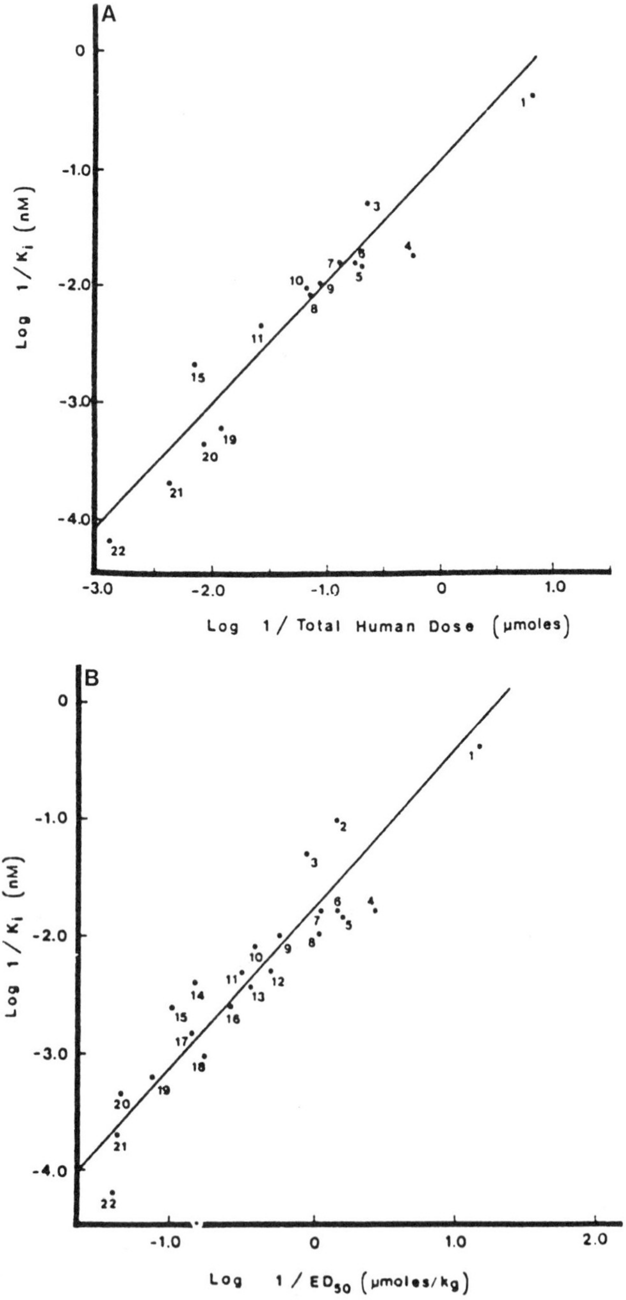

FIGURE 2. Correlation between the affinities of PIA derivatives for ${}^{3}H$-ketanserin–labeled brain $5HT_2$ receptors and **A**) human hallucinogenic potencies ($r = 0.92$;9) and **B**) discrimination-derived ED_{50} values using DOM (2,5-dimethoxyphenlyisopropylamine) as the training drug ($r = 0.94$). The numbers refer to the compounds listed in ref. 9.

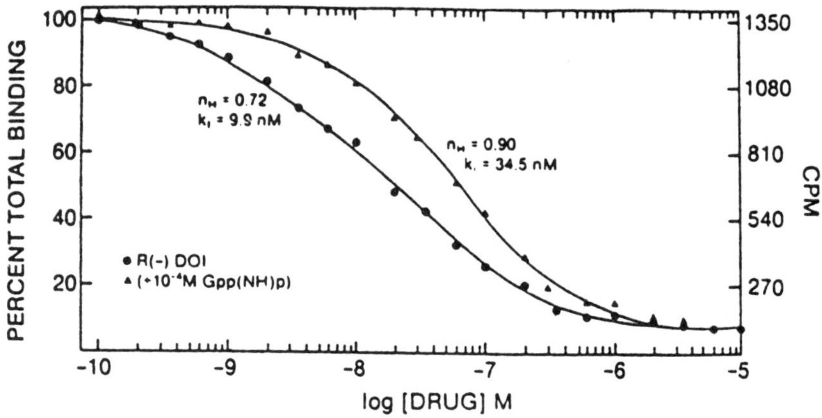

FIGURE 3. Competition curves of (−)DOI for ^{3}H-ketanserin-labeled rat brain $5HT_2$ receptors, in the presence and absence of GppNHp. The lines drawn represent the best fit of the data according to a model for two binding sites (in the absence of GppNHp) and model for one binding site in the presence of GppNHp.

necessitated using very large amounts of rat frontal cortical tissue in order to produce a usable signal. Autoradiographic studies were found to be impossible using ^{3}H-DOB. Therefore ^{125}I-DOI was synthesized by several groups[11,12] in order to take advantage of the extremely high specific activity achievable with iodinated radioligands. In our laboratory ^{125}I-DOI has been found to specifically label sites in human cortical, rat

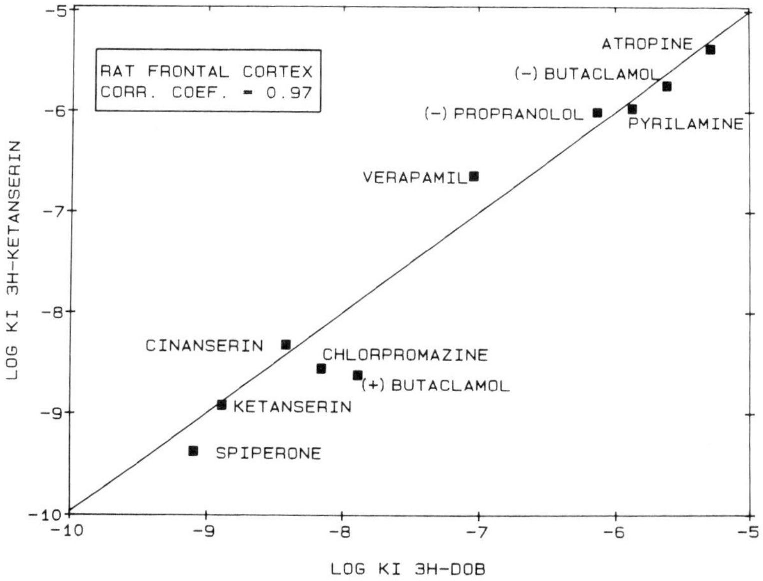

FIGURE 4. Correlation of antagonist K_I values for ^{3}H-DOB–and ^{3}H-ketanserin–specific binding (r = 0.96; ref. 4).

frontal cortical, and bovine frontal cortical tissue with high affinity (FIG. 8; ref. 13). Competition studies indicate that this site is identical to the site labeled by ^{3}H-DOB in rat frontal cortical tissue (FIG. 9).

Autoradiographic studies have also been performed using ^{125}I-DOI (refs. 12, 12A). FIGURE 10 displays the outcome of one such study. As expected from *in vitro* homogenate studies, *in vitro* receptor autoradiography with ^{125}I-DOI revealed a high

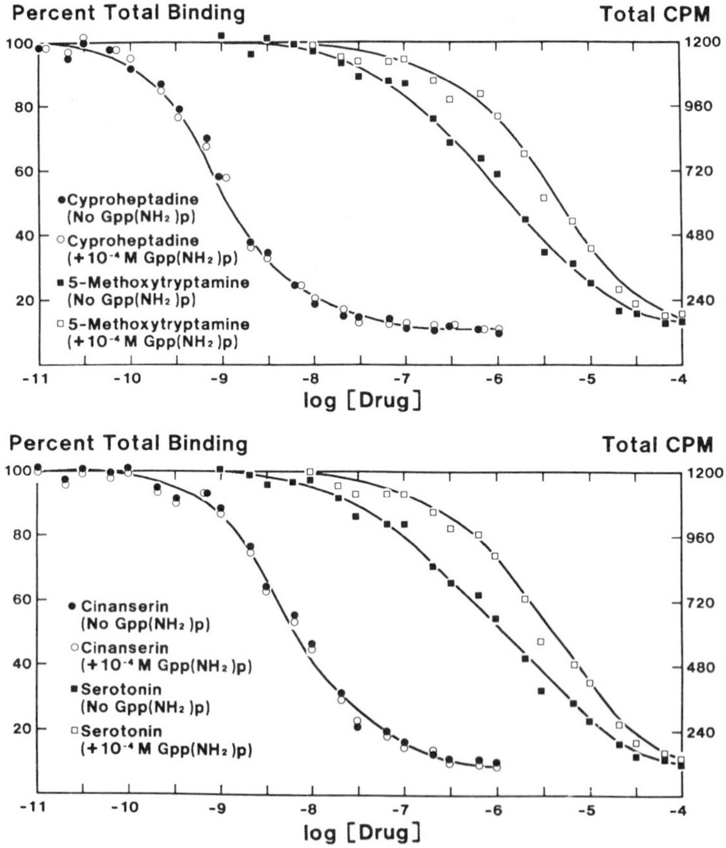

FIGURE 5. Competition curves produced by co-incubating increasing concentrations of nonradioactive antagonists or agonists with 0.4nM ^{3}H-ketanserin in homogenates of rat prefrontal cortex tissue.[5]

density of binding sites in frontal cortex. Highest densities of ^{125}I-DOI-labeled binding sites appear to be concentrated in layer IV of cerebral cortex and extends virtually throughout its length decreasing in density caudally (FIG. 10A). These more caudal receptors were not detectable in homogenate studies using large cortical brain regions. Extremely high levels of binding were also detected in choroid plexus (FIG. 10) lining the ventricles. Results from our preliminary studies support the hypothesis[12] that these sites are $5HT_{1C}$ receptors (FIG. 10C).

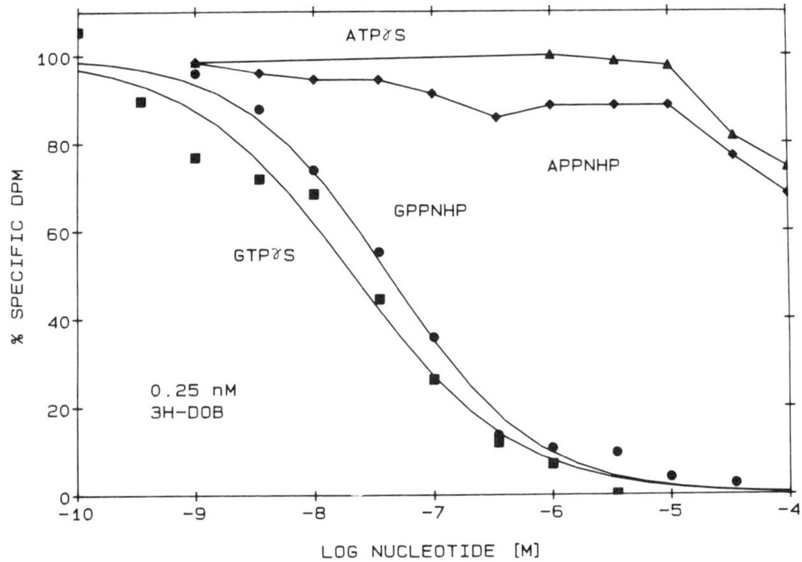

FIGURE 6. Guanyl and adenyl nucleotide inhibition of ^{3}H-DOB–specific binding in rat frontal cortical homogenates.[4]

Stereoselective Interactions of MDA and MDMA with Brain $5HT_2$ Receptors

The perceived therapeutic and/or drug-of-abuse potential of MDMA originally attracted the attention of the psychiatric community and the drug enforcement agencies. Extensive studies have been performed demonstrating the 5HT neurotoxicity

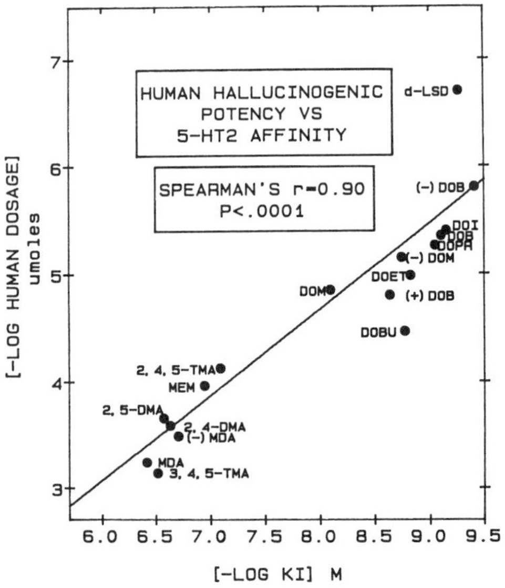

FIGURE 7. Correlation of the affinities of a series of PIA hallucinogens for the ^{3}H-DOB–labeled $5HT_2$ receptor and their potencies as hallucinogens.

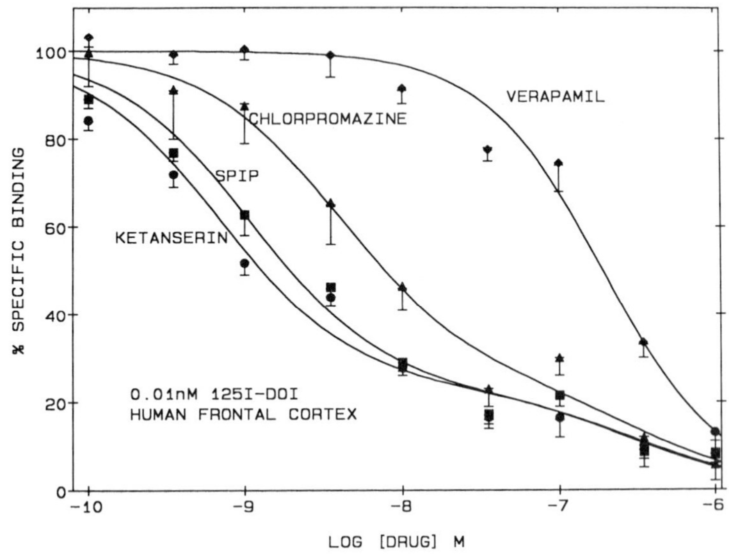

FIGURE 8. Representative antagonist competition curves for ^{125}I-DOI–labeled human cortical $5HT_2$ receptors. Error bars represent SEMs.

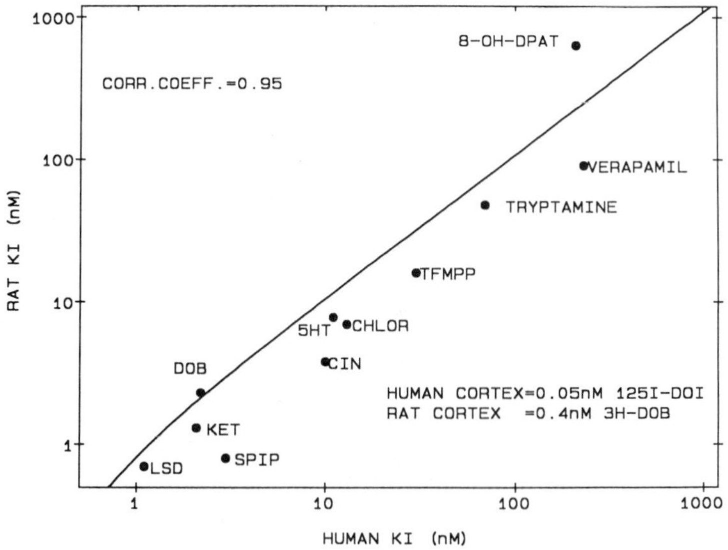

FIGURE 9. Correlation between the affinities of a series of agonists and antagonists in competing for ^{125}I-DOI–labeled human brain $5HT_2$ receptors and ^{3}H-DOB–labeled rat brain $5HT_2$ receptors.

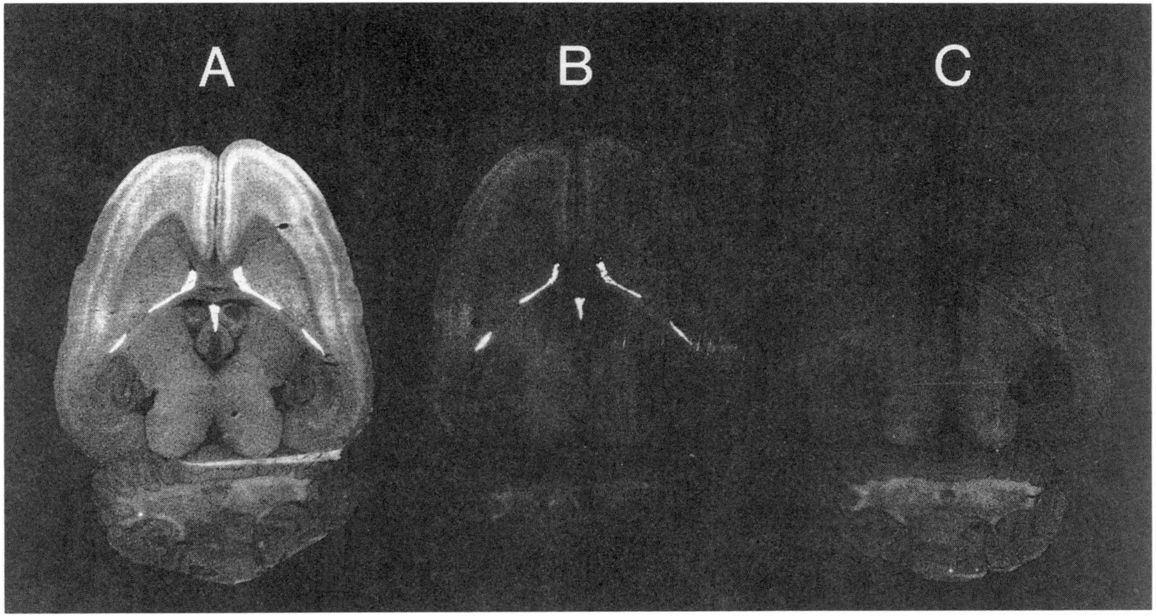

FIGURE 10. Autoradiographic distribution of ^{125}I-DOI binding in rat brain. A–C are darkfield photomicrographs (LKB Ultrofilm) depicting the distribution of autoradiographic grains in horizontal sections of rat brain. In darkfield illumination lighter areas correspond to brain regions displaying highest numbers of binding sites. **A**) Total ^{125}I-DOI binding. **B**) ^{125}I-DOI binding in the presence of 100nM spiperone to block ^{125}I-DOI binding to $5HT_2$ receptors. Note persistence of robust ^{125}I-DOI binding in choroid plexus. **C**) ^{125}I-DOI binding in the presence of 100nM spiperone to block ^{125}I-DOI binding to $5HT_2$ receptors and 1 μM mianserin to block binding to $5HT_{1C}$ receptors. Note disappearance of binding from choroid plexus.

of MDA and MDMA.[14] However, less attention has been paid to the psychoactive effects of these substances. An early study[15] indicated that MDA and MDMA might have different mechanisms-of-action. The reasoning relied on reports that (−)MDA was the more active isomer in man, and that its effects were "LSD-like," while (−)MDMA was reported to be less active than (+)MDMA in man, and had few or no hallucinogenic effects.[16] Glennon and associates had trained one group of animals to recognize the administration of amphetamine and another group of animals to recognize the administration of DOM. It was observed, for those phenylisopropylamines that produced a DOM-like effect, that the *R*(−)-isomers are more potent than their *S*(+)-enantiomers. Conversely, for those phenylisopropylamines that produced amphetamine-like effects, the *S*-isomers are more potent than their *R*-enantiomers. Racemic MDA and *R*(−)MDA (but not (+)MDA) produced DOM-like effects whereas racemic and *S*(+)MDA (but not *R*(−)MDA produced amphetamine-like effects, suggesting that MDA is capable of producing both types of effects but that one type of activity resides primarily with one optical isomer whereas the other type of activity is associated primarily with the other isomer. MDMA, on the other hand, produced only amphetamine-like discriminative effects and not DOM-like effects. *S*(+)MDMA, but not *R*(−)MDMA, also produced amphetamine-like effects and was somewhat more potent in this regard than either racemic MDMA or MDA. Neither *S*(+)- nor *R*(−)MDMA produced DOM-like effects. *N*-monomethylation of MDA, to provide MDMA, effectively eliminates (or at least reduces) the DOM-like character of MDA and increases somewhat the amphetamine-like character of MDA.[15,17] The (−)isomers of MDA and MDMA were found to have higher affinities for the radiolabeled $5HT_2$ receptor than the (+) isomers (TABLE 2). Given the higher potency of (+)MDMA in man, these results would indicate that MDMA may have only weak hallucinogenic effects owing to the relatively weaker interactions of (+)MDMA than (−)MDMA at $5HT_2$ receptors. The major psychoactive effects of MDMA may be through mechanisms similar to amphetamine (i.e. monoamine release). Conversely, since (−)MDA has been reported to be the more potent isomer in man, it is possible that the psychoactive properties of this drug, which include more potent hallucinogenic effects, may be through stimulation of brain $5HT_2$ receptors. Recent studies utilizing ^{3}H-DOB and ^{125}I-DOI as the radiolabel, rather than ^{3}H-ketanserin, have revealed properties supporting the hypothesis that MDA and MDMA are $5HT_2$ agonists (TABLE 2; FIG. 11). MDA and MDMA have higher affinities when an agonist radioligand is used to label the $5HT_2$ receptor rather than an antagonist radioligand; as

TABLE 2. Affinities of the Stereoisomers of MDA and MDMA for Rat Cortical $5HT_2$ Receptors Labeled with Either the Antagonist Radioligand ^{3}H-Ketanserin or the Agonist Radioligands ^{3}H-DOB or ^{125}I-DOI

Drug	^{3}H-DOB[a]	K_I (nM) ^{125}I-DOI	^{3}H-KET.	K_L/K_H
(−)MDA	81	97	3425	35
(+)MDA	257	271	13000	48
(±)MDA	123	110	5330	48
(−)MDMA	192	328	3310	10
(+)MDMA	1380	2784	15800	5.7
(±)MDMA	214	345	8300	24
5HT	8	11	928	84
(−DOB)	0.4	0.9	24	27

[a] Battaglia and De Souza, unpublished data.

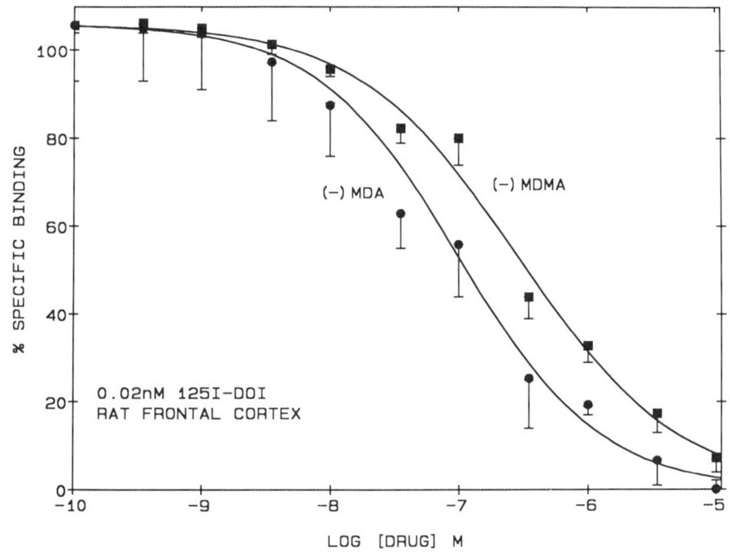

FIGURE 11. Competition curves of (−)MDA and (−)MDMA for ^{125}I-DOI-labeled rat cortical $5HT_2$ receptors.

mentioned above this is a hallmark of receptor agonists. Therefore, although stereochemical arguments indicate that the major mechanism-of-action of MDMA is through a non-$5HT_2$ mechanism, it is possible that high dosages of the drug may induce secondary LSD-like effects.

SUMMARY

The data presented herein appear to strongly implicate the brain $5HT_2$ receptor as the site-of-action of the hallucinogenic PIAs and LSD. If so, this discovery represents a major step in understanding the molecular pharmacology of hallucinogenic drugs. Using radioactive hallucinogenic drugs, detailed properties of brain $5HT_2$ receptors indicating the interaction of $5HT_2$ receptors with GTP-binding proteins have been revealed. Autoradiographic studies have revealed an extensive cortical distribution of brain $5HT_2$ receptors; these studies have also suggested that the PIAs may be $5HT_{1C}$ agonists. Radiolabeling studies in conjunction with drug discrimination studies indicate that MDMA is apparently "amphetamine-like" and not "LSD-like" while MDA is apparently both "LSD-like" and "amphetamine-like." However MDMA does appear to possess the potential to act as a $5HT_2$ agonist at high dosages.

REFERENCES

1. AGHAJANIAN, G. K., W. E. FOOTE & M. H. SHEARD. 1968. Lysergic acid diethylamide; Sensitive neuronal units in midbrain raphe. Science **161:** 706–708.
2. AGHAJANIAN, G. K., W. E. FOOTE & M. H. SHEARD. 1970. Actions of psychotogenic drugs on single midbrain raphe neurons. J. Pharmacol. Exp. Ther. **171:** 178–187.
3. AGHAJANIAN, G. K., H. J. HAIGLER & F. E. BLOOM. 1972. Lysergic acid diethylamide and

serotonin: Direct actions on serotonin-containing neurons in rat brain. Life Sci. **11:** 615–622.

4. LYON, R. A., K. H. DAVIS & M. TITELER. 1987. ^{3}H-DOB (4-bromo-2,5-dimethoxyphenylisopropylamine) labels a guanyl nucleotide sensitive state of cortical 5-HT_2 receptors. Mol. Pharmacol. **31:** 194–199.
5. TITELER, M., G. BATTAGLIA & M. SHANNON. 1984. Guanine nucleotides modulate cortical $5HT_2$ receptors. J. Rec. Res. **4:** 705–712.
6. MCPHERSON, G. A. 1983. A practical computer-based approach to the analysis of radioligand experiments. Comput. Programs Biomed. **17:** 107–114.
7. SHULGIN, A. T. 1978. Psychotomimetic drugs: Structure-activity relationships. *In* Handbook of Psychopharmacology. L. L. Iversen, S. D. Iversen, and S. H. Snyder, Eds. **11:** 243–333. New York: Plenum Press.
8. SHANNON, M., G. BATTAGLIA, R. A. GLENNON & M. TITELER. 1984. $5HT_1$ and $5HT_2$ binding properties of the hallucinogen 1-(2,5-dimethoxyphenyl)-2-aminopropane (2,5-DMA). Eur. J. Pharmacol. **102:** 23–29.
9. GLENNON, R. A., M. TITELER & J. D. MCKENNEY. 1984. Evidence for $5HT_2$ receptor involvement in the mechanism-of-action of hallucinogenic agents. Life Sci. **35:** 2505–2511.
10. GLENNON, R. A., R. YOUNG & J. A. ROSECRANS. 1983. Antagonism of the effects of the hallucinogen DOM and the purported 5HT agonist quipazine by $5HT_2$ antagonists. Life Sci. **35:** 2505–2511.
11. GLENNON, R. A., M. R. SEGGEL, W. H. SOINE, K. H. HERRICK-DAVIS, R. A. LYON, & M. TITELER. 1988. ^{125}I-DOI: an iodinated radioligand that specifically labels the hallucinogen/agonist high affinity state of $5HT_2$ serotonin receptors. J. Med. Chem. **31:** 5–7.
12. MCKENNA, D. J., A. J. NAZARALI, A. J. HOFFMAN, D. E. NICHOLS, C. A. MATHIS & J. M. SAAVEDRA. 1988. Common receptors for hallucinogens in rat brain: A comparative autoradiographic study using ^{125}I-LSD and ^{125}I-DOI, a new psychotomimetic radioligand. Brain Res. **476:** 45–56.

12A. APPEL, N. M., W. M. MITCHELL, R. K. GARLICK, R. A. GLENNOR, M. TEITLER & E. B. DESOUZA. 1990. Autoradiographic characterization of ^{125}I-DOI binding to $5HT_2$ and $5HT_{ic}$ receptors in rat brain. J.P.E.T. In press.

13. LEONHARDT, S. & M. TITELER. 1989. ^{125}I-DOI and ^{3}H-ketanserin binding reveal $5HT_2$ receptor agonist high affinity states in human, rat, and bovine brain. J. Neurochem. **53:** 316–318.
14. O'HEARN, E., G. BATTAGLIA, E. B. DESOUZA, M. J. KUHAR & M. E. MOLLIVER. 1988. Methylenedioxyamphetamine (MDA) and methylenedioxymethamphetamine (MDMA) cause selective ablation of serotonergic axon terminals in forebrain: immunocytochemical evidence for neurotoxicity. J. Neurosci. **8:** 2788–2803.
15. LYON, R. A., R. A. GLENNON & M. TITELER. 1986. 3,4-Methylenedioxymethamphetamine (MDMA): Stereoselective interactions at brain $5HT_1$ and $5HT_2$ receptors. Psychopharmacology **88:** 525–526.
16. ANDERSON, G. M., G. BRAUN, U. BRAUN & D. NICHOLS. 1978. Absolute configuration and pschotomimetic activity. NIDA Research Series. **22:** 8–15.
17. GLENNON, R. A. 1989. Stimulus properties of hallucinogenic phenalkylamines and related designer drugs: formulation of structure-activity relationships. NIDA Research Monographs. **94:** 43–67.

DISCUSSION OF THE PAPER

T. MENNINI (*I.R.F. Mano Negri, Milan*): If I understood correctly, the data shown by your autoradiographic images indicate DOI also labeled 5-HT_{1c} receptors. Therefore, I am wondering if the biphasic inhibition of DOB binding by MDA and MDMA could be interpreted as differential affinity for 5-HT_{1c} and 5-HT_2 receptors.

Teitler: We don't think that the ^{5}H-DOB is labeling the 5-HT_{1c} receptor. Their density relative to 5-HT_2, and the density of these high affinity states are so low that we don't observe any cortical 5-HT_{1c} labeling. We can't see any pharmacological evidence when we use 5-HT_{1c} preferring drugs to suggest 5-HT_{1c} labeling. However, there are other things labeled with ^{3}H-DOB, but exactly what they are, we're not sure. What we try to emphasize is that we are labeling the 5-HT_2 receptor.

Mennini: But doesn't mesulergine inhibit the DOB binding in frontal cortex?

Teitler: Yes. However, mesulergine is a very potent 5-HT_2 and 5-HT_{1c} drug. Possibly a DOB radioligand with higher specific activity may detect 5-HT_{1c} receptors. It is a possibility.

Daniel Hoyer (*Sandoz Ltd., Basel*): Although LSD was discovered in Basel, I still don't know how it works and I think I don't care much how it works. It's not a major issue in our company. The thing is that talking about these differences between 5-HT_{1c} and 5-HT_2 receptors and the labeling of 5-HT_{1c} receptors with DOI, DOM, DOB, in fact we find very little difference for all those compounds on 5-HT_{1c} and 5-HT_2 receptors. Just one point, DOI is the most potent agonist we find on 5-HT_{1c} receptors and I think one has to be aware of this.

Teitler: There are at least three other groups that have reported phenylisopropramines interact potently with 5-HT_{1c} receptors. And we have data that I just didn't have time to show that the phenylisopropramines interact very potently with the 5-HT_{1c}.

Neurotoxicity of MDMA and Related Compounds: Anatomic Studies[a]

MARK E. MOLLIVER,[b] URS V. BERGER,
LAURA A. MAMOUNAS, DEREK C. MOLLIVER,
ELIZABETH O'HEARN, AND MARY ANN WILSON

Departments of Neuroscience & Neurology
The Johns Hopkins University School of Medicine
Baltimore, Maryland 21205

INTRODUCTION

Several recent studies have reported that certain substituted amphetamine derivatives, some of which are abused as recreational drugs, cause acute and lasting decreases in brain levels of serotonin (5-HT). These biochemical changes have been interpreted as resulting from drug-induced degeneration of 5-HT neurons or their projections in the brain. After administering drug treatment regimens that have been shown to produce decreases in biochemical markers for 5-HT, this laboratory has employed anatomic methods in order to detect toxic effects of amphetamine derivatives upon CNS neurons. Immunocytochemistry for the visualization of 5-HT was employed to identify the specific neurons and cellular compartments that are affected by drugs, to detect morphologic evidence of neuronal degeneration, to determine the regional distribution of drug effects, and to assess the potential for neuronal regeneration. The results have been useful in obtaining evidence for degeneration, in identifying the specific structures that are damaged, and in studying the morphologic reaction of cell bodies to injury.

BACKGROUND

One of the most striking anatomic features of 5-HT neurons is the diversity of raphe nuclei, which are located along the midline of the brainstem, extending from the midbrain to the caudal medulla.[1] Ascending 5-HT nerve terminals arise from raphe nuclei in the midbrain,[2-4] and are distributed through widespread areas of the forebrain.[5] The neuronal groups in the midbrain which give rise to ascending 5-HT axons include the dorsal raphe nucleus (DR), which lies primarily in the periaqueductal gray matter, the median raphe nucleus (MR), which lies within the central portion of the midbrain tegmentum, and the B9 cell group, which consists of 5-HT neurons which lie along the dorsal surface of the medial lemniscus. The multiplicity of raphe nuclei in different locations and the heterogeneity of 5-HT cell types suggest that particular raphe cell groups may receive different afferents and have different patterns of integration. Axons which emerge from 5-HT cell bodies in the raphe nuclei ascend

[a]Supported by United States Public Health Service Grants DA04431, NS15199 and NS21011.

[b]Address for correspondence to all authors: Department of Neuroscience, The Johns Hopkins University School of Medicine, 725 N. Wolfe Street, Baltimore, MD 21205.

through the hypothalamus[1-3,5-7] en route to distant parts of the forebrain. Near their destination, the parent axons of passage form pre-terminal branches and then arborize to form axon terminals which are the postulated sites of transmitter release.

All areas of cerebral cortex are densely innervated by 5-HT axons, and there are regional differences in the pattern of 5-HT innervation. Owing to technical limitations, the full extent of cortical 5-HT innervation was not visualized until the development of an immunocytochemical method using antibodies against 5-HT, developed by Steinbusch *et al.*[8,9] A similar antiserum, produced by Lidov, was used to analyze the 5-HT innervation of cerebral cortex.[10] Serotonin-containing axons were found throughout the cerebral cortex of the rat with regional differences in density and laminar pattern. A particularly high density of 5-HT axons was found in frontal cortex, and a distinct laminar pattern of innervation described in several other cortical areas.[10-12] In primates, the 5-HT innervation of cortex is very dense and more highly differentiated than in the rat,[13-15] especially in primary visual cortex where the laminar pattern is most distinctive.[13] Such regional differences in innervation may provide a basis for different functional influences of 5-HT neurons upon separate cortical areas.

The dorsal and median raphe nuclei project widely upon the forebrain, with a smaller contribution from the B9 cell group; the different nuclei form distinct ascending fiber bundles.[7] Recently, axonal transport studies have shown considerable overlap in raphe projections in that most areas of cortex receive convergent projections from both DR and MR, with regional differences in the number of cells projecting from each nucleus.[16,17] Moreover, there is a coarse topography in this projection: cells within different regions of the raphe nuclei project to separate areas of forebrain in rat[16-20] and in primates.[21] These findings indicate that the dorsal raphe nucleus is organized heterogeneously, and that both DR and MR have partially overlapping but dissimilar patterns of forebrain projections.

Several laboratories have described the morphology of 5-HT axon terminals as non-uniform and heterogeneous. In the brainstem, 6 distinct types of 5-HT terminals have been described.[22] At least 2 types of anatomically distinct 5-HT axons have been found in several areas of the forebrain.[10,15,23,24] Using an anterograde transport method (PHA-L) to label dorsal and median raphe neurons separately, Kosofsky discovered that there are consistent morphologic differences between the axons that arise from these two nuclei.[11,25] Axons labeled after MR injections have large, spherical varicosities (typically 2 μm in diameter) which contrast markedly with axons from the DR, which are extremely fine in caliber and have minute, pleomorphic varicosities. These results provided evidence that the dorsal and median raphe nuclei give rise to axons that differ in the morphology of their terminals and in their regional and laminar distribution. Further transport studies are needed to establish the generality of our interpretation that the two types of 5-HT axons consistently arise from separate nuclei within the brainstem. In an independent study in the cat, two corresponding axon types were noted to form mutually distinct axon systems.[24] The functional significance of the two anatomic types of 5-HT axons was not apparent until they were found to exhibit differential vulnerability to psychotropic drugs.

NEUROTOXIC EFFECTS OF DRUGS

Evidence has accumulated that particular amphetamine derivatives may have neurotoxic effects since they produce long-term reductions in biochemical markers for monoamine neurons. For example, repeated administration of methamphetamine in large doses leads to significant decreases in brain markers for 5-HT and dopamine.[26-28] The chlorinated amphetamine derivative, *p*-chloroamphetamine (PCA), also produces a profound and lasting reduction of 5-HT markers, with a characteristic regional distribution.[29-31] A new wave of interest in the toxic effects of amphetamines was stimulated by the report from Ricaurte *et al.*[32] that large doses of the substituted amphetamine, 3,4-methylenedioxyamphetamine (MDA), produced lasting decreases in biochemical markers for 5-HT in forebrain. In rats analyzed 2 weeks after repeated injections of MDA, there were substantial reductions in brain levels of 5-HT, 5-HIAA, and 5-HT uptake.[32] Similar findings were then obtained for the *N*-methyl analog, 3,4-methylenedioxymethamphetamine (MDMA), which showed that this compound produced similar reductions but was somewhat less potent.[33-35] These results demonstrated that the psychotropic amphetamine derivatives, MDA and MDMA, cause initial release of 5-HT followed by apparent degeneration of 5-HT projections. Marked regional differences in 5-HT markers were found after MDA and MDMA, with the most severe reductions in neocortex, striatum and hippocampus, and with much smaller decreases in brainstem and hypothalamus.[35,36] The use of silver stains provided anatomic evidence for axon degeneration in the striatum after treatment with MDA in the rat.[32,33] However, since silver methods have low sensitivity for 5-HT axons, they do not show the full extent of degeneration nor do they indicate which transmitter is associated with a stained axon. In order to further evaluate the cytotoxic effects of MDA and related drugs, it was necessary to conduct additional anatomic studies of 5-HT axons. Therefore, prompted by chemical evidence for neurotoxicity of MDA, we employed immunocytochemical methods to determine whether 5-HT neurons were damaged, and to discern what compartment of the cell may be involved.

In a series of morphologic studies, rats were treated with MDA or MDMA by repeated subcutaneous injections of 20 mg/kg twice daily for 4 days. The rats were sacrificed after different survival times and 5-HT immunocytochemistry was used to visualize 5-HT neurons and their axonal projections. In this study, O'Hearn *et al.*[37,38] demonstrated that MDA and MDMA produced a profound loss of serotonergic axons throughout the forebrain, especially marked in neocortex (FIGS. 1 and 2), striatum and thalamus. Less denervation was seen in hippocampus, hypothalamus (FIGS. 3 and 4), and basal forebrain, while the brainstem was much less affected except for the tectum. The loss of 5-HT axon terminals roughly parallels the reduction in 5-HT levels and uptake sites,[36] and reflects persistent denervation of multiple forebrain structures.

MDA and MDMA produced a selective loss of fine 5-HT axon terminals, while there was intact staining of raphe cell bodies and of straight, thicker axons in deep layers of cortex and in medial forebrain bundle.[38] Dilated pre-terminal axons were intensely immunoreactive at 2 weeks after treatment, presumably owing to accumulation of the neurotransmitter and cytoskeletal components in the proximal stumps of 5-HT axons. Cell bodies in the raphe nuclei remained immunoreactive and did not show cytopathologic changes. The persistence of transmitter in cell bodies and preterminal fibers indicates that these compartments remain functionally intact and that transmitter synthesis and anterograde transport are not evidently impaired. While subtle cytologic changes might not be detected using fixation methods for immunocytochemistry, the sparing of cell bodies and proximal axons suggested that there may be potential for regeneration of 5-HT projections.

In order to detect structural damage, rats were examined at different survival times after fewer doses of MDA. At 24 hours after 3 doses of MDA, cytopathologic changes were frequently seen in surviving 5-HT axons.[38] Typical abnormalities included dilated axons with giant varicosities and fragmentation of axon segments. Markedly swollen varicosities were seen in all areas of cortex in animals that were treated, but not in control rats. The distribution of the enlarged varicosities differed from that of normal varicose 5-HT axons, and were several-fold larger than the largest varicosities found in untreated animals. In addition, swollen axonal stumps were evident in basal forebrain

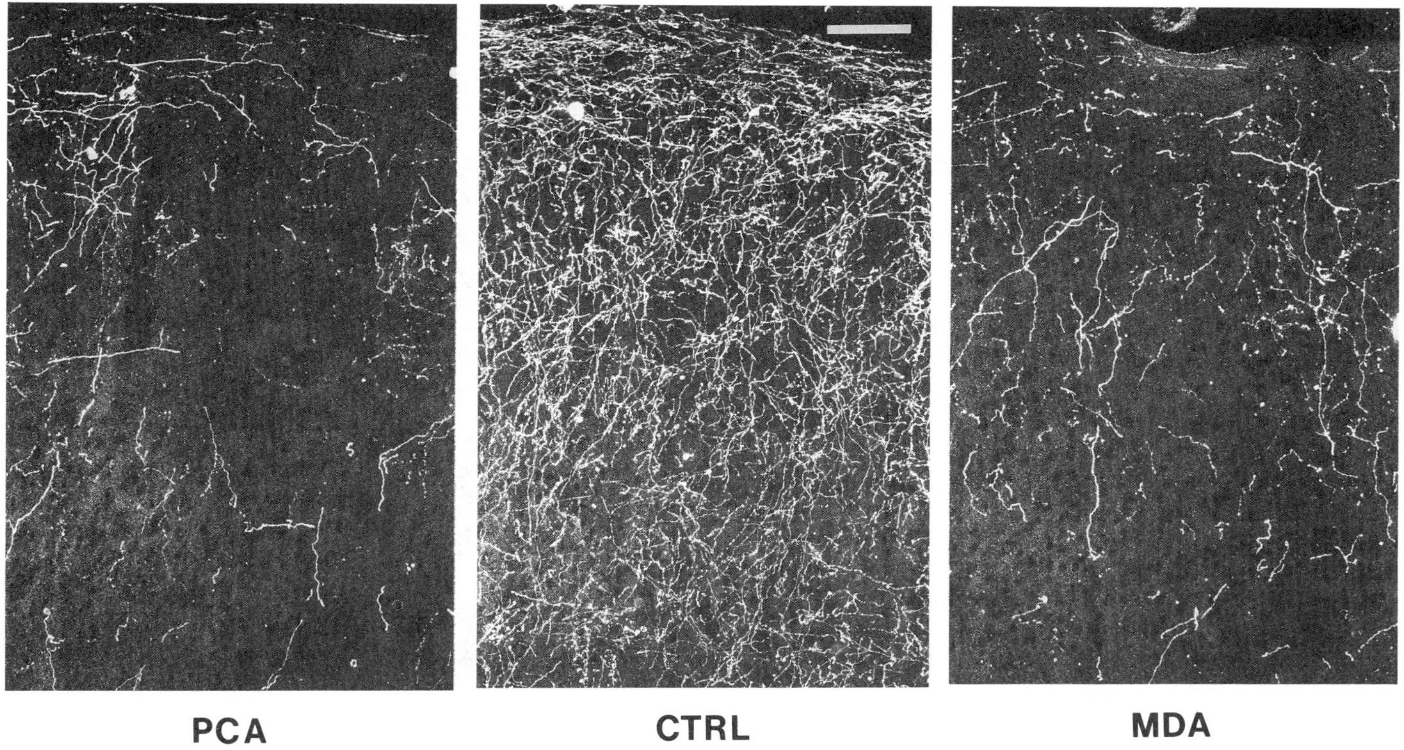

FIGURE 1. 5-HT innervation in parietal cortex of rat. A severe loss of 5-HT axons is apparent 2 weeks after MDA (8 × 20 mg/kg, Q12h) and PCA (2 × 10 mg/kg) as compared with control sections. 5-HT immunocytochemistry, darkfield photomicrograph, Bar = 100 μm.

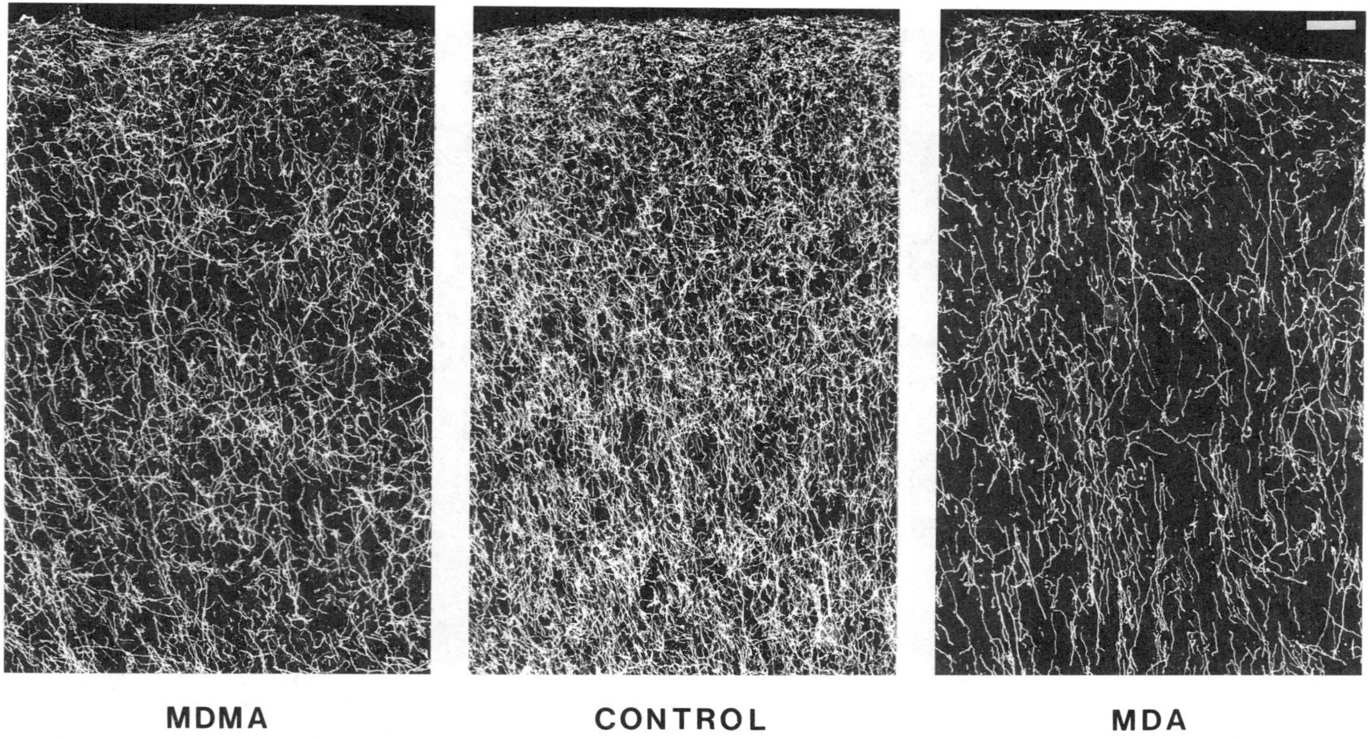

FIGURE 2. Loss of fine 5-HT axons in dorsal frontal cortex of rat 2 weeks after multiple injections of MDA or MDMA (8 × 20 mg/kg). Bar = 100 μm.

and anterior to the genu of the corpus callosum. At survival times of 2 weeks, abnormal axons were not observed in cortex; instead, the persistent loss of fibers reflects denervation of target structures.

The loss of 5-HT axon terminals following treatment with MDA or MDMA is subtotal, and in most cortical regions, there is consistent sparing of some 5-HT axons with a characteristic regional distribution. The spared 5-HT axons have large round varicosities, and are morphologically identical to those reported to arise from the median raphe in normal rats.[25,38] There is a selective decrease in fine axon terminals, which are almost completely ablated. Thus, the two classes of 5-HT axons described above[25] exhibit differential vulnerability to the neurotoxic effects of MDA and MDMA.

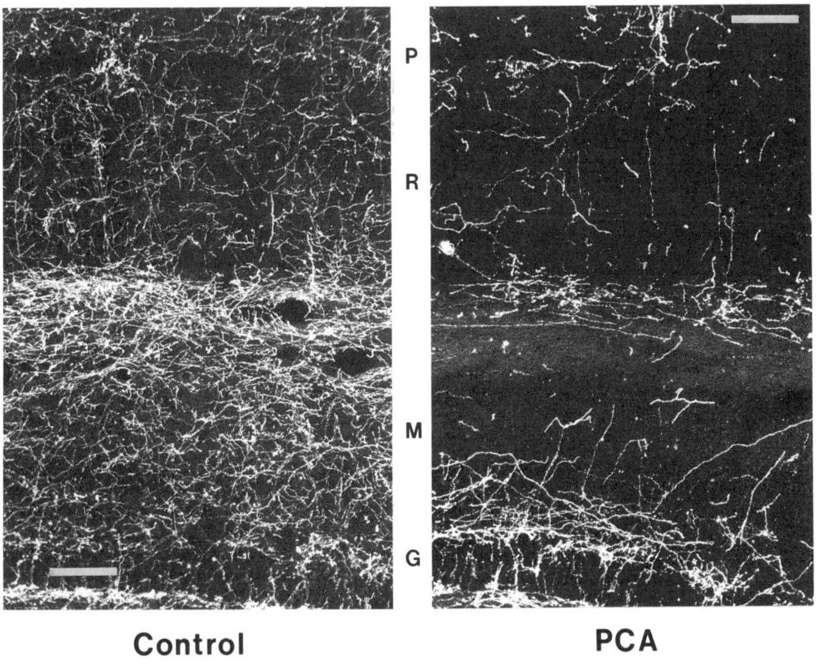

FIGURE 3. Loss of fine 5-HT axons in rat hippocampus (CA_1) 2 weeks after PCA (2 × 10 mg/kg). P–pyramidal cell layer of CA_1; G–dentate gyrus. Bar = 100 μm.

Following drug treatment, surviving axons are consistently found sparsely scattered in the outer layers of cerebral cortex, and more densely concentrated in dentate gyrus of the hippocampus (FIGS. 3 to 5), in the glomerular layer of the olfactory bulb, and in other restricted sites where beaded axons are located in untreated animals.[39] Therefore, the distribution of drug-resistant beaded axons is likely to account for the regional differences in sparing of chemical markers for 5-HT.[36]

In a series of additional studies,[39–41] MDA and MDMA were compared with two other amphetamine derivatives that were previously shown to cause decreased markers for 5-HT, namely PCA and fenfluramine. Each of these compounds produced a reduction of 5-HT axon terminals with the same regional distribution and morphologic

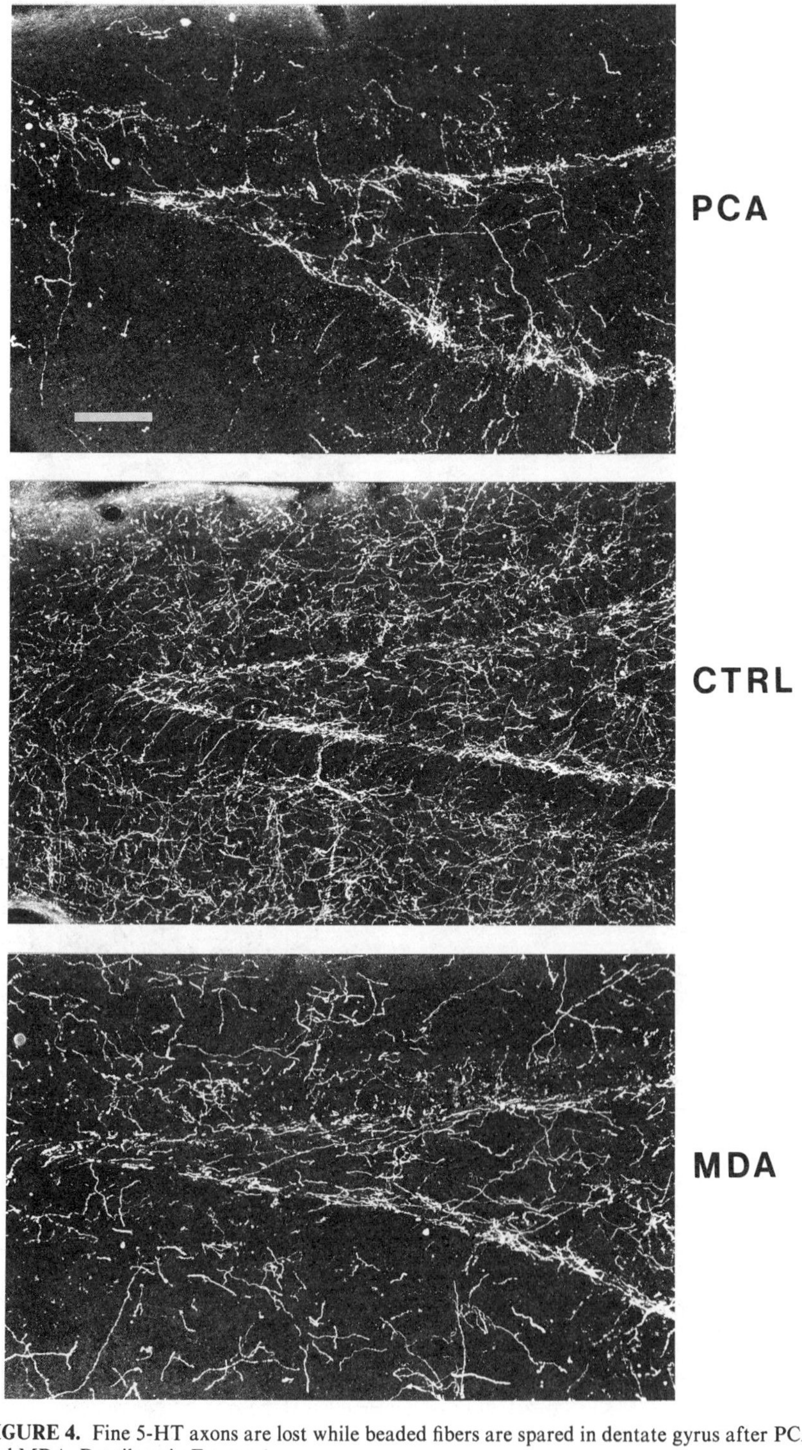

FIGURE 4. Fine 5-HT axons are lost while beaded fibers are spared in dentate gyrus after PCA and MDA. Details as in FIGURE 1.

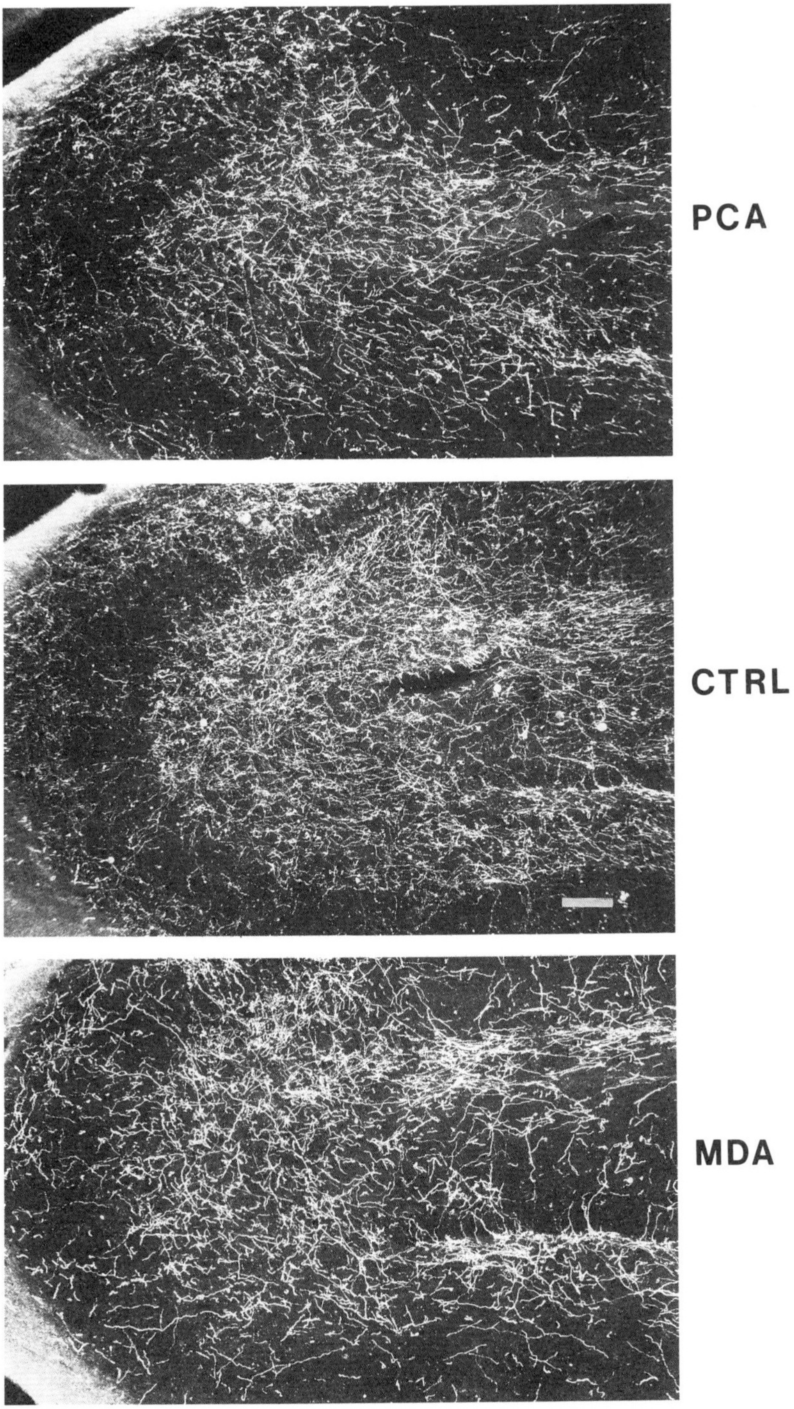

FIGURE 5. Sparing of beaded 5-HT axons in CA_3 of rat hippocampus 2 weeks after PCA or MDA. Reduced axon density reflects the loss of fine axons. Details as in FIGURE 2.

pattern as described for MDA. PCA produced loss of axon terminals throughout cerebral cortex, with a preferential loss of fine terminals while axons with large varicosities were unaffected.[39] The beaded axons which are spared are identical in morphology to beaded axons in control animals, and they have the same distribution, e.g., layers II to III of neocortex, dentate gyrus and CA_1, and the olfactory glomeruli. In addition, each of these drugs spares a unique population of 5-HT axon terminals which line the ependymal surface of the lateral ventricle, third ventricle and aqueduct. This ventricular plexus of drug-resistant 5-HT axons consists of coarse fibers with numerous large varicosities similar to axons arising from the MR. A similar pattern of 5-HT axon loss was obtained after administration of (±)fenfluramine administered as 4 daily doses of 5 or 10 mg/kg.[41,42]

NEUROTOXIC EFFECTS IN PRIMATES

To investigate the neurotoxic effects of these drugs in primates, a series of squirrel and Macaque monkeys were treated with either MDMA or *d*-fenfluramine by repeated subcutaneous injections. Two weeks later, the brains were analyzed for 5-HT markers and for immunocytochemical detection of axons. At doses of 2 to 10 mg/kg repeated 8 times, both drugs produced marked decreases in 5-HT levels in many parts of the forebrain.[15,43,44] As in the rat, the primate cerebral cortex exhibits both fine and beaded 5-HT axon terminals with regional differences in their distribution.[15] Treated monkeys exhibited a profound reduction in 5-HT axon density throughout most cortical areas, characteristic regional differences, and selective vulnerability of fine axons.[15,44,45] This differential vulnerability of fine and beaded axons and their distinct regional distributions suggest that the two ascending 5-HT projections may subserve different functions. The vulnerability of fine 5-HT axons to psychotropic drugs supports the hypothesis that these axons are one of the sites of action for these drugs and may be involved in the control of affective state.[46]

The differential morphology and vulnerability of the two 5-HT axon types led to the hypothesis that the drug-sensitive projections arise from the DR, whereas drug-resistant terminals arise from the MR. Retrograde axon transport studies were performed following drug treatment to verify the differential origin of the two axon types. A fluorescent dye (Fluoro-Gold) was injected into neocortex in treated and control rats, and retrogradely labeled raphe neurons were mapped. In animals that had been treated with PCA, there was greater than 90% reduction in labeled DR neurons;[47] the number of labeled MR cells was unchanged. Since the capacity of DR cells for retrograde axon transport is lost, these results confirm that DR axons which project to neocortex are selectively vulnerable to the neurotoxic effects of PCA in the rat. The retrograde transport studies were based on dye injections into fronto-parietal cortex and are supported by a similar study of the effects of PCA on 5-HT projections to the motor trigeminal nucleus.[48] Ongoing experiments will determine whether fine 5-HT axons in hippocampus and other regions arise exclusively from the DR. The consistent differential vulnerability of particular 5-HT axon types indicates that there are dual 5-HT projections to cortex, which differ in axon morphology, cells of origin, pattern of termination and in pharmacologic properties.

AXON DEGENERATION

One of the goals of these studies was to evaluate whether there is structural evidence for 5-HT axon degeneration following exposure to psychotropic drugs. A limitation of transmitter immunocytochemistry is that visualization of nerve fibers depends upon retention of transmitter within the axon. Axons that are totally depleted of 5-HT cannot be detected by this method, and the loss of axon staining by itself is not sufficient evidence to demonstrate degeneration. Consequently, a series of drug doses and short survival times (24–38 hours) were studied in order to select conditions that would best demonstrate degenerating axon terminals. At the survival times studied, sufficient amounts of 5-HT remain in some degenerating terminals to allow their visualization; while this method provides positive evidence for degeneration, it may underestimate the number of axons that are degenerating at any one time since many axons may be depleted of 5-HT. The results showing axonal damage following MDA are reported elsewhere.[38,46] The largest number of damaged axonal profiles was seen at 36 to 48 hours after treatment with *d*-fenfluramine (3 subcutaneous doses of 5 mg/kg).[41,42] Both rats and monkeys treated with fenfluramine[41,44] exhibited swollen, distorted and fragmented axon terminals in cerebral cortex, as well as swollen proximal axon stumps (FIGS. 6–9). The visualization of enormously swollen axons after administration of MDA or fenfluramine demonstrates that 5-HT nerve fibers are structurally damaged by exposure to these drugs. This positive evidence for degeneration is supported by the damming up of transmitter in swollen pre-terminal fibers, and by the subsequent disappearance of damaged fibers and the loss of fine axon terminals. Similar criteria for degeneration have been previously reported[49,50] based on histofluorescence of monoamine fibers damaged by other neurotoxins.

Unequivocal evidence for axon degeneration is difficult to establish with a single method. To assess axonal integrity, the entire constellation of evidence for axon damage has been taken into account, including biochemical, pharmacologic and morphologic data. Decreases in biochemical parameters have shown that nearly all markers for functional activity of serotonin neurons are reduced (see TABLE 1). Following administration of MDA or PCA, the levels of serotonin in cerebral cortex and other forebrain areas are markedly and persistently decreased.[30,32] A corresponding decrease is found in levels of 5-HIAA, a principal metabolite of 5-HT. A large decrease is seen in the activity of tryptophan hydroxylase, an enzyme required for synthesis of 5-HT.[35,51] Active uptake of tritiated 5-HT into synaptosomal preparations is profoundly decreased[30,32] as is the density of 5-HT uptake sites, as shown by marked reductions in the binding of paroxetine.[36] The morphologic changes[38] observed with immunocytochemistry strongly support the biochemical data for axon loss (see TABLE 2). Following administration of neurotoxic amphetamine derivatives, fine axon terminals in cortex are not detected by immunocytochemistry at survival times of 2 weeks and more.[38] At short survival times, the presence of swollen varicosities and axon fragmentation provide morphologic evidence of acute structural damage to 5-HT axons.[38,41,46] The subsequent disappearance of axon terminals and the presence of swollen, intensely immunoreactive proximal axon stumps reflects damming up of transmitter proximal to the site of axotomy. The intact raphe cell bodies in Nissl stained preparations and persistent 5-HT immunoreactivity in DR neurons[38] reveal that the cell bodies are spared, that 5-HT synthesis is intact and that axonal transport into the proximal axon stump is functional. Axonal transport experiments[47] indicate that, after PCA treatment, dorsal raphe neurons are not capable of retrograde transport from cortex while MR cells retain their projections. Additional morphologic evidence for axon degeneration is derived from the pattern of regeneration, which occurs over several months following drug treatment.[52] Also, axons in striatum exhibited Fink-Heimer-positive staining indicative of degeneration following MDA treatment.[32] The occurrence of drug-induced degeneration of 5-HT axons is supported by the similarity of acute and lasting effects produced by several psychotropic drugs such as MDA,[38] PCA,[30,39,53] and fenfluramine.[41] Taken together, these findings indicate

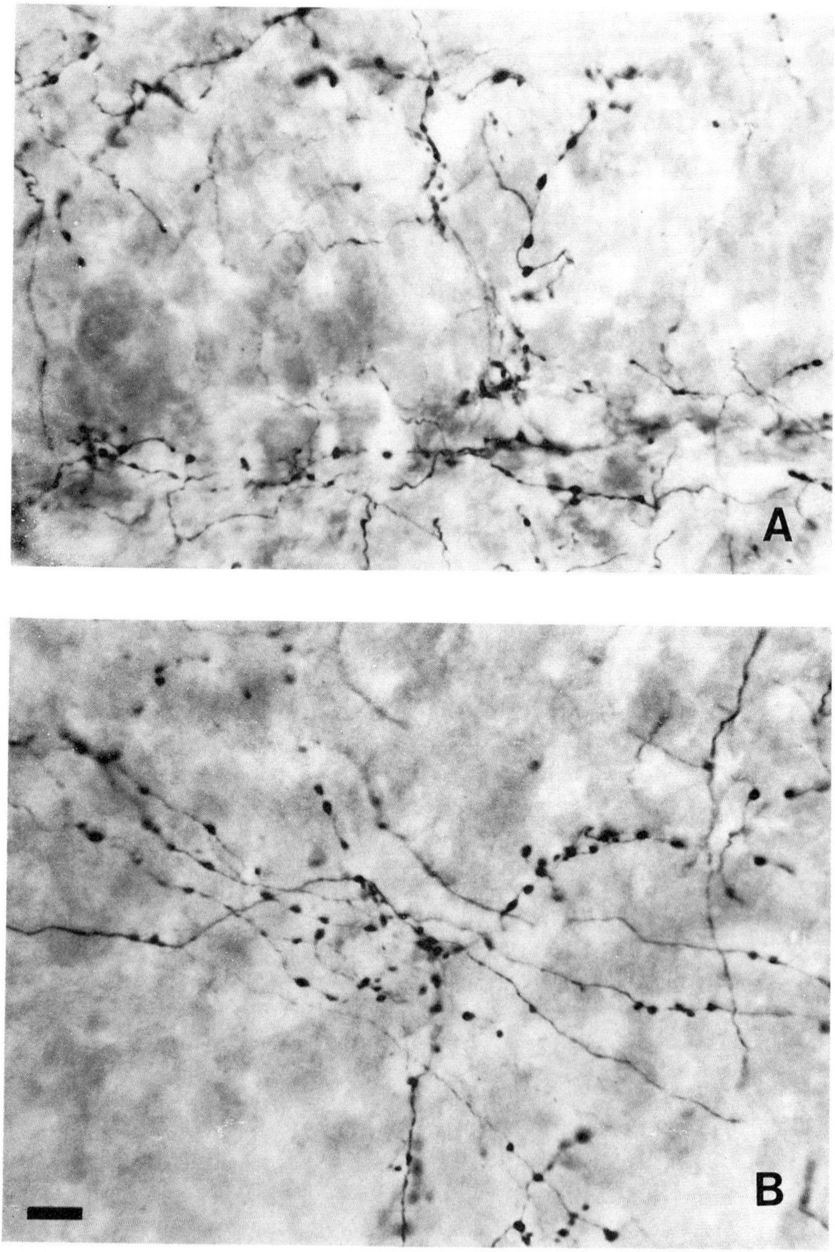

FIGURE 6. Beaded 5-HT axons are spared 2 weeks after MDA (8×20 mg/kg). These axons are identical to the largest 5-HT beaded axons seen in control animals. Bar = 10 μm. A–dentate gyrus; B–parietal cortex.

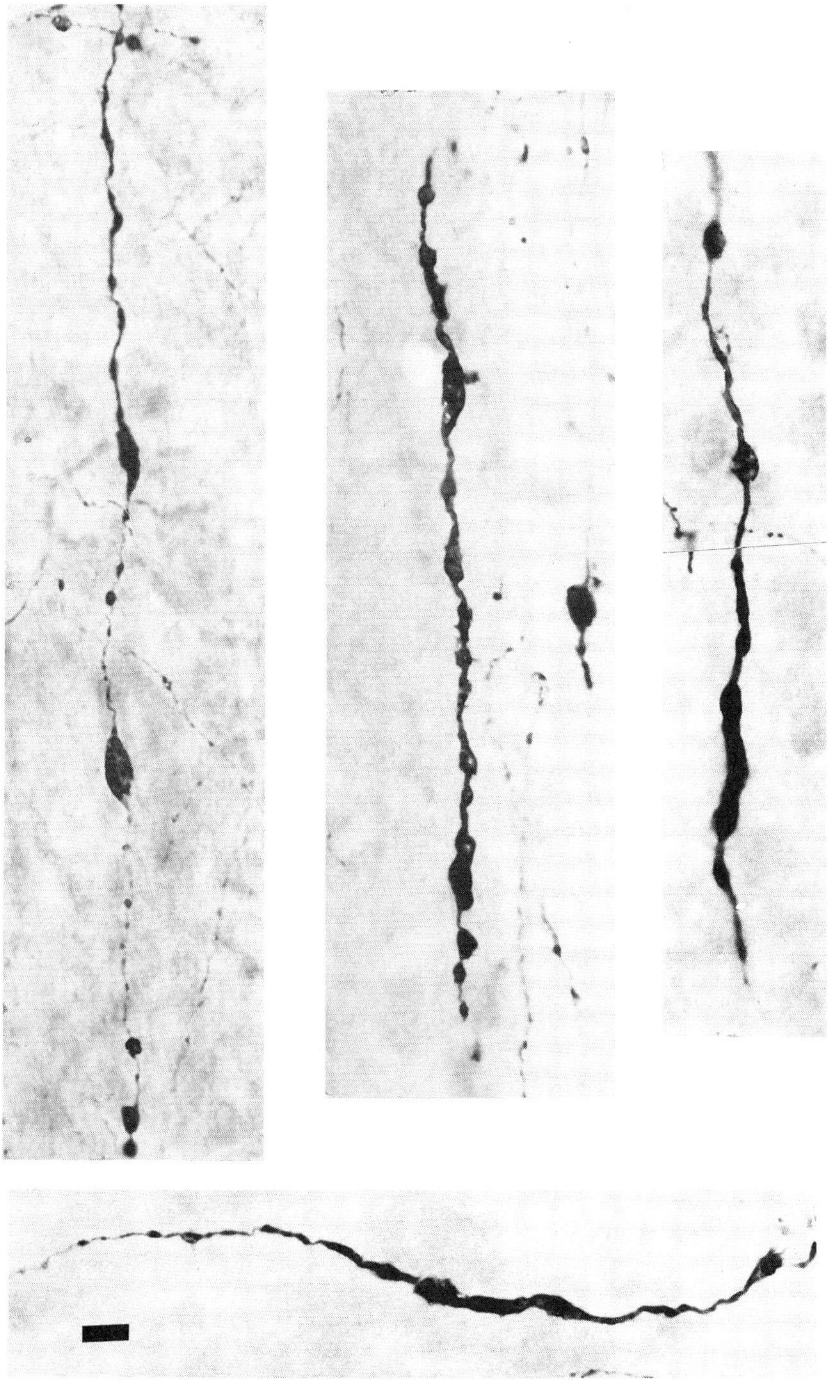

FIGURES 7–9. Examples of irregular, swollen axons within frontal and parietal cortex of rat 48 hours after *d*-fenfluramine (3 × 5 mg/kg). Note the irregular dilatations along axons and the large, swollen varicosities. This type of axonal swelling is never seen in control animals. These cytopathologic changes reflect structural damage and degeneration of axon terminals. Bar = 10 μm.

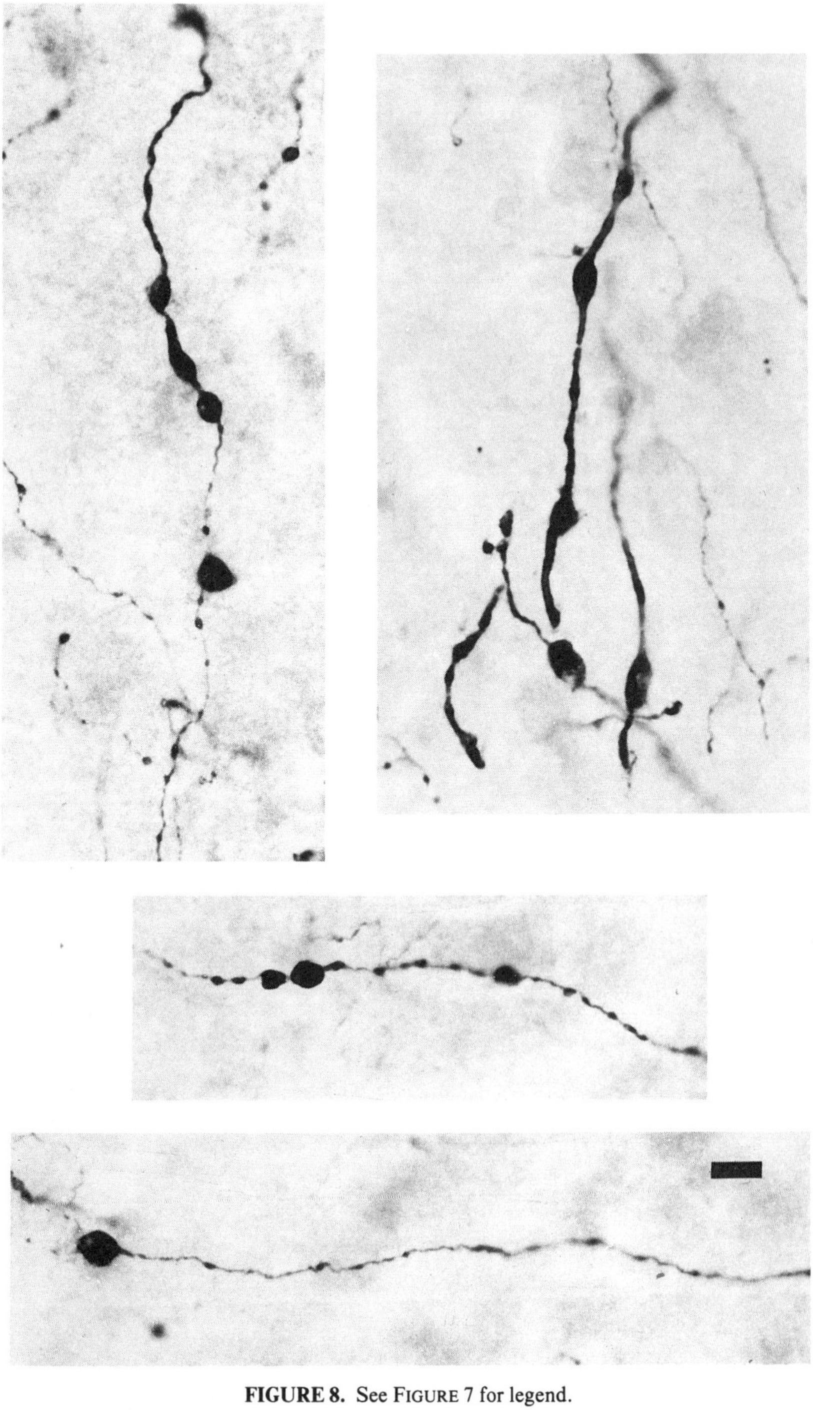

FIGURE 8. See FIGURE 7 for legend.

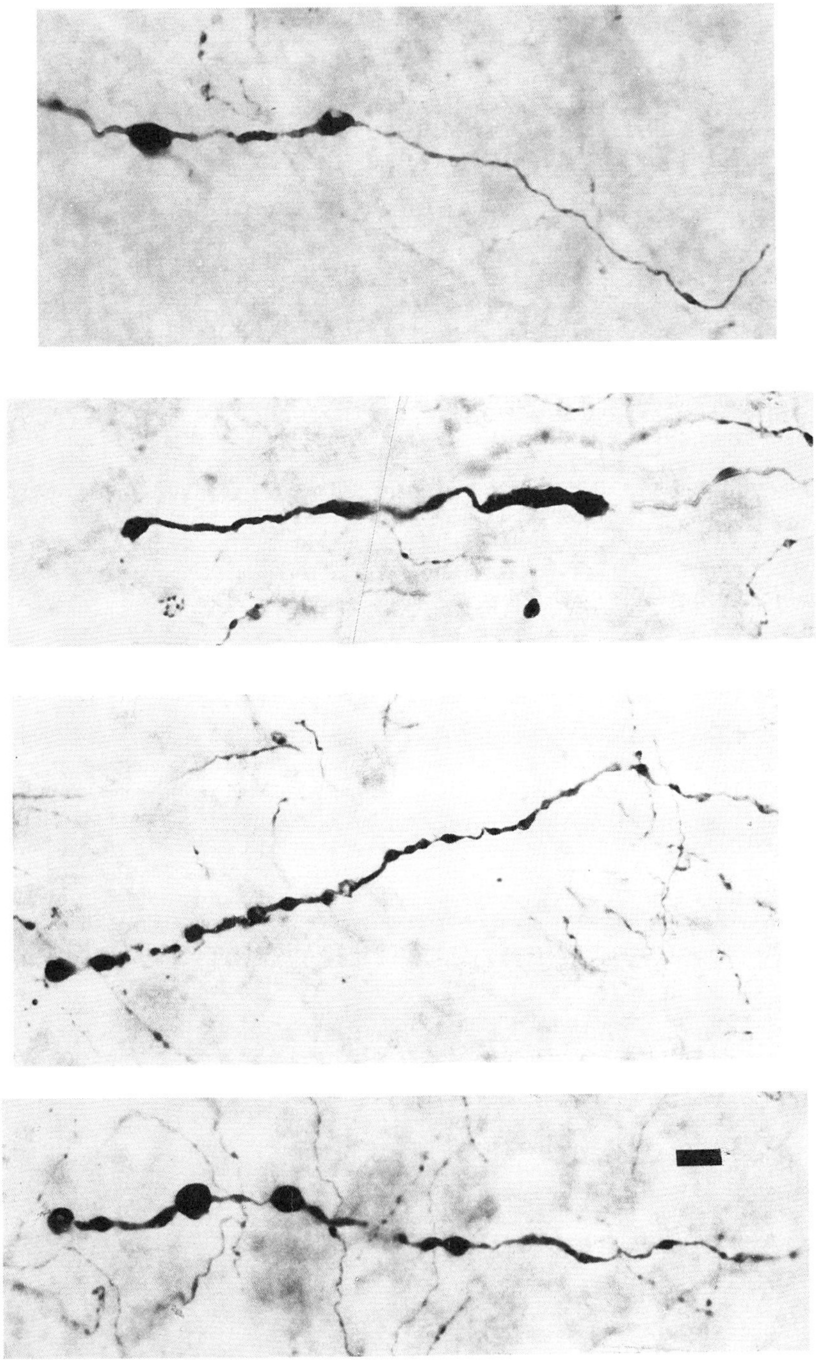

FIGURE 9. See FIGURE 7 for legend.

TABLE 1. Biochemical Evidence for 5-HT Axon Degeneration

- Loss of 5-HT—persistent
- Decreased metabolite levels (5-HIAA)
- Loss of synthetic enzyme activity
- Loss of 5-HT uptake
- Loss of uptake sites for 5-HT

that all properties of axon viability are lost (see TABLE 3). The marked decrease in serotonin uptake and of uptake binding sites results from a loss of functions associated with the plasma membrane of axon terminals. The decrease in terminal 5-HT indicates a loss of transmitter synthesis and storage which is a function of synaptic vesicles. Decreased tryptophan hydroxylase activity reflects a loss of enzyme activity associated with cytoplasmic proteins. The loss of retrograde transport by dorsal raphe cells indicates a severe dysfunction of cytoskeletal components of the axon. This constellation of evidence suggests that virtually all experimentally measurable parameters of axon viability are lost following treatment with large doses of the tested drugs. Based on the positive evidence for structural damage and loss of axon function, we conclude that this data is most consistent with the interpretation that MDA and related compounds produce degeneration of 5-HT axon terminals in forebrain. The lack of any signs of axon viability argues against the possibility that the depleted axon terminals may remain present but undetected.

DUAL SUBTYPES OF MONOAMINE NEURONS

All of the monoaminergic cell groups consist of neuronal subtypes that exhibit differential vulnerability to neurotoxins or other insults. The present studies distinguish two classes of 5-HT neurons that are differentially affected by neurotoxic amphetamines. Evidence for differential vulnerability among dopamine (DA) and norepinephrine (NE) neurons has also been reported. The neurotoxic drug, MPTP, selectively damages DA projections arising from the substantia nigra while sparing axons that arise from the ventral tegmental area.[54] In addition, in cases of Parkinson's disease, a selective loss of DA has been observed in the putamen while projections to the caudate nucleus may be relatively spared.[55] The neurotoxin, DSP-4, causes selective damage to

TABLE 2. Morphologic Evidence for 5-HT Axon Degeneration

- 5-HT Terminals Not Detectable
  - Loss of immunoreactivity (5-HT)
  - Duration—months
- Acute Structural Damage
  - Swollen varicosities and axon fragmentation
- Swollen Proximal Axons
  - Damming up of transmitter in axon stumps
- Spared Raphe Cell Bodies
  - 5-HT synthesis intact
- Loss of Retrograde Axonal Transport
- Pattern of Regeneration
  - Fronto > occipital sprouting
  - Bilaminar axon ingrowth

TABLE 3. Evidence for Axon Degeneration

All properties of axon viability are lost!
Multiple, independent structures are affected.
• Plasma Membrane Function
Loss of 5-HT uptake
Loss of binding sites for uptake carrier
• Synaptic Vesicles
Reduced transmitter synthesis
Loss of 5-HT storage
• Cytoplasmic Proteins
Loss of tryptophan hydroxylase activity
• Cytoskeleton
Loss of axonal transport

NE axons from the locus coeruleus while sparing fibers from other NE cell groups.[56] These data suggest that each monoaminergic cell group contains at least two classes of neurons that are differentially vulnerable to toxic or pathologic agents. Such dual vulnerability may reflect a general principle that would help to understand the causes of neurotoxicity.

MECHANISMS OF DEGENERATION

While the mechanisms by which psychotropic drugs damage 5-HT axons are unknown, the response of 5-HT neurons to neurotoxic amphetamines can be divided into four phases (see TABLE 4). In the first phase, systemic administration of the drug results in release of 5-HT followed by depletion of neurotransmitter from axon terminals.[34,57,58] The time course of this release phase is minutes to hours and the effects are reversible.[58,59] The second phase, at 24 to 48 hours after drug administration, is characterized by axonal swelling, fragmentation, and degeneration.[38] This degenerative phase does not appear reversible. The third phase, from 1 week to several months after drug treatment, is marked by the loss of 5-HT axon terminals with denervation of most forebrain structures. In the fourth phase, over a period of 2 to 8 months following drug treatment, there is progressive sprouting by 5-HT axons, and re-innervation of denervated areas.[52] The effects of psychotropic amphetamines appear to be dependent upon binding to the 5-HT uptake carrier on axon terminals. That view is supported by evidence that the neurotoxic effects of these compounds can be prevented or reversed

TABLE 4. Four Phases in Amphetamine Neurotoxicity

1. Transmitter Release/Depletion
a) Immediate—minutes to hours
b) Reversible
2. Axon Degeneration
a) Sub-acute: 1 to 3 days
b) Not reversible
3. Denervation (regional)
a) Persistent: weeks to months
4. Regeneration
a) Late and slowly progressive
b) 2 to 12 months

by early administration of a 5-HT uptake blocker such as fluoxetine or citalopram.[34,59] However, the subsequent steps which lead to degeneration have not been clarified. Intracerebral injections of MDA or PCA do not produce axonal loss, unlike the neurotoxin 5,7-dihydroxytryptamine (5,7-DHT).[50,60,61] The lack of a direct drug effect led us to postulate that the drugs themselves are not toxic, but that a drug metabolite may be the toxic agent. In one set of studies, 5-HT was depleted by treatment with either reserpine or the 5-HT synthesis inhibitor, *p*-chlorophenylalanine (PCPA) or a combination of both drugs.[61,62] During the phase of 5-HT depletion, animals were protected against the neurotoxic effect of PCA. This result suggests that release of 5-HT is necessary to produce the neurotoxicity, and that metabolism of 5-HT released in the periphery may result in formation of a toxic compound. Other laboratories have previously suggested that amphetamiies may induce the formation of a toxic indoleamine derivative[63] or alternatively, that endogeneous DA release plays a role in producing toxicity.[64]

The mechanism of neurotoxicity and the factors which determine the differential vulnerability of 5-HT axons are unknown at this time. However, several mechanisms may be postulated to explain why the two axon types may be differentially susceptible. Since binding of psychotropic amphetamines to the 5-HT uptake carrier is a prerequisite for these drugs to produce axon damage, the hypothesis is suggested that the two axon types may have distinct uptake carriers for 5-HT. The psychotropic amphetamines may have differential affinities for the uptake binding sites associated with the two types of 5-HT axons. Several other possible differences between the axon terminals should also be entertained. The density of uptake sites may differ on the two axon types, and there are likely to be differences in the ratio of surface area to volume such that a compound taken up by large varicosities may be diluted in a greater volume. Structural or metabolic differences between the two axon types may render them more or less robust or differentially sensitive to particular drugs. Such factors might include mechanisms to bind or inactivate a toxin, or differences in the regulation of intra-axonal calcium concentration. It is not known whether axons arising from DR cells are more divergent than those from the MR, or whether greater divergence would make a cell more vulnerable. These suggested differences between the two axon types are speculative but may serve as a basis for further investigations of the determinants of vulnerability.

REGENERATION

Following MDA or PCA, drug-treated rats were maintained for periods of 2 to 8 months. At long survival times, immunocytochemical studies provide evidence for gradual re-innervation of cortex and other structures by 5-HT axons.[52] A gradual increase in 5-HT axon density is seen over 2 to 8 months in the hippocampus, olfactory bulb, and striatum. In dorsal areas of neocortex, there is a slowly progressive frontal to occipital gradient of recovery. Recovery in frontal cortex occurs more rapidly and shows a moderate increase in axon density at 1 to 2 months after treatment with a slow increase towards normal density at 8 months. Occipital cortex shows the greatest delay in re-innervation. From 2 to 4 months in dorsal neocortex, straight, longitudinally oriented axons gradually extend from rostral to caudal within layers I and VI; these appear to be pre-terminal fibers which form two parallel bands that progressively elongate back to more caudal regions of cortex. The density of fine axons within middle layers of neocortex is initially quite low. From 4 to 8 months, fine terminal branches arise from the longitudinal pre-terminal axons and extend radially towards the middle cortical layers. These newly formed sprouts arborize into the cortex and form fine terminals similar to the fine 5-HT axons seen in controls. There is no evidence to support aberrant re-innervation since the density of beaded 5-HT axons, which survive drug treatment, does not increase. In contrast, there is considerable specificity in the regeneration of fine axonal terminals. The sequence of re-innervation which combines a rostral to caudal gradient with a bilaminar pattern of axon growth simulates the perinatal development of 5-HT projections. This finding suggests that mechanisms which evoke and guide axonal gowth during fetal development may be present in the adult brain. This sprouting response provides a model system for studying axonal regeneration in the mature central nervous system. Further studies are underway to determine the origin and pharmacologic properties of regenerating 5-HT axons and whether the normal pattern of 5-HT projections is re-established.

SUMMARY

The cytotoxic effects of amphetamine derivatives were studied by immunocytochemistry to identify the cellular compartments affected by these drugs, to obtain morphologic evidence of neuronal degeneration, and to assess the potential for regeneration. The substituted amphetamines, MDA, MDMA, PCA, and fenfluramine, all release serotonin and cause acute depletion of 5-HT from most axon terminals in forebrain. (1) Unequivocal signs of axon degeneration were seen at 36–48 hour survivals: 5-HT axons exhibited increased caliber, huge, swollen varicosities, fragmentation, and dilated proximal axon stumps. (2) Fine 5-HT axon terminals were persistently lost after drug administration, while beaded axons and raphe cell bodies were spared. These two types of 5-HT axons, which arise from separate raphe nuclei and form distinct ascending projections, are differentially vulnerable to psychotropic drugs. (3) From 2–8 months after treatment, there was progressive serotonergic re-innervation of neocortex along a fronto-occipital gradient. Longitudinal 5-HT axons grew into layers I and VI from rostral to caudal, before sprouting into middle cortical layers; this bilaminar pattern of growth simulates perinatal development of 5-HT innervation. This study demonstrates differential vulnerability of 5-HT projections, evidence for axonal degeneration, and sprouting of 5-HT axons leading to re-innervation of forebrain. While the sprouting axons are anatomically similar to the type that was damaged, it is not known whether a normal pattern of innervation is re-established.

ACKNOWLEDGMENTS

We appreciate the skilled technical assistance of Patrice Carr and we wish to thank Patricia O'Neill for help in preparation of the manuscript. We also thank Crystal Mullen for helping to conduct many PCA experiments and for several plates.

REFERENCES

1. Dahlstrom, A. & K. Fuxe. 1964. Evidence for the existence of monoamine-containing neurons in the central nervous system. I. Demonstration of monoamines in cell bodies of brain neurons. Acta Physiol. Scand. **62** (Suppl. 232): 1–55.
2. Anden, N. E., A. Dahlstrom, K. Fuxe, K. Larsson, L. Olson & U. Ungerstedt. 1966. Ascending monoamine neurons to the telencephalon and diencephalon. Acta Physiol. Scand. **67:** 313–326.
3. Ungerstedt, U. 1971. Stereotaxic mapping of the monoamine pathways in the rat brain. Acta Physiol. Scand. (Suppl. 367): 1–48.

4. CONRAD, L. C. A., C. M. LEONARD & D. W. PFAFF. 1974. Connections of the median and dorsal raphe nuclei in the rat: Autoradiographic and degeneration study. J. Comp. Neurol. **156:** 179–206.
5. FUXE, K. 1965. Evidence for the existence of monoamine neurons in the central nervous system. IV. Distribution of monoamine nerve terminal in the central nervous system. Acta Physiol. Scand. **64** (Suppl. 247): 39–85.
6. PARENT, A., L. DESCARRIES & A. BEAUDET. 1981. Organization of ascending serotonin neurons in the adult rat brain. A radioautographic study after intraventricular administration of [^{3}H]5-hydroxytryptamine. Neuroscience **6:** 115–138.
7. AZMITIA, E. C. & M. SEGAL. 1978. An autoradiographic analysis of the differential ascending projections of the dorsal and median raphe nuclei in the rat. J. Comp. Neurol. **179:** 641–668.
8. STEINBUSCH, H. W. M., A. A. J. VERHOFSTAD & H. W. J. JOOSTEN. 1978. Localization of serotonin in the central nervous system by immunohistochemistry: Description of a specific and sensitive technique and some applications. Neuroscience **3:** 811–819.
9. STEINBUSCH, H. W. M. 1981. Distribution of serotonin-immunoreactivity in the central nervous system of the rat—Cell bodies and terminals. Neuroscience **6:** 557–618.
10. LIDOV, H. G. W., R. GRZANNA & M. E. MOLLIVER. 1980. The serotonin innervation of the cerebral cortex in the rat—An immunohistochemical analysis. Neuroscience **5:** 207–227.
11. KOSOFSKY, B. E. 1985. The neuroanatomic organization of ascending serotonergic projections to cerebral cortex. Ph.D. Thesis. The Johns Hopkins University School of Medicine. Baltimore, MD.
12. BLUE, M. E., B. E. KOSOFSKY & M. E. MOLLIVER. 1988. Regional differences in the serotonin innervation of rodent cerebral cortex: Differential distribution of two morphologically distinct axon types. Soc. Neurosci. Abstr. **14:** 209.
13. KOSOFSKY, B. E., M. E. MOLLIVER, J. H. MORRISON & S. L. FOOTE. 1984. The serotonin and norepinephrine innervation of primary visual cortex in the Cynomolgus monkey (Macaca fascicularis). J. Comp. Neurol. **230:** 168–178.
14. MORRISON, J. H., S. L. FOOTE, M. E. MOLLIVER, F. E. BLOOM & H. G. W. LIDOV. 1982. Noradrenergic and serotonergic fibers innervate complementary layers in monkey primary visual cortex: An immunohistochemical study. Proc. Natl. Acad. Sci. USA **79:** 2401–2405.
15. WILSON, M. A., G. A. RICAURTE & M. E. MOLLIVER. 1989. Distinct morphologic classes of serotonergic axons in primates exhibit differential vulnerability to the psychotropic drug 3,4-methylenedioxymethamphetamine. Neuroscience **28:** 121–137.
16. O'HEARN, E. & M. E. MOLLIVER. 1984. Organization of raphe-cortical projections in rat: A quantitative retrograde study. Brain Res. Bull. **13:** 709–726.
17. IMAI, H., D. A. STEINDLER & S. T. KITAI. 1986. The organization of divergent axonal projections from the midbrain raphe nuclei in the rat. J. Comp. Neurol. **243:** 363–380.
18. MOLLIVER, M. E. 1987. Serotonergic neuronal systems: What their anatomic organization tells us about function. J. Clin. Psychopharmacol. **7** (Suppl.): 3–23.
19. KOHLER, C. & H. W. M. STEINBUSCH. 1982. Identification of serotonin and non-serotonin-containing neurons of the mid-brain raphe projecting to the entorhinal area and the hippocampal formation. A combined immunohistochemical and fluorescent retrograde tracing study in the rat brain. Neuroscience **7:** 951–975.
20. JACOBS, B. L., S. L. FOOTE & F. E. BLOOM. 1978. Differential projections of neurons within the dorsal raphe nucleus of the rat: A horseradish peroxidase (HRP) study. Brain Res. **147:** 149–153.
21. WILSON, M. A. & M. E. MOLLIVER. 1988. The anatomic organization of serotonergic projections to neocortex in the primate. Soc. Neurosci. Abstr. **14:** 210.
22. CHAN-PALAY, V. 1977. The indoleamine afferent axons to the cerebellum. *In* Cerebellar Dentate Nucleus—Organization, Cytology and Transmitters. V. Chan-Palay, Ed. New York: Springer-Verlag, pp. 390–454.
23. KOHLER, C., V. CHAN-PALAY, L. HAGLUND & H. W. M. STEINBUSCH. 1980. Immunohistochemical localization of serotonin nerve terminals in the lateral entorhinal cortex of the rat: Demonstration of two separate patterns of innervation from the midbrain raphe. Anat. Embryol. **160:** 121–129.

24. MULLIGAN, K. A. & I. TORK. 1988. Serotoninergic innervation of the cat cerebral cortex. J. Comp. Neurol. **270:** 86–110.
25. KOSOFSKY, B. E. & M. E. MOLLIVER. 1987. The serotoninergic innervation of cerebral cortex: Different classes of axon terminals arise from dorsal and median raphe nuclei. Synapse **1:** 153–168.
26. HOTCHKISS, A. J. & J. W. GIBB. 1980. Long-term effects of multiple doses of methamphetamine on tryptophan hydroxylase and tyrosine hydroxylase activity in rat brain. J. Pharmacol. Exp. **214:** 257–262.
27. RICAURTE, G. A., R. W. GUILLERY, L. S. SEIDEN, C. R. SCHUSTER & R. Y. MOORE. 1982. Dopamine nerve terminal degeneration produced by high doses of methylamphetamine in the rat brain. Brain Res. **235:** 93–103.
28. RICAURTE, G. A., C. R. SCHUSTER & L. S. SEIDEN. 1980. Long-term effects of repeated methylamphetamine administration on dopamine and serotonin neurons in the rat brain: A regional study. Brain Res. **193:** 153–163.
29. KOHLER, C., S. B. ROSS, B. SREBRO & S-O. OGREN. 1978. Long-term biochemical and behavioral effects of p-chloroamphetamine in the rat. Ann. NY Acad. Sci. **305:** 645–663.
30. SANDERS-BUSH, E., J. A. BUSHING & F. SULSER. 1975. Long-term effects of p-chloroamphetamine and related drugs on central serotonergic mechanisms. J. Pharmacol. Exp. Ther. **192:** 33–41.
31. SANDERS-BUSH, E. & L. R. STERANKA. 1978. Immediate and long-term effects of p-chloroamphetamine on brain amines. Ann. NY Acad. Sci. **305:** 208–221.
32. RICAURTE, G. A., G. BRYAN, L. STRAUSS, L. SEIDEN & C. SCHUSTER. 1985. Hallucinogenic amphetamine selectively destroys brain serotonin nerve terminals. Science **229:** 986–988.
33. COMMINS, D. L., G. VOSMER, R. M. VIRUS, W. L. WOOLVERTON, C. R. SCHUSTER & L. S. SEIDEN. 1987. Biochemical and histological evidence that methylenedioxymethamphetamine (MDMA) is toxic to neurons in the rat brain. J. Pharmacol. Exp. Ther. **241:** 338–345.
34. SCHMIDT, C. J. 1987. Neurotoxicity of the psychedelic amphetamine, methylenedioxymethamphetamine. J. Pharmacol. Exp. Ther. **240:** 1–7.
35. STONE, D. M., D. C. STAHL, G. R. HANSON & J. W. GIBB. 1986. The effects of 3,4-methylenedioxymethamphetamine (MDMA) and 3,4-methylenedioxyamphetamine (MDA) on monoaminergic systems in the rat brain. Eur. J. Pharmacol. **128:** 41–48.
36. BATTAGLIA, G., S. Y. YEH, E. O'HEARN, M. E. MOLLIVER, M. J. KUHAR & E. B. DE SOUZA. 1987. 3,4-Methylenedioxymethamphetamine (MDMA) and 3,4-Methylenedioxyamphetamine (MDA) preferentially destroy serotonin terminals in rat brain: Quantification of neurodegeneration by measurement of ^{3}H-paroxetine-labeled serotonin uptake sites. J. Pharmacol. Exp. Ther. **242:** 911–916.
37. O'HEARN, E., G. BATTAGLIA, E. B. DE SOUZA, M. J. KUHAR & M. E. MOLLIVER. 1986. Systemic MDA and MDMA, psychotropic substituted amphetamines, produce serotonin neurotoxicity. Soc. Neurosci. Abstr. **12:** 1233.
38. O'HEARN, E., G. BATTAGLIA, E. B. DE SOUZA, M. J. KUHAR & M. E. MOLLIVER. 1988. Methylenedioxyamphetamine (MDA) and methylenedioxymethamphetamine (MDMA) cause selective ablation of serotonergic axon terminals in forebrain: Immunocytochemical evidence for neurotoxicity. J. Neurosci. **8:** 2788–2803.
39. MAMOUNAS, L. A., C. MULLEN & M. E. MOLLIVER. 1988. Morphologically dissimilar serotonergic axon types in rat cerebral cortex are differentially vulnerable to the neurotoxin p-chloroamphetamine (PCA). Soc. Neurosci. Abstr. **14:** 210.
40. STERANKA, L. R. & E. SANDERS-BUSH. 1979. Long-term effects of fenfluramine on central serotonergic mechanisms. Neuropharmacology **18:** 895–903.
41. MOLLIVER, D. C. & M. E. MOLLIVER. 1990. Anatomic evidence for a neurotoxic effect of (±)fenfluramine upon serotonergic projections in the rat. Brain Res. **511:** 165–168.
42. MOLLIVER, D. C. & M. E. MOLLIVER. 1988. Selective neurotoxic effects of (±)fenfluramine upon 5-HT axons in rat brain: Immunocytochemical evidence. Soc. Neurosci. Abstr. **14:** 210.
43. RICAURTE, G. A., L. S. FORNO, M. A. WILSON, L. E. DELANNEY, I. IRWIN, M. E. MOLLIVER & J. W. LANGSTON. 1988. (±)3,4-methylenedioxymethamphetamine

(MDMA) selectively damages central serotonergic neurons in nonhuman primates. J. Am. Med. Assoc. **260:** 51–55.
44. RICAURTE, G. A., M. E. MOLLIVER, J. M. WITKIN, D. C. MOLLIVER, M. A. WILSON & J. L. KATZ. 1989. d-Fenfluramine produces long-term effects on central serotonin neurons in nonhuman primates. Soc. Neurosci. Abstr. **15:** 419.
45. WILSON, M. A., G. A. RICAURTE & M. E. MOLLIVER. 1987. The psychotropic drug 3,4-methylenedioxymethamphetamine (MDMA) destroys serotonergic axons in primate forebrain: Regional and laminar differences in vulnerability. Soc. Neurosci. Abstr. **13:** 906.
46. MOLLIVER, M. E., L. A. MAMOUNAS & M. A. WILSON. 1989. Effects of neurotoxic amphetamines on serotonergic neurons: Immunocytochemical studies. *In* Pharmacology and Toxicology of Amphetamine and Related Designer Drugs. National Institute on Drug Abuse Research Monograph #94.: 270–305. Supt. of Docs., U.S. Govt. Print. Off. Washington, D.C.
47. MAMOUNAS, L. A. & M. E. MOLLIVER. 1988. Evidence for dual serotonergic projections to neocortex: Axons from the dorsal and median raphe nuclei are differentially vulnerable to the neurotoxin p-chloroamphetamine (PCA). Exp. Neurol. **102:** 23–26.
48. FRITSCHY, J-M., W. E. LYONS, M. E. MOLLIVER & R. GRZANNA. 1988. Neurotoxic effects of p-chloroamphetamine on the serotoninergic innervation of the trigeminal motor nucleus: A retrograde transport study. Brain Res. **473:** 261–270.
49. BAUMGARTEN, H. G., L. LACHENMAYER & H. G. SCHLOSSBERGER. 1972. Evidence for a degeneration of indolamine-containing nerve terminals in rat brain, induced by 5,6-dihydroxytryptamine. Z. Zellforsch. **125:** 553–569.
50. BAUMGARTEN, H. G., A. BJORKLUND, L. LACHENMAYER & A. NOBIN. 1973. Evaluation of the effects of 5,7-dihydroxytryptamine on serotonin and catecholamine neurons in the rat CNS. Acta Physiol. Scand. (Suppl. 391): 1–19.
51. JOHNSON, M., A. A. LETTER, K. MERCHANT, G. R. HANSON & J. W. GIBB. 1988. Effects of 3,4-methylenedioxyamphetamine and 3,4-methylenedioxymethamphetamine isomers on central serotonergic, dopaminergic and nigral neurotensin systems of the rat. J. Pharmacol. Exp. Ther. **244:** 977–982.
52. MOLLIVER, M. E., L. A. MAMOUNAS & P. CARR. 1989. Reinnervation of cerebral cortex by 5-HT axons after denervation by psychotropic amphetamine derivatives. Soc. Neurosci. Abstr. **15:** 417.
53. MULLEN, C., L. A. MAMOUNAS, E. O'HEARN & M. E. MOLLIVER. 1987. Dual serotonergic projections to forebrain in the rat: Two classes of axon terminals exhibit differential vulnerability to the psychotropic drugs p-chloroamphetamine (PCA) and 3,4-methylenedioxyamphetamine (MDA). Soc. Neurosci. Abstr. **13:** 907.
54. LANGSTON, J. W., L. S. FORNO, C. S. REBERT & I. IRWIN. 1984. Selective nigral toxicity after systemic administration of 1-methyl-4-phenyl-1,2,5,6-tetrahydropyrine (MPTP) in the squirrel monkey. Brain Res. **292:** 390–394.
55. KISH, S. J., K. SHANNAK & O. HORNYKIEWICZ. 1988. Uneven pattern of dopamine loss in the striatum of patients with idiopathic Parkinson's disease. Pathophysiologic and clinical implications. N. Engl. J. Med. **318:** 876–880.
56. FRITSCHY, J-M. & R. GRZANNA. 1989. Immunohistochemical analysis of the neurotoxic effects of DSP-4 identifies two populations of noradrenergic axon terminals. Neuroscience **30:** 181–197.
57. JOHNSON, M. P., A. J. HOFFMAN & D. E. NICHOLS. 1986. Effects of the enantiomers of MDA, MDMA and related analogues on [^{3}H]serotonin and [^{3}H]dopamine release from superfused rat brain slices. Eur. J. Pharmacol. **132:** 269–276.
58. MOLLIVER, M. E., K. STRATTON, P. CARR, R. GRZANNA & J. BARABAN. 1988. Contrasting in vitro and in vivo effects of p-chloroamphetamine (PCA) on 5-HT axons: Immunocytochemical studies in hippocampal slices. Soc. Neurosci. Abstr. **14:** 210.
59. FULLER, R. W., K. W. PERRY & B. B. MOLLOY. 1975. Reversible and irreversible phases of serotonin depletion by 4-chloroamphetamine. Eur. J. Pharmacol. **33:** 119–124.
60. MOLLIVER, M. E., E. O'HEARN, G. BATTAGLIA & E. B. DE SOUZA. 1986. Direct intracerebral administration of MDA and MDMA does not produce serotonin neurotoxicity. Soc. Neurosci. Abstr. **12:** 1234.

61. BERGER, U. V. 1989. Mechanism of neurotoxicity of substituted amphetamines. Ph.D. Thesis, University of Basel. Switzerland.
62. BERGER, U. V., R. GRZANNA & M. E. MOLLIVER. 1989. Depletion of serotonin using p-chlorophenylalanine (PCPA) and reserpine protects against the neurotoxic effects of p-chloroamphetamine (PCA) in the brain. Exp. Neurol. **103:** 111–115.
63. COMMINS, D. L., K. J. AXT, G. VOSMER & L. S. SEIDEN. 1987. 5,6-Dihydroxytryptamine, a serotonergic neurotoxin, is formed endogenously in the rat brain. Brain Res. **403:** 7–14.
64. STONE, D. M., M. JOHNSON, G. R. HANSON & J. W. GIBB. 1988. Role of endogenous dopamine in the central serotonergic deficits induced by 3,4-methylenedioxymethamphetamine. J. Pharmacol. Exp. Ther. **247:** 79–87.

DISCUSSION OF THE PAPER

R. WURTMAN (*MIT, Cambridge, MA*): I wonder if I could make some counter comments since I don't agree with you about your conclusions. I think that if you take even a second year medical student at Johns Hopkins and ask what drug depletes monoamines, causes axons to swell, and produces pictures like that, he will say "aha, reserpine," and he'll be right. Reserpine is a widely used drug that does all of these things; it continues to be used and nobody would use the term neurotoxicity to describe it. You may have found a selective reserpine for serotonin neurons. I would ask whether you have done a similar experiment using high doses of fluoxetine? We have done similar experiments with fluoxetine and find a marked depletion of serotonin and a prolonged suppression of serotonin release which is comparable to the one you get with these other drugs but it lasts much longer. Have you done studies to determine whether fluoxetine or citalopram produce similar changes, and how do you justify calling your "reserpine-like" action neurotoxicity?

MOLLIVER: Dr. Urs Berger, in our group, has treated animals with large doses of reserpine, and we have not seen evidence of axonal swelling or of degeneration. During the initial 10 days following treatment with reserpine, there is depletion of serotonin from vesicular storage sites. Over the second week, the serotonin levels recover, and the axons appear completely normal within 2 weeks. We have not observed any lasting toxic effects of reserpine.

We have also treated animals with repeated doses of fluoxetine (20–30 mg/kg per day) and found two results. First, there is no axonal swelling, damage or neurotoxicity. In fact, the morphologic quality, caliber and immunoreactive staining of 5-HT axons are highly similar to normal (untreated) animals. Thus, we have seen no evidence of any toxicity caused by fluoxetine. In fact, in some fluoxetine treated animals serotonergic axons may stain more intensely than normal. This may result from the fact that fluoxetine is an uptake blocker and leads to increased serotonin concentrations in the synaptic cleft. As a result, serotonin release may be inhibited, leaving more serotonin remaining in the axon terminals. No swelling or other signs of axonal damage were observed.

A more cogent point is that the administration of fluoxetine, or of other 5-HT uptake blockers, either with or shortly after MDA, MDMA, PCA, or fenfluramine, protects against the neurotoxic effects of these compounds.

RUDY NOBLE (*Cathedral Hill Obesity Clinic, San Francisco, CA*): In the past 21 years we've done about 10 clinical trials with fenfluramine at various dose levels. These have encompassed a total of about 500 patients, all of whom I personally have interviewed. In my experience I have not seen a single case of neurotoxicity. By neurotoxicity I mean complaints from the patients of such things as loss of cognitive function, memory lapses, severe sleep distrubances, or disturbances in sexual function. I really do not believe fenfluramine in any way, at the clinical doses that we prescribe, is neurotoxic. So that's why I'm quite distrubed to see all this basic data coming out. My question is, what level did you give fenfluramine to the experimental animals and how does this relate to the doses that we give in humans at the therapeutic level.

MOLLIVER: Thank you for raising that important issue because our experimental studies were not designed to address the problem of human clinical toxicology. Our use of higher doses, different regimen and route of administration makes it difficult to extrapolate from animal experiments to human effects. The issue of human toxicology requires a completely different type of study. We were using large doses of these drugs as experimental tools in order to identify the types of morphologic effects that might be obtained.

To answer your specific question, the doses of (±)fenfluramine that we found to produce marked levels of axon degeneration in the rat and monkey were 5 mg/kg injected subcutaneously once a day for 2 to 4 days. This dose is considerably higher than the human dose which is taken orally, but the margin of safety is not enormous. What is the dose in mg/kg that you utilized in your patients?

NOBLE: When I was giving DL-fenfluramine we started with 30 or 60 mg/day and ended as high as 120 mg/day. Assuming an average patient weighs 60 kilos, that would be about a 1–2 mg/kg dose.

MOLLIVER: We observed considerable neurotoxicity using 5 mg/kg in the rat. This dose is about 2½ times the maximum dose that you utilized in patients.

NOBLE: Yes, but there may be species differences in effect.

LEWIS S. SEIDEN (*Session Chairman*): Dr. Noble, one thing that puzzles me, if you have 500 patients and you don't have a single incidence of sleep disturbances, sexual dysfunction, or cognitive disorder, I venture to say that's a very special group of patients.

NOBLE: Not at all, as a matter of fact these are very obese patients who had all sorts of problems. However, we specifically asked them when they came back if they had any side effects, which we certainly would have reported to the FDA.

MOLLIVER: It is clear from this morning's proceedings that we do not at this point know how to evaluate what behavioral consequences may result from a serotonin deficiency. There may, first of all, be compensatory mechanisms or there may be subtle changes in mood that would not be easily detected. Alternatively, it is even possible that pruning serotonin neurons may be therapeutic in some cases, and perhaps that's what Dr. Seiden was getting at. Possibly you have done some psychosurgery on your patients and made them better that way.

S. GARATTINI (*I.D.R.F. Mario Negri, Milan, Italy*): I would like to come back a little to the dose of fenfluramine which is used for this toxic effect of the drugs. There was a poster by Neil Roland showing that in fact doses in the range of 0.1 to 0.2 mg/kg were enough to trigger a very obvious anorexic effect of the drug in rats. In my lab, in fact, we looked at the effect of similar doses of fenfluramine to see whether this dose would either produce an inhibition of the uptake or a release of serotonin. And clearly with a low dose, let's say the dose which is required for this behavioral effect, fenfluramine is mainly a serotonin uptake inhibitor and not a releasing agent for serotonin.

So coming back to possible comparisons with other drugs which may have the same profile, if we take for instance fluoxetine, if we increase the dose to 10 times more, maybe much more, 50 times more, we may have a releasing effect and this condition may also trigger some neurotoxic effect regarding 5-HT neurons.

SEIDEN: I think the point has been made that the ratios between therapeutic effect and toxicity are not well understood or known at the present time, especially in human subjects. We can talk about that the rest of the day since in my view, and I think for many of us, that is an open question and a very important issue—to determine the ratio between the long-term effects and the therapeutic or beneficial effects of these compounds.

B. CAMPBELL (*Servier R&D, Ltd., London*): Perhaps the issue could be partially resolved by measuring brain levels of the drug. I didn't hear that today, whether in fact you had measured brain levels.

MOLLIVER: No, we haven't, but as I recall you had made some preliminary determinations in rats. How do those levels compare with the blood or brain level of fenfluramine in the human at an effective dose?

CAMPBELL: The combination, and we have to think about the combination of both fenfluramine and its metabolite norfenfluramine since they're both active, is about 150 nanograms. Now this compares in some of the rat studies to 63,000–122,000 nanograms of the combined drug and metabolite.

MOLLIVER: Is this at the 5 mg level dose in the rat? As I recall you had a level in the rat at 5 mg which was comparable to the human therapeutic level.

CAMPBELL: That's not true. It's the 2 mg per kg level that comparable. At 5 it starts going up non-linearly so we're talking about at least 6–10 times higher. I don't know the actual figures and I'd have to look but it's not a simple dose relationship because you get non-linearity. . .

MOLLIVER: It's a non-linear curve, but I recall there was a sharp change at about the point where the toxicity was occurring.

CAMPBELL: Yes, but the point is you can get enormous differences in the brain levels compared with man.

JOHN BLUNDELL (*University of Leeds, England*): I'm not a pharmaco-kineticist, I'm not a neurochemist and I'm not an anatomist; I am a behavioral biologist and in my view the most fascinating aspect of this morning's work arrives as a paradox. From the psychobiological point of view, the most interesting feature to me is the great disparity between the changes in brain chemistry and the changes that might be observed in behavior. There are huge depletions that apparently occur in 5-HT with marginal, if any, detectable changes in behavior. For example, Stan Lorens mentioned that he'd been quite unable to detect any impairment in fairly sensitive behavioral tests. And I think I'm right in saying that the only behavioral parameter that has been altered was referred to by Dr. Schmidt, and this is the startle response, which was increased, suggesting that they are a bit over responsive to their environment. Now that's a fascinating possibility, far from the possibility of them being impaired, the depletion in 5-HT might actually lead to some marginal advantage. Also, some work from the laboratory of Elaine Sanders-Bush showed that depletion of 5-HT actually led to an increase in acquisition of learning in a maze test. So before I think we start extrapolating and speculating about the effects that any of these drugs might have in man, and the possible relations between the depletions in 5-HT parameters and behavior, we have to sort out what the situation is in the rat. And it seems to me there's nothing at the moment that suggests that these shifts in neurochemical parameters that have been labeled neurotoxic lead to any behavioral imposition. In fact the opposite may be true.

A few moments ago Mark Molliver referred I think rather jocularly to the fact that 5-HT depletion might have a therapeutic effect. I overheard a conversation at coffee time which was suggesting the same thing. I hope this isn't being transcribed by the way.

SEIDEN: The whole thing is.

BLUNDELL: In that case, I hesitate to raise the issue but there's nothing that I've heard at this meeting that suggests that these gross shifts in neurochemical parameters translate into something that you call behavioral toxicity.

SEIDEN: Yes, but I venture to say if you had done the same thing with dopamine 30 years ago and you found 90% dopamine depletions in the brain you wouldn't have been able to predict that a 90% dopamine depletion in the brain would cause Parkinson's disease in the rat. So that has to be tempered somewhat by the fact that we don't know if there's a direct linear relationship between a mean content and behavioral expression.

BLUNDELL: Indeed we don't but even with massive depletions we can't detect any behavioral impairment with the tools that are improved over 25 years. I agree with the first of the argument you're putting forward but it still seems to me to be baffling—baffling and fascinating.

Neurotoxicity of MDMA: Neurochemical Effects

CHRISTOPHER J. SCHMIDT AND JOHN H. KEHNE

Merrell Dow Research Institute
2110 E. Galbraith Road
Cincinnati, Ohio 45215

INTRODUCTION

3,4-Methylenedioxymethamphetamine (MDMA) is one of several amphetamines which can damage serotonergic nerve terminals when administered at high doses to laboratory animals. A single injection of MDMA produces a characteristic biphasic pattern of 5-HT depletion in the rat brain. Remarkably similar patterns are observed following the administration of *p*-chloroamphetamine (PCA), methamphetamine (METH) or methylenedioxyamphetamine (MDA) to rats.[1–3] In our studies we have designated these two phases of 5-HT depletion as the acute or reversible phase and the long-term or neurotoxic phase. The acute neurochemical effects of MDMA begin immediately after drug administration with a rapid decline in the activity of the rate-limiting enzyme for 5-HT synthesis, tryptophan hydroxylase (TPH). During this same period there is an increase in carrier-mediated efflux of 5-HT which is believed to be responsible for the behavioral effects of MDMA. The increase in transmitter release coupled with the inhibition of synthesis results in a significant reduction in brain 5-HT concentrations during the first 3–6 h after drug administration. By 6 h there is a decline in the behavioral effects of MDMA as transmitter efflux ceases. Although TPH activity remains depressed, this is followed by a gradual recovery of 5-HT concentrations over the next 12 hours.[3] The neurotoxic effects of MDMA appear between 24 h and 1 week after drug administration. By 7 days, this second decrease in 5-HT concentrations is associated with a loss of functional 5-HT uptake sites indicating damage to serotonergic nerve terminals.[3]

The reduction in TPH activity is one of the earliest neurochemical events produced by MDMA. This loss of enzyme activity is due to a V_{max} change suggesting that TPH is actually inactivated following MDMA administration.[4] Recently Stone *et al.*[5] demonstrated that the enzyme in cortical homogenates from MDMA-, PCA- or METH-treated rats could be regenerated by incubation with dithiothreitol under N_2. The authors concluded that amphetamines initiate a process resulting in the oxidation of critical sulfhydryl groups on TPH. This suggests that the antioxidant mechanisms which would normally keep TPH in the reduced state must be exhausted. If this is true, we reasoned that early interference with the process leading to this redox imbalance should allow endogenous regeneration of TPH activity. Inhibitors of the 5-HT uptake carrier have previously been shown to block both the acute and neurotoxic effects of MDMA. Administration of the 5-HT uptake inhibitor MDL 27,777 either 1 or 2 h after MDMA blocked further development of the acute effect of MDMA. As shown for hippocampal 5-HT in Figure 1B, the inhibitor prevented any further decline in transmitter concentrations and actually reversed the effect of MDMA. More importantly, TPH activity (FIG. 1A) also recovered indicating reduction of the oxidized enzyme was occurring.[6] An attempt to administer the uptake inhibitor at 3 h post-MDMA indicated regeneration of TPH activity was no longer possible at this time

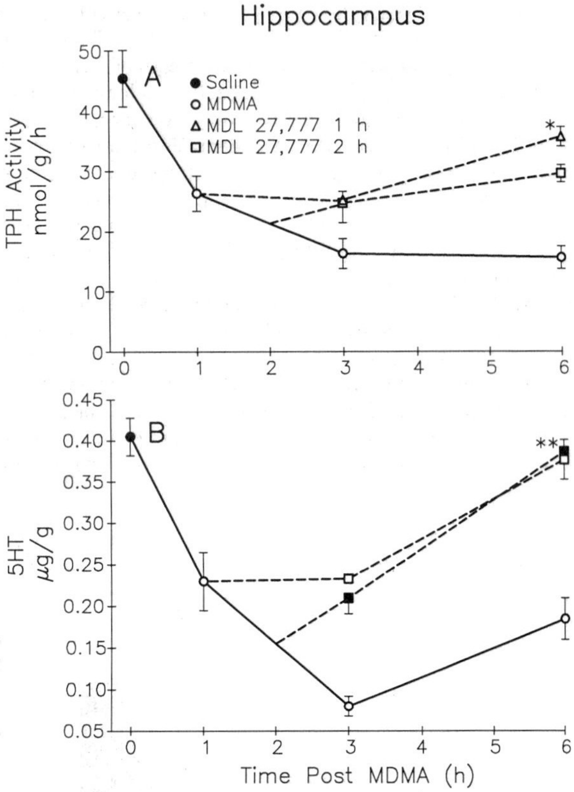

FIGURE 1. Regeneration of TPH activity *in vivo.* The 5-HT uptake inhibitor, MDL 27,777 (5 mg/kg), was administered 1 or 2 h after MDMA (10 mg/kg). Hippocampal 5-HT concentrations recovered in both the 1 and 2 h groups. The recovery of TPH activity reached significance only in the 1 h post MDMA group. *,** $p < 0.05$ and 0.01 vs. activity at 1 or 2 h.

point (unreported observation). Therefore, an antioxidant system(s) exists for maintaining TPH activity in its reduced form and this system(s) becomes severely depleted following MDMA administration. We can infer that MDMA must initiate some process leading to the generation of oxidant species inside the serotonergic nerve terminal.

There is some evidence to suggest that this acute imbalance in the redox state of the serotonergic nerve terminal may be involved in the development of the long-term effects of MDMA. Steranka and Rhind[7] have provided evidence that the sulfhydryl reagent, cysteine, can antagonize the long-term depletion of 5-HT by PCA. We have achieved similar results with cysteine and MDMA. FIGURE 2 displays the results from an experiment in which cysteine (1 g/kg, s.c.) was administered with 20 mg/kg MDMA and the animals killed 1 week later. Cysteine antagonized the MDMA-induced depletion of cortical 5-HT and the loss of cortical TPH activity supporting the hypothesis that the serotonergic neurotoxicity of MDMA and related compounds involves some form of oxidative stress in the affected nerve terminals.

Based on the above discussion we can assume that one early event in the development of MDMA-induced neurotoxicity is the depletion of some critical antioxidants from the nerve terminal. Although the cause of this acute effect is unknown, it can be antagonized by inhibitors of the 5-HT uptake system. Other data suggest that a later uptake event is also necessary for the long-term effects of MDMA and PCA. The administration of the 5-HT uptake inhibitor fluoxetine as long as 3–6 h after MDMA can still antagonize the development of neurotoxicity to varying degrees.[3] This is after the acute oxidation of TPH has occurred. The demonstrated role for dopamine in the serotonergic neurotoxicity of methamphetamine[8,9] suggests dopamine release may be involved in this late event. Using *in vivo* voltametry, Yamamoto and Spanos[10] demonstrated a 2 h lag period between MDMA administration and the peak of dopamine release. Extracellular dopamine was still approximately 200% of control at 3 h, which was the last time point determined. This means that high levels of extracellular dopamine are present well after the point when depletion of cellular antioxidants has advanced to the extent that oxidation of TPH can occur. Spontaneous oxidation of dopamine is known to generate an *o*-quinone, as well as other potentially cytotoxic electrophiles.[11] Furthermore, both *in vitro*[12] and *in vivo*[13] studies have demonstrated that dopamine can be accumulated in serotonergic neurons by the 5-HT uptake carrier. Thus, MDMA produces a condition in which depletion of intracellular antioxidants coincides with the occurrence of high extracellular levels of a pro-oxidant molecule.

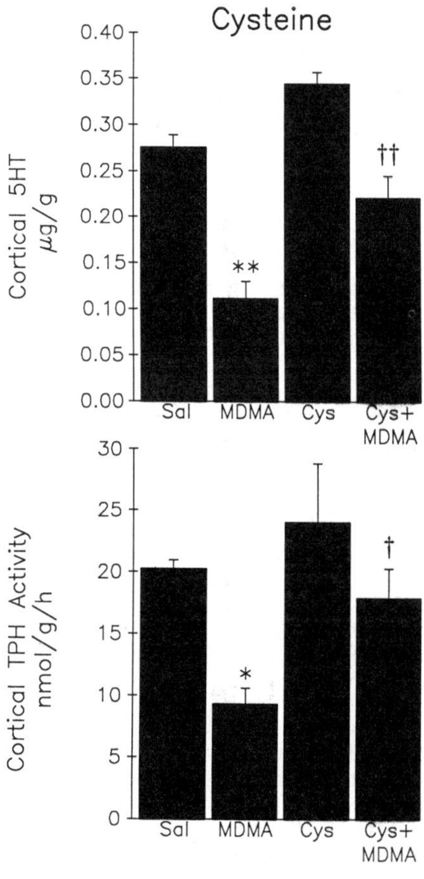

FIGURE 2. Effect of simultaneous administration of cysteine (1 g/kg) on the depletion of cortical 5-HT concentrations and the loss of cortical TPH activity observed 1 week after MDMA (20 mg/kg) administration. *,** $p <$ 0.05 and 0.01 vs. saline; +,++ PL 0.05 and 0.01 vs. MDMA.

Although this speculation is compatible with a great deal of what we know about MDMA, there are several questions remaining. Unlike amphetamine and METH, which produce neurotoxicity in the dopaminergic system, MDMA is a selective serotonergic neurotoxin. Is this difference consistent with dopamine release being involved in the long-term effects of all 3 agents or does this indicate a difference in mechanism? Exactly what conditions are necessary for the neurotoxic effects of MDMA? To examine these questions, we began studies on the role of monoamine release in the neurotoxicity of MDMA.[14]

THE ROLE OF MONOAMINE RELEASE IN THE NEUROTOXICITY OF MDMA

Storage and Synthesis Inhibitors

Pretreatment with reserpine to disrupt vesicular storage of all monoaminergic neurotransmitters was found to significantly antagonize the neurotoxicity of MDMA as measured by the reduction in cortical TPH activity (FIG. 3, *lower left*). In this experiment, the rats were injected with MDMA 2 weeks prior to the measurements to avoid a small elevation of TPH activity produced by reserpine alone. The duration of the effect of reserpine on 5-HT concentrations made evaluation of the effect of the drug combination on this parameter impossible (FIG. 3, *top left*). Protection was also provided if monoamine synthesis was inhibited using the irreversible inhibitor of aromatic L-amino acid decarboxylase, monofluoromethyl-DOPA (MFMD). A 3 day pretreatment with MFMD (100 mg/kg) reduced striatal dopamine and 5-HT to 46 and 32% of control, respectively. This was a transient depletion of monoamines as shown by the recovery of cortical 5-HT by 1 week (FIG. 3, *center*). In this case the protection provided by the inhibitor can be seen both in terms of the blockade of the depression in 5-HT concentrations and the loss of TPH activity produced by MDMA. The results with both reserpine and MFMD indicate that a critical concentration of some monoamine is necessary for MDMA to produce its long-term effects on the serotonergic neuron. It is interesting that both reserpine and MFMD were effective in this regard since each affects a different pool of transmitter in the neuron. Reserpine is generally considered to have little effect on the neurochemical effects of these drugs because amphetamines release transmitters primarily from a cytoplasmic or newly synthesized pool. Reserpine pretreatment has actually been reported to facilitate the neurotoxicity of methamphetamine.[15] In contrast to reserpine, MFMD inhibits new synthesis of transmitter and, thus, depletes the intraneuronal pool from which amphetamines release transmitter.

The role of dopamine was specifically addressed by using α-methyl-p-tryosine (AMPT) to inhibit catecholamine biosynthesis. AMPT (100 mg/kg) was administered 3 h prior to and with MDMA to reduce dopamine levels during the time period when the neurotoxic effects are believed to develop. This treatment was also effective in antagonizing the neurotoxicity of MDMA. FIGURE 3 (*right*) displays the data for both cortical 5-HT and TPH activity. These data provide clear evidence that the mechanism of MDMA neurotoxicity has a catecholaminergic component.

6-Hydroxydopamine Lesions

To verify this apparent requirement for intact levels of dopamine, intranigral injections of 6-hydroxydopamine were used to selectively lesion the primary dopaminergic afferents to the forebrain. Although both unilateral and bilateral lesions were performed, surprisingly only the latter antagonized the long-term serotonergic deficits produced by MDMA administration. FIGURE 4 provides the results for cortical and hippocampal 5-HT concentrations from one such experiment. The basis of the need for the bilateral lesions is unknown. To determine if bilateral nigral inputs to a midline structure, such as the dorsal raphe (DRN), were responsible, we made 6-hydroxydopamine injections into the DRN. In several experiments this treatment had no effect on MDMA-induced neurotoxicity (unreported observation).

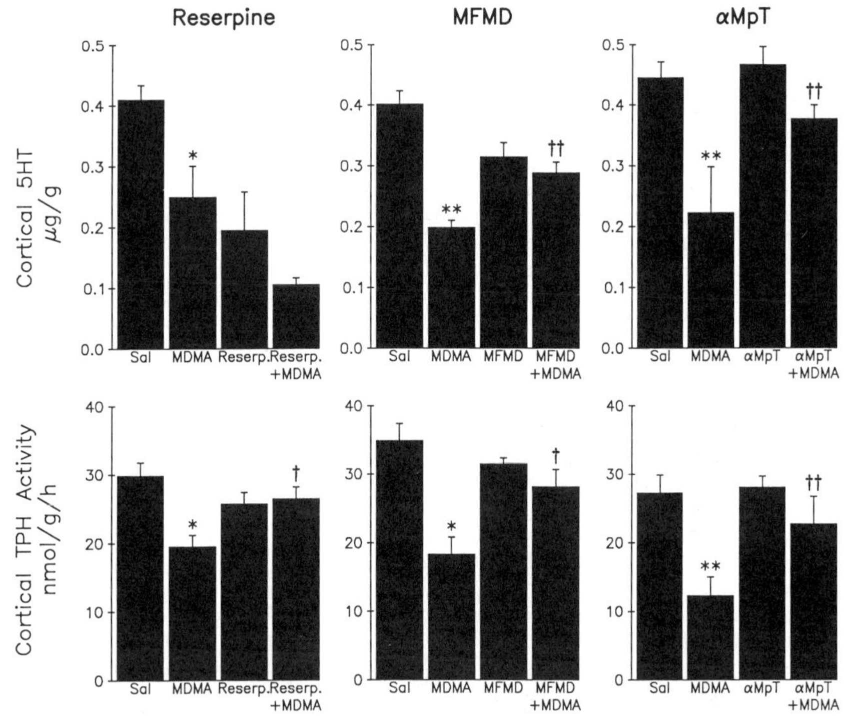

FIGURE 3. Interference with monoamine storage or synthesis antagonizes MDMA-induced neurotoxicity. Reserpine (*left*) was administered 24 h prior to MDMA (20 mg/kg) and the animals were killed 2 weeks later. MFMD (*center*) was given at 100 mg/kg on each of 3 d prior to MDMA (20 mg/kg). Animals were killed 1 week after MDMA administration. αMpT (100 mg/kg) was administered 3 h prior to and with MDMA (20 mg/kg) and the animals were killed 1 week later. *,** $p < 0.05$ and 0.01 vs. saline; +,++ $p < 0.05$ and 0.01 vs. MDMA.

Receptor Antagonists

An obvious consequence of the release of monoamines by MDMA is the activation of the receptors for those transmitters. A variety of receptor antagonists were coadministered with MDMA to determine if receptor activation played any role in the long-term effects of the drug. Complete protection was provided by two selective 5-HT_2 antagonists MDL 11,939 and MDL 26,508.[14] Blockade of the MDMA-induced serotonergic deficits in the cerebral cortex is shown in FIGURE 5 for MDL 11,939. The effect of these agents appeared to be specific since blockade of α_1 receptors with prazosin, or β and 5-HT_1 receptors with pindolol was without effect on the neurochemical response to MDMA. The 5-HT_3 antagonist MDL 73,147 was also without effect.

These data provide convincing evidence that the neurotransmitter release produced by MDMA is important in the development of the neurotoxicity. Although the results with reserpine and the requirement for bilateral 6-hydroxydopamine lesions are

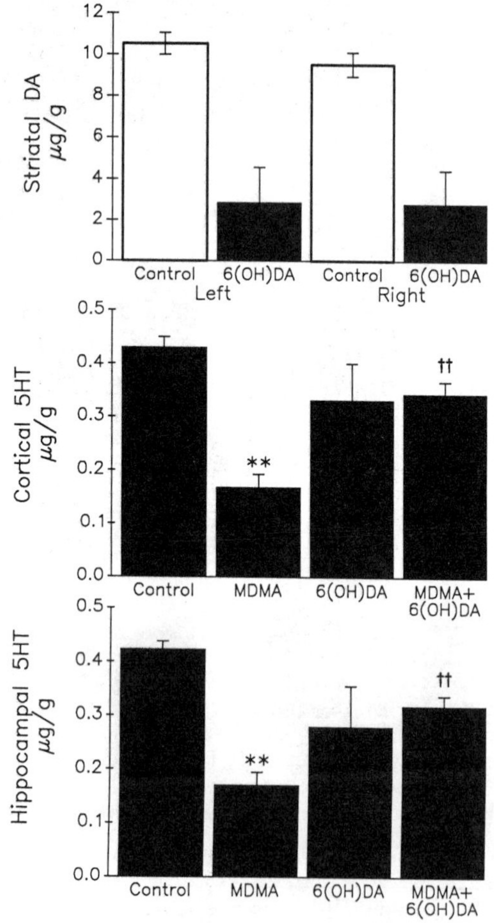

FIGURE 4. Effect of bilateral lesions of the substantia nigra with 6-hydroxy-dopamine on the MDMA-induced depletion of cortical and hippocampal 5-HT at 1 week. The extent of the lesions is indicated in the top panel by the reduction in striatal dopamine concentrations. After lesioning, rats were given 1 week to recover before MDMA (20 mg/kg) was administered.** $p < 0.01$ vs. saline; ++ $p < 0.01$ vs. MDMA.

somewhat surprising, taken with the effects of MFMD and AMPT, the results are consistent with the hypothesis that dopamine may participate in oxidative damage to the serotonergic terminal. The protection provided by the receptor antagonists is more difficult to explain in this context. However, the effectiveness of these agents may be related to the type of neurotoxicity produced by MDMA.

Although several amphetamine analogues produce serotonergic neurotoxicity, the

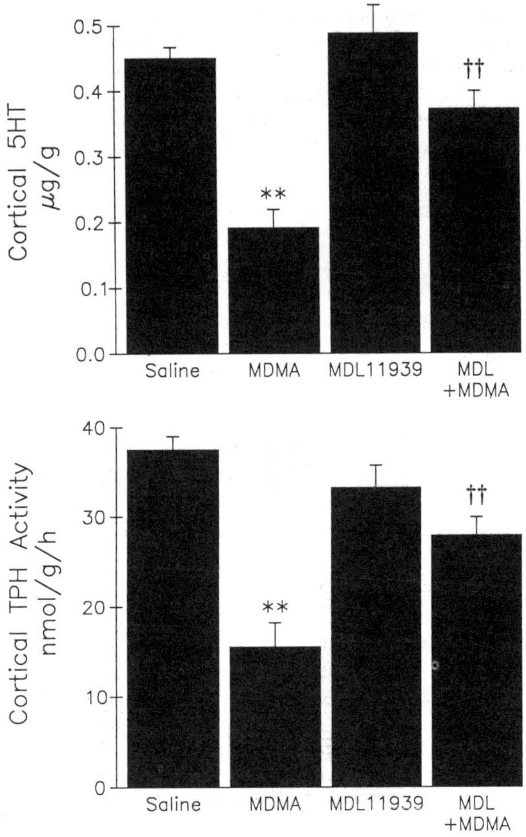

FIGURE 5. Protective effect of the 5-HT_2 antagonist, MDL 11,939 on the serotonergic neurotoxicity produced by MDMA. The antagonist (5 mg/kg) was administered simultaneously with MDMA (20 mg/kg) and the rats were sacrificed 1 week later. ** $p < 0.01$ vs. saline; + + $p < 0.01$ vs. MDMA.

overall pattern of neurochemical effects varies within the class. As diagrammed in FIGURE 6, there is a spectrum of activity which ranges from that of a "pure" dopaminergic neurotoxin, such as amphetamine, to the "pure" serotonergic neurotoxins, such as PCA or MDMA. Between these two extremes is METH which can damage both serotonergic and dopaminergic terminals. The role of dopamine release in the long-term neurochemical effects of amphetamine and METH has already been

	Dopamine Neurotoxin	Dopamine/5HT Neurotoxin	5HT Neurotoxin	nontoxic
example	amphetamine	methamphetamine	MDMA/PCA	MDE
release	DA >> 5HT	DA > 5HT	5HT > DA	5HT >> DA

FIGURE 6. Neurotoxic spectrum of the amphetamines and its correlation with monoamine release. (See text for further discussion).

discussed. The effects of METH on both dopaminergic and serotonergic neurons are susceptible to inhibition of dopamine synthesis[8] or release[16] as are the effects of MDMA on serotonergic neurons. The question remains, if DA release is the basis of the neurotoxicity of methamphetamine, why isn't MDMA a dopaminergic neurotoxin as well? Why does this "spectrum of action" exist? The answer may lay in the pattern of neurotransmitter release. As one progresses from amphetamine to the relatively non-toxic *N*-ethyl-methylenedioxyamphetamine (MDE) the pattern of transmitter release also changes. Amphetamine is a potent dopamine releaser but is a relatively poor 5-HT releasing agent. METH is similar to amphetamine in terms of dopamine release but is a considerably better 5-HT releaser. PCA and MDMA are even less effective at dopamine release but are potent 5-HT releasers. Finally, MDE, although approximately equipotent with MDMA at producing 5-HT release, is the weakest dopamine releaser of the drugs listed. This suggests that the pattern of neurotransmitter release may play a role in determining the pattern of neurotoxicity. The results of the experiments with the 5-HT_2 receptor antagonists support this hypothesis.

Based upon the data and the preceding discussion a revised hypothesis of the mechanism of the neurotoxicity of MDMA must be made. Since agents such as MDE are similar to MDMA at producing the acute phase of 5-HT depletion it is apparent that the oxidative insult which produces the oxidation of TPH is not sufficient to produce the long-term damage to the serotonergic terminal.[17] This again reinforces the importance of that activity by the uptake system which is still occurring after the acute depression of 5-HT concentrations. Since this late effect is believed to require extracellular dopamine, the ultimate cause of MDMA-induced neurotoxicity is the carrier-mediated release of dopamine by MDMA. Neurotoxicity does not occur if sufficient dopamine is not released. This explains the relatively nontoxic nature of MDE and the observation that the long-term effects of MDMA are primarily a property of the (+)stereoisomer since both MDE and (−)MDMA are relatively weak dopamine releasing agents.[17,18] The somewhat greater potency of MDA compared to MDMA as a serotonergic neurotoxin is also consistent with this hypothesis.[17] In spite of this absolute requirement for dopamine it is possible that the release of 5-HT can modify which of the monoaminergic systems are affected.

Further work is obviously required to determine exactly what role 5-HT release plays in the long-term neurochemical effects of MDMA. One possibility being examined in our laboratory arises from the reported inhibition of dopamine-mediated behavior produced by 5-HT.[19] *In vivo* voltammetric studies indicate that there is a 2 h lag between administration of MDMA to rats and the peak of dopamine release.[10] In contrast, results from behavioral studies suggest that 5-HT release increases immediately after drug administration.[20] Therefore, the 5-HT release produced by MDMA could be delaying dopamine release. This inhibition could have the net effect of "postponing" dopamine release until a period when the serotonergic terminal is already under oxidative stress as evidenced by the loss of TPH activity. Activation of 5-HT_2 receptors may therefore be important in delaying the release of dopamine until the terminal is vulnerable to damage by dopamine oxidation products or may extend the period of oxidative stress beyond the point of reversibility. This is consistent with the ability of fluoxetine to block MDMA-induced neurotoxicity for as long as 3 h after MDMA administration.

Understanding the neurochemical mechanisms underlying both the immediate and delayed effects of MDMA and related compounds on brain neurochemistry and behavior could be greatly facilitated by further *in vivo* studies of neurotransmitter release. Recently, several laboratories utilizing the intracerebral dialysis technique to measure PCA-induced release of DA[10] or 5-HT[21] have noted temporally-distinct patterns of release as already outlined for MDMA. Hutson and Curzon[21] found that PCA-induced behavioral effects that were characteristic of the "serotonin syndrome" (head weaving, reciprocal forepaw treading, wet dog shakes) correlated with the peak 5-HT release which occurred in the first 30 min after drug administration. Locomotor stimulation occurred later on, correlating more closely with the time course of DA release.[10] The implication of these early findings are that the behavioral effects and/or the neurotoxicity may be critically dependent upon this exact pattern of release. One has to consider not only the effects of release of one transmitter onto different receptor subtypes, but also interactions between different transmitters.

Behavioral Correlates

The neurochemical selectivity of MDMA and PCA makes them invaluable tools for the study of the serotonergic system. One of the more interesting observations to arise from the use of these agents has been the identification of two subtypes of serotonergic terminals. Molliver and his colleagues[22] have provided evidence that these two populations of axon terminals are structurally distinct and arise from different nuclei in the brain stem. Their data indicate that PCA, MDMA and related amphetamines cause selective degeneration of the fine axons that arise from the dorsal raphe nucleus (DRN). In contrast, the coarse beaded axons which arise from the median raphe nucleus (MRN) are resistant to these neurotoxic effects.

Recently, Blue *et al.*[23] noted a correlation between innervation by the fine axon subtype and the localization of 5-HT_2 receptors. Since 5-HT_2 receptors are believed to mediate the discriminative cue of psychotropic drugs,[24] the authors suggested that the behavioral effects of drugs such as MDMA are mediated by 5-HT release from DRN-derived terminals. Together with the anatomical and pharmacological distinction between the DRN and MRN serotonergic systems these data suggest the possibility that the two populations of neurons may mediate different behavioral functions. To examine this possibility we first defined these two subsystems in neurochemical studies.

Regional 5-HT Depletions Following PCA or MDMA

Lesion studies have shown that destruction of the DRN produces a pattern of forebrain depletion distinct from that produced by MRN destruction[25–29] with the following rank order: corpus striatum, DRN $\gg$ MRN; hippocampus, MRN $\gg$ DRN; cortex, DRN = MRN. A selective effect of PCA or MDMA on DRN terminals should therefore produce a pattern of 5-HT depletion that is similar to that produced by a DRN lesion. FIGURE 7 shows the 5-HT depleting effects of PCA in three supraspinal areas examined (hippocampus, striatum, and cortex) and in one spinal area (ventral lumbar spinal cord). Clearly, in each area, particularly in the spinal cord, the depletion appeared to reach asymptote at a nonzero level. Furthermore, this residual 5-HT was not further depleted when an additional challenge dose of PCA was administered (see FIG. 7), a finding that supports the existence of a PCA-resistant subpopulation. To further extend these findings, radiofrequency lesions of the DRN or MRN were compared to lesions produced by 20 mg/kg PCA or i.c.v. infusions of the 5-HT neurotoxin 5,7-dihydroxytryptamine (5,7-DHT). TABLE 1 shows that the radiofrequency lesions produced region-specific 5-HT depletions characteristic of those described in the literature. In contrast, PCA again produced marked 5-HT depletion in all areas, with no apparent pattern of regional selectivity. The most striking discrep-

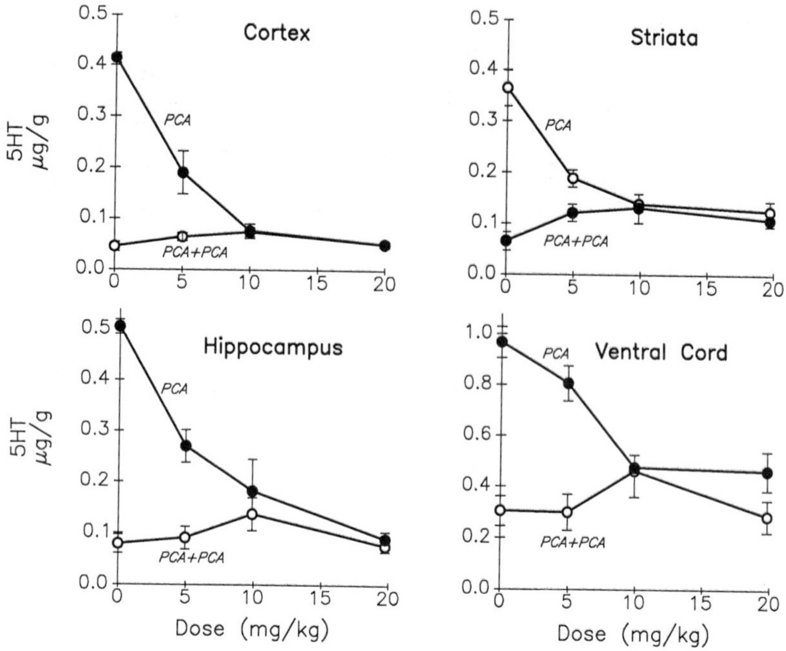

FIGURE 7. Regional dose-dependent depletion of 5-HT concentrations by PCA and the effect of a subsequent 10 mg/kg dose challenge dose of PCA. Note the plateau observed with a single dose of 10 or 20 mg/kg PCA and the inability of the challenge dose to further deplete 5-HT concentrations.

ancy was the large 5-HT depletion produced in the hippocampus by MRN or PCA lesions compared to the minor depletion produced by DRN lesions. In general, the PCA depletions more closely resembled the nonselective 5,7-DHT depletions than the DRN lesions. One explanation for these findings is that PCA destroys 5-HT terminals of ascending pathways other than just those of the DRN, for example those originating in the B9 nuclei.[30] Further studies will be needed to address this possibility.

Immediate and Delayed Behavioral Consequences of PCA and MDMA

Recently, we have initiated investigations of the acute and chronic behavioral correlates of MDMA and PCA administration. Given that serotonergic terminals of

TABLE 1. Effect of Selective Lesions on Regional 5-HT Concentrations

	5-HT (% of Controls)		
Lesion	Hippocampus	Striatum	Cortex
DRN	65	35	46
MRN	27	87	59
PCA	18	24	12
5,7-DHT i.c.v.	7	46	10

the DRN are destroyed by PCA and MDMA, it would be expected that exposure to these neurotoxins would alter the function(s) subserved by these terminals. Furthermore, if 5-HT is released only from terminals that are ultimately destroyed, then any acute, 5-HT mediated effect of PCA on behavior should be disrupted by prior exposure to PCA.

One of the reported behavioral alterations produced by acute administration of PCA is an elevation in the amplitude of acoustic startle reflexes.[31] The startle reflex is a rapid contraction of the body produced by sudden stimulus, such as a loud sound (acoustic startle) or a puff of air delivered to the back (tactile startle). We have assembled an automated apparatus that allows us to measure either acoustic or tactile reflexes in 8 rats. Each rat is placed in a cage that is attached to a floating platform. A stimulus-elicited startle reflex causes a movement of the platform which in turn produces a voltage in a piezoelectric transducer. This voltage is filtered at 10 Hz, rectified, amplified, and fed into a Metrabyte DAS16 I/O board, where it is digitized and sampled over a 250 msec window. Startle amplitude is defined as the average rectified intensity (in arbitrary units) averaged across the recording window. Using this apparatus, we investigated the effects of PCA and MDMA on acoustic and tactile startle reflexes.

Davis[32] previously demonstrated that combined electrolytic lesions of the dorsal and median raphe nuclei produced an elevation in the amplitude of acoustically elicited startle. To determine the relative contributions of the DRN and MRN to this effect, we tested rats with selective radiofrequency lesions of either the DRN or MRN and found that only MRN lesions produced a chronic elevation of acoustic startle amplitude. These data indicated that the acoustic startle reflex was not a suitable functional endpoint for assessing destruction of DRN terminals. Consistent with this conclusion, PCA at neurotoxic doses did not have substantial chronic effects on acoustic startle amplitude.[31,33]

Though not having a consistent long-term effect on acoustic startle amplitude, PCA clearly elevated acoustic startle amplitude when measurements were made at 2.5 h post injection.[31] As seen in FIGURE 8, we replicated this finding and additionally found that tactile startle reflexes were enhanced. Furthermore, we found that MDMA produced a dose-related facilitation of acoustic and tactile startle, though it was less potent than PCA. Note that the MDMA- and PCA-induced increases in startle are rather slow in onset and that they last for the duration of the 3.5 h test session. PCA-enhanced acoustic startle was blocked by the 5-HT synthesis inhibitor PCPA, but not by the catecholamine synthesis inhibitor AMPT,[31] implicating serotonergic involvement. Recently, we found that PCA enhancement of acoustic and tactile reflexes was blocked by the selective 5-HT uptake blockers MDL 27,777 or fluoxetine (tactile results illustrated in FIG. 9). These data support the conclusion that MDMA- or PCA-induced facilitation of startle reflexes is attributable to carrier-mediated release of 5-HT.

As mentioned previously, if 5-HT is released from terminals that are sensitive to the neurotoxic effects of PCA or MDMA, then prior exposure to the neurotoxin should prevent the expression of any acute, 5-HT-mediated behavioral effect. To test this hypothesis, rats previously exposed to a neurotoxic (20 mg/kg) dose of PCA were administered a challenge dose of 10 mg/kg PCA 2 weeks later (FIG. 10). Acoustic and tactile startle reflexes were measured for a 30 min period beginning approximately 45 min after PCA injection. Over this measurement period, prior exposure to PCA did not block the startle enhancing effect of PCA.[34] These preliminary data suggest that the startle-enhancing effects of PCA result from 5-HT release from terminals that are resistant to PCA neurotoxicity, though a firm conclusion awaits the outcome of a more extensive experimental analysis. Further research will be required to ascertain the

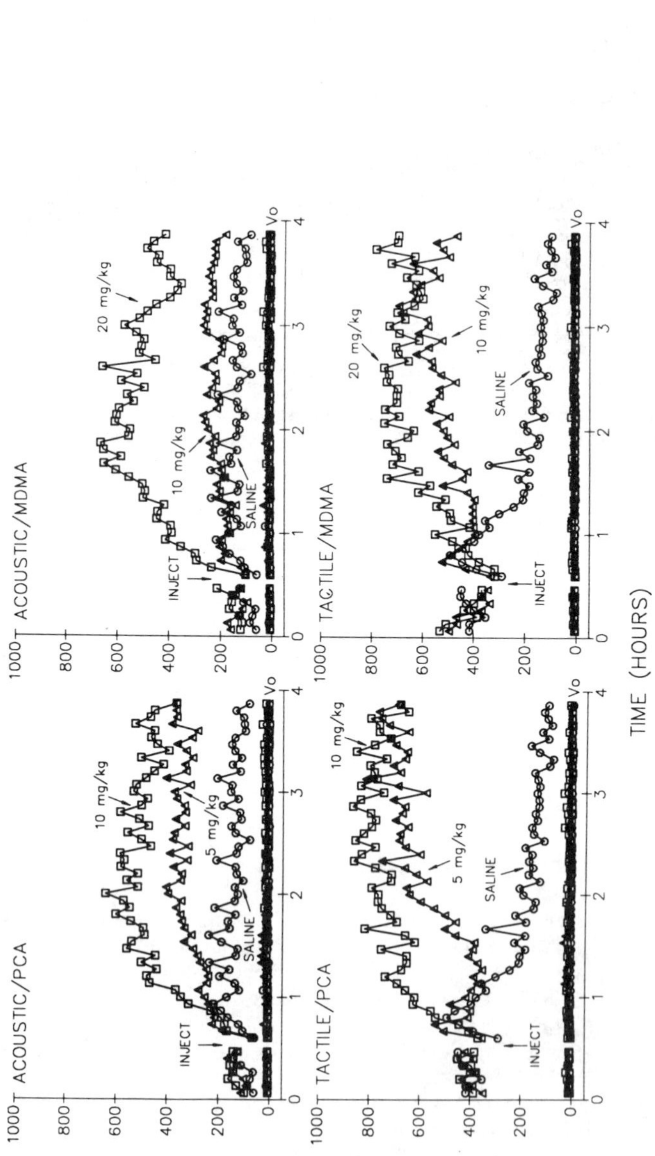

FIGURE 8. Dose-dependent enhancement of acoustic and tactile startle in rats by PCA or MDMA (0–3.5 hours after injection). Each point represents the average of three consecutive stimulus presentations (i.e., 2 minutes of testing). $\mathbf{V_{avg}}$ represents the transducer output for 250 msec following the onset of the acoustic or tactile eliciting stimulus, whereas $\mathbf{V_0}$ represents the output at the beginning of the trial. The very low values of $\mathbf{V_0}$ relative to $\mathbf{V_{avg}}$ support the conclusion that our measure of startle is uncontaminated by any spontaneous (i.e., non-stimulus elicited) movement of the rat within the test cage. Each rat was exposed to alternating acoustic and tactile stimuli within the test session. A 30-minute baseline was taken prior to injection of PCA (*left panels*) or MDMA (*right panels*).

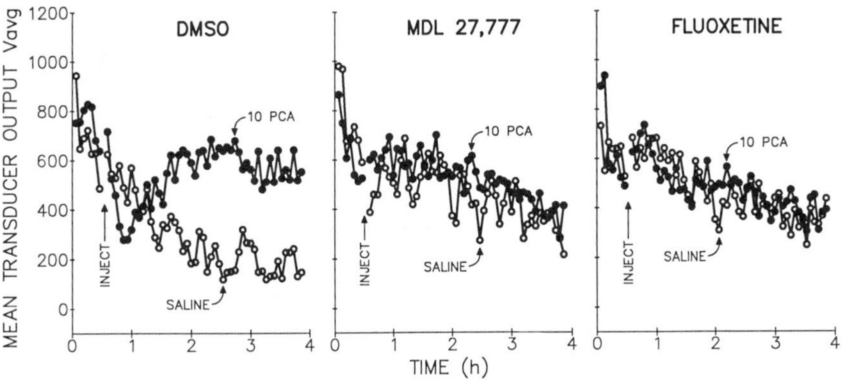

FIGURE 9. The 5-HT uptake inhibitors, MDL 27,777 or fluoxetine (5 mg/kg, P), block the facilitation of tactile startle by MDMA. The uptake blockers were administered at the baseline period (i.e., 30-minutes before injection of MDMA).

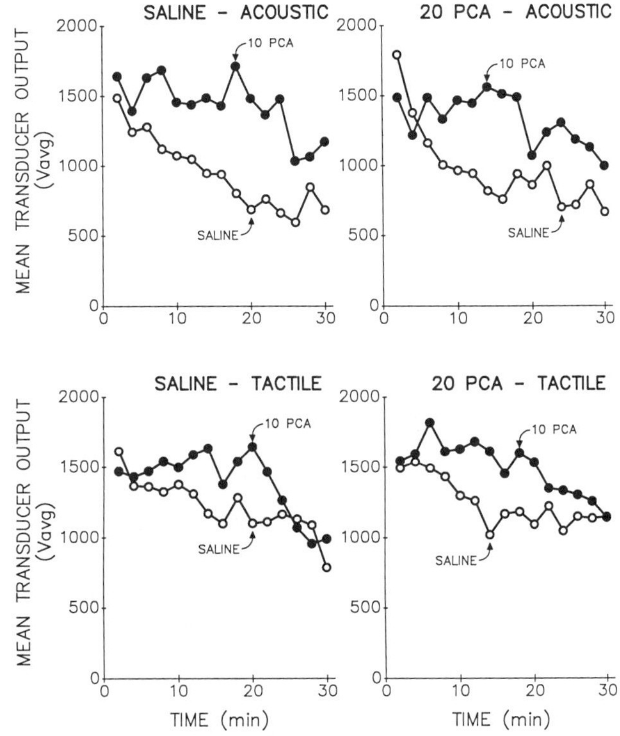

FIGURE 10. Effect of previous exposure to PCA (20 mg/kg) on the enhancement of acoustic and tactile startle by a 10 mg/kg challenge dose of PCA. The neurochemical data for these animals is shown in FIGURE 7. The first dose of PCA was given 2 weeks prior to testing.

identity of the serotonergic pathway(s) that mediate the excitatory effect of PCA on acoustic and tactile startle.

CONCLUSIONS

Evidence from numerous studies has implicated the carrier-mediated release of dopamine in the neurotoxic effects of MDMA and other amphetamines. Although a requirement for dopamine release seems to be a consistent finding in these studies, a mechanism based upon dopamine release alone is insufficient to explain all the available data. We suggest that the ability of the particular amphetamine to also release 5-HT is the final determinant of which monoaminergic system will ultimately be affected. Thus, amphetamine primarily releases dopamine and is a dopaminergic neurotoxin. Serotonergic neurotoxins such as MDMA also release dopamine but must have their primary effect on 5-HT release. This provides us with a slightly modified but very testable hypothesis about the long-term effects of this class of agents.

While the molecular mechanism of amphetamine neurotoxicity remains unknown considerable progress has been made. The disruption of the antioxidant defenses and the rapid oxidation of TPH in serotonergic neurons following MDMA administration certainly suggests oxidative stress may be responsible for initiating damage to the nerve terminal. The protective effect of the antioxidant, cysteine, is consistent with this.

Just as the mechanism of MDMA- or PCA-induced neurotoxicity is unknown so is the explanation for the apparent resistance of some neurons to this effect. The demonstration that PCA can initiate behavior in PCA lesioned rats suggests carrier-mediated release of 5-HT from resistant terminals does occur. If this is true, the results indicate that the immediate action of PCA and MDMA at these terminals is similar regardless of their sensitivity to the neurotoxic effects. Thus, an anatomical or metabolic difference between the sensitive and resistant terminals may be required to explain the difference in sensitivity to PCA and MDMA. Potentially important differences between serotonergic projections are the proximity to dopaminergic terminals or the constitutive levels of antioxidants in the terminal as well as the presence of cotransmitter substances which could modify the effects of released 5-HT.

ACKNOWLEDGMENTS

We wish to acknowledge the excellent technical assistance of V.L. Taylor, C.K. Black and T.C. McCloskey as well as the secretarial assistance of Sue Treadway and Mary Lynn Points.

REFERENCES

1. FULLER, R. W., K. W. PERRY & B. B. MOLLOY. 1975. Reversible and irreversible phase of serotonin depletion by 4-chloroamphetamine. Eur. J. Pharmacol. **33:** 119–124.
2. SCHMIDT, C. J. & J. W. GIBB. 1985. Role of the serotonin uptake carrier in the neurochemical response to methamphetamine: Effects of citalopram and chlorimipramine. Neurochem. Res. **10:** 637–684.
3. SCHMIDT, C. J. 1987. Neurotoxicity of the psychedelic amphetamine, methylenedioxymethamphetamine. J. Pharmacol. Exp. Ther. **240:** 1–7.
4. SCHMIDT, C. J. & V. L. TAYLOR. 1987. Depression of rat brain tryptophan hydroxylase activity following the acute administration of methylenedioxymethamphetamine. Biochem. Pharmacol. **36:** 4095–4102.

5. Stone, D. M., M. Johnson, G. R. Hanson & J. W. Gibb. 1989. Acute inactivation of tryptophan hydroxylase by amphetamine analogs involves the oxidation of sulfhydryl sites. Eur. J. Pharmacol. **172:** 93–97.
6. Schmidt, C. J. & V. L. Taylor. 1990. Reversal of the acute effects of 3,4-methylene-dioxymethamphetamine by 5HT uptake inhibitors. Eur. J. Pharmacol. **181:** 133–136.
7. Steranka, L. R. & A. W. Rhind. 1987. Effect of cysteine on the persistent depletion of brain monoamines by amphetamine, p-chloroamphetamine and MPTP. Eur. J. Pharmacol. **133:** 191–197.
8. Schmidt, C. J., J. K. Ritter, P. K. Sonsalla, G. R. Hanson & J. W. Gibb. 1985. Role of dopamine in the neurotoxic effects of methamphetamine. J. Pharmacol. Exp. Ther. **233:** 539–544.
9. Johnson, M., D. M. Stone, G. R. Hanson & J. W. Gibb. 1987. Role of the nigrostriatal pathway in methamphetamine-induced depression of the neostriatal serotonergic system. Eur. J. Pharmacol. **135:** 231–234.
10. Yamamoto, B. K. & L. J. Spanos. 1988. The acute effect of methylenedioxy-methamphetamine on dopamine release in the awake-behavior rat. Eur. J. Pharmacol. **148:** 195–203.
11. Graham, D. G. 1978. Oxidative pathways for catecholamines in the genesis of neuromelanin and cytotoxic quinones. Mol. Pharmacol. **14:** 633–643.
12. Schmidt, C. J. & W. Lovenberg. 1985. In vitro demonstration of serotonin uptake by neostriatal serotonergic neurons of the rat. Neurosci. Lett. **59:** 9–14.
13. Waldmeier, P. C. 1985. Displacement of striatal 5-hydroxytryptamine by dopamine release from endogenous stores. J. Pharm. Pharmacol. **37:** 58–60.
14. Schmidt, C. J., C. K. Black & V. L. Taylor. 1990. Antagonism of the neurotoxicity due to a single administration of methylenedioxymethamphetamine. Eur. J. Pharmacol. **181:** 59–70.
15. Wagner, G. C., J. B. Lucot, C. R. Schuster & L. S. Seiden. 1982. Alpha-methyltyrosine attenuates and reserpine potentiates methamphetamine induced neuronal changes. Soc. Neurosci. Abstr. **8:** 488.
16. Schmidt, C. J. & J. W. Gibb. 1985. Role of the dopamine uptake carrier in the neurochemical response to methamphetamine: Effects of amfonelic acid. Eur. J. Pharmacol. **109:** 73-80.
17. Schmidt, C. J. 1987. Acute administration of methylenedioxymethamphetamine: comparison with the neurochemical effects of its N-desmethyl and N-ethyl analogs. Eur. J. Pharmacol. **136:** 81–88.
18. Schmidt, C. J., J. A. Levin & W. Lovenberg. 1987. *In vitro* and *in vivo* neurochemical effects of methylenedioxymethamphetamine on striatal monoaminergic systems in the rat brain. Biochem. Pharmacol. **36:** 747–755.
19. Rebee, G. V., K. D. Alloway & S. D. Curtis. 1981. Apparent serotonergic modulation of the dose-dependent biphasic response of neostriatal neurons produced by d-amphetamine. Brain Res. **210:** 277–289.
20. Schecter, M. 1989. Serotonergic-dopaminergic mediation of 3,4-methylene-dioxymethamphetamine (MDMA, "Ecstasy"). Pharmacol. Biochem. Behav. **31:** 817.
21. Hutson, P. H. & G. Curzon. 1989. Concurrent determination of effects of p-chloroamphetamine on central extracellular 5-hydroxytryptamine concentration and behavior. Br. J. Pharmacol. **96:** 801–806.
22. Molliver, M. E. 1987. Serotonergic neuronal systems: What their anatomic organization tells us about function. J. Clin. Psychopharmacol. **7:** 3S–23S.
23. Blue, M. E., K. A. Yagaloff, L. A. Mamounas, P. R. Hartig & M. E. Molliver. 1988. Correspondence between 5-HT_2 receptors and serotonergic axons in rat neocortex. Brain Res. **453:** 315–328.
24. Glennon, R. A., M. Titeler & J. D. McKenney. 1984. Evidence for 5-HT_2 involvement in the mechanism of action of hallucinogenic agents. Life Sci. **35:** 2505–2511.
25. Khanna, J. M., C. Campanelli, A. D. Le & H. Kalant. 1987. Effect of raphe lesions on the development of chronic tolerance to pentobarbital and cross-tolerance to ethanol. Psychopharmacology **91:** 473–478.
26. Robinson, S. E., M. J. F. Austin &D. M. Gibbens. 1985. The role of serotonergic neurons

in dorsal raphe, median raphe and anterior hypothalamic pressor mechanisms. Neuropharmacology **24:** 51–58.
27. NASSIF-CAUDARELLA S., E. KEMPF & L. VELLEY. 1986. Clonidine-induced sedation is not modified by single or combined neurochemical lesions of the locus coeruleus, the median and dorsal raphe nuclei. Pharmacol. Biochem. Behav. **25:** 1211–1216.
28. LORENS, S. A. & H. C. GULDBERG. 1974. Regional 5-hydroxytryptamine following selective midbrain raphe lesions in the rat. Brain Res. **78:** 45–56.
29. JACOBS, B. L. & A. COHEN. 1976. Differential behavioral effects of lesions of the median or dorsal raphe nuclei in rats: Open field and pain-elicited aggression. J. Comp. Physiol. Psychol. **90:** 102–108.
30. HARVEY, J. A., S. E. MCMASTER & L. M. YUNGER. 1975. p-Chloroamphetamine: Selective neurotoxic action in brain. Science **187:** 841–843.
31. DAVIS, M. & M. H. SHEARD. 1976. p-Chloroamphetamine (PCA): Acute and chronic effects on habituation and sensitization of the acoustic startle response in the rat. Eur. J. Pharmacol. **35:** 261–273.
32. DAVIS, M. & M. H. SHEARD. 1974. Habituation and sensitization of the rat startle response: Effects of raphe lesions. Physiol. Behav. **12:** 425–431.
33. KEHNE, J. H., V. L. TAYLOR, C. K. BLACK, P. J. ROBINSON & C. J. SCHMIDT. 1988. Median raphe radiofrequency lesions, but not dorsal raphe lesions or para-chloro-amphetamine, increase acoustic startle responses in rat. Soc. Neurosci. Abstracts. **14:** 373.
34. KEHNE, J. H., T. C. MCCLOSKEY, V. L. TAYLOR, C. K. BLACK & C. J. SCHMIDT. 1989. PCA- or MDMA-enhanced startle reflexes in rats: Effects of the 5-HT uptake blocker MDL 27,777, selective raphe lesions, or prior 5-HT depletion. Soc. Neurosci. Abstracts. **15:** 1185.

DISCUSSION OF THE PAPER

CORNELIUS VANDERWOLF (*University of Western Ontario, London, Ontario, Canada*): I have a question about the recording of the startle response. A few years ago I was involved in some studies which were never published with Peter Osencoff, and we observed that the position the rat has at the time you apply the stimulus has a big effect on the startle. The startle is mainly a sudden flexion of the body. If the animal is standing in a horizontal position on all fours most of the energy of the movement is not transmitted in a vertical dimension. If the animal is rearing then more of the energy of the startle is transmitted to the platform and you get a much bigger deflexion even though the actual behavior under the two circumstances is much the same. And there are a number of other behavorial dependencies as well in the startle response. It's smaller, let's say, if the animal happens to be chewing on something at the time you present it, that kind of thing. So I wonder whether you take any of this sort of thing into account in doing these kinds of tests?

SCHMIDT: Well I think the thing to do would be have John Kehne comment on that, but his apparatus doesn't allow animals to rear for one thing. It's a very confined chamber resting on a platform that is compressed between springs so that none of the animals' other activities really will disturb the transducer. It takes, as you said, an acceleration to produce the signal and this signal is filtered. In the curves shown, the line going across the bottom, represented essentially the background stimulus. As you noticed, it was a flat line indicating that the transducer does not pick up any of the normal increases in locomotion produced by these drugs.

ROSS H. PASTEL (*Walter Reed, Washington, DC*): You've shown that dopamine depletion protects against the neurotoxicity. Have you done any experiments with dopamine uptake blockers and shown a potentiation of the neurotoxicity, you know presuming that dopamine does have part of the neurotoxic effect.

SCHMIDT: I haven't but that's been done in Jim Gibb's lab by Donna Stone I believe. And you really wouldn't expect to see a facilitation necessarily. It's kind of complicated to interpret those experiments. Do you expect the uptake inhibitor to block MDMA-induced release of dopamine or to increase synaptic levels? What probably happens, and what their results would suggest, is that when GBR-12909 was co-administered with MDMA you actually saw a blockade of the neurotoxicity. We did something similar a number of years ago with amfonelic acid and methamphetamine and that also protected. So our conclusion from those experiments is that in the presence of the uptake inhibitor you keep all the transmitter inside the neuron. You have to have that carrier working in reverse to produce the release, of course.

RICHARD WURTMAN (*MIT, Cambridge, MA*): It was shown many years ago that you could deplete tissue catecholamines down to 5–8% of their level and still sustain blood pressure regulation. So the fact that you have difficulty demonstrating a consequence of an 80% depletion of serotonin should not, in itself, be taken as surprising.

SCHMIDT: I agree.

Neurotoxic Effects of MDMA on Brain Serotonin Neurons: Evidence from Neurochemical and Radioligand Binding Studies[a]

ERROL B. DE SOUZA[b]

Laboratory of Neurobiology, Neuroscience Branch
National Institute on Drug Abuse
Addiction Research Center
Baltimore, Maryland 21224

GEORGE BATTAGLIA

Department of Pharmacology
Loyola University Medical Center
Stritch School of Medicine
Maywood, Illinois 60153

THOMAS R. INSEL

Laboratory of Clinical Science
National Institute of Mental Health
Poolesville, Maryland 20837

INTRODUCTION

MDMA (3,4-methylenedioxymethamphetamine, "Ecstasy"), a ring-substituted derivative of methamphetamine, represents one of a number of "designer drugs" which is being increasingly abused, especially among college students. MDMA and its close structural analogue MDA (3,4-methylenedioxyamphetamine) have been reported to exhibit both stimulant and psychotomimetic properties.[1] MDMA has been the subject of continued scientific and legal debate, as several psychiatrists have reported that the drug may "enhance emotions" and "feelings of empathy" and thus serve as an adjunct in psychotherapy.[2] Concern about the clinical use of MDMA has arisen based on data that MDMA is self-administered in nonhuman primates,[3,4] (i.e. has a high potential for abuse in humans) and is a potent neurotoxin that appears to cause selective degeneration of brain serotonin neurons both in rodents[5–9] and in nonhuman primates.[10–14]

This chapter will describe some of the authors' recent data on the neurotoxic effects of MDMA on brain monoamine systems in both rodents and rhesus monkeys. Specifically, studies will be described examining the effects of *in vivo* systemic administration of MDMA on brain monoamine systems with respect to 1) dose dependence and the relative sensitivity of various animal species to neurotoxic effects of

[a]The research upon which this paper is based was supported in part by the U.S. Food and Drug Administration.

[b]Address all correspondence to: Errol B. De Souza, Ph.D., Chief, Laboratory of Neurobiology, NIDA Addiction Research Center, P.O. Box 5180, Baltimore, MD 21224, (301) 550-1462.

the drug; 2) the characteristics and time course of recovery following destruction of serotonin neurons; and 3) the regional differences and morphological specificity of MDMA-induced neurotoxicity.

INDICES OF NEUROTOXICITY

The neurotoxic effects of drugs on monoamine neurons have been typically assessed from long-term reductions in brain levels of monoamines and their metabolites, decreases in the maximal activity of synthetic enzymes, and decreases in the active uptake carrier. However, based on evidence documenting that the methylenedioxy-substituted amphetamines can inhibit the activity of tryptophan hydroxylase,[15,16] the rate-limiting enzyme in serotonin synthesis, it is unclear whether MDMA-induced reductions in brain serotonin and its metabolite 5-hydroxyindoleacetic acid (5-HIAA) are due to suppressed neurotransmission in otherwise structurally intact neurons or are the consequence of the destruction of serotonin neurons.

In the studies to be described below, the authors have used the "traditional" markers described above including the measurement of the content of monoamines and their metabolites in discrete areas of brain and metabolite concentrations in cerebrospinal fluid (CSF). In addition, since monoamine uptake sites are highly concentrated on their respective nerve terminals,[17] we have taken the approach of using selective high-affinity radioligands to directly label these uptake sites in brain and to assess the neurodegeneration of specific monoamine neurons by measuring corresponding reductions in the density of the respective uptake sites. The feasibility of this approach has been validated by recent studies demonstrating decreases in the density of ^{3}H-paroxetine-labeled serotonin uptake sites in homogenates of various brain regions following *in vivo* administration of a variety of amphetamine derivatives including methamphetamine,[18] MDA,[7] MDMA,[7,8,14] and fenfluramine.[19] Immunocytochemical visualization of MDMA-,[9,12] MDA-,[9] and fenfluramine-induced[20] destruction of serotonin axons and terminals using antibodies directed against serotonin and autoradiographic studies demonstrating corresponding decreases in ^{3}H-paroxetine-labeled serotonin uptake sites in slide-mounted brain sections of fenfluramine-,[21] MDA-,[22] and MDMA-treated[23] animals further validate use of changes in the density of serotonin uptake sites as an appropriate index of serotonin neuronal degeneration.

DOSE DEPENDENCE AND SPECIES DIFFERENCES

The studies were designed to assess and quantify both the neurochemical and neurodegenerative effects of subchronic systemic administration of MDMA on brain monoamine neurons in a variety of species including rat, guinea pig, mouse and rhesus monkey. The treatment regimen comprised systemic administration (subcutaneous in rat, mouse, and guinea pig and intramuscular in rhesus monkeys) of MDMA at various doses up to 20 mg/kg. For each dose, the drug was administered twice a day for 4 days. Thus, a dose of 20 mg/kg represents a total of 160 mg/kg MDMA administered over 4 days. There were dose-dependent decreases in a variety of serotonergic markers including serotonin (FIG. 1A), 5-HIAA (FIG. 1B), and the density of ^{3}H-paroxetine-labeled serotonin uptake sites (FIG. 1C) in both rat and monkey frontal cerebral cortex at 18 hours following the last injection. At the lowest dose of MDMA tested in rats (5 mg/kg) and monkeys (2.5 mg/kg), serotonin content was markedly reduced (40–60%), while only a small (14%) decrease in the density of serotonin uptake sites was observed in rats and no change was present in monkeys following administration of 2.5 mg/kg MDMA. At the 5 mg/kg dose in rat, a small decrease in 5-HIAA content was observed while comparable reductions in serotonin and 5-HIAA content were present in monkeys. In rats, the density of serotonin uptake sites decreased progressively as the

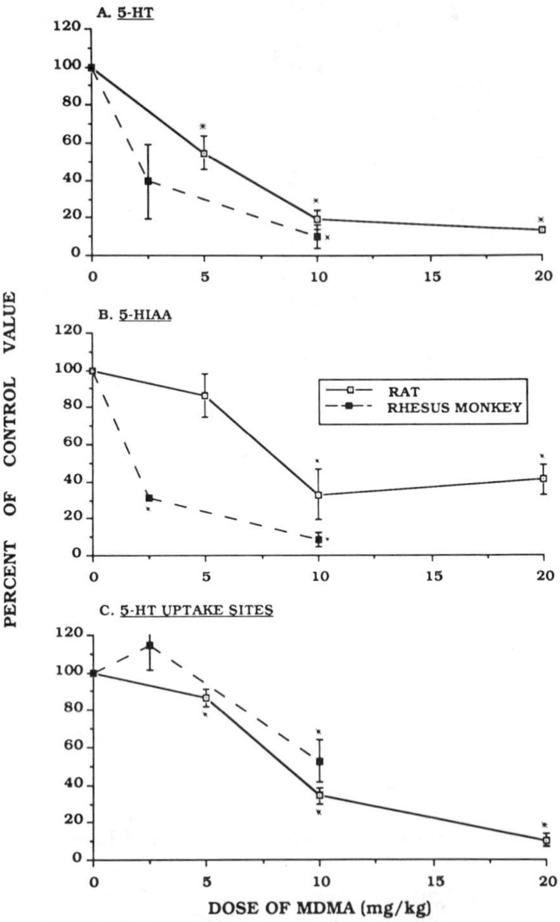

FIGURE 1. Dose-dependent effects of repeated systemic administration of MDMA on A. serotonin (5-HT), B. 5-hydroxyindoleacetic acid (5-HIAA) and C. ^{3}H-paroxetine-labeled 5-HT uptake sites in rat and rhesus monkey frontal cerebral cortex. Data represent the mean ± SEM from 3–5 rats or 3 or 4 monkeys and are expressed as a percent of values in control, saline-injected animals. * indicates significant differences at $p < 0.05$ from saline-injected control animals as determined by ANOVA. (Adapted from Battaglia *et al.*[8] and Insel *et al.*[14]).

dose of MDMA was increased with a maximal reduction of 90% observed at 18 hours following administration of 20 mg/kg of MDMA. In rhesus monkeys, the density of serotonin uptake sites was unaffected by the low-dose treatment (i.e. 2.5 mg/kg) while the 10 mg/kg dose of MDMA decreased the density of serotonin uptake sites by approximately 50% when compared to saline-injected control animals. In both animal species, serotonin levels appeared to be more readily decreased at lower doses of the drug while comparable reductions in serotonin uptake sites were noted only following treatment with higher doses of the drug. This apparent discrepancy between the serotonergic markers measured may relate to effects of lower doses of MDMA to inhibit tryptophan hydroxylase activity whereas the effects of higher doses of MDMA to reduce all three serotonergic markers may relate, in part, to effects on tryptophan hydroxylase activity and, in part, to destruction of serotonin neurons as evidenced by decreases in serotonin uptake sites. MDMA appears to be more potent in rhesus monkey than in rat with respect to its dose-dependent effects on depleting cerebral cortical serotonin and 5-HIAA content (FIG. 1A and B) while its effects on reducing the density of serotonin uptake sites appear to be comparable in the two species. Thus, rhesus monkeys may be more sensitive to the tryptophan hydroxylase inhibitory effects of MDMA than rats while the drug may be equipotent in destroying serotonin terminals (i.e. decreasing the density of serotonin uptake sites) in the two species.

The dose-dependent effects of MDMA treatment in rhesus monkeys on CSF concentrations of monoamines are shown in FIGURE 2. Significant reductions in CSF concentrations of 5-HIAA which were dependent on both the dose as well as the number of injections were seen following MDMA administration. While reductions in CSF concentrations of 5-HIAA were evident by the second day of the treatment regimen (i.e. following the third injection) in the 10 mg/kg treated group, significant changes in the 2.5 mg/kg treated group were observed only on the fifth day (i.e. following the eighth injection). These late decreases in CSF concentration of 5-HIAA were associated with marked (greater than 50%) dose-dependent reductions in the concentration of serotonin and 5-HIAA in the frontal cortex (see FIG. 1) as well as in a variety of other brain areas.

In contrast to the effects of MDMA on serotonergic neurons in rats and monkeys described above, MDMA administration was not associated with widespread or consistent changes in the levels of homovanillic acid (HVA) or 3-methoxy-4-hydroxyphenylethyleneglycol (MHPG) in CSF in rhesus monkeys (FIG. 2) or the concentrations of dopamine, norepinephrine or HVA in most brain regions examined in both rhesus monkeys (TABLE 1) and in rats.[7] The neurotransmitter specificity of the neurotoxic effects of MDMA is further demonstrated by lack of effects of MDMA treatment on the density of ^{3}H-mazindol-labeled norepinephrine uptake sites or ^{3}H-GBR 12935-labeled dopamine uptake sites in monkey (TABLE 1) and rat brain.[7,8]

Since amphetamines have been shown to be metabolized by different pathways in various species,[24] additional studies were carried out to investigate whether MDMA-induced neurotoxicity could be demonstrated in other species such as mouse and guinea pig. Animals in these studies were treated twice daily for four consecutive days with 20 mg/kg MDMA and the levels of serotonin, 5-HIAA and serotonin uptake sites were measured at seven days following the last injection in order to assess the long-term effects of the treatment. MDMA caused comparable and marked decreases in serotonin and 5-HIAA content and in the density of serotonin uptake sites in rat and guinea pig cerebral cortex, but appeared to be without effect on any of these serotonergic markers in the mouse.[8] Other studies have also suggested that mice are less susceptible to the neurotoxic effects of amphetamine derivatives including MDMA[16] and parachloroamphetamine.[25,26] The differential sensitivity may be due, in part, to species-dependent differences in the half-life of the drug.[26] Alternatively, it has been suggested that active neurotoxic metabolite(s) or metabolic intermediate(s) of parachloroamphetamine,[26] MDA,[27] and MDMA[27] may be responsible for their neurotoxic effects on serotonin neurons, although to date no active neurotoxic species has been identified.

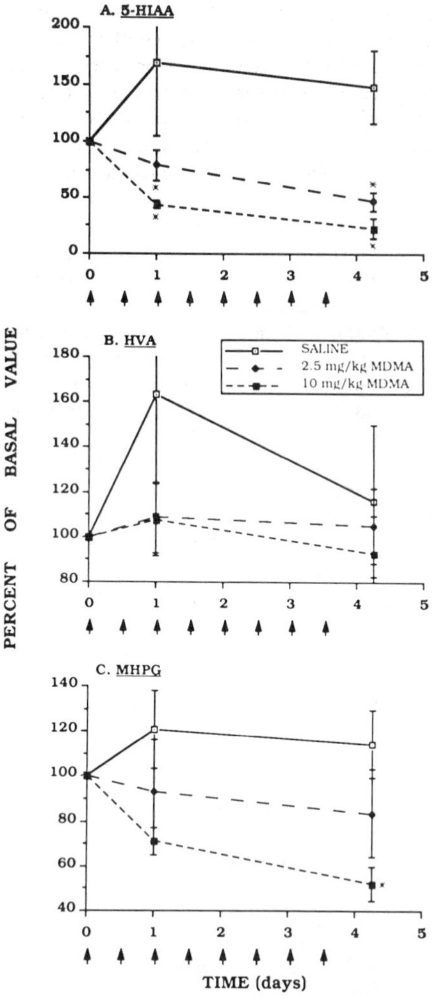

FIGURE 2. Effects of MDMA treatment in rhesus monkeys on cerebrospinal fluid (CSF) concentrations of A. 5-hydroxyindoleacetic acid (5-HIAA), B. homovanillic acid (HVA) and C. 3-methoxy-4-hydroxyphenyleneglycol (MHPG). Data represent the mean ± SEM from 3 or 4 monkeys and are expressed as a percent of basal values (i.e. values prior to beginning of treatment). Data were analyzed by ANOVA with repeated measures and by Scheffe post hoc test. * indicates significant differences at $p < 0.05$ from corresponding monoamine metabolite levels in saline-treated monkeys. Arrows represent time of intramuscular saline or MDMA administration. (Adapted from Insel *et al.*[14]).

REGENERATION OF SEROTONIN NEURONS AND LONG-TERM CONSEQUENCES

The time course of neuronal regeneration was investigated in rats by measuring the recovery of serotonin uptake sites and serotonin levels in frontal cerebral cortex at various time points up to 12 months following repeated systemic administration (i.e. twice daily s.c. injections for 4 days) of 20 mg/kg MDMA. As shown in FIGURE 3, at all timepoints up to six months during the recovery time course, the density of serotonin uptake sites was significantly below the corresponding values in age-matched saline-treated control rats. At 12 months after MDMA treatment, the density of serotonin uptake sites returned to control levels. It was of interest that in spite of the recovery of serotonin uptake sites to control levels, the content of serotonin in the same brain region remained markedly below (40–50%) age-matched controls for as long as one year after MDMA administration in rats. It is unclear from these rodent data whether there is regeneration of axons that have previously undergone degeneration or whether the increased density of uptake sites is a consequence of increased collateral sprouting of neurons unaffected by the drug treatment. It is also possible that axonal regeneration and collateral sprouting are associated with considerably greater densities of serotonin uptake sites per neuron, thereby making it more difficult to assess neuronal recovery from this index.

To assess the long-term consequences of MDMA treatment on brain serotonin neurons in monkeys, the CSF concentrations of 5-HIAA were monitored in saline and MDMA-treated monkeys (10 mg/kg i.m. twice daily for 4 days) for 14 weeks post treatment. In MDMA-treated monkeys, CSF concentrations of 5-HIAA decreased by approximately 75% immediately after treatment (Day 0) and remained decreased by approximately 40% throughout the 14-week post-treatment interval (FIG. 4). There were no apparent effects of MDMA treatment on HVA and MHPG concentrations in CSF.[14] In addition, various monoaminergic markers in brain were measured at time of sacrifice in saline- and MDMA-treated monkeys (14 weeks after the 4-day treatment regimen). As shown in FIGURE 5, brain concentrations of both serotonin and 5-HIAA were dramatically decreased in several cerebral cortical areas with the magnitude of reductions approximating those seen at 18 hours following treatment. Furthermore, the density of serotonin uptake sites in cerebral cortex was reduced to the same extent (i.e. 50%) as that noted at 18 hours following the treatment. Again, the concentration of catecholamine markers were, for the most part, relatively unaffected by MDMA treatment.

TABLE 1. Acute Effects of Systemic MDMA Administration on Catecholamine Markers in Rhesus Monkey Forebrain[a]

	Frontal Cortex	Temporal Cortex	Caudate
	(Percent of Saline-injected Control)		
Dopamine			
MDMA (2.5 mg/kg)	106 ± 70	69 ± 23	90 ± 13
MDMA (10 mg/kg)	106 ± 25	96 ± 31	57 ± 4*
Homovanillic Acid			
MDMA (2.5 mg/kg)	101 ± 28	57 ± 4	122 ± 21
MDMA (10 mg/kg)	67 ± 8	68 ± 9	103 ± 11
Dopamine Uptake Sites			
MDMA (2.5 mg/kg)	94 ± 18	107 ± 11	106 ± 12
MDMA (10 mg/kg)	100 ± 20	65 ± 23	99 ± 14
Norepinephrine			
MDMA (2.5 mg/kg)	109 ± 30	114 ± 6	ND
MDMA (10 mg/kg)	81 ± 22	85 ± 27	ND
Norepinephrine Uptake Sites			
MDMA (2.5 mg/kg)	169 ± 34	92 ± 17	133 ± 14
MDMA (10 mg/kg)	117 ± 26	142 ± 12	124 ± 13

[a]Regional brain levels of dopamine, homovanillic acid and norepinephrine were measured using HPLC; dopamine and norepinephrine uptake sites were measured using ^{3}H-GBR 12395 and ^{3}H-mazindol, respectively. Data are expressed as a percent of values in saline-injected control rats and represent the means ± SEM of determinations in 3 or 4 monkeys. Data were analyzed by factorial ANOVA with dose (0, 2.5 mg/kg, 10 mg/kg) as the main effect. * denotes significant difference at $p < 0.05$ from saline-injected control.

The data in rhesus monkeys parallel the observation in rodents of long-term alterations in various serotonergic parameters in brain[8] and are generally consistent with recent published reports of decreases in serotonin and 5-HIAA content in brain following MDMA administration in primates.[10–13] The reduction in all serotonergic

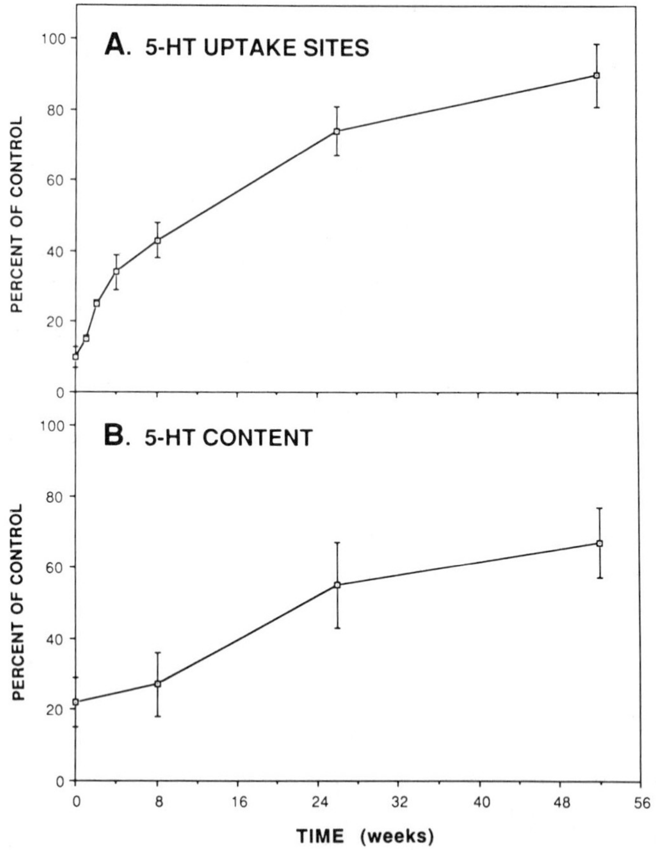

FIGURE 3. Timecourse of recovery of A. serotonin (5-HT) uptake sites and B. 5-HT content in rat cerebral cortex following repeated systemic administration of MDMA. Rats were treated with either saline (1 ml/kg) or MDMA (20 mg/kg) twice daily for 4 consecutive days and then sacrificed at various times up to 12 months later following the last injection. Saline-injected control rats were killed at each of the timepoints, and the data which represent the mean ± SEM of 5 rats per group are plotted as a percent of the values of age-matched saline-injected control rats. (Adapted from Battaglia *et al.*[8]).

markers following high-dose MDMA administration in monkeys appears to be largely a consequence of destruction of serotonin terminals as these alterations persist for at least 14 weeks following treatment. However, the data demonstrate some subtle differences between effects of MDMA in rodents and in primates. First, as described

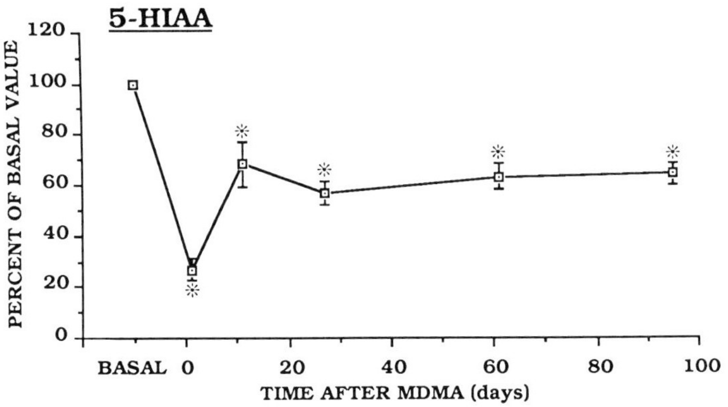

FIGURE 4. Time course of cerebrospinal fluid (CSF) concentration of the serotonin metabolite 5-hydroxyindoleacetic acid (5-HIAA) in rhesus monkeys treated with MDMA. Rhesus monkeys were injected twice daily for 4 days with 10 mg/kg MDMA. CSF was obtained at 2 timepoints prior to start of treatment (Basal) and at various timepoints up to 14 weeks following final drug administration (i.e. time 0). Data are expressed as a percent of basal values in each monkey. A significant overall group difference was noted for 5-HIAA, with * representing timepoints at which group differences exceeded $p < 0.05$ for post hoc comparisons. (Adapted from Insel *et al.*[14]).

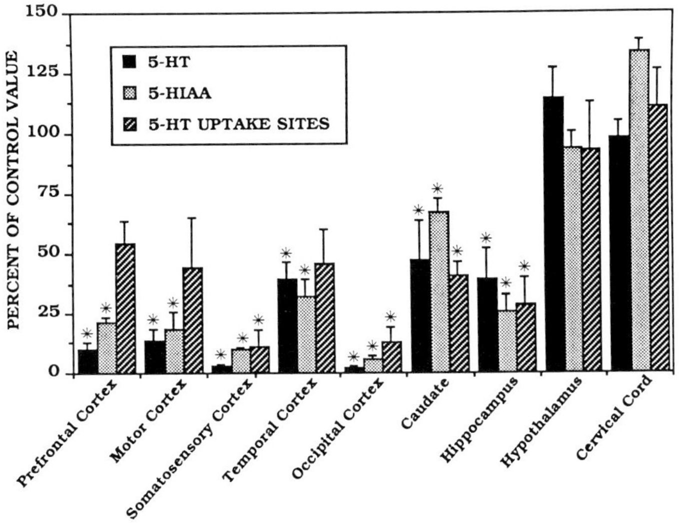

FIGURE 5. Long-term effects of subchronic administration of MDMA on serotonergic markers in discrete areas of monkey CNS. Rhesus monkeys were injected with either saline (n = 4) or MDMA (10 mg/kg; n = 3) twice daily for 4 days and sacrificed at 14 weeks after drug or saline administration. Data represent the mean ± SEM and are expressed as a percent of corresponding concentrations in saline-injected monkeys. * denotes significant differences at $p < 0.05$ as determined by ANOVA. (Adapted from Insel *et al.*[14]).

above, MDMA appears to be more potent in rhesus monkeys than in rats with respect to dose-dependent reductions in brain serotonin and 5-HIAA content. Second, since comparable decreases in brain serotonergic markers are seen at 18 hours and 14 weeks after high-dose MDMA treatment in rhesus monkeys while there is some apparent recovery in rats, the effects of MDMA treatment appear to be longer lasting in primates than in rodents. Evidence from both immunocytochemical studies[9] and autoradiographic studies quantifying changes in the density and distribution of ^{3}H-paroxetine-labeled serotonin uptake sites (see below) indicate that serotonin cell bodies appear to be relatively unaffected following repeated systemic administration of high doses of MDMA in rats. However, some preliminary data in squirrel monkeys of pathological changes in nerve cell bodies in dorsal raphe nucleus[12] suggest that serotonin cell bodies may be affected in this species by MDMA treatment. This apparent species difference in the effects of MDMA may explain, in part, the long-term, persistent decreases in serotonin parameters in rhesus monkeys while there is apparent recovery in rats.

REGIONAL DIFFERENCES AND MORPHOLOGICAL SPECIFICITY

The data described above demonstrate the specific and marked neurotoxic effects of MDMA on serotonin fibers in cerebral cortex. Two approaches were used to further investigate if the effects of MDMA treatment were localized to cerebral cortex or were ubiquitous in the central nervous system (CNS). The first approach involved the measurement of various monoamines, their metabolites and monoamine uptake sites in homogenates of discrete areas of rat and rhesus monkey CNS following treatment with MDMA. Second, we used autoradiographic techniques to visualize the effects of MDMA treatment on the distribution and density of ^{3}H-paroxetine-labeled serotonin uptake sites and ^{3}H-mazindol-labeled norepinephrine and dopamine uptake sites in slide-mounted sections of rat brain.

Neurochemical Studies

The long-term effects of repeated systemic administration of MDMA on various serotonergic parameters were investigated in both rats and rhesus monkeys. In rats, at two weeks following a high-dose MDMA treatment regimen (20 mg/kg, s.c. twice daily for 4 days) dramatic decreases in serotonin, 5-HIAA and 5-HT uptake sites were observed in cerebral cortex, hippocampus, striatum, and hypothalamus.[7] It was apparent that while decreases in the serotonin markers were observed in all of the brain regions examined, the reductions were greater in cerebral cortex and in hippocampus (40–60%) than in striatum and in hypothalamus (20–40%).[7] A similar profile was observed in rhesus monkeys at 14 weeks following a 4-day MDMA treatment regimen (10 mg/kg, i.m. twice daily for 4 days). Specifically, concentrations of both serotonin and 5-HIAA were reduced in a variety of neocortical regions, in hippocampus and in striatum but not in hypothalamus or in spinal cord; the reductions ranged from 50–90% (FIG. 5). Furthermore, the density of serotonin uptake sites was reduced by at least 50% in neocortex, striatum and hippocampus but, consistent with monoamine content, was relatively unaffected in either hypothalamus or spinal cord (FIG. 5). Differences in the sensitivity of various brain regions to the effects of MDMA treatment are not unique as previous data have demonstrated differential sensitivity to the effects of methamphetamine,[28] MDA,[7] fenfluramine,[19] and parachloroamphetamine[25] on serotonergic systems in various brain regions. Biochemical and histochemical data suggest that these compounds affect primarily the ascending serotonin systems whereas the descending pathways are left intact.

In contrast with the marked and consistent effects of MDMA on serotonergic systems, the drug treatment did not produce any widespread or consistent changes in the levels of norepinephrine, dopamine or their metabolites in the various brain regions examined in rhesus monkeys (TABLE 1) and rats.[7] Furthermore, MDMA treatment did not produce any consistent changes in the levels of ^{3}H-mazindol-labeled norepinephrine uptake sites or ^{3}H-GBR 12395-labeled dopamine uptake sites in the brain regions examined (TABLE 1). These data are in keeping with immunocytochemical data which demonstrate no significant changes in the density or morphology of catecholamine axons after subchronic administration of MDMA or MDA in rats.[9]

Autoradiographic Studies

In vitro autoradiographic studies of ^{3}H-paroxetine-labeled serotonin uptake sites in slide-mounted brain sections of saline- and MDMA-treated rats were carried out as previously described[22] in order to assess the extent and neuroanatomic localization of lesions induced by drug treatment. We observed substantial reductions in serotonin uptake sites (50–100% decreases) in all areas of cerebral cortex as early as 18 hours after a 4-day treatment regimen (20 mg/kg MDMA twice daily); the reductions were maintained for at least two weeks. The cerebral cortical regions which showed the most extensive destruction of serotonin uptake sites (i.e. greater than 90%) were the prefrontal, anterior cingulate, entorhinal and parietal cortex (FIG. 6).

As shown in FIGURE 6, marked decreases in serotonin uptake sites were observed in all regions of caudate-putamen, olfactory tubercle, endopiriform nucleus, islands of Calleja, and nucleus accumbens following MDMA administration. Other brain regions which were sensitive to the neurodegenerative effects of MDMA included various thalamic nuclei and regions of hippocampus. On the other hand, the dorsal and medial septal nuclei appeared to be less sensitive to the neurotoxic effects of MDMA as the reductions in serotonin uptake sites in these brain regions (approximately 25%) were not statistically significant (FIG. 6). Likewise, no significant reductions were observed in the indusium griseum which contains primarily serotonin axons of passage.

Within midbrain structures, regions containing serotonin projections appeared to be more dramatically affected by MDMA administration than those containing serotonin cell bodies (FIG. 7). For example, in both the superficial layers of the superior colliculus and profundum, ^{3}H-paroxetine binding was dramatically reduced by approximately 90%, while in dorsal median raphe, central grey and the ventral tegmental region, there was little or no change after MDMA treatment. Likewise, serotonin projections to substantia nigra pars compacta and reticulata were markedly affected, while no changes in serotonin uptake sites were observed in the interpeduncular nucleus and pontine reticular formation.

The neurotransmitter (serotonin) selectivity of the neurodegenerative effects of MDMA in brain were examined by carrying out additional autoradiographic studies of norepinephrine and dopamine uptake sites in brain regions containing catecholamine terminals and cell bodies. Norepinephrine and dopamine uptake sites were labeled with ^{3}H-mazindol in the presence of specific blockers. With respect to norepinephrine uptake sites, no change from control levels of ^{3}H-mazindol binding sites in midbrain regions such as the locus coeruleus, interpeduncular nucleus or substantia nigra was observed following MDMA treatment. Consistent with minimal effects of MDMA treatment on metabolic parameters associated with catecholamine neurons, there were

DA Uptake Sites

5-HT Uptake Sites

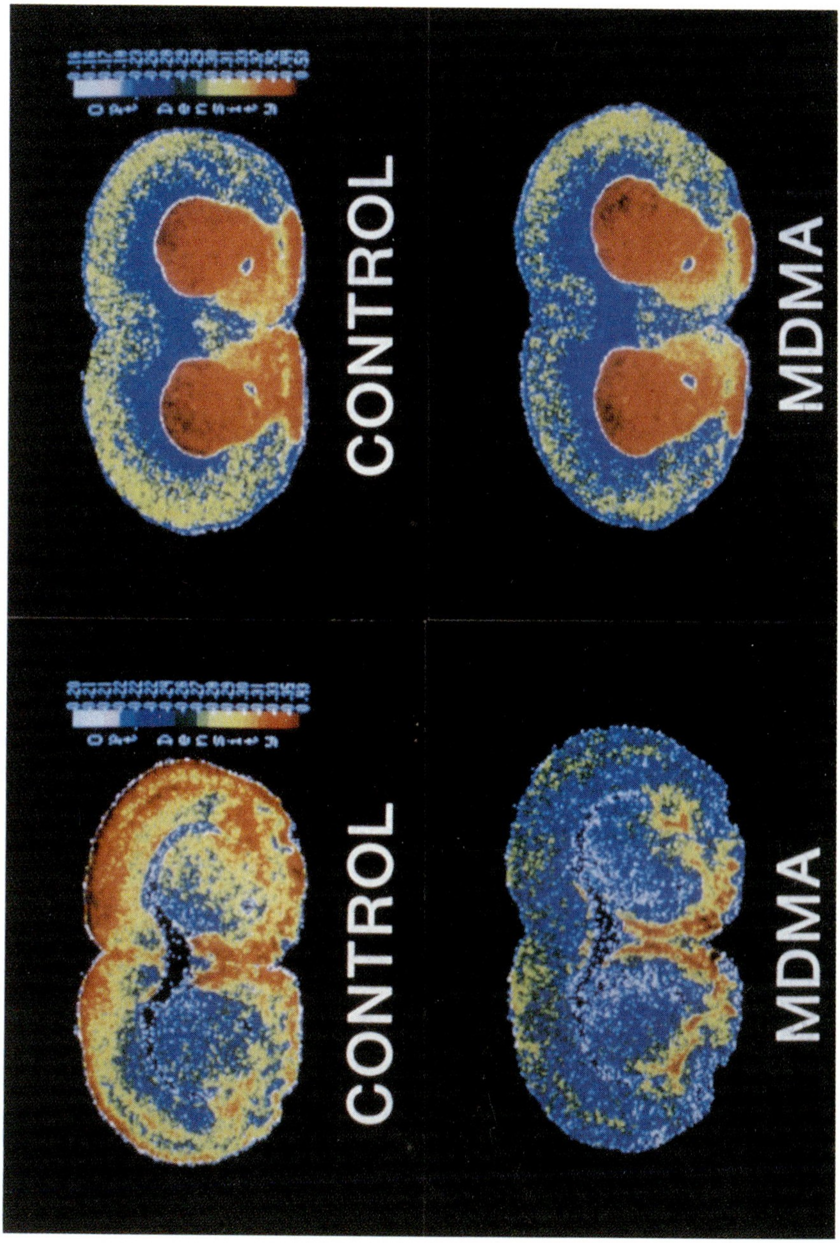

FIGURE 6. Color-coded autoradiograms of the distribution of ^{3}H-paroxetine-labeled serotonin (5-HT) uptake sites (*left-hand panels*) and ^{3}H-mazindol-labeled dopamine (DA) uptake sites (*right-hand panels*) in adjacent forebrain coronal sections of saline- (*upper panels*) and MDMA- (*lower panels*) treated rats. Rats were injected s.c. twice daily for 4 days with saline (1 ml/kg) or MDMA (20 mg/kg) and sacrificed at two weeks after drug administration. In saline-treated animals (control) (*top, left-hand panel*) note the high density of serotonin uptake sites in cerebral cortex, olfactory tubercle, caudate putamen and septum. Following MDMA treatment (*bottom, left-hand panel*), note the dramatic reductions in serotonin uptake sites in regions such as the cerebral cortex, the caudate putamen and olfactory tubercle while the septal nuclei are affected to a lesser extent. A comparable distribution of dopamine uptake sites is observed in saline- (*top right-hand panel*) MDMA- (*bottom right-hand panel*) treated animals, suggesting that MDMA treatment does not affect dopaminergic neurons. Highest levels of binding are shown in red, with progressively lower levels shown in yellow, green, blue and white.

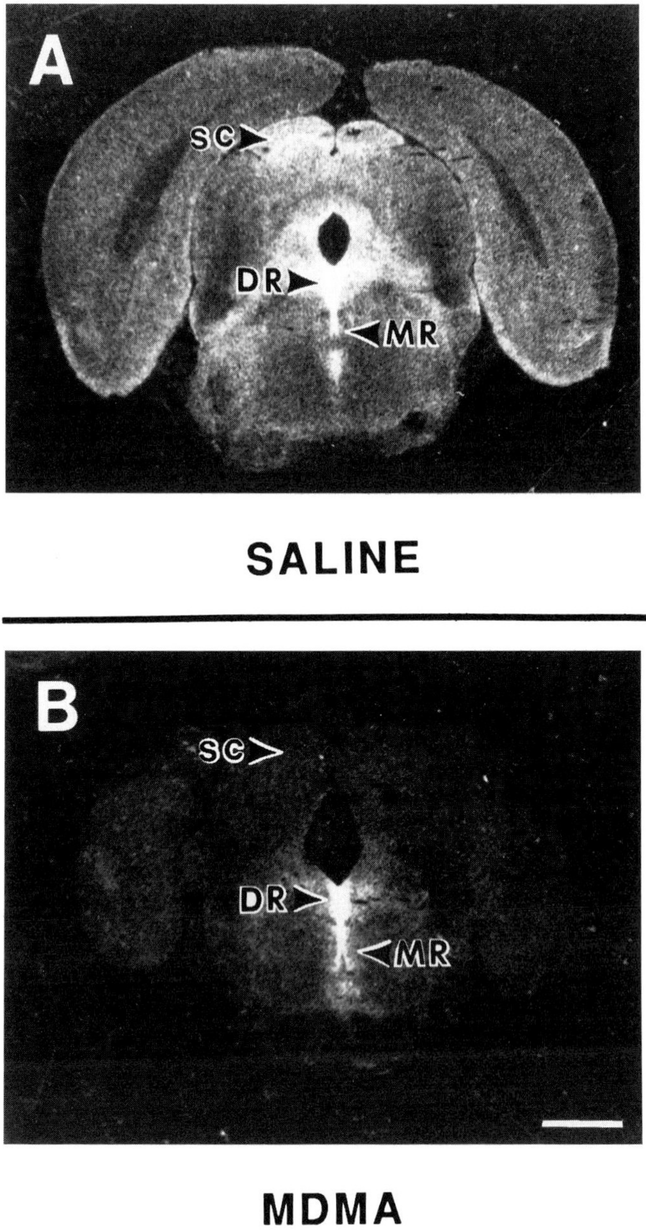

FIGURE 7. Autoradiographic distribution of ^{3}H-paroxetine-labeled serotonin uptake sites in coronal sections at the level of midbrain in (A) saline- and (B) MDMA-treated rats. In A, note the high density of serotonin uptake sites in brain regions containing serotonin projections such as entorhinal cortex, superior colliculus (sc), presubiculum and parasubiculum, as well as in cell body regions such as the dorsal (DR) and median (MR) raphe and central grey. MDMA-treated animals exhibited marked reductions in ^{3}H-paroxetine binding sites in pre- and para-subiculum, entorhinal cortex and superior colliculus, while no change in densities as observed in areas containing primarily serotonin perikarya such as the raphe nuclei (DR and MR).

no changes in the density of dopamine uptake sites in either cell body regions such as the substantia nigra or terminal regions such as the caudate putamen, nucleus accumbens and olfactory tubercle (FIG. 6). Overall, these results are consistent with the neurochemical data in brain homogenates and indicate that the neurodegenerative effects of MDMA appear to be confined primarily to serotonergic pathways since this treatment regimen did not reduce the density of uptake sites associated with catecholamine-containing neurons.

SUMMARY AND CONCLUSIONS

In summary, the data from both neurochemical and neuroanatomical studies demonstrate widespread and long-lasting degeneration of serotonin neurons in brain without any major or consistent effects on catecholamine neurons following *in vivo* administration of MDMA in both rats and rhesus monkeys. A detailed examination of the parameters involved in the neurotoxic and neurodegenerative effects of MDMA on brain serotonin neurons indicate that the severity of the lesion is dependent on the dose of drug administered with the drug being more potent in rhesus monkeys than in rats. Furthermore, the neurodegenerative effects of the drug are long-lasting (up to one year) with respect to neuronal regeneration (i.e. recovery of serotonin uptake sites) while functional recovery may be permanently impaired since serotonin content remains markedly (40–50%) below levels in age-matched controls for as long as one year after drug administration. The neurochemical and autoradiographic data suggest that there are some regional differences and morphological specificity to the neurodegenerative effects of MDMA as demonstrated by greater reductions in serotonin uptake sites in brain regions containing primarily terminals while regions containing axons of passage and cell bodies are relatively unaffected.

ACKNOWLEDGMENTS

Drs. S. Y. Yeh, J. N. Johannessen, J. Sharkey, and M. J. Kuhar contributed to various aspects of the work described. We thank Denise Pearsall, Susan Marra and Steven Culp for technical assistance and Sharon Amos for help in preparation of the manuscript.

REFERENCES

1. SHULGIN, A. T. 1986. The background and chemistry of MDMA. J. Psychoactive Drugs **18:** 291–304.
2. GREER, G. & R. J. STRASSMAN. 1985. Information on "Ecstasy." Am. J. Psychiatry **142:** 1391.
3. BEARDSLEY, P. M., R. L. BALSTER & L. S. HARRIS. 1986. Self-administration of methylenedioxymethamphetamine (MDMA) by rhesus monkeys. Drug Alcohol Depend. **18:** 149–157.
4. LAMB, R. J. & R. R. GRIFFITHS. 1987. Self-administration of *d,l*-methylenedioxymethamphetamine (MDMA) in the baboon. Psychopharmacology **91**: 268–272.
5. SCHMIDT, C. J. 1986. Neurotoxicity of the psychedelic amphetamine, methylenedioxymethamphetamine. J. Pharmacol. Exp. Ther. **240:** 1–7.
6. COMMINS, D. L., G. MOSMER, R. M. VIRUS, W. L. WOOLVERTON, C. R. SCHUSTER & L. S. SEIDEN. 1987. Biochemical and histological evidence that methylenedioxymethamphet-

amine (MDMA) is toxic to neurons in the rat brain. J. Pharmacol. Exp. Ther. **241:** 338–345.
7. BATTAGLIA, G., S. Y. YEH, E. O'HEARN, M. J. KUHAR, M. E. MOLLIVER & E. B. DE SOUZA. 1987. 3,4-Methylenedioxymethamphetamine and 3,4-methylenedioxyamphetamine destroy serotonin terminals in rat brain: Quantification of neurodegeneration by measurement of ^{3}H-paroxetine-labeled serotonin uptake sites. J. Pharmacol. Exp. Ther. **242:** 911–916.
8. BATTAGLIA, G., S. Y. YEH & E. B. DE SOUZA. 1988. MDMA-induced neurotoxicity: Degeneration and recovery of brain serotonin neurons. Pharmacol. Biochem. Behav. **29:** 269–274.
9. O'HEARN, E., G. BATTAGLIA, E. B. DE SOUZA, M. J. KUHAR & M. E. MOLLIVER. 1988. Methylenedioxyamphetamine (MDA) and methylenedioxymethamphetamine (MDMA) cause ablation of serotonin axon terminals in forebrain: Immunocytochemical evidence. J. Neurosci. **8:** 2788–2800.
10. RICAURTE, G. A., L. E. DELANNEY, I. IRWIN & J. W. LANGSTON. 1988. Toxic effects of MDMA on central serotonergic neurons in the primate: Importance of route and frequency of drug administration. Brain Res. **446:** 165–168.
11. RICAURTE, G. A., L. E. DELANNEY, S. G. WIENER, I. IRWIN & J. W. LANGSTON. 1988. 5-Hydroxyindoleacetic acid in cerebrospinal fluid reflects serotonergic damage induced by 3,4-methylenedioxy-methamphetamine in CNS of non-human primates. Brain Res. **474:** 359–363.
12. RICAURTE, G. A., L. S. FORNO, M. A. WILSON, L. E. DELANNEY, I. IRWIN, M. E. MOLLIVER & J. L. LANGSTON. 1988. MDMA selectively damages central serotonergic neurons in the primate. JAMA **260:** 51–55.
13. SLIKKER, W. JR., S. F. ALI, A. C. SCALLET, C. H. FRITH, G. D. NEWPORT & J. R. BAILEY. 1988. Neurochemical and neurohistological alterations in the rat and monkey produced by orally administered methylenedioxymethamphetamine (MDMA). Toxicol. Appl. Pharmacol. **94:** 448–457.
14. INSEL, T. R., G. BATTAGLIA, J. N. JOHANNESSEN, S. MARRA & E. B. DE SOUZA. 1989. 3,4-Methylenedioxymethamphetamine (MDMA; "Ecstasy") selectively destroys brain serotonin terminals in rhesus monkey. J. Pharmacol. Exp. Ther. **249:** 713–720.
15. STONE, D. M., D. C. STAHL, G. R. HANSON & J. W. GIBB. 1986. The effects of 3,4-methylenedioxymethamphetamine (MDMA) and 3,4-methylenedioxyamphetamine (MDA) on monoaminergic systems in the rat brain. Eur. J. Pharmacol. **128:** 41–48.
16. STONE, D. M., G. R. HANSON & J. W. GIBB. 1987. Differences in the central serotonergic effects of methylenedioxymethamphetamine (MDMA) in mice and rats. Neuropharmacology **26:** 1657–1661.
17. KUHAR, M. J. & G. K. AGHAJANIAN. 1973. Selective accumulation of ^{3}H-serotonin by nerve terminals of raphe neurons: An autoradiographic study. Nature **241:** 187–189.
18. ZACZEK, R., G. BATTAGLIA, J. F. CONTRERA, S. CULP & E. B. DE SOUZA. 1989. Methylphenidate and pemoline do not cause depletion of rat brain monoamine markers similar to that observed with methamphetamine. Toxicol. Appl. Pharmacol. **100:** 227–233.
19. ZACZEK, R., G. BATTAGLIA, S. CULP, N. M. APPEL, J. F. CONTRERA & E. B. DE SOUZA. 1990. Effects of repeated fenfluramine administration on indices of monoamine function in rat brain: Pharmacokinetics, dose response, regional specificity and time course data. J Pharmacol. Exp. Ther. In press.
20. APPEL, N. M., J. F. CONTRERA & E. B. DE SOUZA. 1989. Fenfluramine selectively and differentially decreases the density of serotonergic nerve terminals in rat brain: Evidence from immunocytochemical studies. J. Pharmacol. Exp. Ther. **249:** 928–943.
21. APPEL, N. M., W. M. MITCHELL & E. B. DE SOUZA. 1990. Effects of fenfluramine on brain monoamine neurons: Evidence from immunohistochemical and autoradiographic studies. *In* Proceedings of the International Symposium on Serotonin: From Cell Biology to Pharmacology and Therapeutics. P. Paoletti, P. M. Vanhoutte, N. Brunello, and F. M. Maggi, Eds. Boston: Kluwer Academic Publishers, pp. 625–629.
22. DE SOUZA, E. B. & B. L. KUYATT. 1987. Autoradiographic localization of ^{3}H-paroxetine-labeled serotonin uptake sites in rat. Synapse **1:** 488–496.

23. BATTAGLIA, G., R. ZACZEK & E. B. DE SOUZA. 1989. MDMA effects in brain: Pharmacologic profile and evidence of neurotoxicity from neurochemical and autoradiographic studies. *In* MDMA: "Ecstasy" and/or human neurotoxin? S. J. Peroutka, Ed. Boston: Kluwer Academic Publishers, pp. 171–199.
24. CALDWELL, J. 1980. The metabolism of amphetamines and related stimulants in animals and man. *In* Amphetamines and Related Stimulants: Chemical, Biological, Clinical and Sociological Aspects. J. Caldwell, Ed. Boca Raton, FL: CRC Press, pp. 29–46.
25. FULLER, R. W. 1978. Neurochemical effects of serotonin neurotoxins. Ann. N.Y. Acad. Sci. **305:** 178–181.
26. SANDERS-BUSH, E. & L. A. STERANKA. 1978. Immediate and long-term effects of p-chloroamphetamine on brain amines. Ann. N.Y. Acad. Sci. **305:** 208–221.
27. MOLLIVER, M. E., E. O'HEARN, G. BATTAGLIA & E. B. DE SOUZA. 1986. Direct intracerebral administration of MDA and MDMA does not produce serotonin neurotoxicity. Soc. Neurosci. Abstr. 12:1234.
28. RICUARTE, G. A., C. R. SCHUSTER & L. S. SEIDEN. Long-term effects of repeated methamphetamine administration on dopamine and serotonin neurons in the rat brain: A regional study. Brain Res. **193:** 153–163.

DISCUSSION OF THE PAPER

L. S. SEIDEN (*Session Chairman*): Errol you show that the uptake sites are coming back 100%. Going back to Dr. Fuller's remarks early in the meeting where part of the uptake sites at least are attributable to glial cells. Do you have any way to differentiate neuronal from glial uptake sites in any of your techniques?

DE SOUZA: No. we have not . . .

SEIDEN: I think that the glial might be responsible for some of the increase in uptake that you see.

DE SOUZA: Exactly. No we have not differentiated glial versus neuronal uptake sites. Most of the studies that we've done show a good correlation between active serotonin uptake and ^{3}H-paroxetine-labeled serotonin uptake sites. However there was a poster by Dr. Manini at this meeting that showed differentiation in some brain regions, specifically I think it was the hippocampus. But for the most part I think that serotonin uptake sites are a good index of degeneration.

MANFRED GÖTHERT (*University of Bonn*): Did you investigate 5-HT$_{1D}$ binding in non rodents and 5-HT$_{1B}$ bindings in rats, because I think if you have treated rats or the non-rodents with MDMA then loss of these binding sites might be indicative of loss of presynaptic 5-HT receptors.

DE SOUZA: Exactly. I mean that's the reason in my first slide I included presynaptic receptors as another good marker that could be used for looking at presynaptic neuronal loss. Having said that, we have not done the studies. The only receptors that we've looked at in terms of looking for compensatory regulation was the serotonin 2 receptor. In our preliminary studies we didn't see any changes in serotonin 2 receptors in the cortex in rats that were chronically treated with MDMA.

QUESTION: With respect to the difference between the median raphe and dorsal raphe, I think that the comparison between hippocampus and the striatum is quite interesting. In fact from the data by Dr. Molliver I remembered that apparently the striatum was much more affected than hippocampus, which fits with the degeneration of fibers coming preferentially from the dorsal raphe as compared to the median raphe. In all your slides, and for data concerning the rat, you showed clearly an effect which is much more pronounced in the hippocampus than in the striatum. Can you comment on that?

DE SOUZA: In terms of the differences, I think Dr. Steinbusch spoke on some of the discrepancies between studies measuring monoamine content and studies utilizing immunocytochemistry. And maybe some of those differences are related to that but I don't have a really good answer for you.

H. G. BAUMGARTEN (*Free University, Berlin*). Did you try to find a threshold dose that's non-toxic?

DE SOUZA: In terms of a threshold dose, we found in monkeys that 2.5 mg per kg dose of MDMA was showing threshold neurotoxic effects. The 10 mg per kg dose of MDMA clearly produced decreases in all 5-HT markers measured.

BAUMGARTEN: We did the oral dosing in rhesus monkeys, and it's slightly more toxic than the subcutaneous dose.

J. M. PALACIOS (*Sandoz, Basel*): Has somebody looked at the enteric nervous system in these animals?

DE SOUZA: No. The only thing I've looked at in the periphery is the spleen. We didn't see any changes in serotonin uptake sites in spleen following MDMA administration.

MARK MOLLIVER (*Johns Hopkins University, Baltimore, MD*): Very nice presentation Errol, my compliments on that. I had a couple of very brief questions that I think have brief answers. In one of your early slides on regeneration in the rat you showed a time course of recovery but it wasn't clear to me what region of the brain that was from.

DE SOUZA: Oh I'm sorry if I didn't mention it, it was all in the frontal cortex. We haven't looked at a comparison in different brain regions.

MOLLIVER: Frontal cortex, is it just, I mean it has a rather different much more rapid rate of recovery than any other part of the forebrain and there are striking differences as I probably didn't emphasize adequately.

The other point I wanted to mention was that another source of uptake sites in the brain are platelets. Do you rinse out the blood before you do your uptake studies?

DE SOUZA: No we have not, in the autoradiographic study that I did. In the immunocytochemical studies we do repeated washings so I think that would take care of platelet uptake sites. But in the autoradiographic studies we don't do a perfusion as you would in immunocytochemistry, but what we do is a pre-incubation, which may take care of it but I'm not sure.

Aminergic Metabolites in Cerebrospinal Fluid of Humans Previously Exposed to MDMA: Preliminary Observations[a]

G. A. RICAURTE[b]

Department of Neurology
Johns Hopkins University School of Medicine
Baltimore, Maryland 21205

K. T. FINNEGAN

Department of Psychiatry
University of Utah School of Medicine
Salt Lake City, Utah 84148

I. IRWIN AND J. W. LANGSTON

Institute for Medical Research
San Jose, California 95128

INTRODUCTION

Recreational use of the neurotoxic drug (±)3,4-methylenedioxymethamphetamine (MDMA, "Ecstasy") poses a public health problem of uncertain proportions. The reason for this is twofold. First, there is little information regarding the number of individuals that have been exposed to the drug, and even less about how many people are continuing to use it. According to one informal survey,[1] 39% of students on an undergraduate campus reported having experimented with MDMA on at least one occasion. Other informal surveys on different college campuses have yielded lower figures (6–25%). In 1985, Siegel[2] reported that approximately 30,000 dosage units of MDMA were being sold in the United States each month. More recently (1989), Doblin[3] provided an estimate of 100,000 dosage units per month. Whether this represents a true increase in MDMA use, or whether these estimates vary as a function of how they were ascertained is not clear. However, given MDMA's neurotoxic potential (see below), formal epidemiologic efforts to determine the incidence of recreational MDMA use are in order.

The second reason why it is difficult to gauge the seriousness of the threat that MDMA poses to public health is that, although the long-term serotonin-depleting effects of MDMA have been convincingly demonstrated in animals,[4–11] such documentation is lacking in humans. Moreover, the doses, routes, and schedules of MDMA administration used in most (but not all, see reference 11) animal studies have differed significantly from those typically employed by humans. Hence, it is difficult to know to what extent the findings in animals can be generalized to humans.

[a]This work was supported in part by grant DA05938-01 from the National Institute on Drug Abuse, CPHF subcontract No. 019A-701 and NIH grant NS21011.

[b]Address correspondence to: George A. Ricaurte, MD, PhD, Department of Neurology, Johns Hopkins Medical Institutions, 4940 Eastern Avenue, Baltimore, MD 21205.

To the best of our knowledge, thus far only two studies have investigated the neurotoxic potential of MDMA in humans. The first measured the concentration of 5-hydroxyindoleacetic acid (5-HIAA) in the cerebrospinal fluid (CSF) of five individuals with a history of recreational MDMA use.[12] No difference from controls was found. The second study assessed central serotonergic function in nine recreational MDMA users employing the L-tryptophan challenge test.[13] In this study, L-tryptophan induced a rise in serum prolactin in controls, but not in MDMA subjects, suggesting that serotonergic function in MDMA users might be impaired.

The purpose of the present study was to further evaluate the neurotoxic potential of MDMA in humans by measuring CSF 5-HIAA levels in a group of MDMA users who agreed to undergo lumbar puncture two (or more) weeks after their last use of the drug.

SUBJECTS AND METHODS

Demographics

A total of 33 subjects with a history of MDMA use volunteered to participate in this study. The demographics of the MDMA group and the control group are shown in TABLE 1. The mean age for the control group was slightly but significantly higher than for the MDMA group (45 ± 14 years vs. 36 ± 10 years). The relative proportion of males and females in the two groups was approximately the same (58% vs. 64% males in the control and MDMA groups, respectively). Prior to entering the study, all subjects provided a detailed drug history and underwent a complete general physical, neurological, and laboratory examination. Subjects with any significant medical or psychiatric illness (e.g. depression, obsessive compulsive illness) were excluded from the study. To avoid performing lumbar punctures in completely healthy control subjects (in whom there would be no clinical indications for the procedure), the local Institutional Review Board (IRB) requested that controls be recruited from a clinic population undergoing myelography for evaluation of low back pain. Since low back pain is known not to alter the concentration of CSF 5-HIAA, homovanillic acid (HVA), or 3-methoxy-4-hydroxyphenethyleneglycol (MHPG)[14] and this study was exploratory in nature, this recommendation was followed.

MDMA Use

TABLE 2 shows the characteristics of MDMA use by the experimental group. The mean (± SD) number of occasions on which subjects reported using the drug was 52 ± 45 (range 11 to 219). This use had taken place over a mean (± SD) period of 3.4 ± 1.8 years (range 0.5 to 8 years). The typical daily dose ranged from 50 to 200 mg,

TABLE 1. Demographics

	Control Group (N=24)	MDMA Group (N=33)
Age (yrs)		
Mean ± SD	45 ± 14	36 ± 10
Range	21 – 73	19 – 71
Sex (N)		
Males	14	21
Females	10	12

TABLE 2. Characteristics of MDMA Use

Duration of use: 3.5 ± 1.9 years
Frequency of use: 3.4 ± 1.8 weeks
Usual daily dose: 125 ± 44 mg
Cumulative dose: 6.4 ± 5.3 g
Time since last dose: At least 2 wks

corresponding to a dose of 0.98 to 2.6 mg/kg. A common pattern was for subjects to take an initial 100–125 mg dose, followed by a 50–75 mg dose 4–6 hours later. A substantial number of subjects (12 of 33) reported occasional use of higher total daily doses. The highest total daily dose reported was 700 mg, corresponding to approximately 10 mg/kg. The mean (± SD) cumulative dose of MDMA was estimated at 6.4 ± 5.3 g (range 1.4 to 44.2 g). In general, subjects allowed a mean (± SD) period of 3.4 ± 1.8 weeks between MDMA sessions. More frequent use, however, was not uncommon, as numerous (13 of 33) subjects recalled brief periods (5 to 7 days) of daily MDMA use. For the group, the mean (± SD) length of time since they had last used MDMA was 16.7 ± 23.3 weeks (range 2 to 104 weeks). Invariably the drug had been taken orally, although a few (3 of 33) subjects had experimented with MDMA intransally, and one had self-administered the drug intramuscularly. No intravenous use of MDMA was reported.

Other Drugs

Most of the subjects were poly-drug users. The other most commonly used recreational drugs were marijuana, lysergic acid diethylamide (LSD), mescaline, and psylocibin. Occasional use of other ring-substituted amphetamines [3,4-methylenedioxyethylamphetamine (MDE, "Eve")], 4-bromo-2,5-dimethoxyphenethylamine (2-CB)] was also reported, as was sporadic use of cocaine and other psychomotor stimulant drugs. In all subjects, MDMA had been one of the most commonly used drugs. As a prerequisite for participating in the study, subjects agreed to refrain from all recreational drug use for at least two weeks prior to the study.

Analysis of Drug Samples

All subjects were requested to provide a sample of the drug they had been taking and believed to be MDMA. Samples were provided by 16 of the 33 subjects, with one subject providing two samples (only one of which he thought was MDMA; the other he had purchased believing it was MDMA, but had found it to produce different effects). Samples were analyzed for their MDMA content by means of thin layer chromatography (TLC) and gas chromatography/mass spectrometry (GC/MS) with a selective nitrogen phosphorus detector.

Clinical Assessment

In addition to undergoing a complete general physical, neurological, and laboratory examination, all subjects in the MDMA group were specifically asked about changes in functional domains in which serotonin has been implicated, including sleep, pain perception, sexual behavior, aggression, mood regulation, appetite, and food preferences.[15–18] Subjects were also asked if they had noticed any tendency toward impulsive or obsessive compulsive behavior or any evidence of cognitive decline. No formal assessments were carried out in any of these functional spheres.

CSF Examination

After obtaining informed consent, CSF from each MDMA volunteer was obtained by means of lumbar puncture at the L4-L5 intervertebral space using a 22 gauge spinal needle with the subject lying in the lateral decubitus position. Conditions of CSF collection for control and MDMA subjects were closely matched, as neither group was fasted nor kept at bed rest prior to lumbar puncture and lumbar punctures were performed between 0800 and 1700 hours for both groups. Approximately four mls of CSF were collected from each subject. The first ml was sent for analysis of white blood cells and total protein by the clinical laboratory. The second ml of CSF was examined for its content of 5-HIAA, HVA and MHPG as detailed below. The remaining 2 mls were stored frozen in liquid nitrogen.

Determination of CSF 5-HIAA, HVA and MHPG Concentrations

5-HIAA, HVA and MHPG concentrations in CSF were measured by means of high performance liquid chromatography coupled with electrochemical detection (HPLC-EC) using the method of Kilpatrick *et al.*,[19] with minor modification. Briefly, frozen CSF samples were thawed, diluted 1:2 with 0.4 normal perchloric acid, vortexed and centrifuged at 19,600 × *g* for 10 minutes at 4°C. A 1 ml aliquot of the supernatant was immediately returned to liquid nitrogen until HPLC-EC assay. 5-HIAA, HVA and MHPG were separated using a Brownlee C-18 reverse phase column (5 μm particle size) and a mobile phase consisting of 96 parts of an aqueous phase (0.09 M sodium acetate; 0.035 M citric acid; 0.13 mM EDTA and 0.23 mM sodium octane sulfate) and 4 parts methanol. Detection was achieved by means of an ESA coulochem electrochemical dual detector, with detector 1 set at −0.03 volts and detector 2 set at +0.43 volts. Quantitation was accomplished by monitoring the response on detector 2, measuring peak heights of samples, and comparing them with those of standards processed in an identical manner.

Statistical Analysis

Comparison of values between groups was made using a two-tailed Student's *t*-test. Analyses of correlation were performed by calculating the Pearson-product moment correlation coefficient. A p value of less than 0.05 was considered significant.

RESULTS

Concentrations of CSF 5-HIAA, HVA and MHPG

The mean (± SD) concentrations of 5-HIAA, HVA and MHPG in lumbar CSF of control and MDMA subjects are shown in TABLE 3. The control values agree well with

TABLE 3. Concentrations of Monoamine Metabolites in Cerebrospinal Fluid of Control and MDMA Subjects[a]

	5-HIAA	HVA	MHPG
Control (N = 24)	19.1 ± 4.3	39.7 ± 11.3	9.6 ± 2.9
MDMA (N = 33)	14.2 ± 4.8*	33.5 ± 8.9	8.7 ± 2.0

[a]Values shown are the mean (±SD) expressed in ng/ml of CSF. *Significantly different from control; two-tailed Student's t test, $p < 0.05$.

results obtained by other investigators using similar methods.[20–23] Relative to controls, individuals with a history of MDMA use showed a 26% reduction in CSF 5-HIAA. By contrast, no alteration in either CSF HVA or MHPG concentration was observed (TABLE 3).

When analyzed by gender, there was a tendency for CSF 5-HIAA values in females to be higher than those in males in both the control and MDMA groups (TABLE 4). The differences, however, failed to achieve statistical significance.

Correlations

There was no significant correlation between CSF 5HIAA concentration and approximate number of MDMA exposures ($r = 0.18$), duration of MDMA use ($r = 0.28$) or the time since last MDMA exposure ($r = -0.09$) in MDMA users.

Analysis of Drug Samples

In 16 of 16 samples submitttd for analysis, MDMA was identified. As noted above, one subject submitted two samples but suspected that only one contained MDMA. The sample that he identified as containing MDMA did in fact contain the drug. The second sample was found to contain 3,4-methylenedioxyamphetamine (MDA).

Functional Consequences

The majority (30 of 33) of MDMA subjects reported no symptomatic complaints at the time of evaluation. They specifically denied signs or symptoms of depression, sleep disturbance, altered pain perception, or change in sexual function. They also

TABLE 4. CSF 5-HIAA According to Gender in Control and MDMA Subjects[a]

Group	N	CSF 5-HIAA
Controls		
Males	14	18.0 ± 4.2
Females	10	20.7 ± 4.4
MDMA		
Males	21	13.4 ± 5.1*
Females	12	15.6 ± 4.0**

[a]Values shown are the mean (±SD) expressed in ng/ml of CSF. *Significantly different from male control; two-tailed Student's t-test, $p < 0.05$. **Significantly different from female control; two-tailed Student's t-test, $p < 0.05$.

denied changes in appetite or food preference, and none had experienced any substantial weight loss or gain.

Of the three subjects who had functional complaints, one complained of depression and was seeking professional counseling. Another reported "difficulty concentrating" and "poor memory." The third (a musician) reported "diminished creativity" and marked mood swings. All three subjects believed their difficulties were related to their MDMA use, but recognized other factors might be involved.

CSF Protein and Cell Count

The mean (± SD) total protein concentration in the CSF of MDMA subjects was within normal limits (35.3 ± 15.3, normal range 15 to 45 mg/dl). However, in five of the subjects it was slightly elevated (range 50 to 64 mg/dl), and in one, it was twice the upper limit of normal (96 mg/dl). The cell count and differential in the latter individual showed 3 white blood cells (WBC) (2 lymphocytes, 1 polymorphonuclear cell) and the CSF 5-HIAA concentration was 13.8 ng/ml. With one exception, CSF cell counts in MDMA subjects were within normal limits (less than 5 WBC). The one subject with a minimal pleocytocis (9 WBC, predominantly lymphocytes) had no signs or symptoms of meningeal irritation, although his CSF protein was 64 mg/dl and his CSF 5-HIAA concentration was 6.5 ng/ml.

DISCUSSION

The major finding of the present study is that the concentration of 5-HIAA in lumbar CSF of human subjects with a history of MDMA use is reduced by 26%. Since the concentration of 5-HIAA in CSF is thought to reflect the activity of central serotonergic neurons,[24,25] this finding points to a possible impairment of central serotonergic function in recreational MDMA users. While the nature of this impairment cannot be stated with certainty, results from experimental animal studies[4-11] would suggest that damage of central serotonergic axons may be involved.

These findings are in keeping with the recent observation of a blunted prolactin response to L-tryptophan in recreational MDMA users.[13] Both results point to a persistent alteration of central serotonin metabolism in humans repeatedly exposed to MDMA. The present findings, however, are not in accord with those of a prior CSF study.[12] In that study, CSF 5-HIAA levels were found not to be altered in five recreational MDMA users. While the basis for this discrepancy remains to be identified, it could be related to a number of factors including different subjects, differing degrees of MDMA exposure, differences in sample size or to methodological features of the present study (see below). Efforts to shed light on which of these factors is involved are now in progress.

There are a number of reasons why caution should be exercised in interpreting the results of this study. First, almost all of the subjects were poly-drug users. Therefore, it is not possible to know to what extent other drugs might have contributed to the reduction in CSF 5-HIAA. While few recreational drugs other than MDMA are known to damage serotonin neurons, a number of them are known to influence serotonin metabolism (e.g., LSD), and thus could conceivably alter CSF 5-HIAA concentration on a long-term basis. Second, decreased CSF 5-HIAA has been reported in patients suffering from depression and other neuropsychiatric conditions.[26-31] Without pre-MDMA CSF 5-HIAA values, it is difficult to exclude the possibility that the lower CSF 5-HIAA values in MDMA subjects are related to either a pre- or co-existing psychiatric condition. Third, several variables known to influence CSF 5-HIAA concentration (e.g. activity[32] and diet[33]) were not controlled in this study. While this was done in an effort to match conditions of CSF collection for control and MDMA subjects as closely as possible, ideally these variables should have been controlled. Finally, the control group did not consist of entirely healthy human volunteers, but rather was selected from a clinic population undergoing myelography for the evaluation of low back pain. While such subjects have been reported to have normal CSF 5-HIAA levels,[14] a completely healthy control group would have been preferable.

The low degree of correlation between the degree of MDMA exposure and CSF 5-HIAA concentration in recreational MDMA users deserves comment. This could signify that MDMA is not responsible for the lower CSF 5-HIAA levels observed in these individuals. Alternatively, it could be related to the fact that frequency, duration, and pattern of MDMA use were all estimated by means of retrospective accounts, the reliability of which is difficult to ascertain. In addition, since the potency of street samples can vary widely, the number of capsules or tablets taken may not be a true indicator of the degree of exposure. Finally, it is important to note that pre-MDMA CSF 5-HIAA values are not known (making it impossible to calculate true percent decreases for each subject) and that the potential for decreasing 5-HIAA levels in human CSF is limited (theoretically from approximately 20 ng of 5-HIAA per ml of CSF to zero). These factors, coupled with a small coefficient of variation in 5-HIAA determinations, could complicate analyses of correlation.

If decreased 5-HIAA in lumbar CSF of MDMA-exposed humans reflects a lowering of their CNS serotonin content (as it does in the monkey[34]), the question arises as to what extent brain serotonin in these individuals is likely to be depleted. In humans, this question is difficult to answer because the relative contribution of brain and spinal cord to lumbar CSF 5-HIAA is controversial,[35,36] although it is generally believed that CSF 5-HIAA concentrations parallel those in adjacent CNS areas.[37] In MDMA-treated monkeys, a 60% reduction in cervical CSF 5-HIAA is associated with an 80–90% depletion of serotonin (and 5-HIAA) in brain and a 42% depletion of serotonin in spinal cord.[34] Hence, at least in the monkey, the reduction in cervical CSF 5-HIAA appears to underestimate the extent of serotonin depletion in brain and overestimate the extent of serotonin depletion in the spinal cord. Should this also prove to be true in humans, the 26% reduction in lumbar CSF 5-HIAA in MDMA subjects could be associated with a larger depletion of serotonin in the brain of these individuals.

Clearly, one of the more important issues is whether there are any functional consequences that MDMA users experience as a result of their lower CSF 5-HIAA content. As noted above, the majority (30 of 33) of MDMA subjects denied any disturbance in sleep, mood, appetite, or in any functional domain in which serotonin has been implicated. While this paucity of clinical complaints is re-assuring, it is important to bear in mind that formal functional assessments have not been performed. Such assessments could uncover subtle problems not apparent to the subjects themselves. In addition, it is important to keep in mind that partial serotonergic damage could have occurred, but that it is insufficiently large to cause clinical signs or symptoms. To detect such damage, use of specialized imaging techniques (e.g., positron emission tomography) may be required.

In summary, the results of this preliminary study indicate that humans previously exposed to MDMA have lower concentrations of 5-HIAA in their CSF. While the basis for these 5-HIAA decrements remains to be determined, preclinical studies in rodents and primates[4–11] suggest that damage of central serotonin axons and terminals may be involved. In view of this, the public health threat posed by recreational MDMA use may be significant. Therefore, epidemiologic efforts to define the true extent of MDMA use need to be intensified. Finally, in spite of the apparent absence of immediate functional consequences, the possibility that an age-related decline in serotonergic function[38–40] might unmask latent functional deficits in recreational MDMA users needs to be considered.

ACKNOWLEDGMENTS

We acknowledge a donation from the Multidisciplinary Association for Psychedelic Studies, Sarasota, FL. We thank Mr. Rick Doblin for his assistance in recruiting subjects for this study, and Lorrene Davis-Ritchie, David Remmler, ZoAnn McBride and David Rosner for their expert technical assistance.

REFERENCES

1. PEROUTKA, S. J. 1987. Incidence of recreational use of (±)3,4-methylenedioxymethamphetamine (Ecstasy) on an undergraduate campus. N. Engl. J. Med. **317**(24): 1542–1543.
2. SIEGEL, R. K. 1985. Submitted as a preprint to the Drug Enforcement Agency for MDMA hearings.
3. DOBLIN, R. 1989. Risk Assessment, the FDA and MDMA Research. *In* Ecstasy: The MDMA Story. B. Eisner, Ed. Ronin Publishing, Inc., Berkeley, CA: pp. 163–170.
4. SCHMIDT, C. J. 1987. Neurotoxicity of the psychedelic amphetamine, methylenedioxymethamphetamine. J. Pharmacol. Exp. Ther. **240:** 1–7.
5. COMMINS, D. L., G. VOSMER, R. VIRUS, W. WOOLVERTON, C. SCHUSTER & L. SEIDEN. 1987. Biochemical and histological evidence that methylenedioxymethylamphetamine (MDMA) is toxic to neurons in the rat brain. J. Pharmacol. Exp. Ther. **241:** 338–345.
6. STONE, D. M., D. S. STAHL, G. L. HANSON & J. W. GIBB. 1986. The effects of 3,4-methylenedioxymethamphetamine (MDMA) and 3,4-methylenedioxyamphetamine on monoaminergic systems in the rat brain. Eur. J. Pharmacol. **128:** 41–48.
7. BATTAGLIA, G., S. Y. YEH, E. O'HEARN, M. E. MOLLIVER, M. J. KUHAR & E. B. DESOUZA. 1987. 3,4-Methylenedioxymethamphetamine and 3,4-methylenedioxyamphetamine destroy serotonin terminals in rat brain: Quantification of neurodegeneration by measurement of [^{3}H]paroxetine-labeled serotonin uptake sites. J. Pharmacol. Exp. Ther. **242**(3): 911–916.
8. O'HEARN, L., G. BATTAGLIA, E. B. DESOUZA, M. J. KUHAR & M. E. MOLLIVER. 1988. Methylenedioxyamphetamine (MDA) and methylenedioxy-methamphetamine (MDMA) cause ablation of serotonergic axon terminals in forebrain: Immunocytochemical evidence. J. Neurosci. **8:** 2788–2803.
9. RICAURTE, G. A., K. T. FINNEGAN, D. E. NICHOLS, L. E. DELANNEY, I. IRWIN & J. W. LANGSTON. 1987. (±)Methylenedioxyethylamphetamine (MDE) a novel analogue of MDMA, produces long-lasting depletion of serotonin in the rat brain. Eur. J. Pharmacol. **137:** 265–268.
10. RICAURTE, G. A., L. S. FORNO, M. A. WILSON, L. E. DELANNEY, I. IRWIN, M. E. MOLLIVER & J. W. LANGSTON. 1988. (±)3,4-Methylenedioxymethamphetamine (MDMA) selectively damages central serotonergic neurons in non-human primates. JAMA **260**(1) 51–55.
11. RICAURTE, G. A., L. E. DELANNEY, I. IRWIN & J. W. LANGSTON. 1988. Toxic effects of MDMA on central serotonergic neurons in the primate: Importance of route and frequency of drug administration. Brain Res. **446:** 165–168.
12. PEROUTKA, S. J., N. PASCOE & K. F. FAULL. Monoamine metabolites in the cerebrospinal fluid of recreational users of 3,4-methylenedioxymethamphetamine (MDMA; "Ecstasy"). Res. Commun. in Substance Abuse **8:** 125–138.

13. PRICE, L. H., G. A. RICAURTE, J. H. KRYSTAL & G. R. HENINGER. 1988. Neuroendocrine and mood responses to intravenous L-tryptophan in 3,4-methylenedioxymethamphetamine (MDMA) users. Arch Gen. Psych. **46:** 20–22.
14. HYYPPA, M. T., H. SCHEININ, H. ALARANTA, M. HURME, K. LAHTELA & M. SCHEININ. 1985. Neurotransmission and the experience of low back pain: No association between CSF monoamine metabolites and pain. Pain **21**(1): 57–65.
15. BARCHAS, J. & E. USDIN, EDS. 1973. Serotonin and Behavior. New York: Academic Press Inc.
16. LORENS, S. A. 1978. Some behavioral effects of serotonin depletion depend on method. A comparison of 5,7-dihydroxytryptamine, p-chlorophenylalanine, p-chloroamphetamine and electrolytic raphe lesions. Ann. N.Y. Acad. Sci. **305:** 532–555.
17. MESSING, R.B. 1978. Behavioral effects of serotonin neurotoxins: An overview. Ann. N.Y. Acad. Sci. **305:** 480–496.
18. WURTMAN, R. J. 1987. Nutrients affecting brain composition and behavior. Integr. Psychiatry **5:** 226–257.
19. KILPATRICK, I., M. JONES & O. PHILLIPSON. 1986. A semiautomated analysis method for catecholamines, indoleamines, and some prominent metabolites in microdissected regions of the nervous system: An isocratic HPLC technique employing coulometric detection and minimal sample preparation. J. Neurochem. **46**(6): 1865–1876.
20. PAVE S. 1986. Intranasal vasotocin decreases cerebrospinal fluid 5-HIAA levels in man. Brain Res. Bull. **16:** 391–393.
21. BEN MENACHEM, E., L. PERSSON, P. J. SCHECTER, K. D. HAEGELE, N. HUEBERT & J. HARDENBERG. 1989. Cerebrospinal fluid parameters in healthy volunteers during serial lumbar punctures. J. Neurochem. **52:** 632–635.
22. SUNDERLAND, T., P. N. TARIOT, R. M. COHEN, P. A. NEWHOUSE, A. M. MELLOW & E. A. MUELLER. 1987. Dose-dependent effects of deprenyl on CSF monoamine metabolites in patients with Alzheimer's disease. Psychopharmacology **91:** 293–296.
23. POMARA, N., R. SINGH, D. DEPTULA, P. A. LEWITT, G. BISSETTE, M. STANLEY & C. NEMEROFF. 1989. CSF corticotropin-releasing factor (CRF) in Alzheimer's disease: Its relationship to severity of dementia and monoamine metabolites. Biol. Psych. **26:** 500–504.
24. MOIR, A. T., G. W. ASCHCROFT, T. B. CRAWFORD, D. ECCLESTON, & H. GULDBERT. 1970. Cerebral metabolites in cerebrospinal fluid as a biochemical approach to the brain. Brain **93:** 357–368.
25. YOUNG, S. N., E. GARELIS, S. LAL, J. MARTIN, P. MOLINA-NEGRO, R. ETHIER & T. L. SOURKES. 1974. Tryptophan and 5-hydroxyindoleacetic acid in human cerebrospinal fluid. J. Neurochem. **22:** 777–779.
26. MENDELS, J., A. FRAZER, R. G. FITZGERALD, T. A. RAMSEY & J. STOKES. 1972. Biogenic amine metabolites in cerebrospinal fluid of depressed and manic patients. Science **175:** 1380–1382.
27. ASBERG, M., P. THOREN & L. TRASKMAN. 1976. "Serotonin Depression"—a biochemical subgroup within the affective disorders? Science **191:** 478–480.
28. ASHCROFT, G., T. B. CRAWFORD & D. ECCLESTON. 1966. 5-Hydroxyindole compounds in the cerebrospinal fluid of patients with psychiatric or neurologic diseases. Lancet **2:** 1049-1050.
29. BOWERS, M. B. 1974. Lumbar CSF 5-hydroxyindoleacetic acid and homovanillic acid in affective syndromes. J. Nerv. Ment. Dis. **158**(5):325–330.
30. POST, R. M., J. C. BALLENGER & F. K. GOODWIN. 1980. Cerebrospinal fluid studies of neurotransmitter function in manic and depressive illness. *In* Neurobiology of Cerebrospinal Fluid. J. H. Wood, Ed. New York: Plenum Press, pp. 685–717.
31. MELTZER, H. Y. & M. LOWY. 1987. The serotonin hypothesis of depression. *In* Psychopharmacology: The Third Generation of Progress. H. Meltzer, Ed. New York: Raven Press, pp. 513–526.
32. POST, R. M., J. KOTIN, F. K. GOODWIN & E. D. GORDON. 1976. Psychomotor activity and cerebrospinal fluid amine metabolites in affective illness. Am. J. Psychiatry **130:** 67–72.
33. ECCLESTON, D., G. W. ASHCROFT, T. B. CRAWFORD, J. B. STANTON, D. WOOD & P. H.

McTurk. 1970. Effect of tryptophan administration of 5-HIAA administration on 5-HIAA in cerebrospinal fluid in man. J. Neurol. Neurosurg. Psychiat. **33:** 269–272.
34. Ricaurte, G. A., L. E. DeLanney, S. G. Wiener, I. Irwin & J. W. Langston. 1988. 5-Hydroxyindoleacetic acid in cerebrospinal fluid reflects serotonergic damage induced by 3,4-methylenedioxymethamphetamine in CNS of non-human primates. Brain Res. **474:** 359–363.
35. Garelis, E., S. N. Young, S. Lal & T. L. Sourk. 1974. Monoamine metabolites in lumbar CSF: The question of their origin in relation to clinical studies. Brain Res. **79:** 1–8.
36. Weir, R. L., T. N. Chase, L. K. Ng, I. J. Kopin. 1973 5-Hydroxyindoleacetic acid in spinal fluid: relative contribution from brain and spinal cord. Brain Res. **52:** 409–412.
37. Stanley, M., L. Traskman-Bendz & K. Dorovin-Zis. 1985. Correlation between aminergic metabolites simultaneously obtained from human CSF and brain. Life Sci. **37:** 1279–1286.
38. Allen, S. J., J. S. Benton, M. J. Goodhardt, & E. A. Haan, N. R. Sims & C. T. Smith. 1983. Biochemical evidence for selective nerve cell changes in the normal ageing human and rat brain. J. Neurochem. **41:** 256–265.
39. Marcusson, J., L. Oreland, B. Winblad. 1984. Effect of age on human serotonin S1 binding sites. J. Neurochem. **43:** 1669–1705.
40. Goldman-Rakic, P. S. & R. M. Brown. 1981. Regional changes of monoamines in cerebral cortex and subcortical structures of aging rhesus monkeys. Neuroscience **6:** 177–187.

DISCUSSION OF THE PAPER

Pat Stell (*Walter Reed, Washington DC*) Given that Dr. Molliver has shown that fenfluramine has similar neurotoxic effects, and given that fenfluramine has been used therapeutically in humans, have you done, or do you plan to do, any studies with fenfluramine?

Ricaurte: That's a good question. We would be very interested in screening a population of individuals previously exposed to fenfluramine in the same manner that we're screening MDMA subjects. This could give us at least a first indication of whether or not fenfluramine produces long-term changes in human serotonergic systems. In contrast to the MDMA study, however, I think one of the problems with this study will be identifying a cohort of people previously exposed to fenfluramine that is willing to come in for study.

Stell: Well the other thing would be that, since fenfluramine is used therapeutically, you could probably do pre and post studies to show any potential neurotoxicity.

Ricaurte: Let me just say that given what I know currently with regard to the effects of fenfluramine in primates (we've recently been doing some studies with fenfluramine in primates), my personal opinion is that the margin of safety for fenfluramine is too narrow for safe use in humans. At this point, I simply don't know what dose of fenfluramine could be safely administered to humans without running the risk of producing the kind of changes that we're seeing in rodents and non-human primates.

S. Lorens (*Loyola University Medical Center, Chicago*): Dr. Seiden and I recently have been looking for functional consequences of high dose repeated administration of MDMA and also 57-DHT, and we have had major difficulty in discovering any kind of behavorial changes as well as baseline and stress-induced changes in corticosterone release and prolactin. In your work with humans, as well as monkeys, have you seen any indication of any kind of neurological or psychiatric disability?

RICAURTE: We're just beginning such studies, both in humans and in non-human primates. So far we have not seen any changes in either. However, a point that may be pertinent to make at this juncture is that you may need some very detailed cognitive and other forms of testing to see if, in fact, there is any functional consequence. That's precisely what we're attempting to do with the cohort of MDMA subjects in Baltimore.

R. WURTMAN (*MIT, Cambridge, MA*): George, I certainly would agree with your interpretation of your preliminary data so far, but just to make the point as you well know MDMA is a serotonin uptake blocker and there's immense evidence that any serotonin uptake blocker will decrease CSF 5 -HIAA by decreasing the firing of the serotonin neurons and by potentiating the action of presynaptic receptors. So I think the data that you've shown are entirely compatable with an entirely functional effect of MDMA as an uptake blocker, but there's a related point and that is the safety of giving tryptophan to people who have taken MDMA. There are by now at least 10 case reports that I know of published, and perhaps others that I don't know of (if Ray Fuller is here, he may be aware of more) in which people have had very serious toxic reactions after getting tryptophan plus fluoxetine. I think that before I would give a high dose of tryptophan for a prolactin test to anybody taking an uptake blocker like MDMA that I'd worry about that a little bit.

RICAURTE: I think that's a good point and that's one of the reasons we've made great effort to emphasize to the subjects that they remain drug-free prior to the study. Initially we were requesting that they remain drug free for 2 weeks. In our ongoing study in Baltimore, we're requesting that subjects remain drug free for at least 4 weeks.

WILLIAM A. HEWLETT (*Stanford University, Stanford, CA*): Is there any change in the dose required to produce the desired effect in these drug abusers?

RICAURTE: I can only answer that question with some anecdotal data. It's something that I always ask the subjects as they come through the study. In general, they indicate that they have not noted any change in their dose requirements over time. However, I should mention that I have come across one individual who had experimented with high doses of MDMA over a short period of time and he told me that he was finding that he had to remain drug free for at least 2 months if he was going to achieve any of the desired subjective effects of MDMA. So, in this setting (repeated high doses over a short period of time) there is perhaps some evidence for a type of tolerance, if you will. I have also been struck by the number of women who, over time, are reporting that they are experiencing similar effects with lower doses of MDMA then they originally took. These are women who have been using MDMA maybe once a month over a year or two. In some of these women there may be some type of supersensitivity, developing. This is anecdotal data, but it is the only data I have to answer your question.

COMMENT: I think one of the problems that you're going to get is that once the people out there get street wise to the results that are being presented here that say, for example, uptake blockers like fluoxetine will actually decrease the toxicity, we'll find that things like that will start being abused. We'll have drugs of abuse with fluoxetine and other uptake blockers which will then make a big problem for you measuring 5-HIAA because we know that those compounds will also lower them.

S. PEROUTKA (*Stanford University, Stanford, CA*): Just on an anecdotal level I have one patient that was put on fluoxetine for depression and now tells me when she takes MDMA there's no more high, which actually fits with the *in vivo* data. And that's an intelligent PhD-type person. Not that that means anything, obviously.

RICAURTE: From a mechanistic scientific standpoint it's a very intriguing issue. However, I've had subjects report both things, and it just becomes very difficult to know how much one may be coloring or influencing a person's reporting. At times, I get the sense people tell me what they think I want to hear.

COMMENT: I wanted to ask a question. I read a sort of underground report of the use of MDMA several years ago by a psychiatrist and it sounded, subjectively, like a mild stimulant. Then I hear anecdotal reports about it's the most wonderful thing in the world. I was wondering if you could give your integrated subjective feeling from what you've heard, whether it is a mild stimulant or really something entirely different than other ones.

RICAURTE: David Nichols touched on that. However, I can tell you that, after interviewing 40 or so MDMA users, my own impression is that they are describing a unique psychopharmacological effect . I also think that MDMA produces a psychomotor stimulant effect. Subjects clearly tell you that they get a euphoric rush from MDMA. However, they also tell you, that that's not what they find most interesting or rewarding about MDMA.

M. E. MOLLIVER (*The Johns Hopkins University, Baltimore, MD*) Given the increasing rise in the abuse of methamphetamine, and the widespread use of MDMA, wouldn't we be wise as a neurologist or psychiatrist to begin considering evaluating the use of uptake blockers for the acute management of patients who might come into the emergency room with an overdose of such drugs to try to protect their central dopamine or norepinephrine terminals.

RICAURTE: There is certainly a scientific rationale for your suggestion. However, I would be hesitant to embark on such a therapeutic strategy just yet because of data that Lou Seiden and Ray Fuller collected some years ago. What they did is pretreat animals that were going to receive methamphetamine with fluoxetine. The drug did a beautiful job, completely protecting against the serotonergic toxic effect of methamphetamine but it also ended up augmenting the toxic effect of methamphetamine on dopamine neurons. With regard to MDMA while it's true that it's a relatively pure serotonergic neurotoxin if you push the dose, you begin to see effects on the dopamine system. So my concern would be that if you have a patient with an MDMA overdose you can perhaps effectively protect their serotonin system with fluoxetine (if you give the drug in time) but you may be running the risk of causing damage to their dopamine system. Until some of these issues are resolved in experimental animals, it may be a little premature to begin exploring such therapies on overdose patients in emergency rooms.

MDMA and Related Compounds:

Session Summary

LEWIS S. SEIDEN

Department of Pharmacology/Physiological Sciences
The University of Chicago
Chicago, Illinois 60637

SUMMARY OF INDIVIDUAL PAPERS

The historical perspective on drugs as possible neurotoxins was presented by J.W. Gibb. Investigation started in 1964 and continued with the research which noted that halogenated amphetamine derivatives caused prolonged depletion of serotonin in the central nervous system. The halogenated amphetamines although initially developed for use in humans were never used because of prolonged effects on CNS serotonin. After 1971, long-term and perhaps toxic effects of the parent amphetamines that were commonly used by humans (amphetamine, AMPH and methamphetamine, MA) were examined for possible neurotoxicity. These investigations were important because MA and AMPH were abused by humans often in high doses for long periods of time. The finding that MA caused neurotoxicity lead to the investigation of a number of structurally and/or pharmacologically related compounds, and therefore, logically lead to the investigation of methylenedioxyamphetamine (MDA) and methylenedioxymethamphetamine (MDMA).

D. E. NICHOLS presented structure-activity relationships of MDMA analogs. Nichols used two activity measures: first, the ability to be generalized in a drug discrimination pradigm, and second, the ability to cause certain subjective feelings in humans. Nichols expresses the view that MDMA-like drugs may be clinically useful because they increase communication; drugs that have similar discriminative stimulus properties to MDMA may have similar communication facilitatory effects in people. The aim therefore, is to find a drug that is not neurotoxic, subject to abuse, or hallucinogenic, and has clinically useful characteristics. On the basis of their stimulus properties in animals and their reported effects in humans Nichols has proposed a new drug class called enactogens. Based on stimulus properties in animals with a series of structurally related compounds, Nichols has proposed a role for the 5-HT_{1b} site in mediating the effects of the new class and proposed a search for enactogens that do not have hallucinogenic or toxic effects on the 5-HT system. MDA is hallucinogenic and amphetamine-like, but MDMA seems more amphetamine-like. The hallucinogenic properties of MDMA resemble LSD and mescaline. However, MDA, MDMA and other hallucinogens, but not amphetamine can stimulate the 5-HT_2 receptor. The behavioral properties of these drugs are not congruent with the binding site properties. This incongruence raises the possibility that the complete range of relevant binding sites and/or behaviors are not being observed. Nichols uses activity as an hallucinogen in humans and the ability of the various compounds to share stimulus properties with DOM (a powerful hallucinogen in humans) as the criteria for classification of these drugs.

M. E. MOLLIVER has reviewed the differentiation of ascending fibers from the dorsal and medial raphe of the brainstem to rostral terminal fields in the telencephalon; he finds two kinds of 5-HT axons: large, beaded fibers predominantly emanating from the median raphe, and fine fibers arising from the dorsal raphe. Molliver and co-workers have observed that MDMA, MDA and fenfluramine damage fine but not beaded fibers. Between 24 and 36 hours after MDA, MDMA, or fenfluramine administration the fine fibers swell and appear fragmented; at later time points, most of the fine axons are no longer detectable. Molliver interprets the swelling and fragmented nature of the fine 5-HT fibers and the subsequent reduction in visible axons as an indication that the fibers are undergoing changes indicative of axonal degeneration.

The silver-staining methods for degeneration are indicative of cell damage, but it is not known what type of neuron is injured. The immunocytochemical method shows the type of neuron, but since the staining is for the transmitter, an intact neuron devoid of transmitter substance would not be visible; if that were the case, the absence of staining could be misinterpreted as representing a site of neurotoxin-induced damage. In addition to the distorted axons and the subsequent decrease in the density of 5-HT axons, evidence is presented that after several months the damaged 5-HT fine fibers gradually return as a result of sprouting, possibly leading to fairly normal patterns of reinnervation.

C. SCHMIDT presented data indicating that MDMA and related compounds release 5-HT in two phases. The acute phase is marked by a decrease in 5-HT levels in brain and begins immediately after drug administration. The chronic or toxic phase begins 24 hours after drug administration and is associated with a loss of 5-HT, and tryptophan-hydroxylase. Data that could elucidate possible mechanisms of MDMA toxicity were discussed. First, there seemed to be a requirement for dopamine release. Second, the toxicity is blocked by antioxidants; therefore, the neurotoxicity may be the result of an oxidation process that is set in motion by MDMA. Third, 5-HT_2 receptors seem to be involved since antagonism of these receptors prevents the neurotoxicity engendered by MDMA. It was not clear how this interesting data will come together as a unified mechanism.

E. B. DE SOUZA presented evidence that MDMA caused long-term reductions in 5-HT content and 5-HT uptake sites. There appears to be some recovery in 5-HT content and in the number of high affinity uptake sites; recovery in 5-HT levels in rats after varying amounts of MDMA exposure is from a low point of 20% of normal (1 week post-MDMA), to 60% of normal (1 year post-MDMA), but recovery in high affinity uptake sites ranges from about 15% (2 weeks post-MDMA) to 90% (1 year post-MDMA).

G. A. RICAURTE demonstrated that in a human population that had taken MDMA on several occasions, there was a significant decrease in 5-HIAA cerebrospinal levels in the subjects that had used MDMA as compared to controls, while HVA and DOPAC levels were the same for both groups. There were no marked behavioral abnormalties seen in the subjects he studied. The MDMA-using population was similar in sex distribution to the control group, although there was a small but significant age difference.

POINTS OF AGREEMENT

MDMA is considered a drug of abuse by the U.S. Drug Enforcement Agency and is a Schedule I drug. MDMA has long-lasting and perhaps permanent (toxic) effects on 5-HT neurons in the brain. MDMA-increased decreases in 5-HT are more enduring than the reductions in the number of 5-HT uptake sites. Fenfluramine is structurally similar to, but pharmacologically different from MDMA; however, in high doses, fenfluramine, like MDMA, also has long lasting effects on 5-HT neurons.

Nichols states that MDMA might be a useful drug because of its ability to increase communication, if it were not for its toxic effects in humans. Similar drugs without the potential toxicity may be developed in the future for human use. There have not been systematic clinical studies done on MDMA, and the placement of it as a Schedule I drug by the DEA will make such studies difficult in the future. This will make drug development in this class in the future problematical because a "standard" for comparison will be difficult to evaluate. Certain members of the psychiatric community maintain that MDMA is a valuable drug as an adjunct to psychotherapy; others maintain, largely on the basis of the number of illicit drug seizures, that MDMA has abuse potential. Still others find that it is toxic to 5-HT–containing cells in animals and perhaps also in humans. Based on the potential for abuse and on possible harmful effects, as well as a lack of demonstrated therapeutic effects, the DEA has successfully moved to classify the drug as a Schedule I compound. This has been a contentious point because the proponents of the drug argue that its therapeutic value is almost impossible to determine while it remains in a Schedule I classification.

POINTS AT ISSUE

Is fenfluramine uniquely metabolized in humans such that the effective dose would be different from the dose that produces toxic effects to 5-HT neurons? While the largest part of the formal presentation in the session dealt with issues surrounding MDMA, the most intense part of the discussion in this area surrounded the usefulness and potential toxicity of fenfluramine. MDMA has not been demonstrated to have clinical efficacy, it is considered a drug with abuse potential, and it can do damage to the nervous system of animals. On the other hand, fenfluramine has recognized efficacy as a medication in weight reduction and has been widely used by humans for several years without reports of severe adverse side effects or long-term effects after the drug was discontinued. Several investigators have noted that—in experimental animals—fenfluramine causes long-lasting depletions of 5-HT in the CNS. In addition the high affinity uptake sites are reduced, and there appear to be morphological changes in 5-HT neurons immediately after the drug.

There were points of agreement and disagreement regarding fenfluramine. There was general agreement that above certain doses fenfluramine engendered prolonged toxic effects on 5-HT–containing neurons. However, some investigators maintain that fenfluramine causes a toxic effect in 5-HT neurons, while others maintain the effect is a depletion of 5-HT with repletion occurring over time. This issue has not been resolved on the basis of available data. Molliver's data shows morphological changes occurring soon after fenfluramine injection followed by depletion with some evidence that there is at least partial recovery; on the other hand data presented by De Souza indicate that 5-HT depletions last for at least one year. There was a question as to the applicability of data obtained in rats or non-human primates to humans. B. Campbell alluded to evidence that humans do not metabolize fenfluramine in the same way as rats and the major metabolite nor-fenfluramine is not retained in human tissue in the same way that it is retained in rodents and non-human primates; this difference in retention time explains the long-term depletion in animals with high retention times. The only human data on this issue came from a post-mortem case in which death occurred because of an overdose of fenfluramine. The pharmacokinetics may be sufficiently different when death occurs as to raise questions about the relevance of this data. Therefore, based on the small number of these cases and the unique characteristics of an overdose leading to death, further comparative pharmacokinetics will need to be pursued.

Finally, the question of the ratio between the dose that causes long-term 5-HT depletion to the effective dose for suppressing food intake was raised. The disagreement centered mainly on the dose necessary for suppression of food intake, and ultimately on the methodology used to assess the effective anorectic dose. One group used a rodent model in which rats were deprived of food for as long as 23 hours before the anorectic effects of fenfluramine were tested; others measured suppression of food intake in rats with continuous access to food. In the latter case, less drug is required to obtain an effect on overall intake; particularly if one looks at times immediately after fenfluramine administration. In the former case, perhaps the rats are so nutritionally deprived of food that it takes a large dose of fenfluramine to interfere with their feeding behavior. It is difficult to infer from this data what the analog will be in the obese human attempting to lose weight.

POINTS OF QUESTIONS

Schmidt's data strongly suggest that DA release is necessary for the toxicity to 5-HT neurons engendered by MDMA, and there are data to suggest that DA is important to the MA-engendered toxicity to both DA and 5-HT neurons. It is a bit puzzling as to why there is a lack of toxicity to the DA system in the MDMA-treated animals.

One of the most pressing questions raised in the discussion was the consequences of the neurotoxicity. B. Campbell and J. Blundell both noted that in animals who had lost 50% of their DA and/or 5-HT content there does not seem to be any alteration in behaviors. Is it possible to destroy 50% of the 5-HT terminals in the CNS and see no alteration in function? Based on psychopharmacological and neurochemical data the answer would seem to indicate that 5-HT and DA are widely distributed neurotransmitters in the CNS. Drugs acting on uptake sites or receptors have potent effects on behavior and yet a 50% decrease in the transmitter is without effect. 6-OHDA which reduces steady state DA levels by 50% or more has very little effect on behavior; it seems that only with DA depletions of 90% or greater are physiological or behavioral effects observed.

There are altered behavioral states within certain psychiatric illnesses that are not accompanied by known simple alterations in levels of transmitters. Drugs which have their major actions on a given transmitter system are often efficacious in these diseases. These same drugs can alter behavior in animals. From these facts one can infer that there is a role for these transmitters in the maintenance of certain types of behavior and that these transmitter systems may be compromised during disease. However, it is clear that there does not seem to be an isomorphism between behavior and a transmitter system. We have conditions where a transmitter is clearly altered but there are no grossly observable behavioral alterations; the absence of behavioral change with transmitter level alteration seems especially noteworthy.

FUTURE RESEARCH DIRECTIONS

There are several questions on both the chemical and the clinical side that have been raised by the papers that have been presented. First, what is the criterion that shall be used to judge that a drug is neurotoxic. One can draw on chemical, morphological *and* functional criteria to make the definition very strict. That is to say, one would require that there be some evidence of damage to the brain *and* that this damage should have functional consequences. If these criteria are obtained, one would be hard pressed to continue to use the drug except under very extraordinary conditions. This situation is clear-cut, but the situation with MDMA is not quite that clear and the situation with fenfluramine is even less clear. With MDMA there is definitive evidence for long-lasting damage and the efficacy of the drug in therapeutics remains vague. With fenfluramine, the evidence for damage to the CNS is not as convincing as that for MDMA, and the evidence for the usefulness of fenfluramine as a therapeutic agent is more convincing.

The cutoff for using or discontinuing use of a drug is not altogether clear; if we find damage to brain cells that may recover with time and no functional changes, what is the rule to decide whether or not we should continue to use the drug? This is an ethical problem. If one uses the rule to discontinue a drug if it is at all toxic to neurons, then one risks depriving people of a therapeutically effective drug that has minor and/or reversible damaging effects. Therefore, the matter of putative or even established toxicity cannot be the *sole* criterion for the decision; efficacy must enter into the equation. Then one is left with the dilemma of assigning relative weights to each factor.

There are simpler issues that must also be resolved before one gets to the philosophical issues. We must know the ratios of toxicity to efficacy accurately. Some data can best be obtained using animal models and others can be best obtained using retrospective analyses of human data.

Index of Contributors